TECHNIQUES OF CHEMISTRY

ARNOLD WEISSBERGER, *Editor*

VOLUME II

ORGANIC SOLVENTS

Third Edition

TECHNIQUES OF CHEMISTRY

ARNOLD WEISSBERGER, *Editor*

VOLUME I
PHYSICAL METHODS OF CHEMISTRY, in Five Parts
Incorporating Fourth Completely Revised
Edition of Physical Methods of Organic Chemistry
Edited by Arnold Weissberger and Bryant W. Rossiter

VOLUME II
ORGANIC SOLVENTS, Third Edition
John A. Riddick and William B. Bunger

TECHNIQUES OF CHEMISTRY

VOLUME II

ORGANIC SOLVENTS

PHYSICAL PROPERTIES AND METHODS OF PURIFICATION

Third Edition

BY

JOHN A. RIDDICK

Baton Rouge, Louisiana

AND

WILLIAM B. BUNGER

Indiana State University
Terre Haute, Indiana

On the basis of the First Edition
by ARNOLD WEISSBERGER *and* ERIC S. PROSKAUER
 and
the completely revised Second Edition
by JOHN A. RIDDICK *and* EMORY E. TOOPS, *Jr.*

WILEY-INTERSCIENCE

A DIVISION OF JOHN WILEY & SONS, Inc.

New York · London · Sydney · Toronto

Library of Congress Catalogue Card Number: 72-114919

ISBN 0 471 92726 0

Printed in the United States of America

10 9 8 7 6 5 4 3 2 1

INTRODUCTION TO THE SERIES

Techniques of Chemistry is the successor to the Technique of Organic Chemistry Series and its companion—Technique of Inorganic Chemistry. Because many of the methods are employed in all branches of chemical science, the division into techniques for organic and inorganic chemistry has become increasingly artificial. Accordingly, the new series reflects the wider application of techniques, and the component volumes for the most part provide complete treatments of the methods covered. Volumes in which limited areas of application are discussed can be easily recognized by their titles.

Like its predecessors, the series is devoted to a comprehensive presentation of the respective techniques. The authors give the theoretical background for an understanding of the various methods and operations and describe the techniques and tools, their modifications, their merits and limitations, and their handling. The series should contribute to a better understanding and a more rational and effective application of the respective techniques.

Authors and editors hope that readers will find the volumes in this series useful and will communicate to them any criticisms and suggestions for improvements.

Research Laboratories   ARNOLD WEISSBERGER
Eastman Kodak Company
Rochester, New York

FROM THE PREFACE TO
THE FIRST EDITION

In recent years improvements in the methods of organic chemistry and the enhanced interest in the chemical physics of non-aqueous solutions have led to an ever-increasing demand for variety and purity in purification.

For this reason it has seemed desirable to make a collection of physical constants and of methods of purification for these solvents. The aim of this book is to make readily accessible the abundant material which has been accumulated by chemists and physicists in recent years. To this end an organic chemist and a physical chemist have collaborated.

The success of physical and chemical work, both preparation and measurement, is often decided by choice of solvent, and this choice is sometimes not at all simple. We hope that a study of the systematic list of solvents and of the numerical data, classified systematically, contained in this book will make this task easier. The selection of solvents treated is inevitably somewhat arbitrary although we have tried to cater for as many different requirements as possible: we shall be glad of any suggestions for extension or limitation of the list or other amendments. We should be particularly grateful for copies of papers which contain measurements of physical constants of solvents or methods of their organic solvents.

Oxford and Leipzig

A. WEISSBERGER
E. S. PROSKAUER

PREFACE TO THE THIRD EDITION

Since the appearance of the Second Edition of *Organic Solvents* the role of the solvent has become much more significant and better understood. Many more solvents are available for industrial use, generally in a good to high grade of purity, and numerous additional solvents have become available for research.

One hundred solvents have been added to the Third Edition; we could not find justification for deleting any that were listed in the Second Edition. Readers of the Second Edition suggested many compounds and properties to be added in the Third Edition. Of these suggestions those were accepted that were in keeping with the aim of the book provided sufficient information was available. Primary consideration was given to the uniqueness of the solvent property and to present and possible future uses of the solvent, particularly in the newer fields of application such as electrochemical reactions, thin layer chromatography, and preparative solvent extraction. Several solvents were added to complete groups of isomers and to expand homologous series to include a reasonable number of compounds, for instance in the C_nH_{2n+2} and C_nH_{2n} series. The pentane isomers are complete. The *n*-alkyl alcohols through hexanol are given, as are the isomers of the C_3, C_4, and C_5 alcohols.

In most of the functional group classes unsaturated compounds have been added. The amides have been increased from two to nine in keeping with the expanding interest in research and in industrial use. The number of sulfur compounds has been doubled and includes dimethyl sulfoxide and sulfolane. Several heterocyclic solvents have been added. Newly included unique compounds of limited availability are 1,1,2,2-tetramethylurea and hexamethylphosphoric triamide.

Water is a solvent of reference and a standard for calibration and comparison; the physical properties of water have been listed at the beginning of the tables in Chapter III as Tables Oa, Ob, and Oc. We know of no other place where so many of the properties of water are collected in easy-to-use tables.

The number of properties tabulated has been increased by twelve, not including the references to spectra and some properties that have synonymous methods of expression such as the ionization constant, K_i, and the negative logarithm of the acid dissociation constant, pK_a. The newly added properties were selected from a study of the scientific literature beginning in 1960 to determine which properties were used most frequently. Several properties that were of limited academic interest have become useful in newer areas such as electronics and space science.

Eight Index Tables appear in Chapter III, including four new ones— Ebullioscopic Constants, Cryoscopic Constants K_f and A for Equation 2.72 in Chapter II, and Density.

The format of the book remains the same as in the First and Second Editions, as does the aim, "to make readily accessible the abundant material" that has accumulated in the literature on organic solvents.

Chapter II, "Discussion of Properties and Criteria for their Selection," has been completely revised to facilitate the use of the tables in Chapter III. The newer concepts relating to purity have been included in Chapter IV, and the more recent methods for characterizing purity are discussed.

We have calculated many values to make the tables more useful. Some of the calculated values, such as the temperature coefficients of density and of the refractive index, were used during the evaluation of the data prior to inclusion in the tables.

We express our gratitude for their suggestions for the third edition to Dr. Paul D. Bartlett, Harvard University; Professor G. Cauquis, Centre d'Études Nucleaires de Grenoble; Dr. G. L. Covert, Eastman Kodak Company; Dr. Lyman C. Craig, Rockefeller Institute; Mr. P. B. Dalton, General Aniline and Film Corporation; the late Dr. Henry de Laszlo, Koch-Light Laboratories; the late Dr. J. B. Dickey, Tennessee Eastman Company; Mr. Hugh J. Hagemeyer, Texas Eastman Company; Dr. Dan H. Moore, South Jersey Medical Research Foundation; Dr. L. J. Roll and Dr. I. F. Salminen, Eastman Kodak Company; Dr. Dietmar Seyferth, Massachusetts Institute of Technology; Dr. Howard Steinberg, U.S. Borax Research Corporation; Dr. W. West and Mr. C. C. Unruh, Eastman Kodak Company; Dr. D. S. Young, Tennessee Eastman Company; and Dr. C. W. Zuehlke, Eastman Kodak Company.

Our sincere thanks also go to Professor Thomas F. Fagley and Dr. Ray Oglukian of Tulane University, Dr. T. C. Wehman of Michigan State University, and Dr. Robert West of Lehigh University for unpublished information from their research, and to Dr. Arthur and Elizabeth Rose of Applied Science Laboratories for their many helpful suggestions and enlightening discussions. We are grateful to Dr. Arthur and Wanda L. Campbell for their assistance during the first two years of the preparation of this edition. We are especially grateful to Mr. Gene Seidel of Crown Zellerbach Corporation for his generous help in supplying extensive information on dimethyl sulfoxide and dimethyl sulfide and his offer to furnish information on other solvents, and to Commercial Solvents Corporation for providing facilities for our work and for freely allowing the use of previously unpublished information.

To our wives we express our appreciation for their patience during the five years of preparation of this book and for their help in tabulation and proofreading.

Baton Rouge, Louisiana   JOHN A. RIDDICK
Terre Haute, Indiana   WILLIAM B. BUNGER
January 1970

CONTENTS

Chapter I

CLASSIFICATION OF THE SOLVENTS

The choice of compounds not only includes the common solvents but other compounds whose properties might render them suitable as solvents in special cases.

The solvents are arranged in the order of increasing chemical complexity under the following major classes:

Hydrocarbons

Compounds with one type of characteristic atom or group (hydroxy compounds, esters, halogenateds, etc.)

Compounds with more than one type of characteristic atom or group (ether alcohols, amino alcohols, esters of keto acids, etc.)

Subdivisions are made according to the nature of the hydrocarbon residue (saturated, aromatic, unsaturated); the number of times the characteristic atom or group occurs; heterocyclic or open chain structure; and the primary, secondary, or tertiary nature of amines.

Code numbers have been assigned to compounds for easy identification and reference. The same number appears in boldface type with each compound in every tabulation of the solvents, and each relevant use of the name throughout the book.

The names of the solvents included in this book are, in general, those used by *Chemical Abstracts*. The exceptions are mainly those compounds whose *Chemical Abstracts*' name is not well enough known to be recognized readily: for example, **336.** triethanolamine, **320.** sulfolane, and **311.** ε-caprolactam.

Asterisks designate listings in *Chemical Abstracts*. The listing name for many compounds has changed over the years. Former names are followed in parentheses by the last year of use: for example, **272.** 4-Methylvaleronitrile*, isocapronitrile* (through 1952).

WATER

0. Water

HYDROCARBONS

Saturated Aliphatic Hydrocarbons

1. Cyclopentane*, pentamethylene
2. Pentane*, *n*-pentane

   3. 2-Methylbutane*, ethyldimethylmethane
   4. 2,2-Dimethylpropane*, neopentane, tetramethylmethane
   5. Methylcyclopentane*
   6. Cyclohexane*, hexahydrobenzene, hexamethylene
   7. Hexane*, n-hexane
   8. 2-Methylpentane*, dimethylpropylmethane
   9. 3-Methylpentane*, diethylmethylmethane
  10. 2,2-Dimethylbutane*, ethyltrimethylmethane, neohexane
  11. 2,3-Dimethylbutane*, diisopropyl, isopropyldimethylmethane
  12. Methylcyclohexane*, hexahydrotoluene, cyclohexylmethane
  13. Heptane*, n-heptane
  14. 2-Methylhexane*, ethylisobutylmethane
  15. 3-Methylhexane*, ethylmethylpropylmethane
  16. 2,3-Dimethylpentane*, ethylisopropylmethylmethane
  17. 2,4-Dimethylpentane*, diisopropylmethane
  18. Ethylcyclohexane*
  19. Octane*, n-octane
  20. 2,2,3-Trimethylpentane*, tert-butylethylmethylmethane
  21. 2,2,4-Trimethylpentane*, isooctane, isobutyltrimethylmethane
  22. Nonane*, n-nonane
  23. 2,2,5-Trimethylhexane*
  24. Decahydronaphthalene*,   decalin,   bicyclo[4.4.0]decane,   naph-
      thalane, naphthane
  25. cis-Decahydronaphthalene*, cis-decalin, cis-bicyclo[4.4.0]decane
  26. trans-Decahydronaphthalene*, trans-decalin, trans-bicyclo[4.4.0]-
      decane
  27. Decane*, n-decane
  28. Bicyclohexyl*, cyclohexylcyclohexane
  29. Dodecane*, duodecane, bihexyl

Aromatic Hydrocarbons

  30. Benzene*, benzol, benzole, phene
  31. Toluene*, methylbenzene, phenylmethane
  32. o-Xylene*, 1,2-dimethylbenzene, 2-xylene
  33. m-Xylene*, 1,3-dimethylbenzene, 3-xylene
  34. p-Xylene*, 1,4-dimethylbenzene, 4-xylene
  35. Ethylbenzene*, phenylethane
  36. Isopropylbenzene, cumene*, cumol, 2-phenylpropane
  37. Mesitylene*, 1,3,5-trimethylbenzene, sym-trimethylbenzene
  38. Naphthalene*, naphthene
  39. 1,2,3,4-Tetrahydronaphthalene*, naphthalene-1,2,3,4-tetrahydride,
      tetrahydronaphthalene, tetralin

40. Butylbenzene*
41. *sec*-Butylbenzene*, 2-phenylbutane
42. *tert*-Butylbenzene*, 2-methyl-2-phenylpropane, pseudobutylbenzene, trimethylphenylmethane
43. *p*-Cymene*, cymene, 4-isopropyl-1-methylbenzene, *p*-isopropyltoluene
44. Cyclohexylbenzene*, phenylcyclohexane* (through 1950)

Unsaturated Hydrocarbons

45. 1-Pentene*, amylene, α-*n*-amylene, propylethylene
46. 2-Pentene* (mixed isomers), β-*n*-amylene, *sym*-ethylmethylethylene
47. *cis*-2-Pentene*
48. *trans*-2-Pentene*
49. 1-Hexene*, α-hexene
50. 1-Heptene*, α-heptene
51. 1-Octene*, α-octene, caprylene, α-octylene
52. 1-Nonene*, α-nonene, *n*-heptylethylene, 1-nonylene
53. 1-Decene*, α-decene
54. Cyclohexene*, 1,2,3,4-tetrahydrobenzene, tetrahydrobenzene
55. Styrene*, styrol, vinylbenzene, cinnamene, ethenylbenzene, phenylethylene
56. 2-Pinene*, pinene* (through 1956), α-pinene, 2,6,6-trimethylbicyclo[3.1.1]-2-heptene
57. 2(10)-Pinene*, nopinene* (through 1956), β-pinene, 6,6-dimethyl-2-methylenenorpinene

COMPOUNDS WITH ONE TYPE OF CHARACTERISTIC ATOM OR GROUP

HYDROXY COMPOUNDS

MONOHYDRIC ALCOHOLS

Aliphatic Alcohols

58. Methanol*, methyl alcohol, wood alcohol, carbinol
59. Ethanol, ethyl alcohol*, methylcarbinol, alcohol
60. 1-Propanol, propyl alcohol*, *n*-propyl alcohol, ethylcarbinol
61. 2-Propanol, isopropyl alcohol*, dimethylcarbinol, *sec*-propyl alcohol, isopropanol

Butyls

62. 1-Butanol, butyl alcohol*, *n*-butyl alcohol, butanol, butan-1-ol, *n*-propylcarbinol
63. 2-Butanol, *sec*-butyl alcohol*, methylethylcarbinol
64. 2-Methyl-1-propanol, isobutyl alcohol*, isobutanol, isopropylcarbinol
65. 2-Methyl-2-propanol, *tert*-butyl alcohol*, trimethylcarbinol

Pentyls

66. 1-Pentanol, pentyl alcohol*, amyl alcohol* (through 1961), pri-*n*-amyl alcohol, *n*-amyl alcohol, *n*-butylcarbinol
67. 2-Pentanol*, *sec*-amyl alcohol, *sec-n*-amyl alcohol, 1-methyl-1-butanol, methylpropylcarbinol
68. 3-Pentanol*, diethylcarbinol, 1-ethyl-1-propanol
69. 2-Methyl-1-butanol*, *act*-amyl alcohol, *d-pri-act*-amyl alcohol, *d-act*-butylcarbinol
70. 3-Methyl-1-butanol, isopentyl alcohol*, isoamyl alcohol* (through 1951), *pri*-isoamyl alcohol, isobutylcarbinol
71. 2-Methyl-2-butanol* (1952 through 1956), *tert*-pentyl alcohol*, *tert*-amyl alcohol* (through 1951), dimethylethylcarbinol
72. 3-Methyl-2-butanol*, *sec*-isoamyl alcohol, methylisopropylcarbinol
73. 2,2-Dimethyl-1-propanol*, neopentyl alcohol, neopentanol, *tert*-butylcarbinol

Other Aliphatic Alcohols

74. Cyclohexanol*, cyclohexyl alcohol, hexahydrophenol, Hexalin
75. 1-Hexanol, hexyl alcohol*, *n*-hexyl alcohol, amylcarbinol, pentylcarbinol
76. 2-Methyl-1-pentanol*,
77. 4-Methyl-2-pentanol*, methylamyl alcohol, methylisobutylcarbinol
78. 2-Ethyl-1-butanol*, *sec*-hexyl alcohol, 2-ethylbutyl alcohol, *pseudo*-hexyl alcohol

Methylcyclohexanols

79. 1-Methylcyclohexanol*
80. 2-Methylcyclohexanol* (mixed isomers), *o*-methylcyclohexyl alcohol
81. *cis*-2-Methylcyclohexanol*
82. *trans*-2-Methylcyclohexanol*
83. 3-Methylcyclohexanol* (mixed isomers), *m*-methylcyclohexyl alcohol

84. *cis*-3-Methylcyclohexanol*
85. *trans*-3-Methylcyclohexanol*
86. 4-Methylcyclohexanol* (mixed isomers), *p*-methylcyclohexyl alcohol
87. *cis*-4-Methylcyclohexanol*
88. *trans*-4-Methylcyclohexanol*

Other Aliphatic Alcohols

89. 2-Heptanol*, *n-sec*-heptyl alcohol, *n*-heptan-2-ol, *n*-heptanol-2, methylamylcarbinol
90. 1-Octanol, octyl alcohol*, caprylic alcohol, *n*-octyl alcohol, heptylcarbinol
91. 2-Ethyl-1-hexanol*, 2-ethylhexyl alcohol, 1-ethyl-*n*-amylcarbinol

Aromatic Alcohols

92. Benzyl alcohol*, phenylmethyl alcohol, α-hydroxytoluene, phenylcarbinol

Phenols

93. Phenol*, carbolic acid, hydroxybenzene, phenic acid

Cresols

94. *o*-Cresol*, *o*-cresylic acid, *o*-hydroxytoluene, *o*-methylphenol, 2-methylphenol, 1-hydroxy-2-methylbenzene
95. *m*-Cresol*, *m*-cresylic acid, *m*-hydroxytoluene, *m*-methylphenol, 3-methylphenol, 1-hydroxy-3-methylbenzene
96. *p*-Cresol*, *p*-cresylic acid, *p*-hydroxytoluene, *p*-methylphenol, 4-methylphenol, 1-hydroxy-4-methylbenzene

Unsaturated Alcohols

97. 2-Propen-1-ol, allyl alcohol*, propene(1)-ol(3), Δ³-1-propenol, vinylcarbinol
98. *cis*-2-Buten-1-ol*, *cis*-crotonyl alcohol* (1917 through 1926), Δ²-1-butenol* (through 1916), crotyl alcohol, γ-methylallyl alcohol, propenylcarbinol
99. *trans*-2-Buten-1-ol*
100. 2-Propyn-1-ol*, 2-propin-1-ol* (through 1936), propargylic alcohol* (through 1934), propargyl alcohol, acetylenecarbinol

POLYHYDRIC ALCOHOLS

101. *cis*-2-Butene-1,4-diol*
102. *trans*-2-Butene-1,4-diol*
103. 1,2-Ethanediol, ethyleneglycol*, glycol, ethylene alcohol, 1,2-dihydroxyethane
104. 1,2-Propanediol*, propylene glycol, α-propylene glycol, 1,2-dihydroxypropane
105. 1,3-Propanediol*, trimethylene glycol, isopropylene glycol
106. 1,3-Butanediol*, 1,3-butylene glycol
107. Glycerol*, 1,2,3-propanetriol, glycerin, 1,2,3-trihydroxypropane

ETHERS

Aliphatic, Open Chain Ethers

108. Ethylvinyl ether*, vinylethyl ether
109. Ethyl ether*, diethyl ether, ethoxyethane, ethyl oxide, sulfuric ether, 3-oxapentane, ether
110. Propyl ether*, di-*n*-propyl ether, 1-propoxypropane
111. Isopropyl ether*, diisopropyl ether, 2-isopropoxypropane
112. Butylvinyl ether*, vinylbutyl ether
113. Butylethyl ether*, *n*-butylethyl ether, 1-ethoxybutane
114. Butyl ether*, dibutyl ether, di-*n*-butyl ether, 1-butoxybutane
115. Pentyl ether*, amyl ether* (through 1951), *n*-amyl ether, di-*n*-amyl ether, 1-pentoxypentane
116. Isopentyl ether*, isoamyl ether* (through 1951), diisoamyl ether
117. 1,2-Dimethoxyethane*, *sym*-dimethoxyethane* (through 1941), monoglyme, dimethyl Cellosolve, ethylene glycol dimethyl ether
118. Bis(2-methoxyethyl) ether* (after 1940), diglyme, diethylene glycol dimethyl ether

Aliphatic Cyclic Ethers

119. Propylene oxide*, 1,2-epoxypropane, methyloxirane, propene oxide, 2-methyloxacyclopropane
120. 1,2-Epoxybutane*, ethylethylene oxide* (through 1931), ethyloxirane
121. Cineole*, 1,8-epoxy-*p*-menthane* (through 1961), cajeputole, eucalyptole, 1,3,3-trimethyl-2-oxabicyclo(2.2.2)octane
122. Furan*, furfuran, divinylene oxide, oxole, tetrole
123. Tetrahydrofuran*, oxacyclopentane, diethylene oxide, tetramethylene oxide

124. *p*-Dioxane*, diethylene dioxide, 1,4-dioxane
125. Tetrahydropyran*, pentamethylene oxide, oxacyclohexane

Aromatic Ethers

126. Benzylethyl ether*, α-ethoxytoluene
127. Anisole*, methoxybenzene, methylphenyl ether
128. Phenetole*, ethoxybenzene, ethylphenyl ether
129. Benzyl ether*, dibenzyl ether
130. Phenyl ether*, diphenyl ether, diphenyl oxide, phenoxybenzene
131. Veratrole* (through 1961), *o*-dimethoxybenzene*, 1,2-dimethoxy-benzene

ACETALS

132. Dimethoxymethane*, methylal* (through 1961)
133. Acetal* (through 1961), acetaldehyde diethyl acetal*, 1,1-diethoxy-ethane, ethylidine diethyl ether

CARBONYLS

ALDEHYDES

Saturated Aliphatic Aldehydes

134. Acetaldehyde*, ethanal, acetic aldehyde
135. Propionaldehyde*, propanal, methylacetaldehyde
136. Butyraldehyde*, butanal, butaldehyde, butyral, butyric aldehyde
137. Isobutyraldehyde*, 2-methylpropanal, isobutylaldehyde

Aromatic Aldehydes

138. Benzaldehyde*, benzenecarbonal, oil of bitter almond

Unsaturated Aldehydes

139. Acrolein*, 2-propenal, acryl aldehyde, acrylic aldehyde, allyl aldehyde, ethylene aldehyde
140. Crotonaldehyde*, 2-butenal, *trans*-2-butenal, crotonic aldehyde, β-methylacrolein, propenaldehyde, propylene aldehyde

KETONES

Aliphatic Ketones

141. Acetone*, 2-propanone, dimethyl ketone, methyl ketone
142. 2-Butanone*, MEK, ethylmethyl ketone, methylethyl ketone

143. 3-Pentanone*, diethyl ketone, ethyl ketone, *sym*-dimethylacetone
144. Cyclohexanone*, cyclohexyl ketone, hexanon
145. 4-Methyl-2-pentanone*, MIBK, methylisobutyl ketone, isobutyl-methyl ketone
146. Camphor, 2-camphanone, 2-bornanone, d-2-keto-1,7,7-trimethyl-norcamphane

Aromatic Ketones

147. Acetophenone*, methylphenyl ketone, acetylbenzene, hypnone

ACIDS

Saturated Acids

148. Formic acid*, methanoic acid, aminic acid
149. Acetic acid*, ethanoic acid, ethylic acid, methylformic acid
150. Propionic acid*, propanoic acid, methylacetic acid
151. Butyric acid*, butanoic acid, *n*-butyric acid, ethylacetic acid
152. Isobutyric acid*, 2-methylpropionic acid, 2-methylpropanoic acid, dimethylacetic acid, α-methylpropionic acid, isopropylformic acid
153. Valeric acid*, pentanoic acid, *n*-valeric acid
154. Isovaleric acid*, 3-methylbutanoic acid, isopropylacetic acid, 3-methylbutyric acid, β-methylbutyric acid
155. Hexanoic acid*, caproic acid* (through 1950), capronic acid, *n*-hexoic acid, *n*-caproic acid, butylacetic acid
156. Octanoic acid*, caprylic acid* (through 1951)

Unsaturated Acids

157. Acrylic acid*, propenoic acid, acroleic acid, ethylenecarboxylic acid
158. Crotonic acid*, β-methylacrylic acid, *trans*-butenoic acid, α-crotonic acid
159. Methacrylic acid*, 2-methylpropenoic acid, α-methylacrylic acid
160. Oleic acid*, *cis*-octadecen(9)oic acid, *cis*-9-octadecenoic acid

ACID ANHYDRIDES

161. Acetic anhydride*, ethanoic anhydride
162. Propionic anhydride*, propanoic anhydride
163. Butyric anhydride*, butanoic anhydride

ESTERS

Esters of Saturated Aliphatic Monocarboxylic Acids

164. Methyl formate*, methyl methanoate
165. Ethyl formate*, ethyl methanoate
166. Propyl formate*, propyl methanoate
167. Butyl formate*, butyl methanoate
168. Isobutyl formate*, isobutyl methanoate, 2-methyl-1-propyl methanoate
169. Methyl acetate*, methyl ethanoate
170. Vinyl acetate*, ethenyl acetate, ethenyl ethanoate
171. Ethylene glycol diacetate*, glycol diacetate, ethylene acetate, ethylene diacetate
172. Ethyl acetate*, ethyl ethanoate, acetic ester, acetic ether
173. Propargyl acetate*, 2-propyn acetate, propiolic acetate
174. Allyl acetate*, 2-propenyl acetate
175. Propyl acetate*, propyl ethanoate, *n*-propyl acetate
176. Isopropyl acetate*, isopropyl ethanoate, 2-propyl ethanoate
177. Butyl acetate*, butyl ethanoate, butyl acetic ether
178. Isobutyl acetate*, isobutyl ethanoate, 2-methyl-1-propyl ethanoate, β-methylpropyl acetate
179. *sec*-Butyl acetate*, α-methylpropyl acetate, 2-butyl acetate
180. Pentyl acetate*, amyl acetate* (through 1960), pentyl ethanoate, amyl acetic ether, banana oil
181. Isopentyl acetate*, isoamyl acetate* (through 1951), 3-methylbutyl ethanoate, γ-methylbutyl ethanoate, 3-methyl-1-butyl acetate, amyl acetic ether
182. 2-Ethylhexyl acetate*, 2-ethylhexyl ethanoate
183. Benzyl acetate*, benzyl ethanoate
184. Ethyl propionate*, ethyl propanoate
185. Ethyl butyrate*, ethyl butanoate, butyric ether
186. Isobutyl isobutyrate*, 2-methyl-1-propyl 2-methylpropanoate
187. Ethyl isovalerate*, ethyl 3-methylbutanoate, ethyl isopropyl-acetate
188. Isopentyl isovalerate*, isoamyl isovalerate* (through 1951), 3-methylbutyl-3-methylbutanoate, 3-methyl-1-butyl isovalerate
189. Butyl stearate*, butyl octadecanoate

Esters of Unsaturated Aliphatic Monocarboxylic Acids

190. Methyl acrylate*, methyl propenoate, methyl 2-propenoate
191. Ethyl acrylate*, ethyl propenoate

192. Methyl methacrylate*, methyl methyl-2-propenoate, methyl-methylacrylate, methacrylate monomer
193. Methyl oleate*, methyl *cis*-9-octadecenoate

Esters of Aromatic Monocarboxylic Acids

194. Methyl benzoate*, methyl benzenecarboxylate, Niobe oil
195. Ethyl benzoate*, ethyl benzenecarboxylate
196. Propyl benzoate*, propyl benzenecarboxylate
197. Benzyl benzoate*, benzyl benzenecarboxylate, benzyl phenyl-formate, benylate, benzenoate
198. Ethyl cinnamate*, ethyl *trans*-3-phenylpropenoate, ethyl *trans*-3-benzenepropenoate

Lactones

199. γ-Butyrolactone, 4-hydroxybutyric acid γ-lactone*, 4-hydroxy-butanoic acid lactone, butyrolactone

Esters of Dicarboxylic Acids

200. Ethylene carbonate, cyclic ethylene carbonate*, vinylene carbonate, 1,3-dioxolane-2-one, glycol carbonate
201. Ethyl carbonate*, diethyl carbonate
202. Ethyl oxalate*, ethyl ethanedioate, diethyl oxalate, oxalic ester
203. Ethyl malonate*, ethyl propanedioate, diethyl malonate, malonic ester
204. Methyl maleate*, methyl *cis*-butenedioate, dimethyl maleate, methyl *cis*-1,2-ethylenedicarboxylate
205. Ethyl maleate*, ethyl *cis*-butenedioate, diethyl maleate, ethyl *cis*-1,2-ethylenedicarboxylate
206. Butyl maleate*, butyl *cis*-butenedioate, dibutyl maleate, butyl *cis*-1,2-ethylenedicarboxylate
207. Butyl phthalate*, dibutyl phthalate, dibutyl 1,2-benzenedicarbox-ylate
208. Bis(2-ethylhexyl) phthalate*, 2-ethylhexyl phthalate, di-2-ethyl-hexyl phthalate, DOP
209. Butyl sebacate*, butyl decanedioate, dibutyl sebacate

Esters of Polybasic Acids

210. Butyl borate*, tributyl borate
211. Butyl phosphate*, tributyl phosphate

HALOGENATED HYDROCARBONS

FLUORINATED HYDROCARBONS

Aromatic Fluorinated Hydrocarbons

212. Fluorobenzene*, phenyl fluoride
213. o-Fluorotoluene*, 2-fluoro-1-methylbenzene
214. m-Fluorotoluene*, 3-fluoro-1-methylbenzene
215. p-Fluorotoluene*, 4-fluoro-1-methylbenzene
216. Hexafluorobenzene*

Mixed Halogenated Compounds

217. 1,1,2-Trichlorotrifluoroethane*, Freon-113, Genetron-113, F-113
218. 1,1,2,2-Tetrachlorodifluoroethane*, Freon-112, F-112
219. 1,2-Dibromotetrafluoroethane*, Freon-114B2, F-114B2

MONOCHLORINATED HYDROCARBONS

Aliphatic Chlorinated Hydrocarbons

220. Chloroethane*, ethyl chloride
221. 1-Chloropropane*, n-propyl chloride
222. 2-Chloropropane*, isopropyl chloride
223. 1-Chlorobutane*, n-butyl chloride
224. 2-Chlorobutane*, sec-butyl chloride, methylethylchloromethane
225. 1-Chloro-2-methylpropane*, isobutyl chloride, isopropylchloro-methane
226. 2-Chloro-2-methylpropane*, tert-butyl chloride, trimethylchloro-methane
227. 1-Chloropentane*, n-amyl chloride, pentyl chloride

Aromatic Chlorinated Hydrocarbons

228. Chlorobenzene*, phenyl chloride, benzene chloride, monochloro-benzene
229. 1-Chloronaphthalene*, α-chloronaphthalene

POLYCHLORINATED HYDROCARBONS

Aliphatic Chlorinated Hydrocarbons

230. Dichloromethane*, methylene chloride, methylene dichloride
231. Chloroform*, methylene trichloride, trichloromethane

232. Carbon tetrachloride*, tetrachloromethane, perchloromethane
233. 1,1-Dichloroethane*, *asym*-dichloroethane, ethylidene chloride, ethylidene dichloride
234. 1,2-Dichloroethane*, *sym*-dichloroethane, ethylene chloride, ethylene dichloride, EDC
235. 1,1,1-Trichloroethane*, methylchloroform, trichloroethane, Chlorothene
236. 1,1,2,2-Tetrachloroethane*, acetylene tetrachloride, tetrachloroethane
237. Pentachloroethane*, 1,1,1,2,2-pentachloroethane, pentalin

Aromatic Chlorinated Hydrocarbons

238. *o*-Dichlorobenzene*, 1,2-dichlorobenzene, Dowtherm E
239. *m*-Dichlorobenzene*, 1,3-dichlorobenzene
240. *p*-Dichlorobenzene*, 1,4-dichlorobenzene, Dichloricide

Unsaturated Chlorinated Hydrocarbons

241. 3-Chloropropene*, allyl chloride, chlorallylene, 3-chloroprene, 3-chloropropylene
242. 1,1-Dichloroethylene*, vinylidene chloride
243. *cis*-1,2-Dichloroethylene*, *cis*-acetylene dichloride
244. *trans*-1,2-Dichloroethylene*, *trans*-acetylene dichloride
245. Trichloroethylene*, ethinyl trichloride
246. Tetrachloroethylene*, perchloroethylene, ethylene tetrachloride, Perclene

BROMINATED HYDROCARBONS

Aliphatic Monobrominated Hydrocarbons

247. Bromoethane*, ethyl bromide, monobromoethane
248. 1-Bromopropane, *n*-propyl bromide
249. 2-Bromopropane, isopropyl bromide

Aromatic Monobrominated Hydrocarbons

250. Bromobenzene*, phenyl bromide, monobromobenzene
251. 1-Bromonaphthalene*, α-bromonaphthalene

Polybrominated Hydrocarbons

252. Bromoform* (through 1961), tribromomethane*
253. 1,2-Dibromoethane*, *sym*-dibromoethane, ethylene bromide, ethylene dibromide

254. 1,1,2,2-Tetrabromoethane*, acetylene tetrabromide, tetrabromo-ethane, *sym*-tetrabromoethane

IODINATED HYDROCARBONS

Aliphatic Iodinated Hydrocarbons

255. Iodomethane*, methyl iodide
256. Iodoethane*, ethyl iodide
257. 1-Iodopropane*, *n*-propyl iodide
258. 2-Iodopropane*, isopropyl iodide
259. Diiodomethane*, methylene iodide

NITROGEN COMPOUNDS

NITRO COMPOUNDS

260. Nitromethane*, NM, nitrocarbol
261. Nitroethane*, NE
262. 1-Nitropropane*, 1-NP
263. 2-Nitropropane*, 2-NP
264. Nitrobenzene*, nitrobenzol, Oil of Mirbane

NITRILES

Aliphatic Nitriles

265. Acetonitrile*, ethanenitrile, methyl cyanide, cyanomethane, ethyl nitrile
266. Propionitrile*, propanenitrile, ethyl cyanide
267. Succinonitrile*, 1,2-dicyanoethane, ethylene cyanide, ethylene dicyanide
268. Butyronitrile*, butanenitrile, 1-cyanopropane, *n*-propyl cyanide
269. Isobutyronitrile*, 2-cyanopropane, isopropyl cyanide, 2-methyl-propionitrile
270. Valeronitrile*, pentanenitrile, *n*-butyl cyanide
271. Hexanenitrile*, capronitrile* (through 1951), caproic nitrile, *n*-amyl cyanide, pentyl cyanide
272. 4-Methylvaleronitrile*, isocapronitrile* (through 1952), isoamyl cyanide, 4-methylpentanenitrile
273. Octanenitrile*, caprylonitrile* (through 1950), *n*-heptyl cyanide, octyl nitrile
274. α-Tolunitrile*, benzyl cyanide, phenylacetonitrile

Aromatic Nitriles

275. Benzonitrile*, phenyl cyanide, cyanobenzene

Unsaturated Nitriles

276. Acrylonitrile*, vinyl cyanide, cyanoethylene, propenenitrile
277. Methacrylonitrile*, 2-cyanopropene, α-methylacrylonitrile

AMINES
Primary Monoamines

Primary Aliphatic Amines

278. Propylamine*, n-propylamine, monopropylamine, 1-aminopropane
279. Isopropylamine*, 2-propylamine, 2-aminopropane, monoisopropyl-amine
280. Butylamine*, n-butylamine, 1-aminobutane, monobutylamine
281. Isobutylamine*, 1-amino-2-methylpropane, monoisobutylamine
282. sec-Butylamine*, 2-aminobutane, α-methylpropylamine
283. tert-Butylamine*, 2-amino-2-methylpropane, 2-methylisopropyl-amine, 2-aminoisobutane
284. Cyclohexylamine*, aminocyclohexane, hexahydroaniline

Primary Aromatic Amines

285. Aniline*, aminobenzene, phenylamine
286. o-Toluidine*, o-aminotoluene, o-methylaniline, 2-toluidine, 2-aminomethylbenzene
287. m-Toluidine*, m-aminotoluene, m-methylaniline, 3-toluidine, 3-aminomethylbenzene
288. p-Toluidine*, p-aminotoluene, p-methylaniline, 4-toluidine, 4-aminomethylbenzene, tolylamine

Primary Unsaturated Amines

289. Allylamine*, 2-propenylamine, 3-aminopropylene

Primary Polyamines

290. Ethylenediamine*, 1,2-ethanediamine, 1,2-diaminoethane

Secondary Monoamines

291. Ethylenimine*, aziridine, dimethylenimine, vinylimine
292. Diethylamine*
293. Dipropylamine*, di-n-propylamine

294. Diisopropylamine*, 2,4-dimethyl-3-azapentane
295. Dibutylamine*, di-*n*-butylamine
296. Pyrrole*, azole, 1H-pyrrole
297. Piperidine*, hexahydropyridine, pentamethylenimine

Tertiary Monoamines

298. Triethylamine*
299. Pyridine*, azine
300. Quinoline*

AMIDES

301. Formamide*, methanamide
302. N-Methylformamide*
303. N,N-Dimethylformamide*, dimethylformamide, DMF
304. Acetamide*
305. N-Methylacetamide*, monomethylacetamide
306. N,N-Dimethylacetamide*, DMAC
307. N-Methylpropionamide*, monomethylpropionamide
308. 1,1,3,3-Tetramethylurea*, tetramethylurea, temur
309. 2-Pyrrolidinone*, butyrolactam, pyrrolidone
310. 1-Methyl-2-pyrrolidinone*, N-methylpyrrolidone, methylpyrrolidone
311. ε-Caprolactam, caprolactam, hexahydro-2H-azepin-2-one*, 2-oxahexamethylenimine* (through 1956), 2-ketohexamethylenimine

SULFUR COMPOUNDS

Sulfides

312. Carbon disulfide*, carbon bisulfide

Thiols

313. 1-Butanethiol*, *n*-butyl mercaptan, thiobutyl alcohol
314. Benzenethiol*, thiophenol, phenyl mercaptan

Thioethers

315. Methyl sulfide*, dimethyl sulfide, 2-thiapropane
316. Ethyl sulfide*, diethyl sulfide, ethyl thioethane, 3-thiapentane
317. Thiophene*, thiofuran, thiofurfuran, thiole, thiotetrole
318. Tetrahydrothiophene*, thiacyclopentane, diethylene sulfide, thiophane

Oxo-Sulfur Compounds

319. Dimethyl sulfoxide, methyl sulfoxide*, DMSO, methylsulfinyl-methane
320. Sulfolane, tetrahydrothiophene 1,1-dioxide*, 2,3,4,5-tetrahydro-thiophene 1,1-dioxide

COMPOUNDS WITH MORE THAN ONE TYPE OF CHARACTERISTIC ATOM OR GROUP

Ether Alcohols

321. 2-Methoxyethanol*, methyl Cellosolve, methyl glycol, ethylene glycol monomethyl ether
322. 2-Ethoxyethanol*, Cellosolve, hydroxy ether, glycol ethyl ether, glycol monoethyl ether, ethylene glycol monoethyl ether
323. 2-Butoxyethanol*, butyl Cellosolve, butyl glycol, ethylene glycol monobutyl ether
324. Furfuryl alcohol*, furfurol, 2-furancarbinol, 2-furanmethanol, 2-hydroxymethylfuran
325. Tetrahydrofurfural alcohol*, tetrahydro-2-furancarbinol, tetra-hydro-2-furanmethanol
326. Diethylene glycol*, diglycol, di-2-hydroxyethyl ether, 2,2'-di-hydroxyethyl ether, 2,2'-oxydiethanol, 3-oxa-1,5-pentanediol, ethylene diglycol
327. Triethylene glycol*, triglycol, 2,2'-ethylene-dioxydiethanol, glycol bis(hydroxyethyl) ether, 3,6-dioxa-1,8-octanediol, ethylene glycol dihydroxydiethyl ether
328. 2-(2-Methoxyethoxy)ethanol*, methyl Carbitol, methoxydiethylene glycol, methyl diglycol, methoxy diglycol, diethylene glycol mono-methyl ether
329. 2-(2-Ethoxyethoxy)ethanol*, Carbitol, ethoxy diglycol, 2-(β-ethoxyethoxy)ethanol, diethylene glycol monoethyl ether

Carbonyl Alcohols

330. Salicylaldehyde*, o-hydroxybenzaldehyde, salicylic aldehyde
331. 4-Hydroxy-4-methyl-2-pentanone*, diacetone alcohol, diacetone, 4-hydroxy-2-keto-4-methylpentane

Chloro Alcohols

332. 2-Chloroethanol*, ethylene chlorohydrin, 2-chloroethyl alcohol, β-chloroethyl alcohol

Cyano Alcohols

333. 2-Cyanoethanol, hydracrylonitrile*, ethylene cyanohydrin, 2-cyanoethyl alcohol, β-cyanoethyl alcohol, 3-hydroxypropionitrile, glycol cyanohydrin

Amino Alcohols

334. 2-Aminoethanol*, monoethanolamine, ethanolamine, 2-hydroxyethylamine, β-hydroxyethylamine, β-aminoethyl alcohol, ethylolamine

335. Diethanolamine, 2,2'-iminodiethanol*, bis(hydroxyethyl)amine, 2,2'-dihydroxydiethylamine, diethylolamine, 3-aza-1,5-pentanediol

336. Triethanolamine, 2,2',2"-nitrilotriethanol*, trihydroxytriethylamine, tris(hydroxyethyl)amine, triethylolamine, N-(2-hydroxyethyl)-3-aza-1,5-pentanediol

Thioether Alcohols

337. 2,2'-Thiodiethanol*, bis(2-hydroxyethyl)sulfide, thiodiglycol, thiodiethylene glycol

Aldehyde Ethers

338. 2-Furaldehyde*, furfural, furfuraldehyde, fural, furole, 2-furancarbinal

Chloro Ethers

339. Bis(2-chloroethyl)ether*, 2,2'-dichlorodiethyl ether, β,β'-dichlorodiethyl ether, dichlorodiethyl ether, *sym*-dichlorodiethyl ether

340. Epichlorohydrin* (through 1961), 1-chloro-2,3-epoxypropane*, chloromethyloxirane, α-epichlorohydrin, γ-chloropropylene oxide

Nitro Ethers

341. *o*-Nitroanisole*, 1-methoxy-2-nitrobenzene

Amino Ethers

342. Morpholine*, tetrahydro-2H-1,4-oxazine, diethylene oximide, diethyleneimide oxide, tetrahydro-*p*-oxazine

Esters of Hydroxy Acids

343. Ethyl lactate*, ethyl 2-hydroxypropionate

344. Methyl salicylate*, methyl *o*-hydroxybenzoate, oil of wintergreen

Ether Esters

345. 2-Methoxyethyl acetate*, methyl Cellosolve acetate, methoxyethyl ethanoate, methylethylene glycol acetate, glycolmonomethyl ether acetate
346. 2-Ethoxyethyl acetate*, Cellosolve acetate, ethylethylene glycol acetate, ethylene glycol monoethyl ether acetate
347. 2-(2-Ethoxyethoxy)ethyl acetate*, Carbitol acetate, diglycol monoethyl ether acetate, diethylene glycol monoethyl ether acetate

Esters of Keto Acids

348. Methyl acetoacetate*
349. Ethyl acetoacetate*, acetoacetic ester, diacetic ether, ethyl 3-oxobutanoate

Esters of Cyano Acids

350. Methyl cyanoacetate*
351. Ethyl cyanoacetate*, ethyl cyanoethanoate

Fluoro Acids

352. Trifluoroacetic acid*, perfluoroacetic acid, trifluoroethanoic acid

Chloro Amines

353. o-Chloroaniline*, 2-chlorophenylamine

Phosphoramides

354. Hexamethylphosphoric triamide*, HMPA, hexamethylphosphoramide

Chapter II

PHYSICAL PROPERTIES: DISCUSSION OF PROPERTIES AND CRITERIA FOR THEIR SELECTION

The physical properties and numerical values included in the following tables can be grouped into four main types:

1. Values useful for the identification of pure substances and as criteria of purity such as boiling point, melting point, refractive index, and so forth.

2. Values useful for the execution of physicochemical measurements and determinations such as ebullioscopic constants, and so forth.

3. Properties useful for the study of solute systems that make it possible to differentiate between the properties of the solvent and the properties of the solute or the solution such as electrical conductivity, spectra, and so forth.

4. Properties important for the handling and storing of solvents such as toxicity, flash point, and so forth.

With respect to the reliability and accuracy of a particular numerical value, the literature has been used in the following sequence:

1. Papers devoted to precision measurements of physical constants or groups of physical constants of one compound or of a group of compounds. These furnish, as a rule, the best values.

2. Papers devoted to physicochemical measurements in nonaqueous systems in which purification methods for the particular solvents are mentioned. Although papers of this kind often show great precision in the determination of the constants, they may be doubtful with respect to the degree of purity of the solvent.

3. Papers devoted to the synthesis of the solvent or employing the solvent as a medium, for example, for the study of reaction mechanisms. Although these papers are usually reliable with respect to the purity of the solvent, they must be used with care as far as the physical measurements are concerned.

In the absence of such papers, the data given in the literature were evaluated either mathematically or graphically, or simply by inspection, the original sources having been consulted wherever possible. Extensive use has been made of the *Annual Tables of Physical Constants* and the *International Critical Tables*. Data from these sources have, for the most part, been checked back to the original articles, and care has been taken to refer in every case to the original source.

Five groups of workers have been particularly responsible for many reliable physical data and for the development or adaptation of some of the techniques of measurements: Timmermans and coworkers at the Bureau International des Etalons; Swietoslawaki and coworkers of the Institute of Technology, Warsaw; groups at the Bureau of Mines laboratories in Bartles-ville and Laramie; the group at the National Chemical Laboratory in Teddington, Middlesex, England; and Rossini and coworkers at the Bureau of Standards, Carnegie Institute of Technology, and cooperating institutions in the American Petroleum Institute Project 44.

For relatively few compounds, however, have the properties been studied extensively, and often a particular property has been determined at one or a few selected temperatures only. Therefore, in addition to the definitions of the properties listed in the tables and brief references to their determination, in this chapter we emphasize those mathematical relationships that allow the calculation of certain physical properties from values of other properties and the interpolation or extrapolation of data for temperatures not given in the tables.

In some cases, a simpler, although less accurate, relationship has been discussed in preference to a more complex one. For example, the critical temperature has been related to the normal boiling point by several empirical equations. Although these equations are, in general, less accurate than those that involve the parachor as well as the boiling point, they are given in detail because the data necessary for the use of the simpler relations are much more accessible.

While the literature on the correlation and calculation of physical properties and on physical properties–temperature relationships has by no means been completely covered, enough references have been included to guide the reader to the original literature for further information.

Most physical properties are temperature-dependent; a clear concept of temperature and its measurement is needed for the optimum use of tables of physical data or the determination of physical properties. A complete treatment of temperature and its measurement has been presented by the American Institute of Physics [87]. Sturtevant and Weissberger [5110, Chapter 6] and Reilly and Rae [3942] have discussed temperature measurement and control and thermometry.

BOILING POINT

The *boiling point*, bp, is defined as that temperature at which the vapor pressure of the liquid is equal to the opposing pressure. Values listed in the tables refer to an opposing pressure of 760 torr unless otherwise stated. Most routine boiling points for the characterization of substances are measured at "atmospheric pressure"; for comparative purposes these must be corrected to 760 torr. This may be done by using the rule of Crafts [1124].

$$\Delta t = C(273 + t)(760 - p) \tag{2.1}$$

where t is the observed boiling point at p torr pressure, Δt is the correction to be applied to the observed boiling point, and C is a constant that is practically the same for all substances. For alcohols, water, and carboxylic acids, $C = 0.00010$; for very low boiling substances, nitrogen, ammonia, and so forth, $C = 0.00014$; for all other substances, C can be taken as 0.00012.

If a value for dt/dp, the *pressure coefficient of temperature*, has been experimentally determined, it may be used to correct the boiling point to 760 torr pressure.

The boiling point is one of the most useful properties for the characterization of organic compounds. Swietoslawski and Anderson [5110] have presented an excellent discussion of boiling and condensing temperatures and have described several reliable methods for their determination.

An empirical method has been devised by Kinney [2602–2604] for calculating the boiling points of aliphatic and aromatic hydrocarbons and correlating boiling points with molecular structure. The method is useful for predicting the boiling points of new compounds as well as for checking literature values that appear doubtful. Within certain limits, it may also be used to determine the structure of unknown hydrocarbons.

Kinney calculates the boiling point from the boiling point number (bpn) by the equation

$$bp = 230.14 \, (bpn)^{1/3} - 543 \tag{2.2}$$

The boiling point numbers, although empirical, reflect the atomic and structural forces that affect the boiling points of organic compounds. The normal paraffin hydrocarbons and hydrogen were taken as the basis for calculating the boiling point numbers.

Kinney has calculated the boiling point numbers and boiling points for a large number of compounds and has compared the calculated and experimental values. The average deviation for 323 aromatic hydrocarbons was 5°C. The agreement was poorest in the olefin series where the effect of *cistrans* isomerism could not be established with certainty.

VAPOR PRESSURE

The *vapor pressure*, p, of a solvent is the pressure exerted by the vapor when in equilibrium with the liquid or solid phase. The vapor pressure, being a specific property of a compound, is widely used for practical calculations in physical chemistry and chemical engineering. It is also helpful in the characterization of organic compounds.

The error in vapor pressure measurements usually is in the pressure rather than the temperature determination. A discussion of vapor pressure measurements, in which pressure measurements are treated in some detail, has been presented by Thompson [5110].

Experimental vapor pressures usually do not cover the full p-t range; to obtain full usefulness, some type of interpolation formula is necessary. Many different equations have been proposed for representing the vapor pressure–temperature relationship.

One of the most familiar and useful vapor pressure formulas can be obtained by integration of the Clausius-Clapeyron equation

$$\frac{(d \ln p)}{dT} = \frac{\Delta H_v}{RT^2} \tag{2.3}$$

The general integral of (2.3) is

$$\ln p = -\frac{\Delta H_v}{RT} + C$$

$$\log p = A - \frac{B}{T} \tag{2.4}$$

The use of (2.4) implies the following conditions: (1) that ΔH_v, the heat of vaporization, is constant over the temperature range involved; (2) that the volume of the liquid is negligible compared to the volume of vapor; and (3) that the vapor is a perfect gas. Unless the temperature range is very small, these conditions are not fulfilled and the use of (2.4) leads to approximate values.

Antoine [260, 261] first proposed the equation bearing his name for water but later applied it to other liquids. He substituted $(t + C)$ for T in (2.4) to obtain

$$\log p = A - \frac{B}{(t + C)} \tag{2.5}$$

Gutmann and Simmons [1941] have derived the Antoine equation in the form

$$\log p = \frac{(A - B)}{[T - (3\theta/8)]} \tag{2.6}$$

from the theoretical van der Waals and Debye equations. The symbol θ is a characteristic temperature for each compound, and C in (2.5) is equal to $273.15 - (3\theta/8)$. An accurate method for computing θ has been discussed by Dreisbach [1368].

Thomson [4746] has given a comprehensive discussion of the Antoine equation and its use and applicability. He presents several methods for calculating C but states that, for the practical interpretation of data, C may be taken as equal to 230.

The Antoine equation has been found to represent experimental data over a wider range of pressures than the Clausius-Clapeyron equation and is probably the most reliable three-constant equation in the region between the triple point temperature and a reduced temperature of about 0.85. The Antoine equation fails as the temperature approaches the critical temperature. A four-constant equation has been derived by Frost and Kalkwarf [1671] by integrating the Clapeyron equation, $dp/dT = \Delta H_v/T\,\Delta V$, after first expressing ΔH_v as a linear function of T and applying the van der Waals' equation to the solution of the volume of the vapor. The resulting vapor pressure equation (2.7) has been applied by Bond and Thodos [643] and by Perry and Thodos [3721] to data for more than 50 aliphatic and aromatic hydrocarbons. It was found that

$$\log p = A + \frac{B}{T} + C \log T + \frac{Dp}{T^2} \qquad (2.7)$$

accurately described the vapor pressure behavior over the entire liquid range from triple point to critical temperature.

Polynomials in t (or p) have been employed [4382] to represent accurate vapor pressure measurements. These equations usually are of the form

$$p = 760 + A(t - bp) + B(t - bp)^2 + C(t - bp)^3 \qquad (2.8)$$

where bp is the normal boiling point and A, B, and C are constants, and

$$t = bp + A(p - 760) + B(p - 760)^2 + C(p - 760)^3 \qquad (2.9)$$

Equations (2.8) and (2.9) give a very close fit over a limited range of pressures. For example, the average deviation in t for 1,2-dichloroethane [4382] is only $0.001°$ over a pressure range of 660 to 880 torr and the greatest deviation is $0.002°C$. The average deviation in p is 0.01 torr, and the greatest deviation is 0.04 torr from the experimental values.

Many empirical and semitheoretical vapor pressure equations have been proposed. They usually attempt to correct for the bending in the curve that always results when data fitted to (2.4) are plotted. The more common of these are

$$\log p = A - \frac{B}{T} + CT \tag{2.10}$$

$$\log p = A - \frac{B}{T} + \frac{C}{T^2} + \frac{D}{T^3} \tag{2.11}$$

$$\log p = A - \frac{B}{T} - CT + DT^2 \tag{2.12}$$

$$\log p = A - \frac{B}{T} - C \log T \tag{2.13}$$

$$\ln p = A - \frac{B}{T} + C \ln T \tag{2.14}$$

$$\log p = A - \frac{B}{T} + 1.75 \log T + CT \tag{2.15}$$

$$\log p = A - \frac{B}{(T - C)} \tag{2.16}$$

The reader is referred to Thomson [4746, 5110] for well-integrated discussions of vapor pressure–temperature relationships.

In addition to the standard reference works, many articles summarizing and evaluating vapor pressure data have been published. Among these are those of Stull [4535], Dreisbach and Shrader [1373] and Dreisbach and Martin [1372]. The most extensive ones are those of Stull and of Jordan [2444]. The vapor pressure values for many of the compounds in Chapter III have been taken from Stull, who lists over 1200 compounds. Enough values have been listed to enable the reader to calculate a vapor pressure–temperature relationship. Thomson [5110] advises that the critical investigator smooth these data using the Antoine equation with $C = 230$.

In order to give vapor pressure data of maximum usefulness, values for the constants of (2.4) or (2.5) have been included in the tables. When insufficient data were available to permit evaluation of the constants, values of the vapor pressures at several temperatures have been included. It was desirable to give the vapor pressure at 25°C for each solvent. If this value was not available, the vapor pressure nearest 25° was listed.

DENSITY

The *density*, *d*, of a substance is defined as the mass per unit of volume, and is expressed in grams per cubic centimeter in the CGS system. It is one of the

most useful of the physical properties and is widely employed in the characterization of material, both single- and multi-component; it is also a factor
in the calculation of many other properties. Bauer [5110] discusses the determination and uses of density.

The specific gravity of a liquid is the ratio of the weight of any volume of
the liquid to the weight of an equal volume of some standard, usually water.
A dimensionless quantity, it is denoted by the symbol $d_{t_2}^{t_1}$, where t_1 is the
temperature of the liquid and t_2 is that of the water. Water attains its maximum
density of 0.999973 g/cc or 1.00000 g/ml at a temperature of 3.98°C, so that
for nearly all practical purposes the specific gravity of a liquid referred to
water at this temperature may be considered numerically equal to its density;
for this reason, the density is commonly designated by $d_4{}^t$.

Specific gravities are often expressed with reference to the standard at some
temperature other than 3.98°. For example, the specific gravity of ethanol
may be given as $d_{25°}^{25°} = 0.787$, which is the ratio of the weight of some volume
of ethanol at 25° to that of an equal volume of water at 25°. The density of
the alcohol may be obtained by multiplying its specific gravity by the density
of water at 25°:

$$d_{4(\text{alcohol})}^{25} = d_{25(\text{alcohol})}^{25} \times d_{4(\text{water})}^{25} = 0.787 \times 0.997 \, \frac{\text{g}}{\text{ml}} = 0.785 \, \frac{\text{g}}{\text{ml}} \quad (2.17)$$

Conversion of density to specific gravity is accomplished by rearrangement
of (2.17):

$$d_{t_2(\text{alcohol})}^{t_1} = \frac{d_{4(\text{alcohol})}^{t_1}}{d_{4(\text{water})}^{t_2}} \quad (2.18)$$

Often the value of the density or specific gravity of a substance is not
available at the desired temperature. If densities at two temperatures are
known, the *coefficient of density, dd/dt*, may be used to calculate the density
at a particular temperature.

$$\frac{dd}{dt} = \frac{(d_4{}^{t_1} - d_4{}^{t_2})}{(t_2 - t_1)} \quad (2.19)$$

where $t_1 < t_2$. The coefficient varies widely for various liquids and, to some
extent, with temperature for the same liquid. If great accuracy is not desired,
dd/dt can be assumed to be constant from about 0 to 40°, unless there is a
phase change within this range.

The *coefficient of thermal expansion*, α, sometimes called the coefficient of
cubical expansion, is defined as $\alpha = (1/v)(\partial v/\partial t)_p$. For temperatures reasonably
close together, α is expressed as

$$\alpha = \frac{1}{v} \frac{\Delta v}{\Delta t} \quad (2.20)$$

The coefficient, α, may be calculated from two densities, d_1 and d_2 ($d_1 > d_2$), at temperatures t_1 and t_2 ($t_2 > t_1$) as follows: Assume that the liquid has a volume, V_1, equal to unity at temperature t_1. Then the volume, V_2, at t_2 is greater than V_1, and the increase in volume is

$$\Delta v = \frac{d_1}{d_2} - 1 \tag{2.21}$$

and

$$\Delta t = t_2 - t_1 \tag{2.22}$$

Substituting (2.21) and (2.22) in (2.20) gives

$$\alpha = \frac{(d_1/d_2) - 1}{\Delta t} \tag{2.23}$$

Dreisbach [1366] has developed a method for calculating the density at any temperature between 0 and 40° from the specific gravity at 25°/25° and the coefficient of thermal expansion. Dreisbach and Spencer [1375] have also shown that densities of liquids may be calculated from the molecular weight and the boiling point at any pressure.

The application of the Eykman equation for calculating densities from refractive index data is discussed in the following section.

REFRACTIVE INDEX

The *refractive index*, n, is the ratio of the velocity of light in a particular substance to the velocity of light in vacuum. Values usually reported refer to the ratio of the velocity in air to that in the substance saturated with air. The refractive index is useful for characterizing compounds and also for calculating other physical properties. Bauer and Fajans [5110] have discussed the refractive index, including some of its uses and methods of determination. The principles and uses of the several refractometers have been treated by Reilly and Rae [3942].

The refractive index is a function of both temperature and the wavelength of the incident light. Usually the sodium D_1 and D_2 lines are used. They have a weighted mean of 5892.6 Ångström units (Å) and are symbolized by D. Ordinary changes in atmospheric pressure do not produce a change in refractive index detectable with the usual instruments.

The most reliable refractive index data have been determined in the range of 15 to 30°. Between these limits, interpolated values will usually be correct to the fourth decimal place. The calculated temperature coefficient, dn/dt, decreases about 1% per degree of increase in temperature. If dn/dt is calculated from reliable data, the refractive index may be calculated over a small temperature range with an accuracy approaching that of the determined values. Extrapolation or interpolation is not reliable if a transition point lies within the temperature range. When only a single refractive index is available, approximate values at other temperatures may be calculated, the mean value 0.00045 per degree being used for dn/dt.

The empirical equation of Eykman [1530],

$$\frac{(n^2 - 1)}{(n + 0.4)\,d}\cdot\frac{1}{d} = C \qquad (2.24)$$

offers an accurate means not only for checking the accuracy of experimental densities and refractive indices, but also for calculating one from the other. In (2.24), n is the refractive index, d the density at the same temperature, and C a constant. Eykman stated that the constant C is independent of temperature.

In spite of its empirical nature, Eykman's equation is to be preferred to the more familiar Lorenz-Lorentz expression [1797, p. 518]

$$\frac{(n^2 - 1)}{(n^2 + 2)}\cdot\frac{M}{d} = R \qquad (2.25)$$

even though the latter has a theoretical basis.

Kurtz, Amon, and Sankin [2788] have emphasized the usefulness of the Eykman equation over a wide range of temperatures. Their summary of the available data, including the original data of Eykman, shows that the latter's equation accurately represents the experimental data. Their conclusions agree with those previously reached by Drcisbach [1367], Bauer and Fajans [5110], Kurtz and Lipkin [2789], Gibson and Kincaid [1763], and Ward and Kurtz [5079].

Taking Eykman's original data for 42 hydrocarbons, Kurtz and coworkers showed that the observed change in refractive index agreed with the calculated value within an average deviation of $\pm 3 \times 10^{-6}$ per degree. The Lorenz-Lorentz equation showed an average deviation of $+38 \times 10^{-6}$ per degree change in temperature. The minimum temperature interval in the group was $31.0°$ and the maximum $122.3°$. They also stated, ". . . consideration of the currently available data for the change in refractive index and density with temperature leads one to conclude that changes in refractive index calculated from the Eykman equation from experimental data for density at two or more temperatures are likely to be more accurate than experimentally determined changes in refractive index unless the experimental work is very carefully done."

Other equations beside those of Eykman and Lorenz-Lorentz have been proposed for correlating density and refractive index. Since they are, for the most part, of historical interest only, they will not be discussed here.

Relationships have been proposed for specific classes or groups of compounds. Of particular interest are those of Ward and Kurtz [5079]. A careful consideration of existing data led these authors to conclude that the classic Newtonian specific refraction equation

$$\frac{(n^2 - 1)}{d} = r \tag{2.26}$$

accurately represents the relationship between refractive index and density for hydrocarbon isomers.

The equation

$$\Delta n = 0.60 \, \Delta d \tag{2.27}$$

was originally proposed by Ward and Kurtz [5079] to correct refractive indexes of hydrocarbons to 20° with reasonable accuracy. This equation was based on both the Eykman equation and empirical observation. In a later paper Kurtz and coworkers [2788] showed that, for hydrocarbons, this simple equation has a deviation of $\pm 8 \times 10^{-6}$ per degree change in temperature.

VISCOSITY

The *coefficient of viscosity* or *dynamic viscosity*, η, is defined as the force per unit area necessary to maintain a unit velocity gradient between two parallel planes a unit distance apart. The unit of viscosity is the poise (dyne-sec/cm²), but numerical values are usually given in smaller subsidiary units such as the millipoise (10^{-3} poise) or the centipoise, cp, (10^{-2}) poise).

The *kinematic viscosity*, ν, is defined as the ratio of the dynamic viscosity to the density of a fluid. It is used in fluid mechanics and is the only parameter characteristic of the fluid itself to appear in the equations of hydrodynamics. The kinematic viscosity is directly proportional to the time required for the liquid to flow through a capillary-tube viscometer under its own hydrostatic head. The CGS units are called stokes (cm²/sec) and may be converted to poises by means of the equation, poises = stokes × density. Viscosities in this book will be expressed in centipoises, unless otherwise noted.

The absolute viscosity of only a few substances has been measured directly. It is easier to measure the viscosity of a liquid by comparison with a reference standard of known viscosity. The many kinds of viscometers available and the variables and conditions of viscosity measurements have been described by McGoury and Mark [5110].

The primary reference liquid for viscosity measurements is water. In 1931, Bingham [590, 591] evaluated the viscosity data of many workers and concluded that the most probable value for water, at 20°, was 1.005 cp. The *International Critical Tables* (2349) give the viscosity of water, at 20°, as 1.0087 cp. Cannon and Fenske [898–900] used Bingham's evaluation in developing a practical viscometer for routine work.

Coe and Godfrey, whose work at the National Bureau of Standards on the determination of the absolute viscosity of water was interrupted during World War II, published a preliminary report in 1944 [1016]. The final value reported by Swindells, Coe, and Godfrey [4587] is 0.010019 ± 0.000003 poise at 20°. This figure will be rounded off to 0.01002 poise for viscometer calibrations, and the general recognition and use of this value are being encouraged. The values for water at other temperatures between 15 and 60° are best represented by Cragoe's equation [1016]:

$$\log \frac{\eta}{\eta_{20}} = \frac{1.2348(20 - t) - 0.001467(t - 20)^2}{t + 96} \tag{2.28}$$

The lack of consistency in the value used for the viscosity of water prior to the general use of the Swindells, Coe, and Godfrey value has made it difficult to evaluate viscosity data. Few investigators specify the value used for their calibrations. When calibration data are known, viscosities can be corrected to correspond to the accepted value for water by the following simple proportion:

$$\eta = \frac{\eta_1 \eta_2}{\eta_3} \tag{2.29}$$

where η is the corrected viscosity at a particular temperature, η_1 the viscosity of water used in calibration, η_3 the viscosity of water calculated from (2.28), and η_2 the viscosity determined using η_1. Viscosities of water at several temperatures between 0° and 100°C are listed in Table 0b, Chapter III.

Many attempts have been made to correlate viscosity and temperature. One of the earliest, proposed by Arrhenius [298], was of the exponential type:

$$\Phi = A e^{E/RT} \tag{2.30}$$

where A and E are constants for a given liquid and $\Phi = 1/\eta$. Values of A and E have been calculated for 44 hydrocarbons by Kierstead and Turkevich [2588] from viscosity data in the literature. It was found that log Φ did not vary by more than 0.027, an amount that corresponds to an error in Φ of 6.5%. The usual difference between the calculated and observed value was about 1%. Qualitative observations on the effect of structure on the values of A and E were also made.

The values of A and E for the hydrocarbons listed in Chapter III, as far as available, are given in Table 2.1 [2588].

Bingham [590] developed an equation of the form

$$T = A\Phi + C - \frac{B}{\Phi} \tag{2.31}$$

Table 2.1 Viscosity-Temperature Relationship: $\Phi = Ae^{-E/RT}$

Number	Compound	$A \times 10^{-3}$	$E \times 10^{-3}$
1	Cyclopentane	5.554	1.878
2	Pentane	5.520	1.504
3	2-Methylbutane	6.099	1.549
5	Methylcyclopentane	6.603	2.055
6	Cyclohexane	15.21	2.905
7	Hexane	5.598	1.690
8	2-Methylpentane	5.818	1.676
9	3-Methylpentane	5.194	1.638
10	2,2-Dimethylbutane	8.002	1.978
11	2,3-Dimethylbutane	8.563	2.035
12	Methylcyclohexane	2.352	1.678
13	Heptane	5.842	1.858
14	2-Methylhexane	9.122	2.070
19	Octane	6.641	2.088
22	Nonane	8.485	2.394
24	Decahydronaphthalene	7.252	3.034
27	Decane	8.936	2.565
29	Dodecane	11.51	3.010
30	Benzene	9.803	2.428
31	Toluene	5.777	2.059
32	o-Xylene	6.141	2.277
35	Ethylbenzene	5.336	2.084
38	Naphthalene	580.3	5.977
39	1,2,3,4-Tetrahydronaphthalene	9.800	3.112

where A, B, and C are constants, T is the absolute temperature, and Φ is the fluidity or reciprocal of the viscosity. Using this equation, Bingham determined the mean percentage difference between the observed and calculated values for 87 substances to be 0.17. The alcohols had the largest mean percentage difference; if they are omitted, the mean difference for the remaining 70 compounds in 0.09. Values for A, B, and C are given in Table 2.2 [590].

When the fluidity curves vary widely from the simple linear type, (2.31) may be improved by introducing a fourth constant and rewriting it as

$$T = A\Phi + C - \frac{B}{(\Phi + D)} \tag{2.32}$$

Bingham found that, for eight compounds that gave a percentage difference of 0.77 between the observed and calculated value with (2.31), the difference was only 0.07 with (2.32). For example, octane had a difference of 0.16%

Table 2.2 Viscosity-Temperature Relationship: $T = A\Phi + C - (B/\Phi)$; $\Phi = 1/\eta$

Number	Compound	A	B	C
2	Pentane	0.16544	14,937.0	256.61
3	2-Methylbutane	0.18327	10,454.0	234.17
7	Hexane	0.21892	9,137.8	254.03
8	2-Methylpentane	0.20708	9,562.8	252.85
13	Heptane	0.23203	8,681.2	272.70
19	Octane	0.27055	6,332.6	277.25
30	Benzene	0.32052	2,633.1	260.82
31	Toluene	0.32688	4,193.5	262.66
32	o-Xylene	0.37738	3,009.3	271.96
33	m-Xylene	0.34134	4,542.9	266.82
34	p-Xylene	0.32087	5,127.4	277.17
35	Ethylbenzene	0.34180	4,540.8	273.54
58	Methanol	0.24316	4,498.9	279.01
59	Ethanol	0.28395	2,398.6	298.39
60	1-Propanol	0.31496	1,211.7	308.41
61	2-Propanol	0.29810	738.16	300.17
62	1-Butanol	0.33610	877.08	311.94
64	2-Methyl-1-propanol	0.33648	512.18	309.72
65	2-Methyl-2-propanol	0.29657	260.86	305.73
66	1-Pentanol	0.36060	513.18	312.40
69	2-Methyl-1-butanol	0.35020	376.48	311.50
71	2-Methyl-2-butanol	0.29578	259.87	307.20
97	2-Propene-1-ol	0.28815	1,935.7	299.53
109	Ethyl ether	0.16574	14,674.0	256.72
110	Propyl ether	0.23579	6,858.5	266.34
134	Acetaldehyde	0.15205	18,364.0	265.13
141	Acetone	0.23871	8,905.0	247.64
142	2-Butanone	0.27275	5,572.6	252.32
143	3-Pentanone	0.28145	6,179.3	262.41
148	Formic acid	0.52465	963.4	281.57
149	Acetic acid	0.42437	2,716.8	291.81
150	Propionic acid	0.43533	2,908.9	287.53
151	Butyric acid	0.45359	1,951.9	296.43
161	Acetic anhydride	0.40620	2,747.8	273.61
162	Propionic anhydride	0.39705	2,444.6	287.45
164	Methyl formate	0.34444	1,292.0	198.31
165	Ethyl formate	0.29418	4,858.1	239.32
166	Propyl formate	0.29797	4,800.6	260.44
169	Methyl acetate	0.26047	6,475.7	249.48
172	Ethyl acetate	0.27056	5,361.2	257.20
175	Propyl acetate	0.29534	4,262.6	267.50

Table 2.2   (*Continued*)

Number	Compound	A	B	C
184	Ethyl propionate	0.29125	4,846.4	264.51
221	1-Chloropropane	0.25540	7,465.2	246.74
222	2-Chloropropane	0.24993	5,881.9	234.22
225	1-Chloro-2-methylpropane	0.28656	4,973.0	253.00
230	Dichloromethane	0.39806	2,666.5	212.97
231	Chloroform	0.40697	4,400.0	245.73
232	Carbon tetrachloride	0.47337	1,807.5	262.15
233	1,1-Dichloroethane	0.33277	4,651.6	247.99
234	1,2-Dichloroethane	0.44121	2,219.3	258.83
241	3-Chloropropene	0.26292	6,377.2	234.10
247	Bromoethane	0.35853	4,073.2	217.39
248	1-Bromopropane	0.36084	4,950.6	248.95
249	2-Bromopropane	0.33069	5,009.9	248.57
253	1,2-Dibromoethane	0.68897	1,421.6	277.92
255	Iodomethane	0.44916	3,107.1	215.85
256	Iodoethane	0.43828	4,348.2	243.29
257	1-Iodopropane	0.46571	3,474.4	255.85
258	2-Iodopropane	0.41279	4,108.8	262.07
312	Carbon disulfide	0.26901	16,751.0	282.23
316	Ethyl sulfide	0.28517	6,447.9	258.00
317	Thiophene	0.38204	2,967.2	254.95

Table 2.3   Viscosity-Temperature Relationship: $T = A\Phi + C - \{B/(\Phi + D)\}$; $\Phi = 1/\eta$

Number	Compound	A	B	C	D
19	Octane	0.14507	100,745.0	438.10	400
62	1-Butanol	0.23695	4,802.0	349.71	40
64	2-Methyl-1-propanol	0.23700	2,993.7	340.66	30
66	1-Pentanol	0.24191	3,908.7	354.17	35
69	2-Methyl-1-butanol	0.24650	2,942.8	346.82	30
71	2-Methyl-2-butanol	0.22988	2,124.0	328.84	30
151	Butyric acid	0.16154	70,630.0	504.44	250
162	Propionic anhydride	0.23619	52,294.0	425.82	250

Table 2.4 Viscosity-Temperature Relationship:
$\eta = A - B \log t$

Number	Compound	A	B
7	Hexane	2.03	0.83
13	Heptane	2.64	1.08
19	Octane	2.70	1.14
59	Ethanol	4.20	1.16
109	Ethyl ether	2.85	1.25
285	Aniline	14.2	7.10

when the simpler formula was used; with (2.32), this difference was reduced to 0.02%. The constants for (2.32) have been collected in Table 2.3 [590].

Many other equations relating viscosity to temperature have been proposed. Among them are

$$\log \eta = \frac{A}{T} - B \tag{2.33}$$

$$\eta = ae^{b\theta} \qquad \text{(where } \theta = T_{bp}/T) \tag{2.34}$$

$$\eta = A - B \log t \tag{2.35}$$

$$\eta v^{1/2} = Ae^{b/T} \qquad \text{(where } v = \text{specific volume)} \tag{2.36}$$

$$\frac{1}{\eta} = \frac{A}{(T_c - T)^{3/10}} \qquad \text{(where } T_c = \text{critical temperature)} \tag{2.37}$$

Thomson [4746] has stated that viscosity-temperature data for some liquids can be fitted to the Antoine equation (2.5).

The constants in (2.35) for several liquids, between their melting and boiling temperatures, have been determined by Stuart [4532]. These are tabulated in Table 2.4.

Lederer [2866a] proposed the equation

$$\log \eta = \frac{\mathring{q}}{4.571} T - 2.75 \log T + ET + C \tag{2.38}$$

for the variation of viscosity of fatty acids with temperature, where $\eta =$ coefficient of viscosity in poises, $\mathring{q} =$ heat of association at 0°C in cal/mole and E and C are constants. Values of the constants for four fatty acids are given in Table 2.5.

For a discussion of the other equations and a general discussion of viscosity-temperature relationships see Højendahl [2234], Thomas [4721], Kimura

Table 2.5   Viscosity-Temperature Relationship:
$\log \eta = \mathring{q}/4.571T - 2.75 \log T + ET + C$

Number	Compound	$\mathring{q}$	E	C
151	Butyric acid	1960	0.001770	2.9952
153	Valeric acid	2330	0.001663	2.9050
155	Hexanoic acid	2740	0.001401	2.8200
156	Octanoic acid	3475	0.001576	2.4883

[2595, 2596], Telang [4623], Lucatu [3013], Bingham [590], Hatschek [2067], Marschalko and Barna [3159] and Williamson [5201]. See also Table 2.6.

Certain organic solvents form glasses at low temperatures and are suitable for use as trapping matrices in studies of electrons, ions, and free radicals produced by radiation. Viscosity data for these compounds are included in the tables of Chapter III.

Table 2.6   Viscosity-Temperature Relationship: $\log \eta = A/T - B$

Number	Compound	A	B	Range, °K	Reference
117	1,2-Dimethoxyethane	425.	3.773	198–298	926
123	Tetrahydrofuran	393.	3.655	198–298	926
125	Tetrahydropyran	591.	4.10	228–298	3529
127	Anisole	636.8	4.139	283–353	3701
128	Phenetole	763.5	4.451	283–353	3701
320	Sulfolane	1031.3	2.3892	298–323	4924
339	Bis(2-chloroethyl)ether	882.2	4.621	283–353	3701

SURFACE TENSION

The *surface tension*, γ, is defined as the force in dynes acting at right angles to any line 1 cm in length on the surface of a liquid. It has the dimensions of force per unit length. The energy required to extend the area of a surface by 1 cm² is called the surface energy and is numerically equal to the surface tension that opposes the increase. While the surface energy is usually regarded as the fundamental property of a surface, for purposes of calculation the surface tension is more convenient to use. This is justified by the numerical equivalence of these two quantities.

Surface tensions find their widest use in the determination of parachors, though the importance of surface tensions as engineering data is increasing

[3262]. Harkins has described the various methods of measurement, and Thomson has summarized the work done on the parachor [5110]. An interpretation and catalog of parachors of organic compounds have been presented by Quayle [3872a].

Meissner and Michaels [3262] have presented a method for estimating the surface tension of multicomponent liquids based in part on the parachor and molar refraction. The method is also applicable to pure liquids. Empirical relationships that permit the surface tensions of aqueous solutions to be computed have also been derived.

The application of this method to compounds with various functional groups showed an average deviation from the experimentally determined value of 2.8%. The largest deviations were for polyhydric alcohols and halogen and nitro compounds.

Like most physical properties, the surface tension can vary widely with temperature. Vogel [4964–4979], in the course of his work on parachors, has

Table 2.7 Surface Tension-Temperature Relationship: $\gamma = \gamma_0(1 - Bt)$

Number	Compound	γ_0	B	Range °C	Ref.
30	Benzene	30.64	0.00416	0–100	5032
37	Mesitylene	28.98	0.00314	25–165	5032
60	1-Propanol	24.577	0.00318	10–90	2981
103	1,2-Ethanediol	48.48	0.00205	0–132	5032
109	Ethyl ether	18.50	0.00603	0–80	5032
127	Anisole	38.25	0.00323	20–80	3700
		36.43	0.00306	0–153	5032
128	Phenetole	33.79	0.00303	0–152	5032
141	Acetone	25.330	0.00484	0–80	2981
142	2-Butanone	25.964	0.00451	0–45	2981
143	3-Pentanone	25.970	0.00389	0–45	2981
147	Acetophenone	40.34	0.00260	0–171	5032
148	Formic acid	39.35	0.00273	0–80	5032
164	Methyl formate	27.68	0.00553	0–100	5032
172	Ethyl acetate	26.00	0.00463	0–120	5032
228	Chlorobenzene	34.10	0.00320	0–127	5032
232	Carbon tetrachloride	28.17	0.00418	0–100	5032
256	Iodoethane	32.55	0.00411	0–78	5032
264	Nitrobenzene	44.945	0.00262	0–153	5032
265	Acetonitrile	30.62	0.00400	0–73	5032
266	Propionitrile	28.57	0.00385	0–90	5032
275	Benzonitrile	39.45	0.00269	0–150	5032
285	Aniline	44.00	0.00261	0–165	5032
297	Piperidine	31.559	0.00380	0–30	2981

Table 2.8 Surface Tension-Temperature Relationship: $\gamma = A - Bt$

Number	Compound	A	B	Range °C	Ref.
2	Pentane	18.25	0.11021	0–100	2379
7	Hexane	20.44	0.1022	0–100	2379
		20.695	0.1001	0–30	2983
13	Heptane	22.10	0.0980	0–100	2379
19	Octane	23.52	0.09509	0–100	2379
22	Nonane	24.72	0.09347	0–100	2379
27	Decane	25.67	0.09197	0–100	2379
29	Dodecane	27.12	0.08843	0–100	2379
49	1-Hexene	20.47	0.10271	0–100	2379
50	1-Heptene	22.28	0.09908	0–100	2379
51	1-Octene	23.68	0.09581	0–100	2379
52	1-Nonene	24.90	0.09379	0–100	2379
53	1-Decene	25.84	0.09190	0–100	2379
62	1-Butanol	25.348	0.0788	0–40	2982
64	2-Methyl-1-propanol	23.569	0.025	0–40	2982
97	2-Propen-1-ol	26.499	0.0864	0–40	2982
103	1,2-Ethanediol	50.21	0.089	25–150	1688
105	1,3-Propanediol	47.43	0.0903	25–140	1688
118	Bis(2-methoxyethyl)ether	32.47	0.1164	25–80	1688
124	p-Dioxane	36.23	0.1391	15–60	1688
		36.1	0.14	−45–70	2234
147	Acetophenone	40.805	0.1162	30–50	2982
148	Formic acid	38.368	0.1025	0–30	2982
150	Propionic acid	27.797 $+\ 0.000183t^2$	0.1067	10–60	2982
151	Butyric acid	27.762 $+\ 0.000104t^2$	0.0981	0–60	2982
152	Isobutyric acid	26.208	0.0935	0–45	2982
154	Isovaleric acid	26.253	0.0844	20–40	2982
198	Ethyl cinnamate	38.516	0.1042	30–50	2983
202	Ethyl oxalate	33.135	0.1076	10–50	2983
203	Ethyl malonate	33.030	0.1026	10–50	2983
234	1,2-Dichloroethane	33.21	0.1023	25–90	1688
		36.3	0.14	−45–70	2234
260	Nitromethane	40.72	0.1678	0–60	4433
		38.534	0.1423	0–45	2982
261	Nitroethane	35.27	0.1255	0–60	4433
262	1-Nitropropane	32.62	0.1009	0–60	4433
263	2-Nitropropane	32.18	0.1158	0–60	4433
265	Acetonitrile	30.818 $+\ 0.000105t^2$	0.1315	0–50	2983

Table 2.8 (*Continued*)

Number	Compound	A	B	Range °C	Ref.
268	Butyronitrile	28.350	0.1004	30–50	2983
270	Valeronitrile	26.909	0.101	10–50	2983
		$+\ 0.000177t^2$			
275	Benzonitrile	39.469	0.1218	30–40	2983
286	o-Toluidine	41.867	0.1129	0–30	2982
288	p-Toluidine	38.277	0.0924	50–60	2982
301	Formamide	58.306	0.7927	0–40	2982
314	Benzenethiol	40.465	0.1188	10–45	2983
321	2-Methoxyethanol	35.36	0.139	25–80	1688
326	Diethylene glycol	46.97	0.088	20–120	1688
327	Triethylene glycol	47.33	0.088	20–120	1688
338	2-Furaldehyde	44.34	0.1211	0–30	2942
		44.341	0.1221	20–40	2982
339	Bis(2-chloroethyl)ether	40.57	0.1306	25–80	1688
349	Ethyl acetoacetate	33.645	0.1082	30–50	2983
352	Trifluoroacetic acid	15.638	0.08444	24–68	2380

determined the surface tension of many compounds at various temperatures.

The earliest attempt to correlate surface tension and temperature was made by Eötvös [1487] in 1886; it was based on the concept of corresponding states. The integrated form of his equation is

$$\frac{\gamma_1(MV_1)^{2/3} - \gamma_2(MV_2)^{2/3}}{t_2 - t_1} = k \tag{2.39}$$

where γ is the surface tension, M the molecular weight, V the specific volume of liquid, and k the temperature coefficient of molar surface energy.

The original idea was that k is a constant. Ramsay and Shields [3893] found k to be constant and equal to about 2.1 for liquids such as carbon disulfide, carbon tetrachloride, ethyl ether, and ethyl acetate. However, certain liquids such as water, alcohols, and carboxylic acids not only had a temperature coefficient of less than 2.1, but the actual value also varied with temperature. Ramsay and Shields argued that this was a measure of molecular association, though Walden and Swine [5048] later showed that, even for compounds that were neither associated nor dissociated as liquids, the temperature coefficient could vary from about 0.6 to 6.0.

The relationship of Eötvös and other relationships, either derived from it or proposed independently, have not been very useful for the calculation of surface tensions at various temperatures from values at a particular temperature. Frequently, if the temperature range is not large, surface tension-temperature data may be quite accurately fitted to empirical equations of the types

$$\gamma = \gamma_0(1 - Bt) \tag{2.40}$$

and

$$\gamma = A - Bt \tag{2.41}$$

Data for several compounds based on these equations are given in Tables 2.7 and 2.8.

HEAT OF VAPORIZATION

The *heat of vaporization*, ΔH_v, is the amount of heat required to vaporize a definite quantity, for example, one gram or one mole of a liquid. It is often referred to as the *latent* heat of vaporization.

The heat of vaporization decreases with an increase in temperature. It is most often calculated from vapor pressure measurements rather than experimentally determined. Sturtevant [5110] has given a brief discussion of this property and some methods of determination.

Trouton's equation [1797] may be used for a rough approximation of the heat of vaporization of nonpolar substances. This equation is

$$\Delta H_v = 21 T_{bp} \text{ cal mole}^{-1} \tag{2.42}$$

where T_{bp} is the absolute temperature of the boiling point. A modification of (2.42) that gives more reliable estimates has been proposed by Kistiakowsky [2258]:

$$\Delta H_v = 8.75 T_{bp} + 4.571 \log T_{bp}^2 \tag{2.43}$$

More exact values of heats of vaporization may be obtained from vapor pressure data; this is advantageous, since the heats of vaporization may be calculated over as wide a range of temperature as covered by the vapor pressure data.

The simplest calculation of the heat of vaporization can be made from the Clapeyron equation:

$$\Delta H_v = \frac{dp}{dt} T(V_g - V_1) \tag{2.44}$$

where V_g is the molar volume of the vapor and V_1 the molar volume of the liquid. The values calculated from this equation will agree with experimental values provided that dp/dt is determined over a very small interval. For practical calculations, the molar volume of the liquid may be neglected, since it is small compared to that of the vapor.

The accuracy of (2.44) may be increased by using Haggenmacher's correction [1959]

$$\left\{1 - \left(\frac{T_c^3 p}{p_c T^3}\right)\right\}^{1/2}$$

(2.45)

where T_c and p_c are the critical temperature and pressure. Although this factor contains critical constants, which frequently are unknown, Dreisbach and Spencer [1374] have shown that this equation can be applied without knowledge of the values for these constants. They have calculated values for this factor for ten compounds and have found that it is constant for families of compounds and varies only slightly for different families. For detailed information about the application of this correction factor, the reader is referred to the original article [1374].

A method for calculating the heat of vaporization and also the external work of vaporization has been presented by Haggenmacher [1958]. Equations were developed that are applicable over the entire range of the liquid phase, that is, from the triple point to the critical point. The only limit to their application is the availability of exact vapor pressures and the critical temperature and pressure. When exact vapor pressure data are used, the calculated values show a deviation of only a few tenths of a percent from precise experimental measurements. The author includes tables of calculated values of heat and external work of vaporization at various temperatures for 22 hydrocarbons.

The simple relationship

$$\Delta H_v T = k_1$$

(2.46)

where k_1 is a constant and T the absolute temperature, has been proposed by Bowden and Jones [688] for nonassociated liquids. This relationship is applicable over a range to $50°$ above the boiling point. For compounds that do not deviate greatly from Trouton's rule, the relationship has an accuracy of better than 0.5%.

Bowden and Jones [688] also studied the variation of the heat of vaporization with density and found that nonassociated substances conformed to

$$\frac{\Delta H_v}{(D - d) D^{1/3}} = k_2$$

(2.47)

where D is the density of the liquid and d that of the saturated vapor. This equation holds fairly well to within about $50°$ of the critical temperature. The authors had previously shown that normal liquids followed the relationship

$$\frac{\Delta H_v^{4/5}}{D - d} = k_3$$

(2.48)

Table 2.9 Heat of Vaporization-Temperature Relationship:
$\Delta H_v = A + BT + CT^2$ cal mole^{-1}

Number	Compound	A	B	$C \times 10^2$	Range °K	Ref.
1	Cyclopentane	7556.	6.439	−2.990	290–320	3065
5	Methylcyclo-pentane	9701.	−2.100	−1.704	305–345	3065
62	1-Butanol	12347.7	20.159	−6.491	355–390	1109
122	Furan	8854.	−1.882	−1.949	279–303	1940
177	Butyl acetate	16200.	−18.16			4155
210	Butyl borate	12324.	2.1588		390–490	978
212	Fluorobenzene	10689.	−3.850	−1.448	323–373	4208
216	Hexafluoro-benzene	12587.5	−10.3365	−1.0917	315–353	1108
260	Nitromethane	11730.	−4.9977	−1.2400	318–373	3066
286	*o*-Toluidine	7670.3	−2.557		323–473	531
287	*m*-Toluidine	6996.5	−1.673		323–473	531
288	*p*-Toluidine	7145.8	−1.951		323–473	531
296	Pyrrole	12734.	−0.97662	1.8971	362–403	4204
312	Carbon disulfide	7868.	−0.096	−1.419	283–323	5009
313	1-Butanethiol	10655.	−0.4640	−2.0136	330–372	4206
314	Benzenethiol	15808.5	−13.620	−0.1586	373–418	4209
315	Methyl sulfide	9087.	−4.1794	−1.3860	273–300	3064
316	Ethyl sulfide	11298.	−5.0897	−1.3849	325–365	4205

If (2.47) and (2.48) are combined and $D - d$ is eliminated,

$$\frac{\Delta H_v^{3/5}}{D} = k_4 \tag{2.49}$$

where k_4 is a constant. The heats of vaporization calculated from (2.49) agree well with the observed values over a range of temperature to about 100° above the boiling point of the liquid.

Equation 2.50 is frequently used to express the heat of vaporization as a function of temperature. The constants for several solvents

$$\Delta H_v = A + BT + CT^2 \text{ cal mole}^{-1} \tag{2.50}$$

are given in Table 2.9.

The *heat of sublimation*, ΔH_s, is the amount of heat required to transform a definite quantity of a solid, usually one mole, at a specified temperature directly to the vapor phase at the equilibrium vapor pressure.

CRITICAL CONSTANTS

The *critical temperature*, t_c, is defined as the temperature above which a gas cannot be liquefied. The minimum pressure required for liquefaction at this temperature is called the *critical pressure*, p_c. The *critical volume*, v_c, is the volume occupied by one mole of the substance at its critical temperature and critical pressure. F. Brescia [726] presents a good introductory discussion of critical temperature in his proposal for a more accurate concept. A comprehensive review of critical properties has been given by Kobe and Lynn [2651].

The experimental determination of critical temperature and pressure is difficult, and sometimes impossible, because of decomposition at temperatures below the critical points. Many empirical methods for predicting these critical constants have been devised. It is not our purpose to cover the literature completely, but only to point out those methods that are of interest because of their simplicity, or because they require little auxiliary data. An excellent summary of methods for calculating critical constants has been made by Thomson [5110].

The simplest means of estimating the critical temperature is provided by a relationship proposed by Guldberg [1934]. He showed that the following was approximately true:

$$\frac{T_{bp}}{T_c} = k \tag{2.51}$$

where T_{bp} is the absolute temperature of the boiling point and T_c the absolute critical temperature. The values for the constant, k, for 64 compounds ranged from 0.566 to 0.708 and had an average value of 0.651. The average value of k for the various types of aliphatic compounds is given in Table 2.10 [1934].

A number of later investigators have inverted Guldberg's original formula and have given the average value of k as about 1.5 with variations of 1.3 to

Table 2.10 Boiling Point-Critical Temperature Relationship: $T_{bp}/T_c = k$

Compounds	k
Alcohols	0.686
Acids	0.676
Esters	0.677
Amines	0.658
Halides	0.621

1.8. While this relationship should be used cautiously, it has the advantage that only the boiling point is required to estimate the critical temperature.

Many other physical properties have been correlated with critical constants. The equation

$$T_c = K_c \log P - \beta \tag{2.52}$$

has been proposed by Lewis [2930], where P is the parachor and K_c and β are constants for a particular homologous series. Good agreement between experimental and calculated values of T_c are obtained for normal paraffin hydrocarbons with

$$T_c = 551 \log P - 832 \tag{2.53}$$

For aromatic hydrocarbons $K_c = 551$ and $\beta = 716$.

Many other relations have been proposed, some based on surface tensions, parachors, liquid densities, or densities at the boiling point.

Several useful equations have been proposed by Meissner and Redding [3263] which, like Guldberg's formula, require only a knowledge of the boiling point. They are applicable to both polar and nonpolar compounds. For compounds boiling below 235°K, the equation has the form

$$T_c = 1.70 \, T_{bp} - 2 \tag{2.54}$$

Three equations are applicable for compounds boiling above 235°K. For compounds containing sulfur or halogens

$$T_c = 1.41 \, T_{bp} + 66 - 11F \tag{2.55}$$

where F is the number of fluorine atoms in the molecule. The equation

$$T_c = 1.41 \, T_{bp} + 66 - r(0.383 \, T_{bp} - 93) \tag{2.56}$$

applies to halogen- and sulfur-free aromatics or naphthenes. The factor r in (2.56) is the ratio of noncyclic carbon atoms to the total number of carbon atoms in the molecule. All halogen- and sulfur-free compounds other than the aromatics or naphthenes follow the relationship

$$T_c = 1.027 \, T_{bp} + 159 \tag{2.57}$$

When this relationship was tested on compounds boiling below 600°K (with the exception of water), the average deviation between the calculated and observed values was ±5%.

Herzog [2151] has examined various correlations of the critical constants and has evaluated their accuracy. He has proposed four equations involving the parachors and normal boiling points as auxiliary data. Either the experimental or the calculated parachor may be used.

Fewer relations have been proposed for estimating the critical pressure than for the critical temperature. Attempts have been made to correlate the critical pressure with viscosity [3667]. However, the simplest and most useful formulas are based on the critical temperature or critical density.

The relationships

$$p_c = \frac{20.8 T_c}{V_c - 8} \tag{2.58}$$

and

$$p_c = \frac{20.8 T_c}{(M/D_c - 8)} \tag{2.59}$$

have been proposed by Meissner and Redding [3263], where V_c is the critical volume, M the molecular weight, and D_c the critical density. These authors found that V_c could be evaluated from

$$V_c = (0.377P + 11.0)^{1.25} \tag{2.60}$$

where P is the parachor. The authors fully discuss their equations and recommend a procedure for calculation.

Equations developed by Herzog [2151] involving the parachor give somewhat more reliable results in calculating the critical pressure and critical volume than those of Meissner and Redding. Because of the essentially empirical nature of all critical constant equations, careful checks must be made against experimental data whenever possible. Many equations are satisfactory for specific groups of compounds but cannot be applied indiscriminately to all compounds.

Critical densities may be determined from orthobaric densities of liquid and vapor by making use of the law of rectilinear diameter first proposed by Cailletet and Mathias [879a]. This law, (2.61), states that the average density of liquid and vapor is a linear function of the temperature.

$$\tfrac{1}{2}(D + d) = a - bt \tag{2.61}$$

Plots of the density of liquid, D, and density of vapor, d, are made vs. temperature, and extrapolation of the average density line to the point of junction of these two curves gives the critical density.

Hakala [1965a] solved simultaneously the Katayama-Eötvös [2510a] and MacCleod [3057a] equations for surface tension with the Cailletet-Mathias equation to obtain (2.62), from which critical densities can be estimated with good accuracy from data taken far below the critical temperature.

$$D + d = 2d_c - A(D - d)^{10/3} \tag{2.62}$$

The critical density is determined from the intercept of the plot of $D + d$ vs. $(D - d)^{10/3}$. Data for 29 compounds gave results with an average accuracy of 0.5%; the average extrapolation was over a range of 100°C.

HEAT CAPACITY

The *specific heat capacity* or *specific heat* of a substance is the amount of heat required to raise the temperature of one gram of the substance one degree centigrade at constant pressure and without change of phase. Although the specific heat is a function of temperature, average values are frequently quoted for a specified temperature range. The equation

$$c_p = A + Bt + Ct^2 \text{ cal deg}^{-1} \text{ gram}^{-1} \tag{2.63}$$

has been used to express the variation of the specific heat with temperature, and values of the constants for several solvents are given in Table 2.11.

Molar heat capacities are also temperature-dependent and may be precisely defined as $C = dq/dT$, where dq is the differential flow of heat accompanying the infinitesimal temperature change, dT. Conditions of constant pressure or constant volume must be specified and are designated with the symbols, C_p and C_v, respectively. The units of heat capacity are commonly expressed as calories per degree per mole and joules per degree per mole in the CGS System and as Btu per degree F per pound in the FPS System.

Water has been the standard substance for the intercomparison of heat capacity calorimeters because it has characteristics such as high heat capacity and ease of purification to a high degree of purity, which make it suitable for this purpose. However, measurements below 0°C and above 100°C present problems with water and in some calorimeters it is impossible to use water outside of its normal liquid range. The Fourth Congress on Calorimetry [1609a] recommended that three additional substances be used as standards for the intercomparison of heat capacity calorimeters: heptane, benzoic acid, and aluminum oxide. Ginnings and Furukawa [1782] determined the heat capacities of the three proposed substances. They also reported later [1680] on diphenyl ether which was suggested as another additional calorimetric standard (see Chapter V).

Since organic solvents are generally used in the liquid state, the heat capacities given in Chapter III are in calories per degree per mole for the liquid at the stated temperature unless otherwise indicated.

Constants for the equation

$$C_p = A + BT + CT^2 + DT^3 \text{ cal deg}^{-1} \text{ mole}^{-1} \tag{2.64}$$

are given in Table 2.12 for a number of solvents. Kurbatow [2781] has expressed the molar heat capacities of several organic solvents as a linear function of the centigrade temperature (see Table 2.13).

The heat capacity of the ideal gas reference state, C_p°, is a useful thermodynamic property. Values of C_p° may be obtained by a linear extrapolation to

Table 2.11 Specific Heat-Temperature Relationship:
$c_p = A + Bt + Ct^2$ cal deg^{-1} gram^{-1}

No.	Compound	A	B × 10⁴	C × 10⁵	Range °C	Ref.
62	1-Butanol	0.6053	9.76	0.6		3640
63	2-Butanol	0.655	9.20	1.837		3640
66	1-Pentanol	0.5210	15.		40 to 70	2872
67	2-Pentanol	0.592	22.6	0.126		3640
68	3-Pentanol	0.6658	10.		40 to 70	2872
69	2-Methyl-1-butanol	0.5657	13.		40 to 70	2872
70	3-Methyl-1-butanol	0.5277	20.		40 to 70	2872
71	2-Methyl-2-butanol	0.6236	22.		40 to 70	2872
103	1,2-Ethanediol	0.55	12.5		−10 to +195	4611
114	Butyl ether	0.217	93.2	1.98		3640
142	2-Butanone	0.4242	16.2	1.038		3640
148	Formic acid	0.4966	7.09			4143
187	Ethyl isovalerate	0.4539	11.		40 to 70	2872
202	Ethyl oxalate	0.4336	6.5		40 to 70	2872
217	1,1,2-Trichloro-1,2,2-trifluoro-methane	0.2148	2.07		−30 to +63	517
220	Chloroethane	0.3941	7.088		−67 to +15	2782
221	1-Chloropropane	0.3820	7.954		−38 to +43	2782
225	1-Chloro-2-methyl-propane	0.3915	7.334		14 to 58	2782
230	Dichloromethane	0.452	7.448		−76 to +12	2782
231	Chloroform	0.2232	4.4656		−50 to +50	2782
232	Carbon tetra-chloride	0.1933	2.688		−20 to +72	2782
233	1,1-Dichloroethane	0.2953	4.94		−51 to +55	2782
234	1,2-Dichloroethane	0.2812	6.1884		−25 to +82	2782
236	1,1,2,2-Tetra-chloroethane	0.2286	2.957		15 to 145	2782
237	Pentachloroethane	0.2208	4.082		16 to 154	2782
243	1,2-Dichloro-ethylene	0.2618	5.4886		−31 to +54	2782
246	Tetrachloroethylene	0.1937	3.799		16 to 119	2782
247	Bromoethane	0.2088	4.884		−49 to +37	2782
248	1-Bromopropane	0.2589	5.2502		−30 to +67	2782
252	Bromoform	0.1201	1.38322		9 to 147	2782
253	1,2-Dibromoethane	0.1638	3.0994		15 to 127	2782
254	1,1,2,2-Tetra-bromoethane	0.1084	2.436		12 to 132	2782
255	Iodomethane	0.2079	2.50		−56 to +20	2782
256	Iodoethane	0.1677	3.42		−37 to +71	2782
259	Diiodomethane	0.1146	1.6424		15 to 164	2782
286	*o*-Toluidine	0.4706	7.0			4144
292	Diethylamine	0.5604	8.607			2529
298	Triethylamine	0.4987	12.58			2529
338	2-Furaldehyde	0.392	−2.63	1.08	15 to 150	75

Table 2.12 Heat Capacity-Temperature Relationship:
$C_p = A + BT + CT^2 + DT^3$ cal deg^{-1} mole^{-1}

No.	Compound	A	B	$C \times 10^4$	$D \times 10^7$	Range °K	Ref.
13	Heptane	56.582	−0.14490	5.7813	−4.1667	240–370	2299
14	2-Methylhexane	41.850	−0.026750	2.1531	0.10417	160–300	2299
17	2,4-Dimethylpentane	37.649	−0.042300	4.4188	−4.0364	160–310	2299
25	cis-Decahydro-naphthalene	12.68	0.1434			300–350	3350
26	trans-Decahydro-naphthalene	10.95	0.1464			300–350	3350
38	Naphthalene	19.212	0.092572			357–371	3062
39	1,2,3,4-Tetra-hydronaphthalene	44.376	−0.12879	7.4003	−7.467	248–319	4590
40	Butylbenzene	65.158	−0.24196	10.343	−10.108	195–380	3281
107	Glycerol	32.9	0.0761	−0.269		273–523	3594
122	Furan	35.17	−0.14860	5.695	−5.322	191–299	1940
212	Fluorobenzene	39.496	−0.13777	5.7113	−5.3644	235–350	4208
267	Succinonitrile	28.583	0.02922			335–350	5272
290	Ethylenediamine	6.27	0.05871	−0.1829			3493
296	Pyrrole	12.699	0.059825			250–365	4204
314	Benzenethiol	34.985	0.002175	0.6500		258–375	4209

zero pressure of the plots of the heat capacities of the real gas against pressure at constant temperature. For gases obeying the Berthelot equation of state, values of C_p° may be calculated from the critical constants and the measured heat capacity:

$$C_p^\circ = C_{p(g)} + \frac{81RT_c^3 p}{32T^3 p_c} \tag{2.65}$$

Table 2.13 Heat Capacity-Temperature Relationship:
$C_p = A + Bt$ cal deg^{-1} mole^{-1}

No.	Compound	A	B	Range °C
30	Benzene	31.08	0.06342	8 to 80
31	Toluene	35.66	0.08491	−76 to +60
32	o-Xylene	42.31	0.095348	15 to 132
33	m-Xylene	41.37	0.11055	15 to 132
34	p-Xylene	41.43	0.11065	15 to 132
35	Ethylbenzene	41.44	0.1172	15 to 184
36	Isopropylbenzene	48.42	0.10098	16 to 153
37	Mesitylene	48.35	0.10102	14 to 155
43	p-Cymene	55.21	0.1092	10 to 155

Table 2.14 Heat Capacity, Ideal Gas State-Temperature Relationship:
$C_p^\circ = A + BT + CT^2 + DT^3$ cal deg^{-1} mole^{-1}

No.	Compound	A	$B \times 10^3$	$C \times 10^7$	Range $^\circ$K	Ref.
0	Water	7.256	2.298	2.83	300–1500	4457
		7.219	2.374	2.67	300–1500	4458
2	Pentane	5.780	88.843	−273.26	300–1500	4457
		3.140	100.532	−355.60	300–1500	4458
3	2-Methylbutane	2.801	102.820	−367.41	300–1500	4458
4	2,2-Dimethylpropane	6.076	98.954	−353.69	300–1500	4458
5	Methylcyclopentane	−8.096	127.43	−421.1	335–470	3065
7	Hexane	4.298	118.661	−421.30	300–1500	4458
		7.313	104.906	−323.97	300–1500	4457
13	Heptane	8.850	120.974	−374.78	300–1500	4457
		5.401	136.930	−487.71	300–1500	4458
19	Octane	6.231	155.942	−558.57	300–1500	4458
		10.381	137.054	−425.64	300–1500	4457
24	Decahydronaphthalene	−10.48	209.	−823.	300–1500	5003
30	Benzene	−0.283	77.936	−262.96	300–1500	4457
31	Toluene	0.436	94.254	−312.58	300–1500	4457
32	o-Xylene	4.603	104.476	−336.16	300–1500	4457
33	m-Xylene	1.956	109.147	−355.83	300–1500	4457
34	p-Xylene	1.846	108.594	−352.00	300–1500	4457
37	Mesitylene	3.042	124.059	−397.43	300–1500	4457
38	Naphthalene	−7.66	147.11	−620.8	300–1500	5003
39	Tetrahydronaphthalene	−8.90	172.5	−707.	300–1500	5003
59	Ethanol	3.578	49.847	−169.91	300–1000	4458
62	1-Butanol	−0.780	102.9	−456.	400–450	1109
65	2-Methyl-2-propanol	10.836	49.648	240.70	365–435	555
122	Furan	−7.55	93.52	−528.7	320–485	1940
141	Acetone	5.371	49.227	−151.82	300–1500	4457
212	Fluorobenzene	−6.581	115.89	−612.2	345–500	4208
216	Hexafluorobenzene	11.0175	104.91	−584.	340–470	1108
230	Dichloromethane	4.309	31.673	−163.51	250–600	4458
231	Chloroform	7.052	35.598	−216.86	275–770	4458
232	Carbon tetrachloride	22.675	3.274	−3.264[a]	275–770	4458
252	Bromoform	9.356	32.319	−212.72	300–600	4458
254	1,1,2,2-Tetrabromoethane	8.5059	78.346	−789.76	250–750	873
			$+ 3.0734 \times 10^{-8}T^3$			
255	Iodomethane	4.105	24.487	−97.33	300–600	4458
259	Diiodomethane	5.839	32.571	−195.28	250–600	4458
260	Nitromethane	2.352	42.882	−169.4	365–520	3066
265	Acetonitrile	5.018	27.935	−93.02	295–1200	4458
299	Pyridine	−3.016	88.083	−386.65	295–1000	4457
312	Carbon disulfide	7.692	13.426	−91.16	300–1800	4458
313	1-Butanethiol	2.628	97.293	−410.15	360–500	4206
314	Benzenethiol	−5.541	122.12	−657.7	430–500	4209
315	Methylsulfide	5.628	44.990	−155.35	320–500	3064
316	Ethyl sulfide	2.76	94.965	−379.4	350–485	4205
319	Dimethylsulfoxide	6.94	56.	−227.	300–1000	3103

[a] C/T^2

Numerical values of C_p° are given in calories per degree per mole in Chapter III. Constants for the equation

$$C_p^\circ = A + BT + CT^2 + DT^3 \, \text{cal deg}^{-1} \, \text{mole}^{-1} \tag{2.66}$$

are given in Table 2.14.

OPTICAL ACTIVITY

The optical rotatory power of a substance may be expressed as the *specific rotation*, $[\alpha]_\lambda^t$ which is defined by

$$[\alpha]_\lambda^t = \frac{100\alpha}{1c}$$

where α is the observed angle of rotation, 1 the length of the light path in decimeters, c the concentration in grams solute per 100 ml solution, and λ the wavelength of light. Among liquid organic solvents, optical activity is a result of the presence of one or more asymmetric carbon atoms in the molecule. Optical activity may be used in the identification of substances, the elucidation of structure, and the determination of the concentration of solutions.

FREEZING POINT CONSTANT

The *freezing point constant*, or *cryoscopic constant*, K_f, is defined by

$$K_f = \frac{M_1 R T_0^2}{1000\Delta H_m} \tag{2.67}$$

where R is 1.9872 cal deg^{-1} mole^{-1}, T_0 the freezing point of the solvent in degrees Kelvin, M_1 the molecular weight of the solvent, and ΔH_m the molar heat of fusion of the solvent.

The numerical value of K_f is usually expressed as degrees per mole per 1000 g of solvent but is sometimes expressed as degrees per mole per 100 g of solvent. The former unit has been used in Chapter III.

Equation 2.67 is accurate, provided that the solution is dilute enough to obey Raoult's law and ΔH_m is constant between the freezing points of solvent and solution.

The cryoscopic constant is often used in (2.68) for molecular weight determinations.

$$M_2 = \frac{1000 w_2 K_f}{w_1 \Delta T} \tag{2.68}$$

In this equation w_2 is the weight of the solute, w_1 the weight of the solvent, and ΔT the freezing point depression. Molecular weights so determined may vary somewhat with the solute concentration; best values are often obtained by extrapolating a plot of M_2 vs. concentration to zero concentration where the behavior of the solution is most nearly ideal. The determination of molecular weights by (2.68) is accurate to about $\pm 5\%$ under favorable conditions. The usual procedures rarely attain this accuracy but are adequate for checking the correctness of postulated empirical formulas.

Brancker and coworkers [710] have proposed a modification of (2.68) that overcomes the difficulties that arise from variations in K_f and molecular weights with concentration. This modified form is

$$\Delta t = K_f m^b \tag{2.69}$$

where b is a constant and m is the molality of the solution. For benzene as the solvent, the modified equation was found to be

$$\Delta t = 5.229 m^{0.9394} \tag{2.70}$$

The value of K_f in (2.70) was found to vary with different samples of the same cryoscopic solvent. The value of b was constant for all pure solutes but was extremely sensitive to impurities. Table 2.15 illustrates some results.

Using the results of Meldrum, Saxer, and Jones [3266], Brancker and coworkers [710] found the equation for camphor to be

$$\Delta t = 38.73 m^{0.9580} \tag{2.71}$$

Skau and Wakeham [5110] have discussed the theory and determination of molecular weights by the freezing point method, and also the sources of error. The familiar Beckmann procedure [474] and its sources of error have been considered by many investigators, among them Skau and Wakeham [5110], Rall and Smith [3884] and Daniels, Mathews, and Williams [1186].

Table 2.15 Molecular Weights of 1,2,3,4-Tetrahydronaphthalene in Benzene; mw = 132.196

Concentration %w	Molecular Weight (Eq. 2.68)	Molecular Weight (Eq. 2.70)
0.68	111.6	133.2
1.98	118.0	132.4
3.98	127.1	137.0
7.94	131.2	135.5
11.13	134.9	136.6

The unusually high freezing point constant of camphor, about 40, has been utilized by Rast [3921] to develop a simple, rapid procedure for determining the molecular weights of compounds soluble in camphor. This method is described in detail by Kamm [2488], Meyer [3286], and Durand [1417].

A modification of the Rast procedure has been made by Keller and Halban [2546]. Data are given for cyclopentadecanone, diethyldiphenyl, urea, tetrabromomethane, and borneol as solvents.

Rossini [4050]; Glasgow, Krouskop, et al. [1793]; Streiff and Rossini [4525]; and Glasgow, Streiff, and Rossini [1795] have described a method, apparatus, and procedure for the determination of purity by freezing point measurements. This method is adaptable to molecular weight determinations of high accuracy, provided that the fundamental thermodynamic requirements are met by the solvent-solute system.

The basic equation is

$$-\ln N_1 = A(T_0 - T)\{1 + B(T_0 - T) + \cdots\} \qquad (2.72)$$

where N_1 is the mole fraction of the solvent, T the freezing point of the solution, and T_0 the freezing point of the pure solvent; $A = \Delta H_m/RT_0^2$ and $B = (1/T_0) - (\Delta C_p/2\Delta H_m)$. Equation 2.72 may be modified to calculate the mole percent purity, P.

$$\log P = 2.00000 - \frac{A}{2.30259}(T_0 - T)\{1 + B(T_0 - T) + \cdots\} \quad (2.73)$$

It will be noted that the constant, A, is related to K_f through the equation

$$K_f = \frac{M_1}{1000A} \qquad (2.74)$$

Values of A or of A and B for many hydrocarbons have been given in Chapter III. The terms in brackets should be dropped when only the constant A, is given.

$$-\ln N_1 = A(T_0 - T) \qquad (2.75)$$

For dilute solutions, (2.75) becomes

$$N_2 = A(T_0 - T) = \frac{w_2 M_1}{w_1 M_2}$$

from whence a convenient equation for the calculation of molecular weights from freezing point depressions is obtained:

$$M_2 = \frac{w_2 M_1}{w_1 A(T_0 - T)} \qquad (2.76)$$

Witschonke [5227] has tabulated the freezing point of a number of compounds determined by the equilibrium freezing and melting curves. A

Table 2.16 Cryoscopic Constants, K', in Mole % per Degree C

No.	Compound	fp 100%m °C	K'
95	m-Cresol	12.22	0.749
141	Acetone	−94.6	2.14
238	o-Dichlorobenzene	−17.01	2.4
264	Nitrobenzene	5.74	1.87
267	Succinonitrile	58.1	0.42
270	Valeronitrile	−96.20	3.50
274	α-Tolunitrile	−22.0	1.9
275	Benzonitrile	−12.75	1.93
276	Acrylonitrile	−88.55	2.7
285	Aniline	−6.00	1.38
299	Pyridine	−41.6	1.86
300	Quinoline	−14.85	1.95

cryoscopic constant, K', which is described as the mole percent impurity which lowers the freezing point 1°C, was calculated from the equation

$$K' = \frac{100\Delta H_m}{RT_0^2} \qquad (2.77)$$

I, the mole percent impurity, is then calculated from the equation

$$I = K'(T_0 - T_f) \qquad (2.78)$$

where T_0 is the freezing point of the pure solvent and T_f is the freezing point of the sample. Equation 2.78 can be used to calculate the purity of a sample within the range, 95 to 100%. The freezing point of the 100% compound must be known. An example of this calculation may be found in Chapter V, under Criteria of Purity for benzonitrile. Some of Witschonke's values for K' and T_0 are given in Table 2.16.

BOILING POINT CONSTANT

The boiling point constant, K_b, or ebullioscopic constant, is defined as

$$K_b = \frac{M_1 RT_0^2}{1000\Delta H_v} \qquad (2.79)$$

where R is 1.9872 cal deg^{-1} mole^{-1}, T_0 the boiling point of the solvent, and ΔH_v the molar heat of vaporization. The numerical values for K_b in Chapter III are expressed as degrees per mole per 1000 g solvent.

Like the similar equation for K_f (2.67), Eq. 2.79 is valid only for dilute solution for which Raoult's law is at least approximately applicable. Since the boiling point and the heat of vaporization of a solvent vary with pressure, K_b will also vary. In general, the effect is within experimental error for a pressure variation of about 10 torr around atmospheric pressure.

Molecular weights can be determined with the relation

$$M = K_b \frac{1000w_2}{w_1 \Delta T_b} \tag{2.80}$$

where ΔT_b is the elevation of the boiling point brought about by the addition of w_2 grams of solute to w_1 grams of solvent. For the experimental details in determining boiling points, see Swietoslawski and Anderson [5110].

The advantages of the ebullioscopic determination of molecular weights over the cryoscopic method were discussed by Menzies and Wright [3274] in 1921. They also considered the errors encountered in ebullioscopic molecular weight determinations and described the application of a differential thermometric method. The errors that can arise when the gas constant, R, in (2.79) is not corrected by means of the Berthelot equation [1797] have been considered by Hoyt and Fink [2276]. For additional information on the ebullioscopic method, see Rosanoff and Dunphy [4032].

Hill and Brown [2193] have described a stable ebulliometer with highly sensitive differential thermometers to give a wide range of operating temperatures. These authors state that their method represents a tenfold extension of the range and sensitivity of the Menzies and Wright [3274] ebulliometer. This is made possible by the coupling of solvents boiling above 130° with a conventional water thermometer, or by the use of low boiling thermometric liquids with conventional solvents. A precision of 2% was obtained for solutes whose molecular weight ranged from 300 to 13,400.

A thermoelectric procedure for determining the molecular weights of solutes in organic solvents has been described by Taylor and Hall [4613]. It is a modification of the procedure of Hill [2192] for observing the temperature difference between two different solutions or between the solution and the solvent when opposing thermocouples are in an atmosphere saturated with the solvent. Hill's proposal had been used exclusively for water, but Taylor and Hall applied it to organic solvents. For molecular weight determinations, the underlying principles are those of ebulliometry.

ACIDITY AND BASICITY

The characteristics of acids and bases and the mechanisms of their reactions are often explained on the basis of one or both of the two leading theories proposed in 1923. The Lewis theory [2935, first proposed; 2936, amplified] defines a base as an electron-pair donor and an acid as an electron-pair acceptor in the formation of a coordinate covalent bond. Brønsted [753, 754] and Lowry [3010] simultaneously and independently published the proton theory of acids and bases. This theory will be hereinafter referred to as the Brønsted theory. It regards an acid as a proton donor and a base as a proton acceptor. While Brønsted and Lewis bases are the same, Brønsted acids are limited to proton donors; for practical purposes they may be considered as a special, but large, class of Lewis acids.

The acidic or basic character of a solvent often serves as the criterion of its usefulness as a reaction medium, and solvents are sometimes classified on the basis of their acidic or basic properties. Solvents that tend to provide protons are called *protogenic*, while those that act primarily as proton acceptors are said to be *protophilic*. *Amphiprotic* solvents are those that may act either as proton donors or proton acceptors depending on the nature of the solute. Water and the alcohols are well-known examples. Still other solvents are relatively inert with respect to proton activity and are classified as *aprotic* solvents. Most hydrocarbons under conditions usually encountered fall in this class; however, the concept of the aprotic solvent should be employed with discretion because there is no known organic solvent that will not ionize to some extent under some conditions. For example, benzene, usually found to be inert, is a base in hydrogen fluoride having a pK_{BH^+} value of 9.4 [3104]. The influence of the acidic or basic character of the solvent on chemical reactions has been ably discussed by Luder and Zuffanti [3014]. Davis [1218] has lucidly presented the role of the aprotic solvent.

The strengths of acids and bases have generally been determined from measurements made on aqueous solutions of the compounds. Albert and Serjeant [41] discuss the strengths of acids and bases in water using a minimum of mathematics. More extensive and sophisticated presentations including strengths in water and in nonaqueous solvents are given by Bates [436, 2693, Chapter 10], Kolthoff [2693, Chapter 11], Bruckenstein and Kolthoff [2693, Chapter 12], and Kolthoff and Bruckenstein [2693, Chapter 13].

There is no acidity function that expresses the strengths of acids and bases in all different solvent systems. Generally, a substance that is an acid or a base in one solvent is an acid or base in most other solvents, and the measurable strength will vary from solvent to solvent. In the tables of Chapter III, the strengths of acids and bases are expressed as they appeared in the original literature, as equilibrium constants or as functions derived therefrom. Symbols are briefly defined in "Abbreviations and Symbols" and are fully discussed by Bates [436]. Whenever possible, the temperature and the solvent (if different from water) are noted.

Acidity and basicity are chemical properties. Their true measure is their effect on a chemical reaction such as the hydrolysis of sucrose, and so forth.

Since this is not a convenient measure, physical means are used such as the electromotive force of cells, conductivity, or freezing point depressions. The physical methods are reliable within a single solvent system for which they are applicable, and between certain solvent systems. Ionization as a criterion of strengths of acids and bases must be used with caution, particularly from solvent to solvent. Perchloric acid is considered completely ionized in water; it has a relatively low ionization in acetic acid. If ionization were the criterion of strength, sucrose should be hydrolyzed more rapidly by perchloric acid in water than in acetic acid. The reverse is true. Hydrochloric acid also is considered completely ionized in water. It shows no ionization in benzene, yet it is an acid in this solvent, according to the Brønsted and the Lewis concepts. Since it reacts with ammonia in benzene, there is some chemical evidence that it is at least a moderately strong acid. Davis has carried out extensive investigations concerning the effect of the solvent and the nature of acid and base reactions in aprotic solvents. She has presented the behavior of acids and bases in aprotic solvents in a recently published monograph [1218].

The *autoprotolysis constant*, K_s, is the product of the ion activity of the self-ionized ions of a solvent. It need not be limited to ionization involving protons, but its name is so limiting. The general case is

$$2\,HR \rightleftarrows H_2R^+ + R^- \tag{2.81}$$

hence,

$$K_s = \gamma[H_2R^+] \times \gamma[R^-] \tag{2.82}$$

where brackets indicate the molarity of the enclosed entity. Concentrations of the ionic species are generally so small that the activity coefficients may be considered to approach unity and (2.82) becomes

$$K_s = [H_2R^+] \times [R^-] \tag{2.83}$$

The autoprotolysis constant for water is also known as the ion product and is derived from equilibrium considerations thus:

$$H_2O + H_2O \rightleftarrows H_3O^+ + OH^- \tag{2.84}$$

$$K_{aq} = \frac{[H_3O^+] \times [OH^-]}{[H_2O]} \tag{2.85}$$

The relative change in $[H_2O]$ is negligible in comparison to a large change in $[H_3O^+]$ or $[OH^-]$; therefore $[H_2O]$ is considered a constant and (2.85) becomes

$$K_w = K_s = [H_3O^+] \times [OH^-] \tag{2.86}$$

The autoprotolysis constants for only a few amphiprotic solvents have been determined. The specific constants are expressed as indicated in (2.81) and (2.82). Examples from three functional group types are as follows.

Methanol:

$$CH_3OH + CH_3OH \rightleftharpoons CH_3OH_2^+ + CH_3O^- \tag{2.87}$$
$$\text{acid} \qquad \text{base} \qquad \text{acid} \qquad \text{base}$$

$$K_s = [CH_3OH_2^+] \times [CH_3O^-] \tag{2.88}$$

Acetonitrile:

$$CH_3CN + CH_3CN \rightleftharpoons CH_3CNH^+ + CH_2CN^- \tag{2.89}$$
$$\text{acid} \qquad \text{base} \qquad \text{acid} \qquad \text{base}$$

$$K_s = [CH_3CNH^+] \times [CH_2CN^-] \tag{2.90}$$

Acetic Acid:

$$HC_2H_3O_2 + HC_2H_3O_2 \rightleftharpoons H_2C_2H_3O_2^+ + C_2H_3O_2^- \tag{2.91}$$
$$\text{acid} \qquad \text{base} \qquad \text{acid} \qquad \text{base}$$

$$K_s = [H_2C_2H_3O_2^+] \times [C_2H_3O_2^-] \tag{2.92}$$

The H_3O^+, $CH_3OH_2^+$, CH_3CNH^+, and $H_2C_2H_3O_2^+$ and all like protonated ions are *onium* ions, for example, hydronium, methanonium, and so forth. The onium ion is more frequently referred to as the *hydrogen ion*. The composition of the onium ion as H_2R^+ is not universally accepted.

The autoprotolysis constant and the dielectric constant are believed to be the principal influencing factors that determine the nature and extent of the effect on solvates, particularly the acidic or basic nature.

The *ionization constant*, K_i, of an organic solvent is a measure of its strength as an acid or base when it is dissolved in some other solvent, usually water. Acid ionization constants are commonly designated as K_a and base ionization constants as K_b. These constants are the equilibrium constants for the ionization reaction that have been modified to include the concentration of the water that is always present in large excess. For example, for an acid,

$$HC_2H_3O_2 + H_2O \rightleftharpoons H_2C_2H_3O_2^+ + OH \tag{2.93}$$
$$\text{acid} \qquad \text{base} \qquad \text{acid} \qquad \text{base}$$

$$K_a = \frac{[H_2C_2H_3O_2^+] \times [OH^-]}{[HC_2H_3O_2]}$$

and for a base (C_4H_9, *n*-butyl radical = Bu),

$$H_2O + BuNH_2 \rightleftharpoons BuNH_3^+ + OH^- \tag{2.94}$$
$$\text{acid} \qquad \text{base} \qquad \text{acid} \qquad \text{base}$$

$$K_b = \frac{[BuNH_3^+] \times [OH^-]}{[BuNH_2]}$$

The ionization constant, K_a, for acetic acid is 1.754×10^{-5} and K_b for butyl amine is 4.09×10^{-4}. In 1909, Sørensen proposed a more convenient means of expressing the effective hydrogen ion concentration as

$$pH = -\log [H^+]$$

"p" represents the word, "potenz," translated into English as power, the exponent of 10 in the arithmetical expression of the concentration of the hydrogen ion. This reasoning was extended to other expressions such as hydroxyl ion concentration, ionization constants, and autoprotolysis constants. A $[H^+]$ of 0.01 N becomes 10^{-2} and thus a pH of 2; a concentration of 0.00002 N is 2×10^{-5} or $10^{-4.7}$, thus a pH of 4.7.

The autoprotolysis constant for water, $K_w = K_s = [H^+] \times [OH^-]$. Numerically, at 25°C, $K_w = 10^{-14} = 10^{-7} \times 10^{-7}$.

$$pK_w = 14 \quad \text{and} \quad pH = pOH = 7 \quad\quad (2.95)$$

Conversion of the K_a of 1.754×10^{-5} for acetic acid to pK_a is accomplished by subtracting the logarithm of K_a, $\bar{5}.24403$, from zero: $0.00000 - \bar{5}.24403 = 4.75597 = pK_a$. Strengths of bases, expressed as K_b, may be converted to pK_b in the same manner. Conversion of pK_a to K_a is brought about by subtracting the pK_a from zero and taking the antilog of the remainder:

$$0.00000 - 4.75597 = \bar{5}.24403$$

$$\text{antilog } \bar{5}.24403 = 1.754 \times 10^{-5}$$

The strengths of bases are more generally expressed in terms of the protonated base, BH^+. This is derived from K_b by converting to pK_b; for example, for butyl amine in water, $K_b = 4.09 \times 10^{-4}$, $pK_b = 5.389$, and pK_a or pK_{BH^+} is $14.000 - 5.389 = 8.611$.

The subject of acidity and basicity as related to the solvents in this book, of necessity, is brief. The references given herein should be consulted for a more complete presentation of the subject.

ELECTRICAL PROPERTIES

The electrical properties tabulated in Chapter III are the *electrical conductance* or *conductivity*, κ, the *dielectric constant*, ϵ, and the *dipole moment*, μ.

The electrical conductance is the reciprocal of the specific resistance which is the resistance that would be offered by the solvent when placed between plane parallel electrodes 1 cm² in area and 1 cm apart. The units of specific conductance are ohms^{-1} cm^{-1} or mhos cm^{-1}.

In general, pure solvents have a low conductivity. For example, the purest water has a specific resistance at 25° of about 20 million ohm cm. Ordinary water, distilled in the presence of air, has a specific resistance of only about 100,000 ohm cm due to the absorption of carbon dioxide and other gases from the air.

The conductivity of a solvent can be used as a criterion of purity in limited cases. It is an excellent method for determining the purity of water and can be used to determine the completeness of removal of dissociable material from other solvents of high dielectric constant. It is sometimes useful for determining the dryness of solvents. Mixtures of two liquids, each of which is practically nonconducting, may have a conductance of the same magnitude as the individual solvents or they may have a higher conductance.

The dielectric constant is a measure of the relative effect a solvent has on the force with which two oppositely charged plates attract each other. The dielectric constant of a vacuum is defined as unity, but for practical purposes measurements are made in air. It is a unitless number.

Qualitatively, substances of high dielectric constant are good ionizing media. This fact may be used for estimating solubilities. If the solubility of a substance is known in a solvent of known dielectric constant, its solubility in other solvents can often be roughly estimated from their dielectric constants.

The most important use of the dielectric constant is the calculation of *molar polarizations* and from them the dipole moments. The molar polarization, P, is defined by the Clausius-Mosotti equation [1797] as

$$P = \frac{\epsilon - 1}{\epsilon + 2} \frac{M}{d} \tag{2.96}$$

where ϵ = dielectric constant, M = molecular weight, and d = density.

The dipole moment is a measure of the displacement of what might be termed the centers of gravity of positive and negative charges of a molecule. It is the product of a charge and a distance, and the units are expressed either in 10^{-18} esu or in *Debye* units. One Deybe, D, is equal to 10^{-18} esu.

The dipole moment is important in studying the structure of compounds because it provides a means for measuring the polarity, or electrical dissymmetry, of a molecule. Physical properties, such as solubility, the properties of solutions, the influence of solvents on reaction rates, and deviations from the simple gas laws can be quantitatively interpreted more readily if the dipole moments are known.

Many qualitative conclusions about solvents can be drawn from their dipole moments. Perfectly symmetrical molecules are nonpolar. Unsymmetrical compounds, however, are almost invariably polar, particularly when oxygen, nitrogen, or a halogen atom is present. In general, nonpolar substances are volatile and soluble in other nonpolar compounds, whereas polar compounds are less volatile and are soluble only in other polar compounds.

FLASH POINT

The *flash point*, fl pt, is usually not considered a common physical property. It is included in Chapter III because of its widespread use in classifying solvents for storage and shipping. There are many special procedures for measuring the flash point. The more common ones used in the United States are Tag Open Cup (TOC), ASTM D 1310-63; Tag Closed Cup (TCC), ASTM D 56-64; Cleveland Open Cup (COC), ASTM D 92-66; and Pensky-Martens Closed Cup (PMCC), ASTM D 93-66. The American Society for Testing Materials' *ASTM Test Methods* have been standardized and give reliable results within the limits claimed when the standardized apparatus is used in the manner prescribed. The method used often is referred to simply as Open Cup (OC) or Closed Cup (CC). This is not definitive because the several open cup apparatuses and closed cup apparatuses give different values and cannot be converted. The ASTM has defined the use of the several apparatuses as follows:

TOC.   Determination of open cup flash points of volatile flammable materials having a flash point between 0 and 175°F.

TCC.   Determination of the flash points of all mobile liquids flashing below 175°F. (An exception given: not applicable to solvents listed in Chapter I.)

COC.   Determination of flash and fire points of all petroleum products (used for other organic liquids also) except fuel oil and those having an open cup flash below 175°F.

PMCC.   Determination of flash points of such liquids as fuel oil, lubricating oils, viscous materials, and so forth. (Should not be used for common solvents.)

SPECTROSCOPY

In spite of the large volume of work published in spectroscopy, suitable data for some organic solvents are surprisingly scarce. An abundance of papers have appeared for certain other solvents; in these instances, primary consideration has been given to those references that deal with the application of spectra to problems of identification and analysis. Spectroscopic data are not given in the tables of Chapter III but references to *ultraviolet* (uv), *infrared* (ir), *Raman, mass*, and *nuclear magnetic resonance* (nmr) spectroscopy are included.

There are available many excellent books on the possibilities and limitations of the several branches of spectroscopy. Among those of a nonmathematical character are Sawyer [4121], Brode [747], West [5111], Harrison, Lord, and Loofbourow [2044], and Gutowsky [5112].

The ultraviolet is the oldest of the nonvisual spectra that has been used experimentally. The most comprehensive collection of ultraviolet spectra in the region 2100 Å to 3200 Å has been compiled by the American Petroleum

Institute, Research Project 44 [91]. New spectra are added as they become available.

Many analysts consider infrared spectra the most useful of the absorption spectra. With the commonly used spectrometers, many substances do not show a characteristic absorption in the ultraviolet region, but all substances show characteristic absorption in the infrared. Herzberg [2149] has presented a comprehensive discussion on the theory and interpretation of infrared and Raman spectra. The book by Randall, Fowler, Fuson, and Dangl [3898] can be easily used without an extensive background in mathematics. The application of infrared spectra to qualitative and quantitative analysis has been discussed by Barnes, Gore, et al. [406, 407].

Raman spectroscopy has been largely used in the study of molecular structure, but it also has application to problems of analytical chemistry. Braun and Fenske [716] have outlined some of its analytical possibilities and have presented a bibliography on the subject. Several instruments employing continuous gas lasers of the He/Ne type, instead of conventional low pressure mercury arc sources, are now commercially available. These more powerful sources permit the extension of Raman spectroscopy to smaller samples and to dilute solutions. Fluorescence and absorption frequently associated with the mercury arc source are avoided and the narrower excitation line width of the laser source is advantageous in measurements made of the closely spaced rotational spectra of gases.

The mass spectrometer has become an extremely important tool for the analysis of organic compounds and for the elucidation of their structures. The analytical applications encompass hydrocarbons, alcohols, esters, ketones, and aldehydes as well as materials containing appreciable quantities of water. A recent companion to electron bombardment mass spectrometry that results in fragmentation of the molecules is field-ionization mass spectrometry, which gives largely molecular ions. High resolution mass spectrometers ($M/\Delta M$ = ca. 10,000) are becoming common research equipment, and instruments giving resolutions as high as 1,200,000 have been reported. A short, comprehensive survey of the analytical applications and instrumentation has been prepared by Hipple and Shepherd [2196]. They have included an extensive bibliography. The American Petroleum Institute, Research Project 44, and the National Bureau of Standards [90] have issued a catalog of mass spectra data. For recent developments in instrumentation the reader is referred to the review published annually in *Analytical Chemistry*.

Chamberlain [945] and Stehling [4487] present accurate and rapid methods for the determination of the important features of the molecular structures of organic compounds by nuclear magnetic resonance spectroscopy. These methods also may assist in the identification of impurities in solvents.

SAFETY

The brief statements on toxicology in Chapter V are intended only as guides for safe handling and are not meant to give a complete summary of available data. The present maximum allowable concentrations are taken from the list published by the American Conference of Governmental Industrial Hygienists [77]. These are the maximum average atmospheric concentrations to which workers may be exposed for an eight-hour working day without injury to health. They are expressed on a volume/volume basis.

Chapter III

PHYSICAL PROPERTIES: TABULATIONS

EXPLANATION OF THE TABLES

Each property is denoted by an abbreviation or symbol in the first column. The abbreviations and symbols used in this book are defined in the following section. The second column refers to the condition, such as the temperature or the pressure, at which the property was determined. The third column gives the numerical value of the property as it appears in the primary reference.

References to the literature are listed in the fourth column. Sources of the values (primary references) are given by the numbers before the colon; additional references follow the colon. Values calculated by the authors are designated by enclosing in brackets the reference from which the basic data were taken. Additional references are not confined to the condition for the value of the property listed in the second column. Some references have been included because the discussion is of interest, and others are included because the property was determined at other than the usual conditions such as high or low temperature. The references constitute a good basic literature survey.

Temperature is expressed in degrees centigrade unless otherwise indicated and is designated only by the degree sign. The properties are listed in the following sequence and are designated by the indicated abbreviation or symbol.

1. *Molecular weight*, mw
2. *Boiling point*, bp
3. *Pressure coefficient of temperature*, dt/dp, degrees per torr
4. *Vapor pressure*, p, generally at 25°
5. *Constants* for vapor pressure-temperature equations referred to in Chapter II.
6. *Freezing point*, fp
7. *Triple point*, tp
8. *Density*, d, grams per cubic centimeter
9. *Temperature coefficient of density*, dd/dt, grams per cubic centimeter per degree
10. *Coefficient of thermal expansion*, α, cubic centimeters per degree
11. *Refractive index*, n_D, at the average D-line of sodium

12. *Temperature coefficient of refractive index*, dn/dt, change in refractive index per degree
13. *Viscosity*, η, centapoise, or centistoke where indicated by cs
14. *Surface tension*, γ, dynes per centimeter
15. *Heat of vaporization*, ΔH_v, kilocalories per mole
16. *Heat of sublimation*, ΔH_s, kilocalories per mole
17. *Heat of fusion*, ΔH_m, kilocalories per mole
18. *Heat of transition*, ΔH_t, kilocalories per mole
19. *Standard heat of formation*, ΔH_f°, kilocalories per mole
20. *Heat of combustion*, ΔH_c°, standard state; or ΔH_c, constant pressure; or ΔE_c, constant volume, kilocalories per mole
21. *Heat of polymerization*, ΔH_p, kilocalories per mole
22. *Heat capacity for the ideal gas state*, C_p°, cal deg^{-1} mole^{-1}
23. *Heat capacity*, C_p, cal deg^{-1} mole^{-1}
24. *Critical temperature*, t_c
25. *Critical pressure*, p_c, atmospheres
26. *Critical density*, d_c, grams per cubic centimeter
27. *Critical volume*, v_c, liters per mole
28. *Specific rotation*, $[\alpha]_D$, degrees of rotation at the average D-line of sodium
29. *Cryoscopic constant*, A, Eq. 2.72
30. *Cryoscopic constant*, K_f, Eq. 2.67 degrees per mole of solute per 1000 g solvent
31. *Ebullioscopic constant*, K_B, Eq. 2.79 degrees per mole of solute per 1000 g solvent
32. *Ionization constant*, K_i
33. *Acid dissociation constant*, K_a
34. *Base dissociation constant*, K_b
35. *Negative logarithm of the acid dissociation constant*, pK_a
36. *Negative logarithm of the ionization constant of protonated base*, pK_{BH^+}
37. *Negative logarithm of the autoprotolysis constant*, pK_s
38. *Electrical conductance*, κ, ohm^{-1} cm^{-1}
39. *Dielectric constant*, ϵ, limiting values at low frequencies
40. *Dipole moment*, μ, Debye units (10^{-18} esu)
41. *Solubility with water*, mutual, soly in aq and aq in, %w unless otherwise noted
42. *Aqueous azeotrope*, aq az, boiling point and composition (%w)
43. *Flash point*, fl pt, for kind of flash point, see Abbreviations and Symbols, next section
44. *Spectra*, ultraviolet, infrared, Raman, nuclear magnetic resonance; references only

45. *Beilstein reference*, general reference, the system number is in boldface type and the page number of the volume in which the system number appeared in the original work follows in regular type; for example, **59** Ethanol, Beil. **20**, 292. Specific references follow standard Beilstein usage: Beil, **E III 3,** 1182

ABBREVIATIONS AND SYMBOLS

Å	Ångstrom units, 1×10^{-8} cm
A, B, C $\cdots$	Constants
aq	Water
AR	Analytical reagent (grade)
ASTM	American Society for Testing Materials
az	Azeotrope
bp	Boiling point
C	Centigrade or Celsius scale of temperature
C_p	Heat capacity at constant pressure
C_p^o	Heat capacity at constant pressure for the ideal gas state
CC	Closed cup flash point tester, kind not known
COC	Cleveland open cup flash point tester, ASTM D 92-66
cp	Centipoise
cs	Centistoke
d	Density
d	Decomposes, when following a temperature value
d_c	Critical density
dd/dt	Temperature coefficient of density
dn/dt	Temperature coefficient of refractive index at the average D-line of sodium, 5892.6 Å
dt/dp	Pressure coefficient of temperature
eq	Equation
F	Fahrenheit scale of temperature
fl pt	Flash point
fp	Freezing point
g	Gram
(g)	Gaseous or vapor state
gl	Glass or vitreous condition
id	Inside diameter
ir	Infrared
inf	Infinite
K	Kelvin or absolute scale of temperature
K_a	Ionization constant of acid
K_b	Ionization constant of base

K_B	Ebullioscopic constant
K_f	Cryoscopic constant, Eq. 2.67
K_f'	Cryoscopic constant, %m/1°
K_i	Ionization constant
K_s	Autoprotolysis constant
(l)	Liquid state
ml	Milliliter
min	Minute
(ms)	Metastable
mw	Molecular or formula weight
$m\mu$	Millimicron
NBS	National Bureau of Standards
nmr	Nuclear magnetic resonance
n_D	Refractive index at sodium D-line, 5892.6 Å
OC	Open cup flash point tester, kind not known
p	Vapor pressure, torr unless otherwise stated
p_c	Critical pressure, atmospheres
pK	Negative logarithm of ionization constant
pK_a	Negative logarithm of acid ionization constant
pK_b	Negative logarithm of base ionization constant
pK_{BH^+}	Negative logarithm of the ionization constant of the protonated base
pK_s	Negative logarithm of autoprotolysis constant
pK_w	Negative logarithm of the ion product of water
ppb	Parts per billion
ppm	Parts per million
psi	Pounds per square inch
psia	Pounds per square inch absolute
psig	Pounds per square inch gauge
soly	Solubility
sl sol	Slightly soluble
(s)	Solid state
(st)	Stable
t	Temperature, centigrade or Celsius, unless otherwise indicated
T	Temperature, absolute or Kelvin, 0° T = −273.15°
t_c	Critical temperature
TCC	Tag closed cup flash point tester, ASTM D 56-64
TLV	Threshold limit value
TOC	Tag open cup flash point tester, ASTM D 1310-63
tp	Triple point
USP	United States Pharmacopeia
uv	Ultraviolet

(u)	Unstable
V_c	Critical volume
α	Coefficient of thermal expansion
$[\alpha]$	Specific rotation
γ	Surface tension
ΔE_c	Heat of combustion at constant volume
ΔH_c	Heat of combustion at constant pressure
ΔH_c°	Heat of combustion at constant pressure, standard state
ΔH_f°	Standard heat of formation
ΔH_m	Heat of fusion
ΔH_p	Heat of polymerization
ΔH_s	Heat of sublimation
ΔH_v	Heat of vaporization
ϵ	Dielectric constant
η	Viscosity, coefficient of
κ	Specific conductance
μ	Dipole moment, electric moment
μ	Micron
[ ]	Author's remarks; in Chapter III, tables, calculated from designated reference
$^{\circ}$	Degrees temperature, centigrade or Celsius; also degrees of angular rotation
%m	Percent, mole
%v	Percent, volume
%w	Percent, weight
%w/v	Percent, weight/volume, g/100 ml solution
(+)	Dextrorotary
(−)	Levorotary
(±)	Racemic mixture

TABLES OF PHYSICAL PROPERTIES

0a. Water

mw		18.01534			887
bp	760	100°			definition
dt/dp	760	0.03687			4635
p	see	Table 0c			
eq 5		A	B	C	94
	0-29°	8.184254	1791.3	238.1	
	30-39	8.1393986	1767.262	236.29	
	40-49	8.0886767	1739.351	234.10	
	50-59	8.0464202	1715.429	232.14	
	60-69	8.0116295	1695.167	230.41	
	70-79	7.9845588	1678.948	228.97	
	80-89	7.9634288	1665.924	227.77	
	90-99	7.9483960	1656.390	226.86	
	100-150	7.9186968	1636.909	224.92	
fp	760	0°			definition
tp		0.0100			4054
d	see	Table 0b			
n_D	see	Table 0b			
η	see	Table 0b			
γ	see	Table 0b			
$\triangle H_v$	see	Table 0c			
$\triangle H_m$		1.4359			1805
$\triangle H_s$		12.185			1805
$\triangle H_f^\circ$	25°(g)	-57.7979			5013
	25°(l)	-68.315			5358
C_p°	0°	8.001			92
	25	8.025			
	100	8.134			
	Table XIV	Ch II			
Cp	see	Table 0c			
t_c		374.2			4636
p_c		218.3			4636
d_c		0.325			4636
v_c		0.0554			4636
K_f, eq 67	Ch II	1.853			1805
K_B		0.515			1805

0a. Water
(Continued)

pK$_w$	5°	14.734 (molarity scale)	2034
	10	14.535	
	20	14.169	
	25	14.000	
	30	13.837	
	40	13.542	
	50	13.272	
κ, calcd.	25°	5.49×10^{-8}	1980
κ, exp.	15°	3.30×10^{-8}	
	18	3.90×10^{-8}	
	20	4.40×10^{-8}	
	25	5.89×10^{-8}	
	30	7.75×10^{-8}	

0b. Water

t	d	n_D	η	γ
Ref. No.	93	4637, 4759	1805, 4587	1805
0°	0.9998396	1.3339492	1.7702	75.626
1	0.9998987	1.3339475		75.470
5	0.9999641	1.3338835	1.5108	74.860
10	0.9997000	1.3336902	1.3039	74.113
15	0.9991005	1.3333873	1.1374	73.350
15.56	[0.9990003]	[1.3333426]		
20	0.9982058	1.3329880	1.0019	72.583
21	0.9979944	1.3328976	[0.9764]	72.427
22	0.9977726	1.3328037	[0.9532]	72.270
23	0.9975407	1.3327067	[0.9310]	72.113
24	0.9972989	1.3326064	[0.9100]	71.957
25	0.9970474	1.3325029	0.89025	71.810
26	0.9967864	1.3323964	[0.8703]	71.652
27	0.9965160	1.3322869	[0.8512]	71.495
28	0.9962364	1.3321743	[0.8328]	71.339
29	0.9959479	1.3320589	[0.8145]	71.192

0b. Water
(Continued)

t	d	n_D	η	γ
30°	0.9956504	1.3319405	0.79726	71.035
35	[0.9940313]	1.3313076	0.71903	70.230
40	0.9922191	1.3306096	0.65263	69.416
45	0.9902161	1.3298513	0.59716	68.592
50	0.9880382	1.3290364	0.54675	67.799
55	0.9856959	1.3281683	0.50415	66.894
60	0.9831980	1.3272495	0.46688	66.040
65	0.9805524		0.43407	65.167
70	0.9777657		0.40503	64.274
75	0.9748437		0.37918	63.393
80	0.971791		0.35604	62.500
85	0.968613		0.33524	61.587
90	0.965313		0.31647	60.684
95	0,961893		0.29945	59.763
100	0.958357		0.28395	58.802
105	0.954706			

0c. Water

t Ref. No.	p 94	$\triangle H_v$ [3603]	Cp [1782]	ϵ 4953	3129
0°	4.581	10.7669	18.161	87.91	87.740
1	4.925				
5	6.542	10.7165	18.094	85.89	85.763
10	9.208	10.6660	18.051	83.96	83.832
15	12.788	10.6154	18.023	82.11	81.946
15.56	[13.26]				
20	17.536	10.5647	18.006	80.20	80.103
21	18.651		18.004		[79.73]
22	19.828		18.002		[79.38]
23	21.070		18.000		[79.02]
24	22.379		17.998		[78.65]
25	23.758	10.5138	17.997	78.39	78.304
26	25.211		17.996		[77.94]
27	26.741		17.995		[77.60]
28	28.351		17.994		[77.24]
29	30.045		17.993		[76.90]

0c. Water
(Continued)

t	p	$\triangle H_V$	Cp	ϵ	
30°	31.827	10.4629	17.992		76.546
35	42.180	10.4117	17.991		74.828
40	55.33	10.3604	17.992	73.19	73.151
45	71.89	10.3087	17.996		71.512
50	92.55	10.2569	18.001		69.910
55	118.09	10.2048	18.009		68.345
60	149.44	10.1523	18.017		66.815
65	187.62	10.0995	18.028		65.319
70	233.77	10.0463	18.040		63.857
75	289.17	9.99271	18.054		62.427
80	355.26	9.93862	18.069		61.027
85	433.58	9.88406	18.087		59.659
90	525.87	9.82889	18.106		58.319
95	634.00	9.77317	18.129		57.007
100	760.00	9.71671	18.153		55.720
105	906.07				

1. Cyclopentane

$$\underset{\underset{\text{L}\quad\quad\quad\quad\quad\quad\text{J}}{}}{CH_2\ CH_2\ CH_2\ CH_2\ CH_2}$$

C_5H_{10}

mw		70.135	887
bp	760	49.262°	147: 1608, 2370, 5169, 5204, 4781
dt/dp	760	0.04003°	144: 5204, 1608
p	25°	318.	147: 4535, 5204
eq 5	A	6.88676	146: 5204, 2534
	B	1124.162	
	C	231.361	
fp		-93.866°	147: 1608, 2370, 1794, 5169
d	20°	0.74538	144: 460, 2534, 4965, 1732, 1608, 4781
	25	0.74045	5169
n_D	20°	1.40645	144: 460, 4781, 5169, 1608, 4965
	25	1.40363	5170
η	10°	0.492	145: 1732, 4781
	20	0.439	
	25	0.416	
	Table I	Ch II	
γ	10°	23.6	150: 4781, 4965, 2292, 5169
	20	22.42	
	25	21.82	
$\triangle H_v$	25°	6.818	147, 3065, 3006: 4463, 4962
	bp	6.524	
	Table IX	Ch II	
$\triangle H_m$		0.1455	147: 2370
$\triangle H_f^\circ$	25°(l)	-25.23	4462: 149, 2411, 5358
$\triangle H_c^\circ$	25°(l)	-786.62	4462: 2577, 148, 2411, 5358
C_p°	25°	19.82	152: 3065, 4463
C_p	26.84°	30.46	4589: 2370, 4463
t_c		238.60°	151, 964, 1786; 2534, 72
p_c		44.55	151, 2534, 1786
d_c		0.270	151, 1786
v_c		0.260	151
A	eq 72	0.00228	147
ϵ	20°	1.969	4786: 3477
μ		0.00	4906

1. Cyclopentane

(Continued)

soly	in aq, 25°	0.0159%w	3053
	aq in, 20°	0.0142%w	1483
fl pt		− 35°F	3747
uv			3760
ir			89, 2807, 3792, 5078, 386, 3796
Raman			445, 1556, 4294, 1655, 460, 3901, 47
mass			90, 2205, 3997, 3322, 1390
nmr			859, 3098
Beil	452, 19		

2. Pentane

$$CH_3CH_2CH_2CH_2CH_3 \hspace{6cm} C_5H_{12}$$

mw		72.151	887
bp	760	36.074°	100: 1608, 4271, 5234, 2269, 4763
dt/dp		0.03856°	95, 5204: 5234, 1608, 2269
p	25°	512.5	100: 3279
eq 5	A	6.85221	109: 467, 5204, 3279
	B	1064.63	
	C	232.000	
fp		− 129.721°	100: 1947, 3651, 4770, 3280, 3279, 4763
d	20°	0.62624	95: 1345, 2269, 5169, 1732, 1608, 4780
	25	0.62139	
dd/dt	0 − 40°	0.000975	1236: 2269
n_D	20°	1.35748	95: 1345, 2269, 5169, 1608, 3115
	25	1.35472	
dn/dt	20 − 25°	0.00053	2269
η	15°	0.245	106: 4271, 1773, 1732, 4780
	20	0.235	
	25	0.225	
	Tables I, II	Ch II	

2. Pentane
(Continued)

γ	25°	15.48	107: 2378, 3873, 5169, 4780, 4967
	Table VIII	Ch II	
$\triangle H_v$	25°	6.316	100: 3279, 3601, 3776, 3331, 4288, 702
	bp	6.160	
$\triangle H_m$		2.006	100: 3279, 3280, 3651
$\triangle H_f^\circ$	25°(l)	−41.36	112: 3767, 1864, 5358
$\triangle H_c^\circ$	25°(l)	−838.71	3840: 3767, 111, 4052, 2577, 1864, 5358
C_p°	25°	28.73	113: 3722
	Table XIV	Ch II	
C_p	25°	39.96	3280: 2384, 3776, 3279, 3651
t_c		196.34°	98: 64, 3331, 4708, 702, 467, 1100
p_c		33.16	98: 5298, 3331, 702, 467
d_c		0.245	98: 5298, 702, 467
v_c		0.295	98, 467: 4708, 5298, 1100, 5303
A	eq 72	0.04907	100: 3280
κ	19.5°	2×10^{-10}	1723:
ϵ	20°	1.844	3477: 1345
μ		0.00	4906: 2233
soly	in aq, 25°	0.0038%w	3053: 4225, 1064
	aq in, 24.8°	0.0120%w	614
aq az	34.6°	98.6%w	2255
fl pt	TOC	−40°F	3680: 3681
ir			89, 2807, 4547, 388, 3917, 408
Raman			1556, 2210, 4273, 2138, 3901, 1176
mass			90, 3322, 3553, 4434
Beil	10. 130		

3. 2–Methylbutane

Isopentane

$CH_3CH(CH_3)CH_2CH_3$ C_5H_{12}

mw		72.151	887
bp	760	27.852°	110, 5204: 1608, 4780, 4788, 5169
dt/dp	760	0.03815°	105, 5204: 1608, 4780
p	25°	688.	110: 4023, 4194, 4193
eq 5	A	6.78967	109: 4194, 5204
	B	1020.012	
	C	233.097	
fp		− 159.900°	110: 1794, 2645, 3652, 4771, 4857
d	20°	0.61967	105: 1608, 4761, 5169, 1732, 2269, 4788
	25	0.61462	
n_D	20°	1.35373	105: 856, 1608, 5170, 1386, 5169
	25	1.35088	
dn/dt	20 – 25°	0.00058	2269
η	15°	0.237	106: 4783, 1732
	20	0.225	
	25	0.215	
	Table I, II	Ch II	
γ	20°	15.00	107, 2102: 5169
	25	14.46	
$\triangle H_v$	25°	5.937	110: 3331, 4194, 4193, 4207
	bp	5.901	
$\triangle H_m$		1.231	110: 1939, 4193, 3652, 4194
$\triangle H_f^\circ$	25° (l)	− 42.85	112: 3767, 1864, 5358
$\triangle H_c^\circ$	25°	− 837.31	111: 3767, 2645, 2577, 1864, 5358
C_p°	25°	28.39	113: 4207
	Table XIV	Ch II	
C_p	26.84°	39.55	1939: 3652, 4194, 4207, 4193
t_c		187.24°	108: 5302, 3331, 5303, 4553, 72, 4982
p_c		33.66	108: 5302, 4553, 5301, 3331, 4982
d_c		0.236	108: 5302, 5301, 4553, 4982
v_c		0.306	108: 5302, 5301, 5303
A	eq 72	0.04830	110
ϵ	20°	1.843	3477

3. 2–Methylbutane

(Continued)

soly	in aq, 25° aq in, 21.8°	0.0048%w 0.0097%w	3053 614: 1483
fl pt		– 70°F	3747
uv			3817
ir			89, 2429, 3917, 388, 2807
Raman			2479, 3901, 5349, 5292, 3246, 2709
mass			90, 1434, 4434
Beil	**10**, 134		

4. 2,2–Dimethylpropane

Neopentane

$C(CH_3)_4$ C_5H_{12}

		72.151	887
mw			
bp	760	9.503°	110: 310, 5157, 4857
dt/dp	760	0.0361°	105
p	25°	1285.	110: 310, 5157
eq 5	A B C	6.73812 950.84 237.00	109: 464
fp		– 16.58°	110: 310, 5157, 4857
d	0° 20 25	0.613 0.5910 0.5851	5157 105
n_D	0° 6 20	1.3513 1.3476 1.342	5157 105
η	– 10° 0 5	0.391° 0.328 0.303	106
γ	20° 30	12.05 10.98	1370

4. 2,2–Dimethylpropane
(Continued)

$\triangle H_v^*$	25°	5.205	110: 310
	bp	5.438	
$\triangle H_m$		0.752	110: 310
$\triangle H_f^\circ$	25° (l)	– 44.98	112: 3767, 1864, 5358
$\triangle H_c^\circ$	25°	– 835.18	111: 3767, 2645, 2577, 1864, 5358
C_p°	25°	29.07	113
	Table XIV	Ch II	
t_c		160.60°	108: 464, 310, 4708
p_c		31.57	108: 464, 310, 4708
d_c		0.238	108: 464
v_c		0.303	108: 464, 4708
A	eq 72	0.00575	110
ϵ	20°	1.801	1370
soly	in aq, 25°	0.0033%w	3053
fl pt		$– 85^\circ$F	3747
ir			89, 3917, 2429
Raman			2479, 3900, 2664, 3246, 4273, 3231
mass			1301, 3379, 4434, 1434, 90
nmr			3126
Beil	**10**, 141		

* at saturation pressure

5. Methylcyclopentane

$\underline{CH_2 CH_2 CH_2 CH_2 CHCH_3}$ $\hspace{6cm}$ C_6H_{12}

mw		84.162	887
bp	760	71.812°	147, 5204, 1608: 4781, 2172, 5169
dt/dp	760	0.04274°	144: 1608, 5204
p	25°	138.	147: 5204
eq 5	A	6.86283	146: 2809, 5204
	B	1186.059	
	C	226.042	
fp		−142.434°	147: 1608, 2172, 5169, 1794, 4781
d	20°	0.74864	144: 1608, 3391, 5169, 1732, 2534, 4781
	25	0.74394	
n_D	20°	1.40970	144: 1608, 3391, 5169, 2172, 4781, 5170
	25	1.40700	
η	20°	0.507	145: 4781, 1732
	25	0.478	
	Table I	Ch II	
γ	20°	22.19	150: 5169
	25	21.61	
$\triangle H_v$	25°	7.550	147: 3065, 3601, 4962
	bp	6.950	
	Table IX	Ch II	
$\triangle H_m$		1.656	147
$\triangle H_f^\circ$	25°(l)	− 32.92	5358: 2411, 149
$\triangle H_c^\circ$	25°(l)	− 941.28	5358, 2411: 3390, 2577, 148
C_p°	25°	26.24	152, 4203
	Table XIV	Ch II	
t_c		259.61°	151, 1786: 72
p_c		37.36	151, 1786
d_c		0.264	151, 1786, 2534
v_c		0.319	151
A	eq 72	0.04877	147
soly	aq in, 20°	0.0131%w	1483
aq az	63.3°	82%w	3975
fl pt	TCC	−17°F	3747
ir			89, 2807, 3796, 386, 3791, 408
Raman			460, 3901, 1556, 4294
mass			90, 3997
Beil	452, 27		

6. Cyclohexane

$$\underset{\underbrace{\hphantom{CH_2CH_2CH_2CH_2CH_2CH_2}}}{CH_2CH_2CH_2CH_2CH_2CH_2}{}_2 \qquad\qquad C_6H_{12}$$

mw		84.162	887
bp	760	80.725°	156: 1608, 4059, 1791, 5204, 4783
dt/dp	760	0.04376°	153: 1608, 5204, 1900
p	25°	98.	156
eq 5	A	6.84498	155: 5204, 826
	B	1203.526	
	C	222.863	
fp		6.544°	156: 1794, 3288, 4072, 3011, 4158
d	20°	0.77855	153: 1608, 3288, 4125, 1732, 3195, 4059
	25	0.77389	4158, 5169, 4783, 5254
α	25°	0.001217	5254
n_D	20°	1.42623	153: 1140, 4244, 5169, 4109, 4783, 5170
	25	1.42354	
dn/dt		0.00054	1900
η	20°	0.980	154: 4783, 4711, 1732
	25	0.898	
	Table I	Ch II	
γ	20°	24.98	159: 2292, 5169, 4965
	25	24.38	
$\triangle H_v$	25°	7.896	156: 3208, 3601, 2791, 702, 3331, 4463
	bp	7.16	3006, 4962
$\triangle H_m$		0.6399	156: 3652, 4072
$\triangle H_f^\circ$	25°(l)	− 37.34	158, 2411: 4462, 4072, 5358
$\triangle H_c^\circ$	25°(l)	− 936.53	4462: 2577, 3968, 157, 3670, 3390, 2411
C_p°	25°	25.40	161: 731, 4463
C_p	23.39°	37.185	4072: 3745, 4463, 3652
t_c		280°	2651, 5303: 1100, 702, 622, 3331, 964
p_c		40.0	2651, 73, 5301: 702, 5304, 1791, 3331
d_c		0.273	2651, 5301: 702, 5304
v_c		0.3077	[5303]: 1100, 5304, 962
A	eq 72	0.004116	156
K_f		20.0	3189: 3190
K_B		2.75	3187: 3189
κ	25°	$\sim 7\times 10^{-18}$	1603
ϵ	20°	2.023	3477, 3247: 5189, 1140, 4244, 4786, 1037

6. Cyclohexane

(Continued)

soly	aq in, 20° in aq, 25°	0.010%w 0.0055%w	614: 529, 4609, 1483 3053: 2772
aq az	68.95°	91%w	2254: 2255
fl pt	TCC	1°F	3680: 88, 3747
uv			2047, 1237, 4505, 4136, 1238, 3760
ir			89, 3792, 4547, 2454, 1001, 3796, 408
Raman			445, 1854, 47, 1556, 3901, 2395
mass			90, 3997, 1454, 2403, 2205
nmr			2210, 2454, 3098, 859
Beil	**452**, 20		

7. Hexane

$$CH_3(CH_2)_4CH_3 \hspace{6cm} C_6H_{14}$$

mw		86.178	887
bp	760	68.740°	100: 1608, 4166, 4784, 4807, 5234, 1791
dt/dp	760	0.04191°	95: 4166, 4271, 4784, 5234
p	25°	151.3	100:4166
eq 5	A B C	6.87776 1171.530 224.366	117: 2533, 5204
fp		−95.348°	100: 1236, 1608, 1794, 2302, 4784
d	20° 25	0.65937 0.65481	95: 1128, 1345, 1608, 2533, 2901, 1732 3873, 4271, 4784, 4807, 5032, 3392
dd/dt	0−40°	0.000891	1236:
n_D	20° 25	1.37486 1.37226	95: 1128, 1345, 1608, 3115, 4807, 5169 5170
dn/dt	0−40°	0.000542	1236:
η	15° 20 25	0.3275 0.3126 0.2985	96: 1404, 4271, 4784, 4928, 1773, 1732
	Table I, II, IV Ch II		
γ	25°	17.91	[2379], 115: 4967, 5032, 5034, 5169, 2378
	Table VIII	Ch II	

7. Hexane

(Continued)

$\triangle H_v$	25° bp	7.541 6.896	100: 2901, 3331, 3601, 4866, 5007, 2791, 702, 2312
$\triangle H_m$		3.126	100
$\triangle H_f^\circ$	25° (l)	−47.52	120: 1864, 5358
$\triangle H_c^\circ$	25° (l)	−995.01	119, 5358: 2577, 1864, 2398, 3840
C_p°	25°	34.20	121, 3722: 5007
	Table XIV	Ch II	
C_p	25°	46.71	3280: 2302, 3652, 3745, 4536, 1351
t_c		234.15°	98, 72: 622, 2533, 5303, 1100, 1791
p_c		29.71	98: 64, 2533, 3331, 2650, 4708, 5305
d_c		0.233	98: 5301, 4717, 2650, 702
v_0		0.370	98: 4717, 1100, 2650, 4708, 5303
A	eq 72	0.04976	100: 4757
κ	25°	$\langle 10^{-16}$	1603:
ϵ	25°	1.8799	3392: 3477, 1345, 5198, 2048
μ	25° (l)	0.085	3392:
soly	in aq, 25° aq in, 20°	0.00095%w 0.0111%w	3053: 4225, 1677 614: 1483
aq az	61.6°	94.4%w	2255: 2254
fl pt	TCC	−15°F	3680: 3747, 3160, 1370
uv			916
ir			89, 386, 1090, 4547, 3935, 2435, 5278
Raman			1006, 1556, 1854, 1855, 3539, 4040, 4599
mass			90, 3997, 4434, 1434, 471
Beil	**10**, 142		

8. 2-Methylpentane

Isohexane

$(CH_3)_2CHCH_2CH_2CH_3$ $\hspace{6cm}$ C_6H_{14}

mw		86.178	887
bp	760	60.271°	118, 5204: 820, 1608, 4536, 5169
dt/dp	760	0.04141°	114: 1608, 2269
p	25°	212.	118
eq 5	A	6.83910	117: 5204
	B	1135.410	
	C	226.572	
fp		−153.670°	118: 820, 1608, 4536
d	20°	0.65315	114: 820, 1608, 2269, 2901, 3989, 5169
	25	0.64852	1732
n_D	20°	1.37145	114: 820, 1608, 3989, 4806, 5169, 5170
	25	1.36873	
dn/dt	20 − 25°	0.00046	4772: 2269
η	−195.7°gl	2.4 x 10^{15}	2967
	+20	0.310	4772: 1404, 1732
	25	0.295	
	Table I, II	Ch II	
γ	20°	17.38	115: 5169
	25	16.87	
	30	16.37	
$\triangle H_v$	25°	7.138	118: 2901, 3601
	bp	6.643	
$\triangle H_m$		1.498	118: 1351, 4536
$\triangle H_f^\circ$	25°(l)	− 48.82	120: 5358
$\triangle H_c^\circ$	25°(l)	− 993.71	119, 5358
C_p°	25°	34.46	121
C_p	26.84°	46.47	1351: 3601, 4536
t_c		224.30°	116, 72: 2533, 4095
p_c		29.71	116: 2533
d_c		0.235	116
v_c		0.367	116
A	eq 72	0.05287	118
ϵ		1.881	1370

8. 2–Methylpentane

(Continued)

soly	in aq, 25°	0.0014%w	3053: 1064
aq az	55.1°	97 %w	3975
fl pt	TCC	⟨– 20°F	3680: 3747
ir			89, 388, 2807, 2951
Raman			459, 1556, 3246, 4040, 3901, 4294
mass			90, 4434, 1434
Beil	10, 148		

9. 3–Methylpentane

$CH_3CH_2CH(CH_3)CH_2CH_3$ C_6H_{14}

mw		86.178	887
bp	760	63.282°	118, 1608, 5204: 820, 2533, 5169, 5238
dt/dp	760	0.04182°	114, 1608, 5204: 2269, 5238
p	25°	190.	118
eq 5	A	6.84887	117: 5204, 1232
	B	1152.368	
	C	227.129	
fp		gl	1351: 1608, 1794
d	20°	0.66431	114: 820, 1128, 1608, 2269, 2533, 3873
	25	0.65976	3989, 4536, 5007, 5169, 1732
n_D	20°	1.37652	114, 1608: 1128, 3989, 5169, 5170
	25	1.37386	
dn/dt	20 – 25°	0.00054	2269
η	– 195.1°gl	2.2 x 10^{14}	2967
	25°	0.307	4772: 1732
	30	0.292	
	Table I	Ch II	
γ	20°	18.12	115: 5169, 3873
	25	17.60	
	30	17.08	

9. 3-Methylpentane

(Continued)

ΔH_v	25° bp	7.236 6.711	118: 3601, 4965
ΔH_m		forms glass	1351
ΔH_f°	25°(l)	−48.28	120 :5358
ΔH_c°	25°(l)	−994.25	119, 5358
C_p°	25°	34.20	121: 5007
C_p	26.84°	45.77	1351: 3601, 4536, 5007
t_c		231.2°	116: 1232, 2533
p_c		30.83	116: 1232, 2533
d_c		0.235	116
v_c		0.367	116
ϵ		1.895	1370
soly	in aq, 25°	0.0013%w	3053
fl pt	TCC	−25°F	3747
ir			89, 388, 2807, 2951
Raman			459,1556, 3246, 3901, 4040, 4273, 4294
mass			90, 4434, 1434
Beil	10, 149		

10. 2,2-Dimethylbutane

Neohexane

$CH_3C(CH_3)_2CH_2CH_3$ $\qquad\qquad\qquad\qquad\qquad\qquad\qquad\qquad$ C_6H_{14}

mw		86.178	887
bp	760	49.741°	118, 1608, 5204: 759, 2533, 5169, 5238
dt/dp	760	0.04117°	114, 1608, 5204: 759, 2269, 5238
p	25°	320.	118: 759
eq 5	A B C	6.75483 1081.176 229.343	117: 5204
fp		−99.865°	118: 759, 1608, 1794, 4536, 5169

10. 2,2–Dimethylbutane

(Continued)

d	$20°$ 25	0.64916 0.64446	114: 759, 1128, 1608, 2269, 2533, 3989 4536, 5169
n_D	$20°$ 25	1.36876 1.36595	114: 867, 759, 1128, 3989, 5169, 5170
dn/dt	$20 - 25°$	0.00058	2269
η	$25°$ 30	0.351 0.330	4772
	Table I	Ch II	
γ	$20°$ 25 30	16.30 15.81 15.31	115: 5169
$\triangle H_v$	$25°$ bp	6.618 6.287	118: 3601, 5007, 3776
$\triangle H_m$		0.1384	118: 1351, 4536
$\triangle H_f^°$	$25°(l)$	-51.00	120: 5358
$\triangle H_c^°$	$25°(l)$	-991.52	119, 5358
$C_p^°$	$25°$	33.91	121: 5007
C_p	$26.84°$	45.27	1351: 4536, 4095, 3776
t_c		$215.58°$	116, 72: 4708, 4095, 2533
p_c		30.40	116: 4708, 2533
d_c		0.240	116
v_c		0.359	116: 4706
A	eq 72	0.002319	118
ϵ		1.873	1370
soly	in aq, $25°$	0.0018%w	3053
fl pt	TCC	$-25°F$	3747
uv			2700
ir			89, 2807, 4727
Raman			1556, 3246, 3901, 4040, 4599, 4294, 5349
mass			90, 4434, 1434
Beil	**10**, 150		

11. 2,3-Dimethylbutane

Diisopropyl

$(CH_3)_2CHCH(CH_3)_2$ C_6H_{14}

mw		86.178	887
bp	760	57.988°	118, 1608, 5204: 759, 820, 2533, 4536
dt/dp	760	0.04173°	114, 1608, 5204: 759, 5238
p	25°	235.	118: 759
eq 5	A	6.80983	117: 5204
	B	1127.187	
	C	228.900	
fp		−128.538°	118: 759, 820, 1608, 5169, 5238
d	20°	0.66164	114, 1608: 759, 820, 1128, 2533, 2901
	25	0.65702	3989, 4536, 5169, 1732
n_D	20°	1.37495	114, 1608: 759, 820, 1128, 2269, 3989
	25	1.37231	5169, 5170
dn/dt	20 – 25°	0.00049	2269
η	25°	0.361	4772: 1732
	30	0.342	
	Table I	Ch II	
γ	20°	17.37	115: 5169
	25	16.87	
	30	16.37	
$\triangle H_v$	25°	6.961	118: 2901, 3331, 3601
	bp	6.519	
$\triangle H_m$		0.191	118: 1351
$\triangle H_f^\circ$	25° (l)	−42.49	120: 5358
$\triangle H_c^\circ$	25° (l)	−993.05	119, 5358: 2577, 4743
C_p°	25°	33.59	121
C_p	26.84°	45.28	1351: 4207, 4536
t_c		226.78°	116, 72: 5301, 2651, 2602, 359, 5303
p_c		30.86	116: 2602, 5301, 2651, 359, 2533, 3331
d_c		0.241	116: 2602, 5301, 2651, 359
v_0		0.358	116: 359, 5303
A	eq 72	0.00460	118
ϵ		1.890	1370
soly	aq in, 20°	0.0110	1483
fl pt	TOC	⟨−20°F	3680: 3747
ir			89, 2807, 4727, 2951
Raman			1556, 3246, 3901, 4040, 4294, 5349, 5292
mass			90, 4434. 1434
Beil	**10**, 151		

12. Methylcyclohexane

$CH_2CH_2CH_2CH_2CH_2CHCH_3$ C_7H_{14}

mw		98.190	887
bp	760	100.934°	156, 5204: 751, 855, 2172, 5169, 3482
dt/dp	760	0.04671°	153: 1608, 5204
p	25°	46.	156: 1781, 5204
eq 5	A	6.82689	155: 4534, 3482, 2534
	B	1272.864	
	C	221.630	
fp		−126.593°	156: 1608, 1794, 2172, 4771, 5169
d	20°	0.76939	153: 751, 1608, 2172, 3195, 3391, 4783
	25	0.76506	4788, 5169, 1732
n_D	20°	1.42312	153: 326, 751, 1608, 3391, 3901, 4783
	25	1.42058	5169, 5170
η	−195.7°gl	1.2×10^{20}	2967
	20	0.734	154: 4788, 1732
	25	0.685	
	Table I	Ch II	
γ	20°	23.70	159: 2102, 1290a, 2292, 4965, 5169
	25	23.17	
$\triangle H_v$	25°	8.451	156, 3006: 3208, 3601, 4463, 4962
	bp	7.44	
$\triangle H_m$		1.6135	156: 3651
$\triangle H_f^o$	25°(l)	−45.45	158, 2411: 5358
$\triangle H_c^o$	25°(l)	−1091.13	157, 2411, 5358: 2577, 3670
C_p^o	25°	32.27	161: 4463
C_p	21°	43.5	3651: 2384, 4463
t_c		298.97°	72: 2534, 1786, 3482
p_c		34.32	2534, 1786
d_c		0.266	2534, 1786
v_c		0.3445	1369
A	eq 72	0.03780	156
K_f		14.13	2294
κ	25°	$\langle\, 10^{-16}$	1603
ϵ	20°	2.020	3477: 5189
μ		0	5189

12. Methylcyclohexane

(Continued)

soly	in aq, 25°	0.0014%w	3053
	aq in, 20°	0.0116%w	1483
aq az	80.0°	75.9%w	3975
fl pt	TOC	22° F	3747
uv			3817
ir			89, 386, 3791, 3790, 3792, 3796, 408
Raman			460, 1556, 1854, 3901, 3522, 4294, 47
mass			90, 2403, 3997
nmr			245
Beil	**452**, 29		

13. Heptane

$CH_3(CH_2)_5CH_3$ $\hspace{8cm}$ C_7H_{16}

mw		100.206	887
bp	760	98.427°	126: 759, 1657, 1438, 1608, 4380, 5204
dt/dp	760	0.04479°	95: 758, 759, 1608, 4382, 5204, 5234
p	25°	45.7	126: 1546, 4380, 4382, 4413, 5204
eq 5	A	6.90240	125
	B	1268.115	
	C	216.900	
fp		−90.612°	126: 751, 1794, 3115, 3652, 1350, 2299
d	20°	0.68376	95: 55, 758, 759, 822, 1245, 1657
	25	0.67951	1438, 2172, 4271, 4424, 5169, 1732
dd/dt	0 – 50°	0.000840	1236
n_D	20°	1.38764	95, 1606: 751, 759, 822, 1345, 758
	25	1.38511	1438, 1546, 4417, 4486, 4593, 5170
dn/dt	0 – 40°	0.000508	1236
η	−40.01°	1.321 cs	2407: 1404, 1657, 1546, 2939, 4271, 4424
	15	0.4412	96: 4780, 4928, 1773, 1732
	20	0.4181	
	25	0.3967	
	35	0.3592	
	Tables I, II, IV Ch II		

13. Heptane
(Continued)

γ	20° 25 30	20.30 19.80 19.30	123: 22, 1657, 3873, 4780, 4967, 5169 2378
	Table VIII	Ch II	
$\triangle H_v$	25° bp	8.736 7.576	126: 3208, 3331, 3601, 5010, 3775, 3079 126: 4989, 702, 3775
$\triangle H_m$		3.355	126, 3775
$\triangle H_f^\circ$	25° (l)	−53.63	128: 1864, 1204, 5358
$\triangle H_c^\circ$	25° (l)	−1151.27	127, 5358: 2577, 1864, 1204, 2398, 3840
C_p°	25°	39.67	129: 1782, 5010, 1350, 3722, 2299
	Table XIV	Ch II	
C_p	25°	53.76	2299, 3280: 3652, 3745, 5010, 1350, 3601
	Table XII	Ch II	
t_c		266.99°	98: 5301, 5299, 622, 3331, 5303, 72
p_c		27.05	98: 5301, 5299, 1439, 3331, 407, 2602
d_c		0.232	98: 5301, 5299, 2602, 466, 702
v_c		0.432	98: 5301, 5299, 1100, 466, 407, 5303
A	eq 72	0.05067	126: 1439, 4757, 1350, 2299
K_B		3.43	1803
κ	25°	$\langle 10^{-16}$	1603
ϵ	20°	1.924	3477: 1345, 1439, 1621, 4424
μ		0.0	2233: 4424
soly	in aq, 25° aq in, 25°	0.00029%w 0.0091%w	3053: 4225, 1064, 1677, 614, 1483 4134:
aq az	79.2°	87.1%w	2255
fl pt	TOC	30°F	3747: 3680, 3681
uv			3817
ir			89, 386, 2119, 2807, 4547, 408, 2435
Raman			459, 651, 1006, 2138, 3491, 3901, 4273
mass			90, 3997, 1434, 3379, 471
Beil	**10**, 154		

14. 2-Methylhexane

$(CH_3)_2CHCH_2CH_2CH_2CH_3$ C_7H_{16}

mw		100.206	887
bp	760	90.052°	126: 1439, 1438, 4535, 5169
dt/dp	760	0.04431°	122
p	25°	66.	126, 2299: 4535
eq 5	A	6.87318	125: 2299
	B	1236.026	
	C	219.545	
fp		− 118.276°	126: 1439, 1438, 3652, 4522, 5169, 2299
d	20°	0.67859	122: 1439, 1438, 4424, 5169
	25	0.67439	
n_D	20°	1.38485	122, 1605: 1439, 1438, 4424, 5169, 5170
	25	1.38227	
η	20°	0.378	4424: 1439
	Table I	Ch II	
γ	20°	19.29	123: 1439, 5169, 1555, 1370
	25	18.80	
$\triangle H_v$	25°	8.319	126
	bp	7.330	
$\triangle H_m$		2.195	126, 2299: 3652
$\triangle H_f^\circ$	25°(l)	− 54.93	128: 1204, 5358
$\triangle H_c^\circ$	25°(l)	− 1149.97	127, 5358: 1204, 2577
C_p°	25°	39.67	129
C_p	25°	53.28	2299: 3652
	Table XII	Ch II	
t_c		257.16°	124: 1439, 72
p_c		26.98	124: 1439
d_c		0.238	124
v_c		0.421	124
A	eq 72	0.04605	126: 2299
ϵ	20°	1.919	3477: 1439, 3934
μ	20°(l)	0	3056
soly	aq in, 20°	0.0103	1483
ir			89, 2951
Raman			459, 651, 1556, 1914, 4040, 4273, 4294
mass			90, 1434, 2128
Beil	10. 156		

15. 3–Methylhexane

$CH_3CH_2CH(CH_3)CH_2CH_2CH_3$ $\hspace{4cm}$ C_7H_{16}

mw		100.206	887
bp	760	91.850°	126: 1439, 1438, 1792, 4535, 2831, 2096
dt/dp	760	0.04459°	122
p	25°	62.	126: 4535
eq 5	A	6.86764	125
	B	1240.196	
	C	219.223	
fp		−119.4°	4535
d	20°	0.68713	122: 1439, 1438, 1792, 4424, 1433, 2831
	25	0.68295	1841, 2096
n_D	20°	1.38864	122: 1606, 1439, 1438, 1792, 4424, 1433
	25	1.38609	2831, 1841, 2096
η	−195.7°gl	3.2×10^{20}	2967
	20	0.372	4772: 1439, 4424
	30	0.329	
γ	20°	19.79	123: 1439
	25	19.30	
$\triangle H_v$	25°	8.386	126: 4962
	bp	7.359	
$\triangle H_f^°$	25°(l)	−54.35	128: 1204, 5358
$\triangle H_c^°$	25°(l)	−1150.55	127: 5358, 1204, 2577
$C_p^°$	25°	39.67	129
C_p	16.0°	51.2	2301
t_c		262.04°	124: 1439, 4708
p_c		27.77	124, 1439, 4708
d_c		0.248	124
v_c		0.404	124: 4708
$[\alpha]_D$	25°	+9.12°	1841: 1433, 2831, 2096
ϵ	20°	1.927	3477: 1439, 4424
μ	20°(l)	0	3056
fl pt		25°F	3747
ir			89, 2951
Raman			1556, 1914, 4040, 4273, 4294
mass			90, 1434
Beil	**10**, 157		

16. 2,3–Dimethylpentane

$CH_3CH(CH_3)CH(CH_3)CH_2CH_3$ C_7H_{16}

mw		100.206	887
bp	760	89.784°	126: 1439, 1438, 4535, 4593, 5169, 316
dt/dp	760	0.04482°	122
p	25°	69.	126: 4535
eq 5	A	6.85382	125
	B	1238.017	
	C	221.823	
fp		gl	5169
d	20°	0.69508	122: 1439, 1438, 4424, 4593, 5169, 316
	25	0.69091	4851
n_D	20°	1.39196	122: 1127, 1439, 1438, 4424, 4593, 5169
	25	1.38945	5170, 1606, 316, 4851
η	−195.7°gl	3.6 x 10 ^{18}	2967 :1439, 4424
	15	0.436	4772
	30	0.412	
γ	20°	19.96	123: 1439, 5169
	25	19.47	
$\triangle H_v$	25°	8.185	126: 3601
	bp	7.263	
$\triangle H_f^o$	25° (l)	− 55.81	128: 1204, 5112
$\triangle H_c^o$	25° (l)	− 1149.09	127: 5112, 1204, 2577
C_p^o	25°	39.67	129
C_p	18.1°	51.1	[2301]
t_c		264.14°	124: 1439
p_c		28.70	124: 1439
d_c		0.255	124
v_c		0.393	124
$[\alpha]_D$	25°	− 10.39°	2830: 4851
ϵ	20°	1.939	3477: 1439, 4424
μ	20° (l)	0	3056
ir			89, 2951
Raman			651, 1556, 1914, 4040, 4294
mass			90, 1434
Beil	**10**, 157		

17. 2,4–Dimethylpentane

$(CH_3)_2CHCH_2CH(CH_3)_2$ C_7H_{16}

mw		100.206	887
bp	760	80.500°	126: 316, 1438, 1546, 4535, 4767, 5169
dt/dp	760	0.04376°	122
p	25°	98.	126: 1546, 4535
eq 5	A	6.82621	125
	B	1192.041	
	C	221.634	
fp		– 119.242°	126: 1439, 1438, 4522, 4767, 5169, 2299
d	20°	0.67270	122: 1439, 1438, 1546, 4424, 5169, 316
	25	0.66832	
n_D	20°	1.38145	122, 1606: 1439, 4424, 5169, 5170, 316
	25	1.37882	
η	15°	0.375	4772: 1439, 1546, 4424
	20	0.360	
γ	20°	18.15	123: 1439, 5169
	25	17.66	
$\triangle H_v$	25°	7.861	126: 3601, 4962
	bp	7.051	
$\triangle H_m$		1.636	126, 2299
$\triangle H_f^\circ$	25° (l)	– 56.17	128: 1204, 5358
$\triangle H_c^\circ$	25° (l)	– 1148.73	127, 5358: 1204, 2577
C_p°	25°	39.67	129
C_p	25°	53.59	2299
	Table XII	ch II	
t_c		246.58°	124: 1439
p_c		27.01	124: 1439
d_c		0.240	124
v_c		0.418	124
A	eq 72	0.03476	126: 2299
ϵ	20°	1.914	3477: 1439, 4424, 316
μ	20° (l)	0	3056
soly	in aq, 25°	0.0041%w	3053
fl pt		10°F	3747
ir			89, 2951
Raman			651, 1556, 1914, 4040, 4294
mass			90, 1434
Beil	10, 158		

18. Ethylcyclohexane

$\underbrace{CH_2CH_2CH_2CH_2CH_2CHCH_2CH_3}$ C_8H_{16}

mw		112.217	887
bp	760	131.783°	156, 1608, 5204: 460, 4767
dt/dp	760	0.04969°	153: 1608, 5204
p	25°	13.	156: 5204
eq 5	A	6.87041	155: 5204
	B	1384.036	
	C	215.128	
fp		− 111.323°	156: 1608, 1794, 4767, 2303, 4770
d	20°	0.78792	153, 1608: 460, 1732
	25	0.78390	
n_D	20°	1.43304	153, 1608: 460
	25	1.43073	
η	− 195.7°gl	4.4 x 10^{24}	2967, 1732
	20	0.843	154:
	25	0.787	
γ	20°	25.65	159
	25	25.12	
$\triangle H_v$	25°	9.674	156: 3601
	bp	8.20	
$\triangle H_m$		1.9919	156: 2303
$\triangle H_f^\circ$	25° (l)	− 50.72	158, 2413, 2411: 5358
$\triangle H_c^\circ$	25° (l)	− 1248.23	157, 2413, 2411, 5358
C_p°	25°	37.96	161
C_p	25°	50.62	2303
t_c		336.°	160: 1369
p_c		30.	160
d_c		0.25	160: 1369
v_c		0.45	160: 1369
A	eq 72	0.03828	156
ϵ		2.054	1369
μ	100° (g)	0	3056
aq az	90.6°	60.2%w	3975
fl pt		22°C	1369
ir			89, 3791, 3792, 4039, 408
Raman			460, 1556, 3901
mass			90, 3997
Beil	452, 35		

19. Octane

$CH_3(CH_2)_6CH_3$ C_8H_{18}

mw		114.233	887
bp	760	125.665°	100, 3805, 5204: 751, 727, 956, 4407
dt/dp	760	0.04738°	130: 1608, 5204, 5234
p	25°	14.0	100: 2966, 5204
eq 5	A	6.92377	133
	B	1355.126	
	C	209.517	
fp		− 56.795°	100, 3805: 1340, 2302, 4407, 4409, 1574
d	20°	0.70252	130, 3805: 1609: 751, 822, 1340, 4271
	25	0.69849	4784, 458, 1732
dd/dt	0 − 40°	0.000803	1236
n_D	20°	1.39743	130: 3805, 822, 856, 1340, 1345, 3115
	25	1.39505	4271, 4407, 5169, 5170, 458
dn/dt	0 − 40°	0.000483	1236
η	20°	0.5466	96, 4271: 1404, 4784, 4163, 1773, 1732
	25	0.5151	
	Table I, II, III, IV Ch II		
γ	25°	21.14	[2379]: 131, 2102, 3873, 4967, 5169, 2378
	Table VIII	Ch II	
$\triangle H_v$	25°	9.916	100: 3331, 3601, 4962, 1574, 702
	bp	8.225	
$\triangle H_m$		4.957	100: 2302, 3652
$\triangle H_f^o$	25° (l)	− 59.74	136: 1864, 5358
$\triangle H_c^o$	25° (l)	− 1307.53	135, 5358: 2577, 1864, 2398, 3840
C_p^o	25°	45.14	137
	Table XIV	Ch I to Ch II	
C_p	25°	60.74	1574: 2302, 3601, 3652, 3280
t_c		295.54°	98, 5303: 1100, 4708, 72, 1065, 2602
p_c		24.54	98, 5301: 4708, 1065, 5305, 3721, 2602
d_c		0.232	98, 5301: 2602, 702
v_c		0.492	98, 5301: 1100, 4708, 5303
A	eq 72	0.05329	134: 4757, 1574
ϵ	20°	1.948	3477: 1345, 5034
μ	20° (l)	0	1345

19. Octane

(Continued)

soly	in aq, 25$^{\circ}_{\circ}$ aq in, 20	66. x 10^{-6}%w 0.0095%w	3053: 1677 1483
aq az	89.6°	74.5%w	2255
fl pt	TOC	72°F	3618
ir			89, 388, 2807, 408, 3679, 2808, 2435
Raman			651, 1556, 2138, 3396, 3901, 4273, 458
mass			90, 3997, 1568, 471
Beil	**10**, 159		

20. 2,2,3-Trimethylpentane

$(CH_3)_3CCH(CH_3)CH_2CH_3$ C_8H_{18}

mw		114.233	887
bp	760	109.841°	134, 1608, 5204: 4535, 5169
dt/dp	760	0.04755°	130, 1608, 5204
p	25°	32.	134: 4535, 5204
eq 5	A B C	6.82546 1294.875 218.420	133: 5204
fp		-112.27°	134: 1608, 4522, 4535, 5169
d	20° 25	0.71602 0.71207	130: 1608, 2842, 1732
n_D	20° 25	1.40295 1.40066	130, 1608: 2842, 5169, 5170
dn/dt	20 - 25°	0.00046	2269
η	20° 40	0.598 0.471	1732
γ	20° 25	20.67 20.22	131: 5169
$\triangle H_v$	25° bp	8.824 7.65	134: 3601
$\triangle H_m$		2.06	134

20. 2,2,3–Trimethylpentane

(Continued)

$\triangle H_f^\circ$	25°(l)	– 61.44	136, 5358
$\triangle H_c^\circ$	25°(l)	– 1305.83	135, 5358
C_p°	25°	45.14	137
t_c		290.28°	132
p_c		26.94	132
d_c		0.262	132
v_c		0.436	132
A	eq 72	0.0401	134
ϵ	20°	1.962	3971: 3477
ir			89, 2807
Raman			651, 1556
mass			90, 4294
Beil	10, 164		

21. 2,2,4–Trimethylpentane

Isooctane

$(CH_3)_3CCH_2CH(CH_3)_2$ $\qquad\qquad$ C_8H_{18}

mw		114.233	887
bp	760	99.238°	134, 1608, 2535: 751, 758, 759, 3775
dt/dp	760	0.04651°	130, 1608, 5204: 518, 759, 4382
p	25°	49.	134: 465, 2966, 4380, 5204
eq 5	A	6.81189	133: 465, 2535, 4382
	B	1257.840	
	C	220.735	
fp		– 107.388°	134: 758, 1608, 1794, 2172, 3652, 4770
d	20°	0.69193	130: 751, 758, 759, 2535, 2842, 4424
	25	0.68781	4593, 5169, 1732
n_D	20°	1.39145	130, 1608: 751, 758, 759, 1437, 2842
	25	1.38898	4424, 4593, 5170
dn/dt	20 – 25°	0.00047	2269
η	20°	0.504	1732: 1437, 4424
	30	0.447	
	40	0.403	
γ	20°	18.77	131: 5169
	25	18.32	
	30	17.88	

21. 2,2,4–Trimethylpentane

(Continued)

$\triangle H_v$	$25°$	8.397	134: 3601, 3775
	bp	7.411	
$\triangle H_m$		2.198	134: 3652, 3775
$\triangle H_f^°$	$25°$ (l)	-61.97	136: 5358
$\triangle H_c^°$	$25°$ (l)	-1305.29	135, 5358: 2577
$C_p^°$	$25°$	45.14	137
C_p	$25°$	2.0880 int j/g	3601: 3652
t_c		$270.60°$	132: 465, 2535, 4708, 72
p_c		25.34	132: 465, 2535, 4708
d_c		0.244	132
v_c		0.468	132: 4708
A	eq 72	0.04026	134
ϵ	$20°$	1.940	3477: 4424
μ	(l)	0	4424
soly	in aq, $25°$	0.00024%w	3053
	aq in, $20°$	0.0055%w	529: 1483
aq az	$79.4°$	76.7%w	3975
fl pt	TCC	$10°$F	3680: 3747
uv			2428
ir			89, 236, 388, 2119, 408, 4107
Raman			459, 651, 1556, 3900, 3 901, 4599
mass			90, 3997
Beil	10, 164		

22. Nonane

$CH_3(CH_2)_7CH_3$ C_9H_{20}

mw		128.260	887
bp	760	150.798°	142: 956, 3115, 5144, 5204, 2736, 3243
dt/dp	760	0.04967°	138: 1608, 5204, 3243
p	25°	4.3	142: 5204
eq 5	A	6.93513	141: 5204
	B	1428.811	
	C	201.619	
fp		− 53.519°	142: 1340, 1608, 2302, 3243, 4407, 1574
d	20°	0.71763	138: 1340, 1345, 1608, 3873, 4271, 5169
	25	0.71381	2736, 3243
dd/dt	0 − 40°	0.000774	1236: 3243
n_D	20°	1.40542	138, 712: 1340, 1345, 1608, 4271, 4407
	25	1.40311	5169, 5170, 3243
dn/dt	0 − 40°	0.000463	1236: 3243
η	20°	0.7160	96: 4271
	25	0.6696	
	Table I	Ch II	
γ	25°	22.38	2379, 97: 3873, 5169, 2378
	Table VIII	Ch II	
$\triangle H_v$	25°	11.100	142, 2798: 3601, 2712, 702
	bp	8.823	
$\triangle H_m$		3.697	142, 1574: 2265, 4407
$\triangle H_f^o$	25° (l)	− 65.84	102, 2798: 5358
$\triangle H_c^o$	25° (l)	− 1463.80	143, 2412, 2798: 3840, 2398, 5358
C_p^o	25°	50.60	103
C_p	25°	67.97	1574, 3280: 3601, 3652, 4095, 2302
t_c		321.41°	98, 72: 4922, 4708, 4406, 4095, 3721
p_c		22.6	98, 4708: 3721, 4921, 702
d_c		0.234	98: 702
v_c		0.548	98: 4708
A	eq 72	0.03857	142: 4757, 1574
ϵ	20°	1.972	3477: 1345, 3971
soly	aq in, 25°	79 ppm	4134
aq az	95°	60.2%w	2255
fl pt	TCC	88°F	473: 3747
ir			89, 388, 408, 2808
Raman			461, 1556, 2138, 3584, 4273, 3036
mass			90, 3363, 1568, 471
nmr			2664
Beil	**10**, 165		

23. 2,2,5–Trimethylhexane

$(CH_3)_3CCH_2CH_2CH(CH_3)_2$ C_9H_{20}

mw		128.260	887
bp	760	124.084°	142: 2269
dt/dp	760	0.04838°	138
p	25°	16.5	142
eq 5	A	6.83531	141
	B	1324.049	
	C	210.737	
fp		− 105.780°	142
d	20°	0.70721	138
	25	0.70322	
n_D	20°	1.39972	138, 1606: 2269
	25	1.39728	
γ	20°	20.04	139
	25	19.60	
$\triangle H_v$	25°	9.602	142: 2798
	bp	8.07	
$\triangle H_m$		1.48	142
$\triangle H_f^\circ$	25° (l)	− 68.08	2798: 5358
$\triangle H_c^\circ$	25° (l)	− 1459.20	143: 2798, 5358
t_c		294.8°	140
p_c		23.0	140
d_c		0.247	140
v_c		0.519	140
A	eq 72	0.0266	142
fl pt	TOC	55°F	3747
ir			89
Raman			1556
mass			90, 3363, 1568
Beil	10, 167		

24. Decahydronaphthalene

Decalin (mixed isomers)

$C_{10}H_{18}$

mw		138.255	887
bp	760	191.7°	2148: 1703, 3400, 1791
p	23.3°	1.0	1703: 2148, 2966
	66.1	9.7	
	86.8	26.4	
	118.8	87.4	
	150.0	240.0	
fp		− 124 ± 2°	2148
d	20°	0.8865	5116: 1703, 1994, 1927
	25	0.8789	2148
n_D	20°	1.4758	1703: 1994
η	25°	2.415	2148: 1927
	Table I	Ch II	
γ	33.6°	29.36	2148
$\triangle H_v$	bp	9.82	2148
C_p°	25°	44.52	[5003]
	Table XIV	Ch II	
C_p		53.56	3160: 2148, 5116, 1927
t_c		445°	2148
p_c		∼29	2148
K_B		5.762°	2148
ϵ	25°	2.1542	4557: 3400
μ	25° (l)	0	4557: 1539

24. Decahydronaphthalene

(Continued)

soly	in aq 25°	⟨0.02%w	664
	aq in, 20	0.0063%w	1483
fl pt	TCC	136°F	3680
ir			2809, 4076, 2808
Raman			3433, 3229
Beil	453, 92		

25. cis-Decahydronaphthalene

$$H_2C \overset{\displaystyle \overset{H_2}{C}}{} \quad \overset{H}{C} \quad \overset{\displaystyle \overset{H_2}{C}}{} \, CH_2$$

$$H_2C \quad C \quad C \quad CH_2$$
$$\overset{}{\underset{H_2}{C}} \quad \overset{H}{} \quad \overset{}{\underset{H_2}{C}}$$

$C_{10}H_{18}$

mw		138.255	887
bp	760	195.774°	888: 1557, 1646, 1653, 4535
dt/dp	760	0.05710°	888
p	25°	0.78	[888]: 1557, 4247, 4427, 5345
eq 5	A	6.87529	888: 4247
	B	1594.460	
	C	203.392	
fp		−43.01°	4527: 3062, 3649, 4244, 4248, 4535, 4479
d	20°	0.89671	162: 4248, 4273, 1994, 1927
	25	0.89291	888:
dd/dt		0.000760	888

25. cis-Decahydronaphthalene

(Continued)

n_D	20°	1.48098	162
	25	1.47878	888: 1557, 4248, 3352, 1994
dn/dt		0.000440	[888]
η	20°	3.381	4246: 1927
	30	2.723	
γ	20°	32.18	4245
	30	31.01	
$\triangle H_v$	25°	12.271	[1369]
	bp	9.799	4247
$\triangle H_m$		2.268	3062: 3649
$\triangle H_f^{\circ}$	25° (l)	− 52.75	1203: 4459, 5358
$\triangle H_c^{\circ}$	25° (l)	− 1502.4	1203: 4459, 2577, 5358
C_p°	25°	39.84	3350
C_p	25°	55.45	3062: 3649, 4243, 1927
	Table XII	Ch II	
t_c		418.5°	3350: 4247, 964
p_c		27	3350
A	eq 72	0.02154	3062: 4527
B		0.00299	
K_f		19.47°	2294
ϵ	20°	2.197	3477: 4244, 4479
μ	20 – 100° in 30	∿ 0	4244: 1539
soly	aq	ins	2608
aq az	99.1°	22.4%w	3975
ir			89, 4162
Raman			3901, 3709
mass			90
nmr			3473, 2210, 3382
Beil	453, 92		

26. trans-Decahydronaphthalene

$$
\begin{array}{c}
H_2C \\
\end{array}
$$

$C_{10}H_{18}$

mw		138.255	887
bp	760	187.273°	888: 1557, 4247, 4248, 4535
dt/dp	760	0.05656°	888
p	25°	1.23	[888]: 1557, 4247, 4535, 5345
eq 5	A	6.85681	888: 4247
	B	1564.683	
	C	206.259	
fp		– 30.400°	4527: 3062, 3649, 4244, 4248, 4535, 4479
d	20°	0.86971	162: 1329, 4245, 4248, 1994, 1927
	25	0.86592	888:
dd/dt		0.000749	888
n_D	20°	1.46932	162
	25	1.46715	888: 1557, 4248, 1994, 3352
dn/dt		0.000434	[888]
η	20°	2.128	4246: 1329, 1927
	30	1.774	
γ	20°	29.89	4245: 1329
	30	28.87	
$\triangle H_v$	25°	11.92	[1369]
	bp	9.615	4247
$\triangle H_m$		3.445	3062, 3649
$\triangle H_f^o$	25°(l)	– 54.87	1203: 4459, 5358
$\triangle H_c^o$	25°(l)	– 1502.4	1203: 4459, 2577, 5358
C_p^o	25°	40.04	3350
C_p	25°	54.61	3062: 3649, 4243, 1927
	Table XII	Ch II	
t_c		408.3°	3350: 4247, 964
p_c		27	3350

26. trans-**Decahydronaphthalene**

(Continued)

A	eq 72	0.02941	3062: 4527
B		0.00292	
K_f		20.81°	2294
ϵ	20°	2.172	3477: 4244, 4479
μ	20 – 100° in **30**	~ 0	4244: 1539
soly	aq	ins	2608
aq az	98.5°	26.5%w	3975
ir			89, 4162
Raman			3901, 3709
mass			90
nmr			3473, 2210
Beil	**453**, 92		

27. **Decane**

$$CH_3 (CH_2)_8 CH_3 \qquad\qquad\qquad C_{10}H_{22}$$

mw		142.287	887
bp	760	174.123°	100, 1608: 956, 3115, 3400, 4271
dt/dp	760	0.05172°	95, 1608, 5204
p	25°	1.3	100: 5204, 3826
eq 5	A	6.95367	99: 5204
	B	1501.268	
	C	194.480	
fp		– 29.657°	100: 1236, 1340, 3243, 2302, 4524, 1574
d	20°	0.73005	95, 1608: 1345, 3452, 3873, 4271, 4967
	25	0.72625	3243
dd/dt	0 – 40°	0.000751	1236: 3243
n_D	20°	1.41189	95: 1236, 1340, 1345, 3452, 4271, 4967
	25	1.40967	3243
dn/dt	0 – 40°	0.000447	1236: 3243

27. Decane

(Continued)

η	$20°$ 25	0.9284 0.8614	96: 4271, 3452, 4271, 1773
	Table I	Ch II	
γ	$25°$	23.37	[2096]: 97, 2102, 3452, 3873, 4967, 2378
	Table VIII	Ch II	
$\triangle H_v$	$25°$ bp	12.277 9.388	100, 2798: 3601, 702
$\triangle H_m$		6.864	100; 1574, 2302
$\triangle H_f^o$	$25°$ (l)	– 71.95	102, 2798: 5358
$\triangle H_c^o$	$25°$ (l)	– 1620.06	101, 2798, 5358: 3840, 2398, 2577
C_p^o	$25°$	56.07	103: 3652
C_p	$25°$	75.16	1574: 3280, 3601, 2302
t_c		$344.4°$	98, 72: 4922, 4708, 622, 3721, 702
p_c		20.7	98: 4708, 3721, 4921, 702
d_c		0.236	98: 702
v_c		0.603	98: 4708
A	eq 72	0.05826	100: 4757, 1574
ϵ	$20°$	1.991	3477: 1345
μ	– 30 to 170°, liq	$\sim$0	1345
soly	aq in, $25°$	72 ppm	4134
aq az	$98°$	31.5%w	3975
fl pt	TCC	$115°$F	3680: 3747
ir			89, 388, 408, 4304, 2808, 2951, 2435
Raman			1556, 1854, 3584, 2664, 3036
mass			90, 3997, 3964, 1568, 471, 4255
Beil	**10**, 168		

28. Bicyclohexyl

$C_{12}H_{22}$

mw		166.309	887
bp	760	239.04°	3244: 3191, 2455, 4974
dt/dp	760	0.555°	3244
p	25°	0.108	4066
eq 4	A	8.21	1327
	B	2715.	
fp		3.63°	3244: 2455, 4526, 4066, 4974
d	20°	0.88619	3244: 2455, 1927, 4066
	25	0.88249	
dd/dt		0.000740	[3244] : 4066
n_D	20°	1.47995	3244: 1327, 4066
	25	1.47768	
dn/dt		0.000454	[3244]
η	20°	3.75	4066: 2455, 1927, 4974
	25	3.40	
γ	20.5°	32.68	4974: 4066
	41.2	30.56	
$\triangle H_v$	bp	12.4	1327
$\triangle H_c°$	25° (l)	− 912.5	2577
C_p	40°	71.7	[1927]
A	eq 72	0.0110	4526
K_F		14.52	3191: 2294
μ		⟨ 0.4	4066
fl pt	TOC	215°F	3160
ir			89, 1700, 4238, 1698
Raman			3229, 3709
Beil	453, 108		

29. Dodecane

$CH_3(CH_2)_{10}CH_3$ $\qquad\qquad\qquad\qquad\qquad\qquad\qquad$ $C_{12}H_{26}$

mw		170.341	887
bp	760	216.278°	100, 5204: 4271, 1236, 3244
dt/dp	760	0.05528°	95, 5204: 3244
p	25°	0.12	100: 5204, 3826
eq 5	A	6.98059	99: 5204
	B	1625.928	
	C	180.311	
fp		− 9.587°	100: 4271, 1236, 3244, 1574, 2302
d	20°	0.74869	95: 4271, 1236, 3244, 3873, 4967, 733
	25	0.74516	
dd/dt		0.000719	1236
n_D	20°	1.42160	95: 4271, 1236, 3118, 3244
	25	1.41949	
dn/dt		0.000425	1236
η	20°	1.508	96: 4271, 733, 1773
	25	1.378	
	Table I	Ch II	
γ	25°	24.91	[2379]: 97, 3873, 2378
	Table VIII	Ch II	
$\triangle H_v$	25°	14.648	100: 702
	bp	10.43	
$\triangle H_m$		8.804	100, 1574: 2302
$\triangle H_f^\circ$	25° (l)	− 84.16	102
$\triangle H_c^\circ$	25° (l)	− 1932.59	101: 3840, 2398
C_p°	25°	67.00	103
C_p	25°	89.86	1574, 3280: 2302
t_c		385.1°	98: 3721, 4922, 72, 702
p_c		17.8	98: 3721, 4921, 702
d_c		0.239	98: 702
v_c^-		0.713	98
A	eq 72	0.06378	100: 4757, 1574, 4521
ϵ	30°	2.002	1345
μ	− 10 to 210° (l)	0	1345

29. Dodecane

(Continued)

soly	aq in, 25°	65 ppm	4134
aq az	99.6°	11.1%	3975
fl pt	TCC	165°F	3680
uv			2632
ir			89, 408, 2808, 2951, 1699, 2435, 386
Raman			2664, 1855, 3583, 1556, 3036
mass			3801
Beil	10, 171		

30. Benzene

C_6H_6 C_6H_6

mw		78.115	887
bp	760	80.100°	194: 443, 1755, 4381, 4382, 5237, 5350
dt/dp	760	0.04271ᵘ	104, 1608, 5204: 4382
p	25°	95.2	194: 1575, 1779, 2265, 4166, 5350
eq 5	A	6.90565	193: 4125, 4381, 4382, 5204
	B	1211.033	
	C	220.790	
fp		5.533°	194: 3217, 1755, 1794, 3588, 4783
d	10°	0.8895	191: 600, 711, 1028, 1334, 1608, 2832
	20	0.87901	104: 3195, 4125, 4783, 5032, 443, 5350
	25	0.87370	
	30	0.8685	191
α	0 – 80°	0.00138	3160: 1732, 5253
n_D	20°	1.50112	104: 1606, 385, 711, 1028, 1245, 1608
	25	1.49792	1755, 2832, 4783, 5237
η	20°	0.6487	190: 846, 1400, 1717, 2939, 3865, 4783
	25	0.6028	3226, 1732
	30	0.5621	
	40	0.4923	
	Table I, II	Ch II	

30. Benzene

(Continued)

γ	20°	28.88	192: 22, 788, 1334, 2102, 5032, 5035
	25	28.18	5048, 4996
	30	27.49	
	Table VII	Ch II	
$\triangle H_v$	25°	8.090	194: 1575, 3208, 3601, 4866, 1532, 2427
	bp	7.352	702
$\triangle H_s$		10.653	3319
$\triangle H_m$		2.358	194: 3588, 4407, 4601
$\triangle H_f^o$	25°(l)	11.718	196: 4621, 5358
$\triangle H_c^o$	25°(l)	−780.98	195: 4743, 2577, 3968, 5358
C_p^o	25°	19.52	188, 731, 4621
	Table XIV	Ch II	
C_p	25°	32.52	3588: 1028, 2384, 2689, 2781, 3745, 4143
	Table XIII	Ch II	
t_c		288.94°	192a: 622, 3331, 5303, 1100, 1849, 1065
p_c		48.34	192a: 64, 3331, 2602, 73, 643, 1849
d_c		0.302	192a: 2602, 1849, 702
v_c		0.259	192a: 1100, 509, 962, 5303
A	eq 72	0.01528	194: 2745
K_f		5.12°	3672: 491, 1548
K_B		2.53°	2193, 483, 1803: 919, 3274, 5029, 487
pK_{BH+}	0° in HF	9.4	3104
κ	25°	4.43 x 10⁻¹⁷	3160: 1601
ϵ	25°	2.275	2532: 3276, 763, 1621, 3106, 3266, 2048
μ	20 − 60°	0	3733: 2233, 3034, 5198, 65, 66
	(l)		
soly	in aq, 25°	0.1780%w	3053: 529, 636, 2122, 3540, 4225, 1064
	aq in, 25°	0.063%w	5218: 4609, 2445, 1005, 529, 614, 4225
	in aq, (%w)=0.1806 − 0.001095 t + 0.00003179 t²		286
aq az	69.25°	91.17%w	2254
fl pt	TCC	12°F	3680
uv			91, 549, 896, 4175, 4006, 2428, 3541
ir			89, 1001, 1090, 1970, 408, 3766, 3530
Raman			1854, 3899, 3901, 4942, 4252, 1170, 3531
mass			90, 2403, 3997, 4434, 3379, 61
nmr			541, 2210, 4508, 685
Beil	**463**, 179		

31. Toluene

$C_6H_5CH_3$ C_7H_8

mw		92.142	887
bp	760	110.625°	194: 1608, 1755, 4761, 4788, 5204, 5350
dt/dp	760	0.04630°	104: 1608, 4761, 5204
p	25°	28.5	194: 2740, 4166, 4183, 5204
eq 5	A	6.95464	193: 5204
	B	1344.800	
	C	219.482	
fp		– 94.991°	194: 1608, 1755, 1794, 4450, 4536, 4771
d	10°	0.8762	191: 593, 861, 1334, 1608, 1755, 3195
	20	0.86696	104: 4761, 4783, 4788, 4808, 4965, 1732
	25	0.86231	
	30	0.8577	
α	0 – 100°	0.00109	3160
n_D	20°	1.49693	104: 1606, 711, 751, 1608, 1755, 4783
	25	1.49413	4965, 5195
η	20°	0.5866	190: 1324, 2939, 3865, 2151, 1732
	25	0.5516	
	30	0.5203	
	Table I, II	Ch II	
γ	20°	28.53	192, 1334: 2102, 4808, 4965, 5048, 4996
	25	27.92	
	30	27.32	
$\triangle H_v$	25°	9.080	194: 3208, 3601, 5032, 2427, 702
	bp	7.931	
$\triangle H_m$		1.586	194: 4407, 4450, 4536
$\triangle H_f^\circ$	25°(l)	2.867	196: 4621, 5358
$\triangle H_c^\circ$	25°(l)	– 934.50	195: 4743, 1483, 3968, 5358
C_p°	25°	24.80	188: 4621
	Table XIV	Ch II	
C_p	20°	36.09	5195: 2781, 4143, 4144, 4450, 5194
	Table XIII	Ch II	
t_c		318.57°	192a: 64, 622, 1100, 72, 643, 3842
p_c		40.55	192a: 64, 643, 3842, 5305, 702
d_c		0.292	192a: 3160, 702
v_c		0.316	192a: 1100

31. Toluene

(Continued)

A	eq 72	0.02515	194
K_B		3.29°	2193: 3606
pK_{BH+}	0° in HF	6.3	3104
κ	25°	8.0×10^{-16}	1602: 1305, 2569
ϵ	25°	2.379	3477: 2747, 5034, 66, 3733, 1381, 4358
μ	$20 - 60^\circ$ (l)	0.31	3733: 2175, 2747, 4127, 5190, 4754, 66
soly	in aq, 25° aq in, 25°	0.0515%w 0.0334%w	3053: 636, 2122, 1064, 1677, 1912 5218: 4609, 2427, 1483
aq az	85°	79.8%w	2255: 2254
fl pt	TCC	40°F	3680
uv			91, 549, 1347, 2617, 4175, 2428, 3541
ir			89, 3530, 2519, 408, 2764, 5199, 5210
Raman			1854, 2478, 2660, 3901, 5228, 3732, 5210
mass			90, 2403, 3997, 4434, 4618, 61
nmr			937, 954, 4508, 3446, 3126, 685
Beil	**466**, 280		

32. o – Xylene

$o - C_6H_4(CH_3)_2$ C_8H_{10}

mw		106.169	887
bp	760	144.411°	194: 855, 1608, 1755, 5204, 5238, 1791
dt/dp	760	0.04969°	104: 1608, 5204, 5257
p	25°	6.6	194: 2507, 2966, 4535, 5204, 5257
eq 5	A	6.99891	193: 2507, 4534, 1453, 5204
	B	1474.679	
	C	213.686	
fp		– 25.182°	194: 854, 1593, 1755, 1794, 4567, 4051
d	20°	0.88020	104: 852, 1334, 1608, 1755, 3195, 3327
	25	0.87596	1732, 4358
n_D	20°	1.50545	104: 1606, 856, 1608, 1755
	25	1.50295	
dn/dt	8 – 25°	0.00054941	1453
η	20°	0.809	190: 853, 1732
	25	0.756	
	Table I, II	Ch II	
γ	20°	30.03	192: 1334
	25	29.48	
$\triangle H_v$	25°	10.381	194, 3006: 3601, 3208, 782, 702
	bp	8.80	
$\triangle H_m$		3.250	194: 2300, 4051
$\triangle H_f^\circ$	25°(l)	– 5.841	196: 4621, 5358
$\triangle H_c^\circ$	25°	–1088.16	195, 5358: 3968, 2577
C_p°	25°	31.85	197, 4621: 2059
	Table XIV	Ch II	
C_p	30°	43.65	5194: 2300, 2781
	Table XIII	Ch II	
t_c		357.1°	192a: 64, 1453, 643, 73, 1791, 702
p_c		36.84	192a: 643, 73, 1791, 702
d_c		0.288	192a: 702
v_c		0.369	192a:
A	eq 72	0.02660	194

32. o – Xylene

κ	$25°$	6.7×10^{-16}	1602
ϵ	$20°$	2.568	3477: 66, 2036, 1381, 4358
μ	$20 - 60°$ (l)	0.45	3733: 2328, 5190, 65, 66, 2876, 4754
soly	in aq, $25°$	0.0175%w	3053
aq az	$93.5°$	50.1%w	3975: 2255
fl pt	TCC	$63°F$	3680
uv			91, 373, 3418, 4175, 2812, 2428, 1692
ir			89, 403, 406, 4547, 1692, 2357, 2518
Raman			1556, 2137, 4252, 1161, 4944, 1169, 1175
mass			90, 2403, 3997, 61
nmr			4508, 4103, 3929, 1756
Beil	467, 362		

33. m – Xylene

$m - C_6H_4(CH_3)_2$ C_8H_{10}

mw		106.169	887
bp	760	$139.103°$	194: 1608, 4778, 5145, 5204, 5238, 1791
dt/dp	760	$0.04903°$	104: 1715, 5204, 5257,
p	$25°$	8.3	194: 2507, 2966, 4535, 5204, 5257
eq 5	A	7.00908	193: 2507, 4534, 1453, 5204
	B	1462.266	
	C	215.105	
fp		$-47.872°$	194: 1608, 1794, 4567, 4774, 5145, 4770
d	$20°$	0.86417	104: 1334, 1608, 3195, 4778, 5145, 1732
	25	0.85990	
n_D	$20°$	1.49722	104, 1606: 1453, 1608, 5145, 5194
	25	1.49464	
dn/dt	$.5 - 25°$	0.00051262	1453
η	$20°$	0.617	190: 4778, 1732
	25	0.581	
	Table II	Ch II	

33. m – Xylene

(Continued)

γ	$20°$	28.63	192, 1334: 2102, 4778, 5048, 22
	25	28.08	
$\triangle H_v$	$25°$	10.195	194: 3208, 3601, 5032, 3006, 702
	bp	8.69	
$\triangle H_m$		2.765	194: 2300, 4051
$\triangle H_f^°$	$25°$(l)	– 6.075	196: 4621, 5358
$\triangle H_c^°$	$25°$(l)	– 1087.92	195, 5358: 3968, 2577
$C_p^°$	$25°$	30.49	197, 4621
	Table XIV	Ch II	
C_p	$30°$	42.57	5194: 2300, 2781, 4143, 4144
	Table XIII	Ch II	
t_c		$343.82°$	192a: 64, 1453, 5032, 643, 73, 1791
p_c		34.95	192a: 64, 1453, 73, 1791, 643, 702
d_c		0.282	192a: 702
v_c		0.376	192a
A	eq 72	0.02742	194
pK_{BH+}	$0°$in HF	3.2	3104
κ	$25°$	8.6×10^{-16}	1602
ϵ	$20°$	2.374	3477: 66, 2036, 1381
μ	$20 – 60°$	0.30	3733: 4127, 4754, 65, 66, 2876
	(l)		
soly	in aq,$25°$	0.0196%w	646: 2122, 3845
	aq in, $20°$	0.0402%w	1483
aq az	$92.5°$	54.5%w	3975, 2255, 2254
fl pt	TCC	$77°$F	3680
uv			91, 373, 3418, 4175, 2428, 1692, 4603
ir			89, 4817, 4500, 2357, 5199, 2521, 5210
Raman			1556, 1161, 5210, 4944, 1169, 1175, 1176
mass			90
nmr			4508, 3929, 1756, 685
Beil	**467**, 370		

34. p–Xylene

$p - C_6H_4(CH_3)_2$ C_8H_{10}

mw		106.169	887
bp	760	138.351°	194: 1755, 4783, 5145, 5204, 5238, 1791
dt/dp	760	0.04917°	104: 1608, 4783, 5204, 5257
p	25°	8.7	194: 2507, 2966, 4535, 5204, 5257
eq 5	A	6.99052	193: 2507, 4534, 1453, 5204, 3593
	B	1453.430	
	C	215.307	
fp		13.263°	194: 1755, 1794, 2300, 4567, 4783
d	20°	0.86105	104, 1608: 1334, 1755, 3195, 4783, 3226
	25	0.85669	1732, 4358
n_D	20°	1.49582	104, 1606: 1608, 1755, 4783, 5145, 5194
	25	1.49325	
dn/dt	6 – 25°	0.00052667	1453
η	20°	0.644	190: 4783, 1732, 3226
	25	0.605	
	Table II	Ch II	
γ	20°	28.31	192, 1334: 22, 2102
	25	27.76	
$\triangle H_v$	25°	10.128	194, 3006: 3601, 782, 3208, 702, 3593
	bp	8.60	
$\triangle H_m$		4.090	194: 1097, 2300, 4601, 3673
$\triangle H_f^\circ$	25° (l)	– 5.838	196: 4621, 5358
$\triangle H_c^\circ$	25° (l)	–1088.16	195: 2577, 3968, 5358
C_p°	25°	30.32	197, 4621
	Table XIV	Ch II	
C_p	30°	42.18	5194: 1407, 2781, 4143, 4144
	Table XIII	Ch II	
t_c		343.0°	192a: 64, 1453, 643, 73, 1791, 702
p_c		34.65	192a: 64, 1453, 643, 73, 1791, 702
d_c		0.280	192a: 702
v_c		0.379	192a
A	eq 72	0.02509	194
K_f		4.3°	491: 3673
pK_{BH+}	0° in HF	5.7	3104
κ	25°	7.6 x 10⁻¹⁶	1602
ϵ	20°	2.2699	3733, 3477: 3400, 4786, 66, 3994, 1381
μ	20 – 60° (l)	0.02	3733: 4906, 1539, 2328, 4127, 5192, 2876

34. p–Xylene

(Continued)

soly	in aq, 25°	0.019%w	636: 2122, 3845
aq az	92.6°	54.9%w	3975
fl pt	TCC	77°F	3680
uv			91, 373, 3418, 4175, 2428, 1692, 4603
ir			89, 402, 2374, 4500, 5199, 2357, 2520
Raman			1556, 3685, 1161, 4944, 1169, 1175, 1176
mass			90, 61
nmr			4508, 3929, 1756
Beil	467, 382		

35. Ethylbenzene

$C_6H_5CH_2CH_3$ C_8H_{10}

mw		106.169	887
bp	760	136.186°	194: 1608, 1755, 3208, 4783, 5204
dt/dp	760	0.04898°	104: 1608, 4783, 5204, 5257
p	25°	9.6	194: 1966, 4212, 4535, 5204, 5257
eq 5	A	6.95719	193: 4534, 1453, 4212, 5204
	B	1424.255	
	C	213.206	
fp		−94.975°	194: 1608, 1755, 1794
d	20°	0.86702	104: 600, 1334, 1608, 1717, 1755, 3195
	25	0.86264	4783, 4968, 1732
n_D	20°	1.49588	104, 1606: 1608, 1755, 3208, 4783, 4968
	25	1.49320	5194
dn/dt	7 – 25°	0.00051701	1453
η	20°	0.6783	190: 1400, 1717, 4783, 1732
	25	0.6373	
	Table I, II	Ch II	

35. Ethylbenzene

(Continued)

γ	20°	29.04	192, 1334: 4968, 2102
	25	28.48	
$\triangle H_v$	25°	10.098	194: 3208, 3601, 4212, 5032, 3006, 2427
	bp	8.50	4962, 2698
$\triangle H_m$		2.195	194: 2300, 4212
$\triangle H_f^\circ$	25°(l)	−2.977	196: 4621, 5358
$\triangle H_c^\circ$	25°(l)	−1091.03	195: 2577, 3968, 2397, 5358
C_p°	25°	30.69	188, 4621
C_p	30°	43.40	5194: 2300, 2689, 4143, 4144, 4212, 4213
	Table XIII	Ch II	
t_c		343.94°	192a: 64, 622, 72, 643, 3490
p_c		35.62	192a: 64, 643, 5305, 3490
d_c		0.284	192a: 702
v_c		0.374	192a
A	eq 72	0.03479	194
ϵ	20°	2.4042	3733: 3477, 3971, 66
μ	20 − 60°	0.37	3733: 4906, 66, 65, 1154
	(l)		
soly	in aq, 25°	0.0152%w	3053: 3160, 636, 2122, 2631, 1677
	aq in, 25°	0.043%w	[2427], 2254
aq az	92°	67%w	2255: 2254
fl pt	TCC	64°F	3680
uv			91, 4016, 4175, 1751, 3225, 4895, 4603
ir			89, 2554, 4547, 2519, 1879, 2357, 5199
Raman			1556, 1170, 2575, 2573, 1406
mass			90, 2403, 3997, 61
nmr			2461, 63, 954, 3126, 685
Beil	467, 351		

36. Isopropylbenzene

Cumene

$C_6H_5CH(CH_3)_2$ $\hspace{6cm}$ C_9H_{12}

mw		120.196	887
bp	760	152.392°	194: 1608, 1755, 2333, 2652, 5204, 1791
dt/dp	760	0.05074°	104: 4767, 1608, 5204, 5257
p	25°	4.6	194: 2966, 5204, 5257
eq 5	A	6.93666	193: 2652, 5204
	B	1460.793	
	C	207.777	
bp		−96.033°	194: 4767, 1608, 1755, 1794, 2202, 1796
d	20°	0.86179	104, 1608: 1334, 1755, 2202, 2333, 4968
	25	0.85751	1732, 724
n_D	20°	1.49145	104, 1606: 1608, 1755, 2202, 2333, 4407
	25	1.48890	4968
dn/dt	5 − 26°	0.0005186	1453
η	− 195.7°gl	1.7×10^{49}	2967: 1732
	20	0.791	190
	25	0.739	
γ	20°	28.20	192, 1334: 4968, 4996, 2333
	25	27.68	
$\triangle H_v$	25°	10.789	194: 2652, 3601, 3006
	bp	8.97	
$\triangle H_m$		1.861	194
$\triangle H_f^°$	25°(l)	− 9.848	196: 4621, 5358
$\triangle H_c^°$	25° (l)	−1246.52	195, 5358: 3968, 2577
$C_p^°$	25°	36.26	197, 4621
C_p	15 − 30°	47.54	2689: 1453, 2781
	Table XIII	Ch II	
t_c		357.9°	192a: 64, 1453, 643, 73, 1791
p_c		31.67	192a: 64, 643, 73, 1791
d_c		0.28	192a
v_c		0.428	192a
A	eq 72	0.02985	194
ϵ	20°	2.3833	3733: 3971, 66, 3477, 1381, 4358
μ	20 − 60°	0.39	3733: 66, 1154
	(l)		
soly	in aq, 25°	0.0050%w	3053
	aq in, 20°	0.0303%w	1483
aq az	95°	56.2%w	2255
fl pt	TCC	102°F	3680: 88
uv			91, 4006, 3225, 4603
ir			89, 4039, 408, 3679, 3735, 2357, 5199

36. Isopropylbenzene

(Continued)

Raman		1556, 3531, 46
mass		90, 3997
nmr		937, 3126, 685
Beil	**468**, 393	

37. Mesitylene

1,3,5 – Trimethylbenzene

sym–$C_6H_3(CH_3)_3$ C_9H_{12}

		120.196	887
mw		120.196	887
bp	760	164.716°	194: 1579, 4534
dt/dp	760	0.05100°	104
p	25°	2.5	194: 2966, 2507, 2247
eq 5	A	7.07436	193: 2507, 4534
	B	1569.622	
	C	209.578	
fp		– 44.720°	194: 4051, 1794, 2202, 4522
d	20°	0.86518	104: 2202
	25	0.86111	
n_D	20°	1.49937	104: 1606, 1579, 2202
	25	1.49684	
η	20°	1.154	1369
	30	0.936	
γ	20°	28.83	192
	25	28.31	
	Table VII	Ch II	
$\triangle H_v$	25°	11.348	194: 3601, 2074
	bp	9.33	
$\triangle H_m$		2.274	194: 4051
$\triangle H_f^\circ$	25° (l)	– 15.184	196: 4621, 5358
$\triangle H_c^\circ$	25° (l)	– 1241.19	195, 5358: 2577, 3968, 4743

37. Mesitylene

(Continued)

C_p°	25° Table XIV	35.91 Ch II	197, 4621
C_p	25° Table XIII	50.88 Ch II	[2781]
t_c		364.13°	192a: 72; 643
p_c		30.86	192a: 643
d_c		0.278	192a
v_c		0.433	192a
A	eq 72	0.02193	194: 4522
pK_{BH^+}	0°in HF	0.4	3104
κ	25°	$\langle 1 \times 10^{-16}$	1602
ϵ	20°	2.279	66: 3971, 1902
μ	$20 - 30^\circ$ (l)	0	66: 4754, 2875, 65, 2876
soly	aq in, 20°	0.0291	1483
aq az	96.5°		2254
uv			4447, 1751, 4464, 3783, 4466, 1554, 2428
ir			1471, 399, 400, 2951, 3789, 5199, 2357
Raman			4940, 1406, 2139
mass			3997
nmr			1308, 2845, 4508
Beil	468, 406		

38. Naphthalene

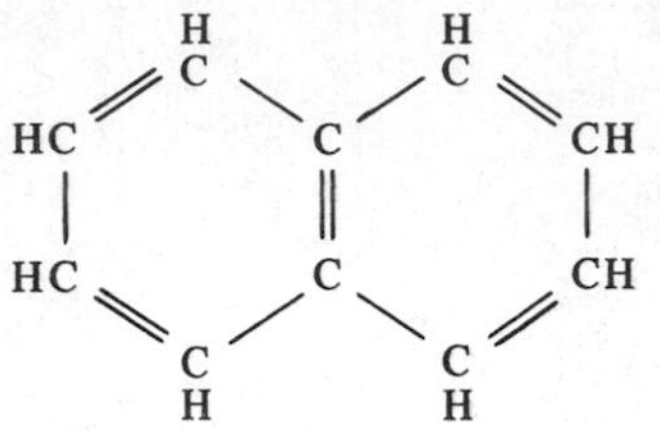

```
     H        H
      C        C
HC  //    \\  //    \\  CH
      C        C
HC  \\    //  \\    //  CH
      C        C
     H        H
```

$C_{10}H_8$

mw		128.175	887
bp	760	217.955°	200, 888: 863, 3118
dt/dp	760	0.05840°	200, 888
p	80.27°	7.48	[1611]: 4218
eq 5	A	10.0896	1611: 4535, 4364, 3324, 703, 1771
$(16 - 80^\circ)$	B	2926.61	
	C	237.332	
$(80 - 218^\circ)$	A	7.01065	1611, 888: 5344, 201
	B	1733.71	
	C	201.859	
fp		80.290°	200: 3062, 1242, 1619, 3118, 4551, 2124
tr pt		80.27°	1611
d	80°	0.9782*	180: 863, 3118, 5032, 5338, 3276, 2101
	90	0.9707	2290, 1857, 200
	100	0.9631	
n_D	85°	1.5898	200, 3118
η	99.8°	0.7802	4174, 863
	Table I	Ch II	
γ	127.2°	27.98	5032, 863
	bp	18.69	
$\triangle H_v$	25°		5032
	bp	10.32	4963: 4962, 3324, 1532, 416, 1126, 2785
$\triangle H_s$	25°	17.50	4459: 3324, 1771, 5247, 4364
$\triangle H_m$	25°	4.27	4459: 3062, 5080, 3207, 4601, 4451, 3203
$\triangle H_f^\circ$	25° (s)	18.57	1039: 4459, 5358
$\triangle H_c^\circ$	25° (s)	-1232.35	1039: 2577, 4459, 5358

* supercooled liquid

38. Naphthalene

(Continued)

C_p^o	$25°$	31.03	416
	Table XIV	Ch II	
C_p	$25°$	39.60	3062: 416, 2279, 2300
	Table XII	Ch II	
t_c		$475.2°$	72: 1100, 962, 643, 964, 5338, 1998
p_c		42.	4190: 643, 1949
v_c		0.408	962: 1100
A	eq 72	0.01827	3062: 4521, 3203
B		0.00261	
K_f		$6.94°$	1531, 323: 3207, 476, 491
K_B		$5.80°$	489: 473, 3606
pK_{BH+}	$0°$ in HF	4.0	3104
κ	$81.8°$	4.35×10^{-10}	3886
ϵ	$80°$	2.54	3477: 936, 2062, 5198, 3276
μ	$25°$	0	5196: 3021, 5198
	in 30		
soly	in aq $25°$	0.344%w	636: 2631
aq az	$98.8°$	16%w	2254
fl pt	TCC	$176°F$	3680
uv			91, 549, 2428, 1560, 1544
ir			89, 1685, 4076, 2809, 275, 408, 5199
Raman			229, 3684, 717, 1567, 1169, 1176, 2704
mass			90, 556
nmr			541, 542, 2415, 4508, 3808
Beil	476, 531		

39. 1,2,3,4 – Tetrahydronaphthalene

Tetralin

$C_{10}H_{12}$

mw		132.207	887
bp	760	207.57°	202, 3119: 1703, 2148, 4535, 1994
dt/dp	760	0.0575°	202
p	25°	0.40	2966: 1703, 2148
eq 5	A	7.9600	[202, 4535]
	B	2442.08	
fp		– 35.790°	202, 2210: 3119, 3062, 2148, 4770
d	20°	0.9702	202, 3119: 1329, 2148, 1994
	25	0.9662	
n_D	20°	1.54135	202, 3119: 1703, 1994
	25	1.53919	
η	20°	2.202	1329: 4711
	25	2.003	2148
	Table I	Ch II	
γ	21.5°	35.46	1329: 2148, 2292
	33.0	32.44	2292
$\triangle H_v$	25°	12.94	[1369]
	bp	10.48	2148
$\triangle H_m$		2.975	3062
$\triangle H_f^\circ$	25° (l)	– 1346.0	2577
C_p°	25°	36.27	[5003]
	Table XIV	Ch II	
C_p	25°	51.97	3062: 2148
	Table XII	Ch II	

39. 1,2,3,4 – Tetrahydronaphthalene
(Continued)

t_c		446°	[1369]
p_c		34.69	[1369]
d_c		0.309	1369
v_c		0.428	[1369]
A B	eq 72	0.02657 0.00248	3062
K_B		5.582°	2148
ϵ	20°	2.773	66: 3477
μ	20 – 30° (l)	0.60	66
aq az	99.4°	19.6%w	3975
fl pt	TCC	172°F	3160
uv			91, 2428, 382, 3417
ir			89, 4076, 4305, 2808, 5199, 1240
Raman			3433, 3229, 1240
mass			90, 2541, 5216
nmr			2461, 541, 4549, 2210
Beil	473, 491		

40. Butylbenzene

$C_6H_5C_4H_9$ $C_{10}H_{14}$

mw		134.223°	887
bp	760	183.270°	199: 2302, 4325, 4535, 4784
dt/dp	760	0.05358°	198
p	25°	1.1	199
eq	A B C	6.98317 1577.965 201.378	198a: 3281
fp	stable metastable	– 87.85° – 88.01	3281, 199: 1794, 3747, 4535, 4784, 2302 3281

40. Butylbenzene

(Continued)

d	$20°$	0.86013	198, 1334: 1453, 3195, 4784, 4965
	25	0.85607	
dd/dt	$0 - 40°$	0.0008130	1453
n_D	$20°$	1.48979	198: 1453, 4784, 4968
	25	1.48742	
η	$-195.7°$gl	3.5×10^{45}	2967: 4711
	20	1.035	184a
	25	0.960	
γ	$20°$	29.23	1334: 2102, 4968, 4996
	30	28.19	
$\triangle H_v$	$25°$	12.202	3281
	bp	9.38	199
$\triangle H_m$	(st)	2.682	3281, 199: 2302
	(ms)	2.691	3281
$\triangle H_f^\circ$	$25°$(l)	-15.28	5358: 4621, 187
$\triangle H_c^\circ$	$25°$(l)	-1403.46	186: 5358
C_p°	$25°$	41.85	188: 4621
C_p	$27°$	58.357	3281: 2302
	Table XII	Ch II	
t_c		$387.3°$	185: 964, 1927
p_c		28.49	185
d_c		0.270	185: 4772
v_c		0.497	185
A	eq 72	0.03931	3281, 199
B	(st)	0.00289	
A	eq 72	0.03951	3281
B	(ms)	0.00264	
ϵ	$20°$	2.359	66
μ	$20 - 30°$(l)	0.36	66: 1154
soly	in aq, $25°$	0.50%w	2631
	aq in, $25°$	0.041%w	2427: 1483
aq az	$98.2°$	30%w	3975
fl pt	TOC	$160°$F	
uv			91, 1751, 3783
ir			89, 3792, 3816, 5199, 408
Raman			1556, 46
mass			90, 3997
Beil	469, 413		

41. sec-Butylbenzene

$C_6H_5CH(CH_3)CH_2CH_3$ $C_{10}H_{14}$

mw		134.223	887
bp	760	173.305°	199: 4767, 841, 1453, 2333, 4329, 4535
dt/dp	760	0.05313°	198
p	25°	1.7	199: 2966, 4535
eq 5	A	6.95097	198a
	B	1540.174	
	C	205.101	
fp		−75.470°	199: 4767, 841, 1794, 4535
d	20°	0.86207	198, 1334: 841, 1160, 1453, 2333
	25	0.85797	
n_D	20°	1.49020	198, 1606: 841, 1160, 2333, 4329
	25	1.48779	
dn/dt	16 – 25°	0.0004846	1453
η	20°	28.53	1334: 2333, 4996
	30	27.53	
$\triangle H_v$	25°	11.83	199
	bp	9.07	
$\triangle H_m$		2.35	199
t_c		383.7°	643
p_c		29.4	643
d_c		0.263	1369
v_c		0.511	[1369]
$[\alpha]_D$	23°	−17.9°	Bell 4th Ed., Vol V, pt 2, p 932
A	eq 72	0.0303	199
ϵ	20°	2.364	66
μ	20 – 30°(l)	0.37	66: 1154
soly	aq in, 20°	0.0317	1483
fl pt	TCC	126°F	3680
uv			91, 3783
ir			89, 3792, 408, 3816, 5199
Raman			1556, 46
mass			90
Beil	469, 414		

42. tert-**Butylbenzene**

$C_6H_5C(CH_3)_3$ $C_{10}H_{14}$

mw		134.223	887
bp	760	169.119°	199: 841, 1453, 2332, 4535
dt/dp	760	0.05269°	198
p	25°	2.1	199: 2966, 4535
eq 5	A	6.92050	198a
	B	1504.572	
	C	203.328	
fp		− 57.850°	199: 841, 1453, 1794, 4535
d	20°	0.86650	198, 1334: 841, 2332, 4593
	25	0.86240	
n_D	20°	1.49266	198, 1606: 841, 2332, 4329, 4593
	25	1.49024	
η	20°	28.13	1334: 2332, 4996
	30	27.14	
$\triangle H_v$	25°	11.73	199
	bp	8.99	
$\triangle H_m$		2.006	199
$\triangle H_c^\circ$	25°(l)	− 1403.2	2577: 3968
t_c		374.1°	643
p_c		29.2	643
d_c		0.274	1369
$\blacktriangledown_c$		0.490	[1369]
A	eq 72	0.02173	199
ϵ	20°	2.366	66: 3971, 2878, 3477
μ	20 – 30°(l)	0.36	66: 4017, 65, 1154, 2878
soly	aq in, 20°	0.0292	1483
fl pt	TOC	140°F	3747
uv			91, 3225
ir			89, 3792, 408, 3816, 5199
Raman			1556, 3629, 412
mass			90
nmr			3126, 1756, 685
Beil	469, 415		

43. p–Cymene

p–CH$_3$C$_6$H$_4$CH(CH$_3$)$_2$ C$_{10}$H$_{14}$

mw		134.223	887
bp	760	177.10°	199: 1453, 1556, 2302, 2652, 4535, 3070
dt/dp	760	0.0528°	198: 5257
p	25°	1.5	199: 2966, 4535, 5257
eq 5	A	6.9260	198a: 2652, 3070
	B	1538.00	
	C	203.10	
fp		– 67.935°	194, 4521: 1755, 2302, 3070, 4770, 1453
d	20°	0.8573	198: 1453, 1755, 2878, 4593
	25	0.8533	
n$_D$	20°	1.4909	198: 1453, 1556, 1755, 2878, 4593
	25	1.4885	
dn/dt	7 – 25°	0.0004830	1453
η	20°	3.402	4773: 396, 846, 1369
	30	1.600	
γ	20°	28.81	1369: 4996
	30	27.74	
△ H$_v$	25°	12.02	199: 2652, 5032, 782
	bp	9.12	
△ H$_m$		2.309	199, 2302
C$_p$	– 23.9°	56.50	2302: 4143
	Table XIII	Ch II	
t$_c$		378.6°	64: 1453, 643
p$_c$		28.6	[64] : 643
d$_c$		0.266	1369
v$_c$		0.505	1369
A	eq 72	0.02759	199: 4521
K$_B$		5.52°	483
κ	25°	2 x 10^{-8}	3676
ε	20°	2.253	66: 2878, 3477
μ	20 - 30°(l)	0	66: 2878, 5210
soly	in aq	0.034%w	[664]
fl pt	TCC	117°F	3680
uv			91
ir			89, 2807, 5199, 4286, 3348
Raman			1556, 1406
mass			90
Beil	469, 420		

44. Cyclohexylbenzene

$C_{12}H_{16}$

mw		160.261	887
bp	760	240.12°	3244: 628, 921
dt/dp	760	0.0525°	3244
p			2966
fp		6.99°	3244: 4328, 616, 4526, 921
d	20° 25	0.94272 0.93874	3244: 628, 1508, 921
n_D	20° 25	1.52633 1.52393	3244: 628, 1508, 921
η	0°	3.681	3488
A	eq 72	0.0295	4526
κ		8.85×10^{-10}	[3160]
fl pt	TOC	210°F	3160
uv			1692
ir			89, 1700, 3816, 3925
Raman			3229
Beil	473, 503		

45. 1–Pentene

$$CH_3CH_2CH_2CH=CH_2 \hspace{4cm} C_5H_{10}$$

mw		70.135	887
bp	760	29.968°	168, 1607: 1233, 1733, 2538, 679, 5183
dt/dp	760	0.03801°	163, 1607
p	25°	638	168: 1233, 4535, 1607
eq 5	A	6.84650	175: 1233, 1607
	B	1044.895	
	C	233.516	
fp		− 165.220°	168, 4528: 4522, 3020, 4798
d	20°	0.64050	163, 1607: 1733, 2538, 679, 5183, 5168
	25	0.63533	917, 4280, 2611
dd/dt	22.5°	0.001034	1607, 1233
n_D	20°	1.37148	163, 1607: 1733, 2538, 679, 5183, 5168
	25	1.36835	67, 917, 2611
η	− 5°	0.25	164
	0	0.24	
γ	0°	18.20	165
	10	17.10	
	20	16.00	
	25	15.45	
$\triangle H_v$	25°	6.088	168: 1233
	bp	6.022	
$\triangle H_m$	25°	1.388	168, 4798
$\triangle H_f^\circ$	25°(l)	− 11.14	5358: 170
$\triangle H_c^\circ$	25°(l)	− 800.68	5358: 4580, 169
C_p°	25	26.19	179, 2589
C_p	25°	37.12	4798
t_c		191.59°	174: 2620, 72
p_c		35.0	174: 2620
d_c		0.230	174: 1370
v_c		0.305	174
A	eq 72	0.05996	168: 4528
ϵ	20°	2.017	67: 4278
μ	20°(l)	0.34	67: 1698, 4906
soly	in aq, 25°	0.0148%w	3053
fl pt	estm	− 60°F	3747
uv			915, 917, 4280, 1718, 2428
ir			89, 4279, 4739, 4275, 3509
Raman			1556, 3901, 1396, 4294, 3925
mass			90, 3997, 1396, 471
Beil	11, 210		

46. 2–Pentene

Mixed Isomers

$CH_3CH_2CH=CHCH_3$ C_5H_{10}

mw		70.135	887
bp	760	36.7°	3805: 1778, 2386, 3651, 917, 679, 5266
dt/dp	760	0.056°	[2444]
p	25°	528	[2444]
eq 4	A	7.7945	[2444]
(–80 – +37°)	B	1512.5	
fp		–138°	1778
d	15°	0.6555	1778: 2386, 3546, 679
	20	0.6545	3805
n_D	20°	1.3798	5266: 2386, 3805, 917, 679
	25	1.3839	1778
η	20°	0.214	5266
	25	0.201	
γ	14.8°	17.68	2386
	23.3	16.79	
$\triangle H_v$	25°	8.42	[2444]
C_p	15.9°	36.1	3651
t_c		202.4°	2651
p_c		40.4	2651
soly	in aq, 25°	0.0203%w	3053
fl pt	estm	–50°F	3747
uv			917, 915
ir			4279, 4739
mass			3997, 3322
Beil	11, 210		

47. cis-2-Pentene

CH$_3$—C=C—CH$_2$CH$_3$ / H   H

C_5H_{10}

mw		70.135	887
bp	760	36.942°	176
dt/dp	760	0.03830°	172
p	25°	495.	176
eq 5	A	6.87274	175: 3747
	B	1067.951	
	C	230.585	
fp		−151.390°	176: 4522, 3020, 4798
d	20°	0.6556	172
	25	0.6504	
n_D	20°	1.3830	172
	25	1.3798	
γ	0°	19.7	173
	10	18.6	
	20	17.38	
	25	16.8	
$\triangle H_v$	25°	6.41	176
	bp	6.24	
$\triangle H_m$		1.6998	176: 4798
$\triangle H_f^\circ$	25°(l)	−12.96	5358: 178, 3841
$\triangle H_c^\circ$	25°(l)	−798.87	5358: 1082, 177, 3841
C_p°	25°	24.32	179, 2589
C_p	25°	36.26	4798
t_c		203.°	174
p_c		36	174
d_c		0.23	174
v_c		0.30	174
A	eq 72	0.05770	176
fl pt	estm	−50°F	3747
uv			1718, 2428
ir			89
Raman			1556, 3901, 4275, 3925
mass			90
Beil	**11**, 210		

48. trans- 2-Pentene

$$CH_3 \diagdown \diagup H$$
$$C=C$$
$$H \diagup \diagdown CH_2CH_3$$

C_5H_{10}

mw		70.135	887
bp	760	36.353°	176
dt/dp	760	0.03824°	172
p	25°	505.	176: 3747
eq 5	A	6.90575	175
	B	1083.987	
	C	232.965	
fp		−140.244°	176: 4522, 4798
d	20°	0.6482	172
	25	0.6431	
n_D	20°	1.3793	172
	25	1.3761	
γ	0°	18.9	173
	10	17.9	
	20	16.90	
	25	16.42	
$\triangle H_v$	25°	6.38	176
	bp	6.23	
$\triangle H_m$		1.9961	176: 4798
$\triangle H_f^\circ$	25° (l)	−14.02	5358: 178
$\triangle H_c^\circ$	25° (l)	−797.81	5358: 1082, 177
C_p°	25°	25.92	179: 2589
C_p	25°	37.52	4798
t_c		202.$^\circ$	174
p_c		36.	174
d_c		0.23	174
v_c		0.30	174
A	eq 72	0.05687	176
uv			1718, 2428
ir			89, 5345, 3509
Raman			1556, 3901, 4275, 3925
mass			90
Beil	11, 210		

49. 1-Hexene

$CH_3(CH_2)_3CH=CH_2$ C_6H_{12}

mw		84.163	887
bp	760	63.485°	168, 1607, 889: 679, 5183, 1733, 4767
dt/dp	760	0.04149°	163, 1607, 889
p	25°	186	168, 889: 1607
eq 5	A	6.86572	167, 1607, 889
	B	1152.971	
	C	225.849	
fp		− 139.819°	168: 4767, 3061, 4917, 4528, 3243
d	20°	0.67317	163, 1607, 889: 679, 5183, 5168, 1733
	25	0.66848	4917, 2700, 3243
dd/dt	25°	0.000943	1607, 889: 3243
n_D	20°	1.38788	163, 1607, 889: 679, 5183, 5168, 67
	25	1.38502	4917, 2700, 3243
dn/dt		0.00057	3243
η	20°	0.26	164: 5266
	25	0.25	
γ	25°	17.90	[2379]: 165, 2377
	Table VIII	Ch II	
$\triangle H_v$	25°	7.32	168
	bp	6.76	889
$\triangle H_m$		2.2341	168, 3061
$\triangle H_f^{\circ}$	25°(l)	− 17.28	5358: 170
$\triangle H_c^{\circ}$	25°(l)	− 956.92	169, 5358
C_p°	25°	31.63	171
C_p	25°	43.81	3061: 2589
t_c		230.83°	72, 166
p_c		31.	166
d_c		0.23	166
v_c		0.37	166
A	eq 72	0.06324	168: 3061, 4528
ϵ	20°	2.051	67
μ	20°(l)	0.34	67: 65, 1731, 3020

49. 1–Hexene

(Continued)

soly	in aq, 25° aq in, 30°	0.0050%w 0.0477	3053 1483
aq az	57.7°	94.3%w	3975
fl pt	estm	– 15°F	3747
uv			915, 1718, 3153, 2428
ir			89, 2567, 389
Raman			717, 2700, 4822, 680, 4876, 4294, 499
mass			1454, 471
nmr			674, 1584
Beil	**11**, 215		

50. 1–Heptene

$CH_3(CH_2)_4CH=CH_2$ $\qquad\qquad\qquad\qquad$ C_7H_{14}

mw		98.190	887
bp	760	93.643°	168, 1607: 679, 5183, 1733, 2611, 917
dt/dp	760	0.04447°	163, 1607
p	25°	56.	168: 1607, 21, 2976
eq 5	A B C	6.90069 1257.505 219.179	167, 1607
fp		– 118.858°	168: 3061, 3661, 4528
d	20° 25	0.69698 0.69267	163, 1607: 679, 5183, 5168, 1733, 2611 917
dd/dt	25°	0.000883	1607
n_D	20° 25	1.39980 1.39713	163: 1607, 679, 5183, 5168, 67, 2611 917
η	20° 25	0.35 0.34	164
γ	25°	19.80	[2379], 165: 2377
	Table VIII	Ch II	

50. 1–Heptene

(Continued)

$\triangle H_v$	25°	8.52	168
	bp	7.43	
$\triangle H_m$		2.964	168, 3061: 3661
$\triangle H_f^\circ$	25°(l)	–23.19	5358: 170
$\triangle H_c^\circ$	25°(l)	–1113.19	169: 4011, 5358
C_p°	25°	37.10	171: 2589
C_p	25°	50.62	3061: 3153
t_c		264.08°	72, 166
p_c		28.	166
d_c		0.22	166
v_c		0.44	166
A	eq 72	0.06266	168: 3061, 4528
ϵ	20°	2.071	67
μ	20°(l)	0.34	67: 1731
soly	aq in, 20.1°	0.1126%w	614: 1483
aq az	77.0°	85.2%w	3975
uv			917, 915, 916
ir			89, 4279, 2807, 4275
Raman			681, 4822, 680, 4275, 4876, 4294, 3925
mass			471
Beil	11, 219		

51. 1–Octene

$CH_3(CH_2)_5CH{=}CH_2$ C_8H_{16}

mw		112.217	887
bp	760	121.280°	168, 1607: 1733, 3805, 5183, 679, 5266
dt/dp	760	0.04711°	163, 1607
p	25°	17	168: 1607
eq 5	A	6.93263	167: 1607
	B	1353.486	
	C	212.764	

51. 1–Octene

(Continued)

fp		$-101.736°$	168: 3805, 3061, 4524
d	$20°$	0.71492	163, 1607: 1733, 3805, 5168, 5183, 679
	25	0.71085	2611, 2700
dd/dt	$25°$	0.000834	1607
n_D	$20°$	1.40870	163, 1607: 67, 3805, 5168, 5183, 679
	25	1.40620	5266, 2611, 2700
η	$20°$	0.470	164: 5266
	25	0.447	
γ	$25°$	21.28	[2379], 165: 2377
	Table VIII	Ch II	
$\triangle H_v$	$25°$	9.70	168
	bp	8.07	
$\triangle H_m$		3.660	168, 3061
$\triangle H_f^\circ$	$25°(l)$	-29.11	5358: 170
$\triangle H_c^\circ$	$25°(l)$	-1269.45	169: 4011, 5358
C_p°	$25°$	42.56	171: 2589
C_p	$25°$	57.65	3061
t_c		$293.4°$	166: 72
p_c		26.	166
d_c		0.22	166
v_c		0.51	166
A	eq 72	0.06268	168: 3061, 4524
ϵ	$20°$	2.084	67
μ	$20°(l)$	0.34	67: 1731
soly	in aq, $25°$	0.00027%w	3053
aq az	$88.0°$	71.3%w	3975
fl pt	TOC	$70°F$	3747
uv			2632, 3784, 5134
ir			89, 3679, 2808, 2567, 4739, 3919, 4275
Raman			681, 1005, 2700, 4822, 680, 4275, 4294
mass			471
Beil	11, 221		

52. 1–Nonene

$CH_3(CH_2)_6CH=CH_2$ C_9H_{18}

mw		126.244	887
bp	760	146.868°	168, 1607: 1733, 679, 5183
dt/dp	760	0.04944°	163, 1607
p	25°	5.3	168: 1607, 4163
eq 5	A	6.95387	167, 1607
	B	1435.359	
	C	205.535	
fp		-81.37°	168, 4527: 4163
d	20°	0.72922	163, 1607: 5168, 1733, 679, 5183, 4163
	25	0.72531	
dd/dt	25°	0.000788	1607
n_D	20°	1.41572	163, 1607: 5168, 679, 5183, 4163, 3133
	25	1.41333	
η	20°	0.620	164
	25	0.586	
γ	25°	22.56	[2379]: 165, 2377
	Table VIII	Ch II	
$\triangle H_v$	25°	10.88	168
	bp	8.68	
$\triangle H_m$		4.3	168
$\triangle H_f^\circ$	25° (l)	-35.58	5358: 170
$\triangle H_c^\circ$	25° (l)	-1425.71	169, 5358: 1082
C_p°	25°	48.03	171: 2589
t_c		319°	166
p_c		23.	166
d_c		0.22	166
v_c		0.58	166
A	eq 72	0.0588	168, 4527
μ	20° in 30	0.59	1731
aq az	94.5°	53.7%w	3975
ir			89, 2808, 1997
Raman			681, 4822, 3925
Beil	11, 223		

53. 1-Decene

$CH_3(CH_2)_7CH=CH_2$ $C_{10}H_{20}$

mw		140.271	887
bp	760	170.570°	168, 1607: 1733, 4756
dt/dp	760	0.05157°	163, 1607
p	25°	1.6	168, 4163
eq 5	A	6.96034	167, 1607
	B	1501.872	
	C	197.578	
fp		−66.310°	168, 2646: 3061, 4163, 4756, 3243, 4527
d	20°	0.74081	163, 1607: 1733, 5168, 4163, 4756, 1731
	25	0.73693	
dd/dt	25°	0.000777	1607: 3243
n_D	20°	1.42146	163, 1607: 5168, 4163, 4756, 3243, 1731
	25	1.41913	5266
dn/dt		0.00051	3243
η	20°	0.805	164: 4163, 5266
	25	0.756	
γ	25°	23.54	[2379]: 165, 2387, 2377
	Table VIII	Ch II	
$\triangle H_v$	25°	12.06	168
	bp	9.24	
$\triangle H_m$		3.300	168: 3061
$\triangle H_f^\circ$	25°(l)	−41.54	5358: 170
$\triangle H_c^\circ$	25°(l)	−1581.97	169: 4011, 5358
C_p°	25°	53.49	171: 2589
C_p	25°	71.78	3061
t_c		342°	166
p_c		22.	166
d_c		0.21	166
v_c		0.65	166
A	eq 72	0.03882	168: 4527, 3061
μ	20° in 30	0.42	1731
aq az	96.7°	35.8%w	3975
uv			89, 1650, 2808, 388, 389, 3750
ir			389
Raman			4822, 2087, 3925
Beil	11, 223		

54. Cyclohexene

C_6H_{10}

mw		82.147	887
bp	760	82.979°	182, 1607: 326, 2292, 3208, 4764, 4965
dt/dp	760	0.04381°	182, 1607
p	25°	88.8	[1607] : 2966
eq 5	A	6.88617	1607
	B	1229.973	
	C	224.104	
fp		− 103.512°	182: 3651, 4764, 4528
d	20°	0.81096	182, 1607: 326, 2292, 3487, 4965, 1732
	25	0.80609	
dd/dt	25°	0.000955	1607
n_D	20°	1.44654	182, 1607: 67, 326, 2292, 3208, 4965
	25	1.44377	1556, 5266
η	20°	0.650	5266: 1732
	30	0.608	
γ	20°	26.54	4772
	30	25.22	
$\triangle H_v$	bp	7.286	3208
$\triangle H_m$			3651
$\triangle H_f^\circ$	25°(l)	− 9.13	5358: 184
$\triangle H_c^\circ$	25° (l)	− 896.62	181: 1483, 3670, 5358
C_p°	25°	25.10	183
C_p			3651
t_c		287.27°	72: 964
d_c		0.288	1369
A	eq 72	0.01376	4528
κ	25°	1.5×10^{-15}	1603
ϵ	20°	2.220	67, 3477: 4786
μ	20°(l)	0.28	67: 3020

54. Cyclohexene

(Continued)

soly	in aq, 25$^{\circ}_{\circ}$	0.0213%w	3053
	aq in, 20	0.0317%w	1483
aq az	70.8°	90.0 %w	2254
fl pt	estm	$-10^{\circ}F$	3747
uv			2047, 2840, 3784, 3818, 2428, 5134, 4860
ir			89, 2809, 3792, 408, 4305, 2567
Raman			1556, 3522, 1406, 3153
mass			90, 3997, 2205
Beil	453, 63		

55. Styrene

$C_6H_5CH=CH_2$ C_8H_8

mw		104.153	887
bp	760	145.14°	203: 1372, 1453, 4535
dt/dp	760	0.049°	203, 1372
p	25°	6.31	3777: 4535, 879, 2966, 829a
eq 5	A	7.22302	1372: 1453, 3777, 829a
	B	1629.2	
	C	230.	
fp		-30.628°	203: 1372, 4535, 3777
d	20°	0.90600	203, 1372: 1453
	25	0.90122	
dd/dt	0–40°	0.0008739	1368
n$_D$	20°	1.54682	203: 1372, 1453, 829a
	25	1.54395	
dn/dt	13–25°	0.0005186	1453
η	20°	0.751	1445
	25	0.696	
	30	0.650	
γ	20°	32.3	1369: 4996
	30	30.98	

55. Styrene

(Continued)

$\triangle H_v$	$25°$	10.500	3777: 879
	bp	9.25	829a: 1369
$\triangle H_m$		2.617	3777
$\triangle H_f^o$	$25°(l)$	24.83	205: 3841, 5358
$\triangle H_c^o$	$25°(l)$	-1050.58	3993: 2577, 172, 328, 3841, 5358
$\triangle H_p$		16.68	3993
C_p^o	$25°$	29.18	206: 470, 4213
C_p	$25.39°$	43.64	1445
t_c		$363.7°$	1369: 4213
p_c		36.3	4213
ϵ	$20°$	2.4257	3733: 3477
μ	$20-60°(l)$	0.13	3733: 3781, 1522, 2139, 2871
soly	in aq, $25°$	0.031%w	2818
	aq in, $25°$	0.066%w	2818
aq az	$93.9°$	59.1%w	2255
fl pt	TCC	$87°F$	3680
uv			91, 549, 4016, 4175, 4006, 4232, 1181
ir			89, 2554, 408, 3679, 4739, 3777
Raman			2573, 499, 670, 3777
mass			90, 2403, 3997, 4618
nmr			801, 3028, 3230
Beil	473, 474		

56. 2-Pinene

α-Pinene
Mixed Isomers

$C_{10}H_{16}$

mw		136.239	887
bp		155.9°	2074: 318, 5093, 4751, 327, 4179, 1929
p	23.64°	4.	1928: 1676, 2966
eq 12	A	26.40174	2074: 4068
	B	3134.525	
	C	6.16045	
fp		−64.°	4047: 318, 1929, 2075
d	20°	0.8582	5093
	25°	0.8539	2074: 1676, 318, 3761, 4751, 4716, 4179
			327, 649, 540, 1062, 1403
n_D	20°	1.4658	5093: 4855, 1676, 318, 3761, 4751, 4716
	25	1.4632	2074: 327, 4179, 649, 1929, 397, 2075
dn/dt		0.00045	1676
η	20°	1.40	396: 1403
γ	30°	25.8	Beil E **III** 5, 368
$\triangle H_v$	3 − 15 torr	11.140	2074: 4855
	215 − 760	9.482	2074
$\triangle H_c°$	25°(l)	−1465.3	2577: 327, 328, 2075

56. 2–Pinene

(Continued)

$[\alpha]_D^{20^\circ}$	$\frac{d}{l}$	$+51.14^\circ$ -51.28°	4751: 327, 4179, 3046, 649, 1062, 397 1929
K_f		16.50°	4047
ϵ	30°	2.2588	1220: 4564, 4716, 3477
μ	30° in 30	0.36	1220: 4564, 3443, 1985, 4563
uv			5134, 4860
ir			89, 408, 3787, 3348
Raman			3523, 563, 649, 4939, 648
mass			90, 4714
nmr			268
Beil	458, 144		

57. 2(10)–Pinene

nopinene
β–Pinene

$C_{10}H_{16}$

mw		136.239	887
bp		166.0°	2074: 4856, 3761, 2075
p	25°	4.6	3761: 1676
eq 12	A	28.77768	2074
	B	3318.845	
	C	6.94243	
fp		-61.54°	2075

57. 2(10)–Pinene

(Continued)

d	$25°$	0.8667	2074: 1676, 3761, 4716, 318
n_D	$15°$	1.4813	1676
	25	1.4768	2074: 4716, 4855, 3761, 318, 2075
dn/dt		0.00045	1676
η	$20°$	1.70	396
$\triangle H_v$	2 – 23 torr	10.390	2074: 4855
	175 – 760	9.610	2074
$\triangle H_c°$	$25°$(l)	– 1485.1	2075: 328
$[\alpha\ ]$	d	$+ 12.6°$	Beil **E II 5**, 102
	l	$– 21.49°$	1676
ϵ		2.4970	4716
uv			5134
ir			89
Raman			2385
mass			4714
nmr			268
Beil	458, 154		

58. Methanol

CH_3OH CH_4O

mw		32.042	887
bp	760	64.70°	4640: 3081, 3184, 4536, 5236, 4635, 4643
dt/dp	760	0.03347°	4643: 4778, 5236, 4635
p	25°	125.03	4640: 2279, 3723, 4535, 4773, 4638, 1292
eq 5	A	7.89750	4639: 4635, 2093
	B	1474.08	
	C	229.13	
fp		−97.68°	4640: 2453, 2547, 3081, 4029, 4483, 4536
d	15°	0.79609	4778: 23, 1717, 2259, 2423, 2543, 2546
	20	0.79129	4644: 3029, 3723, 4761, 4975, 4635, 1895
	25	0.78664	4643
dd/dt		0.00093	[4778]
α		0.0012	2244
n_D	15°	1.33057	2349: 23, 2259, 2543, 3208, 3723, 4778
	20	1.32840	4643: 4975, 4635
	25	1.32652	
η	20°	0.5506	356: 1717, 4778, 4808, 4928, 5022
	25	0.5445	2423
	Table II	Ch II	
γ	20°	22.55	2102, 4778: 22, 2349, 2546, 3723, 4808
	30	21.69	4975
$\triangle H_v$	25°	8.946	4640: 429, 1575, 2346, 3331, 4483, 5123
	bp	8.433	1876, 1873, 4053, 5012
$\triangle H_m$		0.766	4640: 3646, 4483
$\triangle H_f^\circ$	25°(l)	−57.24	949: 4641, 1873, 5358
$\triangle H_c^\circ$	25°(l)	−173.45	949: 1873, 4053, 4743, 2577, 5358
C_p°	25°	10.49	4641: 4393, 5123, 1876
C_p	23.46°	19.48	673: 1575, 2259, 2547, 3745, 4483, 5123
t_c		239.43°	4645: 3331, 5301, 5303, 939, 2093, 1100
p_c		79.9	4645: 3331, 5301, 2651, 2093
d_c		0.272	4645: 5303, 939
v_c		0.118	4645: 5303, 1100
A	eq 72	0.0125	4640
K_B		0.785°	919: 483, 2426, 5029, 3625, 491
pK_a	25° in H_2O	15.5	369: 3787, 3075, 5221
pK_s		16.7	796
κ	25°	1.5×10^{-9}	1525: 2321, 3029, 3723, 4167, 5026

58. Methanol

(Continued)

ϵ	$25°$	32.70	2399, 1188: 1851, 2832, 44, 3802, 2048
μ	$20°$ (l)	2.87	3056: 518, 3113, 3184, 4518, 5243, 2892
soly	in aq, $25°$	inf	1338
aq az		none	2254
fl pt	CC	$54°F$	3680
uv			91, 561, 3772, 4634, 2313, 2042
ir			409, 665, 1112, 1205, 2224, 3089, 4107
Raman			717, 1444, 1969, 2659, 2815, 4943, 2540
mass			90, 3997, 1769, 1883
nmr			3177, 3175, 3821, 3176, 4632, 3126, 951
Beil	**19**, 273		

59. Ethanol

CH_3CH_2OH C_2H_6O

mw		46.070	887
bp	760	$78.29°$	4640: 853, 1776, 3184, 5236, 1373, 811
dt/dp	760	$0.03317°$	4643: 5236, 1372, 811, 2806, 3001
p	$25°$	59.77	4640: 4535, 1562, 4183, 1373, 4123, 4413
eq 5	A	8.32190	4639: 2093, 1372
	B	1718.10	
	C	237.52	
fp		$-114.1°$	4640: 4774, 2548, 1372, 4762
d	$15°$	0.79360	3496: 23, 1492, 1776, 2548, 2835, 4975
	20	0.78937	4644, 4643: 811, 1717, 1851, 2806, 3029
	25	0.78504	5301, 5333
α		0.0011	2244
^nD	$20°$	1.36143	4643: 4425, 811, 1954, 4123, 23, 1492
	25	1.35941	1372, 2806, 4975

59. Ethanol

(Continued)

dn/dt		0.00040	4403
η	25°	1.078	2307: 356, 846, 4808, 384, 1717, 5022
	30	0.991	4793
	Table II, IV	Ch II	
γ	20°	22.32	2102: 22, 2806, 4808, 1492, 2835, 4831
	30	21.48	
$\triangle H_v$	25°	10.112	4640: 5012, 3067, 515, 429, 5032, 3659
	bp	9.260	3208, 3641, 1873, 4962, 2346, 4053
$\triangle H_m$		1.20	4640: 2548, 3646
$\triangle H_f^o$	25°(l)	-66.20	4641, 949: 1873, 1880
$\triangle H_c^o$	25°(l)	-326.86	949: 4743, 2577, 1873, 4053
C_p^o	25°	15.64	4641: 1876, 731, 1880
	Table XIV	Ch II	
C_p	21.15°	26.26	2548: 3641, 1492, 3646, 5030, 1575, 5002
t_c		243.1°	4645: 2093, 5301, 1100, 5303, 3331
p_c		62.96	4645: 2093, 5301, 3331
d_c		0.276	4645: 2651, 5301
v_c		0.167	4645: 1100, 5303
A	eq 72	0.0024	4640
K_B		1.160°	2846: 487, 5029, 2207, 4436, 919, 477
pKa	25° in H_2O	15.9	369: 3075, 420, 3296
pKs		19.1	796
κ	25°	1.35×10^{-9}	1187: 3029, 4167, 2321, 3034, 5026, 2058
ϵ	25°	24.55	2399, 1188: 4425, 2341, 1888, 2747, 5275
μ	20°(l)	1.66	3056: 2893, 2175, 2946, 2176, 4411, 2880
soly	in aq	inf	1338
aq az	78.174°	96.0%w	2254
fl pt	CC	55°F	88
uv			91, 2042, 3268, 560, 1237, 4634
ir			4152, 2950, 4113, 5341, 2812, 3339, 4107
Raman			717, 3132, 4550, 3916, 735, 5017, 4861
mass			3997, 4618, 3378, 1663, 1045, 3269, 1769
nmr			2339, 3177, 3821, 4253, 4861, 2461, 4632
Beil	20, 292		

60. 1–Propanol

$CH_3CH_2CH_2OH$ C_3H_8O

mw		60.097	887
bp	760	97.20°	4640: 4975, 3452, 5236, 4775, 4643, 568
dt/dp	760	0.03475°	4643: 4775, 810, 568, 811, 5236, 3001
p	25°	20.85	4640: 4535, 3846, 2458, 568
eq 5	A	7.84767	4639:.74, 779, 2093, 3206, 568, 826
	B	1499.21	1087
	C	204.64	
fp		−126.2°	4640: 4643, 3650, 4535
tp		−124.40°	1110
d	0°	0.81930	2753: 262, 23, 810, 811, 1717, 2753
	15	0.80749	4775: 2440, 2835, 3029, 3452, 3650, 4829
	20	0.80375	4644: 4979, 5301, 4775, 4643
	25	0.79975	
α		0.00096	2244
n_D	20°	1.38556	4643: 4979, 811, 1954, 3452, 4829
	25	1.38370	810, 23
η	−195.7° gl	3.0×10^{20}	2967: 356, 1404, 1717, 3452, 4808, 4829
	15	2.522	4775
	30	1.722	4772: 4928, 5153
	Table II	Ch II	
γ	20°	23.70	2102: 788, 2835, 3452, 4775, 4808, 4979
	30	22.89	
	Table VII	Ch II	
$\triangle H_v$	25°	11.310	4640, 5012, 1876: 1099, 3001, 1873, 4962
	bp	9.982	5202, 568, 2346, 429, 4989, 4053
$\triangle H_m$		1.284	[1110]: 3650
$\triangle H_f^\circ$	25°(l)	−72.79	949: 4641, 1873
$\triangle H_c^\circ$	25°(l)	−482.64	949: 4435, 4743, 2577, 1873, 4053
C_p°	25°	20.82	4649: 1876, 3206
C_p	25°	34.37	[1110]: 515, 5337, 4829, 2384, 3745, 3001
t_c		263.56°	4645: 74, 3331, 2093, 5303, 5301, 1110
p_c		51.02	4645: 74, 5301, 2093, 3331
d_c		0.275	4645: 74
v_c		0.218	4649: 1110, 5303

60. 1–Propanol

(Continued)

A	eq 72	0.0289	4640
K_B		1.59°	483
κ	18°	9.17×10^{-9}	2321: 3029, 5153
ϵ	25°	20.33	2399, 1188: 3945, 5275, 2321, 3400, 2341
μ	20°(l)	3.09	3056: 5243, 3113, 4409, 1954, 2892, 2880
soly	in aq	inf	1338
aq az	87.65°	71.7%w	2255
fl pt	CC	77°F	88
uv			560, 3785, 3810, 2042
ir			1845, 2756, 3358, 1780, 5354, 3530, 2889
Raman			5105, 717, 3878, 3132, 2761
mass			1769, 90, 1663, 1659, 3269, 1883, 4715
nmr			4632, 3177, 2791
Beil	24, 350		

61. 2–Propanol

$CH_3CHOHCH_3$ C_3H_8O

mw		60.097	887
bp	760	82.26°	4647: 4643, 4775, 4578, 3184, 871
dt/dp	760	0.03294°	4643: 568, 4775, 810, 811
p	25°	45.16	4647
eq 5	A	8.11778	4646: 74, 568, 2754
	B	1580.92	
	C	219.61	
fp		-88.0°	4647: 241, 4775, 3654, 4774, 3650, 4767
d	20°	0.78545	4644: 892, 810, 1717, 2835, 3590, 3648
	25	0.78126	2549, 4643, 3208, 3659, 4808, 4975
n_D	20°	1.37720	4643: 810, 4975, 5194, 871, 5119
	25	1.3752	

61. 2-Propanol

(Continued)

η	15° 30	2.859 1.765	4775: 4373, 1404, 4808, 5153, 1717, 4928
	Table II	Ch II	
γ	15° 30	21.79 20.96	4775: 2102, 4975, 2835
$\triangle H_v$	25° bp	10.88 9.53	4647: 5202, 429, 515, 3648, 3067 3208, 5012, 1968, 532, 4962
$\triangle H_m$		1.292	4647: 241, 3654, 2549
$\triangle H_f^o$	25°(l)	-75.77	949: 4648, 835, 3658
$\triangle H_c^o$	25° (l)	-479.66	949: 4435, 4743, 2577, 3658, 3657
C_p^o	25°	21.21	4649: 532, 1881, 1968
C_p	25.07°	36.950	241: 532, 5194, 2384, 3745, 1968, 3654
t_c		235.17°	3604: 3924, 783
p_c		47.02	3604: 74
d_c		0.273	3604: 74
v_c		0.220	3604: 783
A	eq 72	0.0190	4647
pK_a	25°in aq H_2SO_4	-3.2	420: 3075, 686
κ	25°	5.8×10^{-8}	2058: 4373, 5153
ϵ	25°	19.92	2399, 1188: 2686, 4470, 3309, 2835, 938
μ	30°in 30	1.66	3184: 2177, 4517, 3110, 4411
soly	in aq	inf	
aq az	80.10°	88.0%w	2255
fl pt	CC	53°F	88, 1485
uv			2042, 3810, 4639
ir			4400, 3530, 2203, 1881, 415, 1572, 2926
Raman			1881, 4605, 4943, 304, 5341, 89, 4113
mass			3269, 1663, 1659, 3878, 90, 717, 1769
nmr			4632, 3177, 2223, 3821, 3126, 951
Beil	24, 360		

62. 1–Butanol

$CH_3 CH_2 CH_2 CH_2 OH$ $C_4 H_{10} O$

mw		74.124	887
bp	760	117.66°	4640: 4643, 811, 2422, 3452, 5016, 4764
dt/dp	760	0.03686°	4643: 568, 811, 1372, 4784, 5236, 871
p	25°	6.18	4640: 3846, 1373, 871, 4535
eq 5	A	7.47680	4639: 779, 74, 1372, 871, 2093, 568
	B	1362.39	
	C	178.77	
fp		− 88.62°	4640: 1372, 3646, 4771, 4643, 732
tr pt		− 88.64°	1109
d	20°	0.8097	4644: 4784, 2794, 811, 871, 2440, 4425
	25	0.8060	[779]
α		0.00094	2244
n_D	20°	1.3993	4643: 2968, 23, 817, 3452, 4975, 4425
	25	1.3973	4784, 779, 51, 3208, 4829
dn/dt		0.00039	4784
η	15°	3.379	4784: 811, 1404, 2440, 4829, 846, 2422
	30	2.271	3452
	Table II, III	Ch II	
γ	20°	24.57	2102: 2720, 3452, 2835, 4975
	30	23.75	
	Table VIII	Ch II	
$\triangle H_v$	25°	12.54	4640, 1873: 515, 2346, 1876, 429, 871
	bp	10.310	1109: 568, 3208, 5012, 3001, 3067, 4053
	Table IX	Ch II	
$\triangle H_m$		2.240	1109, 4640: 3646
$\triangle H_f^\circ$	25°(l)	− 78.49	4357: 4641, 949, 1935, 1873
$\triangle H_c^\circ$	25°(l)	− 639.31	4357: 2577, 949, 1935, 1873, 4053
C_p°	25°	26.29	4641: 1876, 1109
	Table XIV	Ch II	
C_p	25°	42.31	1109: 515, 4829, 3745, 3646, 5194
	Table XI	Ch II	
t_c		289.78°	4645: 2093, 74, 4336, 2145, 3924, 3687
p_c		43.55	4645: 74, 4336, 2093, 2145
d_c		0.270	4645: 4336, 74
v_c		0.274	4645

62. 1–Butanol

(Continued)

A	eq 72	0.0331	4640
κ		9.12×10^{-9}	2321: 1190, 2569
ϵ	25°	17.51	1188: 3477, 2835, 1888, 1851, 4425, 5275
μ	25° in 232	1.75	2892: 1954, 4411, 1851, 4409, 4425, 5243
soly	in aq, 25°	7.45%w	4507: 529, 2417, 2191, 1677, 664, 1338
	aq in, 25°	20.5%w	4507: 2191, 2417
aq az	92.7°	57.5%w	2254: 4507
fl pt	CC	84°F	3680
uv			3785, 2812, 560, 4634, 3810
ir			1183, 4107, 2761, 3358, 1699, 5341, 4113
Raman			717, 4943, 3132, 2662, 4942, 2761
mass			3269, 1659, 90, 1769, 4850, 4715
nmr			4632, 3235, 2461
Beil	24, 367		

63. 2–Butanol

sec – Butyl Alcohol

$CH_3CH_2CHOHCH_3$ $C_4H_{10}O$

mw		74.124	887
bp	760	99.55°	4647: 4643, 810, 811, 533, 51, 2260
dt/dp	760	0.03559°	4643: 568, 811, 810, 4784
p	25°	18.29	4647: 568, 871, 4023, 4269, 1085, 4535
eq 5	A	7.47431	4646: 568, 74, 533, 4269, 871
	B	1314.19	
	C	186.55	
fp		-114.7°	4535, 4643
d	15°		4269: 51, 811, 2260, 4784, 810, 871
	20	0.8069	4644: 4643, 2835, 2918
	25	0.8026	

63. 2-Butanol

(Continued)

α		0.00091	2244
n_D	15°	1.39946	4784: 51, 810, 717, 871, 2918
	20	1.3972	4643
	25	1.3950	
η	20°	4.210	4784: 2835, 2794, 4722
	30	3.180	
γ	20°	23.47	2102
	30	22.62	
$\triangle H_v$	25°	11.87	4647, 5012: 871, 3067, 3208
	bp	9.75	
$\triangle H_f^\circ$	25°(l)	−81.90	949: 835, 4648, 4357
$\triangle H_c^\circ$	25°(l)	−635.89	949: 4357
C_p°	25°	27.08	4649: 533
C_p	Table XI	Ch II	
t_c		262.80°	4645: 74
p_c		41.39	4645: 74
d_c		0.276	4645: 74
v_c		0.268	4645
$[\,\alpha\,]_D$	20°	+13.87°	3753, 3755: 2923
κ		⟨ 1.0 x 10⁻⁷	1190
ϵ	25°	16.56	1188: 2835, 1381, 4786, 1190, 3477
soly	in aq, 20°	12.5%w	1338: 2191, 2417
	aq in, 25°	44.1%w	
aq az	87.0°	73.2%w	2255: 68
fl pt	OC	88°F	88
uv			3810, 3785
ir			415, 4856, 1133, 1183, 791, 4533, 4107
Raman			717, 533, 2709
mass			3269, 1659, 90, 1769, 61, 4715
nmr			4632, 3235
Beil	**24**, 371		

64. 2–Methyl–1–propanol

Isobutyl Alcohol

$(CH_3)_2CHCH_2OH$ $C_4H_{10}O$

mw		74.124	887
bp	760	107.66°	4647: 51, 878, 3452, 4643, 3208, 2203
dt/dp	760	0.03613°	4643: 811, 3180, 871
p	25°	10.22	4647: 4535, 2203
eq 5	A	7.22014	4646: 871, 74, 2203
	B	1190.38	
	C	166.67	
fp		– 108°	4535: 4784
tp		– 101.97°	1110
d	20°	0.8016	4644: 4784, 811, 878, 2835, 4829, 4643
	25	0.7978	51, 871, 1717, 3452, 4975
α		0.00095	2244
n_D	15°	1.39768	4784: 51, 2997, 3452, 4975
	20	1.3959	4643: 871, 3208, 4829
	25	1.3939	1190: 811
dn/dt		0.00039	4784
η	15°	4.703	4784: 811, 3452, 4829, 5036, 1717, 4808
	25	3.91	1404: 2794, 4928
	Table II, III	Ch II	
γ	20°	22.98	2102
	30	22.11	
	Table VIII	Ch II	
$\triangle H_v$	25°	12.23	4648: 515, 429, 4962, 5012, 871, 568
	bp	10.07	3208
$\triangle H_m$		1.511	1110
$\triangle H_f^\circ$	25°(l)	– 80.00	4357: 949, 4648
$\triangle H_c^\circ$	25°(l)	– 637.79	4357: 4743, 2577, 949
C_p°	25°	27.	4648
C_p	25°	43.26	1110: 515, 5194, 4829, 5337
t_c		274.58°	4645: 783, 4406, 74, 4095
p_c		42.39	4645
d_c		0.272	4645
v_c		0.272	4645

64. 2-Methyl-1-propanol

(Continued)

K_B		2.166°	2846: 483
κ	25°	1.6×10^{-8}	1190
ϵ	25°	17.93	1188: 2686, 1190, 5034, 1381, 3124, 2835
μ		1.79	3110: 2175, 2177
soly	in aq, 25° aq in, 25°	10%w 16.9%w	1338: 2191, 1677, 664
aq az	89.8°	67%w	2254: 4507
fl pt	CC	82°F	3680
uv			3810
ir			3423, 89, 4113, 304, 5341, 1486
Raman			3629, 3132
mass			1769, 90, 4715
nmr			4632
Beil	24, 373		

65. 2-Methyl-2-propanol

tert-Butyl Alcohol

$(CH_3)_3COH$ $C_4H_{10}O$

mw		74.124	887
bp	760	82.42°	4647: 1372, 871, 4775, 51, 4571, 568
dt/dp	760	0.03348°	4643: 1372, 4775, 871
p	25°	41.98	4647: 1373, 3648, 4535, 555
eq 5	A B C	7.31994 1154.48 177.65	4646: 74, 1372, 871, 555
fp		25.82°	4647, 2579: 1295, 3647, 1252, 4643, 4332
d	25°(a) 30 40	0.7812 0.7757 0.7649	4644: 51, 1372, 3620, 4456, 4571, 4775 4643
n_D	20°(a) 25°(a)	1.3877 1.3851	4643: 1372, 717, 51

(a) Undercooled liquid below normal freezing point

65. 2–Methyl–2–propanol
(Continued)

η	30°	3.316	4775: 4456
	Table II	CH II	
γ	26°	20.02	2102: 4772, 4775, 3620
	35	19.10	
$\triangle H_v$	25°	11.19 (a)	4647: 871, 5012, 3067, 555, 568, 3648
	bp	9.33	1255
$\triangle H_s$	25.00°	12.73	4647: 5245
$\triangle H_m$		1.602	4647, 3579: 4601, 3647, 4357, 1748, 1255
$\triangle H_f^\circ$	25°(l)	– 85.86	4648, 5358: 4357
$\triangle H_c^\circ$	25° (l)	– 631.92	4357: 2577, 4743, 5358
C_p°	25°	27.10	4648: 555
	Table XIV	Ch II	
C_p	25.81°(s)	35.04	3579: 1255, 3647
	25.81° (l)	52.66	
t_c		233.0°	4645: 3687, 74
p_c		39.20	4645: 74
d_c		0.270	4645: 74
v_c		0.274	4645
A	eq 72	0.0090	4647
K_f		8.37°	1748: 3662, 311, 579
K_B		1.745°	311
pK		19	3075: 686 , 888, 1283
κ	27°	2.66 x 10⁻⁸	4419: 2569
ϵ	25°	12.47	1188: 938, 2686, 4419, 4786, 4470, 3400
μ	30°in 30	1.66	3181: 3400
soly	25°	inf	
aq az	79.9°	88.24%w	2254
fl pt	C C	52°F	3680
uv			3810, 3785
ir			2224, 89, 4113, 5341, 4305, 4385, 408
Raman			1854, 717, 3629
mass			1769, 90, 577
nmr			2432, 4256, 3126, 1880
Beil	24, 379		

66. 1–Pentanol

Amyl Alcohol

$CH_3(CH_2)_3CH_2OH$ $C_5H_{12}O$

mw		88.151	887
bp	760	137.8°	4640: 23, 871, 4324, 4948, 3452, 4779
dt/dp	760	0.04069°	4643: 5236, 4779, 871
p	25°	2.35	4640: 4535
eq 5	A	7.17758	4639: 871
	B	1314.56	
	C	168.11	
tp		−77.59°	1110
fp		−78.2°	4640: 3653, 4770, 4324, 4779
d	20°	0.8151	4644: 4779, 871, 2835, 3589, 23
	25	0.8115	4643, 1783, 3452, 4324, 4975
α		0.00092	2608
n_D	20°	1.4100	4643: 4975, 871, 4324, 23, 3452, 4779
	25	1.4079	
η	−195.7°gl	2.1 x 10^{36}	2967: 4722, 4324, 3452
	15	4.650	4779
	25	3.347	2440
	Table II, III	Ch II	
γ	20°	25.60	2102: 2835, 4779, 3452, 4975
	30	24.72	
$\triangle H_v$	25°	13.61	4640, 5012: 1876, 871, 4053, 1873, 2608
	bp	10.8	
$\triangle H_m$		2.510	1110, 3653: 4640
$\triangle H_f^\circ$	25° (l)	−85.85	949: 4641, 1873
$\triangle H_c^\circ$	25° (l)	−794.61	949: 4580, 1873, 4053
C_p°	25°	31.75	4649: 5004, 4095, 1876
C_p	25°	49.78	1110: 3653, 3745, 4648
	Table XI	Ch II	
t_c		312°	2093
p_c		43.8	2093
A	eq 72	0.0311	4640
ϵ	25°	13.9	3477: 2686, 4786, 5275, 1708, 3124, 2835
μ	25° in 232	1.7	2892, 2893: 2175, 3020, 2880, 1494, 3113
soly	in aq, 25°	2.19 %w	1783: 23, 872, 664
	aq in, 25°	7.46%w	
aq az	95.8°	45.6%w	2254

66. 1–Pentanol

(Continued)

fl pt	C C	100°F	3680
uv			3810
ir			4630, 2854, 4533, 5328, 2605, 2077, 4400
Raman			1854, 4943, 1178
mass			90, 1659, 4715
nmr			4632
Beil	**24**, 383		

67. 2–Pentanol

sec–Amyl Alcohol

$CH_3CH_2CH_2CHOHCH_3$ $C_5H_{12}O$

mw		88.151	887
bp	760	119.0°	4640: 5328, 2331, 810, 4779, 719, 1346
dt/dp	760	0.03762°	4643: 4779
p	25°	5.83	4640: 719, 871, 4535
eq 5	A	7.27575	4646: 871
	B	1271.92	
	C	170.37	
fp		gl	4779: 4770
d	20°	0.8094	4644: 4943, 2608, 810, 719, 4779, 4643
	25	0.8054	3772
dd/dt		0.000799	[4772]
α		0.00097	4779
n_D	20°	1.4064	4643: 4943, 766, 719, 2608, 810, 871
	25	1.4044	5328
η	15°	5.130	4779: 4722, 1404
	30	2.780	
γ	20°	23.98	3772: 4779
	30	22.96	

67. 2-Pentanol

(Continued)

$\triangle H_v$	25°	12.7	4640: 871
	bp	10.6	
$\triangle H_f^{\circ}$	25°(l)	– 87.75	949: 4648, 5241
$\triangle H_c^{\circ}$	25°(l)	– 792.41	949: 5241
C_p	Table XI	Ch II	
$[\alpha]_D$	20°	+ 13.901°	719: 3755, 3753
ϵ	22°	13.82	1494: 4786
μ	22° in 30	1.66	1494
soly	in aq, 25° aq in, 25°	4.46%w 11.79	1783: 1338
aq az	91.7°	63.5%w	2254
fl pt	O C	108°F	2608
uv			3810
ir			745, 1183, 5328, 89
Raman			2661
mass			3210, 1659, 472
Beil	24, 384		

68. 3-Pentanol

$(C_2H_5)_2CHOH$ $C_5H_{12}O$

mw		88.151	887
bp	760	115.3°	4647: 810, 3626, 5163, 3990, 1783, 4763
dt/dp	760	0.03765°	4643: 4772
p	25°	8.24	4647
eq 5	A	7.41493	4646
	B	1354.42	
	C	183.41	
fp		gl	4763

68. 3–Pentanol

(Continued)

d	20°	0.8203	4644: 4772, 5163, 1896, 810, 5119, 4643
	25	0.8160	3548
α		0.00088	2244
n_D	20°	1.4103	4643: 3626, 717, 3755, 5163, 5328, 766
	25	1.4079	
η	15°	7.337	4772: 4722
	30	3.306	
γ	20°	24.60	4772: 4779
	30	23.76	
$\triangle H_v$	25°	12.7	4647: 3067
	bp	10.4	
$\triangle H_f^\circ$	25°(l)	– 88.51	949: 4648, 835, 5358
$\triangle H_c^\circ$	25°(l)	– 791.65	949: 5358
C_p	25°(l)	60.0	4648
	Table XI	Ch II	
ϵ	22°	13.02	1494: 2965, 4786
μ	22° in 30	1.64	1494: 2965
soly	in aq, 25°	5.15%w	1783
	aq in, 25°	8.32%w	
aq az	91.7°	64%w	2254: 331
nmr			4632
Beil	24, 385		

69. 2–Methyl–1–butanol

act – Amyl Alcohol

$CH_3CH_2CH(CH_3)CH_2OH$ $C_5H_{12}O$

mw		88.151	887
bp	760	128.7°	4647: 3097, 718, 2261, 343, 4625, 1346
dt/dp	760	0.03898°	4643
p	25°	3.12	4647: 718, 1432
eq 5	A	7.06730	4646
	B	1195.26	
	C	156.83	
fp		$\langle$–70°	1853
d	20°	0.8191	4644: 4722, 4625, 344, 4643, 5160, 718
	25	0.8150	1783, 343, 2918, 1432
n_D	20°	1.4107	4643: 4722, 717, 5160, 718, 2261, 2918
	25	1.4087	
η	19.3°	5.505	4722: 1853
	28.7	3.868	
	Table II, III	Ch II	
γ	25°	25.1	4389
$\triangle H_v$	25°	13.0	4647: 1853, 3067
	bp	10.8	
$\triangle H_f^\circ$	25° (l)	–85.24	949: 5358, 4648
$\triangle H_c^\circ$	25° (l)	– 794.92	949: 5358
C_p	25°	52.6	4648
	Table XI	Ch II	
$[\alpha]_D$	20.7°	– 5.900°	1432: 4625, 343, 5160, 1455, 718, 3320
ϵ	25°	14.7	3477
soly	in aq, 25°	2.97%w	1783
	aq in, 25°	9.19%w	
aq az	93.8°	58.5%w	331
fl pt	O C	122°F	1853
ir			1133, 1183, 4533, 89, 408, 5328
Raman			717, 2661
mass			1659
Beil	**24**, 385		

70. 3–Methyl–1–butanol

Isoamyl Alcohol

$(CH_3)_2CHCH_2CH_2OH$ $C_5H_{12}O$

mw		88.151	887
bp	760	130.5°	4647: 4767, 23, 1346, 5194, 2462, 3208
dt/dp	760	0.03886°	4643: 4777
p	25°	2.37	4647: 4535, 4870. 871
eq 5	A	6.95361	4646: 871
	B	1128.19	
	C	146.47	
fp		-117.2°	4535: 2608
d	20°	0.8104	4644: 4777, 871, 2747, 4975, 5101, 4643
	25	0.8071	23, 1783, 3208
α		0.00092	4777
n_D	20°	1.4071	4643: 4777, 5194, 23, 4975, 871
	25	1.4052	
dn/dt		0.00037	4777
η	-195.7°gl	1.6 x 10^{35}	2967
	15	4.81	4777
	30	2.96	
γ	15°	24.77	4777: 4389
	30	23.44	
$\triangle H_v$	25°	13.3	4647: 3067, 871, 3001, 781, 3208
	bp	10.6	
$\triangle H_f^\circ$	25° (l)	-85.18	949: 5358
$\triangle H_c^\circ$	25° (l)	-794.98	949: 4743, 5358
C_p	25°	50.5	4648: 4144, 5194
t_c		307.$^\circ$	4645: 4406, 783
K_B		2.65°	4864: 483
κ	25°	1.4 x 10^{-9}	2742: 2569
ϵ	25°	14.7	3477: 3309, 3945, 936, 2747, 2686, 2341
μ		1.82	3110
soly	in aq, 25°	2.67%w	2462: 1783, 664, 1677
	aq in, 25°	9.61%w	1783
aq az	95.15°	50.4%w	2254: 331
fl pt	C C	109°F	3680: 2608
uv			3810
ir			89, 1183, 304, 5328, 1572, 4113, 5341
Raman			1178, 4943
mass			90, 1659, 4715
Beil	24, 392		

71. 2–Methyl–2–butanol

tert–Amyl Alcohol

$(CH_3)_2COHCH_2CH_3$ $C_5H_{12}O$

mw		88.151	887
bp	760	102.0°	4647: 3653, 1783, 4763, 312, 4535, 4779
dt/dp	760	0.03727°	4643: 4772
p	25°	16.7	3976: 4535
eq 5	A	6.5193	4646: 871
	B	863.4	
	C	135.3	
fp		-8.8°	4647: 3653, 4763, 4535, 4779
	see Ch V		
d	20°	0.8096	4644: 4722, 871, 3620, 4643, 1783, 4779
	25	0.8050	
α		0.00133	2608: 4779
n_D	20°	1.4049	4643: 4772, 717, 871
	25	1.4024	2608
η	15°	5.480	4779
	30	2.810	
	Table II, III	CH II	
γ	20°	22.77	2102: 4772, 4779, 3620
	30	21.84	
$\triangle H_v$	25°	12.0	4647: 871, 3067, 4601, 782, 13
	bp	9.7	
$\triangle H_m$		1.065	4647, 3653: 4601
$\triangle H_f^\circ$	25°(l)	−90.71	949: 5358
$\triangle H_c^\circ$	25°(l)	-789.45	949: 4743, 2577, 5358
C_p	21.2°	58.4	3653: 2608, 4648
	Table XI	Ch II	
t_c		272.$^\circ$	4645: 4406, 783
A	eq 72	0.0077	4647
K_f		10.4°	2619
K_B		2.255°	4864: 483
pK		19	3075, 686
ϵ	25°	5.82	3477: 4786, 4281, 936
μ		1.7	4281

71. 2-Methyl-2-butanol

(Continued)

soly	in aq, 25° aq in, 25°	11.00%w 23.47%w	1783
aq az	87.35°	72.5%w	2254: 331
fl pt	OC	70°F	2608
uv			3810
ir			1133, 1183, 1486, 89, 5341, 4533
Raman			717, 4943, 3132
mass			90, 1659
nmr			3586, 951
Beil	24, 388		

72. 3-Methyl-2-butanol

sec-Isoamyl Alcohol

$(CH_3)_2CHCHOHCH_3$ $C_5H_{12}O$

mw		88.151	887
bp	760	111.5°	4647: 5328, 2608, 1783
dt/dp	760	0.03781°	4643
p	25°	8.97	4647
eq 5	A B C	6.94210 1090.93 157.15	4646
d	20° 25	0.8184 0.8138	4644: 3754, 2608, 4643, 1783
n_D	20° 25	1.4096 1.4075	4643: 5328, 2608, 3754

72. 3–Methyl–2–butanol

(Continued)

η	$25°$	3.51	1404
	30.8	3.066	4722
γ	$25°$	23.0	4389
$\triangle H_v$	$25°$	12.4	4647: 3067
	bp	10.0	
$\triangle H_f^o$	$25°$ (l)	– 87.63	949: 5358
$\triangle H_c^o$	$25°$ (l)	– 792.53	949: 5358
$[\alpha]_D$	$20°$	+ 4.85	3754
soly	in aq, $25°$	5.55%w	1783
	aq in, $25°$	11.93%w	
aq az	$91.0°$	67%w	2254
ir			5328, 1183, 4856
Raman			2661
mass			1659
nmr			4912
Beil	**24**, 391		

73. 2,2–Dimethyl–1–propanol

Neopentyl Alcohol

$(CH_3)_3CCH_2OH$ $\qquad\qquad\qquad\qquad\qquad\qquad\qquad\qquad$ $C_5H_{12}O$

mw		88.151	887
bp		$113.1°$	4643: 4927, 1783, 4126, 3423, 3810
dt/dp		$0.03676°$	4643
p	$25°$(l) *	10	4646
eq 5	A	7.8753	4646
	B	1604.7	
	C	208.2	
fp		$52°$	4643: 4126, 4792, 5161, 1783, 454
$\triangle H_v$	bp	2.3	[4646]

* Undercooled liquid

73, 2,2–Dimethyl–1–propanol

(Continued)

K_f		11.0	5161
	60°	8.35	4429
soly	in aq, 25°	3.50%w	1783
	aq in, 25°	8.36%w	
uv			3785
ir			3423, 5328
Raman			2661
mass			1659
Beil	24, 406		

74. Cyclohexanol

$$CH_2CH_2CH_2CH_2CH_2CHOH \qquad\qquad C_6H_{12}O$$

mw		100.162	887
bp	760	161.10°	4781: 1256, 1701, 1791, 1140, 2292, 4880
dt/dp	760	0.045°	4781
p	56.0°	10	4535: 1703
eq 5	A	6.80369	3482: 3554
107–160°	B	1199.1	
	C	145.0	
fp		25.15°	4781: 326, 2302, 1140, 2820
			4316
d	25°	0.9684	4063: 1256, 2292, 326, 4965, 5119
	30	0.94155	4781
	45	0.92994	
α		0.00077	2244
n_D	25°	1.46477	4772: 4880, 1703, 4965
	30	1.4629	5119
	37	1.46055	326

74. Cyclohexanol

(Continued)

η	$30°$ 45	41.067 17.194	4772: 4781
γ	$25.5°$ 30	33.91 33.47	4781: 2023, 2292
$\triangle H_v$	$25°$ 158.7	14.82 10.87	5012: 1257 3720
$\triangle H_m$		0.406	2550: 1257
$\triangle H_f^o$	$25°$ (l)	-83.45	4448: 3875, 5358
$\triangle H_c^o$	$25°$ (l)	-890.77	4448: 3875, 4230, 2577, 5358
C_p	$31.9°$	48.38	3745
t_c		$352°$	1791: 3482
p_c		37	1791
K_f		$39.3°$	3313: 1257, 5213
ϵ	$25°$	15.0	3477: 2686, 1140, 5189, 4786, 3400
μ	$25°$in **232**	1.86	2892: 550, 3400, 5189, 1984, 1678
soly	in aq, $24.6°$ aq in, $20°$	3.75%w 11.78%w	4316: 5343, 1256 1338
aq az	$97.8°$	30.5%w	2255: 1850
fl pt	C C	$154°$F	88
uv			2812
ir			2889, 791, 3248, 1275, 5328, 3757, 408 3158
Raman			4943, 3504, 4098, 3522, 4633, 897, 3732
mass			1659, 3804
nmr			4545, 3506, 951, 2866
Beil	**502**, 5		

75. 1–Hexanol

$CH_3CH_2CH_2CH_2CH_2CH_2OH$ $C_6H_{14}O$

mw		102.178	887
bp	760	157.0°	4640: 23, 871, 872, 3452, 4766, 4975
dt/dp	760	0.0406°	4638: 1786
p	25°	0.82	4640: 871, 2268, 4535
eq 5	A	7.86045	4639: 4038, 871, 2268, 2093
	B	1761.26	
	C	196.66	
fp		−44.6°	4650: 1786, 2548, 4535, 4766, 4770
d	20°	0.8196	4651: 1786, 871, 2440, 3589, 4638, 23
	25	0.8159	872, 4975
α		0.00087	2244
n_D	20°	1.4181	4638: 2286, 871, 3452, 4975, 23, 872
	25	1.4161	4335
η	15°	6.293	2268: 1786, 3438, 2440, 3452
	25	4.592	
γ	20°	24.48	2102: 2268, 3452, 4389, 3438, 4975
	30	23.61	
$\triangle H_v$	25°	14.8	4640: 1873, 1876, 5012, 871, 4053, 2268
	bp	11.8	
$\triangle H_m$		3.68	4640: 2548
$\triangle H_f^\circ$	25° (l)	−90.67	949: 1873, 4641, 5358
$\triangle H_c^\circ$	25° (l)	−951.86	949: 1873, 4053, 5358
C_p°	25°	37.2	4641: 5004, 1876
C_p	16.9°	55.56	2548
t_c		337°	2093: 2268, 4095
p_c		40.0	2093
A	eq 72	0.0361	4640
ϵ	25°	13.3	3477: 3945, 3124, 1708, 3577
μ	20° in 30	1.55	2758: 2892, 2880, 3020, 2893, 2959, 2952
soly	aq in, 20°	7.42%w	2609
	in aq, 20°	0.706%w	23: 872, 1677
aq az	97.8°	32.8%w	2255: 2254
fl pt	C C	145°F	88
uv			3810
ir			89, 4475, 2605, 5341
Raman			1133, 1572, 5328, 3132
mass			1659
Beil	24, 407		

76. 2–Methyl–1–pentanol

$CH_3CH_2CH_2CH(CH_3)CH_2OH$ $C_6H_{14}O$

mw		102.178	887
bp	760	148.0°	4650
dt/dp	760	0.0408°	4650
eq 5	A	7.86701	4652
	B	1775.12	
	C	208.00	
d	20°	0.8242	4650: 4651
	25	0.8206	
n_D	20°	1.4190	4650
	25	1.4172	
aq az	97.2°	40%w	2255
Beil	**24**, 409		

77. 4–Methyl–2–pentanol

$(CH_3)_2CHCH_2CH(OH)CH_3$ $C_6H_{14}O$

mw		102.178	887
bp	760	131.7°	4650: 2922, 4327, 766, 1785, 810, 4535
dt/dp	760	0.0413°	4650: 811
p	25°	8.2	2268: 4535
eq 5	A	6.15173	4652: 2268
	B	811.05	
	C	126.60	
d	20°	0.8080	4651: 810, 1785, 2268, 4650, 811, 5119
	25	0.8036	
α		0.00101	2244
n_D	20°	1.4112	4650: 4327, 2268, 811, 810, 5119, 766
	25	1.4090	

77. 4–Methyl–2–pentanol

(Continued)

η	$25°$	4.074	2268
	35	2.715	
γ	$25°$	22.63	2268
	35	21.87	
$\triangle H_v$	bp	10.90	2268
$[\alpha]_D$	$19.5°$	+ 21.19	[2561] : 2922
	$17°$	− 20.80	
soly	in aq, $25°$	1.64%w	1785
	aq in, 25°	6.35%w	
aq az	$94.3°$	56.7	2255
fl pt	TCC	$114°F$	2244
ir			5328, 4856
mass			1659
nmr			4912
Beil	24, 410		

78. 2–Ethyl–1–butanol

sec–Hexyl Alcohol

$CH_3\ CH_2CH(C_2H_5)CH_2OH$ $C_6H_{14}O$

mw		102.178	887
bp	760	$146.5°$	4650: 5159
dt/dp	760	$0.0433°$	4650: 2267
p	$25°$	3.8	2267
eq 5	A	6.84055	4652: 2267
	B	1188.69	
	C	153.70	
fp		$-114.4°$	4650: 2608
d	$20°$	0.8333	4650: 2267, 5119
	25	0.8295	
dd/dt		0.00078	[4650]
α		0.00089	2244

78. 2–Ethyl–1–butanol

(Continued)

n_D	20° 25	1.4224 1.4205	4650: 5159, 2267, 5119
η	15° 25	8.021 5.892	2267
γ	15° 25	25.06 24.32	2267
$\triangle H_v$	bp	10.322	2267
t_c		145.7°	2267
ϵ	90°	6.19	3577
soly	in aq, 20° aq in	0.63%w 4.56%w	1338: 2608
aq az	96.7°	41.3%w	2254
fl pt		145 °F	1338
ir			4475
Beil	**24**, 408		

79. 1–Methylcyclohexanol

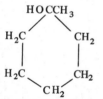

$C_7H_{14}O$

mw		114.189	887
bp		156–158°	5329: 4082
p	70°	25	3419: 3273, 3154, 326, 4083, 3316, 830
eq 4 50–100°	A B	12.08 3677.	3976
fp		26°	3273: 3154, 326, 3316
d	0° 24.65	0.953 0.9251	4082: 5329, 2076, 4083, 3273, 4319 326
n_D	24.65	1.45874	326: 4319, 5329, 3419
$\triangle H_v$	25° bp	19 8.4	3976
mass			**3804**
Beil	**502**, 11		

80. 2–Methylcyclohexanol

Mixed Isomers

$$
\begin{array}{c}
\text{HCOH} \\
\text{H}_2\text{C} \quad\quad \text{CHCH}_3 \\
\text{H}_2\text{C} \quad\quad \text{CH}_2 \\
\text{C} \\
\text{H}_2
\end{array}
$$

$C_7H_{14}O$

mw		114.189	887
bp	760	167.6°	326: 3500, 4360, 4082, 4965
dt/dp	760	0.0621°	3976
p	79°	29	3500
eq 4	A	8.690	3976
	B	2544.	
d	0°	0.9452	4082: 326, 4319, 4360
	20	0.9254	4965
	24.7	0.9215	
n_D	13.4°	1.46585	326: 4965, 4360
	20	1.4610	3500
γ	20°	30.75	4965
	24.7	20.42	
$\triangle H_v$	av	11.6	3976
$\triangle H_c^{\circ}$	25° (l)	– 1048.5	2577
ϵ	20°	13.3	3477: 3400
μ		1.95	5189: 3400
aq az	98.4°	20%w	2254
fl pt	CC	154°F	3680
ir			699
Raman			3522
mass			3804
Beil	**502**, H 6, 11		

81. cis–2–Methylcyclohexanol

$C_7H_{14}O$

mw		114.189	887
bp	760	165°	4360: 1461, 1856, 2291, 1462
dt/dp	760	0.0529°	3976
p	65°	16	1462: 4360, 1856, 2293
eq 4 60 – 165°	A B	8.6543 2528.8	3976
fp		6.8 – 7.3°	2293: 4770, 2291
d	20° 25 30	0.93600 0.93181 0.92750	2293: 2368, 2291, 4360, 1856
n_D	20° 24 30	1.46536 1.46325 1.46074	2293: 4360, 1856
η	25° 30	18.08 13.60	2293: 1856, 2291
$\triangle H_v$	60 – 165°	11.6	3976
$\triangle H_f^\circ$	25° (l)	– 90.9	5358
$\triangle H_c^\circ$	25° (l)	– 1045.7	5358: 1096, 4360
$[\alpha]_D$	20°	– 6.23°	1856
ir			974, 1032
Beil	502, 11		

82. trans–2–Methylcyclohexanol

$C_7H_{14}O$

mw		114.189	887
bp	760	166.5°	4360: 1461, 1462, 2368, 1856, 2291
	760(±)	163–164°	4880
p	72°	20	1462: 1856, 2368, 2293, 4360, 1461
eq 4	A	9.197	3976
80–165°	B	2776.4	
fp		– 4.3 to - 3.7°	2293: 4770, 2291
d	20°	0.92472	2293: 2291, 4360, 2368, 1856
	25	0.92085	
	30	0.91682	
n_D	20°	1.46164	2293: 2368, 4360, 1856
	20(±)	1.4602	4880
	25	1.45972	2293
	30	1.45762	
η	25°	37.13	2293: 1856, 229?
	30	25.14	
$\triangle H_v$	80–165°	12.7	3976
$\triangle H_f^o$	25°(l)	– 97.0	5358
$\triangle H_c^o$	25°(l)	– 1039.6	5358: 1096, 4360
$[\alpha]_D$	20°	(+) 17.19°	1856
		(-) 17.81°	
ir			974, 1032
Beil	502, 11		

83. 3–Methylcyclohexanol

Mixed Isomers

HCOH

H_2C CH_2

H_2C $CHCH_3$

C
H_2

$C_7H_{14}O$

mw		114.189	887
bp	763	172°	4965: 4082, 4360, 1946
p	77°	14	326
d	0°	0.9336	4082: 1946, 4319
	20	0.9168	4965: 326, 4360
n_D	20°	1.45757	4965: 4360, 1946
	25.5	1.45444	326
γ	20°	27.75	4965
	23.1	27.62	
$[\alpha]_D$		– 3.68°	1946
ϵ	20°	12.3	3477: 3400
μ		1.9	5189: 3400, 4361
Raman			3522
mass			3804
Beil	502, 12		

84. cis-3-Methylcyclohexanol

$C_7H_{14}O$

mw		114.189	887
bp	12 760(±)	94° 168°	3096: 1807, 1462, 1461, 3707, 1811 4360
p	65° 60° (±)	5.6 2	1774: 3096 3095
fp		- 6 to - 5°	2295
d	16° 20 30 (±) 30	0.9201 0.9155 0.9065 0.9072	[1807] : 3095, 4360, 1856 2295 3095
n_D	16° 20	1.4589 1.4572	[1807] : 3095, 1856, 1811, 3707, 3096 4360, 2295: 1774
η	16° 25 30	37.74 19.7 17.23	[1807] 1856 2295
γ	30°	29.2	1774
ΔH_f°	25° (l)	- 91.9	5358
ΔH_c°	25° (l)	- 1044.7	5358: 4360
$[\alpha]_D$	20° (-)	3.91°	2295: 1807, 1856, 1774
ϵ	20°	16.47	Beil E III 6, 67
μ		(±) 1.91	4361
ir			974, 1032
Beil	502, 12		

85. trans-3-Methylcyclohexanol

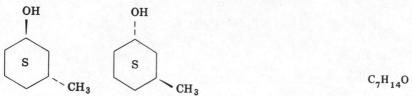

$C_7H_{14}O$

mw		114.189	887
bp	13	84°	3096: 1461, 1807, 1462, 3707, 4360
	(±) 760	166–168°	3556: 3095, 1811, 4880, 1095
p	61°	5.2	1774
eq 4	A	6.890	3976
	B	2062	
fp		−1 to 0°	2295
d	20°	0.9214	2295: 1807, 4360, 1856, 3095, 3096, 1095
	30	0.9138	
n_D	16°	1.4590	1807: 1856, 4360, 2295, 3707, 1810, 3095
	20	1.4580	1774: 1811, 3556, 3096, 1095
	20 (±)	1.4545	4880
η	16°	26.52	1807
	25 (±)	25.1	1856
	30	15.60	2295
γ	30°	28.80	1774
$\triangle H_v$	av	9.44	3976
$\triangle H_f^o$	25° (l)	−97.1	5358
$\triangle H_c^o$	25° (l)	−1039.5	5358: 4360
$[\alpha]_D$	20°	(−) 7.34°	2295: 1807
	21	(+)6.7°	3096
ϵ	20°	8.05	Beil E III 6, 69
μ	(±)	1.75	4361
ir			974, 1032
Beil	502, 12		

86. 4–Methylcyclohexanol

Mixed Isomers

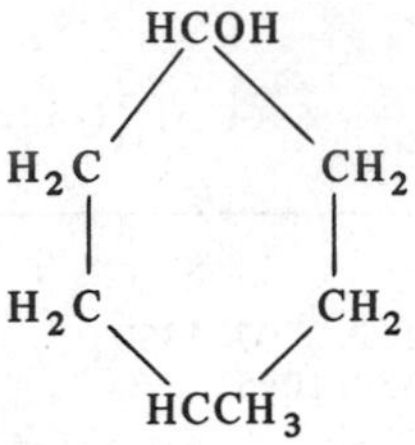

HCOH

H_2C CH_2

H_2C CH_2

HCCH$_3$

$C_7H_{14}O$

mw		114.189	887
bp	763	172°	4965: 4317, 4360, 4082
	12	75°	326
d	0°	0.9328	4082: 4319
	20	0.9122	4965
	22.5	0.9183	326
n_D	20°	1.45647	4965
	22.5	1.45594	326
γ	20°	27.63	4965
ϵ	20°	13.3	3477: 3400
μ		1·9	5189: 3400
Raman			3522
ir			408
Beil	**502, 14**		

87. cis–4–Methylcyclohexanol

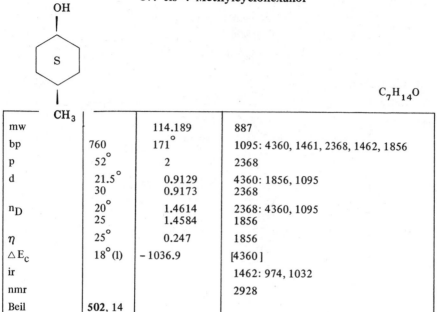

$C_7H_{14}O$

mw		114.189	887
bp	760	171°	1095: 4360, 1461, 2368, 1462, 1856
p	52°	2	2368
d	21.5°	0.9129	4360: 1856, 1095
	30	0.9173	2368
n_D	20°	1.4614	2368: 4360, 1095
	25	1.4584	1856
η	25°	0.247	1856
$\triangle E_c$	18° (l)	– 1036.9	[4360]
ir			1462: 974, 1032
nmr			2928
Beil	**502**, 14		

88. trans–4–Methylcyclohexanol

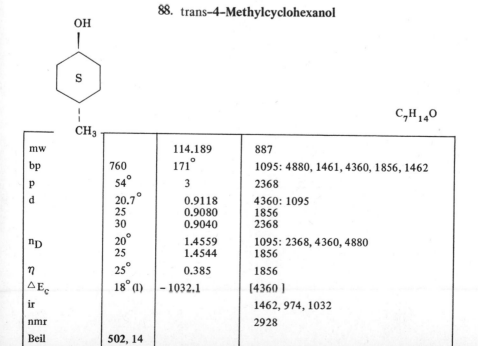

$C_7H_{14}O$

mw		114.189	887
bp	760	171°	1095: 4880, 1461, 4360, 1856, 1462
p	54°	3	2368
d	20.7°	0.9118	4360: 1095
	25	0.9080	1856
	30	0.9040	2368
n_D	20°	1.4559	1095: 2368, 4360, 4880
	25	1.4544	1856
η	25°	0.385	1856
$\triangle E_c$	18° (l)	– 1032.1	[4360]
ir			1462, 974, 1032
nmr			2928
Beil	**502**, 14		

89. 2-Heptanol

sec-Heptyl Alcohol

$CH_3(CH_2)_4CHOHCH_3$ $C_7H_{16}O$

mw		116.205	887
bp	760 (±)	159.7°	4642: 1472, 1346, 1382, 3163, 2331, 4277
			3753, 3842
dt/dp	760 (±)	0.043°	4642
p	20° (±)	0.9	3160: 1472, 4277
eq 4 (±)	A	8.8833	3976
	B	2601.6	
fp		glass	4277
d	20° (±)	0.8171	4642: 3163, 3842, 5119, 4277
	25$_0$ (±)	0.8134	
	20 (−)	0.8184	3753
	35 (+)	0.8050	
α	ca. 20° (±)	0.00094	3160
	20° (±)	1.4210	4642: 4277, 3842, 2331, 3163, 5119, 5328
	25$_0$	1.4190	
	20° (+)	1.4209	3753
η	20° (±)	6.53	3160: 1404
	30	3.321	4722
$\triangle H_v$	65 to 160° (±)	11.9	3976
$[\alpha]_D$	20° (+)	11.45°	3755
	in 59		
	20 (+)	13.71°	
	in 30		
n_D	20 (+)	10.32$_0$°	3753: 3842
	25 (+)	10.21$_0$°	
	17 (−)	10.48°	
ϵ	22°	9.21	1494, 3477: 3304
μ	22 °in 30	1.71	1494
aq az	98.7°	17%w	2254
ir			4293, 5328
mass			4631
Beil	24, 415		

90. 1–Octanol

$CH_3(CH_2)_6CH_2OH$ $\qquad\qquad\qquad\qquad\qquad\qquad\qquad$ $C_8H_{18}O$

mw		130.232	887
bp	760	195.20°	4638: 871, 872, 1248, 1346, 3452, 4766
dt/dp	760	0.0471°	4638: 1372
p	25°	0.075	4640: 871, 4948, 1729, 4535, 1325
eq 5	A	7.08451	4639: 2093, 1372, 871, 4036
	B	1457.76	
	C	151.58	
fp		$-14.97°$	1372: 1248, 4535, 4774, 1346, 4766, 1207
d	20°	0.82555	1372: 871, 1248, 2440, 4425, 872, 1717
	25	0.82209	$\qquad$ 3452, 4975
dd/dt	-60 to 120°	0.00081	[1189]
α	0 to 25°	0.000828	1346
n_D	20°	1.4295	4638: 1372, 872, 3452, 23, 1346, 4975
	25	1.4275	
dn/dt		0.00040	3976
η	15°	10.640	1248: 1717, 3438, 2440, 3452
	30	6.125	
γ	20°	26.06	2102: 1346, 3452, 4389, 3438, 4975
	30	25.21	
$\triangle H_v$	25°	17.2	4640: 871, 1207, 3358, 3450, 4053, 1346
	bp	12.1	$\qquad$ 781, 4963
$\triangle H_s$		25.2	1207
$\triangle H_m$		10.1	1207
$\triangle H_f^°$	25° (l)	-101.62	949: 1873, 5358
$\triangle H_c^°$	25° (l)	-1265.65	949: 2577, 1873, 4053
$C_p^°$	25°	48.2	4641: 5004, 1876
C_p	13°	68.0	1346
t_c		390°	[2093]: 783, 4406
p_c		34.0	[2093]

90. 1–Octanol

(Continued)

κ	$23.1°$	1.39×10^{-7}	4419
ϵ	$20°$	10.34	3477, 3400: 4419, 2421, 4425, 3577
μ	$25°$ in 232	1.76	2892: 4425, 3020, 1851
soly	in aq, $25°$	0.0538%w	1346: 23, 872, 4291
aq az	$99.4°$	10%w	2254
fl pt	TOC	$185°F$	3160
uv			3810
ir			3339, 89, 2605, 1729, 304, 5328
Raman			3132
mass			1659, 4631
Beil	24, 418		

91. 2–Ethyl–1–hexanol

Ethylhexyl Alcohol

$CH_3(CH_2)_3CH(C_2H_5)CH_2OH$ $C_8H_{18}O$

mw		130.232	887
bp	760	$184.34°$	1425: 3691, 2916, 1330, 3058, 3049, 5164
dt/dp	760	$0.049°$	4883
p	$25°$	0.143	[1425]
eq 5	A	6.8889	1425
	B	1325.88	
	C	146.46	
fp	gl	$-76°$	3160: 4883
d	$15°$	0.8435	5164: 2916, 4883, 3049, 2918
	20	0.8332	4795
	25	0.8291	
dd/dt		0.00073	4883: 1189

91. 2–Ethyl–1–hexanol

(Continued)

α	$20°$	0.000875	3160
n_D	$20°$ 25	1.42305 1.4292	[5164] : 3049, 2918 2916
η	$20°$	9.8	4883
$\triangle H_v$	bp	10.8	[4883]
$\triangle H_f^\circ$	$25°(l)$	– 103.46	4795
$\triangle H_c^\circ$	$25°(l)$	– 1263.81	4795
C_p	$25°$	73.45	4883
$[\alpha]_D$	$25°$	(–) 1.53 (+) 1.533	2916 2918
ϵ	$90°$	4.41	3577
μ	$25°$ in **30**	1.74	3691
soly	in aq, $20°$ aq in, $20°$	0.07%w 2.6%w	4883: 1338 4883
aq az	$99.1°$	20%w	2255
fl pt	CC	$185°F$	3160
uv			3049
ir			4533
mass			1659
Beil	**24**, 421		

92. Benzyl Alcohol

$C_6H_5CH_2OH$ C_7H_8O

mw		108.141	887
bp	760	205.45°	4780: 4535, 5008, 1372, 4766, 5130
dt/dp	760	0.050°	4780: 1372, 3001
p	25°	0.11	[4240]
eq 4	A	9.838	4240
	B	3214	
fp		− 15.3°	4780, 4766: 4770, 1372, 3660, 4535
d	15°	1.04927	4780: 1717, 3208, 4975, 5119, 2082, 3973
	25	1.04127	3168: 5032
	30	1.03765	4780
α		0.00075	4780
n_D	15°	1.5426	4772: 1703, 2078, 4780, 4975, 5130, 5119
	20	1.54035	1372: 943
	25	1.53837	
η	15°	7.760	4780: 1400, 3438, 1717, 4709
	30	4.650	
γ	15°	40.41	4780: 3438, 4975
	20	39.96	2899
	30	38.94	
$\triangle H_v$	bp	12.066	3208: 3016, 3001, 4963
$\triangle H_m$		4.074	3660
$\triangle H_f^\circ$	25° (l)	− 38.49	3656: 5358
$\triangle H_c^\circ$	25° (l)	− 893.15	3656: 2577, 5358
C_p	26.8°	52. 37	3660: 3001, 3016
pKa		18	3075
ϵ	20°	13.1	3477: 2686, 1381, 3499, 5034
μ		1.66	5191: 3168, 1539
soly	in aq, 20°	0.08%w	1338
	aq in, 20°	8.37%w	
aq az	99.9°	9%w	2254
fl pt	C C	213°F	88
uv			896, 3237, 3810, 1679, 2812
ir			870, 983, 1209, 1275, 505, 2519, 4107
Raman			2140, 4348, 4934, 4344, 1170
mass			11
nmr			3175, 4632, 951, 3176, 4256
Beil	528, 428		

93. Phenol

C_6H_5OH C_6H_6O

mw		94.114	887
bp	760	181.839°	238, 4677: 4673, 1331, 5198, 4192, 4364
dt/dp	760	0.04788°	238: 1372, 489, 1249, 4673
p	25°	0.41	[4677] : 1373, 489, 1815, 4535
	41	1.39	[238]
eq 5	A	11.5638	238, 570: 370, 4676, 4364, 1372, 1815
9–40°	B	3586.36	
	C	273	
	A	7.13457	238, 570: 4676
110–200°	B	1516.072	
	C	174.569	
fp		40.90°	238: 4673, 2191, 2421, 3203, 2270, 339
	1.37 torr	40.88	4677
d	20°	1.0767(a)	4673: 1372, 3973, 4864, 711, 4781, 2270
	25 (s)	1.132	238
	40 (s)	1.132 (b)	1034
	40	1.0587(a)	4674: 339
	45 (l)	1.05446	4772
n_D	45°	1.54027	4772: 1372, 4781, 711, 4673
	40.6	1.54274	
η	45°	4.076	4781: 1400, 4709
	60	2.578	
γ	50°	37.77	4781: 4772, 339
	55	37.26	
	60	36.69	
$\triangle H_v$	25° (s)	16.410	238: 1198, 1532, 4364, 570, 645, 4962
	25	13.82*	4677
	bp	10.920	
$\triangle H_s$	40.88°	16.36	4677
$\triangle H_m$	tp	2.752	4677: 4491, 1533, 3203, 240, 4601
$\triangle H_f^o$	25° (s)	– 39.46	238: 4678, 1121, 1877, 3556
$\triangle H_c^o$	25° (s)	– 729.80	238: 2577, 4743, 345, 1121
C_p^o	25°	24.75	4678, 1877: 240
C_p	25°	30.460	240: 3041, 3653
	mp (l)	48.14	
t_c		421.1°	4675: 4406, 2145, 71
p_c		60.5	4675: 2145

* (a) Supercooled liquid   (b) Air free

93. Phenol

(Continued)

d_c		0.411	4675
v_c		0.229	4675
A	eq 72	0.01404	4677
K_f		$7.40°$	4864: 4308, 1533, 476, 940, 3973, 3203
K_B		$3.60°$	489: 477, 484
pKa	$25°$ in H_2O	10.02	2126: 596, 3406, 281, 3405, 2841, 3333
κ	$50°$	$1–3 \times 10^{-8}$	5038
ϵ	$60°$	9.78	3477: 3040, 3400, 5198, 1198, 1381
μ	$30°$ in **30**	1.45	2508: 1836, 2946, 1335, 4195, 3168, 2894
soly	in aq, $25°$	8.66%w	2191: 2270, 2417, 4310
	aq in, $25°$	28.72%w	
aq az	$99.6°$	9.2%w	2254
fl pt	C C	$175°F$	3680
uv			303, 635, 2513, 3148, 3224, 3726, 4005
ir			5294, 1877, 3249, 4608, 4630, 1613, 2862
Raman			3040, 3114, 1877, 4550, 5271, 2680, 5228
mass			11, 556
nmr			3177, 3176, 2304, 951, 3674, 1516, 3213
Beil	**512**, 110		

94. o–Cresol

o – $CH_3C_6H_4OH$ C_7H_8O

mw		108.141	887
bp	760	191.004°	4673: 570, 1372, 4535, 238, 3644, 1791
dt/dp	760	0.05021°	4673, 238: 1372, 570
p	25°	0.31	4677: 570, 1373, 4535, 3644, 1815
eq 5 0–30°	A B C	12.7778 3970.17 273	238: 570, 1815, 1372, 4676
110–200°	A B C	7.07055 1542.299 177.110	
fp	 0.52 torr	30.944 30.92	4673: 2461, 908, 1250, 2555, 1796, 3644 4677: 242
d	20°(1) 25 fp 40	1.0460 1.135 1.04 1.0282	4673: 1372, 3972, 3644 3226 4674
n_D	20°(1) 25 (1)	1.5467 1.5442	4673: 1372
η	25° 45 80	7.608 3.506 1.47 cs	2555 4709 3644
γ	40.3° 176	34.8 21.5	2372
$\triangle H_v$	25°(1) bp	15.39 10.801	4677: 570, 238, 645, 4315 238, 4677: 3644
$\triangle H_s$	fp	18.18	4677
$\triangle H_m$		3.781	242: 4677
$\triangle H_f^\circ$	25°(s)	– 48.91	238: 1121, 4678
$\triangle H_c^\circ$	25°(s)	– 882.72	238: 2577, 1121, 1875
C_p°	25°	31.15	4678, 1875
C_p	25°(s) 31.05 (l)	36.94 55.67	242: 3644, 708

(1) Supercooled liquid

94. o–Cresol

(Continued)

t_c		424.4°	4675, 71: 2145, 1791
p_c		49.4	4675: 2145, 1791
d_c		0.384	4675
v_c		0.282	4675
A	eq 72	0.01684	4677
K_f		5.60°	3973: 238, 4315
pKa	25° in H_2O	10.287	583, 584: 2390, 959, 530, 2126
κ	25°	1.27×10^{-9}	2555
ϵ	25°	11. 5	3477: 3040, 3730
μ	25° in 30	1.45	3023: 4420, 1335
soly	in aq, 40°	3.08%w	4315: 4310, 2608
fl pt	C C	178°F	88
uv			3909, 2361, 1238, 1025, 2617, 549, 4001
ir			1657, 4101, 4608, 2630, 5137, 3178, 4500
Raman			3040, 2137, 1875, 2670
mass			11, 1657, 3178, 2630, 5137
nmr			3177, 2304, 3175, 4632
Beil	525, 349		

95. m–Cresol

$m - CH_3C_6H_4OH$ C_7H_8O

mw		108.141	887
bp	760	202.232°	4673: 238, 1815, 4532, 3644, 4781, 570
dt/dp	760	0.05007°	238, 4763: 4781, 570
p	25°	0.143	4677: 4535, 3644, 1815, 570
eq 5	A	9.9653	238: 1815
11–40°	B	3223.45	
	C	273	
	A	7.15904	238, 4676: 570
110–200°	B	1603.811	
	C	172.646	
fp	100%	12.22°	238, 4673: 1249, 1250, 2555, 2461, 908
	triple pt	12.25	242
d	fp	1.05	3226: 4673, 1717, 3644, 4063, 2082, 3973
	20°	1.03410	238: 4781
	25	1.03019	
	30	1.02628	
α	25°	0.000759	238
n_D	20°	1.5414	4673: 3644
	25	1.5396	
η	15°	24.666	4781: 3226, 2555, 4709, 1717, 3644
	30	9.807	
γ	15°	38.01	4781
	30	36.54	
$\triangle H_v$	25°	14.75	238, 4677: 570, 4315, 645, 3644, 3015
	bp	11.329	
$\triangle H_m$		2.559	242: 4063, 4677
$\triangle H_f^\circ$	25° (l)	– 46.38	238: 1121, 4678
$\triangle H_c^\circ$	25° (l)	– 885.25	238: 1121, 1875, 345, 2577
C_p°	25°	29.75	1875, 4678
C_p	25°	53.76	242: 3644, 3015, 708
t_c		432.6°	4675, 71: 1950, 1791
p_c		45.	4675: 1950, 1791
d_c		0.346	4675
v_c		0.312	4675

95. m-Cresol

(Continued)

A	eq 72	0.01340	4677
K_f	1% $\overset{*}{m}$	0.749°	238: 4315
pKa	25° in H_2O	10.09	583, 2126, 584: 959, 530, 668
κ	25°	1.397×10^{-8}	2555
ϵ	25°	11.8	3477
μ		1.543	1335: 4420
soly	in aq, 40°	2.51%w	4315: 4310, 2608
fl pt	C C	187°F	3680
uv			1237, 1238, 908, 2214, 2230, 4954, 4001
ir			4500, 4101, 4608, 3178, 1657, 5137, 4107
Raman			3040, 2136, 1875, 2670
mass			11
nmr			4861, 3175, 1310, 3177, 4632
Beil	526, 373		

* 1 mole percent impurity

96. p-Cresol

p – $CH_3C_6H_4OH$ C_7H_8O

mw		108.141	887
bp	760	201.940°	4673, 238: 955, 1815, 570, 1757, 3644
dt/dp	760	0.04985°	238, 4673: 1372, 570
p	25°	0.13	4677: 1373, 3644, 1815, 4535, 570
eq 5	A	12.0298	238: 1372, 4677, 1817
0–34°	B	3861.98	
	C	273	
110–200°	A	7.11767	238: 570, 4676
	B	1566.029	
	C	167.680	
sublm	0.30 torr	34.70°	4677
fp	100%	34.739°	4677: 4673, 1372, 2555, 238, 1193

96. p–Cresol

(Continued)

tp		34.79°	242
d	20°	1.0348(1)	4673: 1372, 3973, 3644
	25 (s)	1.154	238: 4673
	fp	1.02	3226
	40	1.0185	4674
dd/dt	40 – 60°	0.00078	3976
n_D	25°	1.5391	4673: 3226
	41	1.53115	1372
	46	1.52870	
η	45°	5.607	4709: 2555
	80	2.00 cs	3644
γ	25.6°	34.5	Beil E I 6, 197
	194.5	19.2	
$\triangle H_v$	25°(s)	17.670	238, 4677: 3644, 570, 645, 4962, 238
	25 (1)	16.06	4677
	bp	11.34	
$\triangle H_s$	fp	17.69	4677
$\triangle H_m$		3.037	242: 4315, 4601, 4677, 1533
$\triangle H_f^o$	25°(s)	– 47.64	238: 1121, 4678
$\triangle H_c^o$	25°(s)	– 883.99	238: 2577, 1121, 1875
C_p^o	25°	29.75	1875, 4678
C_p	25°(s)	35.91	242: 3644, 812
	34.79° (l)	54.36	
t_c		431.4°	71: 1791, 4675, 2145
p_c		50.8	4675: 1791, 2145
d_c		0.391	4675
v_c		0.277	4675
A	eq 72	0.01097	4677
K_f		6.96°	1533: 238, 3973, 4315
pKa	25° in H_2O	10.26	583, 584: 2126, 959, 668
κ	25°	1.378×10^{-8}	2555
ϵ	58°	9.91	3477: 3730
μ		1.543	1335: 1836, 4420, 1489

96. p–Cresol

(Continued)

soly	in aq, 40°	2.26%w	4315: 2608, 4310
fl pt	CC	202°F	88
uv			3909, 2361, 1238, 549, 2617, 3903, 908
ir			4500, 4101, 4608, 2630, 5137, 1657, 3178
Raman			3040, 3685, 3396, 2670
mass			11
nmr			4632
Beil	527, 389		

(1) supercooled liquid

97. 2–Propene–1–ol

Allyl Alcohol

$CH_2:CHCH_2OH$ C_3H_6O

mw		58.081	887
bp	760	97.08°	1527, 5062: 5061, 3001, 1333
dt/dp	760	0.040°	4779: 3001
p	25°	28.1	4772
eq 13	A	32.62580	2609
	B	3451.8	
	C	7.94975	
fp	gl	-129°	2609: 4770
d	0°	0.86814	4772: 4264, 5062, 5061, 5119
	15	0.85511	
	30	0.84209	

97. 2–Propene–1–ol

(Continued)

n_D	20°	1.4135	1333
	30	1.4090	5119
η	15°	1.486	4779: 1333, 4711
	30	1.072	
	Table II	Ch II	
γ	15°	26.15	4779
	20	25.68	
	30	24.92	
	Table VIII	Ch II	
$\triangle H_v$	bp	9.550	2609: 3001
$\triangle H_f^o$	25°(l)	– 41.60	5358
$\triangle H_c^o$	25°(l)	– 442.8	2577: 4743, 5358
C_p^o	20°	33.7	[Beil E III 1, 1874]
C_p	20.5 – 95.5°	38.59	3001
t_c		271.9°	3478
pKa	25° in H_2O	15.5	369
ϵ	15°	21.6	3477: 1381, 4752, 2487
μ		1.63	2768
soly	in aq	inf	2609
aq az	88.89°	72.3%w	2254: 4813, 4081, 5061
fl pt	CC	70°F	88
uv			2315, 4604
ir			408, 1267, 3587
Raman			2575, 2612, 3469, 681
mass			4959
nmr			4632, 2424
Beil	25, 436		

98. cis-2-Butene-1-ol

Crotyl Alcohol

HC – CH$_3$
‖
CHCH$_2$OH   C$_4$H$_8$O

		72.108	887
mw		72.108	887
bp	760	123.6°	598: 2066
dt/dp	760	0.0527°	3976
p	63°	60	2066
eq 4	A	8.998	3976
60 – 120°	B	2427	
fp		– 89.44°	598
d	0°	0.8726	2213: 2608, 2066
	20	0.854	
n$_D$	20°	1.4342	598: 2608
△H$_v$	60 – 120°	11.1	3976
soly	in aq	16.6%w	2213
Raman			3469, 1871
Beil	25, 442		

99. trans-2-Butene-1-ol

CH$_3$CH
‖
CHCH$_2$OH   C$_4$H$_8$O

		72.108	887
mw		72.108	887
bp	760	121.2°	598, 2066: 5308, 5309
d	25°	0.8454	2066
n$_D$	20°	1.4289	598: 5308
	25	1.4262	2066
Raman			3469, 1871
Beil	25, 442		

100. 2-Propyn-1-ol

Propargyl Alcohol

HC:CCH$_2$OH   C$_3$H$_4$O

mw		56.065	887
bp	760	113.6°	2105: 4327, 2113, 3959, 2112, 2065, 5061
dt/dp	760	0.0502°	[258]
p	25°	15.5	[258]
eq 4	A	8.656	258
	B	2226.4	
fp		− 51.80	2105: 3959
d	20°	0.9478	2105: 4327, 2065, 2109, 5061
	25	0.9450	4018
n$_D$	20°	1.4320	2105: 4327, 2065, 2109
	25	1.4300	4018
η	20°	1.68	258: 4327
γ	28.5°	35.2	4327
$\triangle H_v$	112°	10.06	258
$\triangle H_c^\circ$	25° (l)	− 423.8	2577, 4743
C$_p$	20°	34.5	258
pKa	25° in H$_2$O	13.55	369
ϵ		24.5	258
μ	25° in 30	1.78	4018: 4327
soly	in aq	inf	258
aq az	97°	45.5%w	2255: 4327, 5061
fl pt	TOC	97°F	258
uv			4607
ir			2366, 5261
Raman			449
nmr			2068
Beil	26, 454		

101. cis-2-Butene-1, 4-diol

cis-Butenediol

HC - CH$_2$OH
‖
HC - CH$_2$OH   C$_4$H$_8$O$_2$

mw		88.107	887
bp	760	235°	4904
dt/dp	760	0.045°	3976
p	97°	2.3	4405: 3828, 1580
eq 4	A	9.6679	3976
	B	3448.7	
fp		11.0°	4405
d	15°	1.082	3828: 4904
	20	1.0740	4405
n$_D$	20°	1.4793	4405: 3828, 4904
$\triangle$H$_v$	100–235°	15.8	3976
K$_b$		2.86°	3976
μ		2.48	4833
ir			4405
Beil	31, 499		

102. trans-2-Butene-1,4-diol

trans-Butenediol

HC - CH$_2$OH
‖
HOCH$_2$- CH   C$_4$H$_8$O$_2$

mw		88.107	887
bp	13	132°	4904
dt/dp	13	1.4°	3976
p	108°	3.8	4405: 1580, 3827, 3847, 4904
eq 4	A	9.9895	3976
100 – 130°	B	3597.3	

102. trans-2-Butene-1, 4-diol

(Continued)

fp		27.3°	4405: 3827, 3847
d	20°	1.0685	4405: 3827, 3847, 4904
n_D	20°	1.4779	4405: 1580, 4904, 3827, 3847
$\triangle H_v$	100–130°	16.5	3976
μ		2.45	4833: 3012
ir			4405
Beil	31, 499		

103. 1,2-Ethanediol

Ethylene Glycol

$(CH_2OH)_2$ $C_2H_6O_2$

mw		62.069	887
bp	760	197.3°	4653: 4141, 677, 1689, 4780, 4611, 3001
dt/dp	760	0.0427°	4653: 3001, 4780
p	25°	0.12	3268: 4535, 3849, 2481, 4832a
eq 5	A	8.7945	4654: 1689, 4832a, 3581
	B	2615.4	
	C	244.9	
eq 4	A	9.2477	4141
130–197°	B	2994.4	
fp*		–13.	4653: 4611, 1687, 677, 4780, 4770
d	0°	1.12763	4780: 1590, 4611, 1762, 5029, 4141, 5119
	15	1.11710	
	20	1.1135	4653: 3983, 1688, 3849
	25	1.1100	
	30	1.10664	[4780]
n_D	15°	1.43312	4780: 1590, 5119, 3849, 3400, 4141
	20	1.4318	4653
	25	1.4306	

103. 1,2–Ethanediol

(Continued)

η	15° 30	26.09 13.55	4780: 1596, 592
γ	20°	46.49	5032, 5034: 1688
	Table VII, VIII Ch II		
$\triangle H_v$	bp	12.06	1254: 3001, 1689, 3016, 3001, 645
$\triangle H_m$		2.778	3654: 4611, 3160, 4601, 1253
$\triangle H_f^\circ$	25° (l)	– 108.73	3663: 5358·
$\triangle H_c^\circ$	25° (l)	– 284.32	3663: 2577, 4743, 3431
C_p	19.8°	35.7	3654: 4611, 3001, 4095
	Table XI	Ch II	
K_i		5.7×10^{-15}	3296: 369, 1847, 2995
κ	25°	1.16×10^{-6}	3449: 3268, 5027, 5035, 2687, 3160
ϵ	25°	37.7	3477: 3268, 2487, 4470, 2687, 2965
μ	20°	2.28	3400: 4815, 2687, 2965
soly	in aq	inf	3268
aq az		none	2254
fl pt	OC	240°F	3268
ir			2787, 836, 2767, 408, 3131
Raman			2021, 2662, 836, 4633
nmr			4253, 5293, 4120, 4632
Beil	**30**, 465		

* The freezing point of 1,2-Ethanediol is in doubt because of the tendency to supercool and form a glass.

104. 1,2-Propanediol

Propylene Glycol

$CH_3CH(OH)CH_2OH$ $\qquad\qquad$ $C_3H_8O_2$

mw		76.096	887
bp	760	187.6°	4653: 1003, 3846, 4151, 4535
dt/dp	760	0.0415°	4653: 3268
p	25°	0.133	3846: 3849
eq 5	A	8.9171	4654: 4141
	B	2645.7	
	C	250.7	
fp		-60°	4653: 1157
d	15°	1.0399	1157: 1003, 1717, 3849, 3846
	20	1.0362	4653
	25	1.0328	
	30	1.0290	1157
n_D	20°	1.4329	4653: 1003, 3400, 1157, 3849
	25	1.4314	
dn_D/dt	20–30°	0.00030	1157
η	0°	243	1157: 2977, 1717
	20	56.0	
	40	18.0	
γ	25°	72.0	1157: 3981
$\triangle H_v$	25°	16.06	[1157]: 645
	bp	12.94	
$\triangle H_f^\circ$	25°(l)	118.10	5358
$\triangle H_c^\circ$	25°(l)	-436.3	2577: 3431, 5358
C_p	20°	45.1	1157
$[\alpha]_D$	20°	-15.37°	2920
ϵ	20°	32.0	3477: 2699, 2965, 3400
μ		2.25	3400: 2965
soly	in aq	inf	3268
aq az		none	2254
fl pt	C O C	210°F	3268
ir			1157
Raman			4633
nmr			3235, 5293, 4632, 951
Beil	30, 472		

105. 1,3-Propanediol

Trimethylene Glycol

$HOCH_2CH_2CH_2OH$ $C_3H_8O_2$

mw		76.096	887
bp	760	214.4°	4653: 4535, 1003, 1689, 4141
dt/dp	760	0.0423°	4653
p	106.2°	10	4654
eq 5	A	9.4710	4654: 4141, 1689
	B	3212.0	
	C	273.0	
fp		-26.7°	4653: 1687
d	20°	1.0538	4141: 1688, 1003, 3412
	25	1.050	4653
α	20°–40°	0.00061	3268
n_D	20°	1.4396	4653, 1687: 3400, 4141, 1003
	25	1.4386	
	40	1.4332	3268
η	20°	46.6	2977
γ	25°	45.17	1688
	Table VIII	Ch II	
$\triangle H_v$	bp	13.83	1689
$\triangle H_c^\circ$	25°(l)	-442.8	2577
ϵ	20°	35.0	3477, 3400: 2487
μ	20°	2.50	3400
soly	in aq	inf	3268
aq az		none	2255
uv			2315
ir			5059
nmr			5293, 951
Beil	30, 472		

106. 1,3-Butanediol

$HOCH_2CH_2CHOHCH_3$ $\qquad\qquad\qquad\qquad\qquad\qquad\qquad\qquad$ $C_4H_{10}O_2$

mw		90.123	887
bp	760	207.5°	4141, 3268: 1003, 527, 2921
dt/dp		0.0415°	4653
p	74°	0.5	1844
eq 5	A	9.8035	4654: 4141
	B	3470.5	
	C	294.8	
fp		⟨-50°	3268
d	20°	1.0053	4653: 4141, 1003, 3268
n_D	20°	1.441	4653: 527, 1003, 1844
	25	1.439	
η	20°	130.3	2977
	25	98.3	3268
	35	89	
γ	25°	37.8	3268
$\triangle H_v$	bp	13.97	645
$\triangle H_c$	(l)	-595.0	3431
$[\alpha]_D$	22°(+)	18.5°	2915: 2923
	25°(-)	18.8	2921
fl pt	TOC	250°F	3268
ir			3131, 4630
nmr			1545, 3219
Beil	30, 477		

107. Glycerol

1,2,3-Propanetriol

$CH_2OHCHOHCH_2OH$ $\qquad\qquad\qquad\qquad\qquad\qquad\qquad\qquad$ $C_3H_8O_3$

mw		92.095	887
bp	760	290.0°	4535, 2932: 2844
p	50°	0.0025	3334: 43, 4535, 2844
eq 4	A	11.27423	4043: 5344
20°-70°	B	4480.5	

107. Glycerol

(Continued)

fp		18.18°	43: 1761, 3863, 4535, 2844, 4988
d	15°	1.26443	4780: 43, 1492, 4167, 672, 2844, 4991
	20	1.26134	
	30	1.25512	
n_D	15°	1.47547	4772: 2844, 1492
	20	1.4746	4519
	25	1.4730	
η	20°	1412	4224: 1596, 2844, 4780, 1492, 3264, 4991
	25	945	1465
	30	612	
	50	142	4224
γ	20°	63.3	3268: 2844, 1492
	90	58.6	
	150	51.9	
$\triangle H_v$	55°	21.06	3334: 483, 2844, 645
	bp	14.59	483
$\triangle H_m$		4.370	1761: 1253, 4601
$\triangle H_f^o$	25°(l)	−159.80	3663: 5358
$\triangle H_c^o$	25°(l)	−395.63	3663: 2577, 5358
C_p	26.2°	53.37	1761: 1492, 4095, 2844
	Table XII	Ch II	
K_f		3.27°–3.69°	3863
K_B		6.52°	483
pKa		14.40	[2995]: 3296
κ	25°	ca 0.6 x 10⁻⁷	4167: 4215
ϵ	25°	42.5	3477: 1184, 3400, 43, 36, 2699, 1381
μ			2487: 3425
soly	in aq	inf	3334
aq az		none	3334, 2254
fl pt	COC, 99%	350°F	3334
uv			4041, 4604
ir			925, 865, 5341, 5187
Raman			1290, 3492, 4096, 3903, 4633, 4909, 4937
mass			3085
nmr			1545, 4120
Beil	38, 502		

108. Ethylvinyl ether

CH$_2$=CHOC$_2$H$_5$ C$_4$H$_8$O

		72.108	887
mw		72.108	887
bp	760	35.72°	1333, 5258, 4997: 2653, 5222
dt/dp	760	0.038°	4888
p	25°	526	[4888] : 2653
eq 4	A	7.41783	[4888]
	B	1400.44	
fp		− 115.8°	1333: 4888
d	9.2°	0.7723	944: 5222
	20	0.7531	5258, 4997
dd/dt		0.00117	4888
n$_D$	20°	1.37542	2653: 1333, 5258, 4997, 4888
	25	1.37288	
η	20°	0.2	4888
γ	20°	19.00	5258: 4997
$\triangle$ H$_v$	25°	6.4	3768
	bp	6.4	[4888]
$\triangle$ H$_f^\circ$	25° (g)	− 33.50	3768
$\triangle$ H$_c^\circ$	25° (g)	− 615.98	3768
t$_c$		202°	2653
p$_c$		40.2	2653
μ	20° in **30**	1.26	1376
soly	in aq, 20°	0.9%w	4888
	aq in, 20°	0.2%w	
aq az	34.6°	98.5%w	2255
ir			3307, 89, 1223
Raman			450, 2612, 2575
nmr			801
Beil	**25**, 433		

109. Ethyl ether

$(C_2H_5)_2O$ $C_4H_{10}O$

mw		74.124	887
bp	760	34.55°	4666: 5235, 3208, 4761, 1776, 3407, 4784
dt/dp	760	0.0372°	4666, 5235: 4761, 4784
p	25°	534.2	4667: 3000
eq 5	A	6.98472	4669
	B	1090.64	
	C	231.20	
fp	(st)	−116.3°	4666, 3650, 4788: 3093, 3301, 4771
	(ms)	−123.3°	3093, 4772
d	15°	0.71925	4784: 1717, 3208, 4486, 4358, 1776, 4165
	20	0.71337	4666: 4788, 4970, 2806
	25	0.70760	
	30	0.70205	[4784]
n_D	15°	1.35555	4784, 2653: 717, 4970
	20	1.35243	4666
	25	1.34954	
	30	1.34682	2653
dn/dt		0.00056	4784
η	15°	0.247	4784: 356, 2939, 4928, 1404, 4808
	20	0.242	1717
	Table II, IV	Ch II	
γ	20°	17.06	2102: 22, 4970, 3407, 5032
	25	16.50	2657
	30	15.95	2102
	Table VII	Ch II	
$\triangle H_v$	25°	6.51	4667: 3650, 3331, 4866
	bp	6.38	4667, 3208: 4962, 4989, 4470
$\triangle H_m$		1.745	4667: 492, 4601
$\triangle H_f^\circ$	25°(l)	−66.83	4668: 930, 3768
$\triangle H_c^\circ$	25°(l)	−651.7	2577: 4743, 4580, 3768
C_p°	25°	28.0	3773: 4172
C_p	15°	39.88	4165: 3650
t_c		193.55°	2771a: 2653, 4968, 5301, 3331, 5032, 1100
p_c		35.90	2771a: 3331, 4968, 5301, 1503, 2653, 5034
d_c		0.265	2771a
v_c		0.280	2771a: 1100, 5303, 2653

109. Ethyl ether

(Continued)

A	eq 72	0.03569	4667
K_f		1.79°	492
K_B		1.824°	919: 483, 475, 2207, 5029, 477
pKa	in aq H_2SO_4 −3.59		280, 283: 5057, 5221
pKb		$\langle 3 \times 10^{-16}$	4990: 1976, 1447, 1446
ϵ	20°	4.335	3477: 998, 4786, 4281, 2747, 3400, 5275
μ	20°	1.15	3400: 1920, 1376, 2946, 2497, 295
soly	in aq, 25°	6.04%w	2462: 2189, 2417, 3540
	aq in, 25	1.468%w	4062: 2462, 3540
	34.15°	98.74%w	2254
fl pt	C C	-49°F	3680
uv			1237, 2131, 2840, 4136, 1238, 2632, 3772
ir			3766, 408, 1469, 3788, 3803, 1275, 4547
Raman			1170, 717, 1008, 4633, 3192, 1172, 735
mass			3086, 90, 4618, 2494, 3997, 4715
nmr			4632, 3559, 937, 3025, 1923
Beil	21, 314		

110. Propyl ether

$(CH_3CH_2CH_2)_2O$ $C_6H_{14}O$

mw		102.178	887
bp	760	89.64°	4666: 5274
dt/dp	760	0.0433°	4666: 5274
p	25°	19.3	4667
eq 4	A	7.821	2444
	B	1791.23	
fp		-123.2°	4666: 4770

110. Propyl ether

(Continued)

d	20°	0.7466	4666
	25	0.7419	
	30	0.737	4358
n_D	20°	1.3805	4666
	25	1.3780	
dn/dt		0.00049	4778
η	15°	0.448	4778: 1404, 1717
	30	0.376	
	Table II	Ch II	
γ	20°	20.53	2102: 4778, 4970
	30	19.35	
$\triangle H_v$	25°	8.53	4667: 1050
	bp	7.91	[2444]
$\triangle H_f^{\circ}$	25°(l)	− 78.58	5358: 4668, 1050
$\triangle H_c^{\circ}$	25°(l)	− 963.93	1050, 5358
C_p°	25°	38	4668
pKa	in aq, H_2SO_4	− 4.40	283
ϵ	26°	3.39	3477: 2371, 4358
μ	25° in 232	1.32	295: 2371, 1920, 4109
soly	in aq, 25°	0.49%w	510
	aq in, 25°	0.45%w	
aq az	72.5°	75%w	3975: 2254
fl pt	O C	− 5°F	3160
ir			1469, 4547, 2181
Raman			717, 1008
mass			3086
nmr			2461, 937, 1923
Beil	**24**, 354		

111. Isopropyl ether

$[(CH_3)_2CH]_2O$ $C_6H_{14}O$

mw		102.187	887
bp	760	68.3°	4670: 1372, 2522, 2653, 1570, 4535
dt/dp	760	0.042°	4670: 1372
p	25°	149.3	746: 3531, 4893, 4671
eq 5	A	7.09712	1372
	B	1257.6	
	C	230	
fp		−85.5°	4670: 1372, 3653, 4705
d	20°	0.7235	4670: 1372, 1570, 2522, 529, 1717, 4970
	25	0.7182	746
n_D	20°	1.3681	4670: 1372, 2522, 2653, 1570, 4970
	25	1.3655	
η	20°	0.329	1717
	25	0.379	1570, 3160
γ	24.5°	17.34	4970: 1570, 746
$\triangle H_v$	25°	7.8	4671: 1050
	bp	7.01	4671: 1570
$\triangle H_m$		2.64	4671
$\triangle H_f^\circ$	25°(l)	−84.01	1050: 5358
$\triangle H_c^\circ$	25°(l)	−958.51	1050, 5358
C_p	19.9°	51.7	3653: 1570
t_c		226.9°	2653: 1503
p_c		28.4	2653: 1503
d_c		0.265	2653
v_c		0.386	2653
A	eq 72	0.0377	4671
pKa	in aq H_2SO_4	−4.30	283
ϵ	25°	3.88	3477: 4744, 1992
μ	20°	1.22	1851: 1503, 4744
soly	in aq, 20°	1.2%w	4893: 529
	aq in	0.57%w	
aq az	62.2°	95.5%w	2254
fl pt	TOC	15°F	3681: 1570
	C C	−18°F	
ir			89, 1648, 1275, 408
Raman			1008, 2181
mass			90, 3086, 1732, 4715
nmr			1923, 937, 3559
Beil	**24**, 362		

112. Butylvinyl ether

$CH_2 = CHOC_4H_9$ $C_6H_{12}O$

mw		100.162	887
bp	760	93.82°	4997, 5258: 4301, 3452, 4888, 4766, 4970
dt/dp	760	0.044°	4888
p	25°	51	4151
fp		-92°	4151
d	20°	0.7792	4997, 5258: 944
	25	0.7727	4151
n_D	20°	1.4007	4888: 4997, 5258, 944
	25	1.3997	4151
η	20°	0.5	4888
	25		
γ	20°	21.99	4997: 5258
$\triangle H_v$	bp	7.71	[4151]
C_p	25°	55.4	4151
μ	25° in 28	1.25	4015
soly	in aq, 20°	0.30%w	4888
	aq in	0.09%w	
aq az	77.5°	88.4%w	2255: 4301
fl pt	COC	30°F	4151, 3681
uv			1670
ir			3307, 1223
Raman			450, 2575, 1670
nmr			801, 1552
Beil	25, 433		

113. Butylethyl ether

$C_4H_9OC_2H_5$ $C_6H_{14}O$

mw		102.178	887
bp	760	92.2°	4670, 4766: 3550, 3452, 4970
dt/dp	760	0.045°	4670
fp		-103°	4670: 4766
d	20°	0.7495	4670, 3452: 4970, 3550
	25	0.7448	
n_D	20°	1.3818	4670, 4970: 3452, 3550
	25	1.3793	
η	20°	0.421	3452
	25	0.397	
γ	20°	20.75	3452, 4970
	25	20.25	
$\triangle H_v$	bp	8.4	4671
C_p°	25°	38	4672
pKa	in aq H_2SO_4	-4.12	283
μ		1.24	2947: 2946
aq az	76.6°	88.1%w	2255
ir			408
mass			3086, 913
Beil	**24**, 369		

114. Butyl ether

$(C_4H_9)_2O$ $C_8H_{18}O$

mw		130.232	887
bp	760	142.2°	4666: 2653, 2771, 4778, 1372, 1656, 3210
dt/dp	760	0.048°	4666: 4778, 1372
p	25°	12.5	4667, 2190: 1373
eq 5	A	7.31540	1372: 2771
	B	1648.4	
	C	230	
fp		−95.2°	4666: 4778, 273, 1372
d	20°	0.7684	4666: 1656, 1920, 4970, 1372, 4778, 3931
	25	0.7641	
n_D	20°	1.3992	4666: 2653, 4970, 1656, 1372
	25	1.3968	
dn/dt		0.00045	4778
η	15°	0.741	4778: 595
	30	0.602	
γ	15°	23.40	4718: 2102, 4970
	30	21.99	
$\triangle H_v$	25°	10.5	4667, 1050
	bp	8.83	4667, 3210
$\triangle H_f^\circ$	25°(l)	−90.34	1050: 4668, 5358
$\triangle H_c^\circ$	25°(l)	−1276.92	1050: 4367, 5358
C_p°	25°	48.8	4668
C_p	Table XI	Ch II	
pKa	in aq H_2SO_4	−5.40	283
ϵ	20°	3.083	3250: 3247, 4744, 2243, 3477
μ	20°	1.18	3250: 295, 1920, 4744, 2947
soly	in aq, 20°	0.03%w	4893: 510
	aq in, 20°	0.19%w	
aq az	92.9°	67%w	2254
fl pt	C C	100°F	3681, 3160
uv			3785
ir			89, 408, 1469
Raman			1007, 1008
mass			3086
nmr			2461, 3559, 1923
Beil	24, 369		

115. Pentyl ether

$(C_5H_{11})_2O$ $C_{10}H_{22}O$

mw		158.286	887
bp	760	186.8°	4666: 4767, 1372, 4784
dt/dp	760	0.052°	4666: 1372
p	105.46°	57.04	1373
eq 5	A	7.45597	1372
	B	1906.7	
	C	230	
fp		− 69.4°	4666: 4766, 4784, 1372, 4767
d	20°	0.7830	4666: 4784, 4970, 1372
	25	0.7790	
n_D	20°	1.4119	4666: 4784, 4970, 1372
	25	1.4098	
η	15°	1.188	4784: 595
	30	0.922	
γ	20°	24.76	2102: 4970
	30	23.78	
$\triangle H_c^{\circ}$	25°(l)	− 1582.0	2577: 4580
C_p°	25°	59.8	4668
ϵ	25°	2.77	3477: 4744, 1381
μ	25° in 232	1.20	295
fl pt	TCC	135°F	3680, 88
ir			408
Raman			1008, 2181
mass			3086
Beil	24, 384		

116. Isopentyl ether

$[(CH_3)_2CHCH_2CH_2]_2O$ $C_{10}H_{22}O$

mw		158.286	887
bp	760	173.4°	4535: 4970, 2852
dt/dp	760	0.050°	[4535]
p	25°	1.4	[4535]
eq 5	A	7.56926	[4535]
	B	1891.32	
	C	230	
d	20°	0.7777	4970
	28	0.7713	
n_D	20°	1.40850	4970
η	11°	1.40	2751
	20	1.012	595
	30	0.866	
γ	21.7°	22.85	2609
	28.0	22.31	4970
$\triangle H_v$	bp	8.40	2785
C_p	24 – 170°	90.56	[2785]
ϵ	20°	2.82	3477: 1496, 5029
μ		1.23	2947: 295, 1496
soly	in aq, 20°	0.02	2609: 664
aq az	97.4°	46%w	2254
ir			408
Raman			1008, 2181
mass			2201, 3086
nmr			3559
Beil	24, 401		

117. 1,2–Dimethoxyethane

Dimethyl Cellosolve

Monoglyme

$CH_3OCH_2CH_2OCH_3$ $C_4H_{10}O_2$

mw		90.123	887
bp	760	93.0°	4535: 5063, 2653, 3635
dt/dp	760	0.049°	[4535]
p	25°	75.2	[4535]
eq 5	A	6.6487	[4535]
	B	1217.03	
	C	230	
fp		−58°	359
d	20°	0.8665	3635: 5063, 904, 5064, 926
	25	0.8621	857
n_D	20°	1.37963	2653: 5063, 3635, 857, 5064, 926, 904
	25	1.37811	
η	25°	0.455	926: 5064, 857
	Table VI	Ch II	
t_c		263°	2653
p_c		38.2	2653
d_c		0.333	2653
v_c		0.270	2653
pKa	in aq H_2SO_4	−2.97	284
ϵ	25°	7.20	926
μ	25° in 30	1.71	2593: 1825
soly	in aq	inf	3975, 857
aq az	77.4°	89.9%w	2255
ir			736, 3349, 3560
mass			2201
nmr			8, 1923
Beil	30, 467		

118. bis(2–Methoxyethyl) ether

Diglyme

$(CH_3OCH_2CH_2)_2O$ $C_6H_{14}O_3$

mw		134.177	887
bp	760	159.76°d	1689: 1687
dt/dp	760	0.066°	[1689]
p	25°	3.4	[1689]
eq 4	A	8.0837	[1689]
	B	2251.5	
d	25°	0.9440	1996*: 1132
dd/dt		0.00106	[1688]
n_D	20°	1.4097	1687
	25	1.4043	1996*
η	25°	0.981	1996*
γ	25°	29.5	[1688]
	bp	14.44	1688
	Table VIII	Ch II	
$\triangle H_v$	bp	10.312	1689
μ	25°in 28	1.97	2593
soly	in aq	inf	3975
aq az	99.8°	∿25%w	3975
fl pt	TCC	145°F	1996*
Beil	30, 468		

* Best commercial grade

119. Propylene oxide

Mixed isomers*

d – form   l – form

C_3H_6O

mw		58.081	887
bp	760	33.9°	1157, 2653: 4888, 4535, 4016
bp (+)	760	36.5 – 38°	4
dt/dp	760	0.036°	1157: 4888
p	20°	445	1157: 4535, 675
eq 5	A	6.96997	3070: 3386, 675
	B	1065.27	
	C	226.283	
fp		∿111.93°	3578: 4535, 3070, 1157
d	0°	0.8598	5347: 4016
	20	0.8287	
d (+)	20/20°	0.8412	4
α	20°	0.00151	1157
n_D	5°	1.3712	5347: 1157, 4016
	20	1.36603	2653
	25	1.36322	
η	0°	0.410	5347
	20	0.327	
	25	0.28	1356
$\triangle H_v$	25°	6.667	4339, 4888
	bp	6.872	[1157]
$\triangle H_m$		1.561	3578
$\triangle H_f^o$	25° (l)	– 28.84	4339: 1874, 5358
$\triangle H_c^o$	25° (l)	– 458.28	4339: 2577, 3431, 5358
C_p^o	25°	17.36	1874
C_p	25°	28.77	3578: 1157
t_c		209.1°	2653: 1157
p_c		48.6	2653
d_c		0.312	2653
v_c		0.186	2653

119. Propylene oxide

(Continued)

$[\alpha]_D$	(+), $18^\circ_\circ$ (−), 18°	$12.72^\circ_\circ$ 8.26°	4: 2920, 2922, 4282 4
μ	25° in **30**	2.00	4016: 2167, 4565
soly	in aq, $20^\circ_\circ$ aq in, 20°	40.5%w 12.8%w	4888: 1157
aq az	None		1157, 2254
fl pt	CC	$-31^\circ F$	88, 3681
ir			2606
Raman			367, 4633, **2606**
mass			1690
nmr			3472
Beil	**2362**, 6		

* Only properties of the d̲ and 1̲ isomers are indicated as such;
mixed d̲l̲ forms not indicated.

120. 1,2-Epoxybutane

$CH_3CH_2CHCH_2O$ C_4H_8O

mw		72.108	887
bp	760	63.2°	2609: 1654
p	20°	141	2609
eq 4	A	7.8461	[2609]
	B	1670.3	
fp		− 150°	2609
d	20°	0.8297	[2609] : 1654
α	55°	0.00139	4886
n_D	20°	1.3840	2609
	25	1.381	1359
η	20°	0.41	2609
	25	0.40	1359
$\triangle H_v$	20 − 63°	7.64	[2609]
$\triangle H_c$	(l)	− 609.4	3431
μ	20° in 30	2.01	3336
soly	in aq, 20°	5.91%w	2609
	aq in, 20°	2.65%w	
aq az	60.1°	92%w	3975
fl pt	TOC	− 25°F	1359
ir			918, 2606
Raman			2606
Beil	2362, 11		

121. Cineole

$$H_3CC \overset{\displaystyle CH_2\,CH_2}{\underset{\displaystyle CH_2\,CH_2}{\diagdown\diagup}} CHC(CH_3)_2$$

$C_{10}H_{18}O$

mw		154.254	887
bp	760	176.0°	4535: 1928, 1929
dt/dp	760	0.053°	[4535]
p	25°	1.9	[4535]: 4274, 1928
eq 5	A	7.27446	[4535]
	B	1783.82	
	C	230	
fp		1.3°	1929: 4535
d	15.5°	0.9284	[1929], 2608
	20	0.9237	1996
	25	0.9192	
n_D	20°	1.4575	1929: 4535, 5065
	25	1.4555	1996
γ	20°	32.1	1996
	25	31.1	
K_f		6.7°	1548
ϵ	23.5°	4.57	3499
soly	in aq, 21°	0.35%w	1428
aq az	97.6°	35.9%w	3975: 2254
fl pt	TCC	118°F	1996
uv			3860
Beil	**2363**, 24		

122. Furan

OCH:CHCH:CH $\qquad$ C_4H_4O

mw		68.076	887
bp		31.36°	1940: 2653, 3210, 1333, 964, 2608
dt/dp	760	0.04131°	1369
p	26.469°	634.00	1940
eq 5	A	6.97523	1940: 3642
	B	1060.851	
	C	227.740	
tp		– 85.61°	1940, 2608: 1333
d	10°	0.95144	1940: 320
	15	0.94467	
	20	0.93781	2608
n_D	20°	1.42140	1940: 2608
	25	1.41871	2653
η	20°	0.380	4773
	25	0.361	
γ	20°	24.10	4773
	25	23.38	
$\triangle H_v$	20°	6.628	1940: 3210, 2608
	bp	6.474	
	Table IX	Ch II	
$\triangle H_m$		0.909	1940
$\triangle H_f^o$	25°(l)	– 14.903	1940: 5358
$\triangle H_c^o$	25°(l)	– 497.97	1940: 1864, 5358
C_p^o	25°	15.64	1940: 355
	Table XIV	Ch II	
C_p	25.93°	27.440	1940
	Table XII	Ch II	
t_c		217.0°	2771a: 964, 1940
p_c		54.3	2771a: 1940
d_c		0.312	1940
v_c		0.218	1940
A	eq 72	0.0130	1940
ϵ	25°	2.942	2036
μ	25° in 30	0.71	2036: 4009, 2879

122. Furan

(Continued)

soly	in aq, 25° aq in	1%w 0.3%w	4274, 2608
fl pt	TCC	− 32°F	3872
uv			3271, 3881, 3832, 3100, 3927, 3759
ir			1940, 355, 2855, 3758, 4736, 3139, 3143
Raman			1940, 38, 7, 653, 2677, 3951, 3950
mass			2205, 90, 2164
nmr			1907, 4130, 6, 1863, 3930, 2845
Beil	2364, 27		

123. Tetrahydrofuran

$\underline{OCH_2CH_2CH_2CH_2}$ C_4H_8O

mw		72.108	887
bp	760	66°	1415, 3871: 634, 3759, 2653, 4349, 964
dt/dp	760	0.044°	[3871]
p	25°	197	[1415] : 2653
eq 5	A	6.75376	[3871] : 930
	B	1146.39	
	C	230	
fp		− 108.5°	3871, 3160, 1415, 4349
d	20°	0.8892	634: 926, 3284, 682
dd/dt		0.00101	[1415]
n_D	20°	1.40716	2653: 634, 3759, 3284, 682
	25	1.40496	
η	20°	0.55	3160: 926, 2794, 3284
	25	0.460	566
	Table VI	Ch II	

123. Tetrahydrofuran

(Continued)

γ	$25°$	26.4	1415
$\triangle H_v$	$\sim 25°$	7.65	3703: 930, 627
	bp	7.07	[1415]
$\triangle H_f^{\circ}$	$25°$ (g)	-44.03	3703: 930
$\triangle H_c^{\circ}$	$25°$ (l)	-752.8	930: 3703
C_p	$20°$	33.8	[1415] : 3703
t_c		$267.0°$	2771a: 964, 2653
p_c		51.2	2653
d_c		0.322	2653
v_c		0.224	2653
pKa	in aq H_2SO_4	-2.08	280, 284: 282
ϵ	$25°$	7.58	3160: 1415, 926, 1152, 566
μ	$25°$ in 30	1.75	1152, 2167: 2381, 1739, 4009, 4429 , 2676
soly	in aq	inf	3871
aq az	$63.4°$	93.3%w	3975: 3957
fl pt	C C	$6°F$	3871
uv			3759
ir			1640, 2766, 4840 2938, 5078, 4304, 89
Raman			2676, 2675, 348, 38
mass			1691, 2205, 1390
nmr			4632, 3559, 2531, 3098
Beil	2362, 10		

124. p–Dioxane

$$C_4H_8O_2$$

mw		88.107	887
bp	760	101.320°	4383: 2745, 443, 2153, 3332, 4504, 4781
dt/dp	760	0.0432°	4383: 4781
p	25°	37.1	[4957] : 4535, 2178, 746
eq 4	A	7.8642	1689: 4957, 2234
	B	1866.7	
fp		11.80°	2153, 4781: 2032, 4198, 1452, 2745, 4182
d	20°	1.03361	4772: 1895, 443, 1565, 3332, 1643, 2144
	25	1.02797	1994: 4056, 4781, 2153, 4426, 4182, 746
dd/dt		0.001128	[4772]
n_D	15°	1.42436	4781: 4182, 1565, 3332, 4504, 2144, 4426
	20	1.42241	2153: 2653
	25	1.42025	1643
η	15°	1.439	4781: 2527, 2144, 1688, 2234
	30	1.087	
γ	15°	34.45	4781: 1688, 2234, 2144, 746
	30	32.20	
	Table VIII	Ch II	
$\triangle H_v$	bp	8.505	[4957] : 1689, 2144
$\triangle H_m$		2.978	4056: 2370, 2745
$\triangle H_f^{\circ}$	25°(l)	– 84.50	5358
$\triangle H_c^{\circ}$	25°(l)	– 564.99	4435: 5358
C_p°	26.84°	22.56	3125
C_p	18°	36.56	4056: 2144, 2370
t_c		314°	2771a: 1791, 2144, 1689, 2234, 2653
p_c		51.4	2653: 1791, 2144, 2234
d_c		0.370	2653
v_c		0.238	2653
K_f		4.63°	2745: 3622, 2144, 4056, 2153, 940
K_B		3.270°	2144: 2193
pKa	in aq, H_2SO_4	– 2.92	284: 2900
κ	25°	5×10^{-15}	2744
ϵ	25°	2.209	3477, 2778: 443, 332, 2744, 550, 938
μ		0.45	4110: 2769, 4923, 4426, 5189

124. p–Dioxane

(Continued)

soly	in aq	inf	2609
aq az	87.82°	82%w	2254
fl pt	C C	54°F	3680
uv			2132, 3759, 2632
ir			3084, 4547, 2100, 5078, 4076, 4400, 408
Raman			3252, 4098, 38, 3896, 3916, 3125
mass			90, 2205
nmr			3559, 3025, 2432
Beil	2668, 3		

125. Tetrahydropyran

$CH_2CH_2CH_2CH_2CH_2O$ $C_5H_{10}O$

		86.135	887
mw			
bp	760	88°	724: 55, 3759, 256, 511
eq 4	A	7.99	930
	B	1660.5	
fp		−45°	3529: 4349
d	20°	0.8814	511: 256, 3529
	25	0.8772	55
α		0.00104	3529
n_D	20°	1.4211	256: 3759
	25	1.4195	55
η	20°	0.826	3529
	25	0.764	
	Table VI	Ch II	

125. Tetrahydropyran

(Continued)

$\triangle H_v$	$25°$	8.35	3703: 930
$\triangle H_f^°$	$25°$ (g)	-53.50	3703
$\triangle H_c^°$	$25°$ (l)	-750.52	4366: 3703, 4435, 930, 4367
C_p	$25°$ (l)	37.4	3703
pKa	in aq, H_2SO_4	-2.79	280, 284
ϵ	$25°$	5.61	3529
μ	$25°$ in **30**	1.55	1152: 55
soly	in aq $25°$	8.02%	[511] : 256
	aq in $25°$	3.14%	
aq az	$71°$	91.5%w	2255: 256
ir			5078, 4304, 348
Raman			348, 1236, 2476
nmr			3098
Beil	**2362**, 12		

126. Benzylethyl ether

$C_6H_5CH_2OC_2H_5$ $C_9H_{12}O$

mw		136.195	887
bp	760	$185.0°$	4535
dt/dp	760	$0.052°$	[4535]
p	$25°$	0.9	[4535]
eq 5	A	7.5247	[4535]
	B	1927.21	
	C	230	
d	$20°$	0.9478	4089: 4842
	25	0.9446	
	40	0.9343	2283
n_D	$20°$	1.4958	4089: 4842
	25	1.4934	

126. Benzylethyl ether

(Continued)

γ	20°	32.82	2283
	25	32.18	
	40	29.97	
ϵ	20°	3.9	3477: 1325
Raman			3461
Beil	528, 431		

127. Anisole

$C_6H_5OCH_3$ C_7H_8O

mw		108.141	887
bp	760	153.80°	1928: 1948, 3017, 3751, 4762, 4761, 4780
dt/dp	760	0.04896°	1372: 3017, 4780, 4761
p	25°	4.2	[1372]: 1373, 4535, 3167, 1948
eq 5	A	7.35950	1372: 3642
	B	1718.7	
	C	230	
fp		−37.5°	4780, 4770: 1372, 4762, 711, 1948, 3591
d	15°	0.99858	4780: 356, 1717, 2082, 1948, 711, 3167
	20	0.99402	1372: 3591, 5032, 4761, 4048, 4358
	25	0.98932	
dd/dt		0.000932	[3701]
n_D	20°	1.51700	1372: 711, 3591, 4970, 1948, 4780, 3992
	25	1.51430	
dn/dt		0.00050	4780
η	15°	1.152	4780: 356, 1728, 4709, 1717, 1948, 3992
	30	0.789	
	Table VI	Ch II	
γ	15°	36.18	4780: 4970, 5048, 5032
	30	34.15	
	Table VII	Ch II	

127. Anisole

(Continued)

$\triangle H_v$	bp	8.802	3017: 5032
$\triangle H_c$	25° (l)	- 905.2	2577: 4743, 345
C_p	31.6°	49.85	3745: 3041, 4144, 3017
t_c		368.5°	1948: 1950
p_c		41.2	2771a: 1950, 1948
K_B		4.502°	1429: 3606
pKa	in aq, H_2SO_4	- 6.54	280, 283
κ	25°	1×10^{-13}	529
ϵ	25°	4.33	3477: 3746, 1198, 5034, 2371, 3400
μ	25° in 232	1.245	293: 2383, 4017, 2946, 1496, 2371, 3769
soly	in aq	1.04%w	[664]
aq az	95.5°	59.5%w	2254
uv			1336, 1347, 2513, 5355, 849, 4384, 4878
ir			3011, 1879, 2361, 4400, 736, 2517, 408
Raman			2660, 2141, 2573, 499, 3685, 2575, 2671
mass			3285, 61
nmr			4512, 5356, 3895
Beil	**514** , 138		

128. Phenetole

$C_6H_5OC_2H_5$ $C_8H_{10}O$

mw		122.168	887
bp	760	170.00°	1372: 1948, 2558, 3591, 4764, 4773
dt/dp	760	0.05055°	1372: 4773
p	25°	2.0	[1372]
eq 5	A	7.40281	1372
	B	1808.8	
	C	230	
fp		- 29.52°	1372: 3591, 4535, 4764

128. Phenetole

(Continued)

d	20°	0.96514	1372: 356, 1717, 1948, 2558, 3591, 4970
	25	0.96049	5032, 4773
dd/dt		0.000934	[3701]
n_D	20°	1.50735	1372: 1948, 4970, 3591, 4773
	25	1.50485	
η	15°	1.364	4773: 2558, 1717, 4709, 356, 1948
	30	1.040	
	Table VI	Ch II	
γ	20°	32.88	4773: 4970
	Table VII	Ch II	
$\triangle H_v$	160°	10.70	3037
	bp	9.83	4963
$\triangle H_c^o$	25°(l)	-1061.5	2577: 345
C_p	20°	54.48	4144
tc		374.0°	1948: 1950, 5032
p_c		33.8	1948: 1950
K_f		7.15°	1531: 3606
K_B		5.0°	3037
κ	25°	$\langle 1.7 \times 10^{-8}$	4106
ϵ	20°	4.22	3477, 2371: 4773
μ	25° in 232	1.36	293: 2667, 2371, 3769, 1496
soly	in aq	0.12%w	664
aq az	97.3°	41%w	2254
uv			373, 1239, 4007, 3910
ir			736, 408, 4100, 2519, 2517, 1879, 3823
Raman			2575, 625, 2573
Beil	514, 140		

129. Benzyl ether

$(C_6H_5CH_2)_2O$ $C_{14}H_{14}O$

mw		198.267	887
bp	760	288.30 $^\circ$	1372: 512, 4564
dt/dp	760	0.06122°	1372
p	25°	0.023	4802
eq 5	A	7.71832	1372
	B	2507.3	
	C	230	
fp		3.60°	512: 1372
d	20°	1.0428	512: 4564, 1372, 5321
	35	1.0341	
n_D	20°	1.54057	1372: 4564, 5321
	25	1.53851	
η	20°	5.333	595
	30	4.062	
	35	3.711	4564
γ	35°	38.2	4564
$\triangle H_m$		4.83	512
K_f		6.27°	512
Soly	in aq, 35°	0.0040%w	4564
uv			3237, 2623
ir			4113, 408
Beil	528, 434		

130. Phenyl ether

$(C_6H_5)_2O$ $C_{12}H_{10}O$

mw		170.213	887
bp	760	258.31°	1372: 1369, 1791, 1070
dt/dp	760	0.05847°	1372: 3786
p	25°	0.0213	1369
eq 5	A	7.09894	1369: 930
(145 – 325°)	B	1871.92	
	C	185.84	
f_p		26.87°	1680: 3786, 3387, 1070, 1372
t_p		26.946°	3786
d	30°	1.06611	1372: 1070, 4834, 3387
	35	1.06117	
dd/dt	30 – 170°	0.00086	[5338]
n_D	30°	1.57625	1372: 4834
	35	1.57387	
	50	1.56681	1369
η	40°	2.4594 cs	1369: 4834
	60	1.7065 cs	
	80	1.2716 cs	
γ	30°	38.82	1369
	40	37.73	
$\triangle H_v$	25°	16.017	[1369] : 930
	bp	11.261	
$\triangle H_m$		4.115	1680: 3786, 1369, 1418
$\triangle H_f^\circ$	25°(l)	– 3.583	[1680]
$\triangle H_c^\circ$	25°(l)	– 1466.63	1680: 930
C_p(s)	25°	51.76	[1680] : 1782
(l)	26.87°	64.15	
t_c		429.5°	1369: 2576, 1791, 5338
p_c		30.8	[1369] : 1791
d_c		0.312	1369
v_c		0.545	[1369]
K_f	A eq 72	0.02177	1369: 3786
K_f		7.88°	1418

130. Phenyl ether

(Continued)

ϵ	20° (s) 20° (l)	2.68 3.686	1496, 2371: 4834
μ	25° in 30	1.16	1832, 2906: 3994, 2946, 2371, 1496, 1080
soly	in aq	0.39%w	[664] : 1356
aq az	99.33°	3.25%w	2254
fl pt		205°	1356
uv			1179, 40 03, 4878, 4634, 2347, 4007, 4879
ir			408, 1179, 3030, 2516, 3136
Raman			2574, 2516, 2575, 2573, 4397, 625
mass			556, 5215
Beil	**514**, 146		

131. Veratrole

$1,2 - C_6H_4(OCH_3)_2$ $C_8H_{10}O_2$

mw		138.168	887
bp	760	206.25°	4772: 3714, 3992
dt/dp	760	0.054°	[3976] : 4772, 1666
p	25°	0.47	[3976]
eq 4	A B	8.1532 2527.58	[3976] : 4772, 1666, 1158
fp	(st) (ms)	22.5° 20.9°	1251: 936, 620, 813, 1158
d	25° 40 60	1.0819 1.0667 1.0470	3992, 1666: 3714, 4772, 1158, 3039
n_D	25°	1.53232	3992: 2608, 1158, 1887

131. Veratrole

(Continued)

η	25°	3.281	3992: 4472, 3039, 1666
	40	2.184	
	60	1.457	
γ	0°	42.5	Beil, E I 6, 383
	131	26.0	3039
	184	22.6	
$\triangle H_v$	bp	11.52	[3976] : 4472, 1666, 1158
$\triangle H_m$		3.834	3003
K_f		6.38°	3286: 3003
ϵ	25°	4.09	3992: 1812, 936
μ	25° in 30	1.32	1304: 2381, 1158, 293, 3992
soly	in aq	sl sol	4519
aq az	99.8°	11%w	3975: 2254
uv			1238, 1824, 1738, 5355, 4998, 3494
ir			736, 2863
Raman			2137, 2671
mass			1123
nmr			3172, 5356, 1812, 1863
Beil	553 , 771		

132. Dimethoxymethane

Methylal

$CH_2(OCH_3)_2$ $C_3H_8O_2$

mw		76.096	887
bp	760	42.30°	4784: 4761, 4762, 3536, 3537
dt/dp	760	0.040°	4761
p	25°	398.7	[3073] : 3537, 3536
eq 5	A	7.10401	3073
	B	1162.58	
	C	−40	
fp		−105.15°	3073: 4784
d	15°	0.86645	4784: 4761, 3635, 4970
	30	0.84745	
dd/dt		0.001267	[4784]
n_D	15°	1.35626	4784: 3536
	20	1.35335	3635: 4970
dn/dt		0.00049	[4784]
η	15°	0.340	4784
	30	0.325	
γ	20°	21.12	2102, 4970, 4969
	30	19.76	
$\triangle H_v$	25°	6.904	3073: 3537, 3536, 545
$\triangle H_m$		1.991	3073: 4769, 4768
$\triangle H_c^\circ$	25°(l)	−455.8	2577: 4743
C_p	25°	38.58	[3073] : 2349
t_c		242.0°	3073
p_c		44.171	[3073]
K_B		2.125°	483: 480, 491
ϵ	20°	2.645	4786: 3477, 5026
μ	34°(g)	0.74	2770
soly	in aq, 16°	24.4%w	[4225]
	aq in, 16°	4.1	
aq az	42.05°	98.6%w	2254: 2255
fl pt	C C	0°	3680
uv			4196
ir			4547, 3561, 5209, 4591, 736, 526, 5054
Raman			717, 5209, 3354, 3562
mass			1660, 2850, 556
nmr			3563, 1624, 1923, 3559, 2682
Beil	75, 574		

133. Acetal

$CH_3CH(OC_2H_5)_2$ $C_6H_{14}O_2$

mw		118.117	887
bp	760	103.6°	[3536] : 4535, 1747, 4300, 3001, 3634, 4970
dt/dp	760	0.046°	3001
p	25°	34.1	3536: 29, 1928, 4535
eq 5	A	7.22012	[3536]
	B	1447.59	
	C	230	
fp		glass	4769
d	20°	0.8264	[4970] : 1747, 29, 4300, 5032
	25	0.8213	
dd/dt		0.000103	[4970]
n_D	20°	1.38054	4970 : 4971, 717, 4300, 4970, 1747
	25	1.3682	29
γ	15°	22.02	4970: 4995, 4969
	21.8	21.26	
$\triangle H_v$	25°	9.05	[3536] : 5032, 5034
	bp	7.819	3001
$\triangle H_c^\circ$	25° (l)	– 924.7	2577
C_p	19.1 – 99°	61.40	3001
t_c		270°	5032
ϵ	25°	3.80	3477: 5034, 1381
μ	in 30	1.38	Beil E III 1, 2643
soly	in aq 25°	5%	2608
aq az	82.6°	85.5%w	2254: 2255
fl pt	CC	97°F	3680: 88
uv			4196, 867
ir			3561, 2746, 1652, 736, 5054
Raman			3563, 717, 2978, 3354
mass			1660, 2850, 3077, 90, 4715
nmr			1923, 1652, 4251
Beil	78, 603		

134. Acetaldehyde

CH_3CHO C_2H_4O

mw		44.054	887
bp		$20.4°$	4656: 1077, 1665, 4404, 4535, 4763
dt/dp		$0.0351°$	4656: 4404
p	$25°$	910	2444: 4535
eq 5	A	7.0565	4655: 4404
	B	1070.6	
	C	236.0	
fp		$-123.°$	4656: 1947, 4546, 4763
d	$15°$	0.7846	4404: 1038, 1077, 1665
	20	0.7780	4656: 4404
dd/dt	$0-30°$	0.001325	[4404]
α	$0-30°$	0.00169	2609
n_D	$20°$	1.3311	4656: 4404
dn/dt	$0-20°$	0.0005635	[4404]: 1077
η	$0.1°$	0.2711	Beil **E III 1**, 2621: 1665
	15	0.2456	2609
	20.3	0.2237	Beil **E III 1**, 2621
	Table II	Ch II	
γ	$20°$	21.2	2609
$\triangle H_v$	bp	6.500	4962: 1038, 5034, 4963
$\triangle H_m$		0.775	2609
$\triangle H_f^o$	$25°$ (g)	-39.67	3780: 5358
$\triangle H_c^o$	$25°$ (l)	-281.90	4743: 2577, 5358
C_p^o	$25°$	15.0	4394
C_p	$24.9°$	14.8	1038: 3780
t_c		$188.°$	2238
p_c		63.2	2609
κ	$0°$	1.20×10^{-6}	5026: 5027
ϵ	$21°$	21.1	3477: 5034, 1381, 4752
μ	$20°$ in **30**	2.51	1077: 2330

134. Acetaldehyde

(Continued)

soly	in aq	inf	4404
aq az		none	2254: 4404
fl pt	TCC	-36°F	3680
uv			91, 2805, 1068, 5256
ir			89, 2599, 1741, 3412, 4732, 1518, 5256
Raman			2384, 3412, 1170, 3732, 2663, 1518
mass			90, 1344, 4715
nmr			937, 4512, 468
Beil	77, 594		

135. Propionaldehyde

CH$_3$CH$_2$CHO   C$_3$H$_6$O

mw		58.081	887
bp		48.0°	4656: 1077, 1346, 1372, 1665, 3081, 4404
dt/dp		0.0380°	4656: 1372, 4404
p	24.65°	314.	4404: 1373
eq 5	A	7.0493	4655: 1372, 4404
	B	1154.8	
	C	229.0	
fp		−80.°	4656: 1372, 3081
d	20°	0.7970	4656: 1077, 1372, 1665, 4404
	25	0.7912	
dd/dt	0−30°	0.001165	[4404]
α	55°	0.00201	4884
n$_D$	20°	1.3619	4656: 1077, 1372, 3081, 4404
	25	1.3593	
dn/dt	0−20°	0.0005180	[4404]
η	15.4°	0.3568	1665
	26.7	0.3167	
△H$_f^o$	25°(l)	−52.65	5358
△H$_c^o$	25°(l)	−429.8	2577: 4743, 4794, 5358
C$_p$	0−30°	30.3	[2608]
κ	25°	9.5 x 10^{-5}	5027
ε	17°	18.5	3477: 770, 1381, 4752
μ	20° in 28	2.54	1077: 2330
soly	in aq, 25°	30.6%w	4404
	aq in, 25°	13.0%w	
aq az	47.79°	98.1%w	4404: 2255
fl pt	TOC	⟨20°F	4884
uv			91, 3100, 611, 1068, 901, 5256, 867
ir			89, 1469, 4946, 4630, 2599, 5256
Raman			1170, 2663
mass			90, 4715
nmr			3575, 4512
Beil	82, 629		

136. Butyraldehyde

$CH_3CH_2CH_2CHO$ C_4H_8O

mw		72.108	887
bp		74.8°	4656: 1346, 1665, 2771, 4404, 4766, 4767
dt/dp		0.0411°	4656: 4404
p	25°	118	[4884] : 4404
eq 5	A	7.0212	4655: 4404, 2771
	B	1233.0	
	C	223.0	
fp		−96.4°	4656, 3655: 4767, 4766
d	20°	0.8016	4656, 4404: 1077, 1665, 4795
	25	0.7964	
dd/dt	0 − 30°	0.001040	[4404]
α	55°	0.00119	4884
n_D	20°	1.3791	4656: 4404
	25	1.3766	
dn/dt	0 - 20°	0.0005045	[4404] : 397, 1077
η	18.35°	0.4575	1665
	38.7	0.3673	
γ	24°	29.9	2609
$\triangle H_v$	bp	7.528	2609
$\triangle H_m$		2.65	[3655]
$\triangle H_f^{\circ}$	25°(l)	− 57.06	4795: 5358
$\triangle H_c^{\circ}$	25°(l)	− 592.42	4795: 5358
C_p	26.85°	39.2	[3655]
ϵ	26°	13.4	3477
μ	40°(l)	2.450	3056: 1077, 2330
soly	in aq, 25°	7.1%w	4404
	aq in, 25°	3.0%w	
aq az	67.8°	93.3%w	2255: 4404
fl pt	TCC	20°F	88, 3680
uv			3100, 1068, 5256
ir			89, 304, 1469, 4630, 641. 2599, 5256
Raman			717, 2663
mass			90, 3194, 4715
nmr			4642, 937, 2454
Beil	87, 662		

137. Isobutyraldehyde

$(CH_3)_2CHCHO$ C_4H_8O

mw		72.108	887
bp		64.1°	4656: 2081, 1077, 3645, 1957, 4237
dt/dp		0.0396°	4656
eq 15	A	6.01614	4237
	B	1946.999	
	C	0.005297	
fp		−65.°	4656: 1957
d	20°	0.7891	4656: 2081, 1077, 803, 1957, 4237, 4795
	25	0.7836	
α	55°	0.00146	4884
n_D	20°	1.3727	4656: 2081, 1077, 803, 1957, 4237
	25	1.3698	
$\triangle H_f^{\circ}$	25°(l)	−34.45	4795
$\triangle H_c^{\circ}$	25°(l)	−546.71	4795: 4743, 4794, 2577
μ	20° in 28	2.58	1077
soly	in aq, 20°	9.1%w	4494: 1957
aq az	64.3°	93.3%w	2255: 2081, 1957
fl pt	O C	⟨20°	4884
uv			611, 1068, 901
ir			641
Raman			2663
nmr			2907
Beil	87, 671		

138. Benzaldehyde

C_6H_5CHO C_7H_6O

mw		106.125	887
bp	760	178.9°	3001: 4535, 1791
dt/dp	760	0.055°	3001
p	26.2°	1	4535: 1077, 2078
fp		−55.6°	4772: 4535, 4764
melts		−26.	4903
d	20°	1.0447	1077
	25	1.0434	1809
n_D	20°	1.5455	1077: 2078
	25	1.5428	1809
η	25°	1.321	1399
γ	15.4°	39.19	2981
	Ch II	Table VI	
$\triangle H_v$	ave	9.47	5034: 3016
	bp	10.2	4963: 3001
$\triangle H_c^o$	25°(l)	−843.1	2577
C_p	15 − 30°	40.93	2689: 3001, 3016
t_c		352°	1791: 5032
p_c		21.5	1791
pK_{BH^+}	25° in aq H_2SO_4	−7.10	5287: 1146
ϵ	20°	17.8	3477: 5027, 5034, 1381, 4752
μ	20°(l)	2.77	3056: 2383, 5287, 1809, 1077, 4178
soly	20° in aq	0.3%w	4494: 664
fl pt	TCC	148°F	88
uv			1347, 3418, 3833, 4555, 5069, 849, 1196
ir			1469, 1646, 4946, 2451, 2519, 4092, 1716
Raman			3732, 3300, 1169, 1176, 412
mass			11, 3194
nmr			4632, 954, 4512
Beil	622, 174		

139. Acrolein

$CH_2=CHCHO$ $\hspace{6cm}$ C_3H_4O

mw		56.065	887
bp		52.69°	4268: 1077, 4535, 4766, 3428
dt/dp		0.00355°	4268
p	25°	265	[4884] : 3428
eq 4	A	7.750	[4535] : 1077, 2078, 3428
	B	1584	
fp		− 86.95°	4268: 4535, 4766, 3428, 4811
d	20°	0.8389	3428: 1077
$d_4^t = 0.86205/(1 + \alpha\,t)$			[3428]
$\alpha\,(0 - 50°) = 0.001318 + 0.0000033t$			3428
n_D	15°	1.4048	4268: 1077
	20	1.4017	
dn/dt		0.00057	4884
$\triangle H_v$	bp	6.77	[3428] : 4268
$\triangle H_c^°$	25°(l)	− 390.7	2577
C_p	17 – 44°	28.65	[3428]
κ		1.55×10^{-7}	3428
μ	25° in 30	2.90	520: 1077
soly	in aq, 20°	20.8%w	2255: 684
	aq in, 20°	6.8%w	
aq az	52.4°	97.4%w	2255: 2616
fl pt	TOC	⟨0°F	4884: 3680
uv			91, 827, 3092, 3100, 611, 4634, 1047
ir			2810, 3766, 548
Raman			681
mass			90
nmr			4512, 1348
Beil	90, 725		

140. Crotonaldehyde

trans -- Crotonaldehyde

CH$_3$- C - H
$\|$
H - C - CHO

C$_4$H$_6$O

mw		70.092	887
bp	760	104.1°	4766: 1333, 1497
p	25°	38.	[4884]
fp		- 76.5°	4766: 1333
d	15.2°	0.8575	Beil E I 1, 379: 1497
	20	0.8516	[4884]
α		0.00123	4884
n$_D$	10°	1.44361	Beil E I 1, 379: 1497, 613, 1333, 3984
	20	1.4373	4884
dn/dt		0.000556	3984
$\triangle$H$_v$	av	8.62	[2609]
$\triangle$H$_c^\circ$	25°(l)	- 547.0	2577: 4580
C$_p$	20°	49.4	[4884]
μ	25°, in 30	3.50	1497: 2330, 520
soly	in aq 20°	15.6%w	4884
	aq in 20°	9.6%w	4884
aq az	84.0°	75.2%w	4884
fl pt	TCC	55°	88: 3680
uv			2072, 3100, 611, 947, 613
ir			2599
Raman			1870, 4117, 2573, 499, 2035
nmr			4632, 3807, 4512, 5176, 1348
Beil	90 , 728		

141. Acetone

CH$_3$COCH$_3$ $\hspace{10cm}$ C$_3$H$_6$O

mw		58.081	887
bp	760	56.29°	4657: 1372, 1491, 4762, 4784, 5307, 5350
dt/dp	760	0.0385°	4657: 1372, 4761, 4784, 5350
p	20°	181.72	3879: 1373, 4165, 4535, 5350
eq 5	A	7.23157	4659: 1372, 3706
	B	1277.03	
	C	237.23	
fp		−94.7°	4657: 1372, 1819, 3093, 3301, 3654, 4762
d	20°	0.78998	4657: 1372, 1491, 1717, 4706, 4761, 4784
	25	0.78440	778
n$_D$	20°	1.35868	4657: 1113, 1372, 1491, 3208, 3628, 4784
	25	1.35596	4829, 5195, 3935, 3424
dn/dt		0.00050	4784
η	15°	0.3371	4784: 3975, 1717, 2557, 2801, 2939
	25	0.3040	4119
	30	0.2954	4784
	Table II	Ch II	
γ	20°	23.32	2102: 22, 1113, 1491, 3621, 4808
	30	22.01	
	Table VII	Ch II	
△ H$_v$	27.26°	7.372	3706: 1048, 3208, 3331, 3654, 5034
	bp	6.952	
△ H$_m$		1.360	3654: 2549
△ H$_f^\circ$	25°(l)	−58.99	5358: 3706, 835
△ H$_c$	25°(l)	−427.77	3321: 4743, 2577, 5358
C$_p^\circ$	65.1°	19.48	3706
	Table XIV	Ch II	
C$_p$	30°	30.87	5195: 515, 1048, 3654, 3745, 4829, 4484
t$_c$		235.0°	2771a: 3331, 2650, 2145, 4658
p$_c$		46.4	2771a: 3331, 5026, 5034, 2650, 2145
d$_c$		0.278	4658: 2650, 2145
v$_c$		0.209	4658: 2650
K$_f$		2.40°	2349: 3599
K$_B$		1.71°	483, 2193: 919, 4864, 5029, 3274, 1803
pK$_{BH+}$	in aq H$_2$SO$_4$	−7.2	894

141. Acetone

(Continued)

κ	25°	4.9 x 10⁻⁹	4119: 2321, 2557, 3034, 5026, 5027
ϵ	25°	20.70	3477: 1888, 2747, 3034, 3389, 5275, 3309
μ	20°(l)	2.69	3056: 3209, 4127, 4178, 5243, 2231, 833
soly	25°	inf	2609
aq az		none	2254
fl pt	TCC	0°F	88
uv			91, 3785, 2812, 1235, 2805, 4505, 3156
ir			4630, 89, 1469, 4107, 1648, 3766, 4547
Raman			717, 3178, 2540, 499, 968, 1009, 4346
mass			90, 3997, 4257, 61, 1344, 3194, 4715
nmr			4512, 3025, 966
Beil	83, 635		

142. 2-Butanone

Methylethylketone, mek

$CH_3CH_2COCH_3$ C_4H_8O

mw		72.108	887
bp	760	79.64°	4657: 1372, 1796, 4265, 4762, 4784, 1043
dt/dp	760	0.0410°	4657: 1372, 4748, 4784, 1043, 3001
p	25°	90.6	4269: 1373, 3879
eq 5	A	7.20867	4659: 4269, 1043
	B	1368.21	
	C	236.50	
fp		−86.69°	4657, 1043: 239, 1947, 4762, 4784, 4340
d	20°	0.8049	4657: 1113, 1372, 1717, 1784, 3208, 4269
	25	0.7997	4761, 4784, 1134
α	20°	0.00076	3160

142. 2-Butanone

(Continued)

n_D	$20°$	1.3788	4657: 717, 941, 1113, 1372, 3208, 4784
	25	1.3764	3424
dn/dt		0.00048	4784
η	$15°$	0.423	4784: 1404, 1717, 1134
	30	0.365	
	Table II	Ch II	
γ	$0°$	26.9	3160: 1113
	24.8	23.97	3621
	Table VII	Ch II	
$\triangle H_v$	$20°$	7.643	4265: 3208, 3001, 3534
	bp	7.570	1043
$\triangle H_m$		2.017	4340: 4769, 3655, 239
$\triangle H_f^°$	$25°$ (l)	− 66.68	3658: 3534, 835, 4340
$\triangle H_c^°$	$25°$ (l)	− 582.80	3658: 4340, 2577
$C_p^°$	$25°$	24.59	4340: 3534
C_p	$25°$	37.98	4340: 515, 2689, 239, 3001, 3655
	Table XI	Ch II	
t_c		$262.4°$	4658: 2650, 1503
p_c		41.0	4658: 2650, 1503
d_c		0.270	4658: 2650
v_c		0.267	4658: 2650
K_B		$2.28°$	2193
pK_{BH+}	in aq H_2SO_4	− 7.2	894
κ		3.6×10^{-9}	2308: 2349, 3449
ϵ	$20°$	18.51	3477: 3389, 5026, 5034, 3309, 1381, 4752
μ	$25°$ in **30**	2.76	1498: 5243, 1503, 3534
soly	in aq, $20°$	24.00%w	2417: 529, 1784, 4269
	aq in, $20°$	10.00%w	2417
aq az	$73.41°$	88.73%w	4269: 2254, 2255
fl pt	TCC	$30°F$	88
uv			91, 573, 597, 3785, 4113, 1235, 3424
ir			89, 1249, 1648, 2865, 4547, 4738, 2837
Raman			597, 717, 4633, 4343, 4822, 2663, 2837
mass			90, 4257, 4850, 61, 913, 4715
nmr			4632, 4512, 3729
Beil	87, 666		

143. 3–Pentanone

$CH_3CH_2COCH_2CH_3$ $C_5H_{10}O$

mw		86.135	887
bp	760	101.99°	4657: 1372, 1113, 3621, 3988, 4767, 1043
dt/dp	760	0.0408°	4657: 1372, 3001, 1043
p	27.9°	20.	3160: 1373, 3988, 4535
eq 5	A	7.30218	4659: 1372, 3988, 1043
	B	1481.17	
	C	233.01	
fp		– 38.97°	4657, 3853: 1372, 4535, 4767, 239
d	20°	0.81430	4657: 1113, 1372, 1717, 3621, 4779
	25	0.80945	
n_D	20°	1.39227	4657: 717, 1113, 1372, 3424
	25	1.39002	
η	15°	0.493	4772: 1717
	25	0.442	4711
	30	0.423	
	Table II	Ch II	
γ	20°	25.26	4772: 1113, 3621
	30	24.37	
	Table VII	Ch II	
$\triangle H_v$	bp	8.060	1043: 4611
$\triangle H_m$		2.771	239
$\triangle H_f^\circ$	25° (g)	– 61.82	835
$\triangle H_c^\wedge$	25° (l)	– 735.6	2577: 4580
C_p	25°	45.63	[239] : 3001
t_c		287.8°	4658, 2650
p_c		36.9	4658, 2650
d_c		0.256	4658, 2650
v_c		0.336	4658, 2650
ϵ	20°	17.00	3477: 3400, 3309, 1381
μ	25° in **232**	2.82	291: 1862, 3400
soly	in aq, 20°	3.4%w	3160: 1911, 664
	aq in, 20°	2.6%w	
aq az	82.9°	86%w	2254
uv			573, 597, 3785, 1235, 3424
ir			89, 1648, 2050, 2865, 4547, 4738, 2438
Raman			597, 717, 968, 4346, 4343, 2663, 2438
mass			90, 4257, 1769, 1344
nmr			2461, 4512, 1923
Beil	87, 679		

144. Cyclohexanone

$CH_2CH_2CH_2CH_2CH_2CO$ (with bracket underneath) $C_6H_{10}O$

mw		98.146	887
bp	760	155.65°	4781: 326, 1140, 4535, 1791
dt/dp	760	0.0048°	4781
p	25°	4.8	59: 3879, 4535
eq 5	A	7.4705	[4535]
1 – 155°	B	1832.2	
	C	244.2	
fp		– 32.1°	4781: 1140, 4770
d	15°	0.95099	4781: 326, 5007
	30	0.93761	
dd/dt		0.00089	[4781]
n_D	15°	1.45203	4781: 326, 4965, 3424
	20	1.45097	1140
η	15°	2.453	4781: 4711
	30	1.803	
γ	20°	34.50	4781: 3577
$\triangle H_v$	29.21°	10.72	3160
	bp	9.62	
$\triangle H_f^\circ$	25° (l)	– 64.84	5358: 3875
$\triangle H_c^\circ$	25° (l)	– 841.04	5358: 4230, 3875
C_p	30.8°	47.89	3745
t_c		356°	1791
p_c		38.	1791
pK_{BH+}	in aq H_2SO_4	– 6.8	894: 503
κ	25°	5×10^{-18}	3160
ϵ	20°	18.3	3477: 1140, 3400, 5189
μ	25° in **30**	3.01	520: 1336, 1984, 3400, 5189, 3971, 520
soly	in aq, 20°	2.3%w	1339
	aq in, 20°	8.0%w	
aq az	96.3°	45%w	2255
fl pt	TCC	147°F	88, 3680
uv			597, 1336, 1604, 4136, 947, 4634, 3424
ir			1648, 4076, 4113, 1148, 408, 1975, 585
Raman			597, 968, 4098
mass			90, 4257, 559, 61
nmr			4512
Beil	**612**, 8		

145. 4–Methyl–2–pentanone

$(CH_3)_2CHCH_2COCH_3$

$C_6H_{12}O$

mw		100.162	887
bp	760	116.5°	4657: 1113, 1784, 4265, 4763, 2503, 1675
dt/dp	760	0.0467°	4657: 4269
p	25°	20.0	4269: 3988, 4535, 1675
eq 5	A	6.9727	4659: 3988, 4269, 1675
	B	1190.69	
	C	195.45	
fp		− 84.°	4657: 4535, 4763
d	20°	0.8008	4657, 4269: 1113, 1784, 4265, 3988, 2503
	25	0.7961	1675
α	20 – 30°	0.00116	3160
n_D	20°	1.3957	4657, 4269: 717, 1113, 2503, 1675
	25	1.3933	
η	25°	0.542	2503
γ	20°	23.64	1113: 1675
	23.7	23.29	
	62.1	19.62	
$\triangle H_v$	20°	10.00	[1675]: 4265, 4269
	bp	8.50	
$\triangle H_c^°$	25° (l)	− 735.6	2577: 4610
C_p	20°	46.	4265
t_c		298.3°	4658, 2650
p_c		32.3	4658, 2650
κ	35°	⟨ 5.2 x 10⁻⁸	2979
ϵ	20°	13.11	3477: 3400, 3309, 2979
soly	in aq, 25°	1.7%w	4269, 1784: 1911
	aq in, 25°	1.9%w	
aq az	87.9°	75.7%	2254
fl pt	TCC	73°F	88
uv			91, 1648, 2812
ir			2861, 2837
Raman			717, 968, 2663, 2837
mass			1769
nmr			4632, 4512, 5176
Beil	87, 691		

146. d–Camphor

$$\begin{array}{c}
\text{CH}_3 \\
| \\
\text{H}_2\text{C} \longrightarrow \text{C} \longrightarrow \text{CO} \\
| \\
\text{H}_3\text{C} - \text{C} - \text{CH}_3 \\
| \\
\text{H}_2\text{C} \longrightarrow \underset{\text{H}}{\text{C}} \longrightarrow \text{CH}_2
\end{array}$$

$C_{10}H_{16}O$

mw		152.238	887
bp	760	207.42°	4772: 1928, 1929, 2608, 2814, 5182
p	24.2°(s)	0.39	4773: 1195, 1928, 4535, 5182
eq 4 0–25°	A B	11.352 3439.2	5344
fp		178.75°	2905: 1417, 1928, 1929, 2608
tr pt	385.8 torr	180.1°	4772
d	0° 20	1.0000 0.9920	1929: 2608, 3454
$\triangle H_v$	bp	14.22	2608: 5182
$\triangle H_m$		1.635	1620: 2608, 4601
$\triangle H_c^\circ$	25°(l)	−1413.2	2577
$[\alpha]_D^{20}$	in 59	+44.22°	2609: 1929, 2608, 4335
K_f		37.7°	1620: 1417, 2608, 3266, 5182
K_B		5.611°	483
ϵ	20°	11.35	1002
μ	25°in 30	3.10	2882: 2382
soly	in aq	ca. 0.01%w	2609, 2608: 1929
uv			1336, 1073
ir			4107, 565
Raman			228, 652, 4750, 2385
mass			4588, 5109
nmr			3420, 4512, 2362
Beil	618, 101		

147. Acetophenone

$C_6H_5COCH_3$ C_8H_8O

mw		120.152	887
bp	760	202.0°	1899: 4535, 4780, 4969
dt/dp	760	0.055°	4780
p	25°	0.37	1369: 4535
eq 5	A	7.15738	1369
	B	1723.46	
	C	201.	
fp		19.62°	4820: 2984, 4780, 5043
d	20°	1.02810	1369: 1809, 4540, 4780, 4820, 4969, 5032
	25	1.02382	2984
n_D	15°	1.53631	4772: 1809, 4780, 4969
	20	1.53423	
dn/dt		0.00045	[4772]
η	15°	2.015	4780
	25.50	1.642	1787
	30	1.511	4780
γ	15°	40.09	4780: 4820, 4969, 5032, 5034
	30	38.21	
	Table VII, VIII Ch II		
$\triangle H_v$	25°	12.76	[1369] : 3015
	bp	9.275	3018: 5032
$\triangle H_f^\circ$	25°(l)	– 34.06	1049
$\triangle H_c^\circ$	25°(l)	– 992.9	2577: 1049
C_p	30°	54.42	3745: 3018, 3015
t_c		456°	5032
K_B		5.65°	1705: 1531
pK_{BH^+}	in aq H_2SO_4	6.15	4503
κ	25°	3.1×10^{-9}	2308: 2557, 2719, 3449, 4540, 3389
ϵ	25°	17.39	4540: 3477, 3389, 5034, 550, 1381
μ	25° in **30**	2.96	520: 1809, 2875, 5287, 886
soly	sl sol		4519
aq az	99.1°	18.5%w	2255
fl pt	TCC	221°F	3680
uv			896, 1347, 3181, 4905, 4461, 1196
ir			969, 3399, 4738, 2451, 2519, 408, 1716
Raman			2660, 3399, 2574, 1170, 499, 3300, 1169
mass			4257, 828
nmr			4632, 3146
Beil	**639** , 271		

148. Formic acid

HCOOH CH_2O_2

mw		46.026	887
bp	760	$100.56°$	4660: 1372, 2447, 4535, 4778, 1145
dt/dp	760	$0.0442°$	4660, 1372: 1249, 4778
p	$25°$	43.1	4516: 602, 4535, 5263, 1076
eq 5	A	7.37790	4661: 1372
	B	1563.28	
	C	247.06	
fp		$8.27°$	5106: 434, 1372, 2453, 4778, 5263, 4516
d	$15°$	1.22647	4772: 1372, 1717, 4778, 4834, 4864, 4868
	25	1.21405	5032
dd/dt		0.00124	4778
n_D	$20°$	1.37140	1372: 4858
	25	1.36938	
dn/dt		0.00038	4778
η	$25°$	1.966	4778: 1404, 1717, 4748, 4868
	30	1.443	
	Table II	Ch II	
γ	$20°$	37.58	4778: 5032, 5034
	30	36.48	
	Table VII, VIII Ch II		
$\triangle H_v$	$25°$	4.754	4516: 1076, 5026, 5032, 5082, 1532
	bp	5.540	781
$\triangle H_m$		3.031	4516
$\triangle H_f^°$	$25°$ (l)	-101.52	4338: 1878, 5358
$\triangle H_c^°$	$25°$ (l)	-60.86	4338: 4743, 2577, 5358
$C_p^°$	$25°$	10.807	1878: 5082
C_p	$26.64°$	23.68	4516: 4144
	Table XI	Ch II	
K_f		$2.77°$	4864: 3911
K_B		$2.4°$	479: 481, 814, 1145
K_i	$25°$	1.772×10^{-4}	2031: 2033, 1201
pK_s		6.2	796
κ		6.08×10^{-5}	5106: 5026, 4868

148. Formic acid

(Continued)

ϵ	$16°$	58.5	3477: 4868, 5026, 5034, 1381
μ	$30°$ in **30**	1.82	3797: 5242, 1490, 5316, 2590
soly	in aq	inf	2609
aq az	$107.65°$	74.5%w	2255: 2254, 4869
uv			573, 1013, 2012, 4546
ir			89, 1075, 2129, 2605, 4152, 4730, 3331
Raman			4633, 447, 448, 659, 3471, 1172, 1168
mass			90, 725, 2013, 61, 4030, 4715
nmr			903, 4512
Beil	**155**, 8		

149. Acetic acid

CH_3COOH $C_2H_4O_2$

mw		60.053	887
bp	760	$117.90°$	4660: 1372, 2813, 4523, 4778, 4975
dt/dp	760	$0.0431°$	4660: 1372, 4778
p	$25.0°$	15.43	2265: 1373, 3071, 4165, 4535, 5264
eq 5	A	7.42728	4661: 1372, 2790, 3814, 3070
	B	1558.03	
	C	224.79	
fp		$16.66°$	4660: 1978, 2154, 2418, 3654, 5264, 3099
d	$20°$	1.04926	4660: 1372, 1717, 2440, 4864, 4975, 4778
	25	1.04366	3476
α	$55°$	0.00111	4881
n_D	$20°$	1.3719	4660: 1372, 3864, 4422, 4975, 5284, 3476
	25	1.3698	

149. Acetic acid

(Continued)

dn/dt		0.00038	4778
η	15°	1.314	4778: 1404, 1717, 2440, 2557
	30	1.040	
	Table II	Ch II	
γ	20°	27.42	2102: 3341, 4778, 4975
	30	26.34	
$\triangle H_v$	25°	5.505	3071: 3331, 5032, 4989, 1532, 3015, 4962
	bp	5.825	781
$\triangle H_m$		2.803	3654: 4601
$\triangle H_f^o$	25° (l)	− 115.7	5112, 5358
$\triangle H_c^o$	25° (l)	− 209.0	5112, 4743, 2577, 5358
C_p	21.5°	29.84	3654: 4143, 4144, 4165, 5301, 3015
t_c		321.30°	2771a, 4662: 5301, 5303, 3924, 1100, 2602
p_c		57.10	4662, 1371: 2602, 3331
d_c		0.3506	1371, 4662
v_c		0.1713	[1371], 4662: 2602, 1100, 5303
K_f		3.90°	4864: 3911
K_B		2.530°	4864: 479, 488, 2152, 475, 3291
K_i	25°	1.754×10^{-5}	2028, 2029: 2033, 1314, 1201, 3099, 5027
pK_s		14.45	796
κ	25°	6×10^{-9}	3476: 3099, 2152, 2557, 3876, 5027, 1790
ϵ	20°	6.15	3477: 4422, 5026, 4786, 4804, 3763, 1381
μ	30° in **30**	1.68	3797: 2176, 5034, 5242, 2497
soly	in aq	inf	2609
aq az		none	2254
fl pt	TCC	104°F	3680
uv			574, 573, 1013, 2012, 2439, 3785, 1196
ir			868, 3766, 4107, 89, 2239, 1093, 2605
Raman			446, 447, 448, 540, 2726, 4938, 1172
mass			90, 2200, 2013, 61, 4715
nmr			2339, 2304, 937, 903, 4289, 4512, 3126
Beil	**158**, 96		

150. Propionic acid

CH_3CH_2COOH $C_3H_6O_2$

mw		74.080	887
bp	760	140.83°	4660: 591, 1372, 1948, 3017, 4767, 4975
dt/dp	760	0.0424°	4660: 1249, 1372, 3017, 4778
p	28.0°	5	4535: 1373
eq 5	A	7.54760	4661: 1372
	B	1617.06	
	C	205.67	
fp		− 20.7°	4660: 501, 1372, 4535, 4767, 4774, 4778
d	20°	0.9934	4660: 589, 1372, 1717, 1948, 3835, 4156
	25	0.9880	4778, 4834, 4975
dd/dt		0.00106	4778
α	20°	0.00110	4881
n_D	20°	1.3865	4660: 1372, 1948, 4834, 4975, 5089, 5139
	25	1.3843	
dn/dt		0.00038	4778
η	15°	1.175	4778: 3975, 1717, 1948
	30	0.958	
	Table II	Ch II	
γ	20°	26.70	2102, 4778: 4975
	30	25.71	
	Table VIII	Ch II	
$\triangle H_v$	25°	13.120	[1371]
	bp	7.716	3208: 781, 3017
$\triangle H_m$		1.800	4769: 3196
$\triangle H_f^\circ$	25° (l)	− 122.12	5358
$\triangle H_c^\circ$	25° (l)	− 364.7	2577: 4743, 4580, 5358
C_p	20°	36.88	4144: 3017, 4143, 3196
t_c		339.5°	[1371]: 4662, 1100, 5077, 1948
p_c		53.0	[1371]: 4662
d_c		0.315	1371: 4662
v_c		0.235	[1371]: 4662, 1100
K_B		3.51°	479
K_i	25°	1.336 x 10^{-5}	2030: 501, 1314, 1372, 4946, 698, 3614
κ	25	$\langle 1 \times 10^{-9}$	2557
ϵ	40°	3.435	3763: 3477, 758, 1381, 4752
μ	30° in 30	1.68	3797: 5242, 2887

150. Propionic acid

(Continued)

soly	in aq	inf	4881
aq az	99.9°	17.7%w	4881: 2254, 1180
fl pt	OC	136°F	4881
uv			574, 573, 2129, 3725
ir			1587, 89, 2050, 4152, 5060, 2239, 1093
Raman			447, 2475, 3471, 4938, 1175, 1168, 1176
mass			90, 2013, 61, 4715
nmr			4632, 937, 903, 4512, 249
Beil	**162**, 234		

151. Butyric acid

$CH_3CH_2CH_2COOH$ $C_4H_8O_2$

mw		88.107	887
bp	760	163.27°	4605: 1314, 1372, 2813, 3452, 4422, 4779
dt/dp	760	0.0426°	4605: 1249, 1372, 4779
p	25°	0.72	1371: 1373, 4535
eq 5	A	7.78434	4661: 1372
	B	1794.04	
	C	202.60	
fp		−5.2°	4605: 501, 1372, 4767, 4779
d	20°	0.9582	4605: 589, 1717, 2440, 3452, 3835, 4422
	25	0.9532	4779, 4859, 2336, 1372
α	20°	0.00103	4881
n_D	20°	1.3980	4605: 1372, 3452, 4859, 4975, 5089, 5139
	25	1.3958	
dn/dt		0.000382	[4772]

151. Butyric acid

(Continued)

η	15° 30	1.814 1.385	4779: 1404, 1717, 2440, 3452
	Table II, III, V Ch II		
γ	20° 30	26.74 25.57	4779: 3452, 4975
	Table VIII Ch II		
$\triangle H_v$	25° bp	14.47 10.040	[1371]: 3647 781
$\triangle H_m$		2.50	4601
$\triangle H_f^o$	25° (l)	– 127.9	5358
$\triangle H_c^o$	25° (l)	– 521.0	2577: 4580, 5358
C_p	20°	41.62	4144: 3647, 4143
t_c		355.0°	1371, 4662: 783, 4406, 5077, 1100
p_c		52.	4662: 1371
d_c		0.302	1371, 4662
v_c		0.2917	[1371], 4662: 1100
K_B		3.94°	479
K_i	25°	1.508×10^{-5}	501: 1314, 1313, 2813, 698, 3360, 5222
ϵ	20°	2.97	3477: 4422, 4804, 3763, 1381, 4752
μ	30° in 30	1.65	3797: 5242, 2887
soly	in aq	inf	4881
aq az	94.4°	18.5%w	2254
fl pt	TCC	170°F	3680
uv			574, 573, 2012, 3785, 3725
ir			4630, 1641, 4113, 89, 2239, 4306, 1093
Raman			447, 2099, 4909, 4938, 1176
mass			2013, 61, 4715
nmr			937, 4289, 4512
Beil	162, 264		

152. Isobutyric acid

$(CH_3)_2CHCOOH$ $C_4H_8O_2$

		88.107	887
mw		88.107	887
bp	760	154.70°	4772, 4775: 3150, 4977
dt/dp	760	0.032°	4761
p	25°	1.39	1371
fp		-46.1°	4772, 4775: 3150
d	20°	0.96815	1371: 4775, 3150, 4975, 4977
	25	0.94288	
n_D	15°	1.39525	Beil E III 2, 638: 3150, 4775
	20	1.39300	4975
η	25°	1.213	[3159]: 4711
	30	1.126	4772
γ	20°	25.55	1371
	30	25.13	
	Table VIII	Ch II	
$\triangle H_v$	25°	13.64	[1371]
	bp	10.620	4962
$\triangle H_m$		1.200	4769
$\triangle H_c^\circ$	25°(l)	-521.0	2577
t_c		336.$^\circ$	4662
p_c		40.	4662
d_c		0.304	4662
v_c		0.2898	4662
K_i	25°	1.38×10^{-5}	1314: 1622, 3614, 1379, 5141, 5179
ϵ	40°	2.73	3477, 3763: 4804, 1381
μ	25°(l)	1.08	3742: 2724
soly	in aq, 20$^\circ_\circ$	22.8%w	2417: 1202
	aq in, 20	44.6%w	
aq az	98.8°	28.2%w	2255
fl pt		170°F	3681
ir			2239, 4306, 1463
Raman			2665, 357
mass			2013, 61
nmr			937, 903, 4289, 3126
Beil	**162**, 288		

153. Valeric acid

$CH_3CH_2CH_2CH_2COOH$ $C_5H_{10}O_2$

mw		102.134	887
bp	760	185.5°	4660: 3452, 4763, 4779, 4975, 2000
dt/dp	760	0.045°	4660: 1249, 4779
p	25°	0.14	2799: 589, 1314, 4535
eq 5	A	7.6569	4661
	B	1777.2	
	C	186.6	
fp		-33.67°	3072: 4535, 4460, 4779
d	20°	0.9390	4660: 589, 1717, 2440, 3452, 4779, 4975
	25	0.9345	
α	55°	0.00101	4881
n_D	20°	1.4080	4660: 3452, 4975, 5089, 2000
	25	1.4060	
η	15°	2.359	4779: 1404, 1717, 2440, 3452
	30	1.774	
	Table V	Ch II	
γ	15°	27.83	4779: 3452, 4975
	30	26.35	
$\triangle H_v$	25°	16.56	[1371]
	bp	10.53	781
$\triangle H_m$		3.385	3072: 4769
$\triangle H_f^\circ$	25°(l)	-131	2000: 5358
$\triangle H_c^\circ$	25°(l)	-677.3	2577: 4580, 2000, 5358
C_p	25°	50.28	[3072]: 4143, 4144
t_c		378.°	4662: 783, 4406, 5077, 1371
p_c		46.1	[1371]
d_c		0.352	1371
v_c		0.290	[1371]
A	eq 72	0.02497	1371
K_i	25°	1.38×10^{-5}	1313: 1379, 3614, 1622, 1314
ϵ	20°	2.66	3477: 1381, 4752

153. Valeric acid

(Continued)

soly	in aq, 20°	2.4%w	4881: 4494
	aq in, 20°	13.0%w	
aq az	99.8°	11%w	4881
fl pt	OC	205°F	4881
uv			3861
ir			1075, 2129, 90, 2239, 4306, 1093
Raman			2099
mass			2013
nmr			4512
Beil	162, 299		

154. Isovaleric acid

$(CH_3)_2CHCH_2COOH$ $C_5H_{10}O_2$

mw		102.134	887
bp	760	176.50°	4779: 1314, 4535, 4762, 4975, 2000
dt/dp	760	0.046°	4779
p	34.5°	1	4535: 1314
fp		−29.3°	4772, 4770: 4762, 4763, 4779, 4535
d	15°	0.93080	4779: 3835, 4834, 4975
	30	0.91708	
n_D	15°	1.4064	4834: 4975
	20	1.4063	
	25	1.4022	2000
η	15°	2.731	4779
	30	1.967	

154. Isovaleric acid

(Continued)

γ	15°	25.78	4779: 4975
	30	24.45	
	Table VIII	Ch II	
$\triangle H_v$	bp	10.32	781
$\triangle H_m$		1.750	4769
$\triangle H_f^o$	25°(l)	− 134	2000
$\triangle H_c^o$	25°(l)	− 678	2000
t_c		361.°	4662: 783
K_i	25°	1.67×10^{-5}	1314: 1622, 1124, 1379, 5179
κ	0 – 80°	$\langle 4 \times 10^{-13}$	3886
ϵ	20°	2.64	3477: 1381
μ	25° in 30	0.63	5242: 3742
soly	in aq 21°	4.1%w	Beil E III 2, 691: 3150
aq az	99.5°	18.4%w	2254
ir			391
Raman			447, 448, 2099
nmr			4289, 4512
Beil	162, 309		

155. Hexanoic acid

caproic acid

$CH_3(CH_2)_3CH_2COOH$ $C_6H_{12}O_2$

mw		116.161	887
bp	760	205.7°	4660: 3450, 3806, 4535, 4324, 4576
dt/dp	760	0.045°	4660
p	61.7°	1	4576

155. Hexanoic acid

(Continued)

eq 5	A B C	7.6245 1790.2 171.7	4661: 639
fp		$-3.95°$	1249: 4576
d	20° 25 30	0.9272 0.9230 0.91832	4660: 1717, 2440, 3452, 4834, 4324, 4576 3835: 1342 4324
n_D	20° 25	1.4168 1.4148	4660: 761, 3452, 4834, 4324, 4035, 639 4576
dn/dt		0.00039	4324, 1342
η	15° 30	3.525 2.511	4324: 1404, 1717, 3452, 2440
	Table V	Ch II	
γ	20° 25	28.05 27.55	3452: 3835, 4576
$\triangle H_v$	94° 190	15.45 13.1	1103 [3150]
$\triangle H_m$		3.60	1712
$\triangle H_c^°$	25° (l)	-833.6	2577: 4580
C_p	$0-23°$	59.30	[4576] : 1712
K_i	25° in H_2O	1.32×10^{-5}	1314: 3614, 1622
κ	mp bp	7×10^{-13} 2.1×10^{-9}	2869 2869
$\ln \kappa = -13.007 + 0.0274\, t - 0.000042\, t^2$			2869
ϵ	71°	2.63	3477
μ	25° (l)	1.13	3056: 3742
soly	in aq, 20°	0.958%w	3886
aq az	99.8°	7.9%w	2255
fl pt		215°F	3681
uv			3887
ir			3571, 1093, 2363
Raman			2099
nmr			4512
Beil	**162**, 321		

156. Octanoic acid

caprylic acid

$CH_3(CH_2)_5CH_2COOH$ $C_8H_{16}O_2$

mw		144.216	887
bp	760	239.9°	4660: 1248, 3452, 3806, 4535
dt/dp	760	0.49°	4660
p	92°	1	3681
eq 5	A	7.3486	4661: 4034
	B	1723.8	
	C	145.9	
fp		16.51°	2218: 4576
d	20°	0.9106	4660: 4576
	25	0.9066	
	80	0.8615	1342
n_D	20°	1.4280	4660: 4035, 4034, 4576
	25	1.4261	
dn/dt		0.00037	4772: 1342
η	20°	5.828	1248: 2440, 846, 1404, 1717, 3452, 4576
	30	4.690	
	Table V	Ch II	
γ	20°	29.2	3452: 3835, 4576
	25	28.7	
$\triangle H_v$	134°	16.73	1126
	bp	13.97	[4576]
$\triangle H_m$		5.11	1713: 4601
C_p	16–36° ca	74	1712: 4576
K_i	25° in H_2O	1.27×10^{-5}	1314: 1622, 5179
κ	mp to 80°	$\langle 3.7 \times 10^{-13}$	3886
ϵ	20°	2.45	3477
μ	25° (l)	1.15	3056: 3742
soly	in aq, 25°	0.0798%w	1452: 1069, 3886, 3150
uv			3785, 4078
ir			3571, 2239, 1093, 2363
Raman			540
nmr			4632
Beil	**162**, 347		

157. Acrylic acid

$CH_2=CHCOOH$ $C_3H_4O_2$

mw		72.064	887
bp	760	141.2°	4881: 3429, 3430, 4875
dt/dp	760	0.042°	4881
p	20°	7.76	4879: 3430, 3429, 3923
eq 4	A	7.6021	[1230]
	B	1954.6	
fp		13.5°	4022: 3430, 3429
d	12°(l)	1.0600	3429: 3430, 1, 3923
	20	1.0511	
α	20°	0.00079	4881
n_D	20°	1.4224	3429: 1, 3923
	25	1.4185	4021
η	25°	1.10 cs	4021
γ	30°	28.1	4881
$\triangle H_v$	50°	11.1	[3429]
	136	8.9	4879
$\triangle H_m$		2.66	4879
$\triangle H_f^\circ$	25°(l)	−91.77	5358
$\triangle H_c^\circ$	25°(l)	−325.6	2577: 5358
$\triangle H_p$		18.49	1505
K_i	25°in H_2O	5.56 x 10^{-5}	1313: 3429, 3614, 720
soly	in aq	inf	4881: 2609
aq az		none	2255
fl pt	COC	155°F	4022
uv			2110
ir			1091, 1587, 1549
Raman			680, 2678, 1549
mass			2013
nmr			3935, 932, 1425, 267, 801, 4512
Beil	163, 397		

158. Crotonic acid

$CH_3CH=CHCOOH$ $C_4H_6O_2$

mw		86.091	887
bp	760	185.0°	4519: 4875
p	30°	0.18	4875: 4519
eq 4	A	**9.08047**	4519
	B	2845.6842	
fp		71.4°	4122: 4879
d	72°	0.973	4879
	77	0.9604	Beil **E III 2**, 1255
n_D	77°	1.4249	Beil **E III 2**, 1255
	80	1.4228	4519, 4879
$\triangle H_m$		3.102	[4879]
$\triangle H_c^\circ$	25°(l)	481.9	2577
C_p	80 – 95°	42.6	2609
K_i	25° in H_2O	1.975×10^{-5}	4122: 1313, 3614, 720, 5222
μ	30° in **30**	2.13	3186
soly	25° in aq	94 g/l	4519
aq az	99.9°	2.2%w	2255
uv			2072, 4078, 3365
ir			1587
Raman			2452
nmr			1259, 4512, 3171, 1047
Bcil	**163**, 408		

159. Methacrylic acid

$CH_2=C(CH_3)COOH$ $C_4H_6O_2$

mw		86.091	887
bp		160.5°	4875: 1684
p	25.5°	1	4022
eq 4	A	9.13473	[4875]
	B	2712.325	
fp		15.0°	4021
d	20°	1.0153	4875: 1684
n_D	20°	1.4314	4875: 5148
	25	1.4288	4021
η	25°	1.30 cs	4021
γ	25°	26.5	4022
$\triangle H_p$		15.84	1505
K_i		3.72×10^{-5}	4022: 2836
μ	0° in 30	1.65	3056: 2724
soly	in aq, 20°°	8.9%w	[4875]
	aq in, 20	28.5%w	
aq az	99.3°	23.1%w	1669
fl pt	COC	170°F	4021
nmr			4438
Beil	**163**, 421		

160. Oleic acid

$CH_3(CH_2)_7CH=CH(CH_2)_6CH_2COOH$ $C_{18}H_{34}O_2$

mw		282.470	887
bp	760	360.0° (d)	4535
p	176.5°	1	4535: 780
	223.0	10	
	264.	50	2737
	286.0	100	4535
	309.8	200	
	334.7	400	
fp		13.38°	2216
mp	α–form	13.38°	2216: 551, 4535, 3152, 4391, 780
	β–form	16.30	
d	15°	0.8939	2543: 3643, 4834
	25	0.8870	
	90	0.8429	
dd/dt	30 – 90°	0.00068	
n_D	20°	1.4599	2216: 3643, 3152, 4391, 780
	35	1.4544	1125
dn/dt		0.000354	[1125]
η	20°	38.80	2543: 846
	25	27.64	
	60	9.41	
	80	4.85	
γ	20°	32.8	42: 4576, 2023
	90	27.94	4231
	180	21.6	42
$\triangle H_v$	bp	16.10	2867: 2868
$\triangle H_v = (26660 + 1.75 \times 1.987T - 0.0304T^2)(1 - p/p_c)$			
$\triangle F_c$	25° (l)	- 2663.9	2543: 1475, 2577, 2544, 2542
C_p	50°	138.	[4576]: 3258, 2868
	100	155.	
	150	180.	
p_c		30.	3150
iodine no	calc'd	89.87	1125: 780
pK_a	25° in aq EtOH	5.02	5146
κ	mp	3 x 10^{-13}	2869, 4576
κ	bp	2.8 x 10^{-9}	
$\ln\kappa = -12.843 + 0.0275t - 0.000042t^2$			
ϵ	20°	2.46	3477: 3643
μ	20° in **232**	1.18	3848: 4492, 4493, 4983, 4984

160. Oleic acid

(Continued)

soly	in aq	ins	1929
uv			410, 4078, 2120, 3861
ir			2437, 4334, 4303, 3570, 35, 2239, 4630
Raman			5312, 4910
nmr			2866, 3856
Beil	**163**, 463		

161. Acetic Anhydride

$(CH_3CO)_2O$ $C_4H_6O_3$

mw		102.091	887
bp	760	140.0°	4778: 2608, 2931, 4535, 4764, 3070, 5266
dt/dp	760	0.044°	4778
p	25°	5.1	[2444] : 2608, 4535
eq 5	A	7.12165	3070
	B	1427.77	
	C	198.037	
fp		−73.1°	4778, 2608: 2418, 4535, 4764
d	15°	1.08712	4778: 2608, 2931, 3476
	30	1.06911	
dd/dt		0.00121	2960
α	20°	0.00112	4881
n_D	15°	1.39299	4778: 5266, 3476
	20	1.3904	2608
dn/dt		0.00041	4778
η	15°	0.971	4778: 2608, 2931, 5266
	30	0.783	
	Table II	Ch II	

161. Acetic Anhydride

(Continued)

γ	20°	32.56	2931: 2102, 2608, 4778
	25	31.90	
$\triangle H_v$	bp	9.13	[4881] : 2608, 5034, 4963, 2785, 2349
$\triangle H_f^\circ$	$25^\circ(l)$	-149.16	5358
$\triangle H_c^\circ$	$25^\circ(l)$	-426.8	2577: 4743, 5358
C_p	30°	45.77	3745: 2608, 2785
t_c		296°	2608
p_c		46.2	2771a: 2608
K_B		3.53°	488
κ	25°	$5. \times 10^{-9}$	3476: 1978, 2608, 5027, 4622
ϵ	19°	20.7	3477: 2608, 5026, 5034, 4622
μ	25° in 312	2.82	3763: 5318
fl pt	TCC	$127^\circ F$	3680
uv			91, 2439
ir			408, 1072
Raman			4628, 2674
nmr			4632
Beil	159, 166		

162. Propionic Anhydride

$(CH_3CH_2CO)_2O$ $C_6H_{10}O_3$

mw		130.145	887
bp	760	169.0°	4881: 2931, 5139, 5266, 4535
dt/dp	760	0.050°	4881
p	20°	0.87	4881

162. Propionic Anhydride

(Continued)

eq 4	A B	8.30 2390.	[4881]
fp		$-43.0°$	4881
d	20° 25	1.0110 1.0057	2931: 4834
α	20°	0.00108	4881
n_D	20°	1.4045	4881: 5139, 4834, 5266
η	20° 25	1.144 1.061	2931: 5266
	Table II, III	Ch II	
γ	20° 25	30.30 29.70	2931
$\triangle H_v$	bp	9.97	[4881]
C_p	25°	56.1	[4881]
ϵ	16°	18.3	3477
fl pt	OC	165°F	4881
ir			411
Raman			2674
Beil	162, 242		

163. Butyric Anhydride

$(CH_3CH_2CH_2CO)_2O$ $C_8H_{14}O_3$

mw		158.199	887
bp	760	199.5°	2609: 2931, 3736, 4767, 5266
dt/dp	760	0.051°	4881
p	20°	0.3	4881
fp		$-65.7°$	2609
d	20° 25	0.96677 0.96199	2931: 3736, 4834

163. Butyric Anhydride

(Continued)

α	20°	0.00100	4881
n_D	9°	1.4148	3736: 5139, 4834, 5266
	20	1.4127	2609
dn/dt	$10 - 60^\circ$	0.000416	3984
η	20°	1.615	3931: 5266
	25	1.486	
γ	20°	28.93	2931
	25	28.44	
$\triangle H_v$	bp	11.95	[4881]
C_p	20°	67.8	4881
ϵ	20°	12.9	3477: 5026
fl pt	OC	$190^\circ F$	4881
ir			2363
Raman			2674
Beil	**162**, 274		

164. Methyl formate

HCOOCH$_3$ C$_2$H$_4$O$_2$

mw		60.053	887
bp	760	31.50°	4778: 3208, 4761, 4762, 5306
dt/dp	760	0.035°	4778: 4761, 5306
p	25°	560.	[2445]
eq 4	A	7.2203	3510
	B	1320.8	
fp		– 99.0°	4778: 2453, 4535, 4762
d	0°	1.00317	4778: 1717, 3009, 3208, 4761, 4834, 5032
	15	0.98149	5301, 5306
	20	0.97421	
n$_D$	15°	1.34648	4778: 4834
	20	1.34332	3208
dn/dt		0.00044	4778
η	15°	0.360	4778: 1717
	25	0.328	
	Table II	Ch II	
γ	20°	24.62	5032
	30	23.09	
	Table VII	Ch II	
△H$_v$	bp	6.749	3208: 781, 3331, 5026
△H$_m$		1.780	5026: 4962
△H$_f^o$	25° (l)	– 90.4	5853
△H$_c^o$	25° (l)	– 233.1	2577: 4743, 5853
C$_p$	12 – 20°	28.7	3442
t$_c$		214.0°	2651, 4663: 4532, 622, 5303, 5306, 1100
p$_c$		59.2	2651: 3331, 5026, 5301, 5306, 4663
d$_c$		0.349	2651, 4663
v$_c$		0.172	2651, 4663: 1100, 5303
K$_B$		1.649°	483
κ	17°	1.92 x 10^{-6}	422
ε	20°	8.5	3477: 5026, 5034, 1381
μ	(g)	1.77	1156: 4028, 3513

164. Methyl formate

(Continued)

soly	in aq, ca 25°	23%w	2608
aq az	None		2254
fl pt	TCC	– 2°F	3680
uv			574, 2007
ir			89, 2050, 4729, 4738, 4812, 3820, 4630
Raman			4712, 1172, 4633
mass			90, 558, 740, 4258, 1769
nmr			4632
Beil	**155**, 18		

165. Ethyl formate

$HCOOC_2H_5$ $C_3H_6O_2$

mw		74.080	887
bp	760	54.15°	4779: 3208, 3452, 3510, 4426, 4535, 4758
dt/dp	760	0.037°	4779: 5306
p	25°	240.	[2444]
eq 4	A	7.8457	3510
	B	1621.6	
fp		– 79.4°	4779: 1947, 4535, 4762, 4770
d	15°	0.92892	4779: 1717, 3009, 3208, 3452, 4426, 4834
	30	0.90958	4868, 5301
α		0.00141	4779
$n_{(5876)}$	15°	1.36253	4779: 3452, 4426, 4834
n_D	20°	1.35994	3208
dn/dt		0.00044	4779
η	15°	0.419	4779: 1717, 3452, 4868
	30	0.358	
	Table II	Ch II	

165. Ethyl formate

(Continued)

γ	15° 30	24.37 22.38	4779: 3452
$\triangle H_v$	bp	7.201	3208: 781, 3331, 5026, 4962
$\triangle H_m$		2.200	4769
$\triangle H_c^o$	25°(l)	– 391.7	2577: 4743
C_p	15 – 30°	35.37	2680: 515
t_c		235.3°	2651, 5301, 5303, 5306, 4663: 622, 1100
p_c		46.8	2651, 5026, 5306: 3331, 5034, 5301, 4663
d_c		0.323	2651, 4663
v_c		0.229	2651, 4663: 1100, 5303
K_B		2.080°	483
κ	20°	1.45 x 10⁻⁹	3160: 422, 4868
ϵ	25°	7.16	3477: 4868, 5026, 5034, 1381
μ	25° in **30**	1.94	4426: 2877
soly	in aq, 25° aq in, 20°	11.8%w 17.0%w	3160: 2608 3160
aq az	52.6°	95%w	2255: 2254
fl pt	TCC	– 4°F	3680
uv			574, 4634
ir			89, 2050, 4738, 3820, 4630
Raman			1175, 1168, 1176, 4633
mass			90, 558, 837, 4258, 1769, 61
Beil	**156**, 19		

166. Propyl formate

$HCOOCH_2CH_2CH_3$ $C_4H_8O_2$

mw		88.107	887
bp	760	80.85°	4773: 4535, 3208, 3510, 4748, 4766, 4971
dt/dp	760	0.040°	5306
p	25°	60.0	[2444]
eq 4	A	7.9925	3510
	B	1806.5	
fp		−92.9°	4766, 4535: 4774
d	15°	0.91109	4773: 600, 601, 1717, 2007, 3009, 3208
	30	0.89406	4971, 5301, 5306, 2992
n_D	15°	1.37898	4773: 2007, 3208, 1630
	20	1.37693	4971
dn/dt		0.00048	[4773]
η	15°	0.544	4773: 1717, 4748
	30	0.460	
	Table II	Ch II	
γ	20°	29.49	4773: 4971
	30	23.28	
$\triangle H_v$	bp	7.766	3208: 781, 3331, 4288, 4962
$\triangle H_c$		−558.80	4743
C_p	9 − 57°	42.7	4143
t_c		264.9°	2651, 5301, 5303, 5306, 4663: 1100, 3331
p_c		40.1	2651: 3331, 5301, 5306, 4663
d_c		0.309	2651, 4663
v_c		0.285	2651, 4663: 1100, 5303
κ	17°	5.5 x 10⁻⁵	422
ϵ	19°	7.72	3477: 1381
μ	22° in 30	1.89	5243
soly	in aq, 22°	2.05%w	4494
aq az	71.6°	97.7%w	2254: 2207
fl pt	TCC	27°F	3680
uv			5346
ir			2050, 2904, 4738, 2860, 3820
Raman			2668
mass			90, 558, 740, 4258, 1769
Beil	156, 21		

167. Butyl formate

HCOOCH$_2$CH$_2$CH$_2$CH$_3$ C$_5$H$_{10}$O$_2$

		102.134	887
mw		102.134	887
bp	760	106.6°	4767: 2007, 3208, 3510, 4535, 4766, 4971
p	25°	28.	[2444]
eq 4	A	8.1232	3510
	B	1983.3	
fp		-90.0°	4766: 4767
d	20°	0.8917	4971: 1717, 2007, 3208, 3453
dd/dt		0.00107	[1664]
n$_D$	20°	1.38903	4971: 2007, 3208, 3453
	25	1.3874	3160
η	16.3°	0.7244	Beil **E III 2**, 39
	20	0.704	1717
γ	21.3°	24.89	4971
	41.3	22.87	
$\triangle$H$_v$	bp	8.861	3208
κ		small	2369
ϵ	ca 80°	2.43	Beil **E II 2**, 31
aq az	83.8°	83.5%w	2254: 2007
fl pt	TOC	169°F	3680
ir			2050, 2857, 2904, 4738
mass			558, 740
Beil	**156**, 21		

168. Isobutyl formate

$HCOOCH_2CH(CH_3)_2$ $C_5H_{10}O_2$

mw		102.134	887
bp	760	98.4°	2007, 4767: 3208, 3510, 4535, 4766, 4971
p	25°	41.	[2444]: 4535, 3209
eq 4	A	7.9060	3510
	B	1863.7	
fp		− 94.5°	4774: 4535, 4766, 4767
d	20°	0.88535	600: 1717, 2007, 3208, 3209, 3453, 4971
	40	0.86360	
n_D	20°	1.38546	4971: 717, 2007, 3208, 3209
η	16.1°	0.688	Beil E III 2, 41
	20	0.680	1717: 3209
	88.0	0.311	Beil E III 2, 41
γ	15.4°	24.47	Beil E III 2, 41
	22.3	23.66	4971
	59.6	19.48	Beil E III 2, 41
$\triangle H_v$	bp	8.018	3208: 781
$\triangle H_c$		− 719.90	4743
C_p	58.2°	50.3	4143
t_c		277.8°	1100: 4406, 3209
p_c		38.29	3209
v_c		0.3545	1100
ϵ	19°	6.41	3477: 1381
μ	22° in 30	1.88	5243: 4028
soly	in aq, 22°	1.0%w	4494
aq az	79.5°	81.1%w	2254
ir			2050, 2857, 4738, 3820
Raman			717
Beil	156, 21		

169. Methyl acetate

CH_3COOCH_3 $C_3H_6O_2$

mw		74.080	887
bp	760	56.323°	5239, 5204: 942, 2007, 3208, 3452, 4762
dt/dp	760	0.0373°	5239, 5204: 5306
p	24.0°	200.	4535: 711, 3879, 4165, 5306
eq 5	A	7.065249	3800
	B	1157.622	
	C	219.724	
fp		− 98.05°	4762: 1947, 4535, 4774
d	20°	0.9342	3452: 942, 1717, 2007, 3009, 3208
	25	0.9279	4971, 5239, 5240
α	20°	0.00139	3160
n_D	20°	1.3614	3452: 2007, 3208, 4971
η	20°	0.385	3452: 942, 1717, 1772, 4793
	25	0.364	
	Table II	Ch II	
γ	20°	24.8	3452: 4971
	25	24.1	
$\triangle H_v$	α	7.73	[3160]: 4989, 4288
	bp	7.268	3208: 781, 3331, 4185
$\triangle H_c^\circ$	25°(l)	− 381.2	2577: 4743
C_p	18 − 42°	37.19	5117: 515, 4165
t_c		233.7°	2651, 5301, 5303, 5306, 4664: 3331
p_c		46.3	2651, 5306: 3331, 5301, 4664
d_c		0.325	2651, 4664
v_c		0.228	2651, 4664: 5303
K_B		2.061°	4185: 483, 3274
κ	20°	3.4 x 10⁻⁶	3160: 422
ϵ	25°	6.68	3477: 2747, 3400, 5026, 4786, 1381
μ	28° in 30	1.61	2760: 2175, 3400, 3445, 5243, 3513

169. Methyl acetate

(Continued)

soly	in aq, 20° aq in, 20°	24%w 8%w	3160: 1677
aq az	56.4°	96.3 – 96.8%w	2254: 2255
fl pt	TCC	14°F	3680
uv			573
ir			1469, 2050, 4113, 4375, 4738, 408, 4107
Raman			1175, 965, 1168, 1176, 4633
mass			90, 558, 740, 4258, 1769, 61, 4715
nmr			4632, 4031
Beil	159, 124		

170. Vinyl acetate

$CH_3COOCH=CH_2$ $C_4H_6O_2$

mw		86.091	887
bp	760	72.5°	4568, 3161, 4535: 1333, 4805, 3205
p	25°	106.	[2444]: 4568, 4535
eq 4	A B	8.091 1798.4	[3161]
fp		– 92.8°	4893: 1333
d	20°	0.9312	1872: 3161, 2881
dd/dt		0.001302	[1872]
n_D	20° 25	1.3959 1.3934	3414: 2881, 1333, 4568, 4481, 3205 3558
η	20°	0.4213	2608: 3558
γ	20° 30	23.95 22.54	1872

170. Vinyl acetate

(Continued)

$\triangle H_v$	bp	8.211	3161
$\triangle H_p$		21.3	2608
C_p	$20°$(g) 1 atm	22.5	2608
t_c		$228.3°$	3161
P_c		22.4	2608
μ	$25°$ in 30	1.79	2881: 3705
soly	in aq, $20°$	2.0%w	4893
	aq in, $20°$	1.0%w	
fl pt	TCC	$18°$ F	3680
uv			2457
ir			408, 2050, 1223, 89, 4630
Raman			2610, 2575, 499, 3300
nmr			932, 4632, 801
Beil	159, 136		

171. Ethylene glycol diacetate

$CH_3COOCH_2CH_2OOCCH_3$ $C_6H_{10}O_4$

		146.144	887
mw		146.144	887
bp	760	$190.9°$	4887: 4535, 4758, 4612, 1157
dt/dp	760	$0.049°$	4887
p	$25°$	0.2	4612: 4535
eq 4	A	8.5945	[4535]
	B	2649.43	

171. Ethylene glycol diacetate

(Continued)

fp		$-41.5°$	4887: 4535, 4612
d	$20°$	1.1043	[4887] : 4612, 4758
dd/dt	$20-30°$	0.00113	4887
α	$55°$	0.00106	4887
n_D	$15°$	1.4183	Beil E I **2**, 66, 4612, 4758
	20	1.4159	1157
η	$20°$	2.9	4887
$\triangle H_v$	bp	10.88	4887
ϵ	$-54°$	~ 10	2699
μ	$30°$ in **30**	2.34	3056
soly	in aq, $20°$	21.3%w	1338
	aq in, $20°$	21.2%w	
aq az	$99.7°$	15.4%w	2255
fl pt	COC	$205°F$	4887
Beil	**159**, 142		

172. Ethyl acetate

$CH_3COOCH_2CH_3$ $C_4H_8O_2$

mw		88.107	887
bp	760	$77.114°$	5240, 5239: 1776, 2558, 3208, 3452, 4778
dt/dp	760	$0.0401°$	5240, 5239: 1372, 4761
p	$25°$	92.	[2444] : 1373, 1779, 3928, 4165, 4535, 5306
eq 5	A	7.102326	3800: 1372
	B	1245.329	
	C	217.911	

172. Ethyl acetate

(Continued)

fp		$-83.97°$	1372: 2556, 3301, 4351, 4535, 4771, 4788
d	$20°$	0.90063	1372: 942, 1776, 3009, 3208, 3452, 4426
	25	0.89455	4761, 4971, 5239, 5240
dd/dt		0.00120	4778
α	$20°$	0.00139	3160
n_D	$20°$	1.37239	1372: 717, 3208, 3452, 4426, 4971
	25	1.36979	
dn/dt		0.00049	4778
η	$15°$	0.473	4778: 1717, 1772, 2557, 2558, 3452
	25	0.426	942
	30	0.400	4778
	Table II	Ch II	
γ	$20°$	23.75	2102, 4778: 1482, 3452, 4971, 5032
	30	22.55	
	Table VII	Ch II	
$\triangle H_v$	$25°$	8.40	5012, 4989, 4962
	bp	7.713	3208
$\triangle H_m$		2.51	1057
$\triangle H_f^°$	$25°$ (l)	-116.3	5358: 5002
$\triangle H_c^°$	$25°$ (l)	-537.5	2577: 4743, 4580, 5358
C_p	$20.4°$	40.4	3653: 4143, 5002
t_c		$250.1°$	2651, 5301, 5303, 5306, 4664: 622, 5032
P_c		37.8	2651: 3331, 5026, 5034, 5301, 5306, 4664
d_c		0.308	2651, 4664
v_c		0.286	2651, 4664: 1100, 5303
K_i	$25°$ in H_2O	$10^{-24.5}$	3695
K_B		$2.583°$	483: 919, 3274, 475
κ		$\langle 1 \times 10^{-9}$	3160: 422, 2557, 2558, 2569
ϵ	$25°$	6.02	3477: 2747, 3400, 4281, 5034, 4786
μ	$25°$ in 30	1.88	4426: 2497, 3400, 3445, 4281, 5243
soly	in aq, $25°$	8.08%w	4225: 2417, 1677
	aq in, $25°$	2.94%w	
aq az	$70.38°$	91.53%w	2254
fl pt	TCC	$24°F$	3680
uv			573, 2012, 748
ir			1469, 2050, 2904, 4113, 4738, 736, 408
Raman			717, 4633, 3468, 3470, 965, 2575, 1168
mass			90, 3997, 558, 740, 4258, 4657, 1769
nmr			4031, 4632
Beil	159, 125		

173. Propargyl acetate

$CH_3 COOCH_2 C\equiv CH$ $C_5 H_6 O_2$

mw		98.102	887
bp	760	121.5°	1925
d	20° 40	0.9982 0.9761	1925: 802
dd/dt		0.0011	1925
n_D	20°	1.41866	1925: 802
γ	20° 40	32.81 30.20	1925
uv			4607
Beil	159, 140		

174. Allyl acctate

$CH_3 COOCH_2 CH\equiv CH_2$ $C_5 H_8 O_2$

mw		100.118	887
bp	760	104.0°	[4000], 421, 2386
p	30°	46.0	421: 419
eq 4	A B	7.908 1893.5	[421]
d	20.9° 28.0	0.9267 0.9190	2386
n_D	20° 24	1.4040 1.3985	2386: 357 419
η	30°	0.2068	421
γ	20.9° 28.0	26.25 25.40	2386

174. Allyl acetate

(Continued)

$\triangle H_c$		− 654.9	4604
aq az	83°	83.3%w	2255
uv			4604
ir			408
Beil	**159**, 136		

175. Propyl acetate

$CH_3COOCH_2CH_2CH_3$ $C_5H_{10}O_2$

mw		102.134	887
bp	760	− 101.548°	5239, 5240: 324, 2007, 3208, 3452
dt/dp	760	0.0430°	5239, 5240: 5306
p	25°	32.3	[2444] : 4535, 3879, 5306
eq 5	A	7.017203	3800
	B	1282.873	
	C	208.664	
fp		− 92.5°	4887: 1947, 4535, 4774
d	15°	0.89377	4773: 324, 600, 1717, 2007, 3009, 3208
	25	0.88303	5240: 3208, 3452, 3453, 4971, 5239, 5301
	30	0.87716	4773
α	20°	0.00131	4887
n_D	15°	1.38656	4773: 324, 2007, 3208, 3209, 3452
	20	1.38442	4971
dn/dt		0.00048	[4773]

175. Propyl acetate

(Continued)

η	20°	0.585	3452: 1261, 2751, 3209, 4748
	25	0.551	
	Table II	Ch II	
γ	20°	24.28	1482: 3452, 4971
	30	23.10	4773
	40	22.39	840
$\triangle H_v$	25°	9.34	5012
	bp	8.200	3208: 781, 3331, 4962, 4288
C_p	20°	46.9	4887: 4143, 4820
t_c		276.2°	2651, 5301, 5303, 5306, 4664: 1100, 3331
p_c		32.9	2651: 3331, 5034, 5301, 5306, 4664
d_c		0.296	2651, 4664
v_c		0.345	2651, 4664: 1100, 5303
κ	17°	2.2 x 10⁻⁷	3160: 422
ϵ	20°	6.002	3746: 3007, 3477, 5034, 1381
μ	22° in **30**	1.78	5243: 3445
soly	in aq, 20°	2.3%w	4887: 2608, 1677
	aq in. 20°	2.9%v	3160
aq az	82.2°	86%w	2254: 2007
fl pt	TCC	58° F	3680
uv			574, 573, 575
ir			2050, 2904, 4113, 4375, 4738
Raman			2229, 965
mass			90, 558, 740, 4258, 1769
nmr			4031, 4632
Beil	**159**, 129		

176. Isopropyl acetate

$CH_3COOCH(CH_3)_2$ $C_5H_{10}O_2$

mw		102.134	887
bp	760	88.2°	1960: 3453, 4535, 4766, 4971
p	25°	60.6	1960: 4535
eq 12	A	14.2517	1960
	B	2170.1	
	C	2.0972	
fp		- 73.4°	4766: 4774, 2608
d	20°	0.8718	4971
	25	0.8668	[4971]
	41.7	0.8502	4971
dd/dt		0.000995	[4971]
n_D	20°	1.37730	4971: 717, 2608, 3453
η	20°	0.569	4571: 1717
γ	22°	22.10	4971
	41.4	20.08	
$\triangle H_v$	25°	8.89	5012
	bp	7.900	1960
C_p	20°	53.2	[3160]
soly	in aq, 20°	2.9%w	4893: 2608, 1677
	aq in, 20°	1.8%w	
aq az	17.6°	89.4%w	2254
fl pt	CC	36°F	88: 3903
ir			2050, 4738
Raman			717, 2229, 965
mass			90, 558, 740, 1769
nmr			4031, 4632
Beil	159, 130		

177. Butyl acetate

$CH_3COOCH_2CH_2CH_2CH_3$ $C_6H_{12}O_2$

mw		116.161	887
bp	760	126.114°	5239: 2007, 3452, 4764, 4971, 5084, 4155
dt/dp	760	0.0456°	5239, 5240: 4155
p	25°	13.	[2444]
eq 5	A	7.0511	2633: 4155
	B	1385.8	
	C	206.1	
fp		−73.5°	4571: 4764
d	15°	0.88652	4773: 2007, 3452, 3453, 4971, 5084, 5228
	25	0.87636	5240
	30	0.87129	4773: 2518
dd/dt	20–30°	0.00102	4887
α	55°	0.00121	4887: 2608, 4155
n_D	15°	1.39636	4773
	30	1.3827	Beil E III 2, 236
dn/dt		0.00047	[4773]
η	15°	0.770	4773: 1772, 3452, 4155
	30	0.628	
γ	20°	25.09	4773: 1482, 3452, 4971, 5084, 840, 4155
	30	23.98	
$\triangle H_v$	25°	10.42	5012
	bp	8.58	781: 2608
	Table IX	Ch II	
$\triangle H_c^{\circ}$	25°	−828.6	4155
C_p	19.41°	57.84	Beil E III 2, 235: 2689
t_c		306°	5077
κ	25°(85%)	1.6×10^{-8}	2569
ϵ	20°	5.01	3477: 3007, 1381
μ	22° in 30	1.84	5243: 3445
soly	in aq, 20°	0.43%w	1338
	aq in, 20°	1.86%w	
aq az	90.2°	71.3%w	2254, 2007
fl pt	TCC	72°F	3680
uv			574, 573, 2812, 748

177. Butyl acetate

(Continued)

ir		1469, 2050, 2904, 4113, 4738
Raman		2229, 965, 2704
mass		90, 558, 740, 4258, 1189, 61
nmr		4632, 4031
Beil	159, 130	

178. Isobutyl acetate

$CH_3 COOCH_2 CH(CH_3)_2$ $C_6 H_{12} O_2$

mw		116.161	887
bp	760	118.0°	4535: 324, 600, 3208, 3452, 4763, 4971
dt/dp	760	0.0444°	3160
p	25°	19.5	[2444]
eq 5	A	7.53858	[2444]
	B	1699.76	
	C	246.9	
fp		−98.85°	4763: 4535
d	20°	0.8745	3452: 324, 600, 1717, 2007, 3208, 3453
	25	0.8695	4971
dd/dt		0.00105	4887
α	55°	0.00126	4887
n_D	20°	1.39018	4971: 324, 717, 2007, 3452, 3453
	25	1.3880	3160

178. Isobutyl acetate

(Continued)

η	20°	0.697	3452: 1717, 2163
	25	0.651	
γ	20°	23.7	3452: 4971
	25	23.15	
$\triangle H_v$	25°	9.38	[4535]
	bp	8.568	3208: 781
C_p	70.2°	58.5	4143: 515
t_c		287.8°	1100
v_c		0.4135	1100
κ	19°	2.55×10^{-4}	422
ϵ	20°	5.29	3477: 3007, 847
μ	22° in 30	1.87	5243: 3445
soly	in aq, 20°	0.67%w	1338: 1677
	aq in, 20°	1.64%w	
aq az	87.4°	83.4%w	2254: 2007
fl pt	CC	64°F	88
ir			2050, 4738, 5124
Raman			717, 965
mass			4850
nmr			4031, 4632
Beil	159, 131		

179. sec–Butyl acetate

$CH_3 COOCH(CH_3)CH_2 CH_3$ $C_6 H_{12} O_2$

mw		116.161	887
bp	760	112.34°	4269: 4915, 3453, 4971, 3756
p	25°	24.	3160
d	17.2° 20 41.9	0.8748 0.8720 0.8478	4971, 3756: 1338, 3453
α	10 – 30°	0.00118	1338: 3160
n_D	20° 25	1.38941 1.3840	4971: 717, 2261, 3756 3453
γ	21.1° 41.9	23.33 21.24	4971
$[\alpha]_D$	20° (+) (–)	25.43° 20.19	3756 2560
soly	in aq, 20° aq in, 20°	0.62%w 1.65%w	1338
aq az	87°	77.5%w	2254
fl pt	OC	88°F	88
ir			2050, 4738
Raman			717, 2229
mass			558, 4850
nmr			558, 4031, 4632
Beil	159, 131		

180. Pentyl acetate

Amyl acetate

$CH_3COOCH_2(CH_2)_3CH_3$ $C_7H_{14}O_2$

mw		130.188	887
bp	760	149.2°	3452: 2007, 4971
p	25°	6.	[1215]
eq 5	A	8.078	1215
	B	2077	
	C	253	
fp	gl	$\leftarrow -100°$	3160
d	20°	0.8753	3452: 2007, 4165, 4971, 840
	25	0.8707	
	70	0.8250	840
dd/dt	40 – 100°	0.000997	[840]
α		0.00115	3160
n_D	20°	1.4028	3452: 2007, 4971
η	20°	0.924	3452: 1772, 2396
	25	0.862	
γ	20°	25.68	4971: 3452, 840
	25.9	25.13	
$\triangle H_v$	bp	9.8	[3160]
$\triangle H_c°$	25° (l)	– 1058.5	2577
C_p	30.1°	66.0	3745: 4143, 4165
t_c		332.°	[4095]
K_B		4.83°	480
κ	25°	1.6×10^{-9}	2569
ϵ	20°	4.75	3477: 2396, 1381
μ		1.91	2396
soly	in aq, 20°	0.17%w	1338
	aq in, 20°	1.15%w	
aq az	95.2°	59%w	2254: 2007
fl pt	TCC	77°F	3680

180. Pentyl acetate

(Continued)

uv		2012, 748
ir		4375, 4630
Raman		965, 1168, 1176
mass		558, 1769
nmr		4031, 4632
Beil	159, 131	

181. Isopentyl acetate

Isoamyl acetate

$CH_3 COOCH_2 CH_2 CH(CH_3)_2$ $C_7 H_{14} O_2$

mw		130.188	887
bp	760	142.0°	4535: 324, 2007, 3453, 4767, 4971
dt/dp	760	0.047°	3160
p	20°	4.5	3160
eq 4	A	8.4163	[2444]
	B	2289.2	
fp		– 78.5°	2608
d	20°	0.8719	4971: 324, 2007, 2608, 3453, 3973, 1540
	25	0.8664	2007
	40	0.8541	4971
α		0.00119	3160
n_D	20°	1.4007	1635, 4156: 324, 4587, 3453, 1540
	25	1.3981	248

181. Isopentyl acetate

(Continued)

η	19.91°	0.872	3323
	25	0.7895	Beil **E II 2**, 60
γ	21.1°	24.62	4971
	30	23.6	Beil **E III 2**, 252
$\triangle H_v$	bp	8.97	781: 5032
C_p	20°	59.73	4143
K_B		4.83°	483
ϵ	30°	4.63	3477: 3007, 4281
μ	22° in **30**	1.82	5243: 4281
soly	in aq, 25°	ca 2%	2608
	aq in, 25°	1.6 %w	[3160]
aq az	93.6°	63.7%w	2254: 2007
fl pt	CC	92°F	88
uv			2012
ir			2050
Raman			965
mass			558
nmr			4632
Beil	**159**, 132		

182. 2–Ethylhexyl acetate

CH$_2$CH$_3$

|

CH$_3$COOCH$_2$CH(CH$_2$)$_3$CH$_3$ C$_{10}$H$_{20}$O$_2$

mw		172.270	887
bp	760	198.6°	3160: 1338, 2063
dt/dp	760	0.053°	4887
p	20°	0.40	3160
eq 4	A	8.4184	[2444]
	B	2521.8	
fp		-93°	3681
d	20°	0.8718	[4887]
α	10 – 30°	0.00099	1338
dd/dt	20 – 30°	0.00087	4887
n$_D$	20°	1.4204	4915, 1338
	25	1.4173	2063
dn/dt		0.00062	[2063]
η	20°	1.5	4887
$\triangle$H$_v$	25°	11.5	[2444]
	bp	10.4	[4887]
C$_p$	20°	85.8	[4887]
soly	in aq, 20°	$\langle$ 0.03%w	3160
	aq in, 20°	0.55%w	
aq az	99.0°	26.5%w	4887: 2255
fl pt	OC	190°F	3160: 1338
Beil	159, 135		

183. Benzyl acetate

$CH_3COOCH_2C_6H_5$ $C_9H_{10}O_2$

mw		150.179	887
bp	760	215.5°	2444: 1703, 2608, 4535
dt/dp	760	0.052°	[2444]
p	60°	1.42	2444: 1703, 4535
eq 4	A	8.1616	[2444]
150 – 220°	B	2583.3	
46 – 150	A	9.1700	
	B	3000.00	
10 – 55	A	9.670	4239
	B	3157.	
fp		– 51.5°	4535, 4764: 2608
d	25°	1.0515	[2527, 1703, 3592, 1067]: 1326, 2608
	30	1.0482	2527: 4834
dd/dt		0.000673	[2527, 1703, 3592, 1067]
n_D	20°	1.5232	2609: 1703, 4834, 4077
η	45°	1.399	4709, 2527
$\triangle H_v$	10 – 55°	14.4	4239
	bp	11.8	[2444]
C_p	32.8°	36.79	3745
ϵ	21°	5.1	3477: 1326, 3499
μ	25° in 30	1.80	3445
soly	in aq	sl sol	2608
aq az	99.60°	12.5%w	2254
fl pt	CC	216°F	3680: 2608
ir			2904, 3918
Raman			3468, 3470, 2670
mass			1476
nmr			4632
Beil	528, 435		

184. Ethyl propionate

$CH_3 CH_2 COOCH_2 CH_3$ $C_5 H_{10} O_2$

mw		102.134	887
bp	760	$99.10°$	4778, 4535: 3208, 3209, 4426, 4748
dt/dp	760	$0.042°$	4778: 4761, 5306
p	$25°$	35.	[2444]
eq 5	A	7.009966	3800
	B	1268.942	
	C	208.301	
fp		$-73.85°$	4774: 1947, 4535, 4762, 4778
d	$15°$	0.89574	4778: 324, 600, 1326, 1717, 3208, 3209
	30	0.87903	3453, 4426, 4761, 4971, 5305, 3970
dd/dt		0.00112	4778
α	$10 - 30°$	0.00125	1338
n_D	$15°$	1.38643	4778: 324, 3208, 3209, 3453, 4426, 4440
	20	1.38394	4971
dn/dt		0.00046	4778
η	$15°$	0.564	4778: 1717, 1772, 3209
	30	0.473	
	Table II	Ch II	
γ	$20°$	24.27	2102, 4778: 1482, 4971, 3970
	30	23.16	
$\triangle H_v$	$20°$	8.92	4057
	bp	8.178	3208: 781, 3331, 4962
$\triangle H_c°$	$25° (l)$	-693.8	2577
C_p	$59.6°$	50.45	4143: 515, 4095
t_c		$272.9°$	4665, 5303: 622, 5301, 5306, 1100, 4095
p_c		33.18	4665: 3331, 5301, 5305, 5306
d_c		0.2965	4665, 5303
v_c		0.345	4665: 1100
κ	$17°$	8.33×10^{-4}	422
ϵ	$19°$	5.65	3477: 1326, 1325, 3007, 1381
μ	$22°, in\ 30$	1.74	5243: 4426, 4410

184. Ethyl propionate

(Continued)

soly	in aq, 20° aq in, 20	1.92%w 1.22%w	1338: 2608, 4440
aq az	81.2°	90%w	2254: 2255
fl pt	TCC	54°F	3680
uv			574, 575, 748
ir			1469, 2050, 4375, 4738, 89
Raman			3468, 1168, 1176
mass			90, 558, 1769, 4258, 3456, 4715
Beil	**162**, 240		

185. Ethyl butyrate

$CH_3 CH_2 CH_2 COOCH_2 CH_3$ $C_6 H_{12} O_2$

mw		116.161	887
bp	760	121.55°	4779: 1928, 2444, 1338, 3209, 3208
dt/dp	760	0.045°	[1928]
p	25°	17.	[2444] : 1928, 1338, 3209
eq 4	A B	8.903 2053.58	2444
fp		-98.0°	4773: 4779
d	15° 25	0.88440 0.87394	4779: 4971, 1338, 3209, 4440, 3969, 3208 [4779]
dd/dt		0.001046	[4779]
α	10 – 30°	0.00116	1338

185. Ethyl butyrate

(Continued)

n_D	$20°$	1.3928	4779: 4971, 3209, 4440, 3208
dn/dt		0.00047	[4779]
η	$15°$	0.771	4779: 3209, 1405
	25	0.6127	1772
	30	0.595	4779
γ	$20°$	24.58	4779: 4971, 3969
	30	23.26	
$\triangle H_v$	$25°$	10.04	5012
	bp	8.679	[3208]
$\triangle H_c^°$	$25°$ (l)	-850.1	2577
C_p	$24.08°$	52.62	2690
t_c		$292.8°$	3209: 4095
p_c		30.24	3209
ϵ	$18°$	5.10	3477: 1381
μ	$22°$ in **30**	1.74	5243: 4410
soly	in aq $20°$	0.49%w	1338: 1677, 4440
	aq in 20	0.75%w	
aq az	$87.9°$	78.5%w	2254
fl pt	TCC	$78°F$	3680
uv			572, 573
ir			89, 4738, 2050
Raman			2666, 1168, 1176
mass			4258, 1769
Beil	**162**, 270		

186. Isobutyl isobutyrate

$(CH_3)_2CHCOOCH_2CH(CH_3)_2$ $C_8H_{16}O_2$

mw		144.216	887
bp	760	147.51°	2444: 366
dt/dp	760	0.061°	[2444]
p	25°	4.8	[2444]
eq 4			
60 – 150°	A	7.9086	[2444]
	B	2111.	
10 – 60	A	8.8522	[2444]
	B	2433	
fp		– 80.7°	4875
d	0°	0.87496	4875
	20	0.8542	[3681]
	148.5	0.7249	4142
n_D	18.2°	1.3986	366
γ	– 76.5°	33.8	2372
	134.5	13.1	
$\triangle H_v$	25°	11.1	[2444]
	bp	9.14	[781] : 4963
t_c		328.74°	783
soly	in aq	0.5%w	3681
aq az	95.5°	60.6%w	2254
fl pt		120°F	3681
uv			3197
ir			4738
Raman			366
Beil	**162**, 291		

187. Ethyl isovalerate

$(CH_3)_2CHCH_2COOCH_2CH_3$ $C_7H_{14}O_2$

mw		130.188	887
bp	760	134.7°	4763: 600, 3453, 4535, 4971, 2000
dt/dp	760	0.047°	[2444]
p	25°	7.9	[2444]
eq 4	A	8.3403	[2444]
	B	2216.8	
fp		−99.3°	4763: 4535
d	20°	0.8652	4971: 3453, 600, 1326, 324, 2024
	40.9	0.8456	
n_D	20°	1.39621	4971: 324
	25	1.3944	2000
γ	21.1°	23.82	4971
	41.4	21.46	
$\triangle H_v$	25°	11.3	[2444]
	bp	8.832	[781]
$\triangle H_f^\circ$	25° (l)	−129	2000
$\triangle H_c^\circ$	25° (l)	−845	2000
	Table XI	Ch II	
t_c		314.87°	783
ϵ	18°	4.71	5034
soly	20° in aq	0.2%w	4494
	aq in 20°	0.3%w	[4494]
aq az	92.2°	69.8%w	2254
ir			2050, 2860
Raman			2666, 4738
Beil	162, 312		

188. Isopentyl isovalerate

$(CH_3)_2CHCH_2COOCH_2CH_2CH(CH_3)_2$ $C_{10}H_{20}O_2$

mw		172.270	887
bp	760	194.0°	4535: 324, 3453
dt/dp	760	0.0521°	[4535]
p	27.0°	1.	4535
eq 4	A	8.13721	[4535]
25 – 100°	B	2440.11	
	A	8.01057	
100 – 200°	B	2394.85	
d	18.7°	0.8583	324: 1325
	25	0.8541	3453
n_D	18.7°	1.41300	324
	25	1.4100	3453
$\triangle H_v$	25°	11.17	[4535]
	bp	10.96	
ϵ	19°	3.62	1325
aq az	98.8°	25.9%w	2254
ir			5124, 2860
Beil	**162**, 312		

189. Butyl stearate

$CH_3(CH_2)_{16}COOCH_2CH_2CH_2CH_3$ $\qquad$ $C_{22}H_{44}O_2$

mw		340.595	887
bp	25	220 – 225°	2608
	760	$\sim 350°$	1055
p	20°	1.27×10^{-6}	704, 705: 3719
eq 4	A	14.88	3719
	B	5220	
fp		26.3°	4390: 2608, 4811, 5139
d	25°	0.8540	4811
	30	0.8501	
a		0.00083	1055
n_D	20°	1.4441	3975
	25	1.4422	
	50	1.4328	5139
η	25°	8.26	4811: 1802
	50	4.9	1802
γ	25°	33.0	4811
	30	32.7	
$\triangle H_v$	75 – 125°	23.9	3719: 4027
$\triangle H_m$		13.6	1055
κ	30°	2.1×10^{-13}	3836
ϵ	30°	3.111	3477: 3400
μ	24° in **30**	1.88	2759: 4028
soly	in aq, 25°	$\sim 0.17\%w$	[1055]
	aq in, 25°	$\sim 0.35\%w$	
fl pt	COC	375°F	4811: 2608, 3680
Beil	**162**, 380		

190. Methyl acrylate

$CH_2=CHCOOCH_3$ $C_4H_6O_2$

mw		86.091	887
bp	760	80.2°	4535: 3978
p	0°	23.4	4021: 4535
	10	40.5	
	20	68.2	
	25	84	[2444]
	30	109.	4021
	50	225.	
	70	530.	
fp		⟨ − 75°	1356: 3978
d	15°	0.9564	3430: 2467
	20	0.9535	3937
n_D	18°	1.4117	3430
	25	1.4003	4021
η	20°	1.3984	2468
$\triangle H_v$	bp	7.9	4021
$\triangle H_p$		20.19	1505
C_p		41.3	[4021]
soly	in aq, 25°	4.94%w	[4021]: 3978
	aq in, 30°	2.65%w	
aq az	71°	92.8%w	2254
fl pt	TCC	26.6°F	4021
uv			377, 1011
ir			4021, 736, 2173, 408, 4812, 4630
Raman			2173, 2673, 4812, 2678, 4633
nmr			932, 801
Beil	163, 399		

191. Ethyl acrylate

$CH_2=CHCOOCH_2CH_3$ $C_5H_8O_2$

mw		100.118	887
bp	760	99.5°	4535: 3978
p	25°	38.	2238: 4021
eq 4	A	8.0854	[4535]
	B	1939.49	
fp		−71.2°	4535: 2238
d	20°	0.9234	3937: 3430
n_D	20°	1.4068	3937: 3430
	25	1.4034	4021
$\triangle H_v$	bp	8.3	4021
$\triangle H_p$		18.6	4021
C_p		47	4021
soly	in aq 25°	1.50%w	[4021] : 3978
	aq in 25°	1.50%w	
aq az	81.1°	15%w	2255: 3978
fl pt	TCC	48°F	4021: 2238
uv			1011
ir			2173, 408, 4630
Raman			2173, 2678, 4633
nmr			4632, 267, 801
Beil	**163**, 399		

192. Methyl methacrylate

$CH_2=C(CH_3)COOCH_3$ $C_5H_8O_2$

mw		100.118	887
bp	760	100.3°	1414
p	25°	38.	[2444] : 2891, 4021
eq 4 0 – 30°	A B	8.71369 2126.21	[4021]
fp		– 48.2°	3239: 3226, 1414
d	20° 30	0.94331 0.93174	1488: 1414, 3936, 2891
dd/dt		0.00116	[1488]
n_D	20° 25	1.4146 1.4120	1488: 1414, 3239, 3936, 2891 4021
η	20°	0.6322	1488
γ	20°	28 – 29	Beil E III 2, 1280: 3278a
$\triangle H_v$	25° bp	9.73 8.6	[4021] 4021: 879
$\triangle H_s$		14.5	4021: 879
$\triangle H_c$		– 645.9	Beil E III 2, 1280 :3278 a
$\triangle H_p$		13.8	4021: 3623
C_p	25°	45.7	3267: 1488
c		2.9	4554
μ	25° in 30	1.675	2891
soly	in aq, 20° aq in, 20°	1.56%w 1.14%w	4021: 4145 4021: 4254
aq az	83°	86%w	2255
fl pt	TCC	50°F	4021
uv			1839, 5291, 1011
ir			4021, 408, 3936, 89, 5291
Raman			3916, 2165
nmr			4632, 5291
Beil	163, 422		

193. Methyl oleate

Methyl cis-9-octadecenoate

$CH_3(CH_2)_7CH{=}CH(CH_2)_7COOCH_3$ $\qquad\qquad\qquad$ $C_{19}H_{36}O_2$

mw		296.498	887
bp	16	$217°$d	63
p	$128.28°$	0.2031	4214: 4151
	185.31	4.709	
eq 5	A	6.8805	4214: 3547
$(128 - 185°)$	B	1963.5	
	C	131.	
fp		$19.9°$	5132
d	$25°$	0.8702	2543: 5132, 3151, 1858
	90	0.8234	
dd/dt	$30 - 90°$	0.00072	[2543]
n_D	$20°$	1.45214	1858: 5132, 1125, 3151
	40	1.44450	
dn/dt		0.000377	[1125]
η	$30°$	4.88	2543
	60	2.62	
	90	1.64	
γ	$25°$	31.3	42
	100	25.4	
	180	19.1	
$\triangle H_v$	1 torr	20.170	4214: 3547
$\triangle E_c$	$25°$(l)	-2837.3	2543: 2542, 2544
K_f		$3.4°$	4214
iodine no	calc'd	85.62	1125: 4214, 2543
ϵ	$20°$	3.211	1858
	40	3.117	
uv			3344, 4370, 1274
ir			1258, 408
mass			1981
nmr			4510, 2052
Beil	163, 467		

194. Methyl benzoate

$C_6H_5COOCH_3$ $C_8H_8O_2$

mw		136.152	887
bp	760	199.50°	4780: 1372, 4535, 4763, 4973
dt/dp	760	0.05328°	1372: 4780
p	25°	0.3944	1369: 1373, 4535
eq 5	A	7.48253	1372
	B	1974.6	
	C	230	
fp	(st)	– 12.10°	1249: 1372, 2556, 4763, 4770
	(ms)	– 13.9°	4780
d	15°	1.09334	4780: 1372, 1717, 3026, 4563, 4973
	30	1.07901	
α		0.000876	3160
n_D	20°	1.51679	1369: 1372, 3026, 4973
	25	1.51457	
	50	1.50298	
dn/dt		0.00046	[4772]
η	15°	2.298	4780: 1717
	30	1.673	
γ	20°	38.14	4780: 4973
	30	36.91	
$\triangle H_v$	25°	13.62	1369: 5034
	bp	10.32	
$\triangle H_m$		2.327	1369
$\triangle H_c^\circ$	25° (l)	– 950.8	2577
C_p	20°	51.92	4144: 4143
t_c		438°	1369
p_c		39.5	[1369]
d_c		0.37	1369
v_c		0.369	[1369]
A	eq 72	0.01722	1369

194. Methyl benzoate

(Continued)

κ	$22°$	1.37×10^{-5}	422
ϵ	$20°$	6.59	3477: 1985, 3007, 5034, 1381
μ	$25°$ in **30**	1.86	1985: 1496, 4563
soly	in aq, $20°$	0.21%w	3160: 664
	aq in, $20°$	0.74%w	3160
aq az	$99.08°$	20.8%w	2254
fl pt	CC	$181°$ F	88
uv			5246, 2257
ir			3918, 4375, 736, 2519
Raman			3467, 2020, 2670, 3853, 1171
mass			11, 1476, 61
nmr			4632
Beil	**900**, 109		

195. Ethyl benzoate

$C_6 H_5 COOCH_2 CH_3$ $C_9 H_{10} O_2$

mw		150.179	887
bp	760	$212.40°$	4780: 2558, 2784, 3452, 4535, 4762, 4973
dt/dp	760	$0.057°$	4780: 2784
p	$20°$	0.18	3160
eq 4			
$90 - 140°$	A	8.6330	[2444]
	B	2750.0	
$140 - 220°$	A	8.0350	
	B	2500.0	

195. Ethyl benzoate

(Continued)

fp		– 34.7°	4780, 4766: 2556, 4535, 4762, 4770
d	15° 30	1.05112 1.03718	4780: 1400, 1712, 2558, 3026, 3452, 4563 4973, 3592
α	20°	0.00089	3160
n_D	15° 20 25	1.50748 1.50570 1.50348	4780: 3026, 3452, 4973, 3592 4772 [4780]
dn/dt		0.00040	[4772]
η	15° 30	2.407 1.751	4780: 1400, 1712, 2557, 2558, 3452
γ	20° 25	35.4 34.75	4773: 3452, 4780, 4973
$\triangle H_v$	av	9.67	2784
$\triangle H_c^o$	25° (l)	– 1107.1	2577
C_p	15 – 30°	57.8	2689: 4143, 4144
κ	25°	$\langle$ 1 x 10⁻⁹	2557: 2558
ϵ	20°	6.02	3477: 1985, 3007, 4281, 5034, 4786, 1381
μ	in 30	1.99	1083: 293, 2760, 2334, 1985, 4281, 4563
soly	in aq, 20° aq in,	0.05%w 0.5%w	3160: 664 3160
aq az	99.40°	16.0%w	2254
fl pt	CC	$\rangle$ 204°F	88
uv			91, 374, 3366, 748
ir			1469, 2275, 4738, 2451, 2959, 2519, 2764
Raman			3467, 3468, 3470, 2020, 3300, 3853, 1171
mass			1476
Beil	900, 110		

196. Propyl benzoate

$C_6H_5COOCH_2CH_2CH_3$ $C_{10}H_{12}O_2$

		164.206	887
mw		164.206	887
bp	760	231.2°	4763: 4535, 4973
dt/dp	760	0.056°	[4535]
p	54.6°	1.	4535
eq 4			
80 – 160°	A	8.30275	[4535]
	B	2712.71	
160 – 230°	A	8.03341	
	B	2597.91	
fp		– 51.6°	4763, 4535
d	15°	1.0274	2947
	20	1.0232	4973
	41.5	1.0046	
n_D	15°	1.50139	3026
	20	1.50031	4973
γ	23.5°	38.87	4973
	41.5	32.04	
$\triangle H_v$	25°	12.41	[4535]
	bp	11.89	
$\triangle H_c^o$	25° (l)	– 1263.4	2577
C_p	20°	65.35	4144: 4143
aq az	99.70°	9.1%w	2254
Beil	**900**, 112		

197. Benzyl benzoate

$C_6H_5COOCH_2C_6H_5$ $C_{14}H_{12}O_2$

mw		212.251	887
bp	760	323.5°	2444: 987, 2558, 2684
dt/dp	760	0.065°	[2444]
p	150°	3.46	2444
eq 4	A	10.115	4240
	B	4057.	
fp		19.4°	2558
d	19.5°	1.1200	1326: 987
	25	1.1121	2558
η	25°	8.292	594: 2557, 2558
	40	5.243	
γ	210.5°	26.6	Beil **E I 9**, 68
$\triangle H_v$	12 – 60°	18.6	4240
κ	25°	$\langle 1 \times 10^{-9}$	2557: 2558
ϵ	20°	4.9	3477: 1326, 3499
μ		1.89 – 1.90	728
fl pt	CC	298°	3680
uv			3366
ir			3030
Raman			3300
mass			11
nmr			4632
Beil	**900**, 121		

198. Ethyl cinnamate

$C_6H_5CH{=}CHCOOCH_2CH_3$ $C_{11}H_{12}O_2$

mw		176.217	887
bp, trans	760	271.0°	4535: 5048, 5200, 524
p, trans	87.6°	1	4535: 2386
cis	131 – 134°	17	524
eq 4	A	8.3708	[4535]
	B	2989.4	
fp, trans		12.°	4535: 5048
d	20°	1.0494	2386
	25	**1.0469**	**5048**
	91.1	0.98815	3712
n_D	20°	1.55983	2386: 5048
	91.1	1.5250	3712
dn/dt		0.000492	3712
η	20°	8.7	4994
γ	19°	37.08	5048: 2386
	24.7	36.57	
	Table VIII	Ch II	
$\triangle H_v$	10 – 100 torr	14.1	[4535]
	400 – 760	14.0	[4535]
κ	curve given		2349: 422
ϵ	18°	6.1	3477: 3007, **1381**
μ , trans	20.2° in **30**	1.84	524
cis	15.4° in **30**	1.77	524
aq az	99.93°	3%w	2254
uv			374
ir			2856
Raman			3466, 3468, 3470, 3853
mass			1476
Beil	**948**, 581		

199. γ–Butyrolactone

$$H_2C \text{———} CH_2$$
$$H_2C \qquad C=O$$
$$O$$

$C_4H_6O_2$

mw		86.091	887
bp	760	204°	3083, 2114
dt/dp	760	0.0477°	[3083]
p	79°	10	1736
eq 4	A	8.60195	[3083]
	B	2728.28	
fp		– 43.53°	3083: 2114
d	25°	1.1254	3083
dd/dt		0.000969	[Beil **XVII**, 234]
n_D	25°	1.4348	3083
η	25°	1.7	1736
$\triangle H_v$	bp	12.48	[3083]
C_p	25°	34.4	[1736]
	60	38.7	
t_c		436°	1736
p_c		34	1736
pK_b		6.12 °	4290
ϵ	20°	39	1736
μ	25° in **30**	4.12	3162: 2998
soly	in aq	inf	1736
aq az		none	2255
fl pt	OC	209°F	1736
ir			644, 1975
mass			3076, 1662
nmr			1736, 2866
Beil	**2457**, 234		

200. Ethylene carbonate

cyclic ethylene carbonate

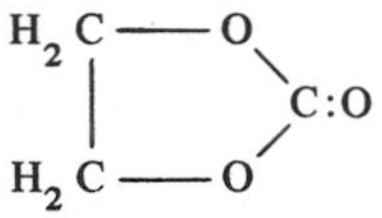

$$\begin{array}{c} H_2C \!\!-\!\! O \\ | \qquad \searrow C\!:\!O \\ H_2C \!\!-\!\! O \nearrow \end{array}$$

$C_3 H_4 O_3$

mw		88.064	887
bp	760	238.°	[4992] : 2552
dt/dp	760	0.043°	[4992]
p	36.4°	0.02	[4992]
eq 4	A	9.8173	[4992]
	B	3546.2	
fp		36.4°	4242: 2552, 3515
d	25°	1.3208	2552
n_D	40°	1.4199	2552
	50	1.4158	3160
$\triangle H_v$		11.98	3160
$\triangle H_m$		2.40	[2553]
$\triangle H_c$		– 199.2	[3160]
C_p	100°	40.6	[3160]
K_f		7.01°	2553
κ		$\langle 1 \times 10^{-7}$	4242
ϵ	40°	89.6	2552: 4242
μ	25° in **30**	4.87	2552: 4242, 2998, 270
soly	in aq, 40°	miscible	3160
aq az		none	2255
fl pt		320°F	3160
ir			247, 1975
nmr			4105, 269
Beil	**2736**, 100		

201. Ethyl carbonate

$(CH_3CH_2O)_2CO$ $C_5H_{10}O_3$

mw		118.134	887
bp	760	126.8°	4778: 687, 3013, 3001, 4978, 2712
dt/dp	760	0.042°	4778, 3001
p	23.8°	10.	4535
eq 4	A	8.2645	[4535]
	B	2148.10	
fp		−43.0°	4778, 4774: 4535
d	15°	0.98043	4778: 4978
	25	0.96926	687, 4745
	30	0.96393	4778
n_D	15°	1.38654	4778: 717, 4978, 2712
	25	1.38287	687: 4745
dn/dt		0.00039	4778
η	15°	0.868	4778: 1404, 2712
	25	0.748	687
γ	20°	26.44	2102, 4778: 539, 4978, 2023
	30	25.47	
$\triangle H_v$	29.53°	9.6	1263
	bp	8.64	3001
$\triangle H_c^{\circ}$	25°(l)	−645.0	2577: 4743, 4580
C_p	15 − 30°	50.38	2689: 4916, 3001
κ	25°	9.1×10^{-10}	2569: 422, 4106
ϵ	20°	2.820	3746, 4786
μ	25° in 30	0.90	4745
soly	in aq, 25°	insol	3681
	aq in, 25°	1.4%w	3160
aq az	91°	70%w	2254
fl pt	CC	77°F	88
uv			3091
ir			4375, 4738, 408, 4029, 4322, 2454
Raman			717, 3354, 4322, 1172
mass			786
nmr			2454
Beil	199, 5		

202. Ethyl oxalate

(COOCH$_2$CH$_3$)$_2$ C$_6$H$_{10}$O$_4$

mw		146.144	887
bp	760	185.4°	4778: 4535, 4762, 4971, 4088, 3001
dt/dp	760	0.052°	4772: 3001
p	25°	0.2	[4535]
eq 4	A	9.3626	[4535]
	B	2974.49	
fp		−40.6°	4778, 4535, 4770: 3637, 4762, 2556
d	15°	1.08426	4778: 687, 1400, 3637, 4426, 4971, 4088
	30	1.06687	4356, 3637
	95.65	0.98829	4772, 4773
dd/dt		0.00116	4778
n$_D$	15°	1.41239	4778: 3637, 4426, 4088, 4356
	20	1.41023	4971
dn/dt		0.00042	4778
η	15°	2.311	4778: 1404, 1400, 733
	30	1.618	
γ	20°	32.22	2102, 4778: 687, 4971
	30	31.03	
	Table VIII	CH II	
△H$_v$	bp	10.04	3001: 3016, 4963
△H$_c$	25°(l)	−723.3	2577
C$_p$	20°	63.29	4144: 3016, 2872, 3001
	Table XI	Ch II	
κ	25°	7.12 x 10^{-7}	2962
ε	21°	1.8	3477: 5034, 1381
μ	25° in 30	2.49	4426: 2877
soly	in aq, 25°	⟨3.6%w	3975
	aq in, 25°	⟨1.2%w	
aq az	hydrolyze s		3975
fl pt	OC	168°F	88
uv			2012, 2133, 748
ir			2857, 2858, 736, 1456
Raman			4097, 4099, 1167, 2452, 3642, 1171
Beil	170, 535		

203. Ethyl malonate

$CH_2(COOCH_2CH_3)_2$ $\qquad\qquad\qquad\qquad$ $C_7H_{12}O_4$

mw		160.171	887
bp	760	199.30°	4775: 3452, 4535, 4762, 4964
dt/dp	760	0.058°	4772
p	40°	1.	4535: 1634
eq 4	A	8.6110	[4535]
	B	2705.20	
fp		− 48.9°	3637: 4535, 4762, 4775
d	15°	1.06040	4775: 1326, 1400, 1634, 3452, 3637
	30	1.04460	3973, 4426, 4964, 4440
dd/dt		0.001053	[4775]
n_D	20°	1.41363	1634: 3452, 3637, 4426, 4440
dn/dt		0.00039	4426
η	20°	2.15	3452: 733, 1404, 1400
	25	1.94	
γ	20°	31.71	2102: 3452, 4964
	30	30.49	
	Table VIII	Ch II	
$\triangle H_v$	bp	13.10	[4535]
$\triangle H_c^\circ$	25° (l)	− 866.6	2577
C_p	21.44°	68.06	1264: 4144
K_a	25° in H_2O	5×10^{-14}	3696
ϵ	25°	7.87	4281: 1326, 1381
μ	25° in 30	2.54	4426: 4281
soly	in aq, 20°	2.7%w	4440
	aq in, 25°	$\langle$ 1.9%w	3975
aq az	99.5°	14%w	3975
fl pt		200°F	3681
uv			3888
ir			1469, 2857, 2858, 4738, 5124, 89, 5087
Raman			4097, 4099, 2452, 1171
mass			694
nmr			9
Beil	171, 573		

204. Methyl maleate

HCCOOCH$_3$
‖
HCCOOCH$_3$ C$_6$H$_8$O$_4$

mw		144.128	887
bp	760	200.4°	4811, 2386, 4535
dt/dp	760	0.046°	[4811]
p	25°	0.30	[4811]
eq 4	A	8.6530	[4811]
	B	2733.74	
fp		− 17.5°	4811
d	20°	1.1513	4811: 2386
	25	1.1462	
α	20 – 30°	0.00089	
n$_D$	20°	1.4422	4811: 2386
	25	1.4405	
	30	1.4385	
η	20°	3.54	4811
	25	3.21	
γ	20°	42.3	4811: 2386, 4541
	25	41.2	
$\triangle$H$_v$	av	12.2	4811
$\triangle$H$_m$		3.52	5088
$\triangle$H$_s$	54°	10.6	5247
$\triangle$H$_c^\circ$	25°(l)	− 674.8	2577
C$_p$	0.41 – 24°	61.98	5088
μ	25° in 232	2.48	2877
soly	in aq, 25°	8.0%w	4811
	aq in, 25°	4.4%w	
aq az	99.3°	12.3%w	3975
fl pt	COC	235°F	4811
uv			4077
ir			408
nmr			4632
Beil	179, 751		

205. Ethyl maleate

HCCOOCH$_2$CH$_3$
$\|$
HCCOOCH$_2$CH$_3$

C$_8$H$_{12}$O$_4$

mw		172.182	887
bp	760	225.3°	4811: 2386, 4535, 5048, 4805
dt/dp	760	0.053°	[4811]
p	30°	2.2	421
eq 4	A	8.4193	[4811]
	B	2760.9	
fp		– 8.8°	4805: 4811
d	20°	1.0687	4811: 739, 2386, 4426, 5048
	25	1.0637	
α	20 – 30°	0.00094	
n$_D$	20°	1.4400	4811: 739, 2386, 4426, 4805
	25	1.4383	
η	20°	3.57	4811: 733, 421
	25	3.14	
γ	20°	37.0	4811: 2386, 5048
	25	36.7	
△H$_V$	av	12.5	4811
[α]			4852
ε	23°	8.58	3477: 739
μ	25° in 30	2.54	4426: 2877
soly	in aq, 30°	ca 1.4%w	4811
	aq in, 30°	ca 1.9%w	
aq az	99.65°	11.8%w	2254
fl pt		250°F	3681
uv			4852
ir			408
Raman			737, 738, 4097, 4099
Beil	179, 751		

206. Butyl maleate

HCCOOCH$_2$ CH$_2$ CH$_2$ CH$_3$
‖
HCCOOCH$_2$ CH$_2$ CH$_2$ CH$_3$

$C_{12} H_{20} O_4$

mw		228.291	887
bp	760	ca 280°d	4811: 2386
dt/dp	150	0.22°	[4811]
p	25°	0.016	3826: 4811
eq 4 140 – 225°	A B	8.5064 3094.71	[4811]
fp	glass	⟨ – 80°	4811: 3681
d	20° 25 86.9	0.9950 0.9907 0.9376	4811: 2386 2386
α	20 – 30°	0.00088	[4811]
n$_D$	20° 25	1.4454 1.4435	4811: 2386
η	20° 25	5.63 4.76	4811
γ	20° 25	29.5 28.7	4811: 2386
△H$_v$	140 – 225°	14.2	4811
soly	in aq, 25° aq in, 25°	⟨ 0.05%w ⟨ 0.05%w	4811
aq az	99.9°	1.6%w	2255
fl pt	COC	285°F	4811
ir			408
Beil	179, 152		

207. Butyl phthalate

$1,2\text{-}C_6H_4(COOCH_2CH_2CH_2CH_3)_2$ $C_{16}H_{22}O_4$

mw		278.351	887
bp	760	340.0°	4535: 1056, 1703, 3290
p	89°	0.1	1703: 704, 827, 2169, 2496, 4535, 4369
eq 11	A	7.065	2444: 4369
134 – 340°	B	1666.	
	C	– 547700.	
fp		– 35°	1056: 733, 3290
d	20°	1.0465	905: 1703, 733, 3948, 3290
	25	1.0426	3975
dd/dt		0.00082	905
α		0.00086	1056
n_D	20°	1.4926	905: 1703, 3948, 3290
	25	1.4901	3975
η	– 195.7°gl	8.7 x 10 66	2967: 905, 733, 3290, 4229
	20	19.91	[3948]
	25	16.47	[733]
	37.8	9.48 cs	[3948]
γ	20°	33.40	271
$\triangle H_v$	bp	18.93	905: 2169
$\triangle H_c$	(l)	– 2058	70
$\triangle H_f^\circ$	(l)	245	70
C_p	21°	119.	4229: 1056
κ	30°	1.8 x 10⁻⁹	3836
ϵ	30°	6.436	3477
soly	in aq, 20°	⟨ 0.01%w	1056: 3296
	aq in, 20°	0.46%w	
aq az		none	3975
fl pt	COC	340°F	1056
uv			91, 4634
ir			408
Raman			452
mass			1322
Beil	**972**, 798		

208. Bis(2-ethylhexyl) phthalate

$$1,2\text{-}C_6H_4[COOCH_2CH(C_2H_5)CH_2CH_2CH_2CH_3]_2 \qquad\qquad C_{24}H_{38}O_4$$

mw		390.568	887
bp	5	231°	4889: 3290
p	183.0°	0.5	2170: 4369
eq 4	A	12.47	2444: 4369
190 – 277°	B	5757	
fp		– 50°	3290
d	20°	0.9843	3948: 3290
	25	0.979	[3773]
α	20°	0.00076	4889
n_D	20°	1.4859	4889: 3948, 3290
	25	1.4845	2170: 4828
η	0°	381.	4889: 3948, 3290
	20	81.4	
	40	22.3	
ϵ	20°	5.3	4028
μ	20°	2.84	4028
soly	in aq, 20°	$\langle$ 0.01%w	4889: 3290
	aq in, 20°	0.20%w	
fl pt	TOC	425°F	4893
Beil	**972**, 798		

209. Butyl sebacate

$C_4 H_9 OOC(CH_2)_8 COOC_4 H_9$ $C_{18} H_{34} O_4$

mw		314.469	887
bp	760	345°	1597: 2608, 3290
p	71°	0.001	3719: 4369
eq 4	A	14.10	3719: 4369
	B	4850	
fp		1°	3290: 733, 2608
d	20°	0.9366	3948: 733, 2608, 3290, 1599
	25	0.9324	3975
n_D	20°	1.4415	3975: 2608, 3948, 3290, 1468, 1597
	25	1.4397	
η	20°	9.03	[3948]: 3290
	25	7.96	[733]
$\triangle H_v$	bp	22.2	3719
κ	30°	1.7×10^{-11}	3836
ϵ	30°	4.540	3477: 1468
μ	20°	2.48	4028
soly	in aq	0.004%w	3290: 2608, 1597
fl pt	C OC	178°	2608: 3948
Beil	178, 719		

210. Butyl borate

$(CH_3 CH_2 CH_2 CH_2 O)_3 B$ $C_{12} H_{27} O_3 B$

mw		230.158	887
bp	760	233.5°	978: 4126, 977, 4811, 1963
dt/dp	760	0.051°	[978]
p	103.8°	10	4811: 272
eq 4	A	8.60626	[978]
	B	2902.14	
fp		⟨ - 70°	4811
d	20°	0.8580	4811: 272, 3618, 4126, 1963
	25	0.8535	
n_D	20°	1.4092	4811: 272, 3618, 4126, 1963
	25	1.4071	
	30	1.4051	
η	20°	1.776	4811: 1963
	25	1.601	
γ	20°	26.2	4811: 272
	25	25.8	
$\triangle H_v$	25°	12.968	[978]: 977, 4811
	bp	13.418	978
	Table IX	Ch II	
$\triangle H_f^\circ$	25°(l)	- 284.7	1686: 952
$\triangle H_c^\circ$	25°(l)	- 1925.2	1686
t_c		470.0°	977
p_c		19.63	977
v_c		0.8633	977
μ	25° in 232	0.77	294: 3618, 1865
soly		hydrolyzes	3975
aq az		hydrolyzes	3975
fl pt	TOC	170°F	4811
ir			5086, 5129
Raman			2401, 2473, 2474
nmr			3748
Beil	24, 369		

211. Butyl phosphate

$(CH_3CH_2CH_2CH_2O)_3PO$ $C_{12}H_{27}O_4P$

mw		266.320	887
bp	760	289°	1507: 858, 4979, 1499
p	114°	0.8	1614
eq 5	A	8.5861	1507
	B	3206.5	
	C	273	
fp		⟨ – 80°	1059
d	25°	0.9760	2016: 1614, 858, 1507, 3371, 4979, 1499
	40	0.9640	
	55	0.9540	
$d\frac{t}{4} = 0.9982(1 - 9.6 \times 10^{-4}t + 2.8 \times 10^{-6}t^2)$			2016
α		0.00093	1507
n_D	20°	1.42496	4979: 1614, 1507, 1499
	25	1.42256	858
dn/dt		0.000480	[4979, 858]
η	25°	3.39	3371: 858, 4853
	35	2.77	
	65	1.61	
γ	20°	27.55	1614
	40	25.96	
	60	24.44	
$\triangle H_v$	bp	14.680	1507
$\triangle H_f^{\circ}$	25° (l)	– 348.6	3542
$\triangle H_c^{\circ}$	25° (l)	– 1905.7	3542
ϵ	30°	7.959	3477: 1499
μ	25° in **30**	3.07	1499: 3298
soly	in aq, 25°	0.039%w	[858]: 2179, 2016, 3311
	aq in, 25°	4.67%w	
aq az	100°	0.6%w	2255
fl pt	COC	295°F	1059
ir			2256, 1727
nmr			4632, 4307, 4919
Beil	**24**, 369		

212. Fluorobenzene

$C_6 H_5 F$ $C_6 H_5 F$

mw		96.105	887
bp	760	84.734°	1352, 4208: 1628, 3388, 4536, 4972, 1100
dt/dp	760	0.040°	4772
eq 5	A	6.95208	4208: 5346
	B	1248.083	
	C	221.827	
fp		−42.21°	4208: 4536, 4770
d	15°	1.03091	4772: 314, 1290a, 1628, 4972, 5301
	30	1.01314	
n_D	15°	1.46837	4772: 314, 1628, 4536, 4972, 1707
dn/dt		0.00050	
η	15°	0.620	4772
	30	0.517	
γ	20°	27.78	4772: 1290a, 4972
	30	26.47	
$\triangle H_v$	45.25°	7.996	4208: 3331, 4962, 5346
	bp	7.457	
	Table IX	Ch II	
$\triangle H_m$		2.702	4208: 4536, 4769
$\triangle H_f^{\circ}$	25° (l)	−34.75	1835, 4208: 5358
$\triangle H_c^{\circ}$	25° (l)	−741.86	1835, 2577, 5358
C_p°	25°	22.57	4208
	Table XIV	Ch II	
C_p	25°	34.98	4208: 4536
	Table XII	Ch II	
t_c		286.92°	1352: 3331, 5301, 5303, 1100, 72, 3843
p_c		44.910	1352: 3843, 2651, 1509
d_c		0.3541	3843: 2651
v_c		0.2688	1352: 3331, 5303, 1100
A	eq 72	0.02549	4208
ϵ	25°	5.42	3477: 3802, 4786, 3247
μ	25° in 30	1.47	2874: 1628, 315, 2328, 2906, 2912, 3388

212. Fluorobenzene

(Continued)

soly	in aq, 30 $^\circ$ aq in, 25 $^\circ$	0.153%w 0.0316%w	1912 5218
uv			91, 302, 4384, 4465, 5250, 4004, 1296
ir			2862, 4547, 2451, 89, 4347, 3793
Raman			2478, 2680, 5228, 2670, 4347, 3434
mass			3087
nmr			4064, 1296, 1513
Beil	**464**, 198		

213. o -Fluorotoluene

o -CH$_3$ C$_6$ H$_4$ F   C$_7$ H$_7$ F

mw		110.132	887
bp	760	114.40°	3815: 3388, 4535
p	21.4°	20.	4535
eq 5	A B C	6.9732 1356.8 217.15	3815
fp		−62.0°	3815: 4535, 2627
d	13.2° 17.3	1.0041 1.00124	Beil **V**, 290 4772
n$_D$	21°	1.4738	4572
η	20° 30	0.680 0.601	4573

213. o-Fluorotoluene

(Continued)

$\triangle H_v$	bp	8.456	[3815]
$\triangle H_c^\circ$	25° (l)	- 903.2	1770
ϵ	30°	4.22	3477: 3464
μ	30° (l)	1.26	3056: 3158, 3388
uv			373, 4384, 4790, 368, 4233, 4603, 2449
ir			4735
Raman			2137, 2660
mass			3087
nmr			4664, 3966, 3446
Beil	**466**, 290		

214. m-Fluorotoluene

$m-CH_3C_6H_4F$ C_7H_7F

mw		110.132	887
bp	760	116.5°	3815: 3388, 4140, 4535
p	23.4°	20.	4535
eq 5	A	7.0095	3815
	B	1382.7	
	C	218.34	
fp		$- 87.7^\circ$	3815: 4535, 4764, 2627
d	20°	0.9974	1290a
n_D	27°	1.46524	4140

214. m–Fluorotoluene

(Continued)

η	20°	0.608	4573
	30	0.534	
γ	20°	27.97	1290 a
ϵ	30°	5.42	3477: 3464
μ	30° (l)	1.66	3056: 3388
uv			373, 368, 4233, 4603
ir			4735, 2863
Raman			1400, 1417
mass			90, 3087
nmr			4064, 3966
Beil	**466**, 290		

215. p–Fluorotoluene

p–$CH_3 C_6 H_4 F$ $C_7 H_7 F$

mw		110.132	887
bp	760	116.6°	3815: 3388, 4140, 4535, 4972
p	24.0°	20.	4535
eq 5	A	7.0574	3815
	B	1410.8	
	C	221.19	
fp		– 56.8°	3815: 2627
d	20°	0.9975	4972: 1430
	26.9	0.9911	
dd/dt		0.00107	[52]
α	30 – 40°	108 x 10^{-5}	1430

215. p–Fluorotoluene

(Continued)

n_D	20°	1.46884	4972: 4140
η	20° 30	0.622 0.522	4573
γ	26.9° 31.5 59.5	27.34 27.2 23.7	4972 52
$\triangle H_f^\circ$	25° (l)	– 44.07	5358
$\triangle H_c^\circ$	25° (l)	– 884.03	5358: 1770
ϵ	30°	5.86	3477: 3464
μ	30° (l)	1.76	3056: 2630, 3388, 4070
uv			373, 368, 4233, 4603
ir			1559, 4735, 3435
Raman			2141, 3685
mass			90, 3087
nmr			4664, 4129, 3966, 3446, 2342
Beil	**466**, 290		

216. Hexafluorobenzene

$C_6 F_6$

mw		186.057	887
bp	760	80.261°	1108: 1509, 3851, 3665, 5325, 3675
dt/dp	760	0.04072°	1108
p	19.75°	65.36	3675
eq 5	A	7.01741	1108: 1509, 3675
	B	1219.410	
	C	214.525	
fp		5.10°	1108: 3851, 1397
d	20°	1.61816	1108: 3851, 5325, 3675
	25	1.60682	
	30	1.59547	
α		0.001412	1108
n_D	20°	1.3781	3851: 3665, 5325
γ	20°	22.6	3851
$\triangle H_v$	25°	8.530	1108: 3675
	bp	7.571	
	Table IX	Ch II	
$\triangle H_m$		2.770	1108
C_p°	25°	37.43	1108
	Table XIV	Ch II	
C_p	25°	52.96	1108
t_c		243.57°	1108: 963, 3675, 1509
p_c		31.25	1509: 963
μ	25° in **30**	0.33	1396
	25° in **6**	0.0	
uv			3665, 368
ir			1268
Raman			1268
mass			1303
nmr			2240, 4508, 2248, 2342
Beil	**464**, 199		

217. 1,1,2–Trichloro–1,2,2–trifluoroethane

$CFCl_2CF_2Cl$ $C_2F_3Cl_3$

mw		187.376	887
bp	760	47.57°	1411: 2987, 2266, 516, 3980, 235, 1536
dt/dp		0.0403°	[3980]
p	25°	330.2	[3980] : 2266
eq 4	A	7.5534	[3980] : 3337, 2266
	B	1499.2	
fp		– 36.4°	2987: 516, 3980, 1536
d	0°	1.6200	2987: 2266, 3980, 1536
	25	1.56354	
	35	1.53982	
dd/dt		0.00229	[2987]
n_D	25°	1.35572	2987: 1536
	35	1.35124	
η	0°	0.925	2266
	20	0.711	
	30	0.627	
γ	0°	19.85	2266
	20	17.75	
	30	16.56	
$\triangle H_v$	26.9°	6.75	517: 3980
	bp	6.528	[1411] : 2266, 516
$\triangle H_f^\circ$	25° (g)	– 166.1	4690
C_p°	25°	30.23	2180
C_p	(g) 1 atm = 27.92 + 0.0374t		517
C_p	25°	40.8	2987: 516, 3980
	Table XI	Ch II	
t_c		214.1°	1411, 516
p_c		33.7	1411, 516
d_c		0.576	1411, 516
v_c		0.325	1411

217. 1,1,2-Trichloro–1,2,2-trifluoroethane

(Continued)

K_B		5.75 °	3960
ϵ	25°	2.41	1411
soly	in aq, 25° aq in, 25°	0.017%w 0.011%w	1411
aq az	44.5 °	99.0%w	1411
ir			2622, 2454, 2482
Raman			2622, 1801
nmr			4539, 2454, 2342
Beil	8, 86		

218. 1,1,2,2-Tetrachloro–difluoroethane

$CFCl_2CFCl_2$ $C_2F_2Cl_4$

mw		203.831	887
bp	760	92.8 °	2987: 2226, 235
p	28.1°	65.8	2226
eq 4	A B	8.1723 1929.27	[2226]
fp		26.0°	1410: 2987
d	25° 35	1.64470 1.62516	2987: 2226
n_D	25° 35	1.41297 1.40831	2987: 2226

218. 1,1,2,2-Tetrachloro–difluoroethane

(Continued)

η	25°	1.21	1411
	30	1.208	2226
	50	0.908	
γ	30°	22.73	2226
	40	21.56	
$\triangle H_v$	bp	8.36	[1410] : 2226
C_p	20°	24.9	[1410]
t_c		278°	1410: 2226
p_c		34.	1410
d_c		0.55	1410
v_c		0.370	1410
K_f		37.7°	543: 2585
ϵ	25°	2.52	1411
soly	in aq, 25°	0.012%w	1411
	aq in, 27.8°	0.0099%w	1410
ir			2465, 2482
Raman			2465
nmr			243, 4753, 3527, 3528
Beil	8, 86		

219. 1,2–Dibromo–tetrafluoroethane

$CF_2 BrCF_2 Br$ $C_2 F_4 Br_2$

mw		259.834	887
bp	760	47.26°	1412: 1801, 2987, 4073, 235
dt/dp	760	0.031°	[4073]
p	25°	434	[1412]
eq 4	A	7.768	4073
	B	1567.0	
fp		−110.5°	1412: 2987, 4073
d	10°	2.209	[1412]
	25	2.163	1411: 2987
dd/dt		0.00304	[4073]
n_D	25°	1.367	1411
η	25°	0.72	1411
	38	0.65	[1412]
γ	20°	18.9	[1412]
	25	18.1	
$\triangle H_V$	bp	7.166	4073
C_p	25°	43.1	1411
t_c		214.5°	1411
P_c		34.0	1411: 1108
d_c		0.790	1411
v_c		0.329	1411
ϵ	25°	2.34	1411
ir			2482
Raman			1801
nmr			2040, 1506, 3528, 3527
Beil	8, 92		

220. Chloroethane

CH_3CH_2Cl C_2H_5Cl

mw		64.515	887
bp	760	12.27°	4680: 1842, 4763, 4781, 3078, 2272
dt/dp	760	0.0350°	4680: 1842, 4781
p	25°	1633.0	1370: 4535, 4679, 2272
eq 5	A	6.94914	4679: 2944, 1842
	B	1012.771	
	C	236.7	
fp		-136.4°	4680: 1842, 4535, 4763, 4781, 4770, 3078
d	0°	0.92390	4781: 3078
	15	0.90280	
α	20°	0.0021	3160
n_D	0°	1.3790	1370: 3078
	10	1.3738	
η	5°	0.292	4781: 3078
	10	0.279	
γ	5°	21.20	4781: 3078
	10	20.64	
$\triangle H_v$	30 torr	6.64	[1370]: 3078
	bp	5.892	1842: 2391
$\triangle H_m$		1.064	1842: 329
$\triangle H_f^\circ$	25° (g)	-26.66	2272: 2817, 5358
$\triangle H_c^\circ$	25° (l)	-322.8	2577: 4743, 5358
C_p°	25°	15.00	4688: 1882, 2272, 2559
C_p	16.90°	24.69	1842: 2782, 3745, 3078
	Table XI	Ch II	
t_c		187.2°	4681, 547: 2651: 3078
p_c		52.	4681: 3078, 547
K_B		1.95°	486
κ	0°	$\langle 3 \times 10^{-9}$	5131
ϵ	20°	9.45	3533: 3477, 3078
μ	20° (l)	1.96	3533: 315, 1674, 2388, 3170, 4411, 3078

220. Chloroethane

(Continued)

soly	in aq, 0°	0.447%w	4494: 3078, 1677
fl pt	TCC	-58° F	3680
uv			2313
ir			1470, 3766, 3068, 89, 1277, 54
Raman			4274, 4986, 5017, 3068, 1170, 54
mass			3087
nmr			3057, 2085, 3126
Beil	8, 82		

221. 1-Chloropropane

$CH_3CH_2CH_2Cl$ C_3H_7Cl

mw		78.542	887
bp	760	46.60°	4778: 3210, 3452, 4535, 4763, 4966
dt/dp	760	0.0391°	4680: 4778
p	21.51°	300	4679: 4535
eq 5	A	6.93111	2944: 4679
	B	1121.123	
	C	230.20	
fp		-122.8°	4763, 4778: 854, 4535
d	20°	0.8909	4680: 1326, 3452, 4778, 4966
	25	0.8850	
dd/dt		0.00123	4778
n_D	20°	1.3879	4680: 3452, 4966
	25	1.3858	
dn/dt		0.00054	4778

221. 1–Chloropropane

(Continued)

η	15°	0.372	4778: 3452, 4711
	30	0.318	
	Table II	Ch II	
γ	20°	21.78	4778, 2102: 3452, 4966
	30	20.48	
$\triangle H_v$	25°	6.81	[1371] : 3210
	bp	6.51	
$\triangle H_m$		1.325	Beil E III 1, 219
$\triangle H_f^\circ$	25° (l)	-38.2	5358
$\triangle H_c^\circ$	25° (l)	-483.0	5358: 4743, 2577
C_p°	25°	20.24	1882: 2559
C_p	25°	31.6	[2782]
	Table XI	Ch II	
t_c		230.0°	547, 4681
p_c		45.18	547, 4681
ϵ	20°	7.7	3477: 1326, 2279
μ	(l)	1.97	3170: 4411, 2884
soly	in aq, 20°	0.271%w	[3680] : 1677
aq az	44°	97.8%w	2255: 2254
fl pt	TCC	$\langle$ 0	3680
uv			2313, 2314
ir			4547, 3068, 3679
Raman			3485, 3878, 3068, 1170
mass			90, 3087
nmr			937, 3146
Beil	10, 104		

222. 2-Chloropropane

$CH_3CHClCH_3$ C_3H_7Cl

mw		78.542	887
bp	760	35.74°	4680: 1919, 2306, 4966, 4784, 2272
dt/dp	760	0.039°	4680: 4784
p	25°	515.3	1371: 4535
fp		− 117.18°	4680: 4763, 4784
d	20°	0.8617	4680
	25	0.8563	
n_D	20°	1.3777	4680: 2272
	25	1.3749	
dn/dt		0.00056	4784
η	15°	0.335	4784
	30	0.286	
	Table II	Ch II	
γ	20°	18.09	1371: 4966
	30	17.13	
$\triangle H_v$	25°	6.42	[1371]
	bp	6.28	
$\triangle H_f^°$	25° (l)	− 39.2	5358: 2272
$\triangle H_c^°$	25° (l)	− 482.0	5358
$C_p^°$	25°	20.93	2272
t_c		212.°	1370
p_c		46.6	[1370]
d_c		0.341	1370
v_c		0.230	[1370]
A	eq 72	0.03661	1370
ϵ	20°	9.82	Beil E III 1, 222
μ	(l)	2.02	3170: 1919, 4411
soly	in aq, 12.5°	0.342%w	1677
aq az	35.0°	99%w	2255: 2254
fl pt	CC	− 26°F	88

222. 2-Chloropropane

(Continued)

uv		2313, 2314
ir		3068, 89
Raman		3878, 3068
mass		90, 3087
nmr		5143, 937, 2085, 3126
Beil	**10**, 105	

223. 1-Chlorobutane

$CH_3CH_2CH_2CH_2Cl$ C_4H_9Cl

mw		92.569	887
bp	760	78.44°	4680: 3210, 3452, 4418, 4857, 4966, 5225
dt/dp	760	0.0426°	4680, 1372: 4478
p	24.47°	100	4679: 905, 1373, 4413, 4535
eq 5	A	6.93790	2944: 1372
	B	1227.433	
	C	224.10	
fp		– 123.1°	4680, 854, 4764, 4478, 4857: 1372
d	20°	0.8862	4680: 315, 1372, 1451, 1919, 3452, 4784
	25	0.8809	4966, 2000
dd/dt	20 – 30°	0.00111	4891
α		0.00080	3160
n_D	20°	1.4021	4680, 1372: 315, 3452, 4413, 4418, 4423
	25	1.4001	4966, 5225

223. 1–Chlorobutane

(Continued)

dn/dt		0.00051	[4772] : 4478
η	$15°$	0.469	4784, 4478: 3452
	30	0.405	
γ	$15°$	24.40	4784, 4478: 2102, 3452, 4966
	30	22.88	
$\triangle H_v$	$25°$	7.96	[1370] : 3210
	bp	7.17	
C_p^o	$25°$	25.71	1882
C_p	$20°$	41.75	3160
t_c		$269.0°$	1370
p_c		36.37	[1370]
d_c		0.207	1370
v_c		0.3119	[1370]
κ	$30°$	10^{-10}	3535: 2000
ϵ	$20°$	7.39	3477: 3400, 4423, 2000
μ	-90 to $70°$ in 13	1.90	4423: 315, 1919, 3668, 4411, 4418, 5225
soly	in aq, $20°$ aq in, 20	0.11%w 0.08%w	905: 1677
aq az	$68.1°$	93.4%w	2254
fl pt	CC	$-18°$ F	3680
uv			2313, 2314
ir			4547
Raman			4986
mass			3087
Beil	**10**, 118		

224. 2-Chlorobutane

sec–Butyl chloride

$CH_3CH_2CHClCH_3$ C_4H_9Cl

mw		92.569	887
bp	760 (active)	68.25°	4784, 4767: 1451, 4426, 5225
dt/dp	760	0.041°	4784: 4680
p	25°	151.6	1371
eq 4	A	7.7223	[4535]
	B	1653.88	
fp	racemic	– 113.3°	4763, 854, 4784
	active	– 140.5	
d	0°, racemic	0.87556	4784: 1451, 4966
	15 , racemic	0.87880	
	20 , racemic	0.87323	
	30°, racemic	0.86210	
	0°, active	0.89497	4784
n_D	20°	1.3971	4680, 4784, 4966: 5225
	25	1.3942	
dn/dt		0.00048	4784
η	15°	0.439	4784: 2794
	30	0.363	
γ	19.7°	21.84	4966
	26.0	21.32	
$\triangle H_v$	25°	7.60	[1371]
	bp	6.98	
t_c		247.5°	4681
$[\alpha]_D$	20°	– 8.48	4784
κ	30°	$\langle 10^{-7}\rangle\,10^{-10}$	3535
ϵ	30°	7.090	4423
μ	25° in 30	2.07	4019: 3668, 4411, 4423, 5225
soly	in aq, 25°	0.1%w	Beil E III 1, 278
	aq in, 25	$\langle$ 0.1%w	3975
aq az	61.1°	94.3%	3975
fl pt	TCC	– 20°F	1996

224. 2-Chlorobutane

(Continued)

uv		2314
ir		3068, 3484
Raman		367, 3068, 1178, 3484
mass		3087
nmr		2461
Beil	**10**, 119	

225. 1-Chloro-2-methylpropane

$(CH_3)_2CHCH_2Cl$ $\qquad\qquad\qquad\qquad$ C_4H_9Cl

mw		92.569	887
bp	760	68.85°	4680, 4783, 4763: 1451, 3452, 5050, 4966
dt/dp	760	0.042°	4680, 4783
eq 4	A	7.6672	[4535]
	B	1641.4	
fp		− 130.3°	4680: 643, 4733, 4783, 4857
d	20°	0.8773	4680: 1451, 3452, 4966, 4783, 2794
	25	0.8717	
n_D	20°	1.3980	4680: 3452, 4783, 4966, 5225
	25	1.3951	
dn/dt		0.00052	4783
η	15°	0.471	4783: 3452, 4748, 2794
	30	0.395	
	Table II	Ch II	

225. 1–Chloro–2–methylpropane

(Continued)

γ	20°	21.99	2102: 3452, 4966, 2024
	30	20.87	
$\triangle H_v$	bp	7.91	[4535]
$\triangle H_c^\circ$	25° (l)	– 635.5	2577: 4743
C_p	20°	41.75	4780: 2782
	Table XI	Ch II	
κ	30°	$\langle 10^{-7} \rangle 10^{-10}$	3535
ϵ	14°	6.49	3477: 3007
μ	$10 - 50^\circ$	2.09	3668, 4411: 5225
	in **30**		
soly	in aq, 12.5°	0.092%w	1677
aq az	61.9°	92.1%w	3975: 2254
uv			3197
ir			1470, 3068
Raman			367, 3068, 2709
mass			3087
nmr			2728
Beil	**10**, 124		

226. 2-Chloro-2-methylpropane

(CH$_3$)$_3$CCl   C$_4$H$_9$Cl

mw		92.569	887
bp	760	50.7°	4680, 4775: 2792
dt/dp	760	0.040°	4680
p	25°	294.8	1371
eq 4	A	7.4644	[4535]
	B	1486.61	
fp		-25.4°	4680: 2792, 4770, 1319, 4775
d	20°	0.8420	4680: 2794, 4775
	25	0.8361	
n$_D$	20°	1.3857	4680: 2792
	25	1.3828	
dn/dt		0.00052	[4772] : 2271
η	15°	0.543	4775: 2794
	30	0.439	
γ	15°	20.06	4775: 4966, 2024
	30	18.35	
$\triangle$H$_v$	25°	7.03	[1371]
	bp	6.55	
$\triangle$H$_m$		0.500	4769: 2792, 1319
$\triangle$H$_f^\circ$	25°(g)	-42.99	2272: 2271, 5358
$\wedge$H$_c^\circ$	25°(g)	-640.6	5358
C$_p^\circ$	25°	27.30	2272
κ	30°	$\langle 10^{-7} \rangle\, 10^{-10}$	3535
ϵ	20°	9.961	314: 3477, 3400, 4786
μ	20° in 30	2.15	3668: 315, 3400, 4411, 4857, 5225, 2878
aq az	47 – 48°	hydrolyzes	3975
fl pt	TCC	23°F	3977
ir			3068, 1519, 2909
Raman			3068, 1170
mass			3087
nmr			2085, 3126
Beil	10, 125		

227. 1–Chloropentane

$CH_3(CH_2)_3CH_2Cl$ $C_5H_{11}Cl$

mw		106.596	887
bp	760	107.76°	4680, 3988: 1919, 3210, 3452, 4324, 4966
dt/dp	760	0.0456°	4680
p	25°	31.07	1371: 4679, 3988
eq 5	A	6.96617	2944: 3988
	B	1332.890	
	C	218.50	
fp		– 99.0°	4680: 4324
d	20°	0.8818	4680: 1919, 3452, 3988, 4324, 4966
	25	0.8769	
n_D	20°	1.4120	4680: 3452, 4324, 4966
	25	1.4100	
η	20°	0.580	3452: 4324
	25	0.547	
γ	20°	25.15	3452: 2102, 4966
	25	24.55	
$\triangle H_v$	25°	9.14	[1371]: 3210
	bp	7.82	
C_p°	25°	31.18	1882
C_p	20°	46.90	[4095]
ϵ	11°	6.6	3477
μ	20° in 30	1.94	3056: 1919, 4411, 5884
soly	in aq, 25°	0.02%w	Beil E III 1, 339
aq az	82°	67.9%w	2254
fl pt	CC	55°F	3680
Raman			1178
mass			3087
Beil	10, 130		

228. Chlorobenzene

C_6H_5Cl C_6H_5Cl

mw		112.560	887
bp	760	131.687°	5350: 1912, 3388, 4536, 4769, 4783, 1509
dt/dp	760	0.0488°	5350: 1372, 4783
p	25°	11.75	1369: 1325, 3167, 3928, 4535, 5350
eq 5	A	7.18473	1372: 5346
	B	1556.6	
	C	230.	
fp		−45.58°	1372: 732, 4351, 4764, 4771, 4774, 4788
d	15°	1.11172	4783: 600, 1372, 1628, 1764, 2134, 2589
	20	1.10630	3208, 3452, 4788, 4930, 4973, 4358
	30	1.09550	
α	20°	0.00098	3160
n_D	15°	1.52748	4271: 711, 1210, 1372, 2158, 3208, 3452
	20	1.52481	4973: 1707
dn/dt		0.00054	4783
η	20°	0.799	4793: 1400, 2158, 2751, 3452, 4783, 1605
	30	0.715	
γ	20°	33.28	2102: 3452, 4973, 5032, 5048
	30	32.11	
	Table VII	Ch II	
$\triangle H_v$	25°	10.17	2427: 3331, 5032, 4962, 5963
	bp	8.735	3208
$\triangle H_m$		2.284	4536: 4769
$\triangle H_f^o$	25°(l)	2.55	2287: 5358
$\triangle H_c^o$	25°(l)	−743.04	2287: 2463, 4743, 5358
C_p^o	25°	23.20	1808
C_p	26.84°	35.90	4536: 3745, 4144, 5195,
t_c		359.2°	2651: 71, 622, 5032, 5301, 5303, 1100
p_c		44.6	2651: 2608, 3331, 5301, 3843, 1509
d_c		0.365	2651: 3843
v_c		0.308	2651: 1100, 5303

228. Chlorobenzene

(Continued)

K_B		$4.15°$	2193: 5120
κ	$25°$	7×10^{-11}	3885: 5130, 4358
ϵ	$25°$	5.621	3477: 998, 2157, 2158, 3400, 5198, 4786
μ	$25°$ in **30**	1.54	.2874: 998, 1577, 2232, 3400, 3445, 4191 3769, 3465, 2876, 1131, 757
soly	in aq, $30°$ aq in, $25°$	0.0488%w 0.0327%w	1912: 664 [5218] : 2427
aq az	$90.2°$	71.6%w	2254
fl pt	CC	$85°F$	3680
uv			91, 301, 1347, 3418, 4465, 4004, 1196
ir			89, 403, 701, 2554, 2862, 3790, 4547 3343, 3393, 2020, 408, 144, 5278, 4347
Raman			1289, 1551, 2478, 2680, 4467, 4942, 5228 4252, 2822, 1178, 3300, 4347, 1169, 1176
mass			3087
nmr			954, 468, 2866, 1513
Beil	**464**, 199		

229. 1–Chloronaphthalene

$C_{10}H_7 Cl$

mw		162.620	887
bp	760	259.3°	2469: 4320, 325, 4535
dt/dp	760	0.062°	[2469]
p	60°	0.39	2444: 2168, 4535, 2469, 2627, 299
fp		– 2.3°	2608: 3669
d	15°	1.1976	[4320] : 325, 2158
	20	1.19382	2469
	40	1.1778	299
dd/dt		0.00077	299
n_D	20°	1.63321	2469: 299
η	25°	2.940	2158: 2108
γ	20°	42.05	299
	40	39.97	
$\triangle H_v$	105 – 140°	13.8	[2469]
	bp	12.44	[2469]
ϵ	25°	5.04	2158: 3477
μ	25° (l)	1.33	2158: 3486, 3669
aq az	99.9°	3.4%w	3975
fl pt	TCC	⟩ 175°F	3977
uv			3859, 1560, 1544
ir			996, 1241, 1561
Raman			3035, 996, 1241, 4222, 4603
mass			2398
Beil	477, 541		

230. Dichloromethane

Methylene chloride

CH_2Cl_2 CH_2Cl_2

mw		84.933	887
bp	760	39.75°	4682, 3185: 3208, 3452, 3720, 4535, 4978
dt/dp	760	0.037°	4682: 4779
p	25°	435.8	1371: 4535, 3720, 4683
eq 5	A	7.0803	4683
	B	1138.91	
	C	231.45	
fp		−95.14°	1371, 4682: 2281, 732, 4762, 4779, 4770
d	15°	1.33479	4779: 393, 1400, 1451, 1717, 3399, 3452
	25	1.31678	[4779] : 4978, 1892, 3078
	30	1.30777	4779
dd/dt		0.00180	[4779]
α		0.00137	4779: 3078
n_D	20°	1.42416	1370: 3185, 3399, 3400, 3452, 4978
	25	1.42115	
dn/dt		0.00055	[4772]
η	15°	0.449	4779: 1400, 1717, 3452, 4748, 1892, 3078
	30	0.393	
	Table II	Ch II	
γ	20°	28.12	4779: 2102, 3452, 4978, 3078, 2024
	30	26.54	
$\triangle H_v$	25°	6.825	[1371] : 4900
	bp	6.688	3208: 3720, 3078
$\triangle H_m$		1.000	4769: 4913
$\triangle H_f^{\circ}$	25° (l)	− 29.03	5358: 3193
$\triangle H_c^{\circ}$	25° (l)	− 133.34	5358: 2577
C_p°	25°	12.22	1735: 538, 1800, 1799
	Table XIV	Ch II	
C_p	19.3°	24.11	3717: 2782, 3078
	Table XI	Ch II	
t_c	237 °		4685, 1427: 3078, 3720
p_c		60.9	4685: 2608, 3078

230. Dichloromethane

(Continued)

K_B		2.60°	5023
κ	25°	4.3×10^{-11}	3876
ϵ	25°	8.93	393: 3112, 3477, 3399, 3400, 3078
μ	(l)	1.14	393: 1917, 3109, 3111, 3399, 3444, 3445
soly	in aq, 25°	1.30%w	2617: 3540, 3078, 664
	aq in 25°	0.198%w	
aq az	38.1°	98.5%w	2608: 2254
fl pt		none	3681
uv			3198, 4634, 4849, 5352
ir			5278, 2536, 1598, 387, 1092, 1477, 3794
Raman			2679, 3539, 4987, 5018, 3632, 4633, 2436
mass			1938, 1344
nmr			3057, 3173, 3174, 3126, 335
Beil	5, 60		

231. Chloroform

$CHCl_3$ $CHCl_3$

mw		119.378	887
bp	760	61.152°	5350: 3185, 3208, 3400, 3972, 4536, 4978
dt/dp	760	0.0403°	5350: 4761, 4783
p	25°	194.8	1371: 3879, 4123, 4165, 4535, 5350
eq 5	A	6.90328	2821
	B	1163.03	
	C	227.4	

231. Chloroform

(Continued)

fp		$-63.55°$	4536: 482, 732, 3301, 4351, 4762, 4771
d	15°	1.49845	4783: 1717, 3208, 3399, 4486, 4761, 4788
	25	1.47988	[4783] : 4978, 3078
	30	1.47060	4783
α		0.00126	3160: 3078
dd/dt		0.001857	[4783]
n_D	15°	1.44858	4783: 3185, 3208, 3400, 3452, 3972, 4978
	25	1.44293	1370: 5194, 5195, 3078
dn/dt		0.00059	4783
η	15°	0.596	4783: 1717, 2939, 3323, 3452, 4793, 3078
	30	0.514	
	Table II	Ch II	
γ	20°	27.16	1541: 22, 2004, 2102, 3452, 4978, 3078
	25	26.53	
$\triangle H_v$	20°	7.97	[4582] : 2791, 3776, 3078
	bp	7.020	3208: 3331, 4866, 5026, 4962
$\triangle H_m$		2.280	4769: 4601
$\triangle H_f^{\circ}$	25°(l)	-32.14	5358: 3193
$\triangle H_c^{\circ}$	25°(l)	-96.07	5358: 2463, 4743, 2577
C_p°	25°	15.71	1735: 538, 1800, 1799
	Table XIV	Ch II	
C_p	30°	27.94	5194: 3745, 3972, 5195, 4484, 3776, 3078
	Table XI	Ch II	
t_c		263.4°	4686, 2651: 3331, 5032, 1100, 622, 3078
p_c		54.	4686: 5026, 2145, 3078, 3331
d_c		0.50	4686, 2651: 2145
v_c		0.24	4686, 2651: 1100
K_f		4.90°	482: 492
K_B		3.62°	483, 491: 487, 4862, 2846, 3606, 580
κ	25°	$\langle 1 \times 10^{-10}$	5037
ϵ	20°	4.806	3477: 2747, 3034, 3399, 3400, 3516, 4786
μ	25° in 30	1.15	2948: 393, 2497, 3185, 3445, 4416, 4421

231. Chloroform

(Continued)

soly	in aq, 20° aq in, 23°	0.815%w 0.072%w	[4225]: 1912, 3078
aq az	56.12°	97.8%w	2254: 2255
fl pt	nonflammable		88
uv			2632, 3198, 4634, 4849, 5352
ir			514, 3178, 4740, 4817, 4740, 4817, 5278 2536, 4630
Raman			1551, 3539, 3899, 3908, 5342, 3916, 2703 1170, 2436, 2704
mass			1344
nmr			3057, 3173, 3174, 2454, 3126
Beil	5, 61		

232. Carbon tetrachloride

CCl_4 CCl_4

mw		153.823	887
bp	760	76.75°	1372, 3452, 4783: 3972, 4761, 5252, 5351
dt/dp	760	0.04320°	1372: 1260, 4761, 4783, 5350
p	25°	115.2	1370: 1373, 4265, 3879, 4413, 4535, 5252
eq 5	A B C	6.89406 1219.58 227.16	2184: 1372
fp		− 22.95°	1249, 4772: 4771, 3301, 4351, 4483, 4762

232. Carbon tetrachloride

(Continued)

d	$15°$	1.60370	4783, 5351: 600, 1118, 1372, 1717, 3208
	25	1.58439	777: 3399, 3678, 4761, 4783, 4978, 3452
	30	1.57480	4783: 778, 3078, 5254
α	$20°$	0.00127	3160: 5254
n_D	$15°$	1.46305	4783: 1210, 1245, 1372, 3208, 3452, 3628
	25	1.45739	777: 3972, 4413, 4486, 4978, 5195, 5351
dn/dt		0.00055	4772
η	$0°$	1.329	4772: 687, 1717, 2939, 3199, 3323, 3452
	15	1.038	4783, 4793, 3078
	30	0.845	
	Table II	Ch II	
γ	$20°$	26.75	3452, 2102: 22, 4978, 5032, 5034, 5048
	25	26.15	
	Table VII	Ch II	
$\triangle H_v$	$25°$	7.746	2184: 3331, 4866, **5026**, 5032, 2712, 4288
	bp	7.161	3208: 2791, 1079, 4962, 3078
$\triangle H_m$		0.581	4536: 2171, 4525
$\triangle H_f^o$	$25°$ (l)	− 32.37	5358: 349
$\triangle H_c^o$	$25°$ (l)	− 61.68	5358: 4743, 2577
C_p^o	$25°$	19.92	1735, 538: 1843, 1799
	Table XIV	Ch II	
C_p	$25°$	31.78	4985: 2171, 2384, 2689, 3745, 5337, 4484
	Table XI	Ch II	
t_c		$283.15°$	4687, 5301, 5303: 622, 3331, 5032, 1100
p_c		44.97	4687, 3331: 2651, 4952, 3078, 5301
d_c		0.558	4687, 2651
v_c		0.275	4687: 1100, 5303
K_f		$29.8°$	492
K_B		$4.88°$	480: 919, 3274, 5029, 491, 1803
κ	$18°$	4×10^{-18}	1723
ϵ	$20°$	2.238	3477: 998, 1326, 1888, 2747, 3399, 4786
			2048, 3247, 3276, 3078
μ	(l)	0	393: 4109, 3078

232. Carbon tetrachloride

(Continued)

soly	in aq, 25°	0.077%w	4225: 1912, 3546, 3078
	aq in, 24°	0.010%w	
aq az	66°	95.9%w	2254
fl pt	nonflammable		88
uv			2439, 2632, 3198, 4634, 4849, 5352
ir			1724, 3666, 3794, 89, 5278, 4630
Raman			976, 1422, 1551, 3408, 3539, 3899, 4942
			5342, 4252, 2540, 2996, 2436, 2704
mass			3997, 3373
nmr			3025, 3126, 468
Beil	5, 64		

233. 1,1-Dichloroethane

ethylidene chloride

CH_3CHCl_2 $C_2H_4Cl_2$

mw		98.960	887
bp	760	57.28°	4682, 3185: 1909, 3452, 3678, 4762, 4978
dt/dp	760	0.040°	4682: 4783
p	25°	227.7	1371: 3928, 4535
eq 5	A	6.9853	4683: 2943
	B	1171.42	
	C	228.12	
fp		−96.98°	2943: 2106, 4535, 4762, 4783

233. 1,1–Dichloroethane

(Continued)

d	15°	1.18350	4783: 1326, 1451, 2106, 3452, 3678, 4978
	25	1.1680	3452: 4358
	30	1.16010	4783
dd/dt		0.00156	[4783]
n_D	15°	1.41975	4783: 2106, 3452, 4978, 427
	20	1.4164	4682
	25	1.4138	
dn/dt		0.00052	4783
η	25°	0.505	4783: 3452, 4748
	30	0.430	
	Table II	Ch II	
γ	20°	24.75	2102, 3452: 4978
	30	23.62	
$\triangle H_v$	25°	7.409	2943: 2712, 1111
	bp	6.83	[1371]
$\triangle H_m$		1.881	2943
$\triangle H_c$	25° (l)	– 267.1	2577: 4743
C_p^o	25°	18.30	4688: 1163, 2943
C_p	25°	30.18	[2943]
	Table XI	Ch II	
t_c		250.0°	601: 4685
p_c		50.0	2651, 4535, 4685
K_B		3.13°	483
κ		2.0×10^{-9}	4358
ϵ	18°	10.0	3477: 1326, 4358
μ	25° in 30	1.98	1910: 4907, 3111, 3185
soly	in aq, 20°	5.03%w	1909: 4908
	aq in, 25°	⟨ 0.2%w	3975
aq az	53.0°	95.3%w	3975
uv			3198, 4634
ir			1477, 3178, 4740, 4741, 3766, 1163, 54
Raman			607, 1170, 2662, 54
nmr			3057, 4934, 3519
Beil	8, 83		

234. 1,2–Dichloroethane

ethylene chloride

CH_2ClCH_2Cl $C_2H_4Cl_2$

mw		98.960	887
bp	760	83.483°	4382: 426, 1909, 3883, 4762, 4784, 581
dt/dp	760	0.047°	4684: 1372, 4761, 4784, 4382
p	20°	83.35	1371: 1570, 3879, 4382, 1213
eq 5	A	7.04636	1372: 1689, 4382
	B	1305.4	
	C	230	
fp		– 35.66°	4684: 426, 1570, 3774, 4762, 1372, 4770
d	0°	1.28164	4784: 354, 426, 1326, 4358, 1451, 1570
	20	1.2531	4684
	25	1.2458	4684: 5357, 3678, 4761, 4951, 4972, 581
	30	1.23831	4784
dd/dt		0.00144	[4784]: 3078, 3452
α	55°	0.00121	4891
n_D	15°	1.44759	4784: 426, 1372, 2158, 3208, 3400, 3452
	20	1.4448	4684
	25	1.4421	4684: 3628, 4972, 581, 1687, 3078, 4772
dn/dt		0.00051	4784
η	15°	0.887	4784: 354, 1400, 1570, 2158, 2234, 3452
	30	0.730	5357, 1605, 3078
	Table II	Ch II	
γ	20°	32.23	2102: 1570, 1688, 2234, 3452, 4972, 3078
	30	30.84	
	Table VIII	Ch II	
$\triangle H_v$	25°	8.19	[1371]: 3078
	bp	7.654	3208: 1570, 1689, 5032
$\triangle H_m$		2.112	3774: 4039, 4769
$\triangle H_f^o$	25° (l)	– 39.6	4341, 5358
$\triangle H_c^o$	25° (l)	– 265.65	4341: 4743, 2577, 5358
C_p^o	25°	18.48	4688: 1951
C_p	26.72 °	30.86	3774: 1570, 2384, 2782, 3883, 4484, 3078
	Table XI	Ch II	

234. 1,2-Dichloroethane

(Continued)

t_c		288.°	4685: 622, 1689, 2235, 2234, 3078, 360
p_c		53.	4685, 2651: 3078, 5034
d_c		0.44	4685, 2651
v_c		0.22	4685: 1100
K_B		3.44°	5053: 483, 491
κ	25°	4×10^{-11}	3885: 2744, 5036, 5357, 4358
ϵ	25°	10.36	3477: 998, 2157, 2158, 4358, 5034, 4786
μ	25° in **30**	1.86	1910: 4907, 998, 2175, 3400, 4486, 5193
soly	in aq, 20° aq in, 20°	0.81%w 0.15%w	905: 1213, 1909, 1912, 3078
aq az	72°	91.8%w	3975: 1213, 1995, 2254, 2255
fl pt	CC	56°F	3680: 1570
uv			3198, 2314
ir			3178, 3790, 4287, 4321, 4375, 89, 3766
Raman			1170, 297, 3353, 3397, 3517, 3241, 5270
mass			90
nmr			3057, 3947
Beil	8, 84		

235. 1,1,1-Trichloroethane

CH_3CCl_3 $C_2H_3Cl_3$

mw		133.405	887
bp	760	74.0°	4772: 428, 2106, 3185, 4857, 4978, 5225
dt/dp	760	0.044°	4772
p	25°	120.7	1371: 4065, 4535
eq 4	A	6.92013	4065
	B	1729	
fp		-30.4°	2106: 428, 4065, 4857, 4913, 1139, 4770
d	15°	1.34587	4772: 428, 2103, 3185, 4978, 581
	20	1.3376	4978
	30	1.32096	4772
dd/dt	$0-30^\circ$	0.001657	[4772]
n_D	20°	1.4379	4857: 4978, 4772, 1139, 581
	25	1.4359	1371
dn/dt		0.00052	[4772]
η	15°	0.903	4772
	25	0.795	1066
	30	0.725	4772
γ	15°	26.17	4772: 4978
	20	25.56	
	30	24.25	
$\triangle H_v$	25°	7.718	[1370] : 4900
	bp	7.692	4065
$\triangle H_m$		1.075	4913, 4065: 1139
$\triangle H_f^\circ$	25° (g)	$-33.$	4689
C_p°	25°	22.39	4688
C_p	26.43°	34.49	4065: 1139
t_c		260°	4065
p_c		50	4065
κ		7.3×10^{-9}	1354
ϵ	20°	7.53	3477: 1066, 4857, 4786, 3802
μ	in 30	1.57	4556: 3185, 5225

235. 1,1,-Trichloroethane

(Continued)

soly	in aq, 20° aq in, 25°	0.132%w 0.034%w	4908 4485
aq az	65.2°	91.7%w	3975
fl pt		None	1354
uv			2314
ir			89, 2454, 54
Raman			2317, 3902, 2662, 1517, 54
nmr			3057, 2454
Beil	8, 85		

236. 1,1,2,2-Tetrachloroethane

Acetylene tetrachloride

CHCl$_2$CHCl$_2$ C$_2$H$_2$Cl$_4$

mw		167.850	887
bp	760	146.20°	4783: 1451, 3208, 3452, 4763, 4978, 3078
dt/dp	760	0.050°	4783
p	25°	5.95	1371: 255, 2147, 3232, 3511, 4535
eq 12	A B C	35.117 3646 8.981	3232: 3511
fp		−43.8°	4783: 255, 2147, 4535, 4763, 3078
d	15° 25 30	1.60255 1.58655 1.57860	4783 [4783]: 255, 3208, 3452, 3678, 4978, 5047 4783: 3078

236. 1,1,2,2-Tetrachloroethane

(Continued)

dd/dt		0.00159	[4783]
α		0.000998	2146
n_D	15°	1.49678	4783
	25	1.4910	1371: 3208, 3452, 4978, 3078
dn/dt		0.00051	4783
η	15°	1.844	4783: 2142, 3452, 5047, 1605, 3078
	30	1.456	
γ	20°	36.04	1451: 3452, 4978, 3078
	40	33.30	
$\triangle H_v$	25°	10.75	1371: 255, 2146, 3232, 3078
	bp	9.236	3208
$\triangle H_f^\circ$	25° (g)	– 35.6	4689
C_p°	25°	23.66	4688
C_p	20°	45.0	2147: 255, 2782, 3078
	Table XI	Ch II	
κ	not measurable		5047
ϵ	20°	8.20	3477: 5051
μ	25° in 30	1.71	3355: 4907, 657, 4718
soly	in aq, 20°	0.287%w	[4908]: 3078
aq az	94.3°	66.0%w	3975
fl pt	nonflammable		255
uv			3200, 2314
ir			89, 3790, 4375, 4547, 4740, 2804, 2454
Raman			230, 3306, 3902, 1170, 603, 412
nmr			1944, 3057, 2454, 4598
Beil	8, 86		

237. Pentachloroethane

$CHCl_2CCl_3$ C_2HCl_5

mw		202.295	887
bp	760	162.00°	4783: 3452, 4763, 3078
dt/dp	760	0.054°	4783: 1249
p	25°	4.4	3511: 255, 4535
eq 4	A	7.80304	3511
	B	2129.6	
fp		-29.0°	4783, 4763: 255, 3078
d	0°	1.71100	4783, 255, 2146, 3452, 3078
	15	1.68813	
	30	1.66530	
n_D	15°	1.50542	4783: 3452, 3078
	20	1.5030	4773
dn/dt		0.00046	4783
η	15°	2.751	4783: 2142, 3452, 3078
	30	2.070	
γ	20°	34.72	2102: 3452
	30	33.58	
$\triangle H_v$	av	9.75	[2444]: 3078
	bp	8.829	2146: 255
$\triangle H_m$		2.710	4913
$\triangle H_f^\circ$	25° (g)	-34.5	4689
C_p°	25°	28.15	4688
C_p	18 – 53°	44.31	5117: 255, 2147, 2782, 2575
	Table XI	Ch II	
t_c		373.0°	622
κ	not measurable		5037
ϵ	20°	3.73	3477: 5037, 5051, 2575
μ	25 – 85°(l)	0.94	2381: 4907, 4718, 2575
soly	in aq, 25°	0.05%w	3185: 2575
	aq in, 25°	0.03%w	
aq az	95.8°	64.9%	3185: 2254
fl pt	nonflammable		255

237. Pentachloroethane

(Continued)

uv		3441
ir		3771, 4547, 4740, 4474, 54
Raman		1742, 3306, 3409, 54
nmr		3057, 3173, 3174
Beil	8, 87	

238. o-Dichlorobenzene

$o\text{-}C_6H_4Cl_2$ $C_6H_4Cl_2$

mw		147.005	887
bp	760	180.48°	1372: 922, 2328, 3388, 3070
dt/dp	760	0.05270 °	1372
p	25°	1.28	1369: 1373, 4535, 3070, 5357
eq 5	A	7.07028	3070: 1372
	B	1649.55	
	C	213.314	
fp		- 17.01°	5227: 1372, 922, 2328, 3070
d	20°	1.30589	1372: 922, 5357, 1892
	25	1.30033	
α		0.00085	3160
n_D	20°	1.55145	1372: 922, 2328
	25	1.54911	
dn/dt		0.00049	[4772]
η	25°	1.324	5357
	31.9	1.2018	1892

238. o–Dichlorobenzene

(Continued)

γ	20°	26.84	1369
	30	35.55	
$\triangle H_v$	25°	12.00	2427
	bp	9.48	2608
$\triangle H_m$		3.01	2608
$\triangle H_f^\circ$	25° (l)	– 4.4	4341: 1212
$\triangle H_c^\circ$	25° (l)	– 707.97	4341: 2577, 1212
C_p	25°	52.98	[2608]
t_c		424.1°	1369
p_c		40.5	[1369]
d_c		0.408	1369
v_c		0.360	[1369]
A	eq 72	0.02215	1369
κ	25°	3×10^{-11}	3885
ϵ	25°	9.93	3477: 3400, 5357
μ	24° in 30	2.27	3056: 4754, 1577, 2328, 3388, 3400, 4420
soly	in aq, 25°	$\langle$ 0.026%w	[664]
	aq in, 25°	0.309%w	5218: 2608, 2427
aq az	98.2°	33.1%w	3975
fl pt	CC	151°F	3680
uv			373, 3833, 4308, 4465, 4603, 254
ir			3790, 408, 89, 2518, 4139, 4474
Raman			2137, 4467, 1178, 4139, 1169
nmr			4632, 49, 4508, 3808, 3172, 1863
Beil	464, 201		

239. m-Dichlorobenzene

m–C$_6$H$_4$Cl$_2$ C$_6$H$_4$Cl$_2$

mw		147.005	887
bp	760	173.00°	1372: 1373, 3407, 4535, 4972
dt/dp	760	0.05206°	1372
p	25°	1.89	1369: 1373, 4535
eq 5	A	7.30364	1372
	B	1782.4	
	C	230	
fp		– 24.76°	1372: 4535
d	20°	1.28844	1372: 4972, 1892
	25	1.28280	
n_D	20°	1.54586	1372: 4972
	25	1.54337	
η	23.3°	1.0450	1892
	32.8	0.9551	
	40.4	0.8807	
γ	20°	36.16	4972
	41.8	33.53	
$\triangle H_v$	25°	11.61	[1369]
	bp	9.23	
$\triangle H_f^\circ$	25° (l)	– 4.89	1212
$\triangle H_c^\circ$	25° (l)	– 706.44	1212
t_c		410.8°	1369
p_c		38.3	[1369]
d_c		0.410	1369
v_c		0.359	[1369]
ϵ	25°	5.04	3477: 3400
μ	24° in 30	1.38	3056: 3400, 4420
soly	in aq, 20°	0.0111%w	Beil E III 5, 542
fl pt		162.°F	1369
uv			373, 3418, 4465, 4603
ir			4547, 408, 2521, 4139, 3793
Raman			2136, 2139, 4467, 1178, 4139
nmr			1583
Beil	464, 202		

240. p–Dichlorobenzene

p-$C_6H_4Cl_2$ $\qquad\qquad\qquad\qquad\qquad\qquad$ $C_6H_4Cl_2$

mw		147.005	887
bp	760	174.12°	1372: 4535
dt/dp	760	0.05217°	1372
p	25°	1.76	1369: 1373, 4535, 5346
eq 5	A	6.99800	3070: 5346, 5074
	B	1575.11	
	C	208.513	
fp		53.13°	5074, 3070: 936, 1912, 3168, 3399, 4601
d	55°	1.24750	1372: 3399, 4063, 1892, 3226
	60	1.24166	
n_D	60°	1.52849	1372
η	55.4°	0.8394	1892: 3226
	69.3	0.7202	
	79.4	0.6678	
γ	68°	30.69	4772
	96	27.58	
$\triangle H_v$	25°	11.7	1369: 5346
	bp	9.27	
$\triangle H_s$	α–form	15.48	5074
	β–form	15.06	
$\triangle H_m$		4.34	2280: 4601, 3782a
$\triangle H_f^\circ$	25°(s)	– 10.12	1212
$\triangle H_c^\circ$	25°(s)	– 701.27	1212: 4397
C_p	(s)	ca 40	2608: 2279
ϵ	50°	2.41	3477: 936, 3399
μ	24° in **30**	0	3056, 2217, 3168, 5192: 4754, 2875, 3876
soly	in aq, 35°	0.010%w	2626: 664, 1912
α–form	in aq, 35°	0.008%w	
β–form	in aq, 55°	0.016%w	
fl pt	OC	153°F	3681
uv			373, 4465, 4603, 254
ir			3967, 4547, 408, 2520, 4139, 1272, 3793
Raman			2667, 3685, 4467, 5001, 1178, 4139, 4797
mass			3085, 4499
nmr			1583,
Beil	**464**, 203		

241. 3–Chloropropene

CH$_2$=CHCH$_2$Cl   C$_3$H$_5$Cl

mw		76.526	887
bp	760	45.10°	4779: 2386, 3452, 4265, 4762, 5266
dt/dp	760	0.042°	4779
p	25°	366.8	1371: 2351, 4265, 4535
eq 4	A	7.848	2444
	B	1577.	
fp		– 134.5°	4774, 4779: 4265, 4535, 4762, 4770
d	15°	0.94419	4779: 2386, 3452, 4265
	30	0.92454	
n$_D$	20°	1.41566	2386: 5266, 3452, 4265
	25	1.4130	2608
dn/dt		0.00069	[4772]
η	15°	0.347	4779: 3452
	25	0.313	5266
	30	0.300	4779
	Table II	Ch II	
γ	15°	28.85	4779: 2386, 3452
	30	21.83	
$\triangle$H$_v$	25°	6.75	[1371]
	bp	6.940	2608: 2351
$\wedge$H$_f^\circ$	25°(g)	– 7.	3193
$\triangle$H$_c^\circ$	25°(g)	– 440.8	2577: 4743
C$_p^\circ$	0°	24.1	4265
t$_c$		ca 241°	2608
ϵ	20°	8.2	3477: 1326, 5034
μ	104°(g)	1.98	2005
soly	in aq, 20°	0.36%w	2608: 4265
	aq in, 20°	0.08%w	
aq az	43.0°	97.8%w	4265
fl pt	CC	– 25°F	3680
uv			506
ir			4741, 3679, 89, 1908
Raman			3732
mass			4959
nmr			2514
Beil	**11**, 198		

242. 1,1–Dichloroethylene

vinylidene chloride

$CH_2{=}CCl_2$ $\qquad\qquad\qquad\qquad\qquad\qquad\qquad\qquad\qquad$ $C_2H_2Cl_2$

mw		96.944	887
bp	760	31.56°	2185: 3944
dt/dp	760	0.037°	[2185]
p	25°	599.0	1501
eq 5	A	6.98200	2185
	B	1104.29	
	C	237.697	
fp		-122.56°	2185: 3944
d	0°	1.2517	1360: 1501, 4015, 3944
	20	1.2132	
	25	1.1747	[1360]
dd/dt		0.00192	
n_D	10°	1.43062	1360: 1501, 3944, 4015, 2800
	15	1.42777	
	20	1.42468	
η	0°	0.422	1501
	20	0.358	
$\triangle H_v$	25°	6.328	2185
	bp	6.257	1360
$\triangle H_m$		1.557	2185
$\triangle H_f^\circ$	25° (l)	-6.0	4341: 3293
$\triangle H_c^\circ$	25° (l)	-261.93	1360, 4341
$\triangle H_p$	25°	-18.0	1360
C_p°	25°	16.02	2185: 1360
C_p	25.15°	26.745	1360: 2185
t_c		222°	2185
p_c		51.3	2185
v_c		0.219	1360
μ	25° in **30**	1.30	4015: 1493, 709

242. 1,1-Dichloroethylene

(Continued)

soly	in aq, 25°	0.021%w	1360
	aq in, 25°	0.035%w	
aq az	30.4°	93.0%w	3975
fl pt	OC	14°F	1501
uv			2799, 3441, 5073
ir			1515
Raman			2996, 1515
nmr			5138
Beil	11, 186		

243. cis-1,2-Dichloroethylene

HCCl
‖
HCCl

$C_2H_2Cl_2$

mw		96.944	887
bp	760	60.63°	2564: 2147, 3185, 3452, 4763, 3078, 2282
dt/dp	760	0.038°	[4535]
p	25°	200	2564: 2147, 4535, 2282
eq 4	A	7.8522	[4535]
	B	1651.52	
fp		- 80.0°	750: 4535, 4763, 4774, 3078
d	15°	1.2917	2565: 1451, 2147, 3452, 5047, 3078
	20	1.2837	2565

243. cis–1,2–Dichloroethylene

(Continued)

dd/dt		0.0016	[2565]
α	15–45°	0.00127	3078
n_D	15°	1.45189	2565: 2142
	20	1.4490	4772
dn/dt		0.0059	2565
η	20°	0.467	2565: 2142, 2147, 3452, 5047, 3078
	25	0.444	
γ	ca 20°	28	2565: 3452
$\triangle H_v$	0°	7.545	2564: 2147
	bp	7.225	2564: 3078
$\triangle H_m$		1.722	750
$\triangle E_c$	18.7°	– 261.06	Beil E III I, 651
C_p°	25°	15.55	3779: 2448
C_p	15°	27.24	3259: 3078
	Table XI	Ch II	
t_c		271.0°	3078
κ	25°	8.5 x 10^{-9}	5047
ϵ	25°	9.20	3477: 2564, 5051, 3078
μ	25° in 30	1.76	3056: 657, 3185, 3445, 3078, 1493, 709
soly	in aq, 25°	0.35%w	3078
	aq in, 25°	0.55%w	3078
aq az	55.3°	96.65%w	2254
fl pt	CC	39°F	88: 3078
uv			2799, 2314, 5071
ir			539, 1477, 3771, 4547, 3779, 4153
mass			90, 1341
nmr			2845, 2278, 5138
Beil	11, 187		

244. trans–1,2–Dichloroethylene

ClCH
‖
HCCl

$C_2H_2Cl_2$

mw		96.944	887
bp	760	47.67°	2564: 2147, 3452, 4763, 3078, 2282
dt/dp	760	0.031°	[2564]
p	25°	340.	[2564]: 2147, 4535, 2282
eq 4	A	7.55657	[2564]
	B	1498.42	
fp		–49.8°	750: 4535, 4763, 4774, 3078
d	15°	1.2631	2565: 1451, 2147, 3452, 3078
	20	1.2547	2565
dd/dt		0.00168	[2565]
α	15–45°	0.00136	3078
n_D	15°	1.45189	2142
	20	1.4462	1451
η	15°	0.423	2565: 2142, 3452, 3078
	20	0.404	
γ	ca 20°	25	2565: 3452
$\triangle H_v$	0°	7.180	2564: 2147
	bp	6.905	2564: 3078
$\triangle H_m$		2.864	750
$\triangle E_c$	18.7°	–261.63	Beil E III 1, 654
C_p°	25°	15.93	3779: 2448, 493
C_p	15°	26.95	3259: 3078
t_c		243.3°	3078
p_c		54.4	3078
ϵ	25°	2.14	3477: 2564, 3078
μ	25° in 30	0.70	3056: 657, 3445, 3078, 1493, 709
soly	in aq, 25°	0.63%w	3978
	aq in, 25°	0.55%w	3078
aq az	45.3°	98.1%w	2254
fl pt	OC	39°F	3160, 3078

244. trans–1,2,–Dichloroethylene

(Continued)

uv		2799, 5070, 2314
ir		539, 1477, 3771, 4547, 3779, 4153
mass		90, 1341
nmr		2845, 2278, 5138
Beil	**11**, 187	

245. Trichloroethylene

CHCl=CCl$_2$ C$_2$HCl$_3$

mw		131.389	887
bp	760	87.19°	3069: 255, 907, 1451, 3208, 3452, 4763
dt/dp	760	0.041°	[1370]
p	25°	47.31	1370: 255, 4535
eq 5	A	7.02808	2821: 3069
	B	1315.04	
	C	230.0	
fp		– 86.4°	4763: 255, 907, 4535, 3078
d	15°	1.4762	907: 255, 1451, 2147, 3069, 3208, 3452
	30	1.4514	4829, 3078
dd/dt		0.001649	[2773]
α	0–40°	0.00117	4482: 2146, 3078
n$_D$	21.4°	1.4767	3069: 3078, 907, 3208, 3452, 4829, 2773
	24.6	1.4750	
dn/dt		0.0005675	[3069]
η	20°	0.566	3452: 907, 2142, 4829, 3078
	25	0.532	
γ	20°	29.5	3452: 3078
	25	28.8	

245. Trichloroethylene

(Continued)

$\triangle H_v$	$25°$	8.190	1370
	bp	7.521	3208: 255, 907, 2146, 3069, 3078
$\triangle H_f^o$	$25°$(g)	1.4	3193
$\triangle E_c$	$18.7°$	-230.01	Beil E III 1, 659
C_p^o	$25°$	19.13	493: 538, 2773, 53
C_p	$20°$	29.3	2147: 255, 907, 2782, 4829, 3078, 2773
t_c		$298.°$	1370: 3078
p_c		48.5	[1370]: 3078
d_c		0.513	1370
v_c		0.256	[1370]
K_B		$4.43°$	5053
κ		8×10^{-12}	2773: 5037
ϵ	ca $16°$	3.42	3477: 907, 5037, 5051, 3078
μ	in 232	0.8	4907: 657, 824, 3078, 709
soly	in aq, $25°$	0.11%w	2608: 907, 3078
	aq in, $25°$	0.32%w	
aq az	$73.6°$	94.6%w	2254
fl pt	nonflammable		4482
uv			2812, 2799, 5072
ir			537, 3771, 3789, 4153
Raman			3902, 4111, 4375, 53
nmr			473, 3174, 5138
Beil	11, 187		

246. Tetrachloroethylene

$CC1_2{=}CC1_2$ C_2Cl_4

mw		165.834	887
bp	760	121.20°	4783, 4778: 3208, 3452, 4763, 3078
dt/dp	760	0.050°	4783, 4778
p	25°	18.47	1370: 255, 1373, 3928, 4535, 4536
eq 5	A	7.02003	2821
	B	1415.49	
	C	221.0	
fp		$-$22.35°	4774, 4763: 4778, 255, 3078
d	15°	1.63109	4783: 255, 2146, 3208, 3452, 4778
	30	1.60640	
dd/dt		0.001646	[4783]
α	0 – 25°	0.00102	4482: 3078
n_D	15°	1.50759	4778: 3452, 3078
	20	150566	3208
dn/dt		0.00053	4778
η	15°	0.932	4783, 4778: 2142, 3452, 3078
	30	0.798	
γ	15°	32.86	4783: 4778, 3452, 3078
	30	31.27	
$\triangle H_v$	25°	9.467	[1370]: 3078
	bp	8.299	3208: 255, 2146
$\triangle H_m$		2.525	4913
$\triangle H_f^\circ$	25° (g)	$-$3.6	5358: 1384
$\triangle H_c^\circ$	25° (l)	$-$162.5	2577: 4743, 5358
C_p°	25°	22.80	538
C_p	20°	35.8	2147: 255
	Table XI	Ch II	
t_c		347.1°	2608: 3078
p_c		44.325	[1370]
d_c		0.573	1370: 3078
v_c		0.2909	[1370]

246. Tetrachloroethylene

(Continued)

K_B		5.50°	460
κ	20°	5.55 x 10^{-4}	3160
ϵ	25°	2.30	3477: 5051, 3078
μ	25° in **232**	0	709: 657, 3078
soly	in aq, 25°	0.015%w	2608: 3078
	aq in, 25°	0.0105%w	
aq az	87.7°	84.2%w	2254
fl pt	nonflammable		3160: 255
uv			3200, 2799, 1852, 2318
ir			1385, 3789, 4800
Raman			3902, 4111, 5229, 1170, 4934, 2436, 412
Beil	**11**, 187		

247. Bromoethane

CH_3CH_2Br C_2H_5Br

mw		108.971	887
bp	760	38.35°	4692: 3452, 4352, 4413, 4761, 4783, 4966
dt/dp	760	0.0381°	5350: 4761, 4783
p	25°	468.6	1371: 4535, 3928, 5350
eq 5	A	6.91995	2499
	B	1090.810	
	C	231.71	
fp		-118.6°	4774, 4770: 732, 4112, 4352, 4535, 4783

247. Bromethane

(Continued)

d	0°	1.50136	4783: 1919, 3399, 3400, 3452, 3678, 4352
	15	1.47080	4783: 4421, 4761, 4777, 4966, 4413
	25	1.4505	4692
	30	1.44030	
n_D	15°	1.42756	4783: 2158, 3452, 4413, 4966, 3025
	20	1.4239	4692
	25	1.4212	
dn/dt		0.00056	4783: 1915
η	15°	0.418	4783: 2158, 2163
	25	0.379	3452
	30	0.348	4783
	Table II	Ch II	
γ	15°	24.83	2102: 3452, 4966, 2024
	20	24.15	
	30	22.83	
$\triangle H_v$	25°	6.506	[1371]: 5034, 4962, 4867
	bp	6.328	
$\triangle H_m$		1.400	4769
$\triangle H_f^o$	25° (l)	− 21.5	5358: 2817
$\triangle H_c^o$	25° (l)	− 335.0	2577: 4743, 5358
C_p^o	25°	15.42	1882: 2559
C_p	25°	24.09	[2782]
	Table XI	Ch II	
t_c		230.7°	4691: 622, 1100, 2145
p_c		61.5	4691, 2145: 5034
d_c		0.507	4691, 2145
v_c		0.215	4691: 2651, 1100
K_B		2.53°	5034: 483
κ	25°	⟨ 2 x 10⁻⁸	3676
ϵ		9.39	3477: 2157, 2158, 3400, 4422, 5275, 4786
μ	25° (l)	1.90	2158: 315, 3400, 4178, 4422, 2895
soly	in aq, 20°	0.91%w	3681: 1677
aq az	37°	99.1%w	2254

247. Bromoethane

(Continued)

fl pt	flammable	
uv		306, 2356
ir		3415, 3788, 4547, 3068, 1277, 784
Raman		2816, 4274, 5017, 3068, 1170
mass		1045, 3087
nmr		937, 3025, 2085, 3126
Beil	8, 88	

248. 1-Bromopropane

$CH_3CH_2CH_2Br$ C_3H_7Br

mw		122.998	887
bp	760	70.97°	4352: 1912, 1919, 3210, 4763, 4783, 4966
dt/dp	760	0.042°	4692: 4783
p	25°	138.3	1371: 4535
eq 5	A	6.91065	2944
	B	1194.889	
	C	225.51	
fp	(st)	-108.1°	Beil **E III 1**, 240: 4352, 4535, 4763, 4783
	(ms)	-109.8°	
d	20°	1.3537	4692: 1119, 1326, 1717, 1919, 4352, 4966
	25	1.3452	4783
n_D	20°	1.4343	4692: 1119, 4966, 2158
	25	1.4317	
dn/dt		0.00052	4783

248. 1–Bromopropane

(Continued)

η	15° 30	0.539 0.459	4783: 1717, 2158, 4711
	Table II	Ch II	
γ	20° 30	25.85 24.59	2102: 4966
$\triangle H_v$	25° 68.8	7.638 7.137	1371: 5012 3210
$\triangle H_m$	(st) (ms)	1.560 2.160	4769: 4768 Beil E III 1, 240
$\triangle H_f^\circ$	25° (l)	– 27.5	5358
$\triangle H_c^\circ$	25° (l)	– 491.54	4773, 609: 2577, 4743, 5358
C_p°	25°	20.66	1882: 2559
C_p	25°	33.46	[2782]
	Table XI	Ch II	
ϵ	25°	8.09	3477: 1326, 2157, 2158
μ	20° in 30	1.93	1119: 1919, 4411, 2895
soly	in aq, 30°	0.230%w	1912: 4534
aq az	62.8°	91.3%	3975
fl pt	T C C	$\langle$ 175°F	3977
uv			2313, 2314
ir			3415, 4547, 2889, 3068, 5296, 4474, 784
Raman			3485, 3975, 3068, 1170, 5296, 412
mass			3087
nmr			937, 3947
Beil	10, 108		

249. 2-Bromopropane

$CH_3CHBrCH_3$ C_3H_7Br

mw		122.998	887
bp	760	59.41°	4352: 1119, 3210, 3452, 4764, 4783, 4966
dt/dp	760	0.0404°	4692: 1371
p	25°	236.3	1371
eq 4	A	7.65582	[4535]
	B	1590.40	
fp		−89.0°	4692: 4764, 4784, 4352
d	20°	1.3140	4692: 4784, 1119, 1919, 4352, 4966, 4783
	25	1.3060	
n_D	15°	1.42847	4783, 4784: 1119, 2158, 3452, 4966
	20	1.4251	4692
	25	1.4221	1371, 4692
dn/dt		0.00057	4783, 4784
η	15°	0.536	4783, 4784: 2158, 3452
	30	0.437	
	Table II	Ch II	
γ	20°	22.90	2102: 3452, 4966
	30	21.74	
$\triangle H_v$	25°	7.21	5012
	58.6	6.788	3210
$\triangle H_f^o$	25° (l)	−29.9	5358
$\triangle H_c^o$	25° (l)	−490.42	609: 5358
ϵ	25°	9.46	3477: 2157, 2158
μ	20° in 30	2.04	1119, 4411: 1919
soly	in aq, 18°	0.286%w	1677
aq az	54.0°	95.3%	3975
uv			2313, 2314
ir			3415, 4547, 3068
Raman			3068, 3975
mass			3087
nmr			937, 2085, 3126
Beil	10, 108		

250. Bromobenzene

C_6H_5Br {{space}} $C_6H_5\,Br$

mw		157.016	887
bp	760	155.908°	5350: 1372, 1912, 3208, 4536, 4783, 1509
dt/dp	760	0.05165 °	5350: 1372, 4783
p	25°	4.182	5350: 5346
eq 5	A	7.25422	1372: 5346
	B	1688.4	
	C	230.	
fp		−30.82°	1372: 4536, 4762, 4783
d	15°	1.50170	4829: 314, 593, 600, 601, 1326, 1372
	25	1.48820	1764, 3208, 4973, 5301
	30	1.48150	
n_D	15°	1.56252	4783: 314, 593, 1372, 2158, 3208, 4973
	25	1.55709	1369
dn/dt		0.00049	4783
η	15°	1.196	4783: 2158, 3323
	30	0.985	
γ	20°	36.34	2102: 4973, 4539
	30	35.09	
$\triangle H_v$	25°	10.87	2427: 3331, 5032. 5034, 5346
	bp	9.049	3208: 4963, 4962
$\triangle H_m$		2.540	4536: 4769
$\triangle H_f^{\circ}$	25° (l)	16.5	5358
$\triangle H_c^{\circ}$	25° (l)	−751.6	5358: 4396, 2463
C_p	29.45°	34.77	2689: 4536, 5195
t_c		397°	5303: 622, 3331, 5301, 1100, 3843, 2651
p_c		44.6	3331: 5301, 3843, 2651, 1509
d_c		0.458	2651, 5301: 3843
v_c		0.343	2651: 1100, 5303
A	eq 72	0.02039	5350
K_B		6.26°	2193
κ	25°	1.2×10^{-11}	3876: 1305
ϵ	25°	5.40	314: 1326, 2157, 2158, 5034, 3802, 4786
μ	25° in 30	1.55	2874: 315, 897, 1922, 2328, 2497, 3445
			4754, 790, 2876, 1153, 3465, 757

250. Bromobenzene

(Continued)

soly	in aq, 30°	0.0446%w	1912
	aq in, 25°	0.0424%w	5218: 2427
aq az	95.5°	51.3%w	3975
fl pt	CC	149°F	3680
uv			91, 301, 1347, 3833, 4004, 1196, 3824
ir			403, 701, 2862, 3415, 3788, 3790, 4547
			2519, 2451, 89, 5278, 4347, 3793
Raman			2478, 2680, 5228, 2670, 3300, 4347, 1169
mass			3087
nmr			954, 3174, 1513
Beil	**464**, 206		

251. 1-Bromonaphthalene

$C_{10}H_7Br$

mw		207.076	887
bp	760	281.1°	4535: 4973, 299
dt/dp	760	0.064°	[2444]
p	84.2°	1.	4535
eq 4	A	7.82728	[2444]
200 – 290°	B	2741.67	

251. 1–Bromonaphthalene

(Continued)

fp	(st)	6.20°	2431: 4535
	(ms)	0.2 – 0.7°	2088
d	20°	1.4834	2431: 4781, 4973, 417, 299
	25	1.4785	
	55	1.4480	2158
dd/dt		0.000934	[299]
n$_D$	20°	1.6580	4559: 2431, 4781, 4973, 299
	40	1.6490	
	60	1.6401	
η	25°	4.52	2158
	40	3.200	
	55	2.330	
γ	27°	44.19	4973: 417, 299
	41	43.03	
	60.4	40.88	
$\triangle H_v$	bp	12.55	[2444]
$\triangle E_c$		– 1200.	4397: 4396
κ	25°	3.66 x 10^{-11}	3876
ϵ	25°	4.83	2158: 1381
μ	25°(l)	1.29	2158: 603
uv			3859
ir			996, 1561
Raman			996, 4222
mass			3087
Beil	477, 547		

252. Bromoform

$CHBr_3$ $CHBr_3$

mw		252.746	887
bp	760	149.55°	4783, 4784: 843, 3400, 4535, 4978
dt/dp	760	0.049°	4783
p	25°	5.9	2444
eq 4	A	7.87725	[4535]
	B	2114.24	
fp		8.05°	4783, 4784: 843, 1912, 4535
d	20°	2.8889	4693: 843, 3400, 3678, 4784, 4978, 4783
	25	2.8758	1290a
n_D	15°	1.60053	4783: 4784, 4978
	20	1.59763	4978
	25	1.5956	4693
dn/dt		0.00055	4783
η	15°	2.152	4783, 4784
	30	1.741	
γ	24.8°	45.10	4978: 1290a
	41.7	42.30	
$\triangle H_v$	bp	2.76	4601
$\triangle H_f^{\circ}$	25°(g)	−4.5	3193
$\triangle E_c$		−90.3	2463
C_p°	25°	17.08	1799, 4498: 1800, 1735
	Table XIV	Ch II	
C_p	17 – 21°	31.75	1029: 2782
	Table XI	Ch II	
K_f		14.4°	226: 5024, 491, 473
κ	25	$\langle 2 \times 10^{-8}$	3676
ϵ	20°	4.39	3477: 936, 1326, 3400, 1381, 4752
μ	10 – 70° in **30**	0.99	4423: 3400, 3445
soly	in aq, 30°	0.318%w	[1912]: 664
aq az	94.2°	66.1%w	3975
fl pt	TCC	$\langle$ 175°F	3977, 3681

252. Bromoform

(Continued)

uv		3199, 2313, 2314
ir		1477, 3666, 3794, 4547, 4817, 2536, 1697
Raman		2679, 4987, 2717, 1170, 2996, 2972
nmr		3057, 3173, 3126
Beil	5, 68	

253. 1,2–Dibromoethane

CH_2BrCH_2Br $C_2H_4Br_2$

mw		187.872	887
bp	760	$131.36°$	4693: 1372, 3208, 3400, 3452, 4414, 4972
dt/dp	760	$0.058°$	4693: 1260, 4783, 1372
p	$25°$	7.79	1371: 885
eq 5	A	7.19807	1372
	B	1560.3	
	C	230	
fp		$9.79°$	4693: 1912, 3774, 3883, 4783, 885, 3301
tp		$9.7°$	885
d	$20°$	2.1791	4693: 1372, 1400, 3208, 3452, 3678, 4414
	25	2.1687	4783, 4972, 1926, 885, 1371, 4358
n_D	$15°$	1.54160	4783: 1372, 2158, 3452, 4414, 4972
	20	1.53874	1926, 885
	25	1.5360	4693
dn/dt		0.00058	4783
η	$15°$	1.880	4783: 1400, 2158, 3452, 4748
	30	1.490	
	Table II	Ch II	

253. 1,2-Dibromethane

(Continued)

γ	20°	38.91	2102: 3452, 4972, 1926, 2024
	30	37.61	
$\triangle H_v$	25°	11.05	[1371]
	bp	8.688	3208: 5032
$\triangle H_m$		2.616	3774: 4039, 4601
$\triangle H_s$		14.075	3572
$\triangle H_c$		-296.5	4396: 3809, 4397
C_p°	110°	23.14	1951
C_p	21.22°	32.48	3774: 2782, 3883, 4789
	Table XI	Ch II	
t_c		309.8°	4694: 4952, 1370
p_c		70.6	4694: 4952, 1370
d_c		0.776	1370
v_c		0.242	[1370]
K_f		12.5°	3372: 491
K_B		6.608°	483: 473
κ	19°	$\langle 2. \times 10^{-10}$	1723: 4358
ϵ	25°	4.78	3477: 1326, 2158, 3400, 2243, 5140, 4358
μ	20° in 30	1.23	3213: 3400, 5193, 4414
soly	in aq, 30°	0.429%w	[1912] : 664
	aq in, 25°	0.071%w	1370
aq az	91.0°	73.7%	3975
fl pt	nonflammable		3681
ir			1477, 3415, 3790, 4321, 5270
Raman			607, 608, 2098, 3353, 3356, 5270, 3572
mass			90
nmr			3057, 3947
Beil	8, 90		

254. 1,1,2,2–Tetrabromoethane

$CHBr_2CHBr_2$ $C_2H_2Br_4$

mw		345.674	887
bp	760	243.5°	4535: 4978
dt/dp	760	0.0555°	1371
p	25°	0.02	1371: 4763
eq 4	A	8.35595	[4535]
	B	2821.95	
fp		0.0°	4763: 2911
d	19.3°	2.9656	4772: 417, 2803, 4978, 3678, 417
	25	2.9529	
	46.0	2.9056	
n_D	20°	1.63533	4978
	25	1.6323	1371
η	10.97°	13.950	3323
	19.91	9.797	
γ	20°	49.44	417: 2024
	25.4	448.65	
$\triangle H_v$	25°	16.73	[1371]
	bp	11.63	
C_p°	25°	25.64	873
	Table XIV	Ch II	
C_p	25°	39.48	[3051]
	Table XI	Ch II	
K_f		21.7°	2911
ϵ	22°	7.0	3477: 5034
μ	20° in 7	1.29	2732
soly	in aq, 30°	0.0651%w	1913: 664
aq az	99.8°	11.4%w	3975
uv			2314
ir			4287, 2454
Raman			230, 1170
mass			90
nmr			3057, 2454
Beil	8, 94		

255. Iodomethane

CH$_3$I   CH$_3$I

mw		141.939	887
bp	760	42.43°	4696: 1119, 1919, 3400, 4747, 4775, 4966
dt/dp	760	0.039°	4696: 4775
p	25°	405.9	1371: 4535, 4733
eq 5	A	6.87991	2944
	B	1093.235	
	C	230.94	
fp		− 66.45°	4696: 4535, 4762, 4775, 4770
d	20°	2.2789	4696: 314, 1119, 1717, 1919, 3399, 3678
	25	2.2649	4747, 4966
n_D	20°	1.5308	4696: 314, 1119, 3400, 4966, 2043
	25	1.5270	
η	15°	0.518	4775: 1717, 4747
	30	0.460	
	Table II	Ch II	
γ	20°	30.14	4775, 2102: 4966
	30	28.61	
$\triangle H_v$	25°	6.579	[1371]
	bp	6.52	5032
$\triangle H_f^\circ$	25° (l)	− 2.9	920: 3193, 1859, 5358
$\triangle H_c^\circ$	25° (l)	− 194.7	2577: 4743, 5358
C_p°	25°	10.55	1440, 1735, 2559
	Table XIV	Ch II	
C_p	30.1°	19.95	2043: 2782
	Table XI	Ch II	
t_c		255°	2651: 2930
K_B		4.19°	5029: 483
ϵ		7.00	3477: 315, 1919, 3399, 3400
μ	25° in 232	1.48	2876: 315, 1119, 3400, 5029, 4411, 5193
soly	in aq, 20°	1.4%w	[2608]: 4908, 1677
aq az	40.0°	99.3%w	3975

255. Iodomethane

(Continued)

uv		2356, 3451, 4135, 2060, 306, 1478, 2314
ir		1111, 2804, 3811, 4547, 4729, 3834
Raman		5016, 2972
mass		3087, 3855, 1883, 4350, 1302
nmr		937, 2085, 3126
Beil	5, 69	

256. Iodoethane

CH_3CH_2I C_2H_5I

mw		155.967	887
bp	760	72.30°	4696: 1912, 1919, 4413, 4784, 4966, 5095
dt/dp	760	0.043°	4696: 4784
p	25°	136.2	1371: 3928, 4413, 4535
eq 5	A	6.83198	2944, 4695
	B	1175.709	
	C	225.26	
fp		−111.1°	4696, 4784: 4535, 4762
d	20°	1.9357	4696: 314, 1717, 1919, 3400, 3452, 4784
	25	1.9244	4966, 5032, 5095, 1119
n_D	20°	1.5133	4696: 314, 1119, 3208, 3452, 4784, 4966
	25	1.5101	5095
dn/dt		0.00063	4784
η	15°	0.617	4784: 1404, 1717, 3452
	30	0.540	
	Table II	Ch II	

256. Iodoethane

(Continued)

γ	$20°$	28.83	2102: 3452, 4966, 5032
	30	27.57	
	Table VII	Ch II	
$\triangle H_v$	$25°$	7.532	[1371]: 5032, 5034
	bp	7.116	3208
$\triangle H_f^o$	$26.85°$	-10.96	2464
$\triangle H_c^o$	$25°$ (l)	-356.0	2577: 4743
C_p^o	$25°$	15.47	2559
C_p	$25°$	27.49	[2782]: 2464
	Table XI	Ch II	
t_c		$281.0°$	622: 5032
K_B		$5.16°$	483: 1803
ϵ	$20°$	7.82	3477: 314, 3400, 4425, 5034, 1381
μ	$20°$ in 30	1.78	1119: 315, 2883, 3111, 3400, 4411
soly	in aq, 30°	3.88%w	[1912]: 1677
aq az	$66°$	96 – 97%v	2254
uv			2356, 4135, 2060, 306, 1478, 2314
ir			3788, 4447, 3068, 1277
Raman			4274, 5017, 3068, 1172
mass			1045, 3087, 3855, 1883
nmr			937, 3025, 2085, 3126
Beil	8, 96		

257. 1-Iodopropane

$CH_3CH_2CH_2I$ C_3H_7I

mw		169.994	887
bp	760	102.45°	4696: 1119, 1912, 1919, 3452, 4775, 4966
dt/dp	760	0.047°	4696
p	25°	43.09	1371, 4535
eq 5	A	6.81603	2944, 4695
	B	1267.062	
	C	219.53	
fp		−101.3°	4696, 4775: 1947, 4535, 4762, 4770
d	20°	1.7489	4696: 1119, 1717, 1919, 3452, 4775, 4966
	25	1.7394	
n_D	20°	1.5058	4696: 1119, 3452, 4966
	25	1.5028	
η	15°	0.837	4775: 1404, 1717, 3452, 4160
	30	0.670	
	Table II	Ch II	
γ	15°	30.11	630: 2102, 3452, 4966
	30	28.47	
$\triangle H_f^\circ$	26.85°	−16.00	2464
$\triangle H_c^\circ$	25°(l)	−514.3	2577
C_p°	25°	20.49	2559
C_p		34.68	4095: 2464
t_c		323°	4095
ϵ	20°	7.00	4477
μ	20° in 30	1.84	1119: 1808, 4411, 2883
soly	in aq, 30°	0.104%w	1912: 1677, 1376
uv			2060, 2062, 1478, 2314
ir			4547
Raman			3485, 3878
mass			3087, 3855
nmr			937, 3947
Beil	**10**, 113		

258. 2-Iodopropane

CH_3CHICH_3 C_3H_7I

mw		169.994	887
bp	760	89.50°	4696: 1119, 2306, 4535, 4763, 4775, 4966
dt/dp	760	0.045°	4696
p	25°	43.	[2444]
eq 4	A	7.75002	[2444]
	B	1765.15	
fp		– 90.0°	4696: 1947, 4763, 4981, 4775, 4535
d	20°	1.7042	4696: 1119, 4775, 4966
	25	1.6946	
n_D	20°	1.4991	4969: 1119, 4966
	25	1.4961	
η	15°	0.732	4775: 1404
	30	0.620	
	Table II	Ch II	
γ	20°	27.42	630: 4966
	30	26.13	
$\triangle H_v$	10 – 100°	8.08	[2444]
$\triangle H_f^\circ$	26.85°	– 18.40	2464
$\triangle H_c^\circ$	25°(l)	– 509.1	2577
C_p	26.85°	21.77	2464
ϵ	20°	8.19	3477
μ	20° in **30**	1.95	1119
soly	in aq 20°	0.140%w	4494
uv			2060, 2314, 1478
ir			4547, 3068
Raman			3878, 3068
mass			4618, 3087, 3855
nmr			937, 2085, 3126
Beil	**10**, 114		

259. Diiodomethane

CH_2I_2 CH_2I_2

mw		267.836	887
bp	760	182.°d	4697
dt/dp	760	0.054°	4697
p	25°	1.2	[4698] : 4978
eq 5	A	6.961	4698
	B	1575.	
	C	204.	
fp	(st)	6.1°	4697, 1249
	(ms)	5.60°	4772, 4779, 1249
d	20°	3.3211	4697: 4779, 1892, 4978
	25	3.3078	
n_D	20°	1.7411	4697
	25	1.7380	
η	15°	3.043	4772: 1892
	30	2.392	
γ	20°	50.68	2022
$\triangle H_v$	25°	12.22	[4698]
	bp	10.01	
$\triangle H_m$	(st)	10.7	478
	(ms)	11.2	
$\triangle H_f^\circ$	25° (l)	16.0	5358: 3193
$\triangle H_c^\circ$	25° (l)	− 178.4	2577, 5358: 2349
C_p°	25°	13.81	1799
	Table XIV	Ch II	
C_p	12.2 – 55.8°	28.83	2782
	Table XI	Ch II	
K_f	(st)	14.4°	478
	(ms)	13.7°	
ϵ	25°	5.316	4423
μ	25 – 50° in 30	1.08	4423: 2758
soly	in aq, 30°	0.124	1912
uv			**2060**, 1478, 2313, 2314
ir			1598, 392, 1477
Raman			1173, 336
nmr			3126, 335
Beil	5, 71		

260. Nitromethane

CH_3NO_2 CH_3O_2N

mw		61.041	887
bp	760	$101.20°$	4810: 1372, 2054, 2211, 3208, 4779, 5225
dt/dp	760	$0.0427°$	4810: 1249, 1372, 4779
p	$25°$	36.66	2442: 1373, 2237, 4535, 534
eq 5	A	7.274170	4810: 2959, 2211, 1372, 3928, 4734, 776
	B	1441.610	902, 3066
	C	226.939	
fp		$-28.55°$	4810: 1372, 4430, 2442, 4764, 4774, 4779
d	$20°$	1.13816	4810: 697, 804, 1372, 1919, 2053, 2324
	25	1.13128	2992, 3208, 4430, 4779, 4978, 5032
	30	1.12439	5265, 776, 777, 4894, 4725, 534
dd/dt		0.001377	
n_D	$20°$	1.38118	4810: 697, 804, 1372, 2237, 2259, 2324
	25	1.37964	2992, 4947, 4430, 4978, 776, 777
	30	1.37738	4894, 4725
dn/dt		0.000450	
η	$20°$	0.647	3977: 2237, 4779, 5022, 5265, 4725
	25	0.610	
	30	0.576	
γ	$20°$	37.48	4433: 697, 2102, 4978, 4725, 4779
	30	35.47	
	Table VIII	Ch II	
$\triangle H_v$	$25°$	9.147	2442: 2237, 3208, 3744, 5032, 4963
	bp	8.225	3778
	Table IX	Ch II	
$\triangle H_m$		2.319	2442
$\triangle H_f^\circ$	$25°(l)$	-27.03	5358: 2237,929
$\triangle H_c^\circ$	$25°(l)$	-169.3	2577: 4743, 2237, 4558, 929, 5358
C_p°	$25°$	13.70	3066
	Table XIV	Ch II	
C_p	$23.97°$	25.28	2442: 2259, 5030
t_c		$315°$	1891: 697
p_c		62.3	[1891]
d_c		0.352	1891

260. Nitromethane

(Continued)

K_B		$1.86°$	3744, 5029
K_i	aci, $25°$	5.5×10^{-4}	4859: 5135, 2460, 4439, 2459
K_a	NO_2 $25°$	6.15×10^{-11}	
κ	$25°$	5×10^{-9}	4894: 5027, 5265, 2979
ϵ	$30°$	35.87	3477: 4430, 5026, 5034, 4786, 2979
μ	$20°$ in 7	3.56	3513: 1919, 2324, 4411, 4415, 5114, 2876
soly	in aq, $25°$	11.1%w	3977: 1053, 513, 664
	aq in, $25°$	2.09%w	
aq az	$23.59°$	76.4%w	4811: 2255, 2254
fl pt	TOC	$112°F$	1053
uv			91, 4734, 5330, 457, 1591, 1722
	aci		5188
ir			3538, 5122, 1275, 4377, 265, 4107, 89
Raman			3216, 3705, 5122, 5230, 4377, 499
mass			61
nmr			937, 1624, 2210, 2226, 5226
	aci		1901
Beil	6, 75		

261. Nitroethane

$CH_3CH_2NO_2$ $C_2H_5O_2N$

mw		75.068	887
bp	760	114.07°	4810: 2053, 2054, 2211, 4978
dt/dp	760	0.0445°	4810
p	25°	20.93	[4810] : 2237
eq 5	A	7.175154	4810: 902, 2211, 2237, 1837
	B	1435.402	
	C	220.184	
fp		-89.52°	4810: 1682, 4532, 2980
d	20°	1.05057	4810: 561, 804, 2053, 2992, 4978, 697
	25	1.04464	2237
	30	1.03870	
dd/dt		0.001187	
n_D	20°	1.39193	4810: 561, 804, 2237, 2329, 697, 2992
	25	1.38973	4978
	30	1.38754	
dn/dt		0.000439	
η	20°	0.677	3977: 2237
	25	0.638	
	30	0.602	
γ	20°	32.66	4433: 697, 4978
	30	31.50	
	Table VIII	Ch II	
$\triangle H_v$	25°	9.94	2237
	bp	9.075	1837
$\triangle H_m$		2.355	2980
$\triangle H_f^\circ$	25° (l)	-33.9	5358: 2237, 929
$\triangle H_c^\circ$	25° (l)	-325.6	2577: 4743, 2237, 929, 5358
C_p	25°	33.1	3303: 2980
t_c		284°	[4095] : 697
K_B		2.60°	5029: 483
K_i	aci, 25°	3.93×10^{-5}	4859: 2458, 5135
K_a	NO_2, 25°	3.50×10^{-9}	
κ	30°	5×10^{-7}	3836
ϵ	30°	28.06	3477: 5026, 5034, 4752
μ	20° in 7	3.60	3513: 1919, 2329, 4411, 2324

261. Nitroethane

(Continued)

soly	in aq, 25	4.68%w	3977: 1053, 664
	aq in, 25	1.05%w	
aq az	87.22°	71%w	4811: 2255, 3975
fl pt	TOC	106 F	1053
uv			5330, 1838, 2061, 4896, 1591, 1722
	aci		5188
ir			89, 3538, 2061, 4377, 265, 837, 4547
Raman			3216, 5230, 2061, 4377, 1730
mass			90
nmr			937, 2226, 5226
	aci		1901
Beil	9, 99		

262. 1-Nitropropane

$CH_3CH_2CH_2NO_2$ $C_3H_7O_2N$

mw		89.095	887
bp	760	131.18°	4810: 1372, 1919, 2211, 4978
dt/dp	760	0.0467°	4810: 1372
p	25°	10.23	[4810]: 1373, 2237, 4535
eq 5	A	7.127539	4810: 1372, 2211
	B	1474.299	
	C	215.986	

262. 1–Nitropropane

(Continued)

fp		$-103.99°$	4810: 1372
d	20°	1.00144	4810: 697, 804, 1372, 1919, 4978
	25	0.99609	
	30	0.99073	
dd/dt		0.001071	
n_D	20°	1.40160	4810: 697, 804, 1372, 2237, 4978
	25	1.39956	
	30	1.39755	
dn/dt		0.000405	
η	20°	0. 844	3977: 2237
	25	0.790	
	30	0.740	
γ	20°	30.64	4433: 697, 4978
	30	29.61	
	Table VIII	Ch II	
$\triangle H_v$	25°	10.37	2237
	bp	9.194	[4810]
$\triangle H_f^°$	25° (l)	-40.15	5358: 2237, 929
$\triangle H_c^°$	25° (l)	-481.9	2577: 2237, 4558, 929, 5358
C_p	25°	41.9	3303
t_c		402.0 °	697
pK_a	NO_2 form	8.98	5135
κ	35°	3.3×10^{-7}	3836
ϵ	30°	23.24	3477
μ	20° in 7	3.59	3513: 1808, 4411, 5225
soly	in aq, 25°	1.50%w	3977: 1053
	aq in, 25°	0.62%w	
aq az	91.63 °	63.5%w	4535: 2255, 3975
fl pt	TOC	120°F	1053
uv			5330, 2061, 4896, 1591
	aci		5188
ir			3766, 2061, 4377, 265, 1730, 837, 3538
Raman			3216, 5230, 2061, 4377, 1730
nmr			937, 2226, 5226
	aci		1901
Beil	10, 115		

263. 2-Nitropropane

$CH_3CH(NO_2)CH_3$ $C_3H_7O_2N$

mw		89.095	887
bp	760	120.25°	4810: 2053, 2211, 4978, 5225
dt/dp	760	0.0460°	4810
p	25°	18.0	[4810]: 1682, 2237, 4535
eq 5	A	7.083240	4810: 2211
	B	1422.898	
	C	218.341	
fp		-91.32°	4810: 575, 4535
d	20°	0.98839	4810: 697, 2053, 2237, 4978
	25	0.98290	
	30	0.97740	
dd/dt		0.001099	
n_D	20°	1.39439	4810: 697, 2237, 4978, 5225
	25	1.39235	
	30	1.39028	
dn/dt		0.000411	
η	20°	0.770	3977: 2237
	25	0.721	
	30	0.677	
γ	20°	29.87	4433: 4978, 697
	30	28.68	
	Table VIII	Ch II	
$\triangle H_v$	25°	9.88	2237
	bp	8.793	4810
$\triangle H_f^\circ$	25° (l)	-43.2	929: 2237, 5358
$\triangle H_c^\circ$	25° (l)	-478.0	5358: 2237, 929, 2477
t_c		344.7°	697
K_i	aci, 25°	7.73×10^{-6}	4859: 5135, 2460
K_a	NO_2, 25°	2.12×10^{-8}	
κ	30°	5×10^{-7}	3836
ϵ	30°	25.52	3477
μ	$120 - 180^\circ$	3.73 (g)	5225

263. 2-Nitropropane

(Continued)

soly	in aq, 25°	1.71%w	3977: 1053
	aq in, 25°	0.53%w	
aq az	88.55°	70.6%w	4535: 2255, 3975
fl pt	TOC	103°F	1053
uv			5330, 2061, 4896, 1591, 1722
	aci		5188
ir			89, 3538, 3766, 408, 2061, 4377, 265
Raman			5230, 2061, 4377, 412
nmr			937, 2226, 5226
Beil	10, 116		

264. Nitrobenzene

$C_6H_5NO_2$ $C_6H_5O_2N$

mw		123.112	887
bp	760	210.80°	4780: 2780, 2963, 3017, 4535, 4761, 4978
dt/dp	760	0.05573°	1369: 3017, 4761, 4780
p	25°	0.284	1369: 800, 3846, 3928, 4535, 3045
eq 5	A	7.545	3045
	B	2064.	
	C	230	
eq 4	A	8.680	4364
10 – 30°	B	2741.	
fp		5.76°	1027: 1538, 1912, 4275, 3242, 5227, 3995

264. Nitrobenzene

(Continued)

d	15°	1.20824	4780: 711, 1027,2963, 2992, 3168, 3202
	25	1.19835	[4780] : 4392, 4761, 4834, 4978, 5032, 5042
	30	1.19341	4780
dd/dt		0.000989	[4780]
n_D	15°	1.55457	4780: 711, 792, 2038, 2469, 2992, 3168
	25	1.54997	[4780] : 4392, 4978, 3051
dn/dt		0.00046	[4772]
η	15°	2.165	4780: 687, 3202, 5022, 5042
	30	1.634	
γ	20°	43.35	2102: 4780, 4978, 5032
	30	42.17	
	Table VII	Ch II	
$\triangle H_v$	10 – 30°	12.54	4364: 1532, 337, 4963
	bp	9.744	3017: 5032
$\triangle H_m$		2.78	4601: 3660
$\triangle H_f^{\circ}$		– 3.9	4574
$\triangle H_c^{\circ}$	25° (l)	– 742.4	2577: 1711, 4574
C_p	30°	42.37	5194: 3017, 3660, 5030
t_c		459°	5032
K_f		6.852°	4864: 490, 3995, 3911
K_B		5.04°	587: 588, 481, 586, 337
κ	25°	2.05×10^{-10}	413: 3876, 5027, 5036, 5044, 5042
ϵ	25°	34.82	3477: 1888, 2038, 5198, 5275, 2048
μ	25° in 30	4.03	2874: 998, 1538, 1851, 1918, 1985, 3168
			3444, 4178, 3769, 1650, 790, 886
soly	in aq, 20°	0.19%w	4225: 1912, 4908, 664, 1912
	aq in, 20°	0.24%w	
aq az	98.6°	12.%w	2254
fl pt	CC	190°F	3680
uv			91, 549, 792, 1347, 1196, 3541, 1544
ir			701, 2554, 2904, 4076, 4547, 2519, 408
Raman			2573, 1169, 625, 5268
mass			61
nmr			954, 1298, 1282, 2866
Beil	**465**, 233		

265. Acetonitrile

CH_3CN C_2H_3N

mw		41.053	887
bp	760	81.60°	4778: 1372, 1948, 2933, 3275, 4761, 5029
dt/dp	760	0.04192°	4702: 4761, 4778, 1372
p	25°	88.81	3866: 1373, 1528, 4535, 1138
eq 5	A	7.12257	1372: 3866
	B	1315.2	
	C	230	
fp		−43.835°	3866: 1372, 4762, 4778, 3217
d	20°	0.7822	4702: 1117, 1192, 1372, 1439, 2933, 4761
	25	0.7766	4702: 4778, 5032, 5040
	30	0.77125	3275
dd/dt		0.001078	[3275]
α	20°	0.00137	
n_D	20°	1.34411	1372: 1117, 1192, 1948, 1439, 2933, 3275
	25	1.34163	4778
dn/dt		0.00045	4778
η	15°	0.375	4778: 1948, 5022, 5040, 1021, 1665
	30	0.325	
γ	20°	19.10	4778, 2102: 2387, 5032
	30	27.80	
	Table VII, VIII Ch II		
$\triangle H_v$	25°	7.941	3866: 3015, 3019, 4962
	80.5	7.127	2470
$\triangle H_m$		1.952	3866: 4769
$\triangle H_f^\circ$	25°(l)	12.3	5358: 3361
$\triangle H_c^\circ$	25°(l)	−302.4	2577, 4581: 4743, 4583, 5358
C_p°	25°	12.48	1526
	Table XIV	Ch II	
C_p	26.84°	21.89	3866: 3019, 1526, 2470, 3018, 5030, 1138
t_c		274.7°	2651: 1950, 1948, 5032
p_c		47.7	2651, 1948: 1950
d_c		0.237	2651
v_c		0.173	2651

265. Acetonitrile

(Continued)

K_B		1.30°	5029: 4864, 3440, 2470, 2471, 2957
pK_{BH+}	in H_2SO_4	-10.12	1281: 1284, 2900
pKs		26.5	1023
κ	25°	6×10^{-10}	2376: 2652, 5026, 5027, 5040, 3455
ϵ	20°	37.5	3477: 2654, 5026, 5034, 550, 4786, 1021
μ	20° in 30	3.44	1117: 2654, 2677, 4411, 5193, 2876, 4911
soly	in aq	inf	4519
aq az	76.7°	84.2%	2449: 2254, 2255
fl pt	CC	$42^\circ F$	88
uv			1854, 4016, 2313, 4197, 4634
ir			341, 4547, 408, 3505, 4911, 1848, 502
Raman			606, 3954, 4941, 720, 3505, 1517, 2436
mass			90, 3087, 232
nmr			4632, 937, 2432, 4850, 1281, 3126
Beil	**159**, 183		

266. Propionitrile

CH_3CH_2CN C_3H_5N

mw		55.080	887
bp	760	97.35°	4702: 1117, 1372, 1808, 1948, 3275, 4535
dt/dp	760	0.0445°	4702: 1372
p	25°	44.63	1371

266. Propionitrile

(Continued)

eq 5	A	7.15217	1372
	B	1398.2	
	C	230	
eq 4	A	8.0269	5102
9 – 25°	B	1894.1	
fp		– 92.78°	5102: 1372, 2453, 4535, 4764, 4775
d	20°	0.78182	1372: 804, 1117, 1118, 1192, 1919, 1948
	25	0.77682	2387, 3275, 5032
	30	0.77196	4775
dd/dt		0.00103	1191
n_D	20°	1.3658	4702: 804, 1117, 1118, 1192, 1372, 1948
	25	1.3636	2387, 5266, 300
dn/dt		0.00045	3275
η	15°	0.454	4775: 1665, 1948, 5022, 5266
	30	0.389	
γ	20°	27.25	2102: 2387
	30	26.19	
	Table VII	Ch II	
$\triangle H_v$	22.85°	8.632	5102: 2090, 3015
	bp	7.40	5032
$\triangle H_m$		1.206	5102: 4769
$\triangle H_c^\circ$	25° (l)	– 459.3	2577: 4583, 4743, 4581
C_p	26.85°	28.61	5103: 5030, 3015
t_c		291.2°	2651: 1950, 1948, 5032, 4952
p_c		41.3	2651, 1948: 1950, 4952
d_c		0.240	2651
v_c		0.230	2651
K_B		1.87°	5029: 483
κ	25°	8.51 x 10^{-8}	5027: 4950, 5035, 3389
ϵ	20°	27.2	3477, 5026: 1118, 3389, 5034, 1191
μ	20° in 30	3.57	1117: 3885, 1919, 2330, 4411, 4837, 5193
soly	in aq, 25°	10.3%w	[664]
aq az	81.5 – 83°	76.%w	2254

266. Propionitrile

(Continued)

uv		1159, 2313
ir		4547, 89, 1395, 1848
Raman		606, 3954, 1395, 1173
mass		90, 3087, 61
nmr		937, 3025, 1281
Beil	**162**, 245	

267. Succinonitrile

$NCCH_2CH_2CN$ $C_4H_4N_2$

mw		80.090	887
bp	760	267°	2608
dt/dp	760	0.06°	[3976]
p(s)	25.14°	7.774μ	5255: 2934
eq 4	A	8.987	[2608, 4519, Beil **E III** 2, 1673]
	B	3259.	
fp		57.88°	5272: 1583, 5255
tp		58.01°	
d	60°	0.98669	4772
	70	0.97860	
	80	0.97059	
n_D	60°	1.41734	4772
dn/dt		0.00037	[4772]

267. Succinonitrile

(Continued)

η	$60°$	2.591	4772
	75	2.008	
γ	$60°$	46.78	4772
	70	45.72	
	80	44.62	
$\triangle H_v$	$25°$	15.29	5255
	bp	11.6	eq 4
$\triangle H_s$	$25°$	16.730	5272: 5255
$\triangle H_m$		0.8851	5272: 5255
$\triangle H_f^°$	$25°$ (g)	54.42	3361
C_p	$61.85°$	38.37	5272
	Table XII	Ch II	
K_f		$18.26°$	2349
κ		5.64×10^{-4}	5026
ϵ	$57.4°$	56.5	3477: 5140
μ	$30°$ in **31**	3.68	2934: 4837, 5193, 4449
soly	in aq, $25°$	11.5%w	2349
ir			1583, 502
Raman			1166
Beil	**172**, 615		

268. Butyronitrile

$CH_3CH_2CH_2CN$ C_4H_7N

mw		69.107	887
bp	760	117.94°	4772: 1919, 1948, 2387, 3017, 4767, 4775
dt/dp	760	0.0463°	4702: 3017
p	25°	19.10	1371
eq 4	A	7.778	2090
	B	1912	
fp		− 111.9°	4775, 4767: 2453, 4763
d	15°	0.79544	4775: 1117, 1192, 1919, 1948, 2387, 3275
	20	0.7911	4702
	25	0.7865	4702
	30	0.78183	4775
n_D	15°	1.38600	3275: 1117, 1192, 1343, 1948, 2387
	20	1.3838	4702
	25	1.3820	
	30	1.37954	3275
dn/dt		0.00043	3275
η	15°	0.624	4775: 1948, 1665
	30	0.515	
γ	20°	27.33	2102: 2387
	30	26.24	
	Table VIII	Ch II	
$\triangle H_v$	25°	8.84	1511
	bp	8.23	1371
$\triangle H_m$		1.200	4769
$\triangle H_f^\circ$	25° (l)	− 1.39	1511
$\triangle H_c^\circ$	25° (l)	− 613.93	1511: 4583, 2577
t_c		309.1°	1948: 1950, 5077
p_c		37.4	1948: 1950
ϵ	21°	20.3	3477, 4160: 5034
μ	20° in 30	3.57	1117: 4411, 4449
soly	in aq	3.3%	1431
	aq in	trace	
aq az	88.7°	67.5%w	2255: 2254
fl pt	TOC	85°F	1431

268. Butyronitrile

(Continued)

uv		2632
ir		2255
Raman		2952, 5268
mass		90, 3087, 61, 2052
nmr		937
Beil	**162**, 275	

269. Isobutyronitrile

$CH_3CH(CH_3)CN$ C_4H_7N

mw		69.107	887
bp	760	103.85°	4775: 300
dt/dp	760	0.0460°	4704, 4775
fp		−71.5°	4775
d	20°	0.7704	4704: 2220
	25	0.7656	
	30	0.76082	4775
n_D	15°	1.37563	4775: 2220, 300
	20	1.3734	4704
	25	1.3712	1451
dn/dt		0.00043	[4775]
η	15°	0.551	4775
	30	0.456	
γ	20°	24.93	4775
	30	23.84	

269. Isobutyronitrile

(Continued)

$\triangle H_v$	$25°$	8.54	1511
$\triangle H_f^°$	$25°$ (l)	-2.92	1511
$\triangle H_c^°$	$25°$ (l)	-612.40	1511
ϵ	$24°$	20.4	4160
μ	$25°$ in **30**	3.61	4014: 4449
aq az	$82.5°$	77%w	2254
Raman			3954
mass			3087, 1585
nmr			937
Beil	**162**, 294		

270. Valeronitrile

$$CH_3CH_2CH_2CH_2CN \hspace{4cm} C_5H_9N$$

mw		83.134	887
bp	760	$141.3°$	4702: 1372, 2387, 3275, 3452, 4767, 4775
dt/dp	760	$0.0489°$	4702: 1372
p	$25°$	7.07	1371: 1373, 2090, 4535
eq 5	A	7.26060	1372
	B	1626.0	
	C	230	
fp		$-96.2°$	4702: 1372, 2453, 4767, 4775
d	$20°$	0.7993	4702: 1117, 1192, 1372, 2387, 3452, 4775
	25	0.7950	
	50	0.7730	380
n_D	$20°$	1.3971	4702: 1117, 1192, 1372, 1343, 2387, 3275
	25	1.3951	3452

270. Valeronitrile

(Continued)

dn/dt		0.00040	3275
η	15°	0.779	4775: 1665, 3452
	25	0.6928	380
	50	0.5084	
γ	20°	27.44	2102: 2387, 3452, 4775
	30	26.33	
	Table VIII	Ch II	
$\triangle H_v$	25°	10.624	[1371]
	bp	7.98	2470
$\triangle H_m$		1.130	4769
$\triangle E_c$		– 772.5	4583
C_p		43.22	[2470]
κ		1.2×10^{-8}	380
ϵ	25°	19.709	1288: 4160, 4786, 380
μ	25° in 30	3.57	1117: 4449, 4411
ir			4547, 1848, 502
Raman			3954
mass			3087, 61
Beil	162, 301		

271. Hexanenitrile

$CH_3(CH_2)_4CN$ $C_6H_{11}N$

mw		97.161	887
bp	760	163.6°	4702, 1372: 1948, 2386, 3275, 4324, 5266
dt/dp	760	0.0508°	4702: 1372
p	25°	2.66	1371: 1373, 2090, 4535

271. Hexanenitrile

(Continued)

eq 5	A	7.32130	1372
	B	1748.0	
	C	230	
fp		-80.3°	4702: 1372, 4324, 4774
d	20°	0.8052	4702: 1192, 1948, 2387, 3275, 4324, 804
	25	0.8012	
n_D	20°	1.4069	4702, 2386: 804, 1192, 1343, 1948, 3275
	25	1.4048	4702, 1372: 4324, 5266
η	15°	1.041	4324: 1665, 1948, 5266
	30	0.830	
γ	20°	27.85	2102: 2387
	30	26.99	
$\triangle H_v$	25°	11.511	[1371]: 2090, 3019
	bp	9.089	
C_p	$18 - 155.5^\circ$	52.63	[3019]
t_c		348.8°	1948: 1950, 4406, 5077
p_c		32.1	1945: 1950
ϵ	25°	17.263	1288: 4160
Raman			3952
mass			3087, 3085
Beil	**162**, 324		

272. 4–Methylvaleronitrile

$(CH_3)_2CHCH_2CH_2CN$ $C_6H_{11}N$

mw		97.161	887
bp		154°	4704, 2387: 5266, 4764
dt/dp		0.050°	4704
p	65.2°	35	4014
fp		– 51.1°	4704, 4774: 4764
d	20°	0.8035	4704: 2387, 4863
	25	0.7993	
n_D	20°	1.4061	4704: 5266
	25	1.4040	
η	20°	0.980	5266
	30	0.843	
γ	20°	26.53	4863: 2387
	30	25.61	
ϵ	22°	15.5	3477
μ	25° in **30**	3.53	4014: 2006
uv			3938
ir			502
Raman			1166, 3952
mass			3087
Beil	**162**, 329		

273. Octanenitrile

$CH_3(CH_2)_6CN$ $C_8H_{15}N$

mw		125.215	887
bp	760	205.2°	4702: 1248, 4535
dt/dp	760	0.0547°	4702
p	25°	0.39	1371: 2090, 4535
fp		-45.6°	4702: 1248, 4774, 4770
d	20°	0.8135	4702: 1192, 1248, 3275
	25	0.8097	
n_D	20°	1.4202	4702: 1343, 3275, 1192
	25	1.4182	
η	15°	1.811	1248
	30	1.356	
γ	20°	28.10	1371
	30	37.96	
$\triangle H_v$	25°	13.27	[1371]
	bp	9.869	
ϵ	25°	13.897	1288
ir			4547, 5260
mass			3087
Beil	**162**. 349		

274. α–Tolunitrile

$C_6H_5CH_2CN$ C_8H_7N

mw		117.152	887
bp	760	233.5 °	4535: 4765, 1899, 5266
dt/dp	760	0.053 °	[4535]
p	60°	1	4535
eq 4	A	8.2285	[4535]
	B	2710.0	
fp		– 23.8 °	4535, 4765
d	20°	1.0155	2387: 1324, 4427, 4539
	25	1.0125	5032
n_D	20°	1.52327	4427: 2387, 5266
	25	1.52086	
η	20°	2.161	5266: 4994, 1324, 5022, 2655
	30	1.779	
γ	20°	41.36	4863: 4539
	30	40.27	
$\triangle H_v$	bp	12.62	[4535]
$\triangle H_c$	25° (l)	– 1029.4	2577
κ	25°	$\langle 0.5 \times 10^{-7}$	5036: 5027
ϵ	27°	18.7	3477: 4160, 1899, 5026, 5034
μ	25° in 30	3.47	4427
uv			91, 1679, 3366, 2763
Raman			2575, 2573, 1167, 3731
Beil	941, 441		

275. Benzonitrile

C_6H_5CN C_7H_5N

mw		103.125	887
bp	760	191.10°	4780: 1747, 1948, 4535, 4761, 4762, 5266
dt/dp	760	0.047°	4780: 4761
p	28.2°	1	4535: 2469, 3167, 3928
eq 5	A	6.74631	2821
	B	1436.72	
	C	181.0	
fp		-12.75°	5227: 4762, 4780, 4535
d	15°	1.00948	4780: 804, 1118, 1400, 1747, 1948, 2873
	25	1.0006	
	30	0.99628	4780: 3166, 3167, 4761, 5032
dd/dt		0.00088	[4780]: 2145
n_D	20°	1.52823	2387: 1118, 1948, 4780 5266
	25.5	1.52570	804
dn/dt		0.00048	4780
η	15°	1.447	4780: 1400, 1948, 3166, 4994, 5022, 5266
	30	1.111	1021
γ	15°	38.65	4780: 5032
	27	38.43	2387
	Table VII, VIII	Ch II	
$\triangle H_v$	25°	13.26	1511: 2470, 5032
	bp	10.98	3744
$\triangle H_m$		2.60	[5227]: 4931
$\triangle H_f^\circ$	25° (l)	39.00	1511
$\triangle H_c^\circ$	25° (l)	-868.15	1511, 2577, 4581
C_p		45.48	3018: 2470
t_c		426.2°	1948: 1950, 5032
p_c		41.6	1948: 1950
K_f		5.34	5227: 4931
K_B		3.87°	3744, 2470: 5029, 5126, 2471, 2957
κ	25°	5×10^{-8}	3166: 5027, 5036, 3389
ϵ	25°	25.20	3166: 1381, 1118, 3389, 5034, 5275, 4786
μ	25° in 30	4.05	1153, 1118, 1522: 2886, 3802

275. Benzonitrile

(Continued)

soly	in aq, 25 aq in, 28	∿0.2%w ∿1%w	3975: 4931 3975
aq az	98.9°	16.7%w	3975
fl pt	COC	none seen	4931
uv			91, 374, 1347, 2204, 3418, 3833, 1196
ir			2862, 4817, 89, 2451, 2519, 408, 2764
Raman			606, 2574, 1170, 3732, 2670, 2573, 499
mass			90, 61, 3265
nmr			1281, 2975
Beil	926, 275		

276. Acrylonitrile

$CH_2=CHCN$ C_3H_3N

mw		53.064	887
bp	760	77.3°	1216: 2330, 4016, 4980
dt/dp	760	0.043°	1216
p	20°	83	4882
eq 4	A B	7.518 1644.7	79
fp		−83.55°	79: 1216
d	20° 25	0.8060 0.8004	1216: 2330, 4016, 4980
dd/dt	0 – 30°	0.001106	[1216]
α	55°	0.00146	4882

276. Acrylonitrile

(Continued)

n_D	$25°$	1.3888	79: 1216, 2330, 4016, 4980
dn/dt		0.000539	[79]
η	$0°$	0.43	4882
	20	0.35	
	25	0.34	79
γ	$17.8°$	27.53	4980
	24	27.3	79
$\triangle H_v$	$0 - 80°$	7.800	1216
	bp	7.78	[4882]
$\triangle H_m$		1.585	79
$\triangle H_f^o$	$25°$ (l)	36.20	79: 3361, 5358
$\triangle H_c^o$	$25°$ (l)	− 420.8	79: 1216, 5358
$\triangle H_p$		17.3	79
C_p^o	$25°$	15.24	1983
C_p	$20°$	26.5	[4882] : 1216
t_c		$246°$	79
p_c		34.9	79
K_f		2.7%m/$°$c	79
ϵ	$20°$	33.0	1191: 1216, 4161
μ	$25°$ in **30**	3.51	4016, 4449: 2330, 2483
soly	in aq, $20°$	7.35%	79: 1216
	aq in, $20°$	3.1%	79
aq az	$69.6°$	86.7%w	4882: 1632, 1216
fl pt	TOC	$32°$F	79: 1216
uv			2621, 79, 1216, 2089
ir			79, 1983, 4737, 408, 89
Raman			79, 1983, 3953, 4760, 2621
mass			79
nmr			1626, 932, 3457
Beil	**163**, 400		

277. Methacrylonitrile

$CH_2=C(CH_3)CN$ C_4H_5N

mw		67.091	887
bp	760	90.3°	3727: 546, 4961
dt/dp	760	0.0433°	3727
p	25°	70	[2444] : 3727, 4535
eq 5	A B C	6.98021 1274.959 220.734	3727
fp		− 35.8°	3727: 4535, 4961
d	20° 30	0.8001 0.7896	3727, 4961
α	20 – 25°	0.00133	4961
n_D	20° 30	1.4007 1.3954	3727: 4961
η	20°	0.392	3727
γ	20°	24.4	3727
$\triangle H_v$	bp	7.6	3727
$\triangle H_p$		15.3	4961
μ	(g)	3.69	4449
soly	in aq, 20° aq in, 20°	2.57%w 1.62%w	3727 3727
aq az	77°	84%w	3727
fl pt	TOC	55°F	3727, 4961
uv			2089, 823
ir			408
nmr			3932
Beil	163, 423		

278. Propylamine

$CH_3CH_2CH_2NH_2$ C_3H_9N

mw		59.112	887
bp	760	48.5°	4700: 804, 4535, 4977, 1791, 1101
dt/dp	760	0.376°	4700
p	25°	307.9	4559: 1085, 4535
eq 5	A	6.92646	3600: 4699
	B	1044.028	
	C	210.833	
fp		− 83.0°	4700: 4535
d	20°	0.7173	4700: 804, 4977, 1115, 1101, 4401
	25	0.7121	
n_D	20°	1.3882	4700: 804, 4977, 1115
	25	1.3851	
η	25°	0.353	3474
γ	19.2°	22.21	4977
	26.4	21.64	
$\triangle H_v$	25°	7.49	4401: 658
	bp	7.106	[1371]
$\triangle H_f^\circ$	25° (l)	− 24.26	4401: 2902, 5358
$\triangle H_c^\circ$	25° (l)	− 565.31	4401: 2902, 4743, 4583, 2577, 4581, 5358
C_p	25°	38.8	4401: 658
t_c		223.8°	547: 1791, 1101
p_c		46.76	547: 1791
pK_{BH+}	25° in H_2O	10.568	437: 1523, 3749, 1974, 1972, 721
ϵ	20°	5.31	1115: 1706
μ	20° in 30	1.33	1115: 4154
soly		inf	3681
aq az	nonazeotropic		2255
fl pt	CC	10°F	88
uv			91
ir			1645, 4113, 3611, 576, 390, 5281, 89
Raman			1444, 2659, 3878, 5249
mass			1813
nmr			937, 3460
Beil	337, 136		

279. Isopropylamine

$CH_3CHNH_2CH_3$ C_3H_9N

mw		59.112	887
bp	760	32.4°	4700: 1101
dt/dp	760	0.033°	4700
p	25°	574.9	1371
eq 5	A	6.89017	3600
	B	985.650	
	C	214.071	
fp		−95.2°	4700
d	20°	0.6875	4700: 1101, 4401
	25	0.6821	
n_D	20°	1.3742	4700
	25	1.3711	
η	25°	0.36	4958
γ	20°	19.53	1371: 2372
	30	18.34	
$\triangle H_v$	25°	6.81	4401: 658
	bp	6.499	1371
$\triangle H_f^\circ$	25° (l)	−26.83	4401
$\triangle H_c^\circ$	25°	−562.74	4401
C_p	25°	39.5	4401: 658
t_c		203.0°	4958: 1101
pK_{BH+}	25° in H_2O	10.63	1972, 1974
ϵ	20°	5.45	4161
soly	in aq	inf	3681
fl pt	OC	−15° F	88
uv			91
ir			2198, 3611, 89
nmr			937
Beil	337, 153		

280. Butylamine

$CH_3CH_2CH_2CH_2NH_2$ $C_4H_{11}N$

mw		73.139	887
bp	760	77.4°	4700: 4014, 4977, 1101, 1791, 1566, 1510
dt/dp	760	0.0419°	4700
p	25°	91.75	1371: 1085
eq 5	A B C	7.213 1308.4 224.2	4699
fp		−49.1°	4700: 1706, 477
d	20° 25	0.7392 0.7346	4700: 1115, 4014, 4977, 1101, 1566
n_D	20° 25	1.4014 1.3987	4700: 1115, 4014, 4977, 1510
η	25°	0.681	3474
γ	19.2° 27.9	24.03 23.17	4977
$\triangle H_v$	25° bp	8.366 7.674	1371: 658, 1510
$\triangle H_f^\circ$	25° (l)	−30.53	1510: 5358, 2902, 4401
$\triangle H_c^\circ$	25° (l)	−721.43	1510: 2902, 4583, 2577, 4581, 5358
C_p		46.325	658
t_c		287.9°	1101: 1791
P_c		41.	1791
pK_{BH+}	25°	10.640	1523: 437, 2217, 3455, 1977, 3697, 1972
ϵ	20°	4.88	1115: 4161, 3477, 1706
μ	20° in 30	1.37	1115: 395, 4014, 4154, 4259, 1004, 2757
soly	in aq	inf	3681
aq az	69°(575 torr)	98.7%w	2255
fl pt	TOC	30°F	4893: 3942, 3440
uv			4932
ir			1645, 408, 1441, 3611, 390, 5281
Raman			2661, 2662, 5249
mass			1813
Beil	337, 156		

281. Isobutylamine

$(CH_3)_2CHCH_2NH_2$ $C_4H_{11}N$

mw		73.139	887
bp	760	67.73°	4700: 804, 4535, 4977, 1101, 4765
dt/dp	760	0.0402°	4700
p	25°	140.7	1371: 1372, 4535
eq 13	A	7.7890	4323
	B	2278.53	
	C	0.007795	
fp		-84.6°	4700: 4323, 4535, 4765
d	20°	0.7346	4700: 804, 4977, 1101
	25	0.7297	
n_D	20°	1.3972	4700: 804, 4977
	25	1.3945	
η	25°	0.553	3474
γ	19.7°	22.25	4977
	25.5	21.71	
$\triangle H_v$	25°	8.006	1371: 658
	bp	7.356	
$\triangle H_f^o$		42.7	2902
$\triangle H_c^o$	25° (l)	-716.4	2577: 2902, 4743, 4583, 4581
C_p		53.92	658
t_c		266.7°	1101
pK_{BH+}	25° in H_2O	10.42	1977: 721, 2217, 720
c	21°	4.43	4161: 3477, 5034
μ	25° in 30	1.27	3056: 395
soly	in aq	inf	3681
aq az	none		3975
fl pt	CC	⟨20°F	3681
ir			1645
Raman			367, 2709
mass			1813
Beil	337, 163		

282. sec–Butylamine

$CH_3CH_2CHNH_2CH_3$ $C_4H_{11}N$

mw		73.139	887
bp	760	62.5°	4700: 4724, 803, 4014, 4977, 1510
dt/dp	760	0.043°	4700
p	25°	172.6	1371
eq 4	A	7.9113	[1371]
	B	1690.8	
d	20°	0.7246	4700: 363, 4724, 804, 4977
	25	0.7201	
n_D	20°	1.3934	4700: 804, 4977, 4014, 1510
	25	1.3907	
γ	21.0°	21.49	4977
	27.4	20.91	
$\triangle H_v$	25°	7.802	[1371] : 1510
	bp	7.236	
$\triangle H_f^°$	25° (l)	– 32.88	1510: 5358, 2902, 4401
$\triangle H_c^°$	25° (l)	– 719.08	1510: 2902, 2577, 4581, 5358
$[\alpha]_D^{20}$	(+) (S)	+ 7.44	4724: 2917
$[\alpha]_D^{20}$	(–) (R)	– 7.40	4724: 363
pK_{BH+}	25°	10.56	1977: 1974, 1972, 6, 721
μ	25° in 30	1.28	4014
aq az	none		3975
fl pt		– 20°F	2609
mass			1259
Beil	**337**, 160		

283. tert–Butylamine

$(CH_3)_3CNH_2$ $C_4H_{11}N$

mw		73.139	887
bp	760	44.4°	4700: 1566, 2499, 4014, 1510
dt/dp	760	0.041°	4700
p	25°	362.1	4700
eq 5	A	6.78204	3600
	B	992.719	
	C	210.423	
fp		−72.65°	2499
d	20°	0.6958	4700: 1566, 2499, 4401
	25	0.6908	
n_D	20°	1.3788	4700: 1556, 2499, 1510
	25	1.3761	
$\triangle H_v$	25°	7.10	4401: 1510
	bp	6.784	1371
$\triangle H_f^\circ$	25°(l)	−36.00	4401: 1510, 2902, 5358
$\triangle H_c^\circ$	25°(l)	−715.94	4401: 2902, 2577, 1510, 4581, 5358
C_p	25°	45.9	[4401]
pK_{BH+}	25°	10.45	1977: 1974, 1972, 720
μ	25° in 30	1.29	4014: 1566
soly	in aq	inf	2609
aq az	none		3975
fl pt		16°F	2609
uv			91
ir			2198, 89
mass			90
Beil	337, 173		

284. Cyclohexylamine

CH$_2$CH$_2$CH$_2$CH$_2$CH$_2$CHNH$_2$ C$_6$H$_{13}$N

mw		99.177	887
bp	760	134.75°	4717: 3554, 923, 2933, 4977, 3996
dt/dp	760	0.050°	4717
p	25°	8.8	[3554] : 923
eq 5	A	6.6272	3554
	B	1203.2	
	C	186.7	
fp		- 17.7°	923
d	15°	0.87128	4773: 1115, 923, 2933, 4977
	30	0.85777	
α		0.001164	1115
n$_D$	20°	1.45926	4977: 1115
	25	1.4565	923
dn/dt		0.00058	4773
η	15°	2.517	4773
	20	1.662	
γ	20°	31.51	4773: 4977
	30	30.29	
$\triangle$H$_v$	25°	10.926	[3554]
	bp	8.688	4958
t$_c$		341.7°	4958
pK$_{BH+}$	25°	10.64	1977: 1974, 1972, 5019
ϵ	20°	4.73	1115: 3477
μ	20° in **30**	1.26	1115: 2933, 4154, 1887
soly	in aq	inf	923
aq az	96.4°	44.2%w	923
fl pt		32°F	923
ir			1467
nmr			3506, 4632, 663
Beil	**1594**, 5		

285. Aniline

$C_6H_5NH_2$ C_6H_7N

mw		93.129	887
bp	760	184.40°	4780: 1372, 2644, 3017, 4761, 4763, 3070
dt/dp	760	0.05042°	1372: 2783, 3017, 4761, 4780
p	25°	0.671	1369
eq 5	A	7.57170	1372: 3070
	B	1941.7	
	C	230.	
fp		-5.98°	266: 4770, 2819, 3301, 4763, 5227, 3070
d	15°	1.02613	4780: 266, 711, 1565, 1764, 1948, 2644
	20	1.02173	1372: 2992, 3167, 4392, 4490, 4761, 4977
	25	1.01750	5032
dd/dt		0.000864	[4780]
n_D	20°	1.58628	1369: 266, 711, 1948, 2259, 2644, 2992
	25	1.58364	4392, 4780, 4977, 1372
η	20°	4.400	4490: 266, 1400, 1948, 4710, 4780, 3474
	25	3.770	
	Table IV	Ch II	
γ	19°	43.38	1555: 4780, 4977, 5032
	25	42.79	246
	32.0	42.01	4772
	Table VII	Ch II	
$\triangle H_v$	25°	13.325	2069: 330, 2783, 3653, 5032, 5000
	bp	10.643	2069: 4962
$\triangle H_f^\circ$	25°(l)	7.47	2069: 2902, 5000, 5358
$\triangle H_c^\circ$	25°(l)	-810.5	2577: 2902, 4583, 4569, 4743, 2069, 4581
C_p°	25°	25.91	2069
C_p	25°	45.90	2069: 612, 1558, 2783, 3653, 4144
t_c		426°	1369: 1950, 1948, 5032
p_c		52.39	1369: 1950, 1948
d_c		0.314	1369
v_c		0.297	1369
K_f		5.87°	227: 1427
K_B		3.22°	484: 5029, 477, 3291, 489
pK_{BH^+}	25° in aq HCl	4.606	582, 3718: 584, 3075, 4531, 755, 1279

285. Aniline

(Continued)

κ	$25°$	2.4×10^{-8}	3690: 3034
ϵ	$20°$	6.89	3477: 4161, 998, 1326, 1985, 3034, 4786
μ	$25°$ in 30	1.51	2888, 2874: 1473, 1496, 4259, 2778, 4127
soly	in aq, $25°$	3.38%w	2608: 664, 3495
	aq in, $25°$	4.76%w	
aq az	$98.6°$(742 torr)	19.2%w	2255: 2254
fl pt	CC	$169°F$	3495: 3680, 88
uv			91, 549, 793, 1347, 2635, 2719, 3418
			3858, 2812, 849, 3541
ir			5294, 2451, 408, 5136, 2764, 2194, 3611
			3495, 1090, 1588, 2605, 2862, 4076, 5286
Raman			1177, 2478, 2680, 4936, 5228, 499, 1174
mass			3377, 61
nmr			2461, 954, 2622, 4632, 3044, 1513
Beil	1598, 59		

286. o-Toluidine

o–CH$_3$C$_6$H$_4$NH$_2$ C$_7$H$_9$N

mw		107.156	887
bp	760	$200.40°$	4780: 531, 1372, 3017, 4763, 1791
dt/dp	760	$0.054°$	4780: 1372, 3017
p	$25°$	0.32	1369: 1373, 2190, 2469, 3928, 4535
eq 5	A	7.60681	1372: 531
	B	2033.6	
	C	230	
fp	stable	$-16.10°$	1249: 1372, 4535, 4763, 4764, 4780, 4770
	metastable	-24.5	

286. o–Toluidine

(Continued)

d	20° 25	0.99843 0.99430	1372: 804, 2992, 4710, 4780
n_D	20° 25	1.57246 1.56987	1372: 804, 2992, 4780
η	15° 25 30	5.195 3.390 3.183	4780: 4710, 3474 246 4780
γ	20° 30	40.03 38.93	4772: 630, 2899
	Table VIII	Ch II	
$\triangle H_v$	25° bp	13.561 10.659	1369: 3017
	Table IX	Ch II	
$\triangle H_c^o$	25° (l)	– 963.3	2577: 4569, 4581
C_p	20°	51.92	4144: 2689, 3017
	Table XI	Ch II	
t_c		421°	1791
p_c		37	1791
pK_{BH+}	25° in aq HCl	4.45	584: 5019, 1977, 755, 1279, 276, 4013
κ	25°	3.792×10^{-7}	3449
ϵ	18°	6.34	3477: 1326, 4161, 3882
μ	25° in **30**	1.60	395: 3882
soly	in aq, 20° aq in, 20°	1.66%w 2.44%w	4225
aq az	99.6°	19%w	3975
fl pt	CC	185°F	3680
uv			373, 2635, 3858
ir			1645, 408, 2518, 5281
Raman			2137, 2670
nmr			2622, 4632
Beil	**1672**, 772		

287. m–Toluidine

$m-CH_3C_6H_4NH_2$ C_7H_9N

mw		107.156	887
bp	760	203.40°	4780: 531, 1372, 4535, 1791
dt/dp	760	0.055°	4780: 1372
p	25°	0.27	1369: 1373, 2190, 2469, 3928, 4535
eq 13	A	18.5043	531: 1372
	B	3200.9	
	C	3.323	
fp		– 30.40°	1372: 4535, 4780, 4770
d	15°	0.99302	4780: 804, 1372, 2469, 4710
	39	0.98096	
n_D	20°	1.56811	1372: 804, 2469
	25	1.56570	
η	15°	4.418	4780: 4710, 595
	30	2.741	
	130	0.506	3474
γ	20°	38.02	630
	30	37.16	
$\triangle H_v$	25°	13.691	1369
	bp	10.719	
	Table IX	Ch II	
$\triangle H_m$		0.930	4769
$\triangle H_c$	25° (l)	– 963.3	2577: 4569
C_p	15 – 30°	51.84	2689
t_c		436°	1791
P_c		41	1791
pK_{BH+}	25° in aq HCl	4.73	584: 1977, 2042, 3359, 755, 276, 4013
κ	25°	5.5 x 10^{-10}	2088
ϵ	18°	5.95	3477: 1326, 4161, 3882
μ	25° in **30**	1.45	395: 3882
soly	in aq	sl sol	2608
aq az	99.6°	17%w	3975
uv			373, 3858
ir			1645, 408, 5136, 2521, 5281
Raman			2670
nmr			2622, 4632
Beil	**1682**, 853		

288. p–Toluidine

p–CH$_3$C$_6$H$_4$NH$_2$ C$_7$H$_9$N

mw		107.156	887
bp	760	200.55°	4781: 531, 2652, 4535, 1791
dt/dp	760	0.054°	4781
p	25°	0.34	1369: 2190, 2469, 3928, 4535
eq 13	A	20.1569	531: 2652
	B	3269.3	
	C	3.877	
fp		43.75°	4781: 531, 936, 1796, 4535, 4864, 3922
d	45°	0.96589	4781: 804, 4710, 4864
	50	0.96155	
dd/dt		0.000803	[4781]
n$_D$	45°	1.55397	1369: 804
	50	1.55348	
η	45°	1.945	4781
	55	1.557	4710
	60	1.425	4781
	130	0.522	3474
γ	45°	36.06	630: 4781
	60	34.10	
	Table VIII	Ch II	
$\triangle$H$_v$	25°	13.431	[1369]: 531, 2652, 2785
	bp	10.581	
	Table IX	Ch II	
$\triangle$H$_m$		4.520	3922: 4569, 4601, 1533
$\triangle$H$_c^o$	25°(l)	− 963.3	2577: 4569, 4581
C$_p$	43°	64.1	2349: 2279
t$_c$		493°	5032: 1791
p$_c$		23.5	1791
K$_f$		5.372°	4864: 1533
K$_B$		4.14°	484
pK$_{BH+}$	25° in aq HCl	5.08	584: 3075, 1977, 720, 755, 3008, 1974
κ	100°	6.2 x 10^{-8}	430
ε	54°	4.98	3477: 936, 3730, 3882
μ	25° in **30**	1.52	395: 4154, 4395, 3882

288. p–Toluidine

(Continued)

soly	in aq, 20.8°	7.35%w	[3008] : 1449
aq az	99.5°	3975	
fl pt	CC	188°F	3680
uv			91, 373, 2635, 4791
ir			1645, 408, 5281
Raman			2141, 3685, 2670
nmr			1767, 4632
Beil	**1683**, 880		

289. Allylamine

$CH_2=CHCH_2NH_2$ C_3H_7N

mw		57.096	887
bp	760	53.3°	3711: 804, 3724, 4977, 2578, 2609
fp		– 88.2°	2609
d	20°	0.7629	[3711], [4977] : 804, 2609, 1789
	25	0.7575	
dd/dt		0.00107	
n_D	20°	1.42051	4977: 804, 2578, 1789
	22	1.41943	2088
η	25°	0.3745	3474
γ	24.5°	24.27	4977: 5033
	31.1	23.57	
$\triangle H_f^{\circ}$		– 2.4	2902
$\triangle E_c$		– 524.8	4583, 4743: 2902
pK_a	25° in H_2O	9.49	1972: 1974, 720, 2609
κ	25°	5.7 x 10⁻⁵	2088
μ	25° in **30**	1.31	4019

289. Allylamine

(Continued)

soly	in aq	inf	2088
aq az		none	2254
fl pt	CC	$10°F$	88, 2609
ir			1645, 1908, 5281
Raman			2477
nmr			50, 1767
Beil	338, 205		

290. Ethylenediamine

$H_2NCH_2CH_2NH_2$ $C_2H_8N_2$

mw		60.099	887
bp	760	$117.26°$	4782: 804, 995, 4977, 5211, 797, 4128
dt/dp	760	$0.045°$	4782
p	$26.51°$	13.1	4773: 4535, 5211
eq 4			
$\langle 50°$	A	9.3539	2444
	B	2468.1	
$50-117°$	A	8.100	
	B	2038.2	
fp		$11.3°$	3738: 5211, 4128, 3829
d	$15°$	0.90001	4780: 804, 2259, 4977, 5211, 4128, 3829
	30	0.88595	
dd/dt		0.0007	5211
n_D	$15°$	1.45887	4782: 804, 2259, 5211, 3829
	20	1.45677	4977
	30	1.4513	995
dn/dt		0.00050	[4782]
η	$15°$	1.722	4782: 623, 4128, 5211
	25	1.54	1400
	30	1.226	4782

290. Ethylenediamine

(Continued)

γ	$20°$	40.77	4782: 4977
	30	39.49	
$\triangle H_v$	$20°$	11.20	4773: 3829, 3493, 2174
	bp	10.0	5211
$\triangle H_m$		4.62	2174: 5211
$\triangle H_c$	$25°$ (l)	-442.8	5211
C_p	$30°$	42.31	2259
	Table XII	Ch II	
t_c		$319.8°$	3493
p_c		62.1	3493
v_c		0.206	3493
K_f		2.43	3738: 3436
pK_{BH+}	$25°$, 1st	6.838	
	2nd	10.960	2160: 3749, 5211, 1524, 634, 1972, 437
κ	$25°$	9×10^{-8}	752: 797, 4128, 3829, 1130
ϵ	$25°$	12.9	3829: 4128, 5140, 5211
μ	$25°$ in **30**	1.90	4837: 2594
soly	in aq	inf	3436
aq az	$119°$	81.6%w	2255: 2254
fl pt	CC	$110°F$	88: 5211
ir			4223
Raman			2662, 1750, 1174, 4633
Beil	**343**, 230		

291. Ethylenimine

CH_2CH_2NH C_2H_5N

mw		43.069	887
bp	760	57°	1358: 844, 2221, 1683
p	20°	160	3681
fp		- 78.0°	1358 , 4770
d	10° 25	0.846 0.832	1358: 1683
n_D	25°	1.4123	1358
η	25°	0.418	1358
γ		32.8	1358
ΔH_v	20°	7.9	1358
ΔH_f°	25° (l)	21.96	3512
ΔH_c°	25° (l)	- 380.86	3512
pK_a	25° in H_2O	8.04	4217: 844, 3596, 4272
κ	25°	8×10^{-6}	2532
ϵ	25°	18.3	2532
μ	25° in **30**	1.77	2532: 2410
soly	in aq	inf	3681: 1683
fl pt		12°F	1358
ir			89, 2221, 4731, 1535, 4262
Raman			2675, 4262
nmr			1945
Beil	**3035**, 1		

292. Diethylamine

$(CH_3CH_2)_2NH$ $C_4H_{11}N$

mw		73.139	887
bp	760	55.45°	4701: 4535, 4764, 4977, 5095, 1101
dt/dp	760	0.0385°	4701
p	25°	233.5	1371: 1085, 3798, 4535
eq 5	A	6.97237	1086
15 – 55°	B	1127.0	
	C	220	
fp		– 49.8°	4701: 3798, 4535, 4764, 1706
d	20°	0.7070	4701: 804, 1665, 2992, 4585, 4977, 5095
	25	0.7016	1101
n_D	20°	1.3854	4701: 804, 2992, 4977, 5095
	25	1.3825	
η	10.2°	0.3878	1665: 3474, 4711
	37.6	0.2732	
γ	15°	20.63	4586: 4977
	25	19.39	
$\triangle H_v$	20°	7.691	[2529]: 4962
	bp	7.130	
$\triangle H_f^\circ$	25° (l)	– 24.79	5358: 2902
$\triangle H_c^\circ$	25° (l)	– 727.15	5358: 2902, 4743, 4583, 4581, 2577
C_p	20°	42.244	[2529]
	Table XI	Ch II	
t_c		223.8°	2145: 547, 1101
p_c		40.0	2145: 547
d_c		0.243	2145
pK_{BH+}	25° in H_2O	10.933	437, 1504: 3455, 1972, 3718, 720, 3697
ϵ	21°	3.58	4161: 1706
μ	25° (l)	1.11	394: 2178
soly		inf	4519
aq az		none	2255
fl pt	CC	⟨0°F	3680

292. Diethylamine

(Continued)

uv		576, 2196, 2632
ir		1090, 1645, 390
Raman		976, 1177, 2478, 1174
mass		1046
nmr		4469, 3460
Beil	**336**, 95	

293. Dipropylamine

$(CH_3CH_2CH_2)_2NH$ $\qquad\qquad\qquad\qquad\qquad\qquad C_6H_{15}N$

mw		101.193	887
bp	760	109.2°	4701: 804, 4764, 4977
dt/dp	760	0.045°	4701
p	25°	24.09	1371
fp		-63.$^\circ$	4701: 4764, 1706
d	20°	0.7375	4701: 1115, 804, 2992, 4977
	25	0.7329	
n_D	20°	1.4043	4701: 1115, 804, 2992, 4977
	25	1.4018	
η	20.1°	0.5335	1665
	37.2	0.4267	
γ	16.9°	23.13	4977
	24.5	22.28	
$\triangle H_v$	25°	9.760	[1371]: 2470
	bp	8.335	
C_p		60.43	2470
pK_a	25° in H_2O	11.00	1972: 1974, 720, 3477
ϵ	20°	3.068	1115: 4161, 721, 1706
μ	20° in **30**	1.03	1115: 2757

293. Dipropylamine

(Continued)

soly	in aq, 25° aq in, 25°	4%w 21%w	[2209]: 1222
aq az	86.7°		2255
fl pt		45°F	3681
uv			576
ir			1645, 408, 390
Raman			4749
mass			1813
nmr			1813
Beil	337, 138		

294. Diisopropylamine

$[(CH_3)_2CH]_2NH$ $C_6H_{15}N$

mw		101.193	887
bp	760	83.9°	4893: 1101, 4977
dt/dp	760	0.042°	[4958]
p	20°	60	4893
eq 4	A B	7.8043 1758.5	[4958]
fp		-96.3°	4893
d	20° 25	0.7153 0.7100	1101: 4977 [1101]
dd/dt		0.00094	
n_D	20°	1.39236	4977
η	25°	0.40	4958
γ	16.0° 26.3	20.04 19.11	4977
$\triangle H_v$	bp	8.24	4958

294. Diisopropylamine

(Continued)

t_c		249.0°	4958
pK_a	25° in H_2O	11.05	1972: 1977, 1974
soly	in aq, 30°	11%	1222
	aq in, 30°	40%	
aq az	74.1°	90.8%w	2254
fl pt	TOC	21°F	4893
ir			89
Raman			4749
mass			1813
nmr			3460
Beil	337, 154		

295. Dibutylamine

$(CH_3CH_2CH_2CH_2)_2NH$ $\qquad\qquad\qquad\qquad\qquad\qquad\qquad$ $C_8H_{19}N$

mw		129.247	887
bp	760	159.6°	4701: 1101, 4977
dt/dp	760	0.051°	4701
p	25°	2.28	1371
eq 5	A	7.26603	1371
	B	1616.4	
	C	209	
fp		$-62.^\circ$	4701
d	20°	0.7619	4701: 1115, 1101, 4977, 1665
	25	0.7577	
n_D	20°	1.4177	4701: 1115, 4977
	25	1.4152	

295. Dibutylamine

(Continued)

η	$20°$	0.95	[4958] : 1665
	30	0.83	
γ	$20°$	24.57	4701
	40.9	22.63	4977
$\triangle H_v$	$25°$	12.08	1371
	bp	9.523	
t_c		$222.6°$	1101
pK_a	$25°$	11.31	1977, 1974: 1972, 3697
ϵ	$20°$	2.978	1115: 1706
μ	$20°$ in 30	1.04	1115: 1004
soly	in aq, $20°$	0.47%w	4893
	aq in, $20°$	6.2%w	
aq az	$97°$	49.5%w	2255
fl pt	OC	$135°F$	88
ir			2645, 408, 390
mass			1813
nmr			3460
Beil	337, 157		

296. Pyrrole

$$C_4H_5N$$

mw		67.091	887
bp	760	$129.76°$	4204: 2097, 4843, 3317
dt/dp	760	$0.0429°$	4204
p	$25°$	8.2	4204: 3317

296. Pyrrole

(Continued)

eq 5	A B C	7.30275 1506.877 210.995	3600: 4204
fp		$-23.4146°$	4204: 3317
tp		$-23.41°$	2097
d	$20°$ 25 30	0.96985 0.96565 0.96133	2097: 4843, 3149, 1116
n_D	$20°$ 25 30	1.51015 1.50779 1.50543	2097: 4843, 3149, 1116
η	$20°$ 25 30	1.352 1.233 1.29	2097
γ	$20°$ 25 30	37.61 37.06 36.51	2097
$\triangle H_v$	$25°$ bp	10.79 9.261	4204: 3317 4204
	Table IX	Ch II	
$\triangle H_m$		1.890	4204
$\triangle H_c°$	$25°$ (l)	-562.07	4204: 2577, 1864
$C_p°$	$25°$	17.05	4204
C_p	$25°$	30.64	4204
	Table XII	Ch II	
t_c		$366.5°$	964: 2254
p_c		56	2254
A	eq 72 B	0.01525 0.00208	4204: 2097
pK_a	$25°$ in aq H_2SO_4	-3.80	970: 3489, 2697, 4602, 2450
ϵ	$25°$	8.13	3249: 4786, 1651
μ	$25°$ in 30	1.80	832: 2658, 4009, 1116, 1831, 2879, 3149
soly	in aq, $25°$	ca 4.5%w	[617]
fl pt	TCC	$102°F$	[2608]
uv			91, 2716, 3489, 3271, 671, 970
ir			89, 2097, 2855, 3933, 2999, 2755, 3394
Raman			654, 656, 3950, 4496, 3142, 2677, 647
mass			2097, 2205, 839, 2393
nmr			1655, 1647, 970, 2443, 1906, 3930, 1317
Beil	3047, 159		

297. Piperidine

$$HNCH_2CH_2CH_2CH_2CH_2$$ $C_5H_{11}N$

mw		85.150	887
bp	760	106.40°	4781: 1948
dt/dp	760	0.044°	4781
p	23.93°	29.1	1528: 3985
eq 5	A	6.85699	3600
	B	1239.577	
	C	205.528	
fp		-10.5°	1528
d	15°	0.86591	4781: 1528, 1948, 2981, 1665
	30	0.85215	
dd/dt		0.001157	[4781]: 1528
n_D	20°	1.4525	3: 800, 1948
dn/dt		0.00048	[4781]
η	15°	1.679	4781: 3474, 4711, 1948, 1665
	25	1.362	2786
	30	1.224	4781
γ	20°	30.05	4781
	30	28.81	
	Table VII	Ch II	
$\triangle H_v$	25°	10.259	[1528]: 4963, 1271
	bp	7.572	[3019]
$\triangle H_f^\circ$	25° (l)	-21.06	474
$\triangle H_c^\circ$	25° (l)	-826.6	2577: 1271, 4743, 474
C_p	20 – 100°	44.56	[3019]: 1271
t_c		320.95°	71: 968
p_c		45.9	1948
K_B		2.84°	Beil **XX**, 7
pK_{BH+}	25° in H_2O	11.123	2159, 2614, 437, 3718: 1974, 1972, 1973
			1976, 1962, 720, 4217, 5126, 2961
ϵ	20°	5.8	4161
μ	25° in **30**	1.19	292: 4259, 4854
soly	in aq	inf	2088
aq az	92.8°	65%w	2254
fl pt		37°F	3

297. Piperidine

(Continued)

ir		390, 1640, 1441, 5078, 89, 2194, 2924
Raman		2924
mass		1164, 1389, 691
nmr		3460
Beil	**3037**, 6	

298. Triethylamine

$(CH_3CH_2)_3N$ $C_6H_{15}N$

mw		101.193	887
bp	760	89.5°	4703, 1107: 4779, 4977, 5095, 4763, 1087
dt/dp	760	0.043°	4703
p	25°	57.07	1371: 1087
eq 5	A	6.9952	1087: 3798, 1086
	B	1300.30	
	C	226.50	
fp		− 114.7°	4703, 4770: 4763, 4779, 2453
d	20°	0.7276	4703: 804, 2992, 4585, 4779, 4977, 5095
	25	0.7230	1665, 1087, 1101
dd/dt		0.00092	[825]
α		0.00126	4779
n_D	20°	1.4010	4703: 804, 2992, 4977, 5095, 825, 1087
	25	1.3980	
η	15°	0.394	4779: 1665
	25	0.363	3474
	30	0.323	4779
γ	20°	20.66	4779: 3278, 4586, 4977
	30	19.62	
$\triangle H_v$	20°	8.571	[2529]
	bp	7.675	4962
$\triangle H_f^\circ$	25° (l)	− 32.07	5358: 2902
$\triangle H_c^\circ$	25° (l)	− 1042.0	2577: 2902, 4743, 4583, 4581, 5358

298. Triethylamine

(Continued)

C_p	$20°$	53.396	2529
	Table XI	Ch II	
t_c		$262.2°$	$2145: 5032, 1101$
p_c		30.0	2145
d_c		0.257	2145
K_B		3.45	$[4703, 4962]$
pK_{BH+}	$25°$ in H_2O	10.715	$437, 1523: 3718, 1972, 720, 1976, 3455$
ϵ	$25°$	2.42	$3477: 5034, 4786$
μ	$25°$ in **30**	0.87	$292: 394, 2178, 4259$
soly	in aq, $20°$	5.5%w	4822
	aq in, $20°$	4.6%w	
aq az	$75°$	90%w	$2254: 800$
fl pt	CC	$20°F$	3680
uv			$576, 2632, 4849$
ir			$1090, 4547, 89, 4107$
Raman			$976, 1177, 2669, 1174, 4633$
mass			1813
nmr			1768
Beil	**336**, 99		

299. Pyridine

$$C_5H_5N$$

		79.102	887
mw		79.102	887
bp	760	$115.256°$	569, 2945, 2127: 2097, 3985, 4781, 4761
dt/dp	760	$0.04535°$	569: 4761, 4781
p	$24.8°$	20.	3160: 2653, 2174, 3928, 3985, 4535
eq 5	A	7.04144	3600: 2127, 2945
	B	1373.990	
	C	215.001	
fp		$-41.55°$	569: 5227, 277, 3660, 4762, 4764, 4781
Triple pt		$-41.67°$	2097
d	$20°$	0.98319	2097: 569, 277, 354, 2898, 3208, 3305
	25	0.97824	3447, 4033, 4761, 4781, 5214
	30	0.97319	
dd/dt		0.00099	[825]
n_D	$20°$	1.51016	2097: 2653, 569, 825, 1900, 2898, 3208
	25	1.50745	3305, 3447, 4033, 4781, 5214
	30	1.50466	
η	$20°$	0.952	2097: 354, 3865, 4781
	25	0.884	
	30	0.815	
γ	$20°$	36.88	2097: 4781
	25	36.33	
	30	35.78	
$\triangle H_v$	$25°$	9.659	2945: 4963, 3019, 1271, 2470, 3208, 5032
	bp	8.697	569
$\triangle H_m$		1.772	2097: 3090, 3660
$\triangle H_f^\circ$	$25°$ (l)	23.95	2945, 1122: 5358
$\triangle H_c^\circ$	$25°$ (l)	-665.00	2945, 1122: 2284, 2577, 4581, 4743
C_p°	$25°$	19.06	2945
	Table XIV	Ch II	
C_p	$17°$	32.4	4584: 708, 1271, 2470, 3660, 3019
t_c		$344.0°$	2145: 2653, 73, 72, 964, 4406
p_c		60.0	2145: 2653, 73
d_c		0.312	2145: 2653
v_c		0.253	2145

299. Pyridine

(Continued)

K_f		$4.75°$	569: 492
A	eq 72	0.01858	2097
K_B		$2.710°$	4864, 5046: 3606, 5126, 5029, 2470
pK_{BH+}	$25°$ in aq HCl	5.25	2969, 3698: 1821, 1976, 3718, 1962, 2213 414, 3359, 24, 1974, 1746
κ	$25°$	4.0×10^{-8}	5039: 3690
ϵ	$21°$	12.4	4160: 550, 4786, 354, 936, 5034, 5275
μ	$25°$ in 232	2.37	4259, 2879: 2898, 3305, 831
soly	in aq	inf	2608
aq az	$93.6°$	58.7%w	2255: 2254
fl pt	CC	$74°F$	3680
uv			91, 2111, 4460, 4468, 2812, 3048, 2097
ir			2859, 4858, 569, 4107, 89, 5278, 2097
Raman			499, 455, 3141, 412
mass			3997, 2097
nmr			4637, 3463, 1770, 1649, 931, 1826, 2843
Beil	3051, 181		

300. Quinoline

C_9H_7N

mw		129.163	887
bp	760	$237.10°$	4781: 4762, 5297, 1899, 1791, 3123
dt/dp	760	$0.057°$	4781: 3123
p	$59.7°$	1	3681: 5297

300. Quinoline

(Continued)

eq 5	A	6.80189	3123: 2444
	B	1656.3	
	C	184.78	
fp		$-14.85°$	5227: 4762, 4770, 3660
d	15°	1.09771	4781: 1326, 804, 4712, 3398, 707
	30	1.08579	
dd/dt		0.000794	[4781]
n_D	15°	1.62928	4781: 804
	20	1.6273	3123
	25	1.62475	5031
η	15°	4.354	4781: 4712, 707
	25	3.37	3474
	30	2.997	4781
γ	20°	45.65	4781: 3398
	30	44.82	
$\triangle H_v$	bp	11.88	[2444] : 1899
$\triangle H_m$		2.58	3660
$\triangle H_c°$	25° (l)	-1123.5	4794: 1270
C_p	26.8°	47.80	3660: 708
t_c		521.3	3398: 71, 1791
p_c		57	1791
K_f'		1.95	5227
K_B		5.84°	4177: 485, 484
pK_{BH+}	25° in aq EtOH	4.94	3698: 276, 40, 4990, 1821, 1976
κ	25°	2.2×10^{-8}	2349
ϵ	25°	9.00	3477: 4160, 1326, 1899, 831
μ	25° in 30	2.184	2890: 2879, 525, 831, 4259
soly	in aq, 20°	0.609%w	39
uv			91, 3048, 1543
ir			89, 408, 1240
Raman			1240, 2972
nmr			615, 1649
Beil	3077, 339		

301. Formamide

HCONH$_2$ CH$_3$ON

mw		45.041	887
bp	760	210.5$^°$(d)	4535, 451: 2372, 3042
p	70.5$^°$	1	4535: 3107, 2350
eq 4	A	10.1366	[4535]
	B	3509.	
fp		2.55$^°$	4387, 4780, 451: 2847, 4614, 2372, 3553
d	20$^°$	1.13339	4387: 804, 1484, 2372, 4614, 4780, 1227
	25	1.12918	
n$_D$	20$^°$	1.44754	4388: 804, 822, 4780, 2350
	25	1.44682	1749
η	20$^°$	3.764	4387: 1484, 2557, 4780, 4994, 1227, 1401
	25	3.302	
γ	20$^°$	58.35	4387: 3147
	25	57.91	
	Table VIII	Ch II	
$\triangle$H$_v$	25$^°$	15.530	451
$\triangle$H$_m$		1.600	451: 4601, 1749, 4446
$\triangle$H$_f^°$	25$^°$(l)	- 60.7	5358
$\triangle$H$_c^°$	25$^°$(l)	- 135.8	5358: 2577
C$_p^°$	25$^°$	10.80	451: 1514
C$_p$	25$^°$	25.72	[4446]: 5030
K$_f$		3.85$^°$	816: 1378, 1749
pK$_a$	20$^°$ in H$_2$O	- 0.48	2309: 2183
κ		$\langle$2 x 10^{-7}	3553: 2557, 2847, 4387, 4947, 5026, 5027
ϵ	20$^°$	111.0	432: 851, 2847, 3477, 5026, 1227, 3042
μ	30$^°$ in 30	3.37	443: 851, 5317, 3042, 296
soly	in aq	inf	3681
fl pt	COC	310$^°$F	2609
uv			2150, 91, 2322
ir			1645, 2903, 4729
Raman			2903, 4107, 1514, 3854, 1170, 3522, 1174
			1177, 2480, 2673, 3904, 428, 3854
mass			1780
nmr			5167, 2484, 2485
Beil	156, 26		

302. N–Methylformamide

HCONH(CH$_3$) C_2H_5ON

mw		59.068	887
bp	760	$180 - 185^\circ$	Beil **4**, 58
p	44°	0.4	2094, 1631: 2350, 1182
	47	0.6	
	55	1.5	
fp		-3.8°	2848: 1182, 2094
d	15°	1.0075	1631: 1182, 2350, 2848, 1840
	25	0.9988	
dd/dt	$0 - 50^\circ$	0.0008674	4342
α	25°	0.000869	4342
n_D	20°	1.4319	2350: 2848
	25	1.4300	1182
η	15°	1.99	1631: 1840
	25	1.65	
γ	30°	37.96	1840
	40	36.50	
	50	35.02	
pK_a	20° in H_2O	-0.04	2309
κ	25°	8×10^{-7}	1631: 2848, 2094
ϵ	25°	182.4	2848
	-40	308	432
μ	25° in **30**	3.86	1826: 296
soly	in aq	inf	Beil **IV**, 58
ir			3351, 2083, 1259, 2433, 3855, 2827
Raman			1259, 3855
mass			1780
nmr			2730, 2731, 2729, 2827
Beil	**335**, 58		

303. N,N-Dimethylformamide

HCON(CH$_3$)$_2$ $\qquad\qquad\qquad\qquad\qquad\qquad\qquad$ C$_3$H$_7$ON

mw		73.095	887
bp	760	153.0°	4075: 1371, 1416, 2350, 1734
dt/dp	760	0.0476°	[3160]
p	25°	3.7	1416: 805, 3826
eq 5	A	7.1085	3299
	B	1537.78	
	C	210.39	
fp		-60.43°	1371: 4075
d	20°	0.94873	1371: 2848, 4075, 1631, 805, 2350, 1734
	25	0.94397	
	30	0.9412	1840
	40	0.9310	
n$_D$	20°	1.43047	1371: 2848, 4075, 1225, 805, 2350, 1734
	25	1.42817	
η	20°	0.9243	1371: 1631, 1840
	25	0.802	1416
	40	0.7386	1371
γ	20°	36.76	1371: 1840
	25	35.2	1416
	40	34.40	1371
$\triangle$H$_v$	25°	11.356	[1371]: 1409, 1734
	bp	9.164	
$\triangle$H$_m$		3.861	1371
$\triangle$H$_c$		-457.5	1409
C$_p^\circ$	100°	26	[1409]
C$_p$	25°	37.45	1734: 1409
t$_c$		323.4°	1416: 1840
p$_c$		51.5	1416: 1734
v$_c$		0.249	1734
pK$_a$	20° in H$_2$O	-0.01	2309: 4627
κ	25°	6 x 10^{-8}	1631: 1225, 3575, 4221
ϵ	25°	36.71	2848
	-40	52.1	432.
μ	25° in 30	3.86	2335: 1740, 4562, 296
soly	in aq	inf	1409
aq az		none	2255
fl pt	TOC	153°F	1409

303. N, N–Dimethylformamide

(Continued)

uv		2322, 5061
ir		2050, 2083, 2433, 3855
Raman		3855
mass		1780
nmr		4632, 3220, 2730, 1615, 2121, 3520
Beil	**335**, 58	

304. Acetamide

CH_3CONH_2 C_2H_5ON

mw		59.068	887
bp		221.15°	4772: 1667
dt/dp	760	0.054°	[4535]
p	65.0°	1.	4535: 4519
eq 4			
150 – bp	A	8.80919	[4535]
	B	2936.07	
65 – 150°	A	9.12026	
	B	3082.80	
mp*		69.5°	3728: 3440, 443, 2779, 1206
fp		80.00°	
d	91.1°	0.9892	1667: 5058, 4863
	111.8	0.9711	
	131.7	0.9538	
n_D	80°	1.4270	3503: 2609
	110	1.4158	4772
	130	1.4079	
η	91.1°	2.182	1667: 1401
	111.8	1.461	
	131.7	1.056	

* see Ch V, amides, general

304. Acetamide

(Continued)

γ	85°	38.96	4863
	95	37.95	
	105	36.96	
$\triangle H_v$	bp	13.4	[4535]
$\triangle H_s$		18.52	1206
$\triangle H_m$	80°	3.754	3728: 2280, 2609
$\triangle H_c^{\circ}$	25°(l)	– 286.5	2577
C_p	80 – 150°	39.67	3728
K_f		4.04°	4128
pK_{BH+}	25° in aq H_2SO_4	0.59	1448: 1976, 2213, 2183, 3695, 1284, 2900 1816, 4990, 2309, 24, 4519, 2572
κ	83.2°	8.8×10^{-7}	2199: 5025, 504
	100	4.3×10^{-5}	5058
ϵ	83°	59	3477: 2848, 1293
μ	30° in 30	3.44	443: 1294, 2779, 2208, 2776, 296
soly	in aq, 20°	40.8%w	4494: 4519
aq az		none	2254
uv			4197, 2133, 2941, 4091, 576
ir			2249, 2903, 2864, 2907, 866, 869
Raman			1170, 2673, 4629, 3855
mass			1780
nmr			1615, 2484
Beil	**159**, 175		

305. N–Methylacetamide

CH$_3$CONH(CH$_3$)   C$_3$H$_7$ON

mw		73.095	887
bp	760	206°	2636: 3479, 496
	90	140.5	1182
p	56°	1.5	2958
fp		30.55°	660: 1182, 2848, 2636, 3479, 1632
d	30°	0.9498	2263: 1228, 2848, 2636, 2958, 1840
	35	0.94604	1632
	45	0.93764	
dd/dt		0.00084	[1632]
n$_D$	28°	1.4286	1182: 2958, 2848, 496, 2263
	35	1.4253	
dn/dt		0.00047	[1182]
η	35°	3.23	1632: 1228, 2636, 1840
	45	2.54	
γ	30°	33.67	1840
	40	32.17	
	50	30.62	
△H$_v$	115 – 205°	14.2	3963
△H$_m$		2.00	660
t$_c$		417°	1840
K$_f$		6.65°	660
pK$_{BH+}$	25° in aq H$_2$SO$_4$	1.25	1448: 1816, 3357, 2309, 2721
κ	40°	2 x 10^{-7}	2636: 1228, 2848, 1632
ε	32°	191.3	660: 1182, 2636, 2958, 1632, 2848, 2263
μ	20.1° in 124	4.39	3357: 3483, 3481
soly	in aq	soluble	3479
uv			3357, 4090
ir			3357, 2515, 2083, 3351, 496, 2433, 3891
Raman			3357, 3891, 2672
mass			1780
nmr			3220, 3221, 1615, 2828, 2484
Beil	335, 58		

306. N,N–Dimethylacetamide

$CH_3CON(CH_3)_2$ C_4H_9ON

mw		87.122	887
bp	760	166.1°	1408: 2779, 1667, 3042, 4979, 2350
dt/dp	760	0.043 $^{\circ}$	[3384]
p	25°	1.3	1408
eq 5	A	7.76228	[3384]
	B	1889.10	
	C	221.0	
fp		-20 $^{\circ}$	1416: 4979
d	15.5 $^{\circ}$	0.9448	3384:1667, 24, 2848, 4979, 2350
	25	0.9366	
	30	**0.9323**	1840
	40	0.9232	
n_D	20 $^{\circ}$	1.4384	2350: 24, 2848, 4979
	25	1.4356	1408
η	20.4 $^{\circ}$	2.141	1667
	30	0.838	1840
	40	0.766	
γ	30°	32.43	1840
	40	30.92	
	50	29.50	
$\triangle H_v$	25°	12.7	[3384]
	bp	10.360	1408: 3384
$\triangle H_m$		2.490	3384
$\triangle H_c$	20°	$-608.$	1408
C_p°	27°	26.	1408
C_p	20°	42	1408
t_c		364°	1840
p_c		38.7	1408
K_f		4.46°	[3384]
K_B		3.22 $^{\circ}$	[3384]
pK_a	$\langle 25$ $^{\circ}$in aq H_2SO_4	-0.19	1816: 2183, 24, 2721
ϵ	25°	37.78	2848: 3042
μ	in 30	3.72	4726: 2779, 3260, 3042
soly	in aq	inf	3384
aq az		none	3384
fl pt	TOC	171°F	3384

306. N,N–Dimethylacetamide

(Continued)

ir		2515, 2903, 1975, 2433, 2257, 2777
Raman		2672
mass		1780
nmr		2730, 1615, 3520
Beil	335, 59	

307. N–Methylpropionamide

$CH_3CH_2CONHCH_3$ C_4H_9ON

mw		87.122	887
bp	100	148.$^{\circ}$	1182, 496, 4049, 2848
dt/dp	100	0.30°	3976
p	10°	94	3976
eq 4	A	8.7813	3976
10 – 100 torr	B	2855.5	
fp		– 30.9°	2245: 1182
d	20°	0.93452	3329: 2848, 1226, 1182, 1840, 2245
	25	0.93050	
	30	0.92650	
	40	0.91850	
n_D	25°	1.4345	1182: 496, 4049, 2848, 3260
η	20°	6.016	3329: 1226, 2561, 2245
	25	5.215	
	30	4.554	
	40	3.533	2561
γ	30°	31.20	1840
	40	30.11	
	50	29.12	
$\triangle H_v$	10 – 100 torr	13.0	5026

307. N–Methylpropionamide

(Continued)

t_c		$412°$	1840
κ	$25°$	8×10^{-8}	2246: 4049, 1226, 2245
ϵ	$25°$	172.2	2848: 1226, 432
	-40	384	432
μ	$110°$ (g)	3.59	3260
soly	in aq	soluble	
aq az		none	
ir			496, 2083, 2827, 3892, 2515, 2434
Raman			3892
nmr			1265
Beil	**335**, 59		

308. 1,1,2,2–Tetramethylurea

$(CH_3)_2NCON(CH_3)_2$ $C_5H_{12}ON_2$

mw		116.164	887
bp	760	$175.2°$	[3042] : 3302, 5324, 2672
dt/dp	760	$0.03°$	[3042]
p	$61.2°$	10	[3042] : 497
eq 4	A	8.182	[3042]
	B	2377.5	
fp		$-1.2°$	3042
d	$15°$	0.972	1616
	20	0.9687	3042
n_D	$25°$	1.4493	5324: 3042
$\triangle H_v$	$60-175°$	10.9	[3042]
$\triangle H_f^o$		-60.8	3042
$\triangle H_c$		-819.3	1316

308. 1,1,2,2–Tetramethylurea

(Continued)

pK_{BH+}	in **260**	0.40	24: 3248, 3042
κ		$\langle 6 \times 10^{-8}$	3042
ϵ		23.06	1725: 3042
μ	$25°$ in **30**	3.47	497: 3043, 498
soly	in aq	inf	3042
fl pt		ca $170°$F	3042
uv			3042
ir			498, 3248, 4132
Raman			2672
Beil	**335**, 74		

309. 2-Pyrrolidinone

C_4H_7ON

mw		85.106	887
bp	760	$245°$	259: 4595
p	$122°$	10	259
fp		$25°$	259
d	$25°$	1.107	259, 1576
	50	1.087	
	100	1.046	
n_D	$25°$	1.486	259, 1576
	30	1.4840	
η	$25°$	13.3	259
$\triangle H_c^°$		-547.1	4530

309. 2–Pyrrolidinone

(Continued)

μ	25° in **30**	3.55	2310: 1293, 567, 1294, 2870, 1576
soly	in aq	inf	259
fl pt	OC	265°F	259
ir			1975, 4250, 2628
Raman			4296
mass			1388
Beil	**3177**, 236		

310. 1–Methyl–2–pyrrolidinone

$C_5 H_9 ON$

mw		99.134	887
bp	760	202°	1737: 4595
p	60°	4	[1737] : 4455
fp		– 24.4°	1737
d	25°	1.0279	4219, 1576: 1737
n_D	25°	1.4680	4219, 1576: 1737
η	25°	1.666	4219: 1737
$\triangle H_p$		0.8	2733
$\triangle H_c^o$	25° (l)	– 715	2733: 1737
κ	25°	$1 - 2 \times 10^{-8}$	1423
ϵ	25°	32.0	4219
μ	30° in **30**	4.09	1576
soly	in aq	inf	4595
fl pt	OC	204°F	1737
ir			2515, 1388, 1736, 3032
Beil	**3177**, 237		

311. ϵ–Caprolactam

$$CH_2(CH_2)_4CO$$
$$\underline{\quad NH \quad}$$

$C_6H_{11}ON$

mw		113.161	887
bp	50	180°	[58]
p	100°	2.9	58
eq 4	A	6.78	3393
80 – 140°	B	2344	
fp		69.207°	3203: 58, 3121, 4211
d	77°	1.02	58
n_D	31°	1.4965	58
	40	1.4935	
η	70°	19.7	1996
	78	9	58
$\triangle H_v$		13.1	58
$\triangle H_m$		3.856	3203
$\triangle H_c^{\circ}$		– 861.6	4530
$\triangle H_p$		20.	58
C_p	25°	37.33	58
A	eq 72	0.01655	3203
B		0.0007	
K_f		7.30°	1996
pK_a	25° in **149**	0.36	2311
μ	25° in **30**	3.88	2310
soly	in aq, 25°	84.0%w	[58]
fl pt	COC	257°F	58
uv			4634, 2311
ir			2515, 1975, 2311, 58, 4250
Raman			4296
nmr			3403
Beil	**3179, 240**		

312. Carbon disulfide

CS$_2$ CS$_2$

mw		76.139	887
bp	760	46.225°	5009: 1776, 3208, 4761, 4762, 5350, 4788
dt/dp	760	0.0407°	1371: 4761, 5350
p	25°	361.6	1371: 3928, 4165, 4535, 5350
eq 5	A	6.94194	5009
	B	1168.623	
	C	241.534	
fp		-111.57°	4536: 732, 785, 3301, 4762, 4784, 4787
d	0°	1.29270	4784: 1118, 1776, 3208, 4486, 4761, 4788
	15	1.27005	523, 842
	30	1.24817	
n_D	15°	1.63189	4784: 2345, 4486, 1971, 2566, 523, 3850
	20	1.62799	[4784]: 842
	25	1.62409	
dn/dt		0.00078	[4784]
η	15°	0.380	4784: 2939, 3323, 4748
	20	0.363	
	Table II	Ch II	
γ	20°	32.25	2102: 5032
	30	30.79	
$\triangle H_v$	25°	6.578	5009: 785, 3208, 3331
	bp	6.390	5009
	Table IX	Ch II	
$\triangle H_m$		1.049	785: 4601, 4769
$\triangle H_f^\circ$	25° (l)	21.37	3102: 1930
$\triangle H_c^\circ$	25° (l)	-394.5	2577: 4743, 1834, 3102
C_p°	25°	10.876	5009
	Table XIV	Ch II	
C_p	24.27°	18.17	785: 1380, 3745, 4165, 4484
t_c		279°	2651: 3331, 5032
p_c		78	2651: 3331
d_c		0.44	2651
v_c		0.170	2651
K_B		2.35°	483: 3274, 475, 3606

312. Carbon disulfide

(Continued)

κ	25°	3.7×10^{-3}	3160
ϵ	20°	2.641	3477: 1118, 1326, 1888, 5034, 5198, 4786
μ	25° in 30	0.06	5198, 2873: 2497, 3444, 734
soly	in aq, 20° aq in, 20°	0.294%w ⟨0.005%w	3540: 4225
aq az	42.6°	97.2%w	2254: 2255
fl pt	CC	− 22°F	3680
uv			225, 549, 2070, 2954, 3831, 4116, 2625
ir			350, 398, 2605, 3795, 4107, 89, 4634
Raman			884, 3223, 3899, 2704
mass			90, 3997, 1344
nmr			2454
Beil	**218**, 197		

313. 1–Butanethiol

Butyl mercaptan

$CH_3CH_2CH_2CH_2SH$ $\qquad\qquad\qquad\qquad\qquad\qquad\qquad\qquad$ $C_4H_{10}S$

mw		90.188	887
bp	760	98.456°	207: 3214, 1285, 4976, 3212, 234
dt/dp	760	0.04470°	207
p	25°	45.5	[4206]
eq 5	A B C	6.92754 1281.018 218.100	4206: 208, 3612, 5147
fp		− 115.67°	207: 1472, 1285
tp		− 115.69°	4206
d	20° 25	0.84159 0.83674	207: 1472, 3214 , 1285, 4976, 3212 2285, 2323
α	25°	1.38×10^{-6}	2285

313. 1–Butanethiol

(Continued)

n_D	$20°$	1.44295	207: 1472, 3212, 1285, 2018, 4976
	25	1.44033	2323
η	$0°$	0.648	592
	20	0.501	[592]
	30	0.450	
	80	0.284	
γ	$21.9°$	25.57	4976
	41.2	23.41	
$\triangle H_v$	$25°$	8.73	212, 209: 3005, 5147
	bp	7.702	212, 209
	Table IX	Ch II	
$\triangle H_m$		2.500	4206, 212, 209
$\triangle H_f^°$	$25°$ (l)	– 29.61	4206, 2285: 211, 3063, 3102, 5358
$\triangle H_c^°$	$25°$ (l)	– 831.77	2285: 3102, 5358
$C_p^°$	$25°$	28.24	4206, 210
	Table XIV	Ch II	
C_p	$25°$	41.18	4206, 2285
A	eq 72	0.05073	212
K_a	$20°$ in aq NaOH	1.65×10^{-11}	5279: 1586
ϵ	$25°$	5.073	3214: 3477
μ	$25°$ in **30**	1.53	3214: 2323, 5068
soly	in aq, $20°$	0.0597%w	[5279]
uv			1679, 2313, 1965, 988
ir			1846, 4836, 5186, 89, 1965
Raman			3212
mass			90, 1965, 2927
Beil	**24**, 370		

314. Benzenethiol

C_6H_5SH $\qquad\qquad\qquad\qquad\qquad\qquad\qquad\qquad\qquad\qquad$ C_6H_6S

mw		110.179	887
bp	760	169.138°	222, 4209: 4535, 4976, 3215, 3413, 2323
dt/dp	760	0.05176°	222: 3215
p	25°	2.98	1369: 4535
eq 5	A	6.99019	223, 4209, 3612
	B	1529.46	
	C	203.05	
fp		−14.94°	222: 3660, 5227, 3413
triple pt		−14.88°	4209
d	20°	1.07755	222: 3215, 711, 4976, 5048, 3413, 2323
	25	1.07273	
n_D	20°	1.59008	222: 3215, 711, 2018, 4976, 3413, 2323
	25	1.58722	
η	20°	1.239	3413
	25	1.144	
γ	20°	39.5	3413: 4976, 5048
	25	38.7	
	Table VIII	Ch II	
$\triangle H_v$	25°	10.84	1369
	bp	8.836	
	Table IX	Ch II	
$\triangle H_m$		2.736	4209: 3660
$\triangle H_f^\circ$	25° (l)	15.02	3102, 4209: 5358
$\triangle H_c^\circ$	25° (l)	−910.78	5358: 3102, 4209
C_p°	25°	25.07	4209
	Table XIV	Ch II	
C_p	25°	41.40	4209: 3660
	Table XII	Ch II	
A	eq 72	0.02064	4209: 3413
B		0.00255	
pK_a	25° in H_2O	6.5	2748: 1199, 4439, 1586, 1955
ϵ	25°	4.382	3215
μ	25° in 30	1.23	3215: 2323

314. Benzenethiol

(Continued)

uv		635, 2656, 4002, 3413, 2618, 3541
ir		1846, 89, 2451, 3413, 4209, 2519
Raman		2478, 2680, 3253, 5239, 4209, 37
mass		3413
nmr		3220, 3221, 1513
Beil	**524**, 294	

315. Methyl sulfide

$(CH_3)_2S$ C_2H_6S

mw		62.134	887
bp	760	37.34°	216: 1149, 4979, 2323, 3567, 3050, 234
dt/dp	760	0.0376°	216
p	25°	484.9	1371
eq 5	A	6.94879	3612
	B	1090.755	
	C	230.799	
fp		-98.27°	216: 3567
d	20°	0.84823	216: 1149, 4979, 2323, 3567
	25	0.84228	
n_D	20°	1.43542	216: 1149, 4979, 2323, 3567
	25	1.43228	
η	20°	0.289	1371
	25	0.279	
γ	20°	24.48	1371: 4979
	30	23.06	

315. Methyl sulfide

(Continude)

$\triangle H_v$	$25°$	6.609	218: 4963, 3064
	bp	6.453	218
	Table IX	Ch II	
$\triangle H_m$		1.9084	218: 3567
$\triangle H_f^°$	$25°$ (l)	-15.64	3064, 3102: 213, 3063, 5358
$\triangle H_c^°$	$25°$ (l)	-521.09	3064, 3102: 2577, 4743, 5358
$C_p^°$	$25°$	17.37	3064: 215
	Table XIV	Ch II	
t_c		$229.0°$	2145
p_c		56.14	2145
d_c		0.301	2145
A	eq 72	0.03140	218: 3567, 3050
K_B		$1.85°$	5126
ϵ	$20°$	6.2	3477
μ	$25°$ in **30**	1.45	1149: 2323, 3764
soly	in aq, $25°$	2. %w	1142
	aq in, $25°$	0.5%w	
fl pt		$-29°$ F	1142
uv			1965, 713, 2072, 1150, 4742, 988
ir			89, 1965, 3708, 2938
mass			1965, 2927
nmr			937, 605, 4105, 4060
Beil	**19**, 288		

316. Ethyl sulfide

$(CH_3CH_2)_2S$ $C_4H_{10}S$

mw		90.188	887
bp	760	92.102°	216: 1149, 4205, 1291, 1964, 4780, 3567
dt/dp	760	0.0439°	216: 4780
p	25°	58.37	1371: 4535
eq 5	A	6.92836	217, 4205, 3612
	B	1257.833	
	C	218.662	
fp		−103.93°	216, 1964: 4780, 4763, 4770, 3567
tp		−103.95°	4205
d	20°	0.83621	216: 1149, 3050, 1291, 332, 3367, 3452
	25	0.83118	4780, 4979, 3567, 2286, 2285
α	25°	1.45×10^{-6}	2285
n_D	20°	1.44294	216: 1149, 3050, 1291, 1964, 332, 2018
	25	1.44015	3452, 4780, 4918, 4977, 3567
η	20°	0.440	1964: 1404, 3452, 4780
	25	0.417	
	Table II	Ch II	
γ	20°	25.2	1964: 3367, 3452, 4780, 840, 4979
	25	24.5	
$\triangle H_v$	25°	8.55	218: 1104, 3101, 2285
	bp	7.591	218, 4205
	Table IX	Ch II	
$\triangle H_m$		2.845	218, 4205: 3567
$\triangle H_f^\circ$	25° (l)	−28.34	2285: 214, 3063, 4205, 2286, 3102, 4552
$\triangle H_c^\circ$	25° (l)	−833.04	2285: 4743, 3102, 2286, 2577
C_p°	25°	27.97	215: 4205
	Table XIV	Ch II	
C_p	25°	40.97	[2285]: 4205
t_c		272.8°	1371, 5032
p_c		32.93	1371
d_c		0.260	1371
v_c		0.3470	1371

316. Ethyl sulfide

(Continued)

A	eq 72	0.05001	218: 4250, 3050, 1964, 3567
K_B		$3.23°$	5126
pK_a	$27°$ in **30**	relative	5221
ϵ	$25°$	5.72	3477: 4786
μ	$25°$ in **30**	1.61	1149: 2323, 4557, 5068
uv			91, 3364, 3368, 713, 2940, 1553, 1150
ir			4836, 1964, 89, 408, 2938
Raman			4345, 4981, 4633
mass			1964, 2927
nmr			937, 605, 4060
Beil	**23**, 344		

317. Thiophene

$C_4 H_4 S$

mw		84.140	887
bp	760	$84.16°$	219: 3567, 1547, 2563, 4535, 5008, 1964
dt/dp	760	$0.0428°$	219: 1547
p	$25°$	79.68	219: 2653, 4535, 5008
eq 5	A	6.95926	220: 5147, 5008
	B	1246.02	
	C	221.35	
fp		$-38.24°$	219, 1964: 3567, 1547, 4535, 5008
d	$20°$	1.06482	219: 1964, 1291, 2036, 3567, 1547, 2563
	25	1.05884	3391

317. Thiophene

(Continued)

n_D	$20°$	1.52890	219, 1964: 1707, 2653, 1291, 3567, 1547
	25	1.52572	2018, 2563, 3391
η	$20°$	0.654	1964: 3567, 1547
	25	0.613	
	Table II	Ch II	
γ	$20°$	32.8	1964
	25	32.0	
$\triangle H_v$	$25°$	8.27	221: 5147, 5008
	bp	7.522	221
$\triangle H_s$		11.199	3319
$\triangle H_m$		1.216	221: 3567, 3648, 2370
$\triangle H_f^o$	$25°$ (l)	19.52	3102, 5008: 4552, 2288, 5358
$\triangle H_c^o$	$25°$ (l)	-675.55	2288: 4743, 2577, 3102, 5008, 5358
C_p^o	$25°$	17.42	2288
C_p	$26.84°$	32.29	3660: 2370, 5008
t_c		$307°$	2653: 3688, 964, 5147, 5032
p_c		56.2	2653: 3688, 5147
d_c		0.385	2653
v_c		0.219	2653
A	eq 72	0.01109	221: 3567, 1964
ϵ	$25°$	2.705	2036: 3477
μ	$25°$ in 30	0.52	2036: 3020, 2879, 2056, 2563, 3022, 4009
soly	in aq	miscible	2608
aq az	minimum	bp	2254
uv			91, 1806, 3318, 1964, 3271, 706, 637
ir			89, 1964, 408, 3532, 2855, 3143, 3450
Raman			653, 650, 2677, 3951, 37, 3950
mass			90, 3997, 1964, 2205
nmr			1863, 6, 4130
Beil	2364, 29		

318. Tetrahydrothiophene

$$H_2C \text{------} CH_2$$

$$H_2C \quad\quad CH_2$$

$$S$$

$$C_4H_8S$$

mw		88.172	887
bp	760	120.9°	1964: 2296
p	25°	18.4	[2296]
eq 5	A	7.39514	2296
	B	1628.8	
	C	240.7	
fp		−96.16°	1964
d	20°	0.99869	1964: 2286
	25	0.99379	
	30	0.98928	
n_D	20°	1.52890	1964: 2296
	25	1.52572	
	30	1.52257	
η	20°	1.042	1964
	25	0.971	
	30	0.914	
γ	20°	35.8	1964
	25	35.0	
	30	34.6	
$\triangle H_v$	25°	9.23	[2286]
$\triangle H_f^\circ$	25° (l)	−17.31	2286: 3063, 5358
$\triangle H_c^\circ$	25° (l)	−758.22	5358: 2286
t_c		358.8°	964
A	eq 72	0.028	1964
pK_a	in aq H_2SO_4	−4.52	282
μ	25° in 30	1.90	1152
uv			1964, 988
ir			1964, 89, 4840, 2938
Raman			4840, 37, 3889
mass			1964, 1691, 1390
nmr			3098
Beil	2362, 10		

319. Dimethyl sulfoxide

$(CH_3)_2SO$ C_2H_6OS

mw		78.134	887
bp	760	189.0°	4157: 1149
p	25°	0.600	1349
eq 12	A	26.49558	1349
20 – 50°	B	3539.32	
	C	6.00000	
fp		18.54°	1714: 1349, 4157, 4408
d	25°	1.0958	2264, 1141: 4220, 2849, 4219
	40	1.0816	1010
	60	1.0616	
α	20°	0.00088	3160
dd/dt		0.000977	[1010, 2264]
n_D	20°	1.4783	1141: 2849, 4219
	25	1.4773	
η	25°	1.996	4220: 2264, 1605, 2849, 4219
	35	1.654	
γ	20°	43.54	1010: 4157
	25	42.86	
	60	38.94	
$\triangle H_v$	25°	12.64	1349
	bp	10.31	1141
$\triangle H_m$		3.33	4355: 2964
$\triangle H_s$		18.4	1141
$\triangle H_f^\circ$	25° (l)	– 47.21	3102: 3103
$\triangle E_c$	25° (l)	– 428.29	1141
C_p°	25°	21.256	3103
	Table XIV	Ch II	
C_p	25°	36.4	1141: 4355
K_f		4.07°	1714: 2964, 4355
pK_a	in 161	1.4	4439: 4489, 2683, 232
pK_s		17.3	2695
κ	25°	2×10^{-9}	823: 3689, 4157
ϵ	25°	46.68	1714: 290, 4219, 2264, 4408
μ	25° in 30	3.9	1102: 4157, 290, 1149

319. Dimethyl sulfoxide

(Continued)

soly	in aq	25.3%w	[1141], 3169
aq az		none	1141
fl pt	OC	203°F	1141
uv			1150, 988
ir			89, 4104, 4157
Raman			4104, 2972
mass			695
nmr			2845, 605, 4105
Beil	19, 289		

320. Sulfolane

$$C_4 H_8 O_2 S$$

mw		120.171	887
bp	760	287.3° d	4267: 4960, 3410, 5102
dt/dp	760	0.065°	4267
p	118°	5.0	4267
eq 13	A	28.6824	4267
118 – 285°C	B	4350.7	
	C	6.5633	
fp		28.45°	1714: 598, 278, 864, 2688, 4960, 3410
d	30°	1.2614	4924: 4960, 3410
	35	1.2568	
	50	1.2447	
	150	1.158	

320. Sulfolane

(Continued)

n_D	30°	1.4820	4924: 4960
	35	1.4798	
	50	1.4747	
η	30°	10.286	4924: 1460, 3410, 864
	35	9.033	
	50	6.312	
	150	1.46	4267
	Table VI	Ch II	
γ	30°	35.5	1996
$\triangle H_v$	100°	15.0	[4267]
	200	14.7	
$\triangle H_m$		0.341	[1714]: 864, 4267
C_p	30°	43	·[4267]
K_f		64.1°	1714: 278, 864
κ	30°	$\langle 2 \times 10^{-8}$	864: 3381
ϵ	30°	43.3	4924: 290, 1460, 1152, 3410, 864
μ	25° in 30	4.81	1152: 2998, 4960, 290
soly	in aq, 30°	miscible	4519
aq az	nonazeotropic		3975
fl pt	OC	$350^\circ F$	4267: 4519
ir			2971, 1328, 32
nmr			1550
Beil	2362, 9		

321. 2-Methoxyethanol

$CH_3OCH_2CH_2OH$ $C_3H_8O_2$

mw		76.096	887
bp	760	124.6°	645: 3752, 1131, 1689, 2228, 4975, 5165
dt/dp	760	0.042°	[3752]
p	25°	9.7	1355: 1200, 4535
eq 5	A	7.7085	3752: 1689
	B	1711.2	
	C	230	
fp		-85.1°	3160
d	20°	0.96459	980: 2231, 1200, 1688, 2228, 3633, 4975
	25	0.96024	5165, 877, 1687, 3752
α	20°	0.00095	3160
n_D	20°	1.4021	980: 2231, 1200, 2228, 4975, 5165, 3752
	25	1.4002	
η	20°	1.72	3160
	25	1.60	
γ	14.9°	31.82	4975, 4216
	41.0	29.28	
	Table VIII	Ch II	
$\triangle H_v$	av	12.90	[3752]: 645, 1689
	bp	9.43	1355
$\triangle H_c$		-440.6	[4326]
C_p	25°	40.	1355
pK_a	25° in H_2O	14.8	369
κ	20°	1.09×10^{-6}	3160: 369
ϵ	25°	16.93	2231
μ	25° in 30	2.04	2231: 1814, 877
soly	in aq	inf	1355: 2228
aq az	99.9°	15.3%	2255: 2254
fl pt	CC	107°F	3160: 2228, 1355
ir			5354, 408, 736, 2618, 2767
Raman			2273
mass			5097
Beil	30, 467		

322. 2-Ethoxyethanol

$CH_3CH_2OCH_2CH_2OH$ $C_4H_{10}O_2$

mw		90.123	887
bp	760	135.6°	4893: 1131, 2228, 3633, 4975, 5165
dt/dp	760	0.042°	[3752]
p	25°	5.3	1355: 1200
eq 5	A	7.8191	3752
	B	1801.9	
	C	230	
fp	gl	$\langle -90°$	4893
d	20°	0.92945	980: 877, 1200, 2228, 3633, 4975, 5165
	25	0.92520	463
α		0.00097	3160
n_D	20°	1.4077	980: 1200, 2228, 4975, 5165
	25	1.4057	
η	20°	2.05	3160
	25	1.85	[1355]
γ	25°	28.2	1355: 4216
	75	23.6	
$\triangle H_v$	25°	11.28	[3752] : 645
	bp	9.688	[1355]
C_p	25°	50.	[1355]
κ		9.3×10^{-8}	463: 2569
ϵ	24°	29.6	463
μ	25° in 30	2.08	877: 1809, 3368
soly	in aq.	inf	2228
aq az	99.4°	28.8%w	2254: 2609
fl pt	CC	112°F	2228
	TOC	120°F	4893
uv			3368
ir			415
Raman			2273
nmr			4632: 4823
Beil	30, 467		

323. 2–Butoxyethanol

$CH_3CH_2CH_2CH_2OCH_2CH_2OH$ $C_6H_{14}O_2$

mw		118.177	887
bp	760	170.2°	4173: 4124, 1131, 4600, 1338, 3526
p	25°	0.852	4124: 4173, 3312, 2903, 1338
eq 4	A	8.6405	4173
	B	2548.0	
d	20°	0.90075	4173: 1338, 3526
	27	0.89460	4124: 2231, 4055, 3312, 1131, 4600
α	20°	0.00092	3160
n_D	20°	1.41980	3638: 4975, 4173, 3312, 2231, 4600
	25	1.4177	3160: 1338
η	25°	3.15	1355
	60	1.51	
γ	25°	27.4	1355
	75	23.3	
$\triangle H_v$	av	11.7	4173
C_p	25°	66.	[1355]
t_c		370°	4173
p_c		38.5	4173
κ	20°	4.32×10^{-7}	3160
ϵ	25°	9.30	2231
μ	25° in **30**	2.08	2231: 877
soly	25°	inf	1120: 1338
aq az	98.8°	20.8%w	2254: 4173
fl pt	CC	141°F	88: 1338
Beil	**30**, 468		

324. Furfuryl alcohol

HC —— CH
 ‖   ‖
HC   C-CH$_2$OH
   O

$C_5 H_6 O_2$

mw		98.102	887
bp	760	170.0°	4535: 4830, 5177
dt/dp	760	0.040°	[4535]
p	25°	0.6	[4535] : 2078, 5177
eq 4	A	9.2086	[4535]
	B	2804.5	
fp	metastable	– 29°	2608
	stable	– 14.63°	3869
d	20°	1.1285	2305: 3869, 2259, 4830, 5177, 3490
	30	1.1238	5119
n_D	20°	1.4868	2305: 2078, 2259, 4830, 5177, 3490
	30	1.4801	5119
dn/dt		0.00041	2305
η	25°	4.62	3869
γ	20°	ca 38	3160
$\triangle H_v$	25°	12.	[4535]
	bp	12.82	[4535]
$\triangle H_m$		3.138	3655: 2608
$\triangle H_f^\circ$	25°(l)	– 66.05	3658
$\triangle H_c^\circ$	25°(l)	– 609.16	3658, 3869
C_p	26.8°	49.0	3655: 2259
ϵ			3490
μ	25° in 30	1.92	3502: 231, 3490
soly	in aq	inf	3869
aq az	98.5°	20%w	2254
fl pt	CC	149°F	[3869]
uv			3271, 3100, 3881, 2305, 4634
ir			789, 2583, 5269
mass			90, 1044, 2164
nmr			6, 3857, 2975
Beil	2382, 112		

325. Tetrahydrofurfuryl alcohol

$$H_2C \text{———} CH_2$$
$$H_2C \diagdown_{}\diagup CHCH_2OH$$
$$O$$

$C_5H_{10}O_2$

mw		102.134	887
bp	760	178.$^\circ$	3870: 5177, 4830, 2530
dt/dp	760	0.049°	[3870]
p	25°	0.8	[3870] : 5177
eq 4 140 – 178°	A B	8.1085 2358.7	[3870] : 1262
eq 4 below 120°	A B	8.9228 2692.9	
fp		$\langle$ 80°	3870
d	20° 24 31	1.0524 1.0483 1.0402	[3870] : 5177, 4830, 2530, 3490
dd/dt		0.00110	3870
α	20 – 37.8°	0.00052	3870
n_D	20° 25	1.4520 1.4499	3870: 5177, 4830, 2530, 3490
η	20°	6.24	3870
γ	25°	37	3870
$\triangle H_v$	bp 25°	10.80 12.32	[3870]
$\triangle H_c$		−709.5	3870
C_p	20 – 27°C	43.3	[3870]
ϵ	23°	13.61	3870: 3490
μ	35° (l)	2.12	3056: 1362, 3490
soly	in aq 25°	inf	3870
fl pt	TOC	183°F	3870
mass			1044
Beil	**2380**, 107		

326. Diethylene glycol

$HOCH_2CH_2OCH_2CH_2OH$ $C_4H_{10}O_3$

mw		106.122	887
bp	760	244.8°	4890: 1200, 1689, 3615, 3986, 4535, 1687
dt/dp	760	0.052°	4890
p	20°	⟨0.01	4890: 3986, 4535, 2685, 3368, 3222
eq 4	A	8.1527	1689
	B	2727.3	
fp		−6.5°	3160: 3986, 3987, 2685, 1687
d	20°	1.1164	4890, 1200: 3986, 3222
dd/dt		0.00075	[1688]
α	20°	0.000635	3160
n_D	15°	1.4490	3986: 2685, 1687, 3222, 4890
	20	1.4475	
	25	1.4461	
η	20°	35.7	4890
	25	30	3986
γ	20°	48.5	3160
	bp	26.28	1688
	Table VIII	Ch II	
$\triangle H_v$	bp	12.491	1689: 3986, 645, 4933
$\triangle H_c$	(l)	−568.5	[3431], 5229, 62
C_p	20°	58.46	[3160]
κ	20°	5.86 x 10⁻⁷	[3160], 2687
ε	20°	31.69	2687
μ	in 30	2.31	3368: 2687
soly	in aq	inf	3986
aq az		none	2254: 2609
fl pt	OC	290°F	4890
uv			3368
Beil	30, 468		

327. Triethylene glycol

$$HOCH_2CH_2OCH_2CH_2OCH_2CH_2OH \qquad\qquad C_6H_{14}O_4$$

mw		150.176	887
bp	760	288.0°	4890: 1689, 1200, 4535, 3222
dt/dp	760	0.057°	4890
p	25°	0.00134	3846: 1200, 1687, 4535, 3222, 2685
eq 4	A	9.6396	1689
	B	3726.2	
fp		– 4.3°	4890: 2685, 1687
d	15°	1.1274	4890: 1200, 1688
	20	1.1235	[4890]
dd/dt		0.00078	4890
α	55°	0.00071	4890
n_D	15°	1.4578	3222: 2685, 1687
	20	1.4561	4890
η	20°	49.0	4890
	60	8.5	3264
γ	20°	45.2	3160
	bp	22.45	1688
	Table VIII	Ch II	
$\triangle H_v$	bp	17.066	1689
$\triangle H_c$	(l)	– 851.1	3431
C_p	20°	78.8	[3160]
κ	20°	8.4 x 10⁻⁸	2687
ϵ	20°	23.69	2687
μ	20° (l)	5.58	2687
soly	in aq	inf	4890
aq az		nonazeotropic	2254
fl pt	OC	330°F	4890
ir			404
Raman			404
Beil	**30**, 468		

328. 2–(2–Methoxyethoxy)ethanol

$CH_3OCH_2CH_2OCH_2CH_2OH$ $C_5H_{12}O_3$

mw		120.150	887
bp	760	194.1°	1355: 5180
dt/dp	760	0.045°	[Beil E III 1, 2095]
p	25°	0.18	1355: 5180, 1844
eq 4	A	8.69464	[Beil E III l, 2095]
	B	2717.39	
fp		– 76°*	5180
d	20°	1.0210	4846: 971
	25	1.0167	
α	20°	0.00086	3160
n_D	20°	1.4264	4846: 971, 1844
	25	1.4245	
	40	1.4188	
η	25°	3.48	[1355]
	60	1.61	
γ	25°	34.8	1355
	75	29.9	
$\triangle H_v$	bp	11.13	[1355]: 645
$\triangle H_c$		– 719.3	[4326]
C_p	25°	64.8	[1355]
soly	in aq	inf	1355
aq az		none	2254
fl pt	OC	200°F	1355
ir			408
Beil	30, 468		

* Supercooled

329. 2-(2-Ethoxyethoxy)ethanol

$CH_3CH_2OCH_2CH_2OCH_2CH_2OH$ $\qquad$ $C_6H_{14}O_3$

mw		134.177	887
bp	760	202.0°	1355: 1200, 2228, 3615, 4227, 4535, 5165
dt/dp	760	0.045°	[4535]
p	25°	0.13	1355: 1200, 1703, 4535
eq 4	A	8.7300	[4535]
	B	2779.0	
d	20°	0.9885	4846: 971, 1200, 1703, 2228, 4227, 5165
	25	0.9841	
n_D	20°	1.4273	4846: 971, 1200, 1703, 4227, 5165
	25	1.4254	
	40	1.4194	
η	20°	3.85	3160
	25	3.71	[1355]
	60	1.72	
γ	25°	31.8	1355
	75	27.2	
$^{\wedge}H_v$	bp	11.34	[1355]
C_p	25°	72	[1355]
κ	25°	2.5×10^{-8}	3160
soly	in aq	inf	2228
aq az		none	2255
fl pt	OC	205°F	1355
ir			408
Beil	30, 468		

330. Salicylaldehyde

$C_7H_6O_2$

mw		122.125	887
bp	760	$196.7°$	3002: 3004, 1158, 3714, 2608, 645, 319
dt/dp	760	$0.055°$	[3002]
p	$33°$	1	4535: 1928, 4720
eq 4	A	7.9407	[1928]
170–200°	B	2360.6	
fp		$-7°$	2608: 924, 2372, 5028
d	$25°$	1.1525	319: 5028, 2608, 924, 3714, 5031, 2372
	50	1.1282	4709, 1534
dd/dt		0.000979	3976
n_D	$20°$	1.5718	1158: 2608, 1534, 2372, 319, 3074
	25	1.57017	5031
η	$20°$	2.90	3439, 5028
	45	1.669	4709
$\triangle H_v$	bp	9.14	645, 3002, 3004: 4963
$\triangle H_c^o$	$25°$ (l)	-794.5	2577: 2608
C_p	$20-195°$	55.4	3002: 3004
K_B		$4.96°$	[1928]
pK_a	$25°$ in H_2O	8.14	3108: 4008, 604, 34, 289
κ	$25°$	1.64×10^{-7}	5026
ϵ	$20°$	13.9	5026: 1381, 4752, 5031
μ	$20°$ in 30	2.86	3020: 5209, 1158, 3245, 3024, 4302
soly	in aq, $85.8°$	1.68%w	4311: 2608
uv			3906, 2092, 10, 1237, 233, 1860
ir			1026, 1419, 408, 973, 4092, 2518
Raman			973, 3300, 3852, 655, 4024
nmr			3812, 1989, 3211, 1266
Beil	744, 31		

331. 4–Hydroxy–4–methyl–2–pentanone

Diacetone alcohol

$(CH_3)_2C(OH)CH_2COCH_3$ $C_6H_{12}O_2$

mw		116.161	887
bp	760	168.1° d	1953: 2228, 4535
dt/dp	760	0.044°	[1675]
p	25°	1.7	[1675]: 4269, 4535, 1338
eq 4	A	8.5552	1675: 4269, 1953
to 115°	B	2482.93	
fp		– 44.0°	2824: 4535
d	20°	0.9387	1675: 2228, 2824
	25	0.9342	[1675]
α	20°	0.00099	3160
n_D	20°	1.4235	2824: 2228, 1953, 1675
	25	1.4213	
η	– 1.8°	7.5	[4269]
	20	2.9	3160
γ	20°	31.0	1675
$\triangle H_v$	30 – 110°	11.4	1675: 645
$\triangle H_c$		– 1000.	[3160]
C_p	20°	52.2	[3160]
ϵ	25°	18.2	3477
μ	20° in 30	3.24	2758: 3477
soly	in aq	inf	4504: 2228
aq az	99.50°	15.7%w	1953: 2254, 4269
fl pt*	CC	136°F	3189
uv			4634
ir			3920, 342
mass			90
nmr			4256, 5282
Beil	113, 836		

* Commercial products may flash lower due to presence of acetone

332. 2-Chloroethanol

HOCH$_2$CH$_2$Cl   C$_2$H$_5$OCl

mw		80.515	887
bp	760	128.6°	4784: 2228, 3208, 4428, 4535, 2490
dt/dp	760	0.040°	4784
eq 15	A	7.83951	2607
	B	2727.40	
	C	0.0067805	
fp		−67.5°	4784: 4535
d	15°	1.20720	4784: 2228, 3208, 4428, 2490
	20	1.20190	
	30	1.19118	
n$_D$	15°	1.44380	4783: 2228, 3208, 4428, 2968, 2490
dn/dt		0.00039	4783
η	15°	3.913	4784: 2968
	30	2.688	
γ	20°	38.9	2607
$\triangle$H$_v$	bp	9.901	3208
$\triangle$E$_c$		−290.2	4397: 3809
pK$_a$	25° in H$_2$O	14.31	369
ϵ	25°	25.8	3477
μ	25°	1.88	4428
soly	in aq	inf	2228
aq az	97.75° at 748 torr	42%w	2254: 5317, 3905
fl pt	OC	105°F	88, 2228
uv			1679, 2840
ir			340, 5348, 5354, 415, 789
Raman			3130, 3485, 2021
nmr			3175, 4253, 4632, 4823
Beil		**22**, 337	

333. 2–Cyanoethanol

NCCH$_2$CH$_2$OH   C$_3$H$_5$ON

mw		71.079	887
bp		220° d	2681: 4037
dt/dp		0.05°	[2681]
p	25°	0.08	4037
eq 4	A	8.822	[2681]
	B	2929.	
fp		– 46°	4037
d	0°	1.059	2088
	25	1.0404	4037
$\triangle$H$_v$	av	13.4	[2681]
soly	in aq	inf	4037
ir			4728, 3227
Raman			2681
nmr			4823
Beil	**222**, 298		

334. 2–Aminoethanol

Monoethanolamine

HOCH$_2$CH$_2$NH$_2$ C$_2$H$_7$ON

mw		61.084	887
bp	760	170.95°	4782: 3949, 3070
dt/dp	760	0.046°	4782
p	20°	0.36	4885, 2896, 3232, 552, 2955
eq 5	A	7.73800	3070: 3232, 1357
	B	1732.11	
	C	186.215	
fp		10.53°	741: 3070, 3949, 3232

334. 2-Aminoethanol

(Continued)

d	$15°$	1.01949	4782: 3949, 2896, 3232, 4847, 2955, 2642
	25	1.01159	[4782]
dd/dt		0.000790	[4782] : 3232
α	$55°$	0.00079	4885
n_D	$20°$	1.4539	4847: 741, 3949, 2896, 3232, 2955, 2642
	25	1.4521	
dn/dt		0.00036	[4782] : 3232
η	$15°$	30.855	4782
	25	19.346	741
	30	14.417	4782
γ	$15.80°$	49.39	4782: 3949
	20	48.89	
	30	47.74	
$\triangle H_v$	$0°$	22.010	3232
	bp	11.910	3232
$\triangle H_m$		4.900	1357
$\triangle H_c$	$25°$ (l)	-220.8	4773
C_p	$30°$	30.4	[1357]
t_c		$341.3°$	1357
P_c		44.1	1357
pK_{BH+}	$25°$ in aq HCl	9.498	437: 1974, 3718, 1972, 441, 1977, 1798
κ	$25°$	11.0×10^{-6}	730
ϵ	$25°$	37.72	741: 2802
μ	$25°$ in 124	2.27	3691
soly	in aq	inf	4885
aq az		none	2255
fl pt	CC	$185°$ F	88
ir			4728, 2766, 4630
Raman			1443, 1750
mass			3085
nmr			1767, 1234
Beil	353, 274		

335. Diethanolamine

$(HOCH_2CH_2)_2NH$ $\qquad\qquad\qquad\qquad$ $C_4H_{11}O_2N$

mw		105.138	887
bp	760	268.39° d	1357: 4885, 2640, 1144, 3691, 2955, 676
p	20°	0.0002	676: 2640, 2955
eq 5	A	8.13968	3070
	B	2328.56	
	C	174.399	
fp		27.95°	3070: 1144, 2640
d	30°	1.0899	[4885] : 2640, 1144, 676, 2642
	40	1.0828	1357
dd/dt		0.00064	4885: 2955
n_D	30°	1.4747	4885: 2640, 2642
	40	1.4720	1357
η	30°	380	4885: 1144
	40	196.4	1357
$\triangle H_v$	160.5°	16.8	[4885]
	bp	15.590	1357
$\triangle H_m$		6.000	1357
C_p	30°	55.8	[1357]
t_c		442.1°	1357
p_c		32.3°	1357
pK_{BH+}	25° in aq HCl	8.883	437: 690, 1977
ϵ	25° in **124**	2.81	3691
soly	in aq 20°	96.4%w	4885
aq az		none	2255
fl pt	OC	280°F	88
ir			4630
mass			3085
Beil	**353**, 283		

336. Triethanolamine

$(HOCH_2CH_2)_3N$ $C_6H_{15}O_3N$

mw		149.191	887
bp	760	335.39°	3070: 1144
dt/dp	760	0.073°	4885
p	20°	⟨0.01	4885: 3070, 3691, 2641, 567, 1144, 5273
eq 5	A	8.55499	3070
	B	2962.73	
	C	186.750	
fp		21.57°	3070: 1144, 2955
d	25°	1.1196	1357: 2641, 1743, 2896, 5273, 676, 2642
	40	1.1116	
dd/dt		0.00054	4885
n_D	25°	1.4835	1357: 2641, 2642
	40	1.4798	
η	25°	613.6	1357: 2896, 1144
	40	208.1	
$\triangle H_v$	bp	16.127	1357
$\triangle H_m$		6.500	1357
C_p	30°	74.1	[1357]: 3258
t_c		514.3°	1357
p_c		24.2	1357
pK_{BH+}	25° in aq HCl	7.762	437, 438: 1972, 1977, 5107, 442
ϵ	25°	29.36	567
μ	25° in 124	3.57	3691
soly	in aq	inf	4885: 2641, 676, 1144
fl pt	OC	355° F	4885: 88
mass			3085
Beil	353, 285		

337. 2,2′-Thiodiethanol

$(HOCH_2CH_2)_2S$ $C_4H_{10}O_2S$

mw		122.187	887
bp	760	$282°$	2609: 1502, 1000, 5156, 3367
dt/dp	25 mm	$\sim 0.9°$	[Beil **E III 1**, 2122] *
p	$175°$	25	[Beil **E III 1**, 2122]
eq 4	A	10.242	[Beil **E III 1**, 2122]
1 – 50 torr	B	3963.	
fp		$-10°$	1000: 1502, 2609
d	$0°$	1.1973	1000: 1502, 2609, 2752
	20	1.1817	3367
	25	1.1793	1000
dd/dt		0.00072	[1000]
n_D	$20°$	1.52031	1502: 2609
	25	1.5146	1000
γ	$20°$	53.8	3367
$\triangle H_v$	av	~ 18	[Beil **E III 1**, 2122]
soly	in aq $20°$	inf	1000: 2609
fl pt	TOC	$320°F$	2609
uv			3368, 2842, 3926, 3364
ir			4188
Beil	**30** , 470		

* Selected values

338. 2–Furaldehyde

Furfural

```
HC ——— CH
 ‖        ‖
HC      C–CHO
    \  /
     O
```

$C_5H_4O_2$

mw		96.086	887
bp	760	161.8°	4535: 3208, 3374
p	25°	2.5	[3868]: 3232, 4766, 4535
eq 4	A	7.959	2942: 3232
	B	2209	
fp		– 36.5°	4535: 4766
d	20°	1.1598	3868: 5032, 3374, 4832, 2305, 3490
	25	1.1545	
dd/dt		0.001057	[2942]
n_D	20°	1.52608	2942: 4832, 3374, 2305, 3490
	25	1.52345	
η	0°	2.48	3868: 5022, 2942
	25	1.49	
	38	1.35	
γ	30°	41.1	3868: 5034
	Table VIII	Ch II	
$\triangle H_v$	bp	10.330	3208: 3232, 5034, 4963
$\triangle H_m$		3.43	2942
$\triangle H_c$	25°	– 560.3	2942: 2577
C_p	25°	37.7	[75]
	Table IX	Ch II	
t_c		397°	2942
p_c		54.4	2942
ϵ	25°	38	2942: 2625, 1381
μ	25° in **30**	3.63	1894: 4302, 2625
soly	in aq, 20°	8.2%w	2417: 4225, 664
	aq in, 20°	6.3%w	
aq az	97.85°	35%w	2254
fl pt	CC	140°F	88
uv			549, 1324, 5069, 3271, 2073, 3100
ir			1162, 408, 3943, 4544
Raman			4544
nmr			4632, 1907, 6, 3857
Beil	**2461**, 272		

339. Bis(2–chloroethyl)ether

$(ClCH_2CH_2)_2O$ $C_4H_8OCl_2$

mw		143.014	887
bp	760	178.75°	1689: 425, 2228, 3368, 3452, 4428, 4970
dt/dp	760	0.0575°	[1689]
p	25°	1.55	[1689] : 1570, 4535, 1765
eq 4	A	8.1040	1689
	B	2359.6	
fp		– 46.8°	4839: 1570, 1760, 3627, 4841
d	20°	1.2192	3452: 425, 1688, 2228, 4428, 4970, 2490
	25	1.2130	
dd/dt		11.70 x 10^{-4}	[3701]
α	10 – 30°	0.00097	1338
n_D	20°	1.45750	4428: 425, 1570, 2228, 3452, 4970, 4839
	25	1.45534	
η	20°	2.41	3452: 1570, 4841, 3627, 1687, 2490
	25	2.14	
	Table VI	Ch II	
γ	20°	37.6	3452: 1570, 1688, 4970
	25	37.0	
	Table VIII	Ch II	
$\triangle H_v$	bp	10.81	1689: 1570, 3627
$\triangle H_m$		2.070	4841: 4839
C_p	30°	52.8	1570: 3627
ϵ	20°	21.2	3477: 4841
μ	25° in 30	2.58	4428: 3368, 4841
soly	20° in aq	1.02%w	1338: 3627, 1570, 2228
	20° aq in	0.1%	
aq az	98°	34.5%w	2255
fl pt	CC	131°F	88
uv			2340: 3368
Raman			2510
nmr			3559
Beil	**22 II**, 337		

340. Epichlorohydrin

ClCH$_2$ CHCH$_2$
\ /
O

C$_3$ H$_5$ OCl

mw		92.526	887
bp		116.11°	4266
dt/dp	760	0.044°	4266
p	25°	18.	4886
eq 5	A	7.4709	4899
	B	1587.9	
	C	230.	
fp		−57.2°	4266: 2372
d	15°	1.18683	4266: 4542
	20	1.18066	
	25	1.17455	
α	55°	0.00104	4886
n$_D$	20°	1.43805	4266
	25	1.4358	
η	0°	1.56	4266: 5023
	25	1.03	
γ	20°	37.00	4266: 4542
	31	35.48	
△H$_v$	bp	9.060	4266: 5034
t$_c$		351°	5032
κ	25°	3.4 x 10^{-8}	4266: 5027
ε	22°	22.6	3477: 1326, 4266, 5034
μ	in 232	1.8	4907
soly	in aq, 20°	6.58%w	4266
	aq in, 20°	1.47%	
aq az	88°	75%w	2254
fl pt	OC	105°F	88
ir			408, 3679
Raman			367, 2912
nmr			4632, 3848, 3941
Beil	2362, 6		

341. o–Nitroanisole

o–CH$_3$OC$_6$H$_4$NO$_2$ C$_7$H$_7$O$_3$N

mw		153.139	887
bp	737	265°	711: 1921, 4312
fp		10.45°	1894: 2009, 711, 1116, 847, 848, 4312
d	20°	1.2527	711: 1921
	25	1.2408	1116
n$_D$	20°	1.56188	711: 4168,
	25	1.5597	1116
γ	26°	45.7	848: 4312
	55	41.93	
μ	20° in **30**	4.83	1116: 2236, 1335, 1921
soly	in aq, 30°	0.169%w	1913
uv			3862, 372, 4365
ir			1988
Raman			5231
mass			2026
nmr			4103
Beil	**523**, 217		

342. Morpholine

OCH$_2$CH$_2$NHCH$_2$CH$_2$ C$_4$H$_9$ON

mw		87.122	887
bp	760	128.94°	4782: 2653
dt/dp	760	0.046°	4885: 4782
p	25°	10.08	4885: 640
eq 5	A	7.71813	2821
0 – 44°	B	1745.8	
	C	235.0	
44 – 170°	A	7.16030	
	B	1447.70	
	C	210.0	

342. Morpholine

(Continued)

fp		$-3.1°$	4885
d	15°	1.00495	4782: 640
	25	0.99547	[4782]
dd/dt		0.000948	[4782]
n_D	15°	1.45733	4782: 2653
	20	1.4542	4885
η	15°	2.534	4782
	30	1.792	
γ	15°	38.27	4782: 640
	20	37.63	
	30	36.24	
$\triangle H_v$	25°	10.508	[1369]
	bp	8.855	[1369]: 640
pK_{BH+}	25° in H_2O	8.492	2159: 5128, 1974, 1972
ϵ	25°	7.42	3576: 4773
μ	25°	1.50	3576
soly	in aq	inf	4885
aq az		none	2255
fl pt	OC	100°F	88
ir			5078, 1658
nmr			2041
Beil	4190, 5		

343. Ethyl lactate

$CH_3CH(OH)COOCH_2CH_3$ $C_5H_{10}O_3$

mw		118.134	887
bp	760	154.5°	4184: 4399, 2228, 5273
dt/dp	760	0.04°	3160
p	30°	5	1338: 1225, 5273
eq 4	A	8.7020	[Beil **221**, 280]
	B	2489.7	
fp		− 26°	3160
d	20°	1.0328	[5251] : 4399, 2228, 5048
	25	1.0272	
dd/dt		0.00112	
α	10 – 30°	0.00098	1338
n_D	20°	1.4124	3375: 2228, 4399, 5048, 1338, 860
η	25°	2.44	3160
γ	17.3°	28.90	5048: 3375, 1482
	38.4	26.80	
$\triangle H_v$	25°	11.8	[Beil **221**, 280]
	bp	11.1	
$\triangle H_c^\circ$	25° (l)	− 654.7	2577
C_p	20 – 30°	60.6	[3160]
$[\alpha]_D$	18.9°	+ 11.26	5251: 4852
κ	25°	1.0×10^{-6}	2569
e	20°	13.1	3375
μ	20° in **30**	2.4	3375
soly	in aq	inf	1338: 2228
fl pt	CC	117 °F	2228
uv			4375, 4852
ir			4375, 638, 3401, 3402
Raman			860
Beil	**221**, 280		

344. Methyl salicylate

o-HOC$_6$H$_4$COOCH$_3$ C$_8$H$_8$O$_3$

mw		152.151	887
bp	760	233.3°	4763: 1928, 3714, 1893
dt/dp	760	0.057°	1893
p	25°	0.11	[3232] : 319, 1928
eq 12	A	36.5929	3232
	B	4410.9	
	C	9.1999	
fp		- 8.6°	4763
d	20°	1.1831	[3714] : 319, 1326
	25	1.1782	
dd/dt		0.00100	[3714]
n$_D$	18.1°	1.53773	319: 3232
	20	1.5365	[1788]
γ	- 19.8°	44.2	2372
	212.2	19.8	
$\triangle$H$_v$	bp	11.155	3232: 645
$\triangle$H$_c^{\circ}$	25°(l)	- 902.2	2577
C$_p$	15 - 30°	59.46	2689
ϵ	30°	9.41	3477: 1326
μ	25° in 30	2.47	2698: 1083
soly	in aq, 30°	0.74%w	4494
fl pt	TCC	210°F	1788
uv			2319, 4634
ir			1653, 2274, 4630, 3401
nmr			3176, 1627, 1655
Beil	1061, 70		

345. 2-Methoxyethyl acetate

$CH_3COOCH_2CH_2OCH_3$ $C_5H_{10}O_3$

mw		118.134	887
bp	760	144.5°	3160: 2228, 1338
dt/dp	760	0.044°	[2444]
p	25°	5	[2444]: 1338
eq 4	A	8.3312	[2444]
	B	2277.68	
fp		−65.1°	4893
d	20°	1.0049	[4893]: 2228, 1338
α	10 – 30°	0.00110	1338
n_D	20°	1.4022	3375
	25	1.4025	2228
$\triangle H_v$	bp	10.5	[2444]
ε	20°	8.25	3375
μ	30° in **30**	2.13	1814: 3375
soly	in aq	inf	4893
aq az	97.0°	48.5%w	2254
fl pt	CC	132°F	88
uv			2812
Beil	**159**, 141		

346. 2-Ethoxyethyl acetate

$CH_3COOCH_2CH_2OCH_2CH_3$ $C_6H_{12}O_3$

mw		132.161	887
bp	760	156.3°	554
p	20°	1.09	554
fp		−61.7°	4893
d	20°	0.9730	[3160]

346. 2-Ethoxyethyl acetate

(Continued)

dd/dt		0.00079	[3743]
n_D	25°	1.4023	[3743]
	30	1.4003	3743
dn/dt		0.00042	[3743]
η	25°	1.025	3160
γ	25°	31.8	3160
ϵ	30°	7.567	3743: 3375
μ	30° in **30**	2.25	1814: 3743, 3375
soly	in aq, 20°	22.9%w	4893
	aq in, 20°	6.5%w	
aq az	97.5°	44.4%w	2255: 2254
fl pt	TOC	134°F	4893
Beil	**159**, 141		

347. 2-(2-Ethoxyethoxy)ethyl acetate

$$CH_3COOCH_2CH_2OCH_2CH_2OCH_2CH_3 \qquad\qquad C_8H_{16}O_4$$

mw		176.214	887
bp	760	217.4°	4893: 2228
dt/dp	760	0.05°	3160
p	20°	0.1	5180, 1338
eq 4	A	8.589	[3160]
	B	2800.	
fp		– 25°	4893
d	20°	1.0096	[4893]: 2228
dd/dp		0.00102	3160
α	20°	0.00101	3160
n_D	20°	1.4213	3160: 2228

347. 2-(2-Ethoxyethoxy)ethyl acetate

(Continued)

η	$20°$	2.8	3160
$\triangle H_v$		21.8	[3160]
C_p	$20°$	94.8	[3160]
soly	in aq	inf	4893
aq az		none	2255: 2254
fl pt	COC	$230°F$	4893
Beil	159, 141		

348. Methyl acetoacetate

$CH_3COCH_2COOCH_3$ $\hspace{4cm}$ $C_5H_8O_3$

mw		116.118	887
bp	760	$171.7°$	Beil E III 3, 1182: 1717, 1932
p	$25°$	6	[Beil E III 3, 1182] : 321
eq 4	A	7.1066	[Beil E III 3, 1182]
	B	1880.4	
fp		$-80°$	4519
d	$15°$	1.0800	[3713] : 321, 1717
	20	1.0747	
	25	1.0724	
n_D	$20°$	1.4186	321
η	$20°$	1.704	1717
$\triangle H_v$	av	8.60	[Beil E III 3, 1182]
$\triangle H_c$		-594.0	1933
fl pt	CC	$170°F$	88
uv			577
mass			692, 691
nmr			850
Beil	280, 632		

349. Ethyl acetoacetate

$CH_3COCH_2COOCH_2CH_3$ $C_6H_{10}O_3$

mw		130.145	887
bp		180.8°	4535
p	keto, 40 – 41°	2	2643: 806, 1402
eq 4	A	8.40179	[4535]
	B	2506.57	
fp	keto	– 39°	2643
	enol	– 44°	2609
d	10° keto	1.0368	3289: 806, 4821, 1402
	enol	1.0119	
	25°	1.02126	4202
dd/dt		0.00097	4202
n_D	10° keto	1.4225	2643: 3289, 806
	enol	1.4480	
	20	1.4192	3424
η	25°	1.5081	1402
γ	14.8°	32.47	3893
	46.4	29.09	
	Table VIII	Ch II	
$\triangle H_c°$	25° (l)	– 755.4	2577
C_p	24.5°	59.8	2609
t_c		400°	883
pK_a	25° in H_2O	10.68	41: 2609, 1457, 1820
κ	25°	4×10^{-8}	2472
ϵ	22°	15.7	1381
μ	– 80° in 312 enol	2.04	554
	18.2° in 30 keto	3.22	
soly	in aq, 25°	12%w	4893
	aq in, 25°	4.9%w	
fl pt	OC	184°	88
uv			577, 3424, 5056
ir			408
Raman			4285
mass			692, 691, 693
nmr			850, 2486
Beil	280, 632		

350. Methyl cyanoacetate

$NCCH_2COOCH_3$ $C_4H_5O_2N$

mw		99.090	887
bp		205.09°	1371: 5026, 2504, 1932
dt/dp	760	0.0490°	1371
	25°	0.14	1371
eq 5	A	7.60624	1371
	B	1914.22	
	C	200.	
fp		−13.07°	1371
d	0°	1.1492	5028: 2504, 1932
	25	1.1225	
n_D	20°	1.41791	1371: 2504
	25	1.41662	
η	20°	2.793	1371: 5028
	40	1.764	
γ	20°	42.32	1371
	30	41.16	
$\triangle H_v$	25°	14.740	[1371]
	bp	11.509	
$\triangle H_c$	25° (l)	−475.8	2577: 1933
A	eq 72	0.02195	1371
κ	25°	4.49 x 10^{-7}	5026: 5035
ϵ	20°	29.30	5052: 5026
soly		sl sol	2537
Raman			1171, 1167
Beil	171, 584		

351. Ethyl cyanoacetate

$NCCH_2COOCH_2CH_3$ $C_5H_7O_2N$

mw		113.117	887
bp	760	206.0°	4535: 5026, 1665, 2504, 3524
p	67.8°	1	4535: 1114, 4782
eq 4	A	9.94022	[4535]
100 – 200°	B	3382.89	
fp		– 22.5°	2537
d	15°	1.06652	4782: 1665, 2504, 804, 1114, 5048
	25	1.0564	5028, 3524
	30	1.05106	4782
dd/dt		0.00103	[4782]
n_D	20°	1.41555	2504: 3524, 804, 1114, 4782
η	0°	5.06	5028, 5022: 1665
	15	3.256	4782
	20	2.63	4994
	25	2.50	2655
	30	2.148	4782
γ	20°	36.48	4782, 1114, 5048
	30	35.42	
$\triangle H_v$	100 – 200°	15.5	[4535]
$\triangle H_c$	25° (l)	– 635.1	2577: 1933
K_a	25°	$\langle 10^{-9}$	3695
κ	25°	6.9×10^{-7}	5026: 5027
ϵ	18°	26.7	1381: 2655, 5026, 5034, 3477, 2802
μ		2.17	4410
soly	in aq	25.9%w	2655
fl pt		230°F	2537
uv			4197
ir			3918
Beil	**171, 585**		

352. Trifluoroacetic acid

CF$_3$COOH   C$_2$HO$_2$F$_3$

mw		114.024	887
bp	760	71.78°	2749: 1758, 2528, 4570, 2104
p	25°	108	[4617] : 2528
eq 4	A	8.389	4617
	B	$-$1895	
fp		$-$15.25°	4331: 4570, 2749, 2104, 2037
d	20°	1.4890	2608: 2528, 4570
	25	1.4785	2749
	30	1.4671	2380
dd/dt		0.002346	2380
n$_D$	20°	1.2850	2608
η	20°	0.926	2528
	25	0.855	
γ	24°	13.63	2380
	Table VIII	Ch II	
$\triangle$H$_v$	av	8.672	4617
	bp	7.949	4617
t$_c$		246°	57
P$_c$		40	57
pK$_a$	25° in H$_2$O	0.23	41: 2104, 2241
ϵ	20°	8.55	2037: 3477, 4331
μ	100°(g)	2.28	1758
soly	in aq	inf	2528
aq az	105.46°	79.4%w	4570: 203, 2107, 2255
ir			508, 4630
Raman			1592
nmr			4064, 903, 2241
Beil	**160**, 194		

353. o–Chloroaniline

o–ClC$_6$H$_4$NH$_2$ C$_6$H$_6$NCl

mw		127.574	887
bp	760	208.84°	1372: 264, 2469, 4535, 1151
dt/dp	760	0.05277°	1372
p	25°	0.2533	1369
eq 5	A	7.63311	1372
	B	2085.5	
	C	230.	
fp		– 1.94°	1372: 264
	α – form	– 11.92	264: 2639
	β – form	– 1.78	
d	20°	1.21251	1372: 264, 2469, 4710
	25	1.20775	
n$_D$	20°	1.58807	1372: 264, 2469, 1151
	25	1.58586	
η	20°	2.9157	1369, 4710
	40	1.8458	
γ	20°	43.66	1369
	30	42.54	
$\triangle$H$_v$	25°	13.565	1369
	bp	10.60	
$\triangle$H$_m$		2.84	4314
A	eq 72	0.01785	1369
pK$_{BH+}$	25°	2.64	582: 3718, 4531, 4947, 4013
ϵ		13.4	5034
μ	25° in **30**	1.77	1151: 2057, 4755, 4395
soly	in aq, 25°	0.876%w	1369: 4314
	aq in, 19°	0.56%w	4494
uv			373, 3862, 5244
ir			1645, 4337
Raman			2137
nmr			1767, 3044
Beil	**1670**, 597		

354. Hexamethylphosphoric triamide

$[(CH_3)_2N]_3PO$ $C_6H_{18}ON_3P$

mw		179.204	887
bp	760	233°	3991: 3544, 4802
dt/dp	25 mm	0.903	[3991]
p	30°	0.07	3545: 3544, 934, 3826
eq 4	A	8.7227	[3991]
	B	2956	
fp		7.20°	3544: 1387, 3991
d	20°	1.027	3991: 4802
n_D	20°	1.4588	666: 4802
	25	1.4570	3545
η	20°	3.47	3991
	60° F	3.5 cs	3545
γ	20°	33.8	3991
$\triangle H_v$	av.	13.53	[3991]
$\triangle H_m$		4.05	[1387]
$\triangle H_f$			3545
K_f		6.93°	3544: 1387
ϵ	20°	30	3544: 3991
μ	25° (l)	4.31	4131: 1383
	25° in 30	5.54	4535
soly	in aq	inf	3991
ir			2008, 1337, 2118
nmr			4919, 5098

BOILING POINT INDEX

Boiling Point 760 torr	Compound Number	Compound
9.503°	4	2,2-Dimethylpropane
12.27	220	Chloroethane
20.4	134	Acetaldehyde
27.852	3	2-Methylbutane
29.968	45	1-Pentene
31.36	122	Furan
31.50	164	Methyl formate
31.56	242	1,1-Dichloroethylene
32.4	279	Isopropylamine
33.9	119	Propylene oxide
34.55	109	Ethyl ether
35.72	108	Ethylvinyl ether
35.74	222	2-Chloropropane
36.074	2	Pentane
36.353	48	trans-2-Pentene
36.7	46	2-Pentene
36.942	47	cis-2-Pentene
37.34	315	Methyl sulfide
38.35	247	Bromoethane
39.75	230	Dichloromethane
42.30	132	Dimethoxymethane
42.43	255	Iodomethane
44.4	283	tert-Butylamine
45.10	241	3-Chloropropene
46.225	312	Carbon disulfide
46.60	221	1-Chloropropane
47.26	219	1,2-Dibromotetrafluoroethane
47.57	217	1,1,2-Trichlorotrifluoroethane
47.67	244	trans-1,2-Dichloroethylene
48.0	135	Propionaldehyde
48.5	278	Propylamine
49.262	1	Cyclopentane
49.741	10	2,2-Dimethylbutane
50.7	226	2-Chloro-2-methylpropane
52.69	139	Acrolein
53.3	289	Allylamine
54.15	165	Ethyl formate
55.45	292	Diethylamine

BOILING POINT INDEX (Continued)

Boiling Point 760 torr	Compound Number	Compound
56.29°	141	Acetone
56.323	169	Methyl acetate
57.	291	Ethylenimine
57.28	233	1,1–Dichloroethane
57.988	11	2,3–Dimethylbutane
59.41	249	2–Bromopropane
60.271	8	2–Methylpentane
60.63	243	cis-1,2–Dichloroethylene
61.152	231	Chloroform
62.5	282	sec-Butylamine
63.2	120	1,2–Epoxybutane
63.282	9	3–Methylpentane
63.485	49	1–Hexene
64.1	137	Isobutyraldehyde
64.70	58	Methanol
66.	123	Tetrahydrofuran
67.73	281	Isobutylamine
68.25	224	2–Chlorobutane
68.3	111	Isopropyl ether
68.740	7	Hexane
68.85	225	1–Chloro–2–methylpropane
70.97	248	1–Bromopropane
71.78	352	Trifluoroacetic acid
71.812	5	Methylcyclopentane
72.30	256	Iodoethane
72.5	170	Vinyl acetate
74.0	235	1,1,1–Trichloroethane
74.8	136	Butyraldehyde
76.75	232	Carbon tetrachloride
77.114	172	Ethyl acetate
77.3	276	Acrylonitrile
77.4	280	Butylamine
78.29	59	Ethanol
78.44	223	1–Chlorobutane
79.64	142	2–Butanone
80.100	30	Benzene
80.2	190	Methyl acrylate
80.261	216	Hexafluorobenzene
80.500	17	2,4–Dimethylpentane

BOILING POINT INDEX (Continued)

Boiling Point 760 torr	Compound Number	Compound
80.725°	6	Cyclohexane
80.85	166	Propyl formate
81.60	265	Acetonitrile
82.26	61	2-Propanol
82.42	65	2-Methyl-2-propanol
82.979	54	Cyclohexene
83.483	234	1,2-Dichloroethane
83.9	294	Diisopropylamine
84.16	317	Thiophene
84.734	212	Fluorobenzene
87.19	245	Trichloroethylene
88.	125	Tetrahydropyran
88.2	176	Isopropyl acetate
89.5	298	Triethylamine
89.50	258	2-Iodopropane
89.64	110	Propyl ether
89.784	16	2,3-Dimethylpentane
90.052	14	2-Methylhexane
90.3	277	Methacrylonitrile
91.850	15	3-Methylhexane
92.102	316	Ethyl sulfide
92.2	113	Butylethyl ether
92.8	218	1,1,2,2-Tetrachlorodifluoroethane
93.0	117	1,2-Dimethoxyethane
93.643	50	1-Heptene
93.82	112	Butylvinyl ether
97.08	97	2-Propen-1-ol
97.20	60	1-Propanol
97.35	266	Propionitrile
98.4	168	Isobutyl formate
98.427	13	Heptane
98.456	313	1-Butanethiol
99.10	184	Ethyl propionate
99.238	21	2,2,4-Trimethylpentane
99.5	191	Ethyl acrylate
99.55	63	2-Butanol
100.000	0	Water
100.3	192	Methyl methacrylate
100.56	148	Formic acid
100.934	12	Methylcyclohexane

BOILING POINT INDEX (Continued)

Boiling Point 760 torr	Compound Number	Compound
101.20°	260	Nitromethane
101.320	124	p–Dioxane
101.548	175	Propyl acetate
101.99	143	3–Pentanone
102.0	71	2–Methyl–2–butanol
102.45	257	1–Iodopropane
103.6	133	Acetal
103.85	269	Isobutyronitrile
104.0	174	Allyl acetate
104.1	140	Crotonaldehyde
106.40	297	Piperidine
106.6	167	Butyl formate
107.66	64	2–Methyl–1–propanol
107.76	227	1–Chloropentane
109.2	293	Dipropylamine
109.841	20	2,2,3–Trimethylpentane
110.625	31	Toluene
111.5	72	3–Methyl–2–butanol
112.34	179	sec–Butyl acetate
113.1	73	2,2–Dimethyl–1–propanol
113.6	100	2–Propyn–1–ol
114.07	261	Nitroethane
114.40	213	o–Fluorotoluene
115.256	299	Pyridine
115.3	68	3–Pentanol
116.11	340	Epichlorohydrin
116.5	145	4–Methyl–2–pentanone
116.5	214	m–Fluorotoluene
116.6	215	p–Fluorotoluene
117.26	290	Ethylenediamine
117.66	62	1–Butanol
117.90	149	Acetic acid
117.94	268	Butyronitrile
118.0	178	Isobutyl acetate
119.0	67	2–Pentanol
120.25	263	2–Nitropropane
120.9	318	Tetrahydrothiophene
121.2	99	trans–2–Buten–1–ol

BOILING POINT INDEX (Continued)

Boiling Point 760 torr	Compound Number	Compound
121.20°	246	Tetrachloroethylene
121.280	51	1–Octene
121.5	173	Propargyl acetate
121.55	185	Ethyl butyrate
123.6	98	cis–2–buten–1–ol
124.084	23	2,2,5–Trimethylhexane
124.6	321	2–Methoxyethanol
125.665	19	Octane
126.114	177	Butyl acetate
126.8	201	Ethyl carbonate
128.6	332	2–Chloroethanol
128.7	69	2–Methyl–1–butanol
128.94	342	Morpholine
129.76	296	Pyrrole
130.5	70	3–Methyl–1–butanol
131.18	262	1–Nitropropane
131.36	253	1,2–Dibromethane
131.687	228	Chlorobenzene
131.7	77	4–Methyl–2–pentanol
131.783	18	Ethylcyclohexane
132.	102	trans–2–Butene–1,4–diol
134.7	187	Ethyl isovalerate
134.75	284	Cyclohexylamine
135.6	322	2–Ethoxyethanol
136.186	35	Ethylbenzene
137.8	66	1–Pentanol
138.351	34	p–Xylene
139.103	33	m–Xylene
140.0	161	Acetic anhydride
140.83	150	Propionic acid
141.2	157	Acrylic acid
141.3	270	Valeronitrile
142.0	181	Isopentyl acetate
142.2	114	Butyl ether
144.411	32	o–Xylene
144.5	345	2–Methoxyethyl acetate
145.14	55	Styrene
146.20	236	1,1,2,2–Tetrachloroethane
146.5	78	2–Ethyl–1–butanol
146.868	52	1–Nonene

BOILING POINT INDEX (Continued)

Boiling Point 760 torr	Compound Number	Compound
147.51°	186	Isobutyl isobutyrate
148.	307	N-Methylpropionamide
148.0	76	2-Methyl-1-pentanol
149.2	180	Pentyl acetate
149.55	252	Bromoform
150.798	22	Nonane
152.392	36	Isopropylbenzene
153.0	303	N,N-Dimethylformamide
153.80	127	Anisole
154.	272	4-Methylvaleronitrile
154.5	343	Ethyl lactate
154.70	152	Isobutyric acid
155.65	144	Cyclohexanone
155.9	56	2-Pinene
155.908	250	Bromobenzene
156.3	346	2-Ethoxyethyl acetate
157.	79	1-Methylcyclohexanol
157.0	75	1-Hexanol
159.6	295	Dibutylamine
159.7	89	2-Heptanol
159.76 d	118	Bis(2-methoxyethyl) ether
160.5	159	Methacrylic acid
161.10	74	Cyclohexanol
161.8	338	2-Furaldehyde
162.00	237	Pentachloroethane
163.27	151	Butyric acid
163.6	271	Hexanenitrile
164.716	37	Mesitylene
165	81	cis-2-Methylcyclohexanol
166.0	57	2(10)-Pinene
166.1	306	N,N-Dimethylacetamide
166.5	82	trans-2-Methylcyclohexanol
ca 167.	85	trans-3-Methylcyclohexanol
167.6	80	2-Methylcyclohexanol (mixed isomers)
ca 168.	84	cis-3-Methylcyclohexanol
168.1 d	331	4-Hydroxy-4-methyl-2-pentanone
169.0	162	Propionic anhydride
169.119	42	tert-Butylbenzene
169.138	314	Benzenethiol
170.0	324	Furfuryl alcohol
170.00	128	Phenetole

BOILING POINT INDEX (Continued)

Boiling Point 760 torr	Compound Number	Compound
170.2°	323	2-Butoxyethanol
170.570	53	1-Decene
170.95	334	2-Aminoethanol
171.	87	cis-4-Methylcyclohexanol
171.	88	trans-4-Methylcyclohexanol
171.7	348	Methyl acetoacetate
172.	83	3-Methylcyclohexanol (mixed isomers)
172.	86	4-Methylcyclohexanol (mixed isomers)
173.00	239	m-Dichlorobenzene
173.305	41	sec-Butylbenzene
173.4	116	Isopentyl ether
174.12	240	p-Dichlorobenzene
174.123	27	Decane
175.2	308	1,1,3,3-Tetramethylurea
176.0	121	Cineol
176.50	154	Isovaleric acid
177.10	43	p-Cymene
178.	325	Tetrahydrofurfural alcohol
178.75	339	Bis(2-chloroethyl) ether
178.9	138	Benzaldehyde
180.48	238	o-Dichlorobenzene
180.8	349	Ethyl acetoacetate
181.839	93	Phenol
182. d	259	Diiodomethane
ca 182.	302	N-Methylformamide
183.270	40	Butylbenzene
184.34	91	2-Ethyl-1-hexanol
184.40	285	Aniline
185.0	126	Benzylethyl ether
185.0	158	Crotonic acid
185.4	202	Ethyl oxalate
185.5	153	Valeric acid
186.8	115	Pentyl ether————
187.273	26	trans-Decahydronaphthalene
187.6	104	1,2-Propanediol
189.0	319	Dimethyl sulfoxide
190.9	171	Ethylene glycol diacetate
191.004	94	o-Cresol
191.10	275	Benzonitrile
191.7	24	Decahydronaphthalene, mixed

BOILING POINT INDEX (Continued)

Boiling Point 760 torr	Compound Number	Compound
194.0°	188	Isopentyl isovalerate
194.1	328	2-(2-Methoxyethoxy)ethanol
195.2	90	1-Octanol
195.774	25	cis-Decahydronaphthalene
196.7	330	Salicylaldehyde
197.3	103	1,2-Ethanediol
198.6	182	2-Ethylhexyl acetate
199.30	203	Ethyl malonate
199.5	163	Butyric anhydride
199.50	194	Methyl benzoate
200.4	204	Methyl maleate
200.40	286	o-Toluidine
200.55	288	p-Toluidine
201.940	96	p-Cresol
202.	310	1-Methyl-2-pyrrolidinone
202.0	147	Acetophenone
202.0	329	2-(2-Ethoxyethoxy)ethanol
202.232	95	m-Cresol
203.40	287	m-Toluidine
204.	199	γ-Butyrolactone
205.09	350	Methyl cyanoacetate
205.2	273	Octanenitrile
205.45	92	Benzyl alcohol
205.7	155	Hexanoic acid
206.	305	N-Methylacetamide
206.0	351	Ethyl cyanoacetate
206.25	131	Veratrole
207.42	146	Camphor
207.5	106	1,3-Butanediol
207.57	39	1,2,3,4-Tetrahydronaphthalene
208.84	353	o-Chloroaniline
210.5 d	301	Formamide
210.80	264	Nitrobenzene
212.40	195	Ethyl benzoate
214.4	105	1,3-Propanediol
215.5	183	Benzyl acetate
216.278	29	Dodecane
217.4	347	2-(2-Ethoxyethoxy)ethyl acetate
217.955	38	Naphthalene
220. d	333	2-Cyanoethanol

BOILING POINT INDEX (Continued)

Boiling Point 760 torr	Compound Number	Compound
221.15°	304	Acetamide
225.3	205	Ethyl maleate
231.	208	Bis(2-ethylhexyl) phthalate
231.2	196	Propyl benzoate
233.	354	Hexamethylphosphoric triamide
233.3	344	Methyl salicylate
233.5	210	Butyl borate
233.5	274	α-Tolunitrile
235.	101	cis-2-Butene-1,4-diol
237.10	300	Quinoline
238.	200	Ethylene carbonate
239.04	28	Bicyclohexyl
239.9	156	Octanoic acid
240.12	44	Cyclohexylbenzene
243.5	254	1,1,2,2-Tetrabromoethane
244.8	326	Diethylene glycol
245.	309	2-Pyrrolidinone
258.31	130	Phenyl ether
259.3	229	1-Chloronaphthalene
265.	341	o-Nitroanisole
267.	267	Succinonitrile
268.39 d	335	Diethanolamine
271.0	198	Ethyl cinnamate
ca 280 d	206	Butyl maleate
281.1	251	1-Bromonaphthalene
282.	337	2,2'-Thiodiethanol
287.3 d	320	Sulfolane
288.0	327	Triethylene glycol
288.30	129	Benzyl ether
289.	211	Butyl phosphate
290.0	107	Glycerol
323.5	197	Benzyl benzoate
335.39	336	Triethanolamine
340.0	207	Butyl phthalate
345.	209	Butyl sebacate
ca 350.	189	Butyl sterate
360.0 d	160	Oleic acid

FREEZING POINT INDEX

Freezing Point	Compound Number	Compound
− 165.220°	45	1-Pentene
− 159.900	3	2-Methylbutane
− 153.670	8	2-Methylpentane
− 151.390	47	cis-2-Pentene
− 150.	120	1,2-Epoxybutane
− 142.434	5	Methylcyclopentane
− 140.5 (active)	224	2-Chlorobutanol
− 140.244	48	trans-2-Pentene
− 139.819	49	1-Hexene
− 138.	46	2-Pentene
− 136.4	220	Chloroethane
− 134.5	241	3-Chloropropene
− 130.3	225	1-Chloro-2-methylpropane
− 129.721	2	Pentane
− 128.538	11	2,3-Dimethylbutane
− 126.593	12	Methylcyclohexane
− 126.2	60	1-Propanol
− 124.$\pm$ 2	24	Decahydronaphthalene, mixed
− 123.8 (ms)	109	Ethyl ether
− 123.2	110	Propyl ether
− 123.1	223	1-Chlorobutane
− 123.	134	Acetaldehyde
− 122.8	221	1-Chloropropane
− 122.56	242	1,1-Dichloroethylene
− 119.4	15	3-Methylhexane
− 119.242	17	2,4-Dimethylpentane
− 118.858	50	1-Heptene
− 118.6	247	Bromoethane
− 118.276	14	2-Methylhexane
− 117.2	70	3-Methyl-1-butanol
− 117.18	222	2-Chloropropane
− 116.3 (st)	109	Ethyl ether
− 115.8	108	Ethylvinyl ether
− 115.67	313	1-Butanethiol
− 114.7	63	2-Butanol
− 114.7	298	Triethylamine
− 114.4	78	2-Ethyl-1-butanol
− 114.1	59	Ethanol
− 113.3 (racemic)	224	2-Chlorobutanol
− 112.27	20	2,2,3-Trimethylpentane

FREEZING POINT INDEX (Continued)

Freezing Point	Compound Number	Compound
− 111.93°	119	Propylene oxide
− 111.9	268	Butyronitrile
− 111.57	312	Carbon disulfide
− 111.323	18	Ethylcyclohexane
− 111.1	256	Iodoethane
− 110.5	219	1,2-Dibromotetrafluoroethane
− 109.8 (ms)	248	1-Bromopropane
− 108.5	123	Tetrahydrofuran
− 108.1 (st)	248	1-Bromopropane
− 108.	64	2-Methyl-1-propanol
− 107.388	21	2,2,4-Trimethylpentane
− 105.780	23	2,2,5-Trimethylhexane
− 105.15	132	Dimethoxymethane
− 103.99	262	1-Nitropropane
− 103.93	316	Ethyl sulfide
− 103.512	54	Cyclohexene
− 103.	113	Butylethyl ether
− 101.736	51	1-Octene
− 101.3	257	1-Iodopropane
− 99.865	10	2,2-Dimethylbutane
− 99.3	187	Ethyl isovalerate
− 99.0	164	Methyl formate
− 99.0	227	1-Chloropentane
− 98.85	178	Isobutyl acetate
− 98.27	315	Methyl sulfide
− 98.05	169	Methyl acetate
− 98.0	185	Ethyl butyrate
− 97.68	58	Methanol
− 96.98	233	1,1-Dichloroethane
− 96.4	136	Butyraldehyde
− 96.3	294	Diisopropylamine
− 96.2	270	Valeronitrile
− 96.16	318	Tetrahydrothiophene
− 96.033	36	Isopropylbenzene
− 95.348	7	Hexane
− 95.2	114	Butyl ether
− 95.2	279	Isopropylamine
− 95.14	230	Dichloromethane

FREEZING POINT INDEX (Continued)

Freezing Point	Compound Number	Compound
- 94.991°	31	Toluene
- 94.975	35	Ethylbenzene
-94.7	141	Acetone
- 94.5	168	Isobutyl formate
- 93.866	1	Cyclopentane
- 93.	182	2-Ethylhexyl acetate
- 92.9	166	Propyl formate
- 92.8	170	Vinyl acetate
- 92.78	266	Propionitrile
- 92.5	175	Propyl acetate
- 92.	112	Butylvinyl ether
- 91.32	263	2-Nitropropane
- 90.612	13	Heptane
- 90.0	167	Butyl formate
- 90.0	258	2-Iodopropane
- 89.52	261	Nitroethane
- 89.44	98	cis-2-Buten-1-ol
- 89.0	249	2-Bromopropane
- 88.62	62	1-Butanol
- 88.2	289	Allylamine
- 88.01 (ms)	40	Butylbenzene
- 88.0	61	2-Propanol
- 87.85 (st)	40	Butylbenzene
- 87.7	214	m-Fluorotoluene
- 86.95	139	Acrolein
- 86.69	142	2-Butanone
- 86.4	245	Trichloroethylene
- 85.61	122	Furan
- 85.5	111	Isopropyl ether
- 85.1	321	2-Methoxyethanol
- 84.6	281	Isopropylamine
- 84.0	145	4-Methyl-2-pentanone
- 83.97	172	Ethyl acetate
- 83.55	276	Acrylonitrile
- 83.0	278	Propylamine
- 81.37	52	1-Nonene
- 80.7	186	Isobutyl isobutyrate
- 80.3	271	Hexanenitrile
- 80.	135	Propionaldehyde
- 80.	348	Methyl acetoacetate

FREEZING POINT INDEX (continued)

Freezing Point	Compound Number	Compound
- 80.0°	243	cis–1,2–Dichloroethylene
- 79.4	165	Ethyl formate
- 78.5	181	Isopentyl acetate
- 78.0	291	Ethylenimine
- 77.2	66	1–Pentanol
- 76.5	140	Crotonaldehyde
- 76.	91	2–Ethyl–1–hexanol
- 76.	328	2–(2–Methoxyethoxy)ethanol
- 75.470	41	sec–Butylbenzene
- 73.85	184	Ethyl propionate
- 73.5	177	Butyl acetate
- 73.4	176	Isopropyl acetate
- 73.1	161	Acetic anhydride
- 72.65	283	tert–Butylamine
- 71.5	269	Isobutyronitrile
- 71.2	191	Ethyl acrylate
- 69.4	115	Pentyl ether
- 67.935	43	p–Cymene
- 67.5	332	2–Chloroethanol
- 66.45	255	Iodomethane
- 66.310	53	1–Decene
- 65.7	163	Butyric anhydride
- 65.	137	Isobutyraldehyde
- 65.1	345	2–Methoxyethyl acetate
- 64.	56	2–Pinene
- 63.55	231	Chloroform
- 63.	293	Dipropylamine
- 62.0	213	o–Fluorotoluene
- 62.	295	Dibutylamine
- 61.7	346	2–Ethoxyethyl acetate
- 61.54	57	2(10)–Pinene
- 60.43	303	N,N–Dimethylformamide
- 60.	104	1,2–Propanediol
- 58.	117	1,2–Dimethoxyethane
- 57.850	42	tert–Butylbenzene
- 57.2	340	Epichlorohydrin
- 56.8	215	p–Fluorotoluene
- 56.795	19	Octane

FREEZING POINT INDEX (Continued)

Freezing Point	Compound Number	Compound
− 55.6°	138	Benzaldehyde
− 53.519	22	Nonane
− 51.80	100	2–Propyn–1–ol
− 51.6	196	Propyl benzoate
− 51.5	183	Benzyl acetate
− 51.1	272	4–Methylvaleronitrile
− 50.	208	Bis(2–ethylhexyl) phthalate
− 49.8	244	trans–1,2–Dichloroethylene
− 49.8	292	Diethylamine
− 49.1	280	Butylamine
− 48.9	203	Ethyl malonate
− 48.2	192	Methyl methacrylate
− 47.872	33	m–Xylene
− 46.8	339	Bis(2–chloroethyl) ether
− 46.1	152	Isobutyric acid
− 46.	333	2–Cyanoethanol
− 45.6	273	Octanenitrile
− 45.58	228	Chlorobenzene
ca − 45.	125	Tetrahydropyran
− 44.720	37	Mesitylene
− 44.6	75	1–Hexanol
− 44.0	331	4–Hydroxy–4–methyl–2–pentanone
− 44. (enol)	349	Ethyl acetoacetate
− 43.835	265	Acetonitrile
− 43.8	236	1,1,2,2–Tetrachloroethane
− 43.55	199	γ–Butyrolactone
− 43.01	25	cis–Decahydronaphthalene
− 43.0	162	Propionic anhydride
− 43.0	201	Ethyl carbonate
− 42.21	212	Fluorobenzene
− 41.55	299	Pyridine
− 41.5	171	Ethylene glycol diacetate
− 40.6	202	Ethyl oxalate
− 39. (keto)	349	Ethyl acetoacetate
−38.97	143	3–Pentanone
− 38.24	317	Thiophene
− 37.5	127	Anisole
− 36.5	338	2–Furaldehyde
− 36.4	217	1,1,2–Trichlorotrifluoroethane

FREEZING POINT INDEX (Continued)

Freezing Point	Compound Number	Compound
− 35.8°	277	Methacrylonitrile
− 35.790	39	1,2,3,4−Tetrahydronaphthalene
−35.66	234	1,2−Dichloroethane
− 35.	207	Butyl phthalate
− 34.7	195	Ethyl benzoate
− 33.67	153	Valeric acid
− 32.1	144	Cyclohexanone
− 30.9	307	N−Methylpropionamide
− 30.82	250	Bromobenzene
− 30.628	55	Styrene
− 30.400	26	trans−Decahydronaphthalene
− 30.40	287	m−Toluidine
− 30.4	235	1,1,1−Trichloroethane
− 29.657	27	Decane
− 29.52	128	Phenetole
− 29.3	154	Isovaleric acid
− 29.0	237	Pentachloroethane
− 29. (ms)	324	Furfuryl alcohol
− 28.55	260	Nitromethane
− 27.6	105	1,3−Propanediol
− 26. melts	138	Benzaldehyde
− 26.	343	Ethyl lactate
− 25.4	226	2−Chloro−2−methylpropane
− 25.182	32	o−Xylene
− 25.	347	2−(2−Ethoxyethoxy)ethyl acetate
− 24.76	239	m−Dichlorobenzene
− 24.5 (ms)	286	o−Toluidine
− 24.4	310	1−Methyl−2−pyrrolidinone
− 23.8	274	α−Tolunitrile
− 23.4146	296	Pyrrole
− 22.95	232	Carbon tetrachloride
− 22.5	351	Ethyl cyanoacetate
− 22.35	246	Tetrachloroethylene
− 20.7	150	Propionic acid
− 20.	306	N,N−Dimethylacetamide
− 17.7	284	Cyclohexylamine
− 17.5	204	Methyl maleate
− 17.01	238	o−Dichlorobenzene
− 16.58	4	2,2−Dimethylpropane
− 16.10 (st)	286	o−Toluidine

FREEZING POINT INDEX (Continued)

Freezing Point	Compound Number	Compound
− 15.3°	92	Benzyl alcohol
− 15.25	352	Trifluoroacetic acid
− 14.97	90	1–Octanol
− 14.94	314	Benzenethiol
− 14.85	300	Quinoline
− 14.63 (st)	324	Furfuryl alcohol
− 13.9 (ms)	194	Methyl benzoate
− 13.07	350	Methyl cyanoacetate
ca − 13.	103	1,2–Ethanediol
− 12.75	275	Benzonitrile
− 12.10 (st)	194	Methyl benzoate
− 11.92 (α–form)	353	o–Chloroaniline
− 10.5	297	Piperidine
− 10.	337	2,2′–Thiodiethanol
− 9.587	29	Dodecane
− 8.8	71	2–Methyl–2–butanol
− 8.8	205	Ethyl maleate
− 8.6	344	Methyl salicylate
− 7.	330	Salicylaldehyde
− 6.5	326	Diethylene glycol
− 5.98	285	Aniline
ca − 5.5	84	cis–3–Methylcyclohexanol
− 5.2	151	Butyric acid
− 4.3	327	Triethylene glycol
ca − 4.0	82	trans–2–Methylcyclohexanol
− 3.95	155	Hexanoic acid
− 3.8	302	N–Methylformamide
− 3.1	342	Morpholine
− 2.3	229	1–Chloronaphthalene
− 1.78 (β–form)	353	o–Chloroaniline
− 1.2	308	1,1,3,3–Tetramethylurea
ca− 0.5	85	trans–3–Methylcyclohexanol
0.00	0	Water
0.0	254	1,1,2,2–Tetrabromoethane
ca 0.4 (ms)	251	1–Bromonaphthalene
1.	209	Butyl sebacate
1.3	121	Cineole
1.94	353	o–Chloroaniline
2.55	301	Formamide

FREEZING POINT INDEX (Continued)

Freezing Point	Compound Number	Compound
3.60°	129	Benzyl ether
3.63	28	Bicyclohexyl
5.10	216	Hexafluorobenzene
5.533	30	Benzene
5.60 (ms)	259	Diiodomethane
5.76	264	Nitrobenzene
6.1 (st)	259	Diiodomethane
6.20	251	1-Bromonaphthalene
6.544	6	Cyclohexane
6.99	44	Cyclohexylbenzene
ca 7.0	81	cis-2-Methylcyclohexanol
7.20	354	Hexamethylphosphoric triamide
8.05	252	Bromoform
8.27	148	Formic acid
9.79	253	1,2-Dibromoethane
10.45	341	o-Nitroanisole
10.53	334	2-Aminoethanol
11.0	101	cis-2-Butene-1,4-diol
11.3	290	Ethylenediamine
11.80	124	p-Dioxane
12. (trans)	198	Ethyl cinnamate
12.22	95	m-Cresol
13.263	34	p-Xylene
13.38 (α-form)	160	Oleic acid
13.5	157	Acrylic acid
15.0	159	Methacrylic acid
16.30 (β-form)	160	Oleic acid
16.51	156	Octanoic acid
16.66	149	Acetic acid
18.18	107	Glycerol
18.54	319	Dimethyl sulfoxide
19.4	197	Benzyl benzoate
19.62	147	Acetophenone
19.9	193	Methyl oleate
20.9 (ms)	131	Veratrole
21.57	336	Triethanolamine
22.5 (st)	131	Veratrole
25.	309	2-Pyrrolidinone

FREEZING POINT INDEX (Continued)

Freezing Point	Compound Number	Compound
25.15°	74	Cyclohexanol
25.82	65	2-Methyl-2-propanol
26.	79	1-Methylcyclohexanol
26.0	218	1,1,2,2-Tetrachlorodifluoroethane
26.3	189	Butyl stearate
26.87	130	Phenyl ether
27.3	102	trans-2-Butene-1,4-diol
27.95	335	Diethanolamine
28.45	320	Sulfolane
30.55	305	N-Methylacetamide
30.944	94	o-Cresol
34.739	96	p-Cresol
36.4	200	Ethylene carbonate
40.90	93	Phenol
43.75	288	p-Toluidine
52.	73	2,2-Dimethyl-1-propanol
53.13	240	p-Dichlorobenzene
57.88	267	Succinonitrile
69.207	311	ϵ-Caprolactam
71.4	158	Crotonic acid
80.00	304	Acetamide
80.290	38	Naphthalene
178.75	146	Camphor

DENSITY INDEX AT 25°

d	Compound Number	Compound
0.5851	4	2,2–Dimethylpropane
0.61462	3	2–Methylbutane
0.62139	2	Pentane
0.63533	45	1–Pentene
0.6431	48	trans–2–Pentene
0.64446	10	2,2–Dimethylbutane
0.64852	8	2–Methylpentane
0.6504	47	cis–2–Pentene
0.6535	46	2–Pentene
0.65481	7	Hexane
0.65702	11	2,3–Dimethylbutane
0.65976	9	3–Methylpentane
0.66832	17	2,4–Dimethylpentane
0.66848	49	1–Hexene
0.67439	14	2–Methylhexane
0.67951	13	Heptane
0.6821	279	Isopropylamine
0.68295	15	3–Methylhexane
0.68781	21	2,2,4–Trimethylpentane
0.6908	283	tert–Butylamine
0.69091	16	2,3–Dimethylpentane
0.69267	50	1–Heptane
0.69849	19	Octane
0.7016	292	Diethylamine
0.70322	23	2,2,5–Trimethylhexane
0.70760	109	Ethyl ether
0.7100	294	Diisopropylamine
0.71085	51	1–Octene
0.71207	20	2,2,3–Trimethylpentane
0.7121	278	Propylamine
0.71381	22	Nonane
0.7182	111	Isopropyl ether
0.7201	282	sec–Butylamine
0.7230	298	Triethylamine
0.72531	52	1–Nonene
0.72625	27	Decane
0.7297	281	Isobutylamine
0.7329	293	Dipropylamine
0.7346	280	Butylamine

DENSITY INDEX AT 25° (Continued)

d	Compound Number	Compound
0.73693	53	1-Decene
0.74054	1	Cyclopentane
0.7419	110	Propyl ether
0.74394	5	Methylcyclopentane
0.7448	113	Butylethyl ether
0.74561	29	Dodecane
0.7473	108	Ethylvinyl ether
0.7575	289	Allylamine
0.7577	295	Dibutylamine
0.7641	114	Butyl ether
0.76506	12	Methylcyclohexane
0.7656	269	Isobutyronitrile
0.7727	112	Butylvinyl ether
0.7737	116	Isopentyl ether
0.77389	6	Cyclohexane
0.7766	265	Acetonitrile
0.77682	266	Propionitrile
0.7780	134	Acetaldehyde
0.7790	115	Pentyl ether
0.7812	65	2-Methyl-2-propanol
0.78126	61	2-Propanol
0.7836	137	Isobutyraldehyde
0.78390	18	Ethylcyclohexane
0.78440	141	Acetone
0.78504	59	Ethanol
0.7865	268	Butyronitrile
0.78664	58	Methanol
0.7912	135	Propionaldehyde
0.7948	277	Methacrylonitrile
0.7950	270	Valeronitrile
0.7961	145	4-Methyl-2-pentanone
0.7964	136	Butyraldehyde
0.7978	64	2-Methyl-1-propanol
0.7993	272	4-Methylvaleronitrile
0.7997	142	2-Butanone
0.79975	60	1-Propanol
0.8004	276	Acrylonitrile
0.8012	271	Hexanenitrile
0.8026	63	2-Butanol
0.8036	77	4-Methyl-2-pentanol

DENSITY INDEX AT 25° (Continued)

d	Compound Number	Compound
0.8050	71	2–Methyl–2–butanol
0.8054	67	2–Pentanol
0.8060	62	1–Butanol
0.80609	54	Cyclohexene
0.8071	70	3–Methyl–1–butanol
0.80945	143	3–Pentanone
0.8097	273	Octanenitrile
0.8115	66	1–Pentanol
0.8134	89	2–Heptanol
0.8138	72	3–Methyl–2–butanol
0.8150	69	2–Methyl–1–butanol
0.8159	75	1–Hexanol
0.8160	68	3–Pentanol
0.8194	120	1,2–Epoxybutane
0.8206	76	2–Methyl–1–pentanol
0.8209	119	Propylene oxide
0.8213	133	Acetal
0.82209	90	1–Octanol
0.8291	91	2–Ethyl–1–hexanol
0.8295	78	2–Ethyl–1–butanol
0.83118	316	Ethyl sulfide
0.832	291	Ethylenimine
0.8361	226	2–Chloro–2–methylpropane
0.83674	313	1–Butanethiol
0.8389 (20°)	139	Acrolein
0.84228	315	Methyl sulfide
0.8454	99	trans–2–Buten–1–ol
0.8456	140	Crotonaldehyde
0.84643	97	2–Propen–1–ol
0.8533	43	p–Cymene
0.8535	210	Butyl borate
0.85378	132	Dimethoxymethane
0.8539	56	2–Pinene
0.8540	189	Butyl sterate
0.8541	188	Isopentyl isovalerate
0.8542 (20°)	186	Isobutyl isobutyrate
0.85607	40	Butylbenzene
0.8563	222	2–Chloropropane

DENSITY INDEX AT 25° (Continued)

d	Compound Number	Compound
0.85669	34	p–Xylene
0.85751	36	Isopropylbenzene
0.85794	297	Piperidine
0.85797	41	sec–Butylbenzene
0.85990	33	m–Xylene
0.8603	187	Ethyl isovalerate
0.86111	37	Mesitylene
0.8621	117	1,2–Dimethoxyethane
0.86227	284	Cyclohexylamine
0.86231	31	Toluene
0.86240	42	tert–Butylbenzene
0.86264	35	Ethylbenzene
0.86592	26	trans–Decahydronaphthalene
0.8664	181	Isopentyl acetate
0.8667	57	2(10)–Pinene
0.8668	176	Isopropyl acetate
0.867	224	2–Chlorobutane, active
0.8674	182	2–Ethylhexyl acetate
0.86766	224	2–Chlorobutane, racemic
0.8695	178	Isobutyl acetate
0.8702	193	Methyl oleate
0.8707	180	Pentyl acetate
0.8717	225	1–Chloro–2–methylpropane
0.8720 (20°)	179	sec–Butyl acetate
0.8772	125	Tetrahydropyran
0.87370	30	Benzene
0.87596	32	o–Xylene
0.87636	177	Butyl acetate
0.8769	227	1–Chloropentane
0.8789	24	Decahydronaphthalene, mixed isomers
0.87991	168	Isobutyl formate
0.87394	185	Ethyl butyrate
0.8809	223	1–Chlorobutane
0.88249	28	Bicyclohexyl
0.88303	175	Propyl acetate
0.8842	123	Tetrahydrofuran
0.88463	184	Ethyl propionate
0.8850	221	1–Chloropropane
0.8863	167	Butyl formate
0.8870	160	Oleic acid

DENSITY INDEX AT 25° (Continued)

d	Compound Number	Compound
0.89064	290	Ethylenediamine
0.89291	25	cis-Decahydronaphthalene
0.89455	172	Ethyl acetate
0.89634	323	2-Butoxyethanol
0.89974	166	Propyl formate
0.90122	55	Styrene
0.90280	220	Chloroethane
0.9066	156	Octanoic acid
0.9080	88	trans-4-Methylcyclohexanol
0.9110	84	cis-3-Methylcyclohexanol
0.9126	83	3-Methylcyclohexanol, mixed isomers
0.9129	87	cis-4-Methylcyclohexanol
0.91593	165	Ethyl formate
0.9176	85	trans-3-Methylcyclohexanol
0.9183 (22.5°)	86	4-Methylcyclohexanol, mixed isomers
0.9192	121	Cineol
0.92058	82	trans-2-Methylcyclohexanol
0.9215	80	2-Methylcyclohexanol, mixed isomers
0.92165	154	Isovaleric acid
0.9223	174	Allyl acetate
0.9230	155	Hexanoic acid
0.9234 (20°)	191	Ethyl acrylate
0.9247	170	Vinyl acetate
0.9251 (24.65°)	79	1-Methylcyclohexanol
0.92520	322	2-Ethoxyethanol
0.9279	169	Methyl acetate
0.93050	307	N-Methylpropionamide
0.93095	122	Furan
0.93109	241	3-Chloropropene
0.93181	81	cis-2-Methylcyclohexanol
0.9324	209	Butyl sebacate
0.9342	331	4-Hydroxy-4-methyl-2-pentanone
0.9345	153	Valeric acid
0.9366	306	N,N-Dimethylacetamide
0.93758	192	Methylmethacrylate
0.93874	44	Cyclohexylbenzene
0.94206	144	Cyclohexanone
0.94288	152	Isobutyric acid
0.94397	303	N,N-Dimethylformamide
0.9440	118	Bis(2-methoxyethyl) ether

DENSITY INDEX AT 25° (Continued)

d	Compound Number	Compound
0.9446	126	Benzylethyl ether
0.9450	100	2-Propyn-1-ol
0.9498 (30°)	305	N-Methylacetamide
0.9506	190	Methyl acrylate
0.9532	151	Butyric acid
0.96024	321	2-Methoxyethanol
0.96049	128	Phenetole
0.96199	163	Butyric anhydride
0.9640 (77°)	158	Crotonic acid
0.96565	296	Pyrrole
0.96589 (45°)	288	p-Toluidine
0.9662	39	1,2,3,4-Tetrahydronaphthalene
0.96693	164	Methyl formate
0.9684	74	Cyclohexanol
0.9687 (20°)	308	1,1,3,3-Tetramethylurea
0.9691	346	2-Ethoxyethyl acetate
0.96926	201	Ethyl carbonate
0.9760	211	Butyl phosphate
0.9782 (80°)	38	Napthalene
0.97824	299	Pyridine
0.979	208	Bis(2-ethylhexyl) phthalate
0.98290	263	2-Nitropropane
0.9841	329	2-(2-Ethoxyethoxy)ethanol
0.98498	287	m-Toluidine
0.98669 (60°)	267	Succinonitrile
0.9880	150	Propionic acid
0.9892 (91.1°)	304	Acetamide
0.98932	127	Anisole
0.9900	146	Camphor
0.9907	206	Butyl maleate
0.9921	215	p-Fluorotoluene
0.9927	173	Propargyl acetate
0.99379	318	Tetrahydrothiophene
0.99430	286	o-Toluidine
0.99547	342	Morpholine
0.99609	262	1-Nitropropane
0.99705	0	Water
0.9974 (20°)	214	m-Fluorotoluene
0.9988	302	N-Methylformamide

DENSITY INDEX AT 25° (Continued)

d	Compound Number	Compound
1.0006	275	Benzonitrile
1.00124	213	o–Fluorotoluene
1.0046	347	2-(2-Ethoxyethoxy)ethyl acetate
1.0049	345	2-Methoxyethyl acetate
1.0053	106	1,3-Butanediol
1.0057	162	Propionic anhydride
1.01159	334	2-Aminoethanol
1.0125	274	Tolunitrile
1.0153 (20°)	159	Methacrylic acid
1.0167	328	2-(2-Methoxyethoxy)ethanol
1.01750	285	Aniline
1.0190	196	Propyl benzoate
1.01906	212	Fluorobenzene
1.02 (77°)	311	ε–Caprolactam
1.02126	349	Ethyl acetoacetate
1.02382	147	Acetophenone
1.027 (20°)	354	Hexamethylphosphoric triamide
1.0272	343	Ethyl lactate
1.0279	310	1-Methyl-2-pyrrolidinone
1.02797	124	p-Dioxane
1.03019	95	m–Cresol
1.0328	104	1,2-Propanediol
1.0399	129	Benzyl ether
1.0404	333	2-Cyanoethanol
1.04127	92	Benzyl alcohol
1.04183	195	Ethyl benzoate
1.0426	207	Butyl phthalate
1.0434	138	Benzaldehyde
1.04366	149	Acetic acid
1.04464	261	Nitroethane
1.0469	198	Ethyl cinnamate
1.0472	325	Tetrahydrofurfural alcohol
1.04987	203	Ethyl malonate
1.050	105	1,3-Propanediol
1.0511 (20°)	157	Acrylic acid
1.0515	183	Benzyl acetate
1.0564	351	Ethyl cyanoacetate
1.05884	317	Thiophene
1.0637	205	Ethyl maleate
1.06611 (30°)	130	Phenyl ether

DENSITY INDEX AT 25° (Continued)

d	Compound Number	Compound
1.0685 (20°)	102	trans–2–Butene–1,4–diol
1.0724	348	Methyl acetoacetate
1.07267	202	Ethyl oxalate
1.07273	314	Benzenethiol
1.0740 (20°)	101	cis–2–Butene–1,4–diol
1.07511	161	Acetic anhydride
1.0819	131	Veratrole
1.08376	194	Methyl benzoate
1.08976	300	Quinoline
1.0899	335	Diethanolamine
1.0958	319	Dimethyl sulfoxide
1.0987	171	Ethylene glycol diacetate
1.10090	228	Chlorobenzene
1.107	309	2–Pyrrolidinone
1.1100	103	1,2–Ethanediol
1.1121	197	Benzyl benzoate
1.1126	326	Diethylene glycol
1.1196	327	Triethylene glycol
1.1196	336	Triethanolamine
1.1225	350	Methyl cyanoacetate
1.1254	199	γ–Butyrolactone
1.1261	324	Furfuryl alcohol
1.12918	301	Formamide
1.13128	260	Nitromethane
1.132 (st)	93	Phenol
1.135 *	94	o–Cresol
1.1462	204	Methyl maleate
1.1525	330	Salicylaldehyde
1.154 (st)	96	p–Cresol
1.1545	338	2–Furaldehyde
1.1680	233	1,1–Dichloroethane
1.17455	340	Epichlorohydrin
1.1747	242	1,1–Dichloroethylene
1.1782	344	Methyl salicylate
1.1793	337	2,2'–Thiodiethanol
1.1902	229	1–Chloronaphthalene
1.19654	332	2–Chloroethanol
1.19835	264	Nitrobenzene
1.20775	353	o–Chloroaniline
1.2130	339	Bis(2–chloroethyl) ether
1.21405	148	Formic acid

* Supercooled liquid

DENSITY INDEX AT 25° (Continued)

d	Compound Number	Compound
1.2408	341	o–Nitroanisole
1.2458	234	1,2–Dichloroethane
1.2463	244	trans–1,2–Dichloroethylene
1.24750 (55°)	240	p–Dichlorobenzene
1.25546	312	Carbon disulfide
1.25823	107	Glycerol
1.2614 (30°)	320	Sulfolane
1.2757	243	cis–1,2–Dichloroethylene
1.28280	239	m–Dichlorobenzene
1.30033	238	o–Dichlorobenzene
1.3060	249	2–Bromopropane
1.31678	230	Dichloromethane
1.3208	200	Ethylene carbonate
1.32924	235	1,1,1–Trichloroethane
1.3452	248	1–Bromopropane
1.4505	247	Bromoethane
1.4597	245	Trichloroethylene
1.4785	251	1–Bromonaphthalene
1.4785	352	Trifluoroacetic acid
1.47988	231	Chloroform
1.48820	250	Bromobenzene
1.56354	217	1,1,2–Trichlorotrifluoroethane
1.58439	232	Carbon tetrachloride
1.58655	236	1,1,2,2–Tetrachloroethane
1.60682	216	Hexafluorobenzene
1.61463	246	Tetrachloroethylene
1.64470	218	1,1,2,2–Tetrachlorodifluoroethane
1.67291	237	Pentachloroethane
1.6946	258	2–Iodopropane
1.7394	257	1–Iodopropane
1.9244	256	Iodoethane
2.163	219	1,2–Dibromotetrafluoroethane
2.1687	253	1,2–Dibromoethane
2.2649	255	Iodomethane
2.8758	252	Bromoform
2.9529	254	1,1,2,2–Tetrabromoethane
3.3078	259	Diiodomethane

EBULLIOSCOPIC CONSTANT INDEX

K_B	Compound Number	Compound
0.515°	0	Water
0.785	58	Methanol
1.160	59	Ethanol
1.30	265	Acetonitrile
1.59	60	1-Propanol
1.649	164	Methyl formate
1.71	141	Acetone
1.745	65	2-Methyl-2-propanol
1.824	109	Ethyl ether
1.85	315	Methyl sulfide
1.86	260	Nitromethane
1.87	266	Propionitrile
1.95	220	Chloroethane
2.061	169	Methyl acetate
2.080	165	Ethyl formate
2.125	132	Dimethoxymethane
2.166	64	2-Methyl-1-propanol
2.255	71	2-Methyl-2-butanol
2.28	142	2-Butanone
2.35	312	Carbon disulfide
2.4	148	Formic acid
2.53	30	Benzene
2.530	149	Acetic acid
2.53	247	Bromoethane
2.583	172	Ethyl acetate
2.65	70	3-Methyl-1-butanol
2.60	230	Dichloromethane
2.60	261	Nitroethane
2.710	299	Pyridine
2.75	6	Cyclohexane
2.84	297	Piperidine
2.86	101	cis-2-Butene-1,4-diol
3.13	233	1,1-Dichloroethane
3.22	285	Aniline
3.22	306	N,N-Dimethylacetamide
3.23	316	Ethyl sulfide
3.270	124	p-Dioxane
3.29	31	Toluene

EBULLIOSCOPIC CONSTANT INDEX (Continued)

K_B	Compound Number	Compound
3.43°	13	Heptane
3.44	234	1,2-Dichloroethane
3.45	298	Triethylamine
3.51	150	Propionic acid
3.53	161	Acetic anhydride
3.60	93	Phenol
3.62	231	Chloroform
3.87	275	Benzonitrile
3.94	151	Butyric acid
4.15	228	Chlorobenzene
4.19	255	Iodomethane
4.43	245	Trichloroethylene
4.48	232	Carbon tetrachloride
4.502	127	Anisole
4.83	180	Pentyl acetate
4.83	181	Isopentyl acetate
4.96	330	Salicylaldehyde
5.0	128	Phenetole
5.04	264	Nitrobenzene
5.16	256	Iodoethane
5.50	246	Tetrachloroethylene
5.52	43	p-Cymene
5.582	39	1,2,3,4-Tetrahydronaphthalene
5.611	146	Camphor
5.65	147	Acetophenone
5.75	217	1,1,2-Trichlorotrifluoroethane
5.762	24	Decahydronaphthalene (mixed isomers)
5.80	38	Naphthalene
5.84	300	Quinoline
6.26	250	Bromobenzene
6.52	107	Glycerol
6.608	253	1,2-Dibromoethane
7.01	200	Ethylene carbonate

CRYOSCOPIC CONSTANT INDEX

K_f eq. 67 Ch. II

K_f	Compound Number	Compound
1.79°	109	Ethyl ether
1.853	0	Water
2.40	141	Acetone
2.77	148	Formic acid
3.4	193	Methyl oleate
3.48	107	Glycerol
3.85	301	Formamide
3.90	149	Acetic acid
4.04	304	Acetamide
4.07	319	Dimethyl sulfoxide
4.3	34	p-Xylene
4.46	306	N,N-Dimethylacetamide
4.63	124	p-Dioxane
4.75	299	Pyridine
4.90	231	Chloroform
5.12	30	Benzene
5.34	275	Benzonitrile
5.60	94	o-Cresol
5.87	285	Aniline
6.27	129	Benzyl ether
6.38	131	Veratrole
6.65	305	N-Methylacetamide
6.7	121	Cineole
6.852	264	Nitrobenzene
6.94	38	Naphthalene
6.96	96	p-Cresol
7.15	128	Phenetole
7.30	311	ε-Caprolactam
7.40	93	Phenol
7.88	130	Phenyl ether
8.37	65	2-Methyl-2-propanol
10.4	71	2-Methyl-2-butanol
11.0	73	2,2-Dimethyl-1-propanol
12.5	253	1,2-Dibromoethane
13.7 (ms)	259	Diiodomethane
14.13	12	Methylcyclohexane
14.4	252	Bromoform
14.4 (st)	259	Diiodomethane
14.52	28	Bicyclohexyl

CRYOSCOPIC CONSTANT INDEX (Continued)

K_f	Compound Number	Compound
18.26	267	Succinonitrile
16.50	56	2–Pinene
19.47	25	cis–Decahydronaphthalene
20.81	26	trans–Decahydronaphthalene
21.7	254	1,1,2,2–Tetrabromoethane
29.8	232	Carbon tetrachloride
37.7	146	Camphor
39.3	74	Cyclohexanol
64.1	320	Sulfolane

CRYOSCOPIC CONSTANT INDEX, A
Eq. 72, Ch. II

A	Compound Number	Compound
0.00228	1	Cyclopentane
0.0024	59	Ethanol
0.004116	6	Cyclohexane
0.00460	11	2,3–Dimethylbutane
0.00575	4	2,2–Dimethylpropane
0.0077	71	2–Methyl–2–butanol
0.0090	65	2–Methyl–2–propanol
0.01097	96	p–Cresol
0.0110	28	Bicyclohexyl
0.01109	317	Thiophene
0.0125	58	Methanol
0.0130	122	Furan
0.01340	95	m–Cresol
0.01376	54	Cyclohexene
0.01404	93	Phenol
0.01525	296	Pyrrole
0.01528	30	Benzene
0.01655	311	ε–Caprolactam
0.01684	94	o–Cresol
0.01722	194	Methyl benzoate
0.01785	353	o–Chloroaniline
0.01827	38	Naphthalene
0.01858	299	Pyridine
0.0190	61	2–Propanol
0.01940	300	Quinoline
0.02039	250	Bromobenzene
0.02064	314	Benzenethiol
0.02154	25	cis–Decahydronaphthalene
0.02177	130	Phenyl ether
0.02178	42	tert–Butylbenzene
0.02193	37	Mesitylene
0.02195	350	Methyl cyanoacetate
0.02215	238	o–Dichlorobenzene
0.02319	10	2,2–Dimethylbutane
0.0295	44	Cyclohexylbenzene
0.02497	153	Valeric acid
0.02509	34	p–Xylene
0.02515	31	Toluene
0.02549	212	Fluorobenzene

CRYOSCOPIC CONSTANT INDEX, A (Continued)

A	Compound Number	Compound
0.02657	39	1,2,3,4-Tetrahydronaphthalene
0.0266	23	2,2,5-Trimethylhexane
0.02660	32	o-Xylene
0.02742	33	m-Xylene
0.02759	43	p-Cymene
0.028	318	Tetrahydrothiophene
0.0289	60	1-Propanol
0.02941	26	trans-Decahydronaphthalene
0.02985	36	Isopropylbenzene
0.0303	41	sec-Butylbenzene
0.0311	66	1-Pentanol
0.03140	315	Methyl sulfide
0.0331	62	1-Butanol
0.03476	17	2,4-Dimethylpentane
0.03479	35	Ethylbenzene
0.03569	109	Ethyl ether
0.0361	75	1-Hexanol
0.03661	222	2-Chloropropane
0.0377	111	Isopropyl ether
0.03780	12	Methylcyclohexane
0.03828	18	Ethylcyclohexane
0.03851	22	Nonane
0.03882	53	1-Decene
0.03931 (st)	40	Butylbenzene
0.03951 (ms)	40	Butylbenzene
0.0401	20	2,2,3-Trimethylpentane
0.04026	21	2,2,4-Trimethylpentane
0.04605	14	2-Methylhexane
0.04830	3	2-Methylbutane
0.04877	5	Methylcyclopentane
0.04907	2	Pentane
0.04976	7	Hexane
0.05001	316	Ethyl sulfide
0.05067	13	Heptane
0.05073	313	1-Butanethiol
0.05287	8	2-Methylpentane
0.05329	19	Octane
0.05687	48	trans-2-Pentene

CRYOSCOPIC CONSTANT INDEX, A (Continued)

A	Compound Number	Compound
0.05770	47	cis–2–Pentene
0.05826	27	Decane
0.0588	52	1–Nonene
0.05996	45	1–Pentene
0.06266	50	1–Heptene
0.06268	51	1–Octene
0.06324	49	1–Hexene
0.06378	29	Dodecane

DIELECTRIC CONSTANT INDEX

Dielectric Constant	Temp	Compound Number	Compound
1.8	21°	202	Ethyl oxalate
1.801	20	4	2,2-Dimethylpropane
1.843	20	3	2-Methylbutane
1.844	20	2	Pentane
1.873		10	2,2-Dimethylbutane
1.8799	25	7	Hexane
1.881		8	2-Methylpentane
1.890		11	2,3-Dimethylbutane
1.895		9	3-Methylpentane
1.914	20	17	2,4-Dimethylpentane
1.919	20	14	2-Methylhexane
1.924	20	13	Heptane
1.927	20	15	3-Methylhexane
1.939	20	16	2,3-Dimethylpentane
1.940	20	21	2,2,4-Trimethylpentane
1.948	20	19	Octane
1.962	20	20	2,2,3-Trimethylpentane
1.969	20	1	Cyclopentane
1.972	20	22	Nonane
1.991	20	27	Decane
2.002	30	29	Dodecane
2.017	20	45	1-Pentene
2.020	20	12	Methylcyclohexane
2.023	20	6	Cyclohexane
2.051	20	49	1-Hexene
2.054		18	Ethylcyclohexane
2.071	20	50	1-Heptene
2.084	20	51	1-Octene
2.14	25	244	trans-1,2-Dichloroethylene
2.1542	25	24	Decahydronaphthalene (mixed isomers)
2.172	20	26	trans-Decahydronaphthalene
2.197	20	25	cis-Decahydronaphthalene
2.209	25	124	p-Dioxane
2.220	20	54	Cyclohexene
2.238	20	232	Carbon tetrachloride
2.253	20	43	p-Cymene
2.2588	30	56	2-Pinene
2.2699	20	34	p-Xylene
2.275	25	30	Benzene
2.279	20	37	Mesitylene
2.30	25	246	Tetrachloroethylene

DIELECTRIC CONSTANT INDEX (Continued)

Dielectric Constant	Temp	Compound Number	Compound
2.359	20°	40	Butylbenzene
2.34	25	219	1,2–Dibromotetrafluoroethane
2.364	20	41	sec–Butylbenzene
2.366	20	42	tert–Butylbenzene
2.374	20	33	m–Xylene
2.379	25	31	Toluene
2.3833	20	36	Isopropylbenzene
2.4042	20	35	Ethylbenzene
2.41	50	240	p–Dichlorobenzene
2.41	25	217	1,1,2–Trichlorotrifluotoethane
2.42	25	298	Triethylamine
2.4257	20	55	Styrene
2.43	80	167	Butyl formate
2.45	20	156	Octanoic acid
2.46	20	160	Oleic acid
2.4970	25	57	2(10)–Pinene
2.52	25	218	1,1,2,2–Tetrachlorodifluoroethane
2.54	80	38	Naphthalene
2.568	20	32	o–Xylene
2.63	71	155	Hexanoic acid
2.64	20	154	Isovaleric acid
2.641	20	312	Carbon disulfide
2.645	20	132	Dimethoxymethane
2.66	20	153	Valeric acid
2.68 (s)	20	130	Phenyl ether
2.705	25	317	Thiophene
2.73	40	152	Isobutyric acid
2.77	25	115	Pentyl ether
2.773	20	39	1,2,3,4–Tetrahydronaphthalene
2.82	20	116	Isopentyl ether
2.820	20	201	Ethyl carbonate
2.9		192	Methyl methacrylate
2.942	25	122	Furan
2.97	20	151	Butyric acid
2.978	20	295	Dibutylamine
3.068	20	293	Dipropylamine
3.083	20	114	Butyl ether
3.111	30	189	Butyl sterate
3.211	20	193	Methyl oleate
3.39	26	110	Propyl ether
3.42	16	245	Trichloroethylene

DIELECTRIC CONSTANT INDEX (Continued)

Dielectric Constant	Temp.	Compound Number	Compound
3.435	40°	150	Propionic acid
3.58	21	292	Diethylamine
3.62	19	188	Isopentyl isovalerate
3.686 (l)	20	130	Phenyl ether
3.73	20	237	Pentachloroethane
3.80	25	133	Acetal
3.88	25	111	Isopropyl ether
3.9	20	126	Benzylethyl ether
4.09	25	131	Veratrole
4.22	30	213	o-Fluorotoluene
4.22	20	128	Phenetole
4.33	25	127	Anisole
4.335	20	109	Ethyl ether
4.382	25	314	Benzenethiol
4.39	20	252	Bromoform
4.41	90	91	2-Ethyl-1-hexanol
4.43	21	281	Isobutylamine
4.540	30	209	Butyl sebacate
4.57	23.5	121	Cineole
4.63	30	181	Isopentyl acetate
4.71	18	187	Ethyl isovalerate
4.73	20	284	Cyclohexylamine
4.75	20	180	Pentyl acetate
4.78	25	253	1,2-Dibromoethane
4.806	20	231	Chloroform
4.83	25	251	1-Bromonaphthalene
4.88	20	280	Butylamine
4.9	20	197	Benzyl benzoate
4.98	54	288	p-Toluidine
5.01	20	177	Butyl acetate
5.04	25	229	1-Chloronaphthalene
5.04	25	239	m-Dichlorobenzene
5.073	25	313	1-Butanethiol
5.1	21	183	Benzyl acetate
5.10	18	185	Ethyl butyrate
5.29	20	178	Isobutyl acetate
5.3	20	208	Bis(2-ethylhexyl) phthalate
5.31	20	278	Propylamine
5.316	25	259	Diiodomethane
5.40	25	250	Bromobenzene

DIELECTRIC CONSTANT INDEX (Continued)

Dielectric Constant	Temp	Compound Number	Compound
5.42	25°	212	Fluorobenzene
5.42	30	214	m–Fluorotoluene
5.45	20	279	Isopropylamine
5.61	25	125	Tetrahydropyran
5.621	25	228	Chlorobenzene
5.65	19	184	Ethyl propionate
5.72	25	316	Ethyl sulfide
5.8	20	297	Piperidine
5.82	25	71	2-Methyl–2–butanol
5.86	30	215	p–Fluorotoluene
5.95	18	287	m–Toluidine
6.002	20	175	Propyl acetate
6.02	25	172	Ethyl acetate
6.02	20	195	Ethyl benzoate
6.1	18	198	trans–Ethyl cinnamate
6.15	20	149	Acetic acid
6.19	90	78	2–Ethyl–1–butanol
6.2	20	315	Methyl sulfide
6.34	18	286	o–Toluidine
6.41	19	168	Isobutyl formate
6.436	30	207	Butyl phthalate
6.49	14	225	1–Chloro–2–methylpropane
6.59	20	194	Methyl benzoate
6.6	11	227	1–Chloropentane
6.68	25	169	Methyl acetate
6.89	20	285	Aniline
7.0	22	254	1,1,2,2–Tetrabromoethane
7.00		255	Iodomethane
7.00	20	257	1–Iodopropane
7.090	30	224	2–Chlorobutane
7.16	25	165	Ethyl formate
7.20	25	117	1,2–Dimethoxyethane
7.39	20	223	1–Chlorobutane
7.42	25	342	Morpholine
7.53	20	235	1,1,1–Trichloroethane
7.567	30	346	2–Ethoxyethyl acetate
7.58	25	123	Tetrahydrofuran
7.7	20	221	1–Chloropropane
7.72	19	166	Propyl formate
7.82	20	256	Iodoethane
7.87	25	203	Ethyl malonate

DIELECTRIC CONSTANT INDEX (Continued)

Dielectric Constant	Temp	Compound Number	Compound
7.959	30°	211	Butyl phosphate
8.05	20	85	trans-3-Methylcyclohexanol
8.09	25	248	1-Bromopropane
8.13	25	296	Pyrrole
8.19	20	258	2-Iodopropane
8.2	20	241	3-Chloropropene
8.20	20	236	1,1,2,2-Tetrachloroethane
8.25	20	345	2-Methoxyethyl acetate
8.35	60	73	2,3-Dimethyl-1-propanol
8.5	20	164	Methyl formate
8.55	20	352	Trifluoroacetic acid
8.58	23	205	Ethyl maleate
8.93	25	230	Dichloromethane
9.00	25	300	Quinoline
9.20	25	243	cis-1,2-Dichloroethylene
9.21	22	89	2-Heptanol
9.30	25	323	2-Butoxyethanol
9.39		247	Bromoethane
9.41	30	344	Methyl salicylate
9.45	20	220	Chloroethane
9.46	25	249	2-Bromopropane
9.78	60	93	Phenol
9.82	20	222	2-Chloropropane
9.91	58	96	p-Cresol
9.93	25	238	o-Dichlorobenzene
9.961	20	226	2-Chloro-2-methylpropane
ca 10.	-54	171	Ethylene glycol diacetate
10.0	18	233	1,1-Dichloroethane
10.34	20	90	1-Octanol
10.36	25	234	1,2-Dichloroethane
11.35	20	146	Camphor
11.5	25	94	o-Cresol
11.8	25	95	m-Cresol
12.3	20	83	3-Methylcyclohexanol, (mixed isomers)
12.4	21	299	Pyridine
12.47	25	65	2-Methyl-2-propanol
12.9	20	163	Butyric anhydride
12.9	25	290	Ethylenediamine
13.02	22	68	3-Pentanol
13.1	20	92	Benzyl alcohol

DIELECTRIC CONSTANT INDEX (Continued)

Dielectric Constant	Temp	Compound Number	Compound
13.1	20°	343	Ethyl lactate
13.11	20	145	4-Methyl-2-pentanone
13.3	25	75	1-Hexanol
13.3	25	80	2-Methylcyclohexanol (mixed Isomers)
13.3	20	86	4-Methylcyclohexanol (mixed isomers)
13.4	26	136	Butyraldehyde
13.4		353	o-Chloroaniline
13.6	23	325	Tetrahydrofurfural alcohol
13.82	22	67	2-Pentanol
13.897	25	273	Octanenitrile
13.9	25	66	1-Pentanol
13.9	20	330	Salicylaldehyde
14.7	25	69	2-Methyl-1-butanol
14.7	25	70	3-Methyl-1-butanol
15.0	25	74	Cyclohexanol
15.5	22	272	4-Methylvalcronitrile
15.7	22	349	Ethyl acetoacetate
16.47	20	84	cis-3-Methylcyclohexanol
16.56	25	63	2-Butanol
16.93	25	321	2-Methoxyethanol
17.00	20	143	3-Pentanone
17.263	25	271	Hexanenitrile
17.39	25	147	Acetophenone
17.51	25	62	1-Butanol
17.8	20	138	Benzaldehyde
17.93	25	64	2-Methyl-1-propanol
18.2	25	331	4-Hydroxy-4-methyl-2-pentanone
18.3	20	144	Cyclohexanone
18.3	16	162	Propionic anhydride
18.3	25	291	Ethylenimine
18.5	17	135	Propionaldehyde
18.51	20	142	2-Butanone
18.7	27	274	α-Tolunitrile
19.709	25	270	Valeronitrile
19.92	25	61	2-Propanol
20.3	21	268	Butyronitrile
20.33	25	60	1-Propanol
20.4	24	269	Isobutyronitrile
20.7	19	161	Acetic anhydride
20.70	25	141	Acetone
21.1	21	134	Acetaldehyde

DIELECTRIC CONSTANT INDEX (Continued)

Dielectric Constant	Temp	Compound Number	Compound
21.2	20°	339	Bis(2–chloroethyl) ether
21.6	15	97	2–Propene-1- ol
22.6	22	340	Epichlorohydrin
23.06		308	1,1, 3,3–Tetramethylurea
23.24	30	262	1–Nitropropane
23.69	20	327	Triethylene glycol
24.5	25	100	2–Propyn-1-ol
24.55	25	59	Ethanol
25.20	25	275	Benzonitrile
25.52	30	263	2–Nitropropane
25.8	25	332	2–Chloroethanol
26.7	18	351	Ethyl cyanoacetate
27.2	20	266	Propionitrile
28.06	30	261	Nitroethane
29.30	20	350	Methyl cyanoacetate
29.36	25	336	Triethanolamine
29.6	24	322	2–Ethoxyethanol
30.	20	354	Hexamethylp hosphoric triamide
31.69	20	326	Diethylene glycol
32.0	20	104	1,2–Propanediol
32.0	25	310	1–Methyl-2-pyrrolidinone
32.70	25	58	Methanol
33.0	20	276	Acrylonitrile
34.82	25	264	Nitrobenzene
35.0	20	105	1,3–Propanediol
35.87	30	260	Nitromethane
36.71	25	303	N,N–Dimethylformamide
37.5	20	265	Acetonitrile
37.7	25	103	1,2–Ethanediol
37.72	25	334	2–Aminoethanol
37.78	25	306	N,N–Dimethylacetamide
38	25	338	2–Furaldehyde
39	20	199	γ–Butyrolactone
42.5	25	107	Glycerol
43.3	30	320	Sulfolane
46.68	25	319	Dimethyl sulfoxide
52.1	– 40	303	N,N–Dimethylformamide
56.5	57.4	267	Succinonitrile

DIELECTRIC CONSTANT INDEX (Continued)

Dielectric Constant	Temp	Compound Number	Compound
58.5	16°	148	Formic acid
59	83	304	Acetamide
78.39	25	0	Water
89.6	40	200	Ethylene carbonate
111.0	20	301	Formamide
172.2	25	307	N–Methylpropionamide
182.4	25	302	N–Methylformamide
191.3	32	305	N–Methylacetamide
308	– 40	302	N–Methylformamide
384	– 40	307	N–Methylpropionamide

DIPOLE MOMENT INDEX

μ	Temp	Condition	Compound Number	Compound

"Condition" refers to the state of the compound for which the dipole moment was determined; e.g., "(g) or (l)" – purified compound in gaseous or liquid state respectively; "in 30" signifies a solution in solvent corresponding to the number given, in the case cited the solvent is benzene.

μ	Temp	Condition	Compound Number	Compound
0.00	25°		1	Cyclopentane
0.00	25		2	Pentane
0.0	25		13	Heptane
0.0	25	in 6	216	Hexafluorobenzene
0.	25		12	Methylcyclohexane
0.	25	(l)	14	2-Methylhexane
0.	25	(l)	15	3-Methylhexane
0.	25	(l)	16	2,3-Dimethylpentane
0.	25	(l)	17	2,4-Dimethylpentane
0.	100	(g)	18	Ethylcyclohexane
0.	25	(l)	19	Octane
0.	25	(l)	21	2,2,4-Trimethylpentane
0.	25	(l)	24	Decahydronaphthalene (mixed isomers)
0.	20 – 60	(l)	30	Benzene
0.	20 – 30	(l)	37	Mesitylene
0.	25	in 30	28	Naphthalene
0.	20 – 30	(l)	43	p-Cymene
0.	25	(l)	232	Carbon tetrachloride
0.	24	in 30	240	p-Dichlorobenzene
0.	25	in 232	246	Tetrachloroethylene
ca 0.	20 – 100	in 30	25	cis-Decahydronaphthalene
ca 0.	20 – 100	in 30	26	trans-Decahydronaphthalene
ca 0.	– 30 – 170	(l)	27	Decane
ca 0.	– 10 – 210	(l)	29	Dodecane
0.02	20 – 60	(l)	34	p-Xylene
0.06	25	in 30	312	Carbon disulfide
0.085	25	(l)	7	Hexane
0.13	20 – 60	(l)	55	Styrene
0.28	20	(l)	54	Cyclohexene
0.30	20 – 60	(l)	33	m-Xylene
0.31	20 – 60	(l)	31	Toluene
0.33	25	in 30	216	Hexafluorobenzene
0.34	20	(l)	45	1-Pentene
0.34	20	(l)	49	1-Hexene
0.34	20	(l)	50	1-Heptene
0.34	20	(l)	51	1-Octene

DIPOLE MOMENT INDEX (Continued)

μ	Temp	Condi-tion	Compound Number	Compound
0.36	20 – 30°	(l)	40	Butylbenzene
0.36	20 – 30	(l)	42	tert–Butylbenzene
0.36	30	in 30	56	2-Pinene
0.37	20 – 60	(l)	35	Ethylbenzene
0.37	20 – 30	(l)	41	sec–Butylbenzene
0.39	20 – 60	(l)	36	Isopropylbenzene
⟨ 0.4	25		28	Bicyclohexyl
0.42	20	in 30	53	1-Decene
0.45	20 – 60	(l)	32	o-Xylene
0.45			124	p-Dioxane
0.52	25	in 30	317	Thiophene
0.59	20	in 30	52	1-Nonene
0.60	20 – 30	(l)	39	1,2,3,4-Tetrahydronaphthalene
0.63	25	in 30	154	Isovaleric acid
0.70	25	in 30	244	trans—1,2-Dichloroethylene
0.71	25	in 30	122	Furan
0.74	34	(g)	132	Dimethoxymethane
0.77	25	in 232	210	Butyl borate
0.8		in 232	245	Trichloroethylene
0.87	25	in 30	298	Triethylamine
0.90	25	in 30	201	Ethyl carbonate
0.94	25 – 85	(l)	237	Pentachloroethane
0.99	10 – 70	in 30	252	Bromoform
1.03	20	in 30	293	Dipropylamine
1.04	20	in 30	295	Dibutylamine
1.08	25	(l)	152	Isobutyric acid
1.08	25 – 50	in 30	259	Diiodomethane
1.11	25	(l)	292	Diethylamine
1.13	25	(l)	155	Hexanoic acid
1.14		(l)	230	Dichloromethane
1.15	20		109	Ethyl ether
1.15	25	(l)	156	Octanoic acid
1.15	25	in 30	231	Chloroform
1.16	25	in 30	130	Phenyl ether
1.18	20		114	Butyl ether
1.18	10	in 232	160	Oleic acid
1.19	25	in 30	297	Piperidine
1.20	25	in 232	115	Pentyl ether
1.22	20		111	Isopropyl ether
1.23	25	in 232	110	Propyl ether

DIPOLE MOMENT INDEX (Continued)

μ	Temp	Condi-tion	Compound Number	Compound
1.23			116	Isopentyl ether
1.23	20°	in 30	253	1,2-Dibromoethane
1.23	25	in 30	314	Benzenethiol
1.24			113	Butylethyl ether
1.245	25	in 232	127	Anisole
1.25	25	in 30	112	Butylvinyl ether
1.26	20	in 30	108	Ethylvinyl ether
1.26	30	(l)	213	o-Fluorotoluene
1.26	20	in 30	284	Cyclohexylamine
1.27	25	in 30	281	Isobutylamine
1.28	25	in 30	282	sec-Butylamine
1.29	25	(l)	251	1-Bromonaphthalene
1.29	20	in 7	254	1,1,2,2-Tetrabromoethane
1.29	25	in 30	283	tert-Butylamine
1.30	25	in 30	242	1,1-Dichloroethylene
1.31	25	in 30	289	Allylamine
1.32	25	in 30	131	Veratrole
1.33	25	(l)	229	1-Chloronaphthalene
1.33	20	in 30	278	Propylamine
1.36	25	in 232	128	Phenetole
1.37	20	in 30	280	Butylamine
1.38		in 30	133	Acetal
1.38	24	in 30	239	m-Dichlorobenzene
1.45	30	in 30	93	Phenol
1.45	25	in 30	94	o-Cresol
1.45	25	in 30	287	m-Toluidine
1.45	25	in 30	315	Methyl sulfide
1.47	25	in 30	212	Fluorobenzene
1.48	25	in 232	255	Iodomethane
1.50	25		342	Morpholine
1.51	25	in 30	285	Aniline
1.52	25	in 30	288	p-Toluidine
1.53	25	in 30	313	1-Butanethiol
1.54	25	in 30	228	Chlorobenzene
1.543			95	m-Cresol
1.543			96	p-Cresol
1.55	20	in 30	75	1-Hexanol
1.55	25	in 30	125	Tetrahydropyran
1.55	25	in 30	250	Bromobenzene

DIPOLE MOMENT INDEX (Continued)

μ	Temp	Condition	Compound Number	Compound
1.57		in 30	235	1,1,1–Trichloroethane
1.60	25°	in 30	286	o–Toluidine
1.61	28	in 30	169	Methyl acetate
1.61	25	in 30	316	Ethyl sulfide
1.63			97	2–Propene–1–ol
1.64	22	in 30	68	3–Pentanol
1.65	30	in 30	151	Butyric acid
1.65	0	in 30	159	Methacrylic acid
1.66	20	(l)	59	Ethanol
1.66	30	in 30	61	2–Propanol
1.66	30	in 30	65	2–Methyl–2–propanol
1.66	22	in 30	67	2–Pentanol
1.66			92	Benzyl alcohol
1.66	30	(l)	214	m–Fluorotoluene
1.675	25	in 30	192	Methyl methacrylate
1.68	30	in 30	149	Acetic acid
1.68	30	in 30	150	Propionic acid
1.7	25	in 232	66	1–Pentanol
1.7			71	2–Methyl–2–butanol
1.71	22	in 30	89	2–Heptanol
1.71	25	in 30	117	1,2–Dimethoxymethane
1.71	25	in 30	236	1,1,2,2–Tetrachloroethane
1.74	25	in 30	91	2–Ethyl–1–hexanol
1.74	22	in 30	184	Ethyl propionate
1.74	22	in 30	185	Ethyl butyrate
ca 1.75			85	trans–3–Methylcyclohexanol
1.75	25	in 232	62	1–Butanol
1.75	25	in 30	123	Tetrahydrofuran
1.76	25	in 232	90	1–Octanol
1.76	30	(l)	215	p–Fluorotoluene
1.76	25	in 30	243	cis–1,2–Dichloroethylene
1.77		(g)	164	Methyl formate
1.77	15.4	in 30	198	cis–Ethyl cinnamate
1.77	25	in 30	291	Ethyleneimine
1.77	25	in 30	353	o–Chloroaniline
1.78	25	in 30	100	2–Propyn–1–ol

DIPOLE MOMENT INDEX (Continued)

μ	Temp	Condition	Compound Number	Compound
1.78	22°	in 30	175	Propyl acetate
1.78	20	in 30	256	Iodoethane
1.79			64	2-Methyl-1-propanol
1.79	25	in 30	170	Vinyl acetate
1.8		in 232	340	Epichlorohydrin
1.80	25	in 30	183	Benzyl acetate
1.80	25	in 30	296	Pyrrole
1.82	22	in 30	181	Isopentyl acetate
1.82			70	3-Methyl-1-butanol
1.82	30	in 30	148	Formic acid
1.84	22	in 30	177	Butyl acetate
1.84	20.2	in 30	198	trans-Ethyl cinnamate
1.84	20	in 30	257	1-Iodopropane
1.86	25	in 232	74	Cyclohexanol
1.86	25	in 30	194	Methyl benzoate
1.86	25	in 30	234	Dichloroethane
1.87	22	in 30	178	Isobutyl acetate
1.88	25	in 30	172	Ethyl acetate
1.88	22	in 30	168	Isobutyl formate
1.88	24	in 30	189	Butyl stearate
1.88	25		332	2-Chloroethanol
1.89	22	in 30	166	Propyl formate
1.89 – 1.90			197	Benzyl benzoate
1.9			83	3-Methylcyclohexanol (mixed isomers)
1.9			86	4-Methylcyclohexanol (mixed isomers)
1.90	– 90 – 70	in 13	223	1-Chlorobutane
1.90	25	(l)	247	Bromoethane
1.90	25	in 30	290	Ethylenediamine
1.90	25	in 30	318	Tetrahydrothiophene
1.91			84	cis-3-Methylcyclohexanol
1.91			180	Phenyl acetate
1.92	25	in 30	324	Furfuryl alcohol
1.93	20	in 30	248	1-Bromopropane
1.94	25	in 30	165	Ethyl formate
1.94	20	in 30	227	1-Chloropentane

DIPOLE MOMENT INDEX (Continued)

μ	Temp	Condition	Compound Number	Compound
1.95			80	2–Methylcyclohexanol (mixed isomers)
1.95	20°	in 30	258	2–Iodopropane
1.96	20	(l)	220	Chloroethane
1.97	25	in 30	118	Bis(2–methoxyethyl) ether
1.97		(l)	221	1–Chloropropane
1.98	25	in 30	233	1,1–Dichloroethane
1.98	104	(g)	241	3–Chloropropene
1.99		in 30	195	Ethyl benzoate
2.00	25	in 30	119	Propylene oxide
2.01	20	in 30	120	1,2–Epoxybutane
2.02		(l)	222	2–Chloropropane
2.04	20	in 30	249	2–Bromopropane
2.04	25	in 30	321	2–Methoxyethanol
2.04	– 80	in 312	349	Ethyl acetoacetate, enol
2.07	25	in 30	224	2–Chlorobutane
2.08	25	in 30	322	2–Ethoxyethanol
2.08	25	in 30	323	2–Butoxyethanol
2.09	10 – 50	in 30	225	1–Chloro–2–methylpropane
2.12	35	(l)	325	Tetrahydrofurfuryl alcohol
2.13	30	in 30	158	Crotonic acid
2.13	30	in 30	345	2–Methoxyethyl acetate
2.15	20	in 30	226	2–Chloro–2–methylpropane
2.17			331	Ethyl cyanoacetate
2.184	25	in 30	300	Quinoline
2.25			104	1,2–Propanediol
2.25	30	in 30	346	2–Ethoxyethyl acetate
2.27	24	in 30	238	o–Dichlorobenzene
2.27	25	in 124	334	2–Aminoethanol
2.28	20		103	1,2–Ethanediol
2.28	100	(g)	352	Trifluoroacetic acid
2.31		in 30	326	Diethylene glycol
2.34	30	in 30	171	Ethylene glycol diacetate
2.37	25	in 232	299	Pyridine—
2.4	20	in 30	343	Ethyl lactate
2.45			102	trans–2–Butene–1,4–diol
2.450	40	(l)	136	Butyraldehyde
2.47	25	in 30	344	Methyl salicylate
2.48			101	cis–2–Butene–1,4–diol

DIPOLE MOMENT INDEX (Continued)

μ	Temp	Condition	Compound Number	Compound
2.48	25°	in 232	204	Methyl maleate
2.48	20		209	Butyl sebacate
2.49	25	in 30	202	Ethyl oxalate
2.50	20		105	1,3-Propanediol
2.51	20	in 30	134	Acetaldehyde
2.54	20	in 30	135	Propionaldehyde
2.54	25	in 30	203	Ethyl malonate
2.54	25	in 30	205	Ethyl maleate
2.58	20	in 30	137	Isobutyraldehyde
2.58	25	in 30	339	Bis(2-chloroethyl) ether
2.69	20	(l)	141	Acetone
2.76	25	in 30	142	2-Butanone
2.77	20	(l)	138	Benzaldehyde
2.81	25	in 124	335	Diethanolamine
2.82	25	in 232	143	3-Pentanone
2.82	25	in 312	161	Acetic anhydride
2.84	20		208	Bis(2-ethylhexyl) phthalate
2.86	20	in 30	330	Salicylaldehyde
2.87	20	(l)	58	Methanol
2.90	25	in 30	139	Acrolein
2.96	25	in 30	147	Acetophenone
3.01	25	in 30	144	Cyclohexanone
3.07	25	in 30	211	Butyl phosphate
3.09	20	(l)	60	1-Propanol
3.10	25	in 30	146	Camphor
3.22	18.2	in 30	349	Ethyl acetoacetate, keto
3.24	20	in 30	331	4-Hydroxy-4-methyl-2-pentanone
3.37	30		301	Formamide
3.44	20	in 30	265	Acetonitrile
3.44	30	in 30	304	Acetamide
3.47	25	in 30	274	α-Tolunitrile
3.47	25	in 30	308	1,1,2,2-Tetramethylurea
3.50	25	in 30	140	Crotonaldehyde
3.51	25	in 30	276	Acrylonitrile
3.53	25	in 30	272	4-Methylvaleronitrile
3.55	25	in 30	309	2-Pyrrolidinone
3.56	20	in 7	260	Nitromethane
3.57	20	in 30	266	Propionitrile
3.57	20	in 30	268	Butyronitrile
3.57	25	in 30	270	Valeronitrile

DIPOLE MOMENT INDEX (Continued)

μ	Temp	Condi-tion	Compound Number	Compound
3.57	25 °	in 124	336	Triethanolamine
3.59	20	in 7	262	1–Nitropropane
3.59	110	(g)	307	N–Methylpropionamide
3.60	20	in 7	261	Nitroethane
3.61	25	in 30	269	Isobutyronitrile
3.63	25	in 30	338	2–Furaldehyde
3.68	30	in 31	267	Succinonitrile
3.69		(g)	277	Methacrylonitrile
3.72		in 30	306	N,N–Dimethylacetamide
3.73	120 – 180		263	2–Nitropropane
3.86	25	in 30	302	N–Methylformamide
3.86	25	in 30	303	N,N–Dimethylformamide
3.88	25	in 30	311	ϵ–Caprolactam
3.9	25	in 30	319	Dimethyl sulfoxide
4.03	25	in 30	264	Nitrobenzene
4.05	25	in 30	275	Benzonitrile
4.09	30	in 30	310	1–Methyl–2–pyrrolidinone
4.12	25	in 30	199	γ–Butyrolacetone
4.31		(l)	354	Hexamethylphosphoric triamide
4.39	20.1	in 124	305	N–Methylacetamide
4.81	25	in 30	320	Sulfolane
4.83	20	in30	341	o–Nitroanisole
4.87	25	in 30	200	Ethylene carbonate
5.58	20	(l)	327	Triethylene glycol

Chapter IV

CRITERIA OF PURITY.
DRYING AND DETERMINATION OF WATER

The purity of a substance can be defined either thermodynamically or kinetically. Purity in a thermodynamic sense implies that the substance behaves in a system of more than one phase as one and only one independent component, so that it possesses a sharply defined chemical potential which is the same in all operations. According to the kinetic definition, the substance is pure when it consists of molecules of one type only. These definitions are not rigorous and must be modified in certain cases, as in the thermodynamic definition for racemates and the kinetic definition for isotopes. The concept of purity and the difficulties that arise in its definition are fully discussed by Timmermans [4785] and in the Symposium on Purity and Identity of Organic Compounds [405].

These definitions set an ideal that can only be experimentally approached more or less closely. There are, in fact, no criteria of ideal purity. Ostwald considered a pure substance to be the point of convergence of various methods of purification. In practice, the substance is obtained from as many different sources as possible, subjected to a variety of purification processes and the constancy of the physical properties determined. Timmermans [4785, p. 64] gives the following general rules: (1) The starting material should be the best specimen of a substance commercially available; for precision work this must always be subjected to some process of purification; (2) impurities may often be removed more simply by transforming them chemically than by separating them in the form in which they are initially present; and (3) if a product of a high degree of purity is to be obtained synthetically by several distinct operations, each intermediate product should be purified as far as possible.

The term *pure* has a relative connotation and, for practical use, must be defined in explicit terms. The explicit definition must be adequate to define the purity to the extent desired. This definition must be based upon the specific use for which the substance is intended. For instance, water for the calibration of thermometers at the steam point must boil at 100° at 760 torr. This water may contain impurities that make it entirely unsuited for conductivity measurements, and water completely satisfactory for conductivity measurements may contain so much particulate matter that it is not suitable as a solvent for light scattering photometry. The definition of purity for practical

552

purposes may be stated as: *A material is sufficiently pure if it does not contain impurities of such nature and in such quantity as to interfere with the use for which it is intended.*

CRITERIA OF PURITY

The criteria of purity are standards by which the purity of a substance may be judged. The standards are based on results of tests of known reliability. Some tests have proven thermodynamic and stoichiometric accuracy, and some are empirical. The kind and amount of information that may be deduced from chemical and physical tests vary. These tests must be sensitive enough to characterize the purity to the desired accuracy.

Recent advances in electronic instrumentation have resulted in a great improvement in the sensitivity of test methods and in the ability to separate impurities bearing close structural resemblance to the main compound. Gas chromatography is the greatest, and certainly the most spectacular test method to be developed for the characterization of organic solvents. It has been followed in importance by significant advances in mass spectrometry and in spectrophotometry.

The extension of the sensitivity of tests has made it possible to prepare purer substances; this has given rise to a new grade of purity called *ultrapure*. Although the degree of purity implied by the new term has been the subject of some confusion, Wilcox, et al. [5181], and Janke, et al. [4260] have presented lucid discussions that serve as excellent guides. The former group proposed 500 ppm for the upper limit for impurities in organic compounds classed as ultrapure.

It is sometimes desirable to express the composition as the percentage of the major component present instead of as concentration values of the impurities. Since the percentage figure for a substance containing, for example, 50 parts per billion, ppb, of impurities would be somewhat cumbersome, 99.99995%, the "nine" or "N" method of expression has been proposed and apparently has been well received. This method gives the number of successive nines in the percentage as an Arabic figure followed by an "N," which in turn is followed by the Arabic figure(s) following the nines. Thus, 50 ppb becomes 99.99995% or 6N5, 99.992% becomes 4N2 and 99.9% becomes 3N.

Caution should be exercised when composition is expressed on the basis of results from impurity tests. A purity of 99.992%, for example, implies that the purity of the principal component is 99.992% within the accuracy of the test method. This situation rarely exists when the figure is based on impurity analyses, particularly for organic solvents. The expression "effective purity" is compatible with the stated definition of purity. It implies that the objectionable impurities are present to no greater extent than 0.008%.

Chemical Tests

Chemical tests are usually based on the reaction of a functional group and have certain limitations as criteria of purity. Generally, it is not feasible to analyze a substance of high purity for the principal component. Functional group analysis usually has an accuracy of about $\pm 0.2\%$, and it is therefore desirable to determine at least one physical property, and preferably two, to use in conjunction with the chemical results.

Empirical composition obtained from elemental or functional group analyses is no guarantee of purity. For instance, isomers that may be present will give the same elemental, and possibly functional, analysis; in high molecular weight compounds, adjacent homologues will have to be present in fairly large amounts to be detected by the usual chemical methods.

The application of chemical methods of analysis in the testing of high purity organic compounds presupposes a knowledge of the types of impurities present. This is a broad field; specific information may be obtained from Rosin [4042], ACS *Reagent Chemicals* [76], and Chapter V.

Physical Tests

The physical tests may be used as a quantitative or qualitative indication of the purity, and most of them are not specific for the chemical nature of the impurity present.

Test methods furnish four types of standards for judging purity:

1. Total amount of impurity present, but no differentiation as to kind.
2. The amount of a general class of impurities present.
3. The amount of a specific impurity present.
4. Empirical tests based on some artificial standard.

The selection of the type of test to be used depends on the material being tested, the accuracy desired to define the required purity, knowledge of impurities present and available equipment. The more common standards and tests are discussed in the following pages.

Cryoscopy

Cryoscopy, according to the *Random House Dictionary of the English Language*, is "... the determination of the freezing points of liquids or solutions or the lowering of the freezing point by dissolved substances." Glasgow and Ross [2694, Chapter 88] "... have included melting points because the freezing and the melting point designate the same temperature according to thermodynamic definition." The cryoscopy concept of Glasgow and Ross is logical and brings together a number of methods derived from the same basic considerations. Cryoscopy is based on the assumption that the solid phase is in equilibrium with the liquid phase and that the temperature measured depends on the change in composition of the liquid phase.

The several cryoscopic procedures are alike in that the temperatures for solid-liquid equilibria are measured. They differ in the method used to determine the relative amounts of the liquid and solid phases present and in the precision obtained. The procedures now widely used are:

1. thermometric—relates temperature and time
2. calorimetric—relates temperature and heat input
3. dilatometric—relates temperature and volume change

Recently Ross and Frolen [4045] developed a method based on the determination of the dielectric constant during melting.

Glasgow and Ross [2694, Chap. 88], point out that "Notwithstanding the wide variation in apparatus and procedures for cryoscopic measurements as to cost, complexity, accuracy, sample size, and time of determination, each cryoscopic procedure finds important application as an analytical tool." The procedure and apparatus that is used must characterize the solvent to the desired precision.

METHODS BASED ON THE FREEZING POINT

The *thermometric method*, or time-temperature method, is used to designate the open method of White [5149] as improved and simplified by Rossini and coworkers [1795, 3117, 4620] and the variations that are now in common use. The less precise method in which a mercury or alcohol-filled thermometer is generally used will be referred to as the *freezing curve method*.

The thermometric method is one of the most effective and accurate means of evaluating the purity of materials, and particularly of high purity substances. It is a thermodynamic method based on the phase rule interpretation of Raoult's law. For an ideal or sufficiently dilute solution, it measures the equilibrium that exists between a single crystalline phase consisting of the major component and a liquid phase consisting of one or more components, one of which is the major component. It is applicable to all substances that form ideal solutions, or whose solutions are so dilute that the departure from the ideal state is not significant. It is concerned only with the amount of impurity and not the kind. Rossini [4050] gives a relationship for those cases in which a significant departure from the ideal state occurs. The method is applicable when the purity is above 95%m, and its reliability increases as the purity increases. It has an accuracy in some cases of 0.002%m; the sensitivity for some systems has been reported to be in the 0.0002–0.0005%m range.

Hüttner and Tammann [2336] in 1905 observed that it was possible, from the behavior during crystallization, to draw conclusions as to the amount of impurities. White [5149], fifteen years later, proposed a method for estimating the amount of impurity by means of the melting curve. Rossini and coworkers [3117] in 1941 developed the method to its present high sensitivity and also expanded it to enable the true freezing point to be calculated for the major component, even though it contained an appreciable amount of impurity.

When only information about the degree of purity is desired, irrespective of the nature of the contaminants, this method has the advantage that no other knowledge of the system is necessary except that the substance to be tested have a high purity. The temperature-indicating device need not be calibrated but must measure precisely throughout the temperature range. The values are uncertain if a solid solution is formed or if there is a eutectic point near 100%m. It also has the disadvantage that the equipment required for the most precise measurements is expensive. Where the high precision of resistance thermometry is not required, the apparatus of White [5149] should find wide use.

The *closed thermometric method* was developed somewhat later than the open method. Glasgow and Ross [2694, Chap. 88] give the following advantages of the closed cell: There is less chance of sample contamination during transfer and subsequent measurement. Noxious or highly flammable substances may be measured without discomfort or danger. The error due to dissolved gases (air or other blanketing gases) may be eliminated. The error due to dissolved gases becomes appreciable in the ultrapure range, particularly >99.99%m.

The cell described by Ross and Dixon [4044] permits cleaning the system of adsorbed gases and water, effective outgassing of the sample and transferring of the liquid from the sample vial to the freezing chamber. Substances with a purity range of 99.0 to 99.999%m were used for the study.

A *simplified freezing point method* was introduced recently by Ross and Glasgow [4046]. It utilizes a semi-closed system apparatus, which makes it possible to follow a purification or other operation depending on change of composition with a precision of about 0.01°. This precision in a typical purity analysis corresponds to an estimated standard deviation of 0.01%m. The apparatus is inexpensive, rugged, and easy to operate. It is easily cleaned and has a cell capacity of 8 ml.

The *freezing curve method* includes various procedures common in the laboratory for determining the freezing point with a precision of 0.05° or less, usually less. The sample is generally placed in a double-walled test tube. A thermometer may be used as a stirrer or held in place with a stopper that also serves as a guide for a stirrer. The bath may be any convenient medium that may be held 5–10° below the temperature of the sample. The temperature-time data are plotted and the section where Δ-temp/Δ-time $= 0$ is taken as the freezing point. This is a convenient, rapid, and inexpensive test that is used for characterizing many commercially pure substances. It is also used with specific tests for impurities to characterize a high functional purity.

The *freezing point* was early recognized as a specific property of a pure substance. There are many methods that have been reported, and many of these have been extensively used that have not been mentioned.

A differential thermometer may be used to follow the purification of a substance. If a reference sample is available, the purity may be calculated from the difference in freezing points. The use of the freezing point constant has been extensively studied; see references for K_f (Eq. 2.67) and particularly Brancker et al. [710], Meldrum et al. [3266], Kraus and Vingee [2745], Rall and Smith [3884], Menzies and Wright [3274], Adams [14], and Washburn and Read [5085].

METHODS BASED ON THE MELTING POINT

The *adiabatic calorimetric method* is the most accurate and the most sensitive of the cryoscopic methods. The basic technique was proposed by Giauque and Wiebe [1754] and was further improved by Aston and Messerly [310]. Aston and Fink [307] describe the apparatus and give typical data. The method is based on the measurement of the temperatures of the liquid-solid equilibrium at known liquid-to-solid ratios, precisely evaluated from the total heat input, heat of fusion, and heat capacity. The equipment is costly and the experiment is time-consuming, but there is no other method of characterization that gives as much information about the system being studied. The method has an accuracy of at least 0.001 %m and a sensitivity of at least 0.0001 %m.

The *simple melting curve method* has been developed by Gunn [1936] to a sensitivity of the order of 10^{-5} mole fraction impurity and an accuracy of 10 to 20 % using about 0.3 g sample. The apparatus is simple and a determination requires about 2 hr. The data may be used to calculate not only purity but also heat of fusion and heat capacity.

The *capillary tube* and *melting block techniques* are approximate methods for the determination of melting points, and are not suitable for characterizing a very pure substance. These techniques serve a purpose when their possibilities and limitations are considered. They are dependent on such factors as the state of transition taken as the melting point, the rate of heat transfer from the bath medium to the substance, the rate of response of the thermometer to change of temperature, and the rate of heating. When the freezing point of a solvent is high enough to permit the use of either of these techniques, they are useful devices to approximate the degree of purity.

There is a voluminous literature on capillary tube and melting block melting points. One of the more significant studies is that of Francis and Collins [1619]. They observed that a single observer can determine a capillary melting point to within 0.1–0.2°. Using the same apparatus and the same materials, different observers may differ by 0.2–0.3°. Freezing points can be determined by different observers to within 0.02°. The melting and freezing points of the same material should be the same, but, when determined by this technique, the former are invariably higher. These authors further observed that capillary melting points determined with different apparatus may differ by 2°.

Ebulliometry

Ebulliometry is defined by Leslie and Kuehner [2694, Chapter 89] as the science of measuring and using boiling points. They continue, "properly measured, the boiling point has definite thermodynamic significance and can be used to identify materials, determine or indicate their purity, determine molecular weights, study azeotropes and the association of molecules. . . ."

The boiling point of a liquid is that temperature at which the vapor pressure of the liquid equals the pressure on the system at the vapor-liquid equilibrium.

The *differential ebulliometric method* proposed by Swietoslawski is a sensitive test that is easy to make with inexpensive apparatus. A comprehensive discussion of this method is presented by Swietoslawski in [4577], and an excellent summary suitable for general use was prepared by Swietoslawski and Anderson [5110]. Swietoslawski has set up an arbitrary scale of purity based on the difference between the boiling temperatures, t_b, and the condensing temperatures, t_c; see Table 4.1.

Table 4.1

Degree of Purity	$\Delta t = t_b - t_c$
I	1.00–0.10°
II	0.10–0.05°
III	0.050–0.020°
IV	0.020–0.005°
V	0.005–0.000°

It is possible, when this scale is used, for two samples of liquid to have the same amount of different impurities but to exhibit different degrees of purity. The effect of an impurity is dependent on the vapor pressure curve of the mixture. This does not detract from the usefulness of the method. If an impurity forms an azeotrope, it is the effect of the azeotrope that is manifest and not that of the impurity itself. If the impurity is known, the amount present may be estimated from the vapor pressures, provided that the concentration is low enough for Raoult's law to be valid.

This is the only ebullioscopic method by which the purity may be estimated with any degree of certainty. It is an indication of the amount, and is fairly independent of the kind, of impurity. It is considered possible to detect 0.002% of impurity by the differential ebulliometer. Zepalova-Mikhailova [4577, p. 94] found almost a linear relationship between the amount of impurity present and Δt for several systems. This method has the further advantage that only the temperature-measuring instrument need be precise. The actual temperature is of no importance, only the difference between the boiling and condensing temperatures.

The *equilibrium boiling point* has long been regarded as a criterion of purity, although two possible circumstances may negate its usefulness for this purpose. An impurity whose boiling point lies very close to that of the principal component or which forms an azeotrope having a proximate boiling point will probably go undetected. The equilibrium boiling point must therefore be used in conjunction with one, and preferably two, other tests such as refractive index and density. The accuracy of this method is dependent upon how closely the pressure may be controlled or measured. The precision of the measurements of the auxiliary properties must be comparable.

The boiling point may be used in one of two ways as a test of purity. If the boiling point of the substance has been established to an accuracy suitable for the particular characterization, the substance may be purified until it has this boiling point. If the boiling point of the substance is unknown, or has not been determined with sufficient accuracy, the substance may be purified until there is no further change in its boiling point.

It should be noted that the boiling points of relatively few substances have been accurately determined. This may be attributed to several factors, namely, impure material, inaccurate temperature-measuring devices, inaccurate pressure measurements, and improperly designed and operated ebulliometers.

A brief description and diagrams of the several ebulliometers are presented by Swietoslawski and Anderson [5110]. The Quiggle, Tongberg, and Fenske [3874] modification of the Cottrell [1103] apparatus offers a convenient and rapid means of determining the boiling point, and to a limited extent the distillation range.

The determination of the *distillation range* (boiling range) is an arbitrary test that is suitable for characterizing commercial grade materials. The results are qualitative but will show the presence of low and high boiling impurities. The sensitivity depends on the amount of impurities present and the difference between their boiling points and those of the principal constituents. It requires as much as, or more material than, the differential ebulliometric method, about the same amount of time, and is less informative. ASTM D 86-67 and D 1078-67 are the most precise modifications of this technique and are recommended when the distillation range is desired for characterization.

The distillation range is a useful and convenient test for many solvents for general laboratory use. It is a criterion used by Rosin [4042] and in the ACS *Reagent Chemicals* [76]. It is one of the standard industrial tests. In conjunction with other data, such as density and refractive index, it may be used to characterize a liquid but not to define its purity.

Fractional distillation is a convenient means of detecting and estimating impurities. For instance, 10 μg of benzene per liter of ethanol may be determined readily with an accuracy of ± 1 μg by removing the lower boiling azeotrope and analyzing for benzene.

Gas Chromatography

Gas chromatography was invented by A. J. P. Martin in 1952. The growth and development of this versatile analytical tool may be characterized as phenomenal. It is the most useful method of analysis for solvents that exists and may be used to analyze both the major component and the volatile impurities in a solvent system. Allen [56] and his associates have analyzed most of the solvents in this book by gas chromatography, either as a major component or as an impurity using a thermal conductivity detector.

Gas chromatography is the simplest and usually the fastest means of analyzing a solvent, and is a convenient and rapid means of following purification. When the purification has reached the 3N to 4N range, or better, the impurities are best analyzed with a high sensitivity detector. Some impurities may be determined in the low nano-range.

The sensitivity obtained from a gas chromatograph for a given solvent system depends on the characteristics of the instrument and on the operating conditions. Such factors as the sensitivity of the recorder, the length of column, the kind of column packing and the method used in packing the column, sample port and column temperatures, and the carrier gas and carrier gas pressure play important roles.

It has been reported by Allen [56] that a Teflon solid phase with either Carbowax 20-M, Carbowax 1540, or diisodecyl phthalate liquid phase is the most versatile packing used at the present time. Good separation is obtained for polar compounds, low boiling hydrocarbons, and mixtures of polar and nonpolar compounds. A cross-linked polystyrene with no liquid phase makes a good packing for intermediate boiling compounds. These packings may or may not be the best for a particular solvent, but they are excellent for general use.

The detection or sensitivity characteristics of several of the detectors are given.

Thermal conductivity is the most commonly used detection system for routine and research analyses of solvents when a solvent analysis of 100% is desired, that is, major and minor components. The lowest response that has been reported is about 3×10^{-4} mv mg^{-1} ml^{-1} with a noise level of <2 μV, or about 50 ppb. With a lower noise level, the limit of detection can be reduced to about 20 ppb, all other factors being optimal. A sensitivity of at least 0.01% may be expected for routine handling of most solvent systems.

Gas density is reported to have a sensitivity probably in the 1 to 10 ppm range.

Hydrogen flame ionization has a detection limit of about 8×10^{-14} gram-atoms of carbon sec^{-1} (nearly 10^{-14} mol heptane sec^{-1}) with a clean system.

Electron capture has about the same or better sensitivity for favorable samples as the hydrogen flame ionization detector.

Spectrometry

Spectrometric methods are widely used in the detection and determination of impurities in solvents; criteria for certain grades of solvents are defined in terms of absorption in the infrared and ultraviolet regions (see Chapter V, General section). Some manufacturers supply individual absorption curves and gas chromatograms showing mole per cent purity for their best grades of the more common solvents such as acetone, benzene, carbontetrachloride, methanol, and so forth. Mass spectrometry and nuclear magnetic resonance spectrometry are also useful for establishing the purity of organic solvents.

Other Methods

The determination of *density* (specific gravity) is one of the standard means of characterizing a substance. Bauer [5110] defines and discusses density and several methods for its determination. Density is easy to determine, but the simplicity of the determination is deceptive—considerable care and experience are required to secure accurate data. Bauer [5110] gives the accuracy of the thermometer-type pycnometer, without thermostating, as ± 0.001. This is the maximum accuracy that can be expected; usually the error is nearer ± 0.005. Several pycnometers are capable of a sensitivity of $\pm 5 \times 10^{-6}$ if properly handled. The specific gravity or density balance is capable of an accuracy of ± 0.001 if the thermometer plummet is used. The density balance modified similarly to that described by Forziati, Mair, and Rossini [1609] has a sensitivity of ± 2 to 3×10^{-5} and an overall uncertainty of $\pm 5 \times 10^{-5}$ g/ml.

That the density does not change on successive purifications is no indication that a pure material has been achieved. If the successive purifications are made by different methods, such as fractional distillation followed by fractional crystallization, and there is no change in density, it is probable that the substance is as pure as may be detected by the sensitivity of the density determination. This property has the same limitations, when used alone, for characterizing purity as the boiling point and refractive index.

The density-composition relationship for dilute solutions may be considered additive. The most convenient method of determining the relationship is to determine the density of the solution and plot the two points on suitable graph paper. The calculation may be made algebraically:

$$d_{\text{soln}} = (C_1 \times d_1) + (C_2 \times d_2)$$

where d_{soln} is the density of the solution, d_1 and d_2 are the densities of pure solvent and solute, respectively, and C_1 and C_2 are weight fractions of each present.

The sensitivity of density as a quantitative measure depends upon the sensitivity of the measurement and the difference in the densities of the solvent and the solute. For example, the densities of cyclohexane and benzene differ by 0.09917 g/ml at 30°. If the method used to determine the density is sensitive to 5×10^{-5} g/ml, benzene can be determined in cyclohexane with an accuracy of slightly less than 1 part per 1000. The density of hexane and that of 3-methylpentane differ by 0.00495 g/ml at 25°. A density method sensitive to 5×10^{-5} g/ml would detect only 1 part of 3-methylpentane in 50 parts of n-hexane, or 2%. The density of water and that of methanol differ by 0.21033 g/ml at 25°. Density will determine as little as 0.05%w if the method is sensitive to 5×10^{-5} g/ml.

The *refractive index* affords one of the easiest and most rapid tests that can be made. Bauer and Fajans [5111] discuss the theory and practice in a concise and practical manner. If the refractive index of the substance has been previously established with sufficient accuracy, a material may be purified until it has the proper refractive index. If the refractive index has not been determined, or has not been determined with sufficient reliability, the substance may be purified to a constant refractive index. The refractive index alone is not a sufficient criterion of purity but, together with the density and equilibrium boiling point, it characterizes the purity of a liquid quite closely.

The refractive index is a convenient test to control the progress of a purification. For instance, the refractive index of the distillate, together with the reflux temperature, is very useful in fractional distillations but its value depends upon the sensitivity of the refractometer and the purity desired.

Refractive indexes may be considered additive for dilute solutions. An ethanol, n_D 25° 1.35941, and benzene, n_D 25° 1.49790, mixture can be analyzed accurately with an Abbé or dipping refractometer; a 1% benzene solution of alcohol could be analyzed to ±0.1%. Ethanol, as an impurity in iodoethane, can be determined with considerably greater accuracy. The refractive index of iodoethane differs from that of ethanol by 0.56512. One part ethanol in 2500 parts of iodoethane can be detected with an Abbé refractometer. On the other hand, refractometry is almost useless for measuring ethanol in ethyl ether. Systems that do not adhere fairly closely to Raoult's law may offer some anomalies that make the refractive index useless when the concentration increases beyond a few percent. The data of Carr and Riddick [914] show that little information can be obtained from the refractive index of the methanol-water system.

The small Abbé refractometer, precise to $\pm 2 \times 10^{-4}$ units, is satisfactory for characterizing solvents for the majority of laboratory uses. The Precision Abbé Refractometer has a precision of $\pm 2 \times 10^{-5}$ and the dipping refractometer of about $\pm 3.5 \times 10^{-5}$ units. The differential refractometer has a precision of $\pm 3 \times 10^{-6}$ units, but has the disadvantage that the refractive index of the sample is compared to the refractive index of a standard substance whose refractive index must be within 0.01 unit of the sample.

The *specific conductance* has been used widely as a criterion of purity. If the impurity and the solvent are both nonconductive, the specific conductance is of little, if any, use. For example, benzene cannot be detected in toluene by this method, but ethanol can be detected because its specific conductance is much greater than that of toluene. Water, salts, inorganic and organic acids, and basic impurities may be readily detected by the conductance of the solution.

The specific conductance is a most useful criterion for the presence or absence of ionizable impurities. It is a sensitive test for water in hydrocarbons and in most halogenated hydrocarbons, and for acids and alkaline substances in alcohols, ketones, aldehydes, and some esters. Its reliability as a means of detecting decomposition products during the purification of formamide [4947] illustrates its usefulness. The increased interest in electrochemical studies in nonaqueous systems has brought about an increase in the use of specific conductance as a rapid and sensitive method for the detection and determination of water and other ionic substances; for example, ethylenediamine [1130], nitromethane [4894], acetonitrile [2691], and acetamide [2199].

The *critical solution temperature* is again being recognized as a convenient criterion that has wide application. The amount of impurities that will be detected by this method depends on the system used and the exactness of the determination. A simple application of the critical solution temperature is the "cloud point" test for water in many solvents. This is considered in more detail later in the chapter.

A. W. Francis [1617] has compiled a table of more than 6000 critical solution temperatures. The book includes an introduction to the subject, description of methods of determination, and other pertinent information. It is well documented.

Specifications are criteria set up, generally between the seller and the buyer, based on specific test methods to assure a uniform product. Certain organizations and individuals have published specifications for certain types of materials. The United States Pharmacopoeia lists purity and limits of impurities with prescribed tests of determined reliability that are considered safe for substances to be used as pharmaceuticals. It is not the intention of the USP Committee to set standards and devise tests to characterize a material as suitable for other than pharmaceutical use, although many solvents meeting USP specifications are suitable for general laboratory use. Each group of specifications and tests is generally designed to fit the substance for a specific use. The *ACS Reagent Chemicals* [76] lists purity, limits of impurities and tests to qualify the substance as a general analytical reagent. Rosin [4042] includes many reagents in *Reagent Chemicals and Standards* that are in the ACS list, and many that are not, together with specifications and tests.

The American Society for Testing Materials lists specifications and tests for a variety of solvents for industrial use in the *1968 Book of ASTM Standards*, Part 20.

Manufacturers have developed specifications and test methods to characterize their products. Some specifications are as stringent as those of the American Chemical Society, while others characterize a "technical" or "commercial" product. Unless the solvent is produced by a number of manufacturers, for example, methanol and acetone, the specifications and tests generally characterize the product of the particular manufacturer. The impurities present often vary from manufacturer to manufacturer, particularly if the product is made by different processes.

Standard or organizational specifications are excellent criteria of purity for a solvent for a specific type of use. The manufacturer or seller does not guarantee that the solvent will be suitable for all uses. For instance, "absolute alcohol" is guaranteed to meet certain minimum requirements. Some manufacturers remove the water from 95% ethanol by the benzene-water-ethanol azeotrope. A very small amount of benzene remains in the absolute alcohol. This alcohol is not suitable for a solvent in ultraviolet spectroscopy. The manufacturers or venders have not stated, nor implied, that the dry ethanol meets ultraviolet clarity as one of the criteria of purity.

The level of impurities sometimes lies in a range above the maximum permitted by the criteria and below the minimum determinable by direct application of available analytical methods. In such cases it is convenient to carry out a *concentration of impurities* prior to the analysis of the sample. An example is the concentration of benzene in ethanol, which is mentioned under fractional distillation and discussed under **59** ethanol in Chapter V.

Concentration by fractional distillation has many applications. It is most useful when the impurity and the principal component, *C*, form a positive azeotrope boiling several degrees below C; or when another component that can be added to the system forms a positive ternary azeotrope with the impurity and C or forms a positive binary azeotrope with the impurity and not with C. The distilled system permits a greater accuracy if the azeotrope is homogeneous. Generally, impurities that are higher boiling than C and do not form an azeotrope are difficult to concentrate unless the distillation alpha is large; microamounts usually distill with C.

Concentration by fractional solidification, see Zief and Wilcox [5340], offers a general, more positive means than fractional distillation. Fractional freezing and zone melting are two techniques that have been frequently used.

The following references are recommended for additional information: *Cryoscopy:* Glasgow and Ross [2694, Chapter 88], Skau, Arthur, and Wakeham [5110, Chapter 7], Smit [4374]; *Ebulliometry:* Swietoslawski [4577], Swietoslawski and Anderson [5110, Chapter 8], Lestie and Kuehner [2694, Chapter 89], Hála, Pick, Fried, and Vilím [1966]; *Temperature measurements:* Sturtevant [5110, Chapter 6], Corruccini [2694, Chapter 87].

DRYING OF SOLVENTS AND DETERMINATION OF WATER

Water is the universal impurity in all organic solvents because it usually remains in the solvent during its preparation and because practically all organic solvents are hygroscopic to some extent. Small amounts of water are often difficult to detect because of the limited number of reliable and dilicate chemical tests and because the presence of water does not cause an appreciable change in many physical properties. The literature concerning the drying of organic liquids and the detection of water and its effect on the physical proportion is voluminous.

Drying of organic liquids may be accomplished by either chemical or physical means or both. The method selected depends on the solvent, on the material or equipment available, and the degree of dryness desired.

The chemical method of drying depends on the use of substances generally classed as *drying agents*. These agents remove water by either chemical or physical combination. The efficiency of drying agents depends upon the vapor pressure of the hydrated compound, the physical state of the agent, and the technique of use. Other factors being equal, the lower the vapor pressure, the more efficient is the drying agent. However, it is seldom possible to fulfill the basic requirement of "other factors being equal." Drying agents usually are rated by the amount of residual water left in air, under specified test conditions, when moist air is passed over the agent. A number of comparative investigations are listed in Table 4.2 (cf. the recent work of Trusell and Diehl [4838] and Bower [689].

The *residual water* criterion provides an excellent method for rating drying agents for removing moisture from gases. There is assumed to be a

Table 4.2 Comparative Efficiency of Some Drying Agents (micrograms residual water per liter of dried air)

Alumina, Al_2O_3	2.9
Barium perchlorate	599.
Barium oxide	0.6
Calcium bromide	200.
Calcium chloride, anhydrous	360.
Calcium chloride, anhydrous, technical	1250.
Calcium chloride, granular	1500.
Calcium oxide	3.
Calcium sulfate, Drierite	5.
Magnesium perchlorate, anhydrous	2.
Magnesium perchlorate, Anhydrone	1.5
Magnesium perchlorate, $Mg(ClO_4)_2 \cdot 3H_2O$	4.4
Molecular sieves, 5A	3.9
Sodium hydroxide, sticks	513.
Phosphorus pentoxide	$<0.2^a$
Silica gel	30.
Sulfuric acid, 100%	8.
Sulfuric acid, 95.1%	300.

[a] The reported values vary from 0.02 to 2.9.

correlation of a drying agent's efficiency for drying gases and its ability to dry organic solvents. Pearson and Ollerenshaw [3693] used a rapid, near-infrared method for the rating of the efficiency of drying agents for organic liquids. The rate and the extent of drying for five agents with three solvents have been reported. Small samples were withdrawn at intervals and analyzed to determine the rate. The data are given in Table 4.3.

The data of Pearson and Ollerenshaw point out that the several drying agents have different drying efficiencies for the same solvent and for different solvents. These differences presumably are due to such factors as hydrogen bonding, the rate of contact of the water molecule with the drying agent, and the rate of diffusion of the water from the surface to the interior of the drying agent particle.

Bower [689] classified drying agents in five groups according to the type of reaction with water:

1. $A(H_2O)_{y(s)} + xH_2O \rightarrow A(H_2O)_{x+y(s)}$. (e.g., calcium chloride, calcium sulfate, and copper sulfate.)
2. $A_{(s)} + H_2O \rightarrow$ saturated solution. (e.g., sodium chloride. This type of drying is used primarily for removing large amounts of water from lower molecular weight compounds such as alcohols.)

Table 4.3 Comparative Efficiency of Five Drying Agents for Three Solvents

	$CaCl_2$	*Drierite*	Molecular Sieves	$MgSO_4$	Na_2SO_4
Benzene					
C_1, %w	0.054	0.028	0.048	0.023	0.030
C_2, %w	0.000	0.009	0.000	0.019	0.021
T_{50}, min	2.	3.	30.	38.	30.
T_{95}, min	40.	20.	9 hr	75.	4 days
Ethyl ether					
C_1, %w	1.468	1.580	1.468	1.570	2.070
C_2, %w	0.011	0.320	0.009	0.220	1.830
T_{50}, min	0.5	1.	3.5	1.	0.5
T_{95}, min	25.	20.	60.	30 hr	2.
Ethyl acetate					
C_1, %w	0.370	0.515	0.370	0.495	1.490
C_2, %w	0.042	0.195	0.039	0.346	1.280
T_{50}, min	5.	1.5	26.	9 hr	6 days
T_{95}, min	2 hr	25.	4 hr	7 days	12 days

C_1 is the initial water concentration.

C_2 is the concentration of water after the solution had been in contact with the drying agent for three to four weeks.

T_{50} is the time required to reduce the concentration, C_1, by 50% based on the final concentration, C_2.

T_{95} is the time required to reduce C_1 by 95% based on C_2, i.e., to reduce C_1 to $[0.05(C_1 - C_2) + C_2]$.

3. $A_{(1)} + H_2O \rightarrow$ solution. (e.g., sulfuric acid.)
4. Adsorption. (e.g., silica gel, alumina.)
5. $M + (H_2O)_x \rightarrow M(OH)_x + xH_2$. (e.g., sodium.)

A chemical classification of types is also possible:

1. Direct union. (e.g., phosphorus pentoxide and calcium oxide.)
2. Double decomposition. (e.g., lithium aluminum hydride, calcium carbide, and alkali alkoxides.)
3. Displacement. (e.g., sodium.)
4. Hydrate formation. (e.g., calcium sulfate, magnesium perchlorate, and sulfuric acid.)
5. Adsorption. (e.g., silica gel and alumina.)

This system does not account for the removal of water by concentrated solutions of calcium chloride and potassium carbonate.

Phosphorus pentoxide has long been the standard for drying agents. It is among the best of the agents but it has numerous disadvantages. It is hard to handle; it reacts with many solvents such as amines; it dehydrates many compounds or catalyzes reactions in others. Very dry substances may be obtained using phosphorus pentoxide. The most successful method is to reflux the material with the oxide and distill. It is particularly recommended for hydrocarbons, acids, anhydrides, ethers, and halohydrocarbons.

Metal hydrides such as lithium hydride, calcium hydride, and lithium-aluminum hydride are excellent drying agents for solvents with which there is no deleterious reaction. Riddick [3975] has found calcium hydride to be particularly effective. The reaction products are not a problem in many solvents; hydrogen escapes as a gas and calcium hydroxide is insoluble. Methanol, ethanol, and several hydrocarbons have been dried to the available lower limit of detection, about 0.0005% water.

Magnesium perchlorate is one of the most efficient drying agents. However, it is soluble in many solvents and may cause explosions if improperly used.

Molecular sieves, particularly 4A and 3A, are the most universally useful and efficient of the drying agents. Riddick [3975] has reduced the water in alcohols to 0.5% or less and then slowly passed them through long columns of molecular sieves. The water content was found to be equal to, or less than, 0.001%.

The *alkali hydroxides* are good agents for drying organic bases to 0.1% or less water. They are inexpensive and easy to use.

Calcium chloride, long a favorite drying agent, has several excellent characteristics. It is inexpensive, easy to use, and it is insoluble in most substances. It is an excellent agent for removing water to 0.1%, or in some instances to 0.05% or lower.

Potassium carbonate finds its primary use in drying alcohols. The carbonate is added until it remains solid on prolonged shaking.

Magnesium sulfate as a drying agent does not have any outstanding characteristics. It is included because it is used in several methods of purification given in Chapter V. An anhydrous grade is available commercially. Riddick [3975] studied the effectiveness of magnesium sulfate, together with several other drying agents for the drying of solvents. The heptahydrate must be dehydrated at reduced pressure to prevent fusion of the particles at the dehydration temperature of 300°. The particles are free-flowing, porous, and somewhat fragile. If the temperature of dehydration is below 300°, the water is not removed completely; if it is heated much above 300°, its effectiveness as a drying agent is decreased. It is soluble in many organic solvents.

Silica gel and *alumina* are effective drying agents. Finely divided silica gel was used to dry some API hydrocarbons to a very low water content; see

Chapter V. The effectiveness depends upon the amount of water to be removed, the fineness of the agent, and the length of time and effectiveness of the contact.

Wymore [5276] found that the *sulfonic-type, cation-exchange resins* are excellent desiccants for drying organic liquids. The resins used were production grade and, after conversion to the desired form, were dried at 110° in vacuo. The potassium form of Dowex 50W-X8 resin, 20–50 mesh, was the best tested. The rate of flow depends on the polarity of the solvent and the amount of water in the liquid. The rate for nonpolar substances may be quite high, and the resin swelling is small. Nonpolar solvents are easily dried to less than 5 ppm of water. Polar solvents may be dried to 10 ppm or less of water but the flow rate per unit of resin is less.

Drierite, a form of anhydrous *calcium sulfate*, was the most practical laboratory drying agent until the advent of molecular sieves. It still is a convenient and efficient agent.

Fractional distillation is a convenient and generally applicable method for the drying of solvents. If the solvent and water do not form a positive azeotrope, the greater the difference in boiling points between the water and the organic liquid and the more efficient the fractionating column, the drier will be the distilled compound, other factors remaining constant. The xylenes, nitrobenzene, aniline, and *N*-methylacetamide may be dried to at least 1–2 ppm of water by removing the water and distilling a small amount of the solvent to clear out the distillation apparatus. Methanol containing 2–5 ppm of water can be prepared by fractional distillation. Generally, about one-quarter of the charge must be kept in the still pot to retain the water if a particularly dry product is desired. Acetone could not be dried to less than about 0.02% water in a 100-plate still at a reflux ratio of 50:1 and better.

Many organic compounds form azeotropes with water. A large number of these boil several degrees below the boiling point of the solvent and the latter may be dried by distilling the azeotrope. If the boiling points of the compound and the azeotrope are too close together to permit separation by fractional distillation, a third substance sometimes may be found that forms a lower boiling ternary azeotrope with the solvent and water. For instance, the boiling point of the ethanol-water azeotrope (78.174°) is too near that of ethanol (78.325°) for efficient removal of water by fractional distillation. If benzene is added, it forms a ternary azeotrope (bp 64.86°) that is easily removed. The alcohol may be dried to less than 0.01% water.

Barabanov [383] describes an electrolytic method for the removal of water and other ionic impurities from organic solvents. The apparatus is a modification of the Hittorf tube. Electrical conductance is the criterion of ionic impurities present.

THE DETERMINATION OF WATER

A critical review of the methods for the determination of water has been prepared by Tranchant [4819]. He presents the critique in three main sections: review of the principal methods for the determination of water; precautions to take for the determination of water; and the elements of choice of the method of determination.

Cloud Point

Generally, the solubility of water in solvents partially miscible with water decreases with decrease in temperature. In those solvents in which water is only slightly soluble, the amount of water may be determined quite accurately by lowering the temperature of the liquid until a faint cloudiness is observed. This method may be used to within a few degrees of the freezing point of the liquid. Knowledge of the solubility curve for water in the solvent is necessary. Many of these temperature-composition systems are cited by Francis [1617].

Babko and Schevchenko [334] found that cobalt (II) iodide was the most sensitive of the cobalt halides for the determination of water in alcohols, ketones, nitriles, and some esters.

Knight and Weiss [2638] developed a *calcium carbide-gas chromatography* method for the determination of traces of water in hydrocarbons. The precision at 3 and 20 ppm levels is about 15 and 7%, respectively. Meda and Bertino [3251] describe the apparatus and method for the determination of water in glycerol, phenol, ethanol, essential oils, syrups, and so forth, based on measurements of the amount of acetylene liberated from calcium carbide. The inadequacies of other methods are pointed out.

Pearson and Ollerenshaw [3693] point out that water determination in many organic solvents by means of the overtone band of water near 5300 cm^{-1} has become a simple matter. The method is rapid and only a small volume of sample is required. An accuracy of $\pm 0.02\%$ has been suggested for the range, 0.02 to 1.00% water, but probably can be improved. Good agreement has been reported between this and other methods that have been proven reliable; a comparison with the Karl Fischer Method has been made by Meeker and coworkers [3255] for several functional-group types. Streim and coworkers [4529] have compared results with those obtained by gas chromatography for amines and alcohols. Cordes and Tait [1088] report the results of a study of the determination of water in hydrazines, and Keyworth [2571] reports the application for alcohols.

Forbes [1594] devised an infrared method for the determination of water in liquids in the few parts per million range; examples are given from 2.37–16.8 ppm. The liquid was treated with calcium carbide, the acetylene collected in a dry gas carrier, transferred to a cell, dissolved in carbon tetrachloride, and then measured.

The *Karl Fischer Method* is the most universal and accurate method for the determination of water. Mitchell and Smith incorporated a study of accuracy, precision, and application of the Karl Fischer method in [3343].

Swensen and Keyworth [4575] found that iodine can be coulometrically generated from potassium iodide dissolved in a solvent containing pyridine, formamide, and sulfur dioxide, it can be used in a Karl Fischer-type reaction for the determination of water below the 10-ppm range in benzene and related-type solvents.

Water cannot be determined in vinyl ethers by the regular Karl Fischer method. The reagent reacts rapidly with the ethers. It is believed by Barnes and Parvlak [401] that the methanol present in the reagent as the sample solvent takes part in the iodoacetal reaction:

$$ROCH{=}CH_2 + I_2 + CH_3OH \rightarrow ROCH(OCH_3)CH_2I + HI$$

When methanol is eliminated, the water in vinyl ethers can be determined. A commercially available, stabilized Karl Fischer reagent was used containing 2-methoxyethanol instead of methanol as the solvent.

The *acetyl pyridinium chloride method* developed by Smith and Bryant [4378, 4379] is a convenient method for determining water in liquids that do not react with the reagent. It may be preferred to the modified Karl Fischer method, which uses hydrocyanic acid for aldehydes and ketones.

Electrical conductivity offers a rapid and convenient means of determining water in many organic liquids. The electrical conductivity is very sensitive to small amounts of water in most solvents. Schmidt and Jones [4167], Lund and Bjerrum [3029], and de Brouckère and Prigogine [1244] have used electrical conductivity as a criterion of purity. Some factors that increase the precision of the method for measuring the conductivity of dilute solutions have been reported by de Brouckère and Prigogine [1243].

Other methods, such as the measurement of density, refractive index, viscosity, boiling point, freezing point, infrared spectroscopy, and mass spectroscopy, are convenient for the determination of the amount of water in some organic liquids. The physical property difference caused by the water must be of sufficient magnitude to give the accuracy desired.

PURIFICATION METHODS

General

Introduction
Solvents of Known Quality Available

Hydrocarbons **1–57**

General
  Methods of Preparation
  Preparation of Carbonyl-Free Solvents
Saturated Aliphatic Hydrocarbons, **1–29**
  General
    Methods of Preparation
    Drying and Criteria of Purity
    Standards for Raman Spectroscopy
    Purification for Ultraviolet Spectroscopic Solvents
  Specific Solvents, **1–29**
Aromatic Hydrocarbons, **30–44**
  General
    Preparation of Alkyl Derivatives of Benzene
    Purification for Specific Conductance
  Specific Solvents, **30–44**
Unsaturated Hydrocarbons, **45–57**
  General
    Methods of Preparation
    Methods of Purification
    Safety
  Specific Solvents, **45–57**

Compounds with One Type of Characteristic Atom or Group, **58–319**

Hydroxy Compounds, **58–107**
  Monohydric Alcohols, **58–100**
    Aliphatic Alcohols, **58–91**
      General
        Methods of Preparation
        Methods of Purification

GENERAL

Introduction

The compounds are discussed in this chapter in the order established in Chapter I.

General information concerning an entire class of compounds is given, when available, following the title of the class or subgroup class. Purification and preparative methods applicable to several members of a functional group or subgroup are given in the General section. Sections pertaining to specific compounds follow the General section for each group and contain information on preparation, purification, criteria of purity for specific applications, and safety. When the number of compounds prepared or purified by a method is small, or if compounds from several functional groups or subgroups are involved, the method is given under one compound and reference is made thereto under the other compounds; for example, purification of ethyleneimine, Searles and coworkers [4217] and piperidine. The primary entry is under ethyleneimine.

Reference should be made to related compounds when using the directions, since one method of purification is applicable to several compounds.

Since the publication of the second edition of *Organic Solvents*, interest in the formation, presence, and determination of *peroxides* has greatly increased. Formerly associated principally with ethers, peroxide formation is now known to occur among such functional-group types as hydrocarbons (particularly unsaturated ones), alcohols, ethers, esters, and nitriles.

The general discussion of peroxides is in the *Ether* section. When a method or procedure is confined, either chemically, or by the author, to one compound or one functional-group type, the information appears under the compound or group.

The sections on *safety* have been included as a convenient resumé of available information on the safe handling of the several solvents. Patty's *Industrial Hygiene and Toxicology*, [3680, 3681], the *Handbook of Organic Industrial Solvents*, [88], and *Documentation of Threshold Limit Values*, [4509] have been drawn upon for most of the information presented.

The modes of exposure given primary consideration are the inhalation of vapors and absorption of liquid through the skin. Eye damage from some liquids and vapors is an important consideration.

The *threshold limit value* (TLV) often referred to as maximum permissible concentration, MPC, or maximum allowable concentration, MAC, is generally accepted to mean that the average person can be exposed to the stated concentration of the vapors for 8 hr per day, 5 days per week without harm. These values should be considered as guides to good practice.

The concentration of vapors in the air that is fatal to 100% and 50% of the test animals in the stated time is represented by LC_{100} and LC_{50}, respectively. The maximum concentration in the air that was not fatal to any of the test animals is represented by LC_0. LD_{100}, LD_{50}, and LD_0 represent lethal dosages, the subscript having the same meaning as for lethal concentrations.

The *flammable limits* represent the concentration range in air in which explosions occur.

The *minimum ignition temperature* is the minimum temperature at which rapid combustion becomes independent of externally supplied heat.

While distillation remains the most important general method of purification of organic solvents, newer techniques involving solid-liquid equilibria deserve wider consideration. Specific applications include the purification of solvents for use as reference standards and as high purity materials for electrical, physical, and thermal studies. The method of *zone melting* has several ramifications of which *zone refining*, the purification technique developed by Pfann in 1952, is but one. Zone refining is the phase of zone melting of particular interest for solvent purification. The general practice of using the less specific term, zone melting, will be followed in most instances. Zeif and Wilcox [5340] introduced the term *fractional solidification* as a general term for a group of processes including zone melting.

There are excellent books and reviews available, for example, *Fractional Solidification* [5340], *Zone Melting of Organic Compounds* [2123], *Zone Melting* [3739], *Zonenschmelzen* [4148], and "Zone Melting of Organic Compounds" [5181].

The principal limitation of zone melting and related techniques is that mixtures of eutectic composition present difficulties in separation. The equipment becomes expensive to obtain and operate when the temperature required to maintain the substance in the solid state is lower than that obtainable from commercially available refrigeration units.

The technique has shown itself capable of removing impurities to the part per billion range. The need of ultrapure germanium for transistors brought zone refining into being. It is feasible to extend zone refining to achieve a purity in many solvents comparable to that which has been reached with metals.

Dickinson and Eaborn [1306] found that *purification by progressive freezing* is superior to the zone melting technique involving a single zone.

Dugacheva and Anikin [1391] describe a zone melting method for the purification of solvents crystallizing between 0° and −100° in a hermetically sealed chamber without contact with air. Five-fold melting results in 8 to 9-fold purification of the sample. Examples are: acetone, mp −56.798° and hexane, mp −95.347°.

Solvents of Known Quality Available

There are a number of solvents available in laboratory quantities that are of good purity and have been precisely and/or accurately characterized. Some of these materials are designed for specific end use. The small amount of impurities generally present permit them to be used for any purpose except: (1) where a very pure material is required (Research grade excepted), (2) when a certain impurity, or type of impurity, is present that is deleterious to the intended use. These solvents may be further purified to the desired degree of purity relatively easily.

Some of the solvent grades available follow.

American Chemical Society Reagent grade is often designated as ACS Reagent grade and sometimes as ACS when the meaning is obvious. The requirements that serve as the set of minimum standards for this grade of solvents may be found in *Reagent Chemicals* [76]. Some solvents, for example, hexane, meet the Reagent grade requirements, and a second grade has, in addition, an optical absorbance requirement. Rosin's *Reagent Chemicals and Standards* [4042] is similar to *Reagent Chemicals* but includes more compounds. In general, if a solvent will pass the requirements of one, it will pass those of the other.

Research grade is believed to have been first used by Phillips Petroleum Company to designate its high purity research and reference standards characterized by the thermodynamic freezing curve [3747]. It is used herein to include solvents of similar purity from other sources. A Research Grade solvent usually has a purity of 99.9+ % m. Often the impurities, or at least some of them, will be given.

Spectrophotometric grade was originally conceived as suitable for the ultraviolet region of the spectrum. It has been extended by several suppliers to include the visible, near-infrared and infrared regions. Some suppliers will furnish copies of the spectrum of the batch shipped. Generally, these solvents are also of Reagent grade quality. Eastman's Distillation Products Industries' brochure, *Spectrophotometric Solvents*, and Matheson Coleman & Bell's brochure, *Spectroquality Solvents*, list 30+ solvents each and give typical spectra and related information.

Chromatographic Quality solvents are 99+ % m determined by gas chromatography and are available from a limited number of sources. A chromatogram may be obtained with each unit that will also give information for the conditions of analysis.

Pesticide grade is also known by the registered trademark name *Nanograde*. These solvents are suitable for pesticide residue analysis. They meet ACS Reagent grade requirements and, in addition, do not contain more than 10 nanograms per liter of a chlorinated pesticide or an equivalent amount of

Table 5.1 Specification for Electronic Grade Solvents

	Acetone	Methanol	Trichloroethylene
Assay, min, %	99.5	99.5	99.5
Color, APHA	5	5	5
Boiling point, °C, 760 torr	56.1	64.6	87.0
Boiling range, °C, 760 torr	±0.5	±0.5	±0.5
Specific gravity, 25°C	0.7871	0.7886	1.460 to 1.464
Specific conductance, max, 25°C, ohm^{-1} cm^{-1}	1.0	4.0	0.08
Residue on evaporation, %	0.001	0.001	0.001
Water, Karl Fischer, %, max	0.5	0.10	0.01
Solubility in water, ACS	PT[1]	PT[1]	—
Acidity as acetic acid, %	0.002	0.002	0.0 as HCl
Alkalinity as NH$_3$, %	0.001	0.0003	PT[2]
Methanol, %	0.05	—	—
Acetone, %	—	0.001	—
Aldehydes, %	PT[1]	—	—
Substances reducing KMnO$_4$, ACS	PT[1]	PT[1]	—
Substances darkened by H$_2$SO$_4$	—	PT[1]	—
Heavy metals, as lead, %	—	—	0.001
Free halogen	—	—	none
Al, Fe, Cu, As, Sb, Pb, Ni, Ag, Au, and B, each, max, ppm	0.1	0.1	0.1

PT[1], to pass test; PT[2], to pass test as NaOH

Table 5.2   Solvents Available for Special Purposes

			Code		
A	American Chemical		C	Chromatographic grade	
	Society Reagent grade		P	Pesticide grade	
R	Research grade		N	Nonaqueous Titration	
S	Spectrophotometric grade			Solvent grade	
		E	Electronic grade		

No.	*Compound*	A	R	S	C	P	N	E
1	Cyclopentane	.	×	×	.	.	.	.
2	Pentane	.	×	×	×	.	.	.'
3	3-Methylbutane	.	×	×	×	.	.	.
4	2,2-Dimethylpropane	.	×	.	.	.	.	.
5	Methylcyclopentane	.	×	.	×	.	.	.
6	Cyclohexane	.	×	×	×	×	×	.
7	Hexane	×	×	×	×	.	.	.
8	2-Methylpentane	.	×	.	×	×	.	.
9	3-Methylpentane	.	×	.	×	.	.	.
10	2,2-Dimethylbutane	.	×	.	×	.	.	.
11	2,3-Dimethylbutane	.	×	.	×	.	.	.
12	Methylcylcohexane	.	×	×	×	.	.	.
13	Heptane	.	×	×	×	×	.	.
15	3-Methylhexane	.	×	.	.	.	.	.
17	2,4-Dimethylpentane	.	.	×	.	.	.	.
19	Octane	.	×	.	×	.	.	.
21	2,2,4-Trimethylpentane	×	×	×	×	×	.	.
22	Nonane	.	×	.	×	.	.	.
23	2,2,5-Trimethylhexane	.	×	.	×	.	.	.
24	Decahydronaphthalene	.	.	×	.	.	.	.
25	*cis*-Decahydronaphthalene	.	×	.	.	.	.	.
26	*trans*-Decahydronaphthalene	.	×	.	.	.	.	.
27	Decane	.	×	.	.	.	.	.
29	Dodecane	.	×	.	.	.	.	.
30	Benzene	×	×	×	×	×	×	.
31	Toluene	×	×	×	×	×	×	.
32	*o*-Xylene	.	×	.	×	.	.	.
33	*m*-Xylene	.	×	×	×	.	.	.
34	*p*-Xylene	.	×	.	×	.	.	.
35	Ethylbenzene	.	×	.	.	.	.	.

Table 5.2 (Continued)

No.	Compound	A	R	S	C	P	N	E
36	Isopropylbenzene	.	×	.	.	.	.	.
38	Naphthalene	.	×	.	.	.	.	.
40	Butylbenzene	.	×	.	.	.	.	.
41	*sec*-Butylbenzene	.	×	.	.	.	.	.
42	*tert*-Butylbenzene	.	×	.	.	.	.	.
45	1-Pentene	.	×	.	.	.	.	.
46	2-Pentene, mixed isomers	.	×	.	.	.	.	.
47	*cis*-2-Pentene	.	×	.	.	.	.	.
49	1-Hexene	.	×	.	.	.	.	.
51	1-Octene	.	×	.	.	.	.	.
54	Cyclohexene	.	×	.	.	.	.	.
58	Methanol	×	.	×	×	×	×	×
59	Ethanol	×	.	.	.	×	.	.
60	1-Propanol	.	.	.	.	.	×	.
61	2-Propanol	×	.	×	×	×	×	.
62	1-Butanol	×	.	×	×	×	×	.
63	2-Butanol	.	.	.	×	.	.	.
64	2-Methyl-1-propanol	×	.	×	×	×	.	.
65	2-Methyl-2-propanol	.	.	.	×	.	×	.
66	1-Pentanol	.	.	.	×	.	.	.
70	3-Methyl-1-butanol	×	.	×	×	.	.	.
74	Cyclohexanol	.	.	.	×	.	.	.
93	Phenol	.	.	.	.	.	×	.
97	2-Propen-1-ol	.	.	.	×	.	.	.
100	2-Propyn-1-ol	.	.	.	×	.	.	.
103	1,2-Ethanediol	.	.	.	×	.	×	.
104	1,2-Propanediol	.	.	.	×	.	×	.
106	1,3-Butanediol	.	.	.	×	.	.	.
107	Glycerol	×	.	×	.	.	×	.
109	Ethyl ether	×	.	×	×	×	×	.
111	Isopropyl ether	.	.	.	×	.	.	.
113	Butylethyl	.	.	×	.	.	.	.
118	Bis(2-methoxyethyl) ether	.	.	×	.	.	.	.
122	Furan	.	.	.	×	.	.	.
123	Tetrahydrofuran	.	.	.	×	.	×	.

Table 5.2 (Continued)

No.	Compound	A	R	S	C	P	N	E
124	*p*-Dioxane	.	×	×	×	×	×	.
141	Acetone	×	.	×	×	×	×	×
142	2-Butanone	.	.	.	×	×	×	.
144	Cyclohexanone	.	.	.	×	.	.	.
145	4-Methyl-2-pentanone	.	.	.	.	.	×	.
147	Acetophenone	.	.	.	×	.	.	.
148	Formic acid	×	.	.	.	.	×	.
149	Acetic acid	×	.	.	.	.	×	.
150	Propionic acid	.	.	.	.	.	×	.
161	Acetic anhydride	×	.	.	.	.	×	.
162	Propionic anhydride	.	.	.	.	.	×	.
164	Methyl formate	.	.	×	.	.	.	.
165	Ethyl formate	.	.	×	.	.	.	.
169	Methyl acetate	.	.	.	×	.	.	.
172	Ethyl acetate	×	.	×	.	×	×	.
177	Butyl acetate	.	.	×	×	.	.	.
180	Pentyl acetate	.	.	.	×	.	.	.
184	Ethyl propionate	.	.	×	.	.	.	.
191	Ethyl acrylate	.	.	.	×	.	.	.
192	Methyl methacrylate	.	.	.	×	.	.	.
200	Ethylene carbonate	.	.	×	.	.	.	.
217	1,1,2-Trichlorotrifluoroethane	.	.	×	.	.	.	.
222	2-Chloropropane	.	.	.	×	.	.	.
228	Chlorobenzene	.	.	.	.	.	×	.
230	Dichloromethane	.	.	×	×	×	×	.
231	Chloroform	×	.	×	×	×	×	.
232	Carbon tetrachloride	×	.	×	×	×	×	.
234	1,2-Dichloroethane	.	.	×	×	.	.	.
236	1,1,2,2-Tetrachloroethane	.	.	.	×	.	.	.
240	*p*-Dichlorobenzene	.	×	.	.	.	.	.
245	Trichloroethylene	.	.	.	×	.	.	×
246	Tetrachloroethylene	.	.	×	.	.	.	.
249	2-Bromopropane	.	.	.	×	.	.	.
252	Bromoform	.	.	×	.	.	.	.
255	Iodomethane	.	.	.	.	.	×	.

Table 5.2 (Continued)

No.	Compound	A	R	S	C	P	N	E
260	Nitromethane	.	.	×	.	.	×	.
264	Nitrobenzene	.	.	.	.	.	×	.
265	Acetonitrile	.	.	×	×	×	×	.
275	Benzonitrile	.	.	×	.	.	×	.
276	Acrylonitrile	.	.	.	×	.	.	.
280	Butylamine	.	.	.	.	.	×	.
285	Aniline	×	.	.	×	.	×	.
292	Diethylamine	.	.	.	.	.	×	.
298	Triethylamine	.	.	.	.	.	×	.
299	Pyridine	×	.	×	×	.	×	.
302	N-Methylformamide	.	.	.	×	.	.	.
303	N,N-Dimethylformamide	.	.	×	.	×	×	.
306	N,N-Dimethylacetamide	.	.	×	.	.	.	.
310	1-Methyl-2-pyrrolidinone	.	.	×	.	.	.	.
312	Carbon disulfide	×	.	×	×	.	.	.
319	Dimethyl sulfoxide	.	.	×	.	×	×	.
320	Sulfolane	.	.	.	.	.	×	.
321	2-Methoxyethanol	.	.	×	×	×	×	.
322	2-Ethoxyethanol	.	.	.	×	.	×	.
323	2-Butoxyethanol	.	.	.	×	.	.	.
324	Furfuryl alcohol	.	.	.	×	.	.	.
325	Tetrahydrofurfuryl alcohol	.	.	.	×	.	.	.
326	Diethylene glycol	.	.	.	.	.	×	.
329	2-(2-Ethoxyethoxy)ethanol	.	.	.	×	.	.	.
332	2-Chloroethanol	.	.	.	×	.	.	.
334	2-Aminoethanol	.	.	.	.	.	×	.
338	2-Furaldehyde	.	.	.	×	.	.	.
342	Morpholine	.	.	.	.	.	×	.

an impurity that would produce a comparable chromatographic peak at similar operating conditions.

Nonaqueous Titration Solvent grade solvents are low in water and in substances that are acidic or basic in the solvent. They are suitable for normal use as received. They often will have to be dried for high precision titrations. The Eastman Kodak Company lists 54 solvents suitable for normal titrimetric use in their 1968 brochure *Titrants, Indicators, Solvents for Non-Aqueous Titrimetry.*

Electronic grade solvents are used for precleaning and drying metal parts of electric circuitry. They have low boron and metallic impurities and low specific conductance. The specifications for three Electronic grade solvents are given by the Fisher Scientific Co. in *Technical Data Sheet*, TD-143, 10/63. The specifications are given in Table 5.1.

There are other specialized grades of solvents, but those cited are the most common and easily obtained.

Table 5.2 lists the solvents available in 1968 in several categories. This list has been compiled from the following sources: Matheson Coleman & Bell, *Laboratory Chemical Catalogue*, 1967 and MC/B *Spectroquality Solvents*, 1965; *Phillips 66 Hydrocarbons and Petro-Sulfur Compounds*, Bulletin 522, 6th ed.; *Fisher Chemical Index 67-C;* E. H. Sargent & Co., *Chemical Reagents;* Eastman's Distillation Products Industries, *Spectrophotometric Solvents;* Eastman Organic Chemicals, *Titration, Indicators, Solvents for Non-Aqueous Titrimetry;* Mallinckrodt, *Product List*, 1965. Matheson Coleman & Bell was the primary reference source because of the large number of solvents tabulated in the "Appendix" of their Chemical Catalogue according to the several grades. Phillips 66 was the primary reference for hydrocarbons and petro-sulfur compounds, and Eastman's Distillation Products for nonaqueous titration solvents.

HYDROCARBONS

General

Aromatic hydrocarbons were available in England about the middle of the nineteenth century. The discovery of the first synthetic dye, *mauve*, by Perkins in 1856 ushered in the golden age of the aromatics. Stillman's analysis of the surface Titusville petroleum was made in 1859. About three-quarters of a century passed before scientific technology became sufficiently sophisticated to utilize the potentials of petroleum hydrocarbons. W. Nelson Axe believes that "As a chemical class, hydrocarbons must rank close to air, water, carbohydrates, and protein in basic importance to the human race."

Improved technology in the petroleum industry has provided additional aromatic and many paraffinic and naphthenic hydrocarbons in several grades of purity at a reasonable cost. The impurities present in the different grades may be characterized and evaluated by gas chromatography and mass spectroscopy. The amount of impurities present, but not the kind, is most accurately evaluated in high purity materials by the thermometric freezing or the adiabatic calorimetric method (see Chapter IV).

The American Petroleum Institute Research Project 6 was begun in 1927 as a comprehensive investigation of hydrocarbons in petroleum [4050]. This project was the real start of the intensive and systematic separation and study of the several components of petroleum. The success of this study was made possible by the development of needed methods and equipment to solve the many problems associated with such a complex system. The methods used for separating the hydrocarbons have been mainly the physical processes of distillation, crystallization, extraction, adsorption, and the group of proc- esses characterized as fractional solidification.

Several variations of distillation such as regular, reduced pressure, and azeotropic, have been used. Regular distillation has normally been the first step in the purification. In the absence of azeotropic mixtures, it removes all except the very close boiling impurities.

The primary purposes of azeotropic distillation are: (1) to remove impurities of a type different from the main component; (2) to remove isomeric or closely related impurities; (3) to reduce the column hold-up by selecting a suitable low boiling azeotrope-forming compound; and (4) to permit complete distillation of the main component by using an excess of the azeotropic- forming compound. The distillation equipment and operational techniques used by the API-NBS group have been discussed by Willingham and Rossini [5203].

Adsorption has been used to remove water and nonhydrocarbon impurities. It is also applicable for removing aromatic hydrocarbons from paraffinic and cycloparaffinic hydrocarbons and will, in some cases, remove isomeric impurities from hydrocarbon compounds. The chromatographic method for separating the aromatic hydrocarbons from paraffinic and cycloparaffinic compounds has been described by Mair and Forziati [3116a]. This method has been extended by Mair [3115a] to allow the separation of aromatic, paraffinic, and cycloparaffinic hydrocarbons from a mixture of the three. It may be used analytically to determine the amount of each type of hydrocarbon in a mixture [3116].

The results in Table 5.3 summarize the purification of paraffins and mono- olefins as carried out by the API-NBS group [4522]. In connection with the purification, the authors state: "(1) A logical, simple purification following the original synthesis or other preparation of a given hydrocarbon concen- trate will usually remove all impurities except more or less close boiling isomers, (2) an impurity of several percent of close-boiling isomers will have relatively little effect on the boiling point, refractive index, or density, but will normally affect the freezing point appreciably, (3) in a series of fractions obtained from a distillation at a high efficiency of a hydrocarbon containing close-boiling isomers as impurity, the fraction of highest purity will be beyond or ahead of the middle portion of the distillate as frequently as in the middle

Table 5.3 Purification of Paraffins and Mono-olefins by Fractional Distillation

No.	Compound	Initial Purity, %m	Kind of Distillation	Azeotrope-forming Substance	Final Purity, %m
14	2-Methylhexane	—	Azeotropic	Methanol	99.82 ± 0.07
15	3-Methylhexane	—	Azeotropic	Ethanol	99.80 ± 0.15
16	2,3-Dimethylpentane	—	Azeotropic	Ethanol	99.80 ± 0.15
17	2,4-Dimethylpentane	99.35	Azeotropic	Ethanol	99.88 ± 0.05
19	Octane	98.78	Regular	—	99.95 ± 0.04
20	2,2,3-Trimethylpentane	91.4	Regular	—	99.68 ± 0.20
45	1-Pentene	95.0	Regular	—	99.49 ± 0.40
47	cis-2-Pentene	—	Azeotropic	Methanol	99.80 ± 0.15
48	trans-2-Pentene	99.90	Regular	—	99.93 ± 0.05

Table 5.4 Purification of Hydrocarbons by Fractional Distillation

No.	Compound	Kind of Distillation	Azeotrope-forming Substance	Final Purity, %m
2	Pentane	Azeotropic	Methanol	99.86
3	2-Methylbutane	Regular	—	99.46
5	Methylcyclopentane	Azeotropic	Methanol	99.87
6	Cyclohexane	Regular	—	99.997
7	Hexane	Azeotropic	Methanol	99.91
8	2-Methylpentane	Azeotropic	Methanol	99.89
9	3-Methylpentane	Azeotropic	Methanol	—
10	2,2-Dimethylbutane	Azeotropic	Methanol	99.94
11	2,3-Dimethylbutane	Azeotropic	Methanol	99.90
12	Methylcyclohexane	Azeotropic	Methanol	99.90
13	Heptane	Azeotropic	Methanol	99.93
19	Octane	Azeotropic	Ethanol	99.96
20	2,2,3-Trimethylpentane	Azeotropic	2-Methoxyethanol	99.3
21	2,2,4-Trimethylpentane	Azeotropic	Ethanol	99.88
22	Nonane	Azeotropic	2-Ethoxyethanol	>99.7
27	Decane	Azeotropic	2-Butoxyethanol	>99.7
30	Benzene	Regular	—	99.96
31	Toluene	Regular	—	99.90
32	o-Xylene	Azeotropic	2-Ethoxyethanol	99.90
33	m-Xylene	Azeotropic	2-Ethoxyethanol	99.71
34	p-Xylene	Azeotropic	2-Ethoxyethanol	99.92
35	Ethylbenzene	Regular	—	99.54
36	Isopropylbenzene	Azeotropic	2-Ethoxyethanol	99.96

portion, and (4) for producing material of highest purity, the blending of the fractions of a distillate can be done safely only on the basis of the freezing points of the selected fractions."

Additional purification data from API Project 6 [1608] are summarized in Table 5.4. For further information, see Glasgow, Murphy, and coworkers [1794] and Streiff, Murphy, and coworkers [4523].

Mair and coworkers [3120] have described a method of purification and sealing in vacuum of standard samples of hydrocarbons. The technique and apparatus described generally are useful for storing pure solvents.

The freezing point of a pure substance can be calculated from the freezing point of the impure substance by the thermometric freezing point method developed by Streiff and Rossini [4525], Taylor and Rossini [4620], and Glasgow, Streiff, and Rossini [1795]. The purity must be above 95%m for the best accuracy, and the impurity must be a similar substance or substances. The purity of the material is determined at the same time. This method has been used as the criterion of purity for all hydrocarbons isolated and purified by the API-NBS group.

Fenske, Quiggle, and Tongberg [1558b] isolated from Pennsylvania straight-run gasoline the normal paraffins as high as tridecane in a pure state. The gasoline was distilled in a 27-ft column, 3 in. in diameter [1558a]. The cuts containing the normal paraffins were treated with chlorosulfonic acid according to the method of Shepard and Henne [4270]. The remaining hydrocarbon was washed, dried, and fractionally distilled. Tongberg and Fenske [4806] isolated cyclohexane from the hexane fraction of gasoline, after removing benzene, by fractional distillation in a 52-ft, $\frac{3}{4}$-in.-diameter column [1558a].

Vondráček and Dostál [4989a] found that the high solubility of aromatic hydrocarbons in phenol was useful in the purification of paraffins; pure hexane could be fractionally distilled from a petroleum fraction after it had been extracted with phenol. 2-Methylpentane, methylcyclohexane, heptane 3-methylpentane, and cyclohexane were also obtained in a high state of purity.

Bruun, Hicks-Bruun, and Faulconer [820] isolated 2-methylpentane, 3-methylpentane, and 2,3-dimethylbutane from the 55–65° commercial cut from natural gas of the Clendener Gas Field of West Virginia by fractional distillation in a Bruun-type column [818a] constructed by Brunn and Faulconer [819a].

Nonoxygenated extractive solvents have been prepared *carbonyl-free* by a simple process that is adaptable to laboratory or production quantities by Hornstein and Crowe [2252]. A uniform blend was made of 60 ml of concentrated sulfuric acid and 100 g of Celite 545 and packed into a chromatographic tube as follows: some of the solvent was poured into the tube and approximately 10 cm of granular Reagent grade sodium sulfate added. Then

160 g of the Celite mixture was added in portions, tamping each portion when added and the column topped with 7–8 cm of crystalline sodium sulfate. A flow rate of 3–5 ml per min was found satisfactory. The results for one pass of six solvents, hexane, Skelly F, benzene, dichloromethane, chloroform, and carbon tetrachloride were reported. The carbonyl content of the un-purified solvent varied from very small to 280 micromoles per liter; that of the purified solvent from 0.000 to approximately 0.45 micromoles per liter. Celite impregnated with 2,4-dinitrophenylhydrazine will also remove carbonyl compounds [4201].

SATURATED ALIPHATIC HYDROCARBONS

General

METHODS OF PREPARATION

Howard and coworkers [1564] prepared 65 paraffinic hydrocarbons in pure form, many by chemical means. The procedures included several techniques for chemical preparation in large laboratory quantities of intermediates and final compounds. Several paraffins were prepared by hydrogenating alkenes. The alkenes were prepared by dehydrating the alcohols made by the Grignard synthesis.

Aschan [303a] observed that chlorosulfonic acid reacted rather rapidly with branched-chain hydrocarbons. Young [5299a] reported that the acid reacted much more readily with branched-chain than with straight chain hydrocarbons. The acid has long been known to react readily with unsaturated hydrocarbons. Shepard, Henne, and Midgley [4271] isolated the normal saturated aliphatic hydrocarbons, pentane through decane, by removing the unsaturated and branched-chain components by treating an American gasoline fraction with chlorosulfonic acid until there was no change in density. The normal paraffins were separated from the naphthenes by fractional distillation. The distillation and freezing curves were used as the criteria of purity.

Isomeric hexanes were prepared by Cramer and Mulligan [1128] by the general alcohol-olefin-paraffin method. For example, hexane was prepared by passing 3-hexanol over an aluminum oxide catalyst at 350° and hydrogenating the olefin in the presence of platinum black, according to the method of Adams and Shriner [59].

A general method for the preparation of branched-chain paraffinic hydro-carbons having one carbon atom less than the parent compound is given under 2,2-dimethylpentane [1956]. Another general method for preparing branched-chain compounds is described in [830].

Pines [3769a] has reported a general method for alkylation, isomerization, autodestructive alkylation, and dealkylation. Some examples cited are: methylcyclohexane to dimethylcyclopentane or ethylcyclopentane; methyl-cyclopentane to cyclohexane; pentane to methylpropane, 2-methylbutane and some unsaturates; benzene and 2-chloro-2-methylbutane to *tert*-butyl-benzene, *p-tert*-dibutylbenzene, and other *tert*-butylbenzenes. Van Pelt and Wibaut [4915] prepared esters by Spassow's method [4453]: The esters were decomposed by passing them through a tube filled with glass wool at 500–520°. Alkenes were formed that could be hydrogenated to paraffins.

Fischer and Klemm [1579] prepared aliphatic hydrocarbons by the Grignard synthesis and by decomposing nitriles with sodium.

The Wurtz synthesis has been used to prepare several compounds; see, for example, hexane, octane, and decane.

Cyclopentane, cyclohexane, and methylcyclohexane appear to be gaining status as standards in Raman spectroscopy. Aleksanyan and Sterin [47] determined the spectra of these compounds at 25–35° under precise conditions.

Hesse and Schildknecht [2155] prepared hydrocarbons for ultraviolet spectrographic study by treating them with sulfuric acid and filtering through a suitable grade of alumina. Some technical grade substances, for example, cyclohexane can be purified by successive passage through alumina or silica gel.

DRYING AND CRITERIA OF PURITY

All of the drying agents listed in Table 4.1 have been used to dry hydro-carbons. Phosphorus pentoxide, sodium, molecular sieves, and metallic hydrides are all very efficient. Silica gel has been used by Glasgow and co-workers [1794] to dry aliphatic hydrocarbons. The freezing point of the dried materials indicates that the amount of water remaining was less than 0.005%m. For the usual laboratory work, calcium chloride, sodium hydroxide, and most of the other common drying agents can be used.

The most accurate criterion of purity of hydrocarbons is the adiabatic calorimetric method. The thermometric freezing point method is somewhat more rapid and has an accuracy for some systems to ±0.002%m. Boiling point, refractive index, and density are inadequate when isomeric impurities are present, but these impurities usually have an appreciable effect on the freezing point.

The boiling point, refractive index, and density, in combination, are satisfactory for detecting impurities in the 0.01%w and higher range, particularly for nonhydrocarbons. The ebullioscopic technique of Swietoslawski [4577] is a rapid and convenient method with good sensitivity.

SAFETY

"Pharmacologically, the hydrocarbons above ethane can be grouped with the general anesthetics . . . . The vapors of these hydrocarbons are mildly irritating to the mucous membranes, the irritation increasing in intensity from pentane to octane. The liquid paraffin hydrocarbons are fat solvents and primary skin irritants." The *threshold limit value* is available only for the first eight members of the normal, or straight chain, homologous series. The value for pentane is 1000 ppm; for hexane, heptane, and octane it is 500 ppm [3681].

Specific Solvents

1. CYCLOPENTANE

Vogel [4965] reduced pure cyclopentene in ethanol with hydrogen in the presence of Adams' catalyst [1778, p. 430] and purified the product; bp 48.4–48.6 at 763 torr. For preparation from dicyclopentadiene, see Jacobs and Parks [2370].

SAFETY

The toxicity resembles that of pentane [3681]. The *lower flammable limit* in air is about 1.4%v [224].

2. PENTANE

Directions for the preparation of pentane from 2-bromopentane are given in Marvel [3182, p. 84] and Nollar [616].

Pitzer [3776] purified a commercial product for *thermodynamic study* by fractionally freezing about 40% on a copper coil through which cold air was circulating. It was then washed with concentrated sulfuric acid and fractionally distilled. The middle 60% was used.

SAFETY

Fairhall [4509] concluded that narcosis and irritation are effects of inhaling pentane; see also Stoughton and Lamson [4514].

The *threshold limit value* is 1000 ppm or 2950 mg/m^3 [3681].

The *flammable limits* in air are 1.42 and 7.80%v. The vapor is 2.49 times as heavy as air [3681]. The *minimum ignition temperature* in air is 287° [3680].

3. 2-METHYLBUTANE

See Table 5.4.

Purified 2-methylbutane was found by Potts [3817] to have remarkable transmission in the *far ultraviolet*. He studied the transmission of four promising solvents as received and treated by the following purification methods:

1. Vigorous stirring with concentrated sulfuric acid for four hours; washing twice with water and drying with anhydrous calcium sulfate. Longer or repeated treatment with sulfuric acid was ineffective.

2. Passing through a water-jacketed, 1-in.-diameter column of Davidson 200-mesh silica gel 18-in. long. The method was most effective when the hydrocarbon and the adsorbant were absolutely dry. The silica gel was dried in place at 350° for 12 hr before use. The hydrocarbon was refluxed with sodium for 1 hr and fractionally distilled from sodium through a 40 theoretical plate Podbielniak column directly into the silica gel column.

3. Same as No. 2 except the one-hour reflux period with sodium was omitted.

Numbers 2 and 3 were about equally effective. 2-Methylbutane has a useful transmission to 1790 Å in a 1-cm cell and to 1720 Å in a 0.13-cm cell.

SAFETY

The *flammable limits* in air are 1.2 and 8.2%v [4501]. The *minimum ignition temperature* in air is 420° [3680].

4. 2,2-DIMETHYLPROPANE

Whitmore and Fleming [5157] prepared 2,2-dimethylpropane by reacting methylmagnesium chloride and 2-chloro-2-methylpropane in toluene at 45–50°; yield 42–50%. The crude product was freed from olefins, fractionally distilled, scrubbed with 85% sulfuric acid and then 25% potassium hydroxide, dried with phosphorus pentoxide, vaporized and condensed, and finally fractionally distilled. The freezing point indicated a high order of purity.

The compound has also been prepared in larger yields from zinc dimethyl and 2-chloro-2-methylpropane in toluene at 5° [2269].

Haensel and Ipatieff [1956] developed a method for the preparation of hydrocarbons having one less carbon atom than the parent compound. Branched-chain hydrocarbons were reacted with hydrogen in the presence of a nickel or cobalt catalyst [2352] to yield hydrocarbons of one less carbon atom. The carbon attached to the secondary carbon is removed more rapidly than those attached to a tertiary or quaternary carbon. Conditions are given for the production of 2,2-dimethylpropane 2,3-dimethylbutane, and 2,3-dimethylpentane.

Buck and coworkers [830] have described a method for preparing any branched-chain hydrocarbon including those containing a quaternary carbon atom by starting with the appropriate ketone or iso-ester.

Heat capacity measurements were made on a high purity 2,2-dimethyl-propane purified by Aston and Messerly [310]. A partially purified sample from Whitmore and Fleming [5157] was fractionally sublimed, twice distilled in the absence of air in an all-glass system, and the middle fraction collected in a glass bulb. The material was distilled from the bulb for use, cooled to liquid air temperature, and the air pumped out to less than 10^{-5} torr, melted, cooled, and repumped.

SAFETY

The *lower flammable limit* in air is 1.4%v [5314].

5. METHYLCYCLOPENTANE

See Table 5.4.

Tooke [4809] separated a mixture of hexane and methylcyclopentane. The procedure is described under hexane.

6. CYCLOHEXANE

Cyclohexane is produced in large quantities by the hydrogenation of synthetic benzene.

Palfray [3631] hydrogenated commercial, thiophene-free benzene, raising the temperature every 15 min and obtained 97.5% yield of a very pure cyclohexane. The hydrogen adsorption is rapid at 155°. The pressure for benzene was not given but toluene was hydrogenated at 115 kg.

Seyer and coworkers [4249] also prepared cyclohexane by hydrogenating benzene. The hydrogenated product was fractionally distilled in a Penn-State-type column 4-meters long filled with No. 18 jack chain. The distillate of 3 liters was taken off in 30 fractions at a rate of 5 ml/hr at a reflux ratio of 35:1. The density, refractive index, and freezing point were in good agreement with API values. See also [358, 4965, 2579].

Cyclohexane as a *solvent for spectroscopy in the ultraviolet* was prepared by Ashmore [305] by passing the liquid through a 1-meter long, 10-mm-inside-diameter column packed with silica. This appeared to be an application of the extensive study of Mair and Forziati [3116] on the analytical determination of hydrocarbons by adsorption on silica gel.

De Pauw and Limido [1286] prepared cyclohexane containing less than 0.01% benzene for *spectrographic use* by chlorinating in the dark at room temperature catalyzed by anhydrous iron (III) chloride. It was washed with sodium carbonate and distilled from the benzene-iron salt complex.

Samsanova and coworkers [4108] purified cyclohexane as a solvent for *ultraviolet and infrared spectroscopy* by passing it through three-11-mm-diameter, 50-ml burets filled with 40 g of silica gel. This removed benzene, paraffinic hydrocarbons, and carbonyl compounds. The treated solvent was fractionally distilled in a 25-plate column and the first 15% of the distillate discarded.

Zaugg and Schaefer [5326] used four purified solvents to study the *effect on the ultraviolet spectra* of salts of phenols and enols. Spectrographic grade cyclohexane was stored over lithium aluminum hydride and twice distilled from the hydride in an atmosphere of nitrogen just before use.

Timmermans and Martin [4783] treated cyclohexane containing benzene with nitrating acid. Cyclohexane is practically insoluble, whereas the nitrobenzene produced is appreciably soluble. After this treatment, any remaining benzene can be removed by the method of Mair and Forziati [3116].

Crowe and Smyth [1140] purified cyclohexane for *dielectric constant measurements*. It was washed several times in the cold with a mixture of concentrated nitric and sulfuric acids to nitrate the benzene that may have been present. After repeated washings with distilled water, it was fractionally distilled over sodium; bp 80.6°, mp 6.5°, n_D 20° 1.42650.

The use of *ultraviolet spectrophotometry for the comparison of the purity of* different samples of several solvents (cyclohexane among them), for the *detection of impurities* originally present or formed by deterioration and for the testing of purification procedures has been described by Maclean, Jencks, and Acree [3105].

Forster [1603] purified cyclohexane in a similar manner as he did benzene [1601] for *electrical conductance* studies.

Cyclohexane was purified by Brown [790] as a solvent for *dipole moment studies* by passing it through a 60-cm column of acid-washed alumina followed by careful fractional distillation from phosphorus pentoxide. It was stored over sodium wire; d 24° 0.77410.

Aston and Mastrangelo [309] designed the apparatus and purified several compounds by *fractional melting*. Cyclohexane of 85% was purified to 99.22%m with a 37.8% yield. Another pass increased the purity to 99.75%m. The results are of doubtful validity because about 1%m methylcyclopentane was present, which forms solid solutions with cyclohexane.

Staudhammer and Sayer [4479] purified cyclohexane as a solvent for *dielectric constant measurements* by crystallizing 13 times; fp 6.51°. Before use, the cyclohexane was allowed to stand over silica gel with the hope that any polar compounds would be removed.

A spectrographic grade cyclohexane was purified by Cheng [962] for *critical constant study* by fractional crystallization until the refractive indices of the melted crystals and the liquor were the same. It was dried by storing over sodium wire; bp 80.7–80.8°, n_D 20° 1.4261, t_c 280.2°.

Mascarelli and coworkers did considerable work with cyclohexane as a *cryoscopic and ebullioscopic solvent*. Mascarelli and Musatty's [3187] value is given for K_B. They concluded that it is not reliable as a solvent for the determination of molecular weight because it caused association with solutes containing the hydroxyl, carboxyl, nitro, and carbonyl groups [3189].

Some molecular weight values are given in [3188]; see also [3190]. References to several more articles relating to this subject may be found in [957a].

The *nature of the hydrogen bonded ion-pair* was studied by Yerger [5290] by using cyclohexane that was stirred with sulfuric acid, washed, dried, and distilled.

Brown and Ives [762] prepared cyclohexane for *dielectric constant measurements* by recrystallizing three times followed by fractional distillation; bp 80.8°, d 25° 0.77379.

A rapid test for cyclohexane is given in [1137].

Determination in air; range 0.01–100 mg/liter with $\pm 3\%$ error [2649, p. 63].

SAFETY

Cyclohexane is absorbed by inhalation [3681]. The *threshold limit value* is 300 ppm or 1050 mg/m³ [4509].

The *flammable limits* in air are 1.33–8.35%v [3681]. The *minimum ignition temperature* in air is 260° [3680].

7. HEXANE

Petroleum ether, bp 60–70°, is principally hexane. This product may be used for hexane for many solvent purposes. The aromatics that absorb in the ultraviolet may be removed by one of the procedures described below. Fuchs [1673] describes a procedure for purifying "benzin" or hexane contaminated with benzene.

Kelso and Felsing [2551] prepared hexane by treating iodopropane successively with quantities of sodium. The hexane was washed, dried, and repeatedly treated with fresh sodium until an excess of the metal remained unchanged. The solvent was allowed to stand over sodium for 24 hr and then distilled. The distillate was heated repeatedly with concentrated sulfuric acid, washed with aqueous sodium carbonate, and then dried and distilled.

Timmermans and Martin [4784] prepared hexane by the Wurtz reaction from 1-bromopropane and sodium. They removed unsaturated compounds from the crude product by treating with nitrating acid. After being boiled with potassium carbonate until the halide was completely decomposed, the solvent was fractionated to a common density and a constant critical solution temperature in nitrobenzene. See also Vogel [4967].

Shepard, Henne, and Midgley [4271] treated crude hexane with chlorosulfonic acid for 20 days, fractionally distilled the mixture, and obtained a product distilling with a constant density to ± 0.00001 g/cm³.

Tongberg and Johnston [4807] prepared pure hexane from petroleum ether, bp 60–70°, by fractional distillation in a 6-ft column. Benzene was removed by treating with an equal volume of nitrating mixture (58%w concentrated sulfuric acid, 25%w concentrated nitric acid and 17%w water) and shaking for 8 hr. The hydrocarbon layer was washed with concentrated sulfuric acid, then with water, and dried. It was distilled over sodium and the distillate dried over sodium. See also Castille and Henri [933].

Tooke [4809] separated hexane and methylcyclopentane by azeotropic distillation with methanol in a 35-plate batch still at a pressure of 10–50 psi and a reflux ratio of 20:1.

The *heat capacity* of hexane was measured by Waddington and Douslin [5007] on hexane purified as follows: The aromatics were first removed by passing through silica gel. It was fractionally distilled through a 100-plate column at a reflux ratio of 50:1. The product had a purity of 99.77%m calculated from the freezing curve.

Castille and Henri [933] prepared hexane, suitable for *optical measurements*, from petroleum ether. See also Weigert [5108]. Ley and Hünecke [2941a] purified hexane for *optical purposes* by shaking hexane repeatedly with fuming sulfuric acid until the acid was only slightly colored.

Nikuradse [3543] purified hexane to obtain a product of low *specific conductance*. After being dried with phosphorus pentoxide and filtered, the suspended and electrolytic impurities still present were removed by the application of a high potential. See also Jaffé [2373].

Forster [1603] purified hexane in the same manner as he did benzene [1601] for *specific conductance studies*.

For purification by *zone melting*, see Dugacheva and Anikin [1391].

Hartmann and coworkers [2048] purified hexane for *dielectric constant studies;* see benzyl alcohol.

For use as a solvent for *dipole measurements*, Morgan and Lowry [3399] treated hexane by shaking it several times with concentrated sulfuric acid, then with a 0.1 N solution of potassium permanganate in 10% sulfuric acid and finally with a 0.1 N solution of permanganate in 10% sodium hydroxide. The hexane was washed with water, dried over sodium wire, and distilled. The fraction boiling between 65–70° was used.

For *optical measurements*, hexane must have a transparency down to 1900 Å, according to Castille and Henri [933]. The copper 1944 Å line must appear without sensible reduction in intensity through a 10 to 15-mm layer and must be detectable through a 40-mm layer. As a solvent for ordinary ultraviolet absorption measurements, hexane need only be optically transparent to about 2100 Å in a 10-cm cell. The procedure described by Maclean, Jencks, and Acree [3105] provides a means of determining the suitability of the hexane and of evaluating the effectiveness of the purification procedure.

The initial temperature of thermal decomposition of hexane is 230–235° [4865].

SAFETY

Hexane is three times as toxic to mice as pentane. The *threshold limit value* is 500 ppm or 1800 mg/m³ [3681, 3508].

The *flammable limits* in air are 1.18–7.43%v [3681]; see also [31]. The *minimum ignition temperature* in air is 260° [3680].

8. 2-METHYLPENTANE

Cramer and Mulligan [1128] prepared 2-methylpentane by heating 2-methyl-1-pentanol with iodine at 122°. The product was hydrogenated and purified.

SAFETY

The *flammable limits* in air are 1.18–6.95%v [3680]; see also [5314]. The *minimum ignition temperature* in air is 306° [3680].

9. 3-METHYLPENTANE

Howard and coworkers [2269] prepared 3-methyl-3-pentanol from ethyl-magnesium chloride and ethyl acetate. The alcohol was dehydrated by refluxing with 0.2% β-naphthalene sulfonic acid. The alkenes were hydrogenated to alkanes. The mixture was fractionally distilled, filtered through silica gel, and refractionated. The criteria of purity were the boiling point and the density. See also Cramer and Mulligan [1128].

10. 2,2-DIMETHYLBUTANE

Brooks and coworkers [759] dehydrated pinacolyl alcohol with alumina to 3,3-dimethyl-1-butene, 2,3-dimethyl-1-butene, and 2,3-dimethyl-2-butene. After separation by fractional distillation, the 3,3-dimethyl-1-butene was hydrogenated in the presence of Raney nickel to 2,2-dimethylbutane. See also Cramer and Mulligan [1128].

Waddington and Douslin [5007] measured the *heat capacity* of 2,2-dimethylbutane purified by fractional distillation through a 77-plate column at a reflux ratio of 60:1; the purity was found to be 99.70% from the freezing curve.

SAFETY

The *lower flammable limit* in air is 1.2%v [5314]. The *minimum ignition temperature* in air is 425° [3680].

11. 2,3-DIMETHYLBUTANE

Kay [2533] prepared 2,3-dimethylbutane by catalytic alkylation of 2-methylpropane and ethylene. The product was purified by fractional distillation through a 15-plate column at a high reflux ratio. It was used to determine the *critical temperature* and *critical pressure*.

Kotlyarevskiĭ and coworkers [2727] found the optimum conditions for alkylation to be 1.5% aluminum chloride catalyst and a temperature of 50–55°; yield 42–47%. Knight and Kelly [2637] studied various catalysts and found iron (III) pyrophosphate tetrahydrate ($Fe_4(P_2O_7)_3 \cdot 4H_2O$) and boron trifluoride in a 1:1 mole ratio to be the most efficient. See also Haensel and Ipatieff [1956] and Brooks and coworkers [759].

SAFETY

The *lower flammable limit* in air is 1.2%v [5314]. The *minimum ignition temperature* in air is 420° [3680].

12. METHYLCYCLOHEXANE

Palfray [3631] hydrogenated toluene at 150° and 115 kg pressure for 5 hr and obtained practically pure methylcyclohexane. Vogel [4965] prepared methylcyclohexane from the 2-, 3-, and 4-methylcyclohexanols.

The ultraviolet absorption spectrum was used by Maclean, Jencks, and Acree [3105] to characterize the effectiveness of purification.

Forster [1603] purified methylcyclohexane in the same manner as he did benzene [1601] for *specific conductance studies.*

Methylcyclohexane was one of the four solvents studied by Potts for their transmission in the far ultraviolet; see 2-methylbutane [3817].

SAFETY

Methylcyclohexane is absorbed by inhalation. Rabbits exposed to 15,000 ppm (1.5%) of methylcyclohexane in air died in 70 min. The symptoms were conjunctival congestion, salivation, labored breathing, narcosis, and convulsions [4825]. All rabbits died after exposure to 10,000 ppm (1.0%) for two weeks for 6 hr per day and five days a week. Four weeks' exposure to about 0.5% caused no deaths [3681].

The *threshold limit value* is 500 ppm or 2000 g/m³ [4509].

The *lower flammable limit* in air is 1.15%v. The *minimum ignition temperature* in air is 285° [3680].

13. HEPTANE

Heptane has been gaining recognition as a standard reference substance. It is relatively easy to obtain or prepare and to maintain in a high state of purity. It has many attributes to recommend it as a standard substance. Many of its physical and thermodynamic properties have been accurately determined.

Vogel [4967] prepared heptane by reducing 4-heptanone with zinc and hydrochloric acid. The hydrocarbon was steam distilled, washed successively with water, 10% sodium carbonate solution and water. After a preliminary drying with anhydrous magnesium sulfate, it was distilled from sodium. Toussaint [4816] obtained heptane by hydrogenating 2-ethyl-2-hexenal.

The dipole moment was measured by Smyth and Rogers [4423] on heptane obtained from the Ethyl Gasoline Corporation. No improvement was observed when the material was subjected to purification.

Heptane for *optical purposes* was purified by Herold and Wolf [2134]. Petroleum heptane was shaken twice with concentrated sulfuric acid (12 hr each time); for 12 hr more with sulfuric monohydrate; and then for 1 hr each with water, dilute potassium hydroxide, and again with water. After standing over potassium hydroxide for 24 hr, it was fractionally distilled.

Forster [1603] purified heptane in the same manner as he did benzene [1601] for *specific conductance studies*.

Heptane was one of the four solvents that Potts [3817] studied for the transmission in the far ultraviolet; see 2-methylbutane.

Aston and Mastrangelo [309] purified heptane by *fractional melting*. Four successive meltings raised the purity from 99.5 to 99.996%m in 70% yield.

The initial temperature of thermal decomposition of heptane is 210–215° [4865].

SAFETY

Slight dizziness develops in man after breathing air containing 1000 ppm (0.1%) of heptane for 6 min and for 4 min at 2000 ppm [3681].

The *threshold limit value* is 500 ppm or 2000 mg/m³. It is sufficiently low to protect against unpleasant sensory effects [4509].

The flammable limits in air are 1.10–6.70%v. The *minimum ignition temperature* in air is 223° [3680].

14. 2-METHYLHEXANE

Streiff and coworkers [4522] purified 2-methylhexane by azeotropic distillation with methanol at a reflux ratio of 290:1. Azeotropic distillation with ethanol in a 135-plate column at a 145:1 reflux ratio was also used. The purity was 99.82%m.

15. 3-METHYLHEXANE

Optically active 3-methylhexane was prepared by Gordon and Burwell [1841] from optically active 4-methyl-2-hexanol, which had been prepared from 2-methyl-2-bromobutane and acetaldehyde by the Grignard reaction; $[\alpha]_D$ 25° +9.12°, d 25° 0.6824 and n_D 25° 1.3889. Heller [2096] converted d(−) 2-methyl-1-butanol to the d(+) 2-methyl-1-bromobutane. The yield of the optically active hydrocarbon by the Grignard reagent was low. He also prepared the d(+) hydrocarbon by the malonic ester synthesis; bp 91.9°, n_D 20° 1.3886, d 23° 0.6855 and $[\alpha]_D$ 23° +8.02°.

Easton and Hargreaves [1433] prepared *optically active* 3-methylhexane by three methods starting with optically active 2-methyl-1-butanol. One method was a slight modification of that of Heller [2096]. Another method involved the use of ethyl lithium and 1-bromo-2-methylbutane. The third method involved the preparation of pentyl lithium and its reaction with dimethyl sulfate to yield 44.4% purified product with a rotation of $[\alpha]_D$ 19.9° +9.43°, n_D 19.9° 1.3892, d 19.9° 0.6898.

Lardicci and coworkers [2831] prepared a number of *optically active* paraffins and alcohols by the addition of ethylene to (D)(+)-2-methyl-1-butyl lithium. The lithium reagent was prepared from (L)(+)-1-chloro-2-methylbutane in a petroleum ether solution in 80–90% yield. The reaction product of the lithium alkyl and acetylene was hydrogenated and the optically active paraffin purified; bp 91–92°, at 760 torr, n_D 25° 1.3861, d 25° 0.6826 and $[\alpha]_D$ 25° +9.43°.

16. 2,3-DIMETHYLPENTANE

Streiff and coworkers [4522] purified 2,3-dimethylpentane by azeotropic distillation with ethanol in a 130-plate column at a reflux ratio of 145:1; 99.80%m from the freezing curve.

For preparation, see Haensel and Ipatieff [1956].

Lardicci and coworkers [2830] reacted (−)-(s)-3-methylpentanal with formaldehyde and dimethylammonium chloride at 80° and obtained 60% (+)-(s)-2-methylene-3-methylpentanal. This was reduced with lithium aluminum hydride to (+)-(s)-2-methylene-3-methylpentanol, which was transformed into (−)-(s)-2,3-dimethylpentane; $[\alpha]_D$ 25° −10.39°.

SAFETY

The *explosive limits* in air are 1.12–6.75%v. The *minimum ignition temperature* in air is 337° [3680].

17. 2,4-DIMETHYLPENTANE

Fawcett [1546] prepared 2,4-dimethylpentane by alkylating 2-methylbutane with propylene in the presence of aluminum chloride. The reaction product was distilled, filtered through silica gel to remove halogen-containing compounds, and then fractionally distilled; bp 80.7°, d 20° 0.6738, n_D 20° 1.3821.

18. ETHYLCYCLOHEXANE

Glasgow and coworkers [1794] purified ethylcyclohexane by azeotropic distillation with 2-ethoxyethanol. The purity determined from the freezing curve was 99.90%m. See also Forziati et al. [1608].

SAFETY

The *flammable limits* in air at elevated temperature are 0.95–6.60%v. The *minimum ignition temperature* in air is 262 [3680].

19. OCTANE

Doolittle and Peterson [1340] prepared octane for the determination of *physical properties*. Constant density octanol, obtained by fractional distillation, was converted to the bromo compound by saturation with anhydrous hydrogen bromide at 100° and the reaction mixture distilled. Constant density fractions were converted to octane by the Grignard reaction. The product was fractionally distilled twice, and the heart cuts of identical densities were used for the measurements.

Vogel [4967] prepared octane for *parachor studies* by two methods: (1) the Wurtz synthesis from sodium and 1-bromobutane, and (2) the Clemmensen reduction of 2-octanone. The boiling point, refractive index, and density were used as the criteria of purity. Several references to other methods of preparation are given.

According to Lewis and coworkers [2937], large quantities of octane can be obtained from 1-bromobutane by the Wurtz synthesis; 450 g yield.

Pomerantz and coworkers [3805] prepared a high purity 1-octene by fractional distillation of a commercial product and reduced it to octane by hydrogenation. The octane was passed through silica gel and fractionally distilled through a 2.5 × 127-cm glass-Heli-grid packed column. The hydrocarbon was 99.7%m and was used for *physical property studies*.

Purification by zone melting; see Dugacheva and Anikin [1391].

The initial temperature of thermal decomposition is 195–200° [4865].

SAFETY

The *threshold limit value* has been established at 500 ppm, or 2350 mg/m^3, based on analogy with heptane, pentane, and gasoline [4509].

The *flammable limits* in air are 0.96–4.66%v [3681]; see also [31, 5314]. The *minimum ignition temperature* in air is 220° [3680].

20. 2,2,3-TRIMETHYLPENTANE

Howard and coworkers [2269] prepared this compound in a 31% yield from isopropylmagnesium chloride and 2-chloro-2-methylbutane

21. 2,2,4-TRIMETHYLPENTANE

Samsanova and coworkers [4108] freed 2,2,4-trimethylpentane from isomeric compounds and a small amount of toluene by passing it through a 11-mm diameter, 50-ml buret packed with 40 g silica gel and found it *spectrographically pure* as a solvent for *ultraviolet* spectroscopy. It was found suitable for *infrared* use when passed through a second column.

A 99.6%m 2,2,4-trimethylpentane was purified by one pass of fractional melting by Aston and Mastrangelo [309] to a purity of 99.99%m in a 78.0%

yield. An improved apparatus [3204] permitted a purification from 94.4 to 99.998%m in six passes in 80% yield.

The initial temperature of thermal decomposition is 295–300° [4865].

SAFETY

The *flammable limits* in air at elevated temperature are 1.00–6.03%v [3680]; see also [5314]. The *minimum ignition temperature* in air is 434° [3680].

22. NONANE

Nonane was first synthesized by Krafft [2736] in 1882 by heating pelagonic (nonanoic) acid with hydroiodic acid and phosphorus in a sealed tube at 240°.

Bazhulin and coworkers [461] prepared nonane from 5-nonanol, ethyl formate, and butylmagnesium bromide by dehydration and hydrogenation over a platinum catalyst.

For *physical property measurements*, Doolittle and Peterson [1340] isolated 1-decanol, fp > 6.5°, by fractionally distilling du Pont "Antifoam" LF and hydrogenating to nonane; yield 85%. It was fractionally distilled twice and the final center cuts of identical density were used.

Mears [3243] and coworkers prepared 2.38 kg of high purity nonane in three steps: 5-nonanol was prepared by a modification of the method of Coleman and Craig [616, p. 179] by reacting methyl formate with butyl-magnesium chloride. The 5-nonanol was dehydrated with aluminum oxide catalyst at 285–325° to nonenes, which were hydrogenated with nickel-on-kieselguhr at 140–160° at an initial pressure of 1350 psi. The nonane was purified by fractional distillation in a Podbielniak Hypercal Heligrid packed column 2.5 × 250-cm. The freezing point was the criterion of purity.

The initial temperature of thermal decomposition is 185–190° [4865].

SAFETY

The *minimum ignition temperature* in air is 206° [3680].

24. DECAHYDRONAPHTHALENE (MIXED ISOMERS)

25. *cis*-DECAHYDRONAPHTHALENE

26. *trans*-DECAHYDRONAPHTHALENE

Commercial decahydronaphthalene is a mixture of the *cis*- and *trans*-isomers produced by the hydrogenation of naphthalene. The purity of the hydrogenated product depends on the purity of the naphthalene and the purification process of the manufacturer.

Several processes for the purification of the mixed isomers differ slightly from one another [2148, 4191, 5205, 5341]. The following method is based on those processes cited and the experience of the authors in purifying a similar material. Shake with several portions of about 7% sulfuric acid until no appreciable darkening is observed. Wash with water, dilute sodium hydroxide, and three portions of water; and then dry with Drierite. Fractionally distill under reduced pressure. The boiling points of the *cis*- and *trans*-isomers at a particular pressure may be calculated from the vapor pressure-temperature equation. The purity of the product will depend on the efficiency of the column used, the reflux ratio, and the closeness of the cut to the calculated boiling point of the isomers.

The washed material may be dried with calcium chloride and then allowed to stand over sodium wire for several days with occasional shaking, or the last traces of water may be removed from the distillate by passing it through a column of silica gel.

Decahydronaphthalene from different sources will show considerable variation in composition of the *cis*- and *trans*-isomers. Seyer and Walker [4248] found that the material decomposed when distilled at atmospheric pressure. The solvents from various sources were combined and fractionally distilled into five fractions. Fractions one and two were mixed and refractionated as were fractions four and five. The respective distillates were then fractionally crystallized to a constant melting point. Nine solutions of *cis*- and *trans*-isomers were prepared by mixing weighed amounts. It was found that density and refractive index were linear functions of composition. Other *physical property measurements* have been reported [4245, 4246].

For preparation by hydrogenating naphthalene, see Baker and Schuetz [358] and Palfray [3631].

Musser and Adkins [3475] studied the selective hydrogenation of naphthalene and biphenyl with Raney nickel and copper-chromium oxide catalyst, respectively, to 1,2,3,4-tetrahydronaphthalene, decahydronaphthalene, cyclohexylbenzene and bicyclohexyl. Raney nickel is the more active catalyst, requiring the lowest temperature and the shortest time of hydrogenation.

Commercial decalin was fractionally distilled by Allinger and Coke [60] through an all-glass Penn-State-type still with a 4-ft column packed with glass helices. The fractions distilling at 192.0–193.0° at 740 torr were refractionated in the same equipment and gave *cis*- and *trans*-isomers that did not show impurities by gas chromatography.

Hydrogenation of *cis*- or *trans*-decahydronaphthalene with Raney nickel catalyst at 250–300° and 200 atm was reported by Boelhouwer and coworkers [631] to give an equilibrium mixture containing 10% of the *cis*-isomer. At 275° and 15% catalyst, the *cis*-isomer was transformed into the equilibrium mixture in 4 hr and the *trans*-isomers in 1 hr.

Zelinsky and Turowa-Pollak [5331] found that aluminum chloride at room temperature converts the *cis*- into the more stable *trans*-form.

Fenske, Myers, and Quiggle [1557] separated the *cis-* and *trans-*isomers from decalin by fractional distillation through a 75-plate column at a reflux ratio of 40:1. *Physical properties* and *vapor-liquid equilibrium data* were determined on the center cut of each isomer. The boiling point and refractive index were used as the criteria of purity.

The isomeric decahydronaphthalenes were purified by Staudhammer and Sayer [4479] for *dielectric constant study.* The materials were fractionally distilled, followed by fractional crystallization. The *freezing point* of the *cis-*isomer after 12 successive crystallizations was −43.15°, that of the *trans-*isomer after 11 successive crystallizations was −30.56°.

Streiff and coworkers [4527] purified *cis-*decahydronaphthalene for *physical and thermodynamic property studies* by fractionally distilling in a 130 theoretical plate column at an approximate reflux ratio of 145:1; purity 99.93%m calculated from the freezing curve. The *trans-*isomer was distilled in a 200 theoretical plate column at a reflux ratio of 160:1; purity 99.97%m.

Seyer's [4243] study of *heat capacity* of *cis-*decahydronaphthalene definitely suggests a *transition point* at 50.1–50.5°. He did not report a similar irregularity for the *trans-*isomer. No anomalous behavior of the *trans-*isomer was detected from the temperature-viscosity relationship [4897] by Urazovskii and Chernyavskii.

SAFETY

Decahydronaphthalene vapor is irritating to the eyes, nose, and throat. Colored urine has been reported by workers exposed to a mixture of decalin and tetralin. Decalin and each isomer is absorbed through the skin. Dermatitis, but no systemic poisoning, has been reported by painters using decalin. Prolonged application to the skin caused the death of guinea pigs [3861].

The *threshold limit value* of 25 ppm, 140 mg/m³, has been suggested by Stokinger [4509].

The *minimum ignition temperature* in air is 262° [3680].

27. DECANE

Vogel [4967] prepared decane by the Wurtz synthesis from 1-bromopentane and sodium. The material was used for *parachor studies.*

Chu [979] purified "olefin-free"-grade decane and bis(2-chloroethyl) ether for *critical opalescence studies* by essentially the same method. The decane was dried over sodium. Both distilled liquids were then purified by preparative gas chromatography on an Aerograph Model A-700 Autoprep Vapor Fractometer. A 20-ft × ⅜-in. o.d. column packed with 30% SE-30 (General Electric methylsilicone rubber) on 42/60 chromosorb P, operated at 150° for decane (165° for the ether) and 40 psig was used. The helium carrier gas rate was about 200 ml per min.

Decane containing 1.40% impurity was given two passes of zone refining and attained a purity of 99.73% in an automatic zone melting apparatus capable of operating between 0 and −170°, designed by Kieffer and co-workers [2586].

Mears and coworkers [3243] prepared decane by hydrogenating 1-decene at 170° with nickel-on-kieselguhr catalyst at a hydrogen pressure of 500 psi. The decane was fractionally distilled in a 2.5 × 250-cm Podbielniak Hypercal Heligrid column. The freezing point was the criterion of purity.

Karabinos [2498] prepared decane by the Kolbe synthesis by the electrolysis of hexanoic acid. The reduction product was shown to be essentially free from branched-chain hydrocarbons.

SAFETY

The *flammable limits* in air are 0.72–5.0%v [31]; see also [5314]. The *minimum ignition temperature* in air is 208° [3680].

28. BICYCLOHEXYL

Ipatiew [2355] reported the preparation of bicyclohexyl in 1907 from diphenyl at 191 atm and 260° with a reduced nickel catalyst. In 1912, Sabatier and Murat [4084] reduced diphenyl in two stages at 180° and 160° with nickel and hydrogen.

Bicyclohexyl was prepared in 40% yield by Gardner and Borgstrom [1704] from cyclohexylmagnesium bromide and an equivalent amount of dry silver bromide in dry ethyl ether.

A commercially available bicyclohexyl was purified by Vogel [4974] for *physical property and chemical constitution studies* by shaking with one-half of its volume of concentrated sulfuric acid. The acid was separated and the hydrocarbon was washed repeatedly with water, dried with calcium chloride, heated 5 hr with sodium at 110°, filtered, and distilled.

Mears and coworkers [3244] prepared bicyclohexyl for *physical property study* to a purity of 99.96%m by catalytic hydrogenation of selected fractions of cyclohexylbenzene with nickel-on-kieselguhr catalyst at 180° at an initial hydrogen pressure of about 1850 psig. The hydrogenated material was filtered and percolated through silica gel. The portion of material boiling at 148.5–150° at 58 torr with a refractive index at 20° of 1.4796–1.4797 was collected and again percolated through the silica gel. See also Musser and Adkins [3475] under decahydronaphthalene.

Schrauth and Gorig [4187] present data to support the existence of three stereoisomers of bicyclohexyl.

The initial thermal decomposition temperature is >240° [1327].

29. DODECANE

Garach [1699] prepared saturated and unsaturated hydrocarbons to study their properties, and especially the *infrared spectra*, for optical analysis. The unsaturated compounds were prepared by reacting heptanal with the proper Grignard reagent, hydrolyzing and dehydrating the alcohol over kaolin just below 300°. The saturated compounds were made by hydrogenation over nickel at 170–180° and purifying with sulfuric acid.

Deanesly and Carleton [1236] prepared and purified dodecane for precise *physical property study*. Laurel alcohol (dodecanol) was purified by fractional distillation. It was converted to iododecane [3543a, p. 29] which was filtered, washed, and reduced to dodecane [3543a, p. 27]. The alkane was extracted with ethyl ether and fractionally distilled into 100 ml cuts. Those whose n_D 20° was 1.42156 were combined and crystallized five times without solvent. The freezing point rose only 0.05° after the fifth crystallization. It was washed with 98% sulfuric acid containing 3% silver sulfate. It was crystallized twice, and the temperature rose less than 0.002°. The freezing point was the principle criterion of purity.

Mears and coworkers [3244] prepared dodecane for *physical property studies* by dehydrating 1-dodecanol over alumina at 350–375°; 80% yield per pass. The mixed dodecenes were washed, dried, and fractionally distilled through a short helices-packed column at 210–215°. They were hydrogenated with nickel-on-kieselguhr catalyst at 140° with an initial pressure of 1800 psig. The dodecane was freed from nickel, percolated through silica gel and fractionally distilled; purity 99.92%m calculated from the freezing curve.

Vogel [4967] prepared dodecane by Wurtz synthesis for *physical property studies*. See also Timmermans and Martin [4784, under hexane].

SAFETY

The *flammable limits* in air at 25° are 0.61–4.7%v [31]; see also [5314]. The *minimum ignition temperature* in air is 204° [3680].

AROMATIC HYDROCARBONS

General

Benzene and several of its simple alkyl and alkene derivatives are now prepared synthetically from petroleum. The impurities in the synthetic compounds differ from those present in products from the destructive distillation of coal. This is an important consideration when the nature of the impurity is a factor.

Alkyl derivatives of benzene can be prepared from one mole each of the corresponding alcohol and benzene and 20–65 g of boron trifluoride, depending on the alcohol used. Normal, iso-, secondary, and tertiary alcohols react.

Simons and Archer [4328] prepared the corresponding alkyl benzene from the olefin and benzene with hydrogen fluoride as the condensing agent. Some of the compounds prepared were isopropylbenzene, *tert*-butylbenzene, and cyclohexylbenzene.

A method for preparing *alkyl derivatives of benzene* by catalyzing a mixture of benzene and an alkyl bromide with aluminum was reported by Turova-Pollack and Maslova [4865]. The mixture of benzene and the alkyl bromide is heated with aluminum shavings 1.5–2 hours until the reaction starts, followed by cooling to 5–10° until the reaction ceases. The reaction mixture was heated to 70–80° and let stand at room temperature for 2 hr. The reddish complex was broken with dilute acid, and the alkyl benzene was separated and purified. Examples are: 96% yield of ethylbenzene from 8 moles of benzene, 1 mole bromoethane and 0.01 gram-atom of aluminum; isopropyl benzene was obtained from the same ratio.

Forster [1601, 1602] purified saturated and unsaturated hydrocarbons and methyl-substituted benzenes for *specific conductance studies* from 99% commercially available material by fractional distillation in a nitrogen atmosphere. The fraction that distilled within ±0.1° of the reported boiling point was fractionally crystallized thrice and degassed twice. No impurities were detected using a gas chromatograph at the highest sensitivity.

Specific Solvents

30. BENZENE

Benzene, because of its ease of purification, is one of the secondary standards for physical measurements and is a common solvent for *dipole moment* determinations. It has been used extensively as a solvent in acid-base titrations in nonaqueous solvents [3974].

Benzene, sufficiently pure for most purposes, may be purchased in quantity. ACS *Reagent Chemicals* [76] and Rosin [4042] list specifications and tests for benzene for use as an analytical solvent or test medium; boiling range 79.5–81.0°, minimum freezing point 5.2°. Some sulfur and traces of other hydrocarbons remain. It is not sufficiently pure for a physical property standard or for the preparation of solutions for most accurate physical measurements.

A Research grade, 99.94%m, is available from Phillips Petroleum Co. Other high purity grades are available.

An anomalous behavior was deduced from a viscosity study; see *p*-dioxane [4897].

Anderson and Engelder [237] isolated and identified the nonbenzenoid hydrocarbons and toluene from a sample believed to be typical of the type called "nitration grade" from the destructive distillation of coal. The benzene met ACS *Reagent* grade specifications for boiling range and freezing point. The saturated nonbenzenoid hydrocarbons were obtained by fractional crystallization, fractional adsorption, and fractional distillation. It was found that the impurities were a complex mixture that is predominantly naphthenic.

The estimated impurities were: saturated nonbenzenoid hydrocarbon, 0.6%v; toluene, 0.004%v; unidentified, 0.1%v. The results indicated a correlation between the freezing point depression of 0.67° and a paraffin content of 1%. Another investigation showed that European nitration grade benzene from coke-plant production contained mainly naphthenes and methylcyclohexane.

Forziati and coworkers [1608] distilled benzene, after preliminary treatment, to a purity of 100.00%m. A National Bureau of Standards' Standard Samples of benzene were prepared by Mair and coworkers [3120] by a combination of crystallization, filtration through silica gel, and distillation. The crystallization was performed as follows: "A volume of about 200 ml of benzene was mixed with 50 ml of ethanol in a cylindrical brass container (5 cm in diameter and 20 cm long), which was placed in a cooling bath of ice and salt (temperature about −10°). With vigorous hand-stirring and scraping, a thick mush, or slurry, of hydrocarbon and alcohol was produced. This slurry was then transferred to the basket of a centrifuge having a jacket cooled to nearly −10° with a mixture of ice and salt. The centrifuge was operated for about 5 min, leaving in the basket about half of the benzene in crystalline form. The crystals were removed from the basket of the centrifuge and allowed to melt. The purified liquid benzene was then washed three times with distilled water and then filtered through silica gel to remove any alcohol and water remaining." The purity was determined from the freezing curve [1795].

Swietoslawski [4579] stated that benzene cannot be purified, after removal of thiophene and unsaturated hydrocarbons, by fractional distillation and crystallization alone. Crystallization from methanol or methanol-water produces 99.998% purity. The purity test was carried out in a dilatometric cryometer. The benzene was frozen in the apparatus and the decrease in volume due to partial solidification determined. The decrease in volume was then plotted against the equilibrium temperature. Once the curve had been established, the purity of a sample of benzene could be determined by obtaining only one point on the curve.

Stull [4536] purified thiophene-free benzene for the determination of the *thermodynamic properties*. The material was crystallized six times; one-fourth was discarded each time as unfrozen liquid. It was then dried over phosphorus pentoxide for two weeks and finally fractionally distilled through a 5-ft column packed with glass helices. The purity was above 99.95% determined from the freezing curve.

Washburn and Read [5085] purified thiophene-free benzene by allowing it to stand in contact with pure sulfuric acid, with frequent shaking, for four weeks. The treated benzene was then distilled from sulfuric acid into a distilling flask containing metallic calcium, and from there into the apparatus in which it was to be used. The freezing point was 5.43°.

The determination of sulfur compounds in crude and refined benzene from coke has been thoroughly studied by Claxton and Hoffert [997a]. It was shown that crude benzene contained about half of its sulfur content as carbon disulfide. While the remainder of the sulfur content is usually considered to be thiophene and its derivatives, other sulfur compounds such as mercaptans, sulfides, and disulfides are also present. The reader is referred to the original article for the recommended analytical procedure. See also French and Claxton [1637].

Wojciechowski [5237] purified ACS Reagent grade benzene by three different methods.

1. two liters were fractionally distilled through a Swietoslawski-type, vacuum-jacketed column. Five middle fractions of 200 ml each were collected.

2. a second portion was recrystallized and fractionally distilled.

3. a third portion was subjected to azeotropic distillation with ethanol and water.

The alcohol was removed from the distillate by washing with water and the benzene redistilled. The maximum deviation in the boiling point between samples purified by the three methods was 0.006°.

Bender and coworkers [509] purified benzene by two methods.

1. Research grade benzene was treated with concentrated sulfuric acid until thiophene-free. After repeated water washes, it was dried first with calcium chloride, then with sodium wire. It was recrystallized by the second method of Schwab and Wiehers [4199] after distillation from sodium wire.

2. Thiophene-free, Reagent grade benzene was dried with sodium wire and distilled at a reflux ratio of 20:1 through a 30-plate Oldershaw column. The constant boiling portion was retained.

Leonard and Sutton [2906] purified an analytical grade of benzene for use as a solvent for determining *dielectric constants*. The benzene was frozen three times, boiled under reflux over phosphorus pentoxide, and finally distilled in a stream of dry air. The purified material was stored under dry air.

Thiophene-free benzene was dried with sodium sulfate by Brown and Ives [762] for *dielectric constant studies*. It was fractionally crystallized three times, dried with sodium, and fractionally distilled in a 4-ft, vacuum-jacketed column packed with helices; bp 80.1°, d 25° 0.87365.

Meighan and Cole [3260] used reagent grade benzene in *dielectric studies* after refluxing with calcium hydride for 48 hr and distilling. Subsequent handling was in a closed system.

Pearce and Berhenke [3691] made benzene thiophene free and purified it in the usual way for *dielectric moment measurements*. It was finally dried over sodium wire and fractionally crystallized; bp 760 torr 80.1°, fp 5.4°.

A spectrographic-quality, reagent grade benzene was used as such or further purified by Forster [1601] for study of *specific conductance* in liquid hydrocarbons. It was further purified by fractional distillation in an atmosphere of nitrogen. The fraction distilling at 80.00°–80.20° was collected and fractionally crystallized followed by two degassing cycles. The highest sensitivity available for gas chromatography did not show any impurities. Thallium shavings removed traces of oxygen without impairing the purity of the benzene. The material was stored under vacuum in the freezing compartment of a refrigerator to prevent decomposition.

For *specific conductance measurements*, Walden [5037] dried benzene with sodium and distilled in an all-glass apparatus closed with a calcium chloride-soda lime tube. The middle fraction was used for measurements.

Bruckenstein and Saito [798] purified ACS Reagent grade benzene for *acid-base equilibria studies* by storing the solvent over 4A molecular sieves ($\frac{1}{16}$ in. pellets), filtering and finally fractionally distilling. The water by Karl Fischer method was 5×10^{-3} M.

Maruyama [3180] describes a method for purifying benzene for use in *polarography*.

Röck [4010] refined benzene containing about 0.2% impurity by zone melting and obtained a fraction containing <0.008%m impurity.

Thiophene in benzene was reduced from 0.1%–0.01% in 15 *zone refining* passes [1273].

Dickinson and Eaborn [1306] studied *progressive freezing* purification of benzene and concluded that it was as effective as the more complex zone melting technique.

Molinari [3373] converted benzene feed stock, mp 4.7°, containing 3.8% paraffins and 0.3% total sulfur, into 20% benzene, mp 5.53° and 55% nitration grade, mp 5.2°, by using the Probad process.

Graul and Karabinos [1866] developed a rapid method for removing *thiophene from benzene*. They refluxed 100 ml of benzene containing 1% thiophene for 15 min with 10 g of Raney nickel prepared according to the method of Pavlic and Adkins [3686]. The isatin-sulfuric acid test was negative, indicating the removal of thiophene. When 5 g of Raney nickel was used, not all of the thiophene was removed.

Cherkasova and Gorin [967] reported thiophene can be removed from benzene by condensing with formaldehyde in the presence of sulfuricacid.

Mathieu [3218] prepared benzene by the thermal decomposition of calcium benzoate and purified it by fusion and crystallization for direct measurement of *pressure coefficient of temperature, dt/dp, and change of volume on melting* in a Swietoslawski-type dilatometric cryometer; purity 99.997%m, tp 5.690°.

Hartmann and coworkers [2048] purified benzene; see benzyl alcohol.

Total sulfur in benzene; combustion, oxidation to sulfate and determination by conductometric titration is given in [1636].

Benzene sufficiently pure for all except measurements of the highest accuracy is most conveniently prepared by shaking a *good commercial grade* successively with concentrated sulfuric acid until free from thiophene, then with water, dilute sodium hydroxide, and finally with two portions of water. The benzene may be dried by first shaking with anhydrous calcium chloride and removing the last traces of water by distilling over phosphorus pentoxide, shaking with molecular sieves or passing through a column of silica gel. Kraus and Vingee [2745] used a final washing of only one portion of water, distilling from phosphorus pentoxide when preparing benzene for use as a *cryoscopic solvent*. The purified benzene was stored over sodium-lead alloy and distilled just prior to use.

CRITERIA OF PURITY

The *freezing point* is, perhaps, the most easily determined property and the most sensitive to impurities. The amount of impurity may be calculated from the freezing curve [1795].

The *dilatometric cryometric method* of Swietoslawski [4579] is sensitive enough for use as a criterion of high purity; see Mathieu [3218] who characterized benzene to 99.997 %m.

The *boiling point*, *density*, and *refractive index* offer a means of estimating the purity.

The following specifications and tests, which are easy to make, characterize benzene sufficiently well for general solvent and laboratory use:

Freezing point: not less than 5.2°. Fill a double-walled test tube half full of benzene, immerse in an ice-water slurry, and determine the freezing point. See Skau and coauthors [5110, p. 287 ff.] or [76, p. 98].

Nonvolatile material: transfer 100 ml of benzene to a tared evaporating dish, evaporate on a steam bath, dry at 110° for 30 min and weigh. Maximum increase in weight must not exceed 0.001 g.

Substances darkened by sulfuric acid: shake 25 ml of benzene with 15 ml of ACS Reagent grade sulfuric acid about 20 sec and allow the phases to separate. There should be no darkening in either phase [76, 4042].

Thiophene: add about 3 mg of isatin mixture from the preceding test and shake well. A blue-green color should not develop in the acid phase within 1 hr.

Water: the water content is most conveniently determined by the Karl Fischer method.

Gas chromatography may be used to characterize any grade of benzene.

SAFETY

Benzene may be regarded as a cumulative poison due to the slow build-up in the body tissues and fluids because of its low solubility in the circulating blood. After exposure the depletion is equally slow. Benzene poisoning results almost exclusively from breathing the vapors. Small quantities of benzene may be absorbed through the skin, but Gerarde [3681] does not believe that systemic poisoning can arise from immersing the hands in benzene. However, skin defatting, erythema, scaling, and even secondary infection may occur.

The oral LD_{50} for rats has been reported as 4080 mg/kg [2392]; see also [4615].

The danger inherent in inhaling benzene has been controversial [756, 1377, 1466]. The *threshold limit value* of 100 ppm was established in 1946. It has been successively reduced to the present 25 ppm, 80 mg/m³ [4509].

Many methods have been proposed for the determination of benzene in air; see Patty [3681].

The *flammable limits* in air are given by Huff [2298] as 1.4–7.1 %v; see also [646]. The *minimum ignition temperature* in air is 562° [3680].

31. TOLUENE

Toluene has a limited use as a secondary standard for physical and thermal measurements. It may be purchased in quantities sufficiently pure for most purposes and in limited quantity with a purity of 99.97%m. The specifications for Analytical grade are given in *Reagent Chemicals* [76] and by Rosin [4042].

Toluene, like benzene, is available from two different basic manufacturing sources, the coking of coal and synthesis from petroleum. The toluene from each source has its own characteristic impurities. There is little published information concerning the impurities in the synthetic material. The impurities in coke toluene are of the same type as in benzene [237], and they are more difficult to remove completely.

Toluene may be purified by a combination of chemical treatments, fractional distillation, and fractional crystallization, but the freezing point of toluene, −95°, is too low for convenient crystallization; see Ramsay and Steele [3894]. The sulfur impurities may be removed by methods similar to those described for benzene.

Vogel [4965] prepared toluene for *physical property measurements* as follows: 100 g of purified and redistilled benzaldehyde, bp 179° at 757 torr, was reduced with 200 g of amalgamated zinc and hydrochloric acid; yield 43 g toluene boiling at 109–110° at 748 torr. It was washed twice with 10% of its volume of concentrated sulfuric acid, then with sodium carbonate solution, and finally with water. After being dried over anhydrous magnesium chloride, it was twice distilled from sodium and passed over constantly at 110° at 763 torr. The middle fraction was used.

Ramsay and Steele [3894] (see also Mitsukuri and Nakatsuchi [3346]) prepared pure toluene from *p*-toluidine by diazotization; Perkin [3714] (see also Orton and Jones [3597]) purified the potassium salt of *p*-toluenesulfonic acid by recrystallization and obtained pure toluene by hydrolysis with steam in a sulfuric acid solution.

Hardy [2017] studied the synthesis of toluene from benzene and chloromethane by the Friedel-Crafts reaction. The best yields were obtained at moderate pressures of 4–10 atmospheres and temperatures of 4–15°. The best benzene conversion to toluene under these conditions was 63%, corresponding to a chloromethane conversion of 43%.

Schwable [4200] recommended the following process for the removal of thiophene and its homologues: Nitrous oxide and nitric oxide are passed into the toluene until the concentration is 0.2–0.5%; after standing for 2–3 hr, the mixture is shaken with 2% of its volume of concentrated sulfuric acid for 5 min, allowed to stand for 15 min, and separated. The operation is repeated two or three times until the acid is only slightly yellow in color; the toluene is then shaken with water and dilute sodium hydroxide until fresh portions no longer are colored.

According to Timmermans and Martin [4783], to obtain toluene pure it is necessary to convert it into a crystalline substance from which the hydrocarbon can be recovered after purification, or to prepare it in the first place from a substance purified by crystallization.

Mathews [3208] purified commercial toluene for the determination of *heat of vaporization* by successive shaking with sulfuric acid, sodium hydroxide, and mercury, followed by drying over phosphorus pentoxide and then fractionally distilling. A similar process was used by Williams and Krchma [5197] for purifying toluene for the determination of the *dielectric constant*. See also Richards and Wallace [3972].

Forster [1602] purified toluene in the same manner as he did benzene for *specific conductance* studies. For *conductance* measurements, Walden [5037] purified a good grade of Kahlbaum's toluene by the same method that he used for benzene.

For *thermoregulators*, Beal and Souther [462] washed toluene first with sulfuric acid, then with water, and then boiled it under reflux with 1% sodium amalgam. After decanting, the solvent was washed with water and distilled.

For an additional method of purification, see Duncan and coworkers [1396, under mesitylene].

Kuss and coworkers [2795] prepared "very pure" toluene by reducing benzyl alcohol at 200–300° with a chromium-copper catalyst.

CRITERIA OF PURITY

The freezing curve method has been successfully used [1795, 1608, 3120].

Timmermans and Martin [4783] found that the boiling point should not be used as a criterion of purity. Fractions having the same boiling point were found to have different densities.

Lumsden [3027] recommended the following criteria: (1) density; (2) nitration followed by the determination of the amount of nitration in a nitrometer; (3) miscibility with acetic acid and the determination of the critical solution temperature. See also Orton and Jones [3597].

Rosin [4042] lists the following specifications for Reagent grade toluene:

Distillation range: 100-ml sample to distill between 110° and 111°.

Nonvolatiles: 0.001 % maximum.

Acidity: none.

Alkalinity: none.

Substances darkened by sulfuric acid: shake 15 ml of toluene with 15 ml of sulfuric acid for 15 min. There should be no darkening of the toluene phase, the acid phase should be no darker than a mixture of 2 parts of water and 1 part of a solution containing 5 g of cobalt chloride, 40 g iron [III] chloride and 20 ml of hydrochloric acid per liter of solution.

Sulfur compounds (as S): 0.003 % maximum.

Water: no cloudiness should be produced at 0°.

There are no simple tests for homologues, naphthenic, and paraffinic hydrocarbons except gas chromatography.

The water is most conveniently determined by the Karl Fischer method [3343].

SAFETY

Gerarde [3681] states that toluene is a more powerful narcotic and is more acutely toxic than benzene. Controlled exposure of humans to concentrations of 50 to 800 ppm indicates that 200 ppm for a period of 8 hr produces mild fatigue, weakness, confusion, and paresthesia of the skin. Toluene, like benzene, is most dangerous by inhalation. It is irritating to the skin, and contact should be avoided when possible.

The *threshold limit value* has been set at 200 ppm, 750 mg/m³ [4509].

The *flammable limits* in air at elevated temperature are 1.17–7.10%v. The *minimum ignition temperature* in air is 536° [3680].

32. *o*-XYLENE

33. *m*-XYLENE

34. *p*-XYLENE

Less than 2% of the xylene produced today comes from coal tars. The remainder is obtained by refining petroleum naphtha. The aromatics are extracted and the benzene, toluene, and C₈-aromatics separated by fractional distillation. The *o*-xylene and ethylbenzene are separated by distillation, and *p*-xylene is fractionally crystallized from the residue. The *m*-xylene is separated from part of the residue by distillation and/or clathration. The mixed xylenes usually contain about 40% *meta-* and 20% each of the *ortho-* and *para-*isomers and ethylbenzene. It is estimated that the annual xylene production in 1970 will be *para-* 1,500,000,000 lb, *ortho-* 425,000 lb and *meta-* 80,000,000 lb.

The composition of coal tar xylene varies in about the same range as the petroleum product. The principal impurities are ethylbenzene and paraffins [4577, p. 141].

Analytical Reagent grade xylene, boiling range of 137 to 140° [76], is available from most supply houses. Research grade *o-*, *m-*, and *p-*xylenes are available in 99.9+ %m purity. Other grades of the isomers are available in 99+ %m purity. A 95%w minimum *p-*xylene is available in quantity.

The xylenes were purified by Forster [1602] for *specific conductance studies* in the same manner as was benzene [1601].

A method for separating the xylenes has been developed by Arnold [287]. The mixture of *m-* and *p-*xylenes is diluted with a relatively volatile, inert solvent having a freezing point below −58.5°, the eutectic point of the xylenes, and cooled until a large part of the *p-*xylene has crystallized. The diluent is removed by distillation and the *m-*xylene crystallized from the remaining liquid. Suitable solvents are methanol, ethanol, 2-propanol, acetone, butanol, toluene, the pentanes, and the pentenes.

Aromatic isomers were separated by Radzitzky and Hanotier [3880] by clathration with Werner complexes of the form $Ni(SCN)_2(RC_6H_4-CHR'NH_2)_4$.

Timmermans and Hennaut-Roland [4778] sulfonated mixed xylenes and separated the alkali salts of the sulfonic acids by fractional crystallization. The purification was attended with considerable loss.

An extraction method was developed by McCaulay and coworkers [3055] for separating *m-*xylene from commercial xylene; they used a mixture of hydrogen fluoride and boron trifluoride. The reactions involved are rapid and sufficiently selective, so that 95% *m-*xylene can be separated from mixed xylenes. An inert hydrocarbon diluent further improves the selectivity of the extraction. The authors state that, when their method is used in conjunction with fractional distillation or crystallization, each of the xylene isomers can be separated in a high state of purity.

Clarke and Taylor [993] sulfonated 4400 g of technical *o-*xylene, 95% of which boiled between 143 and 144°, by stirring for 4 hr with 2.5 liters of concentrated sulfuric acid at 95°; after cooling the mixture the product was separated from unsulfonated material, diluted with 3 liters of water, and neutralized with 40% sodium hydroxide. The precipitate that formed on cooling was separated and crystallized from half its weight of water. A further quantity was obtained by concentrating the mother liquor to a third of its volume and cooling; this was recrystallized from half its weight of water and purified with the first fraction. A further crystallization of the whole material produced a homogeneous product, and the mother liquor was worked up as before. The purified sodium salt was dissolved in the necessary quantity of cold water, mixed with the same volume of concentrated sulfuric acid and steam distilled after heating to 110°. The yield of *o*-xylene was 1980 g or 43%. The boiling point was 144–145°.

Pure *o*-xylene was prepared by Skita and Schneck [4363] from *o*-toluidine (purified by means of the acetyl compound) by conversion into *o*-bromo-toluene, followed by the Grignard reaction with dimethyl sulfate.

Moldavskiĭ and Turetskaya [3370] prepared *o*-xylene by hydrogenating phthalic anhydride or phthalide at 300–330° and 120 atm initial pressure with 10% molybdenum disulfide as catalyst. The phthalic anhydride was dissolved in 1,2,3,4-tetrahydronaphthalene and gave a 32% yield. The phthalide was dissolved in two volumes of 1,2,3,4-tetrahydronaphthalene and hydrogenated for 5 hr; the yield was 85%.

Clarke and Taylor [993] started with a technical product boiling within 1° and boiled it with dilute nitric acid (one part acid to three parts water). The *o*- and *p*-xylenes were oxidized, but the *m*-xylene is unattacked. After washing with water and alkali, the product was distilled, first with steam and then through a column, and finally sulfonated by stirring at 95° with half of its volume of concentrated sulfuric acid. The resulting solution was separated from the unsulfonated material, and an equal volume of concentrated sulfuric acid and twice its volume of water was added. Pure *m*-xylene passes over from 110 to 120° during steam distillation.

Cole [1035] selectively sulfonated *m*-xylene in a mixture of xylenes. The mixture was refluxed with the theoretical amount of 50–70% sulfuric acid at 85–95° *in vacuo*. The water-xylene vapors were condensed, the xylene returned to the reaction vessel, and the water withdrawn until the amount present in the sulfuric acid and the theoretical amount formed from the reaction were collected. Water was then added to the reaction mixture and the unreacted xylenes removed at reduced pressure; *m*-Xylene monosulfonic acid was hydrolyzed by steam distillation up to 140°. The unreacted xylenes, chiefly *ortho* and *para*, may be separated into material of fair purity by fractional distillation.

Mathews [3208] purified *m*- and *p*-xylenes for the determination of the *heat of vaporization* by successive shaking with sulfuric acid, sodium hydroxide and mercury, followed by drying over phosphorus pentoxide and fractional distillation. The *p*-xylene was fractionally crystallized six times.

The *meta*- and *para*-xylenes form a binary eutectic of 87% *meta*. It is not possible to get a pure *m*-xylene by *fractional solidification* of mixtures containing less than 87% *meta*. Laconte [5340] states that Egan and Luthy took advantage of the solid compound formed between carbon tetrachloride and *p*-xylene. When the mixture was cooled, the solid complex separated and the mother liquor contained 98% *meta*.

Skita and Schneck [4363] prepared pure *m*-xylene from asymmetric *m*-xylidine (purified by means of the acetate) by reduction of its diazo compound with tin in an alkaline solution. By the same method, they prepared pure *p*-xylene from *p*-xylidine which had been purified by means of the benzylidene compound.

According to Timmermans and Martin [4783] *p*-xylene can be obtained pure quite easily by fractional crystallization. *p*-Xylene was purified in this way until the freezing point was constant (13.35 ± 0.03°).

Greenburg [1884] isolated *p*-xylene from mixtures containing at least 16% *para* from close-boiling constituents such as *o*- and *m*-xylenes or ethylbenzene by fractional melting. About equal volumes of the hydrocarbon mixture were mixed with methanol or ethanol and partially solidified. The solid was fractionally melted into steps not exceeding 40°. The liquid was withdrawn at each step. The last fraction, melting above 10°, consisted substantially of *p*-xylene.

Cole and Burtt [1036] developed a method for separating *m*- and *p*-xylenes. A mixture of coke-oven xylenes was fractionally distilled to remove the *o*-isomer. The mixture (100 parts) of the *m*- and *p*-isomers was treated with 120 parts of 26% fuming sulfuric acid. The mixture was partially hydrolyzed by steam distillation. The first distillate contained paraffinic hydrocarbons, and ethylbenzene was discarded. It was followed by very pure *m*-xylene. The residue was cooled to 10°, the crystals of *p*-xylenesulfonic acid were collected by filtration, washed with aqueous sulfuric acid, and then steam distilled. Pure *p*-xylene was obtained. The separated *m*- and *p*-isomers may be washed with dilute sodium carbonate and further purified.

McArdle and Mason [3052] developed a crystallization procedure similar to that of Arnold [287] for separating *m*- and *p*-xylenes. It was stated that 56.6% of the *m*-xylene was crystallized from a mixture with C_8-hydrocarbons.

Sue and coworkers [4543] effectively removed *o*-xylene from *p*-xylene by *zone melting*.

CRITERIA OF PURITY

ACS Reagent Chemicals [76] list the following specifications for Reagent grade xylene (mixed isomers):

Boiling range: from 137 to 140°.
Residue after evaporation: not more than 0.002%.
Sulfur compounds (as S): not more than 0.003%.
Water (H_2O): not more than 0.05%.

Mair and coworkers [3120] used the freezing curve method to determine the purity of the NBS Standard Samples of the xylenes.

According to Timmermans and Martin [4783], density and melting point are the most satisfactory criteria of purity. Swietoslawski [4577] found that differential ebulliometric measurements gave a lower accuracy for the xylenes than for other compounds.

All of the known hydrocarbon impurities that have been found in the xylenes have been resolved by gas chromatography.

SAFETY

The acute toxicity of the xylenes probably is greater than that of benzene or toluene. The isomers differ in their acute toxicity, but their relationship has not been definitely established. The chronic toxicity appears to be less than that of benzene. The absorption of xylene takes place chiefly by breathing the vapors. Skin irritation is more serious than for benzene or toluene [3681].

There is definite irritation to the eyes, nose, and throat at 200 ppm and olfactory fatigue is rapid. The *threshold limit value* has been established at 100 ppm or 435 mg/m³ [4509]. The *flammable limits* in air for the *ortho-*, *meta-*, and *para-*isomers respectively are 1.0–6.0, 1.1–7.0 and 1.1–7.0%v [88]; at elevated temperature the corresponding values are 1.09–6.40, 1.09–6.4, and 1.08–6.60%v. The respective *minimum ignition temperatures* in air are 464°, 528°, and 529° [3680].

35. ETHYLBENZENE

Most of the ethylbenzene is obtained from petroleum (see the introduction to the xylenes). It is used principally to make styrene. It is available in Research grade, 99.90%m, and in several special grades.

Ethylbenzene was prepared by O'Kelly and coworkers [3585] by passing benzene and ethylene over a silica-alumina catalyst at temperatures of 448–496° and pressures of 57–75 psig; conversion was as high as 81.1% based on ethylene.

Hanai [1999] established the best condition for the formation of ethylbenzene from benzene and ethylene. The yield was 30–40%. The conditions were: 100 moles of benzene, 40–60 moles of ethylene, 8–10% aluminum chloride (benzene = 100%), saturated with hydrogen chloride. The mixture was kept under a slight pressure.

Hammick and Roberts [1993] catalyzed benzene and ethylene at atmospheric pressure to give predominantly ethylbenzene. Although the yield is low per pass of gas, the method is of interest because pressure equipment is not necessary.

Buu-Hoï and Janicaud [875] synthesized aromatic ethyl compounds by the Friedel-Crafts reaction using ethyl chloroformate. The alkylating agent is used in the usual manner in the presence of an aluminum or iron (III) chloride catalyst.

Passino [3671] reacted benzene (two parts), ethylene (one part) and hydrogen fluoride, equivalent to 12% of the benzene, in a shaking autoclave for 16 hr at 500 psi and 70–80°F. Ethylbenzene equivalent to 10% of the benzene was obtained.

Vogel [4968] reduced 100 g of acetophenone with 200 g of zinc amalgam and hydrochloric acid; yield 53 g, bp 134.5–135° at 758 torr. The ethylbenzene was purified by shaking with 6-ml portions of concentrated sulfuric acid until the acid layer was colorless, then with sodium carbonate solution, then with water and finally drying twice with anhydrous magnesium sulfate. After two distillations over sodium, the middle fraction of the second distillate was used for *physical property determinations.*

For preparation from bromoethane and benzene, see Aromatic Hydrocarbons, General [4865].

Kuss and coworkers [2795] prepared "very pure" ethylbenzene by reducing acetophenone with a chromium-copper catalyst at 180–200°.

Scott and Brickwedde [4212] further purified a sample of ethylbenzene obtained from M. R. Fenske as follows: The material was partially frozen three times, one-fourth of the liquid being discarded each time. It was passed over finely divided silica gel to remove water. No cloudiness was observed near the freezing point. It was used to determine *thermodynamic properties.*

SAFETY

The acute toxic effects in mice are comparable to toluene and *m*-xylene. Men exposed to 1000 ppm experienced eye irritation, which gradually decreased on continued exposure. Animals exposed to 400–2200 ppm 7 to 8 hr a day for 5 days a week for as long as 6 months were not affected, judged by the usual criteria of injury. Absorption is chiefly by inhalation [3681].

The *threshold limit value* in air has been set at 100 ppm or 435 mg/m^3 [4509].

The *explosive limits* in air at elevated temperature are 0.99 and 6.70%v. The *minimum ignition temperature* in air is 432° [3680].

36. ISOPROPYLBENZENE

Isopropylbenzene is a significant commercial petroleum solvent in the boiling range of 150–160°. It is readily synthesized from benzene and propylene or benzene and 2-propanol. See, for example, O'Kelly and coworkers [3585], Huston and Kaye [2333], Simons and Hart [4329], Brun [808] and Vermillion and Hill [4949].

Schmerling [4162] developed a method for the alkylation of aromatic hydrocarbons by olefins in a solution of aluminum chloride with a nitroparaffin. The temperature was from −10 to 100° and pressured from 1 to 100 atm. Suggestions are given for batch or continuous operations.

Rueggeberg and coworkers [4071] used chlorosulfonic acid to catalyze the reaction between an aryl hydrocarbon and an aliphatic alcohol. The method was not thoroughly investigated, but it is simple and the yields are fair.

Two moles of benzene and 1 mole of 2-chloropropane that were reacted 4 hr at 25° with 75% iron (III) chloride saturated with hydrogen chloride gave 66% isopropylbenzene [2394]. For a preparation from benzene and 1-bromopropane, see Aromatic Hydrocarbons, General [4865].

Eastman Kodak isopropylbenzene was fractionally distilled by Hentz [2115] in a 100-theoretical plate column operating at 85% efficiency. Approximately one-fifth of the middle cut fraction was used for *γ-radiation studies*, n_D 20° 1.4913. Not more than two days prior to each experiment, 20 ml was passed through 10 ml of Alcoa F-20 alumina in a 50-ml buret with a Teflon stopcock to remove the peroxides.

Thizy [4707] reports that olefins can be removed from isopropylbenzene by the addition of chlorine for 5–7 min. The liquid was washed with water and then steam-distilled in a glass column in a 95% yield.

For purification, see mesitylene, Duncan and coworkers [1396].

Vogel [4968] purified a commercial product by washing it five times with concentrated sulfuric acid in a ratio of 10:1; then with water, sodium carbonate solution, and water again. This was followed by drying with anhydrous magnesium sulfate. The isopropylbenzene was then fractionally distilled in a three-section Young and Thomas column and about 90% collected. It was refractionated over sodium and the middle fraction collected for *physical property measurements*.

SAFETY

Isopropylbenzene is a depressant to the central nervous system; its narcotic action is slowly induced and of long duration. Considered a primary eye and skin irritant, it is absorbed through the skin more rapidly than is benzene, toluene, the xylenes, or ethylbenzene.

A *threshold limit value* of 50 ppm or 250 mg/m^3 has been recommended [4509]. Gerarde [3681] states that, if this concentration elicits no mucous membrane effects, it should be a safe working atmosphere.

The *flammable limits* in air are 0.9 and 6.5%v [88]. An upper limit of 8.8%v has been reported [874]. The *minimum ignition temperature* in air is 424° [3680].

A method for the determination of isopropylbenzene in air has been reported [49].

37. MESITYLENE

Adams and Hufferd gave a method for preparing mesitylene from acetone and concentrated sulfuric acid in 13–15% yield in Conant [1060]. Ipatiew and coworkers [2354] studied the acetone to mesitylene reaction, using hydrogen chloride as the catalyst, and found the optimum conditions for maximum yield of 43%. Sucharda and Kuczyński [4538] found that lower pressures gave a 47% conversion in 48 hr.

Mesitylene was prepared by Norris and Ingraham [3549] from toluene and aluminum chloride cooled to 10–15°, to which a mixture of toluene and methanol had been added. The reaction mixture was heated to 110° for 3 hr and poured over chipped ice and purified by chemical and physical means; purity 99+%. Norris and Sturgis [3551] heated aluminum chloride, toluene, and methanol in a 2, 2.5, and 1 mole ratio, respectively, at 100° for 3.5 hr and obtained 53% mesitylene.

Duncan and coworkers [1396] purified reagent or commercial grade mesitylene, cyclohexene, toluene, p-xylene, isopropylbenzene, and cyclohexane for *thermodynamic studies*. The substances were twice fractionally distilled through a 40-theoretical plate column retaining the middle one-third of each distillate. The purity was found to be better than 99% by gas chromatography for all except mesitylene. The latter compound was "extremely difficult to purify," and was further fractionally crystallized and distilled. The final analysis was: mesitylene 98.52%, m- and/or p-ethyltoluene 0.65%, ethylbenzene 0.03%, and nonaromatic compounds 0.80%.

Mesitylene was purified by Streiff and coworkers [4522] for *physical and thermodynamic property studies* by distilling about 8 liters through a 125-plate column at a 125:1 reflux ratio. The criterion of purity was the freezing curve, 99.8%m.

Spectral studies in the 1700 to 2300 Å range were made by Platt and Klevens [3783] using mesitylene, butylbenzene, sec-butylbenzene, and tert-butylbenzene by crystallizing several times, then distilling twice through a 50-plate Podbielniak column taking the center cut each time. The boiling point and refractive index were used as the criteria of purity.

Forster [1602] purified mesitylene in the same manner as he did benzene [5295] for *specific conductance studies*.

38. NAPHTHALENE

The commercial source of naphthalene until 1961 was the high temperature coking of coal. Since then, naphthalene has been recovered from petroleum. About 43% of the U.S. production in 1965 was from the new source.

The impurities in coke oven naphthalene have been extensively studied. Thianaphthene, which forms mixed crystals with naphthalene, is the most universally objectionable and is difficult to remove. The methyl- and dimethylnaphthalenes are usually present, as in anthracene. Koptyug and coworkers [2710] studied the impurities in coke oven naphthalene by *infrared spectroscopy* and *gas chromatography;* they found thianaphthene; α- and β-methylnaphthalene; indene; indane; indole; o-, m-, and p-xylene;

phenol; 2,4-, 2,5-, 3,5-, and 2,3-xylenol; and unidentified higher boiling components. Weissgerber and Kruber [5118] found thianaphthene in a pure commercial product.

Coke oven naphthalene was purified by Garcia [1702] to 99.9% and in 95% yield with 10% sodium hydroxide to remove phenols, with 50% sodium hydroxide to remove nitriles, with 10% sulfuric acid to remove organic bases, and with 0.8 g aluminum chloride per 100 g naphthalene to remove thianaphthene and alkyl derivatives of cumarin and indene. All extractions were done at 85°. The material was then treated with 20% sulfuric acid, then with 15% sodium carbonate, and finally distilled.

Armstrong and coworkers [275] found thianaphthene to be the main impurity in commercially pure naphthalene from coke. Their purification consisted in heating at 145° for 18 hr with 5% sodamid. The residual liquid was fractionally distilled through a 15-in. glass helices-packed column jacketed with naphthalene vapors at the boiling point. The sulfur content was reduced to 0.015%. A second similar treatment reduced the sulfur to less than 0.005%. The product was recrystallized twice from redistilled methanol and fractionally distilled again; mp 80.32°. An *infrared study* suggested the presence of small amounts of other impurities believed to be methylnaphthalenes.

The sulfur content of naphthalene was reduced from 0.15 to 0.031% by one crystallization from alcohol [5206].

Herington and coworkers [2124] report that recrystallization and "simple" distillation do not remove thianaphthene very effectively from naphthalene. They purified commercial crystals for *freezing point determination* by keeping the naphthalene overnight under a stream of nitrogen to remove water. It was stirred 24 hr at 140° with 5% sodamid, decanted, stirred 24 hr at 180° with 1% sodamid, and filtered through glass wool. It was fractionally distilled in a 10-plate adiabatic column. The middle fraction was twice crystallized from methanol and fractionally distilled again. The sulfur content was less than 0.002%; fp $80.278 \pm 0.002°$, purity $99.978 \pm 0.10\%$m.

Mair and Streiff [3118] isolated naphthalene from a kerosene and purified it by systematic crystallization from ethanol. The residues rich in naphthalene were dissolved in a minimum of hot alcohol, the solution cooled to about 10°, and the crystals separated by filtration with suction. Second crops of crystals were obtained by removing some of the alcohol in a fractional distillation unit. Before determining the freezing point, the naphthalene was fractionally distilled at 56 torr; mp 80.24°, n_D 85° 1.5898.

Washburn and Read [5085] purified naphthalene by recrystallization from the melt, discarding 75 ml of the liquid from 500 g of the original melt. Distillation at reduced pressure gave a middle fraction melting at 80.09°.

The *thermodynamic properties* of naphthalene were determined by Barrow and McClellan [416] on resublimed material that had been repeatedly recrystallized from ether and dried by vacuum.

Rosanoff and Dunphy [4032] purified naphthalene by twice subliming, twice recrystallizing from ethanol, washing with water, and drying over phosphorus pentoxide.

Cheng and coworkers [964] purified naphthalene for *critical temperature studies* by chromatography on alumina using benzene as the eluent. It was then crystallized from benzene, which was removed by vacuum sublimation; mp 80.3°.

For the determination of the *absorption spectrum*, Baly and Tuck [375] purified naphthalene by heating with concentrated sulfuric acid and manganese dioxide (see also Lunge [3033]), distilling with steam, and, after a repetition of the process, several recrystallizations of the picrate. The regenerated naphthalene was then distilled with steam and recrystallized from dilute ethanol.

Mastrangelo [3203] purified refined liquid naphthalene for *cryoscopic studies* by slow cooling in an insulated Dewar flask and discarding the center core. The procedure was repeated four times in successively smaller flasks.

Naphthalene was purified as the picrate for *hydrogen bonding studies* [5295].

More work has been done on the purification of naphthalene by *zone melting* and related techniques than on any other organic compound. Anthracene is often used as the criterion of efficacy for purification because of its undesirableness as an impurity in naphthalene, for some uses even in the low parts per billion range, and its ability to fluoresce at parts per billion concentrations. Wolf and Deutsch [5241] reduced the concentration of anthracene from 0.1% to 1 ppm in two zone passes. It was suggested that further purification be done *in vacuo* or a nitrogen atmosphere. Herington and coworkers [2125] reduced the anthracene concentration from 0.2% to <200 ppb in seven zone passes. Some of the other references that have made significant contributions to the refining of naphthalene by zone melting are [2001, 557, 5277, 4147, 4149, 2416, 1672].

Paûlopoulos and El-Sayed [3684] purified naphthalene by crystallization followed by 40-zone melting passes.

A commercial naphthalene containing 1% thianaphthene was recrystallized from ethanol by Miller [3324] and purified by zone melting for *vapor pressure measurements*. A freezing point of 80.1° from the cooling curve corresponded to 99.7%m.

Mattox [3233] prepared naphthalene by passing the vapors of ethylbenzene over a catalyst of silica and alumina, thoria, or zirconia at 525–700°.

Naphthalene is available from the Fisher Scientific Co. [2798a] as a *temperature standard* defined by the triple point and certified accurate to ±0.05°. Naphthalene decomposes 1%m per hour at 480° [2402].

SAFETY

Naphthalene has a characteristic well-known odor. A concentration of 25 ppm in air is easily detectable by the odor. The inhalation of the vapors may cause headache, confusion, nausea, and profuse perspiration. Severe exposures to vapors of naphthalene may cause vomiting, optic neuritis, and hematuria. It is irritating to the skin, and hypersensitivity of certain individuals has been reported [3681].

It is reported that concentrations in excess of about 15 ppm result in noticeable irritation of the eyes. The *threshold limit value* was, therefore, set at 10 ppm or 50 mg/m³ [4509].

The *flammable limits* in air at elevated temperature are 0.88 and 5.90%v [3680]; see also [2425]. The *minimum ignition temperature* in air is 515° [3680].

39. 1,2,3,4-TETRAHYDRONAPHTHALENE

Mair and Streiff [3119] purified 1,2,3,4-tetrahydronaphthalene for *physical property studies* by five fractional crystallizations from a mixture of 50% dichlorodifluoromethane. The freezing curve was the criterion of purity; 99.985%m.

1,2,3,4-Tetrahydronaphthalene was purified by Morton and de Gouveia [3417] for *spectral studies* by preparing the barium sulfate salt by the Willstätler and Seitz [5207] method and recrystallizing several times. It was then converted to the sodium salt and recrystallized. The solvent was regenerated by adding 66% sulfuric acid and fractionally distilling with superheated steam at 160–180°. The principal impurities found were dihydronaphthalene, hexahydronaphthalene, decahydronaphthalene, and naphthalene. They are the impurities that are normal for the tetrahydro compound prepared from a high purity naphthalene.

Bass [431] purified 1,2,3,4-tetrahydronaphthalene by extracting 2.0 moles of commercially available material with 3–30 ml of concentrated sulfuric acid, washing with 250 ml of water, and drying with sodium sulfate. It was then distilled from sodium through a short, lagged Vigreux column; yield 230 ml. The partially purified material was vigorously shaken with 150 ml of concentrated sulfuric acid and then heated on a water bath for about 2 hr with stirring. The warm mixture was poured into a solution of 120 g ammonium chloride in 400 ml of water. The crystals of ammonium 1,2,3,4-tetrahydronaphthalene-6-sulfonate were separated, recrystallized, washed with 50% aqueous ethanol, and dried at 100°. Traces of naphthalene were removed by steam distillation. The salt was recrystallized from water until it was pure. One mole of salt was mixed with 5 moles of concentrated sulfuric acid and steam distilled from an oil bath at 165–170°. The distillate was extractèd with ether, washed with sodium carbonate solution, then with water. The oil layer was then dried and the ether removed by distillation. The liquid was finally refluxed over sodium and distilled through a Vigreux column.

Tanaka [4606] hydrogenated a mixture of 96 g of phenol and 120 g of naphthalene at 96 atm in the presence of nickel. Normally, the hydrogenation of naphthalene gives decahydronaphthalene, but phenol limits the reaction to the addition of four hydrogen atoms. Naphthalene can also be reduced to tetrahydronaphthalene in the presence of sodium hydroxide with activated aluminum-nickel catalyst. Some other metal catalysts are also effective [347].

Palfray [3631] hydrogenated naphthalene in the presence of Raney nickel at 125° and 115 kg pressure for 30 min. The tetrahydronaphthalene was separated from the decahydronaphthalene by fractional distillation. (See also under decahydronaphthalene [3475].)

SAFETY

The odor of 1,2,3,4-tetrahydronaphthalene is similar to that of naphthalene, and 110 ppm in the air is irritating and offensive. Dermatitis has been reported by painters working with tetralin. They also complained of headache; malaise; and irritation of the eyes, throat, and mucous membranes of the nasal passage. Severe exposure has resulted in the elimination of green urine [3681]. Smyth and coworkers [4431] report LD_{50} for rats.

The *threshold limit value* of 25 ppm or 135 mg/m^3 has been recommended [3681].

40. BUTYLBENZENE

Read and coworkers [3926] prepared butylbenzene as follows: A mixture of 411 g of 1-bromobutane and 471 g of bromobenzene were added to 300 ml of dry ether containing 161 g sodium over a period of about 2.5 hr. The temperature was maintained at about 20°. The mixture was allowed to stand for two days, the liquid decanted, 300 ml of methanol added, and the mixture refluxed for 4 hr. Eight hundred milliliters of water was added and the aqueous layer extracted with ether. The decanted liquid, the hydrocarbon layer, and the ether extract were combined, dried, and fractionally distilled; yield 65–70%.

Vogel [4968] prepared butylbenzene by reducing two ketones to the hydrocarbon.

1. Seventy-five grams of butyrophenone was reduced with 150 g zinc amalgam and concentrated hydrochloric acid; yield, 50 g crude product upon

steam distillation. The crude material was distilled from sodium and washed with 7-ml portions of concentrated sulfuric acid until the acid was no longer colored. After additional washings with water and sodium carbonate, it was dried with anhydrous magnesium sulfate. After it was fractionally distilled twice from sodium, the middle fraction from the second distillation was used for *physical property measurements*.

2. Seventy-five grams of benzylethyl ketone was used for the second preparation. The same procedure was used as described for (1).

Platt and Klevens [3783] purified butylbenzene for *ultraviolet studies;* see mesitylene.

SAFETY

The *minimum ignition temperature* in air is 412° [3680].

41. sec-BUTYLBENZENE

Huston and Kaye [2333] prepared 20 secondary aliphatic derivatives of benzene by condensing benzene with secondary alcohols. The crude *sec*-butylbenzene was dried over anhydrous sodium sulfate and the remaining benzene removed by vacuum. The residue was distilled through a 10-in Vigreux column; yield 71%. A much purer product may be prepared by drying with Drierite or molecular sieves and fractionally distilling at a high reflux ratio in an efficient fractionation column. A second fractional distillation should give a product approaching 100% purity if the starting materials were of good purity.

Simons and Hart [4329] prepared *sec*-butylbenzene by condensing 1-chlorobutane and benzene with hydrogen chloride at 100 psi and 195°; yield 30%.

Platt and Klevens [3783] purified *sec*-butylbenzene for *ultraviolet studies;* see mesitylene.

SAFETY

The *minimum ignition temperature* in air is 418° [3680].

42. tert-BUTYLBENZENE

Huston and coworkers [2332] prepared *tert*-butylbenzene by condensing benzene and 2-methyl-2-propanol. The crude product was purified by fractional distillation through a 30-cm column; bp 740 torr was 168–170°, yield 65–70%.

Simons and Hart [4329] prepared *tert*-butylbenzene by condensing 2-chloro-2-methylpropane and benzene in the presence of hydrogen chloride at 100 psi and 150°; yield 45.5%, bp 168.0–168.8°.

Norris and Sturgis [3551] prepared *tert*-butylbenzene in 84% yield by allowing 0.6 moles of aluminum chloride, 0.5 moles of 2-methyl-2-propanol, and 3 moles of benzene to react at room temperature for 24 hr.

Bowman and coworkers prepared *tert*-butylbenzene [696] by direct alkylation of benzene with 2-methylpropene in the presence of boron trifluoride hydrate ($BF_3 \cdot H_2O$).

Platte and Klevens [3783] purified *tert*-butylbenzene for *ultraviolet studies;* see mesitylene.

SAFETY

The *minimum ignition temperature* in air is 450° [3680].

43. *p*-CYMENE

Le Fèvre and coworkers [2878] purified *p*-cymene as follows: Several liters of the technical product was refluxed over powdered sulfur for two days. It was then shaken with successive quantities of concentrated sulfuric acid until the latter was no longer colored, then twice with small amounts of chlorosulfonic acid and, after a water wash, with potassium permanganate solution, followed by dilute aqueous sodium hydroxide. The material was dried over sodium sulfate and fractionally distilled.

Lombard [2993] obtained *p*-cymene from the fraction of pine-tar oil of the cluster pine boiling at 200°. For additional natural sources, see Katasura [2509] and Hasabe [2051].

Simons and Hart [4329, 4330] prepared isopropyltoluenes by condensing 2-chloropropane and toluene in the presence of hydrogen chloride at 300 psi and 235°; yield 67% mixed isomers.

Plate and Tarasova [3782] converted turpentine sulfate into *p*-cymene at 380–520° in contact with three catalysts: chromium oxide on aluminum oxide, 1:3; molybdenum oxide on aluminum oxide, 1:3; and synthetic Houdry aluminosilicate catalyst. The first two catalysts gave as much as 66% *p*-cymene at 400° calculated from turpentine, and 77% calculated on the basis of the pinene-carene content. The aluminosilicate promoted the formation of lower aromatic compounds.

SAFETY

The *minimum ignition temperature* in air is 436° [3680].

44. CYCLOHEXYLBENZENE

The alkylation of benzene was studied by Carson and Ipatieff [921]. Fifty-eight grams of cyclohexylbenzene was prepared by stirring an ice-cooled mixture of 176 g benzene and 60 g of anhydrous aluminum chloride, adding 246 g cyclohexane over a 2.8-hr period and stirring an additional hour. The catalyst was decomposed with dilute hydrochloric acid. The hydrocarbon was purified by fractional distillation and fractional crystallization. They also prepared cyclohexylbenzene by using sulfuric acid as the catalyst. Ipatieff and Gross [2353] used aluminum chloride and hydrogen chloride.

Bodroux [628] used only aluminum chloride. Pajeau [3630] used aluminum chloride, cyclohexanol, and benzene.

The reaction of chlorocyclohexane and benzene to give cyclohexylbenzene was studied by Neunhoeffer [3521]. It was found that the larger the ratio of benzene to chlorocyclohexane, the greater the yield. The yield was 0.8 moles at a ratio of 12:1.

Metallic chlorides were found to be alkylating catalysts for aromatic compounds. One mole of chlorocyclohexane, 2 moles of benzene, and 100 ml of 83.5% zinc chloride saturated with hydrogen chloride at 75° for 6 hr gave 50% cyclohexylbenzene; the substitution of cyclohexane gave 43% and cyclohexanol 70%.

For compound from bicyclohexyl, see [458] under decahydronaphthalene; from benzene and cyclohexene, see Aromatic Hydrocarbons, General [4328].

Mears and coworkers [3244] purified commercially available cyclohexylbenzene for *physical property studies* by fractionally distilling at 60 torr in a Heli-grid column. Fractions of about 50 ml were collected and classified according to refractive index. The purest material was 99.83%m. The less pure fractions were hydrogenated to bicyclohexyl.

Unsaturated Hydrocarbons

General

METHODS OF PREPARATION

A general method of preparing alkenes by the decomposition of xanthates was developed by Tschugaeff [4844] and applied to the preparation of several compounds [4845]. Whitmore and Simpson [5162] have studied this method and have made several improvements.

Kirrmann [2611] prepared 1-alkenes from Grignard compounds of suitable alkyls and bromopropene in ethyl ether. Propyl ether was used for 1-pentene. The yields are reported to be excellent.

Wilkinson [5183] treated 3-bromopropene with the appropriate Grignard reagent and obtained high purity 1-alkenes. Moisture and carbon dioxide were excluded from the system. The five straight chain 1-alkenes of C_5, C_6, C_7, C_8, and C_9 were prepared.

Normant [3546] found that aliphatic, cyclic or vinyl-magnesium halides are obtained in good yields by reacting magnesium with the corresponding bromo compounds (chloro compounds will work) in tetrahydrofuran, 2-methyltetrahydrofuran or tetrahydropyran. The reaction of bromo compounds with magnesium is promoted by iodine. The vinylmagnesium halides apparently form oxonium salts with the solvent. Hydrolysis of the Grignard compound gives the corresponding alkene.

Swallen and Boord [4566] found that β-bromoalkylethyl ethers react with magnesium to form alkenes. The method was further developed by Dykstra and coworkers [1424]. The latter quote an earlier author concerning the preparation of olefins, "High temperature and many chemical reagents, particularly acids, cause such α-olefins to rearrange or the double bond to shift position. Only reactions employing low temperatures and absence of isomerization agents can be expected to produce α-olefins in any degree of purity."

Bourguel [679] reported that the reduction of alkynes to the corresponding alkenes with hydrogen and colloidal platinum at room temperature gave purer compounds than previously described methods.

Table 5.5 Conditions for van Pelt and Wibaut Preparation of Alkenes

Alkene	Acetate Ester	Temp. °C	Feed Rate Ester, g/hr	N₂, l./hr	Yield %
1-Hexene	Pentyl	500°	22	4	66
1-Heptene	Heptyl	515°	17	—	72
1-Octene	Octyl	500°	14.7	—	77
2-Pentene	3-Pentyl	430°	20.4	—	71

Van Pelt and Wibaut [4915] reported a method for the preparation of alkenes from the corresponding alkyl acetate by passing through a 50-cm tube 2.5 cm in diameter packed with glass wool in a tube furnace heated to the temperature most suited to the particular decomposition. A preheater and a source of nitrogen were provided to prevent undesirable decomposition and polymerization. Information for the preparation of pertinent alkenes is given in Table 5.5. See also [4453].

Olefins can be prepared by catalytically hydrogenating acetylenes. Campbell and Eby [895] studied this procedure and the sodium-liquid ammonia method. During catalytic hydrogenation, it is difficult to stop precisely at the olefin stage; therefore the product usually contains traces of saturated hydrocarbons that are difficult, if not impossible, to separate by fractional distillation. When the sodium-ammonia method is used, there is no danger of forming saturated hydrocarbons and a much purer product will be obtained. 1-Hexene, 1-heptene and 1-octene were prepared by both methods and *physical properties* were determined.

A simple general method for preparing alkenes was reported by Brandenberg and Galat [712]. Boric acid or anhydride will dehydrate primary, secondary, or tertiary alcohols. The boric ester, $B(OCH_2CH_2R)_3$, which is formed is decomposed by heat, regenerating the catalyst, and producing an alkene. A typical example is: 1 mole of 1-octanol and 1 mole of boric acid were gradually heated to 350°. Water was continually separated from the azeotrope and the alcohol returned to the reaction vessel. When 1 mole of water was removed, the water-octene azeotrope started to distill. After the alkene azeotrope had all distilled, a second mole of the alcohol was added, then the third, and so forth; 90% yield.

No proof has been obtained that 1-alcohols give only 1-alkenes.

PURIFICATION

Several alkenes are now produced commercially. Riddick [3975] obtained samples from the pilot plants of the C_6–C_{10} 1-alkene cut of the Gulf Oil Corp., Petrochemical Department; and the California Chemical Company. The individual 1-alkenes were separated and purified to 99.7–99.9+% for *azeotrope studies* as follows: The primary separation was made in an 8-ft, Penn-State-type column packed with 0.16 × 0.16 inch protruded stainless steel packing and fitted with an automatic take-off head. The pressure was maintained at 760 torr under nitrogen. The desired product was taken at a reflux ratio of 25:1 at the beginning and end of the fraction, decreasing to 10:1 during the take-off of the main part. The partially purified (98+%) product was fractionally distilled in the same or a similar still. The reflux ratio at the beginning and end of the distillation was high. The main cut was taken at either a 10:1 or a 15:1 reflux ratio, depending on the difficulty of separating the impurities. The reflux ratios and the cut points were determined by gas chromatographic analysis of the product stream. The final analysis of the individual 1-alkenes was verified by mass spectroscopy.

SAFETY

Only a few unsaturated aliphatic hydrocarbons have been available in commercial quantities until recently. Consequently, there is little safety information available other than a modicum for the C_2, C_3, and C_4's. The lower members are simple asphixiants or weak anesthetics. Anesthetic potency increases with increasing chain length in the ethylene series. Pentene has been used for surgical anesthesia [3861]. Caution is recommended when extending conclusions based on C_1–C_4 and C_5 to higher homologs.

Specific Solvents

45. 1-PENTENE

Whitmore and Simpson [5162] made 1-pentene from 1-pentanol by the Tschugaeff method [4844].

Preparing 1-alkenes for standards for the Soviet oil industry, Kazanskiï and coworkers [2538] obtained 1-pentene from the Grignard reaction in 40–45%

yield; boiling point, density, and refractive index indicated a product of high purity.

Streiff and coworkers [4528] purified 1-pentene from Phillips Petroleum Co. by fractionally distilling in a 200 theoretical plate column at a reflux ratio of 160:1 to a purity of 99.82%m.

Research, Pure, and technical grades are available.

46. 2-PENTENE (MIXED ISOMERS)

47. *cis*-2-PENTENE

48. *trans*-2-PENTENE

Directions for the preparation of 2-pentene by the dehydration of 2-pentanol in 65–80% yield are given in Gilman and Blatt [1778, p. 430].

Jeffery and Vogel [2386] prepared 2-pentene for *physical property measurements* from the low boiling product obtained during the preparation of 3-bromopentane.

Norris [1778] prepared 2-pentene by dehydrating 2-pentanol with sulfuric acid and water mixed 1:1.

Alkenes were prepared by thermal decomposition of alkyl acetates by Wibaut and van Pelt [5171]. 2-Pentene was obtained from 2-pentyl acetate by heating at 440°; heptyl acetate gave 1-heptene in 75% yield. See also [4915] under Unsaturated Hydrocarbons, General.

Pomerantz and coworkers [3805] prepared 2-pentene from 3-pentanol for *physical property studies* by passing the alcohol over alumina at 300–320° with an 86% yield. The pentene layer was washed with water, dried over anhydrous sodium carbonate and fractionally distilled in a 2.5 × 250 cm glass-Heligrid packed column. The center cut had a purity of 99.4%m by mass spectrometry analysis.

Aston and Mastrangelo [308] purified 95.7% *trans*-2-pentene by two passes of fractional melting to a purity of 99.99%m in 60% yield.

49. 1-HEXENE

Van Risseghem [4917] prepared 1- and 2-hexene in several different ways to study the possibility of *stereoisomer* existence.

Mears and coworkers [3243] prepared 1-hexene by dehydrating 1-hexanol over alumina at 400°. Thirty-two liters of commercial hexanol was fractionally distilled in a 2.5 × 183-cm total reflux, variable take-off column packed with $\frac{3}{16}$-in. glass helices to remove 2-ethyl-1-butanol which would dehydrate to form hexenes boiling close to 1-hexene. The crude olefin was distilled in a column similar to the one described and 4.2 liters of 1-hexene boiling at 63.47° was obtained.

Waterman and de Kok [5090] prepared pure 1-hexene by reacting par-aldehyde, ethanol, and hydrogen chloride to 1-chloroethylethyl ether which was brominated to 1,2-dibromoethylethyl ether. The ether was reacted with butylmagnesium bromide to 1-butyl-2-bromoethylethyl ether. The latter was reduced with zinc in ethanol to 1-hexene "in a high state of purity." They also prepared 1-hexene from 3-bromopropane and butylmagnesium bromide. It was stored over sodium for several months and fractionally distilled from sodium.

Komarewsky and coworkers [2700] prepared the 1-alkene from 1-hexanol without other unsaturated compounds being present; see 1-octanol.

See also [895, 4915] under Unsaturated Hydrocarbons, General.

Streiff and coworkers [4528] purified 1-hexene for API Research Project 45 for *physical property* and *thermodynamic studies* by fractional distillation in a 135 theoretical plate column at a reflux ratio of 165:1 to a purity of 99.87%m.

50. 1-HEPTENE

A pure 1-heptene was prepared by Waterman and de Kok [5091] from 3-bromo-1-propene and butylmagnesium bromide. The crude material was kept over sodium for several months until the white deposit no longer formed. It was then fractionally distilled from sodium.

For preparations by heating heptyl acetate at 525°, see 2-pentene [3549] and Unsaturated Hydrocarbons, General [4915], also by hydrogenation and sodium-ammonia reduction of acetylenes in the same section [895].

Streiff and coworkers [4528] purified 1-heptene from the same source and by the same method they used for 1-hexene, to a purity of 99.84%m.

Sherrill and coworkers [4278] purified 1-heptene by azeotropic distillation with ethanol.

51. 1-OCTENE

Whitmore and Herndon [5158] showed that octene prepared by the dehydration of either 1- or 2-octanol (capryl alcohol) was a mixture of 1- and 2-octene. Komarewsky and coworkers [2700] reported that they passed 1-octanol over aluminum oxide catalyst in a 12-mm-diameter, Pyrex-brand glass tube at 350° at 35 ml per hr and obtained 1-octene and about 2% of an impurity they surmised to be 2-, 3-, or 4-*trans*-octene. The criteria of purity were: bp 121–121.4° at 750 torr, n_D 20° 1.4095, d 20° 0.7152, and the Raman spectra. Shchekin [4261] reported that 1- and 2-octanols may be dehydrated to the corresponding octenes over aluminum silicate or platinized aluminum silicate without accompanying polymerization or hydrogen redistribution.

Waterman and de Kok [5092] prepared a high purity 1-octene from 3-bromopropene and butylmagnesium bromide and studied the *molecular refraction*.

Pomerantz and coworkers [3805] prepared this alkene to a purity of 99.2%
for *physical property studies* by fractionally distilling 1-octene in the same
still used to purify 2-pentene.

Streiff and coworkers [4524] purified 99.70%m 1-octene by fractional
distillation in a 130 theoretical plate column at a reflux ratio of 155:1 to a
purity of 99.77%m determined from the freezing curve.

See Unsaturated Hydrocarbons, General, for preparation from octyl
acetate [4915], hydrogenation and sodium ammonia reduction of acetylenes
[895], and Grignard preparation [5183].

52. 1-NONENE

Anomalous behavior was found in the melting region during *low temperature thermal studies* of 1-nonene [3061].

Adamson and Kenner [21] prepared 1-nonene from nonyl-ammonium
nitrate in 83.4% yield.

Streiff and coworkers [4527] purified 1-nonene for *physical and thermodynamic property studies* by fractional distillation in a 200 theoretical plate
column at a reflux ratio of 160:1 to a purity of 99.76%m calculated from the
freezing curve.

53. 1-DECENE

1-Decene was isolated from cracked petroleum in 1949 [4756].

The 1-alkene was purified in the same column that was used for 1-hexanol,
see 1-hexene, by Mears and coworkers [3243] for *physical property studies*
and the preparation of *n*-decane by hydrogenation.

Thermodynamic and physical property studies were made on 1-decene
purified by Streiff and coworkers [4527] in a 125 theoretical plate column at a
reflux ratio of 125:1 to a purity of 99.91%m calculated from the freezing
curve.

54. CYCLOHEXENE

Most of the commercially available cyclohexene has been separated from
petroleum. It contains small amounts of other hydrocarbons that are difficult
to remove. If a purity of better than 99.9% is desired, it is advisable to prepare
the cyclohexene from a high purity cyclohexanol. Directions are given in
Marvel [3181] for the preparation of large quantities of cyclohexene by
dehydrating cyclohexanol with concentrated sulfuric acid. See also Senderens
[4234]. Dehn and Jackson [1260] prepared cyclohexene from cyclohexanol
and phosphorus pentoxide with a 96% yield.

Mousseron and coworkers [3432] reported that cyclohexene may be prepared by heating cyclohexylmethyl and butyl ethers, gradually adding
sulfuric acid.

Kharasch and Urry [2579] prepared a mixture of cyclohexane and cyclo-hexene (35%) by a modification of the Grignard reaction.

Waterman and Van Westen [5094] purified cyclohexene prepared from cyclohexanol by fractional distillation through a 60-cm Vigreux tube, refluxing the cyclohexene cut over sodium for 7 hr and redistilling. Bromine and hydrogen bromide absorption was used as a test for purity.

Bodroux [628] dehydrated cyclohexanol with phosphoric acid and obtained cyclohexene.

Balandin and coworkers [361] obtained cyclohexene in 100% yield by passing cyclohexanol over anhydrous magnesium sulfate with a space velocity of 0.2–0.4.

Mathews [3208] dried commercial cyclohexene over phosphorus pentoxide and fractionally distilled for the determination of *heat of vaporization*.

Cyclohexene was washed several times with iron (II) sulfate by Chen and coworkers [958], dried over magnesium perchlorate, and distilled for the investigation of *molecular polarizability* and *dipole moment*. The portion boiling at 83–83.4° at 760 torr was collected and stored under nitrogen until ready to be used. Air was excluded to prevent oxidation. The *critical temperature* was determined with material that was washed with aqueous iron (II) sulfate and fractionally distilled at 83.0° by Cheng and coworkers [964].

Streiff and coworkers [4528] of API Research Project 44 purified com-mercially available cyclohexene by fractional distillation in a 135 theoretical plate column at a reflux ratio of 165:1 to a purity of 99.97%m for *physical property thermodynamic studies*.

Forster purified cyclohexene in the same manner as he did benzene [1603]. See also Duncan and coworkers [1396] under mesitylene.

55. STYRENE

Styrene is difficult to purify and to keep pure. Purified styrene must be used immediately, unless it contains an inhibitor.

Pitzer and coworkers [3777] added a few milligrams of hydroquinone to about 80 g of styrene to prevent polymerization for *thermodynamic studies*.

Adams [16, p. 84] gives a method for the preparation from cinnamic acid; yield 40–42 g. Paul and Tchelitcheff [3683] prepared styrene in 81% yield by heating benzaldehyde, malonic acid, and quinoline in a water bath. Styrene may be prepared in about 25% yield by reacting acetophenone with form-amide. The reaction product, 1-phenyl-1-ethylamine, is converted to the hydrochloride and the salt heated to give styrene [2358].

Cheney and McAllister [961] dehydrogenated isopropylbenzene by passing it through an unpacked column at 750° with a residence time of 0.1 sec at atmospheric pressure. A dehydrogenation of 68% per pass of feed was reported with a yield of 33% styrene. The styrene was separated by fractional distillation. When activated alumina was used as the catalyst in the dehydrogenation column, operating temperatures could be reduced to 450–650°, but dehydrogenation per pass was lower.

A number of methods were used by Hippel and Wesson [2197] to prepare highly purified styrene for *dielectric loss measurements*. The best results were obtained by drying Dow Chemical Company's N-100 with Drierite for two days, followed by vacuum distillation at 25° in the presence of an inhibitor (*p-tert*-butylcatechol, 0.005%). The dielectric loss increases if the inhibitor is removed before drying and distilling, or if the liquid is overheated, or if air is admitted. The removal of all traces of water was found to be difficult as determined by the Karl Fischer method.

Kienitz [2587] studied the purification of organic substances by fractional crystallization. The method essentially consisted of partially crystallizing the liquid, withdrawing the liquid, melting the solid, and then repeating the process. A mathematical treatment shows the determination of absolute purity from the change in the equilibrium temperature with time and the cryoscopic constant. Examples of purification of 99.9_6–99.9_9%m styrene and 98.4_0–99.9_0%m pyridine are given with their time-temperature curves.

For the determination of less than 0.01% of *o*-xylene in styrene, see Marquardt and Luce [3157].

SAFETY

Guinea pigs exposed to 650 ppm of styrene repeatedly for a long period of time appeared as well in all respects as the controls. The vapors in concentrations of 200–400 ppm have a transient irritating effect on the eyes and mucous membranes of the nose. The initial odor is easily noticeable and diminishes to faint or very faint [3681, 4509].

The *threshold limit value* has been set at 100 ppm or 420 mg/m³ [4509].

The *flammable limits* in air at elevated temperature are 1.10 and 6.10%v. The *minimum ignition temperature* in air is 490° [3680].

Ivanov [2364] describes a rather fast method for the determination of styrene in air. Another method [49] gives the sensitivity, specificity, and likely interferences.

56. 2-PINENE

57. 2(10)-PINENE

The nomenclature of the pinenes is somewhat confused in the literature. The most commonly used names for these compounds appear to be α-*pinene* and β-*pinene*, respectively. See also synonyms in Chapter I.

d-2-Pinene is the principal constituent of Greek turpentine oil, Russian turpentine oil, and hinoki wood oil. American turpentine oil contains a small amount. *l*-2-Pinene is the principal constituent of Spanish, Austrian, French, and American turpentine oils. It is also found in oil of lemon, American peppermint oil, coriander seed oil, and other essential oils.

The pinenes are isomerized by heat, acids, and certain catalysts. It is advisable to distill pinenes in an atmosphere of nitrogen and store them in the dark. The material should be used shortly after preparation if the presence of isomerization or oxidation products is objectionable.

2-Pinene and 2(10)-pinene can be separated by fractional distillation. 2(10)-Pinene has a constant optical rotation, regardless of the source and time of collection.

Svirbely and coworkers [4563] obtained chemically pure 2-pinene from the de Haen Company and used it without further purification for the determination of *dipole moment*.

Guenther [1929] describes several methods for the isolation of 2-pinene and its separation into optically active isomers. 2-Pinene can be isolated from essential oils by fractional distillation and subsequent purification.

Mirov [3340] has reported that turpentine from *Pinus muricata* contains 98–99% *d*-2-pinene. The physical properties of the turpentine fractions obtained by distillation are given.

2-Pinene and 2(10)-pinene were purified from commercial products by fractional distillation in a Lecky and Ewell column [2853] at a reflux ratio of about 100:1. The refractive index was used as the criterion of purity; 2-pinene 1.4634 and 2(10)-pinene 1.4768.

Fuguitt and coworkers [1676] purified commercial 2- and 2(10)-pinenes for *physical property studies* by fractional distillation in a Lecky and Ewell column [2853] at 20 torr pressure and a reflux ratio of 40:1, 75-ml fractions were collected. All fractions of 2-pinene having a refractive index of 1.4631–1.4633 at 25° were redistilled under the same conditions and 50-ml fractions collected. Those fractions that have constant refractive indices and optical rotation were considered pure. The purified 2(10)-pinene was prepared in the same manner.

2-Pinene was purified from American turpentine oil by fractional distillation, followed by purification through nitrosochloride by Waterman and coworkers [5093].

Widmark [5175] found the most efficient way to purify 2-pinene is by fractional distillation at reduced pressure.

Hawkins and Eriksen [2075] purified 2- and 2(10)-pinene for *heat of combustion studies* in a 120-plate column. The purity was calculated from the freezing point.

Schorger [4179] isolated *d*-2-pinene from the Port Oxford cedar, *Chamaecyparis lawsoniana* (Murr.) *Parlatore*, by rectifying, followed by fractionation over sodium. He found the highest specific rotation $[\alpha]_D$ 51.21° to that date; Grecian turpentine oil gives a *d*-2-pinene with $[\alpha]_D$ of 48.7°.

Lynn [3046] *purified the optically active form* of 2-pinene by mixing equal volumes of pinene, ethyl nitrate, and absolute ethanol, and then cooling in a freezing mixture. An amount of hydrogen chloride in dry ethanol equivalent to the pinene was added at such a rate that the temperature of the mixture was maintained below −5°. It was allowed to stand for about ½ hr and the crystals separated by filtration. The filtrate was diluted with one or two volumes of ethanol and the temperature reduced to below −10°. Crystals separated from the filtrate; they were washed with cool ethanol.

The crystals were dissolved in ethanol and aniline added, and the resulting red solution was distilled with steam. The isolated pinenes had an optical rotation of ±53.75°. This was slightly higher than $[\alpha]_D$ 51.21° for d-2-pinene reported by Schorger [4179].

Gasopoulos [1719] reports that 2-pinene immediately precipitated mercury (I) acetate from an ethanolic solution, whereas the mixture containing 2(10)-pinene remains clear for two or three days.

Thomas and Hawkins [4716] stored purified 2(10)-pinene for *physical and thermodynamic property studies* under nitrogen at −20° until ready to use.

2-Pinene is available commercially. The Hercules Powder Company specifications are:

Specific gravity, 15.6/15.6°: 0.863.
Refractive index, 20°: 1.466.
Unpolymerized residue: 0.8%.
Distillation range, 5%: 156°; 95%: 158°.

SAFETY

It is suggested that the safety practices observed for turpentine be followed for the pinenes until specific information becomes available.

The *minimum ignition temperature* of pinene in oxygen is 275° [3860].

COMPOUNDS WITH ONE TYPE OF CHARACTERISTIC ATOM OR GROUP

HYDROXY COMPOUNDS

Aliphatic Monohydric Alcohols

General

PREPARATION

Brown and Subba Rao [767, 769] and Brown and Zweifel [773] prepared primary alcohols from the appropriate olefins using sodium borohydride and aluminum chloride, first obtaining the trialkylborane. The latter was hydrolyzed in ethanol with sodium hydroxide and hydrogen peroxide. The boron is preferentially attached to the terminal carbon in the olefin.

Pohoryles and coworkers [3799] prepared ethanol, 2-butene-1-ol and 1,3-and 1,4-butanediol by lithium aluminum hydride reduction.

Sladkov and coworkers [4368] and Bolotiv and Dolgov [642] prepared various alcohols from ethylene and other unsaturated hydrocarbons by reaction with alkyl aluminums, followed by oxidation and acid hydrolysis.

Olefins [1495] were condensed with perfluorocarboxylic acids to produce esters. These were hydrolyzed to alcohols.

PURIFICATION

Biddiscombe and coworkers [568] purified and characterized 1- and 2-propanol, 1- and 2-butanol, 2-methyl-1-propanol, and 2-methyl-2-propanol for *physical and thermodynamic property studies.* Commercially available

Table 5.6

	Gas Chromatography	fp
1-Propanol	99.94%	—
2-Propanol	99.96	99.96%
1-Butanol	99.94	99.92
2-Butanol	99.95	—
2-Methyl-1-propanol	99.93	—
2-Methyl-2-propanol	—	99.96

materials were fractionally distilled twice in a column of 50–75 theoretical plates and the best fractions were selected by gas chromatographic analysis. Purity was evaluated by comparison with samples containing known amounts of impurities. Further purification of 2-methyl-2-propanol was made by fractional crystallization. Water was removed by treatment with calcium hydride. The purity of 2-propanol, 1-butanol, and 2-methyl-2-propanol were checked by the freezing curve. The other alcohols did not freeze but turned to glasses. The purities found are shown in Table 5.6.

Ballinger and Long [369] purified a number of alcohols for *ionization studies*, starting with the best material available. They were treated with anhydrous sodium carbonate to remove acidic materials, then with a suitable drying agent, and fractionally distilled. Gas chromatography was used whenever possible during the purification procedure to establish the purity. Traces of any materials stronger than the alcohol being studied would lead to spurious values for the pK, more particularly those greater than 14. The *criterion of acceptability* was specific conductance.

McCurdy and Laidler [3067] purified 14 Certified Reagent grade C_1–C_5 alcohols by refluxing with freshly ignited calcium oxide for 4 hr and then fractionally distilling through a 40-cm column packed with 4-mm Raschig rings. The middle fraction was used to determine the *heat of vaporization*.

Purification of alcohols, particularly the optically active compounds, often can be most satisfactorily accomplished by the methods of Ingersoll [17]. A pure acid anhydride, usually phthalic anhydride, that reacts rather readily with the alcohol is used to form the ester or acid ester in an appropriate solvent such as pyridine. The acid ester is converted to the sodium salt with sodium carbonate. The salt is filtered and washed with immiscible solvents to remove the nonreactive impurities, then reprecipitated by acidification with a mineral acid.

Resolution of optically active alcohols may be accomplished by forming the alkaloid salts of the acid ester, crystallizing, washing, and so forth. After reaching the desired state of purity, the salt is decomposed with hydrochloric acid and the ester saponified with aqueous alkali. The alcohol is recovered and purified by customary methods.

Campbell [890] has prepared the acid ester by heating the alcohol and anhydride gently to fusion. In some cases, this simplifies the preparation.

McEwen [3075] determined the *ionization constants* of alcohols that were carefully fractionally distilled from their sodium salts and kept in a sealed bottle in a desiccator over phosphorus pentoxide. Different samples of the same compound were used in the various runs to guard against any one of the samples being contaminated.

Halpern and Westley [1982] found that many optically active mixtures of secondary alcohols could be resolved by esterification with L-amino acids in the presence of p-toluenesulfonic acid. The crystal sulfonates of the ester are produced in 60–80% yield.

Pfenninger [3741] found that all aliphatic alcohols to four carbon atoms can be separated by gas chromatography using dioctyl sebacate on Celite.

Nagai [3480] reports that the R_f values of alcohols used for paper chromatography generally increase with the length of the carbon chain. Gasparič [1720] presented the R_f values of a large number and variety of alcohols.

Peltonen [3704] reported a method for the determination of sulfur in alcohols.

SAFETY

Zagradnik and coworkers [5315] determined the toxicity of saturated aliphatic alcohols to C_8 and tabulated the LD_{50}s.

Specific Solvents

58. METHANOL

Practically all methanol available today is synthetic. The commercial grade is sufficiently pure to meet all except the most exacting requirements. The impurities that have been identified in a representative manufacturer's product are carbon dioxide, methyl ether, methylal, methylol, methyl formate, methyl acetate, formaldehyde, acetaldehyde, acetone, ethanol, 2-propanol, 1-butanol, 2-butanol, 2-methyl-2-propanol, and water. The concentration of all impurities is small. Methanol as it comes from the still usually contains from 0.01–0.04% water, the impurity normally present in the largest amount. Each time the methanol is handled, unless extraordinary care is exercised, the water content increases. The acetone content of synthetic methanol is low. There are numerous articles in the literature concerning the removal of acetone from methanol. This was a major problem when methanol was obtained by the destructive distillation of wood. Water is the most objectionable substance present for most users. Recently, it has been found that ethanol, in normal impurity concentrations, poisons certain catalysts. A grade of methanol is available that contains 10 ppm or less ethanol.

The following manufacturer's specifications are typical of United States production:

Purity, min: 99.85%.
Specific gravity at 20/20°, max 25/25°: 0.7926–0.7889.
Acidity as acetic acid, max: 0.003%.
Acetone, max: 0.003%.
Alkalinity as ammonia, max: 0.0003%.
Distillation range including 64.5°, max: 1.0°.
Nonvolatile matter, g/100 ml, max: 0.0010.
Water, max: 0.1%.
Color, max: 5 APHA.
Permanganate time at 15°, max: 50 min.
Carbonizable substances (H_2SO_4 test) max: 50 APHA.
Chlorides: to pass test.
Sulfur: to pass test.
Appearance: clear and free from suspended matter.
Odor: characteristic, free from foreign odor, nonresidual.
Water solubility: no turbidity after one hour at 25° when one volume of methanol is mixed with three volumes of distilled water.

The purity is calculated from the specific gravity.

Gas chromatography could be used to determine the impurities present in methanol, but it would be slow because a high precision method, such as flame ionization, would be required for those components in the low ppm range.

The acidity of commercial methanol is mainly due to carbon dioxide and methyl formate. The alkalinity is due to ammonia or perhaps methylamines.

Purified methanol must be kept in an atmosphere free of oxygen to prevent the formation of formaldehyde.

The removal of impurities from methanol was studied by Morton and Mark [3416]. The alcohol was refluxed with 2-furaldehyde and aqueous sodium hydroxide and fractionally distilled. A yield of 95 % of the methanol treated gave no test with Nessler's reagent. The reagent is sensitive to 6 ppm (0.0006 %) acetone. Tests with aniline acetate indicated that 2-furaldehyde has been removed to at least 0.0001 %. Kretschmer and Wiebe [2754] used the Morton-Mark treatment, fractionally distilled through a 50-plate column, and dried the distillate with magnesium; d 25° 0.78653.

Evers and Knox [1525] purified methanol for *specific conductance studies.* Approximately 4.5 liters of synthetic methanol was refluxed for 24 hr over 50 g of magnesium. Four liters were removed by distillation and refluxed for 24 hr over silver nitrate, excluding atmospheric moisture and carbon dioxide. After distilling, the solvent was shaken for 24 hr with activated alumina and filtered through a sintered glass funnel into a pot containing 1 atm of pure nitrogen. The methanol was carefully distilled in an all-glass distilling column having a conductivity cell sealed between the column condenser and the receiver. Methanol with a low specific conductance may be stored for several weeks without noticeable effect.

Lund and Bjerrum [3029] dried methanol for *specific conductance studies* by means of magnesium activated with iodine. To a flask provided with a reflux condenser are added 0.5 g iodine, 5 g magnesium, and 50–75 ml of methanol. The flask is warmed until the iodine disappears. If vigorous evolution of hydrogen does not occur, add 0.5 g more of iodine and heat until all of the magnesium has been converted into methylate. An additional 900 ml of methanol is then added and the mixture boiled for $\frac{1}{2}$ hr under reflux; the product is distilled with the exclusion of moisture and redistilled over tribromobenzoic acid to remove basic impurities.

Maryott [3184] purified methanol for measurement of the *dielectric constant* by drying over magnesium ribbon and then fractionally distilling through a 180-cm Dufton column; bp 64.51°, critical solution temperature in carbon disulfide 35.2°.

Thermal properties of carefully purified methanol were determined by Fiock and coworkers [1575].

Bruun [819] collected a 1000-ml center fraction from 2500 ml of chloroform-free methanol by distilling through a 30-plate, bubble-cap column. The center fraction was refractionated in an atmosphere of dry carbon dioxide.

Pesce [3723] stated that the best way to purify methanol is to start with a product containing no acetone and fractionally distill through an efficient column, dehydrate with calcium hydride, and repeat the distillation and drying three times. This process will yield about 50 % of the starting material.

Gillo [1775] found that one distillation of methanol over sodium reduced the water content to 0.003%; after the second distillation, it was 0.00005%. Acetone cannot be removed satisfactorily by distillation; chemical means must be used. Bates and coworkers [435] removed 0.2% acetone from methanol by treating with sodium hypoiodite. Twenty-five grams of iodine was dissolved in 1 liter of methanol and the solution slowly poured, with constant stirring, into 500 ml of 1 N sodium hydroxide. The iodoform was precipitated upon the addition of 150 ml of water. After standing overnight, the solution was filtered and the filtrate boiled under reflux until the smell of iodoform disappeared. A single fractional distillation produced 800 ml of acetone-free methanol.

Bredig and Bayers [722] obtained methanol *free from aldehydes and ketones* by treatment with sodium hypoiodite and silver oxide. See also Pearce and Mortimer [3692] and the methods described under ethanol.

Weigert [5108] recommended that a sodium hydroxide solution containing iodine be added to the alcohol; that after standing for one day the mixture be poured slowly into about one-quarter of its volume of a 10% silver nitrate solution. After being shaken for several hours, it is distilled.

For *electrochemical measurements*, Hartley and Raikes [2046] used methanol freed from acetone with sodium hypoiodite. The alcohol was dried with aluminum amalgam, and ammonia and other volatile impurities were removed by boiling for 6 hr with freshly dehydrated copper sulfate (2 g/liter) while passing through a current of dry air. Sulfanilic acid was found to be less efficient for removing basic impurities. Distillation over silver nitrate to remove reducing impurities did not seem to be necessary. See also Walden and coworkers [5049].

For *optical purposes*, Herold and Wolf [2134] dried methanol with magnesium or calcium and distilled over sulfanilic acid.

Methanol free from impurities may be prepared from carefully purified methyl esters. The formation of formaldehyde may be reduced materially by working in an atmosphere of nitrogen. Haller [1979] used potassium methyl phthalate to prepared pure methanol. See also Wohler [5233], Carius [906], and Kraemer and Grodzki [2735].

REMOVAL OF WATER

Appreciable amounts of water are most easily removed by fractional distillation. Efficient distillation can reduce the water to 0.01%. Methanol absorbs water rather readily from the air. Hampton and Riddick [1996] found that the convenient way to prepare a methanol containing not more than 0.0007% water, the sensitivity limit of the test method, was by a combination of techniques. If a purity of components other than water, greater than that of a good commercial methanol, is desired, a suitable purification is made and the water content reduced to at least 0.04%. The purified material, or a commercial product containing not more than 0.04% water, is slowly percolated through a column of Type 3A or 4A molecular sieves. The effluent methanol is stored over calcium hydride. The hydride is effective as long as it remains in a lump. The solubility of calcium hydride or calcium oxide in methanol is very small. *Drierite is not suitable for drying methanol.*

Hartley and Raikes [2046] concluded that the use of calcium oxide for drying methanol is tedious and wasteful. Timmermans and Hennaut-Roland [4778] found that calcium and barium oxides do not remove all of the water; calcium also leads to contamination with ammonia. Sodium is not efficient because the hydrolysis of sodium methylate is reversible. Goldschmidt and Thuesen [1823] dried methanol repeatedly with calcium and then removed ammonia with sulfanilic acid. Bjerrum and Zechmeister [610] recommended magnesium for the preparation of absolute methanol. Calcium carbide was recommended as a drying agent for alcohols by Cook and Haines [1071].

CRITERIA OF PURITY

ACS *Reagent Chemicals* [76] lists the following requirements for Reagent grade methanol:

Appearance: clear and colorless.
Water: not more than 0.20%.
Boiling range: 1 ml–95 ml, not more than 1°; 95 ml to dryness, not more than 1°.
Residue after evaporation: not more than 0.001%.
Solubility in water: to pass test.
Acetone, aldehydes: to pass test, limit about 0.001% acetone.
Acidity (as HCOOH): not more than 0.002%.
Alkalinity (as NH_3): not more than 0.0003%.
Substances darkened by sulfuric acid: to pass test.
Substances reducing permanganate: to pass test.

The test methods are given.

A comparison of the Reagent grade specifications with the typical industrial specifications reveals that the latter are "more demanding."

Gas chromatography is a convenient method for determining the purity of methanol for impurities present in the range of 0.01–0.02% or greater. High sensitivity detectors may be used for impurities in the ppm or ppb range.

SAFETY

Methanol does not have suitable warning or irritating properties except at high concentrations.

The literature on the toxicology of methanol is voluminous, indicating a wide variation in tolerance to methanol among human beings. Apparently there is not much danger in breathing moderate concentrations of methanol vapor for a short time.

Treon [3681] concludes that industrial exposures to 200 ppm are not considered hazardous, even if the variation in individual response is kept in mind. In contrast to the many cases of blindness and fatalities from ingestion of methanol, little difficulty has been encountered in industry from inhalation or skin contact.

A comprehensive study of the toxicological effects of methanol is reported by Cooper and coworkers [1081, 2600, 2601] and Chang [948].

Alekseeva [48] developed a method for the determination of methanol and formaldehyde in air that is sensitive to 0.008 μg/ml.

The *threshold limit value* has been established as 200 ppm or 260 mg/m^3 [4509].

The *flammable limits* in air are 6.72 and 36.50%v [3680]; the upper limit is at elevated temperature. The *minimum ignition temperature* in air is 867° [3680].

59. ETHANOL

Commercial ethanol is manufactured either by chemical synthesis or by fermentation. The major portion manufactured in the United States is synthetic. It may be obtained as the water azeotrope containing about 5% water or as absolute alcohol which contains 0.1% or less water. Internal Revenue Department restrictions on the use of alcohol, make it more convenient for private laboratories to use Specially Denatured Alcohols (SDA). The formulas listed in Table 5.7 have been found [3975] to meet practically all laboratory solvent requirements of ethanol. The major producers of ethanol will supply, upon request, booklets on industrial alcohol containing a variety of useful information, the SDA formulas, and a synopsis of the United States Internal Revenue Department regulations.

NATURE AND DETECTION OF IMPURITIES

The usual impurities in fermentation alcohol are fusel oil, aldehydes, ketones, esters, and water. The exact composition and amounts of each

Table 5.7

SDA Number	Formulation
2-B	100 gal ethanol + 0.5 gal benzene
3-A	100 gal ethanol + 5 gal methanol
13-A	100 gal ethanol + 10 gal ethyl ether
23-A	100 gal ethanol + 10 gal USP acetone

component depend on the composition of the mash and the rectifying process. Synthetic alcohol contains acidity, aldehydes, ketones, and other oxygenated compounds and possibly organic sulfur and some cyclic compounds, olefinic hydrocarbons, and "unknown" polymerics.

The most satisfactory method of analysis for the impurities is gas chromatography with a sensitivity adequate for the end use of the solvent.

Deckenbrock [1246] states that "grain spirits" can be identified by the methanol content. Mohr [3369] found that 0.05% methanol could be detected by dimethyldihydroresorcinol. See also [4042, 76]. Zapletálek [5322] developed a polarographic method for the determination of methanol (1–10%) in ethanol.

Fusel oil is the trade name for a mixture that includes alcohols boiling higher than ethanol, principally the pentyl alcohols, which cause the optical activity of fusel oil. If present in more than trace amounts, fusel oil can be detected by its distinctive odor by this procedure: Wet a piece of filter paper with the alcohol and allow the alcohol to evaporate. If fusel oil is present, it can be detected as a distinctive odor on the filter paper. See also [76, 4042]. The Komarowsky color reaction [2701] has been studied by Coles and Tournay [1041]. This method is sensitive enough to detect 1 g of higher alcohols in 1000 liters of ethanol. The higher alcohols may be removed by filtering through a bed of Fuller's earth [1042]. Aldehydes may be removed by the method of Stout and Schuette [4515].

Megaloikonomos [3257] developed a method using Tollins and Deniges' reagent for the determination of acetone in ethanol in the presence of aldehydes.

Orchin [3605] has reported a practical test for detecting ethanol, 1-propanol, and acetone in the presence of each other.

Raposo and coworkers [3914] studied the minor components of fermentation alcohol and found formaldehyde, acetaldehyde, methanol, ethyl acetate, 1-propanol, and 3-methyl-1-butanol.

PURIFICATION

Ethanol is available primarily in two grades based on water content. The 190 proof, or about 95%v, is the composition of the ethanol-water azeotrope. The 200 proof is commonly referred to as *absolute alcohol*. It is termed "anhydrous" to the trade and contains not more than 0.01%v water. The water from 190 proof usually is removed by azeotropic distillation.

If it is not expedient to purchase anhydrous ethanol, or if a dryer material is desired than is available, there are a number of ways whereby a dry alcohol may be prepared in the laboratory.

The removal of the major portion of the water from 190 proof is easily accomplished by distilling the water as one of the components of a ternary heteroazeotrope. Many azeotroping agents have been used. The use of benzene to form the ternary azeotrope, benzene-ethanol-water, bp 64.48°, is the basis of a widely used commercial process for the production of anhydrous ethanol [5300]. It is also a convenient means for drying ethanol in the laboratory. A 10-plate, Penn-State column has been found adequate for laboratory use. About 5% benzene is added to the alcohol and the column adjusted so that the take-off of the lower aqueous phase is less than the amount of water condensing.

After most of the water has been removed, the rate of the aqueous phase distillation will decrease. When this stage has been reached, the lower aqueous phase is withdrawn intermittently as it collects in the head. After the water has been removed, the homogeneous benzene-ethanol azeotrope (bp 68.24°, 32.4%w ethanol) is distilled until benzene no longer separates upon dilution with distilled water. If the ethanol dried by the benzene-azeotrope process is to be used as a solvent for *ultraviolet measurements*, the column should be operated at intermediate take-off until the distillate shows no absorption in ultraviolet. About 10 ml of distillate is taken off every $\frac{1}{2}$ hr per liter of pot charge.

Hexane, cyclohexane, and 2,2,4-trimethylpentane are also convenient to use as ternary azeotroping agents.

Ethanol may be dried to a water content approaching 0.005% if adequate precautions are used to exclude water during the distillation. It has been found more convenient to take only normal precautions and get an ethanol containing not more than 0.05% water. The final drying may be accomplished with Type 3A or 4A molecular sieves or Drierite; see methanol.

The classical method for the preparation of dry ethanol is to reflux over calcium oxide, then distill or fractionally distill. The literature for this method is especially voluminous. See, for example, Noyes [3557] and Danner and Hildebrand [1187], and others cited later in the text. Sometimes cloudiness is observed in absolute ethanol prepared from calcium oxide. This is probably due to the hydrolysis of calcium ethylate, which can be prevented by careful distillation to remove all calcium salts.

Smyth and Stoops [4425] determined the *dielectric constant* of ethanol that had been refluxed over freshly ignited calcium oxide. The calcium oxide was changed once during the 24 hr of refluxing. The anhydrous ethanol was then fractionally distilled.

For *electrical measurements*, Danner and Hildebrand [1187] distilled 95% ethanol with 5 ml of concentrated sulfuric acid and 20 ml of water per liter. Acetaldehyde was removed by boiling the distillate for several hours under reflux with 10 g of silver nitrate and 1 g of potassium hydroxide per liter. After distillation, the ethanol was boiled for 8 hr with 600–700 g per liter of the best commercial lime, and the mixture was shaken vigorously at room temperature for 24–36 hr. It was boiled for a further 4–6 hr with 100–150 g per liter of special lime (see below) and distilled into a special all-glass vacuum distillation apparatus connected to the conductivity vessel where it was subjected to its final purification. Preparation of calcium oxide: A mixture of calcium hydroxide and carbonate obtained from the hydroxide by drying in air was heated until completely converted to the oxide without sintering.

Thermal properties of ethanol were determined on carefully purified material by Fiock and coworkers [1575]. Low aldehyde alcohol was refluxed over freshly burned lime and distilled. The remainder of the water was removed as the ternary benzene azeotrope in a modified Bruun still [819].

Locquin [2988] states that commercial ethanol may be dried with aluminum foil slightly amalgamated with mercury. The alcohol thus dried was used to *dehydrate biological specimens*. It was found that amalgamated aluminum foil in the bottom of the jar kept the ethanol anhydrous.

Maryott [3184] purified absolute ethanol for *dielectric constant* determinations by drying over magnesium ribbon and distilling in a 180-cm Dufton column, bp 78.34°. The *critical solution temperature* in carbon disulfide was −23.5°, indicating less than 0.1% water.

According to Winkler [5219] 98–99% ethanol is best dried with metallic calcium. Technical calcium is cleaned with a wire brush under 70% ethanol and rasped; the raspings are added to the alcohol (20 g/liter), which is then warmed on a water bath for several hours until the evolution of hydrogen is almost over. The alcohol obtained by distillation is 99.9%; further distillation over a few grams of calcium filings, the first part of the distillate being discarded, gives absolute alcohol.

Konek [2706] obtained absolute ethanol by boiling commercial absolute ethanol for 1–2 hr with 2–10% of magnesium amalgam; see also Evans and Fetsch [1521] and Andrews [244]. Lund and Bjerrum [3029] dried ethanol with magnesium activated with iodine by the method described for methanol.

Walden and coworkers [5049] used amalgamated aluminum chippings for the preparation of dry ethanol for *specific conductance measurements*. For this purpose, the aluminum chippings were degreased, treated with alkali until there was a vigorous evolution of hydrogen, washed with a little water until the washings were weakly alkaline; then they were treated with 1% mercury (II) chloride solution and after 2 min washed quickly with water, with alcohol, with ether, and then dried with filter paper. If the product is satisfactory, it will become warm during this treatment. The amalgam was then added to the alcohol to be dried, gently warmed until the evolution of hydrogen ceased (requiring a few hours), then distilled and pure air passed through the distillates for a long time. See also Thomas and Maxum [4719], Kraus and Callis [2743], Wislicenus and Kaufmann [5224], Osborne and coworkers [3602], Brunel and coworkers [811] and Müller and coworkers [3448].

According to Leighton and coworkers [2897], the best method for the preparation of ethanol for *optical measurements* is the following: 95% ethanol is distilled over a period of several hours with 25 ml of 12 N sulfuric acid per liter, and the distillate heated under reflux with 20 g of potassium hydroxide and 10 g of silver nitrate per liter and distilled. After standing for one week over activated aluminum amalgam, it is filtered and distilled again. The material so obtained showed a smaller absorption than ethanol dried over lime. See also Lüthy [3038].

Goldschmidt [1817] prepared absolute ethanol from ethanol dried with lime by boiling several hours with a quantity of nitride-free calcium, in pieces the size of a pea, corresponding to ten times the water content. For removing traces of ammonia, a stream of dried air, dried by passing over sodium wire, was led into the vapor during the boiling.

According to Riiber [3982], the last traces of water can be removed from 99.9% ethanol, obtained by drying with lime, by means of calcium hydride. This, he claims, is more suitable for the rapid, convenient, and certain preparation of absolute ethanol than calcium oxide or metallic calcium. The process is carried out as follows: Three and one-half grams of calcium hydride is crushed to a powder and dissolved in 200 ml of absolute alcohol by gently boiling. Two-thirds of the alcohol is then distilled off to remove ammonia. The calcium ethylate solution so obtained is poured into 2 liters of 99.9% alcohol in a distillation apparatus fitted with a double condenser. The mixture is boiled for 20 hr so slowly that only about 100 ml, which contains all aldehydes, distills; at the same time a slow stream of pure and carefully dried hydrogen is bled in. Slow distillation, in a current of hydrogen, the first and last 100 ml of the distillate being discarded, gives pure ethanol; d 20° 0.789334 ± 0.000003.

Smith and Bonner [4403] purified absolute alcohol for *refractive index studies*. It was first tested for methanol, 2-propanol, 2-methyl-2-butanol, pentanol, carbonyls, acids, benzene, water, and nonvolatile matter. The only appreciable impurity found was water. The alcohol was dried for three days with calcium sulfate and distilled in a dry nitrogen atmosphere; bp 78.32°.

Othmer and Wentworth [3616] prepared anhydrous ethanol by using ethyl ether as the entrainment agent in the distilling column. The column was operated under pressure, and under these conditions the azeotrope contained only ether and water. No ternary azeotrope was formed. See also Wentworth [5125].

The use of dichloromethane for drying ethanol was proposed in [333]. It forms an azeotrope with water boiling at 38.1° and containing 1.8% water.

Arnold [285] found C_8–C_{12} hydrocarbons more effective for dehydrating ethanol than benzene; 2,2,4-trimethylpentane is preferred. When it is used, 3–5% of the aqueous phase is entrained in the hydrocarbon phase, up to 25% in benzene. The stratification is faster when aliphatic hydrocarbons are used. The data indicate a saving in heat requirement of 9–19% for the hydrocarbons over benzene.

Several chloroparaffins have been proposed to remove water as the ternary azeotrope. The ternary azeotropes of ethanol, water, and other compounds are given by Horsley [2254, 2255].

Maruyama [3180] describes a method for purifying alcohol for use in *polarography*. Ethanol is discussed as a *polarographic solvent* [4814]. Maruta and Matubora [3179] determined as little as 0.0007% acetaldehyde and 0.0005% diacetal furfural in ethanol *polarographically*.

Hassion and Cole [2058] prepared alcohol for *electrical measurements* in the same manner as they did 2-propanol, collecting the fraction distilling at 78.3°, n_D 25° 1.3594, and specific conductance 8.2×10^{-8} ohm^{-1} cm^{-1} at room temperature.

Teramoto and coworkers [4624] purified alcohol from sweet potatoes, barley, molasses, and so forth by passing over an acid and then an alkaline ion exchange resin. Up to 95% of the aldehyde and 40% of the fusel oil were removed.

For purification for *dielectric constant measurements*, see [3802], also [3309, 2058].

Izergin [2367] purified ethanol by zone melting.

CRITERIA OF PURITY

The criteria of purity for an analytical Reagent grade are given by Rosin [4042] and ACS *Reagent Chemicals* [76]. Water is easily determined by the Karl Fischer method. If water is the principal impurity, as it is in any good grade of ethanol, it may be determined from density measurements and density tables [3496].

Gas chromatography of the required precision to characterize ethanol for the desired end use is to be recommended.

Gillo [1776] of the Bureau des Etalons Physico-Chemiques proposed a method for the purification and criteria of purity for ethanol as an organic standard. Five hundred milliliters of ethanol and 75 ml of carbon disulfide are placed in an all-glass distilling column and distilled in an atmosphere of nitrogen. Three hundred milliliters of ethanol can be obtained, which will meet the following specifications: bp 78.30°, density 0° 0.80624, ethyl acetate 0.003%, acetaldehyde 0.004%, water 0.001%, 2-propanone 0.001%, methanol 0.0005%, and carbon disulfide 0.0001–0.0002%. Careful fractional distillation will remove the ethyl acetate; the acetaldehyde and methanol are difficult to remove.

Dreisbach and Martin [1372] determined the purity from the freezing curve. Swietoslawski [4577] used the differential ebulliometric method for determining both volatile and nonvolatile impurities.

SAFETY

It is practically impossible to produce any toxic effects by inhalation of ethanol vapors under usual industrial conditions. The minimum identifiable odor is about 350 ppm. Concentrations of 6000–9000 ppm have an intense odor that may be practically intolerable at first, but one becomes acclimated soon. Concentrations of about 1000 ppm cause slight irritation of the mucous membranes, and other symptoms [3681, 4509].

The *threshold limit value* has been set at 1000 ppm or 1900 mg/m³ [4509]. The *flammable limits* in air are 3.28 and 18.95%v; the upper limit is at elevated temperature. The *minimum ignition temperature* in air is 439° [3680].

60. 1-PROPANOL

1-Propanol is produced commercially by the hydration of propylene. It is available as either the anhydrous alcohol or as the water azeotrope that contains 72% 1-propanol. Some manufacturers produce a product comparable in purity to the best grade of ethanol. One of the principal impurities is 2-propene-1-ol.

1-Propanol is also produced by the partial oxidation of propane [5178].

Water may be removed from the 72% alcohol by azeotropic distillation similar to the procedures described for ethanol. Ternary azeotropes suitable for this purpose are given by Horsley [2254, 2255].

De Brouckère [1244] purified technical 1-propanol by refluxing over lime for 5 hr and then distilling through a 1-meter column. The fraction boiling at 97–97.3° was redistilled in a stream of hydrogen at 40–50-mm pressure. The best fraction had a specific conductance of 4.4 × 10⁻⁹ ohm⁻¹ cm⁻¹.

Brunel [810] purified the alcohol by means of the phthalate ester.

Brown and Ives [762] prepared 1-propanol for *dielectric constant* measurements. Some trouble was experienced in obtaining material that did not give a slight haze with water. This was overcome by bromination to remove 2-propene-1-ol, refluxing with sodium hydroxide treatment with, and distillation from, Hewett's reagent, and a final fractional distillation through a 4-ft column; bp 97.1°, d 25° 0.79995.

Kretschmer [2753] found that 1-propanol contained as its principal impurity about 1.5% 2-propene-1-ol. He carefully purified the alcohol for the determination of the *thermal expansion* as follows: About 1.5 ml of bromine was added to a liter of alcohol. It was fractionally distilled from a small amount of potassium carbonate through a 75-plate column. The middle fraction of 600 ml was dried with 1 g of magnesium ribbon, freshly cleaned with steel wool, in a storage flask attached to a vacuum system. The dried sample was then withdrawn as needed by vacuum distillation at room temperature. Before the flask was sealed, 1 g of 2,4-dinitrophenylhydrazine was added to react with propionaldehyde that had not been removed by distillation.

Mathews and McKetta [3206] studied the *thermodynamic properties* of 1-propanol purified by fractional distillation twice through a 1.2 meter column packed with glass helices. The product has a boiling range of less than 0.02°.

Purification to 99.9%m for *physical and thermodynamic* property studies, see Biddiscombe [568] in Alcohols, General.

For *specific conductance measurements*, Keyes and Winninghoff [2570] dried with metallic sodium and fractionally distilled. See also Kraus and Bishop [2741]. Goldschmidt and Thomas [1822] dried 1-propanol with aluminum amalgam and, to remove the basic impurities, distilled over sulfanilic or tartaric acid.

For *refractive index measurements*, Berner [536] boiled 1-propanol with lime for 6 hr and, after distilling, warmed the middle fraction of about 80% with calcium hydride in a stream of hydrogen for 3 hr. He finally fractionated in a stream of hydrogen. The treatment with calcium hydride and the fractional distillation were repeated until the density and refractive index were constant.

For additional purifications, see Brunel and coworkers [811] and Lund and Bjerrum [3029].

SAFETY

1-Propanol is one of the less toxic and one of the least irritating solvents as indicated by the almost complete absence of ill effects reported from industrial usage. Two of six rats died when exposed for 6 hr to a concentration of the alcohol of 4000 ppm. Some of the symptoms that may become apparent if the concentration in the air is sufficient are intoxication, ataxia, and irritation of the mucous membranes [3681]. The LD_{50} for rats, single dose, is 1.87 g/kg [4432]; see also [4615].

The *threshold limit value* of 200 ppm or 490 mg/m³ has been recommended [4509, undated supplement].

The *flammable limits* in air are 2.15 and 13.50%v at elevated temperature. *The minimum ignition temperature* in air is 439° [3680].

61. 2-PROPANOL

2-propanol is manufactured by absorption of propylene in sulfuric acid and subsequent hydrolysis of the mono- and diisopropyl sulfate to propanol and sulfuric acid. The crude mixture is distilled as the 91% alcohol-water azeotrope. The process is basically the same as reported by Bertholet in 1855. The ternary heteroazeotrope is used to produce anhydrous 2-propanol commercially. Isopropyl ether is often used as the heteroazeotroping agent but ethyl ether, ethyl-*tert*-butyl ether, and several other agents are also used [1485].

An especially high purity 2-propanol for such uses as perfumes is made by the hydrogenation of acetone.

The availability of 2-propanol of a high purity and at a reasonable price has resulted in the use of large volumes as a solvent to replace ethanol for many of its uses. The anhydrous alcohol from several sources, including several grades such as spectrographic, reagent and essence, were analyzed by gas chromatography and by the Karl Fischer method for water. The gas chromatograph had a sensitivity for most substances in the range of 0.01–0.02%. Only four impurities, including water, were detected. Water was present in the greatest amount, 0.05–0.06% [3975].

Specifications are available for several grades of 2-propanol. Some of these are included in Table 5.8.

For purification to 99.9+ %m for *physical and thermodynamic properties*, see Biddiscombe [568], Alcohols, General.

Itakura [2360] prepared 2-propanol by the catalytic hydration of propylene with sulfuric acid. The maximum yield was 76% obtained by using 99% sulfuric acid at 15°. At 100° with 85% acid, hardly any propylene was absorbed.

Reynolds and Grudgings [3962] obtained an increased yield of 2-propanol and less polymer formation by reducing propylene with tungsten oxide catalyst. When the catalyst was pelletized at 62 tons psi pressure, the conversion at 250° was 16%. If the pellets were formed at only 32 tons pressure, the conversion at the same temperature was 10%.

For additional preparations, see de Melle [1276], Hunter [2325], and Brunel [810].

Ozol and Masterson [3624] purified 2-propanol, produced from olefins such as propylene, by first treating with aqueous sodium hydroxide solution, followed by preliminary distillation. Before final distillation, the alcohol was stabilized and the odor improved by adding a small amount of copper (I) chloride (0.5% or less).

Frejacques [1629] prepared anhydrous 2-propanol by saturating an aqueous solution with ammonia and carbon dioxide. Two layers were formed. The water-rich layer was distilled, yielding ammonia, carbon dioxide, and the azeotrope. The water-poor layer was fractionally distilled to give the azeotrope and anhydrous alcohol.

Maryott [3184] purified a good commercial grade by drying over magnesium ribbon and fractionally distilling in a 180-cm Dufton column. The best fraction, boiling at 82.33 to 82.39°, was used for *dipole moment studies*.

Table 5.8 Specifications for Several Grades of 2-Propanol

	ACS Reagent Grade	ASTM D 770-59	Commercial 99%	Commercial 91%	Essence[1] Grade	British Standard
Density, 25°C	0.781–0.783	—	—	—	—	—
Sp. Gr., 20°/20°C	—	0.785–0.790	0.7862–0.7867	0.8175–0.8185	0.785–0.789	0.785–0.789
Color, not greater than	Co-Pt 10	Colorless	Co-Pt 10	Co-Pt 10	Water-white[10]	Colorless
Distillation Range, °C	See [76]	1.5°C[8]	0.5°C[8]	1.0°C[9]	95% 80–82	95% 81–83%
Nonvolatile matter, max	0.001%	0.005 g/100 ml	0.001 g/100 ml	0.002 g/100 ml	0.001%	0.01%
Odor, characteristic	—	Nonresidual	Nonresidual	Nonresidual	—	—
Water, max %	0.50[3]	0.5[2]	0.1	—	1.0	0.5
Miscibility with water	PT[4]	PT[4]	Clear	Clear	Complete	Complete
Acidity, as H_2SO_4 max %	PT[4]	0.002[5]	0.002[5]	0.0024[5]	None	0.002
Alkalinity[6]	PT[4]	—	—	—	None	None
Carbonyls,[7] max	—	—	—	—	—	0.2
Permanganate time, 15°C	—	—	20 minute	30 minute	—	—
2-Propanol, min %	—	99%	99.8%w	87.50%w	98	—
Bibliography No	76	See title above	4892	4892	3160	3160

1. Essence or Cosmetic grade. 2. Based on cloud test. 3. Karl Fischer method. 4. Pass test. 5. Acetic acid. 6. To phenolphthalein. 7. As acetone. 8. To include 82.3°. 9. To include 80.4 ± 0.1°C. 10. This is an indefinite term. One of the definitions of (Water-white) is a Hazen or Co-Pt value of 25–30.

Mathews [3208] purified 2-propanol for the determination of the *heat of vaporization* by drying first with calcium chloride and then with barium oxide, and then three careful fractional distillations.

Lebo [2851] prepared "absolute" 2-propanol by distillation over freshly burned lime through an efficient column. The fraction boiling at 82–82.4° was collected and shaken for 2 days with anhydrous copper sulfate. It was then fractionally distilled several times until the boiling point remained constant. The alcohol so obtained contained less than 0.010% water.

For drying with calcium oxide or aluminum amalgam, see Brunel and co-workers [811].

Aerov and Motina [30] report that isopropyl ether is economically advantageous for removing the water from 2-propanol. They rectified alcohol containing 45.6%m water in a 44-sieve plate column at a reflux ratio of 8.6:1.

Aluminum isopropylate, $Al(OCH(CH_3)_2)_3$, was found by Bezborodko [562] to be effective for drying 2-propanol and removing traces of catalyst.

Catterall [935] dried 91%v 2-propanol to 99.9% by continuous azeotropic distillation with ethyl ether in a column of about 50 plates at 150–170 psig.

Berman and coworkers [532] purified commercial 2-propanol by storing over magnesium ribbon, followed by distillation through a 1.2-m, glass-ring packed column. The middle cut boiling over a range of 0.04° was stored under vacuum.

The *solubility* of normally gaseous paraffinic hydrocarbons was studied by Kretschmer and Wiebe [2754] in a 2-propanol that was purified by adding a small quantity of 2,4-dinitrophenylhydrazine and fractional distillation through a 50-plate column; d 25° 0.7881.

For *optical measurements*, Herold and Wolf [2134] dried with magnesium and distilled from sulfanilic acid.

A laboratory method for purifying 91% 2-propanol may be found in [1781].

Métra and coworkers [3283] developed a method for the determination of 2-propanol in the presence of other alcohols. An oxidimetric method for the determination of 2-propanol in a mixture of methanol, ethanol, 1-propanol, and acetone is described by Wehle [5105].

For use as a polarographic solvent, see [4814].

CRITERIA OF PURITY

Test methods are available for all of the specifications given in Table 5.8.

Swietoslawski [4577] characterized 2-propanol by differential ebullio-metric measurements.

Gas chromatography procedure of suitable sensitivity for the end use of the solvent is common practice.

SAFETY

The effect of exposure to 800 ppm of 2-propanol in air is not severe but most people find it objectionable. Mild irritation to the eyes, nose, and throat is experienced by most human subjects to 400 ppm for 3–5 min. The minimum identifiable concentration is about 200 ppm [3681].

The *threshold limit value* has been established at 400 ppm. This is low enough to prevent narcosis but slight irritation may occur [4509].

The *flammable limits* in air are 2.02 and 11.80%v at elevated temperature [3680]. The *minimum ignition temperature* in air is 853°F [3681].

The Butyl Alcohols

62. 1-BUTANOL

Bremner [723] prepared 1-butanol in 70% yield by the hydrogenation of furan at 100 atm and 220° in the presence of copper-barium oxalate catalyst. The production of tetrahydrofuran or its derivatives is suppressed if a small amount of an acid, or acid salt, is added.

For *dielectric constant measurements*, Smyth and Stoops [4425] refluxed 1-butanol, prepared by fermentation, over freshly ignited lime and fractionally distilled the product.

Jones and Christian [2422] refluxed 1000 g of a good commercial grade of 1-butanol over 50 g of freshly ignited calcium oxide for 4 hr. The outlet tube was protected from atmospheric moisture. The alcohol was decanted from the lime, refluxed with magnesium turnings, and then distilled through a 75-cm column. A small, low boiling, fraction was discarded. The bulk of the material distilled at 117.70° ± 0.01°, d 0° 0.82460, d 25° 0.80572.

Williams and Daniels [5194] digested 1-butanol over unslaked lime and distilled. The center portion was further digested over barium oxide, then treated with sodium and fractionally distilled. The purified alcohol was used for *specific heat measurements.*

Clarke and coworkers [994] washed 1-butanol with dilute sulfuric acid and with sodium bisulfite solution to remove bases, aldehydes, and ketones. Esters were removed by boiling for 1.5 hr with 20% sodium hydroxide. The alcohol was dried with potassium carbonate, followed by barium oxide, and finally distilled through an efficient column.

Harking and Wampler [2025] treated 1-butanol with sodium bisulfite solution, then boiled for 4 hr with 10% sodium hydroxide solution, washed the separated alcohol with water, and neutralized with hydrochloric acid. The alcohol was dried overnight with lime and boiled three times with fresh lime for 3 hr. It was finally fractionally distilled.

The usual purification did not appear satisfactory to Orton and Jones [3597]. They purified by means of the sodium salt of the salicylic ester, which separates out immediately when the ester is dropped, with vigorous stirring, into a slight excess of sodium hydroxide. It is pressed out and converted back to the alcohol by treatment with very dilute hydrochloric acid. The purification process is then repeated. The 1-butanol is dried with potassium carbonate and distilled.

Reagent grade 1-butanol was refluxed for several hours with lime by Venkatasetty and Brown [4935] and distilled three times. The fraction boiling between 117.3 and 117.5° was collected from the final distillation.

For purification to >99.9%m for *physical and thermodynamic properties*, see Biddiscombe [568], under Alcohols, General.

1-Butanol containing 9% water can be dried by fractional distillation. The water passes over in the first fraction as the binary azeotrope containing about 27% water [3975].

For *optical measurements* Herold and Wolf [2134] dried 1-butanol with magnesium ribbon and distilled over sulfanilic acid. See also Berner [536].

For *specific conductance measurements*, Goldschmidt and Mathiesen [1819] used 1-butanol, boiling between 116 and 117°, dried by Walden's method with aluminum amalgam. The dried alcohol was distilled from tartaric acid.

For drying with barium oxide or with calcium and aluminum amalgam, see Brunel and coworkers [811].

For 1-butanol as a polarographic solvent, see [4814].

CRITERIA OF PURITY

The specifications for Reagent grade 1-butanol are given by Rosin [4042] including distillation range, nonvolatiles, and free acidity.

Orton and Jones [3597] used the critical solution temperature in hydrochloric acid to characterize the purity.

Dreisbach and Martin [1372] determined the purity by the Swietoslawski differential ebulliometric method [4577].

SAFETY

1-Butanol is potentially more toxic than any of the lower alcohols but the actual hazard is reduced by its low vapor pressure. Humans briefly exposed to concentrations of 25 ppm showed mild irritation of the nose, throat, and eyes. A concentration of 50 ppm is objectionable because it produces pronounced irritation of the throat of all persons and mild headaches in some. The minimum concentration with identifiable odor is 15 ppm [3681].

The *threshold limit value* has been set at 100 ppm or 300 mg/m^3 [4509].

The *flammable limits* in air are 1.45 and 11.24%v; the upper limit is at elevated temperature. The *minimum ignition temperature* in air is 345° [3681].

63. 2-BUTANOL

Whitman [5152] prepared 2-butanol by hydrogenation of 3-hydroxy-1-butene in the presence of titanium hydride at 175° and pressures of 1000–2000 psi. The reaction mixture contained 26% 2-butanol.

Houtman and coworkers [2261] prepared 2-butanol by hydrogenating butanone at 200° in the presence of a nickel catalyst. See also Brunel [810].

Dannhauser and Cole [1190] prepared 2-butanol for *electrical measurements* by drying over calcium hydride and distilling; $\kappa = 1.0 \times 10^{-7}$ ohm^{-1} cm^{-1}.

For purification to 99.9 + %m for *physical and thermodynamic properties*, see Biddiscombe [568], Alcohols, General.

Berman and McKetta [533] purified 2-butanol for *thermodynamic study* by repeated fractional distillation through a 1-inch, 90-plate, Oldershaw column to a purity of 99.92% by gas chromatography analysis.

Pickard and Kenyon [3753] have described a method for *resolving racemic 2-butanol* into its optically active isomers and have recorded some of their *physical properties*.

The method is further refined and described [3755] wherein ethylmagnesium bromide and acetaldehyde are employed in large scale preparations. The resolution employs the 2-butyl hydrogen phthalate and brucine. The salt was recrystallized from acetone, then methanol, followed by hydrolysis.

2-butanol and 2-pentanol were resolved into their *optical isomers* by Adembri [25] by slowly reacting the recemate with phosgene at −60°, allowing the phosgene and hydrogen chloride to escape, and then purging with air. The chlorocarbonates were reacted with 1-asparagine and the isomers separated by crystallization. The salts were hydrolyzed with 25% potassium hydroxide and the alcohols isolated and purified. The optical rotation was taken in ethanol and was found to be; L(+)-2-butanol 20° [α]$_D$ +13.8°, D(−)-2-butanol 20° [α]$_D$ −13.5°, (−)-2-pentanol 20° [α]$_D$ −16.8°, and (+)-2-pentanol 20° [α]$_D$ +16.6°.

Hargreaves [2019] has purified 2-butanol by conversion to the alkyl hydrogen phthalate and recrystallization to constant values.

Levene and coworkers [2918] describe a method for preparing the optical isomers of 2-butanol.

Optically active 2-butanol was prepared from material obtained from a chemical supply house by Kantor and Hauser [2495]. The alcohol was converted to the acid phthalate ester and 2.01 moles thoroughly mixed with 2.0 moles of brucine. The mix was charged into a 6-liter Erlenmeyer flask with 2-liters of acetone and refluxed for 24 hr. The solid changed in appearance during the reflux period and was kept broken up so that the acetone could come in contact with all of the solid. The mixture was filtered hot and recrystallized two or three times from methanol. The brucine salt was hydrolyzed with sodium hydroxide to give a high purity (+)2-butanol; bp 98–99.5°, d 30° 0.799, n_D 23° 1.3955, [α]$_D$ 27° +13.28° corresponding to an optical purity of at least 98.2% based on the highest reported value, +13.52, found by Timmermans and Martin [4784].

Synthetic 2-butanol may have traces to 1–2% of one or more of the following substances depending on the method of manufacture and purification: 2-propanol, C_8 (mostly) hydrocarbons, butanone, sec-butyl ether, 2-methyl-2-propanol, and other alcohols.

SAFETY

Limited data for the acute toxicity of 2-butanol indicate that is is less toxic than the normal homologue. The physiological response of animals to the vapors of the alcohol in air is about the same. There does not appear to be any difference in the toxicity of the optical isomers of 2-butanol [3681; 4509, undated supplement].

The *threshold limit value* has been set at 150 ppm or 450 mg/m³. This limit has been set on meager data mainly to prevent narcotic and irritative effects [4509, undated supplement].

The *minimum ignition temperature* in air is 406° [3680].

64. 2-METHYL-1-PROPANOL

Itakura [2360] prepared 2-methyl-1-propanol by the catalytic hydration of isobutylene with sulfuric acid. A maximum yield of 25% was obtained by using 50% sulfuric acid at 105°.

The preparation of 2-methyl-1-propanol and 3-methyl-1-butanol is easily accomplished by fractional distillation in the presence of sodium chloride, according to Chatskii [4548]. Further purification is obtained by extraction with benzene and carbon tetrachloride.

Sarancha and Dubovikova [4114] studied the products of synthesis of 2-methyl-1-propanol from carbon monoxide and hydrogen on a 2-meter chromatograph column of diatomite saturated with dinonyl phthalate. The by-products were identified as 1-butanol, 3-methyl-1-butanol, 1-propanol, and methanol.

Brunel [809] boiled commercial 2-methyl-1-propanol with lime for 5 hr and then with calcium shavings for 1.5 hr. The dried alcohol was fractionally distilled with care several times through an efficient column. See also Brunel and coworkers [811].

Michael and coworkers [3295] fractionally distilled a good commercial grade through a long Hempel column until the boiling point was constant to within 0.15° and obtained a 40% yield. After being shaken with dilute sulfuric acid to remove basic substances, the alcohol was washed with water and dried, first with ignited potassium carbonate and then with lime. The alcohol was further purified by conversion to the borate ester by heating for 6 hr in an autoclave at 160–175° with a quarter of its weight of boric acid. The ester was fractionally distilled several times in a vacuum, then hydrolyzed by heating for a short time with aqueous alkali. After drying with lime, it was distilled.

For purification to 99.9 + %m for *physical and thermodynamic* properties, see Biddiscombe [568], under Alcohols, General.

Hückel and Ackermann [2289] converted 2-methyl-1-propanol into the acid phthalate by heating with phthalic anhydride. The ester was recrystallized from petroleum ether until the melting point was constant at 65° and then hydrolyzed with 15% potassium hydroxide. The alcohol was distilled as the water azeotrope, dried with potassium carbonate, anhydrous copper sulfate, and finally with magnesium turnings. It was then fractionally distilled.

Mathews [3208] dried commercial 2-methyl-1-propanol repeatedly with barium oxide and carefully distilled. The dried material was used for the determination of the *heat of vaporization.*

For *specific conductance measurements,* Goldschmidt [1818] dried and purified the alcohol by repeated treatment with calcium or by a single treatment with aluminum amalgam. See Walden [5049]. The dried material was then fractionally distilled from sulfanilic or tartaric acid.

SAFETY

2-Methyl-1-propanol, generally, may be considered of the same order of toxicity as 1- and 2-butanol. Its primary action is narcotic. No evidence of eye irritation was noted with repeated 8-hr exposures. It may be a skin irritant [3681].

The *lower flammable limit* in air is 1.68%v. The *minimum ignition temperature* in air is 434° [3681].

65. 2-METHYL-2-PROPANOL

Remiz and Frost [3955] prepared 2-methyl-2-propanol by hydrating iso-butylene with solutions containing 10–30%w of sulfuric acid and 2–3% silver sulfate at 85–95°. When the gas was passed over the catalytic solution rapidly, the yield of the alcohol was 2–3 kg per kilogram of the sulfuric acid-silver sulfate mixture. An increase in the temperature or acid concentration caused an increase in the polymerization processes. See also Katsuno [2523].

Getman [1748] determined the *cryoscopic constant* of 2-methyl-2-propanol that was dried over lime and fractionally distilled.

Bigelow [579] recommended 2-methyl-2-propanol as a solvent for *cryoscopic molecular weight determination.* It has a convenient melting point at, or just below, room temperature and is available in adequate purity at a reasonable cost. The use of this alcohol, in general, is limited to substances that do not associate extensively with it such as hydrocarbons, carbonyls, tertiary amines, ethers, and nitro- and halogen compounds. A simple, rapid method is described that utilizes common laboratory equipment.

Fritz and Marple [1668] found that 2-methyl-2-propanol was a solvent that was well suited for the differentiating of titration-weak acids and weakly acidic compounds.

For purification to 99.9+%m for *physical and thermodynamic properties*, see Biddiscombe [568], under Alcohols, General.

Parks and Anderson [3647] purified a good grade of a commercial alcohol by distilling from lime and fractionally crystallizing the distillate eight times; mp 25.4°.

Simonsen and Washburn [4332] reported evidence of a second form of 2-methyl-2-propanol melting at 25.0°.

Natradze and Novikova [3498] extracted 2-methyl-2-propanol from its aqueous solution with chloroform. The extracted mixture was fractionated several times to free it from water. The chloroform was distilled and 2-methyl-2-propanol containing 0.1–0.15% water was obtained.

Maryott [3184] distilled 2-methyl-2-propanol from sodium and fractionally crystallized from its own melt until no change in melting point occurred; mp 24.85–25.0°.

Parks and coworkers [3662] purified 2-methyl-2-propanol melting at about 20° for *cryoscopic studies* by two fractional distillations, followed by two fractional crystallizations; mp 25.4°.

DeVries and Soffer [1295] proposed the use of 2-methyl-2-propanol as a *thermometric standard*. A commercial grade of the alcohol, melting at 24.08°, was first crystallized in an open beaker. This resulted in the lowering of the freezing point due to the absorption of moisture from the air. The alcohol was then dried over lime, decanted, and distilled into a cell, to the inlet tube of which a short length of rubber tubing fitted with two pinch clamps was then attached. The alcohol in the cell was frozen until only 5–10 ml of liquid remained. The cell was inverted and the liquid was withdrawn by alternately opening and closing the pinch clamps. This crystallization procedure was repeated 13 times. Thirty-seven cooling curves were made with an NBS calibrated platinum resistance thermometer, or with one calibrated at the ice and steam points. After the addition of a small amount of calcium hydride, the freezing point was increased by 0.06° to 25.66°.

Smyth and McNeight [4419] purified 2-methyl-2-propanol for *electrical measurement studies* by washing with a saturated solution of sodium bicarbonate, refluxing with 10% sodium hydroxide, distilling, and drying with sodium sulfate or potassium carbonate. It was then refluxed over lime and distilled. It was again distilled after standing over calcium turnings for two days. A final distillation was made from tartaric acid; bp 8.23°.

A Certified Reagent grade 2-methyl-2-propanol was purified by Cocivera [1014] for *kinetic measurements* by refluxing for 24 hr with 1 g calcium hydride per liter of alcohol and then distilling. The center portion was redistilled with about 25 mg benzoic acid per liter. The center fraction contained less than 10^{-5} M acid and less than 0.002 M water.

CRITERIA OF PURITY

Dreisbach and Martin [1372] determined the purity from the freezing curve.

SAFETY

Animals exposed to vapors of 2-methyl-2-propanol show symptoms of intoxication similar to other butyl alcohols. It has a stronger narcotic action on mice than the normal- or iso-homologues. Skin contact by humans resulted in only slight erythemia and hyperemia [4509].

The *threshold limit value* has been set at 100 ppm or 300 mg/m³ to prevent narcosis [4509].

The *lower flammable limit* in air at 25° is 2.35%v; the *upper flammable limit* in air at 55° is 8%v. The *minimum ignition temperature* in air is 478° [3681].

The Pentyl Alcohols

All of the eight pentyl alcohols, $C_5H_{11}OH$, have been included in the Third Edition.

There are three commercial sources of the pentyl alcohols. The earliest of these sources was fusel oil; see 3-methyl-1 butanol. All of the pentyl alcohols are obtained, except 2,2-dimethyl-1-propanol, when a mixture of pentane and 2-methylbutane is chlorinated and the chloropentanes hydrolyzed to the alcohols [331]. These alcohols are available in varying degrees of purity; their composition may be determined by gas chromatography. The pentyl alcohols are also produced by the Oxo process. The impurities are other monomeric alcohols, dimeric alcohols, acetals, and several other miscellaneous substances. These impurities vary from producer to producer.

Several references in this section discuss the impurities in the pentanols from different sources.

SAFETY

The physiological action of the several pentyl alcohols does not differ to an appreciable extent. 3-Methyl-1-butanol (isoamyl) appears to have been studied the most. The most important effect to be expected from vapor inhalation is narcosis. The narcotic action for rabbits decreases from 2-pentanol to 2-methyl-2-butanol to 3-methyl-1-butanol; the toxicity of the same isomers decreases in the following order: tertiary, secondary, primary. The toxicity for rats decreases from 2-methyl-2-butanol to 2-pentanol to 1-pentanol. The pentanols may be irritating to the skin [3681, 4509].

66. 1-PENTANOL

1-Pentanol can be prepared in 56–61% yields from ethyl valerate [19].

Brown and Subba Rao [769] prepared primary alcohols from internally unsaturated hydrocarbons through hydroboration and oxidation.

Table 5.9 Safe Air Concentration and Flammability of the Pentyl Alcohols

Isomer	Threshold Limit Value	Minimum Ignition Temperature	Lower Flammability Limit
66. 1-Pentanol	—	391°	1.19%v
67. 2-Pentanol	—	343	—
69. 3-Methyl-1-butanol	100 ppm[a]	343	1.2
71. 2-Methyl-2-butanol	—	437	6.0

[a] 360 mg/m³.   [3681, 4509]

Medoks and Ozerskaya [3254] prepared 1-pentanol by the Grignard reaction between butylmagnesium bromide and trioxymethylene.

Olivier [3589] purified 1-pentanol by esterifying with *p*-hydroxybenzoic acid, crystallizing the ester from carbon disulfide, and saponifying with alcoholic potassium hydroxide. Further purification by drying with Drierite and fractionally distilling will give a high purity product.

1-Propanol, 2-methyl-2-propanol, and 1-pentanol were separated from purified fermented sulfite liquor [845].

67. 2-PENTANOL

Brauns [719] prepared 4.5 kg of the inactive 2-pentanol from bromopropane and acetaldehyde by the Grignard method. The directions are given by Pickard and Kenyon [3753]. The alcohol was kept for a few months over dry potassium carbonate and fractionally distilled; bp 119.4°, yield 40%. The racemic mixture was then separated into the *optically active* forms. Physical properties are given.

Brunel [810] prepared 2-pentanol from the corresponding ketone and purified it by fractional distillation.

Brown and Subba Rao [767, 769] reported a method for the preparation of primary alcohols from olefins by aluminum borohydride. See General, Aliphatic Monohydric Alcohols.

Timmermans [4772] recommended the preparation of 2-pentanol from acetaldehyde and propylmagnesium bromide. Paraldehyde, bp 124°, forms an azeotrope boiling at 118.5°, which is difficult to separate. The alcohol is reported to be unstable at its boiling point. Brown and Nakagawa [766] obtained 82% yield from methylmagnesium iodide and butyraldehyde.

Norton and Hass [3552] prepared 2-pentanol by reacting 1,2-epoxypropane with (1) diethylmagnesium, yield 23%, or (2) ethylmagnesium bromide, yield 11.7%.

68. 3-PENTANOL

Sabatier and Senderens [4085] prepared 3-pentanol by the reduction of 3-pentanone with hydrogen on nickel catalyst at 130–140°.

Grignard [1897] prepared 3-pentanol from ethylmagnesium bromide and ethyl formate; see also Brown and Nakagawa [766].

Hamelin [1987] found that ethylmagnesium bromide and propionaldehyde gave 3-pentanol, propanol, and *tert*-heptanol; the distribution depended on the reaction conditions.

For separation of the isomers, see 2-butanol [25].

69. 2-METHYL-1-BUTANOL

Houtman and coworkers [2261] prepared 2-methyl-1-butanol from formaldehyde and 2-butylmagnesium bromide. Brown and Zweifel [771] obtained the alcohol from internally unsaturated hydrocarbons by hydrobromination.

Birun [602] found that the inactive 2-methyl-1-butanol reacted 2.5 times as fast with hydrogen chloride as the active form. A mixture was enriched with the active form in this manner.

Kortuem and Faltusz [2718] found that 2,3-dichloro-1-propanol was the most effective entrainer tested for the separation of 2-methyl- and 3-methyl-1-butanols by extractive distillation. The average *alpha* for the mixture could be increased from 1.078 to 1.2 with 92% of the entrainer.

2-Methyl-1-butanol may be prepared in a high state of purity by fractional crystallization of a suitable ester such as the trinitrophthalate [4772]. Terry and coworkers [4625] prepared the 3-nitrophthalic ester of the alcohol by the method of Marckwald and McKenzie [3144]. The acid ester was converted to the cinchoidine salt in acetone and then recrystallized from chloroform by adding pentane. The salt was saponified, extracted with ether, and fractionally distilled; bp 128.5°, n_D 25° 1.4082, d 25° 0.8161, and $[\alpha]_D$ 25° −9.50°. d-2-Methyl-1-butanol was prepared by Ehrlich [1455] from *d*-isoleucine.

Easton and coworkers [1432] found that fusel oil contained up to 20% dextrorotatory 2-methyl-1-butanol. They isolated the compound to measure the *optical rotation* by first drying a commercial product with calcium oxide and then fractionally distilling in a 73–117 theoretical plate Stedman column operating at 70:1 to 120:1 reflux ratio, depending on the separation required. The distillate was converted to the (1) 3-nitrophthalate acid ester and to (2) the acid phthalate ester. It was then crystallized, hydrolyzed, and the active alcohol separated and dried; $[\alpha]_D$ 20.7° −5.900°.

Brauns [718] and Whitmore and Olewine [5160] prepared the *optically active isomer* from fusel oil by fractional distillation. Brauns used a column packed with glass Raschig rings. Whitmore and Oilwine used a Penn State packing. Apparently, both prepared a high purity optical product.

70. 3-METHYL-1-BUTANOL

3-Methyl-1-butanol is the principal component of *fusel oil*, a by-product of alcoholic fermentation of starches and sugars. The composition of fusel oil varies considerably with the fermentation organism, the fermentation medium, and the process and equipment used for separating it from crude spirits. The composition shown in Table 5.10 is an average.

2-Furaldehyde may be detected by shaking the alcohol with an equal quantity of concentrated sulfuric acid. No more than a pale yellow or reddish color should be produced. The alcohol should not become colored on shaking with an equal quantity of potassium hydroxide solution. For a

Table 5.10 Average Composition of Fusel Oil

3-Methyl-1-butanol	60–65%
l-2-Methyl-1-butanol	8–10
Ethanol and 1-propanol	4
Isobutyl alcohol	20
Water	0.1–1
1-Butanol, pentanols, and higher	traces
Aldehydes, acids, pyridines, pyrazines, 2-furaldehyde	traces

sensitive test for 2-furaldehyde with β-naphthol and sulfuric acid, see Udránsky [4871].

Refined fusel oil is produced by chemical treatment and rectification of the crude material.

Refined amyl alcohol is produced by further chemical treatment and more careful rectification. It contains principally about 85% 3-methyl-1-butanol and 15% 2-methyl-1-butanol. The 2-methyl-1-butanol may be removed from the higher boiling isomer by fractional distillation in an efficient column of at least 70–100 plates at a high reflux ratio, fractional crystallization, or preparative gas chromatography.

For preparation from fusel oil, see Sukhodol and Chatskii [4548] under 2-methyl-1-propanol.

Hargreaves [2019] prepared 3-methyl-1-butanol through the alkyl hydrogen phthalate.

Brown and Zweifel [771] obtained the alcohol by hydroboration of an internally unsaturated hydrocarbon.

For separation of 3-methyl-1-butanol from 2-methyl-1-butanol, see 2-methyl-1-butanol [2718].

Timmermans and Hennaut-Roland [4777] prepared 3-methyl-1-butanol by allowing the magnesium derivative of 1-bromo-2-methylpropane to react with pure, dry trioxymethylene in dry air. After distilling in a Crismer column, the alcohol had a boiling point of 132.00°. See also Veibel et al. [4927] and Timmermans [4772].

Udránsky [4872] purified 3-methyl-1-butanol by recrystallizing potassium amyl sulfate several times. Alcohol practically free from 2-furaldehyde can be obtained by heating fermentation 3-methyl-1-butanol with half its volume of concentrated sulfuric acid for 8 hr on a steam bath, separating the acid, shaking with calcium carbonate, and steam distilling. The process is repeated several times.

Kablukov and Malischeva [2462] dried 3-methyl-1-butanol over anhydrous copper (II) sulfate and distilled twice from calcium for *solubility studies.*

For *electrical measurements*, Müller et al. [3448] boiled 3-methyl-1-butanol with concentrated potassium hydroxide solution, washed it with a dilute solution of phosphoric acid, dried first over potassium carbonate and then over anhydrous copper sulfate, and finally fractionated.

Williams and Daniels [5194] purified 3-methyl-1-butanol for *specific heat measurements* by digesting over unslaked lime and fractionally distilling. The center portion of the distillate was further digested over barium oxide and redistilled. See also Meyer [3287] and Kraus and Bishop [2742].

71. 2-METHYL-2-BUTANOL

Adams and coworkers [18] prepared 2-methyl-2-butanol in high yields from amylene.

A relatively pure material can be prepared from the mixed synthetic pentanols by fractional distillation. Fractional crystallization, zone melting, or preparative gas chromatography may be used for further purification.

Ginnings and Baum [1783] prepared 2-methyl-2-butanol for *solubility studies* from ethylmagnesium bromide and acetone.

72. 3-METHYL-2-BUTANOL

The preparation of 3-methyl-2-butanol from 600 g of 2-bromopropane, 146 g magnesium turnings, and 200 g of acetaldehyde has been described in Whitmore [5155, p. 48]. The yield is 210–215 g. It may also be prepared from methylmagnesium halides and isobutyraldehyde, by the reduction of 3-methyl-2-butanone [5154, p. 130], or from 2-chloropropane and acetaldehyde [3754].

3-Methyl-2-butanol can be separated from synthetic pentanols by careful fractional distillation [2500].

73. 2,2-DIMETHYL-1-PROPANOL

This alcohol has been known for about a century. Tissier [4792] reported its preparation from 2,2-dimethyl propionyl chloride and sodium amalgam.

Despite the fact that this compound has some unique characteristics, there has been only casual interest shown in its physical properties. However, the literature for its preparation is quite extensive. This alcohol is a solid at ambient temperatures; the crystalline material sublimes and has an odor of peppermint.

Purification by fractional distillation is difficult unless the entire still is in an atmosphere whose temperature is greater than 55°. Fractional crystallization, zone melting, and related methods are well suited for the purification of this alcohol.

Whitmore and Rothrock [5161] reported that 2,2-dimethyl-1-propanol is stable to heat and that 5% water liquifies the solid at room temperature.

Trimethylacetaldehyde reacts with *tert*-butylmagnesium chloride in an ether solution to give 2,2-dimethyl-1-propanol [1063]. Greenwood et al. [1886] obtained an almost theoretical yield from these reagents. Beattie [5161], Ginnings and Baum [1783], and Conant and coworkers [1063] used formaldehyde with the magnesium alkyl reagent.

Hoffman and Boord [2222] reported the preparation of 2,2,-dimethyl-1-propanol from 30% hydrogen peroxide and diisobutylene in cold 95% sulfuric acid solution. The alcohol and acetone peroxide are formed; the latter is removed by suction filtration. The acid filtrate is poured over ice and the alcohol separated and fractionally distilled. The heart cut was redistilled and sublimed twice; mp 54.5–55.5°.

D'Adams and Kienle [1165] reduced 2,2-dimethylpropionic acid with lithium aluminum hydride to 2,2-dimethyl-1-propanol in 85% yield; mp 51–52°; Kornblum and Iffland [2713] obtained 85–90% yield.

Ethyl 2,2-dimethylpropionate was reduced with hydrogen at 350° and 220 atm in the presence of copper chromate catalyst in 88% yields [3564, 28].

Sommer and coworkers [4444] reacted *tert*-butylmagnesium chloride and methyl formate for a 72% yield of 2,2,-dimethyl-1-propanol.

74. CYCLOHEXANOL

Cyclohexanol is produced principally by the catalytic air oxidation of cyclohexane. An appreciable amount of cyclohexanone is produced at the same time. A small amount of the alcohol is produced by the catalytic hydrogenation of phenol. Richards and Shipley [3971] report that cyclohexanol is very hygroscopic and state " . . . exposure to air for a few minutes lowers the freezing point several tenths of a degree." Kelley [2550] reports two crystalline forms of cyclohexanol.

Brown [764] gave a general method for preparing alcohols from carbonyls with diborane.

Foresti [1599] hydrogenated cyclohexanone in an acid medium in the presence of platinized pumice and obtained cyclohexanol along with small quantities of cyclohexene. In an alkaline medium, only the alcohol was obtained. Ferrier [1564] hydrogenated either cyclohexylaldehyde or cyclohexanone with aluminum *sec*-butoxide in 75% yield.

Zal'kind and Markov [5320] mixed 1 mole of cyclohexene and 1–1.5 moles of 70–75% sulfuric acid at 55° for 1 hr, diluted the mixture with five volumes of distilled water, and steam distilled. The cyclohexanol was salted out with sodium chloride, dried with potassium carbonate, and fractionally distilled; yield 55%.

The literature for preparing cyclohexanol from phenol is voluminous; a few methods are cited. Sasa [4115] reports that the catalyst obtained by boiling nickel (II) formate in biphenyl, phenyl ether, or a mixture of the two, was most active. Phenol was completely hydrogenated at 160° and 15 atm in 6.3 hr to cyclohexanol. Yeh and coworkers [5289] quantitatively hydrogenated 940 g of phenol to cyclohexanol over 60 g of nickel-kaolin catalyst at 140° and 80 atm. Bag et al. [347] reported the reduction of phenol at 63° in 95% yield in the presence of fragments of sodium hydroxide, activated nickel-aluminum and other metal catalysts. A more detailed description was given the following year [346].

Vogel [4965] purified cyclohexanone through the bisulfite complex and reduced the ketone to the alcohol with sodium for the *determination of physical properties.*

Cyclohexanol was purified for *dielectric constant measurements* by fractional crystallization in the absence of moist air [3971].

Wilson and Heron [5213] state that cyclohexanol of sufficient purity for a *cryoscopic solvent* may be prepared by fractional distillation. The melting point may be used as the *criterion of purity.*

Mikulak and Runquist [3313] found cyclohexanol to be an ideal *cryoscopic solvent* for molecular weight determinations because of its large freezing point constant, fairly general solvent properties, and a melting point slightly above room temperature. They have devised a simple and rapid method that overcomes the difficulty encountered due to hygroscopicity. The method is accurate to $\pm 2\%$ using a thermometer graduated by 0.1° and is suitable as an experiment in general chemistry, or for use in an analytical or physical chemistry laboratory. The commercially available cyclohexanol was distilled at reduced pressure at the rate of 10 ml per minute. The material that freezes at 24° or higher was collected and placed in vials that were stoppered and sealed with paraffin. A found K_f of 39.3 $\pm$ 0.5° compared favorably with the value calculated from the heat of fusion, 39.6°.

Crowe and Smyth [1140] purified commercial cyclohexanol by refluxing 24 hr over freshly ignited lime, followed by fractional distillation. The distillate was heated with small pieces of sodium and fractionally distilled a second time. Since cyclohexanol is extremely hygroscopic, the purified material was kept in a desiccator until needed. The *dielectric constant* was determined.

For purification by vacuum distillation and fractional crystallization, see Schreiner and Frivold [4189], Sidgwick and Sutton [4316], Herz and Bloch [2143], Kelley [2550], and Lange [2820].

CRITERIA OF PURITY

The freezing point is the most convenient means of determining the purity of cyclohexanol. Aldehydes, ketones, and other common impurities may be determined by tests in Rosin [4042] or ACS *Chemical Reagents* [76]. For a cryoscopic test of purity, see Lange [2820]. Gas chromatography is a convenient means for determining the number and amount of the impurities.

SAFETY

Cyclohexanol has a low vapor pressure at room temperature. It has a faint odor that has been described as resembling menthol or camphor. Nelson et al. [3508] found the highest concentration tolerable in air for an 8-hr day to be less than 100 ppm. Treon and coworkers [4826] showed that 1000 ppm caused narcosis but 100 ppm has no effect on rabbits and monkeys.

The *threshold limit value* has been chosen as 50 ppm or 200 mg/m^3 to reduce objectionable irritation [4509].

The *minimum ignition temperature* in air is 300° [3680].

75. 1-HEXANOL

1-Hexanol is available commercially. Generally it has been separated from a mixture of alcohols; it contains other alcohols that are quite difficult to separate. If a material of high purity is required, synthesis that does not produce homologues is preferred.

Hovorka et al. [2268] prepared 1-hexanol by the Grignard synthesis. Preliminary purification was carried out in a 40-cm column packed with glass beads. A final fractional distillation was performed in a similar column. Aluminum amalgam was used as the drying agent. Some decomposition to the corresponding olefin occurred with the drying agent. The final distillation was done in the absence of the amalgam. The purified alcohol had a boiling range not greater than 0.04°.

The preparation of 1-hexanol by the Grignard synthesis is given in [1777, p. 54].

Brown and Subba Rao [769] prepared 1-hexanol; see 1-pentanol.

Decker and Halz [1247] prepared 2-ethylbutanol and 1-hexanol from butyraldehyde and acetaldehyde. Detailed preparation and recovery are given.

Brown and Subba Rao [770] added diborane to hexane and then oxidized the trihexylboron produced with alkaline hydrogen peroxide to produce hexanol.

Olivier [3589] purified 1-hexanol by esterification with hydroxybenzoic acid, recrystallization of the ester, followed by saponification.

Adkins and Billica [26] reduced hexaldehyde with a specially prepared Raney nickel catalyst. For additional information, see 2-propanol.

SAFETY

The *minimum ignition temperature* in air is 300° [3680].

77. 4-METHYL-2-PENTANOL

Brown and Nakagawa [766] prepared 4-methyl-2-pentanol in 84% yield by sodium borohydride reduction in methanol of 4-methyl-2-pentanone.

Brown and Zweifel [772] prepared 4-methyl-2-pentanol in 57% yield from *trans*-isopropylmethylethylene and *bis*-3-methyl-2-butylborane.

Levene and Walti [2922] *prepared optically active* 4-methyl-2-pentanol from 0.6 moles of *dextro*-propylene oxide dissolved in 100 ml of dry ethyl ether which had been added dropwise to a cooled solution of 0.6 moles of isopropylmagnesium bromide in 250 ml of dry ethyl ether with stirring. The mixture was allowed to stand seven days, and the ether was removed by distillation. To the residue was added 70 ml of dry pyridine and 40 g of phthalic anhydride. The purified phthalate was steam-distilled in the presence of sodium hydroxide. The distillate was extracted with ether and the extract dried and fractionated; bp 65–66° at 60 torr.

Hovorka and coworkers [2268] prepared 4-methyl-2-pentanol for *thermodynamic and physical property studies* from 4-methyl-2-pentanone and sodium in ethanol.

78. 2-ETHYL-1-BUTANOL

Matsui [3228] fractionally distilled the hydrogenated polymer of acetaldehyde. From 82 g of volatile distillate, he obtained 15.6 g of ethanol, 16.2 g of 1-butanol, 7.5 g of 2-ethyl-1-butanol, and 1.2 g of 1-octanol. The polymerization of the acetaldehyde was carried out with cooling in the presence of barium hydroxide, which was then neutralized by passing carbon dioxide through the polymerization product. The reduction was carried out in the presence of a nickel catalyst at pressures below 80 atm and temperatures up to 200°.

Hovorka et al. [2267] allowed 2-ethyl-1-butanol to stand over Drierite for at least two weeks, filtered, and then fractionally distilled. A fraction boiling over a range of 0.02° was used for *physical property measurements*.

The Methylcyclohexanols

There are four methylcyclohexanols, three of which exist in *cis*- and *trans*-configurations; in the case of 2-methylcyclohexanol and 3-methylcyclohexanol, each possesses an asymmetric carbon atom and exhibits optical activity. A disagreement in the literature concerning the correct identification of the several isomers apparently has been resolved by Noyce and Denney [3556] who identified *cis*-3-methylcyclohexanol as the isomer of lower refractive index and lower density in contradiction to the prediction of the von Auwers-Skita rule. Discussions of *cis*- and *trans*-methylcyclohexanols have been presented by Skita and Faust [4360], Goering and Serres [1811] and Jackman and coworkers [2368]. No attempt has been made in this chapter to reevaluate configurational assignments.

Gough et al. [1856] have studied extensively the various forms of the methylcyclohexanols and present considerable data. They express the belief that their α-form is *trans*- and the β-form is *cis*-.

Practically all references to the stereoisomers give methods for their preparation and separation. Only a few of these methods are presented here or under the appropriate compound because of the complexity. Some unique reduction methods with lithium aluminum hydride, sodium borohydride, and aluminum isopropoxide are reported by Noyce and coworkers [3555, 1197] and Siegel [4318].

Skita [4359] and Skita et al. [4362] concluded from their work that *cis*- and *trans*-alcohols could be produced in predominating amounts depending on the method of preparation. Peppiatt and Wicker [3707] and Wicker [5172, 5173] have shown that the Skita rule is not too reliable for the methylcyclohexanols. The three articles [3707, 5172, 5173] constitute a good reference source up to 1957.

Vogel [4965] prepared 2- and 3-methylcyclohexanol from the corresponding hexanones. Palfray [3631] hydrogenated the corresponding cresol in the presence of Raney nickel at 95° and 120 kg pressure.

The cresols were reduced by Brode and Van Dolah [749] under hydrogen using copper chromate catalyst at 500–2600 psi at 250–300°. Shigeru and coworkers [4284] employed 1:1 nickel-aluminum catalyst to reduce the same compounds.

Calas and coworkers [880] reduced 2-, 3-, and 4-methylcyclohexanone with trichlorosilane to the respective *trans*-methylcyclohexanols in 60, 30, and 75% yields.

Eliel and Haber [1461] separated the isomers from commercially available 2-, 3-, and 4-methylcyclohexanols by fractional distillation in a 4-ft Podbielniak column at a reflux ratio of 100:1 to 120:1. A charge of 300–500 g gave about one-third to one-half quite pure epimer by infrared examination. It was pointed out that the alcohols must be free from the corresponding cyclohexanone, since it is not satisfactorily separated from the lower boiling alcohol.

Komers and coworkers [2702] have achieved good separation of the *cis*- and *trans*-2-methyl, 3-methyl, and 4-methylcyclohexanols using gas-liquid partition where the stationary phase is capable of hydrogen bond formation.

Brownstein and Miller [794] have used nuclear magnetic resonance to determine the *cis*- and *trans*-methylcyclohexanols.

CRITERIA OF PURITY

The freezing curve may be used with its usual precision and accuracy. The *cis*- and *trans*-isomer distribution may be determined by gas chromatography and by nuclear magnetic resonance.

The British Standards Institute [745] lists the standards for methylcyclohexanols.

SAFETY

Most of the toxicological and safety information available is for a commercial product usually listed as "methylcyclohexanol," made by hydrogenating *m*- and *p*-cresols. The isomers may occur as the *cis*- and *trans*-forms. The vapors of methylcyclohexanol are recognizable at 500 ppm. They are capable of causing upper respiratory irritation at this concentration. It has been concluded that the 2-compound is more toxic than the other two isomers [3681, 4509].

The *threshold limit value* has been set at 100 ppm or 470 mg/m³ [4509].

The *minimum ignition temperatures* in air of the 2- and 4-isomers are, respectively, 296° and 295° [3680].

79. 1-METHYLCYCLOHEXANOL

Markownikaw and Tscherdynzew [3154] prepared 1-methylcyclohexanol from 1-amino-1-methylcyclohexanol and sodium nitrite.

Zelinsky [5329], Sabatier and Mailhe [4083], and Wallach [5066] prepared 1-methylcyclohexanol from cyclohexanone and methylmagnesium iodide.

Grignard and Vegnon [1898] treated 1,5-bis-pentamethylenemagnesium bromide with ethyl acetate and hydrolyzed the product to 1-methylcyclohexanol.

80. 2-METHYLCYCLOHEXANOL (MIXED ISOMERS)

81. *cis*-2-METHYLCYCLOHEXANOL

82. *trans*-2-METHYLCYCLOHEXANOL

Anziani and coworkers [263] found that reduction of 2-substituted cyclohexanones with 2 moles of isobutylmagnesium bromide at 35° gave the corresponding alcohols, mostly the *cis*-isomer. Similar reduction with *tert*-butylmagnesium chloride, *tert*-pentylmagnesium chloride or *tert*-butylmagnesium bromide gave mainly the *trans*-isomer. Methylcyclohexanone reduced with isobutylmagnesium bromide gave a 37% yield of 87% of *cis*-isomer.

Cornubert and coworkers [1094] produced chiefly *trans*-isomers with mild reducing agents.

McQuillin and Ord [3094] reduced 1-methylcyclohexene oxide in ethyl acetate containing perchloric acid and obtained 58% *trans*-isomer of 2-methylcyclohexanol.

Brown and Zweifel [774] report that hydroboration of 1-methylcyclohexene, followed by oxidation with alkaline hydrogen peroxide produces pure *trans*-2-methylcyclohexanol.

83. 3-METHYLCYCLOHEXANOL (MIXED ISOMERS)

84. *cis*-3-METHYLCYCLOHEXANOL

85. *trans*-3-METHYLCYCLOHEXANOL

Ebersole [1436] prepared 3-methylcyclohexanol by hydrogenating *m*-cresol in the presence of nickel-on-kieselguhr catalyst. A temperature of 160° was used with maximum pressure of 1800 psi. The theoretical yield was obtained in 12 hr.

Peppiatt and Wicker [3707] prepared the *trans*-isomer as 77% of the mixture by hydrogenating 3-methylcyclohexanone using a nickel catalyst. The mixed isomers were converted to the 3,5-dinitrobenzyl ester. Repeated crystallizations produced a pure *trans*-ester, which on hydrolysis gave the *trans*-alcohol; bp 170–177° at 772 torr, n_D 20° 1.4581. They also reduced *m*-cresol and obtained a product containing 55% of the *cis*-isomer, which was converted to the hydrogen phthalate ester, then to the potassium salt. The salt was refluxed with excess potassium hydroxide until completely hydrolyzed. The *cis*-alcohol was steam distilled: bp 175–177° at 777 torr, n_D 20° 1.4577.

Goering and Serres [1811] and Noyce and Denney [3556] worked through *p*-toluenesulfonic acid to isolate the *cis*- and *trans*-compounds.

Macbeth and Mills [3095] prepared *dl-cis*-3-methylcyclohexanol, d 30° 0.917, n_D 20° 1.4583, and *dl-trans*-3-methylcyclohexanol, d 30° 0.9022, n_D 20° 1.4573. Included are the derivatives such as hydrogen phthalate, *p*-nitrobenzoate, phenylurethane and α-naphthylurethane.

Hückel and Kurz [2295] isolated (−)*cis*- and (+)*trans*-3-methylcyclohexanol but gave no data for the physical constants.

86. 4-METHYLCYCLOHEXANOL (MIXED ISOMERS)

87. *cis*-4-METHYLCYCLOHEXANOL

88. *trans*-4-METHYLCYCLOHEXANOL

Schmidt and Seydel [4169] report 58% yield of 4-methylcyclohexanol by heating with stirring for 5 hr at 260–280° and 200–250 atm 1 part of chromium (2%) activated catalyst, 15 parts 4-aminotoluene, and 10 parts of water.

Wheeler and Mateos [5133] reduced 4-methylcyclohexanone with lithium tri-*tert*-butoxyaluminum hydride (Li(tert-$C_4H_9O)_3$AlH); LiAlH$_4$ in excess *tert*-butyl alcohol and obtained 84% of the *trans*-alcohol.

Kuivila and Beumel [2774] reduced 4-methylcyclohexanone with organo-tin hydrides and obtained an 84% yield of 4-methylcyclohexanol, containing $70 \pm 2\%$ *trans*- and $30 \pm 2\%$ *cis*-isomer.

Stork and White [4511] report for *cis*-4-methylcyclohexanol a bp of 183–184° and 174° for the *trans*-isomer. There is uncertainty which compound is *cis*- and which is *trans*-.

Other Aliphatic Alcohols

89. 2-HEPTANOL

Directions for the preparation of 2-heptanol from 228 g of 2-heptanone, 600 ml of 95% ethanol, and 130 g of sodium are given in [992, p. 60]. The yield is 145–150 g of distilled alcohol. Farkas and Stribley [1542] prepared 2-heptanol by controlled oxidation of methylcyclohexane with air. Fractionation of the oxidation product gave nearly equal parts of methylcyclohexanol and 2-heptanone. The ketone was separated by extraction with sodium bisulfite and hydrogenated to the alcohol.

Malinovskiĭ and coworkers [3127] prepared 2-heptanol in 30% yield by the Grignard reaction.

Thomas and Meatyard [4722] used bromopentane and acetaldehyde to prepare the alcohol for *molecular association studies.*

Prout and Spikner [3842] reacted pentylmagnesium bromide with acet-aldehyde and obtained 85% yield of *dl*-2-heptanol. Proceeding through the acid phthalate brucine salt, they obtained (+)-2-heptanol; bp 157–159°, n_D 25° 1.4178, d 25° 0.835, and $[\alpha]_D$ 25° 10.11°.

90. 1-OCTANOL

Deffet [1248] decomposed 1-heptylmagnesium iodide with formaldehyde to obtain 1-octanol.

Smyth and Stoops [4425] purified a commercial product by fractional distillation for determination of the *dielectric constant.*

Ueno and Komori [4874] fractionally distilled commercial 1-octanol and treated the distillate with boric anhydride. The treated alcohol was fractionally distilled at 5 torr pressure and the portion distilling at 195–205° was collected. The alcohol was neutralized with sodium hydroxide and again fractionally distilled. The purified alcohol boiled at 98° at 19 torr pressure.

Brown and Subba Rao [769] prepared the alcohol, see 1-pentanol.

Kondo [2705] converted butanol to octanol and octanoic acid by heating with metallic sodium at 320° and 100 atm.

Takao and Kumamoto [4596] prepared 1-octanol from octyl caprylate in a 97.1% yield by catalytic reduction with a copper chromate catalyst at 220°.

Rose and coworkers [4036] purified 1-octanol for *vapor-liquid equilibria studies*. It was precision vacuum-distilled through a 35-in. long, 2-in.-diameter column packed with 0.16 × 0.16-in. protruded packing at 20–40 torr pressure and a reflux ratio of 40:1. Criterion of purity: distillation at several pressures to give product of constant refractive index at 25°, 1.42727.

CRITERIA OF PURITY

Dreisbach and Martin [1372] determined the purity from the freezing curve.

O'Connor and Norris [3569] report a colorimetric method for the analysis of acetals and aldehydes in C_8–C_{10} alcohols below the 100 ppm range.

91. 2-ETHYL-1-HEXANOL

Kobayashi and coworkers [2647] prepared 2-ethyl-1-hexanol by the condensation of 1-butanol, using sodium or potassium hydroxides as the principal catalyst and magnesium as the cocatalyst. Two moles of butanol were condensed by Guerbet's reaction [1639] with 1 mole of tetrabutoxy silicone, $(C_4H_9O)_4Si$, with 1–2 moles of alkaline catalyst and various cocatalysts at 200–290° in 80–93% yield [2648]; see also [3328].

Some other condensations involving 1-butanol are obtained by: refluxing butanol in toluene with sodium hydroxide, heating butanol at high temperature and pressure with sodium or potassium butyrate [3940] or with copper and potassium carbonate [4210], and heating butanol in the presence of alkaline earth alcoholates [1638]. Condensations are obtained in 50% yield from butanol with sodium and zinc chloride at 195°. Yields of 71–72% were obtained from butyric acid, sodium butyrate, and butanol with a nickel oxide catalyst [5310].

Pearce and Berhenke [3691] purified commercial 2-ethyl-1-hexanol for *dipole moment studies* by treating with sodium to remove water and lower alcohols. It was fractionally distilled, and the middle fraction boiling at 185.5° at 760 torr was retained.

Levene and Taylor [2919] reduced 2-ethylhexylate with sodium and absolute alcohol by the method of Levene and Cretcher [2914] to prepare 2-ethyl-1-hexanol in 78% yield; bp 743 torr 181–183°, d 20° 0.8328, n_D 20° 1.4328.

SAFETY

Large quantities of 2-ethyl-1-hexanol have been used in industry without reports of injury [477]. Hodge [2212] found the median lethal dose for rats to be about 0.8 ml/kg interperitoneally or about 3.9 ml/kg by stomach tube.

Aromatic Alcohols

92. BENZYL ALCOHOL

Palfray [3631] hydrogenated benzaldehyde to benzyl alcohol at 105° and 115 kg pressure for 20 min in the presence of Raney nickel. Adkins and Billica [26] reduced benzaldehyde with a specially prepared Raney nickel catalyst. For additional information, see 2-propanol.

Hazlet and Callison [2078] prepared benzyl alcohol and furfuryl alcohol as follows: 53 g of benzaldehyde and 48 g of 2-furaldehyde were treated, with cooling, with 22.5 g of sodium hydroxide and 45 ml of water, shaken for 2 hr, and then allowed to stand at room temperature for 24–48 hr. Sufficient water was added to dissolve the solid and the solution extracted with ether for 8–12 hr. After acidification of the extracted alkaline solution, the solution was again extracted with ether for an additional 8–10 hr. The average mole ratio of benzyl to furfuryl alcohols was 0.56:1. The ether extract was dried over anhydrous magnesium sulfate, the ether removed by distillation, and the alcohols distilled under reduced pressure in an atmosphere of nitrogen.

Nystrom and Brown [3565] reduced benzoic acid with lithium aluminum hydride in a dry ether solution to benzyl alcohol. The yield was 81%. Chaikin and Brown [943] reduced benzoyl chloride to benzyl alcohol with sodium borohydride.

For benzyl alcohol from benzaldehyde and diborane, see cyclohexanol, Brown [764] for a general method.

The general method for purification is careful fractional distillation of the commercial alcohol, which, according to Mathews [3208] is best carried out at reduced pressure with the exclusion of air. Martin and George [3168] purified the alcohol by shaking with aqueous potassium hydroxide and extracting with ether that had been freed from peroxides with silver nitrate and sodium hydroxide. After being washed, the extract was treated with saturated sodium hydrogen sulfide solution, filtered, washed, and then dried over potassium carbonate. After removal of the ether, the alcohol was distilled under reduced pressure and the middle fraction dried over lime that has been burned in an atmosphere of nitrogen. Schiff's reagent indicated a barely detectable quantity of benzaldehyde.

Benzaldehyde was tested for purity by gas chromatography by Hartmann and coworkers [2048] and purified for *dielectric constant studies* by fractional distillation in a 25 theoretical plate column at a reflux ratio of 30:1 to 20:1. The last traces of water were removed with molecular sieves. Gas chromatography was used as the criterion of purity.

CRITERIA OF PURITY

Rosin lists the following specifications: boiling range, 202–206°; nonvolatile matter, not more than 0.005%; chlorine, about 0.02%; 2 ml of alcohol shaken with 20 ml of water should give a clear solution.

Dreisbach and Martin [1372] determined the purity from the freezing curve.

Rees and Anderson [3934] determined benzaldehyde in benzyl alcohol by ultraviolet measurements at 283 mμ.

Hartmann and coworkers [2048] determined the purity by gas chromatography.

SAFETY

Animal experiments indicate that benzyl alcohol is 2.5 to 3 times as toxic as 2-propanol or 1-butanol. Inhalation of the vapors at a concentration of 200–300 ppm, about 880–1325 mg/m^3, by rats gave a LC_0 for 2 hr and a LC_{33} for 4 hr with a LC_{100} for 8hr. It is believed that benzene and benzyl alcohol that were present in a poorly ventilated area were the cause of violent headaches, vertigo, nausea, and other symptoms [3681].

The *minimum ignition temperature* in air is 426° [3681].

The Phenols

General

Keading and coworkers [2539] report an air oxidation of benzoic acid to phenol in the presence of copper [II] benzoate. Of particular interest is the production of pure *meta*-cresol by the air oxidation of either *p*- or *o*-toluic acids. *m*-Toluic acid gives about 40% *ortho*-cresol.

Andon and coworkers [238] prepared high purity phenols for the determination of *high precision physical and thermodynamic properties*. All specimens were similarly purified. Steam was passed into a boiling solution containing 1 mole of the phenol and 1.5 to 2.0 moles of sodium hydroxide in 5 liters of water until no nonacidic material distilled. The cooled residue was made acidic with 20% sulfuric acid. The phenol was separated and dried over calcium sulfate. The aqueous layer was extracted with ether, and the recovered phenol was again treated with steam. The dried phenol was distilled at approximately 10 torr pressure. It was then fractionally distilled at a pressure of 100–212 torr through a 45 theoretical plate column at a reflux ratio of 50:1. Batches of 1200 g were distilled; when the column had reached a steady state of operation, fractions of 150 ml were collected. A series of fractions with an overall boiling range of 0.5° was examined by infrared spectroscopy, and the best fractions were combined. The fractional distillation of the best material was repeated if it seemed desirable. The final purification was done by fractional crystallization or zone melting, or both, several times. In the latter stages of the purification, all material was protected from light and manipulations made under dry nitrogen. The final colorless material was stored in the dark in a vacuum.

Chen and Laidler [959] purified phenols for the determination of *ionization constants* by starting with the highest purity product obtainable and then vacuum distilling or crystallizing from water. The water was prepared in an all-glass still in a carbon dioxide-free atmosphere.

Winters and Shelmerdine [5220] report that sulfur can be removed from phenolic materials by passing the vapors and hydrogen over a cobalt-molybdate catalyst. For example, m-cresol containing 0.08% sulfur was treated with hydrogen at atmospheric pressure at 238° over the catalyst at a liquid space velocity of 0.5 vol/vol of catalyst per hour.

Phenols and alkylated phenols were identified by Gasparic [1721] through their R_f values as 3,5-dinitrobenzoates.

Specific Solvents

93. PHENOL

The majority of phenol is synthetically produced. Natural sources, that is, degeneration of natural products such as carbonization of coal, accounts for only a small portion of the United States' total production. Four sources account for practically all phenol in the United States: carbonization of coal, alkali fusion of benzene sulfonic acid, hydrolysis of chlorobenzene, and the cumene process which involves the cleavage of cumene hydroperoxide.

Phenol is a hygroscopic solid at normal room temperature. Synthetic material usually is at least 99.5% phenol, and the majority of the 0.5% is water. In storage, phenol tends to acquire a yellow, pink to red, or brown discoloration. The formation of the color is promoted by trace amounts of substances acting as catalyst such as iron or copper. Commercial phenol may contain a preservative.

Boorman and coworkers [662] found that phenol from different sources contained characteristic impurities: (1) that prepared from alkali fusion of benzene sulfonic acid contained o- and p-hydroxydiphenyls, (2) that from high temperature hydrolysis of chlorobenzene contained diphenyl ether, and (3) phenol obtained from coal tar contained naphthalene. The following methods are given for testing the purity and identifying the source: (a) "Distill the sample at 1–1.5 torr in a bath at 58–60° until the distillation ceases. Take up the residue from 1 kg in ether, filter, evaporate, and heat the residue to 100° in a current of air to remove traces of phenol." Class (1) yields 0.3–10 g of residue, class (2) or (3) yields a trace of residue. The weighed residue from class (1) was identified by infrared analysis. (b) "Dissolve 500 g in 800 ml of 30% w/v sodium hydroxide and distill in a current of steam until 200 ml has been collected. Add 80 ml of the alkali to this, and repeat the steam distillation. Collect 100 ml of the distillate allowing the condenser to become warm toward the end of the operation. Extract the final distillate with 20 ml of carbon tetrachloride and examine the infrared spectrum of the solution.

Filar [1571] states that phenol prepared from cumene contains generic impurities not separable by distillation. He recommended that the phenol be treated with nonpolar adsorbents such as activated carbon, talc, or precipitated calcium carbonate, the solids separated, and the final purification made by fractional distillation. Impurities in phenol from cumene are reported to be mesityl oxide, methylstyrene, cumene, and benzofuran. One method of purifying phenol consists of treating with hydrogen peroxide neutralizing with sodium hydroxide, and steam distilling [2242]. Another method recommends neutralizing with sodium hydroxide to a pH 7, heating to 108° for 90 hr and fractionally distilling [1318].

Yamamoto and Ohara [5284] prepared phenol by sulfonating commercial chlorobenzene (bp 127–133°) with sulfuric acid. The product was then treated with sodium hydroxide, acidified with sulfuric acid, and heated to 230° On cooling, the crude phenol appeared at the surface. It was extracted with ether and distilled. The yield was 54–58%.

Cresols [360] were demethylated when passed over nickel-on-aluminum catalyst in steam at 410–470°; see also [4284].

Timmermans [4772] states that phenol is purified by fractional distillation in vacuo and by fractional fusion. The criteria of purity are the consistency of the melting point and the critical solution point in water for successive fractions.

Commercial grade phenol, mp 40.3°, was purified by Wille and Rappen [5184] by mixing 7380 g with 2620 ml of water. The mixture was cooled to 11° and seeded with $C_6H_5OH \cdot \frac{1}{2}H_2O$. After crystallization was complete, the crystals were centrifuged and washed at 5° with 500 ml of water saturated with phenol at 0–2°; yield 6590 g with a mp of 40.9°.

Davison [1221] purified phenol, o- and p-cresols, and nitrobenzene by distilling through a small column and then fractionally crystallizing until a constant melting point was obtained. The material was stored in a desiccator and distilled, just prior to use, to remove water and any oxidation products that might have accumulated.

Draper and Pollard [1364] purified phenol by adding 12% water, 0.1% aluminum, and 0.05% sodium bicarbonate. The mixture was distilled at atmospheric pressure until the azeotrope was removed. The pure phenol was then distilled at 25 torr until 20 ml of the black residue remained in the distilling flask.

Williams and Ogg [5198] determined the *dielectric constant* and *dipole moment* on phenol that was purified by crystallization and fractional distillation; bp 179.5–180.0°.

Judson and Kilpatrick [2456] purified a good commercial product by fractional distillation for *dissociation constant* measurements; bp 182.0–182.5°, mp 39.8–39.9°.

Phenol was purified of catalytic poisons with mixed sodium and potassium salts of ethylenediamine tetracetic acid [1393].

Phenols were purified by distillation with steam at 6 and 60 torr. The recovery was better than that obtained at atmospheric distillation [2629].

The purification of phenol by *zone melting* has been studied. Deluzarche and coworkers [1273] used phenol containing methyl red to test the operation and efficiency of an automatic zone melting apparatus. Schildknecht and Vetter [4150] also used phenol to test the efficiency of a zone melting apparatus. Sorensen [4448] studied the speed of zone travel and number of passes to remove impurities from phenol.

CRITERIA OF PURITY

ACS Reagent grade phenol meets the following specifications [76]:
Water: to pass test (limit about 0.5%).
Insoluble matter: to pass test.
Residue on evaporation: not more than 0.05%.

[Note: Phenol that conforms to this specification may contain preservatives. If a preservative is present, it should be stated on the label.]

Water in phenol may be determined by the Karl Fischer method [3343].

Dolique [1332] proposed the critical solution temperature in water as a criterion of purity; see also Timmermans [4772].

The freezing point is the most exact criterion of purity. Dreisbach and Martin [1372] used the freezing curve to determine the purity of a high purity (99.96%m) material; see also Timmermans [4772].

SAFETY

The minimum detection by odor is 0.022 mg/m³; the minimum concentration affecting light sensitivity of the eye is 0.0155 mg/m³; the minimum concentration affecting electrocortical activity is 0.0156 mg/m³; the maximum concentration not affecting cholinesterase activity, porphyrin metabolism and motor chronoxy of rats during continuous exposure for 61 days is 0.01 mg/m³. On the basis of these findings, the recommended threshold limit value for single or continued exposure was 1 ppm or 3.84 mg/m³ [3437].

Phenol vapors or liquid is readily absorbed through the skin or any mucous membrane. A concentrated solution causes bad acid burns. Acute and chronic poisoning may result from skin exposures. Repeated skin contamination by low concentrations of phenol, either vapor or solution, will cause chronic poisoning. Many individuals are allergic to phenol. Phenol is not sufficiently volatile to constitute a respiratory hazard under normal conditions [4509, 3681].

The *threshold limit value* of 5 ppm or 19 mg/m³ contains a sufficiently large safety factor to prevent systemic poisoning, provided that there is no skin contact [4509].

The *minimum ignition temperature* in air is 715° [3680].

The Cresols

The cresols, which are by-products of the coking and petroleum industries, are available as crude or refined materials containing varying amounts of the three isomers. *o*-Cresol is easily separated and purified. There is a large literature on the separation and purification of the isomeric cresols.

Gluzman and coworkers [1804] obtained 96–99% *m*-cresol from a mixture of the *meta*- and *para*-isomers by forming a complex between sodium acetate and the *m*-cresol.

Cislak and Otto [984] found that the *m*- and *p*-cresols selectively formed crystalline addition compounds with cyclic nitrogen compounds. *m*-Cresol forms addition compounds with 2,4-xylenol and 2,6-lutidine; *p*-cresol crystallizes with 4-picoline or 2,3,6-collidine. Both addition compounds, when regenerated, give cresols of about 80% purity. Additional directions are given for obtaining products in high states of purity.

o-Cresol may be isolated from cresol mixtures by fractional distillation. *m*-Cresol and *p*-cresol must be separated by chemical, or a combination of chemical and physical means. The data available indicate that the cresol mixtures follow Raoult's law fairly well. Usually, three distillations are required with efficient equipment and a high reflux ratio to produce material approaching 99.9+% *o*-cresol.

Several techniques have been developed for separating *m*- and *p*-cresols by differences of their sulfonates. Brückner [799] states that a convenient way to carry out the sulfonation is as follows: Mix 100 g of the cresol with an equal volume of sulfuric acid and stir with a glass rod until no more streaking is noticeable. Heat for 3 hr at 103–105°. Dilute carefully with 200–300 ml of water, heat to the boiling point, and steam-distill until all of the unsulfonated cresol has been removed. Cool and extract the resins with ether. Evaporate the solution until the boiling point reaches the decomposition temperature of the sulfonate (133–135°), and then introduce steam. Shorygin and coworkers [4297] tried several methods for separating *m*- and *p*-cresols and found that Brückner's method gave the purest *meta* compound.

Stevens [4497] isolated 250 g of mixed cresols with a boiling range of 1° as a middle cut from 450 g of a commercial mixture of cresols by fractional distillation. To this fraction 12.5 g of concentrated sulfuric acid was added and isobutylene was passed into the mixture, with rapid agitation, as fast as it could be absorbed. The excess isobutylene was washed from the mixture with natural gas. The mixture was neutralized with sodium carbonate and washed. The washed material was fractionally distilled through a 15-plate column at 20 torr pressure and a reflux ratio of 20:1. The tertiary butylated cresols boil 20° apart at this pressure. Debutylation was accomplished by heating with 0.2% of concentrated sulfuric acid, neutralizing with sodium carbonate, and again fractionally distilling.

Kotake [2723] treated a mixture of m- and p-cresols with concentrated phosphoric acid or with a phosphate and strong acid, and treated the resulting tolyl phosphate with petroleum ether; the para-compound is insoluble, the meta-compound soluble.

Golumbic [1829] developed a countercurrent distribution method suitable for partial separation of the m- and p-cresols.

Widiger [5174] found that impurities in alkyl phenols prepared by alkaline hydrolysis of aryl chlorides may be removed by treating with aqueous alkali and air at 60–120°.

CRITERIA OF PURITY

The freezing point is the most convenient and sensitive measurement for the determination of the purity of cresol isomers. Dreisbach and Martin [1372] determined the purity of o- and p-cresols from the freezing curve. The infrared spectrum was used by Whiffen and Thompson [5137] for the analysis of cresols.

SAFETY

The hazards for the cresols are about the same as described for phenol. Deichmann and Witherup [1261] found m-cresol the least toxic of the isomers. p-Cresol did not seem to be any more toxic than phenol.

94. o-CRESOL

Carney and Sanford [908] purified o-cresol for use in developing an an *ultraviolet method* for the isomeric cresols. It was recrystallized several times from a petroleum solvent with a distillation range of 80–120°, followed by a distillation. The fp was 30.937° corresponding to 99.99%m and a 100% mp of 30.994°.

95. m-CRESOL

A high purity m-cresol was prepared by Carney and Sanford [908] for *ultraviolet analytical method studies* of the isomeric cresols from commercial 12° mp material. It was fractionally distilled through a Podbielniak high temperature distillation analyzer and a heart cut collected; mp 11.67° corresponding to 99.93%m purity and a 100% mp of 12.20°.

m-Cresol can be prepared nitrogen free [5185] by forming the m-cresol-phenol complex, washing, distilling, and regenerating; purity 99.6%.

Millikan [3330] reports a method for the separation of m-cresol from p-cresol by azeotropic distillation with benzyl alcohol. A mixture of 29 parts of p-cresol, 29 parts of m-cresol, and 42 parts of benzyl alcohol by volume was fractionally distilled at 30 torr and a reflux ratio of 6:1 in a 10 theoretical plate still; 70–75% *m*-cresol was obtained. A more efficient column and a higher reflux ratio will produce up to 98% *m*-cresol.

The best grades available of *m*-cresol and benzyl acetate were dried over anhydrous sodium sulfate for long periods and fractionally distilled for *physical property studies* of mixtures by Katti and Chaudhri [2526]; the density was used as the criterion of purity.

Chivate and Shah [975] reported the separation of *m*- and *p*-cresols by extractive crystallization of the eutectic with acetic acid.

Keicher and Klopfer [2545] removed *m*-cresol from other phenols by conversion to the urea additives, washing with solvents, then decomposing the additive compounds by treatment with chlorinated hydrocarbons at elevated temperatures. The liberated urea is insoluble.

Rigamonti and Schiavina [3981] developed a process for separating *m*- and *p*-cresols based on the ternary system *m*-cresol-ligroin-methanol (aq) and *p*-cresol-ligroin-methanol (aq). A complicated column operation was employed.

m-Cresol was treated [4873] with *o*-toluidine, 1:4, at 90–100°, cooled to room temperature, and allowed to stand. The crystals of the addition compound were filtered, washed with benzene, and distilled. The purity was 95–97%.

Engel [1479] added phenol to a mixture containing at least 35% *m*-cresol and *p*-cresol. A paraffinic petroleum solvent, boiling at 90–150°, was added, and the resulting mixture cooled. The phenol-*m*-cresol addition compound crystallized out and was separated.

96. *p*-CRESOL

The phenol *p*-Cresol was purified by Carney and Sanford [908] for *ultraviolet studies* in the same manner described for *o* cresol except that it was not distilled. The bp was 34.734°, corresponding to 99.99%m and a 100% bp of 34.739°.

Erić and coworkers [1489] purified *p*-cresol by distillation and recrystallization from light petroleum cut, bp 40–60°.

A mixture of *m*- and *p*-cresols can be separated by crystallizing from a paraffinic hydrocarbon containing not more than 5%w aromatics and not more than 10%w of olefins. Engel [1480] obtained *p*-cresol in 98.6% purity from a mixture containing 85% *p*-cresol. The remainder was principally the *meta*-isomer with a small amount of mixed xylenols. See also Fox and Barker (1612).

Golumbic [1829] developed a counter current distribution method suitable for the partial separation of *m*- and *p*-cresol.

Kendall and Beaver [2555] prepared *p*-cresol by diazotizing pure *p*-toluidine and boiling the diazo solution. The product was steam-distilled, dried, and fractionally distilled. However, Hartman and coworkers [991, p. 37] state that the quality of p-cresol prepared by diazotizing p-toluidine is almost always poor.

For *optical measurements*, Savard [4118] converted p-cresol to sodium p-cresoxyacetate, which was crystallized from water. The purified salt was decomposed by heating with hydrochloric acid in an autoclave. The yield was 60%.

Gibbs [1757] describes a method for purifying technical p-cresol, which depends on the fact that o- and m-cresols react with 2,6-dichloroquinone-chlorimide to form indophenols, while the p-isomer does not.

Unsaturated Monohydric Alcohols

Specific Solvents

97. 2-PROPENE-1-OL

Directions are given in Adams [15, p. 15] for the preparation from glycerol and formic acid. The yield is 845 g, 45–47% of theory, of 2-propene-1-ol. It may be dried to 98–99% by refluxing with successive portions of fused potassium carbonate until no further action is observed. An azeotropic method for producing anhydrous material is described. See also Coffey and Ward [1024].

Biedermann and Raichle [571] heated 1,3-propanediol with ethyl carbonate and potassium carbonate to 200–210° until evolution ceased; yield 80.2%.

Bharuchka and Weedon [564] reduced butyne-1-ol with lithium aluminum hydride to *trans*-2-propene-1-ol.

2-Propene-1-ol may be prepared by the high temperature chlorination of propylene [1537]. 2-Propene-1-ol is produced from the chloride by alkaline hydrolysis. The process is used commercially.

Ballard and coworkers [365] prepared 2-propene-1-ol as follows: Vapors of ethanol and propenealdehyde were passed over uncalcined magnesium oxide and zinc oxide in a mole ratio of 2.3 : 1 and at a rate of 0.099 moles per 100 ml of catalyst per minute. The yield of 2-propene-1-ol at 391–401° was 77.5%.

Shokal and Evans [4295] heated an aqueous solution of acetic acid, propylene, and selenium dioxide almost to the boiling point. An excess of aqueous sodium hydroxide was added and the ester saponified. The 2-propene-1-ol was distilled as the azeotrope.

Rapean and coworkers [3912] found that pure commercial grade 2-propene-1-ol is relatively stable on heating. Less than 1% decomposes per hour at 225° in stainless steel. Iron, iron salts, and copper have no significant effect on the decomposition rate. It is stable when heated in the presence of sodium carbonate. Extensive decomposition takes place in the presence of sodium hydroxide, starting as low as 100°. At about 135°, the reaction becomes exothermic and becomes explosive at about 190°.

A commercial 2-propene-1-ol was fractionally distilled by Kepner and Andrews [2562] through a 1.5 × 120-cm, vacuum-jacketed column packed with 3-mm glass helices to study its *complex with copper:* bp 97.1° at 761 torr.

Azeotropic distillation may be used to dry 2-propene-1-ol [2002].

SAFETY

2-Propene-1-ol has a sharp pungent odor and is a powerful lachrymator. The threshold odor concentration is reported to be less than 0.78 ppm. Nasal irritation starts in the same range as odor perception. It is absorbed through the skin [3681]. Dunlap and coworkers [1398] found that the LD_{50} for rabbits is 45–105 mg/kg. Chronic exposure of 150 ppm was fatal to four animals.

Torkelson et al. [4799] suggest that the average exposure concentration of 2-propene-1-ol not exceed 2 ppm and that concentrations always be kept below 5 ppm.

The *threshold limit value* of 1 ppm, or 3 mg/m³, has been suggested [4509].

The *flammable limits* in air are 2.52 and 18.00%v; the upper value at elevated temperature. The *minimum ignition temperature* in air is 378° [3680].

98. *cis*-2-BUTEN-1-OL

99. *trans*-2-BUTEN-1-OL

Lieben and Zeisel [2953] reduced crotonaldehyde with iron and acetic acid to 2-buten-1-ol in 1881.

Yada and coworkers [5280] selectively reduced crotonaldehyde with hydrogen over Raney copper-cadmium catalyst at 225–245°.

Clarke and Crombie [989] reduced 2-butyn-1-ol to *cis*-2-butene-1-ol (3,5-dinitrobenzoate, mp 52–53°) using a copper-zinc couple in boiling ethanol. Apparently this is a specific *cis*-reduction. Ingold and Ingold [2348] used the same method of preparation. They purified and dried the alcohol to a constant refractive index. See also Charon [953].

Petrov [3734] saturated 1,3-butadiene with hydrogen bromide in acetic acid at −10° and obtained 60% of the bromobutane. The latter was digested 1.5–2 hr with 10% sodium hydroxide. A mixture of 2-buten-1-ol and 3-buten-1-ol was obtained.

Hatch and Nesbitt [2066] prepared *cis*-2-buten-1-ol from 25 g of 2-butyn-1-ol stirred with 8 g of Schmidt [4164] catalyst (palladium on barium sulfate) in 290 ml of methanol and treated with 0.357 moles of hydrogen for 3 hr in a 1-liter suction flask. The methanol was removed by fractional distillation, and the *cis*-2-buten-1-ol was fractionally distilled at reduced pressure; yield 76%. It was also prepared from isocrotonic acid. *trans*-Buten-1-ol was prepared from *trans*-crotonaldehyde by reduction with lithium aluminum hydride.

Young and Andrews [5308] prepared *trans*-2-buten-1-ol from the corresponding chloro-compound by hydrolysis with silver oxide.

Smets [4372] prepared *cis*-ethylenic alcohols from acetylenic magnesium halide and formaldehyde. The triple-bonded alcohol was reduced to the ethylenic alcohol by mild reduction with Bourgel's catalyst. The *trans*-alcohols are formed from Grignard reagent and an unsaturated aldehyde.

Meerwein and coworkers [3256] reduced 2-butenal with $Mg(OC_2H_5)_2 \cdot Mg(OC_2H_5)Cl$ in 60% yield to 2-buten-1-ol. No butanol was found.

Young and coworkers [5309] studied methods to 1936 for the preparation of 2-buten-1-ol and found that all gave low yields. They report a method that gave 60% theoretical of 93% unsaturated alcohol. Aluminum isopropoxide is first prepared in the flask and then reacted with crotonaldehyde.

100. 2-PROPYN-1-OL

2-Propyn-1-ol has a mild geranium-like odor. Commercial material is of good purity and contains a stabilizer. It is miscible with water and a large number of organic compounds such as benzene, chloroform, 1,2-dichloroethane, ethyl ether, acetone, p-dioxane, tetrahydrofuran, and pyridine. It is partially miscible with carbon tetrachloride and immiscible with aliphatic hydrocarbons.

2-Propyn-1-ol is considered to be a stable compound. Manufacturers' experience indicates a decrease in assay of about 1% per six months' storage. Alkalies appear to cause polymerization and development of color on heating or storage over a period of time. Acids stabilize against polymerization during the heating of aqueous solutions [258].

2-Propyn-1-ol has been prepared from acetylene and formaldehyde. Yamamoto and coworkers [5283] used methanol and acetone as the reaction medium in the presence of copper chromate; Suzuki [4561] used tetrahydrofuran, and Hanford and Fuller [2003] used a copper-bismuth catalyst.

Hatch and Moore [2065] studied the preparation of 2-propyn-1-ol from *cis*- and *trans*-3-chloro-2-propene-1-ol. The lower boiling chloroalcohol was refluxed with 10% excess 12.5% sodium hydroxide for 3 hr and a 69.3% yield of the alcohol was obtained. The water-alcohol azeotrope was obtained. It was found that repeated extraction with ether was the most satisfactory way of separating the alcohol.

Henry [2112] in 1872 described the preparation of 2-propyn-1-ol by the action of potassium hydroxide on monobromoalkyl alcohols.

Johnson [2404] prepared 2-propyn-1-ol by heating 2-butyn-1,4-diol with potassium carbonate at 140–150° and Kreimeier [2750] reacted sodium acetylide with paraformaldehyde. See also Smets [4372] under 2-buten-1-ol.

Shiga and Tsuruta [4283] concentrated 2-propyn-1-ol by feeding the aqueous solution to the top of an extraction tower and two volumes of water saturated ethyl acetate into the bottom in counter-current flow. Distillation of the acetate phase collected at the top gave 99% 2-propyn-1-ol.

Lespieau [2910] purified 2-propyn-1-ol by freezing the aqueous solution, forming the monohydrate, mp 16.52°. The triiodide was used to characterize the alcohol.

An aqueous solution of 2-propyn-1-ol can be concentrated by azeotropic distillation with either butanol or butyl acetate [3582].

McKinley [3082] found that 0.25–0.5% of succinic acid or aqueous hydrogen chloride prevents decomposition of 2-propyn-1-ol during distillation.

SAFETY

2-Propyn-1-ol is primarily a skin irritant, but not a skin sensitizer. The LD_{50} for white rats and guinea pigs, respectively, is 0.07 and 0.06 ml/kg [258].

The *minimum ignition temperature* in air is 374° [3680].

Polyhydric Alcohols

Specific Solvents

101. *cis*-2-BUTENE-1,4-DIOL

102. *trans*-2-BUTENE-1,4-DIOL

Reppe [3959] partially reduced *cis*-2-butyn-1,4-diol with hydrogen and palladium on calcium carbonate to *cis*-2-butene-1,4-diol. He also prepared the alcohol from acetylene and aqueous formaldehyde at 100° under pressure.

Clarke and Crombie [989] reduced 2-butyne-1,4-diol to *cis*-2-butene-1,4-diol, mp 8–10°, using a copper-zinc couple in boiling ethanol. Apparently this is a specific *cis*-reduction.

Valette [4904] prepared *trans*-2-butene-1,4-diol by lengthy heating of *trans*-1,4-dibromo-2-butene with water. He also prepared the *cis*-alcohol by partial reduction of 2-butyn-1,4-diol in ethanol with hydrogen and Raney nickel catalyst.

Kato and Sakuma [2511] used polarographic reduction of 2-butyn-1,4-diol to 2-butene-1,4-diol on platinum, palladium, silver, and copper cathodes. No 1,4-butanediol was found.

Kato et al. [2512] reduced 2-butyne-1,4-diol on spongy silver and obtained almost pure *cis*-2-butene-1,4-diol.

103. 1,2-ETHANEDIOL

Most of the 1,2-ethanediol is produced by the hydration of ethylene oxide by one of two processes. Diethylene glycol and triethylene glycol are by-products. The principal impurities found in 1,2-ethanediol are 1,2-propanediol, butanediol, and water. It is very hygroscopic.

Pukirev [3849] prepared 1,2-ethanediol in a 34% yield by saponifying 1,2-dibromoethane with potassium carbonate.

Ghosh and coworkers [1753] prepared the diol and ethylene chlorohydrin by the electrolytic reduction of ethylene on porous carbon anodes. Ethylene was fed to a porous carbon tube anode at the rate of 57 ml per hr per cm² of anode surface, with a current density of 0.023 amp per cm² in 10% sodium chloride solution: 1,2-ethanediol and ethylene chlorohydrin were formed. The current efficiency of the chlorohydrin was 91% at 1° and 1.1% at 91°. The glycol efficiency was 5.4% at 1° and 16.5% at 91°.

Fogg and coworkers [1590] purified technical grade 1,2-ethanediol, 99.5%, by fractional distillation at 7–10 torr at a reflux ratio of 10:1 in a nitrogen-blanketed, adiabatically operated, $\frac{1}{16}$-in. glass helices packed 1 × 36-in. column. Only the middle third distilling at constant temperature was collected. Eleven lots were purified for *density and refractive indices studies of aqueous solutions;* d 25° varied from 1.1098 to 1.1099, n_D 25° varied from 1.43179 to 1.43182.

Koizumi and Hanai [2687] purified commercial 1,2 ethanediol for *electrical property studies* by drying over dehydrated sodium sulfate and then running three successive fractional distillations at 8 torr pressure, retaining only the middle fractions. It was then fractionally crystallized twice.

Smyth and Walls [4426] fractionally distilled 1,2-ethanediol in a vacuum, dried the middle fraction over sodium sulfate, and, after decanting, fractionated repeatedly. The water was removed by fractional distillation, the first fraction containing all of the water.

Taylor and Rinkenbach [4612] concluded, after studying the results of careful investigators and comparing their own experience, that the true melting point of 1,2-ethanediol is intermediate between −17.4 and −11.5° due to its tendency to supercool.

Kanbayashi and Nukada [2492] *studied the infrared* and *Raman spectra* of 1,2-ethanediol as liquids and in solutions and their temperature response. A *rotational isomer* was found that exists only in the gauche form in the liquid state.

1,2-Ethanediol and 1,2-propanediol were simultaneously determined by Warshowsky and Elving [5083] by oxidation with periodic acid and then analyzing polarographically for the formaldehyde in the presence of acetaldehyde by the technique of Whitnack and Moshier [5166]. See also Francis [1618], Reinke and Luce [3946] and Hoepe and Treadwell [2215].

CRITERIA OF PURITY

The boiling point, density, and refractive index may be used as a qualitative measure of purity for 99–99.8% material. Water can be determined by the Karl Fischer method [3343]. Gas chromatography may be used for the analysis of 1,2-ethanediol.

SAFETY

1,2-Ethanediol is a colorless, odorless liquid with a bittersweet taste. It has a low vapor pressure at normal temperature. It presents negligible hazards to health except, possibly, when being used at elevated temperature. It has a low, acute oral toxicity. Bornmann [669] found the LD_{50} for mice to be 13.79 mg/kg. It was fed to monkeys at a dietary level of 0.2 and 0.5% for three years without any toxic effects [621]. See also [3411]. There does not appear to be any significant irritation from skin contact. Toxic amounts can be absorbed through the skin [3681]. Bornmann [669] reports that all glycols have a narcotic effect and, in toxic amounts, cause a central paralysis of respiration and the heart.

The *threshold limit value* of 100 ppm, 274 mg/m³, appears to be generally acceptable [88, 3681].

The *minimum ignition temperature* in air is 413° [3680].

Several methods have been published for the determination in air. Mandric [3135] oxidized the glycol to formaldehyde and determined with chromatropic acid. He also reports the determination in air to a sensitivity of 0.001 mg per 7 ml of solution. See also [3160].

104. 1,2-PROPANEDIOL

Pukirev [3849] converted 2-propanol (free of tertiary alcohols) to propylene with sulfuric acid and aluminum sulfate. The propylene was passed into cold bromine to yield dibromopropane, which, upon heating with dry silver acetate in 100% acetic acid for 75 hr, gave the diacetate. The diacetate was redistilled and saponified with barium hydroxide in water to give 1,2-propanediol.

The conditions for the production of 1,2-propanediol and glycerol from propylene, air, and steam were studied by Newitt and Mene [3525].

Curme and Johnston [1157] hydrated propylene oxide and obtained 1,2-propanediol.

Levene and Walti [2920] prepared *levo*-1,2-propanediol from *levo*-1-amino-2-hydroxypropane.

SAFETY

Rowe [3681] states, "The hazards to health in industrial handling and use of propylene glycol would seem to be negligible. Its systemic toxicity is especially low and, since 1942, it has been considered a proper ingredient for pharmaceutical products. The Food and Drug Administration does not object to its use in food products or cosmetics. The inhalation of atmospheres containing propylene glycol vapor presents no health hazard." Rowe further finds it hard to visualize how a 1,2-propanediol fog in the air would be a health hazard.

Walther [5075] found that the liquid diol was not irritating to the skin of animals or of human beings. When taken orally, vomiting results; but even single large doses cause no ill effects. It has been firmly established, however, that this glycol is no more toxic than glycerol.

A *threshold limit value* is not considered necessary [3681].

The *flammable limits* in air are 2.62 and 12.55%v at elevated temperature [3680]. The *minimum ignition temperature* in air is 216° [2297].

105. 1,3-PROPANEDIOL

Hatch and Evans [2064] hydrated acrolein in dilute sulfuric acid, followed with reduction over Raney nickel at 32–40°. 1,3-Propanediol constituted about 25% of the product and may be separated by fractional distillation.

Senkus [4236] recovered 1,3-propanediol by treating with formaldehyde to form the 1,3-dioxane.

SAFETY

Van Winkle [4920] found 1,3-propanediol to be about twice as toxic as the 1,2-isomer.

A *threshold limit value* does not seem necessary because of the low vapor pressure at room temperature and the relatively low toxicity [3681].

106. 1,3-BUTANEDIOL

1,3-Butanediol is a relatively new commercial product. It is produced by essentially the same type of hydration process of the corresponding oxide as the commercial synthetic glycols.

Kling and Roy [2634] prepared 1,3-butanediol and other glycols similarly by the action of magnesium amalgam on acetaldehyde in benzene solution. Hibbert [2166] condensed acetaldehyde in the presence of potassium carbonate and reduced with aluminum amalgam. See also [743].

Carothers and Adams [910] reduced aldol to the butanediol with hydrogen and platinum oxide in 86% yield. Palfray [3631] used nickel as the catalyst. Levene and coworkers [2923] reduced freshly prepared aldol by biochemical means by the method of Neuberg and Kerb [3518] to *dextro*-1,3-butanediol. The electrolytic reduction of aldol has been reported [680]. Acetaldol has also been used as the starting material for reduction [744].

SAFETY

Ingestion of 1 ml/100 g body weight twice weekly for 45 to 185 days: no damage to vital organs, teeth, and bones. It did not cause globinuria or proteinuria in man or experimental animals [2708]. Loeser found butanediol to be no more toxic than glycerol on oral administration. It was not harmful to the skin or mucous membranes [2991].

There does not seem to be a necessity for a *threshold limit value* due to the low vapor pressure at room temperature and the low toxicity.

107. GLYCEROL

All glycerol was obtained from glycerides of fats and oils until 1949, when the synthetic product was first manufactured. The synthetic material now accounts for more than 60% of the total production. It is made by several processes.

Milas [3315] prepared glycerol in 60% yield by hydroxylating 2-propene-1-ol with substantially anhydrous solutions of hydrogen peroxide. The preferred catalysts were oxides of osmium, rubidium, vanadium, molybdenum, or chromium. The temperature was an important variable. The formation of glycols in high yields was favored by a temperature of 0–21°.

2-Propene-1-ol is converted to glycerol in 70% yield when treated with hydrogen peroxide in the presence of phosphotungstic acid at 80–82° for 2 hr [3240].

Hass and Patterson [2055] purified glycerol, prepared by hydrogenolysis, by dissolving it in an equal volume of 1-butanol. The solution was placed in a watertight bottle, cooled, seeded, and slowly revolved in an ice-water slurry until crystallization had taken place. The impurities and most of the solvent were removed by centrifuging. The crystals were washed with cold acetone or isopropyl ether. Over 60% of the fraction boiling at 290° was isolated as crystalline glycerol. The crystalline material passes the USP specifications. Other solvents that may be used are 1-propanol, the pentanols, or liquid ammonia.

Eeckelaers [1450] purified glycerol by passing it through a cation-exchange resin followed by an ion-exchange resin and then concentrating in vacuum.

Glycerol was purified for *viscosity studies* by Segur and Oberstar [4224] by distilling C.P. material from a 5-liter flask through a 25-cm Vigreux tube. The middle portion, weighing 4330 g and distilling at 140° at 1 torr, was used. The specific gravity at 25/25° was 1.26192, which corresponded to a purity of 99.97%.

Hoyt [2277] purified by digesting with hot concentrated sulfuric acid, saponifying with lime paste, acidifying with sulfuric acid, filtering, treating with an anion absorbent resin, and finally fractionally distilling. The recovery was 91%.

Wallerstein and coworkers [5067] purified glycerol, produced by fermentation of carbohydrates, by distilling. Before distillation, the impure alcohol was treated with a small quantity of formaldehyde and kept for approximately 1 hr in an alkaline condition.

Evans [1520] removed the color from distilled and substantially dry glycerol by extracting with 2,2,4-trimethylpentane.

The 1.93 μ band is characteristic of water and has been used to determine water in glycerol [950].

Glycerol containing ionic and/or nonionic materials may be purified by ion exchange or ionexclusion or a combination of both [3268, p. 149].

Glycerol is very hygroscopic and can absorb water up to 50% of its weight from air [3268, p. 150].

Glycerol decomposes at its boiling point.

CRITERIA OF PURITY

Requirements and test methods for ACS Reagent grade are given in [76]. A dynamite grade is available. It has a higher density than the ACS Reagent grade, that is, less water. A USP and BP grade is available in the United States and the British Commonwealth, respectively.

Glycerol is difficult to keep dry. After other impurities have been removed, the density and the critical solution point in 3-methyl-1-butanol are good criteria of the amount of water present [4772].

SAFETY

Glycerol is the standard of comparison for substances that are considered to be safe for use in medicinals, cosmetics, and similar products; for example, see *Safety* for 1,2-propanediol and 1,3-butanediol.

ETHERS

General

PREPARATION AND PURIFICATION

Reppe [3958, p. 28 ff.] gives methods, or references thereto, for a number of methods for the preparation of vinyl ethers.

Kobe et al. [2653] purified 20 compounds by the same general procedure for use in the determination of *critical properties and vapor pressures*. The compounds were dried with Drierite and those suspected of forming peroxides were treated with iron (II) sulfate. The compounds were fractionally distilled three times and 80% heart cuts were taken. Reflux ratios of 60 to 80:1 were used. The criterion of purity was the boiling range measured with a double-junction thermopile sensitive to about $1/80°$.

PEROXIDES

Ethers form peroxides under normal storage conditions. Unless a test shows peroxides to be absent, ethers should not be evaporated or distilled to near dryness. The rate of formation of peroxides in a given ether depends upon the storage conditions. Peroxide formation is accelerated by heat, light, and air. A small amount of moisture will accelerate peroxide formation in most ethers. Ethers kept in the laboratory should be in a dark bottle in an atmosphere of nitrogen in contact with iron filings or copper powder. Storage in a refrigerator is further recommended.

Removal or Destruction of Peroxides.   Bodea and Silberg [626] prepared peroxide-free ethers by heating 1 liter of the ether with about 50 g of phenothiazine, refluxing for 5 min and distilling.

Peroxides may be removed from water-insoluble ethers by shaking the ether with a solution of iron (II) sulfate acidified with sulfuric acid until the ether phase no longer gives a test for peroxides.

Fierz-David [1569] freed ethyl ether from peroxides by shaking 1 liter with 10 g of copper-zinc couple. The couple was prepared by suspending 10 g zinc dust in 50 ml of hot water, adding 5 ml of 2 N hydrochloric acid and decanting after 20 sec. The dust was then washed twice with water and covered with 50 ml of water and 5 ml of 5% copper (I) sulfate with swirling. The liquid was decanted and the residue washed three times with a small amount of water, three times with 20 ml of ethanol, and twice with 20 ml of ethyl ether. The treated ether was stabilized by adding 2 mg of pure diphenylamine. It is not necessary to remove the zinc dust before stabilizing.

A summary [3140] of methods of destroying peroxides has been prepared. The peroxide content is lowered markedly by passing the ether over an alumina or Dowex 1 column.

Dasler and Bauer [1194] removed peroxides from ethyl ether, butyl ether, and *p*-dioxane by passing through a column containing activated aluminum oxide. This treatment also reduced the aldehyde content. The ratio of aluminum oxide to the volume of ether varied with the amount of peroxides present; 82 g of aluminum oxide in a column 1.9×23 cm was sufficient for the complete purification of 700 ml of ethyl ether containing 127 micromoles of peroxides per liter.

Peroxides may be destroyed by sodium bisulfite, or more slowly by an alkali.

Fisher and Baxter [1581] report that peroxides in *p*-dioxane can be decreased by shaking with tin (II) chloride or by refluxing with lead (IV) oxide and filtering through a slow filter paper. Isopropyl, ethyl, and probably isopentyl ethers can be freed from peroxides by lead (IV) oxide. No lead was found in the ethers after treatment.

Ramsey and Aldridge [3897] efficiently removed peroxides from ethers with cerium (III) hydroxide. The reagent has the advantage that it is a solid and is insoluble in ethers. It apparently is quite safe. Some reagents, such as lithium aluminum hydride, are reported to have caused fire [3362]. Cerium (III) hydroxide is recommended and is prepared in the following manner: Sodium hydroxide is added to a cerium (III) salt solution until the supernatant liquid is slightly alkaline. The white precipitate is separated and washed by centrifugation. It is used undried. The amount of the hydroxide to use is not given. Water is required in the ratio of 1 part to 15 parts of ether. The cerium (III) hydroxide presumably is shaken or stirred vigorously with the ether. The peroxide removal is complete in about 15 min. The hydroxide changes color from white to reddish brown, depending upon the amount of peroxide present. The solid is removed from the ether by decantation and centrifugation. The clear ether can be tested by the benzidine test [1555].

Nineteen ethers were tested and only two, allylethyl and benzyl-*n*-butyl, gave positive tests for cerium. Starch-iodide paper for testing for the presence of peroxides is not as reliable as a slightly acidified solution of potassium iodide and starch.

Detection and Determination of Peroxides.   Rinse a small glass-stoppered cylinder with the ether; then add 10 ml of the ether to the tube and 1 ml of a 10% solution of freshly prepared potassium iodide. No yellow color should be observed in either layer at the end of 1 min [76, 4042]. This test may be made more sensitive by adding a drop of starch indicator and shaking.

Prepare an approximately 1% solution of iron (II) ammonium sulfate. Transfer 5 ml to each of two test tubes and add 0.5 ml of 1 N sulfuric acid and 0.5 ml of 0.1 N thiocyanate solution to each tube. Add 10 ml of ether to one tube and shake well. The aqueous phase in the ether tube should not have a redder color than the blank. When testing water soluble ethers, such as *p*-dioxane, add 10 ml more of water to the blank.

A rapid method for the detection of microgram quantities of peroxides in ethers was reported by Dugan [1392]. N,N-Dimethyl-*p*-phenylenediamine was dissolved in 10 ml of water and brought to 100.0 ml with methanol. Two milliliters of the reagent was mixed with 2 to 5 ml of the ether to be tested and a like amount of ether that had been distilled from sodium and allowed to stand for 5 min in subdued light. A red-blue reaction product develops when peroxides are present. The reagent is an amine that will form Schiff bases with aldehydes. If the ether contains aldehydes in concentrations of 1000 mg/liter, or higher, interference will develop for the test.

Five different methods for determining the amount of peroxides in ethers were studied by Rothenbach [4058]. He found that the dichromate method was the least sensitive, that the potassium iodide test was somewhat better, and that the method using reduced alkaline phenolphthalein (sensitive to 0.1 mg/liter) was the most sensitive. The basic titanium sulfate-sulfuric acid test was found suitable for both general qualitative and quantitative determinations, although the sensitivity was only one-hundredth that of reduced phenolphthalein.

A spectrographic method for the determination of hydroperoxides in ethers using acidic titanium tetrachloride was reported by Wolfe [5248]. The agreement with the iodometric method is good.

Wagner and coworkers [5014] report a modification of the iodometric method for peroxides that has a wide application. The iodine does not react with olefinic compounds. Add 40 ml of dry 2-propanol, 2 ml of acetic acid, and the sample (5 to 10 ml, depending on the peroxide content) to a 250-ml Erlenmeyer flask. Heat until reflux starts, add 10 ml of saturated 2-propanol-potassium iodide solution, reflux 5 min, add 5 ml of water, and titrate with 0.1 or 0.01 N sodium thiosulfate. Prepare the potassium iodide solution by refluxing 25 g of the salt with 100 ml of 2-propanol. Blank determination on all reagents will be nil unless oxidizing impurities are present in the alcohol. A blank determination on each new lot of 2-propanol is sufficient.

Banerjee and Budke [378] describe a sensitive spectrophotometric method for the determination of traces of organic peroxides in organic solvents. Iodine is liberated from potassium iodide and measured at 470 mμ. Active oxygen in the range of 5–80 ppm can be detected in a 1.0 cm cell; this range can be lowered to 0 to 5 ppm in a 1.5 cm cell at 410 mμ. They extended their method, so that it is applicable to the parent method of Wagner and coworkers [5014] and, in addition to hydrocarbons, ethers, esters, alcohols, and so forth, include α, β-conjugated unsaturates such as acrolein, isoprene, and sorbic acid.

Inhibition of Peroxide Formation. Bailey and Roy [353] found that peroxide formation can best be inhibited by storing ethers over sodium amalgam in brown bottles.

In France, du Pont has patented a stabilizer for cyclic ethers, 4,4'-thiobis(6-*tert*-butyl-*m*-cresol) effective in concentrations of 0.01 to 1 % [1413].

Hamstead and Van Delinder [1998] present an excellent discussion of autoxidation and inhibition of peroxides in ethers. They studied the inhibition of isopropyl ether and present information that should be valuable for the storage of all ethers.

Robertson and Jones [3998] state that aldehydes, ethers, and esters of unsaturated acids are stabilized from peroxide formation by the addition of 0.0001–1 % of 2,5-bis(dimethylaminomethyl)hydroquinone.

Jones [2419] found 3,3',5,5'-tetralkylstilbenequinone to be a stabilizer for aldehydes, ethers, and esters of unsaturated acids.

N-Benzyl-*p*-aminophenol, 16 ppm; diethylenetriamine, 50 ppm; triethylenetetramine, 50 ppm; tetraethylenepentamine, 50 ppm; morpholine, 300 ppm: and ethylenediamine, 50 ppm were found effective for peroxide inhibition is isopropyl ether [998]. The last two amines were the least effective.

Specific Solvents

Aliphatic Open Chain Ethers

108. ETHYLVINYL ETHER

Dolliver and coworkers [1333] passed acetal over 5 % platinized asbestos at 280–290°. Repeated fractional distillation, followed by drying with calcium chloride, gave a pure ethyl-vinyl ether.

Böhme and Bentler [633] prepared ethylvinyl ether in 49% yield by refluxing 21.7 g of 2-ethoxy-2-chloroethane with 36 g of N,N-dimethyl-aniline on a water bath until the reaction starts and fractionally distilling the upper layer.

Wacker-Chemie [5006] reported the preparation of ethylvinyl ether by stirring for 30 min at −15 to −8° 340 g of vinyl acetate, 92 g ethanol, 3 g of mercury acetate, and 3.05 g concentrated sulfuric acid. The mixture was cooled to −15°, and 128 g of ethylvinyl ether was distilled at 8 torr over a 3-hr period; 96% yield, 90% conversion.

Twenty-four grams of 1-chloroacetyl ether was added to 40 ml of pyridine at room temperature. Fourteen grams distilled below 100° and was redistilled through a 12-cm Vigreux column to yield 6.8 g of ethylvinyl ether [2326].

Pilcher and coworkers [3768] purified ethylvinyl ether for the determination of *heat of combustion* by fractional distillation from sodium in a still with a 50-cm column packed with glass helices. The center fraction was redistilled using a 1 in. × 5 ft column packed with stainless steel Knitmesh Multifil. The center fraction of 100 ml from a charge of 700 ml had a purity of 99.99%m determined in a melting point calorimeter. Portions were distilled from calcium hydride before use.

Water may be determined in ethylvinyl ether by a modified Karl Fischer method.

SAFETY

Ethylvinyl ether has a narcotic action similar to, but slightly weaker than, ethyl ether. Its action disappears faster and apparently does not produce any aftereffects [4298].

109. ETHYL ETHER

The principal impurities found in commercial ethyl ether are ethanol, water, aldehydes, peroxides, and sometimes acetone. The following method of purification has been found to produce an ether of high purity satisfactory for exacting *analytical techniques*. Shake the ether with one-tenth its volume of 10% sodium bisulfite solution intermittently for 1 hr, withdraw the aqueous phase, and wash the ether with a saturated sodium chloride solution containing 0.5% sodium hydroxide. Then wash with a saturated sodium chloride solution containing a small amount of sulfuric acid, then with two portions of saturated sodium chloride solution, and finally distill in an atmosphere of nitrogen.

A substantially dry and alcohol-free ether may be prepared by fractional distillation. The distillate generally contains some aldehydes.

Rabinowitsch [3876] prepared ether for *specific conductance* measurements by distilling the ether over calcium chloride and then several times over sodium. Fresh distillate was always used for the measurements. Vogel [4970] determined some *physical properties* on ethyl ether purified in the same manner, but distilled only once from sodium.

Bruce [795] prepared absolute ethyl ether by the action of flaked, technical grade sodium hydroxide on USP ether for a period of two weeks. One pound of sodium hydroxide was used for each 3 liters of ether. Comparison with commercial ether by the method of Forbes and Coolidge [1593], gave the following reaction times with strongly alkaline potassium permanganate: USP, 3 sec; absolute ether, 12 sec; and the product prepared with sodium hydroxide, 20 sec.

Lamond [2811] states that ethyl ether may be rendered completely free of alcohol by washing with 10% sodium chloride solution. One liter of ether may be conveniently treated in a 2-liter separatory funnel with three successive 300-ml portions of salt solution. The ether is then dried and distilled. Lassar-Cohn [2838] recommends drying the ether and refluxing for 24 hr with sodium potassium amalgam to remove alcohol. Aldehydes were removed by shaking with an alkaline solution of a mercury (II) salt for several days.

Ziebell [5339] freed ethyl ether from aldehydes by adding hydrazine hydrogen sulfate, phenylhydrazine or $NH_2CSNHNH_2$ and distilling the ether from the nonvolatile hydrozone compounds.

It is reported that ethanol may be removed by repeated washing with water. Wade and Finnemore [5011] washed 700 ml of commercial ether 16 times with 50-ml portions of water before the iodoform test became negative.

For *optical measurements*, Castille and Henri [933] treated ether for 12 hr with a 10% solution of sodium carbonate, dried over calcium chloride, and distilled slowly over sodium through a long column in an apparatus with ground joints; the first and last portions of the distillate were discarded. Ether purified in this manner was transparent in a 10-mm layer to 2050 Å.

Strecker and Spitaler [4520] treated ether with potassium permanganate and potassium hydroxide, with water and then with 50% sulfuric acid, and dried over calcium chloride. The dried ether was shaken with a few grams of mercury, further dried over sodium, and fractionally distilled.

Scheibe and coworkers [4137] recommended shaking several times with sulfuric acid, sodium hydroxide solution, and water. After drying with sodium, the ether is distilled through an efficient column.

Werner [5127] added 4 g of silver nitrate dissolved in 30 ml of water to 500 ml of ether in a liter bottle. Fifty ml of a 4% solution of sodium hydroxide was then introduced and the mixture shaken for 6 min. The ether was removed and found to be free of peroxides, aldehydes, and unsaturated substances.

A *reference standard* of ethyl ether for *infrared absorption* studied was prepared by Colon and Frediani [1051]. Residual ethanol and water were removed from absolute ether by drying with anhydrous calcium chloride and then adding sodium ribbon. The ether was repeatedly decanted and fresh sodium added until evolution of hydrogen ceased. It was filtered and distilled in an essentially dry system.

Gillo [1776] proposed a method for the purification and the criteria of purity for ethyl ether as a *standard*. Two methods of purification were recommended.

1. Ether is shaken with 50% sulfuric acid, distilled, dried 8–15 days over calcium chloride and for one month over sodium, and then distilled.

2. This method is the same as the first except that the ether is washed with a saturated potassium permanganate solution containing 5% sodium hydroxide. The final drying period over sodium is two months in the absence of light and air.

The products of these two methods differ in boiling point by 0.006–0.014°. The water and acetaldehyde content varied from 0.0008–0.0010%. The peroxide content of ether prepared by method 1 was about 1 mg/100 ml, that of the ether prepared by method 2 about 0.01–0.1 mg/100 ml. Ether is best preserved by purifying by the second method and storing in a sealed container in the absence of both air and light. Phosphorus pentoxide is a good preservative.

The preparation of ethyl ether as a polarographic solvent is described [4957].

Kablukov and Malischeva [2462] state that the presence of a third substance, even in insignificant amounts, may have a decided influence on the *mutual solubility* of ethyl ether and water. They washed ether five times with water, dried over calcium chloride, and distilled from sodium. A portion of 100–200 ml was redistilled from sodium for each determination; the first fraction being rejected.

CRITERIA OF PURITY

ACS Reagent Chemicals [76] and Rosin [4042] list specifications and test methods for both anhydrous ether and that containing 2–3% ethanol and 0.5% water.

Lamond [2811] modified the method of Duke and Smith [1394] for the determination of ethanol in ether. The results are accurate to 0.01 g of ethanol in 100 ml of ether.

An infrared absorption method for determining ethanol in ethyl ether has been reported by Colon and Frediani [1051]. The water is first determined by the Karl Fischer method, and then ethanol and water are determined at 2.83 μ. Once the calibration curve has been established, a determination can be made in a short time with a relative accuracy of $\pm 2\%$.

Pesez [3724] has described a method for detecting acetaldehyde in ethyl ether that is sensitive to 10 ppm in 1 hr and 2 ppm in 12 hr.

Gas chromatography is a convenient method for the determination of impurities in ethyl ether.

SAFETY

The vapor pressure of ethyl ether at normal temperatures is high, the flammable limits in air are wide, and the minimum ignition temperature is low. Therefore the principal hazard with ethyl ether is considered its explosibility with air. Concentrations of ethyl ether in air of 3.6–6.5%v produce anaesthesia in humans. Concentrations of over 10%v are usually fatal. Repeated-dose inhalation is often the result of intentional exposure to an "ether jag," which results in loss of appetite, dizziness, headache, exhaustion, psychic disturbances, and damage or malfunction to some body organ [3681].

A *threshold limit value* of 400 ppm, 1200 mg/m³, is suggested [4509].

Ethyl ether has wide *flammable limits* of 1.85 and 36.50%v in air. The *minimum ignition temperature* in air is low, 174° [3680].

110. PROPYL ETHER

SAFETY

Marsden and Mann [3160] state that detailed safety data are not available. Propyl ether is more toxic than ethyl ether. Exposure to vapors of 1%v do not seem to produce any ill effect other than an occasional intoxication.

Propyl ether is a highly flammable substance.

111. ISOPROPYL ETHER

Nekrasov and Krentsel [3507] prepared isopropyl ether by passing 2-propanol vapor over an ascarite clay catalyst. Under optimum conditions, the yield was 52%w.

Tabuteau and Gautier [4592] absorbed propylene in concentrated sulfuric acid in the presence of a copper and copper-sulfate catalyst. The ether was obtained by adding either 2-propanol or water.

Calderazzo [882] dried tetrahydrofuran and isopropyl ether over sodium and distilled over lithium aluminum hydride and stored under nitrogen for *carbonylation of amines studies.*

For purification, see Ethers, General [2653].

Peroxide formation is isopropyl ether is rapid. It is advisable to test for peroxides, even though it has been freed of them only a short time before. Katsuno [2524] found that the addition of 0.001%w of such dihydric phenols as pyrocatechol, resorcinol, and hydroquinone prevented the formation of peroxides in isopropyl ether for six months.

SAFETY

The vapors of isopropyl ether in air are an explosive hazard. The physiological properties of isopropyl ether are similar to those of ethyl ether but the former is more toxic. It is not absorbed through the skin in harmful amounts. A concentration of 500 ppm in air produced no complaints after 15-min exposure of human volunteers, but 35% of the subjects objected to the odor at 300 ppm [3681, 4509].

A *threshold limit value* has been set at 500 ppm, 2100 mg/m³, to prevent complaints of irritation and to provide a level free from general narcotic action [4509].

The *flammable limits* in air are 1.38 and 7.90%v. The *minimum ignition temperature* in air is 189° [3680].

112. BUTYLVINYL ETHER

Amagasa and coworkers [69] prepared butylvinyl ether by mixing 5 g of 1-butanol and 1 g of potassium hydroxide, and adding 40 g of a solution of acetylene in liquid ammonia. The mixture was heated for 8 hr at 100° and extracted with chloroform.

Alkylvinyl ethers were prepared by a continuous vapor phase reaction by Kozlov and Chumakov [2734]. The reactor contained a zinc acetate or copper acetalide catalyst and was fitted with an acetylene meter and a device so that the alcohol could be dropped on the catalyst. The alkylvinyl ethers were prepared from 1-butanol, 2-methyl-1-propanol, and 3-methyl-1-butanol. The yield of butylvinyl ether at 200–220° was 47.6%.

Reppe [3957] reported on the reaction of acetylene and alkyl alcohols over soda lime in the vapor phase.

Schildknect and coworkers [4151] developed an improved method of purification and determined more precisely additional *physical properties* of butylvinyl ether. The ether was washed with water at room temperature to remove alcohols and the stabilizing alkali until no further volume decrease was noted. The pH was kept below 8. It was dried with a neutral or alkaline drying agent and finally, by standing over sodium wire. The finished product was obtained by fractional distillation over sodium wire through a 10–20 plate column packed with glass helices. The distillate should be stored over potassium hydroxide or with triethanolamine to prevent hydrolysis or polymerization.

Water can be determined by a modified Karl Fischer method.

SAFETY

The oral LD_{50} for rats is 3.10 g/kg; the vapor exposure is 4 hr for no deaths, when the animal is exposed to vapors obtained by passing air through the ether at room temperature. A LD_{50} of 3.00 ml/kg was found for absorption through the skin of rabbits [3681]. Butylvinyl ether tested on frogs showed a similar reaction to ethyl ether [4298].

113. BUTYLETHYL ETHER

Vogel [4970] prepared butylethyl ether for *physical property measurements* from 1-butanol, sodium, and iodoethane. The ether was purified by repeated refluxing over sodium and then distilling.

Norris and Rigby [3550] prepared butylethyl ether from sodium butylate and bromoethane; Kranzfelder and coworkers [2738] used ethyl sulfate instead of bromoethane.

SAFETY

Six rats survived inhaling air containing 1000 ppm for 4 hr [3681].

Butylethyl ether may be determined in air with a sensitivity of 5 mg in 2.7 ml of solution [2797].

114. BUTYL ETHER

Vogel [4970] prepared butyl and higher homologous ethers for *physical property measurements* by removing the theoretical amount of water as the alcohol-water azeotrope from the alcohol-sulfuric acid mixture, and then steam distilling the ether.

Technical grade butyl ether was purified by Kusama and Koike [2790] by treating with carbon disulfide and sodium hydroxide. The excess sulfide was expelled by heating. The ether layer was washed with water, dried with sodium hydroxide and vacuum distilled.

Dasler and Bauer [1194] freed the ether of peroxides and aldehydes by adsorption of the impurities on alumina.

For purification; see Ethers, General [2653].

SAFETY

Butyl ether seems to have a higher toxicity by inhalation than the lower members of the series. It is more irritating to the skin, eyes, and nose. Rats survived air saturated with the ether for 30 min. It has been estimated that 100 ppm, 533 mg/m^3 should be the highest concentration to which workmen should be exposed for an 8-hr day [3681].

115. PENTYL ETHER

Vogel [4970] prepared pentyl ether in the same manner as he did butyl ether.

Kranzfelder and coworkers [2738] prepared pentyl ether from 1-pentanol and pentyl sulfate.

SAFETY

No specific data are available. The low vapor pressure of pentyl ether makes it a relatively safe solvent at ambient temperatures with normal handling care. One should avoid breathing the vapors at elevated temperature over a period of time [3160].

116. ISOPENTYL ETHER

Vogel [4970] prepared isopentyl ether in the same manner as he did butyl ether.

Schorigin and Makaroff-Semljanski [4181] freed isopentyl ether of the alcohol with boric acid.

Underwood and Toone [4877] decomposed the peroxides in isopentyl ether by shaking with potassium iodide solution.

Späth [4454] prepared alcohol-free isopentyl ether by boiling the commercial product for 5 hr under reflux with sodium followed by distillation. See also Zappi and Degiorgi [5323]. For the removal of 3-methyl-1-butanol from isopentyl ether with barium, see [4181].

For *electrochemical measurements*, Bruus [821] dried isopentyl ether over calcium chloride, purified by careful fractional distillation, and then dried over sodium or phosphorus pentoxide.

117. 1,2-DIMETHOXYETHANE

Capinjola [904] prepared 1,2-dimethoxyethane by refluxing 1830 g of 2-methoxyethanol and 138 g sodium until the latter dissolved. Chloromethane was passed into the liquid only as fast as it would react. The weight of the flask was used to determine when sufficient chloromethane was added. The portion that distilled below 123° was collected and fractionally distilled.

1,2-Dimethoxyethane was prepared in 81 % yield by Kranzfelder and Vogt [2739] by treating the monomethyl ether in the presence of sodium with methyl sulfate.

Commercial 1,2-methoxyethane was treated with lithium aluminum hydride by Wallace and Mathews [5063] and fractionally distilled immediately before making aqueous solutions to study some of their *physical properties*.

For the *study of solvent effect* of alkali salts of phenols and enols, 1,2-dimethoxyethene was purified by Zaugg and Schaefer [5326] in the same way they purified hexane.

Agami [33] discusses 1,2-dimethoxyethane as a solvent for inorganic, organometallic, and organic substances, and as a medium for several types of reactions.

The glycol diethers may contain inhibitors such as N-methylpyrrolidine, N-butylpyrrole, and similar compounds [5332].

118. BIS(2-METHOXYETHYL) ETHER

Hamamoto and Tone [1986] prepared bis(2-methoxyethyl) ether by a general method for the preparation of bis-ethers of mono-, di-, or triethylene glycols. Ethylene oxide reacts readily with methanol, or ethanol, in the presence of sodium methoxide, or ethoxide, with the generation of heat (to about 90°). The reaction is complete in about an hour.

Gallaugher and Hibbert [1687] prepared bis(2-methoxyethyl) ether by the action of bis(2-chloroethyl) ether on sodium methylate in methanol at 25°. Thomas and Weisse [4723] used sodium hydroxide to bring about the same reaction.

Bis(2-methoxyethyl) ether starts to decompose at 60° [1689].

Aliphatic Cyclic Ethers

Specific Solvents

119. PROPYLENE OXIDE

Propylene oxide was purified by Oetting [3578] for *heat capacity measurements* by drying with Drierite and fractionally distilling in a 100-plate still. Melting curves showed the purity to be 99.94%m. The sample was transferred to a high vacuum line, degassed, and loaded into the sample container by a one-plate distillation. Helium was admitted to the container before sealing.

Price and Osgan [3830] purified commercial propylene oxide by refluxing for 2 hr over potassium hydroxide, then fractionally distilling, and discarding the first and last 15%.

Bartlett [418] described a method for purifying epoxides containing 2–12 carbon atoms by treating with ammonia, amines, hydrazine, a hydroxylamine, or a semicarbazide in such amounts that not more than 10% of the epoxide will be consumed.

Levine and Walti [2920] prepared *dextro*-propylene oxide: 40 ml of water and 19 g of sodium hydroxide were put in a double-necked distilling flask immersed in a 40° water bath. The flask was provided with a condenser through which 0° water circulated. 3-Bromo-1,2-propanediol was added and the temperature was raised to 60°. The flask was shaken vigorously, while the propylene oxide was distilled.

Rogers [4016] fractionally distilled a commercial sample of propylene oxide through an efficient column for determination of the *dipole moment*. The boiling point, refractive index, and density were used as the criteria of purity.

Abderhalden and Eichwald [4] prepared *dextro*-propylene oxide from *d*-2-chloro-1-propanol.

A crystalline hydrate, $C_3H_6O\cdot16H_2O$, mp −3°, has been reported by Zimakov and Sokolova [5347].

SAFETY

Propylene oxide is a highly reactive substance. The liquid is relatively stable. It reacts exothermally with any substance that has a labile hydrogen such as water, alcohols, amines, and organic acids. Certain substances, such as acids, alkalies, and certain salts, act as catalysts. It also polymerizes exothermally. No acetylide-forming metals, such as copper or copper alloys should be in touch with propylene oxide.

Physiologically, it is primarily an irritant, a mild protoplasmic poison, and a mild depressant of the central nervous system. Contact with the skin, even in dilute solutions, may cause irritation and narcosis. Excessive exposure to the vapor may cause irritation to the eyes, respiratory tract, and lungs [3681].

All animals tolerate 100 ppm without adverse effects.

A survey [624] of the several methods for the determination of propylene oxide in air indicates that the method based on oxidation to formaldehyde is the most rapid and sensitive.

The *threshold limit value* of 100 ppm, 240 mg/m^3, has been suggested [4509].

The *flammable limits* in air are 2.1 and 38.5%v. The *minimum ignition temperature* in air is 421° [3680].

120. 1,2-EPOXYBUTANE

Houtman and coworkers [2262] describe a procedure for straight chain epoxybutane substantially free of isobutylene oxide where the oxygen is attached to the adjacent carbon. Agitate to create an intimate depression of the butylene oxide and add 30–50% aqueous alkali hydroxide, allow the phases to separate, draw off the aqueous phase, and fractionally distill the oxide. The small amount of water and isobutylene oxide remaining will distill as the azeotrope.

De Montmollin and Matile [1278] prepared 1,2-epoxybutane for *infrared studies* by treating 2-chloro- or 2-bromo-1-butanol with sodium hydroxide; bp 58.5–59°.

SAFETY

The *explosive limits* in air are 3.0 and 15.3%v [1359].

121. CINEOLE

Cineole is a syrupy oil with a camphor-like odor and a pungent taste. It occurs in many essential oils. The oils from eucalyptus species contain from 30–92% cineole; hence the synonym, *oil of eucalyptol*. It is also found in oils of ginger, cajuput, niouli, cardamom, spike lavender, laurel leaf, rosemary, and certain of the *Ocimum artemesia* and *O. alpina* [1929]. It has been reported in oil from leaves of *Chamaelyparis obtusa formosana* [2509].

If the cineole content of the oil is high, it can be separated by fractional distillation. The 170–180° fraction is cooled and the cineole separated by fractional crystallization. The first crystallization is reported to be of high purity. The authors experienced some difficulty in fractionally distilling California bay tree oil at atmospheric pressure, but cineole and other components were separated without difficulty at reduced pressure. Johnson [2406] has described a method for separating cineole from hydrocarbons by fractional azeotropic distillation with phenol.

Cineole forms addition compounds with hydrogen halides. The hydrochloride compound is quite soluble in the other constituents of the essential oil, but the hydrobromide is less soluble and therefore suitable for isolation. The procedure is as follows: Dilute refined cineole, bp 175–180°, with an equal volume of petroleum ether, and saturate the solution with dry hydrogen bromide. Filter the white precipitate, wash with several small portions of petroleum ether, and regenerate the cineole by stirring the crystals with water.

If the cineole content of the oil is high, it may be isolated by forming the addition compound with resorcinol. Low cineole oils first must be fractionally distilled. The details of this and other methods are given by Guenther [1929].

CRITERIA OF PURITY

See Guenther [1929].

SAFETY

Cineole is used widely in pharmaceuticals for both internal and topical application.

122. FURAN

Furan can be prepared by treating 2-furaldehyde with steam at elevated temperatures in the presence of a catalyst containing a dehydrogenating oxide, a mixture of dehydrogenating oxides, or a chromite of a dehydrogenating oxide [1407]. One volume of 2-furaldehyde and five volumes of steam were passed over manganese chromite pellets containing potassium chromate at about 400°. After stripping, drying, and redistilling, one-half of the theoretical amount of pure furan was obtained.

Katsuno [2525] prepared furan by converting 2-furaldehyde to 2-furoic acid and furfuryl alcohol by the Cannizzaro reaction. The furoic acid was decarboxylated to furan in 83–88% yield. See also Whitmore [5155, p. 40].

A method has been described by Wilson [1778] for the preparation of furan by the decarboxylation of 2-furancarboxylic acid.

Guthrie and coworkers [1940] distilled technical furan through a 70-plate column packed with steel helices under an atmosphere of nitrogen. Hydroquinone was added to each cut as an inhibitor. The heart cuts were 99.98%m. Mass spectra indicated 0.01%m isopentane and a trace of an unidentified impurity.

For purification, see Ethers, General [2653].

Cheng and coworkers [964] purified furan for *critical temperature studies* by distilling commercially available material, treating with 5% potassium hydroxide, drying over anhydrous sodium sulfate, and distilling *in vacuo* over sodium; bp 31.33° at 765 torr.

Cass [928] reported that 0.01 to 1.0% hydroquinone will stabilize furan against polymerization and color formation.

SAFETY

The Quaker Oats Company [3872] states that the concentration of furan vapors in the air always should be kept to a minimum both as a health and as a fire precaution. Chronic exposure of dogs to concentrations in air of 10 ppm produces noticeable circulatory disturbances. Furan is readily absorbed through the skin. Continual exposure of man to low concentrations may develop fatigue, headache, and gastrointestinal disturbances.

The *explosive limits* in air are 2.3 and 14.3%v [3872].

123. TETRAHYDROFURAN

Ivanskii and Dolgov [2365] developed two catalysts: (1) 60% Cu, 15% Al_2O_3, 25% SiO_2, and (2) 60% NiO, 15% Al_2O_3, 25% SiO_2 for converting 1,4-butanediol to tetrahydrofuran and γ-butyrolactone. The catalysts were made by adding sodium hydroxide and sodium silicate to a hot solution of the metal nitrates, drying the precipitate, heating to 250–450° and reducing in a stream of hydrogen. At 190° catalyst 1 converts 1,4-butanediol quantitatively to γ-butyrolactone. At 190–260°, catalyst 2 converts the diol quantitatively to tetrahydrofuran.

Cyanuric chloride and 1,4-butanediol will form tetrahydrofuran, according to Matuszko and Chang [3234]; see also tetrahydropyran.

Starr and Hixon [2408, p. 77] prepared tetrahydrofuran by catalytic hydrogenation of furan using palladium oxide.

Mikhailov [3310] prepared furan by catalytic decarboxylation of furaldehyde in hydrogen or steam. Hydrogenation of the furan over nickel, palladium, or osmium gave tetrahydrofuran.

Kirrman and Hamaide [2613] prepared tetrahydrofuran in high yields by heating a *gamma-*, *delta-* or *epsilon-*methoxylated alkyl bromide with a few milligrams of iron (III) chloride.

Furan containing 0.01% dibutylamine was hydrogenated in the vapor phase to tetrahydrofuran at 85° in almost quantitative yield by Manly [3138].

Arnett and Wu [284] report that, despite the widely different nature of acidic systems, the same order of basicity is always found for the cyclic ethers. When 2 pK's are plotted against any other function of basicity, linear correlations are nearly always obtained.

The cyclic ethers are generally five to ten times more basic than the simple acyclic ones. Tetrahydrofuran is very basic in all systems, which probably explains, at least in part, its superior solvent ability among the ethers for organometallic systems.

Commercial tetrahydrofuran was distilled from sodium for *vapor-liquid equilibrium studies* [1726].

For purification, see Ethers, General [2653]; for *carbonylation of amines studies*, see Calderazzo [882] under isopropyl ether.

Bhattacharyya and coworkers [566] purified tetrahydrofuran for *conductance studies* by refluxing overnight with sodium-potassium alloy and fractionally distilling. In some cases, the ether was further purified by stirring with the alloy, to which benzophenone had been added, and then vacuum-distilling.

Commercial tetrahydrofuran was prepared for electrochemical studies by Perichon and Buvet [3710] by allowing it to stand for 15 days over potassium hydroxide and then over naphthalene sodium for several hours. It was distilled in an inert atmosphere, refluxed over lithium aluminum hydride, and distilled directly into the electrolytic cell in an argon atmosphere. They discuss the *possibilities and limitations of electrochemical reactions* in this ether.

Tetrahydrofuran was purified for the study of the *solvent effect on ultraviolet spectra* of alkali phenols and enols by Zaugg and Schaefer [5326] in exactly the same way that they purified cyclohexane.

Pickett and coworkers [3759] purified tetrahydrofuran from three sources for *vacuum ultraviolet absorption studies* by drying over calcium sulfate and metallic sodium, and then fractionally distilling through a 30-cm Penn State glass-helices packed column. Because of the unique appearance of the spectrum and the possibility that impurities might be present, two samples were prepared by the catalytic hydrogenation of furan.

Cheng and coworkers [964] purified tetrahydrofuran for *critical temperature studies* by distilling commercial material, allowing it to stand 48 hr over freshly fused sodium hydroxide and 24 hr over sodium wire, over which it was refluxed. It was fractionally distilled in an atmosphere of dry nitrogen, and finally vacuum-distilled from lithium aluminum hydride; bp 66.2–66.4°at 766 torr.

Paul and coworkers [3682] purified tetrahydrofuran as a *solvent for sodium metal resection study* with aromatic hydrocarbons as follows: It was distilled from a mixture of activated alumina and sodium hydroxide pellets into a receiver containing sodium chips and anthracene. It was allowed to stand at least 1 hr and distilled into a receiver containing the reactants in active form.

Hernandez [2130] purified tetrahydrofuran for *ultraviolet studies* by repeated vacuum distillation, retaining the middle cut each time. Samples were stored in vacuum until ready for use.

Morikawa and Yoshida [3404] studied the purification and aging of tetrahydrofuran, and identified eight aging products: namely, tetrahydrofuran α-hydroperoxide, butyraldehyde, 4-hydroxybutyric acid, 4-hydroxyper-butyric acid, butyric acid, 4-hydroxybutyraldehyde, α-oxotetrahydrofuran, and water.

A sample of du Pont tetrahydrofuran was analyzed by gas chromatography for *azeotrope studies* [3975]. It was found to be 99.9% with less than 0.1% water and traces of two impurities that eluted prior to the ether.

Some tetrahydrofurans contain impurities such as 2- or 3-*tert*-butyl-4-hydrofuran, or 0.025% butylated hydroxytoluene, introduced to inhibit peroxidation.

Bordner [667] found 0.01 to 1.0% *p*-cresol will inhibit peroxide formation in tetrahydrofuran. Hinegardner [2195] found 0.05 to 0.10% hydroquinone effective.

A photometric method [5353] is given for the determination of peroxides in tetrahydrofuran that compares favorably with the iodometric method; range 0.002 to 0.8%.

SAFETY

The Quaker Oats Company [3471] states that tetrahydrofuran is only moderately toxic; normal precautions recommended for volatile solvents are satisfactory.

Tetrahydrofuran has a hypnotic effect at 6%. The onset is slower than ethyl ether and the effect is longer lasting. Jockmann reported 20 to 50% aqueous solutions irritated the skin of rabbits. However, it has been concluded that it is not irritating to the skin and is not a skin-sensitizer [4509].

The *threshold limit value* of 200 ppm, 590 mg/m^3, has been recommended [4509]; see also [78].

The *explosive limits* in air are 1.8 and 11.8%v at 25°. The *minimum ignition temperature* in air is 321° [3871].

A TLV of 100.0 mg/m^3 was suggested in 1964 for Russia [2711].

Deyanova [1297] reported a method for the determination of tetrahydro-furan in the air; sensitivity, 1 μg/ml solution; on a sample collected by adsorption on silica gel.

124. *p*-DIOXANE

Urazovskii and Chernyavskii [4897] determined the viscosity of benzene, *cis*- and *trans*-decahydronaphthalene, and *p*-dioxane over a wide temperature range and plotted the data as log η vs. $1/T$. Only the data for *trans*-deca-hydronaphthalene gave a straight line. Data for the other compounds gave two intersecting straight lines. Explanations are offered. Urazovskii and Ezhik [4898] continued the study of the behavior of *p*-dioxane and found that the temperature dependence curves of several properties exhibit an anomaly at about 26°. The dipole moment is negligible at 20° but is about 0.30 at 30°.

They conclude that the *p*-dioxane molecule started to assume the boat form above this temperature.

Hess and Frahm [2153] found that the chief impurities in commercial *p*-dioxane are acetic acid, water, and glycol acetal (CH$_3$CH—O—CH$_2$CH$_2$O).

Depending upon the composition of the dioxane, heating with hydrochloric acid may be sufficient to remove the acetal. When the acetal is present in only small amounts, prolonged heating with sodium is adequate. Additional purification is effected by careful fractional distillation and crystallization. Once purified, *p*-dioxane is stable indefinitely when protected from atmospheric oxygen.

According to Hepworth [2117], *p*-dioxane prepared from 1,2-ethanediol by heating with concentrated sulfuric acid contains aldehydes. These can be removed by repeated boiling with silver oxide. After several distillations over freshly fused potassium hydroxide, it is fractionally distilled over sodium.

Rieche [3979] and Milas [3314] state that *p*-dioxane is auto-oxidized, but according to experiments of Eigenberger [1458] it is not auto-oxidized when it is pure and free from acetal. As the best method for the preparation of acetal-free *p*-dioxane, they recommended that ethylene acetal be decomposed by acid, whereby a better yield is obtained than by fractional distillation. Crude *p*-dioxane is boiled under reflux for 7 hr with 10% of its volume of 1 N hydrochloric acid, a slow stream of air being passed in through the condenser to remove the acetaldehyde formed. To remove water, the distilled *p*-dioxane is treated with potassium hydroxide, and the aqueous layer removed. It is then allowed to stand over pellets of potassium hydroxide for one day and dried over sodium for several hours. It is then distilled from the sodium.

p-Dioxane was purified for *dielectric constant studies* by the method of Hess and Frahm [2153], except that recrystallization was omitted. It had a freezing point of 11.73° and a purity of not less than 99.95%m from the freezing curve [1135].

Vinson and Martin [4957] allowed commercial grade *p*-dioxane to stand over calcium chloride for several weeks for *heat of vaporization* and *vapor pressure studies*. The boiling temperature showed no rising or declining trend over several consecutive determinations; mass spectral analysis showed 0.04% water.

Smith and Wojciechowski [4383] purified a good commercial grade of *p*-dioxane by distilling through a 40-bulb Swietoslawski-type column of about 30 theoretical plates. McKinney and coworkers [3084] purified by refluxing over sodium and distilling through a 24-in. column packed with Raschig rings.

Kraus and Vingee [2745] purified *p*-dioxane as a solvent for a *study of nonpolar solvent solutions of electrolytes* as follows: The *p*-dioxane was refluxed over sodium hydroxide, distilled, and refluxed for an extended period over fresh sodium, fresh sodium being added from time to time.

Kraus and Fuoss [2744] boiled technical *p*-dioxane with sodium hydroxide, dried over barium oxide, and fractionally distilled from a sodium-lead alloy. The middle fraction was dissolved in liquid ammonia and the ammonia boiled off. The *p*-dioxane was redistilled, stored over sodium-lead alloy, and distilled only when needed.

Oglukian [3580] discovered, when *following kinetic reactions by ultra violet spectroscopy* in *p*-dioxane as the solvent, that decomposition of the solvent was causing an increase in absorbency in the spectral region of interest. A study of the purification was made, including more than 50 published methods. The following method was evolved.

Step 1.   *p*-Dioxane, any good grade, was let stand over iron (II) sulfate for at least two days.

Nitrogen was used continually in all subsequent operations at a light bubbling rate unless otherwise specified. It was water-pumped grade and passed through a concentrated sulfuric acid drying tower, then through a 1 × 24-in. tube of sodium hydroxide pellets.

Step 2.   (a) 200 ml of water and 28 ml of concentrated hydrochloric acid were added to 2 liters of *p*-dioxane. A pale yellow color developed.

(b) It was refluxed not less than 8 nor more than 12 hr with heavy nitrogen bubbling.

(c) Potassium hydroxide pellets were added to the solution, while still warm, until two layers formed. The yellow color disappeared. The contents of the flask were cooled, often rapidly in an ice bath.

(d) It was magnetically stirred and potassium hydroxide pellets added until no more dissolved. The solution was cooled when the flask became hot to the touch.

(e) The flask was allowed to sit for 4–12 hr until the two phases were sharply separated. If the lower phase was black, the lot was discarded and the purification started again with Step 1.

(f) The upper phase was decanted rapidly into a clean flask and then into a still pot containing sodium.

Step 3.   The *p*-dioxane was refluxed in a still with an 18-plate Oldershaw column until the sodium remained very shiny for some time. After the system had been under total reflux for an hour, 30–50 ml were slowly distilled until the head temperature reached 101° with the nitrogen bubbling shut off during the reading. The spectrum was checked every 4–6 hr after the appearance of the shiny sodium to assure that a minimum absorbency below 2500 Å had been attained. The material was then fractionally distilled into a round bottom flask and stoppered.

Step 4.   The distillate was recrystallized three times, discarding 15–20% each time. The flask was placed in an 8° refrigerator and shaken frequently or magnetically stirred. Nitrogen was not bubbled through the p-dioxane during the freezing. It was bubbled through during the melting. If the spectral cut-off was higher than 2170 Å, a crystallization would not help and the process was started over. By the third crystallization, there was little or no spectral difference between the mother liquor and the melted solid therefrom.

The material was stored in a refrigerator until ready to use. It was thawed, refluxed over sodium for 48 hr, and distilled as needed directly into the container in which it was to be used. The first 5–10 ml were discarded from each lot distilled.

All joints were teflon-tape-clad and p-dioxane was its own lubricant for the stopcocks.

The p-dioxane has a zero absorbency at 2200 Å.

The density at 25° was 1.02803. It was very hygroscopic, so that any operation requiring exposure to air was difficult.

Katti and Chaudhri [2526] purified AR grade p-dioxane for *physical property of mixtures studies* by refluxing with hydrochloric acid, treating with potassium hydroxide, allowing to stand over sodium and fractionally distilling; d 30° 1.0232.

For purification, see Ethers, General [2653].

Palyi and Peter [3639] prepared *polarographically pure p*-dioxane from a commercial product. It was mixed with concentrated hydrochloric acid in a ratio of 5:2 ether to acid and refluxed at 120–130° for 10 hr under nitrogen. The resinous material was separated by filtration when cool and the filtrate neutralized with solid sodium carbonate and 25 g of sodium hydroxide per liter. The aqueous phase was allowed to separate; 30 g iron (II) sulfate or chloride was added per liter of ether. The mixture was shaken mechanically for 3–5 hr. The solids were filtered and the two phases of the filtrate allowed to separate. The p-dioxane was separated from the aqueous phase. It was distilled twice from solid sodium hydroxide and once from sodium under an oxygen and water free atmosphere. The purified material was stored in a glass bottle.

p-Dioxane was purified by Brown [790] as a solvent for *dipole moment studies* by drying with anhydrous magnesium sulfate, refluxing with sodium, and fractionally distilling. It was stored over sodium wire; d 24° 1.0292.

Sideri and Osol [4309] reported that p-dioxane that had been shaken with 20 g per liter of Powminco brand asbestos (used for filtering mats) did not develop a color when used as a *solvent for standard perchloric acid titrant*. Riddick [3975] checked this purification and found one lot of Powminco brand asbestos out of four was not effective. No other brand of asbestos was found to be effective.

Levi and coworkers [2925] found that Amberlite IRC 50 was effective for removing the objectionable impurities from p-dioxane that caused discoloration when used as a solvent to prepare *standard perchloric acid solutions*.

Fisher and Baxter [1581] found that the peroxide content of p-dioxane could be decreased either by shaking with tin (II) chloride and distilling, or by refluxing with lead (IV) oxide and filtering through a slow filter paper.

SAFETY

The liquid is painful to the eyes and irritating to the skin upon prolonged contact. It can be absorbed through the skin in toxic amounts. p-Dioxane is insidious. Its vapors have poor warning properties; they are faint and inoffensive. Concentrations in air of 300 ppm cause irritation of the eyes, nose, and throat of humans. The vapors can be inhaled in amounts that cause serious systemic injury, particularly to the liver and kidneys [3681]. Johnstone [2414] reports a death that was supposedly due to p-dioxane after exposure during working hours for one week to concentrations of 208–650 ppm in an unventilated room. It had been used to dissolve glue from the victim's hands.

The *threshold limit value* has been set at 100 ppm, 360 mg/m³, and is believed to be low enough to prevent systemic poisoning with a good margin of safety [4509].

The *flammable limits* in air are 1.97 and 22.25%v [3681]. The *minimum ignition temperature* in air is 266° [3680].

125. TETRAHYDROPYRAN

Dihydropyran containing 0.01% aniline was hydrogenated to tetrahydropyran in the vapor phase at 85° over a highly active nickel catalyst in quantitative yield by Manly [3138]. Andrus and Johnson [4398] present a laboratory method for hydrogenation in the liquid phase that is almost quantitative.

Allen and Hibbert [55] prepared tetrahydropyran in 90% yield for *electrical moment measurements* by heating at 150° for 4 hr in a sealed tube 25 g of purified 1,5 dibromopentane, 25 g of distilled water, and 8 g of pure zinc oxide. The reaction mixture was extracted three times with ether and dried over sodium sulfate. The fractions from three preparations boiling at 85–90° were further purified by fractional distillation.

Matuszko and Chang [3234] studied the reaction of cyanuric chloride and diols. Tetrahydropyran was prepared in 77% yield from 1,5 pentanediol; see also tetrahydrofuran.

Kirrmann and Hamaide [2613] prepared tetrahydropyran in high yields by heating *gamma-*, *delta-*, or *epsilon-* methoxylated alkyl bromide with a few milligrams of iron (III) chloride.

Pickett and coworkers [3759] purified two samples of tetrahydropyran for *vacuum ultraviolet absorption studies* by drying over calcium sulfate; and refluxing over, and distilling from, sodium. Both samples were found to contain benzene (0.0005% estimated) which was removed by passing through a graphon column.

Hernandez [2130] purified tetrahydropyran for *ultraviolet studies* by repeated vacuum distillation retaining the middle cut each time. Samples were
stored in vacuum until ready for use.

Aromatic Ethers

Specific Solvents

126. BENZYLETHYL ETHER

Braun [715] prepared benzylethyl ether by the action of ethanol on benzyl
bromide and removed the unchanged bromide by adding a small quantity of
a base, for example, piperidine, warming for a short time, and shaking with
dilute hydrochloric acid. After drying over calcium chloride, the ether was
fractionally distilled.

White and coworkers [5142] prepared the ether by the action of potassium
ethylate on benzyl chloride. The ether solution was dried with sodium
hydroxide and fractionally distilled.

The unchanged ester in benzylethyl ether, prepared by electrolytic reduction
of ethyl benzoate, can be removed by boiling with alcoholic potassium
hydroxide [4594].

For *optical measurements*, Baly and Collie [371] purified the ether by
fractional distillation until the boiling point was constant.

According to Schorigin [4180] benzylethyl ether is decomposed by sodium
at 140°.

127. ANISOLE

King and Wright [2598] prepared anisole in 75% yield by adding 3.91 g
of potassium to 30 ml of ice cold methanol and then adding 9.4 g of phenol
and 19.4 g of methyl phthalate. The excess methanol was removed by
warming at reduced pressure and the syrupy liquid was refluxed with an air
condenser in an oil bath (a heating mantle may be used) at 190–200° for 3 hr
and allowed to cool. The solid mass was dissolved in water and ethyl ether.
The ether soluble portion was washed with 2 N sodium hydroxide and fractionally distilled.

Hiers and Hager [1778] prepared anisole from phenol, sodium hydroxide,
and methyl sulfate.

Anisole was prepared in the same manner as phenetole in 83% yield [1363];
see phenetole.

Cullinane and Davies [1147] treated phenol with a methylating agent, such
as methyl ether or methanol, in the presence of activated alumina at 175–225°.
By recycling the phenol, the yield of anisole was increased to 43%. See also
Conant [1061] and Vogel [4970].

Vaughn and coworkers [4925] purified anisole for *dielectric relaxation studies* by distilling from sodium and storing over sodium sulfate. Roberti and Smyth [3992] distilled from barium oxide.

SAFETY

Data are not available to suggest a *threshold limit value*. The absence of reported adverse effects indicates that there probably is little need for this information [3681].

128. PHENETOLE

Phenetole was prepared by Drahowzal and Klamann [1363] in 79% yield by heating phenol and ethyl *p*-toluenesulfonate in the presence of sodium hydroxide. The organic product was extracted with ether, washed with dilute sodium hydroxide, dried, and distilled.

Phenetole may be prepared in 66% yield by the same method used for anisole [2598] by substituting an equivalent amount of ethyl phthalate for methyl ether.

For preparation from sodium phenolate and bromoethane in liquid ammonia, see [5142].

SAFETY

The limited information that is available indicates that phenetole is quite low in acute toxicity. It is not believed that this ether would present any unusual hazards [3681].

129. BENZYL ETHER

Bennett and Willis [512] refluxed 30 g of benzyl alcohol and 25 g of α-chlorotoluene with 25 g potassium hydroxide for 5 hr and fractionally distilled to prepare benzyl ether for *freezing point studies*. Repeated distillation at reduced pressure did not reduce the odor of benzaldehyde. It was concluded that the ether decomposed even at temperatures below 200°. It was further purified by fractional freezing. The purified substance had a faint odor quite distinct from that of the aldehyde.

It was found necessary to seed benzyl ether when using it as a cryoscopic solvent.

130. PHENYL ETHER

Phenyl ether has been suggested by Furukawa and coworkers [1680] of the National Bureau of Standards as a *calorimetric standard*. They accurately determined the *thermal properties* of a high purity material. It is a relatively stable substance and can be obtained in a high state of purity by fractional distillation, followed by fractional crystallization. It melts close to room temperature, 26.87°.

The phenyl ether used in the thermal studies was fractionally distilled and then fractionally crystallized 25 times. Dissolved air and water were removed by slowly freezing in a glass bulb during evacuation. The material was remelted and the freezing-evacuation process repeated. It was then distilled into another bulb and the freezing-evacuation process repeated. The purity was determined from melting point studies.

Phenyl ether decomposes 1 %m per hr at 538° [2402].

Merz and Weith [3277] prepared phenyl ether by heating phenol with zinc chloride.

SAFETY

Hake and Rowe [3681, pp. 1698 ff.] state that information regarding the physiological effects of phenyl ether is not extensive. It appeared to be relatively nonirritating to the skin and of a low order of toxicity. Dowtherm A is an eutectic mixture of phenyl ether and biphenyl. It is used as a heat transfer medium below 750°F (399°). Industrial experience with this eutectic is a good criterion of the hazards of phenyl ether. Its odor is disagreeable and it sometimes causes nausea. Pecchiai and Saffotti [3699] investigated the oral toxicity.

It is not an explosion or fire hazard due to its low vapor pressure.

The *threshold limit value* probably should be 1.0 ppm, 6.79 mg/m^3, or perhaps less than 0.1 ppm for some persons. The exact level will depend on the individual concerned [3681].

The *spontaneous ignition temperature* is 1195° (646°) [3681].

131. VERATROLE

Veratrole was prepared in 95% yield by Perkins and Weizmann [3715] by dissolving 100 g of 1,2-dihydroxybenzene in 200 ml of methanol containing 75 g of dimethyl sulfate at −50°. A solution of 150 g of potassium hydroxide in 350 ml of water was added rapidly. The reaction was fast. The mixture was diluted and extracted with ether

Curran [1158] prepared veratrole for *dipole moment studies* by treating catechol with sodium hydroxide and methyl sulfate. The organic phase was washed with successive portions of sodium hydroxide solution to remove guaiacol before drying and distilling.

Roberti and Smyth [3992] purified veratrole for *dielectric relaxation studies* by fractional distillation from barium oxide. Vaughn and coworkers [4925] distilled from sodium and stored over anhydrous sodium sulfate for the same purpose.

Cauwood and Turner [936] purified veratrole for *dielectric constant studies* by repeated crystallization from a low boiling petroleum fraction to a constant melting point.

Veratrole was distilled over calcium hydride and the fraction boiling at 206–207° was used for the study of the *variation of the dipole moment* with temperature by Di Bello and coworkers [1304].

SAFETY

The oral LD_{50} for rats is reported to be 1360 mg/kg [2392].

Acetals

General

Croxall and Neher [1143] have described a general method for the preparation of acetals. Acetals are formed when a vinyl ester and a primary or secondary alcohol are heated at 20–50° with a mercury compound and a strongly acidic catalyst. Mercury (II) sulfate will serve as a combined catalyst. An excellent combination is mercury (II) oxide and boron trifluoride. A ratio of mercury compound to acid catalyst of 1 g to 10 per mole of ester and an excess of alcohol are preferred. See also Shostakovskiĭ and Gershteĭn [4300].

Another general method for the preparation of acetals has been described by Heinemann [2091]. It is applicable to alcohols from methanol to the pentanols and also to higher homologues that are liquid at the reaction temperature; suitable aldehydes are the lower aliphatics and some aromatics. A catalyst bed, prepared from a clay absorbent, such as bauxite, fuller's earth or bentonite, is activated at 600–1200° for a sufficient time to reduce the water to 5% or less. The alcohol-aldehyde mixture in a mole ratio of 2:1 to 4:1 is passed through the bed, which has been heated to a suitable temperature between 0 and 50°, at a velocity of 0.01 to 0.1 volume per volume of catalyts per hour. The yield per pass is reported to be about 50% of the theoretical.

Renault [3956] described a general method for preparing acetals from alcohols, aldehydes, and hydrochloric acid in the presence of calcium chloride.

Helferich and Hausen [2095] reported a general method for the preparation by treating aldehydes or ketones with the appropriate tetralkyl silicate, using dry hydrogen chloride as the catalyst.

Specific Solvents

132. DIMETHOXYMETHANE

According to Bourgom [678], the chief impurity in dimethoxymethane is methanol, which can be removed by treatment with sodium. Fractional distillation over sodium gives a 50% yield of a product boiling at 42.3°. See also the methods of purification given for acetal.

Rambraud and Besserre [3890] describe a laboratory method for dimethoxymethane in 84% yield from trioxymethylene and methanol in the presence of calcium chloride and hydrogen chloride. Emde and Hornemann

[1474] prepared dimethoxymethane by distilling a formaldehyde solution of methanol containing ammonium chloride.

Buehler and coworkers [2251] boiled formaldehyde and methanol with calcium chloride and a trace of hydrogen chloride to produce dimethoxymethane.

Vinokurov [4956] prepared a technical formaldehyde in 95% yield and gave three methods for further purifying.

1. The technical dimethoxymethane, containing about 8% methanol, was allowed to stand with paraformaldehyde and a few drops of sulfuric acid 24 hr at room temperature and then distilled.

2. The technical material was shaken with an equal volume of 20–30% sodium hydroxide at 20°. The phases were separated at the end of 30 min and the acetal distilled.

3. Methanol free formaldehyde was added to the middle of the distilling column during the preparation of the technical product.

Haarer and Rühl [1952] produced methanol free dimethoxymethane by introducing water into the lower third of a fractionating column. The product contained 0.5% water.

McEachern and Kilpatrick [3073] purified commercial material for *thermodynamic study* by fractional distillation. The distilled liquid was dried over sodium; purity 99.9% by gas chromatography.

A commercial grade should meet the following specifications: ASTM distillation range 42.0–43.5°, sp. gr. 20/20° 0.863–0.865, n_D 20° 1.353–1.355, purity 97%w min, color APHA 15 max, aldehydes 0.1% max, water 1.5%w max, acidity as acetic acid 0.1%w max [3160].

SAFETY

The acetals only recently are becoming industrially important. There is only a meager amount of toxicological and other safety data available.

Dimethoxymethane has a chloroform-like odor and a pungent taste. The lethal concentration in air for rats is approximately 15,000 ppm. It may cause liver and kidney damage in humans; in higher concentration, narcosis [4519].

Fassett [3681, pp. 1984 ff.] has summarized the data that has been reported. Dimethoxymethane has been studied as a possible anaesthetic. The LD_{50} for mice was found to be about 18,000 ppm. The investigators concluded that the threshold toxic concentration for mice and guinea pigs is of the order of 11,000 ppm.

The *threshold limit value* has been set at 1000 ppm, 3100 mg/m³ [4509].

The *flammable limits* in air are 2.95 and 17.40%v. The *minimum ignition temperature* in air is 237° [3680].

133. ACETAL

Reichert and coworkers [3939] prepared acetal by passing acetylene into a mixture of ethanol, sulfuric acid, and mercury (II) sulfate.

Boxel-Maletra [700] prepared acetal with a purity of 98.5% by stirring acetaldehyde, ethanol, and ammonium chloride for 6 hr at room temperature.

For preparation from alkylvinyl ethers, see [4299, 3501, 1778].

For *physical property measurements*, Walden [5032] dried acetal over sodium and purified by distillation. Adams and Adkins [13], who studied catalysis in acetal formation, purified acetal by shaking with hydrogen peroxide at 65° to remove aldehydes. The acetal was then washed with water, dried first with anhydrous potassium carbonate, and then with sodium and distilled. Fougue and Cabanac [1610] allowed acetal to stand over sodium for several weeks to remove alcohols and water and to polymerize aldehyde.

Commercial acetal was purified by Vogel [4970] for *physical property measurements* by treating with alkaline hydrogen peroxide solution at 40–45°, saturating the solution with sodium chloride, drying with anhydrous potassium carbonate, and distilling over sodium.

SAFETY

Acetal produces only minor symptoms of weakness on rats; even at high dose levels, no typical anaesthesia was noted [3681].

The *explosive limits* in air are 1.6 and 10.4%v. The *minimum ignition temperature* in air is 230° [3680].

CARBONYLS

Aldehydes

General

Aldehydes are quite reactive and their use as solvents is, therefore, limited. They are easily oxidized by oxygen in the air.

Zakharkin and Khorlina [5319] obtained aldehydes by the reduction of nitriles with diisobutylammonium hydroxide.

Brown and Rao [768] prepared aldehydes of widely varying structure by reduction of acid chlorides with lithium tri-*tert*-butoxyaluminum hydroxide in bis(2-methoxyethyl) ether solution at −78°. Aromatic aldehydes were obtained in better yields than aliphatic aldehydes but the experimental simplicity of the method makes it preferable when there is no advantage in yield. Stabilization of aldehydes from peroxide formation; see Ethers, General, Robertson and Jones [3998]

Specific Solvents

Saturated Aliphatic Aldehydes

134. ACETALDEHYDE

Mita [3342] prepared acetaldehyde from acetylene and a special copper catalyst. Three grams of 1-chloroethylene (vinyl chloride) and 40 g of acetaldehyde were obtained by passing 300 liters of acetylene through a solution at 78° containing 100 g of copper (I) chloride, 50 g of ammonium chloride, 32 g leucine hydrochloride, 150 ml of water, and 30 ml of concentrated hydrochloric acid

Smith and Bonner [4404] purified acetaldehyde by fractional distillation through a 100-cm column packed with glass helices. During the distillation, the column was kept under a positive pressure of dry nitrogen (1 in. of water pressure or 2.5 cm). The heads were removed at a reflux ratio of 20:1, the main fraction at 8:1.

Coleman and De Vries [1038] purified by distillation through a 4-ft Fenske-type column. The distilled fraction was sealed in Pyrex-brand glass containers and stored in the dark until used.

Muroi and coworkers [3459] developed a modification of the Karl Fischer method for the determination of water in acetaldehyde and other aldehydes. Pyridine and 1,2-propanediol in 3:1 volume ratio were used as the sample solvent. The titrant was Karl Fischer Reagent SS from Mitsubishi.

SAFETY

The vapor pressure of acetaldehyde is above 1 atm at normal laboratory temperature; therefore the material must be handled with caution because of the explosion hazard. The odor of the vapor can be detected well below 50 ppm; some persons can detect it below 25 ppm. Volunteers were exposed to vapors of 50 ppm for 15 min and some showed signs of eye irritation. At 200 ppm, all subjects had red eyes and conjunctivitis. The LD_{50} for rats is about 20,000 ppm for a 30-min exposure. At this level, the animals develop pronounced excitement, followed by an anaesthetic-like state after about 15 min; survivors recover rapidly. The principal finding at autopsy is pulmonary edema [3681].

Although acetaldehyde is irritating in higher concentrations, it is rapidly metabolized [4513].

The *threshold limit value* is 200 ppm, 360 mg/m³ (4509).

The *explosive limits* in air are 3.97 and 57.00%v [3680]; 4.1 and 55%v have also been reported [88]. The *minimum ignition temperature* in air is 176° [3680].

135. PROPIONALDEHYDE

Hurd and Meinert [616] oxidized 1-propanol with acid dichromate to obtain propionaldehyde; see also [4466, p. 64].

Hoffmann [2225] passed propylene oxide and steam over silica gel at 300° and obtained a yield of 70% propionaldehyde.

McKenna and coworkers [3081] purified propionaldehyde by distillation under nitrogen. The distillate was collected in glass containers that had been boiled with water for 20–30 min and dried.

Smith and Bonner [4404] purified both propionaldehyde and butyraldehyde by carefully drying three times with Drierite and then distilling under nitrogen in a Podbielniak column. The heads were removed at a 50:1 reflux ratio, the main fraction at 20:1.

SAFETY

Propionaldehyde has about the same skin and eye irritation as acetaldehyde [4884].

136. BUTYRALDEHYDE

Purification; see Smith and Bonner [4404] under propionaldehyde.

SAFETY

Butyraldehyde appears to have no primary skin irritation for rabbits but it is a severe eye irritant [4884].

The *threshold limit value* of 1.0 mg/m³ has been suggested for Russia in 1964 [2711].

The *lower explosive limit* in air is 2.5%v [88].

137. ISOBUTYRALDEHYDE

Pfeiffer [3740] oxidized 2-methyl-1-propanol with acid dichromate, distilled, and purified through the acid bisulfite derivative; see also Lipp [2970].

SAFETY

It did not cause skin irritation to rabbits but did cause moderate eye injury [4884].

Aromatic Aldehydes

138. BENZALDEHYDE

Highet and Wildman [2182] prepared benzaldehyde by oxidation of benzyl alcohol in chloroform, ether, or hexane with manganese dioxide; yields 61–89% were obtained depending upon the solvent used and the reaction time.

Yamashita and Matsumura [5285] oxidized primary alcohols with *p*-benzoquinone to obtain aldehydes. Two moles of *p*-benzoquinone and $\frac{1}{3}$ mole of aluminum phenolate were reacted with 1 mole of alcohol in benzene; benzyl alcohol gave benzaldehyde and furfuryl alcohol gave 2-furaldehyde.

Hesse and Schrödel [2156] prepared benzaldehyde in 92% yield by the reduction of benzonitrile with sodium hydrogen aluminum ethoxide (NaHAl-$(C_2H_5O)_3$), in tetrahydrofuran at room temperature.

Meyers [3294] reported that potassium hypochlorite oxidized benzyl alcohol to the corresponding aldehyde without any side reactions in 77% yield.

For additional preparations see [5336, 5311].

Benzaldehyde may be purified through its bisulfite addition compound or by fractional distillation at reduced pressure [2078].

CRITERIA OF PURITY

Rosin [4042] describes the Reagent grade as containing at least 98% benzaldehyde when analyzed by the hydroxylammonium chloride method and not more than 0.01% chlorine compounds. It should give no reaction for nitrobenzene. He cautions to keep the aldehyde protected from the light. The *National Formulary* [3497] also specifies 98% and no hydrocyanic acid.

SAFETY

The *minimum ignition temperature* in air is 192° [3680].

Unsaturated Aldehydes

139. ACROLEIN

Acrolein polymerizes easily. The dimer, 2-formyl-3,4-dihydro-2H-pyran, is present in all except freshly prepared and stabilized monomers.

Gilman [1777, p. 1] gives directions for the preparation of acrolein from 1 kg of potassium hydrogen sulfate, 200 g potassium sulfate, and 600 g dry glycerol. Hydroquinone is used as a stabilizer; yield 240–350 g.

Malinovskii et al. [3128] prepared acrolein by the gas phase aldol condensation of formaldehyde and acetaldehyde. A yield of 56.5% was obtained using a silica gel catalyst saturated with an aqueous solution of sodium and potassium silicates, tungstates, and titinates in the temperature range of 250–320°.

Bremner and Jones [725] and Wilson [5112] prepared acrolein from tetrahydrofurfuryl alcohol over a catalyst in two stages.

Parry [3664] stabilized the monomer by the synergistic action exerted by copper and phenolic antioxidants in the presence of water. The rate of polymer formation in monomeric acrolein containing 0.1%w in contact with copper metal and in the presence of 0.001–0.002% hydroquinone decreased by a factor of 100 or more with respect to the same system without copper. Hydroquinone has been reported to be effective as a monomer stabilizer [2609] and

Union Carbide includes hydroquinone in their specifications, 0.10–0.25% [4884].

Blacet and coworkers [613] purified acrolein by distilling in a nitrogen-filled, 90-cm column packed with glass rings. To avoid the formation of diacryl, the vapor was passed through an ice-cooled condenser into a receiver containing 0.5 g catechol in an ice-salt bath. The acrolein was twice distilled from anhydrous copper sulfate at low pressure. During the distillation, catechol was placed in both the distilling flask and the receiver to avoid polymerization.

SAFETY

The committee on Threshold Limit Values [4509] has summarized the pertinent information concerning the toxicity of acrolein: Exposure of cats to 10 ppm for 3.5 hr caused respiratory irritation, salivation, lacrimation, and mild narcosis; 1 ppm is immediately detectable and produces marked irritation to the eyes and nose; and 5.5 ppm is intensely irritating and 10.0 ppm and over is lethal in a short time. Di Macco [1311] reviewed the general toxic effects and states that acrolein inhibits the action of the enzymatic oxidation process and causes disturbances in the endocrine and neuro-regenerative mechanisms.

The *threshold limit value* is 0.1 ppm, 0.25 mg/m³ [4509].

The *flammable limits* in air are 2.85 and 30.5%v. The *minimum ignition temperature* in air is 234° [3680].

140. CROTONALDEHYDE

Commercial crotonaldehyde is more than 95% *trans*-form [2609]. It polymerizes rather readily and becomes colored. Water and hydroquinone are two of the substances that have been found to inhibit polymerization, which is really an oxidation and resinification, according to some reports.

Horsley [2253] excluded oxygen and carried out the aldol condensation in a faintly alkaline solution buffered with sodium acetate and acetic acid at about 30°. A British patent [4441] describes a method for making croton-aldehyde by mixing equal quantities of acetaldehyde and water, cooling, gradually adding sodium hydroxide solution, and then acidifying. The aldehyde is distilled from the mixture. There have been other reported modifications of the two methods cited but apparently their principal novelty was mostly to evade existing patents and not to change significantly the yield or cost.

Raphael and Sondheimer [3913] hydrogenated 3,3-diethoxypropyne using a palladium-calcium carbonate catalyst to obtain *cis*-crotonaldehyde diethyl acetal. Hydrolysis of the latter with hot aqueous oxalic acid gave *trans*-crotonaldehyde.

Blacet and coworkers [613] converted the *cis*-portion of a commercial crotonaldehyde to the *trans*-isomer by exposing to bright sunlight a sample containing hydrochloric acid. This was distilled and the portion distilling between 102.6 and 102.8° was collected.

Hünig [2320] prepared acid-free crotonaldehyde to study its *condensation with secondary amines* by distilling a technical product over sodium pyrogallate and collecting the fraction boiling at 95–105°. It was redistilled in a Widmer column in a nitrogen atmosphere.

The *hydrogen bonding properties* of crotonaldehyde were studied by Chandra and Sannigrahi [947] by the ultraviolet spectra. The aldehyde was purified by drying with anhydrous calcium chloride and then distilling.

Dolliver and coworkers [1333] carefully fractionated commercial crotonaldehyde, saving the fraction boiling at 102.29–102.30° to study the *heat of hydrogenation*.

SAFETY

The Committee on Threshold Limit Values [4509] summarized pertinent information concerning the toxicity of crotonaldehyde. It produces the same symptoms as acrolein. The LD_{50} for rats for a 30-min exposure was 1500 ppm, 135 ppm for acrolein, and 825 ppm for formaldehyde. A case of apparent sensitization of a person handling small amounts of the aldehyde has been reported. Union Carbide [4884] states that crotonaldehyde is not a primary skin irritant to rabbits but causes severe eye injuries.

The threshold limit value has been set as 2 ppm, 5.6 mg/m^3 [4509].

The *flammable limits* in air are 2.12 and 15.5%v, the latter value at elevated temperature. The *minimum ignition temperature* in air is 232° [3680].

Ketones

General

Harrison and Eisenbraun [2045] pointed out that steam volatile ketones may be purified through their bisulfite addition product, semicarbazone, hydrazones of *p*-hydrazinobenzenesulfonic acid [4824] or 2,4-dinitrophenyl-hydrazones. The bisulfite or semicarbazone products are preferred if they form a good yield, since they are readily cleaved to the ketone. The bisulfite addition product is limited to aldehydes and methyl-ketones but they are readily split with dilute acid or alkali. The semicarbazones are hydrolyzed by steam distillation in the presence of oxalic acid [1323] or by exchange with pyruvic acid [2135]. 2,4-Dinitrophenylhydrazones are impractical because the carbonyl is not readily recovered. Several procedures have been reported in which pyruvic acid or levulinic acid is used as an acceptor for 2,4-dinitro-phenylhydrazine, thereby freeing the ketone [1321, 1287].

A method [2045] has been specifically designed for steam volatile ketones using the steam nonvolatile α-ketoglutaric acid as the acceptor for 2,4-dinitrophenylhydrazine. The procedure consists of steam-distilling the 2,4-dinitrophenylhydrazone in the presence of an equivalent amount of α-ketoglutaric acid and an excess of 30–50% sulfuric acid.

Ketones exist in the tautomeric *keto-* and *enol-*forms, both forms being present in the normal material. The amount of the enol form present in several ketones has been reported and is tabulated in Table 5.11.

Table 5.11 Amount of Enol-Form Normally Present in Some Ketones

No.	Ketone	Enol Present %	Reference
141	Acetone	1.5×10^{-4}	1744
142	2-Butanone	1.2×10^{-1}	1744, 1745
143	3-Pentanone	7×10^{-2}	1744, 1745
144	Cyclohexanone	1.18	1744, 1745
145	4-Methyl-2-pentanone	7.4×10^{-2}	1744
146	Camphor	0.123	1744
147	Acetophenone	3.5×10^{-2}	1744
203	Ethyl Malonate	7.7×10^{-3}	1744
348	Methyl acetoacetate	5.0	1744
349	Ethyl acetoacetate	13.2	1744, 1307
350	Ethyl cyanoacetate	2.5×10^{-1}	1744

A general route to ketones from esters was discovered by Corey and Chaykovsky [1089]. The esters are treated with the sodium salt of dimethyl sulfoxide to form a β-keto-sulfoxide which was reacted with aluminum amalgam in 90% tetrahydrofuran-10% water mixture to give ketones. Acetophenone, for example, was formed from ethyl benzoate in 77% yield.

Specific Solvents

Aliphatic Ketones

141. ACETONE

More than 95% of the acetone produced in the United States is from four synthetic processes. The remainder is from fermentation, and a small amount is still produced from pyroligneous acid. A good grade of commercial acetone should contain less than 0.1% organic impurities. The water content is generally below 0.4% and often below 0.2%. Acetone is much more reactive than is generally supposed. Such mildly basic material as alumina gel induces the aldol condensation to 4-hydroxy-4-methyl-2-pentanone, and an appreciable quantity is formed in a short time if the acetone is warm. Small amounts of acidic material, even as mild as anhydrous magnesium sulfate, causes acetone to condense [3975].

Acetone is available in a high state of purity at such a reasonable cost that it is impractical to prepare it in the laboratory. References to methods of preparation may be found in [433, 1382, 1481, 2316].

Livingston [2985] purified acetone as follows: The acetone is saturated with dry sodium iodide at 25–30°, the solution decanted from the excess solid, cooled to about −10° and the mother liquor removed from the crystals by filtration. The cold salt is transferred to a flask and warmed to above 26° (about 30° is optimum). The resulting liquid is transferred to a distilling flask and distilled, the last 10 % being rejected. This is a modification of the Shipsey-Werner method of purification [4292].

Timmermans and Gillo [4776] found that it was difficult to produce a dry acetone; they questioned the possibility of preparing a very pure material and of storing it in a high state of purity if it were produced. They were working with acetone from pyroligneous acid, which contained relatively large amounts of various impurities, some of which are also present in synthetic and fermentation acetone. Since it was impossible to free acetone completely from certain hydrocarbons, homologues, methanol, acetals, and water by simple distillation, other methods of purification were necessary if a very pure product was to be produced. They suggested precipitation as the water-insoluble complex with sodium bisulfite or sodium iodide.

Methanol may be eliminated as its bromomethane azeotrope boils at 35°. The last traces of methanol may be eliminated by treating with acetyl chloride [1642] or by heating with acidified permanganate. The last traces of aldehydes may be removed by treatment with ammoniacal silver nitrate solution. Timmermans and Gillo [4776] found that water was the most difficult impurity to remove. They could not separate water by distillation, and the dehydrating agents such as sodium, phosphorus pentoxide, and calcium chloride caused condensation of the acetone. Dehydration with such salts as calcium nitrate and potassium carbonate does not dry to better than 0.1 % water. The following procedure was finally adopted. The acetone was treated with calcium chloride, distilled, treated with phosphorus pentoxide, and fractionally distilled. The distillate contained 0.01–0.02 % water. A second distillation from phosphorus pentoxide reduced the water content to 0.001 % or lower.

An analytical grade acetone was dried for *heat of mixing studies* by fractional distillation from anhydrous copper sulfate. Molecular sieves 4A were also used. The same refractive index was obtained with both drying agents [3395].

A number of methods involve a treatment with potassium permanganate, distilling, drying, and then usually fractional distillation. Some of these methods may be found in [707, 4087, 4093, 5104].

Werner [5127] has recommended the following method: To 700 ml of acetone in a liter bottle is added 3 g of silver nitrate dissolved in 20 ml of water. Twenty milliliters of 1 N sodium hydroxide is introduced and the mixture shaken for about 10 min. The acetone is then dried with calcium chloride and distilled.

Duclaux and Lanzenberg [1386] purified acetone by azeotropic distillation with carbon disulfide.

Lannung [2823] purified acetone for *solubility determinations* by drying for several days over potassium carbonate and distilling in a vacuum over fresh potassium carbonate.

Frankforter and Cohen [1623] dried over calcium chloride, treated with sodium amalgam, and fractionally distilled. For *electrical measurements*, acetone was distilled over anhydrous copper sulfate [599]. It was dried by treating for a long time with anhydrous potassium carbonate and distilling several times [5041].

For *optical measurements*, Scheibe and coworkers [4137] allowed acetone from the bisulfite compound to stand for three days over potassium permanganate, then boiled for 2 hr; after distillation it was dried over potassium carbonate and then fractionally distilled through an efficient column.

According to Lannung [2823], the most suitable drying agent is anhydrous potassium carbonate. Calcium chloride forms a compound with acetone and limits its efficiency, resulting in aa appreciable loss of acetone. Phosphorus pentoxide results in contamination by condensation products.

Maruyama [3180] describes a method for purifying acetone for use as a *polarographic solvent;* see also [4814].

Smith and coworkers [4402] allowed Reagent grade acetone to percolate slowly through a 2-ft column packed with $\frac{1}{16}$-inch pellets of Linde-type 4A molecular sieves. The dried acetone was distilled from 4A sieves through an 18-in. Vigreux column and was used for *solvolysis rate studies*. Savedoff [4119] found that acetone prepared for *conductivity studies* by this method had a specific conductance of 1.66×10^{-8} ohm^{-1} cm^{-1} and contained less than 0.005% water.

For purification for *dielectric constant studies*, see [2048], benzyl alcohol.

An extensive study has been made by Riddick [3975] on the preparation of high purity acetone with a water content of 0.05% or less. The acetone used was a special middle cut, containing less than 0.25% water, from a 10,000 gallon distillation of fermentation acetone. The acetone contained 0.01% or less methanol, no aldehydes as tested by a modified Tollen's reagent, and no detectable hydrocarbons. The acidity and alkalinity was 0.002% or less, and the permanganate time was 30 min or more.

The acetone was fractionally distilled through all-glass-type, Penn State-type columns packed with $\frac{1}{8}$-in. single-turn glass helices. See Table 5.12.

Silica gel and alumina increased the water content of the acetone, presumably through aldol condensation and subsequent dehydration. The water content of acetone was increased from 0.24–0.46% by one pass over alumina. All other drying agents tried, including copper sulfate, potassium carbonate, calcium chloride, sodium sulfate, and phosphorus pentoxide, caused some condensation. Drierite and anhydrous magnesium sulfate were the most

Table 5.12 Purification of Acetone by Fractional Distillation

Column	Reflux Ratio	Distilled %	Water Content of Distillate, %
100 plates	25:1	6–16	0.002
	25:1	16–25	0.012
35 plates	10:1	47	0.04
	2:1	94	0.05
25	2:1	94	0.16

satisfactory. Magnesium sulfate prepared by heating in a furnace to 400° was unsatisfactory. The most satisfactory method of preparation was to heat Epsom salt crystals gradually to 300° at approximately 10 torr pressure. The dried salt contained about 0.2% water. Weight for weight, Drierite was found to be 2–4 times as efficient as magnesium sulfate in drying acetone to about 0.05% water. A large excess of the magnesium salt will not dry acetone below 0.04% water.

The water content of acetone can be reduced to 0.001%, or less, with Drierite. It is not practical to reduce the water below the 0.01% range, except in relatively small amounts, because the drying agent required is large. The acetone should first be distilled to reduce the water content to 0.25% or less water. Any chemical treatment to remove impurities should be done before the drying operation.

Acetone dried with Drierite is suitable for most purposes without distillation. If a product 99.95+% is required, the acetone must be distilled. There is a trace of 4-hydroxy-4-methyl-2-pentanone in all dried acetone. Acetone distilled from a vessel whose wall temperature is greater than 20° above the boiling point of acetone will result in a distilled product that has a greater color and a shorter permanganate time than if the wall temperature were kept within 20° of the boiling point. The most satisfactory way to control the wall temperature is to immerse the distillation flask in a water bath. The following procedure has been found to produce a distillate from dry 99.95% acetone that had a color no greater than water distilled from alkaline permanganate and a permanganate time of 12+ hr: The previously dried acetone is distilling from an all-glass still fitted with a 15-in. spray trap containing $\frac{1}{4} \times \frac{1}{4}$-in. Raschig rings. The still pot is immersed in a water bath and a slow stream of dry nitrogen is passed into the receiver. The temperature of

Table 5.13 Specifications for a Commercial and ACS Reagent Grade Acetone

	Commercial	ACS Reagent
Purity, min	99.5%	—
Density, 25°	0.784–0.786	0.7857, max
Specific gravity, 25/25°	0.786–0.788	—
Distillation range	1° Including 56.1°	1 ml to 95 ml, not more than 0.5°; 95 ml to dryness, not more than 0.5°
Acidity, max	0.002% as CO_2	0.002% as acetic acid
Alkalinity, max	None to p-nitrophenol	0.001% as ammonia
Aldehydes, max	—	0.002%
Permanganate time at 25°	2 hr	15 min
Methanol, not more than	—	40 mg methanol/100 ml
Nonvolatile matter, max	0.001 g/100 ml	0.001% max
Water solubility	Miscible with no turbidity	to pass test
Color, max	5 APHA[a]	—

[a] ASTM now includes this test as D 2108-64.

the water bath is brought to 70–75°, the still flushed with acetone vapors and the condenser water started. The receiver is washed with several portions of distilled acetone before the main fraction is collected. The distillation is continued until about 100 ml remains in the pot. The temperature of the water bath must be reduced as the volume of acetone in the still pot decreases to keep the volume of distillate within the capacity of the condenser.

A more recent investigation indicates that 3A or 4A molecular sieves may be used instead of Drierite without causing condensation of the acetone. Weight for weight the sieves are more efficient than Drierite.

CRITERIA OF PURITY

The specifications for a good commercial grade (1054) and ACS Reagent grade [76] acetone are given in Table 5.13.

Swietoslawski [4577] used the differential ebulliometer to detect impurities.

Timmermans and Gillo [4776] studied the determination of the several impurities present in acetone and recommended the following procedures.

Aldehydes: Schiff-Villavecchia reaction, sensitive to 0.001 %.

Acidity: titration with barium hydroxide to phenolphthalein end point, sensitivity to better than 0.001 %.

Action of permanganate: sensitive to less than 0.003 % methanol.

Total impurities: add known amounts of water, methanol, and acetic acid to pure acetone and observe the differential boiling-condensing temperature, sensitivity to about 0.001 %.

Water: differential boiling-condensing temperature, sensitive to 0.001 %; infrared absorption at 3 μ range, sensitive to better than 0.001 %.

Dreisbach and Martin [1372] determined the purity from the freezing curve.

SAFETY

Acetone presents a relatively low degree of health hazard. Acetone has a high vapor pressure at normal room temperatures [3681].

The method of Alekseeva [49] for the determination in air gives the sensitivity, specificity, and likely interferences.

The *threshold limit value* has been set at 1000 ppm or 2400 mg/m³ [4509].

The *flammable limits* in air are 2.55 and 12.80 %v. The *minimum ignition temperature* in air is 561° [3680].

142. 2-BUTANONE

Bewley [553] prepared 2-butanone by the continuous dehydration of 2,3-butanediol. See also Robertson [3999].

2-Butanone generally can be purified by the methods described for acetone.

Newman and coworkers [3526] dried 2-butanone by extractive distillation and solvent extraction. 2-Butoxyethanol was found suitable for extractive distillation. Suitable solvents for solvent extraction were 1,1,2-trichloroethane, trichloroethylene, and chlorobenzene.

Bresler and Drasky [727] give directions for purifying 2-butanone, which is contaminated and discolored by diketones. The color-forming diketones can be removed by agitation with an aqueous solution of ammonia or a primary amine.

Lochte [2986] purified 2-butanone, like acetone and other ketones, by means of the addition compound with sodium iodide. The impure product was saturated with sodium iodide by boiling under reflux, and filtering while hot. It was then cooled and the white needles of the addition compound separated and were removed by filtration. The crystals melt at 73–74° and yield the ketone quantitatively upon distillation.

For *electrical measurements*, Müller and coworkers [3449] treated 2-butanone several times with potassium carbonate solution to remove acidic impurities, separated the two layers, and distilled to remove most of the water. The ketone was then dried for several days over sodium sulfate and anhydrous potassium carbonate and, after decanting, fractionally distilled several times. Walden and Birr [5041] purified the ketone by means of the bisulfite compound. Commercial 2-butanone was fractionally distilled and the bisulfite compound precipitated from the middle fraction with concentrated sodium bisulfite solution. After filtering, the crystals were dissolved in water and precipitated by the addition of sodium bisulfite. The precipitate was filtered, pressed out, and decomposed with concentrated potassium carbonate solution. The nonaqueous layer was separated (ketones remaining in the solution were separated by distillation), dried over potassium carbonate, distilled, and air passed through the liquid for 24 hr. After drying twice more with potassium carbonate, it was poured off and distilled very carefully, great care being taken to insure that evaporation took place only from the surface, so that splashing through bubble formation was avoided.

Collerson and coworkers [1043] purified 2-butanone to determine *thermodynamic properties* as follows: Commercial material was fractionally distilled at 700 torr in a 100 theoretical plate still at a reflux ratio of 100:1. The fractions of highest purity by gas chromatographic analysis were combined. Boiling range and gas chromatography indicated adequate purity; 99.95%m from the freezing curve.

For *optical measurements*, Rice [3965] treated 2-butanone with a saturated solution of sodium bisulfite until no more reacted. The mixture was cooled to 0°, filtered, washed twice with ether, and dried on a porous plate. The dried addition compound was added to a sodium bicarbonate solution and steam-distilled. The ketone was salted out of the distillate with potassium carbonate, separated, and treated with potassium hydroxide to remove sulfur dioxide or carbon dioxide. After standing for 4 hr over calcium chloride, it was distilled. See also Mathews [3208].

Sinke and Oetting [4340] purified a commercial 2-butanone for *low temperature studies* by drying with calcium sulfate and fractionally distilling in a 100-plate still; 99.78%m shown by variation in melting point with fraction melted.

CRITERIA OF PURITY

Tests for acetone may be applied to 2-butanone. Dreisbach and Martin [1372] determined the purity from the freezing curve.

SAFETY

2-Butanone presents a low degree of hazard with normal handling. It has a low order of toxicity. The liquid may produce moderate skin irritation when exposures are frequent and prolonged. It may be classed as a mild eye irritant. The threshold odor detection concentration is less than 25 ppm. There have been no recorded instances of human illness due to the use of 2-butanone [3681].

The *threshold limit value* has been set at 200 ppm, or 590 mg/m³, to prevent objectionable irritation [4509].

The *flammable limits* in air are 2.05 and 11.00%v. The *minimum ignition temperature* in air is 505° [3680].

143. 3-PENTANONE

Collerson and coworkers [1043] distilled and tested 3-pentanone as described for 2-pentanone to determine *thermodynamic properties*. Further purification was necessary. Fractional freezing gave a satisfactory product; 99.95%m from the freezing curve.

SAFETY

The LD_{50} for rats, single dose, was found to be 2.14 g/kg [4432].

The *minimum ignition temperature* in air is 452° [3680].

144. CYCLOHEXANONE

Joris and Vitrone [2446] heated phenol containing 0.1% palladium on charcoal at 140–145° under hydrogen at atmospheric pressure for 30 hr to obtain cyclohexanone in 79% yield.

For the preparation of cyclohexanone from phenol through cyclohexanol see cyclohexanol, Yeh and coworkers [5289].

According to v. Auwers and coworkers [326], commercial cyclohexanone can be freed from cyclohexanol with chromic acid. It can be purified by means of the semicarbazone. Herz and Bloch [2143] dried a good grade of commercial material over anhydrous sodium sulfate for two days and fractionally distilled. Garland and Reid [1709] purified by means of the bisulfite addition complex, which they decomposed with sodium carbonate and then steam-distilled.

Crowe and Smyth [1140] purified cyclohexanone for *dielectric constant measurements;* it was dried for 24 hr over anhydrous sodium sulfate and fractionally distilled twice.

Cyclohexanone was purified for use in *ultraviolet studies of hydrogen bonding* by Chandra and Sannigrahi [947] by preparation of the bisulfite derivative and subsequent release with 80% sodium hydroxide solution. The ketone layer was dried with anhydrous sodium sulfate and distilled.

Takamitsu and Sumida [4597] identified 1-butanol, 1-pentanol, cyclohexanol, caproaldehyde, 3-pentanone, 2-hexanone, 2-heptanone, 3-heptanone, 4-heptanone, cyclopentanone, cyclohexane, and benzene as impurities in cyclohexanone made by the autoxidation of cyclohexane in the liquid phase.

SAFETY

Cyclohexanone is considered to be of a low degree of hazard under normal use conditions. It has strong warning properties at low concentrations. It defats the skin, and frequent contact of the skin and the liquid should be avoided. Contact of the liquid with the eyes causes marked irritation and some transient corneal injury. The principal health hazard is the inhalation of the vapors [3681]. Cyclohexanone was found [535] to cause vascular and degenerative lesions of the lungs and liver. See also [1104].

The *threshold limit value* is 50 ppm or 200 mg/m³ [4509].

The *lower flammable limit* in air at elevated temperature is 1.11%v. The *minimum ignition temperature* in air is 453° [3680].

145. 4-METHYL-2-PENTANONE

Karr and coworkers [2503] purified a special commercial material (99.5%) for *phase equilibrium studies* by taking a 70% middle cut from a fractional distillation at a reflux ratio of 25:1 from a 33 theoretical plate still.

SAFETY

4-Methyl-2-pentanone is not considered to be hazardous to health under normal conditions of handling. Inhalation does not normally constitute a health problem because of its warning properties. Workers exposed to about 100 pm complained of headache and nausea [3681].

The *threshold limit value* is 100 ppm or 410 mg/m³ [88].

The *flammable limits* in air are 1.35 and 7.60%v, the latter value at elevated temperature [3680]; see also [3160]. The *minimum ignition temperature* in air is 465° [3680].

146. CAMPHOR

Camphor occurs naturally in its optically active forms and as the racemate. Synthetic camphor is optically inactive. The commercial nomenclature is confusing. The *d*-form is called Japan camphor, Formosa camphor, laurel camphor, or gum camphor and is obtained principally from the camphor tree, *Cinnamomum camphora* Nees and Ebermaier (fam. Lauracea) [1929]. *l*-Camphor has been reported in the oil from *Salvia triloba*, *Blumea balsamfera*, *Artemisia austrachanica*, *Lippa adoensis*, and *Lavendula pedunculata*. The *l*-camphor is generally called "matricaria camphor." *dl*-Camphor occurs in the oil of *Chrysanthemum sinense* var. *japonicum* and in oil of sage.

Although *d*- and *l*-camphor are optical antipodes, the melting points are the same within 0.2°, the boiling points of the *d*- and *l*-forms as given in the literature differ by 3°. The density of the *l*-form has been reported to be about

0.001 g/cm^3 lower than that of the d-form. Synthetic camphor is just as suitable for the Rast melting point method for molecular weights as d-camphor. The l-form is not used as a solvent.

Camphor shows most of the characteristic reaction of the ketones. The dl-form, which is less reactive in some instances than the separate optically active forms, does not give an addition compound with sodium bisulfite.

Commercial camphor is a material of high purity. It may be further purified by sublimation, fractional crystallization, or crystallization from 50% ethanol-water solution or from a low boiling petroleum solvent. A French patent [3385] states that the impurities can be removed from camphor by dissolving it in concentrated acetic acid and then adding water until the camphor precipitates.

CRITERIA OF PURITY

The optical rotation and freezing point have been used most frequently as criteria of purity.

Aromatic Ketones

147. ACETOPHENONE

Synthetic acetophenone is prepared by the Friedel-Crafts reaction from benzene and acetic anhydride [1903, 1904, 4102]. Smeets and Verhulst [4371] report that continuous elimination of hydrogen chloride with a stream of dry air in the Friedel-Crafts reaction using acyl chlorides gives a yield of 97%, or better, ketone with mild temperature conditions in about 0.5 hr with no side reactions.

Acetophenone occurs naturally as the main constituent of oil of *Stirlinga latifolia*.

Webb and Webb [5100] prepared acetophenone by the catalytic decomposition of methyl or ethyl benzoate and acetic acid over thorium oxide at an elevated temperature. See also Porter and Cosby [3813].

Petrov and Lagucheva [3737] describe a method for preparing aromatic and alkaryl ketones using silicon tetrachloride and aluminum chloride. An acid is reacted with silicon tetrachloride to form $(RCO_2)_4Si$, which is reacted with the aromatic hydrocarbon in the presence of aluminum chloride to form the ketone.

Hauser and coworkers [2071] prepared acetophenone in 33% yield by the Grignard reaction with acetonitrile.

The criteria of purity and methods of purification of acetophenone have been studied by Livingston and coworkers [2984]. An extensive bibliography is given to 1924. They recommend subjecting commercial acetophenone to filtration, followed by careful, slow, and repeated crystallization with exclusion of light and moisture. Since fairly large quantities of water can be removed more efficiently by distillation than by crystallization, the mother liquors were frequently distilled at atmospheric pressure through a Vigreux column, and further purified by crystallization.

Sudgen [4540] froze acetophenone until the ternary freezing point in contact with water was constant to 0.2°. He then dried over calcium chloride and repeatedly distilled at 2 torr.

CRITERIA OF PURITY

Chemical purity as the ketone can be determined by the hydroxylammonium chloride method. The freezing point is the most sensitive and accurate physical criterion.

SAFETY

Acetophenone has been used as a hypnotic [4519]. A method for the determination in air has been reported [49]. The sensitivity, specificity, and likely interferences are given.

The *minimum ignition temperature* in air is 571° [3680].

ACIDS

Specific Solvents

Saturated Acids

148. FORMIC ACIDS

Formic acid usually is produced from sodium formate, which is made by reacting carbon monoxide and sodium hydroxide under pressure. It is also obtained as a by-product of the pentaerythritol process and the acetic acid process from butane. In the acetic acid process, the formic acid is most efficiently separated from the acetic acid by azeotropic distillation with an aliphatic chlorohydrocarbon. A dark color develops in the formic acid in the subsequent process. Höfermann [2219] found that the coloration was due to an ill-defined polymerization product of isopropenyl ketone, which boils only 3° from formic acid.

It is available in several concentrations and grades of purity. ACS Reagent grade is not less than 88% formic acid; high purity commercial grades are available in 85 and 90%. An anhydrous grade is available in limited quantities. Some chemical supply houses list a 97% grade.

Formic acid slowly decomposes into carbon monoxide and water at room temperature. The decomposition can be prevented by keeping the acid frozen [1075]. Hammett and Dietz [1990] found it a stronger "super acid" than acetic acid. Its solvent use is limited because of its rapid decomposition. Formic acid generally is referred to as "very" or "extremely" hygroscopic.

Hammett and Dietz [ibid.], however, exposed a sample freezing at 8.05° to air for 13 min and could not detect a change in the freezing point.

Coolidge [1075] purified the best grade of acid available for *vapor density measurements*. The acid was distilled in vacuum at room temperature to avoid decomposition. The vapors were condensed in a bulb packed with ice. Five fractional distillations yielded a product having a vapor pressure of 11.15 torr at 0°. It was then fractionally crystallized and sublimed three times. The vapor pressure at 0° (11.16 torr) and the melting point of 8.26° were used as the criteria of purity. It was found that dehydrating agents were unsatisfactory for preparing an anhydrous product.

Hammett and Dietz [1990] distilled in vacuum the best commercial grade of formic acid and then followed the method of Coolidge [1075]. The purified acid was used as a solvent for *potentiometric studies*. Precautions were taken to exclude moist air. See also Ewins [1529], Auerbach and Zeglin [317], Beckmann [479], and Timmermans and Hennaut-Roland [4778].

Wehman [5106] purified formic acid by bulk fractional freezing with temperatures ranging from 0° to 20° for freezing and melting, respectively. The limiting purity was checked by means of conductance, 6.08×10^{-5} ohm^{-1} cm^{-1}. The autoprotolysis constant was determined to be 2.19×10^{-7}.

Stout and Fisher [4516] prepared a high purity formic acid for *thermodynamic studies*. An 85% solution of phosphoric acid was concentrated by heating to 250°. It was then cooled and added to ammonium formate and the formic acid distilled; the temperature of the flask was kept below 50°. The formic acid-water mixture was distilled into a bulb containing Drierite. The dried formic acid was fractionally distilled through a small column and the middle fraction used. The apparatus was an all-glass system from which air had been eliminated. The sample was kept frozen except during transfer and during the measurement of melting. The criteria of purity was the heat capacity and the change in melting point with fraction melted.

Boric anhydride and anhydrous copper sulfate have been used as drying agents but phosphorus pentoxide and calcium chloride are unsuitable, since they react with formic acid. Schlesinger and Martin [4159] dried formic acid in an all-glass apparatus for several days with powdered boric anhydride and distilled under reduced pressure at 22–25°. For the preparation of boric anhydride, boric acid was melted in an oven at a high temperature, poured on an iron plate, cooled in a desiccator, and powdered. Garner and coworkers [1710] distilled under reduced pressure over anhydrous copper sulfate.

Formic acid of greater than 99% purity can be produced from a dilute aqueous solution by a two-step distillation process involving butyric acid [742]. The first distillation removes the bulk of the water and leaves a residue containing about 77% of the acid. The residue is then distilled with 3–6 times its amount of butyric acid as an azeotroping agent.

Sinke [4338] purified formic acid for *heat of combustion* determination by fractional crystallization to a purity of better than 99.9%m determined from the freezing curve.

Beckmann [479] froze the available formic acid, centrifuged, dried over boric anhydride, and fractionally distilled. The distillation system was protected against moisture; mp 5.6°.

Harned and Embree [2031] twice fractionally distilled the best formic acid available for the determination of the *ionization constant*.

CRITERIA OF PURITY

Specifications and test methods are given for the minimum 88% Reagent grade formic acid by Rosin [4042] and ACS Reagent grade [76].

Coolidge [1075] used the vapor pressure at 0° and the freezing point as the criteria of purity. Hammett and Dietz [1990] and Dreisbach and Martin [1372] used the freezing point.

Stout and Fisher [4516] used the heat capacity and the change in melting point with fraction melted.

SAFETY

Formic acid is very corrosive to the skin, eyes, and mucous membranes. The vapors are irritating to the eyes, respiratory tract, and the skin [3681, 4509].

The *threshold limit value* has been set at 5 ppm or 9 mg/m³ [4509].

Care should be exercised when opening formic acid containers. When the acid has been stored for a long period of time or subjected to elevated temperatures, considerable pressure can develop as a result of decomposition.

149. ACETIC ACID

Acetic acid is available at a reasonable cost and in good purity in quantities as desired. The ACS Reagent grade is 99.7% based on the freezing point. The adjective *glacial* usually used to characterize the better grades of the acid is an indefinite description. Monnier [2609] states, "*glacial* is a term applied to acetic acid of high purity (above 99%) which congeals to ice-like crystals at 58–60°F."; (14.4–15.6°). The Random House *Dictionary of the English Language*, defines it as at least 99.5% and solidifying at 16.7°. *USP* XVII lists the assay based on total acid titration as 99.4% and the freezing point is not a requirement. Marsden and Mann [3160] list a 99% minimum assay and a minimum crystallizing point of 15.5°.

Acetic acid is the most extensively used of the nonaqueous solvents for the titration of bases [3974].

Othmer and White [3617] developed a method for drying acetic acid by removing the water as the butyl acetate-water azeotrope. The method is applicable to other low molecular weight aliphatic acids.

Kahane [2466] determined the heat of mixing of perchloric acid ($HClO_4$·-$2.5H_2O$) and acetic anhydride, and found that the mixture was perchloric acid and acetic acid. He cautioned that care should be exercised when using mixtures of perchloric acid and acetic acid.

MacInnes and Shedlovsky [3099] purified acetic acid for *ionization constant study* by fractionally distilling from 2% potassium permanganate, then from "a little chromic anhydride," followed by a final fractional distillation. The first quarter of the final distillation was discarded. The remaining fractions were kept or discarded on the basis of their freezing points.

Hutchinson and Chandlee [2334] determined the *activity coefficient* of sulfuric acid in anhydrous acetic acid purified as follows: USP acetic acid was refluxed for 10 hr with an excess of chromium (III) oxide to remove any easily oxidizable material. The acid was distilled from the solid residue and then fractionally distilled at a reflux ratio of 10:1 in a Marshall column [3163] packed with small Raschig rings. The distillate was protected from atmospheric moisture. McDougall [3071] purified acetic acid by the above method. A fraction of the distillate was further purified by fractional crystallization; mp $16.60 \pm 0.01°$.

According to Orton and coworkers [3609], impurities that react with chlorine or bromine, if present in an acid not melting below 16°, can be removed by a single distillation over phosphorus pentoxide. The distillate will contain a little acetic anhydride due to the action of phosphorus pentoxide. According to Orton and Bradfield [3607], technical acetic acid is best purified by distillation with chromium (III) oxide, preferably with the addition of a quantity of acetic anhydride corresponding to the water content of the acetic acid.

Harned and Ehlers [2028] purified acetic acid for *dissociation constant studies* by fractionally distilling three times in an all-glass still from 2% chromium (III) oxide. The first and last quarters were discarded. A fourth like distillation was made without the chromium (III) oxide.

Vogel [4975] found that acetic acid that had been distilled from potassium permanganate was not suitable for *physical property measurements*. The acid used was prepared by partially freezing and rejecting about one-half as unfrozen liquid, then fractionally distilling from permanganate. The portion distilling between 116.5 and 117.5° at 765 torr was collected, partially frozen and about one-half of the acid discarded as liquid. The solid was melted and again fractionally distilled from permanganate.

Turner and Pollard [4864] purified commercial glacial acetic acid for use as a solvent for *molecular weight determinations*. The acid was distilled from potassium permanganate and the middle portion collected and dried over phosphorus pentoxide and redistilled. The middle portion from the second distillation was subjected to fractional freezing. See also Hess [2152].

Acetic acid was dried over phosphorus pentoxide for 24 hr, then refluxed over phosphorus pentoxide for 5 hr, and finally distilled. About 450 ml were distilled and then 2–10-ml portions collected for study of the *conductance behavior* of the acid in N-methylacetamide. It was shown that small quantities of acetic acid and/or the acetate remained with persistent tenacity in the amide even after it had been subjected to a rather extensive purification procedure [1229].

For *kinetic investigation*, Dimroth [1312] used glacial acetic acid distilled in a quartz apparatus. Commercial glacial acetic acid always contains traces of copper, which may act as a catalyst.

For *specific conductance measurements*, Rabinowitsch [3876] dried acetic acid before distilling by shaking for 20 min with phosphorus pentoxide. The acid was filtered into the distillation flask through a layer of glass wool and phosphorus pentoxide. Schall and Thieme-Wiedtmarckter [4133] treated the purest commercial acetic acid, free from known impurities, with a considerable quantity of boron triacetate (Picket and Gelenzoff [3762]), allowed it to stand for three days in a sealed flask at 30–40°, distilled under reduced pressure, and purified by freezing once. The acid was kept in a desiccator over sulfuric acid, particular care being taken when samples were withdrawn.

Myers [3476] purified acetic acid for *electrical property studies*. Reagent grade acid was analyzed for water by a colorimetric method similar to that of Greathause and coworkers [1869]. A solution of 6 ml of perchloric acid in 39 ml of acetic acid was used as the catalyst to avoid rise in temperature. See [3476] for details of the colorimetric method. The amount of acetic anhydride plus 2% was added to the acetic acid to react with the water present. The reaction was catalyzed by 1 g of anhydrous 5-sulfosalicylic acid per liter. The solution was kept overnight at near 100° and distilled through a 1-meter column packed with glass helices. The distillate had a *specific conductance* of 6×10^{-9} ohm^{-1} cm^{-1}.

Lundin [3031] purified acetic acid by mixing a commercial acid with its own volume of water and adding 1% of activated carbon or 2% of charcoal. The mixture was boiled 10–15 min, filtered, and distilled.

Suitable drying agents for acetic acid are phosphorus pentoxide, boron triacetate, magnesium perchlorate [2152], anhydrous copper sulfate, and chromium (III) acetate. Acetic acid may be dried with acetic anhydride and used without further treatment if the small amount of excess anhydride is not objectionable. For the determination of acetic anhydride in acetic acid, see Benson and Kitchen [519].

CRITERIA OF PURITY

The specifications and test methods for Reagent grade acetic acid are given by Rosin [4042] and for ACS Reagent grade in *Reagent Chemicals* [76]. The

Reagent grade "Glacial" acetic acid from most suppliers meets more stringent specifications than those of ACS Reagent grade. The melting point is the most critical criterion of purity. Jones and Betts [2418] suggest the critical solution temperature with benzene; Bousfield and Lowry [683] suggest the specific conductance. The freezing curve was used by Driesbach and Martin [1372].

SAFETY

The odor of acetic acid has been described as sharp and penetrating. It can be detected by odor below 10 ppm. Normally, the chief hazard from acetic acid is the liquid on the skin or in the eyes. Concentrations greater than 80% produced severe burns on the skin of guinea pigs. Concentrations below 50% produce relatively mild injuries [3681].

The *threshold limit value* is 10 ppm or 25 mg/m^3 [4509].

The *minimum ignition temperature* in air is 550° [3680].

150. PROPIONIC ACID

Child [972] states that propionic acid is thermally unstable at the normal boiling point.

Weizmann [5121] prepared propionic acid by the air oxidation of propionaldehyde in the presence of a catalyst containing lead, manganese, and cobalt oleates. The air was passed through a stirred mixture containing 288 parts of the aldehyde and 5.76 parts of the catalyst for 8 hr. No external heat was used, but the temperature increased to 45° over the first 2 hr and remained there for the remaining reaction time. The propionic acid was isolated by distillation; yield 62.5%.

Vogel [4975] purified commercial propionic acid for *physical property determinations*. About 1 liter was dried over anhydrous sodium sulfate and fractionally distilled through a Young and Thomas column. The fraction distilling at 139–141° was collected and refractionated over potassium permanganate. The middle fraction distilling at 140.7° at 760 torr was collected.

CRITERIA OF PURITY

Rosin [4042] gives the specifications and test methods for a Reagent grade propionic acid.

Dreisbach and Martin [1372] determined the purity from the freezing curve,

SAFETY

The chief hazard from propionic acid is the local damage of the liquid on the skin or in the eyes. No accumulative effects are known from industrial exposures [3681].

151. BUTYRIC ACID

Vogel [4975] purified a commercial grade of butyric acid to determine some *physical properties*. About 250 ml of redistilled acid was mixed with 5 g of potassium permanganate and refractionated. The first one-third was discarded and the remainder distilled constantly at 162.5° at 767 torr.

CRITERIA OF PURITY

Driesbach and Martin determined the purity from the freezing curve [1372].

SAFETY

Stasenkova and Kochetkova [4476] concluded, from their test with animals, that butyric acid is not very toxic. Lethal vapor concentrations could not be obtained but the vapors are irritating to the skin and eyes. The chronic threshold concentration was found to be 0.1 mg/liter and a permissible trial concentration of 0.01 mg/liter was recommended. A *threshold limit value* of 10.0 mg/m³ was recommended for Russia in 1964 [2711]. For oral toxicity, see [4432].

152. ISOBUTYRIC ACID

Isobutyric acid is prepared by the oxidation of 2-methyl-1-propanol or from 2-propanol through the chloride or nitrile. It has the same characteristic odor of rancid butter as butyric acid.

Lippincott and Hass [2974] prepared isobutyric acid by the hydrolysis of 1-nitro-2-methylpropane. Groll and Tamele [1905] prepared the acid from isobutyraldehyde by air oxidation at less than 45° in the presence of iron isobutyrate.

153. VALERIC ACID

Valeric acid has a putrid odor. It is made by the action of carbon dioxide and water with olefins, by the electrolytic oxidation of 1-pentanol, and by the reaction of formic acid and butene.

Valeric acid can be prepared by treating the nitrile with sodium hydroxide, removing the excess alcohol and freeing the acid with sulfuric acid [19]. About 300 g with an 81% yield was prepared.

Vogel [4975] prepared valeric acid from 1-bromobutane. A large sample was distilled from the middle fraction, bp 184° at 768 torr, and collected for *physical property measurements.*

Two samples of valeric acid were purified by McDougall and Kilpatrick [3072] for *entropy and related thermodynamic properties study* using a Nester and Faust spinning band column at reduced pressure. The refinements in collection and storage of the samples reduced the impurity from 1.79%m in the first sample to 0.41%m ± 0.01 in the second, based on the sharpness of the melting under 1 atm of helium. The samples analyzed 99.6 and 99.8% by gas chromatography.

154. ISOVALERIC ACID

Isovaleric acid has a disagreeable rancid cheese odor. A limited amount is obtained from pyroligneous acid, which is one of the compounds formed during the high pressure reaction in the manufacture of alcohols. It is made by the electrolytic oxidation of 3-methyl-1-butanol and the catalytic oxidation of the alcohol by carbon dioxide.

Vogel [4975] dried about 500 g of a pure commercial acid over anhydrous sodium sulfate and fractionally distilled; bp 176.5° at 762 torr.

155. HEXANOIC ACID

Hexanoic (caproic) acid has the characteristic odor of goats, L. *caper*. It is manufactured by the oxidation of 1-hexanol and by the hydrolysis of hexanenitrile.

Butyl malonic ester can be saponified to hexanoic acid in a 74% yield [19].

Vogel [4975] dried a pure grade of hexanoic acid and fractionally distilled. The fraction collected at 203° at 756 torr was used for *physical property measurements*.

Rose and coworkers [4035] purified hexanoic and octanoic acids for *liquid-vapor equilibrium studies*. Commercial grade materials with a purity of 85 to 95% were fractionally distilled at a high reflux ratio in a 30-plate packed column at a pressure of about 50 torr. The heart cuts were used. The criteria of purity were the acid number, the iodine number, carbon and hydrogen analysis, water by Karl Fischer, and partial crystallization, followed by comparison of fractions.

Garner and coworkers [1712] studied some *heat values* of hexanoic acid purified by fractional distillation or crystallization at a constant setting point of −3.6°.

SAFETY

Skin and eye irritation are relatively severe [3681]. The LD_{50} single dose for rats is 5.19 g/kg [4432].

156. OCTANOIC ACID

Vogel [4975] carefully fractionally distilled commercial octanoic acid for *physical property studies;* bp 236° at 769 torr.

See Rose and coworkers [4035] under hexanoic acid for purification.

SAFETY

Octanoic (caprylic) acid is a rather mild irritant. It can be detected by odor at about 0.008 ppm [3681].

The LD_{50} for rats is 10,080 mg/kg [2392].

Unsaturated Acids

General

Ratchford and coworkers [3923] found that unsaturated acids could be prepared in high yields by passing a suitable corresponding ester through a Pyrex brand glass tube heated to 500 to 580°. The ethyl ester of acrylic and methacrylic acids pyrolized to 92 and 87%, respectively. The secondary esters gave the highest yields. Butyl crotonate at 537° and a residence time of 7.2 sec gave 84% crotonic acid. The purification of the acids is discussed.

Specific Solvents

157. ACRYLIC ACID

Rheberg [2251] described a method for the preparation of acrylic acid by acidolysis of ethyl acrylate with formic acid.

Moureau [3426] prepared acrylic acid by refluxing one mole of 3-bromo-propionic acid with two moles of aqueous potassium hydroxide. The acrylic acid was liberated with sulfuric acid and distilled.

Rohm & Haas [4022] lists 0.020% p-methoxyphenol or 0.10% N,N'-diphenyl-p-phenylenediamine as polymerization inhibitors. Ababi and Mihailia [1] used p-methoxyphenol as a stabilizer and removed it by distilling at 15 torr.

SAFETY

Acrylic acid has an acrid odor. It is a severe skin, eye, and respiratory irritant as a liquid or in concentrated solutions. It should be handled with the usual precautions used for acetic or propionic acids [3681].

158. CROTONIC ACID

Dutt [1420] states that malonic acid easily condenses with aldehydes in the presence of piperidine in pyridine solution to alkylidine- and arylidenemalonic acids, which when heated lose carbon dioxide to give the α-unsaturated carboxylic acid. Acetaldehyde gave 75% crotonic acid; see also [4138].

Crotonic acid may be purified by crystallization either by partial freezing or from a solvent such as hexane. It may be sublimed.

159. METHACRYLIC ACID

Crawford and McGrath [1129] heated acetone cyanohydrin and concentrated sulfuric acid at 130°. The methacrylamide was hydrolyzed in boiling aqueous solution to the acid.

Rohm & Haas [4022] lists 0.025% p-methoxyphenol, 0.10% hydroquinone, and 0.05% N,N'-diphenyl-p-phenylene diamine as polymerization inhibitors.

160. OLEIC ACID

Oleic acid is considered the most prevalent acid in nature. It comprises 50%, or more, of the total acids of many fats and oils. Few fats are known to contain less than 10% of this acid [3150]. The fatty acids of olive oil are 80% oleic. Markley [3152] states that all natural fats contain unsaponifiable matter such as hydrocarbons, long-chain aliphatic alcohols, sterols, and so forth. The concentration may vary from 0.2% in lard and tallow to 20% in shark liver oil. These impurities tend to separate in the alcoholic fractions during the lead salt purification.

Markley [3152, Chapter 20] has an excellent 140-page discussion, including 569 references, concerning the techniques of separating naturally occurring fatty acids. The discussion includes consideration of hydrolysis and esterification for preparation from natural fats, oils, and other material. The separation and purification of the acids are discussed during the presentation of the three principal separation techniques, namely, distillation, salt solubility, and low temperature crystallization.

See methyl oleate for preparation and purification of the ester which is used as part of the purification of oleic acid.

Moreno [3396] separated oleic acid by extracting olive oil with 96% ethanol in a tower packed with Raschig rings or pieces of solid glass. The latter was more effective. On passing through the tower, the mixture separated into an oil phase and an alcohol phase, the latter containing 75–90% oleic acid.

Hartsuch [2049] separated oleic acid from linoleic and saturated acids by successive crystallization from acetone at −20, −40, and −60°. Part of the remaining saturated acids were precipitated in ethanol as lead soaps. The product was then distilled at 1 torr in a 4-ft column packed with glass helices. The oleic acid had a purity of 97.8% as determined by the iodine and thiocyanogen numbers.

Ware and Dunell [5080] purified a commercial "purified grade" oleic acid by crystallization from acetone at −11°. The crystals were pulverized and dried *in vacuo* at −11° for one week for *proton magnetic resonance studies.*

Keffler and McLean [2543] made an extensive study of the *purification* and *thermochemical properties* of oleic acid. The "solid" acids were separated from the "liquid" acids by the lead salt method repeated two or three times. The "liquid" acids were converted into the lithium salts and recrystallized from 80% ethanol as many as 12 times, until the iodine value became and remained identical within 0.2% for the successive precipitates and corresponding mother liquors. The oleic acid was free of linoleic (9,10-octadecadienoic) and linolenic (9,12,15-octadecatrienoic) acids but contained about 4% saturated acids. The acid was fractionally distilled at 0.1 torr several times. It was found that a less laborious method was to convert the oleic acid containing the 4% saturates to the methyl or ethyl ester and fractionally distill at a pressure obtained with two mercury vapor pumps in series.

Two lots of oleic acid were prepared, one from the methyl ester and the other from the ethyl ester. The iodine value and the density were the criteria of purity.

Wheeler and Riemeneschneider [5132] prepared 99.6% methyl ester. The acid was obtained from the ester. The methyl esters from the acids from olive oil were distilled. The unsaturated esters were crystallized from acetone at different concentrations and temperatures at −60, −35, and −60°. The material was again fractionally distilled and crystallized twice from a petroleum naphtha at about −65°.

Brown and Shinowara [780] concluded from previous work that high purity oleic acid could be obtained by direct crystallization from acetone. Their method involved the removal of the saturated acids by a −20° acetone precipitation, crystallization of the unsaturated acids at −60° four or more times, followed by a partial crystallization from acetone at −35° to remove the small amount of palmitic acid not removed in the initial −20° crystallization. Three specimens melted at 13.0°, mean molecular weight 282.4, mean iodine number 89.90 (calculated value 89.93), mean n_D 20° 1.4585.

Smith [4391] carefully prepared oleic acid from methyl oleate that was fractionally distilled. The acid was crystallized by the Brown and Shinowara [780] method to a constant melting point. Smith is also of the opinion that pure oleic acid may be obtained by crystallizing from acetone only but he recommends a preliminary esterification and fractional distillation. He prepared and measured the *freezing points* of the α- and β-forms.

The methyl and ethyl esters are generally used in the separation of naturally occurring fatty acids by fractional distillation. Metcalf and Schmitz [3282] converted the acid to the ester by boiling 2 min with boron trifluoride-methanol reagent and obtained practically quantitative conversion. The ester was hydrolyzed and converted to the lead salt and crystallized from ethanol or ethyl ether.

Lithium salts are crystallized from acetone and barium salts from benzene [3152].

Hoerr and Harwood [2216] prepared a high purity oleic acid from olive oil for *solubility studies* by acetone crystallization and fractional distillation. They concur with, and explain, Hartsuch's [2049] conclusion that oleic acid exceeding 98% purity had not been prepared because stearic and oleic acids form an eutectic at 98% oleic acid. Hartsuch's statement, therefore, applies only to oleic acid prepared by crystallization. The solubility of the acid was determined in 23 solvents of various functional group types at seven temperatures from −40 to 20°. This article is a good reference source for pertinent work done on oleic acid prior to 1952. Hoerr and Harwood are of the opinion that the freezing point (not the melting point) is the best criterion of purity.

SAFETY

There are no known hazards in the use of oleic acid in industry [3681].

ACID ANHYDRIDES

General

The acid anhydrides are hygroscopic and slowly react with water to form acids.

Allen and coworkers [2251] found that an acid chloride in a dry solution of pyridine in benzene reacts with an acid to form the anhydride. Mixed anhydrides may be prepared by this method.

Kostyuk and coworkers [2722] report a method for preparing the anhydrides of carboxylic acids by dehydrating the acids with acetic anhydride. An equimolar mixture of the anhydride desired and acetic anhydride (an excess may be used) was refluxed in a 20 theoretical plate column until the head temperature came to the boiling point of acetic acid. The mixture was distilled at a reflux ratio, so that the distillation took 2.4 hr. Propionic, iso-valeric, hexanoic, undecanoic, and benzoic anhydrides were prepared.

Lewis [2931] purified acetic, propionic, and hexanoic anhydrides and determined their *boiling points*, *surface tensions*, *densities*, and *viscosities*. The anhydrides were fractionated through a glass bead column until they gave a fairly constant boiling point and were relatively free from the lower boiling acid. They were refractionated twice over the fused sodium salt of the corresponding fatty acid and twice under reduced pressure. The higher anhydrides decompose slowly when distilled at atmospheric pressure.

Cockerille [1015] described a method for purifying aliphatic acid anhydrides. About 0.2–2% of sulfuric acid was added to the impure anhydride in two stages. In the first stage, the reaction mixture was heated to between 60° and the boiling point; in the second stage, the mixture was heated to 50–55° for some time. An alkali or alkaline earth metal salt of the corresponding acid was added to the anhydride to neutralize all traces of the sulfuric acid; the mixture was then distilled.

Specific Solvents

161. ACETIC ANHYDRIDE

Acetic anhydride may be purified by fractional distillation through an efficient column; see Orton and Jones [3610] and Jones and Betts [2418]. To remove acid, Walton and Withrow [5075] allowed 97% anhydride to stand several days over thin slices of sodium, boiled it in a vacuum under reflux for several hours, and finally distilled it over a mixture of sodium and sodium acetate. It was further purified by fractional distillation. Since sodium reacts vigorously with acetic anhydride at 65–70°, Calcott et al. [881] treated 95% anhydride (1450 g) with coarse magnesium filings (200 g) under reflux at 80–90° for five days. After boiling moderately for 17 hr, it was fractionally distilled through a 30-cm Hempel column.

Myers [3476] purified acetic anhydride for *electrical properties study* by adding toluene to the anhydride and removing the toluene-acetic acid azeotrope in an efficient distillation column; bp 100.6°. The toluene was removed and the anhydride distilled at a reflux ratio of 20:1. The distillate reached the lowest conductivity when about one-half of the charge had distilled; 5×10^{-9} ohm^{-1} cm^{-1}.

Rollett and Leimüller [4025] freed acetic anhydride from most of the acetic acid present by refluxing with calcium carbide and distilling the anhydride. The calcium acetate showed no tendency to decompose. Aluminum chloride also could be used; however, it decomposes slightly during distillation.

McClure and coworkers [3059] report two methods for the direct determination of acetic acid in acetic anhydride. One is a colorimetric and the other a thermometric titration method. Neither method is applicable if amines are present in the anhydride.

A method for the determination of acetic acid in acetic anhydride was reported by Mitra and coworkers [3345].

CRITERIA OF PURITY

ACS Reagent grade acetic anhydride is not less than 97% anhydride by alkali titration of the acidity in aqueous solution [76]. Rosin's specifications [4042] are essentially the same as those for the ACS Reagent grade. A 99+% grade is available.

Timmermans and Hennaut-Roland [4778] used the consistency of the density of fractions as a criterion of purity.

SAFETY

Acetic anhydride is a severe eye and skin irritant. It has a pungent odor and is a lachrymator. Bronchial and lung injuries are likely to occur from inhalation of the vapor [4509, 3681].

The *threshold limit value* is 5 ppm or 20 mg/m^3 [4509].

The *flammable limits* in air at elevated temperature are 2.90 and 10.30%v [3680]; see also [88]. The *minimum ignition temperature* in air is 392° [3680].

162. PROPIONIC ANHYDRIDE

Smith and Hunter [4386] reduced crotonic anhydride in the presence of nickel, cobalt, or Raney nickel at 100 psi pressure at 80°.

The effects are similar to those of acetic anhydride [3681].

163. BUTYRIC ANHYDRIDE

Butyric anhydride is a skin and eye irritant [3681]. A *threshold limit value* of 0.001 ml/liter has been recommended [4477].

ESTERS

General

The same methods of preparation and purification are usually applicable to a large number of esters, particularly to members of a homologous series.

Spassow [4453] developed a method for preparation of esters from aliphatic primary, secondary, and tertiary alcohols, and aromatic alcohols including phenol (except triphenyl methanol). The method is simple and rapid and produces esters in good yield. The general directions are: The alcohol and magnesium are mixed with ether in the ratio of 0.1 mole of alcohol to 0.1 atom weight of magnesium to 10–15 g ether; some tertiary alcohols require a ratio of ether to alcohol of 15–20:0.1 mole. Acid chloride in ether (0.1 mole: 5–10 g) is slowly dropped, with moderate cooling if necessary, into the alcohol mixed in slight excess (1.25–1.5 equivalents). The mixture is allowed to stand for 1 hr, heated 1–2 hr in a water bath, or poured into water or dilute sodium carbonate solution, extracted several times with ether, dried, and fractionally distilled to the desired purity.

Lippincott [2973] devised a method of preparing esters by heating acetals with an aliphatic acid of low molecular weight with, or without, 10% sulfuric acid. For example, 56 parts of dimethoxymethane, 92 parts of formic acid and 30 parts of 10% sulfuric acid were placed in a reactor equipped with an efficient fractionating column. Upon heating, 120 parts of methyl formate were obtained by distillation.

Gonzalez [1833] prepared a series of esters of saturated and unsaturated mono- and dicarboxylic and iso- and heterocyclic acids. In general, 1 mole of the acid, 3.5 moles of the alcohol, and 0.02 moles of chlorosulfinic acid were heated for 2 hr at the boiling point of the mixture. The esters were neutralized, washed with sodium carbonate solution and then with water, and dried over anhydrous sodium sulfate. Solid esters were purified by crystallization. Typical yields were: ethyl acetate 82.5%; isopentyl acetate, 78.8%; ethyl propionate, 77.7%; ethyl benzoate, 81.8%, and ethyl oxalate, 14.9%.

Extensive work on purifying low molecular weight fatty acid esters has been done by Young and Thomas [5306]. The esters were obtained from

Kahlbaum and also prepared from the acid or acid anhydride and the alcohol. Materials from both sources were carefully purified and the physical properties compared. It was assumed that the impurities present were alcohols, acids, water, and homologous esters. The acids were removed by repeated agitation with strong aqueous solutions of potassium carbonate; it was observed that solid potassium carbonate was useless for this purpose. The esters were then washed several times with water, except those compounds with an appreciable water solubility. The lower alcohols were largely removed during the water wash. Drying was necessary before fractional distillation. Potassium carbonate was useful only for clearing the esters when turbid with suspended water. The final drying was made with phosphorus pentoxide, which combined with the water and any residual alcohols. If appreciable amounts of water and alcohols were present, the oxide would hiss and liquify rapidly. The ester was repeatedly decanted and more phosphorus pentoxide added until the oxide merely became pasty; the ester was then distilled. Phosphorus pentoxide was added to the distillate; by the next day a jelly had formed with the ester. This is a common phenomenon when phosphorus pentoxide is allowed to stand with a dry, alcohol-free ester. The ester was again distilled in a dry atmosphere.

Esters, alcohols, and water are noted for azeotrope formation. For example, ethanol and water, ethanol and ethyl acetate, and ethyl acetate and water form binary azeotropes; the three compounds form a ternary azeotrope. Esters, as a rule, do not form azeotropes with esters, particularly with those of the same homologous series. Esters boiling below 100° usually can be distilled at atmospheric pressure. Some esters decompose above 100°, but many can be distilled as high as 150° without decomposition.

SAFETY

The simple aliphatic esters produce irritation and a slight to marked anesthetic action when inhaled in sufficiently high concentration. Some irritate the upper respiratory tract. They produce a considerable fatigue of the olfactory sense, so that the sense of smell is effective only in detecting initial exposure.

The formates are the least toxic of the simple esters, the toxicity increasing slightly from formates to acetates to propionates, and so forth. Within each series, the toxicity increases with molecular weight and boiling point, with the exception of the methyl derivatives. The formates, especially the methyl and ethyl esters, are quite irritating to the eyes and respiratory tract, probably because they hydrolyze readily in contact with water to formic acid. The benzyl fatty acid esters are more toxic than the aliphatic compounds.

The chronic effects of the simpler aliphatic esters are relatively slight.

The *threshold limit value*, the *flammable limits* in air, and the *minimum*

Table 5.14 Safety Data for Esters of Aliphatic Monocarboxylic Acids

Ester	TLV ppm	mg/m³	FL in Air, %v Lower	Upper	MIT in Air
164. Methyl formate	100	250	5.90	21.70	456°
165. Ethyl formate	100	300	3.15	16.40	455
166. Propyl formate	100	360	—	—	455
167. Butyl formate	—	—	1.73	8.15	334
168. Isobutyl formate	—	—	2.00	8.90	—
169. Methyl acetate	200	610	3.85	15.20	502
170. Vinyl acetate	—	—	—	—	427
172. Ethyl acetate	400	1400	2.18	11.40	484
175. Propyl acetate	200	840	1.77	8.00ᵃ	450
176. Isopropyl acetate	250	950	1.78	7.80ᵃ	476
177. Butyl acetate	150	700	1.40	8.00ᵃ	421
178. Isobutyl acetate	—	—	—	—	423
180. Pentyl acetate	100	525	1.03	7.13ᵃ	358
181. Isopentyl acetate	—	—	1.16	6.96ᵃ	—
184. Ethyl propionate	—	—	1.85	11.05	440
185. Ethyl butyrate	—	—	—	—	463
189. Butyl stearate	—	—	—	—	355
190. Methyl acrylate	10	35	—	—	—
191. Ethyl acrylate	25	100	—	—	—
192. Methyl methacrylate	100	410	—	—	—

TLV is threshold limit value; FL is flammable limits in air; and MIT is minimum ignition temperature. TLV reference [4509, 88]; FL and MIT reference [3680].

ᵃ At elevated temperature

ignition temperature in air for aliphatic esters of some saturated and unsaturated monocarboxylic acids are given in Table 5.14.

Specific Solvents

Esters of Saturated Aliphatic Monocarboxylic Acids

164. METHYL FORMATE

For the determination of *physical constants*, Timmermans and Hennaut-Roland [4778] treated methyl formate with sodium carbonate and distilled several times in a water bath from phosphorus pentoxide.

165. ETHYL FORMATE

Bodroux [629] prepared ethyl formate by distilling a mixture of ethanol, formic acid, and water. The yield depends on the proportion of the reactants.

Schiff [4143] found that ethyl formate, like methyl acetate, could not be dried with calcium chloride because it reacts rapidly with the ester to form a crystalline compound.

166. PROPYL FORMATE

Vogel purified propyl formate for *physical property measurements*. The distillate boiling below 85° was washed successively with saturated sodium chloride solution and saturated sodium bicarbonate solution in the presence of solid sodium chloride, dried with anhydrous magnesium sulfate, and distilled [4971].

167. BUTYL FORMATE

Vogel [4971] prepared and purified butyl formate for *physical property measurements* by refluxing 46 g of formic acid and 37 g of 1-butanol for 26 hr. The reaction mixture was successively washed with a solution saturated with sodium bicarbonate and sodium chloride until effervescence ceased, and then dried and distilled. The yield was 40 g of ester boiling at 105–107° at 765 torr.

168. ISOBUTYL FORMATE

Isobutyl formate was prepared and purified by Vogel [4971] in a similar manner to butyl formate.

169. METHYL ACETATE

Hurd and Strong [2327] purified methyl and ethyl acetates by the following procedure: One liter of the ester was refluxed for 6 hr with 85 ml of acetic anhydride and then distilled through a Vigreux column. The distillate was shaken with 20 g of anhydrous potassium carbonate and redistilled.

Vogel [4971] purified methyl acetate for *physical property measurements*. The ester was washed with saturated sodium chloride solution, dried with anhydrous magnesium sulfate, and distilled.

To remove methanol, Schiff [4143] treated methyl acetate with acetyl chloride, washed with concentrated sodium chloride solution, and dried over calcium oxide. It must not be dried with calcium chloride, since it reacts rapidly to form a crystalline compound.

170. VINYL ACETATE

Mitsutani and Kominami [3347] studied the vinyl acetate reaction product and identified several compounds from the mass spectra. The following compounds may be present in the commercial product as impurities: acetaldehyde, acetone, 2-butenal, ethyl acetate, isopropyl acetate, 1-butene, butenyne ($CH\equiv CCH\equiv CH_2$), 1,3-butadiene, 1-hexene, 1-octene, benzene, toluene, and 1,7-octadiene.

Ushakov and Feinstein [4901] prepared vinyl acetate by reacting acetylene and acetic acid in the vapor state in the presence of zinc or cadmium acetate on activated carbon at 180–250°.

Acetylene and glacial acetic acid react to form the ester in the presence of mercury (II) sulfate [957].

Nozaki and Bartlett [3558] fractionally distilled vinyl acetate through a 100+ plate column with stainless steel Lecky-Ewell packing and a total reflux, intermittent take-off head designed to avoid contact with mercury, grease, air, or metal other than the packing. The fraction distilling at 72.3° at 756 torr was used to study the *rate constant* in addition polymerization.

Vinyl acetate was found to be stabilized by 0.05% anthracene [528].

171. ETHYLENE GLYCOL DIACETATE

Taylor and Rinkenbach [4612] dried a commercial product boiling at 188–189.6° at 739 torr with calcium chloride, filtered with the exclusion of moist air, and fractionally distilled at reduced pressure.

SAFETY

Rowe [3681] believes that ethylene glycol diacetate should be considered like ethylene glycol, based on available limited information.

172. ETHYL ACETATE

This ester is available commercially with a composition corresponding to the water azeotrope of about 85–92% ester. It is also available in an anhydrous, or absolute grade, usually guaranteed to be at least 99.5% ester by saponification.

Vogel [4971] prepared ethyl acetate for *physical property measurements*. The preparation and purification were similar to those described for butyl formate.

Gillo [1776] proposed a method of purification and criteria of purity for ethyl acetate as an *organic standard*. The ethyl acetate was dried over potassium carbonate for two months, filtered, distilled, and the first and last portions of the distillate discarded. The entire center fraction was distilled over phosphorus pentoxide (10–20 g/kg) from an all-glass apparatus. The first 10 ml was discarded before the main fraction was collected. The instability and decomposition of the ester stored in the absence of air for 4, 6, and 13 years are discussed.

Hurd and Strong [2327] purified ethyl acetate in the same manner as they did methyl acetate.

CRITERIA OF PURITY

The ACS Reagent grade [76] ethyl acetate is characterized by density and distillation range, plus tests for impurities. Rosin [4042] bases the characterization on a saponification assay and a distillation range. Dreisbach and

Martin [1372] determined the purity from the freezing curve. Maclean and Jencks [3105] characterized the effectiveness of their purification procedure by the ultraviolet absorption spectra.

173. PROPARGYL ACETATE

Yields of propargyl acetate up to 50% were reported by Golendeev and Okrokova [1827] by esterification of acetic acid and 2-propyl-1-ol in the presence of concentrated sulfuric acid and with benzene as the solvent.

174. ALLYL ACETATE

Wislicenus [5223] prepared allyl acetate from 3-iodopropene and sodium acetate. Palomaa and Juvala [3636] reacted allyl alcohol and acetyl chloride to obtain the ester.

Bartlett and Altschul [419] prepared allyl acetate for *polymerization studies* by heating for 2 hr under reflux 1.5 moles of allyl alcohol and 0.75 moles of acetic anhydride to which 5 ml of 96% sulfuric acid had been added. The mixture was poured slowly into ice water with stirring. It was then extracted with ice-cold sodium carbonate and cold saturated calcium chloride solutions. The ester was dried and then distilled in an all-glass apparatus in a nitrogen atmosphere. The middle fraction was collected.

175. PROPYL ACETATE

Vogel [4971] prepared and purified propyl acetate for *physical property measurements* by refluxing 40 g of 1-propanol, 120 g of glacial acetic acid, and 2 g of sulfuric acid for 12 hr. An equal volume of water was then added to the reaction mixture and the ester separated and purified as described for butyl formate.

176. ISOPROPYL ACETATE

Haggerty and Weiler [1960] purified isopropyl acetate for *vapor pressure determinations*. It was shaken with potassium carbonate (50 g/100 ml of ester) to remove acids, and then with a strong solution of sodium chloride to remove alcohols. It was allowed to stand over calcium chloride overnight and then carefully distilled. A fraction boiling within 0.1° was collected for measurements. See also Vogel [4971].

177. BUTYL ACETATE

Butyl acetate can be purchased in a relatively pure state. For laboratory preparation Vogel [4971] passed dry hydrogen chloride into 37 g of 1-butanol until 1 g was absorbed; 120 g of acetic acid was then added and the mixture refluxed 8 hr. The reaction mixture was diluted with an excess of water saturated with sodium chloride to separate the ester. It was washed with aqueous saturated sodium bicarbonate solution until neutral. It was dried with anhydrous magnesium sulfate and fractionally distilled.

180. PENTYL ACETATE

Vogel [4971] prepared and purified pentyl acetate for *physical property measurements* by the procedure described for propyl acetate.

181. ISOPENTYL ACETATE

Vogel [4971] prepared and purified isopentyl acetate for *physical property measurements* by the procedure described for propyl acetate.

182. 2-ETHYLHEXYL ACETATE

Van Pelt and Wibaut [4915] prepared 2-ethylhexyl acetate from 2-ethyl-1-hexanol and acetyl chloride; Hatch and Adkins [2063] prepared it with acetic anhydride.

183. BENZYL ACETATE

Three methods of preparation of benzyl acetate were studied by Su and P'an [4537]. Potassium acetate in an excess of α-chlorotoluene in acetic acid gave a 71.9% yield. Potassium or sodium acetate with α-chlorotoluene and trimethylamine as a catalyst gave 80% ester. Benzyl alcohol, acetic acid, and phosphoric acid gave 84% ester.

Benzyl acetate was prepared in large batches by Bogdanov and Antonov [632] in 98.2 to 99.9% yield of high purity material. They reacted benzyl alcohol, acetic acid, and acetic anhydride catalyzed by sulfuric acid.

Joshi and Merchant [2448] heated 256 g of benzyl alcohol, 1200 ml of water, and 640 g of sodium acetate for 8 hr in an oil bath at 115° with continuous stirring. The top oil layer was washed with twice its volume of water, dried over calcium chloride, and distilled to give 68% yield of benzyl acetate; bp 208–215°.

Katti and Chaudhri [2526] purified benzyl acetate for the study of *physical properties of solutions;* see *m*-cresol.

SAFETY

The oral LD_{50} for rats was found to be 2490 mg/kg [2392].

184. ETHYL PROPIONATE

Vogel [4971] purified ethyl propionate for *physical property measurements* in the same manner that he used for ethyl formate.

186. ISOBUTYL ISOBUTYRATE

Pierre and Puchot [3765] prepared isobutyl isobutyrate by partial oxidation of 2-methyl-1-propanol with potassium dichromate and strong sulfuric acid.

Grünzweig [1924] prepared the ester from the alcohol and the acid with sulfuric acid as the catalyst.

SAFETY

Isobutyl isobutyrate is considered to be practically nontoxic orally and only slightly toxic intraperitoneally. Only slight skin irritation results from prolonged contact with the liquid [3681].

187. ETHYL ISOVALERATE

Vogel [4971] prepared and purified ethyl isovalerate for *physical property measurements*. A mixture of 51 g of isovaleric acid, 11 g of absolute ethanol, and 1 g of sulfuric acid was refluxed for 20 hr and then fractionally distilled to a temperature of 155°. After being washed with saturated sodium bicarbonate and sodium chloride solutions, the ester was dried and distilled.

189. BUTYL STEARATE

Whitby [5139] prepared butyl stearate from silver stearate and 1-iodobutane by heating for 1.5 hr at 100° and extracting with hot ethanol.

Butyl stearate has been purified by Riddick [3975] by fractional freezing and fractional crystallization from solvents whose boiling points are below 100°. Acidic impurities were removed by shaking with 0.05 N sodium hydroxide or a 2% sodium bicarbonate solution, followed by several water washes. The acidity may be determined by titration of an aqueous wash with sodium hydroxide to the phenolphthalein end point. A more sensitive and accurate method is to titrate with a quarternary ammonium hydroxide in methanol-benzene solution. Butyl stearate may be distilled at 1–2 torr.

SAFETY

Two years of feeding rats at 1.25 and 6.25% produced no effects on growth, mortality, or fertility. There are no ill effects known from industrial handling.

Esters of Unsaturated Aliphatic Monocarboxylic Acids

General

Stabilization of esters of unsaturated acids from peroxide formation; see Ethers, General, Robertson and Jones [3998].

Specific Solvents

191. ETHYL ACRYLATE

Ethyl acrylate has a persistent acrid odor and usually contains polymerization inhibitors.

Moureau and coworkers [3430] warmed acrylic acid, dry ethanol, and 10% sulfuric acid to prepare the ester.

Ethyl acrylate is moderately toxic and highly irritating by several routes. It is absorbed through the skin [2338].

192. METHYL METHACRYLATE

Methacrylate esters were prepared by Ruzicka [4080] by heating 50 g of the corresponding γ-bromoisobutyric ester with 50 g of quinoline to boiling under reflux until the heat of reaction was sufficient to maintain the reflux. Heat was again applied at the end to complete the reaction. The methacrylate ester was distilled in 1 atm of hydrogen.

Methyl methacrylate was prepared by Bruson and Washburne [818] by leading methyl α-hydroxyisobutyrate over phosphoric acid-wetted alumina.

Methyl methacrylate of 0.2% impurities was purified to 0.01% impurities by Shelepin and coworkers [4263] by washing with 25% alkali solution, the alkali washed out, and the material distilled in *vacuo* in a stream of nitrogen. The carbonyl compounds were removed by boiling 12–14 hr with powdered potassium hydroxide, filtering, and distilling at 90 torr in oxygen-free nitrogen. The middle fraction was collected and repeatedly distilled until carbonyl-free to 2,4-dinitrophenylhydrazine. The peroxide content was reduced to less than 1×10^{-4} moles/liter by iodometric determination. The freezing point was the criterion of purity.

Methyl methacrylate was purified by zone melting and the purity determined cryoscopically by Anikin and Dugacheva [251]. The monomer was heated above its melting point in a hermetically sealed vessel and then subjected to the action of a cooling agent. The purity was determined from a cooling curve; see also [253, 252, 249].

Makarov and coworkers [3122] stabilized methyl methacrylate by passing it through a sorbent of silica gel or KIL clay. The addition of either sorbent to the stabilized ester increases its storage time.

Yates and Ihrig [5288] have studied the retardation of methyl methacrylate by aromatic amines. They conclude that the amines (like phenols) act as antioxidants by conserving the supply of oxygen inhibitor present.

SAFETY

The odor of methyl methacrylate is readily detectable at 50 ppm. A person can become tolerant to lower concentrations.

A method for the determination of methyl methacrylate in air is described and gives the sensitivity, specificity, and likely interferences.

193. METHYL OLEATE

Methyl oleate is an unusual compound in several respects [3150]. The use of the methyl ester is an important part of the preparation of high purity oleic acid. A combination of fractional distillation and crystallization at different temperatures appears to be the most satisfactory method of preparing a high purity methyl oleate as well as oleic acid. Much of the information concerning the purification of this ester is given under oleic acid.

It appears that Wheeler and Riemeneschneider [5132] developed a method for preparing a high purity methyl oleate; see oleic acid.

Markley [3151, pp. 778 ff.] discusses "Esterification Practices" on both industrial and laboratory scales. The more recent methods for laboratory esterification of fatty acids commonly used are modifications of the Fischer-Speier method [1578] of 1895. Their method consists of heating the acid to be esterified with an excess of alcohol containing 3% dry hydrogen chloride.

There is no generally accepted method for esterifying a specific acid with a specific alcohol. However, Markley [3151, pp. 778 ff.] states that, for methanolic esterification of fatty acids, the concentration of the hydrogen chloride catalyst is usually 3–5%w of the alcohol. Apparently, it makes little difference if the hydrogen chloride is added to the methanol or to the mixture of alcohol and oleic acid.

Senderens and Aboulenc [4235] demonstrated that sulfuric acid could be used as a catalyst.

Hilditch [2186] gives three variations for esterification:

1. About twice the weight of oleic acid to methanol was added in the presence of 2% concentrated sulfuric acid. Then add ethyl ether and wash with dilute potassium carbonate solution to remove unesterified acids. Conversion usually is 97–98%.

2. The acid is converted to the methyl ester by boiling with four times its weight of methanol in the presence of 1% concentrated sulfuric acid. About 70–80% of the methanol is distilled from the reaction mixture, ethyl ether is added, and the free acid removed by shaking with dilute potassium carbonate [2187].

3. To esterify acids separated by the lead salt method (see oleic acid), add about four times the weight of methanol containing 2% concentrated sulfuric acid and reflux 2 hr. Remove as much of the alcohol as possible by distillation from a water bath. Dissolve the residue in ethyl ether, wash with water, and then cautiously with dilute sodium carbonate to remove unesterified acids [2188].

Other workers use hydrogen chloride with equal success.

The conditions for preparing methyl oleate may be varied widely without sacrificing the yield.

One precaution must be observed. The unesterified acids must be removed to a quite low concentration. If they are present in appreciable amount, they will emulsify with the alkali carbonates and hydroxides during the washing out of the acids. Buxton and Kapp [876] report a more satisfactory procedure for removing substantially all of the unreacted fatty acid. Distill the unreacted

Table 5.15   Solubility of Methyl Oleate in Acetone

t	Solubility[a]	t	Solubility[a]
−27.5°	208.0	−35.3°	12.0
−29.4	98.0	−40.0	5.42
−31.0	44.9	−55.8	0.60
−33.0	27.1		

[a] g ester per 100 g acetone.

alcohol from the reaction mixture, dissolve the residue in 2 to 5 parts of such a solvent as 1,2-dichloroethane to one part of ether. Determine the free acid in the mixture by titration. Add the calculated equivalent amount of 38% potassium hydroxide slowly with constant stirring. Allow the mixture to stand: the potassium soap will rise to the top as a floc and the mineral acid salt will settle to the bottom. Filter without suction and wash the precipitate with a small amount of 1,2-dichloroethane. Distill the filtrate.

The ester may be purified by a combination of fractional distillation at reduced pressure and crystallization by, or similar to, that of Wheeler and Riemenschneider [5132].

Craig [1125] prepared methyl oleate for *refractive index studies* from olive oil in the usual manner and distilled the ester. The C_{18} fraction was recrystallized repeatedly from acetone until the precipitate and filtrate had the same refractive index.

The purification of methyl oleate usually is done by a combination of fractional distillation and crystallization from acetone. The solubilities of the ester in acetone are given in Table 5.15 [3837].

Althouse and Triebold [63] determined that methyl oleate decomposes at 217° at 16 torr.

CRITERIA OF PURITY

The iodine number, refractive index, and melting or freezing point are usually used to characterize the purity. The capillary tube melting point is reported to be −19.9 to −19.6° [5132].

Keffler and McLean [2543] point out that the density of methyl oleate, and other unsaturated fatty acid esters, is not very sensitive to the presence of saturated fatty acid ester impurities; for example, for the 97.4% ester the density is 0.8773, and for the 99.8% ester the density is 0.8774.

Esters of Aromatic Monocarboxylic Acids

General

The toxicological properties of benzoates have not been studied extensively. The information available indicates a low order of toxicity. Their low vapor pressure at ambient temperature greatly reduces any danger from inhalation of the vapors. They may be slight or moderate skin irritants.

Specific Solvents

194. METHYL BENZOATE

Vogel [4973] prepared methyl benzoate for *physical property measurements* from 1 mole of benzoic acid and 10 moles of absolute methanol and concentrated sulfuric acid (5%w of the methanol). The boiling point was 199° at 775 torr.

CRITERIA OF PURITY

Dreisbach and Martin [1372] determined the purity from the freezing curve.

195. ETHYL BENZOATE

Vogel [4973] prepared ethyl benzoate for *physical property measurements* in the same manner as described for methyl benzoate.

Cohen and Mier [1030] reacted benzoic acid and triethyl orthoformate for 45 min at 165–192° and obtained an 80–94% yield of ethyl benzoate.

196. PROPYL BENZOATE

Vogel [4973] prepared propyl benzoate by refluxing for 35 hr 30.5 g of benzoic acid, 30 g 1-propanol, and 50 ml of dry benzene and concentrated sulfuric acid. The yield was 37 g; bp 229.5° at /66 torr. Some *physical constants* were determined.

197. BENZYL BENZOATE

Benzyl benzoate is reported to decompose slightly when distilled at atmospheric pressure [594].

Kamm and Mathews [2491] recommend that benzyl benzoate be prepared if a high purity material is desired. They prepared it by dissolving 3 g of sodium in 70 g of benzyl alcohol and, after cooling, gradually added 454 g of benzaldehyde, the temperature being kept at 50–60° by cooling. After warming on a water bath for 1 hr, the product was washed with water and distilled in vacuum. The benzyl alcohol and benzaldehyde passed over in the first portion of the distillate; yield 410–420 g ester boiling at 183–185° at 15 torr. The ester had a purity of 99%.

Guerin [1931] heated and stirred 80 g of sodium benzoate with 125 ml of 1,2-ethanediol for a few minutes, then added 63 g of α-chlorotoluene, refluxed for 15 min, and then cooled. A little water was added to dissolve the salt and the benzyl benzoate upper phase separated from the 1,2-ethanediol. The ester was washed with dilute ammonium hydroxide and distilled.

198. ETHYL CINNAMATE

Jeffery and Vogel [2386] washed commercial ethyl cinnamate with 10% sodium carbonate, then with water; the product was dried with anhydrous magnesium sulfate and distilled. The purified ester was saponified with aqueous potassium hydroxide and the acid isolated, washed, and dried. A mixture of 25 g of the purified cinnamic acid, 23 g of absolute ethanol, 4 g of concentrated sulfuric acid, and 100 ml of dry benzene was refluxed for 15 hr. The ester was isolated and purified. The yield of pure ester was 23.5 g; bp 127° at 6 torr. See [4471] for a slight modification of this method.

Cinnamic acid when reacted with triethyl orthoformate at 180–222° for 1 hr yielded 94% ethyl cinnamate.

Lactones

199. γ-BUTYROLACTONE

This lactone is a powerful solvent. Its viscosity is low for its volatility and it has good heat stability. γ-Butyrolactone is stable at pH 7; it hydrolyzes rapidly in alkaline solution and less rapidly in acid solution. The hydrolysis is reversible and the lactone is reformed when the pH is restored to 7 [1736].

For the quantitative preparation of γ-butyrolactone from 1,4-butanediol; see tetrahydrofuran, Ivanskii and Dolgov [2365].

Fittig and Chanlarow [1582] studied the formation of γ-butyrolactone from γ-hydroxybutyric acid. The lactone forms slowly at room temperature and goes to completion rapidly when distilled.

SAFETY

γ-Butyrolactone is stated to be neither a primary irritant nor skin-sensitizer [1736]. It probably has a low toxicity. It appears to be absorbed readily through the skin. Inhalation of the vapors and absorption of the liquid through the skin should be avoided because of possible delayed effects [3681].

Esters of Dicarboxylic Acids

200. ETHYLENE CARBONATE

Ethylene carbonate exhibits some unusual solvent properties for inorganic compounds; for instance, the alkali metal chlorides are almost insoluble and the chlorides of iron (II), mercury (II), and other "heavy metals" are soluble. The carbonate ester dissolves electrolytes and forms electrolytic solutions as expected from its large dielectric constant.

Vorländer [4992] and Nemirowsky [3515] prepared ethylene carbonate by reacting equimolecular amounts of 1,2-ethanediol and phosgene in a closed tube at room temperature.

Kempa and Lee [2552] purified ethylene carbonate for *dielectric constant* and *dipole moment studies* by fractional distillation at reduced pressure and fractional crystallization from dry ethyl ether; mp 36.2°, n_D 40° 1.4199.

Cislan and Cornilescu [985] studied a process for the continuous purification of cyclic ethylene carbonate by continuous crystallization in a column with thermal gradient; the product was 99.93%.

SAFETY

Inhalation by rats of the concentrated vapor for 8 hr resulted in no deaths. It is only a slight irritant to the skin of rabbits and produces moderate irritation to their eyes [3681].

202. ETHYL OXALATE

Dutt [1421] obtained good yields of ethyl oxalate by passing alcohol vapor through oxalic acid until water ceased to distill.

Vogel [4971] described a general method for preparing dialkyl oxalates. Finely ground AR grade oxalic acid dihydrate was spread thinly on a large watch glass and heated to 105° for 6 hr to produce the anhydrous acid. One mole of the anhydrous acid, 2.5–3.5 moles of alcohol, dry AR grade benzene (twice the volume of the alcohol used), and concentrated sulfuric acid (60%w of the oxalic acid) were refluxed for 6–12 hr and poured into a large excess of water. For the actual preparation cited, the reflux period was 20–34 hr. The benzene phase was separated, the aqueous phase extracted with ether, and the extracts added to the benzene phase. The benzene-ether extract was washed first with aqueous saturated sodium bicarbonate until free from acid and then with water. It was dried with anhydrous magnesium sulfate and distilled.

Ethyl oxalate was prepared by Clarke and Davis [1778] from oxalic acid and 95% ethanol by removing the water as the carbon tetrachloride azeotrope.

Palomaa and Mikkila [3637] determined the *melting point* of ethyl oxalate that was purified by distillation under vacuum.

204. METHYL MALEATE

Toops [4811] fractionally distilled 1500 ml of technical ester (99.26% ester by saponification) in a Podbielniak Hyper-Cal column at 6 torr. The column was preflooded and operated under total reflux for 1 hr. A head fraction of 200 ml was removed at a reflux ratio of 45:1. One liter of distillate was removed at a reflux ratio of 30:1 for *physical property measurements*.

205. ETHYL MALEATE

Toops [4811] fractionally distilled 3 liters of diethyl maleate (99.25% ester by saponification) in a 20-plate, Penn-State-type column packed with ⅛-in. single turn glass helices. The column was operated at total reflux for

1 hr. The first one-third was removed at a 10:1 reflux ratio and discarded. The center fraction was distilled at a reflux ratio of 5:1 and used for the determination of *physical constants*. The distilled ester was 99.53% by saponification.

206. BUTYL MALEATE

Toops [4811] determined some *physical constants* on a technical grade of this ester. The purity was 99.98% by saponification; 0.01% acidity as maleic acid; 0.08% water.

207. BUTYL PHTHALATE

Butyl phthalate is easily purified due, largely, to its relative chemical inertness, its insolubility in water and its high boiling point. It may be freed from alcohol by water washing and from acids and any butyl hydrogen phthalate with dilute caustic. A careful fractional distillation at 10 torr, or less, will produce a product approaching 100%, as nearly as can be determined by several analytical approaches [3975].

Kuskov and Zhukova [2793] refluxed 30 g of phthalic anhydride and 25 g of 1-butanol for 2 hr at 200°, cooled it to 100°, added 2.5 g boric acid, heated it to 200° and added 27 g of 1-butanol during 12 hr; yield, 80% butyl phthalate.

SAFETY

Butyl phthalate is quite toxic when breathed. Spasovski [4452] states that it is not absorbed through the skin; Elizarova [1464] reports that it is. The Committee on Threshold Limit Values [4509] considers it a substance of low acute and chronic oral toxicity under conditions of normal use.

The *threshold limit value* of 5 mg/m³ has been recommended more to control air-born mist rather than as a health measure [4509].

208. BIS(2-ETHYLHEXYL) PHTHALATE

Hickman and Trevoy [2170] purified bis(2-ethylhexyl) phthalate by high vacuum evaporation. They describe the apparatus in the article cited and discuss the technique and apparatus in articles immediately following. Trevoy and Torpey [4828] purified the ester by high vacuum fractional distillation in a falling-stream still.

SAFETY

Fassett [3681] concluded, after studying all available toxicological information, that he sees no reason for predicting that the use of this ester in industry would be associated with any health hazard. See Carpenter and coworkers [911] for a study of animal feeding of the ester.

Esters of Polybasic Acids

210. BUTYL BORATE

The chief impurities in butyl borate are 1-butanol and boric acid from hydrolysis. The ester hydrolyzes readily on contact with atmospheric moisture. Purification must be carried out in a closed system and transfers of the dry ester made in a dry box [3975].

Butyl borate undergoes some decomposition at its boiling point [5086].

Carothers [909, p. 16] gives directions for the preparation from 124 g of boric acid and 666 g of 1-butanol. The yield is 400–425 g of vacuum-distilled product.

Toops [4811] fractionally distilled 1500 ml of commercial butyl borate in a Podbielniak Hyper-Cal column at 25 torr. A center fraction of the distillate was removed at a reflux ratio of 20:1 and used for *physical property measurements*.

Lappert [2829] gives a good review of the methods of preparation. Washburn and coworkers [5086] present an extensive discussion of the methods of preparation. Kuskov and Zhukova [2793] prepared butyl borate in 95.6% yield. O'Brien [3568] prepared the trialkyl orthoborates by removing the water formed during esterification as the benzene azeotrope. Butyl borate was prepared in 80% yield. See also [1963].

SAFETY

Little information has been published on the toxicity of butyl borate. The toxicity of boric acid is considered to be the gauge [5086].

211. BUTYL PHOSPHATE

Commercial butyl phosphate is generally made by reacting 1-butanol with phosphorus oxychloride. The ester is washed with water and then with sodium hydroxide or sodium carbonate, and finally with water. The ester may be dried at a reduced pressure of about 600 torr and a temperature of not more than 125° with good agitation. The likely impurities are phosphates of the alcohol impurities present in the 1-butanol, sodium dibutyl phosphate, disodium butyl phosphate, and traces of butanol and water. Butyl phosphate may be distilled at less than 1 torr if an inert gas, such as nitrogen, is bled into the flask. Decomposition will be quite small under these conditions [3975]. Dyrssen [1426] reported that this ester formed a complex with dibutyl phosphate. A more recent study has been by Liem [2956].

Commercial butyl phosphate has been purified by stirring with 5% sodium carbonate or a 1% sodium hydroxide solution for several hours, washing well with water and drying at reduced pressure. The ester is then vacuum-distilled, discarding the first and last 10% [3975].

Evans and coworkers [1507] prepared butyl phosphate from sodium butoxide and phosphorus oxychloride. Forty-six grams of sodium were added to 600 ml of 1-butanol and the excess alcohol distilled off under reduced pressure. The butoxide was mixed with 200 ml of dry ether, and 60 ml of phosphorus oxychloride in 100 ml of ethyl ether was added with cooling. The mixture was boiled gently for 1 hr, water was added and the ether layer was dried over anhydrous sodium sulfate. The ether was removed and the ester distilled at 50 torr.

Butyl phosphate was purified for *solubility studies* [45] by boiling 100 ml of the ester with 500 ml of 0.4% sodium hydroxide solution, followed by distillation of 200 ml to remove steam volatile impurities. The remaining ester was washed repeatedly with water and dried under vacuum. See also Hardy and coworkers [2016].

Butyl phosphate was purified by Liem [2956] for *distribution studies* by washing with an equal volume of 0.1 M sodium hydroxide, then with 0.1 M nitric acid, and finally with water. It was dried with the heat of an infrared lamp at about 120° for $\frac{1}{2}$ hour at reduced pressure in 1 atm of nitrogen.

The 7170^{-1} and 6960^{-1} absorption bands were used to determine the water in butyl phosphate in the range of 0.5 to 5.0% with an error of ≥ 3 to 4%. This is as accurate as the Karl Fischer method and much faster [2505].

SAFETY

Vapors from the steam distillation and from the hot ester are respiratory irritants. Available information indicates that butyl phosphate should be handled with caution [3681].

HALOGENATED HYDROCARBONS

General

In their study of the kinetics of dehydrochlorination of substituted chlorohydrocarbons, Barton and Howlett [426] developed a general procedure for purifying chlorohydrocarbons. The compound is repeatedly shaken with concentrated sulfuric acid until no further color develops in the acid. The chloro compound is then washed with a solution of sodium bicarbonate and then with water, dried with calcium chloride, and fractionally distilled through an efficient column. The final purification is made by fractional crystallization until the freezing range is constant to 0.1°. In each crystallization only half of the liquid is frozen, the remainder being discarded.

Barton [424] has shown that trace impurities have a marked influence on the thermal dehydrochlorination of chlorohydrocarbons. This was further demonstrated by Barton and Howlett [426] and Barton and coworkers [425].

For the preparation of *carbonyl-free* nonoxygenated extractive solvents, see Hornstein and Crowe [2252], Hydrocarbons, General.

SAFETY

The toxicity of halogenated hydrocarbons varies from dichlorodifluoromethane, which is among the most physiologically inert organic substances to 3-chloropropene (allyl chloride), which is quite toxic, both acutely and chronically. It seems necessary to consider the halogenateds individually, instead of grouping them together or even dividing them into subgroups [3681].

Fluorinated Hydrocarbons

Aromatic Fluorinated Hydrocarbons

Specific Solvents

212. FLUOROBENZENE

Balz and Schiemann [376] prepared fluoroaryl compounds from diazonium fluoroborates. The method involves the diazotization of the appropriate aryl amine in as concentrated hydrochloric acid as possible, precipitation of the fluoroborate with 40% fluoroboric acid (HBF$_4$), washing and drying, and decomposition of the solid to the corresponding fluoro aryl compound. The yields are high, some being almost theoretical. It is stated that there are no side reactions. The preparation of fluorobenzene and p-fluorotoluene were reported.

Carothers (909, p. 46) gives directions for preparing 780–870 g of fluorobenzene from 16 moles of aniline.

Vogel [4972] prepared fluorobenzene for *physical property measurements* according to the method described by Flood [909]; bp 84.5° at 760 torr. The sodium borofluoride can replace the fluoroboric acid.

Stull [4536] fractionally distilled a commercial product to 99.9+%m. Moore and Hobbs [3388] fractionally distilled a commercial grade through a 60-cm Widmer column and used the middle fraction for *dipole moment measurements;* bp 84.74–84.76° at 760 torr.

CRITERIA OF PURITY

Stull [4536] determined the purity from the freezing curve.

SAFETY

Lapik [2826] reported the median lethal concentration for mice for a 2-hr exposure to be 45 mg/liter.

213. *o*-FLUOROTOLUENE

214. *m*-FLUOROTOLUENE

215. *p*-FLUOROTOLUENE

For preparation, see Balz and Schiemann [376] under fluorobenzene.

Vogel [4973] prepared the *para*-isomer for *physical property measurements* from pure p-toluidine; bp 111.5° at 756 torr.

Moore and Hobbs [3388] purified the three isomers for *dipole moment measurements* by the method described for fluorobenzene.

The *vapor pressure* and *freezing point* of the three isomers were determined by Potter and Saylor [3815] on the best commercial grades, which had been dried over Drierite and fractionally distilled at a high reflux ratio. The material distilling within 0.01° was used.

CRITERIA OF PURITY

Potter and Saylor [3815] determined the purity from the freezing curve.

SAFETY

Lapik [2826] reports the median lethal concentration for mice for a 2-hr exposure to be 55 mg/liter.

216. HEXAFLUOROBENZENE

An easy, though low yield, method for preparing hexafluorobenzene was reported by Vorozhtsov and coworkers [4999]. Dry hexachlorobenzene and potassium fluoride were heated in an autoclave at 450–500°; yield 21%.

Evans and Tiley [1509] purified a relatively high purity material by de-aerating by a three-stage distillation at 10^{-5} torr; purity 99.8% by gas chromatography on a dodecylphthalate column. The material was used to determine *critical* and *other physical properties*.

Duncan and Swinton [1397] obtained a commercial product that was about 98% hexafluorobenzene. The main impurities were incompletely fluorinated benzenes, primarily pentafluorobenzene. The material was purified for *thermodynamic property studies* of systems containing hexafluorobenzene. The commercial material was contacted with oleum for 4 hr at room temperature. The treatment was repeated until the oleum was not discolored. It was washed several times with distilled water and dried with phosphorus pentoxide. The hexafluorobenzene was fractionally crystallized 15 times. The criterion of purity was the melting point, 5.082 ± 0.005°, which indicated a purity of 99.95 ± 0.05%m.

Hexafluorobenzene decomposes at a temperature >650° in the vapor phase [2402].

SAFETY

Lapik [2826] reports the median lethal concentration for mice for a 2-hr exposure to be 95 mg/liter compared to 37 mg/liter for benzene.

Mixed Halogenated Compounds

General

The mixed fluorohalogenated aliphatic compounds in this section, in general, are stable to a degree not ordinarily found in organic compounds.

Each compound has, of course, its own degree of stability. The presence of fluorine atoms in the molecule is responsible for the stability; as a general rule, the more fluorine the greater the stability [1411].

Specific Solvents

217. 1,1,2-TRICHLOROTRIFLUOROETHANE

Locke and coworkers [2987] purified laboratory-prepared material to study its *physical properties* by washing with water and then with a weak caustic solution. The compound was dried with calcium chloride or sulfuric acid, and finally fractionally distilled.

1,1,2-Trichlorotrifluoroethane shows the first trace of decomposition in quartz at about 300° [1411].

SAFETY

Michaelson and Huntsman [3297] found the oral LD_{50} for rats to be 43.0 ± 4.8 g/kg or about 2.5 times less toxic then ethanol.

The compound is not flammable or explosive in air [1411].

218. 1,1,2,2-TETRACHLORODIFLUOROETHANE

For purification for *physical properties*, see Locke and coworkers [2987], 1,1,2-trichlorotrifluoroethane.

SAFETY

Toxicity studies by Clayton and coworkers [999] indicate a low order of acute, dermal, and inhalation toxicity. The approximate lethal dose for rats by inhalation for a single 4-hr exposure is approximately 15,000 ppm. Ten repeated daily exposures of 4-hr each to 3000 ppm and 31 4-hr exposures of rats, guinea pigs, and mice to 1000 ppm were not lethal.

A *threshold limit value* of 500 ppm has been recommended [999].

1,1,2,2-Tetrachlorodifluoroethane is nonflammable and nonexplosive [1410].

219. 1,2-DIBROMOTETRAFLUOROETHANE

For purification for *physical properties*, see Locke and coworkers [2987], 1,1,2-trichlorotrifluoroethane.

SAFETY

The LD_{50} for rats exposed for 15 min is approximately 126,000 ppm [1412].

1,2-Dibromotetrafluoroethane is nonflammable [1412].

Monochlorinated Hydrocarbons

Aliphatic Chlorinated Hydrocarbons

Specific Solvents

220. CHLOROETHANE

Commercial chloroethane is generally prepared by one of three methods: the addition of hydrogen chloride to ethylene, chlorination of ethane, and the reaction of ethanol and hydrogen chloride. The first method is the most widely used. Technical chloroethane has a purity of at least 99%.

Chloroethane was prepared in 97.6% yield by Norris and Sturgis [3551] by mixing equal molecular amounts of ethanol and aluminum chloride and heating to 110°, then to 120° at which temperature the reaction was rapid, and finally to 150°. The chloro compound was purified by distillation.

Groves [1916] passed chloroethane vapors through sulfuric acid to obtain a dry and alcohol-free product.

Gordon and Giauque [1842] purified a commercial material by twice fractionally crystallizing, pumping free from air, fractionally distilling, and retaining the middle fraction. From the amount of premelting in *heat capacity measurements*, the purity was estimated to be 99.98%m.

SAFETY

The vapors of chloroethane produce narcosis and slight symptoms of irritation. Less toxic than chloroform, it has been used extensively as an anesthetic [3681]. Troshina [4835] reported LC_{100}, LC_{50} and LC_{min} of 180, 152 and 145 mg/liter, respectively, for 2-hr exposure for rats.

The *threshold limit value* has been set at 1000 ppm or 2600 mg/m³ [4509, 3681].

The *flammable limits* in air are 3.85 and 15.40%v. The *minimum ignition temperature* in air is 494° [3680].

221. 1-CHLOROPROPANE

SAFETY

1-Chloropropane may be considered relatively safe. Liver and kidney injury are likely if the time of administration is long or repeated [3681].

The *flammable limits* in air are 2.60 and 11.10%v, the latter value at elevated temperature [3680].

222. 2-CHLOROPROPANE

Norris and Sturgis [3551] prepared 2-chloropropane in 70% yield in a similar way to chloroethane.

Hughes and Shapiro [2306] dried 2-chloropropane over phosphorus pentoxide and fractionated from a small amount of anhydrous sodium carbonate for *kinetic studies;* bp 36.5° at 760 torr.

SAFETY

2-Chloropropane has been used as an anesthetic. Some histological changes were observed in livers and kidneys of animals exposed 127 times 7 hr a day and 5 days a week over a period of 181 days [3681].

The threshold limit value should be well below 1000 ppm or 3210 mg/m³ [3681].

2-Chloropropane is highly flammable [3681].

223. 1-CHLOROBUTANE

1-Chlorobutane is manufactured from 1-butanol and hydrogen chloride. It is available at reasonable cost with a 1° distillation range.

Deb [1239, p. 27] gives directions for the preparation of about 60 g of 1-chlorobutanol from 1-butanol; see also [1778].

Smyth and McAlpine [4418] purified a commercial product for the determination of *dipole moment* by gently refluxing with concentrated sulfuric acid, washing several times with water, drying over two portions of calcium chloride, and distilling.

See Barton and coworkers [425] and Barton and Howlett [426] under Halogenated Hydrocarbons, General, for additional methods of purification.

CRITERIA OF PURITY

Smyth and Rogers [4423] used the boiling point and the refractive index as the criteria of purity.

SAFETY

The *flammable limits* in air are 1.85 and 10.10%v, the latter value at elevated temperature. The *minimum ignition temperature* in air is 460° [3680].

225. 1-CHLORO-2-METHYLPROPANE

SAFETY

The *flammable limits* in air are 2.05 and 8.75%v, the latter value at elevated temperature [3681].

226. 2-CHLORO-2-METHYLPROPANE

Adams (16, p. 50) describes a simple method of preparation by shaking 2-methyl-2-propanol with concentrated hydrochloric acid. The 2-chloro-2-methylpropane is separated, shaken with 5% sodium bicarbonate and then with water until it is neutral to litmus, dried with calcium chloride, and fractionally distilled.

Turkevich and Smyth [4857] washed a commercial material with ice water, dried over fused calcium chloride, and fractionally distilled; bp 50.5°, fp $-25.4°$, n_D 20° 1.38786.

Howlett [2271] determined *equilibrium constants* on a commercial product that was distilled. The main fraction, bp 50.3–50.5°, was fractionally crystallized six times. The freezing point was constant at $-25.4°$ for the last three crystallizations until almost the entire bulk was frozen. The material was finally dried over calcium chloride and a small amount of calcium oxide bp 50.4°, fp $-25.4°$, n_D 20° 1.3852.

CRITERIA OF PURITY

Turkevich and Smyth [4857] and Howlett [2271] used the boiling point, freezing point, and refractive index as criteria of purity.

Aromatic Chlorinated Hydrocarbons

228. CHLOROBENZENE

The aromatic chloro compounds are, in general, relatively stable compounds and are much less reactive than the aliphatic ones.

Ralph and Gilkerson [3885] purified chlorobenzene in the same manner as they did o-dichlorobenzene; $\kappa = 7 \times 10^{-11}$ ohm^{-1} cm^{-1}.

Chlorobenzene was purified by Bayles and Chetwyn [456] for use as a solvent for the determination of *acid-base function in nonaqueous media*. It was distilled and then fractionally distilled and the middle fraction dried over phosphorus pentoxide for two to nine days and again fractionally distilled. The portion distilling at $132 \pm 0.5°$ was collected for use. The solvent was used only if the aqueous extract showed a pH of 7 with a British Drug House Universal indicator.

For purification, see Hartmann and coworkers [2048] under benzyl alcohol.

Vogel [4973] prepared chlorobenzene from AR grade aniline. The crude chlorobenzene was washed with sodium hydroxide, cold concentrated sulfuric acid, and then with water; dried and distilled; bp 131.5° at 765 torr.

McAlpine and Smyth [3051] determined the *dipole moment* of commercial chlorobenzene that was purified by shaking repeatedly with portions of sulfuric acid until the acid was no longer colored. It was washed with water and with dilute potassium bicarbonate solution, dried with calcium chloride and fractionally distilled; bp 131.1°, n_D 20° 1.52459. After redrying over phosphorus pentoxide and again distilling, the boiling point dropped to 130.7°, while the refractive index did not change.

Stull [4536] fractionally distilled a commercial product through a 5-ft column and obtained a purity of 99.9+%, calculated from the freezing curve.

Moore and Hobbs [3388] purified the best commercial grade, as described under fluorobenzene, for *dipole moment measurements;* bp 131.24–131.25° at 745.9 torr.

Variations of the method of McAlpine and Smyth have been described. Often, chemical treatment is omitted and the commercial material, usually a good grade, is dried with either calcium chloride or phosphorus pentoxide and fractionally distilled; see also Timmermans and Martin [4783], Bramley [707], Sudgen [4540], and Williams and Ogg [5198].

CRITERIA OF PURITY

Stull [4536] and Dreisbach and Martin [1372] determined the purity from the freezing curve.

Timmermans and Martin [4783] purified by fractional distillation until the physical properties of the fractions were constant. McAlpine and Smyth

[3051] used the boiling point and the refractive index as their criteria of purity.

Swietoslawski [4577] recommended the differential ebulliometric method.

Maclean and coworkers [3105] characterized the effectiveness of their purification by ultraviolet absorption spectra.

SAFETY

Voluntary acute exposure by inhalation to chlorobenzene is unlikely because eye and nasal irritation begins at about 200 ppm, at which concentration the odor is unpleasant. There are dangers of chronic exposure. Repeated skin contact should be avoided.

The *threshold limit value* has been established at 75 ppm or 345 mg/m³ [81].

The *flammable limits* in air are 1.35 and 7.05%v at elevated temperature. The *minimum ignition temperature* in air is 638° [3680].

229. 1-CHLORONAPHTHALENE

1-Chloronaphthalene is prepared commercially by the chlorination of naphthalene in the presence of a catalyst. Some isomeric monochloronaphthalenes are formed together with small amounts of dichloronaphthalenes.

A material of high purity may be prepared by the Sandmeyer reaction from 1-naphthylamine.

Satisfactory purification can be attained by fractional distillation of commercial 1-chloronaphthalene. A higher purity may be achieved by fractionally crystallizing the middle fraction from the distillation.

SAFETY

Some information is available for chloronaphthalenes as a group but very little is available specifically for 1-chloronaphthalene. The following information is from the *Hygienic Guide Series* [85]: Nothing is known for man for short exposure tolerance. There is no recommended threshold limit value but data for tri- and pentachloronaphthalenes, would indicate the value for the monochloro compounds should be 1 to 2 ppm or greater. The chloronaphthalenes cause dermatitis. The areas most frequently affected are the face and neck. Probably liver damage would result from repeated inhalation of 1-chloronaphthalene. Fortunately, its vapor pressure is quite low.

Polychlorinated Compounds

General

Carbonyl-free halogenated compounds were prepared by Hornstein and Crowe [2252] for *extractive studies;* see Hydrocarbons, General.

STABILITY AND STABILIZATION

Chlorinated hydrocarbons, particularly chlorinated aliphatic hydrocarbons, are unstable under normal storage conditions. They generally react with oxygen in the presence of moisture to give hydrogen chloride, phosgene and other reaction products, depending on the chloronated compound. Light, heat, and common metals catalyze the reaction. Skeeters [4354] believes that the products of oxidation catalyze the oxidation reaction, and the reaction once initiated is self-catalyzed and self-sustaining. Some of the impurities in the commercial material often are more reactive than the principal component and hence initiate the reaction.

High purity chlorinated compounds are more stable than the commercial grade that have not been stabilized. The most convenient means of purifying most chlorinated hydrocarbons is to neutralize the acid impurities, wash well with water, and fractionally distill. Wash the chlorinated compounds with one-fourth their volume of 1 % aqueous sodium carbonate solution and then three times with the same amounts of distilled water. Thorough agitation is necessary for efficient washing. Select a still and operating conditions that will give a product of the purity desired. Practically all chlorinated hydrocarbons form positive azeotropes with water, as do many of their stabilizers; therefore these impurities are generally removed as heads. Flush the still and continually bleed nitrogen into the receiver to keep out oxygen and moisture. Remove at least 10% heads and collect the 80% middle fraction. Store the product in a nitrogen atmosphere in the dark, preferably in a refrigerator or freezer. Most chlorinated hydrocarbons can be obtained with a purity of 99.8–99.9+% [3975].

The manufacturer's problem is a two-phase one; to retard or prevent the oxygen-moisture reaction and to retard or prevent the reaction of the commercial polychlorinateds with metals, particularly aluminum.

Rapp [3915] stated that various metals are attacked by polychlorinated hydrocarbons and that aluminum is the most reactive of all of the common metals.

A large amount of work has been done to develop stabilizers for the large volume industrial chlorinated solvents. The use and purification of these solvents usually is facilitated if some knowledge of the stabilizers is available.

Stabilizers usually consist of two and sometimes three substances. It has been found that certain substances exhibit synergistic properties. Most stabilizers are synergistic combinations. Some of the stabilizers are included in the following list.

1. For chlorinated hydrocarbons: an amide and an epoxide [1084].
2. Polychlorinated aliphatic hydrocarbons: nitromethane and p-dioxane [3915].
3. 1,1,1-Trichloroethane; nitromethane and amines or acetylenic alcohols [4333]; nitromethane and 1,3-dioxolane and epichlorohydrin [1861]; tetrahydrofuran and N-methylpyrrole [893].

4. Trichloroethylene; tetrahydrothiophene and an amine or phenol or a 1,2-epoxide [1299].

5. Tetrachloroethylene; lower molecular weight, C_1–C_3, nitroalkanes and chloro derivatives thereof (purification greatly enhances the effect).

Sims [4333] found ". . . that the discoveries relative to the stabilization of particular chlorinated hydrocarbon solvents are neither applicable nor translatable to the problem of stabilizing other chlorinated hydrocarbon solvents." Rapp [3915], however, studied the inhibition of dichloromethane and translated the findings to other hydrocarbon solvents; he states,"Hydrocarbon solvents which are inhibited in the same manner are, in general, any polychlorinated aliphatic hydrocarbon, such as, for example, carbon tetrachloride, 1,2-dichloroethane, 1,1,1-trichloroethane tetrachloroethylene, chloroform, and trichloroethylene." The works and conclusions of Copelin [1084] and Dial [1299] seem to support the findings of Rapp.

Bromo- and Iodoalkanes. It was reported [2839] that the iodoalkanes are less stable than the bromoalkanes. Copper and zinc are effective for stabilizing the iodo-compounds and copper silicate for the bromo-compounds.

Aliphatic Chlorinated Hydrocarbons

Specific Solvents

230. DICHLOROMETHANE

Dichloromethane is manufactured commercially by the chlorination of methane or chloromethane, generally the former. All possible chlorine substitution products are formed. The purity of the methane determines the amount of C_2 and higher chlorine compounds present in the crude product.

Dichloromethane appears to be stable to the common laboratory reagents. It reacts slowly with oxygen and must be stored in a colored container out of direct light. There is no appreciable hydrolysis due to dissolved water in the compound, but excess water causes hydrolysis, the rate being rather rapid at elevated temperatures. At low temperature it forms a hydrate that is stable to 2°.

Treszczanowicz and Bakowski [4827] describe a method and apparatus for the continuous preparation of dichloromethane from methane and chlorine and from chloromethane and chlorine.

Mathews [3208] purified dichloromethane by washing with water and sodium carbonate solution, drying over calcium chloride, and fractionally distilling. For *dielectric constant measurements*, Morgan and Lowry [3399] repeatedly fractionally distilled the commercial product until the conductivity of the middle fraction was constant. Maryott and coworkers [3185] washed a commercial product with concentrated sulfuric acid, then with dilute sodium hydroxide, and finally with water. The washed material was left standing overnight over sodium hydroxide and calcium chloride and fractionally distilled in a 60-cm Widmer column; bp 39.93–40.12°, n_D 20° 1.4249.

Vogel [4978] fractionated a commercial product and collected the fraction boiling at 40–41°. This was washed with 5% sodium bicarbonate solution, then with water, dried, and fractionally distilled in a Widmer column.

CRITERIA OF PURITY

Maryott and coworkers [3185] characterized the purity by the boiling point and refractive index. Morgan and Lowry [3399] fractionally distilled to a minimum conductivity. Massol and Faucon [3198] used the optical clarity as a criterion.

SAFETY

Irish states [3681] "Methylene chloride is by far the least toxic of the four chlorinated methanes. The toxic effect is predominantly narcosis. . . . It does not cause significant organic injury." It is mildly irritating to the skin on repeated contact if free to evaporate. It is painful to the eyes but no permanent damage may be expected [84].

The *threshold limit value* is 500 ppm, 1750 mg/m³, which is considered low enough to prevent anesthetic effects and hazard to health [4509].

The *minimum ignition temperature* in air is 642° [3680].

Dichloromethane has no flash point but does give an effect that has been called "a halo," particularly in the TCC Flash Point Tester [3975].

231. CHLOROFORM

Chloroform reacts slowly with oxygen, or oxidizing agents, when exposed to the air and light. The principal products of this decomposition are carbonyl chloride (phosgene), chlorine, and hydrogen chloride.

Serious accidents have been reported from phosgene poisoning from chloroform that had stood in a warm place for a long time. Potassium dichromate and sulfuric acid oxidizes chloroform to phosgene and chlorine. Concentrated nitric acid and chloroform produce a small amount of chloropicrin. Alkalies react with chloroform in a number of ways. It is not advisable to heat chloroform with alkali when purifying.

Chloroform formerly was made almost entirely from acetone and ethanol. It is made now principally by the chlorination of methane. A limited amount is made by the reduction of carbon tetrachloride [2609].

The many procedures for the purification of chloroform in the literature can be divided, in general, into those that use concentrated sulfuric acid for the removal of alcohols and those that do not. There are slight variations in the individual methods. Several modifications for each group are given.

REMOVAL OF CARBONYL CHLORIDE

This impurity reacts slowly with water, alcohols, and alkalies and rapidly with sodium phenolate even in small concentrations. Gillo [1775] found chloroform hard to purify because of its instability and sensitivity to light and oxygen. He found that the concentration of carbonyl chloride could not be reduced below 0.0003 %.

REMOVAL OF ETHANOL

For *preparative and analytical use*, ethanol may be removed by washing five or six times with about one-half its volume of water and then drying.

Budde [838] recommends shaking with concentrated acid as the best method for removing alcohol. Morgan and Lowry [3399] washed several times with concentrated sulfuric acid, then with dilute sodium hydroxide, and then with ice water; then dried over potassium carbonate, stored in a completely filled brown flask, and distilled shortly before use. Williams and Daniels [5194] washed with dilute sodium hydroxide and then with water, dried over fused calcium chloride and then over phosphorus pentoxide, and fractionally distilled for *specific heat measurements*. Vogel [4978] washed with water until neutral to litmus, dried over calcium sulfate, and distilled in an all-glass Widmer column for *physical property measurements*. See also Maryott and coworkers [3185] under dichloromethane, Ball [364], Gibby and Hall [1759], Hantzsch and Hofmann [2011], Mathews [3208], Peddle and Turner [3702], and Cohen [1031].

Yerger [5290] washed commercial chloroform with sulfuric acid, dilute sodium carbonate solution, and water. It was dried with calcium chloride and distilled in an atmosphere of nitrogen in a 30-cm column. Freshly distilled material was used to study the *nature of hydrogen bonded ion-pairs*.

Gross and Saylor [1912] washed chloroform for 1 hr with sodium hydroxide solution and twice with distilled water. It was then treated three times with concentrated sulfuric acid, twice with distilled water, once with mercury, and finally with distilled water. It was dried with calcium chloride and distilled through a 1-meter column for use in *solubility measurements*. According to Richards and Wallace [3972], washing with sulfuric acid, sodium hydroxide, and water leads to partial oxidation to carbonyl chloride. They fractionally distilled over calcium chloride until the boiling point and refractive index were constant for *specific heat measurements*.

For *electrical measurements*, Walden and coworkers [5050] allowed chloroform to stand several weeks over calcium chloride, then distilled and dried a number of times over either sodium sulfate or potassium carbonate, distilling after each drying. They recommended passing carbon dioxide-free air through the liquid, preferably during a slow distillation. Walden [5037] made *specific conductance measurements* on chloroform that was washed with water, dried with calcined sodium carbonate or phosphorus pentoxide, and fractionally distilled over calcined sodium carbonate.

Scheibe and coworkers [4137] purified chloroform for *optical measurements* by the method reported by Anschütz [257].

Chloroform was refluxed by Brown [790], simply distilled, and stored over sodium for *dipole moment studies;* d 25° 1.4754.

Coomber and Rose [1078] found that chloroform prepared from acetone and bleaching powder does not contain the impurities present in commercially available material. It should be used in the assay of strychnine and other organic bases.

Technical grade chloroform may be stabilized by absolute ethanol, methylated spirits, *tert*-butylphenol, *n*-octylphenol, or thymol [2609]. Reagent grade chloroform contains about 0.75% ethanol [76]. Anesthetic grade contains 0.5 to 1.0% alcohol (USP XVI).

The impurities in a typical technical grade chloroform are: 300 ppm of water, 2 ppm of acid as HCl, 200 ppm of dichloromethane, 555 ppm of bromochloromethane, 1500 ppm carbon tetrachloride, 20 ppm of residue on evaporation, and no detectable free halogen [2609].

CRITERIA OF PURITY

ACS *Reagent Chemicals* [76] lists two grades: the regular Reagent grade and one suitable for spectrophotometry. Requirements and test methods are given.

Timmermans and Martin [4783] consider consistency of the density of distilled fractions a sufficient criterion of purity.

Swietoslawski [4577] recommends the differential ebulliometric method.

Maryott and coworkers [3185] used the boiling point and the refractive index as the criteria.

SAFETY

The USP XVI cautions "Care should be taken not to vaporize chloroform in the presence of a naked flame, because of the production of noxious gases."

Rosin [4042] states "Keep in tightly closed containers, protected from light, in a cool place." ACS *Reagent Chemicals* [76] adds "Chloroform should be supplied and stored in amber containers . . ."

Chloroform can produce kidney and liver injuries and heart irregularities; its industrial use has not caused many difficulties.

The *threshold limit value* has been set at 50 ppm or 250 mg/m³ [4509].

232. CARBON TETRACHLORIDE

Most carbon tetrachloride produced in the United States is made by direct chlorination of methane. The impurities present are similar to those found in chloroform. The chlorination of carbon disulfide is still used extensively in other countries, and to a limited extent in the United States.

A good portion of the literature concerning the purification of carbon tetrachloride deals with the removal of carbon disulfide from the carbon disulfide process, since this is the most objectionable impurity. Carbon disulfide is removed by boiling with dilute alkali. Schmitz-Dumont [4170] recommended that carbon tetrachloride be shaken with $1\frac{1}{2}$ times the amount of potassium hydroxide (dissolved in an equal volume of water and 100 ml ethanol) necessary to react with the carbon disulfide. The mixture is shaken thoroughly for $\frac{1}{2}$ hr at 50–60°. Klein [2624] refluxed for $\frac{1}{2}$ hr with 5% sodium hydroxide. Günther and coworkers [1937] treated carbon tetrachloride with three successive portions of 100 ml each of alcoholic sodium hydroxide; the carbon disulfide-free solvent was then washed several times with water, dried, and distilled.

For *thermal measurements*, Williams and Daniels [5194] purified carbon tetrachloride by refluxing over mercury to remove sulfides, followed by washing with dilute sodium hydroxide and water. It was dried over fused calcium chloride and distilled. Hicks and coworkers [2171] purified two lots by the same method. They were fractionally distilled twice in a vacuum-jacketed column, the first and last quarters being discarded each time. The second sample was protected from light. See also Williams and Krchma [5197].

Zimm [5348] purified a Reagent grade for *critical point studies*. It was fractionally distilled through a 20-plate column. The first one-fifth was discarded and the remainder collected; bp 76.84°.

Stull [4536] fractionally distilled a good grade twice through a 5-ft column and obtained a purity of 99.98%, presumably calculated from the freezing curve.

Purification for *hydrogen bonded ion-pair studies;* see Yerger [5290] under chloroform. See also Hartmann and coworkers [2048] under benzyl alcohol.

Bonner and coworkers [660] purified carbon tetrachloride for *freezing point depression* measurements by passing it over alumina and then storing it for several days over alumina that has been heated to 1000°. It was distilled prior to use.

Taylor [4619] reports that carbon tetrachloride reacts with water on silica gel at 110° to produce phosgene and hydrogen chloride; see also Rao and Rao [3907].

CRITERIA OF PURITY

Carbon tetrachloride is available in a number of different grades. It may be purchased in a purity that is adequate for most solvent uses.

Dreisbach and Martin [1372] and Stull [4536] determined the purity from the freezing curve.

Maclean and coworkers [3105] characterized the effectiveness of their purification by the ultraviolet absorption spectra.

Swietoslawski [4577] recommended the differential ebulliometric method for estimating the purity.

Ingold and Powell [2347] distilled commercial carbon tetrachloride and tested for chloroform in the constant boiling middle fraction by warming to 60° with Fehling's solution. The latter is reduced by chloroform but not by the tetrachloride.

SAFETY

Carbon tetrachloride is a toxic substance, both acutely and chronically. It is an excellent solvent for many purposes and it is easy to obtain in the purity desired. Therefore, since it is widely used, one tends to become careless, which increases the toxic hazard. Carbon tetrachloride may enter the body by inhalation, ingestion, and absorption through the skin. The threshold concentration of detection by odor is about 80 ppm, and the odor is strong at about 175 ppm. There have been several reports of illness from breathing air containing 25 ppm. The concentration, therefore, is too high when the odor is detectable [3681, 4509].

The *threshold limit value* is 10 ppm, 64 mg/m³ [4509].

Carbon tetrachloride is not flammable; it has no flash or fire point; it will neither explode nor support combustion.

233. 1,1-DICHLOROETHANE

Barton [302] prepared and purified 1,1-dichloroethane for kinetic studies. Paraformaldehyde was treated with phosphorus pentachloride, the reaction product shaken with concentrated sulfuric acid or aqueous permanganate solution and then fractionally distilled; bp 58.0–58.7°.

Maryott and coworkers [3185] dried a good commercial grade with anhydrous potassium carbonate and distilled over Drierite through a Dufton still.

Vogel [4978] washed a commercial product with a saturated sodium bicarbonate solution until effervescence ceased, then with water; dried and fractionally distilled.

Timmermans and Martin [4783] purified 1,1-dichloroethane by fractional distillation until the density and boiling point were constant. About 200 g were obtained from a 500-g commercial product.

Li and Pitzer [2943] purified a commercially available 1,1-dichloroethane for *thermodynamic property study* by fractional distillation and crystallization three times. The criterion of purity was the change in melting point as a function of the fraction melted.

SAFETY

Heppel and coworkers [2116] concluded that 1,1-dichloroethane is one of the most toxic of the commonly used chlorinated hydrocarbons. It has a chronic toxicity somewhat less than carbon tetrachloride, and its most important effect seems to be on the liver.

The *threshold limit value* of 100 ppm, 400 mg/m³, may be low enough to prevent injury, considering the distinctive odor and irritating properties [4509].

The *minimum ignition temperature* in air is 458° [3681].

234. 1,2-DICHLOROETHANE

1,2-Dichloroethane is a very useful solvent with a favorable boiling point for crystallization and extraction. It is more reactive than many of the commonly used laboratory chloro solvents. It is generally prepared commercially by chlorinating ethylene; beside the principal addition product, some substitution takes place. The other impurities present depend on the impurities in the ethylene. Generally, there are several other chlorinated hydrocarbons present in varying amounts. It is available in a variety of purities.

Vogel [4972] prepared 1,2-dichloroethane for *physical property measurements* from dry 1,2-ethanediol and thionyl chloride. The crude material was purified by washing with 5% sodium hydroxide solution and water; after drying it was distilled; bp 83° at 756 torr. A commercial product was purified in a similar manner; bp 83° at 762 torr.

Pitzer [3774] washed 1,2-dichloroethane with water and repeatedly fractionally distilled for *heat capacity* and *related physical property studies*. The impurities were estimated from the rise in the heat capacity below the melting point; 0.058% impurities; see also Railing [3883].

The spectra of *trans*-1,2-dichloroethane prepared by fractional distillation, followed by several crystallizations just below the melting point was studied by Tokuhiro [4801].

This compound has been purified by Götz [1853], Smyth and coworkers [4412] and Sudgen [4540] by washing with dilute potassium hydroxide and water, drying over calcium chloride or phosphorus pentoxide, and fractionally distilling through an efficient column.

For *electrical measurements*, Walden and coworkers [5050] alternately dried and distilled over sodium sulfate for several days, finally dried for one day over phosphorus pentoxide, and collected the fractions distilling within 0.5° of their measurement. Walden and Busch [5045] fractionally distilled commercially pure material, dried the middle fraction for about four weeks with anhydrous potassium carbonate, and distilled in a distillation apparatus painted black.

Barton [423] purified by fractional distillation to a boiling point of 83.5–83.7° for *kinetic studies*.

A good commercial grade of 1,2-dichloroethane was fractionally distilled by Zwolenik and Fuoss [5357] in a Todd still at a reflux ratio of 5:1. The first 10% was discarded and the main fraction distilled sharply at 83.7°. It was used for *specific conductance measurements*.

Hammond [1995] purified the crude compound by shaking with sodium hydroxide solution to remove the acid, heating to vaporize and remove material boiling below 71.5°, adding water, and distilling the azeotrope at 71.5°. The lower phase was dried by distilling over potassium or sodium hydroxide.

Railing [3883] purified by the procedure described for 1,2-dibromo-ethane; premelting indicated a purity of 99.6%.

Davis and Schumann [1219] found that 1,2-dichloroethane stored in a colorless glass container in direct sunlight acquired an acid reaction within a few weeks. Deterioration was very slow when the solvent was protected from light.

1,2-Dichloroethane was purified by Coetzee and Lok [1022] as a *non-hydrogen-bonding solvent* to study *self-association of acids and bases* as follows: Reagent material was shaken with Brockman Activity 1 basic alumina, 1 g per liter, to remove traces of hydrochloric acid. It was refluxed over a like amount of alumina and fractionally distilled at a reflux ratio of 8:1; bp 83.5°.

Ralph and Gilkerson [3885] purified 1,2-dichloroethane in the same manner that they used for o-dichlorobenzene; $\kappa = 4 \times 10^{-11}$ ohm^{-1} cm^{-1}.

Beckers [469] found that purified 1,2-dichloroethane could be stabilized more than 207 days by the addition of 0.05–0.10% diisopropyl amine.

SAFETY

The odor of 1,2-dichloroethane is definitely detectable at 100 ppm and pronounced at 200 ppm. One can become adapted to the odor at low concentrations, therefore it cannot be considered as a reliable warning [83]. The acute and chronic effects of 1,2-dichloroethane can be significant [3681].

The threshold limit value has been set at 50 ppm, 200 mg/m³ [4509].

The *flammable limits* in air are 6.20 and 15.90%v, the value at elevated temperature. The *minimum ignition temperature* in air is 413° [3680].

235. 1,1,1-TRICHLOROETHANE

1,1,1-Trichloroethane reacts with calcium hydroxide and sodium ethoxide at ordinary temperatures.

Barton and Onyon [428] purified a commercial material for *pyrolysis studies*. It was washed with concentrated hydrochloric acid, 10% potassium carbonate solution, and 10% sodium chloride solution, dried over calcium chloride, and distilled through an 18-in., Penn-State-type column. The main fraction boiling at 73.9° was fractionally crystallized three times, the lower freezing fraction being rejected each time.

Hardies [2014] stabilized 1,1,1-trichloroethane with 0.1–1.0%w di-methoxyethane, 1.0–5%w 2-methyl-2-propanol, 0.5–5%w nitromethane and 0.1–1%w butylene oxide. Hardies and Pray [2015] found 0.5%w tetrahydro-furan, 1.5%w 2-methyl-2-butanol, and 1.5%w 3-hydroxy-3-methylbutyne to by effective. p-Dioxane, nitromethane, and 2-butanol and a vicinyl C_4–C_6 monoepoxide were found effective [1353].

Air oxidation of tri and tetrachloroethane can be prevented, or delayed for 22 hr or longer, during the purification by fractional distillation by the addition of 100 mg of phenol or 25 mg of pyrocatechol. When both were present the chloro compound was stabilized for ten days [4443].

SAFETY

The American Industrial Hygiene Association introduced a new toxico-logical factor in 1964 [82], the *Emergency Exposure Limit* (EEL). This is the concentration of a substance in air that can be tolerated without adversely affecting health, but not necessarily without acute discomfort or other evidence of irritation or intoxication. These values are intended to give guidance in the management of single brief exposures. Only a few substances have had their EEL's established. The EEL's for 1,1,1-trichloroethane are: 5 min at 2500 ppm (13,500 mg/m^3); 15 min at 2000 ppm (10,800 mg/m^3); 60 min at 1000 ppm (5400 mg/m^3). Exposure to 2–3%v for more than a few minutes will cause complete incoordination, unconsciousness, and possible death.

The *threshold limit value* in air is 350 ppm, 1900 mg/m^3, [4509].

The *flammable limits* in air are 10 to 15.5%v by hot wire ignition. It will not sustain combustion [3681].

236. 1,1,2,2-TETRACHLOROETHANE

Thomas and Gwinn [4718] repeatedly washed a technical grade with con-centrated sulfuric acid, steam distilled, and then fractionally distilled in a 30-plate column. The center portion boiling within 0.20° was stored over anhydrous calcium sulfate until it was ready for use in determining *rotational configuration* and *dipole moment*.

Physical properties were determined by Vogel [4978] on technical material that was purified by stirring 135 ml with 17 ml of concentrated sulfuric acid for 10 min at 80–90°. The discolored acid was removed and the acid washing repeated two more times. The 1,1,2,2-tetrachloroethane was washed with water, steam distilled, again washed with water, dried with potassium carbonate, and fractionally distilled through a Widmer column.

For *kinetic studies*, Barton [423] purified by shaking with concentrated sulfuric acid and fractionally distilling; bp 146–147°.

Walden and Werner [5051] determined the *dielectric constant* on material that was shaken with potassium carbonate solution, dried over anhydrous potassium carbonate, fractionated under reduced pressure at 60°, and the middle fraction again fractionated at atmospheric pressure. The distillate was further dried over potassium carbonate for one day before use.

Walden and Gloy [5047] washed commercial material with potassium carbonate solution, dried, and distilled; bp 145.6°. The *specific conductance* was found to be immeasurably low.

Stabilization during fractional distillation; see 1,1,1-trichloroethane [4443].

SAFETY

It is reported that 1,1,2,2-tetrachloroethane has a noticeable odor at 3 ppm; however, the odor is not particularly striking and the acceptable level is 5 ppm. Neither level is of any significance as a warning property. This compound is considered to be highly toxic [3681, 4509].

The *threshold limit value* is 5 ppm, 35 mg/m³ [4509].

237. PENTACHLOROETHANE

Pentachloroethane is manufactured by the chlorination of trichloroethylene. The crude product contains about 5% unreacted trichloroethylene.

This compound cannot be distilled at atmospheric pressure without some decomposition. Water in excess of that dissolved by the compound causes appreciable hydrolysis, even at room temperature. Sodium ethylate, alcoholic potassium hydroxide, and aqueous calcium hydroxide slurry split out hydrogen chloride. Sodium hydroxide in an aqueous-alcoholic solution is reported to form an aldehyde resin. Ammonia and primary and secondary amines also react [2608].

Walden and Werner [5051] treated pentachloroethane with potassium carbonate solution, dried over potassium carbonate, fractionally distilled at reduced pressure, and finally fractionated the middle portion at atmospheric pressure. The fraction boiling at 157.5–158.5° was used for *dielectric constant measurements* after being dried over potassium carbonate.

Thomas and Gwinn [4718] repeatedly washed a commercial product with concentrated sulfuric acid, steam-distilled, and then fractionally distilled through a 30-plate column. The center portion distilling within 0.2° was stored over anhydrous calcium sulfate and used for *rotational configuration* and *dipole moment measurements*.

Barton [423] purified pentachloroethane for *kinetic studies* by shaking with concentrated sulfuric acid and fractionally distilling.

SAFETY

Pentachloroethane has a strong narcotic effect, perhaps even greater than chloroform. Exposure to this compound may result in injury to the liver, lungs, and kidneys. It is irritating to the eyes and upper respiratory tract.

Cats exposed to 121 ppm, 8–9 hr daily for 23 days showed significant patho-
logical changes in the liver, lungs, and kidneys [3681].

Aromatic Chlorinated Hydrocarbons

Specific Solvents

238. o-DICHLOROBENZENE

This hydrocarbon is a by-product in the manufacture of chlorobenzene and
p-dichlorobenzene. The technical grade approximated the composition of the
ortho-para eutectic, about 84% *ortho* and 15–16% *para*, plus as much as 1%
trichlorobenzene. The purified product may be obtained with a purity of
98–99+%. A 99.9%m material is available but it is expensive.

o-Dichlorobenzene appears to be stable to many of the common reagents
and has some excellent characteristics as a solvent.

Moore and Hobbs [3388] purified Eastman's best grade, as described under
fluorobenzene, for *dipole moment measurements;* bp 179.7° at 746.2 torr.

A technical o-dichlorobenzene was passed through Alcoa grade F-20
alumina in a 30 × 2-cm column by Ralph and Gilkerson [3885] and then
fractionally distilled through a 35-cm column packed with glass helices. The
middle fraction, bp 48° at 5 torr, was stored over alumina which had been
heated to about 800° for 1 hr. Just before use, the solvent was passed through
a 30 × 2-cm column, the bottom of which was heated alumina and the top
one-half of which was 5A molecular sieves. The specific conductance was
3–8 × 10⁻¹¹ ohm⁻¹ cm⁻¹. See also [660].

The isomeric dichlorobenzenes were separated by Radzitzky and Hanotier
[3880] by clatheration with Werner complexes of the form

$$Ni(SCN)_2(RC_6H_4CHR'NH_2)_4.$$

SAFETY

The physiological effect of o-dichlorobenzene is primarily injury to the liver,
secondarily to the kidneys. Short exposures to high concentrations may
result in depression of the central nervous system. The odor becomes strong
and irritating at concentrations of 100 ppm.

The threshold limit value is 50 ppm, 300 mg/m³ [4509].

The minimum ignition temperature in air is 648° [3680].

239. m-DICHLOROBENZENE

Vogel [4972] further purified a pure commercial product for *physical
property measurements.* It was washed successively with 10% sodium
hydroxide solution, then with water until the washings were neutral, dried,
and distilled; bp 172.5° at 760 torr.

The compound attacks rubber stoppers.

For separation of dichlorobenzene isomers, see *o*-dichlorobenzene, Radzitzky and Hanotier [3880].

240. *p*-DICHLOROBENZENE

The principal impurity is *o*-dichlorobenzene. The presence of 0.5%w of the *ortho*-compound depresses the freezing point 0.25°.

p-Dichlorobenzene is stable to the common laboratory reagents. It reacts with concentrated sulfuric acid and other reagents that react with benzene.

For separation of the dichlorobenzene isomers, see *o*-dichlorobenzene, Radzitzky and Hanotier [3880].

SAFETY

p-Dichlorobenzene has a very distinctive aromatic odor detectable at 15–30 ppm in air. It is painful to the eyes and nose at concentrations of 80–160 ppm. A person may become accustomed to and tolerate high concentrations. The physiological symptoms are similar to those of the *o*-isomer but this isomer is somewhat less toxic [3681, 4509].

The *threshold limit value* is 75 ppm, 450 mg/m³ [4509].

Unsaturated Chlorinated Hydrocarbons

Specific Solvents

241. 3-CHLOROPROPENE

This solvent is manufactured by the chlorination of propylene. The commercial product is likely to contain 2-chloropropene, 2-chloropropane, 1-chloropropane, 3,3-dichloropropene and 1,3-dichloropropene. It usually has a purity of about 97%.

3-Chloropropene is more reactive than the saturated chloroparaffins and less active than propene. Alcoholic potassium hydroxide reacts to form small amounts of allene (propadiene) and large amounts of allylethyl ether. Aqueous alkalies are reported to cause some hydrolysis.

SAFETY

Allyl chloride is a highly toxic and irritating material. It is very irritating to the eyes and upper respiratory tract. Lung irritation and kidney injury are the primary responses to acute exposure. Similar injuries are observed for chronic exposures but liver injuries become more important. It is readily absorbed through the skin. The threshold odor concentration for all persons is considered to be 25 ppm; for one-half of the people from 3 to 6 ppm.

The *threshold limit value* is 1 ppm, 3 mg/m³ [4509].

The flammable limits in air are 2.90 and 11.30%v. The *minimum ignition temperature* in air is 487° [3680].

242. 1,1-DICHLOROETHYLENE

Rogers [4015] prepared 1,1-dichloroethylene by dehydrohalogenation of 1,1,2-trichloroethane with aqueous calcium hydroxide. It was purified by fractional distillation for *dipole moment studies*.

Hildenbrand and coworkers [2185] purified 1,1-dichloroethylene for *thermodynamic and spectrographic studies* by fractional distillation; purity 99.97%m by calorimetric melting studies. Precautions were taken to avoid elevated temperatures and contact with air.

1,1-Dichloroethylene must be stored away from light and air. When stored in the presence of oxygen, a peroxide is formed [1501].

The Dow Chemical Company [1360] states "The safe handling of vinylidine chloride monomer is dependent on knowledge of the characteristics of the material." The observance of this principle contributes to the safe handling of all solvents.

The Ethyl Corporation [1500] reports that 1 ppm of *p*-methoxyphenol is an effective stabilizer for the monomer; it is about 50 times as effective as phenol under similar experimental conditions.

Dow Chemical Company [1360] furnishes the monomer inhibited either with 0.6–0.8%w phenol or with 200 ppm of monomethoxyhydroquinone. The latter inhibitor usually is preferred for most purposes because it is present in smaller amount and does not interfere with many uses.

SAFETY

The odor is the sweet smell characteristic of chloroform. Most people can detect it at 1000 ppm but some can detect the odor at 500 ppm [3681]. Ethyl Corporation [1500] states that 1,1-dichloroethylene is irritating to the skin and eyes and may cause burns. Overexposure to the vapors may cause anesthesia.

Irish [3681] suggests that the *threshold limit value* be less than 25 ppm, 99 mg/m^3.

The flammable limits in air are 7–16%v [1501]. The *minimum ignition temperature* in air is 570° [3681].

243. *cis*-1,2-DICHLOROETHYLENE

244. *trans*-1,2-DICHLOROETHYLENE

Commercial 1,2-dichloroethylene is a mixture of the *cis*- and *trans*-isomers. The individual isomers are available at a reasonable cost. There is some confusion in the literature as to which is the *cis*- and which is the *trans*-form. Some of the properties listed in the tables may be interchanged because it was not possible to identify positively the isomer in all publications.

Knox and Riddick [2646] separated a commercial product into the isomers by fractional distillation. The *cis*-fraction contained about 0.01% *trans*-isomer and 0.01% of an unidentified substance. The *trans*-fraction contained less than 0.0001% *cis*-isomer. The purified material was stored in dark bottles with a small amount of hydroquinone to inhibit oxidation and polymerization.

Maryott and coworkers [3185] purified Eastman's *cis*-product for *dielectric constant* and *dipole moment measurements* by shaking with mercury, drying over anhydrous potassium carbonate, and distilling from Drierite in a Dufton still; bp 60.33–60.38°.

Walden and Gloy [5047] separated the *cis*-isomer from technical 1,2-dichloroethylene by fractional distillation from fused potassium carbonate; bp 59–61°; the specific conductance at 25° was 8.5×10^{-9} ohm^{-1} cm^{-1}.

SAFETY

Apparently 1,2-dichloroethylene does not have a high toxicity. The major response to both isomers is one of central nervous system depression. Liver and kidney injury does not appear to be a major factor [3681].

The *threshold limit value* is 200 ppm, 790 mg/m^3 [4509].

245. TRICHLOROETHYLENE

Like chloroform, trichloroethylene undergoes autoxidation, forming, among other components, hydrogen chloride, carbon monoxide, and carbonyl chloride.

It has been reported that trichloroethylene and potassium or sodium hydroxide may explode. Trichloroethylene reacts with 90% sulfuric acid, and dropwise addition of nitric acid produces, among other substances, tri-chloronitromethane (chloropicrin).

Bruyne and coworkers [824] treated commercial trichloroethylene containing carbonyl chloride with potassium carbonate solution and then with water. After drying over potassium carbonate and calcium chloride, it was fractionally distilled through a Widmer column in an atmosphere of dry carbon dioxide.

McDonald [3069] distilled a commercial material in an Ewell column of 20 theoretical plates. The 40% middle fraction was used for vapor *pressure measurements*.

Commercial trichloroethylene was purified by Carlisle and Levine [907] for *physical property measurements* and *stability studies*. It was steam-distilled from a 10%w calcium hydroxide slurry. The oil phase was cooled to −30 to −50° and the ice removed by filtration. The filtrate was fractionally distilled at 252 torr at a reflux ratio of 3:1. The middle fraction was collected in a receiver covered with black paper; bp 86.7°, fp −88°. Neither purified nor commercial material showed any evidence of decomposition in one year when stored in glass in the dark, or in steel. It behaved like chloroform toward light and air.

For *conductivity measurements*, Walden [5037] distilled technical material several times, dried over sodium sulfate, distilled again, and dried the fraction distilling at exactly 87° for a long time over potassium carbonate.

Trichloroethylene may be stabilized against oxidation by 0.01–0.02% 1-ethoxy 2-iminoethane [3838]; against heat and oxidation by mixtures of pyrrole or its derivatives and an epoxy compound [4442]; 0.05–1% methacrylonitrile is claimed to be an effective stabilizer [3165]; and 0.001–1.0% acetylenic alcohol such as 3-methyl-3-hydroxypentyne and a light stabilizing substituted phenol such as isoeugenol is claimed to be effective [4353].

SAFETY

The predominant physiological response to trichloroethylene is that of depression of the central nervous system, particularly from acute exposures. Visual disturbance and mental confusion accompanied by incoordination may lead to accidents.

The *threshold limit value* is 100 ppm, 520 mg/m^3 [4509].

The *minimum ignition temperature* in air is 463° [3680].

246. TETRACHLOROETHYLENE

Tetrachloroethylene is an excellent solvent and is less reactive than many of the other halogenated hydrocarbons. It reacts with concentrated sulfuric and nitric acids. It is oxidized to phosgene and trichloroacetic acid [352] under the same conditions of storage given for chloroform. Bailey [351] found that ethanol, ethyl ether, or thymol could be used to retard the oxidation, the effectiveness increasing in the order given. Little oxidation was observed in material containing 2–5 ppm thymol when stored in a dark glass container. The pure material is more stable than the commercial material.

Timmermans and Hennaut-Roland [4778] treated a commercial product with sodium carbonate and fractionally distilled until the density was constant. Walden and Werner [5051] dried a commercial product over sodium sulfate or calcium chloride and fractionally distilled. See also Mathews [3208].

Bretscher [728] distilled the material in a vacuum in order to avoid phosgene formation and stored it in the dark out of contact with air. Greenwald [1885] removed 1,1,2-trichloroethane and 1,1,1,2-tetrachloroethane by countercurrent extraction with an ethanol-water mixture. The purified product is more stable to light, heat, moisture, and oxidation.

Tetrachloroethylene is stabilized against heat and oxidation by mixtures of pyrrole and its derivatives and an epoxy compound [4442]. It has also been stabilized by the addition of 0.005–1 % of diallylamine or tripropyleneamine. Baldridge [362] found 0.01–1.0% acetylenic halides effective as a stabilizer such as 3-chloro- or 3-iodo-1-propyne, 4-bromo-1-butyne, or 1,4-dichloro-1-butyne.

SAFETY

Tetrachloroethylene is a central nervous system depressant at high concentrations. Although tetrachloroethylene is considered a relatively safe solvent, the available information indicates that the word *relatively* should be remembered [3681].

The *threshold limit value* is 100 ppm, 670 mg/m³ [4509].

It is nonflammable, nonexplosive, and will not support combustion.

Brominated Hydrocarbons

Specific Solvents

Aliphatic Monobrominated Hydrocarbons

247. BROMOETHANE

Adams [15, p. 6] gives directions for preparing about 1 kilogram (kg) from hydrobromic acid, ethanol, and concentrated sulfuric acid. Kamm and Marvel [2489] prepared about 1 kg in about 90–95% yield from ethanol and bromine. Rodionov [4012] claims good yields from ethyl-*p*-toluene sulfonate, and potassium bromide.

Smyth and Engel [4413] purified a commercial product by washing several times with cold concentrated sulfuric acid and then with water until all the sulfate was removed. The washed material was dried with calcium chloride and fractionally distilled; bp 38.3–38.4°.

The principal impurities in bromoethane are ethanol and water, both of which form azeotropes with bromoethane; the three form a ternary azeotrope. Material prepared according to Adams [15] can be further purified by rapidly agitating the interphase of cold concentrated sulfuric acid and bromoethane, 1:5 by volume, for 2 hr [3975]. The alcohol-free material is washed with water and then agitated as described with 5% sodium bicarbonate solution, washed several times with water, shaken with calcium chloride, and dried overnight with Drierite. It is then fractionally distilled in a 25-plate, all-glass, Penn-State-type column. The column is operated under total reflux for 2 hr, 5%v removed at a reflux ratio of 25:1 and 90%v removed at a reflux ratio of 10–15:1. The middle 90% distills within 0.1° at a constant density.

SAFETY

The *threshold limit value* has been set at 200 ppm, 890 mg/m³. It is believed that this is sufficiently low to prevent systemic effects and narcosis [4509].

The *flammable limits* in air are 6.75 and 11.25%v. The *minimum ignition temperature* in air is 511° [3680].

248. 1-BROMOPROPANE

249. 2-BROMOPROPANE

Goshorn and coworkers [1778] give directions for the preparation of the isomeric bromopropanes from the corresponding alcohols, red or yellow phosphorus, and bromine. The yield for 1-bromopropane is 143–152 g or 69–73%; for the 2-isomer the yield is 170–186 g or 82–90%.

Maccoll and Thomas [3097] purified 2-bromopropane for *pyrolysis studies* by fractional distillation through a 1 meter glass helices-packed column bp 59.32° at 760 torr, n_D 15° 1.4285.

Aromatic Monobrominated Hydrocarbons

250. BROMOBENZENE

Bromobenzene was purified as a *solute for freezing point depression measurements* by passing through alumina and then storing over alumina that had been heated to 1000°. It was distilled prior to use [660].

Vogel [4973] prepared bromobenzene for *physical property measurements* from AR grade aniline, according to the method in Marvel [3181] for bromotoluene. The crude material was purified in the manner described for chlorobenzene; for specific directions, see Timmermans and Martin [4783], Mathews [3208], Rabinowitsch [3876], Williams and Krchma [5197] and Hurdis and Smyth [2328].

Stull [4536] fractionally distilled a commercial product through a 5-ft column and obtained a purity of 99.9+%m, as calculated from the freezing curve.

CRITERIA OF PURITY

Stull [4536] and Dreisbach and Martin [1372] determined the purity from the freezing curve.

Swietoslawski [4577] reports that the differential ebulliometric method may be used as the criterion of purity.

251. 1-BROMONAPHTHALENE

Wright and Mura [5267] describe a classroom experiment for preparing 1-bromonaphthalene; 6.4 g (0.05 moles) of naphthalene is mixed with 100 ml of ethyl acetate and 4–5 g of anhydrous iron (III) chloride. A solution of 8 g (0.05 moles) of bromine (apparently in ethyl acetate) is added to the mixture with efficient stirring. The reaction mass is refluxed 3–4 hr during the release of hydrogen bromide. The residue is poured into 200 ml of ice water after removing 125 ml of ethyl acetate by distillation. The product is steam distilled. No 2-bromonaphthalene was detected by the authors by gas chromatography.

Polybrominated Hydrocarbons

252. BROMOFORM

The stability on storage and reactions of bromoform are similar to those of chloroform.

For *dielectric constant measurements*, Smyth and Rogers [4423] washed bromoform with concentrated sulfuric acid, dilute sodium hydroxide, and water. After drying over anhydrous potassium carbonate, it was distilled at reduced pressure. The density to the third decimal place is given as the criterion of purity. Cauwood and Turner [936] fractionally crystallized eight times; fp 7.5°.

Vogel [4978] washed bromoform containing 4% ethanol with saturated calcium chloride solution, dried with calcium chloride, and fractionally distilled. The colorless product, after standing several days, developed a yellow cloudiness at the surface.

SAFETY

Relatively little is known of the toxicity of bromoform. It has been shown to cause liver damage [3680, 4509].

The *threshold limit value* has been set at 0.5 ppm, 5.0 mg/m³ [4509].

253. 1,2-DIBROMOETHANE

Vogel [4972] prepared 1,2-dibromoethane from 48% hydrobromic acid, concentrated sulfuric acid, and 1,2-ethanediol. The crude material was washed twice with concentrated hydrochloric acid, water, sodium bicarbonate solution, and finally with water. It was dried and distilled; bp 130.5° at 758 torr.

Smyth and Kamerling [4414] washed commercial 1,2-dibromoethane with concentrated sulfuric acid, sodium carbonate solution and water, dried over calcium chloride, and fractionally distilled.

Meisenheimer and Dorner [3261] dried 1,2-dibromoethane over phosphorus pentoxide and fractionated repeatedly under reduced pressure until the melting point was 9.98°. Repeated fractional distillation, followed by fractional crystallization several times, and a final distillation produced a material melting at 10.06°, but the melting point began to fall after a short time.

For *cryoscopic measurements* Moles [3372] dried 1,2-dibromoethane over calcium chloride, fractionally distilled, and dried over sulfuric acid in a desiccator for six days, and finally fractionally crystallized.

Railing [3883] purified a commercial product for *specific heat* and *heat of fusion measurements* by fractionally distilling twice through a 5-ft column, retaining a portion boiling over a 0.02° range. This narrow boiling range material was fractionally crystallized twice and then fractionally distilled. The amount of premelting indicated a purity of 99.92%m.

Timmermans and Martin [4783] purified the material by repeated distillation and fractional crystallization; see also Gross and Saylor [1912].

Biron [601] stated that fractional crystallization must be carried out with the exclusion of light, because even in diffused light a small quantity of a deep red substance is formed.

CRITERIA OF PURITY

Railing [3883] used the amount of premelting, and Meisenheimer and Dorner [3261] the melting point as their criteria of purity.

SAFETY

Injury to the lungs, liver, and kidneys is generally observed following acute exposure. The difference between the concentration that is tolerable, 25 ppm and that which causes severe injury and death, 50 ppm, is not great. 1,2-Dibromoethane is injurious and irritating to the skin. Any clothing upon which the liquid has been spilled should be removed immediately. It will penetrate through several kinds of protective clothing, particularly neoprene-type rubber and several types of plastic [3681, 4509].

The *threshold limit value* is 25 ppm, 200 mg/m³ (4509).

254. 1,1,2,2-TETRABROMOETHANE

Vogel [4978] washed a commercial product three times with 12% of its volume of concentrated sulfuric acid and three times with water, dried with potassium carbonate and calcium sulfate, and distilled.

SAFETY

The vapor pressure is low at room temperature, and the material may be handled with reasonable precautions and ordinary ventilation [3681].

The *threshold limit value* is 1 ppm, 14 mg/m³ [4509].

Iodinated Hydrocarbons

Aliphatic Iodinated Hydrocarbons

Specific Solvents

255. IODOMETHANE

Carothers [909, p. 60] gives directions for the preparation from iodine, red and yellow phosphorus, and methanol. The method is suitable for the preparation of iodoethane, 1-iodopropane, 2-iodopropane, and the iodobutyl and iodopentyl compounds, excluding the tertiaries.

Adams and Voorhees [20] give a general method for preparing alkyl iodides. They prepared 3.15 kg of iodomethane in 94% yields, 3.4 kg of iodoethane in 92% yield, and 0.48 kg of 1-iodopropane in 96% yield.

Cowley and Partington [1119] purified the lower members of the aliphatic bromides and iodides by shaking with dilute sodium carbonate solution and washing repeatedly with water. After a preliminary drying with calcium chloride, the material was allowed to stand for a day over phosphorus pentoxide and fractionally distilled twice. Iodomethane, iodoethane, iodopropane, and bromopropane so purified were used for *dipole moment measurements*.

Iodomethane is slightly unstable. A drop of mercury, copper turnings, or silver powder generally will keep the compound clean. Discolored material may be decolorized by shaking with mercury, sodium sulfite, or other reducing agents. It may be kept for an extended period of time if it is fractionally distilled and collected in 1 atm of oxygen-free nitrogen in a dark bottle, preferably a Corning Low Actinic or similar type. The bottle must be stoppered and sealed without admitting oxygen and stored in a dark cool place.

A method for purifying alkyl iodides is described by Gand [1693, 1694, 1695, 1696]; see iodoethane.

CRITERIA OF PURITY

Rosin [4042] gives specifications and test methods for Reagent grade.

SAFETY

Iodomethane should be considered as a toxic substance, more toxic than carbon tetrachloride and chronically dangerous [4509, 3681].

The *threshold limit value* is 5 ppm, 28 mg/m^3 [4509].

256. IODOETHANE

Taylor and Grant [4616] discovered that anhydrous hydrogen iodide reacts immediately with ethyl ether at room temperature to produce ethanol and iodoethane as a two phase system that can be readily separated.

Smyth and Stoops [4425] purified iodoethane for *dielectric constant determinations* by repeatedly washing CP material with water, drying over calcium chloride, and fractionally distilling. The boiling point was used as the criteron of purity. See also Cowley and Partington [1119] under iodomethane.

For *vapor pressure measurements*, Smyth and Engel [4413] purified a commercial product by washing with very dilute potassium hydroxide solution until decolorized. Several drops of the caustic solution were added in excess and the iodoethane separated, thoroughly washed with water, dried over fused calcium chloride, and fractionally distilled. The fraction distilling without any appreciable change in temperature at 73.2° was collected.

Gand [1693] developed a method for removing free iodine from alkyl iodines. The iodides are distilled through a LeBel-Henninger-type column containing copper turnings. Purification by shaking with alkaline solutions and storing over silver powder are adversely criticized. Storing over silver is unsatisfactory for preserving iodoethane because surface films of silver iodide are formed. The best preservative is metallic calcium [1696].

Further experiments [1694] indicated that distillation through copper turnings removes all traces of hydrogen iodide. Investigations of drying agents showed that calcium chloride does not remove all of the water. Calcium carbide removes all of the water but makes the iodoethane impure, and phosphorus pentoxide is unsatisfactory because of several physical difficulties. The iodoethane also becomes contaminated with ethyl phosphate. Sodium wire was found to be effective but some sodium iodide was formed. Iodoethane, distilled after being dried with sodium, was reported to be very pure. The advantage of calcium for drying was stressed.

Gand concluded [1695] that, after iodoethane had been purified by distilling through copper turnings, photolysis can be prevented by suspending metallic calcium in the liquid. No photolysis was detected even after a period of several months.

SAFETY

Iodoethane causes central nervous system depression. It may affect the kidneys, thyroid, lungs, and liver. A concentration of 150 ppm, 960 mg/m³, for 24 hr was fatal to mice [3681].

257. 1-IODOPROPANE

258. 2-IODOPROPANE

Gross and Saylor [1912] decolorized 1-iodopropane by shaking with sodium thiosulfate solution. After being washed twice with distilled water, it was allowed to stand for ten days over aluminum oxide and then fractionally distilled. A fraction whose distillation range was 102.28–102.58° was collected for *solubility measurements*. See also Adams and Voorhees [20] and Cowley and Partington [1119] under iodoethane.

For *kinetic studies*, Hughes and Shapiro [2306] treated 2-iodopropane with mercury to remove iodine, dried over phosphorus pentoxide, and fractionated twice; bp 89.5° at 760 torr. No color was observed on storing in a brown bottle in the dark.

See Gand's studies on purification of iodoalkanes under iodoethane [1693, 1694, 1695, 1696].

259. DIIODOMETHANE

Adams and Marvel [1778] describe a method for converting iodoform to diiodomethane with alkaline arsenic (III) oxide.

NITROGEN COMPOUNDS

Nitro Compounds

Specific Solvents

260. NITROMETHANE

261. NITROETHANE

262. 1-NITROPROPANE

263. 2-NITROPROPANE

The four nitroalkanes (nitroparaffins) nitromethane, nitroethane, 1-nitropropane, and 2-nitropropane are available commercially. They are manufactured by the vapor phase nitration of propane, purification of the reaction product, and separation of the nitrocompounds. The principal impurities in the nitroalkanes are isomers or homologues. Water is present to 0.1% or less, usually less. The nitroalkanes may be analyzed by mass spectrometry or gas chromatography. The latter is more rapid and offers a wide range of sensitivities. Allen [56] found that a thermistor or hot wire detector is sensitive to 0.005–0.05% impurity. All known possible impurities are detectable in the 0.005–0.01% range except the carboxylic acids. He has summarized the impurity analyses of the nitroalkanes on a water-free basis as follows.

260. Nitromethane: Nitroethane, 4–5%; 2-nitropropane and acrylonitrile, less than 0.1%; acetonitrile and propionitrile, traces.

261. Nitroethane: 2-Nitropropane, 3–4%; nitromethane, about 1%; 1-nitropropane and 1- and 2-propanol, traces.

262. 1-Nitropropane: 2-Nitropropane, 3–4%; nitroethane, trace to 0.5%; 2-methyl-2-nitropropane, 1- and 2-nitrobutane, 1- and 2-propanol, traces.

263. 2-Nitropropane: 1-Nitropropane, 4–5%; nitroethane, 0.5–1.0%; acetone, trace to 0.1%; 2-methyl-2-nitropropane, trace to 0.5%.

"Trace" is used to express an amount near the limit of detectability.

These solvents are quite reactive. The primary and secondary nitroalkanes exist in equilibrium with their tautomeric aci-form, a pseudo acid. They are sparingly soluble in water; the aci-form is more soluble and reacts with alkalies to form a water soluble salt that may be regenerated with a weak acid, particularly carbonic acid. Riddick [3975] has quantitatively titrated the nitroalkanes, except nitromethane, with alkali in nonaqueous systems. The best titration that was obtained for nitromethane was 99%.

MUTUAL SOLUBILITY WITH WATER

The mutual solubility of nitroethane, 1-nitropropane, and 2-nitropropane with water was determined by Hampton and Riddick [1996] from 0°–80° and that of nitromethane and water from 0° to the critical solution temperature.

The nitroalkanes used for this study were prepared by Toops [4810] and were at least 99.9% the principal component. The solubilities were determined

Table 5.16 Mutual Solubilities of Nitroalkanes and Water

$t°$	Nitromethane		Nitroethane		1-Nitropropane		2-Nitropropane	
	NM in aq, %w	aq in NM, %w	NE in aq, %w	aq in NE, %w	1-NP in aq, %w	aq in 1-NP, %w	2-NP in aq, %w	aq in 2-NP, %w
0	8.9	1.0	4.6	0.5	1.4	0.2	1.7	0.2
10	9.6	1.3	4.6	0.6	1.4	0.4	1.7	0.3
20	10.5	1.8	4.6	0.9	1.5	0.6	1.7	0.5
25	11.1	2.1	4.7	1.0	1.5	0.6	1.7	0.5
30	11.6	2.5	4.8	1.2	1.5	0.7	1.7	0.6
40	13.1	3.6	5.0	1.7	1.6	0.9	1.8	0.8
50	14.8	4.7	5.4	2.0	1.8	1.1	1.9	1.1
60	16.9	6.0	5.9	2.4	1.9	1.4	2.1	1.3
70	19.3	7.6	6.6	3.0	2.2	1.7	2.3	1.6
80	22.1	9.4	7.3	3.5	2.4	2.0	2.6	1.9
90	29.	14.	—	—	—	—	—	—
100	40.	25.	—	—	—	—	—	—
105	58.	42.	—	—	—	—	—	—

by a precision cloud point technique. Selected points were checked by: (1) water in the nitroalkane by the Karl Fischer method, and (2) nitroalkanes in water in concentrations less than 10% by the appropriate colorimetric method of Jones and Riddick [2430]. The data were plotted and a smooth curve was drawn through the points. The solubilities, read from the curve are given in Table 5.16.

SYNTHESES

Nitromethane is best prepared from chloroacetic acid and sodium nitrite, see Clarke [991, p. 83]. It may be prepared by the Walden synthesis [5020] from sodium nitrite and dimethyl sulfate. Toops [4811] prepared it from chloroacetic acid and sodium nitrite, distilled in a 25-plate, Penn-State-type column at 100 torr and obtained a 25% yield with a purity of 99.98%m from the freezing curve.

The other nitroalkanes may be prepared by the Victor Meyer [3293] synthesis. An alkyl halide is reacted with silver nitrite; the bromo compound has been found to be the most satisfactory. Average yields for replicate preparations are: nitroethane, 51%; 1-nitropropane, 57%; and 2-nitropropane, 13%. McCombie and coworkers [3060] prepared nitroethane in 46% yield by the Walden synthesis.

Kornblum and coworkers [2714] found that primary and secondary alkyl nitro compounds could be prepared in a better yield, 55–62%, in dimethylformamide than in water as the reaction medium. A bromo- or iodoalkyl or cycloalkyl was reacted with sodium nitrite in dimethylformamide. Chloro compounds react very slowly. Hardie, an associate of Kornblum, found that the reaction took place in dimethyl sulfoxide and that the latter is a better solvent for the sodium salts.

PURIFICATION

The purification of commercial nitroalkanes has been studied by Riddick [3975]. Nitromethane, 1-nitropropane and 2-nitropropane may be prepared in 99.9%m purity by fractional distillation in a 35–50 theoretical plate column at a pressure sufficient to give a pot temperature not greater than 90°; 100 torr is a convenient pressure if the pressure drop in the column is not too great. There is some decomposition of nitroalkanes above 90°. The purity depends on the reflux ratio used during the distillation.

Wright and coworkers [5265] made conductivity measurements on commercial nitromethane that was dried over calcium chloride and the remaining water removed by azeotropic distillation. Volatile impurities were removed by boiling under reflux in a stream of dry air for several hours.

Unni and coworkers [4894] purified 5 liters of commercial nitromethane by distillation through a 3.6 × 300-cm column packed with glass helices. The distillation was made at 4–6 torr and a 10:1 reflux ratio. The first and last 1.5 liters were discarded. The middle two liters were fractionally crystallized. After repeated distillations and crystallizations, the specific conductance was 5. × 10^{-9} ohm^{-1} cm^{-1}.

Coetzee and Cunningham [1018] studied the *conductivity of single ions* in nitromethane purified as by the following modification of Unni and coworkers [4894]. A spectrographic grade material was fractionally distilled four times at reduced pressure in the presence of Drierite. It was fractionally crystallized six times. It was their opinion that fractional crystallization was more effective than fractional distillation as a means of purification.

Mathews [3208] purified nitromethane by drying over phosphorus pentoxide and fractionally distilling. Smyth and McAlpine [4415] purified nitromethane for the determination of the *dipole moment* from a commercial material that was dried over calcium chloride and fractionally distilled. The fraction boiling at 100.8–101.1° was further dried over phosphorus pentoxide and fractionally distilled. The fraction distilling at 100.5° at 748 torr, n_D

1.38195 was used. Smyth and Walls [4430] determined that the *dipole moment* on commercial material was distilled from phosphorus pentoxide twice and fractionally distilled; mp $-28.6°$, bp $101.5°$ at 762 torr, n_D 25° 1.37970, d 25° 1.1314. For the effects of phosphorus pentoxide, see Wright and co-workers [5265].

Thompson and coworkers [4725] purified nitromethane essentially according to Riddick and determined the *physical properties;* 99.9+%m from the freezing curve. McCullough and coworkers [3066] dried the Thompson material by passing the vapors through magnesium perchlorate. The purity was redetermined by the freezing curve and found to be unchanged. Impurities of different volatility were not present in significant amount because the difference in the boiling and condensing temperatures was only 0.006° at 760 torr. The material was used to study *thermodynamic* properties.

Aplin and coworkers [265] purified the nitroalkanes for mass spectral studies by preparative gas chromatography in a Wilkins Aerograph instrument with a polybutylene column operating at 30 psi helium pressure.

Nitromethane for *voltammetric reaction studies* was purified [2833] by passing 1 liter of a "Certified Reagent" grade, boiling range 100–102°, through a 2.5-cm i.d. × 35-cm column of Woelm chromatographic neutral alumina, activity grade 1, protected from air by a calcium chloride tube. Three passes over fresh alumina reduced the water to 3×10^{-3} M. Nitromethane treated in the same manner passed through a fine porosity fritted glass was used for *dielectric constant measurements* [2688].

Nitromethane is a differentiating solvent for the titration of weakly basic substances. It was purified by Clarke and Sandler [990] using a strongly acidic cation exchange resin in the hydrogen form, washing several times, first with anhydrous methanol and then with two-300 ml portions of nitromethane, and discarding the eluate. Fifteen to twenty-five liters of nitromethane can be treated with 75 g of the resin.

Nitroethane was purified by Pearson [3694] for *kinetics of neutralization studies* by drying with magnesium sulfate, refluxing with urea, and drying with phosphorus pentoxide. It was finally distilled through a 10-plate column.

The purification of nitromethane by progressive freezing was studied by Dickinson and Eaborn [1306]. The 4% impurities, other than water, were removed by eight passes.

Wright and coworkers [5265] found phosphorus pentoxide unsuited for drying nitromethane. When it is dried with the pentoxide and distilled, a white residue collects in the condenser. (This is characteristic of nitroalkanes containing an appreciable amount of formaldehyde or when a reaction takes place forming formaldehyde.)

Turnbull and Maron [4859] purified nitromethane, nitroethane, and 2-nitropropane before determining the *ionization constants* of the aci- and nitro-forms. The individual nitroparaffins were carefully purified by fractional distillation. Water and possible oxides of nitrogen were eliminated by treating first with urea and then with anhydrous sodium sulfate, and finally separating from the drying agent by low temperature distillation.

SAFETY

The nitroalkanes act chiefly as moderate irritants when inhaled. The nitropropanes are somewhat more toxic than nitromethane and nitroethane. There is no evidence of sufficient absorption through the skin to produce systemic injuries [3681]. Prolonged exposure to concentrations of vapors greater than the threshold limit values may produce warning symptoms such

Table 5.17   The Threshold Limit Values and Lower Explosive Limits of the Nitroalkanes

	Threshold Limit Values		Lower Explosive Limits
Nitromethane	100 ppm	250 mg/m^3	7.3 %v
Nitroethane	100	310	3.4
1-Nitropropane	25	90	2.2
2-Nitropropane	25	90	2.5

as irritation of the respiratory tract, headache, and nausea, which clear promptly when the concentration is reduced by proper ventilation.

Avoid evaporation of alkaline mixtures containing nitroalkanes because of the formation of nitronate salts, some of which are explosive. Observe the special precautions recommended by the manufacturer [1058] for handling nitromethane.

The *threshold limit values* [4509] and the *lower explosive limits* in air [5313] are given in Table 5.17.

264. NITROBENZENE

Nitrobenzene is made by the direct nitration of benzene. The impurities likely to be present are the nitro derivatives of the cyclic impurities that were present in the benzene. The most common are nitrotoluene and dinitrothiophene. Free acidity may be present in some improperly purified material.

For *electrical measurements*, McAlpine and Smyth [3051] purified a good commercial material crystallizing five times from itself, drying with phosphorus pentoxide and distilling at reduced pressure. The criteria of purity were the freezing point and the refractive index; see also Roberts and Bury [3995], Murray-Rust and coworkers [3462], and Walden [5036].

The aim of a considerable amount of investigation has been the production of nitrobenzene free from nitrotoluene, dinitrobenzene, and dinitrothiophene. Cohen and Boekhorst [1027] made an exhaustive study of purification. A "pure" commercial product was repeatedly fractionally crystallized, then fractionally distilled at 2 torr several times, dried with phosphorus pentoxide, and finally fractionally crystallized in a dry atmosphere. The melting point in an atmosphere of dry hydrogen was $5.76 \pm 0.01°$. The material was kept in the dark in a sealed vessel for five months. Its melting point changed less than $0.01°$.

Masson [3201] fractionally crystallized thiophene-free benzene several times, nitrated, and purified. The melting point was the same as that reported by Cohen and Boekhorst. No allotropy was observed.

Sidgwick and Ewbank [4313] and Davy and Sidgwick [1224] nitrated thiophene-free benzene with the theoretical amount of nitric acid in the presence of a large excess of sulfuric acid. The temperature was kept below $40°$ to avoid the formation of dinitro compounds. It was distilled at reduced pressure and fractionally crystallized.

Hantzsch [2009] observed that nitrobenzene prepared from pure benzene and purified by repeated fractional distillation and crystallization is colorless. Impure nitrobenzene rapidly becomes colored when standing over phosphorus pentoxide, but a pure product shows hardly any color even after standing in contact with the pentoxide for several months; see also Roberts and Bury [3995] and Sidgwick and Ewbank [4313].

For nitrobenzene as a polarographic solvent, see [4814].

Catalyst poisons were removed from nitrobenzene with mixed sodium and potassium salts of EDTA.

Fractional distillation followed by fractional crystallization and a final distillation was used by Brown and Ives [762] to purify nitrobenzene for *dielectric constant measurements;* bp $210.8°$, d $25°$ 1.19817.

Nitrobenzene was purified by Barreira and Hills [413] to study the *kinetics of ion migration.* It was considered necessary to remove two kinds of impurities, water and dissolved ionic impurities. The main impurities, other than water, found in nitrobenzene are believed to be *o*- and *p*-nitrophenols. These cannot be removed by fractional distillation because of the closeness of the boiling points to the boiling point of nitrobenzene. They state, "White claims to have purified nitrobenzene to 10^{-12} ohm^{-1} cm^{-1} by a succession of distillations and fractional freezings." They purified A. R. grade nitrobenzene by drying over calcium chloride for five days and distilling slowly at 2 torr to avoid local heating and consequent decomposition. The middle half of the distillate was collected and kept in 1 atm of dry nitrogen. It was passed through a column of dried chromatographic alumina 3 cm × 20 cm long. After about 50 ml of the nitrobenzene had passed through the column, a yellow ring was clearly discernible at the top of the column. The final product was a straw color but it darkened on exposure to light. The solvent was stored in the dark to avoid darkening. It had a specific conductance of 2.05×10^{-10} ohm^{-1} cm^{-1}.

For *cryoscopic measurements*, Meisenheimer and Dorner [3261] dried a commercial nitrobenzene with calcium chloride and fractionated at reduced pressure. The nitrobenzene was distilled fresh for each series of measurements. Roberts and Bury [3995] purified a commercial product by fractional distillation and fractional crystallization five times. The middle fraction of the last crystallization was preserved with phosphorus pentoxide. Just before it was to be used, it was distilled into a carefully cleaned freezing apparatus in a vacuum, only the middle fraction being used.

Table 5.18

Distillation range	210–212°
Freezing point, min	5°
Acid (as nitric), max	0.003%
Water, max	0.05%

For *solubility determinations*, Smith and coworkers [4392] purified nitrobenzene by vacuum distillation and drying over calcium chloride. The refractive index was used as the criterion of purity.

Williams and Ogg [5198] fractionally crystallized nitrobenzene repeatedly and then fractionally distilled; bp 210.6–210.8°; see also Davison [1221] under phenols.

For *optical measurements*, Brand and Kranz [711] nitrated carefully purified benzene, steam-distilled the nitrobenzene, dried with calcium chloride, and distilled several times under reduced pressure. Scheibe and coworkers [4137] prepared nitrobenzene from thiophene-free benzene and distilled and recrystallized from ethanol to remove the dinitro compounds. See also Möller [3376] and Hehlgans [2086].

CRITERIA OF PURITY

Rosin [4042] lists the accompanying specifications for a reagent chemical grade. The acidity is determined by shaking 16 ml of nitrobenzene with 50 ml of water for 1 min, separating the water phase and adding 2 drops of bromophenol blue indicator solution. The yellow color of the solution is changed to a bluish violet by not more than 0.5 ml of 0.02 N sodium hydroxide, see Table 5.18.

The freezing point established by Cohen and Boekhorst [1027] has been used as the criterion of purity by many investigators.

SAFETY

Nitrobenzene is a powerful methemoglobin former [3681]. The effects of overexposure are grave, resulting in methemoglobinemia, central nervous system injury, and effects on the spleen and liver. It is readily absorbed through the skin [4509].

The *threshold limit value* is 1 ppm, 5 mg/m³ [4509].

The *minimum ignition temperature* in air is 482° [3680].

Nitriles

General

Nitriles are relatively inert to many of the common reagents. They can be hydrolyzed under certain conditions; the ease of hydrolysis depends upon the nature of the R-group attached to the CN. Nitriles should not be used with strong alkalies or boiled with dilute acids. Strong dilute acids, and weaker acids, such as carboxylic acid, slowly hydrolyze nitriles. This can be demonstrated by periodically analyzing a 0.1 N solution of perchloric acid in acetonitrile.

A general method for the preparation of the lower molecular weight aliphatic nitriles was given by Pyryalova and Zil'berman [3867]. Adiponitrile $NC(CH_2)_4CN$, was heated with the acid corresponding to the nitrile desired in the presence of a catalyst; for example, 54 g of adiponitrile and 64 g of valeric acid was refluxed for 3 hr with 1.7 g p-toluenesulfonic acid and then distilled during a 4-hr period through a 35-cm packed column. The distillate was washed with sodium carbonate solution, dried with phosphorus pentoxide and redistilled at 93° at 20 torr; yield, 91% valeronitrile.

The refluxing of amides with sulfamic acid at 180–220° is a generally applicable method for the preparation of nitriles reported by Kirsanov and Zolotov [2615].

Friedman and Stechter [1661] found that primary and secondary chloro compounds react with sodium cyanide in dimethyl sulfoxide rapidly and efficiently. Advantages are also realized for the reaction of the corresponding bromo compounds to nitriles. Rearrangement does not occur. Typical procedures illustrating the various experimental methods are given.

For the preparation of *carbonyl-free*, nonoxygenated, extractive solvents, see Hornstein and Crowe [2252], Hydrocarbons, General.

Aliphatic Nitriles

Specific Solvents

265. ACETONITRILE

Acetonitrile was first produced commercially in 1952. The unusual solvent potential and high selectivity make it outstanding for liquid-liquid extraction, selective recrystallization, and extractive distillation. Its high solubility parameter and low hydrogen bonding ability make it an excellent polymer solvent. Acetonitrile is the preferred solvent for many reactions [4926].

The favorable dielectric, solvent, and optical properties of acetonitrile have permitted it to be used widely in spectrophotometric and electrochemical experiments and in peptide chemistry. The impurities generally present in the commercial grade material are water, unsaturated nitriles, toluene, and some aldehydes and amines. If the nitrile were subjected to hydrolytic conditions, it would contain acetic acid and ammonia. O'Donnell and coworkers [3574] found that the Reagent grade material has a higher chemical purity than the commercial grade but may actually have poorer optical properties and is not suitable for electrochemical use. The spectrographic grade does not show optimum optical or electrochemical properties.

Coetzee [1017] prepared a report for the Analytical Chemistry Division, Commission on Electroanalytical Chemistry, International Union of Pure and Applied Chemistry. This report included the effect of impurities, purification, handling and storage of purified material, and the detection and determination of impurities and purification of other nitriles.

Acetonitrile, a relatively inert material, is readily available in several purity grades that are suitable for much of its solvent use. It has a favorable boiling point for general use and is relatively easy to purify to a higher state of purity. It is a weaker base and a much weaker acid than water. Coetzee [1017] points out that many possible impurities are sufficiently reactive to modify its properties significantly, even when present in extremely low concentrations.

The most widely used methods of purification prior to the late fifties were to treat the commercial material with a saturated aqueous solution of potassium hydroxide or with anhydrous potassium or sodium carbonate and subsequent distillation from phosphorus pentoxide. Water removal by azeotropic distillation slightly improved the purity. The final purification was by fractional distillation.

Coetzee [1017] states, "There can be little doubt that the pretreatment with potassium hydroxide does more harm than good, since undesirable base hydrolysis may occur . . . . Repeated distillation of acetonitrile from phosphorus pentoxide has the disadvantage that it usually causes extensive polymerization of the solvent."

O'Donnell and coworkers [3574] in their comments on the sodium carbonate-phosphorus pentoxide treatment state, "MeCN prepared by us with this method . . . contains less than 0.05% water, and is a satisfactory voltammetric solvent. GLC analysis does reveal, however, a substantial amount (ca. 0.5%) of an impurity which apparently forms an azeotrope with MeCN and which is not removed until the greater part of the charge has been distilled over."

Walden and Birr [5040] purified a commercial product for *conductivity measurements* by distilling from phosphorus pentoxide in an all-glass apparatus protected from air by a phosphorus pentoxide tube. The same distillation procedure was repeated until the phosphorus pentoxide in the still pot no longer became colored. The acetonitrile was then distilled over anhydrous potassium carbonate to remove traces of the phosphorus pentoxide and finally distilled without a drying agent.

Wawzonek and Runner [5096] shook a good grade of acetonitrile with a saturated solution of potassium hydroxide. It was decanted and dried over anhydrous sodium carbonate. It was then fractionally distilled from phosphorus pentoxide in a nitrogen atmosphere through a 3-ft, Penn-State-type column. The first 150–200 ml of distillate from a 2-liter batch was discarded. The nitrile was used as a *polarographic solvent* in the study of alkali and alkaline earth and heavy metal ions. Kolthoff and Coetzee [2692] used essentially the same method of purification for further *polarographic studies* of metal ions. The water content was always less than 2 millimolar. Coetzee and Kolthoff [1020] used the same purification method for further polarographic studies: *amperometric titration* of amines with perchloric acid.

Putnam and coworkers [3866] purified a commercial product for *thermodynamic studies* by treatment with anhydrous potassium carbonate, followed by several distillations from phosphorus pentoxide. It was then fractionally distilled in a 90-cm glass helices packed column. The center cut was redistilled twice and stored over anhydrous calcium sulfate. Apparently the purified acetonitrile attacked slightly the inside of the copper sample container of the calorimeter.

The *ionization processes* of hydrogen halides were studied by Janz and Danyluk [2376] in acetonitrile that had been purified to a water content of less than 0.01 % and free from such contaminants as acetamide, acetic acid, and methyl isonitrile. A Reagent grade material was allowed to stand over anhydrous calcium chloride for several days. It was distilled from a Hyper-Cal distillation unit, and the middle fraction stored over phosphorus pentoxide. A bright orange coloration was noted. This had been reported previously by French and Roe [1633]. Coetzee [1017] stated that the orange-colored material is a polymer. The nitrile was decanted and redistilled. It was necessary to repeat the phosphorus pentoxide treatment-distillation process at least six times before coloration no longer developed. The nitrile was distilled into a vessel containing fused potassium carbonate and then distilled from the fused salt. It was then fractionally distilled; $\kappa = 0.6$–2.5×10^{-9}.

Lewis and Smyth [2934] dried acetonitrile over calcium chloride, refluxed over phosphorus pentoxide repeatedly until no color appeared on the oxide, distilled onto freshly fused potassium carbonate, distilled, and finally fractionally distilled. The middle portion was used for *dipole moment measurements*.

Pratt [3825] purified acetonitrile by azeotropic distillation. The azeotrope can be dried by using a third component such as benzene or trichloroethylene. At 776.5 torr, the ternary azeotrope, benzene-acetonitrile-water, boils at about 66°; trichloroethylene-acetonitrile-water boils at 67.5°.

Levi and coworkers [2925] found that Amberlite IRC-50 would remove the interfering basic substances from acetonitrile when it was used as a solvent in *nonaqueous titrimetry*.

Kolthoff and coworkers [2691] purified acetonitrile, which contained traces of water, acetamide, ammonium acetate, and ammonia, by treating with activated alumina, then distilling through a 1-meter column packed with stainless steel wool. The picric acid test indicated that the ammonia had been reduced to 10^{-6} molar. The specific conductance was $0.7–1.5 \times 10^{-7}$ ohm^{-1} cm^{-1}.

Coetzee and coworkers [1019] discussed the purification of acetonitrile as a solvent for exact measurements. A number of methods of purification are given.

Coetzee and McGuire [1021] purified acetonitrile by preliminary drying with silica gel, then with calcium hydride, followed by distillation from phosphorus pentoxide and again from calcium hydride; see [1019].

Muney and Coetzee [3455] studied the *properties of bases by electrical measurements* in acetonitrile that was purified as follows: It was shaken successively with two lots of silica gel for 30–45 min each to remove most of the water, then successively with two lots of Fisher 80–200 mesh alumina to remove acetic acid. It was finally shaken with phosphorus pentoxide to remove most of the remaining water and then fractionally distilled from 5–10 g of phosphorus pentoxide per liter at a high reflux ratio through a 30-in. column packed with glass helices. It was noted that large amounts of phosphorus pentoxide caused polymerization of the nitrile during distillation. Usually the solvent was fractionally distilled twice with no appreciable improvement from the second distillation. The middle portion of the distillate was stored in a black painted container from which it was siphoned as needed. The liquid was protected from atmospheric contamination. The main reason for the extreme weakness of ionization in acetonitrile as a solvent is discussed; see also [1023].

Coetzee and Padmanabhan [1023] purified acetonitrile by three methods to study the *properties of bases* therein and to determine the *protolysis constant*.

Method A. The acetonitrile was dried with silica gel and then phosphorus pentoxide, and fractionally distilled twice from phosphorus pentoxide.

Method B. The nitrile was dried with silica gel and then with calcium hydride. It was fractionally distilled from phosphorus pentoxide and then from calcium hydride.

Method C. This method removed traces of unsaturated nitriles from the solvent. It was refluxed with 1 ml of 1% aqueous potassium hydroxide per liter, fractionally distilled, dried with calcium hydride, then fractionally distilled from phosphorus pentoxide and finally from calcium hydride.

The water content of acetonitrile prepared by Method A was generally between 1 and 2 millimoles; from Methods B and C it was below 1 millimole.

O'Donnell and coworkers [3574] describe a relatively rapid procedure that produces an acetonitrile that is more generally satisfactory for *electrochemical experimentation*, as a *spectrometric solvent* and in *peptide chemistry*. Reflux 800 ml of acetonitrile containing less than 0.2% water with benzoyl chloride for 1 hr. Distill into a receiver containing 10 ml of water at a rate of 5–10 ml per min. Add 20 g of sodium carbonate, reflux 2 hr, and distill into a receiver with a drying tube. Add 10 g of anhydrous sodium carbonate and 15 g of potassium permanganate and distill at 5–10 ml per min into a receiver protected from moisture. Make the distillate slightly acidic with concentrated sulfuric acid. Decant from the crystals and distill from a 30-plate column at a reflux ratio of 20:1 at a rate of 10 ml per hr, discarding the first 40 ml of the distillate.

The purified solvent has a transmittancy of at least 85% at 200 mμ and a UV cutoff at 189 mμ. Only one impurity was discernible in the gas chromatogram using a flame ionization detector. Material prepared using only the latter part of the purification procedure starting with the permanganate treatment is recommended for *voltammetric* and *coulometric* purposes.

Coetzee [1017] recommends two methods of purification of acetonitrile based on extensive investigations with purifications and use. The method used depends on the use of the nitrile. For the majority of uses, Procedure I is recommended. When the presence of traces of unsaturated nitriles would be objectionable, Procedure II is recommended. The procedures are given in the proper sequence of operations.*

"*Procedure I*. (1) If a considerable amount of water is present (more than approx. 0.1 M), preliminary drying by shaking with silica gel or molecular sieves (e.g. Linde Type 3A) is advisable. This step can often be omitted.

(2) Shake or stir (a magnetic stirrer is particularly effective) with sufficient calcium hydride to remove most of the remaining water, until hydrogen evolution ceases (relieve pressure). Decant.

(3) Distill in an all-glass apparatus under a high reflux ratio from phosphorus pentoxide (not more than 5 g per liter, otherwise gel formation may be excessive). Protect the distillate from atmospheric moisture by means of a

---

* Reprinted by permission of the International Union of Pure and Applied Chemistry and Butterworths Scientific Publications.

drying tube packed with anhydrous magnesium perchlorate. If extensive gel formation occurs, take care not to overheat the distillation flask. Discard the first 5 percent of the distillate, collect the next 80 to 85 percent (if possible).

(4) Reflux the distillate over calcium hydride (5 g per liter) for at least an hour, and then distill slowly under maximum reflux ratio (b.p. 81.6°C). Again collect the 5–90 percent fraction. (*Note:* Careful fractionation is required to reduce the acrylonitrile content.)

"*Procedure II.* (1) Reflux the acetonitrile with a small amount of potassium hydroxide (1 ml of 1 percent solution per liter) for at least an hour, longer if it proves necessary.

(2) Distill fractionally. Collect the 10–90 percent fraction.

(3) Follow Procedure I, steps 2 through 4.

"*Alternate Procedure II.* 1. Reflux over sodium hydride (1 g per liter) for 10 min.

(2) Distill rapidly.

(3) Reflux over potassium bisulfite (2 g per liter) for 20 min. in order to remove amines produced in step 1.

(4) Distill rapidly.

(5) Follow Procedure I, Step 4."

Sherman and Olson [4276] report that acetonitrile is a difficult solvent to purify for *electrochemical applications*. The most bothersome impurities are unsaturated nitriles, particularly acrylonitrile, which requires drastic treatment for removal. They found that the method of Coetzee and coworkers [1019] is long and involved and may introduce ammonia into the nitrile. The ammonia can be removed by the second step of Forcier and Olver [1595]. They have developed a method that is relatively short and simple for purifying large batches to approximately the same unsaturated concentration, 10^{-3} M, as the longer and more involved method of Coetzee [1017]. A weak acid remains at about 3×10^{-5} M, which will have to be considered for certain uses.

They charged 3.8 liters of a purified grade of acetonitrile (such as Mallinckrodt's Nanograde or Matheson, Coleman, and Bell's Pesticide quality) into a 5-liter round-bottom flask fitted with a calcium chloride tube. Six milliliters of nitrogen tetroxide was added. The solution was heated for about 2 hr at 50°. Thirty-eight grams of calcium hydride (equal parts of Metal Hydride's −40 and −4 + 40 mesh) was added and the mixture was purged with nitrogen at the boiling point to remove the nitrogen tetroxide. The nitrile was rapidly distilled through a short column in an atmosphere of nitrogen and a 250-ml forecut and a 200-ml pot residue were discarded. The distillate was refluxed with calcium hydride and the acetonitrile fractionally distilled from calcium hydride through a Nester/Faust (Model NF-136) 36-in. spinning band column. The 1-in. delivery tube for the distillation apparatus was packed for 20 in. with F-20 chromatographic alumina which was freshly activated at 225° for 12 hr. The first 400 ml distilled at a reflux ratio of 15:1 was discarded. The remainder was distilled at a reflux ratio of 5:1, leaving a 200-ml pot residue.

Addison and coworkers [12] purified acetonitrile as a solvent for the *determination of infrared and Raman spectra of metal salts* by shaking with 5 g per liter of phosphorus pentoxide, followed by distillation from calcium hydride. The distilled solvent was free from fluorescent impurities.

Table 5.19   Specifications for Reagent Grade Acetonitrile

Specification	ACS^a	Rosin
Distillation range	1-95 ml 1°	81–82°
95 ml to dryness	95 ml to dryness	1°
Boiling point	81.6°	—
Density, 25°	0.775–0.780	—
Residue on evaporation, max	0.005%	0.01%
Neutrality to red and blue litmus	—	neutral
Acidity as CH$_3$COOH, max	0.05%	—
Alkalinity as NH$_3$, max	0.001%	—
Water, max	0.3%	0.2%
Permanganate time, max	—	30 min
Hydrogen cyanide, max	—	0.005%
Miscibility with water	—	to pass test^b
Color, APHA	10	—
Appearance and odor	to pass test	—

^{a} From *Fisher Chemical Index*, 67-C.
^{b} 5 ml acetonitrile and 15 ml of water; remains clear for 30 min.

The impurities in acetonitrile were reduced from 0.43 to 0.03% by ten zone melting passes in a closed tube [250].

Acetonitrile is a strongly differentiating solvent for acids and bases [3455, 2692, 1020]. It is an excellent solvent for many inorganic substances [2376].

CRITERIA OF PURITY

Lewis and Smyth [2934] used the physical properties such as the boiling point, density, and refractive index as the criteria of purity; Timmermans and Hennaut-Roland distilled until the density was constant [4778].

Dreisbach and Martin [1372] determined the purity from the freezing curve.

Rosin [4042] and ACS Reagent Chemicals give specifications and test methods for Reagent grade acetonitrile. The specifications are listed in Table 5.19.

O'Donnell and coworkers used the transmittancy at 200 mμ and the UV cutoff and the gas chromatographic examination with a flame ionization detector for spectrographic solvent use [3574]. An electrolytic test was used for voltammetric and coulometric purposes. The specific conductance is a criterion for all acetonitrile used for conductometric investigations.

Coetzee [1017] gives the methods for the detection and determination of important impurities in acetonitrile. These are water, acetic acid, ammonia, and unsaturated nitriles.

Nemes and Orville-Thomas [3514] determined the purity of acetonitrile by examining the infrared spectrum for impurities.

SAFETY

Animal studies indicate that acetonitrile is only moderately toxic, necessitating the use of ordinary precautionary measures. The odor is easily detectable at 40 ppm [48]. Human volunteers were studied at levels of 40, 80, and 160 ppm of acetonitrile vapors for periods of 4 hr. No specific subjective responses were noted. Several accidents have been reported from extended exposure to undetermined concentrations. Marked delay of the onset of symptoms appears to be characteristic [3681].

The *threshold limit value* is 40 ppm, 70 mg/m^3 [4509].

266. PROPIONITRILE

Teter and Merwin [4626] found that acetonitrile and propionitrile, when prepared by high temperature ammoniation of olefins, contained hydrocarbon impurities that were difficult to remove because of the formation of azeotropes with the nitriles. The hydrocarbons may be removed from propionitrile by adding acetonitrile, which forms lower boiling azeotropes with the hydrocarbons than does propionitrile. After the removal of the excess acetonitrile, the higher nitrile may be obtained in substantially pure condition.

Jeffrey and Vogel [2387] prepared propionitrile for *physical property measurements* by an elaboration of the Walden method [5020]. The upper phase was distilled from the reaction mixture, treated with saturated calcium chloride solution, and cooled in ice to remove cyanide impurities. The isocyanides were destroyed by shaking twice with 5-ml portions of concentrated hydrochloric acid for 5 min. It was shaken with saturated potassium carbonate solution and then with calcium chloride solution, dried with anhydrous magnesium sulfate, and fractionally distilled three times; bp 97.5° at 765 torr.

Gresham [1890] prepared propionitrile in a 91% yield by dehydrating propionamide at 320–450° over a catalyst of iron, nickel, or manganese chromite.

Propionitrile was purified by Duncan and Janz [1395] for *molecular structure studies* by fractional distillation of a commercial product after drying over phosphorus pentoxide; n_D 25° 1.3633, d 25° 0.755.

CRITERIA OF PURITY

Dreisbach and Martin [1372] determined the purity from the freezing curve.

SAFETY

Propionitrile is quite toxic. It is believed that the toxicity is produced through the metabolism of the cyanide ion. The LD_{50} oral dose for rats is reported to be about 39 mg/kg; 500 ppm in air was fatal to one-third of the rats in 4 hr [3687, see also 3681].

267. SUCCINONITRILE

Succinonitrile was purified by Wulff and Westrum [5272] for *thermo-dynamic property studies* by two vacuum sublimations. The purity was determined by fractional fusion.

Lewis and Smyth [2934] determined the *dipole moment* of succinonitrile that was twice recrystallized from acetone and distilled four times at about 1 torr; bp about 108°.

Succinonitrile may be heated for 72 hr at 200° without decomposition [4519].

SAFETY

The acute toxicity appears to be somewhat lower than that of propio- or butyronitriles. Adequate precautions should be taken against skin and eye contact. The inhalation hazard is uncertain [3681].

268. BUTYRONITRILE

Jeffery and Vogel [2387] prepared butyronitrile for *physical property measurements* from sodium cyanide and 1-bromopropane. The product was purified in a manner similar to propionitrile.

Teter and Merwin [4626] used the method of purification described for propionitrile; see also acetonitrile.

SAFETY

Eastman Chemical Products [1431] states that butyronitrile is an extremely toxic compound producing symptoms similar to these caused by other nitriles such as isobutyronitrile. Results of screening studies in their Laboratory of Industrial Medicine show this compound to be extremely hazardous on oral ingestion, skin contact, or inhalation.

The *lower explosive limit* in air is 1.65%v. The *auto-ignition temperature* is 502° [1431].

269. ISOBUTYRONITRILE

SAFETY

Eastman Chemical Products [1430] states that isobutyronitrile is capable of being absorbed by the intact skin and in full strength can cause systemic toxicity and even death. Fassett [3681] cautions about the inhalation of the vapors.

270. VALERONITRILE

Jeffery and Vogel [2387] prepared valeronitrile for *physical property measurements* by a modification of the method given by Adams and Marvel [19]. The crude nitrile was washed twice with half its volume of concentrated hydrochloric acid and then with saturated sodium bicarbonate solution, dried with anhydrous magnesium sulfate, and distilled; bp 141° at 764 torr.

Valeronitrile was prepared in 54–75% yield by Ferguson [1563] by heating 51 g of 96% sodium cyanide, 2 g of potassium iodide, and 65 ml of water; then adding 130 ml of 2-ethoxyethanol and 118 g of 1-bromobutane and refluxing vigorously for 2 hr. Then 200 ml of water was added and distilled until a single phase was reached. The upper phase was washed with 40% calcium chloride solution, water, cold 50% sulfuric acid, and saturated sodium bicarbonate solution; then dried with anhydrous calcium chloride and distilled. The yield dropped to 30–40% when 1-chlorobutane was used.

Banewicz and coworkers [380] purified valeronitrile that was prepared by the method of Ferguson [1563] by drying with anhydrous calcium chloride and distilling. The distillate stood over phosphorus pentoxide and was fractionally distilled. The fraction distilling at 140–141° was used as a solvent for *specific conductance studies* of tetraethylammonium perchlorate. Some *physical properties* were also determined.

CRITERIA OF PURITY

Dreisbach and Martin [1372] determined the purity from the freezing curve.

271. HEXANENITRILE

Jeffery and Vogel [2387] prepared hexanenitrile for *physical property measurements* from sodium cyanide and 1-iodopentane. The material was purified in the manner described for valeronitrile.

Vvedenskiĭ and coworkers [5005] prepared hexanenitrile and octanenitrile in 97–100% yield by treatment of the respective acids with ammonia at 350–370° over activated alumina.

CRITERIA OF PURITY

Dreisbach and Martin [1372] determined the purity from the freezing curve.

272. 4-METHYLVALERONITRILE

Jeffery and Vogel [2387] prepared 4-methylvaleronitrile *for physical property measurements* by refluxing 30 g of sodium cyanide, 40 ml of water, and 77 g of 1-bromo-3-methylbutane in 150 ml of methanol for 27 hr. It is assumed that the material was purified in the same manner as described for propionitrile; bp 153.5° at 756 torr.

273. OCTANENITRILE

Vvedenskiĭ and coworkers prepared octanenitrile; see hexanenitrile [5005].

274. α-TOLUNITRILE

This aliphatic nitrile may be purchased at a reasonable cost, or it may be easily prepared [1060, p. 9]. Further purification is possible by fractional distillation.

Gomberg and Buchler [1830] prepared the nitrile from benzyl chloride and aqueous sodium cyanide in a 50–60% yield as compared to 80–90% yield when the compound was prepared according to Conant [1060].

Jeffery and Vogel [2387] prepared α-tolunitrile by a method similar to the one used by Conant [1060] for *physical property measurements;* bp 108.5° at 15 torr.

Benzyl isocyanide, which gives the product a disagreeable odor, may be removed after the first distillation by shaking vigorously with an equal volume of 50%w sulfuric acid at 60°, washing with saturated sodium bicarbonate solution, and then with half-saturated sodium chloride solution. The material is then dried and fractionally distilled under reduced pressure. The product is colorless and will remain so for some time [2408, p. 89].

SAFETY

The LD_{50} for rats was found to be less than 350 mg/kg body weight [1309].

Aromatic Nitriles

275. BENZONITRILE

Benzonitrile is a strongly polar aromatic solvent that dissolves a wide variety of organic substances. It also dissolves many anhydrous inorganic salts and organometallic compounds. Its low vapor pressure, low mutual solubility with water, and wide solvent ability makes it a uniquely versatile solvent.

Dangyan [1185] prepared benzonitrile in 85% yield by refluxing dicyandiamide and benzoic acid in a mole ratio of 1:3 for 15 hr.

Jeffery and Vogel [2387] prepared benzonitrile for *physical property measurements* by heating 45 g of benzamide and 75 g of phosphorus pentoxide. The product was washed with potassium carbonate solution, dried, and distilled; bp 189° at 758 torr.

Everard and coworkers [1522] found that fractional distillation of two commercial samples of benzonitrile gave unsatisfactory material for *dipole moment determinations.* They prepared the nitrile from benzamide (mp

127–128°) by heating with phosphorus pentoxide for 15 min and distilling. Before use, four-fifths was slowly frozen and the rest discarded.

Martin [3166] purified commercial material by steam distillation to remove small quantities of carbylamine, washed the distillate with sodium carbonate solution to remove benzoic acid, dissolved in ether, washed with water, and dried with calcium chloride. The solution was filtered and the ether removed by distillation. The residue was fractionally distilled under reduced pressure. The middle fraction that distilled at 81° and 22 torr was shaken with a mixture of potassium carbonate and calcium chloride, and then with calcium chloride, and allowed to stand several days over fresh calcium chloride. After each of the drying operations, the material was distilled. It was used for *electrical measurements*.

Benzonitrile was fractionally distilled three times through a 2-ft column by Brown and Ives [762] for *dielectric constant measurements;* bp 191.0°, d 25° 1.00061.

The *solvent effects* on the *polarographic reduction* of metal ions in benzo-nitrile-acetonitrile system was studied by Larson and Iwamoto [2834]. A practical grade benzonitrile was dried over calcium sulfate for several days. It was transferred to a distilling flask and distilled from fresh calcium sulfate. The distillate was repeatedly redistilled from phosphorus pentoxide in an all-glass apparatus, the center portions only being collected, until almost no black residue remained in the distilling flask; bp 188 ± 0.5° at 735 torr. The distillate was kept in the receiving flask under a nitrogen atmosphere.

Walden and Werner [5052] fractionated a commercial product in vacuum several times and dried for several days with phosphorus pentoxide between distillations. Müller and coworkers [3448] dried with lime and fractionated several times. Sudgen [4540] fractionated repeatedly at atmospheric pressure in an all-glass apparatus.

Coetzee and McGuire [1021] purified benzonitrile to study the *relative basicity of solvents* by first shaking with silica gel, then stirring with calcium hydride for 2 hr, carefully decanting and vacuum-distilling from a small amount of phosphorus pentoxide.

CRITERIA OF PURITY

Velsicol [4931] uses the freezing point and calculates the mole per cent impurity by the method of Witschonke [5227]. The assay method is: Add about 25 ml of the benzonitrile sample to a large test tube. Place the test tube in a dry ice bath and agitate continuously with a thermometer (range −36 to +54°, graduated in 0.2°). Take point just before supercooling begins, where the temperature remains constant the longest, as the freezing point. Calibrate the thermometer by determining the freezing point of the water.

The assay is based on the calculated freezing point of 100%m benzonitrile of −12.75° and a freezing point constant, $K_f' = 1.93$ %m̄/1° [5227]. This

Table 5.20

Purity, %m	fp
100.0	−12.75°
99.6	−13.0
97.9	−14.0
96.1	−15

method is discussed in Chapter II; see eq. 2.77 and 2.78 and Table 5.16. Plot the data in Table 5.20 and read the purity from the chart.

SAFETY

The Velsicol Company in its *Technical Data Bulletin* [4931] states that the physiological properties of benzonitrile have not been completely established. It may be toxic to some persons. The symptoms of exposure are headache, nausea, weakness, increased heart beat and dizziness. There have been no indications of chronic toxic effects in handling and use where the normal precautions have been observed.

Unsaturated Nitriles

276. ACRYLONITRILE

Acrylonitrile was first prepared by Moureu [3427] in 1893 from acrylamide and twice its weight of phosphorus pentoxide at 150–250° in 30% yield. Cyanamid reports that the preparation is also published in [3425a].

Present industrial production starts with either ethylene cyanohydrin or acetylene [79]. Ethylene cyanohydrin decomposes spontaneously and rapidly in the presence of a suitable catalyst and operating conditions. The wet nitrile may be dried by binary azeotropic distillation of the two components or by the use of a ternary azeotrope using dichloromethane or chloroform. Acrylonitrile is formed when acetylene and hydrogen cyanide are passed over a catalyst at 400–500°. The reaction also takes place rapidly in an aqueous solution of ammonium chloride and copper (I) chloride.

Cyanamid [79] states that highly purified acrylonitrile may polymerize spontaneously, particularly in the absence of oxygen or on exposure to visible light. Traces of ammonia, ammonium carbonate, or one of the products resulting from the reaction of ammonia and acrylonitrile will stabilize it. Other materials that appear in the literature as stabilizing agents are: alkylamines, hydrazines, mercaptans and sulfides, anthracene, aromatic nitro compounds, benzoquinone, chlorine, bromine, iodine, ethylene cyanohydrin, metals (such as copper, zinc, lead, mercury), pyridine, and unsaturated amines. It was found to be stabilized effectively by 0.01% of methoxyhydroquinone, 2-aminoethanol, diethanolamine, triethanolamine and monoethanoldiethylamine.

Acrylonitrile polymerizes violently in the presence of concentrated alkali. Caution is urged if it is necessary to treat acrylonitrile with strongly alkaline materials.

It is miscible with most organic liquids, including many from the following functional group classes: aliphatic and aromatic hydrocarbons, alcohols, ethers, and esters.

Butskus [829] prepared acrylonitrile in 60% yield by treating 22 g of β-bromopropionitrile with 9.4 g of potassium hydroxide in 30 ml of water.

Baum and Hermann [453] prepared acrylonitrile by reacting acetylene and hydrogen cyanide in the vapor phase in the presence of a catalyst. Heuser [2162] reacted acetylene and hydrogen cyanide in an aqueous solution of ammonium chloride and copper (I) chloride.

Sevryugova and coworkers [4241] purified crude acrylonitrile by a three-stage process. The low boiling components, such as hydrogen cyanide, were removed by fractional distillation in a 50-plate column at 150 torr. The acrylonitrile was distilled in a 100-plate column at 50–100 torr. Four volumes of water and 0.001% hydroquinone were added to the distilled acrylonitrile; the azeotrope, which contained water, acrylonitrile, acrolein, and hydrogen cyanide was fractionally distilled from a 50-plate column at 50–100 torr. The acrylonitrile residue contained 5×10^{-4}% hydrogen cyanide.

Hurdis and Smyth [2330] purified 200 g of commercial acrylonitrile for *structural studies* by *dipole moment* by drying over calcium chloride and fractionally distilling; bp 77.3° at 750 torr.

Thompson and Torkington [4737] used a purified sample of acrylonitrile for *infrared studies* that was furnished by the manufacturer and was stabilized by copper powder. It was distilled before use.

CRITERIA OF PURITY

Union Carbide's [4882] test method for acrylonitrile is mass spectrometry using the 41 mass-to-charge ratio as a measure of the impurity.

Cyanamid [79] describes an assay method based on the addition of 1-dodecanethiol to the double bond, followed by the titration of the excess mercaptan with bromate-iodide solution.

SAFETY

Acrylonitrile is toxic by inhalation, oral ingestion, and skin penetration. The toxicity of acrylonitrile results from the release of the cyanide radical [4882]. There is no evidence to indicate that repeated exposures produce a cumulative effect [79]. Acrylonitrile has a characteristic unpleasant odor resembling that of pyridine. Fatigue develops rapidly, and the odor cannot be used as a reliable exposure index.

The *threshold limit value* is 20 ppm, 45 mg/m³ [4509].

The *explosive limits* in air are 3.05 and 17.0%v. The *minimum ignition temperature* in air is 481° [3681].

277. METHACRYLONITRILE

Peters and coworkers [3727] describe a process for the manufacture of methacrylonitrile by passing methallylamine, air and steam over a silver oxide catalyst at 450–600°. The reaction product was slightly acidified, and the nitrile and other volatile products removed. The methacrylonitrile distilled in this manner was about 96%; it contained about 2% of other nitriles. Further purification may be accomplished by fractional distillation. The acidification of the reaction product is important because, in the absence of acid, by-product ammonia and unreacted amines react rather rapidly with unsaturated nitriles to form high boiling nitrogen bases.

SAFETY

The toxicity of methacrylonitrile is similar to that of acrylonitrile. There is no appreciable or easily recognizable odor, and it is only mildly irritating to the eyes, nose, and skin. Odor and irritation, therefore, cannot be relied upon as warning devices [3681, 4961].

There has been no threshold limit established. Osborn [4691] believes it should be less than 20 ppm, 34.8 mg/m³. Pozzani and coworkers [3822] in a later and more extensive study, including human volunteers, believe the threshold limit value should not be greater than 3 ppm. They found that "it is very likely that most people will not be able to detect the odor of dangerous concentrations of methacrylonitrile vapors after 2 or 3 min."

Amines

General

The primary aliphatic amines, C_1–C_5, are soluble in alcohol, ether, and water. As the molecular weight increases from *n*-hexylamine, the solubility in water decreases. The secondary amines, R_2NH, have a limited solubility in water, the solubility decreasing rapidly above C_3, but are generally soluble in ether and alcohol. Trimethylamine is the only tertiary amine that is completely soluble in water. The lower molecular weight amines form stable hydrates with water, which make some of them difficult to purify. Aromatic amines are only slightly soluble in water [2609].

Amines are bases and form coordinated covalent compounds with acids. The complexes with the strong and moderately strong acids are quite stable.

Most amines, C_6 or less, can be titrated in water with a strong acid, provided that the proper indicator is used or the titration made potentiometrically; the latter gives more accurate results. Amines are strong bases in acetic acid and may be titrated in this medium with a precision of $\pm 0.05\%$ [3155, 3974].

When an alcohol and ammonia are passed over a dehydration catalyst, such as nickel on keiselguhr, at 200°, the corresponding nitrile is formed. The mixture, when passed over a hydrogenation catalyst, such as copper at 180° or nickel at 100°, gives an equilibrium mixture of primary, secondary, and tertiary amines, nitriles, alcohol, and water. The desired amine can be recovered by rectification and the other components recycled. The yield is nearly 100% [4902].

Primary amines can be prepared uncontaminated with secondary or tertiary amines by the method of Brown and Jones [787].

$$2RMgX + CH_3ONH_2 \rightarrow RNH_2MgX + RH + CH_3OMgX$$

$$X = Cl, Br \text{ but not } I$$

The reaction mixture is acidified with hydrochloric acid and the amines separated by adding potassium hydroxide. (Sodium hydroxide is equally effective.) The purification depends on the amine prepared and the purpose for which it will be used.

Delépine [1269] prepared primary amines by converting the halide of the desired alkyl amine to a quaternary salt with hexamethylenetetramine. The salt was hydrolyzed and the amine isolated.

SAFETY

The physiological action of, and response to, aliphatic and alicyclic amines varies only little among the individual amines. The toxicological information available for amines is not great. Acute oral, cutaneous, and eye toxicities of the amines do not show marked change with increasing chain length because these are influenced to a major degree by the acute local irritating properties. Amine vapors are irritating to the eyes, producing tears, inflammation, and a halo around lights. The vapors irritate the mucous membranes of the respiratory tract. Polyamines may cause asthmatic symptoms and perhaps respiratory tract sensitization. The vapors may cause skin irritation and dermatitis. Liquid amines splashed in the eye will cause severe pain and may result in permanent eye damage. The liquid will produce burns on the skin if allowed to remain long enough. Amines may be absorbed through the skin and mucous membranes either as vapor or as liquid [3681].

The *threshold limit values* that have been established or suggested, the *flammable limits*, and the *minimum ignition temperatures* are tabulated in Tables 5.21 and 5.22.

Table 5.21   Threshold Limit Value for Aliphatic Amines

Compound Number	Compound Name	TLV ppm	TLV mg/m³	Reference Number
279	2-Propylamine	5	12	4509
280	Butylamine	5	15	4509
284	Cyclohexylamine	20	81	3681
290	Ethylenediamine	10	30	4509
292	Diethylamine	25	75	4509
294	Diisopropylamine	10	41	3681
298	Triethylamine	25	100	4509

Table 5.22   Flammable Limits and Minimum Ignition Temperatures of Aliphatic Amines in Air [3680]

Compound Number	Compound Name	Flammable Limits, %v Lower	Flammable Limits, %v Upper	Minimum Ignition Temp.
278	Propylamine	2.01	10.35	318°
279	2-Propylamine	—	—	402°
280	Butylamine	1.70	9.75ᵃ	312°
281	Isobutylamine	—	—	378°
289	Allylamine	2.20	22.05	374°
291	Ethylenimine	3.60	45.80	322°
292	Diethylamine	1.77	10.10	312°
298	Triethylamine	1.25	7.90ᵃ	232°

ᵃ Determination made at elevated temperatures.

Primary Aliphatic Amines

Specific Solvents

280. BUTYLAMINE

Butylamine was dried over potassium carbonate for one day and fractionally distilled. The portion distilling at 77.0° was collected by Bayles and Chetwyn [456] to study *acid-base function in nonaqueous systems.*

Cluett [1012] purified a good commercial butylamine by distilling over potassium hydroxide, collecting the middle 80% and storing under dry carbon dioxide-free nitrogen. The material was used to study the *differentiating titration* of phenylureas as acids.

Muney and Coetzee [3455] purified butyl-, diethyl- and triethylamines obtained from supply houses for *conductivity studies* in acetonitrile as follows: The amines were dried over sodium hydroxide, then over sodium, followed by a 2-hr refluxing over sodium, and finally fractional distillation. The criteron of purity was the boiling point.

283. *tert*-BUTYLAMINE

Karabinos and Serijan [2499] reported a convenient method for the preparation of a relatively large quantity of *tert*-butylamine by catalytic hydrogenation of 2,2-dimethyleneimine; see also Horning [2251, p. 148].

Pearson et al. [2251] prepared *tert*-butylamine by the hydrolysis of *tert*-butylurea in 71–78% yields.

284. CYCLOHEXYLAMINE

Auwers [322] prepared and purified cyclohexylamine for *optical measurements* by reducing cyclohexanone oxime with sodium in boiling alcohol. The reaction product was steam-distilled, acidified with hydrochloric acid, evaporated to dryness, made alkaline, and extracted with ether. The ether extract was dried over potassium hydroxide and rectified in a stream of hydrogen; a Linnemann column was used.

Sabatier and Senderens [4086] separated this amine from aniline by distillation at 30 torr. Baker and Schuetz [358] prepared cyclohexylamine from aniline in the same manner as cyclohexane.

Palfray [3631] hydrogenated aniline at 150° and 60 kg pressure in the presence of Raney nickel. Fractional distillation of 199 g of product gave 96 g cyclohexylamine for a 61% yield.

Adkins and Billica [26] reduced cyclohexanone oxime with a specially prepared Raney nickel catalyst; see 2-propanol.

Carswell and Morrill [923] determined the *physical properties* of cyclohexylamine purified by crystallizing the hydrochloride several times from water, liberating the amine with alkali, and fractionally distilling under nitrogen.

Lewis and Smyth [2933] distilled a commercial product; dried it with calcium chloride, and then with sodium in an atmosphere of nitrogen; and redistilled; bp 133.8°, d 25° 0.86253.

Primary Aromatic Amines

Specific Solvents

285. ANILINE

Aniline, the simplest of the aromatic amines, exhibits the chemical properties of both the amine and the aromatic nucleus. It has some amine reactions not common to aliphatic amines. It forms salts readily, and the amine group is easily oxidized, as shown by darkening on exposure to air. It is a weak base in water and a strong base in some solvents such as acetic acid, benzene, and carbon tetrachloride.

Aniline is manufactured by: (1) vapor-phase reduction of nitrobenzene with hydrogen at temperatures up to 350° in the presence of a catalyst, (2) reduction of nitrobenzene with hydrochloric acid and iron filings or powder, and (3) heating under pressure chlorobenzene, aqueous ammonia, and copper (I) oxide. Distillation of the "aniline oil" obtained from the three reactions yields a product of high purity because of its high boiling point and scarcity of side reactions.

Bag and coworkers [347] reported reducing nitrobenzene almost stoichiometrically to aniline in the presence of fragments of sodium hydroxide activated with nickel and aluminum plus other metals as a catalyst.

Palfray [3631] rapidly hydrogenated quantitatively 1 mole of nitrobenzene at 75° and 95-kg pressure (the temperature rose to 115°) in the presence of Raney nickel.

In 1920, Knowles made a critical study [2644] of the more common properties of aniline reported in the literature. He concluded that it was not possible to evaluate the data because of the wide variation in the reported values. He found it a "matter of considerable difficulty" to purify aniline. He concluded that the following method would produce a high purity product. Distill 2 liters of a colorless aniline four times under reduced pressure, rejecting the first and last portions each time. Allow the product to drop slowly into a solution of recrystallized oxalic acid. Filter the oxalate, wash several times with water, and recrystallize three times from 95% ethanol. Regenerate the aniline from the salt with a saturated solution of recrystallized sodium carbonate and distill the aniline from the solution. Distill the aniline three times under reduced pressure. When possible, use the aniline immediately; otherwise keep it in a dark place in a vacuum desiccator; bp 184.32–184.39°, fp −6.24°, d 15°/15° 1.0268, n_D 20° 1.5850. Aniline is extremely hygroscopic, and precautions must be taken to keep moisture from an anhydrous product.

Few and Smith [1565] purified aniline by dissolving in 40% sulfuric acid. Nonbasic materials were removed by passing steam through the solution for 1 hr. The base was liberated by the addition of sodium hydroxide pellets, steam-distilled, and dried over potassium hydroxide. It was distilled twice from zinc dust at 20 torr, dried over freshly prepared barium oxide, and finally distilled from barium oxide in an all-glass apparatus.

Smith and coworkers [4392] distilled aniline under vacuum and dried over sodium hydroxide. The purified aniline was used for *solubility studies*.

Keyes and Hildebrand [2568] dissolved a pure commercial material in hydrochloric acid, steam-distilled to remove traces of nitro compounds, made alkaline to free the base, and steam-distilled. The aniline was separated from the water, dried with potassium hydroxide, and fractionally distilled at reduced pressure.

Jones and Kenner [2420] reported that aromatic amine hydrochlorides form double salts with copper (I) chloride that may be used for purification.

Oddo and Tognacchini [3573] crystallized aniline several times from ethyl ether and distilled at reduced pressure.

Hantzsch and Freese [2010] offered an explanation for the rapid coloration of aniline in air. They found that *sulfur-free* aniline remained colorless on standing exposed to air for weeks, whereas technical aniline became brown. The sulfur-free aniline was prepared by refluxing aniline with 10% acetone for 10 hr, the mixture was acidified with hydrochloric acid to congo red, and extracted with ethyl ether until it no longer showed any color. The hydrochloride was purified by repeated crystallizations; the aniline freed, dried, and distilled.

Weissberger and Strasser [5115] found that the tendency of aromatic amines to become colored by oxidation from the air was greatly reduced by treatment with tin (II) chloride, which precipitated the sulfur.

Aniline has been stabilized by the addition of mercaptobenzimidazoline [3623].

LaPaglia [2825] purified aniline for *fluorescent studies* by distilling twice in an argon atmosphere and then freezing and degassing in successive cycles. No impurities were found by gas chromatography.

Pearce [3690] allowed a commercial product to stand several weeks over potassium hydroxide, distilled and treated the portion distilling at 181–182° by the Hantzsch-Freese procedure (see above). The middle fraction was distilled twice over zinc. It was used for *electrical measurements*. See also Pound [3819] and Muller and coworkers [3448] for other purifications for *electrical measurements*.

Aniline was purified by Armanet and Pouyet [274] by fractional distillation, crystallization and zone melting. The improvement in purity was followed by the melting point, specific conductance, and near-ultraviolet and visual spectroscopy.

CRITERIA OF PURITY

At least two grades are available from most supply houses but most list three: a 99+%m, a Reagent grade to meet, at least, ACS Reagent grade specifications, and a purified grade. Fischer Scientific Company's specifications for their reagent grade are given in Table 5.23.

SAFETY

The toxicology of aniline in animals and its effect on man have been extensively studied. It is considered a hazardous substance. It is an acute and a chronic poison. It is absorbed through the respiratory tract and through

Table 5.23 Specifications for Reagent Grade Aniline[a]

Boiling point	184.4°
Distillation range, max	3°
Residue after ignition, max	0.005%
Chlorobenzene, max	0.010%
Hydrocarbons	pass test
Nitrobenzene, max	0.0010%
Appearance	pass test

[a] From *Fisher Chemical Index* 67-C.

the skin [3681]. Tkachev [4796] found the threshold odor concentration to be 0.37 mg/m³.

The *threshold limit value* is 5 ppm, 19 mg/m³ [4509].

286. *o*-TOLUIDINE

287. *m*-TOLUIDINE

288. *p*-TOLUIDINE

The toluidines undergo about the same reactions as aniline. They form salts with most acids, are weak bases in water and strong bases in acetic acid and other organic solvents. The methods of purification for aniline are generally applicable to the toluidines.

Adkins and Billica [26] reduced *p*-nitrotoluene with a specially prepared Raney nickel catalyst; see under 2-propanol.

Berliner and May [531] purified the three isomeric toluidines as follows for *vapor pressure studies:*

1. *o-Toluidine.* The material was twice distilled and then dissolved in four times its volume of ethyl ether. An equivalent amount of oxalic acid to form the dioxalate was added to the ether solution and the crystals of *p*-toluidine dioxalate removed by filtration. The ether solution was evaporated and the crystals of *o*-toluidine dioxalate formed were separated by filtration. They were recrystallized five times from water containing a small amount of oxalic acid to prevent hydrolysis. The purified crystals were treated with a dilute solution of sodium carbonate and the amine separated, dried with calcium chloride, and distilled three times at reduced pressure; bp 199.84°.

2. *m-Toluidine.* The amine was twice distilled and heated with a slight excess of hydrochloric acid. The hydrochloride was fractionally crystallized five times from 95% ethanol and twice from distilled water. In each case, the

material that crystallized first was discarded. The amine was regenerated and distilled as described above; bp 202.86°.

3. *p-Toluidine*. The amine was distilled three times and sublimed twice at 30°. It was then dissolved in five volumes of ethyl ether containing an equivalent amount of oxalic acid. The filtered and washed crystals were recrystallized three times from hot distilled water and the base regenerated with sodium carbonate solution. The base was recrystallized three times from distilled water; mp 43.5°. Seven recrystallizations from ethanol did not produce any change in the melting point.

Rastogi and coworkers [3922] purified *p*-toluidine for *thermodynamic studies* by fractional crystallization.

o-Toluidine has been stabilized by the addition of mercaptobenzimidazoline [3623].

SAFETY

Hamblin [3681] states that *o*- and *p*-toluidines have a toxicity similar to that of aniline, but in addition cause a transient hematuria which, in observed experience, cleared completely upon removal from further exposure.

The *threshold limit value* has been set at 5 ppm, 22 mg/m³ [4509].

The LD_{50} for rats, intraperitoneal injection for the toluidines has been reported as: *ortho* 0.083, *meta* 0.1166, and *para* 0.150 g/kg [986].

Primary Unsaturated Amines

289. ALLYLAMINE

Tollens [4803] reported the preparation of allylamine in 1872.

Clavier [997] prepared allylamine in 90% yield from 3-bromopropene and hexamethylenetetramine and in 25% yield from 3-chloropropene. A modification raised the yields for the chloro compound to 55–60%.

Fuson [1681, p. 5] gives directions for preparation from allyl isothiocyanate, hydrochloric acid, and potassium hydroxide.

Primary Polyamines

290. ETHYLENEDIAMINE

Ethylenediamine is a water-soluble, colorless liquid with an ammoniacal odor. It is very hygroscopic and forms a carbamate with carbon dioxide. It is commonly available in two grades: a "practical" grade of 91–93% and a 98–100% grade often referred to as "anhydrous."

The amine has found a variety of solvent uses such as an ionizing solvent for electrochemical investigation, a reaction medium particularly for ionic reactions, and as an enhancing solvent for acids in nonaqueous titrations.

Ethylenediamine always should be protected from atmospheric moisture and carbon dioxide during purification. It should be protected during use if the moisture and carbon dioxide absorbed from the air would cause a greater error in the results than should be tolerated.

The "anhydrous" grade ethylenediamine is suitable for some solvent uses. It is recommended that this grade be used as the starting material for purification. The water and carbon dioxide may be reduced by shaking the amine with sodium or potassium hydroxide pellets for several hours, transferring the liquid to a still, and distilling the amine through a spray trap. The amount of water may be reduced by treating with a suitable drying agent, preferably 3A or 4A molecular sieves, or with alumina. The amine is then distilled from the sieves or alumina.

The water may be removed by azeotropic distillation. Ethylenediamine and water form a negative azeotrope that has a boiling point less than 2° above that of the amine. Therefore, it is not feasible to try to separate the amine from the azeotrope by fractional distillation. A third substance that will form an azeotrope with water but not with ethylenediamine, and that boils sufficiently below the amine to permit the two to be easily separated by fractional distillation, may be used. There has not been much information published concerning azeotropes of ethylenediamine; however, Creamer and Chambers [1130] have used both benzene and isopropyl ether as azeotropic separating agents. Benzene is preferable because it removes about twice the amount of water on a weight basis.

The preparation of high purity ethylenediamine requires an additional step, treatment with a substance that will react quantitatively with water. The alkali metals, sodium hydride or boron hydride, and subsequent efficient fractional distillation have been successfully used. The following procedure has given ethylenediamine that analyzed 99.9+%w by mass spectrometry and gas chromatography: Stir commercially "anhydrous" amine for 2–4 hr with 10%w freshly opened sodium hydroxide pellets. Transfer the amine to a still pot and add about 2%w of sodium wire, ribbon or fine slices, connect to an efficient distillation column and reflux for about 3 hr. Fractionally distill at a reflux ratio sufficient to give the purity desired. A 7-ft (2.1 m) × 25 mm i.d., Penn-State-type column packed with 0.16 × 0.16-in. protruded packing and fitted with an automatic vacuum-jacketed head has been used to obtain a 99.9+% amine when operated as follows. It was refluxed for 3 hr just under flooding, and 50 ml was distilled at a reflux ratio of 50:1; about 50 ml more was removed at a reflux ratio of 35:1 and the remainder of a 10%v heads fraction volume of a 3500-ml pot charge was taken at a reflux ratio of 25:1. The middle 80% was distilled at a reflux ratio of 10:1. Samples were analyzed from time-to-time to be sure the desired purity was distilling.

Several methods of purification are given that have some unique features, or that have been used for specific purposes requiring special criteria.

Heumann and coworkers [2161] purified "anhydrous" grade for *ion exchange studies* in the amine. The amine was dried for several days with porous barium oxide (Barium and Chemicals, Inc.). It was decanted into a vacuum distillation apparatus fitted with Teflon-sleeved ground-glass joints and Teflon stopcocks. Liquid potassium (75%)-sodium alloy was added until a persistent dark blue color developed. The amine was distilled at reduced pressure at a temperature not greater than 40°.

Windwer and Sundheim [5217] purified the amine by drying over calcium oxide, then filtering in a cellophane bag enclosure in an atmosphere of nitrogen. It was refluxed overnight with sodium and fractionally distilled through a 10-ft column, and the middle fraction was collected. Sodium-potassium alloy was introduced under vacuum and then repeatedly cycled through shaking, freezing, evacuating, and remelting until a permanent blue color developed. This color is evidence of a high purity.

For removal of acidic impurities as a solvent for *titration of acids*, see Harlow et al. [2027] under piperidine.

Bruckenstein and Mukherjee [797] extensively studied the purification of ethylenediamine to obtain the best purity as a solvent for *equilibrium studies*. The 114–117° fraction was shaken with either Alcoa F-20 or Kaiser KA 101 activated alumina, the liquid decanted and fractionally distilled in a dry carbon dioxide-free atmosphere at a reflux ratio of 20:1. This procedure was as satisfactory as sodium treatment. The fraction boiling at 117.2° was used.

Creamer and Chambers [1130] prepared anhydrous ethylene diamine for *electrochemical investigation* by removing the water as the benzene azeotrope in a fractionating still with a phase-separating stillhead; the specific conductance was 1×10^{-5} ohm^{-1} cm^{-1}.

Schaap and coworkers [4128] found that they obtained the best product for *polarographic studies* by twice refluxing over sodium and vacuum-distilling.

Mukherjee and Bruckenstein [3436] continued their study of the purification of ethylenediamine and developed a method for producing "pure" amine suitable for *electroanalytical chemistry*. In addition to the purification method, it is an excellent review and discussion article. Their recommended procedure is: Shake 98% "anhydrous" commercial ethylenediamine with 70 g per liter of activated Linde Type 5A molecular sieves for about 12 hr. Decant the amine from the molecular sieves and shake approximately 12 hr with 50 g per liter of calcium oxide and 15 g per liter of potassium hydroxide. Keep in an atmosphere free of carbon dioxide and water during the remainder of the purification. Fractionally distill the supernatant liquid at a reflux ratio of 20:1 from freshly activated molecular sieves and collect the material that distills at 117.2° at 760 torr. Fractionally distill the 117.2° cut from sodium.

SAFETY

The vapors of ethylenediamine are irritating to the eyes, mucous membranes, and respiratory tract. This liquid causes severe skin and eye injuries. Dermatitis is a common result of repeated exposure to the vapors [3681]. It is both irritating and allergenic, and some individuals are hypersensitive [4509]. The *threshold limit value* is 10 ppm, 30 mg/m^3 [4509].

Secondary Monoamines

Specific Solvents

291. ETHYLENIMINE

Ethylenimine is the simplest of the heterocyclic nitrogen compounds. It is a colorless mobile liquid with an amine-like odor. Traces of acid cause exothermic polymerization to a water soluble polyimine [1358]

$$n \; H_2C \text{———} CH_2 \xrightarrow{H^+} \text{—}(\text{—}CH_2\text{—}CH_2\text{—}NH\text{—})_n\text{—}$$

A few pellets of sodium hydroxide will react with any acid absorbed by ethylenimine and prevent ring opening and polymerization.

Gray and Jones [1868] purified ethylenimine by distilling twice in an atmosphere of nitrogen and retaining the constant boiling fraction each time. A gas chromatogram showed the distillate to be pure.

Ethylenimine and piperidine were purified by Searles and coworkers [4217] for *hydrogen bonding and basicity studies* by drying over barium oxide and sodium and distilling through a 5-plate column in a nitrogen atmosphere. See also Kay and coworkers [2532].

Allen and coworkers [3877] prepared ethylenimine from β-aminoethyl-sulfuric acid.

SAFETY

The Dow Chemical Company [1358] recommends special precautions when handling and using ethylenimine. It can polymerize violently in the presence of acids or acid-forming substances. The liquid is corrosive to the skin and is readily absorbed in toxic amounts. Short contact will cause marked irritation and longer contact is likely to result in severe burns that heal slowly. Repeated exposure to vapors not concentrated enough to cause immediate discomfort may result in delayed lung injury.

The *threshold limit value* is 0.5 ppm, 1 mg/m³ [4509].

The *flammable limits* in air are 3.60 and 45.80%v. The minimum ignition temperature in air is 322° [3680].

292. DIETHYLAMINE

Swift [4585] converted a commercial diethylammonium chloride into *p*-toluenesulfonamide and recrystallized it three times from pure dry ligroin (bp 90–120°) until a sharp melting point was obtained. The amide was hydrolyzed with hydrochloric acid, excess sodium hydroxide added, and the amine distilled through a tower of activated alumina. The diethylamine was then distilled through a 5-ft adiabatic column at a reflux ratio of 20:1. A fraction distilling over a range of about 0.01° was collected under nitrogen and sealed under vacuum. It was dried with activated alumina before use.

For purification for *conductivity studies*, see butyl amine, Muney and Coetzee [3455].

295. DIBUTYLAMINE

Dibutylamine was dried over sodium hydroxide and fractionally distilled. The portion distilling from 158–159° was collected by Bayles and Chetwyn [456] to study *acid-base functions* in nonaqueous systems.

296. PYRROLE

Pyrrole has unusual solvent properties that have been investigated only slightly because of its ease of reacting with atmospheric oxygen. It should be handled, used, and stored in an inert atmosphere. Its existence as an entity was first surmised by Runge in 1834. Anderson isolated it from coal tar in 1857. It is a good solvent for donor molecules of high molecular weight. This is illustrated by the strong tendency for the formation of —N—H ← O bonds [3183]. It is an extremely weak base but strong proton donor acids form

protonated coordinate covalent compounds. The protonated ion of pyrrole is highly unstable and undergoes polymerization readily [3489].

Ciamician [982] presented a historical summary of pyrrole in 1905 and gave several methods of preparation. The original method of Schwanert [982] published in 1860 of heating dry ammonium mucate is still practical. Goldschmidt in 1867 found the addition of glycerol helpful; Khotinsky [2584] added ammonia; Blicke et al. [618, 617] further modified the method for the preparation of large or small amounts in yields of 40–52%. McElvain and Bolligen [1778] have further refined the basic method of Blicke and Powers.

Physical properties were determined by Helm and coworkers [2097] on a commercially available pyrrole that was fractionally distilled in a 1-in., 80-plate (65 theoretical plates) Oldershaw all-glass column. The product was fractionally redistilled in a 200 theoretical plate Helipac still. The highest purity material was 99.994%m determined from the freezing curve.

They found that pyrrole was stable for at least one year in sealed glass ampules, whether stored exposed on the roof or in the dark at room temperature. A sample of the pyrrole in a loosely capped vial was exposed to air and light in the laboratory. The sample darkened with exposure, and a decrease in purity was found from the freezing curve. The change was followed by infrared spectroscopy. There was an increased absorbance at $5.87\ \mu$, the carbonyl region.

Scott and coworkers [4204] used a portion of the pyrrole prepared by Helm et al. [2097] for their *thermodynamic studies*. Scrupulous care was taken to avoid exposing the material to atmospheric oxygen. It was stored and handled *in vacuo* or blanketed with dry, oxygen-free helium.

SAFETY

Scant evidence indicates that pyrrole can be considered of low toxicity, both systemically and locally [3681].

297. PIPERIDINE

Pyridine is readily reduced to piperidine by catalytic reduction in acid solution with a platinum catalyst, in neutral solution with Raney nickel catalyst, electrolytically or by sodium and ethanol. All the reduction products contain some impurities. Davies and McGee [1214] studied the impurities in commercial piperidine. Their initial study consisted of fractionally distilling a "pure" 98–99% material in a 46 theoretical plate still at a reflux ratio varying from 30:1 to 15:1. A 6% fraction that distilled at 92° was supposed to have been "tetrahydropyridine." However, it was shown to be the piperidine-water azeotrope. The fraction distilling at 106° and comprising 70% of the charge was piperidine. A 20% fraction distilling at 107° had been assumed to be 1,2,3,6-tetrahydropyridine but it was shown to be 1,2,5,6-tetrahydropyridine. They found that not all commercial compounds contained the 1,2,5,6- partial reduction compound. None was found in a laboratory-prepared piperidine with nickel catalyst. Available information indicated that the electrolytically reduced product contains the 1,2,5,6-impurity.

Schnildknecht [4146] reduced the pyridine in piperidine from 0.1 to 0.001 mole fraction by zone melting.

Werner and coworkers [5126] purified piperidine as a solvent for the *study of constitution of inorganic compounds* by removing the water and carbon dioxide with potassium hydroxide. It was warmed over sodium, allowed to stand, and distilled; bp 104.5–105.0°. The *ebullioscopic constant* was found to be 2.842.

Bates and Bower [439] purified "CP" and "Purified" grades of piperidine in the following ways to determine the *dissociation constant* of the piperidinium ion and *related thermodynamic quantities:* (1) piperidine was distilled from sodium hydroxide pellets and redistilled, rejecting the first and the last tenths of the distillate; (2) a portion of the redistilled base was frozen and the last 5% of the liquid rejected; the main portion was redistilled; and (3) a portion of the finished (2) sample was twice redistilled and the first and last fractions rejected.

Searles and coworkers [4217] purified piperidine and ethylenediamine for *hydrogen bonding studies:* see ethylenediamine.

Harlow and coworkers [2027] purified piperidine, pyridine, and ethylenediamine as solvents for the *potentiometric titration study* of very weak acids by passing them through a column of activated Alcoa F-20 alumina.

Hudson and coworkers [2296] allowed "purified" piperidine to stand over freshly fused barium oxide for three days and then fractionally distilling, as did Davies and McGee [1214] for *vapor diffusion coefficient and collision parameter studies.*

SAFETY

The oral toxicity of the salts of piperidine is very low; for instance, the LD_{50} of piperidinium citrate for mice is 11 g/kg. Piperidine is a strong base and has the inherent hazards of such compounds [3681]. First degree burns have been reported by Linch [2961] when it was in contact with the skin for 3 min. The vapors are irritating to the respiratory tract [3].

Tertiary Monoamines

Specific Solvents

298. TRIETHYLAMINE

Swift [4585] distilled commercial anhydrous triethylamine from acetic anhydride to remove traces of primary and secondary amines, dried with activated alumina, and distilled three times under reduced pressure.

Counsell and coworkers [1107] purified triethylamine for *phase diagram studies* by refluxing a high purity commercial product containing a small amount of lower molecular weight amines with p-toluenesulfonyl chloride and potassium hydroxide to remove primary and secondary amines. It was then fractionally distilled from metallic sodium at a reflux ratio of about 15:1; bp 89.50° at 760 torr.

Muney and Coetzee [3455] purified triethylamine for *conductivity studies;* see butylamine.

Commercial triethylamine was purified by Bryant and Wardrop [825] by converting to the triethylammonium chloride and recrystallizing from ethanol to a sharp and constant melting point, 254°. The amine was regenerated with aqueous sodium hydroxide, dried over potassium hydroxide, and twice fractionally distilled over sodium in an atmosphere of nitrogen. It was used for *conductimetric studies* in a nonaqueous system.

Copp and Findlay [1087] purified triethylamine for *thermodynamic studies of binary systems* by fractionally distilling at a reflux ratio of 8:1 in a 30-plate column that was functioning adiabatically. The amine was first distilled from potassium hydroxide pellets. A middle fraction was dried with sodium and again distilled.

SAFETY

See also Amines, General. Kulagina and Kochetkova [2775] found the lethal concentrations of triethylamine in air to be: LC_{100} 7.5, LC_{50} 6, and LC_{min} 5 mg/liter. Concentrations as low as 0.03 g/liter caused chronic effects.

299. PYRIDINE

Pyridine is obtained from the destructive distillation of coal and from petroleum. A limited amount is found in "bone oil." It is available commercially in three regular grades and several grades for special purposes. Pyridine is a weak base in water and a strong base in acetic acid. It forms pyridinium salts with a wide variety of acids. It is very hygroscopic.

Coulson and Ditcham [1105] studied the impurities occurring in pyridine. *o*-Xylene and tetrahydrothiophene were both isolated and identified. No other sulfur compounds were positively identified, although others were present. Methods of purification were given, particularly for reducing the sulfur-bearing impurities.

Whitford [5151] separated pyridine from some of its homologues by utilizing the difference in the solubilities of the oxalates. The impure pyridine was slowly added to a stirred solution of oxalic acid in acetone. The pyridinium oxalate was precipitated, filtered, washed with cold acetone, regenerated, and isolated. The oxalate has a low density and consequently a large bulk; the method is convenient only for small quantities.

Arndt and Nachtwey [277] found that pyridinium perchlorate could easily be crystallized to a high purity. Unlike the perchlorates of pyridine homologues, it is sparingly soluble in water. Technical pyridine was dried with potassium hydroxide and fractionally distilled. The distillate was neutralized with 6 N hydrochloric acid until the odor of pyridine could not be detected.

A slight excess of 6 N perchloric acid, based on the hydrochloric acid used, was added and the mixture allowed to stand for several hours in the cold. The crystals were removed by filtration, washed with cold water or 20% perchloric acid solution, and dried; mp 288°. The base was liberated with 50% sodium hydroxide or ammonia; the latter is preferable because no water is involved. A slow stream of ammonia was passed into the cooled flask containing the dry pyridinium perchlorate until only a fraction of the salt remained. Dry air was passed through the warmed flask to remove the ammonia. The product was distilled at reduced pressure. Zecherl [5327] stated that this method yielded a very dry pyridine *but is very dangerous.*

Wilson and Hughes [5214] purified 800 ml of technical pyridine by stirring for 24 hr with a mixture prepared by gently grinding 20 g of technical cerium (IV) sulfate with 15 g of anhydrous potassium carbonate. Heating was not necessary. The mixture was filtered through a Buchner funnel and fractionally distilled. A column of at least 12–14 theoretical plates was recommended.

Heap and coworkers [2079] dried crude pyridine over sodium hydroxide, fractionally distilled several times, and purified by means of the coordinate compound with zinc chloride. The double salt was prepared as follows: One liter of pyridine was added to a mixture of 845 g of zinc chloride dissolved in a mixture of 600 ml of water, 345 ml of concentrated hydrochloric acid, and 690 ml of absolute alcohol. The crystalline mass was filtered with suction, recrystallized twice from dry ethanol, dried, and decomposed with 26.7 g of sodium hydroxide per 100 g of dry precipitate. The pyridine was dried over sodium hydroxide and distilled. Bryant and Wardrop [825] used the same method of purification except that the freed pyridine was dried with potassium hydroxide and distilled from barium oxide. They commented that the method was "tedious and wasteful." The pyridine was used for *specific conductance studies.*

Ralph and Gilkerson [3885] distilled a Reagent grade pyridine from potassium hydroxide and then from barium oxide for *ion-solvent interaction studies.* Stewart and O'Donnell [4502] distilled a like-grade material from barium oxide for use as a solvent for the *determination of the H-function* of strongly basic systems.

Leis and Currans [2898] purified AR grade pyridine for the *determination of dipole moments* by repeated fractional freezing, refluxing over freshly heated barium oxide, and distilling through a column at least 1 meter long packed with glass helices. The middle fraction, with a constant refractive index, was used for *physical measurements;* n_D 20° 1.5094, d 20° 0.9832. Another portion, from which the lower boiling material had been removed by fractional distillation, was purified as described; bp 114.5° at 760 torr, n_D 25° 1.5067, d 25° 0.9786.

Müller and coworkers [3448] purified pyridine for *specific conductance studies* by allowing it to stand over freshly fused potassium hydroxide for a long time and by repeated fractional distillation.

The principal acidic impurity present in most basic solvents used in non-aqueous titrimetry is carbon dioxide. The solvent is adequate for precision titrations, in most cases, if this impurity is removed [3975]. A column of alumina such as that used by Harlow and coworkers [2027] is convenient (see piperidine). If the column is connected to a reservoir of the solvent, an adequate supply of low acid solvent is available at all times.

Pyridine and pyridine mixed with another solvent are widely used as the sample solvent in acid-base titrimetry because certain of the mixtures are such excellent differentiating solvents for acids. The technical 2° material is usually used with or without a blank correction, depending on the accuracy desired. Often a purified solvent with a low, or negligible, blank is desired or is necessary. Cundiff and Markunas [1155] purified technical grade pyridine by letting it stand overnight over sodium hydroxide pellets and then flash-distilling for their study of the titration of strong acids. They purified the solvent by flash distillation from barium oxide for their study of the determination of the alkoxyl group.

Banick [381] developed a simple and rapid procedure for purifying large volumes of pyridine to a blank of less than 0.05 ml of 0.1 N titrant. Shake intermittently for 5 min in a 1000-ml, iodine-type flask 500 ml of pyridine and 20 g of alumina and filter by suction through a 9.0-cm, glass-fiber filter paper containing a layer of 5.0 g of diatomaceous earth (Celite grade is satisfactory).

Brown and Barbaras [765] purified a Reagent grade pyridine by fractional distillation through a 15 theoretical plate column. The middle portion boiling at 114.9° at 750 torr was collected and used for the study of the *steric nature of the ortho effect*.

Vapor diffusion coefficients and *collision parameters* were studied [2296] with pyridine purified as follows: It stood over freshly fused potassium hydroxide for two months and was distilled. It was fractionally frozen three times, each time rejecting one-third of the mother liquor and remelting the crystals. The final material was dried over freshly fused potassium hydoxide and distilled.

For purification for *critical properties*, see Ethers, General [2653].

Kienitz [2587] partially froze 98.4% pyridine and obtained 99.9%m material after three cycles.

For purification by progressive freezing, see General, Chapter V, Dickinson and Eaborn [1306].

Trachuk [4818] reported that pyridine could be effectively dried to less than 0.05% water by distilling as the ternary pyridine-benzene-water azeotrope; bp 69.3°. Berg and coworkers [521] found that toluene could be used successfully as a ternary azeotroping agent for drying pyridine.

The water may be reduced to less than 0.01% with Linde 3A or 4A molecular sieves [3975].

For special techniques of drying, see Zerewitinoff [5334], Müller and Brenneis [3447], and Timmermans [4762].

Eberius and Kowalski [1435] describe a method for the dehydration of a methanol-pyridine mixture for the preparation of Karl Fischer reagent. Coulson and coworkers [1106] developed an infrared procedure for the

Table 5.24 ACS Specifications for Reagent Grade Pyridine

Distillation range, max	2°[a]
Solubility in water	to pass test
Residue on evaporation, max	0.002%
Water, max	0.10%
Chloride (Cl), max	0.001%
Sulfate (SO_4), max	0.001%
Ammonia (NH_3), max	0.002%
Copper (Cu), max about	0.0005%
Reducing substances	to pass test

[a] Some supply houses now list a Reagent grade with a 1° distillation range.

determination of water in pyridine down to 0.002% v/v that is rapid and requires only a 0.1-ml sample. The 3450 cm^{-1} band depends on water concentration. The disadvantages of the other methods for the determination of water in pyridine are discussed.

Waldron [5055] reported a good color stability for pyridine that was refluxed for 0.5 hr with 20% of its volume of 28% sodium hydroxide and then fractionated.

Rosin [4042] and ACS Reagent Chemicals [76] give the specifications and test methods for Reagent grade pyridine. The ACS requirements are given in Table 5.24.

Cox and coworkers [1122] characterized pyridine for *heat of combustion studies* by the freezing curve.

The titratable purity may be determined potentiometrically to ±0.05% in an acetic acid solvent system by the method of Markunas and Riddick [3155].

SAFETY

The vapor is irritating to the mucous surfaces and causes eye and nasal irritation. The inhalation of low concentrations that produce chronic poisoning is considered the greatest hazard. It affects the kidneys, liver, and bone marrow. Pyridine is absorbed through the skin and can cause acute or chronic conditions [4509].

The *threshold limit value* is 5 ppm, 15 mg/m^3 [4509].

The *explosive limits* in air are 1.81 and 12.40%v, the latter value at elevated temperature. The *minimum ignition temperature* in air is 482° [3680].

300. QUINOLINE

Quinoline is produced from coal tar. It is made by several synthetic methods, one of the best known is Skraup's synthesis from aniline, glycerol, nitrobenzene, and sulfuric acid [1778].

Mastrangello [3203] purified refined quinoline by repeated precipitation of the phosphate, followed by washing the crystalline mass three times with methanol after filtering and drying overnight at 55°. The quinoline was liberated with sodium hydroxide. The wet base was fractionally distilled and the first 10% and the last 20% rejected. The material used for *calorimetric studies* was distilled *in vacuo* to remove the last traces of water.

Malanowski [3123] purified several quinoline bases for precise *vapor pressure studies*. They were distilled through a 30-plate adiabatic column with a Swietoslawski ebulliometric head. One-half of each fraction distilling within ±0.1° was converted to the chloride and twice recrystallized from the remaining base. The quinoline base was regenerated with sodium hydroxide and again fractionally distilled. The distillate condensing within 0.05° was collected. The ebulliometric purity by a Swietoslawski differential ebulliometer was 0.002°.

Quinoline decomposes 1% per hr at 510–535° [2402].

SAFETY

Limited animal tests indicate that quinoline has a moderate degree of toxicity by ingestion, skin penetration, or inhalation [3681].

Amides

General

The amides are excellent solvents both for organic and inorganic compounds with properties that make them of particular interest to both research and production. The Second Edition of *Organic Solvents* contained two amides; this Edition contains eleven, distributed among several structural types.

Peterson [3728] found that aliphatic amides are characterized by two melting points (different melting and freezing points). If the melt is allowed to cool slowly and is seeded just below the lower melting point, crystallization takes place at a sharply defined temperature. The melting point on remelting is higher. The curve for the melting point for acetamide was extrapolated to 69.5°. The slopes of the freezing and melting curves are different.

Gopal and Rizvi [1840] purified six amides as follows: Freshly ignited calcium oxide was added to the liquid and it was distilled at reduced pressure. The middle fraction was collected. The process was repeated until the specific conductance of the distillate was reduced to 10^{-5} to 10^{-6} ohm^{-1} cm^{-1}. They were stored in an amber bottle in a dry nitrogen box and used, when possible, immediately after distillation. *Physical properties* at different temperatures were determined on the amides. Five of the six amides that were used are included in this section: N-methylformamide, N,N-dimethylformamide, N-methylacetamide, N,N-dimethylacetamide, and N-methylpropionamide.

Campbell [891] developed a method for the vapor-phase reaction at atmospheric pressure of carboxylic acids with a boiling point such that it can be vaporized without difficulty, not over 250°, and primary or secondary amines to give the corresponding amide with a high purity. The reaction is

$$RCOOH + HNR_1R_2 \xrightarrow{190-300°} RCONR_1R_2 + H_2O$$

R and R_1 may be H or an aliphatic group. R_2 is an aliphatic group.

There are little if any side reactions. The reaction takes place in a heated Pyrex-brand, glass tube 100 cm long by 22 mm i.d. with an 8-mm o.d. glass tube thermowell running the length of the reactor. The reactor is packed with Alcoa E-10 alumina or Davis Chemical Co. catalyst SMR-55-5138. The preferred amine-carboxylic acid ratio is 1.05. The optimum temperature and feed rate depends upon the reactants. If the reaction effluent is fed into, or near, the middle of a fractionating column, the lower section of which is maintained between the boiling point of water and that of the amide, conditions can be maintained to give an amide with a purity of 99+% in the still pot.

For a method of preparing N-methylamides of C_1–C_8 carboxylic acids, see D'Alelio and Reid [1182] under N-methylpropionamide.

Loder [2990] describes a process for the preparation of N-alkylformamides in which an alcohol reacts with ammonia and carbon monoxide at 900-atm pressure and 210–260°.

Hobbs and Bates [2208] conclude from dielectric and dipole moment studies that N-unsubstituted amides exist principally in the *cis*-configuration and dimerize similarly to carboxylic acids:

Carpenter and Donohue [912] studied the crystal structure of peptides and concluded that N-monosubstituted amides, CONH, existed principally in the keto form. Shortly thereafter, Mizushima and Simanouti [3357] surmised from infrared and dipole moment studies that CONH existed mainly in the *trans* configuration. Worsham and Hobbs [5259] inferred from dipole moment that a chain association of the N-monosubstituted amides exists in the following form:

$$
\begin{array}{ccc}
 & & R \\
 & & \diagup \\
C{=}O\cdots H{-}N & \\
\diagup & & \diagdown \\
H{-}N & & C{=}O \\
| & & \\
R & &
\end{array}
$$

Kotera and Shibata [2725] confirmed the *trans* configuration from ultraviolet and dipole moment studies. Davies and Thomas [1208] studied energies and entropies of association of amides in benzene solution and stated that the chain association of CONH amides is a consequence of the *trans* configuration of the monomer. The *cis* configuration of monomeric amides, for steric reasons, would be essentially confined to cyclic dimerization. Suzuki [4560] questioned whether amides, such as N-methylformamide, contained any of the *cis* form. LaPlanche and Rogers [2827] continued the configuration study with nuclear magnetic resonance; they believe that the CONH amides exist substantially as the *trans* configuration.

Specific Solvents

301. FORMAMIDE

Röhler [4020] in 1910 pointed out that formamide resembles water in that both have a high dielectric constant. From his study of the amide as a *solvent for inorganic salts* and as an *electrolysis solvent*, he concluded that the salts were solvated in a manner similar to water. Walden [5021] studied formamide as an ionizing solvent and concluded that it imitates, in a remarkable manner, the physical characteristics and constants of water. He found that, when binary salts were used as solutes, the salts may exhibit a degree of ionization greater than in aqueous solution. Strong organic acids do not become markedly ionized in this solvent.

Formamide is an excellent ionizing solvent soluble in water, lower alcohols, and glycols and insoluble in hydrocarbons, chlorinated hydrocarbons, and nitrobenzene. It dissolves casein, gelatin, zein, animal glue, and related water soluble gums and resins. The chlorides and some sulfates and nitrates of copper, lead, zinc, tin, nickel, cobalt, iron, aluminum, and magnesium are soluble. Toops [4811] found that Drierite could not be used as a drying agent. The salt dissolved and the solution set to a gel on standing overnight.

Formamide is a very weak base, even weaker than water. It is easily hydrolyzed to the parent acid and ammonia by acids, bases, and enzymes. It reacts with peroxides, and the amide hydrogen may be replaced by acid halides, acid anhydrides, and esters. On heating with alcohols, it gives formic esters. Strong dehydrating agents remove a molecule of water to form the corresponding nitrile. Maxim and Mavrodineanu [3236] reported that it reacts with Grignard compounds. Berthelot and Gaudechon [544] found that formamide was decomposed by exposure to an ordinary incandescent lamp.

Smith [4387] found that neither fractional distillation nor fractional crystallization would purify formamide to a high state of purity. He recommended that it be purified by fractional distillation to a freezing point of 2.25° and then by fractional crystallization to a freezing point of 2.55°. Formamide is very hygroscopic, and precautions must be taken to exclude water so that the purified product will have a freezing point of 2.55° and a specific conductance of 1.98×10^{-6} ohm^{-1} cm^{-1}.

Practically all of the simpler amides possess the above described properties to a greater or lesser degree.

Berger and Dawson [522] studied the *determination of chlorides* in formamide solutions using a commercial material that had been purified by the same general method of Smith [4387] above, except that the order of operations was reversed. The material was fractionally frozen several times in a specially constructed flask. The impure mother liquor was drained and the solid melted. The liquid was dried overnight with calcium oxide and then slowly distilled at about 1 torr. The middle 80% was collected and stored in "low actinic" glass-stoppered bottles that had been thoroughly purged with nitrogen.

Verhoek [4947] found that commercial formamide often contains acids and ammonium formate. The latter is so largely solvated that the ammonia distills, leaving the formic acid, which cannot be separated satisfactorily from the amide by distillation. Water present during the distillation hydrolyzes the formamide. He found the following method of purification to be the most satisfactory for preparing a material for a solvent for the *determination of the strength of acids*.

A few pieces of bromothymol blue were added to the formamide and the acidity exactly neutralized with sodium hydroxide. The neutral liquid was heated to 80–90° under reduced pressure, the ammonia and water pumped off, and the amide again neutralized. The procedure was repeated four or five times until the liquid remained neutral when distillation began. Sodium formate was added to the still pot and the formamide distilled between 80 and 90°. The distillate was neutralized and distilled as before and the last four-fifths collected; mp ca. 2.2°, $\kappa = 5 \times 10^{-5}$ ohm^{-1} cm^{-1}. The distillate was then fractionally crystallized in a water- and carbon dioxide-free atmosphere; $\kappa = 1-2 \times 10^{-6}$ ohm^{-1} cm^{-1}. The liquid from the crystallization was again neutralized, distilled, and fractionally crystallized.

The pure material is not stable. If a stream of air, purified by passing through soda-lime, calcium chloride, phosphorus pentoxide, and cotton, is bubbled rapidly through the formamide, the stability is greatly improved. The lowest *specific conductance* found was 6.2×10^{-7} ohm^{-1} cm^{-1}. The solvent was considered sufficiently pure when the specific conductance was 3.5×10^{-6} ohm^{-1} cm^{-1}.

A measure of the hygroscopicity was made by weighing 5.6 g into a small beaker with a free surface of about 9 cm^2. The beaker was exposed to the air in the laboratory (humidity not given) and the following gains in weight noted: 1 hr 0.2%, 5.5 hr 1%, 141 hr 10%. It was found that water up to 1% had no appreciable effect on conductimetric or potentiometric experiments.

"Chemically pure" grade formamide was found by Letaw and Gropp [2913] to be *polarographically pure* in the range they studied, after being held at a pressure of 1 torr, or less, for 2 hr.

Conductimetric studies of solutions of hydrogen chloride and potassium chloride in formamide were made by Dawson and coworkers [1227] purified from a commercial material by distilling several times from calcium oxide at a pressure of less than 0.1 torr between 51 and 58°. Decomposition of the vapors was held to a minimum by using a fractionating column in the region where the vapors were carried over a short path and condensed immediately at room temperature. The specific conductance ranged from 3×10^{-5} to 1×10^{-5} ohm^{-1} cm^{-1} at 20°.

Bates and Hobbs [443] determined the *dipole moment* of a good commercial grade purified as follows: The amide was dried over anhydrous sodium sulfate and vacuum-distilled. The middle portion of the distillate was fractionally crystallized three times, about one-half of the material being discarded in each crystallization; fp $2.50 \pm 0.06°$.

Leader [2847] purified a commercial material for *dielectric constant measurements*. The material was treated with 5 g of calcium oxide per liter and distilled at a pressure of 1 torr in a still, affording essentially no fractionation. Two additional distillations of the middle fractions were made with lime treatment each time. The final distilled product was statically fractionally crystallized three times in a container that was protected from carbon dioxide and water; fp 2.3-2.4°; $\kappa = 4 \times 10^{-6}$ ohm^{-1} cm^{-1}. The conductivity increased to 1×10^{-5} in 5-6 hr.

Notley and Spiro [3553] developed a method to produce formamide of low water and ion content in large laboratory amounts for *electrochemical investigation*. The chief impurities in high-grade commercial formamide were found to be about 0.04 M water and an electrolyte concentration of about 0.01 M.

Removal of Water.   A 5-cm column was packed with 1 kg of 3A molecular sieves in $\frac{1}{16}$-in. pellets. The column was heated electrically to about 60° to increase the rate of water removal. Formamide was passed through at the rate of 0.2–0.25 liters/hr. The water content was found to be about 0.008 to 0.01 M. The column will dry about 20 liters of the liquid. The sieves were washed with water, dried at 100°, and heated at 360° in a column through which a stream of nitrogen was flowing.

Removal of Ions.   Ions were removed by a mixed bed resin that had been especially treated. Two hundred-fifty grams of Amberlite IR-120 cation exchange resin was well washed with water and taken through several regenerations with 2 M hydrochloric acid and sodium hydroxide. It was washed with ethanol until the effluent was colorless, taken through a further cycle and let soak in water. Amberlite IRA-400 anion exchange resin was similarly treated. The IR-120 was drained of water, put in a column, and washed with formamide (ca. 7 bed-volumes) until the effluent contained less than 0.05 M water. One bed-volume of 2 M sulfuric acid in formamide was allowed to stand one day on the resin. Two liters of 2 M sulfuric acid in the solvent was passed through the resin. The anion resin was similarly treated with 2 M sodium formamide in formamide. The two resins were washed with formamide until the respective effluents were only slightly acidic or basic. The resins were mixed and about six liters of formamide were passed through: $\kappa = 2 \times 10^{-7}$ ohm^{-1} cm^{-1}, fp 2.3°. (It is recommended that one read the article if the use of the procedure is considered. Only the essential steps are in the above description.)

Cole and coworkers [1033] purified formamide for use as a solvent for the paper *chromatography of steroids* by shaking 4 hr with a 20% charcoal suspension and filtering. Taylor and Davis [4614] used formamide as a solvent for the study of the *velocity of esterification*. A commercial material was dried over anhydrous sodium sulfate for several days and distilled 15 times at 30–40 torr. It was slightly alkaline and odorless; fp 2.20°.

CRITERIA OF PURITY

Rosin [4042] gives specifications and test methods for Reagent grade formamide. The requirements are listed in Table 5.25.

Smith [4387] found a freezing point of 2.55° to characterize a very pure substance with a specific conductance of 1.98×10^{-6} ohm^{-1} cm^{-1}. Notley and Spiro [3553] found a specific conductance of less than 2×10^{-7} ohm^{-1} cm^{-1} for a material especially prepared for electrochemical investigation; however, its freezing point was 2.3°.

Table 5.25   Rosin's Specifications for Reagent Grade
Formamide

Specific gravity, 25°/25°	1.132–1.134
Freezing temperature, not less than	2°
Miscibility with water	to pass test
Ammonia, amines, etc.	to pass test
Evaporation residue, max	0.1%
Water, max	0.3%

SAFETY

No human injuries have been reported for formamide. It was absorbed through the skin of guinea pigs. The LD_{50} was found to be <5 ml/kg with only slight irritation [3681].

302. N-METHYLFORMAMIDE

Leader and Gormley [2848] led gaseous methylamine into 98–99% formic acid until its proper weight had been absorbed. The salt solution was heated to remove water. The resulting amide was vacuum-fractionated through a 1×35-cm glass-helices packed column and successively crystallized to a constant melting point. Further purification using dried Amberlite IR-120 resin in the hydrogen form gave a satisfactory product *for conductivity measurements;* $\kappa = $ ca. 5×10^{-5} ohm^{-1} cm^{-1}.

Held and Criss [2094] studied the *heat of solution* of selected alkali metal halides in anhydrous N-methylformamide purified as follows: It was treated with sodium hydroxide pellets and barium oxide for at least 4 hr. It was decanted into a distilling flask containing barium oxide and distilled at 1 torr through a 35-cm Vigreux tube. The specific conductance was $3–5 \times 10^{-6}$ ohm^{-1} cm^{-1}. Water by the Karl Fischer method was found to be 0.003%.

LaPlanche and Rogers [2827] purified a commercial material for *nuclear magnetic resonance studies* by drying with anhydrous sodium sulfate and fractionally distilling *in vacuo.*

French and Glover [1631] purified N-methylformamide for the study of *specific conductance* in solvents of high dielectric constant. It was shaken with phosphorus pentoxide, filtered through glass wool, and distilled at about 1 torr; bp 51°. This treatment was repeated three times, followed by two distillations without treatment. The specific conductance normally was in the range of $1.0–2.0 \times 10^{-6}$ ohm^{-1} cm^{-1}. This method of purification reduced the change of conductance with time to a very small amount.

303. N,N-DIMETHYLFORMAMIDE

N,N-Dimethylformamide is a colorless mobile liquid miscible with water and organic solvents [3160, 4075]. The slight amine odor results from hydrolysis by absorbed water. It is a uniquely versatile and powerful solvent that has a wide liquid range, good chemical and thermal stability, a high polarity, and a wide solubility range for both organic and inorganic compounds. It is only slightly basic. It is considered to be thermally stable at its boiling point. Above 350°, decomposition may occur to dimethylamine and carbon monoxide [1409]. Thomas and Rochow [4713] found that it decomposes slightly at the normal boiling point. Acidic and basic materials catalyze the decomposition at lower temperatures. It will decompose at room temperature when in contact with basic substances such as solid sodium or potassium hydroxide. It is somewhat sensitive to ultraviolet radiation; dimethylamine and formaldehyde are the main decomposition products.

N,N-Dimethylformamide was prepared by Campbell [891] using an alumina catalyst; see Amides, General.

Dawson and coworkers [1225] used N,N-dimethylformamide as a solvent for studying *limiting equivalent conductance*. They purified by distilling under reduced pressure; $\kappa = 1.83 \times 10^{-6}$ ohm^{-1} cm^{-1}, n_D 25° 1.4294.

Chan and Valleau [946] observed that it was the usual practice when preparing water-free N,N-dimethylformamide to treat it with a drying agent and distill under reduced pressure. They shook a Reagent grade (water below 0.05%) for 12 hr with Linde-type 4A molecular sieves. The liquid was decanted and distilled in a nitrogen atmosphere in a 50-plate bubble cap column with a head pressure of 20 torr; $\kappa = 3.5 \times 10^{-7}$ ohm^{-1} cm^{-1} at 25°.

Specific *conductance studies* of quaternary ammonium bromides were made by Sears and coworkers [4221] in N,N-dimethylformamide that had been purified by allowing it to stand in contact with potassium hydroxide for several days, taking a large center cut from an atmospheric distillation, and finally retaining a cut from a 10 torr fractional distillation. Criterion of purity; $\kappa = 0.6-2.0 \times 10^{-7}$ ohm^{-1} cm^{-1}.

Brummer [807] purified commercial N,N-dimethylformamide as a solvent for *temperature and pressure coefficient of ionic conductance*. It was dried for three days over type 4A molecular sieves. During this time, several changes of sieves were made; finally the N,N-dimethylformamide was run slowly down a column of fresh sieves into a still that had been flushed with dry nitrogen. It was fractionally distilled through a 1-meter, Penn State-type column packed with "Fenske helices" and collected as soon as the specific conductance had reached 0.8×10^{-7} ohm^{-1} cm^{-1}, usually after 200 ml had distilled from a 2000-ml charge. The conductance steadily decreased during the course of the distillation, and the final liter had a specific conductance of $0.2-0.5 \times 10^{-7}$ ohm^{-1} cm^{-1}. Dry nitrogen was passed through the apparatus during the distillation.

Thomas and Rochow [4713] studied the *conductance of ionophoric compounds* in the solvent and state ". . . it is very difficult to purify." Trace-amounts of proton-releasing impurities have a profound effect on the observed conductimetric behavior of any solute that yields a hydrogen halide as a solvolysis product. Every effort was made to eliminate such impurities or to estimate their concentrations, so that their effects could be taken into account in the interpretation of the results. Alcohols, water, and primary and secondary amines are deleterious impurities in this respect.

It is doubtful that distillation alone can remove water from N,N-dimethylformamide. When 20% water is added to the amide, a chemical reaction takes place above 20°. It is recommended that a chemical method be used to remove protonic impurities, and necessarily one that does not cause decomposition.

Technical grade N,N-dimethylformamide was first dried by adding about 10%v of benzene, which had been dried over calcium hydride. The benzene-water azeotrope was removed by distillation at atmospheric pressure. (*Author's note:* slow heating recommended to prevent superheating.) N,N-dimethylformamide dried in this manner is referred to as "benzene-dried."

ACS Reagent grade magnesium sulfate was heated overnight at 300–400° and 25 g was shaken with each liter of benzene-dried N,N-dimethylformamide for one day. A comparable amount of the heated magnesium sulfate was added and the amide was distilled at 15–20 torr in a 3-ft, vacuum-jacketed column packed with "steel" helices. The middle fraction of the distillate was used; $\kappa = 0.9-1.5 \times 10^{-7}$ ohm^{-1} cm^{-1}. The specific conductance was reduced to $0.3-0.9 \times 10^{-7}$ when a middle fraction was treated with 50 g per liter of Merck Chromatographic grade alumina powder that had been heated overnight at 500–600°. The solvent was distilled from an additional portion of the heated alumina at 5–10 torr and the middle fraction retained.

It is pointed out that specific conductance is no criterion of purity with respect to freedom from covalent impurities. Methods are given for estimating these impurities.

Prue and Sherrington [3844] removed the major part of the water from Reagent grade N,N-dimethylformamide as the benzene azeotrope [4713]. The product was stored in glass bottles with glass stoppers. About 800-ml portions were shaken with phosphorus pentoxide for three days; about 10 g of fresh oxide was added each morning. The amide was decanted and shaken with potassium hydroxide pellets to neutralize the formic acid. It was fractionally distilled at 50–64° at 15–30 torr with a slow stream of dry nitrogen passing into the liquid. The middle fraction was collected for use. The water was estimated to be 4×10^{-3} M by reaction with triphenylsilyl chloride. The purified material was used for *conductance studies to test the Fuoss-Onsager equation* and to determine the *size of ions* in N,N-dimethylformamide.

The high solvent capacity of N,N-dimethylformamide combined with its transparency in the visible and near ultraviolet regions of the spectrum has accounted for its serviceability in spectrographic work. Zaugg and Schaefer [5325] purified N,N-dimethylformamide by the modification of a combination of the methods of Thomas and Rochow [4713] and Prue and

Sherrington [3844] for *spectrophotometric studies*. Two liters of Reagent grade N,N-dimethylformamide was mixed with 400 ml of Reagent grade benzene and distilled at atmospheric pressure until the vapor temperature reached 130°. The undistilled liquid was transferred to two 1-liter bottles that had been flushed with dry nitrogen. About 25 g of phosphorus pentoxide was added to each bottle and the bottles closed with Teflon-covered rubber stoppers, then shaken for about 3.5 hr. After the solid had settled, the liquid phase was transferred to two other dry nitrogen-filled bottles and 50 g of potassium hydroxide pellets was added to each. They were shaken for about 1.5 hr. The liquid was decanted into a 3-liter, round-bottom flask with side arm, filled with nitrogen and distilled at 20–25 torr in 1 atm of dry nitrogen. The dry nitrogen was bled into the flask through a fine capillary tube during the distillation. The distillation was repeated and the center cut taken for use.

The *vacuum ultraviolet spectrum* (1500–2340 Å) was determined by Hunt and Simpson [2322] on a "quite pure" N,N-dimethylformamide that was dried over magnesium sulfate for 24 hr and then over potassium hydroxide to remove water and formic acid. It was distilled at atmospheric pressure and the middle fraction used; bp 152.6° at 760 torr, n_D 25° 1.4281.

Crompton and Buckley [1136] used N,N-dimethylformamide as a *solvent for polarographic studies*. The high purity material obtained from the chemical supply houses is sufficiently pure. A blank polarogram will show whether or not a purification is necessary. The method of Sears and coworkers [4221] will produce satisfactory material.

N,N-Dimethylformamide has been purified as a *solvent for nonaqueous titrations* by treating with air-dried Dowex 1-X10(OH form) resin [3421].

Bender and coworkers [507] purified N,N-dimethylformamide, bp 148–150°, for *solvent use in thin layer chromatography* by distilling over sodium bicarbonate.

The du Pont de Nemours & Co. [1409] found that molecular sieves can be used to lower the water content from 0.5–1.0% to a few parts per million. Ion exchange resins have been used successfully to remove ionic impurities. Calcium hydride has been used successfully as a drying agent.

Water in N,N-dimethylformamide may be determined by the Karl Fischer method [1409]. Infrared absorption, generally at 9.2 microns, provides an accurate and fast method of analysis.

SAFETY

N,N-Dimethylformamide was found by Martelli [3164] to affect the kidneys, liver, and digestive tract of guinea pigs when given orally or subcutaneously. Liver disturbances were found in workers subjected to vapors. There were numerous complaints of stomachache, headache, loss of appetite, and nausea from workers exposed to less than 20 ppm [4509]. The Russians apparently have done considerable toxicological investigation of N,N-dimethylformamide. In 1962, 0.001 mg/liter was recommended as the MPC (TLV) and 10.0 mg/m³ was recommended in 1964 [2493, 2711, respectively]. An extensive study of the toxicological properties has been made by du Pont [1409]. N,N-Dimethylformamide is capable of producing both acute and chronic effects. It is absorbed through the skin and by breathing the vapors.

The *threshold limit value* is 10 ppm or 30 mg/m³ [4509].

The *flammable limits* in air are 2.2 and 15.2%v. The *minimum ignition temperature* in air is 445° [1409].

304. ACETAMIDE

Acetamide is a colorless deliquescent crystalline substance. It is odorless when pure but otherwise has a mousy odor. It is an excellent solvent for both organic and inorganic compounds.

Walker and Johnson [5058] in 1905 apparently were the first to recognize the unusual solvent and ionizing ability of acetamide. They *studied the compound as a solvent and as an ionizing solvent.* It was recrystallized twice to remove the chief impurities, ammonium acetate and water. The specific conductance at 100° was 4.3 × 10⁻⁵ ohm⁻¹ cm⁻¹. They were of the opinion that the conductivity could be lowered comparable to that of other liquids by additional crystallizations. See also Walden [5025]. Menschutkin [3272] in 1907 followed with additional information of acetamide as a solvent for inorganic compounds and showed that acetamidates are similar in nature to hydrates. Stafford [4473] determined the approximate solubility of about 400 organic and 200 inorganic compounds and reported his findings in 1933. He concluded that acetamide has a wider solvent ability than any other substance that had been reported.

An extensive study of liquid acetamide as a solvent was made by Jander and Winkler [2375] in 1959. There is not much new that could not have been predicted by considering earlier work and the general characterization of amides. The study does present a useful panorama of acetamide as an ionizing solvent, showing its similarity to water in many respects. The amide ionizes and solvates, therefore, giving rise to a system of acids and bases that permits conductimetric, potentiometric, and similar types of measurements.

Müller [3440] studied the *stable and unstable forms* of acetamide and *determined the heats of fusion and heats of crystallization.* Many of the melting points that had been reported were low because acetamide absorbs water from the air. See also Peterson [3728], Amides, General.

A refinement of the classical method for the preparation of acetamide is presented by Coleman and Alvarado [1040] who reacted acetic acid with the proper amount of ammonium carbonate. The product was fractionally distilled and the desired fraction recrystallized from benzene-ethyl acetate solvent.

Kirsanov and Zolotov [2614] heated acetic acid and sulfamide in pyridine 3 hr at 100°. It was treated with sodium carbonate and extracted with acetone; yield 84%.

Wagner [5015] is of the opinion that complete purification of acetamide by distillation at atmospheric pressure probably is impossible. Hitch and Gilbert [2206] observed that acetamide distilled at atmospheric pressure had a lower melting point than that purified by crystallization. The high temperature causes some dehydration to acetonitrile and water, which lowers the melting point of the distillate. Wagner agrees and presents a purification procedure by crystallization that yields a product of crisp, well-formed, needle crystals that are dry and without odor. He discusses the disadvantages of the several solvents that are used. Ethyl ether gives the best material but the solubility is so low at the solvent's boiling point that the preparation of only small amounts of the amide is practical. The solubility in chloroform, benzene, and a mixture of benzene and ethyl acetate is so great that their saturated solution "sets" and the crystals and mother liquor cannot be separated satisfactorily. He combined the solvents of excessive and sparse solubilities to produce the following simple purification procedure.

Dissolve the acetamide in hot methanol in the ratio of 1 g of amide to each 0.5 ml of solvent. If the solution is not clear, filter rapidly with suction or through cotton into a vessel that will permit a 10-fold dilution. Wash the flask and filter with an amount of warm methanol to bring the ratio to 1 g amide per 0.8 ml of alcohol. Measure a volume of ethyl ether of 8–10 ml for each gram of acetamide. Add the ether slowly at first and with stirring, but avoid splashing. Crystallization usually begins when about one-fourth, or slightly more, of the ether has been added. Seed at this point, if necessary, to start crystallization. When crystallization starts, stop stirring and stop ether addition; allow the liquid to rest until the crystallization appears to be about complete. Add the remainder of the ether rapidly with vigorous stirring and cool in an ice bath. Filter the crystals rapidly with suction. Wash once with cold ether. Continue the suction only long enough to remove most of the ether. Finish the solvent removal and drying *in vacuo* or over sulfuric acid.

Reagent grade acetamide was purified by Bates and Hobbs [443] for *dipole moment studies* by crystallization from methyl acetate, an equal volume of ethyl acetate and benzene, and finally from benzene. It was then sublimed and stored over phosphorus pentoxide for two weeks. The very dry material was difficult to handle during weighing because a high static electrical charge had accumulated.

Acetamide was purified by Kumler and Porter [2779] for *electrical measurements* by distillation, once crystallizing from ethyl acetate, subliming in vacuum, crystallizing once from ethyl acetate, and three times from chloroform. It was dried at 70° and kept in a desiccator over phosphorus pentoxide. The product was odorless and melted at 80.0–80.1°. Belladen [504] measured the *conductance of some electrolytes* in liquid acetamide. The amide was prepared from ammonium acetate and acetic acid [2454a]. It was distilled repeatedly at 220 torr, shaking the middle fraction with anhydrous ether and vacuum-drying; specific conductance at 90° was 1.5×10^{-5} ohm^{-1} cm^{-1}.

Hirano [2199] measured the specific *conductance* of acetamide that had been fractionally distilled, recrystallized three times from benzene by the Lewin and Vance [2929] method, and dried in an Abderhalden drier for 5 hr; 83.2° $\kappa = 8.8 \times 10^{-7}$, 87.5° $\kappa = 1.0 \times 10^{-6}$ ohm^{-1} cm^{-1}.

Friend and Hargreaves [1667] purified acetamide for *structural studies by viscosity function* from material that was crystallized twice from a mixture of benzene and dry ethyl acetate.

Davis [1217] recommends acetamide as the solvent for laboratory experiments in general and physical chemistry for freezing point depression for *molecular weight determination* and for *solubility experiments*. Reasonably accurate results have been obtained by high school and college freshmen chemistry students with simple equipment and mass measurements to 1 mg and temperature to 0.1°. The curve for acetamide characteristically exhibits a short supercooling period and a well-defined freezing plateau. He points out that acetamide can serve as a solvent for research-oriented experiments.

Acetamide was successfully purified by Schildknecht and Vetter [4150] by zone-refining.

305. N-METHYLACETAMIDE

N-Methylacetamide like all N-alkyl amides of the lower fatty acids, has wide solvent application. It is easy to prepare and to purify. It is characterized by a high dielectric constant (165.5 at 40°) and extensive solubility and dissociating ability [1231]. The characteristics of solutions of a number of salts in this solvent have been studied. Most soluble salts are completely ionized.

Dawson and coworkers [1228] prepared N-methylacetamide by reacting methylamine and acetic acid and subsequently heating to split out and distill water. The product was purified by fractional distillation, followed by five, or more, fractional freezing cycles using the procedure of Berger and Dawson [522]; see formamide.

Knecht and Kolthoff [2636] found that commercially available N-methylacetamide was not sufficiently pure for *polarographic studies*. At best, it contains methylamine and acetic acid. They studied methods of preparation and purification that would yield large volumes of a solvent sufficiently pure as a polarographic solvent. It was prepared by allowing an excess of acetic acid to react with 40% methylamine under reflux for 12–15 hr. A preliminary purification gave a "crude product." Two purification procedures were developed.

Purification Procedure 1.   The crude product contained only acetic acid and methylamine. It was fractionally distilled under an atmosphere of nitrogen through an 18-in. heated column packed with porcelain saddles. The major fraction distilling at 204–206° was found to contain three impurities: methylamine, acetic acid, and impurity "X." These impurities were present in small amounts and were removed by extracting "X" with petroleum ether. Water was added to separate the phases. The water and dissolved petroleum ether were removed by simple distillation to 130°. The pot temperature was then allowed to cool to room temperature. About 10 ml of concentrated sulfuric acid was added and the mixture distilled at about 80° at 3–5 torr. The acetic acid and most of the remaining water were removed by allowing the mixture to stand several hours over calcium oxide, after which it was filtered and vacuum-distilled.

Purification Procedure 2.   The crude product was distilled slowly to 130° to remove as much of the methylamine, acetic acid, and water as possible. The methylamine was removed by adding sulfuric acid and distilling as in Procedure 1. There was too much acetic acid to remove with calcium oxide. Potassium oxide was added to a known excess and the mixture allowed to stand several hours, filtered, and vacuum-distilled. When potassium acetate started to precipitate, the mixture was cooled and filtered. The distillation of the filtrate was continued.

Pure N-methylacetamide has no detectable odor.

CRITERIA OF PURITY

The criteria of purity were: (1) the polarographic range, determined with 0.1 M $(C_2H_5)_4NClO_4$ supporting electrolyte, between $+0.35$ and -2.75 volts, (2) water content by the Karl Fischer method, and (3) specific conductance less than 3×10^{-7} ohm^{-1} cm^{-1}.

It is recommended that the original article be consulted if this method is to be used.

N-Methylacetamide was vacuum-distilled from calcium oxide and fractionally crystallized through several freezing cycles by Dawson and coworkers [1229] as a *solvent for the determination of the dielectric constant of acetic acid;* $\kappa = 5 \times 10^{-8}$ ohm^{-1} cm^{-1} and n_D 32° 1.4282.

Lin and Dannhauser [2958] made *dielectric constant studies* with N-methylacetamide that was purified by distilling at reduced pressure through a 1.5×80-cm glass helices packed column at a high reflux rate.

French and Glover [1632] purified N-methylacetamide as a *solvent for the conductance studies of salts* as follows: The liquid was shaken with phosphorus pentoxide, filtered through glass wool, and vacuum-distilled. This treatment was repeated three times, followed by two distillations without the phosphorus pentoxide treatment. The normal specific conductance was 6×10^{-7}; the lowest found was 4.2×10^{-7} ohm^{-1} cm^{-1}. Their criteron of purity, other than low electrical conductance, and the one that determined the selection of the method of purification, was the constancy of the conductance with time. The electrical conductance of the purified material rose with time by all other methods tried.

Commercial N-methylacetamide was purified by Bonner and coworkers [660] by vacuum distillation, followed by zone refining in an apparatus designed especially for this research. Atmospheric contaminations were avoided during transfer by the use of a 50-ml hypodermic syringe. N-methylacetamide is unstable in the presence of air. The purified material was used to determine the *freezing point, freezing point constant, heat of fusion, and dielectric constant.* Appreciable changes are believed to occur in the hydrogen bonding of the solvent when small quantities of impurities are present.

Rey-Lafon [3961] *determined the infrared spectra* of N-methylacetamide in 33 inert solvents, including proton donors and acceptors. The bands for the CO and NH groups were similar to those in cyclohexanone and pyrrol in dilute solutions of the same solvents.

306. N,N-DIMETHYLACETAMIDE

High molecular weight resins and polymers are soluble in N,N-dimethyl-acetamide. The lower paraffinic hydrocarbons have only a limited solubility at ambient temperatures. It is thermally stable at its atmospheric boiling point and in the presence of acidic and alkaline substances. It decomposes above 350° to dimethylamine and acetic acid. It is not very hygroscopic. It acts as a dehydrohalogenating agent with halogen compounds. The heat of reaction with highly halogenated substances is large and the reaction may become violent, particularly in the presence of iron [1408].

Campbell prepared N,N-dimethylacetamide (see Amides, General) over alumina catalyst at 230° [891]. The crude amide was fractionally distilled to remove the impurities; yield 99.5% containing very small amounts of acetic acid and water.

Friend and Hargreaves [1667] prepared N,N-dimethylacetamide for *structural studies through viscosity* by heating dimethylammonium chloride with an excess of alkali, drying with solid potassium hydroxide. The dimethyl-amine was passed into a solution of acetyl chloride in benzene. The precipi-tated amide was filtered and fractionally distilled.

Schmulbach and Drago [4171] found that Eastman white label N,N-dimethylacetamide contained two impurities after shaking with barium oxide and fractionally distilling through a 2.0 × 40-cm, vacuum-jacketed Podbielniak column at reduced pressure. The center fraction boiled at

58.0–58.5° at 11.4 torr and was retained. Chromatography indicated that the impurities were below 0.01 %.

Drago and coworkers [1361] purified N,N-dimethylacetamide for *solvolysis studies of iron (III) complexes* by shaking with barium oxide and distilling at reduced pressure from barium oxide. The middle fraction was distilled at reduced pressure with calcium hydride.

Gutmann and coworkers [1942] purified N,N-dimethylacetamide as a *polarographic solvent* by fractionally distilling twice from calcium hydride at 10 torr under nitrogen in an 80-cm Vigreux column, and finally without the hydride.

SAFETY

Toxicity by dermal contact and by chronic inhalation was investigated by Horn [2250]. It was found that the amide was readily absorbed into the bloodstream by both routes and that repeated exposure to relatively low levels resulted in liver damage. In [1408] du Pont cautioned that N,N-dimethylacetamide should be handled with care, that contact of the liquid with the skin and the eyes should be avoided. Spills on the skin should be immediately and thoroughly flushed with water; if splashed in the eye, irrigate for 30 min, and get medical care. Breathing N,N-dimethylacetamide, vapors should be avoided.

The *threshold limit value* of 10 ppm, 35 mg/m³, has been recommended by du Pont [1408].

The *flammable limits* in air are: at 100°, 1.70–18.5%v; at 200°, 1.45–15.2%v; and the upper limit at 160° and 740 torr, 11.5%v. The *minimum ignition temperature* in air is 420° [1408].

307. N-METHYLPROPIONAMIDE

Only the *trans* isomer of N-methylpropionamide was found by *configurational studies by infrared* made by La Planche and Rogers [2827] in a sample of the amide that had been purified by drying over anhydrous sodium sulfate and fractionally distilling *in vacuo*.

Leader and Gromley [2848] passed gaseous methylamine into propionic acid until an equivalent weight had been absorbed. It was heated at 160–170° to split out water. The pressure was gradually reduced and it was fractionally distilled *in vacuo* through a 1 × 35-cm glass helices packed column. It was further purified by successive crystallizations for *dielectric constant studies*. Hoover [2245] used the same preparation and purification for *conductance studies*, except that the final purification was by repeated fractional distillation at 5 torr in a 4 × 1000-cm column packed with 3-mm helices.

D'Alelio and Reid [1182] prepared the N-methyl amides of the first nine members of the homologous series of n-RCOOH acids by slowly dropping the corresponding acid chloride into three moles of concentrated aqueous solution of methylamine at −10 to −20° during the addition. Solid potassium hydroxide was added and the amide separated. It was then dried with solid potassium hydroxide and distilled. Ross and Labes [4049] prepared N-methylpropionamide by the D'Alelio and Reid method and distilled from calcium oxide before use for *solvolyses studies;* $\kappa = 8.3 \times 10^{-5}$ ohm^{-1} cm^{-1}, 0.002–0.008% water by Karl Fischer.

N-Methylpropionamide was prepared for Beer and coworkers [496] for *infrared spectra studies* by the reaction of propionic anhydride with aqueous methyl amine. The amide was salted out, dried, and distilled; bp 104° at 16 torr, n_D 24.1° 1.4313.

Dawson and coworkers [1226] prepared N-methylpropionamide as a solvent for the *study of solutions of potassium halides* by preparing the methylammonium propionate. Xylene was added and the mixture heated to split out the water. The water and unreacted propionic acid were removed as their respective xylene azeotropes. The amide was fractionally distilled several times at 5 torr through an efficient column; specific conductance at 30° was 3×10^{-6} ohm^{-1} cm^{-1}.

The amide was prepared by Meighan and Cole [3260] for *dielectric property studies* from 40% aqueous methylamine and propionic acid. Water was removed by distillation. It was dried over calcium oxide and vacuum-distilled three times. Bass and coworkers [432] prepared the amide in the same manner but distilled five times.

308. 1,1,3,3-TETRAMETHYLUREA

1,1,3,3-Tetramethylurea is unique among the solvents. It is an amide with a relatively low dielectric constant. It is an excellent solvent for organic substances, particularly aromatics. It is a suitable reaction medium for the base catalyzed isomerization, alkylation, acylations, and other condensation reactions. It is inert toward Grignard compounds and makes a useful solvent for these reactions and for compounds sparingly soluble in ethers [3042]. It is a solvent for acetylene and for polyacrylonitrile. It dissolves inorganic compounds to about the same extent as does acetone. 1,1,3,3-Tetramethylurea has the dissolving power and miscibility of pyridine. It has a more convenient boiling point and can be made practically anhydrous by simple distillation. It has a mild and pleasant odor.

Michler and Escherich [3302] first reported the preparation of 1,1,3,3-tetramethylurea in 1897 from dimethylamine and phosgene. Lüttringhaus and Dirksen [3042] reported that the reaction proceeded via N,N-dimethylcarbamyl chloride (bp 164–165°), which is easy to isolate. This procedure is not recommended [3042] for large scale preparative purposes because of the amount of heat evolved and the formation of a troublesome by-product salt. There have been several modifications of the original method that would be of interest to anyone using the phosgene method. Of greatest interest is circumventing the formation of dimethylammonium chloride [4069, 4488].

Dirksen [1315] has developed a method that overcomes the troublesome features of the phosgene method. He charged 1.17 moles (135 g) of 40% aqueous dimethylamine into a 500-ml four-neck flask, equipped with two dropping funnels and a stirrer, and immersed into a cooling bath. One mole (107 g) of dimethylcarbamyl chloride was added drop by drop, so that the temperature did not get above $-10°$. After the first 20 ml of the chloride was added, 100 ml of 40% sodium hydroxide was added at such a rate that the two dropping funnels were emptied at about the same rate. Sixty grams of sodium hydroxide was added and two layers formed. The upper layer was separated and freed from salts by reduced pressure distillation. The lower layer was extracted three times with benzene and combined with the distillate. The benzene, water, and dimethylamine were removed by distillation. The residue was fractionally distilled at 10–20 torr through a short column; yield 78–85% based on the acid chloride.

Lüttringhaus and Dirksen [3042] describe a method of Farbenfabriken Bayer: Introduce gaseous dimethylamine into 10 moles (2140 g) of diphenyl carbonate externally cooled with water until the increase in weight is 25 moles (1150 g). Heat the mixture 4 hr at 200° in an autoclave at about 20 atm. Add 7 liters of 5 N sodium hydroxide after cooling and extract the mixture with ether for at least 24 hr. Fractionally distill through a short column and collect the fraction boiling at 60–66° at 13 torr; yield 850 g. Further distillation of the residue yields 345 g of the urethane, which can be reused.

1,1,3,3-Tetramethylurea was purified by Foerster and coworkers [1589] to study its *solvent properties* of inorganic substances by drying over porous barium oxide and then distilling in an atmosphere of nitrogen.

SAFETY

Dixon and coworkers [1320] made a study of the toxicity of tetramethylurea to mice and monkeys. Monkeys tolerated 150 mg/kg/day for five days without showing signs of toxicity. The LD_{50} was found by Hotovy [3042, ref. 56] to be 1.1 g/kg for rats.

309. 2-PYRROLIDINONE

Sakai and Tsunakawa [4094] prepared 2-pyrrolidinone by mixing 53.2 g γ-butyrolactone, 53.5 g ammonium chloride, and 40 g of sodium hydroxide in an autoclave, heating to 210°, filtering, and distilling *in vacuo;* yield 69.4 g, bp 139° at 22 torr.

Lynn [3047] prepared 2-pyrrolidinone in 91% yield by hydrogenating 254 g of ethyl 3-cyanopropionate (prepared by a combination of the Kurtz method [Ann. Chem 572, 52(1951)], 750 ml methanol, 50 g wet Raney nickel, and

117 g of ammonia at 90–95° and 100 psi for 1 hr. It was purified by distillation.

The *dipole moments* were determined by Fischer [1576] on commercial 2-pyrrolidinone and 1-methyl-2-pyrrolidinone by vacuum distillation immediately before use.

SAFETY

Antara Chemicals [259] describes 2-pyrrolidinone as a mild primary skin irritant with definite sensitizing properties. Prompt removal from the skin with soap and water is recommended.

The oral toxicities for white rats and guinea pigs are LD_0 4.0 ml/kg, LD_{50} 6.5 ml/kg, and LD_{100} 9.5 ml/kg [259].

310. 1-METHYL-2-PYRROLIDINONE

1-Methyl-2-pyrrolidinone is a stable, hygroscopic liquid completely miscible with water. It can be hydrolyzed by 4% aqueous sodium hydroxide to 50–70% in 8 hr; the lactam ring can be split by concentrated hydrochloric acid when exposed for an extended period of time.

It is a lactam with good stability that is a versatile solvent with a wide utility. The Antara Division of National Aniline [1737] further characterizes it as a maximum safety solvent with a high flash point, low vapor hazard, high boiling and low freezing points, and chemical and thermal stability.

Tafel and Wassmuth [4595] prepared 1-methyl-2-pyrrolidinone in benzene solution from the sodium salt of pyrrolidinone and iodomethane by shaking for 2 hr at 60–70°. The sodium iodide was removed by filtration. The filtrate was distilled and the portion distilling at 197–202° and 736 torr was saved.

Sears and coworkers [4219] purified 1-methyl-2-pyrrolidinone as one of the solvents for studying the *dipole moment and solution viscosity* of sulfamic acid. The lactam was dried by removing the water as the benzene azeotrope. It was fractionally distilled at 10 torr through a 100-cm column packed with glass helices and the middle 60% retained.

Fischer [1576] purified for *dipole moment;* see 2-pyrrolidinone.

SAFETY

Stasenkova and Kochetkova [4478] reported the LD_{100} and LD_{min} as 5 and 3 g/kg, respectively. General Aniline [1737] gives LD_{100} and LD_0 as 10 and 3 ml/kg, respectively, for albino rats for acute oral toxicity. They also reported the inhalation of saturated vapor to be nontoxic and suggested a threshold limit value of 0.05 mg/liter. A TLV of 100.0 mg/m³ was suggested for Russia in 1964 [2711].

The *minimum ignition temperature* in air is 346° [4472].

311. ε-CAPROLACTAM

ε-Caprolactam, hereinafter called caprolactam, is a high melting, 69.2°, solvent that slowly polymerizes at its melting point. It is very hygroscopic. It is an excellent solvent, particularly for high molecular weight polymeric materials that are difficult to make soluble, or that are insoluble, in other solvents.

Scott and coworkers [4211] prepared caprolactam by rearranging cyclohexanone oxime. The caprolactam-acid phase was neutralized with ammonia and the oily upper layer separated, and then dried and purified by distillation at 2.5 torr. The distillate was crystallized from a mixture of 185 ml of petroleum ether, bp 70°, and 30 ml of 2-methyl-2-propanol; yield 88%, mp 70.5–71.5°.

Either nitromethane, nitroethane, or 1-nitropropane was treated by Kimura and coworkers [2592] in sulfuric acid in a mole ratio of 1:5 at 125–130° and hydroxylammonium hydrogen sulfate was formed in about 90% yield. Cyclohexanone was added at 120–125°. The oxime was formed and the rearrangement proceeded smoothly.

Caprolactam deteriorates on long standing; in the liquid state it turns yellow in contact with air. The yellow color formation may be suppressed by the addition of basic substances such as sodium hydroxide, carbonate, or tetraborate; or lithium acetate in concentrations of about 0.2 g per liter [2344]. Indest and coworkers [2343] found that caprolactam may be stabilized with 0.001–1% of sodium trichloroacetate. Vereinigte Glanzstoff-Fabriken [4945] found 0.07% hexamethylenetetramine to be effective.

A purification procedure was reported [661] that gives a product that does not discolor on exposure to the air and light: Dissolve 250 g of discolored caprolactam in 750 g of nitromethane and 6 ml of 30% hydrogen peroxide, heat 1 hr and evaporate. Distill the lactam *in vacuo* in the presence of 0.5% sodium hydroxide; yield 288 g.

Kornev and coworkers [2715] purified caprolactam by mixing 600 g of the amide with 21 ml of 40% sodium hydroxide and 6 g of paraformaldehyde. Light ends and water were distilled at 65–80 torr until foaming ceased. About 10% of the caprolactam was distilled at 5–10 torr at 125–135° and added to the next purification batch. The distillation was continued rapidly to produce pure caprolactam in good yield.

Commercial Solvents Corporation [1052] substantially freed caprolactam from volatile bases by refluxing 1 hr in a stream of nitrogen at 30 torr or for a shorter time with 0.1% sodium hydroxide present, and then distilling at 10 torr.

Snider and coworkers [4437] reported that oxidizable impurities may be removed from caprolactam by heating at 50° with about 15% water and 25% sodium hydroxide solution. The aqueous phase was separated and the lactam distilled *in vacuo* at 145° and crystallized.

Grekov [1889] purified caprolactam by zone melting.

Hampton and Riddick [1996] found caprolactam to be an excellent solvent for the determination of molecular weights by the freezing point method for substances that are not sufficiently soluble in other cryoscopic solvents and for many substances that are difficult to make soluble, where the degree of polymerization is not large. It was found suitable for routine laboratory use for molecular weight determination for polymers in the range of $(M)_2-(M)_{20}$. It was not tried for polymers above $(M)_{20}$. Caprolactam was the only solvent that would dissolve some polymers.

The freezing point constant was determined using naphthalene and biphenyl. The freezing point apparatus was essentially the freezing tube and stirrer described by Glasgow and coworkers [1795] actuated by a windshield wiper motor. The freezing tube was unsilvered and the inner tube of uniform diameter. A differential thermometer was used.

The freezing point constant was determined in the following manner: The melting point tube was heated to 80–100°, the heat turned off, and the tube stoppered with a cork. It was allowed to cool to about 60° and 25–30 g of caprolactam, accurately weighed, was added. The heat was turned on, so that the rate of heating was about 1° per min to about 70°. The heat was then reduced and the lactam melted. When all of the solid was melted, the heat was shut off, the stirrer started, 0.5 g of 20–40 mesh 3A or 4A molecular sieves added. The apparatus was evacuated to about 5 torr and the sample added. There was only about 0.1° supercooling. The temperature-time data were taken, from which the freezing point was determined. The freezing point of the solvent was determined in the same manner without the addition of the sample.

The range and standard deviation of the melting points of 10 replicates were 0.332° and 0.110°, respectively, for caprolactam without molecular sieves. The same values for eight replicas with molecular sieves added were 0.084 and 0.029°.

The average freezing point constant for 28 replicates, using naphthalene and biphenyl as the solute, was 7.3° with a standard deviation of 0.2°. Some values were determined by adding only one increment of solute to the solvent, and others by adding several increments. The precision of replicates is not nearly as good with caprolactam as it is with such solvents as benzene and camphor, for instance. The standard deviation may be reduced to below 0.1° if the system is kept under 1 atm of nitrogen.

The K_f calculated from the heat of fusion is 6.8 [3203].

SAFETY

Caprolactam has a low order of toxicity and presents no appreciable health hazard if handled properly [58].

SULFUR COMPOUNDS

Specific Solvents

Sulfides

312. CARBON DISULFIDE

Prior to about 1950, carbon disulfide was manufactured by the high temperature reaction of carbon and sulfur vapors. The natural gas-sulfur process is used almost exclusively in the United States at the present time.

Carbon disulfide is an extremely volatile and flammable liquid. The pure substance has a weak odor and keeps well in the dark. Gillo [1776] proposed a method in 1939 for the purification of carbon disulfide and criteria of purity for its use as an *organic standard*. Waddington and coworkers [5009] reported in 1962 that carbon disulfide met the requirements as a *reference standard* for vapor flow calorimetry. They purified C. P. carbon disulfide in an efficient fractional distillation column. The purity was determined from the freezing curve to be 99.98%m. The Swietoslawski ebulliometric Δt was 0.001°.

Brown and Manov [785] purified carbon disulfide for *heat capacity measurements*. It was dried with calcium chloride and fractionally distilled several times; then two liters was fractionally distilled in an atmosphere of helium through a 60-cm column packed with brass shoe eyelets. About half of the distillate was collected as the middle fraction. The fractionation procedure was repeated twice. The carbon disulfide contained less than 0.001% impurities as determined by fractional melting.

Obach [3566] purified carbon disulfide by first distilling over lime and then treating with potassium permanganate, 5 g/liter. The carbon disulfide was allowed to stand until the hydrogen sulfide was completely removed. After separating, the liquid was shaken for some time with mercury to remove sulfur, poured off and shaken with 25 g/liter of mercury (II) sulfate until the unpleasant smell of carbon disulfide had disappeared. The liquid was poured off and distilled from calcium chloride, bright daylight being excluded. The purified material was stored in the dark. See also Hammick and Howard [1991].

Chenevier [960] allowed carbon disulfide to stand with 0.5 ml of bromine per liter for 3–4 hr, removed the bromine by shaking with potassium hydroxide solution or copper turnings, and dried over calcium chloride. McKelvy and Simpson [3080] purified 12 liters of a technical product by distilling from 900 g of ceresin and distilling the middle fraction from fused calcium chloride.

For the *measurement of vapor pressure*, Stock and Seelig [4506] agitated carbon disulfide with mercury, distilled, dried with phosphorus pentoxide, and fractionally distilled in vacuum, avoiding all greased joints. The first fraction always had a rather high vapor pressure, probably because of the presence of carbon dioxide or carbonyl sulfide.

For *complete removal of hydrocarbons*, Ruff and Golla [4074] mechanically agitated a solution of 130 g of sodium sulfide and 150 g of water with 45–50 g of technical carbon disulfide for 24 hr at 35–40°. The sodium thiocarbonate solution was separated from the excess carbon disulfide, precipitated with 140 g of copper sulfate in 350 g of water with cooling; after the copper thiocarbonate was separated, it was decomposed with steam. The distillate was separated from water and distilled from phosphorus pentoxide.

According to Staudinger [4480] shock can cause an explosive reaction between alkali metals and carbon disulfide.

CRITERIA OF PURITY

McKelvy and Simpson [3080] recommended the critical solution point in ethanol as a criterion of purity. MacLean and coworkers [3105] characterized the effectiveness of their purification by the ultraviolet spectrum. Gillo [1776] reported the boiling point and refractive index as criteria for the material they studied as an organic standard. Waddington and coworkers [5009] characterized the material they studied as a reference standard for vapor flow calorimetry by the freezing curve and the Swietoslawski Δt.

Table 5.26 Specifications for Reagent Grade Carbon Disulfide

Distillation range	46–47°
1 ml to 95 ml, not more than 0.5°; 95 ml to dryness not more than 0.5°.	
Residue on evaporation, max	0.002%
Foreign sulfides and dissolved sulfur	TPT[a]
Sulfites and sulfates	0.002%
Water, max	0.05%

By Karl Fischer method or cool 10 ml in a test tube to 0°; no turbidity or drops of water develop.

[a] To pass test.

ACS Reagent Chemicals [76] and Rosin [4042] give the requirements and test method for *Reagent* grade carbon disulfide. They are very nearly the same. The combined requirements are given in Table 5.26.

SAFETY

The high vapor pressure of carbon disulfide together with its low minimum ignition temperature, wide flammable limits, and toxicity makes it a hazardous substance. Reagent grade carbon disulfide has a somewhat unpleasant, slightly etheral odor that does not offer adequate warning for unsafe concentrations. Carbon disulfide is readily absorbed through the lungs. It has a limited absorption through the skin. Liquid or vapors may cause dermatitis or blistering of the skin or mucous membranes [3681].

The *threshold limit value* is 20 ppm, 60 mg/m^3 [4509]. Kashin [2506] studied workers under actual conditions and concluded that the then threshold limit value (Russian) of 10 mg/m^3 was too high.

The *flammable limits* in air are 1.25 and 50.00%v. The *minimum ignition temperature* in air is 120° [3680].

Thiols

313. 1-BUTANETHIOL

Denyer and coworkers [1285] isolated 1-butanethiol from mixed thiols obtained from petroleum naphtha purification. The mixture was fractionally distilled through a 25-plate column into fractions of definite boiling ranges. The 70–110° fraction was refractionated in a 100-plate still. The 1-butanethiol was distilled as a series of azeotropes with various hydrocarbons. The thiol was separated from the hydrocarbons by extractive distillation with aniline. The purity was 99.1%m calculated from the freezing curve.

1-Butanethiol was prepared by Mathias [3212] for *refractivity studies* by the method of Backer and Dijkstra [338]. It was purified by converting to the lead salt, regenerating, and thrice fractionally distilling in a 22 theoretical plate still.

Ellis and Reid [1472] prepared aliphatic thiols (ethyl through heptyl) by adding the appropriate alkyl bromide to alcoholic potassium hydroxide solution saturated with hydrogen sulfide. The solution was gently stirred by a stream of hydrogen sulfide. The reaction mixture was maintained at 50–60° for 2 hr or at reflux for 1 hr. The alkyl thiol separated when the solution was diluted with a large volume of water. The impurities present were: bromoalkane, ethanol, and alkyl sulfide. The crude thiol was separated, dissolved in 20% sodium hydroxide, extracted with a small portion of benzene, and steam-distilled until the alkaline solution became clear. The solution was slightly acidified with 15% sulfuric acid when cooled and the thiol distilled.

The material used by Scott and coworkers [4206] in their *thermodynamic studies* between 12 and 500°K was prepared by the Laramie Station of the Bureau of Mines. The purity was 99.990 ± 0.005%m determined by the calorimetric melting point.

SAFETY

It has been reported that butanethiol caused the death of some animals in 2 hr at a concentration of 60 mg/liter [619].

314. BENZENETHIOL

Adams [15, p. 71] gives directions for preparation from 7200 g cracked ice, 2400 g concentrated sulfuric acid, 600 g crude benzene sulfonyl chloride, and 1200 g zinc dust. The yield is 359 g of crude benzenethiol or 340 g of purified material. The crude product is purified by drying over calcium chloride and distilling at 15 torr.

Morris and coworkers [3413] purified Eastman Kodak Co. benzenethiol for *physical property study* by twice fractionally distilling at 90 torr. The best fractions were combined and had a purity of 99.98 %m determined from the freezing curve.

Scott and coworkers [4209] used a portion of the standard sample prepared at the Laramie Station of the Bureau of Mines for *thermodynamic property study*. The purity was found to be 99.98 ± 0.01 %m by the calorimetric melting point as a function of the fraction melted. The sample was dried by passing the vapors through magnesium perchlorate.

A commercial product was purified by Mathias and coworkers [3215] for *physical and electrical property study* by distillation at reduced pressure in a Claisen apparatus, dried with calcium chloride, and successively fractionally distilled.

Thioethers

315. METHYL SULFIDE

Methyl sulfide is a solvent for most classes of organic compounds, for many organic resins, and for certain inorganic compounds. It also is a good solvent for a number of reactions. It is mutually only slightly soluble in water [1142].

Cumper and coworkers [1149] prepared methyl sulfide for *dipole moment studies* by slowly adding an aqueous solution of sodium sulfide to a boiling solution of iodomethane and 95% ethanol, refluxing the mixture for 2 hr and distilling. Methyl sulfide separated as a yellow oil when water was added to the mixture. The oil was allowed to stand over sodium hydroxide pellets for 24 hr, washed until neutral, and dried. It was fractionally distilled and the middle fraction repeatedly washed with 15% sodium hydroxide, then with water and dried. It was fractionally distilled from sodium.

The *preparation, purification, and physical properties* of 14 dialkyl sulfides were studied by McAllan and coworkers [3050]. Two general methods for the preparation are given.

Method I. Dissolve 1.5 moles of sodium sulfide, $Na_2S \cdot 9H_2O$, in 250 ml water in a 3-necked flask fitted with a condenser, a nonreactive sealed stirrer, and a dropping funnel. Add slowly 2 moles of the desired alkyl halide, or 1 mole each of the corresponding alkyl halides for a mixed thioether, to the vigorously stirred sodium sulfide solution. Regulate the addition to maintain a reasonable reflux. Apply external heat to continue the reflux for 3 hr after the alkyl halide has been added. Steam-distill the thioether, wash well with water, then with 10% sodium hydroxide, and finally with water; then dry with calcium chloride.

Method II. Dissolve 1.025 moles of sodium hydroxide in 250 ml of water in the same apparatus described in Method I. Slowly add 1 mole of mercaptan with vigorous stirring. When the reaction has ceased, add 1 mole of alkyl halide as rapidly as the reflux will permit. Continue as described for Method I.

Purification. Fractionally distill in an efficient still. Purify the desired alkyl sulfide fraction through the mercury (II) chloride complex. Dissolve 1 mole of mercury (II) chloride in 1250 ml of ethanol and slowly add the boiling alcoholic solution of the proper ratio of sulfide to give the complex: $2(CH_3)_2S \cdot 3HgCl_2$ or $(C_2H_5)_2S \cdot 2HgCl_2$. Recrystallize the complex to a constant melting point. Heat 500 g of the purified complex with 250 ml of concentrated hydrochloric acid in 750 ml water. Separate the sulfide, wash with water, and dry with calcium chloride; purity, methyl sulfide 99.995%m, ethyl sulfide 99.8%m from the melting curve.

The equilibrium $R_1R_2S(l) \rightleftharpoons R_1R_2S(s)$ apparently is difficult to establish during freezing. The melting curves were used to calculate the purity.

SAFETY

There is little toxicity data available for methyl sulfide. The fire hazard may be comparable to that of ethyl ether [1142].

316. ETHYL SULFIDE

Ethyl sulfide for *physical property and chemical constitution studies* was prepared by Vogel and Cowan [4979] by distilling an aqueous solution of sodium ethyl sulfate with an excess of sodium sulfide until the temperature of the reaction mixture reached 120°. The sulfide layer of the distillate was separated, shaken several times with 20% sodium hydroxide solution, and kept over sodium hydroxide pellets for 24 hr. It was washed with water until neutral, dried over calcium chloride, and distilled over sodium.

Drefahl and Schick [1365] prepared ethyl sulfide from thiourea and bromoethane in 85% yield. Eighty-six grams of thiourea was refluxed 6 hr with 50 ml of ethanol and 109 g bromoethane. The solution was concentrated, 100 ml water added drop by drop, made alkaline with 4 moles of sodium hydroxide in water, and refluxed 1 hr. It was cooled, 164 g bromoethane added drop by drop, and refluxed 5 hr.

For preparation and purification, see McAllan and coworkers methyl sulfide [3050].

Cumper and coworkers [1149] prepared ethyl sulfide by the reaction

$$Na_2S + C_2H_5I \rightarrow 2NaI + (C_2H_5)_2S$$

but could not free it from iodoethane; see Cumper et al., under methyl sulfide. It was prepared by the method of Gray and Gutekunst [1867] for butyl sulfide from fuming sulfuric acid, anhydrous ethanol, and sodium sulfide.

Approximately 3 liters of commercially available ethyl sulfide was fractionally distilled by Haines and coworkers [1964]. Fractions were selected on the basis of refractive index and composited. The freezing curve of the 1.68 liters showed a purity of 99.94%m. Samples of purified ethyl sulfide and thiophene were sealed in vacuum in clear glass ampules and exposed to sunlight for one year. Other samples were refluxed in an all-glass apparatus at atmospheric pressure for 240 hr. The apparatus was vented through traps to collect any decomposition products. Decomposition or decrease in purity was not noted for either compound in either test.

31.7 THIOPHENE

Whitmore [5156, p. 72] gives directions for preparation from sodium succinate and phosphorus trisulfide; yield 20–30% after fractional distillation; see also [616].

Keswani and Freiser [2563] washed 99%m thiophene with hydrochloric acid and then with water, dried, and carefully distilled in a 3-ft column for *dipole moment and structure studies.*

Fawcett and Rasmussen [1547] purified a commercial grade material for *physical property studies* by both physical and chemical means. The thiophene was washed successively with dilute hydrochloric acid, sodium hydroxide, and distilled water, then dried over calcium chloride. About 2 liters were fractionally distilled at atmospheric pressure at a 50:1 reflux ratio through a 235-cm column packed with 2.4-mm stainless steel helices. The first and last quarters of the distillate were discarded. Mass spectra analysis indicated 0.37%m benzene. The distillate was then fractionally crystallized six times. The purified thiophene was degassed and sealed in Pyrex-brand glass flasks. A portion of the material rejected from the fractional crystallization was treated with mercury (I) chloride in an ethanol-sodium acetate solution. The solid obtained was heated under reflux with dilute hydrochloric acid and the thiophene extracted from the cooled liquid with pentane. The pentane solution was dried over calcium chloride and fractionally distilled through a 28-plate column; fp −38.5°.

Thiophene was purified of its hydrogen sulfide impurity by Brady [706] to use in the development of a *spectrophotometric (uv) method* for its determination in producer gas by adding piperidine to the sulfur compound in the ratio of 1:50. The mixture was fractionally distilled and the heads and residue discarded.

For purification for *critical properties*, see Ethers, General [2653].

Haines and coworkers [1964] purified commercial thiophene by fractional distillation through a 1-in. × 9-ft column packed with stainless steel helices. The product was 99.99%m and shown to be benzene-free by its mass spectrum. The freezing curve failed to reveal benzene because benzene and thiophene form a solid solution [1547]. Stability tests are given under Haines and coworkers; see ethyl sulfide.

SAFETY

Mikhailets [3308] obtained a LC_{50} of 9.5 mg/liter for mice by inhalation for 2 hr for thiophene.

318. TETRAHYDROTHIOPHENE

Whitehead and coworkers [5150] prepared 1,4-dibromobutane from tetrahydrofuran in 71% yield in an estimated purity of 99.85%m. A solution of 5 moles (1203 g) of sodium sulfide, $Na_2S \cdot 9H_2O$, was made with 1100 ml water and 1350 ml ethanol. An 1100-ml portion of this solution was refluxed in a 5-liter, 3-necked flask with a reflux condenser and two dropping funnels. The remainder of the solution was placed in one of the funnels and 3.5 moles of dibromide in the other. The liquids were added to the flask at such a rate that both funnels were emptied at the same time, about 30 min. The crude material was purified by crystallization of the mercury (II) chloride complex to a constant melting point; see methyl sulfide, McAllan and coworkers [3050]. The regenerated tetrahydrothiophene was washed, dried, and fractionally distilled in a 20-plate, glass-packed still at reduced pressure. There was decomposition when the distillation was done in a 100-plate still at atmospheric pressure.

Pipparelli et al. [3770] prepared tetrahydrothiophene from tetrahydrofuran and hydrogen sulfide using an aluminum oxide gel catalyst at 400°. Yields of 88–90% were obtained under optimum conditions.

Lawson and coworkers [2908, p. 89] describe a method of preparation from 1,4-dichlorobutane and sodium sulfide; yield 160–170 g or 73–78%.

Commercially available material from the Oronite Chemical Co. was fractionally distilled by Haines and coworkers [1964]. Several 30-ml cuts were made. The freezing points and mass spectra were used to select succeeding cuts that were composited; purity 99.95%m.

Oxo-Sulfur Compounds

319. DIMETHYL SULFOXIDE

Dimethyl sulfoxide has been known for more than a century, but only recently has it been available commercially. It is a versatile solvent. Schläfer and Schaffernicht [4157] discuss it as a solvent for inorganic compounds; see also [2]. Martin and others [3169] discuss its use as a solvent for organic reactions and for organic compounds. It is a useful solvent for a number of gases [2], and a variety of polymers [1766, 1961, 3137]. Kononenko and Herstein [2707] found dimethyl sulfoxide probably to be the best non-protogenic solvent for sucrose. Gutman and Schober [1943] used dimethyl sulfoxide as a *solvent in polarographic studies* for inorganic ions. It has favorable attributes as a cryoscopic solvent for molecular weight determination and has been used as the stationary phase in paper chromatographic separation.

Dimethyl sulfoxide is a colorless, odorless, hygroscopic liquid with a slightly bitter taste. It has an equilibrium moisture content of 10% with air at 20°. The partial decomposition observed during boiling at atmospheric pressure is dependent on the kind and amount of impurities present. Thermal decomposition is catalyzed by acids. Many weak bases, and basic and neutral salts inhibit decomposition.

LeBel and Goring [2849] purified commercial material for *physical property studies* of the dimethyl sulfoxide-water system. The solvent was kept in contact with Drierite or barium oxide for several days and shaken from time-to-time. It was distilled near 90° under reduced pressure.

Menashi and coworkers [3270] purified dimethyl sulfoxide for *iron (II) and iron (III) isotope exchange studies*. Commercial material from the manufacturer, or a 99.9% analyzed material from a chemical supply house, gave identical products when purified. It was fractionally distilled at 2–3 torr and about 50°. The center two-thirds was collected and shaken overnight with alumina, followed by fractionation under the previously given conditions. The center one-half was saved for use. It was stored at 5° in a flask fitted with a magnesium perchlorate guard tube. The purified dimethyl sulfoxide contained 0.015%m water by Karl Fischer analysis.

Johnson and coworkers [2405] purified dimethyl sulfoxide as a solvent for the study of the *polarographic reduction of oxygen* by holding a commercially purified material over sodium hydroxide for 3 hr at 90°, distilling at reduced pressure, and using the center fraction.

Kolthoff and Reddy [2696] studied the *polarography and voltammetry* in dimethyl sulfoxide as a solvent. The sulfoxide was shaken overnight with freshly heated and cooled Woelm chromatographic grade alumina in a sealed container. The liquid was siphoned into a distilling flask with suction. It was fractionally distilled at a few torr from alumina through a 20-in. column packed with porcelain saddles. The middle 60% of the distillate was saved. The purified fraction was subjected to the same procedure twice more or until the specific conductance was 2×10^{-8} ohm^{-1} cm^{-1}. The water was 0.01%, or less, determined by the Karl Fischer method.

Kolthoff and Reddy [2695] also *determined acid-base strengths* in dimethyl sulfoxide purified as described above. They found the autoprotolysis constant to be of the order of 5×10^{-18}, that dimethyl sulfoxide is a base of strength comparable to water but it is a much weaker acid, and that water is an extremely weak base in dimethyl sulfoxide.

Stewart and O'Donnell [4502] purified Reagent grade dimethyl sulfoxide as a *solvent for determining the H-function* of strongly basic systems by distilling from sodium hydroxide pellets before use.

Franzen and Driesen [1625] purified dimethyl sulfoxide as a solvent to study the *reaction of sulfonium ylides with polar double bonds* by shaking for 24 hr with calcium sulfate, cooling to 5° until about 75% had crystallized, and filtering. The molten crystals were fractionally distilled at 12 torr. It was finally shaken 24 hr with 4A molecular sieves.

Commercial dimethyl sulfoxide was purified as a solvent for the study of the *chemical shift of the hydroxyl proton of phenols* by distilling from calcium hydride under reduced pressure and storing over molecular sieves by Ouellette [3619].

A spectrographic grade dimethyl sulfoxide was purified by Smyrl and Tobias [4408] for lithium chloride *solution thermodynamics* by distilling under reduced pressure at 80°; mp 18.58° and water <50 ppm.

Dimethyl sulfoxide and sulfolane were both purified in the same manner by Garnsey and Prue [1714] for *cryoscopic determination of osmotic coefficients*. The solvent was fractionally distilled through a column packed with "small glass rings" at 20 torr with dry nitrogen bled into the pot. The middle 70% of the distillate was collected over granular calcium hydride. The collected material was fractionally distilled two or three times from calcium hydride at 0.05 torr. The middle 80% was collected each time, the equilibrium temperature with 50% of the solid phase present was only 0.01° lower than when it finally disappeared for dimethyl sulfoxide. This corresponds to a 0.012%m impurity or 0.002%w if the impurity is water.

Dimethyl sulfoxide was purified by Krueger and Johnson [2765] as a *solvent for hydrolysis studies* by slowly heating a liter of commercial material containing 1 g of potassium hydroxide for 1 hr at 125°. It was distilled at about 72° at 10 torr through a 16-in. column packed with glass helices. The 70% center cut was used.

Johnson [2409] purified dimethyl sulfoxide by treating with B.D.H. 5A molecular sieves, followed by drying with barium oxide, and fractionally distilling at 12 torr as solvent for the study of the reaction of the *bromide and p-nitrobenzenediazonium ions*.

Willson and coworkers [5205] purified commercial dimethyl sulfoxide for a *toxicological study* by filtering it through solvent-resistant Millipore filters, type OH, 1.5 μ pore size.

SAFETY

Crown Zellerbach [1141] states that preliminary toxicity studies show a low degree of toxicity. All reports confirm their statement, for example, [5205, 2505]. Buckley [834] points out that dimethyl sulfoxide dehydrates and defats the skin, but seems to be relatively free from toxic effects. Schläfer and Schaffernicht [4157] point out that dimethyl sulfoxide penetrates the skin and that toxic solutes are carried with it into the body fluid.

Several substances have been reported to have produced an explosion when mixed with dimethyl sulfoxide. A violent pressure explosion occurred at Mount Zion Hospital, Cancer Chemotherapy Department, after about 1 hr when 4.5 moles of sodium hydride was added to 18.4 moles of dimethyl sulfoxide [956a]. Rowe and coworkers [4061] describe an explosion that occurred 5–10 min after about 70 ml of dimethyl sulfoxide was added to 34.0 g of periodic acid. According to [957a] an explosion of dimethyl sulfoxide with magnesium perchlorate occurred at Washington State University following a procedure given in [2385a] and another with 70% perchloric acid following a procedure reported in *Organic Sulfur Compounds* [2499a]. Patricia Able [5] reported a mild explosion during the addition of 70% perchloric acid to dimethyl sulfoxide to prepare a 0.05 N solution.

320. SULFOLANE

Sulfolane is an aprotic solvent with a moderately high dielectric constant. It possesses very weak acidic and basic characteristics and Morman and Harlow [3410] state that this makes it suitable for both very weak acid and very weak base titrations, while at the same time it is non-levelling for strong acids and bases. They found that the commercial material contains an acidic impurity and had to be purified before it could be used as a solvent for *acid-base titrimetric studies*. It could have been purified by fractional distillation at reduced pressure. It is more easily purified for acid-base titrimetry by passing through a column of freshly activated alumina (Alcoa F-20 chromatographic).

Sulfolane is a good solvent for most classes of organic compounds and many common polymers. It is also soluble in most solvents except paraffins. It is stable to 220° and decomposes slightly at 285° [4267]; see also [5102].

It has a very high cryoscopic constant [1714].

A commercial grade sulfolane was purified by Stewart and O'Donnell [4502] as a *solvent for determining H-function of strongly basic systems* by distilling from sodium hydroxide.

Arnett and Douty [278] vacuum-distilled sulfolane from sodium hydroxide pellets repeatedly until 1 ml did not develop a visible color within 5 min after the addition of an equal part of 100% sulfuric acid. The purified material was used as a *weakly basic aprotic solvent for high dielectric constant study.*

Monica and coworkers [3381] purified sulfolane for *transport number and ion conductivity studies* by distilling from solid potassium hydroxide at 10^{-4} torr until the specific conductance reached 2.0×10^{-8} ohm^{-1} cm^{-1}.

Monica and coworkers [3380] purified commercial sulfolane for *cryoscopic studies* by repeated fractional distillation from phosphorus pentoxide at 10^{-4} torr through a 1.5-m Podbielniak column packed with glass helices.

See Garnsey and Prue [1714] under dimethyl sulfoxide for purification of sulfolane for *cryoscopic determination of osmotic coefficients*. They doubt if the fraction melted *versus* temperature is a reliable criterion of purity for sulfolane because the solid and the liquid phases are probably so similar that any impurity not removed by distillation is likely to be soluble in the solid phase. Batch samples of about 500 ml were redistilled until the melting point was not raised further by distillation. The Karl Fischer method is of limited sensitivity because of a solvent-iodine complex, but it indicated that the water content was below 0.007%w.

SAFETY

No vapor toxicity has been reported because of the low vapor pressure of sulfolane. Shell Chemical Corp. [4267] classifies it as a slightly toxic compound. It is nonirritating to the skin.

COMPOUNDS WITH MORE THAN ONE TYPE OF CHARACTERISTIC ATOM OR GROUP

Specific Solvents

Ether Alcohols

321. 2-METHOXYETHANOL

Ruch and Critchfield [4067] employed 2-methoxyethanol as the solvent of choice in the *titration of small amounts of tertiary amine* in primary and secondary amines.

Chu and Thompson [980] purified 2-methoxyethanol and 2-ethoxyethanol by fractional distillation in a closed system in an adiabatically operated, 36-in. column packed with $\frac{3}{16}$-in. glass helices at a reflux ratio of 25:1. The 20–90% constant boiling fraction was used for *density and refractive index study* of aqueous solutions.

Ballinger and Long [369] determined the *ionization constant* of 2-methoxyethanol that had been treated with anhydrous sodium carbonate to remove mineral or carboxylic acids. It was dried and fractionally distilled. Gas chromatography and specific conductance were criteria of purity.

SAFETY

2-Methoxyethanol in concentrations estimated to be as low as 25 ppm caused neurological and hematological changes in 19 exposed workers.

Postmortem findings of hemorrhagic gastritis and kidney and liver changes were reported following a death by ingestion. It does not cause skin irritation but toxic amounts are readily absorbed through the skin. It causes immediate pain when introduced into the eye [3681, 4509].

The *threshold limit value* is 25 ppm, 80 mg/m³ [4509].

The *flammable limits* in air are 2.50 and 19.80%v at elevated temperature. The *minimum ignition temperature* in air is 383° [3680].

322. 2-ETHOXYETHANOL

Beal and Mann [463] purified 2-ethoxyethanol as a solvent for the study of the *electrolysis of metal perchlorates* by drying with Drierite, filtering, and distilling *in vacuo* from fresh calcium turnings. It was fractionally distilled and the fraction distilling at 133.8° at 738 torr was used; $\kappa = 9.3 \times 10^{-8}$ ohm^{-1} cm^{-1}.

For purification for *physical properties of aqueous solutions*, see Chu and Thompson under 2-methoxyethanol [980].

Miller [3326] purified C.P. 2-ethoxyethanol for *boiling point-composition studies* by twice fractionally distilling from a 3-ft, all-glass column; the center cut was retained.

SAFETY

2-Ethoxyethanol is less toxic than its methyl homologue, 2-methoxyethanol.

The *threshold limit value* is 200 ppm, 740 mg/m³ [4509]. Rowe [3681, p. 1550] believes this to be a reasonable figure and the margin of safety to be small.

The *minimum ignition temperature* in air is 238° [3680].

323. 2-BUTOXYETHANOL

This ether alcohol is soluble in water in all proportions at 25° but exhibits partial miscibility between 48.6 and 128° [1120].

Commercial 2-butoxyethanol was purified by Scatchard and Wilson [4124] by repeated distillation at 45 torr in a 4-ft glass helices-packed column with nitrogen bubbling into the pot. The midfraction of the fourth distillate, d 27°/4° 0.89473, was used for *vapor pressure studies* of the binary water system.

Mikhant'ev and Pyrakhina [3312] added 70 g of sodium to 500 g 1,2-ethanediol at 50° and then added 300 g of 1-chlorobutane. The product was heated on the steam bath for 3–4 hr to obtain a 61% yield of 2-butoxyethanol.

SAFETY

2-Butoxyethanol is moderately toxic orally, appreciably irritating and injurious to the eyes, and not significantly irritating to the skin. It is readily absorbed through the skin in toxic amounts. It is moderately toxic when inhaled Rowe [3681, pp. 1552–53]. He cautions [3681, p. 1558] concerning the possible absorption of toxic quantities through the skin.

The *threshold limit value* is 50 ppm, 240 mg/m³ [4509].

The *flammable limits* in air are 1.1 and 10.6%v [88]. The *minimum ignition temperature* is 245° [3680].

324. FURFURYL ALCOHOL

Furfuryl alcohol is stable in the presence of alkaline substances. It is extremely sensitive to strong acids and resinifies in their presence [3869].

Brown and Hixon [775] prepared furfuryl alcohol by the continuous vapor-phase hydrogenation of 2-furaldehyde, using a calcium-stabilized copper chromite catalyst prepared according to the directions of Burnette and coworkers [862]. Yields of furfuryl alcohol up to 95% were recovered after one pass. In a laboratory model, at least 68 g of the aldehyde could be hydrogenated for each gram of catalyst in the reactor.

Hazlet and Callison [2078] prepared benzyl and furfuryl alcohols by the crossed Cannizzaro reaction of benzaldehyde and 2-furaldehyde. For details see benzyl alcohol. The purification was carried out in the same manner as for benzyl alcohol.

Nystrom and Brown [3565] reduced furoic acid with lithium aluminum hydride in dry ether solution to furfuryl alcohol with an 85% yield.

Issoire [2359] prepared furfuryl alcohol in 80% yield by the drop by drop addition of an aqueous solution of 2.5 M sodium hydroxide to a stirred mixture of one mole of 2-furaldehyde and 2 moles of formaldehyde at 25–30°.

SAFETY

Furfuryl alcohol apparently is only moderately toxic. Rats and mice exposed to 19 ppm were restless the first 5–10 min and drowsy the remainder of the 6 hr period. Changes were observed only in the respiratory tract of rats repeatedly exposed to 19 ppm. Treon [3681] states that a threshold limit value of 5 ppm was established in 1958.

The Committee on Threshold Limits of American Conference of Governmental Industrial Hygienists established the *threshold limit value* at 50 ppm in 1961, Treon [3681] and [4509].

The *flammable limits* in air are 1.8 and 16.3%v in the temperature range of 72.5–122°. The *minimum ignition temperature* in air is 391° [3869].

325. TETRAHYDROFURFURYL ALCOHOL

SAFETY

The Quaker Oats Company [3870] states that there has been no explosion reported associated with the heating or distilling of tetrahydrofurfuryl alcohol due to peroxide formation. This does not preclude the possibility that there is an inherent danger from peroxide formation.

The Quaker Oats Company also reports that this alcohol has been in use in substantial quantities for over 20 years without evidence of harmful effect to the health. They caution to provide good ventilation and avoid contact with the skin; good safety practices should be observed.

The *flammable limits* in air are 1.5 and 9.7%v. The *minimum ignition temperature* in air is 282° [3870].

326. DIETHYLENE GLYCOL

Rinkenbach [3986] purified diethylene glycol for *physical property determinations* by distilling 1650 ml at reduced pressure. The first fraction of 480 ml was discarded. The next fraction of 1000 ml was fractionally crystallized and gave a final volume of 700 ml.

Jen [2390] reacted 1,2-ethanediol, 2-chloroethanol, and sodium hydroxide for 1 hr at 120° and 3 hr at 130°, obtaining a 74.8% yield of diethylene glycol. He [2389] prepared it in a 71.6% yield by condensing 1,2-ethanediol with ethylene oxide in the presence of a small amount of sulfuric acid. The most favorable conditions were found to be a mole ratio of 1,2-ethanediol to ethylene oxide of 16, a temperature of 120–130°, and a reaction time of 3 hr.

Koizumi and Hanai [2687] purified diethylene glycol in the same manner as they did 1,2-ethanediol for *dielectric property study*.

SAFETY

Rowe [3681, pp. 1502] states that diethylene glycol presents negligible hazards to health in industrial handling except possibly when it is being used at elevated temperature. There does not appear to be a need to establish a threshold limit value because of the low vapor pressure.

The *minimum ignition temperature* in air is 229° [3680].

327. TRIETHYLENE GLYCOL

Koizumi and Hanai [2687] purified triethylene glycol in the same manner as they did 1,2-ethanediol for *dielectric property studies*.

SAFETY

Rowe [3681, pp. 1507 ff.] believes that the industrial handling of triethylene glycol should present no significant problem from ingestion, skin contact, or vapor inhalation. He does not consider a hygienic standard necessary for triethylene glycol.

The *flammable limits* in air at elevated temperature are 0.89 and 9.20% v. The *minimum ignition temperature* in air is 371° [3680].

328. 2-(2-METHOXYETHOXY)ETHANOL

SAFETY

2-(2-Methoxyethoxy)ethanol is low in single dose oral toxicity, moderate in repeated dose oral toxicity, and appreciably irritating to the skin. It is fairly irritating to the mucous membranes. It is toxic when inhaled and is readily absorbed through the skin in toxic amounts [3681].

The *threshold limit value* is 25 ppm [3681].

329. 2-(2-ETHOXYETHOXY)ETHANOL

Seikel [4227] found 1,2-ethanediol difficult to remove from 2-(2-ethoxyethoxy)ethanol by fractional distillation alone. It could best be removed by extracting 250 g of alcohol in 750 ml benzene with 5-ml portions of water, allowing 10 min for separation of the phases. The volumes of the aqueous extracts were accurately measured, the increase in volume calculated, and the extraction continued until the increase in volume became constant.

SAFETY

Rowe [3681, pp. 1561 ff.] states that it is generally agreed that the relatively pure ether does not present any serious industrial hazard. Reasonable precautions are adequate to insure safe handling. Impurities, such as 1,2-ethanediol, in appreciable amounts increase the toxicity.

A hygienic standard is not considered necessary because of the low toxicity and low vapor pressure.

Carbonyl Alcohols

330. SALICYLALDEHYDE

Casnati and coworkers [927] found that phenoxymagnesium halide heated with an excess of triethyl orthoformate, $HC(OC_2H_5)_3$, gave salicylaldehyde in 42% yield.

Salicylaldehyde was prepared in 92% yield by Davies and Hodgson [1211] by heating under pressure salicylic acid, titanium dioxide, and formic acid at a temperature of 250–256°. They [1212] also reduced salicylic acid with sodium amalgam in 64% yield.

Carswell and Pfeifer [924] fractionally distilled a commercial product, fp −7°, in a 4-ft column packed with Raschig rings. The midcut has a freezing point of 1°. Three hundred grams of the 1° cut was slowly added, with agitation, to an ethanolic solution of sodium bisulfite such that a slight excess of the bisulfite was present. When the yellow bisulfite addition product settled, it was filtered, washed with ethanol, and recrystallized from 10% ethanol. The bisulfite complex was mixed with sodium carbonate. The yellowish-white precipitate was filtered, stirred with water, and treated with hydrochloric acid to the congo-red end point. The yellow oil was fractionally distilled *in vacuo* and the middle cut refractionated at atmospheric pressure. The distillate was used to *determine physical properties.*

Bratus, Voronin, et al. [714] precipitated the sodium salt of salicylaldehyde from the benzene, toluene, or carbon tetrachloride solution of a technical grade material by adding sodium carbonate and mixing. The insoluble sodium salt was filtered, washed with solvent, and added with cooling to 50% sulfuric acid. It was steam-distilled and gave an 86% yield of 92–95% purity.

Perkins [3716] purified a crude salicylaldehyde that contained phenol by dissolving in water and adding the hydroxide of an alkaline earth metal, separating the metal salt and regenerating the aldehyde by acidification.

331. 4-HYDROXY-4-METHYL-2-PENTANONE

The preparation of 4-hydroxy-4-methyl-2-pentanone is a long procedure. It can be purchased commercially at reasonable cost.

Adams [15, p. 45] gives directions for preparation from 1190 g acetone and barium hydroxide; yield 850 g. See also Locquin [2989], Edmonds [1442] and Conant and Tuttle [1778, p. 199].

4-Hydroxy-4-methyl-2-pentanone is very reactive and loses water when heated. It is doubted that a high purity material has been prepared and kept in such a state long enough to determine highly accurate physical properties. It is best purified by fractional distillation at reduced pressure. It can be dried with Drierite.

Commercial grade material was fractionally distilled three times at 20 torr and the middle 80% retained each time by Hack and Van Winkle [1953] for *vapor-liquid equilibria studies.*

Fuge and coworkers [1675] three times fractionally distilled the purest obtainable 4-hydroxy-4-methyl-2-pentanone at reduced pressure for *physical property studies;* bp 61.7° at 13 torr. They reported that the vapor pressure indicated that decomposition starts to become evident at about 130°.

SAFETY

4-Hydroxy-4-methyl-2-pentanone is only slightly toxic by skin absorption; however, it defats the skin and may cause dermatitis on prolonged or frequent contact. The liquid causes irritation of the eyes. It is readily absorbed through the lungs, but a relatively high concentration is required to be injurious to health. A much lower concentration is uncomfortable to the eyes, nose, and throat. It is a low health hazard; Rowe and Wolf [3681, pp. 1750 ff.].

The *threshold limit value* of 50 ppm, 238 mg/m^3, has been suggested [3681, p. 1752].

Chloro Alcohols

332. 2-CHLOROETHANOL

Ghosh et al. [1753] prepared 2-chloroethanol and 1,2-ethanediol by the electrolytic reduction of ethylene on porous carbon anodes; see 1,2-ethanediol.

Heard [2080] treated a mixture of ethylene and 1-chloro-2-propanol (propylene chlorohydrin) at 25° with an amount of calcium hydroxide equivalent to the 1-chloro-2-propanol. After 5 min, the mixture was stripped to remove the propylene oxide. Almost pure 2-chloroethanol was obtained.

Ballinger and Long [369] used the best 2-chloroethanol available, treated it with anhydrous sodium carbonate and fractionally distilled for *ionization constant determination*. The specific resistance was used as a criterion for freedom from electrolytes. A unique method for determining the specific resistance is given.

Mathews [3208], de Laszlo [2840], and Smyth and Walls [4428] used fractional distillation for the purification of 2-chloroethanol.

SAFETY

2-Chloroethanol is considered quite toxic. The toxicity by skin absorption has been found to be greater than by inhalation. Inhalation by animals has resulted in nasal irritation, incoordination, convulsions, prostration, and respiratory failure.

The *threshold limit value* has been established at 5 ppm or 16.5 mg/m³ [3681].

The *minimum ignition temperature* in air is 425° [3680].

Cyano Alcohols

333. 2-CYANOETHANOL

Sekino prepared 2-cyanoethanol [4228] of 80–95% concentration in 95–96% yield by the reaction of liquefied ethylene oxide and hydrogen cyanide using alkali or alkaline earth oxides as catalysts.

Kendall and Mackenzie [1778] described the preparation of 2-cyanoethanol by the reaction of sodium cyanide and 2-chloroethanol.

SAFETY

Fassett [3681, pp. 2018–19] believes 2-cyanoethanol to be of a low order of toxicity compared to some nitriles. There has been little evidence of skin irritation and no significant skin absorption. A hygienic standard does not seem warranted.

Amino Alcohols

334. 2-AMINOETHANOL

Reitmeier and coworkers [3949] purified 2-aminoethanol for *physical property measurements* as follows: The alcohol was given a preliminary distillation, followed by repeated washings with ether and then recrystallization from ethanol. This was followed by fractional distillation, care being taken to prevent the absorption of carbon dioxide. Two successive crystallizations and distillations gave a product with a constant melting point of 10.51°. The purity was determined by potentiometric titration with standard hydrochloric acid using a quinhydrone electrode, and by weight titration using methyl orange as the indicator. The results were 99.97 and 99.96%, respectively.

2-Aminoethanol was purified by vacuum distillation by Tseng and Thompson [4847] for *physical property measurements of aqueous solutions.* Water was determined by the Karl Fischer method.

Arnold and Churms [288] distilled 2-aminoethanol at 82° and 20 torr and stored it under dry carbon dioxide-free nitrogen. The water content could be reduced only to 0.05%. The aminoalcohol was used to study the *swelling and exchange equilibria of cation exchange resins.*

Brewster and coworkers [729] found that there was a slight decomposition of 2-aminoethanol when distilled at atmospheric pressure, resulting in the formation of conducting substances. They did not find a suitable drying agent. The drying and purification was done by vacuum distillation at 5 torr.

2-Aminoethanol was purified by Bates and Pinching [441] for the study of the *acid dissociation constant and related thermodynamic quantities* by twice fractionally distilling and collecting the middle third.

SAFETY

2-Aminoethanol can cause severe injuries to the eyes and skin and the vapors are irritating to the eyes, skin, and mucous membranes. Its low vapor pressure reduces the hazard by inhalation [3681].

The *threshold limit value* is 3 ppm, 6 mg/m³ [4509].

336. TRIETHANOLAMINE

Pearce and Berhenke [3691] purified commercially available triethanolamine for *dipole moment studies* by fractional distillation; bp 175° at 2 torr.

Bates and Schwarzenbach [442] gave the thermodynamic acidity constants, pK, of triethanolamine at 20°, 25°, and 30° as 7.87, 7.77, and 7.68, respectively. They *recommend it as a buffer substance* in the pH-range of 7–8.5.

Triethanolamine is considered to have a low acute and chronic toxicity. Any toxic effects probably would be due to the alkalinity. The greatest hazard is from spills and splashes on the skin and in the eyes [3681].

Thioether Alcohols

337. 2,2'-THIODIETHANOL

Victor Meyer [3292] prepared 2,2'-thiodiethanol from 2-chlorethanol and concentrated aqueous potassium sulfide.

Faber and Miller [5156] prepared 2,2'-thiodiethanol by adding sodium sulfide to aqueous 2-cyanoethanol. Water was removed by distillation at 30–40 torr after the reaction was completed and the sulfide extracted from the residue with hot ethanol.

Clayton and Reid [1000] prepared 2,2'-thiodiethanol from cyanohydrin. The crude material was freed from dithiane, $S(CH_2CH_2)_2S$, and its polymers by distillation at 8 torr. The distillate was diluted with sufficient water to bring the boiling point to 165°, and superheated steam was passed through the mixture. After evaporation of the water, the 2,2'-thiodiethanol was distilled at 147.5° at 6 torr.

Ross [4048] distilled a commercial product and used the middle cut boiling at 137° at 5 torr to *study the role of oxidation.*

Aldehyde Ethers

338. 2-FURALDEHYDE

This aldehyde ether, available commercially at a reasonable cost, is unstable and must be purified by fractional distillation under reduced pressure before use. 2-Furaldehyde turns dark red in the presence of air, light, and acids. It undergoes oxidation by absorbing oxygen that stops when the aldehyde decomposition proceeds to 7–8%. Formic acid, β-formylacrylic acid, and furan-α-carboxylic acid have been found in the oxidized material. Decomposition and apparent polymerization increase with increasing temperature. Polymerization takes place at 230° regardless of time.

Several reviews on its preparation and properties are available; see, for example, Monroe [3383], Miner and coauthors [3335], and van Os [4914].

Adams [15, p. 49] gives directions for the preparation from 1500 g corn cobs, 5 liters of 10% sulfuric acid, and 2000 g sodium chloride; yield 180–220 g.

For preparation from furfuryl alcohol, see Zetzsche and Zala [5336] and Yamashita and Matsumura [5285].

2-Furaldehyde may be separated from impurities other than carbonyl compounds by the bisulfite addition compound [1929].

Mayland [3238] purified 2-furaldehyde that was contaminated with close boiling hydrocarbons by water extraction in the presence of a C_3—C_8 hydrocarbon. The amount of light hydrocarbon added corresponded to 50–100 times the weight of the contaminating hydrocarbon present in the 2-furaldehyde. The amount of water for the extraction was regulated to give a solution of 10–50% of the aldehyde in water. The extraction temperature was from 32–65°. The aqueous solution was distilled and the 2-furaldehyde separated from the azeotrope.

Miller [3325] recovered 2-furaldehyde from a mixture with its polymer by treatment with a large volume of water containing a surface active agent that caused coagulation and settling of the polymer. The aldehyde was separated from the supernatant aqueous solution by distillation.

2-Furaldehyde was purified by Harris and Zoch [2039] for *analytical spectrophotometric studies* by saving the middle one-third from successive fractional distillations. It was stored in sealed glass ampules in a freezer for four years. It was originally colorless but developed a pale yellow color during storage.

Chueh and Briggs [981] purified 2-furaldehyde by distilling twice *in vacuo* and storing *in vacuo* for no more than four days.

Impurities that develop during storage of 2-furaldehyde may be removed by passing through a chromatographic grade alumina [2581]; the resulting product is reported to be nearly equivalent to that produced by vacuum distillation.

Furamide, 0.08%, was used to stabilize 2-furaldehyde [4442].

Khol'kin and Chernyaeva [2582] reduced resin formation by storing technical 2-furaldehyde under an inert atmosphere. The amount of resin after 40 days under oxygen, air, carbon dioxide, and nitrogen was 1.7, 0.3, 0.1, and 0.05%, respectively. The acidity was lower in the inert atmosphere. p-Hydroxydiphenylamine, diphenylamine, cadmium iodide, hydroquinone, pyrogallol and β-naphthol were found to be the most effective oxidation inhibitors at 0.1% level of 160 compounds tested.

Oosterhout and Roddy [3595] stabilized 2-furaldehyde by the addition of 0.001–0.1% of N-phenyl-substituted guanidines, thioureas, or naphthalamines. Polymer formation was prevented at 60–170°. Stabilized 2-furaldehyde *is useful for the extraction of unsaturated materials from hydrocarbon mixtures.*

The presence of formaldehyde in 2-furaldehyde may be detected spectrophotometrically [4929]. For the determination of unsaturated aldehydes, see Wearn and coworkers [5099].

SAFETY

The odor threshold of 2-furaldehyde has been reported to be 0.25–0.38 ppm. Some hygienists believe that the odor becomes readily noticeable near

5 ppm. Nasal irritation would indicate that the concentration in the air is too high. Human eye injury has been reported from 2-furaldehyde vapors. Full strength liquid resulted in immediate injury to the cornea of a rabbit's eye. Contact with the skin should be avoided [2337].

The *threshold limit value* is 5 ppm, 20 mg/m³ [4509].

The *lower flammable limit* in air at 125° and 740 torr is 2.1%v. The *minimum ignition temperature* in air is 315° [2337].

Chloro Ethers

339. BIS(2-CHLOROETHYL) ETHER

Barton and coworkers [425] purified commercial bis-(2-chloroethyl) ether for *kinetic studies* by washing several times with concentrated hydrochloric acid and then fractionally distilling at reduced pressure.

Kamm and Waldo [2490] heated 2-chloroethanol and concentrated sulfuric acid to produce bis(2-chloroethyl) ether.

For purification for *critical opalescence studies*, see Chu [979] under decane.

SAFETY

Most people can detect the odor of bis(2-chloroethyl) ether at its *threshold limit value* of 15 ppm, 90 mg/m³. Hake and Rowe [3681, pp. 1673 ff.] caution that the odor should not be relied upon for monitoring purposes. Brief exposures of human volunteers to concentrations above 550 ppm resulted in severe irritation to the eyes and nasal passages and was considered intolerable; coughing, retching, and nausea were also reported. Concentrations from 100–260 ppm were not considered intolerable but the irritating effects were still present to some extent. Although it is not injurious to the intact skin, it is rapidly absorbed in lethal amounts through the skin. Its vapors can cause delayed response resulting in lung lesions.

The *minimum ignition temperature* in air is 369° [3680].

340. EPICHLOROHYDRIN

Epichlorohydrin was prepared by Clarke and Hartman [1778] from glycerol-1,3-dichlorohydrin by treatment with solid, powdered sodium hydroxide in ether; see also [991, p. 47; 2408, p. 30].

Epichlorohydrin may be purified by fractional distillation.

SAFETY

Systemically, epichlorohydrin is moderately toxic by whatever way it enters the body. It is very irritating to the eyes, nose, throat, mucous membranes, and skin [3681].

The *threshold limit value* is 5 ppm, 19 mg/m³ [4509]. Two recommended threshold limit values from Russia are 1.0 mg/m³ [2711] in 1964 and 0.26 ppm [2493] in 1962.

Nitro Ethers

341. o-NITROANISOLE

According to Hantzsch [2009], the purest nitroanisole is colorless. It can be purified by repeated vacuum distillation of a freshly prepared sample as free from air as possible, followed by distillation in a high vacuum in a sealed apparatus.

Amino Ethers

342. MORPHOLINE

Morpholine has a penetrating odor and has been described as a mild base. Its double function as both an ether and an amine makes it an unusually interesting solvent.

Kobe and coworkers [2653] purified morpholine for *critical property and vapor pressure studies*. It was dried with Drierite and fractionally distilled three times at a reflux ratio of 60 to 80:1 and the 80% heart cut taken. The first two distillations were in a 12 mm × 36 in.-column packed with 0.25-in. glass helices. The third distillation was in a 20 mm × 48 in.-column packed with the same size helices.

Friedel and McKinney [1658] purified morpholine for *infrared studies* by refluxing with sodium for 1 hr and distilling onto fresh sodium in an atmosphere of pure dry nitrogen. The middle fraction was redistilled from sodium and collected in a nitrogen atmosphere.

SAFETY

Morpholine has the usual amine hazards to the eyes, skin, and mucous membranes. Concentrated morpholine penetrates the skin. The hazards diminish as the base is diluted to less than 25% with water [3681].

The *threshold limit value* is 20 ppm, 70 mg/m^3 [4509].

Esters of Hydroxy Acids

343. ETHYL LACTATE

Ethyl lactate was prepared by D'Ianni and Adkins [1300] by refluxing 250 g of 85% lactic acid, 500 ml dry ethanol, and 250 ml carbon tetrachloride for 24 hr using an automatic separator that returned the heavier liquid to the reaction flask. Additional ethanol was added and the reaction continued as long as water was formed. The ester was purified by distillation at 50 torr.

Wood and coworkers [5251] resolved commercial lactic acid into the *l*-form by the means found most suitable from the studies of Patterson and Forsyth [3677] and converted it to the zinc ammonium *l*-lactate. A mixture of 36 g of the zinc ammonium salt, 195 g of anhydrous ethanol, and 27.8 g of concentrated sulfuric acid were heated under reflux for 4 hr. The excess sulfuric acid was neutralized with potassium carbonate and the salts filtered off. The filtrate was extracted with dry ethyl ether. The ether was evaporated and the ester fractionally distilled four times at 15 torr; yield 72%.

Smith and Claborn [4399] studied methods of preparing 18 lactate esters.

SAFETY

Fassett [3681, p. 1865] states that little toxicity data are available and there does not seem to be any problem in industrial handling. No cumulative effects are expected.

344. METHYL SALICYLATE

Copeland and Rigg [1083] determined the *dipole moment* of methyl salicylate that had been fractionally distilled under reduced pressure as needed. A middle, constant boiling fraction, free from water and acid, was used.

SAFETY

Methyl salicylate long has been used as a flavoring agent as "oil of wintergreen." It is a counterirritant used in some ointments and linaments and formerly was considered helpful for "rheumatism." It is absorbed through the skin. The oral lethal dose is about 0.5 g/kg [3681].

The *minimum ignition temperature* in air is 454° [3680].

Ether Esters

345. 2-METHOXYETHYL ACETATE

SAFETY

This ether ester is not significantly irritating to the eyes or skin. It is poorly absorbed through the skin and is moderately toxic when inhaled [3681].

The *threshold limit value* is 25 ppm, 120 mg/m³ [4509].

The *flammable limits* in air are 1.75 and 8.20%v at elevated temperature [3680].

346. 2-ETHOXYETHYL ACETATE

SAFETY

High concentrations of the vapors are irritating to the eyes and nose. It is not appreciably irritating to the skin and is poorly absorbed through it. 2-Ethoxyethyl acetate is not especially toxic when inhaled in concentrations likely to be encountered. It is capable of causing central nervous system depression and lung and kidney damage [3681].

347. 2-(2-ETHOXYETHOXY)ETHYL ACETATE

SAFETY

No adverse human experience has been reported, nor would any be expected under ordinary conditions with usual care [3681].

Esters of Keto Acids

348. METHYL ACETOACETATE

Several high boiling impurities in methyl acetoacetate, which was made from diketene and methanol, were reported by Motodo and Yoshie [3422] from gas chromatographic studies. The following were found: methyl-acetoxy-3-butenoate, methyl diacetoacetate, *trans*-enolacetate and *cis*-enolacetate of methyl acetoacetate and the methyl ether of the subject compound.

SAFETY

No deaths were reported from rats exposed to the concentrated vapors of methyl acetoacetate for 8 hr [3681].

349. ETHYL ACETOACETATE

Ethyl acetoacetate from a chemical supply house was purified for *dissociation constant determinations* by fractional distillation at 18 torr by Eidinoff [1457] and the portion distilling at 79.5–80.0° was collected. The narrow range fraction was refractionated under the same conditions.

SAFETY

No deaths were reported from rats exposed to the concentrated vapors of ethyl acetoacetate for 8 hr [3681].

Esters of Cyano Acids

SAFETY

Fassett [3681, p. 2029] gives the oral LD_{50} to be 400 mg/kg or greater for these two esters. They are toxic from skin contact. Skin contact and inhalation of the vapors should be avoided.

350. METHYL CYANOACETATE

Cowan and Vogel [1114] purified a commercially available pure product for *physical property and chemical constitution studies* by shaking with 10% sodium carbonate solution, washing well with water, drying with anhydrous sodium sulfate and distilling from a fractionating Claissen flask at 6 torr.

Methyl cyanoacetate may be prepared by heating methyl monochloro-acetate with sodium cyanide in the presence of acetic acid at 80–85°.

351. ETHYL CYANOACETATE

Inglis [1778] reported the preparation of cyanoacetic acid from chloroacetic acid, followed by esterification with ethanol.

Clarke [991, p. 53] gives the directions for preparation from 625 g chloroacetic acid and 940 g cracked ice. The mixture is neutralized to litmus with about 825 ml of sodium hydroxide, 333 g/liter. A solution of 387.5 g sodium cyanide in 780 ml water is gently heated and boiled. The sodium chloroacetate solution is added to the hot cyanide solution in portions of 300 ml at such a rate that, without the application of heat, a gentle reaction is maintained. After purification, the yield is about 490–550 g.

Stephens [4495] prepared the ester from ethyl chloroacetate and sodium cyanide and by esterifying cyanoacetic acid with ethanol.

Newitt and coworkers [3524] purified ethyl cyanoacetate for *hydrolysis studies* by distilling at 46 torr.

Fluoro Acids

352. TRIFLUOROACETIC ACID

Trifluoroacetic acid is a strong acid that is not easily oxidized or reduced. It has a high thermal stability and can be heated in borosilicate glass to 400° without significant decomposition [2528]. It is as highly ionized in water as hydrochloric acid and forms salts and esters readily. The CF_3 group is not hydrolyzed by acids or bases [1573]. It is very hygroscopic [57].

Yerger [5290] distilled trifluoroacetic acid in a nitrogen atmosphere and a constant boiling center cut was used for studying the *nature of the hydrogen bond ion-pair.*

SAFETY

Trifluoroacetic acid fumes strongly in air and has a sharp biting odor [4570]. When spilled or splashed on the skin, it can cause painful burns that are slow to heal. The vapors are irritating to the eyes, nose, and mucous membranes. Vapors should not be breathed [3681]. Kheilo and Kremneva [2580] studied the toxicity of trifluoroacetic acid with mice, rats, and humans. They considered that 0.025–0.05 mg/liter would be a safe threshold concentration range.

Chloro Amines

353. o-CHLOROANILINE

Sidgwick and Rubie [4314] freed o-chloroaniline from small amounts of the *p*-isomer by the method of Beilstein and Kurbatow [500]; see also Jones and Orton [2441]. The amine was dissolved in an equivalent quantity of sulfuric acid and steam-distilled; the *para* compound remained behind as the sulfate.

For the separation of o-chloroaniline from 2,4-dichloroaniline, see Jones and Orton [2441].

For o-chloroaniline that contains only a small amount of the p-isomer, King and Orton [2597] recommended dissolving in warm 10% hydrochloric acid (11 ml per gram of the amine). On cooling, the hydrochloride of o-chloroaniline separates out and is further purified by recrystallization until the melting point of the acetyl compound is constant. The yields of pure product obtained in this way are better than those obtained by recrystallization of the picrate from ethanol or of the acetyl compound from petroleum ether.

Hale and Cheney [1967] prepared o-chloroaniline by heating o-dichlorobenzene with alcoholic ammonia in the presence of copper and copper (I) chloride at 150–250°.

SAFETY

o-Chloroaniline is readily absorbed through the skin and may cause kidney, and to a lesser extent, liver damage. Avoid breathing the vapors [3681].

Phosphoramides

354. HEXAMETHYLPHOSPHORIC TRIAMIDE

Normant [3544] characterized hexamethylphosphoric triamide as the most remarkable of the polar aprotic solvents. It is a colorless mobile liquid, miscible in all proportions with water. It is miscible with many polar and nonpolar solvents, but not with saturated hydrocarbons. It can be extracted from aqueous solution with chlorinated solvents with which it forms a complex. It has been studied as a solvent for gases, organic and inorganic salts, and for polymers. It has some unique properties as a reaction medium and shows promise for electrochemical processes. The symmetrical distribution of the positive charge over the N_3P grouping and the high electron density on the oxygen give rise to a large dipole moment and a high basicity. Normant [3544, 3545] and Robert [3991] present reviews of the physical and chemical properties and uses of hexamethylphosphoric triamide.

Catterall and coworkers [934] purified hexamethylphosphoric triamide as a solvent for *spectral study of solutions of alkali metals*. The amide was refluxed over calcium oxide for 24 hr and then distilled from sodium at 10^{-3} torr.

Brooks and Dewald [760] purified hexamethylphosphoric triamide supplied by Dow Chemical Company as a *solvent* for studying *absorption of alkali metals*. An all-Pyrex brand glass purification assembly with only break-seals and seal-offs was used. The system could be evacuated to 5×10^{-6} torr after flaming. The triamide was stored over sodium for one week, transferred to the stillpot and refluxed at about 0.5 torr in an atmosphere of nitrogen for 24 hr. The charge was fractionally distilled through a 3-ft column packed with helices. The middle 450 ml was collected on a potassium mirror. It immediately formed a blue solution. The excess metal was allowed to react (about two days). The distillation and collection on a potassium mirror was repeated, allowing the excess metal to react. The solvent was distilled into the reservoir and transferred to proper size break-seal ampules.

They found it difficult to purify hexamethylphosphoric triamide in large quantities. The solvent was transferred without exposure to the atmosphere for recovery.

Wayland and Drago [5098] obtained hexamethylphosphoric triamide commercially and distilled it at reduced pressure from barium oxide. The middle fraction that distilled at 127° at 20 torr was used for *spectra studies*.

Hanson and Bouck [2008] purified a number of phosphoryl compounds by fractional distillation. The hexamethylphosphoric triamide contained less than 5% $[(C_2H_5)_2N]_2POCl$ and decomposed in attempts at further purification.

Robert [3991] states that hexamethylphosphoric triamide is generally prepared by the reaction of an excess dimethylamine with phosphorus oxychloride in a solvent such as isopropyl ether. The dimethylammonium chloride is removed by filtration and the phosphorus amide is obtained by distilling the solvent from the filtrate.

Hexamethylphosphoric triamide was purified by Drago and coworkers [1362] for *chloride-36 exchange studies* by distilling through a micropath column just prior to use and storing in a desiccator containing activated alumina or Drierite in the absence of light.

Basic impurities were found to be present in hexamethylphosphoric triamide from three suppliers by Dietrich and coworkers [1310]. Reduced pressure distillation gave a neutral material of low specific conductance. This displayed some basicity toward phenol by widening the nuclear magnetic resonance peaks. The purified amide was stored over 4A molecular sieves before use.

Ducom and Normant [1387] purified hexamethylphosphoric triamide by two vacuum distillations over lithium aluminum hydride with argon bleeding into the flask by capillary. The distillate was kept in an inert atmosphere and was used to determine the *freezing point* and *cryoscopic constant*.

SAFETY

Kimbrough and Gaines [2591] found the LD_{50} for male and female rats by oral administration to be 2.650 and 3.360 mg/kg, respectively. Dermal administration for both sexes was about 4000 mg/kg. Testicular atrophy in male rats was noted at a single dose of 1000 mg/kg. Prolonged dietary intake of 40–80 mg/kg produced partial or complete atrophy.

Chapter **VI**

BIBLIOGRAPHY

A

1. Ababi, V. and G. Mihaila, *Studia Univ. Babes-Bolyia, Ser. Chem.* **8,** 429 (1963); *C. A.* **61,** 12694 (1964).
2. Abadie-Maumert, F. A., *Papeterie* **81,** 187 (1959); from 3049.
3. Abbott Laboratories, Technical Information Bulletin No. 325, "Piperidine," North Chicago.
4. Abderhalden, E. and E. Eichwald, *Ber. Deut. Chem. Gesell.* **51,** 1312 (1918).
5. Abel, Patricia, St. Mary's-of-the-Woods College, Private communication.
6. Abraham, R. J. and H. J. Bernstein, *Can. J. Chem.* **37,** 1056, 2095 (1959).
7. Abraham, R. J. and H. J. Bernstein, *Can. J. Chem.* **39,** 216 (1961).
8. Abraham, R. J. and K. G. R. Pachler, *Mol. Phys.* **7,** 165 (1963–1964); *C. A.* **60,** 14034 (1964).
9. Abramovitch, R. A., J. B. Rajan, et al., *J. Chem. Eng. Data* **12,** 594 (1967).
10. Achyuta, I. and V. R. Rao, *Indian J. Phys.* **34,** 196 (1960); *C. A.* **54,** 21996 (1960).
11. Aczel, T. and H. E. Lumpkin, *Anal. Chem.* **32,** 1819 (1960); **34,** 32 (1962).
12. Addison, C. C., D. W. Amon, et al., *J. Chem. Soc.* **1968A,** 2285.
13. Adams, E. W. and H. Adkins, *J. Amer. Chem. Soc.* **47,** 1358 (1925).
14. Adams, L. H., *J. Amer. Chem. Soc.* **37,** 481 (1915).
15. Adams, R., *Organic Synthesis*, Vol. 1. Wiley, New York, 1921.
16. Adams, R., *Organic Synthesis*, Vol. 8. Wiley, New York, 1928.
17. Adams, R. and co-editors, *Organic Synthesis Collection*, Vol. 2. Wiley, New York, 1943.
18. Adams, R., O. Kamm, et al., *J. Amer. Chem. Soc.* **40,** 1950 (1918).
19. Adams, R. and C. S. Marvel, *J. Amer. Chem. Soc.* **42,** 310 (1920).
20. Adams, R. and V. Voorhees, *J. Amer. Chem. Soc.* **41,** 789 (1919).
21. Adamson, D. W. and J. Kenner, *J. Chem. Soc.* **1934,** 838.
22. Addison, C. C., *J. Chem. Soc.* **1943,** 535.
23. Addison, C. C., *J. Chem. Soc.* **1945,** 98.
24. Adelman, R. L., *J. Org. Chem.* **29,** 1837 (1964).
25. Adembri, G., *Ann. Chim.* (Rome) **46,** 62 (1956); *C. A.* **50,** 13714 (1956).
26. Adkins, H. and H. R. Billica, *J. Amer. Chem. Soc.* **70,** 695 (1948).
27. Adkins, H. and H. R. Billica, *J. Amer. Chem. Soc.* **70,** 3121 (1948).
28. Adkins, H. and K. Folkers, *J. Amer. Chem. Soc.* **53,** 1095 (1931).
29. Adkins, H. and B. J. Nissen, *J. Amer. Chem. Soc.* **44,** 2749 (1922).
30. Aerov, M. E. and G. L. Motina, *Khim. Prom.* **1962,** 368; *C. A.* **58,** 2356 (1963).
31. Affens, W. A., *J. Chem. Eng. Data* **11,** 197 (1966).
32. AF Materials Laboratory, Research and Technology Division, RTD-TDR-63-4016, "Infrared Spectra of Organosulfur Compounds between 2000 and 250 cm^{-1}," Wright-Patterson Air Force Base, 1963.
33. Agami, C., *Bull. Soc. Chim. Fr.*, **1968,** 1205.
34. Agren, A., *Acta Chem. Scand.* **9,** 39 (1955); *C. A.* **49,** 13008 (1955).
35. Ahlers, N. H. E., R. A. Brett, et al., *J. Appl. Chem.* (London) **3,** 433 (1953).

36. Åkerlöf, G., *J. Amer. Chem. Soc.* **54**, 4125 (1932).
37. Akishin, P. A., N. G. Rambidi, et al., *Sbornik Trudov Mezhvuz. Soveshchaniya po Khim. Nefti, Moscow* **1956**, 146; *C. A.* **55**, 17218 (1961).
38. Akishin, P. A., N. G. Rambidi, et al., *Vestn. Mosk. Univ.* **10**, No. 12, *Ser. Fiz. Mat. i Estestven. Nauk.* No. 8, 103 (1955); *C. A.* **50**, 8329 (1956).
39. Albersmeyer, W., *Gas u. Wasserfach* **99**, 269 (1958); *C. A.* **52**, 10553 (1958).
40. Albert, A. and R. Goldacre, *Nature* **153**, 467 (1944).
41. Albert, A. and E. P. Serjeant, *Ionization Constants of Acids and Bases*, Wiley, New York, 1962.
42. Albert, O., *Osterr. Chem.-Ztg.* **41**, 287 (1938); *Beil* 3rd Supp., Vol. 2, pt. 2, *p.* 1388.
43. Albright, P. S., *J. Amer. Chem. Soc.* **59**, 2098 (1937).
44. Albright, P. S. and L. J. Gosting, *J. Amer. Chem. Soc.* **68**, 1061 (1946).
45. Alcock, K., S. S. Grimley, et al., *Trans. Faraday Soc.* **52**, 39 (1956).
46. Aleksanyan, V. T., K. E. Sterin, et al., *Izv. Akad. Nauk SSSR* **19**, 225 (1955); *C. A.* **50**, 3892 (1956).
47. Aleksanyan, V. T. and K. E. Sterin, *Opt. Spektrosk.* **2**, 562 (1957); *C. A.* **51**, 16097 (1957).
48. Alekseeva, M. V., *Predel'no Dopustimye Kontsentratsii Atm. Zagryaznenii* **1960**, No. 4, 150.
49. Alekseeva, M. V., *Predel'no Dopustimye Kontsentratsii Atm. Zagryaznenii* **1962**, No. 6, 187; *C. A.* **58**, 8344 (1963).
50. Alexander, S., *J. Chem. Phys.* **32**, 1700 (1960).
51. Allen, B. B., S. P. Lingo, et al., *J. Phys. Chem.* **43**, 425 (1939).
52. Allen, E. C. and S. Sudgen, *J. Chem. Soc.* **1932**, 76.
53. Allen, G and H. J. Bernstein, *Can. J. Chem.* **32**, 1044 (1954).
54. Allen, G. and H. J. Bernstein, *Can. J. Chem.* **32**, 1124 (1954).
55. Allen, J. S. and H. Hibbert, *J. Amer. Chem. Soc.* **56**, 1398 (1934).
56. Allen, J. T., private communication.
57. Allied Chemical, General Chemical Division, "Trifluoroacetic Acid (3FA*)," Product Data Sheet, New York, 1962.
58. Allied Chemical Corporation, National Aniline Division, "ε-Caprolactam," Tech. Bull. I-14R; "Molten Caprolactam."
59. Allied Chemical Corporation, National Aniline Division, "Tech. Bulletin I-19."
60. Allinger, N. L. and J. L. Coke, *J. Amer. Chem. Soc.* **81**, 4080 (1959).
61. Alpin, R. T., H. Budzikiewicz, et al., *J. Amer. Chem. Soc.* **87**, 3180 (1965).
62. Al'shits, I. M. and I. E. Flis, *Zh. Prik. Khim.* **34**, 644 (1961); *C. A.* **55**, 16115 (1961).
63. Althouse, P. M. and H. O. Triebold, *Ind. Eng. Chem., Anal. Ed.* **16**, 605 (1944).
64 Altschul, M., *Z. phys. Chem.* (Leipzig) **11**, 577 (1893).
65. Altschuller, A. P., *J. Phys. Chem.* **57**, 538 (1953).
66. Altschuller, A. P., *J. Phys. Chem.* **58**, 392 (1954).
67. Altschuller, A. P., *J. Phys. Chem.* **59**, 32 (1955).
68. Altsybeeva, A. I., V. P. Belousov, et al., *Zh. Fiz. Khim.* **38**, 1242 (1964); *C. A.* **61**, 7757 (1964).
69. Amagasa, M., I. Yamaguchi, et al., Japanese Patent 954, Apr. 26, 1962; *C. A.* **58**, 3399 (1963).
70. Ambler, R. H., *J. Soc. Chem. Ind.* **55**, 291T (1936).
71. Ambrose, D., *Trans. Faraday Soc.* **59**, 1988 (1963).
72. Ambrose, D., J. D. Cox, et al., *Trans. Faraday Soc.* **56**, 1452 (1960).
73. Ambrose, D. and D. G. Grant, *Trans. Faraday Soc.* **53**, 771 (1957).
74. Ambrose, D. and R. Townsend, *J. Chem. Soc.* **1963**, 3614.

75. Amel'chenko, F. S., *Izv. Vyssh. Ucheb. Zaved., Pishch. Tekhnol.* **1962**, 151; *C. A.* **57**, 10591 (1962).
76. American Chemical Society, *Reagent Chemicals*, American Chemical Society, Washington, D.C. 1960.
77. American Conference of Governmental Industrial Hygienists, *Arch. Ind. Hyg. and Occupational Med.* **8**, 296 (1953); or **9**, 530 (1954).
78. American Conference of Governmental Industrial Hygienists, *Threshold Limit Values for 1963*, Cincinnati, Ohio.
79. American Cyanamid Company, *The Chemistry of Acrylonitrile*, 2d ed., Cyanamid, New York, 1959.
80. American Industrial Hygiene Association, *Amer. Ind. Hyg. Assoc. J.*, **25**, 94 (1964).
81. American Industrial Hygiene Association, *Amer. Ind. Hyg. Assoc. J.*, **25**, 97 (1964).
82. American Industrial Hygiene Association, *Amer. Ind. Hyg. Assoc. J.*, **25**, 578 (1964).
83. American Industrial Hygiene Association, *Amer. Ind. Hyg. Assoc. J.* **26**, 435 (1965).
84. American Industrial Hygiene Association, *Amer. Ind. Hyg. Assoc. J.* **26**, 633, 636, 640 (1965).
85. American Industrial Hygiene Association, *Amer. Ind. Hyg. Assoc. J.* **27**, 89 (1966).
86. American Industrial Hygiene Association, *Amer. Ind. Hyg. Assoc. J.* **27**, 571 (1966).
87. American Institute of Physics, *Temperature, Its Measurement and Control in Science and Industry*, Reinhold, New York, 1941.
88. American Mutual Insurance Alliance (formerly National Association of Mutual Casualty Companies), *Handbook of Organic Industrial Solvents*, 3rd ed., Chicago, 1966

    American Petroleum Institute Research Project 44, Thermodynamics Research Center, College Station, Texas A & M University: references 89–223.
89. *Catalog of Infrared Spectral Data.*
90. *Catalog of Mass Spectral Data.*
91. *Catalog of Ultraviolet Spectral Data.*

Selected Values of Properties for References 92–223: When data were used from tables bearing the old designation, they were changed to what the new designation will be when the table is revised and reissued.

92. Table Ov-G, June 30, 1949.*
93. Table 2-1-d, Apr. 30, 1967.
94. Table 2-1-(1.02)-k, Oct. 31, 1964.
95. Table 23-2-(1.101)-a, Apr. 30, 1956.
96. Table 23-2-(1.101)-c, Oct. 31, 1952.
97. Table 23-2-(1.101)-e, Oct. 31, 1955.
98. Table 23-2-(1.101)-i, Apr. 30, 1966.
99. Table 23-2-(1.101)-k, Apr. 30, 1954.
100. Table 23-2-(1.101)-m, Oct. 31, 1965.
101. Table 23-2-(1.101)-n, Oct. 31, 1954.
102. Table 23-2-(1.101)-p, Oct. 31, 1954.
103. Table 23-2-(1.101)-v, Dec. 31, 1952.
104. Table 23-2-(33.1110)-a, Apr. 30, 1956.
105. Table 23-2-(1.200)-a, Oct. 31, 1952.
106. Table 23-2-(1.200)-c, Oct. 31, 1952.
107. Table 23-2-(1.200)-e, Oct. 31, 1952.
108. Table 23-2-(1.200)-i, Apr. 30, 1966.

109. Table 23-2-(1.200)-k, Dec. 31, 1952.
110. Table 23-2-(1.200)-m, Oct. 31, 1965.
111. Table 23-2-(1.200)-n, Dec. 31, 1952.
112. Table 23-2-(1.200)-p. Dec. 31, 1952.
113. Table 23-2-(1.200)-v, Dec. 31, 1952.
114. Table 23-2-(1.201)-a, Oct. 31, 1952.
115. Table 23-2-(1.201)-e, Oct. 31, 1952.
116. Table 23-2-(1.201)-i, Apr. 30, 1966.
117. Table 23-2-(1.201)-k, Oct. 31, 1967.
118. Table 23-2-(1.201)-m, Oct. 31, 1965.
119. Table 23-2-(1.201)-n, Apr. 30, 1962.
120. Table 23-2-(1.201)-p, Dec. 31, 1952.
121. Table 23-2-(1.201)-v, Dec. 31, 1952.
122. Table 23-2-(1.202)-a, Oct. 31, 1952.
123. Table 23-2-(1.202)-e, Oct. 31, 1952.
124. Table 23-2-(1.202)-i, Apr. 30, 1966.
125. Table 23-2-(1.202)-k, Dec. 31, 1952.
126. Table 23-2-(1.202)-m, Oct. 31, 1964.
127. Table 23-2-(1.202)-n, Apr. 30, 1960.
128. Table 23-2-(1.202)-p, Dec. 31, 1952.
129. Table 23-2-(1.202)-v, Dec. 31, 1952.
130. Table 23-2-(1.203)-a, Apr. 30, 1956.
131. Table 23-2-(1.203)-e, Oct. 31, 1952.
132. Table 23-2-(1.203)-i, Apr. 30, 1966.
133. Table 23-2-(1.203)-k, Oct. 31, 1967.
134. Table 23-2-(1.203)-m, Oct. 31, 1965.
135. Table 23-2-(1.203)-n, Apr. 30, 1960.
136. Table 23-2-(1.203)-p, Dec. 31, 1952.
137. Table 23-2-(1.203)-v, Dec. 31, 1952.
138. Table 23-2-(1.204)-a, Apr. 30, 1958.
139. Table 23-2-(1.204)-e, Oct. 31, 1953.
140. Table 23-2-(1.204)-i, Apr. 30, 1966.
141. Table 23-2-(1.204)-k, Apr. 30, 1966.
142. Table 23-2-(1.204)-m, Oct. 31, 1965.
143. Table 23-2-(1.204)-n, Apr. 30, 1960.
144. Table 23-2-(3.1030)-a, Apr. 30, 1956.
145. Table 23-2-(3.1030)-c, Dec. 31, 1948.
146. Table 23-2-(3.1030)-k, Oct. 31, 1967.
147. Table 23-2-(3.1030)-m, Oct. 31, 1965.
148. Table 23-2-(3.1030)-n, Apr. 30, 1962.
149. Table 23-2-(3-1030)-p, Dec. 31, 1952.
150. Table 23-2-(3.1032)-e, Oct. 31, 1953.
151. Table 23-2-(3.1032)-i, Apr. 30, 1967.
152. Table 23-2-(3.1032)-v, Dec. 31, 1952.
153. Table 23-2-(3.1110)-a, Apr. 30, 1956.
154. Table 23-2-(3.1110)-c, Dec. 31, 1948.
155. Table 23-2-(3.1110)-k, Apr. 30, 1954.
156. Table 23-2-(3.1110)-m, Oct. 31, 1965.
157. Table 23-2-(3.1110)-n, Apr. 30, 1962.
158. Table 23-2-(3.1110)-p, Dec. 31, 1952.

159. Table 23-2-(3.1112)-e, Oct. 31, 1953.
160. Table 23-2-(3.1112)-i, Apr. 30, 1967.
161. Table 23-2-(3.1112)-v, Apr. 30, 1947.
162. Table 23-2-(3.5212)-a, Oct. 31, 1967.
163. Table 23-2-(5.1101)-a, Apr. 30, 1956.
164. Table 23-2-(5.1101)-c, Aug. 31, 1949.
165. Table 23-2-(5.1101)-e, Oct. 31, 1955.
166. Table 23-2-(5.1101)-i, Apr. 30, 1967.
167. Table 23-2-(5.1101)-k, Sept. 30, 1951.
168. Table 23-2-(5.1101)-m, Oct. 31, 1965.
169. Table 23-2-(5.1101)-n, Oct. 31, 1962.
170. Table 23-2-(5.1101)-p, Dec. 31, 1952.
171. Table 23-2-(5.1101)-v, Dec. 31, 1952.
172. Table 23-2-(5.1200)-a, Apr. 30, 1954.
173. Table 23-2-(5.1200)-e, Oct. 31, 1955.
174. Table 23-2-(5.1200)-i, Apr. 30, 1967.
175. Table 23-2-(5.1200)-k, Dec. 31, 1952.
176. Table 23-2-(5.1200)-m, Oct. 31, 1965.
177. Table 23-2-(5.1200)-n, Oct. 31, 1956.
178. Table 23-2-(5.1200)-p, Dec. 31, 1952.
179. Table 23-2-(5.1200)-v, Dec. 31, 1952.
180. Table 23-2-(5.200)-a, Oct. 31, 1967.
181. Table 23-2-(9.11100)-n, Apr. 30, 1960.
182. Table 23-2-(9.1112)-a, Oct. 31, 1961.
183. Table 23-2-(9.1112)-v, Dec. 31, 1952.
184. Table 23-2-(9.1112)-w, Dec. 31, 1952.
184a. Table 23-2-(33.1100)-c, Oct. 31, 1948.
185. Table 23-2-(33.1100)-i, Apr. 30, 1967.
186. Table 23-2-(33.1100)-n, Apr. 30, 1962.
187. Table 23-2-(33.1100)-p, Dec. 31, 1952.
188. Table 23-2-(33.1100)-v, Dec. 31, 1952.
189. Table 23-2-(33.1110)-a, Apr. 30, 1956.
190. Table 23-2-(33.1110)-c, Nov. 30, 1949.
191. Table 23-2-(33.1110)-d, Oct. 31, 1952.
192. Table 23-2-(3.31110)-e, Oct. 31, 1952.
192a. Table 23-2-(33.1110)-i, Apr. 30, 1967.
193. Table 23-2-(33.1110)-k, Apr. 30, 1954.
194. Table 23-2-(33.1110)-m, Apr. 30, 1966.
195. Table 23-2-(33.1110)-n, Apr. 30, 1960.
196. Table 23-2-(33.1110)-p, Nov. 30, 1945.
197. Table 23-2-(33.1110)-v, Nov. 30, 1945.
198. Table 23-2-(33.1111)-a, Oct. 31, 1957.
198a. Table 23-2-(33.1111)-k, Dec. 31, 1952.
199. Table 23-2-(33.1111)-m, Apr. 30, 1966.
200. Table 23-2-(33.5210)-a, Oct. 31, 1955.
201. Table 23-2-(33.5210)-k, Dec. 31, 1952.
202. Table 23-2-(35.5214)-a, Oct. 31, 1952.
203. Table 23-2-(37.1110)-a, Oct. 31, 1955.
204. Table 23-2-(37.1110)-n, Apr. 30, 1960.
205. Table 23-2-(37.1110)-p, Feb. 28, 1949.

206. Table 23-2-(37.1110)-v, Sept. 30, 1947.
207. Table 23-14-2-(1.1000)-a, Oct. 31, 1964.
208. Table 23-14-2-(1.1000)-k, Apr. 30, 1961.
209. Table 23-14-2-(1.1000)-m, Oct. 31, 1966.
210. Table 23-14-2-(1.0000)-v, Oct. 31, 1963.
211. Table 23-14-2-(1.1000)-w, Oct. 31, 1963.
212. Table 23-14-2-(1.1020)-m, Dec. 31, 1965.
213. Table 23-14-2-(1.2100)-w, Oct. 31, 1963.
214. Table 23-14-2-(1.2101)-w, Oct. 31, 1963.
215. Table 23-14-2-(1.2111)-v, Oct. 31, 1963.
216. Table 23-14-2-(1.2120)-a, Oct. 31, 1964.
217. Table 23-14-2-(1.2120)-k, Oct. 31, 1964.
218. Table 23-14-2-(1.2120)-m, Oct. 31, 1965.
219. Table 23-14-2-(9.2020)-a, Oct. 31, 1964.
220. Table 23-14-2-(9.2020)-k, Oct. 31, 1953.
221. Table 23-14-2-(9.2020)-m, Oct. 31, 1965.
222. Table 23-14-2-(33.1020)-a, Oct. 31, 1964.
223. Table 23-14-2-(33.1020)-k, Oct. 31, 1954.
224. American Society for Testing Materials, ASTM Special Technical Publication No. 109A, "Physical Constants of Hydrocarbons C_1 to C_{10}," Philadelphia, 1963.
225. Amiot, O., and H. Marsac, *J. phys. radium* **8**, 498 (1937); *C. A.* **32**, 2834 (1938).
226. Ampola, G. and C. Manuelli, *Gazzetta* **25 B**, 91 (1895); *J. Chem. Soc.* **1896**, Aii, 238.
227. Ampola, G. and C. Rimatori, *Gazzetta* **27 I**, 35 (1897); *J. Chem. Soc.* **1897**, Aii, 306.
228. Anand, B. M. and S. Narain, *Indian J. Phys.* **13**, 159 (1939); *C. A.* **33**, 9140 (1939).
229. Ananthakrishnan, R., *Proc. Indian Acad. Sci.* **5A**, 200 (1937); *C. A.* **31**, 4596 (1937).
230. Ananthakrishnan, R., *Proc. Indian Acad. Sci.* **5A**, 285 (1937); *C. A.* **31**, 4596 (1937).
231. Anantakrishnan, S. V. and S. Soundararajan, *Proc. Indian Acad. Sci.* **59A**, 365 (1964); *C. A.* **63**, 114 (1965).
232. Andersen, K. K., W. H. Edmonds, et al., *J. Org. Chem.* **31**, 2859 (1966).
233. Andersen, L., *Suomen Kemistilehti* **29B**, No. 3, 94 (1956); *C. A.* **51**, 5551 (1957).
234. Anderson, H. H., *J. Chem. Eng. Data* **9**, 272 (1964).
235. Anderson, H. H., *J. Chem. Eng. Data* **11**, 117 (1966).
236. Anderson, J. A., *Anal. Chem.* **20**, 801 (1948).
237. Anderson, J. R. and C. J. Engelder, *Ind. Eng. Chem.* **37**, 541 (1945).
238. Andon, R. J. L., D. P. Biddiscombe, et al., *J. Chem. Soc.* **1960**, 5246.
239. Andon, R. J. L., J. F. Counsell et al., *J. Chem. Soc.* **1968A**, 1894.
240. Andon, R. J. L., J. F. Counsell, et al., *Trans. Faraday Soc.* **59**, 830 (1963).
241. Andon, R. J. L., J. F. Counsell, et al., *Trans. Faraday Soc.* **59**, 1555 (1963).
242. Andon, R. J. L., J. F. Counsell, et al., *Trans. Faraday Soc.* **63**, 1115 (1967).
243. Andrew, E. R. and D. P. Tunstall, *Proc. Phys. Soc.* **81**, 986 (1963); *C. A.* **59**, 2299 (1963).
244. Andrews, L. W., *J. Amer. Chem. Soc.* **30**, 353 (1908).
245. Anet, F. A. L., *Can. J. Chem.* **39**, 2262 (1961).
246. Angelescu, E. and C. Eustatiu, *Z. Phys. Chem.* (Leipzig) **177A**, 263 (1936).
247. Angell, C. L., *Trans. Faraday Soc.* **52**, 1178 (1956); *C. A.* **51**, 5555 (1957).
248. Angus, W. R. and W. K. Hill, *Trans. Faraday Soc.* **39**, 190 (1943).
249. Anikin, A. G., G. M. Dugacheva, et al., *Russ. J. Phys. Chem.* **36**, 1115 (1962); *Chem. Zentr.* **1964**, No. 19, 0768; see also 6247, 6244.
250. Anikin, A. G. and G. M. Dugacheva, *Vestn. Mosk. Univ. Ser. II, Khim.* **20**, 16 (1965); *C. A.* **63**, 5528 (1965).

251. Anikin, A. G. and G. M. Dugacheva, *Zh. Fiz. Khim.* **38,** 1372 (1964); *C. A.* **61,** 7101 (1964).

252. Anikin, A. G., Y. I. Gerasimov, et al., *Plast. Massy* **1962,** No. 12, 13; *Chem. Zentr.* **1964,** No. 8, 2231; see also 6247.

253. Anikin, A. G. and H. V. Ormanetz, *Dokl. Akad. Nauk. SSSR* **142,** 817 (1962); *Chem. Zentr.* **1965,** No. 4, 0895; see also 6247.

254. Anno, T. and I. Matsubara, *J. Chem. Phys.* **23,** 796 (1955).

255. Anonymous, *Chem. Age* (London) **18,** 606 (1928).

256. Anonymous, *Plastics* (London) **10,** 579 (1946); *C. A.* **41,** 2269 (1947).

257. Anschütz, R., *Ber. Deut. Chem. Gesell.* **25,** 3512 (1892).

258. Antara Chemicals Division, General Aniline & Film Corp., "Propargyl Alcohol," brochure, New York, 1961.

259. Antara Chemicals, "2-Pyrrolidone," General Aniline & Film Corp., New York, 1957.

260. Antoine, C., *Compt. rend.* **107,** 681 (1888).

261. Antoine, C., *Compt. rend.* **107,** 836 (1888).

262. Antonoff, G., *J. Phys. Chem.* **48,** 80 (1944).

263. Anziani, P., A. Aubry, et al., *Bull. Soc. Chim. Fr.* **1955,** 408; *C. A.* **50,** 2449 (1956).

264. Aoyama, S. and I. Morita, *J. Pharm. Soc. Jap.* **53,** 1089 (1933); *C. A.* **29,** 7303 (1935).

265. Aplin, R. T., M. Fischer, et al., *J. Amer. Chem. Soc.* **87,** 4888 (1965).

266. Applebey, M. P. and P. G. Davies, *J. Chem. Soc.* **1925,** 1836.

267. Arata, Y., *J. Phys. Soc. Jap.* **15,** 2119 (1960); *C. A.* **55,** 10076 (1961).

268. Arbuzov, B. A., Z. G. Isaeva, et al., *Dokl. Akad. Nauk. SSSR* **137,** 589 (1961); *C. A.* **55,** 18311i (1961).

269. Arbuzov, B. A., Y. Y. Samitov, et al., *Izv. Akad. Nauk SSSR, Ser. Fiz.* **27,** No. 1, 89 (1963); *C. A.* **59,** 147 (1963).

270. Arbuzov, B. A. and T. G. Shavsha, *Dokl. Akad. Nauk SSSR* **68,** 1045 (1949); *C. A.* **44,** 886 (1950).

271. Arbuzov, B. A. and Z. Z. Valeeva, *Zhur. Fiz. Khim.* **27,** 713 (1953); *C. A.* **48,** 13 (1954).

272. Arbuzov, B. A. and V. S. Vinogradova, *Compt. rend. acad. sci. U.R.S.S.* **55,** 411 (1947); *C. A.* **41,** 7183 (1947).

273. Archibald, R. C., *J. Amer. Chem. Soc.* **53,** 4452 (1931).

274. Armanet, G. and B. Pouyet, *Bull. Soc. Chim. Fr.* **1966,** 1931; *C. A.* **65,** 13581 (1966).

275. Armstrong, W. E., A. B. Densham, et al., *J. Chem. Soc.* **1950,** 3359.

276. Arnall, F., *J. Chem. Soc.* **1920,** 835.

277. Arndt, F. and P. Nachtwey, *Ber. Deut. Chem. Gesell.* **59B,** 448 (1926).

278. Arnett, E. M. and C. F. Douty, *J. Amer. Chem. Soc.* **86,** 409 (1964).

279. Arnett, E. M. and C. Y. Wu, *Chem. Ind.* (London) **1959,** 1488.

280. Arnett, E. M. and C. Y. Wu, *J. Amer. Chem. Soc.* **82,** 4999 (1960).

281. Arnett, E. M. and C. Y. Wu, *J. Amer. Chem. Soc.* **82,** 5660 (1960).

282. Arnett, E. M., C. Y. Wu, et al., *J. Amer. Chem. Soc.* **84,** 1674 (1962).

283. Arnett, E. M. and C. Y. Wu, *J. Amer. Chem. Soc.* **84,** 1680 (1962).

284. Arnett, E. M. and C. Y. Wu, *J. Amer. Chem. Soc.* **84,** 1684 (1962).

285. Arnold, C., British Patent 590,311, July 14, 1947; *C. A.* **42,** 200 (1948).

286. Arnold, D. S., C. A. Plank, et al., *Ind. Eng. Chem., Chem. Eng. Data Series* **3,** 253 (1958).

287. Arnold, J. C., British Patent 585,076, Jan. 29, 1947; *C. A.* **41,** 5897 (1947).

288. Arnold, R. and S. C. Churms, *J. Chem. Soc.* **1965,** 325.

289. Arnold, R. T. and J. Sprung, *J. Amer. Chem. Soc.* **61**, 2475 (1939).
290. Aroney, M. J., L. R. Fisher, et al., *J. Chem. Soc.* **1963**, 4450.
291. Aroney, M., D. Izsak, et al., *J. Chem. Soc.* **1961**, 4148.
292. Aroney, M. and R. J. W. Le Fèvre, *J. Chem. Soc.* **1958**, 3002.
293. Aroney, M., R. J. W. Le Fèvre, et al., *J. Chem. Soc.* **1960**, 3173.
294. Aroney, M., R. J. W. Le Fèvre, et al., *J. Chem. Soc.* **1961**, 4140.
295. Aroney, M. J., R. J. W. Le Fèvre, et al., *J. Chem. Soc.* **1962**, 2886.
296. Aroney, M. J., R. J. W. Le Fèvre, et al., *J. Chem. Soc.* **1965**, 3179.
297. Aronov, O. L., V. M. Tatevskii, et al., *Dokl. Akad. Nauk SSSR* **60**, 387, 1177, 1343 (1948); *C. A.* **42**, 7235 (1948).
298. Arrhenius, S., *Medd. Vetensk. Nobel.* **3**, [20](1916).
299. Arrowsmith, G. B., G. H. Jeffery, et al., *J. Chem. Soc.* **1965**, 2072.
300. Arthur, P., D. C. England, et al., *J. Amer. Chem. Soc.* **76**, 5364 (1954).
301. Asagoe, K. and Y. Ikemoto, *Proc. Phys.-Math. Soc. Jap.* **22**, 677, 685 (1940); *C. A.* **35**, 1317 (1941).
302. Asagoe, K. and Y. Ikemoto, *Proc. Phys.-Math. Soc. Jap.* **23**, 829 (1941); *C. A.* **41**, 6150 (1947).
303. Asagoe, K., Y. Simokawa, et al., *Jap. J. Phys.* **14**, 11 (1940–1941); *C. A.* **35**, 7827 (1941).
303a. Aschan, O., *Ber. Deut. Chem. Gesell.* **31**, 1801 (1898).
304. Ashdown, A. and T. A. Kletz, *J. Chem. Soc.* **1948**, 1454.
305. Ashmore, S. A., *Analyst* **72**, 206 (1947).
306. Astoin, N., J. Granier, et al., *J. Phys. Radium* **19**, 507 (1958); *C. A.* **53**, 8805 (1959).
307. Aston, J. G. and H. L. Fink, *Anal. Chem.* **19**, 218 (1947).
308. Aston, J. G. and S. V. R. Mastrangelo, *Anal. Chem.* **22**, 197 (1950).
309. Aston, J. G. and S. V. R. Mastrangelo, *Anal. Chem.* **22**, 636 (1950).
310. Aston, J. G. and G. H. Messerly, *J. Amer. Chem. Soc.* **58**, 2354 (1936).
311. Atkins, W. R. G., *J. Chem. Soc.* **1911**, 10.
312. Atkins, W. R. G. and T. A. Wallace, *J. Chem. Soc.* **1913**, 1461.
313. Auclair, M. and N. Hameau, *Compt. Rend. Soc. Biol.* **158**, 245 (1964); *C. A.* **61**, 12535 (1964).
314. Audsley, A. and F. R. Goss, *J. Chem. Soc.* **1941**, 864.
315. Audsley, A. and F. R. Goss, *J. Chem. Soc.* **1942**, 497.
316. Audsley, A. and F. R. Goss, *J. Chem. Soc.* **1951**, 120.
317. Auerbach, F. and H. Zeglin, *Z. Phys. Chem.* (Leipzig) **103**, 178 (1923).
318. Austerweil, G., *Chim. Ind.* (Milan), Special No. 603 (Sept. 1926); *C. A.* **21**, 818 (1927).
319. Auwers, K. v., *Ann. Chem.* **408**, 212 (1915); *Chem. Zentr.* **1915**, I, 936.
320. Auwers, K. v., *Ann. Chem.* **408**, 270 (1915); Beil **E177**, 16.
321. Auwers, K. v., *Ber. Deut. Chem. Gesell* **46**, 494 (1913).
322. Auwers, K. v., *Ges. Wiss. Marburg* **62**, No. 4, 113 (1927).
323. Auwers, K. v., *Z. Phys. Chem.* (Leipzig) **18**, 595 (1895).
324. Auwers, K. v. and F. Eisenlohr, *Z. Phys. Chem.* (Leipzig) **83**, 429 (1913).
325. Auwers, K. v. and A. Frühling, *Ann. Chem.* **422**, 192 (1921).
326. Auwers, K. v., R. Hinterseber, et al., *Ann. Chem.* **410**, 257 (1915); *C. A.* **10**, 458 (1916); *Chem. Zentr.* **1916**, I, 99.
327. Auwers, K., W. A. Roth, et al., *Ann. Chem.* **373**, 267 (1910); *C. A.* **4**, 2143 (1910).
328. Auwers K., W. A. Roth, et al., *Ann. Chem.* **385**, 102 (1911); *C. A.* **6**, 604 (1912).
329. Awbery, J. H., *Phil. Mag.* **31**, 247 (1941); *C. A.* **35**, 3887 (1941).
330. Awbery, J. H. and E. Griffiths, *Proc. Phys. Soc.* **36**, 303 (1924).

331. Ayres, E. E., *Ind. Eng. Chem.* **21**, 899 (1929).
332. Ayers, G. W. and M. S. Agruss, *J. Amer. Chem. Soc.* **61**, 83 (1939).

B

333. B., S., *J. Soc. Chem. Ind.* (London) **61**, *Chem. Ind.* (London) **20**, 120 (1942).
334. Babko, A. K. and L. L. Shevchenko, *Dopov. Akad. Nauk Ukr. RSR* **1958**, 1212; *C. A.* **53**, 9890 (1959).
335. Babushkina, T. A. and G. K. Semin, *Zh. Strukt. Khim.* **7**, 114 (1966); *C. A.* **64**, 18730 (1966).
336. Bacher, W. and J. Wagner, *Z. Phys. Chem.* (Leipzig) **B43**, 191 (1939); *C. A.* **33**, 6719 (1933).
337. Bachmann, P. and K. Dziewònski, *Ber. Deut. Chem. Gesell.* **36**, 971 (1903).
338. Backer, H. J. and N. D. Dijkstra, *Rec. Trav. Chim.* Pays-Bas **51**, 290 (1932); Backer, H. J., *ibid.*, **54**, 215 (1935); from [3212].
339. Badachhape, R. B., M. K. Gharpurey, et al., *J. Chem. Eng. Data* **10**, 143 (1965).
340. Badger, R. M., *J. Chem. Phys.* **8**, 288 (1940).
341. Badger, R. M. and S. H. Bauer, *J. Amer. Chem. Soc.* **59**, 303 (1937).
342. Badger, R. M. and S. H. Bauer, *J. Chem. Phys.* **5**, 839 (1937).
343. Badin, E. J. and E. Pacsu, *J. Amer. Chem. Soc.* **66**, 1963 (1944).
344. Badin, E. J. and E. Pacsu, *J. Amer. Chem. Soc.* **67**, 1352 (1945).
345. Badoche, M., *Bull. Soc. Chim. Fr.* **8**, 212 (1941); *C. A.* **36**, 1840 (1942).
346. Bag, A., T. Egupov, et al., *Masloboĭno Zhirovoe Delo* **13**, No. 2, 27 (1937); *C. A.* **31**, 7410 (1937).
347. Bag, A., T. Egupov, et al., *Org. Chem. Ind.* (USSR) **2**, 141 (1936); *C. A.* **31**, 1006 (1937).
348. Baggett, N., S. A. Barker, et al., *J. Chem. Soc.* **1960**, 4565.
349. Baibuz, V. F., *Dokl. Akad. Nauk SSSR* **140**, 1358 (1961); *C. A.* **56**, 10999 (1962).
350. Bailey, C. R., *Nature* **140**, 851 (1937).
351. Bailey, K. C., *J. Chem. Soc.* **1939**, 767.
352. Bailey, K. C. and W. E. S. Hickson, *J. Chem. Soc.* **1941**, 145.
353. Bailey, W. G. and A. Roy, *J. Soc. Chem. Ind.* (London) **65**, 421 (1946).
354. Bair, E. J. and C. A. Kraus, *J. Amer. Chem. Soc.* **73**, 2459 (1951).
355. Bak, B., S. Brodersen, et al., *Acta Chem. Scand.* **9**, 749 (1955); *C. A.* **50**, 2296 (1956).
356. Baker, F., *J. Chem. Soc.* **1912**, 1409.
357. Baker, R. H., *J. Amer. Chem. Soc.* **60**, 2673 (1938).
358. Baker, R. H. and R. D. Schuetz, *J. Amer. Chem. Soc.* **69**, 1250 (1947).
359. Baker, W. and F. B. Field, *J. Chem. Soc.* **1932**, 86.
360. Balandin, A. A., T. A. Slovokhotova, et al., *Dokl. Akad. Nauk SSSR* **141**, 839 (1961); *C. A.* **56**, 14138 (1962).
361. Balandin, A. A., M. B. Turova-Pollak, et al., *Dokl. Akad. Nauk SSSR* **114**, 773 (1957); *C. A.* **52**, 3701 (1958).
362. Baldridge, J. R., British Patent 795,891, June 4, 1958; *C. A.* **53**, 290 (1959).
363. Baldwin, W. C. G., *Proc. Roy. Soc.* **162A**, 215 (1937); *Chem. Zentr.* **1938**, II, 1567.
364. Ball, A. O., *J. Chem. Soc.* **1930**, 570.
365. Ballard, S. A., H. de V. Finch, et al., British Patent 619,014, Mar. 2, 1949; *C. A.* **43**, 7497 (1949).
366. Ballaus, O., *Monatsh. Chem.* **74**, 88 (1943); *C. A.* **37**, 6191 (1943), Beil **162**, 291.
367. Ballaus, O. and J. Wagner, *Z. Phys. Chem.* (Leipzig) **45B**, 272 (1939); *C. A.* **34**, 4991 (1940).
368. Ballester, M., J. Riera, et al. *J. Amer. Chem. Soc.* **86**, 4276 (1964).

369. Ballinger, P. and F. A. Long, *J. Amer. Chem. Soc.* **82,** 795 (1960).
370. Balson, E. W., *Trans. Faraday Soc.* **43,** 48 (1947); *C. A.* **41,** 5346 (1947).
371. Baly, E. C. C. and J. N. Collie, *J. Chem. Soc.* **1905,** 1332.
372. Baly, E. C. C., W. H. Edwards, et al., *J. Chem. Soc.* **1906,** 514.
373. Baly, E. C. C. and E. K. Ewbank, *J. Chem. Soc.* **1905,** 1347, 1355.
374. Baly, E. C. C. and F. G. Tryhorn, *J. Chem. Soc.* **1915,** 1058.
375. Baly, E. C. C. and W. B. Tuck, *J. Chem. Soc.* **1908,** 1902.
376. Balz, G. and G. Schiemann, *Ber. Deut. Chem. Gesell.* **60B,** 1186 (1927).
377. Bandow, F., *Biochem. Z.* **298,** 81 (1938); *Chem. Zentr.* **1939,** I, 2389.
378. Banerjee, D. K. and C. C. Budke, *Anal. Chem.* **36,** 792 (1964).
379. Banerjee, D. K. and C. C. Budke, *Anal. Chem.* **36,** 2367 (1964).
380. Banewicz, J. J., J. A. Maguire, et al., *J. Phys. Chem.* **72,** 1960 (1968).
381. Banick, W. M., *Anal. Chem.* **34,** 296 (1962).
382. Bapat, R. N., *Proc. Indian Acad. Sci.* **48A,** 119 (1958); *C. A.* **53,** 10954 (1959).
383. Barabanov, V. P., *Zh. Fiz. Khim.* **37,** 710 (1963); *C. A.* **59,** 3551 (1963).
384. Barak, M. and H. Hartley, *Z. Phys. Chem.* (Leipzig) **165A,** 272 (1933).
385. Barbaudy, J., *Bull. Soc. Chim. Fr.* **39,** 371 (1926).
386. Bârcă-Gălăteanu, D., *Bull. Soc. Roumaine Phys.* **38,** No. 69, 109 (1938); *C. A.* **32,** 4433 (1938).
387. Bârcă-Gălăteanu, D., *Bull. Soc. Roumaine Phys.* **46,** No. 83, 9 (1945); *C. A.* **41,** 6812 (1947).
388. Barchewitz, P., *Ann. Combustibles Liquides* **13,** 501 (1938); *C. A.* **33,** 2412 (1939).
389. Barchewitz, P., *Ann. Office Nat. Combustibles Liquides* **30,** 501 (May/June, 1938), Paris, Sorbonne; *Chem. Zentr.* **1939** I, 1743.
390. Barchewitz, P., *Compt. Rend.* **204,** 246 (1937); *Chem. Zentr.* **1937,** II, 3591.
391. Barchewitz, P., *Compt. Rend.* **208,** 807 (1939); *C. A.* **33,** 3694 (1939).
392. Barchewitz, P. and M. Parodi, *J. Phys. Radium* **10,** 143 (1939); *Chem. Zentr.* **1939,** II, 827.
393. Barclay, G. A. and R. J. W. Le Fèvre, *J. Chem. Soc.* **1950,** 556.
394. Barclay, G. A., R. J. W. Le Fèvre, et al., *Trans. Faraday Soc.* **46,** 812 (1950).
395. Barclay, G. A., R. J. W. Le Fèvre, et al., *Trans. Faraday Soc.* **47,** 357 (1951); *C. A.* **45,** 8828 (1951).
396. Bardyshev, I. I., *Zh. Prik. Khim.* **21,** 1019 (1948); *C. A.* **43,** 5249 (1949).
397. Bardyshev, I. I., A. L. Piryatinskiĭ, et al. *Zh. Prikl. Khim.* **20,** 1308 (1947); *C. A.* **42,** 9201 (1948).
398. Barker, E. F., *Rev. Mod. Phys.* **14,** 198 (1942); *C. A.* **37,** 1331 (1943).
399. Barnes, J. and W. H. Fulweiler, *J. Amer. Chem. Soc.* **49,** 2034 (1927).
400. Barnes, J. and W. H. Fulweiler, *J. Amer. Chem. Soc.* **51,** 1750 (1929).
401. Barnes, L. and M. S. Parvlak, *Anal. Chem.* **31,** 1875 (1959).
402. Barnes, R. B., *Nature* **124,** 300 (1929).
403. Barnes, R. B., *Phys. Rev.* **35,** 1524 (1930).
404. Barnes, R. B., L. G. Bonner, et al., *J. Chem. Phys.* **4,** 772 (1936).
405. Barnes, R. B., H. Eyring, et al., *Anal. Chem.* **20,** 96 (1948).
406. Barnes, R. B., R. C. Gore, et al., *Anal. Chem.* **20,** 402 (1948).
407. Barnes, R. B., R. C. Gore, et al., *Infrared Spectroscopy,* Reinhold, New York, 1944.
408. Barnes, R. B., U. Liddel, et al., *Ind. Eng. Chem. Anal. Ed.* **15,** 659 (1943).
409. Barnes, R. B., R. S. McDonald, et al., *J. Appl. Phys.* **16,** 77 (1945).
410. Barnes, R. H., I. I. Rusoff, et al., *Ind. Eng. Chem. Anal. Ed.* **16,** 385 (1944).
411. Barr, E. S. and E. K. Plyler, *J. Chem. Phys.* **4,** 90 (1936).
412. Barrett, J. J. and M. C. Tobin, *J. Opt. Soc. Amer.* **56,** 129 (1966).

413. Barreira, F. and G. J. Hills, *Trans. Faraday Soc.* **64**, 1359 (1968).
414. Barron, E. S. G., *J. Biol. Chem.* **121**, 313 (1937).
415. Barrow, G. M., *J. Phys. Chem.* **59**, 1129 (1955).
416. Barrow, G. M. and A. L. McClellan, *J. Amer. Chem. Soc.* **73**, 573 (1951).
417. Bartell, F. E. and A. D. Wooley, *J. Amer. Chem. Soc.* **55**, 3518 (1933).
418. Bartlett, P. D., U.S. Patent 3,149,131, Sept. 15, 1964; *C. A.* **61**, 14639 (1964).
419. Bartlett, P. D. and R. Altschul, *J. Amer. Chem. Soc.* **67**, 812 (1945).
420. Bartlett, P. D. and J. D. McCollum, *J. Amer. Chem. Soc.* **78**, 1441 (1956).
421. Bartlett, P. D. and K. Nozaki, *J. Amer. Chem. Soc.* **68**, 1495 (1946).
422. Bartoli, A., *Gazzetta* **24**, II, 156 (1894).
423. Barton, D. H. R., *J. Chem. Soc.* **1949**, 148.
424. Barton, D. H. R., *Nature* **157**, 626 (1946).
425. Barton, D. H. R., A. J. Head, et al., *J. Chem. Soc.* **1951**, 2039.
426. Barton, D. H. R. and K. E. Howlett, *J. Chem. Soc.* **1949**, 155.
427. Barton, D. H. R. and K. E. Howlett, *J. Chem. Soc.* **1949**, 165.
428. Barton, D. H. R. and P. F. Onyon, *J. Amer. Chem. Soc.* **72**, 988 (1950).
429. Bartoszewicz, E., *Rocz. Chem.* **11**, 90 (1931); *C. A.* **25**, 5343 (1931).
430. Baskov, A., *J. Russ. Phys. Chem. Soc.* **45**, 1604 (1913).
431. Bass, K. C., *J. Chem. Soc.* **1964**, 3498.
432. Bass, S. J., W. I. Nathan, et al., *J. Phys. Chem.* **68**, 509 (1964).
433. Bataafsche, de N. V., Petroleum Maatschappij, Dutch Patent 63,335, June 15, 1949; *C. A.* **43**, 7502 (1949).
434. Bateman, L. C. and E. D. Hughes, *J. Chem. Soc.* **1937**, 1187.
435. Bates, H. H., J. M. Mullaly, et al., *J. Chem. Soc.* **1923**, 401.
436. Bates, R. G., *Determination of pH, Theory and Practice*, Wiley, New York, 1964.
437. Bates, R. G., *Ann. N. Y. Acad. Sci.* **92**, Art. 2, 341 (1961).
438. Bates, R. G. and G. F. Allen, *J. Res. Nat. Bur. Stand.* **64A**, 343 (1960).
439. Bates, R. G. and V. E. Bower, *J. Res. Nat. Bur. Stand.* **57**, 153 (1956).
440. Bates, R. G. and H. B. Hetzer, *J. Phys. Chem.* **65**, 667 (1961).
441. Bates, R. G. and G. D. Pinching, *J. Res. Nat. Bur. Stand.* **46**, 349 (1951).
442. Bates, R. G. and G. Schwarzenbach, *Helv. Chim. Acta* **37**, 1437 (1954).
443. Bates, W. W. and M. E. Hobbs, *J. Amer. Chem. Soc.* **73**, 2151 (1951).
444. Batisheheva, M. G., *Inzh.-Fiz. Zh.* **2**, No. 5, 107 (1959); *C. A.* **54**, 5245 (1960).
445. Batuev, M. I., *Bull. Acad. Sci. URSS, Classe Sci. Chim.* **1947**, No. 1, 3; *C. A.* **42**, 4463 (1948).
446. Batuev, M. I., *Compt. Rend. Acad. Sci. URSS* **52**, 401 (1946); *C. A.* **41**, 1936 (1947)
447. Batuev, M. I., *Compt. Rend. Acad. Sci. URSS* **53**, 507 (1946); *C. A.* **41**, 2995 (1947).
448. Batuev, M. I., *Dokl. Akad. Nauk SSSR* **59**, 1117 (1948); *C. A.* **42**, 6664 (1948).
449. Batuev, M. I., A. A. Akhrem, et al., *Izv. Akad. Nauk SSSR, Otdel. Khim. Nauk* **1960**, 2201; *C. A.* **55**, 13056 (1961).
450. Batuev, M. I., E. N. Prilezhaeva, et al., *Bull. Acad. Sci. URSS, Classe Sci Chim.* **1947**, 123; *C. A.* **42**, 4464 (1948).
451. Bauder, A. and H. H. Günthard, *Helv. Chim. Acta* **41**, 670 (1958).
452. Baudler, M., *Z. Elektrochem.* **59**, 173 (1955); *C. A.* **49**, 11418 (1955).
453. Baum, E. and W. O. Hermann, German Patent 559,734, May 20, 1930; *C. A.* **27**, 735 (1933).
454. Bauveault, L., *Compt. Rend.* **138**, 1108 (1904); Beil I, 406.
455. Bayard, P., *Bull. Soc. Roy. Sci. Liege* **12**, 179 (1943); *Chem. Zentr.* **1943**, II, 409; *C. A.* **38**, 5147 (1944).
456. Bayles, J. W. and A. Chetwyn, *J. Chem. Soc.* **1958**, 2328.

457. Bayliss, N. S. and C. J. Breckenridge, *J. Amer. Chem. Soc.* **77,** 3959 (1955).
458. Bazhulin, P. A., A. V. Koperina, et al., *Izv. Akad. Nauk SSSR, Otd. Khim. Nauk* **1954,** 709; *C. A.* **49,** 10866 (1955).
459. Bazhulin, P. A., A. F. Plate, et al., *Bull. Acad. Sci. URSS, Classe Sci. Chim.* **1941,** 13; *C. A.* **37,** 5315 (1943).
460. Bazhulin, P. A., K. E. Sterin, et al., *Izv. Akad. Nauk SSSR, Otd. Khim. Nauk* **1946,** No. 1, 7; *C. A.* **42,** 6238 (1948).
461. Bazhulin, P. A., S. A. Ukholin, et al., *Izv. Akad. Nauk SSSR, Otd. Khim. Nauk* **1949,** 481; *C. A.* **44,** 1331 (1950).
462. Beal, G. D. and B. L. Souther, *J. Amer. Chem. Soc.* **49,** 1994 (1927).
463. Beal, J. L. and C. A. Mann, *J. Phys. Chem.* **42,** 283 (1938).
464. Beattie, J. A., D. R. Douslin, et al., *J. Chem. Phys.* **19,** 948 (1951).
465. Beattie, J. A. and D. G. Edwards, *J. Amer. Chem. Soc.* **70,** 3382 (1948).
466. Beattie, J. A. and W. C. Kay, *J. Amer. Chem. Soc.* **59,** 1586 (1937).
467. Beattie, J. A. and S. W. Levine, *J. Amer. Chem. Soc.* **73,** 4431 (1951).
468. Becconsall, J. K. and P. Hampson, *Mol. Phys.* **10,** 21 (1965); *C. A.* **64,** 13578 (1965).
469. Beckers, N. L., U.S. Patent 2,903,488, Sept. 8, 1959; *C. A.* **54,** 2165 (1960).
470. Beckett, C. W. and K. S. Pitzer, *J. Amer. Chem. Soc.* **68,** 2213 (1946).
471. Beckey, H. D. and P. Schulze, *Z. Naturforsch.* **20A,** 1329, 1335 (1965); *C. A.* **64,** 4411 (1966).
472. Beckey, H. D. and P. Schulze, *Z. Naturforsch.* **21A,** 214 (1966); *C. A.* **65,** 4816 (1966).
473. Beckmann, E., *Arch. Pharm.* (Weinheim) **245,** 211; *Chem. Zentr.* **1907, II,** 191.
474. Beckmann, E., *Z. Phys. Chem.* (Leipzig) **2,** 638 (1888); 715 (1888).
475. Beckmann, E., *Z. Phys. Chem.* (Leipzig) **6,** 437 (1890).
476. Beckmann, E., *Z. Phys. Chem.* (Leipzig) **7,** 323 (1891).
477. Beckmann, E., *Z. Phys. Chem.* (Leipzig) **8,** 223 (1891).
478. Beckmann, E., *Z. Phys. Chem.* (Leipzig) **46,** 853 (1903).
479. Beckmann, E., *Z. Phys. Chem.* (Leipzig) **57,** 129 (1907).
480. Beckmann, E., *Z. Phys. Chem.* (Leipzig) **58,** 543 (1907).
481. Beckmann, E., *Z. Phys. Chem.* (Leipzig) **63,** 177 (1908).
482. Beckmann, E. and O. Faust, *Z. Phys. Chem.* (Leipzig) **89,** 247 (1915).
483. Beckmann, E., G. Fuchs, et al., *Z. Phys. Chem.* (Liepzig) **18,** 473 (1895).
484. Beckmann, E. and W. Gabel, *Ber. Deut. Chem. Gesell.* **39,** 2611 (1906).
485. Beckmann, E. and W. Gabel, *Z. Anorg. Allg. Chem.* **51,** 236 (1906); *Chem. Zentr.* **1906, II,** 1804.
486. Beckmann, E. and F. Junker, *Z. Anorg. Allg. Chem.* **55,** 371 (1907); *J. Chem. Soc.* **1907,** Aii, 927.
487. Beckmann, E. and O. Liesche, *Z. Phys. Chem.* (Leipzig) **88,** 23 (1914).
488. Beckmann, E. and O. Liesche, *Z. Phys. Chem.* (Leipzig) **88,** 419 (1914).
489. Beckmann, E. and O. Liesche, *Z. Phys. Chem.* (Leipzig) **89,** 111 (1915).
490. Beckmann, E. and G. Lockemann, *Z. Phys. Chem.* (Leipzig) **60,** 385 (1907).
491. Beckmann, E. and A. Stock, *Z. Phys. Chem.* (Leipzig) **17,** 107 (1895).
492. Beckmann, E. and P. Waentig, *Z. Anorg. Allg. Chem.* **67,** 17 (1910); *Chem. Zentr.* **1910, II,** 277.
493. Beckwith, W. F. and R. W. Fakien, *Chem. Eng. Progr. Symp. Ser.* **44,** Thermodynamics 75 (1963).
494. Bedford, A. F., A. E. Beezer, et al., *J. Chem. Soc.* **1963,** 2039.
495. Bednenko, P. F., *Vrachebnoe Delo* **27,** 85 (1947); *C. A.* **42,** 4675 (1948).
496. Beer, M., H. B. Kessler, et al., *J. Chem. Phys.* **29,** 1097 (1958).

497. Béguin, Cl. and T. Gaümann, *Helv. Chim. Acta* **41**, 1971 (1958).
498. Béguin, Cl. and Hs. H. Günthard, *Helv. Chim. Acta* **42**, 2262 (1959).
499. Behringer, J., *Z. Elektrochem.* **62**, 544 (1958); *C. A.* **52**, 16872 (1958).
500. Beilstein, F. and A. Kurbatow, *Ann. Chem.* **176**, 27 (1875).
501. Belcher, D., *J. Amer. Chem. Soc.* **60**, 2744 (1938).
502. Bell, F. K., *J. Amer. Chem. Soc.* **57**, 1023 (1935).
503. Bell, R. P. and P. W. Smith, *J. Chem. Soc., Phys. Org.* **1966 B**, 241.
504. Belladen, L., *Gazz. Chim. Ital.* **57**, 407 (1927); *C. A.* **21**, 3298 (1927).
505. Bellanato, J. and E. D. Schmid, *An. Real Soc. Espan. Fis. Quim.* Ser. A**57**, No. 11/12, 319 (1961); *C. A.* **57**, 5481 (1962).
506. Bencze, K., *Chem. Zvesti* **15**, 571 (1961); *C. A.* **57**, 53 (1962).
507. Bender, D. F., E. Sawicki, et al., *Anal. Chem.* **36**, 1011 (1964).
508. Bender, M. L., *J. Amer. Chem. Soc.* **75**, 5986 (1953).
509. Bender, P., G. T. Furukawa, et al., *Ind. Eng. Chem.* **44**, 387 (1952).
510. Bennett, G. M. and W. G. Philip, *J. Chem. Soc.* **1928**, 1930.
511. Bennett, G. M. and W. G. Philip, *J. Chem. Soc.* **1928**, 1937.
512. Bennett, G. M. and G. H. Willis, *J. Chem. Soc.* **1928**, 2305.
513. Bennett, O. F., *Anal. Chem.* **36**, 684 (1964).
514. Bennett, W. H. and F. Daniels, *J. Amer. Chem. Soc.* **49**, 50 (1927).
515. Bennewitz, K. and W. Rossner, *Z. Phys. Chem.* (Leipzig) **39B**, 126 (1938); *C. A.* **32**, 4868 (1938).
516. Benning, A. F. and R. C. McHarness, *Ind. Eng. Chem.* **31**, 912 (1939).
517. Benning, A. F., R. C. McHarness, et al., *Ind. Eng. Chem.* **32**, 976 (1940).
518. Benoit, J. and G. Ney, *Compt. Rend.* **208**, 1888 (1939); *C. A.* **33**, 6103 (1939).
519. Benson, G. and R. M. Kitchen, *Can. J. Res.* **27F**, 266 (1949); *C. A.* **43**, 6947 (1949).
520. Bentley, J. B., K. B. Everard, et al., *J. Chem. Soc.* **1949**, 2957.
521. Berg, L., J. M. Harrison, et al., *Ind. Eng. Chem.* **37**, 585 (1945).
522. Berger, C. and L. R. Dawson, *Anal. Chem.* **24**, 994 (1952).
523. Berghoff, V., *Z. Phys. Chem.* (Leipzig) **15**, 422 (1894).
524. Bergmann, E., *J. Chem. Soc.* **139**, 402 (1936).
525. Bergmann, E., L. Engel, et al., *Ber. Deut. Chem. Gesell.* **65**, 446 (1932).
526. Bergmann, E. D. and S. Pinchas, *Rec. Trav. Chim. Pays-Bas* **71**, 161 (1952); *C. A.* **47**, 6397 (1953).
527. Bergmann, M., A. Mickeley, et al., *Ber. Deut. Chem. Gesell.* **62**, 1467 (1929).
528. Bergmeister, E. and G. Hübner, German Patent 1,005,953, Apr. 11, 1957; *C. A.* **53**, 16597 (1959).
529. Berkengeïm, T. I., *Zavodskaya Lab.* **10**, 592 (1941); *C. A.* **40**, 6961 (1946).
530. Berkman, Y. P. and L. A. Polonskaya, *Nauchn. Zapiski L'vov. Politekhn. Inst.* **22**, 49 (1956); *Referat. Zhur. Khim.* **1957**, Abstr. No. 756; *C. A.* **53**, 4867 (1959).
531. Berliner, J. F. T. and O. E. May, *J. Amer. Chem. Soc.* **49**, 1007 (1927).
532. Berman, N. S., C. W. Larkam, et al., *J. Chem. Eng. Data* **9**, 218 (1964).
533. Berman, N. S. and J. J. McKetta, *J. Phys. Chem.* **66**, 1444 (1962).
534. Berman, H. A. and E. D. West, *J. Chem. Eng. Data* **12**, 197 (1967).
535. Bernard, B., S. Pallade, et al., *Prom. Toksikol. i Klinika Prof. Zabolevanii Khim. Etiol. Sb* **1962**, 133; *C. A.* **61**, 3600 (1964).
536. Berner, E., *Z. Phys. Chem.* (Leipzig) **141A**, 91 (1929).
537. Bernstein, H. J., *Can. J. Res.* **28B**, 132 (1950).
538. Bernstein, H. J., *J. Chem. Phys.* **24**, 911 (1956).
539. Bernstein, H. J. and D. A. Ramsay, *J. Chem. Phys.* **17**, 556 (1949).

540. Bernstein, H. J., R. G. Romans, et al., *Trans. Roy. Soc. Can.* **III, 33,** 49, 57 (1936); *C. A.* **31,** 1701 (1937).

541. Bernstein, H. J. and W. G. Schneider, *Proc. Roy. Soc.* **A236,** 515 (1956); *C. A.* **51,** 76 (1957).

542. Bernstein, H. J., W. G. Schneider, et al., *J. Chem. Phys.* **26,** 957 (1957).

543. Bernstein, J. and W. T. Miller, *J. Amer. Chem. Soc.* **62,** 948 (1940).

544. Berthelot, D. and H. Gaudechon, *Compt. Rend.* **151,** 478 (1910); *C. A.* **5,** 2629 (1911).

545. Berthelot, D. and J. Ogier, *Ann. Chim. Phys.* **23,** 201 (1881).

546. Berthet, M., *Bull. Classe Sci. Acad. Roy. Belges* **27,** 212 (1941); *Chem. Zentr.* **1942,** I, 2115; *C. A.* **37,** 3400 (1943).

547. Berthoud, A., *J. Chim. Phys.* **15,** 3 (1917); *C. A.* **11,** 1922 (1917).

548. Bertoluzza, A., G. Battista, et al., *J. Chim. Phys.* **63,** 395 (1966); *C. A.* **65,** 1595 (1966).

549. Berton, A., *Ann. Chim.* **19,** (Paris) 394 (1944); *C. A.* **40,** 1733 (1946).

550. Bertram, R. and K. Cruse, *Ber. Bunsenges. Phys. Chem.* **67,** 98 (1963); *C. A.* **58,** 12038 (1963).

551. Bertram, S. H., *Rec. Trav. Chim. Pays-Bas* **46,** 397 (1927).

552. Beskov, S. D., L. I. Kochetkova, et al., *Uchenye Zapiski Mosk. Gos. Ped. Inst. im V. I. Lenina* **99,** 147 (1957); *C. A.* **53,** 20993 (1959).

553. Bewley, T., British Patent 584,788, Jan. 23, 1947; *C. A.* **41,** 5145 (1947).

554. Beyaert, M., *Natuurwetensch. Tijdschr.* **19,** 197 (1937); *Chem. Zentr.* **1938,** I, 1768.

555. Beynon, E. T. and J. J. McKetta, *J. Phys. Chem.* **67,** 2761 (1963).

556. Beynon, J. H., G. R. Lester, et al., *J. Phys. Chem.* **63,** 1861 (1959).

557. Beynon, J. H. and R. A. Saunders, *Brit. J. Appl. Phys.* **11,** 128 (1960); *C. A.* **54,** 14805 (1960); Ref. [2123].

558. Beynon, J. H., R. A. Saunders, et al., *Anal. Chem.* **33,** 221 (1961).

559. Beynon, J. H., R. A. Saunders, et al., *Appl. Spectrosc.* **14,** 95 (1960); *C. A.* **54,** 20464 (1960).

560. Beynon, W. J. G., *Phil. Mag.* **25,** 443 (1938); *C. A.* **32,** 4077 (1938).

561. Beynon, W. J. G. and E. J. Evans, *Phil. Mag.* **25,** 476 (1938); *C. A.* **32,** 4076 (1938).

562. Bezborodko, G. L., *Plast. Massy* **1961,** No. 4, 1963; *C. A.* **58,** 5496 (1963).

563. Bhagvantam, S. and S. Venkateswaran, *Nature* **125,** 237 (1930); *C. A.* **24,** 2051 (1930).

564. Bharucha, K. R. and R. C. L. Weedon, *J. Chem. Soc.* **1953,** 1584.

565. Bhatnagar, V. M., *Chem. Ind.* (London) **1966,** 731.

566. Bhattacharyya, D. N., C. L. Lee, et al., *J. Phys. Chem.* **69,** 608 (1965).

567. Bhattacharyya, S. K. and S. H. Nakhati, *J. Indian Chem. Soc.* **24,** 1 (1947); Beil E III **4,** 697.

568. Biddiscombe, D. P., R. R. Collerson, et al., *J. Chem. Soc.* **1963,** 1954.

569. Biddiscombe, D. P., E. A. Coulson, et al., *J. Chem. Soc.* **1954,** 1957.

570. Biddiscombe, D. P. and J. F. Martin, *Trans. Faraday Soc.* **54,** 1316 (1958).

571. Biedermann, W. and K. Raichle, German Patent 964,044, May 16, 1957; *C. A.* **53,** 11223 (1959).

572. Bielecki, J. and V. Henri, *Ber. Deut. Chem. Gesell.* **45,** 2819 (1912).

573. Bielecki, J. and V. Henri, *Compt. Rend.* **155,** 456 (1912); *Chem. Zentr.* **1912,** II, 1330.

574. Bielecki, J. and V. Henri, *Compt. Rend.* **155,** 1617 (1912); *Chem. Zentr.* **1913,** I, 587.

575. Bielecki, J. and V. Henri, *Compt. Rend.* **156,** 550 (1913).

576. Bielecki, J. and V. Henri, *Compt. Rend.* **156,** 1861 (1913); *Chem. Zentr.* **1913,** II, 649.

577. Bielecki, J. and V. Henri, *Compt. Rend.* **156,** 1322 (1913); *Chem. Zentr.* **1913,** II, 238.

578. Bien, G. S., C. A. Kraus, et al., *J. Amer. Chem. Soc.* **56,** 1860 (1934).

579. Bigelow, M. J., *J. Chem. Educ.* **45**, 108 (1968).
580. Bigelow, S. L., *Amer. Chem. J.* **22**, 280 (1899).
581. Bigg, D. C., S. C. Banerjee, et al., *J. Chem. Eng. Data* **9**, 17 (1964).
582. Biggs, A. I., *J. Chem. Soc.* **1961**, 2572.
583. Biggs, A. I., *Trans. Faraday Soc.* **52**, 35 (1956).
584. Biggs, A. I. and R. A. Robinson, *J. Chem. Soc.* **1961**, 388.
585. Billeter, E., T. Bürer, et al., *Helv. Chim. Acta* **40**, 2046 (1957).
586. Biltz, H., *Ber. Deut. Chem. Gesell.* **36**, 1110 (1903).
587. Biltz, H., *Monatsh. Chem.* **22**, 627 (1901); *J. Chem. Soc.* **1903 Aii**, 411.
588. Biltz, H., *Z. Phys. Chem.* (Leipzig) **19**, 385 (1896).
589. Biltz, W., W. Fischer, et al., *Z. Phys. Chem.* (Leipzig) **151A**, 13 (1930).
590. Bingham, E. C., *Fluidity and Plasticity*, 1st ed., McGraw-Hill, New York, 1922.
591. Bingham, E. C., *J. Rheology* **2**, 403 (1931).
592. Bingham, E. C. and H. J. Fornwalt, *J. Rheology* **1**, 372 (1929–1930).
593. Bingham, E. C. and J. A. Geddes, *Physics* **5**, 42 (1934).
594. Bingham, E. C. and L. A. Sarver, *J. Amer. Chem. Soc.* **42**, 2011 (1920).
595. Bingham, E. C. and L. W. Spooner, *J. Rheology* **3**, 221 (1932).
596. Binns, E. H., *Trans. Faraday Soc.* **55**, 1900 (1959).
597. Biquard, D., *Bull. Soc. Chim. France* **7**, 894 (1940); *C. A.* **36**, 2207 (1942).
598. Birch, S. F. and D. T. McAllan, *J. Chem. Soc.* **1951**, 2556.
599. Birkenstock, W., *Z. Phys. Chem.* (Leipzig) **138A**, 432 (1928).
600. Biron, E., *J. Russ. Phys. Chem. Soc.* **42**, 135, 167; *C. A.* **5**, 609 (1911); *Chem. Zentr.* **1910, I**, 1914.
601. Biron, E., *Z. Phys. Chem.* (Leipzig) **81**, 590 (1913).
602. Birun, A. M., *Nauch. Zapiski Inst. Narodnogo Khoz. im. Plekhanova* **1938**, No. 1, 120; *C. A.* **34**, 1969 (1940).
603. Bisanz, T., *Rocz. Chem.* **37**, 133 (1963); *C. A.* **59**, 7348 (1963).
604. Bisanz, T. and B. Dybowska, *Rocz. Chem.* **33**, 975 (1959); *C. A.* **54**, 6656 (1960).
605. Biscarini, P., F. Taddei, et al., *Boll. Sci. Fac. Chim. Ind. Bologna* **21**, 169 (1963); *C. A.* **60**, 2464 (1964).
606. Bishui, B. M., *Indian J. Phys.* **22**, 167 (1948); *C. A.* **42**, 8651 (1948).
607. Bishui, B. M., *Indian J. Phys.* **22**, 253 (1948); *C. A.* **43**, 2514 (1948).
608. Bishui, B. M. and S. B. Senyal, *Indian J. Phys.* **21**, 233 (1947); *C. A.* **42**, 6238 (1948).
609. Bjellerup, L., *Acta Chem. Scand.* **15**, 121 (1961); *C. A.* **55**, 19446 (1961).
610. Bjerrum, N. and L. Zechmeister, *Ber. Deut. Chem. Gesell.* **56**, 894 (1923).
611. Blacet, F. E., *J. Phys. Colloid Chem.* **52**, 534 (1948).
612. Blacet, F. E., P. A. Leighton, et al., *J. Phys. Chem.* **35**, 1935 (1931).
613. Blacet, F. E., W. E. Young, et al., *J. Amer. Chem. Soc.* **59**, 608 (1937); *C. A.* **31**, 4595 (1937).
614. Black, C., G. G. Joris, et al., *J. Chem. Phys.* **16**, 537 (1948).
615. Black, P. J. and M. L. Heffernan, *Aust. J. Chem.* **17**, 558 (1964); *C. A.* **61**, 1735 (1964).
616. Blatt, A. H., ed., *Organic Syntheses, Collective Volume II*, 1st ed., Wiley, New York, 1943.
617. Blicke, F. F. and E. S. Blake, *J. Amer. Chem. Soc.* **52**, 235 (1930).
618. Blicke, F. F. and J. L. Powers, *Ind. Eng. Chem.* **19**, 1334 (1927); Additional notes, *J. Amer. Chem. Soc.* **52**, 235 (1930).
619. Blinova, E. A., *Toksikol. Seraorgan. Soedinenii, Ufa, Sb.* **1964**, 43; *C. A.* **63**, 4854 (1965).
620. Block, H., *Z. Phys. Chem.* (Leipzig) **78**, 385 (1912); **82**, 403 (1913).

621. Blood, F. R., G. A. Elliott, et al., *Toxicol. Appl. Pharmacol.* **4,** 489 (1962).
622. Boas, A., *Ind. Eng. Chem.* **40,** 2202 (1948).
623. Boas-Traube, S. G., E. M. Dresel, et al., *Nature* **162,** 960 (1948).
624. Bobbio, G., L. Pozzoli, et al., *Lavoro Umano* **15**(12), 600 (1963); *C. A.* **61,** 6255 (1964).
625. Bobovitch, Y. S., *Opt. Spektrosk.* **20,** 252 (1966); *C. A.* **65,** 2105 (1966).
626. Bodea, C. and I. Silberg, Belgian Patent 625,714, Mar. 29, 1963; *C. A.* **59,** 10066 (1963).
627. Bodor, E., G. Bor, et al., *Veszprémi Vegyipari Egyetem Közleményei* **1,** 89 (1957); *C. A.* **55,** 3176 (1961).
628. Bodroux, D., *Ann. Chim.* (Paris) **11,** 511 (1929); *C. A.* **23,** 4936 (1929).
629. Bodroux, F., *Compt. Rend.* **160,** 204 (1915); *C. A.* **9,** 1309 (1915).
630. Bodson, H., *Annual Tables of Physical Constants*, Sec. 700 (c), 1941.
631. Boelhouwer, C., G. A. M. Diepen, et al., *Brennst.-Chem.* **39,** 173 (1958); *C. A.* **52** 17193 (1958).
632. Bogdanov, K. A. and A. P. Antonov, *Masloboĭno-Zhirovaya Prom.* **18,** No. 11, 19 (1953); *C. A.* **49,** 932 (1955).
633. Böhme, H. and H. Bentler, *Chem. Ber.* **89,** 1468 (1956).
634. Böhme, H. and W. Schürhoff, *Chem. Ber.* **84,** 28 (1951).
635. Böhme, H. and J. Wagner, *Ber. Deut. Chem. Gesell.* **75,** 606 (1942).
636. Bohon, R. L. and W. F. Claussen, *J. Amer. Chem. Soc.* **73,** 1571 (1951).
637. Boig, F. S., G. W. Costa, et al., *J. Org. Chem.* **18,** 775 (1953).
638. Bolard, J., *J. Chim. Phys.* **62,** 887 (1965); *C. A.* **64,** 160 (1966).
639. Bolle, J. and L. Bourgeois, *Compt. Rend.* **233,** 1466 (1951); *C. A.* **46,** 7514 (1952).
640. Bolling, J. M., A. R. Collett, et al., *Proc. W. Virginia Acad. Sci.* **19,** 61 (1947); *C. A.* **45,** 8314 (1951).
641. Bologna, L. and F. Albonico, *Chim. Ind.* (Milan) **33,** 141 (1951); *C. A.* **45,** 6475 (1951).
642. Bolotiv, B. A. and B. N. Dolgov, *Kataliz v Vysshei Shkole, Min. Vysshego i Srednego Spets. Obrazov. SSSR, Tr l-go (Pervogo) Mezhuuz. Soveshch. po Katalizu* **1958** (1) Pt. 2, 248 (1962); *C. A.* **59,** 8579 (1963).
643. Bond, D. L. and G. Thodos, *J. Chem. Eng. Data* **5,** 289 (1960).
644. Bond, R. P. M., T. Cairns, et al., *J. Chem. Soc.* **1965,** 3958
645. Dondi, A. and D. J. Simkin, *A. I. Ch. E. J.* **3,** 473 (1957).
646. Bondroit, C., *Rev. Universelle Mines* **15,** 197 (1939); *C. A.* **34,** 6076 (1940).
647. Bonino, G. B., *Gazz. Chim. Ital.* **66,** 316 (1936); *C. A.* **31,** 2099 (1937).
648. Bonino, G. B. and L. Brüll, *Gazz. Chim. Ital.* **59,** 728 (1929); *C. A.* **24,** 839 (1930).
649. Bonino, G. B. and P. Cella, *Mem. R. Accad. Italia, Classe di Chimica* **2,** No. 4, 5 (1931); *Chem. Zentr.* **1932** I, 2684.
650. Bonino, G. B. and R. Manzoni-Ansidei, *Atti Accad. Naz. Lincei, Cl. Sci. Fis. Mat. Nat.* **24,** 207 (1936); *C. A.* **31,** 4207 (1937).
651. Bonino, G. B. and R. Manzoni-Ansidei, *Proc. Indian Acad. Sci.* **8A,** 405 (1938); *C. A.* **33,** 4874 (1939).
652. Bonino, G. B. and R. Manzoni-Ansidei, *Ric. Sci.* **6,** II, No. 5–6 (1935); *C. A.* **31,** 3389 (1937).
653. Bonino, G. B. and R. Manzoni-Ansidei, *Ric. Sci.* **7,** I, No. 7–8 (1936); *C. A.* **31,** 3388 (1937).
654. Bonino, G. B. and R. Manzoni-Ansidei, *Ric. Sci.* **7,** I, No. 11–12 (1936); *C. A.* **31,** 3388 (1937).

655. Bonino, G. B., R. Manzoni-Ansidei, et al., *Ric. Sci.* **8, III** No. 5–6 (reprint 1937); *C. A.* **32,** 4846 (1938).

656. Bonino, G. B., R. Manzoni-Ansidei, et al., *Z. Phys. Chem.* (Leipzig) **22B,** 21 (1933); *Chem. Zentr.* **1933 II,** 1149.

657. Bonino, G. B. and P. Mascherpa, *Arch. Ital. Sci. Farmacol. Vol. Giubilare A. Bene-dicenti* **15,** (1937); *C. A.* **32,** 4662 (1938).

658. Bonnefoi, J., *Ann. Chim. Phys.* **23,** 317; *Chem. Zentr.* **1901, II,** 396.

659. Bonner, L. G. and J. S. Kirby-Smith, *Phys. Rev.* **57,** 1078 (1939); *C. A.* **35,** 4688 (1941).

660. Bonner, O. D., C. F. Jordan, et al., *J. Phys. Chem.* **68,** 2450 (1964).

661. Boon, J. W. P. and L. W. F. Kampschmidt, U.S. Patent 2,752,336 (Cl. 260-239.3), June 26, 1956.

662 Boorman, E. J., C. G. Daubney, et al., *Analyst* **72,** 246 (1947).

663. Booth, H., N. C. Franklin, et al., *Tetrahedron* **21,** 1077 (1965).

664. Booth, H. S. and H. E. Everson, *Ind. Eng. Chem.* **40,** 1491 (1948).

665. Borden, A. and E. F. Barker, *J. Chem. Phys.* **6,** 553 (1938).

666. Borden Chemical Company, The, *Preliminary Data Sheet No. 5482, Hexamethyl-phosphoramide*, Philadelphia, 1964.

667. Bordner, C. A., U.S. Patent 2,489,260, Nov. 29, 1949; *C. A.* **44,** 1543 (1950).

668. Bordwell, F. G. and G. D. Cooper, *J. Amer. Chem. Soc.* **74,** 1058 (1952).

669. Bornmann, G., *Arzneim.-Forsch.* **4,** 643, 710 (1954); **5,** 38 (1955); *C. A.* **49,** 7311 (1955).

670. Bortfeld, D. P. and M. Geller, *J. Chem. Phys.* **40,** 1770 (1964).

671. Borynice, A. and L. Marchlewski, *Bull. Inter. Acad. Polonaise* **1931A,** 392; *C. A.* **27,** 229 (1933).

672. Bosart, L. W. and A. O. Snoddy, *Ind. Eng. Chem.* **19,** 506 (1927).

673. Bose, E. and A. Müller, *Z. Phys. Chem.* (Leipzig) **58,** 586 (1907).

674. Bothner-By, A. A. and C. Naar-Colin, *J. Amer. Chem. Soc.* **83,** 231 (1961).

675. Bott, T. R. and H. N. Sadler, *J. Chem. Eng. Data* **11,** 25 (1966).

676. Bottomo, R. R., *Ind. Eng. Chem.* **23,** 501 (1931).

677. Bourchardat, G., *Compt. Rend.* **100,** 452 (1885); *Ber. Deut. Chem. Gesell.* **18R,** 178 (1885).

678. Bourgom, A., *Bull. Soc. Chim. Belges.* **33,** 101 (1924); *C. A.* **18,** 1814 (1924).

679. Bourguel, M., *Bull. Soc. Chim. Fr.* **41,** 1475 (1927); *Chem. Zentr.* **1928I,** 673.

680. Bourguel, M., *Compt. Rend.* **194,** 1736 (1932); *C. A.* **26,** 4252 (1932).

681. Bourguel, M. and L. Piaux, *Bull. Soc. Chim. Fr.* (5), **2,** 1958 (1935); *C. A.* **30,** 962 (1936).

682. Bourguignon, A., *Bull. Soc. Chim. Belges* **22,** 87 (1908); *J. Chem. Soc.* **1908Ai,** 280.

683. Bousfield, W. R. and T. M. Lowry, *J. Chem. Soc.* **1911,** 1432.

684. Boutaric, A., *J. Chim. Phys.* **18,** 126 (1920); *C. A.* **15,** 790 (1921).

685. Bovey, F. A., F. P. Hood, et al., *J. Amer. Chem. Soc.* **87,** 2060 (1965).

686. Bowden, K., *Can. J. Chem.* **43,** 2624 (1965).

687. Bowden, S. T. and E. T. Butler, *J. Chem. Soc.* **1939,** 75, 79.

688. Bowden, S. T. and W. J. Jones, *J. Chem. Ind.* (London) **66,** 342 (1947).

689. Bower, J. H., *J. Res. Nat. Bur. Stand.* **12,** 241 (1934).

690. Bower, V. E., R. A. Robinson, and R. G. Bates; from R. G. Bates and H. B. Hetzer, *J. Phys. Chem.* **65,** 667 (1961).

691. Bowie, J. H., R. Grigg, et al., *Chem. Commun.* **1965,** 403; *C. A.* **63,** 16183 (1965).

692. Bowie, J. H., R. Grigg, et al., *J. Amer. Chem. Soc.* **88,** 1699 (1966).

693. Bowie, J. H., S. O. Lawesson, et al., *J. Amer. Chem. Soc.* **87,** 5742 (1965).

694. Bowie, J. H., D. H. Williams, et al., *J. Org. Chem.* **31,** 1792 (1966).
695. Bowie, J. H., D. H. Williams, et al., *Tetrahedron* **22,** 3515 (1966).
696. Bowman, R. S., D. R. Stevens, et al., *J. Amer. Chem. Soc.* **79,** 87 (1957).
697. Boyd, G. E. and L. E. Copeland, *J. Amer. Chem. Soc.* **64,** 2540 (1942).
698. Boyd, R. H. and C.-H. Wang, *J. Amer. Chem. Soc.* **87,** 430 (1965).
699. Boyer, M., M. M. Claudon, et al., *Bull. Soc. Chim. Fr.* **1966,** 1139; *C. A.* **65,** 3180 (1966).
700. Bozel-Maletra, British Patent 607,130, Aug. 26, 1948; *C. A.* **43,** 2222 (1949).
701. Barchewitz, P., *Compt. Rend.* **203,** 930 (1936); *C. A.* **31,** 319 (1937).
702. Bradford, M. L. and G. Thodos, *J. Chem. Eng. Data* **12,** 373 (1967).
703. Bradley, R. S. and T. G. Cleasby, *J. Chem. Soc.* **1953,** 1690.
704. Bradley, R. S. and M. G. Evans, *Proc. Roy. Soc.* **186A,** 368 (1946).
705. Bradley, R. S., M.G. Evans, et al., *Proc.Roy.Soc.* **186A,** 368 (1946); Beil **EIII2,** 1016.
706. Brady, L. J., *Anal. Chem.* **20,** 512 (1948).
707. Bramley, A., *J. Chem. Soc.* **1916,** 10, 434.
708. Bramley, A., *J. Chem. Soc.* **1916,** 496.
709. Bramley, R., C. G. Le Fèvre, et al., *J. Chem. Soc.* **161,** 1183 (1959).
710. Brancker, A. V., S. J. Leach, et al., *Nature* **153,** 407 (1944).
711. Brand, K. and K. W. Kranz, *J. Prakt. Chem.* **115,** 143 (1927); *Chem. Zentr.* **1927,** I, 2726.
712. Brandenberg, W. and A. Galat, *J. Amer. Chem. Soc.* **72,** 3275 (1950).
713. Brandt, G. R. A., H. J. Emeléno, et al., *J. Chem. Soc.* **1952,** 2549.
714. Bratus, I. N., V. G. Voronin, et al., *Tr. Vses. Nauchn.-Issled. Inst. Sintetich. Natural'n. Dushistykh Veschchestv* **1961,** No. 5, 111; *C. A.* **57,** 8398 (1962).
715. Braun, J. v., *Ber. Deut. Chem. Gesell.* **43,** 1350, 2594 (1910).
716. Braun, W. G. and M. R. Fenske, *Anal. Chem.* **21,** 12 (1949).
717. Braun, W. G., D. F. Spooner, et al., *Anal. Chem.* **22,** 1074 (1950).
718. Brauns, D. H., *J. Res. Nat. Bur. Stand.* **18,** 315 (1937).
719. Brauns, D. H., *J. Res. Nat. Bur. Stand.* **31,** 83 (1943).
720. Bredig, G., *Z. Phys. Chem.* (Leipzig) **13,** 191 (1894).
721. Bredig, G., *Z. Phys. Chem.* (Leipzig) **13,** 289 (1894).
722. Bredig, G. and R. Bayers, *Z. Phys. Chem.* (Leipzig) **130A,** 15 (1927).
723. Bremner, J. G. M., British Patent 586,222, Mar. 11, 1947; *C. A.* **41,** 6892 (1947).
724. Bremner, J. G. M., D. G. Jones, et al., British Patent 619,976, Aug. 14, 1946; *C. A.* **43,** 5650 (1949).
725. Bremner, J. G. M. and D. G. Jones, U.S. Patent 2,451,712, Oct. 19, 1948; British Patent 573,507, Nov. 23, 1945; *C. A.* **43,** 1795 (1949).
726. Brescia, F., *J. Chem. Educ.* **24,** 123 (1947).
727. Bresler, F. and A. Drasky, U.S. Patent 2,204,956, June 18, 1940; *C. A.* **34,** 6944 (1940).
728. Bretscher, E., *Phys. Z.* **32,** 765 (1931).
729. Brewster, P. W., F. C. Schmidt, et al., *J. Amer. Chem. Soc.* **81,** 5532 (1959).
730. Brewster, P. W., F. C. Schmidt, et al., *J. Phys. Chem.* **65,** 990 (1961).
731. Brickwedde, F. G., M. Moskow, et al., *J. Res. Nat. Bur. Stand.* **37,** 263 (1946).
732. Bridgman, P. W., *J. Chem. Phys.* **9,** 794 (1941).
733. Bried, E. M., H. F. Kidder, et al., *Ind. Eng. Chem.* **39,** 484 (1947).
734. Briegleb, G., *Z. Phys. Chem.* (Leipzig) **16B,** 276 (1932); *Chem. Zentr.* **1932,** II, 1272.
735. Briegleb, G. and W. Lauppe, *Z. Phys. Chem.* (Leipzig) **37B,** 260 (1937); *C. A.* **32,** 48 (1937).
736. Briggs, L. H. and L. D. Colebrook, *Anal. Chem.* **29,** 904 (1957).

737. Briner, E., E. Perrottet, et al., *Helv. Chim. Acta* **19**, 558 (1936).
738. Briner, E., E. Perrottet, et al., *Helv. Chim. Acta* **19**, 1163 (1936).
739. Briner, E., E. Perrottet, et al., *Helv. Chim. Acta* **19**, 1354 (1936).
740. Brion, C. E. and W. J. Dunning, *Trans. Faraday Soc.* **59**, 647 (1963).
741. Briscol, H. T. and T. P. Dirkse, *J. Phys. Chem.* **44**, 388 (1940).
742. British Celanese Ltd., British Patent 597,078, Jan. 16, 1948; *C. A.* **42**, 4603 (1948).
743. British Patent 17,259, July 28, 1911; *C. A.* **7**, 417 (1913).
744. British Patent 853,266, Feb. 11, 1960; *Chem. Zentr.* **1962**, 13941.
745. British Standards Institute, Brit. Stand. 2713 (1956).
746. Broad, D. W. and A. G. Foster, *J. Chem. Soc.* **1946**, 446.
747. Brode, W. R., *Chemical Spectroscopy*, 2d ed., Wiley, New York, 1943.
748. Brode, W. R., *J. Phys. Chem.* **30**, 56 (1926).
749. Brode, W. R. and R. W. Van Dolah, *Ind. Eng. Chem.* **39**, 1157 (1947).
750. Broers, G. H. J., J. A. A. Ketelaar, et al., *Rec. Trav. Chim. Pays-Bas* **69**, 1122 (1950); *C. A.* **45**, 2761 (1951).
751. Bromiley, E. C. and D. Quiggle, *Ind. Eng. Chem.* **25**, 1136 (1936).
752. Bromley, W. H. and W. F. Luder, *J. Amer. Chem. Soc.* **66**, 107 (1944).
753. Brønsted, J. N., *Chem. Rev.* **5**, 231 (1928).
754. Brønsted, J. N., *Rec. Trav. Chim. Pays-Bas* **42**, 718 (1923).
755. Brønsted, J. N. and H. C. Duus, *Z. Phys. Chem.* (Leipzig) **117**, 299 (1925).
756. Brooks, A. L., *Ind. Med.* **6**, 299 (1937); *C. A.* **31**, 8064 (1937).
757. Brooks, C. S. and M. E. Hobbs, *J. Amer. Chem. Soc.* **62**, 2851 (1940).
758. Brooks, D. B., *J. Res. Nat. Bur. Stand.* **21**, 847 (1938).
759. Brooks, D. B., F. L. Howard, et al., *J. Res. Nat. Bur. Stand.* **24**, 33 (1940).
760. Brooks, J. M. and R. R. Dewald, *J. Chem. Phys.* **72**, 2655 (1968).
761. Broughton, G., *Trans. Faraday Soc.* **30**, 367 (1934); *Chem. Zentr.* **1934**, II, 2382.
762. Brown, A. C. and D. J. G. Ives, *J. Chem. Soc.* **1962**, 1608.
763. Brown, A. S., P. M. Levin, et al., *J. Chem. Phys.* **19**, 1226 (1951).
764. Brown, H. C., U.S. Patent 2,709,704, May 31, 1955; *C. A.* **50**, 7156 (1956).
765. Brown, H. C. and G. K. Barbaras, *J. Amer. Chem. Soc.* **69**, 1137 (1947).
766. Brown, H. C. and M. Nakagawa, *J. Amer. Chem. Soc.* **77**, 3614 (1955).
767. Brown, H. C. and B. C. Subba Rao, *J. Amer. Chem. Soc.* **78**, 5694 (1956).
768. Brown, H. C. and B. C. Subba Rao, *J. Amer. Chem. Soc.* **80**, 5377 (1958).
769. Brown, H. C. and B. C. Subba Rao, *J. Amer. Chem. Soc.* **81**, 6423, 6428, 6434 (1959).
770. Brown, H. C. and B. C. Subba Rao, *J. Org. Chem.* **22**, 1136 (1957).
771. Brown, H. C. and G. Zweifel, *J. Amer. Chem. Soc.* **82**, 1504 (1960).
772. Brown, H. C. and G. Zweifel, *J. Amer. Chem. Soc.* **82**, 3222 (1960).
773. Brown, H. C. and G. Zweifel, *J. Amer. Chem. Soc.* **83**, 1241 (1961).
774. Brown, H. C. and G. Zweifel, *J. Amer. Chem. Soc.* **83**, 2544 (1961).
775. Brown, H. D. and R. M. Hixon, *Ind. Eng. Chem.* **41**, 1382 (1949).
776. Brown, I. and F. Smith, *Aust. J. Chem.* **8**, 62 (1955).
777. Brown, I. and F. Smith, *Aust. J. Chem.* **8**, 501 (1955).
778. Brown, I. and F. Smith, *Aust. J. Chem.* **10**, 423 (1957).
779. Brown, I. and F. Smith, *Aust. J. Chem.* **12**, 407 (1959).
780. Brown, J. B. and G. Y. Shinowara, *J. Amer. Chem. Soc.* **59**, 6 (1937).
781. Brown, J. C., *J. Chem. Soc.* **1903**, 987.
782. Brown, J. C., *J. Chem. Soc.* **1905**, 265.
783. Brown, J. C., *J. Chem. Soc.* **1906**, 311.
784. Brown, J. K. and N. Sheppard, *Trans. Faraday Soc.* **50**, 535 (1954).
785. Brown, O. L. I. and G. G. Manov, *J. Amer. Chem. Soc.* **59**, 500 (1937).

786. Brown, P. and C. Djerassi, *J. Amer. Chem. Soc.* **88**, 2469 (1966).
787. Brown, R. and W. E. Jones, *J. Chem. Soc.* **1946**, 781.
788. Brown, R. C., *Phil. Mag.* **13**, 578 (1932).
789. Brown, T. L., *J. Amer. Chem. Soc.* **80**, 6489 (1958).
790. Brown, T. L., *J. Amer. Chem. Soc.* **81**, 3232 (1959).
791. Brown, T. L. and M. T. Rogers, *J. Amer. Chem. Soc.* **79**, 577 (1957).
792. Brown, W. G. and H. Reagan, *J. Amer. Chem. Soc.* **69**, 1032 (1947).
793. Brownlie, I. A., *J. Roy. Tech. Coll.* (Glasgow) **5**, 161 (1950); *C. A.* **45**, 4476 (1951).
794. Brownstein, S. and R. Miller, *J. Org. Chem.* **24**, 1886 (1959).
795. Bruce, W. F., *Science* **87**, 171 (1938).
796. Bruckenstein, S. and I. M. Kolthoff, *Treatise on Analytical Chemistry*, Part I, Vol. 1,
     p. 475 ff, Interscience, New York, 1959.
797. Bruckenstein, S. and L. M. Mukherjee, *J. Phys. Chem.* **64**, 1601 (1960).
798. Bruckenstein, S. and A. Saito, *J. Amer. Chem. Soc.* **87**, 698 (1965).
799. Brückner, H., *Z. Anal. Chem.* **75**, 289 (1928); *C. A.* **23**, 1738 (1929).
800. Brückner, H., *Z. Anorg. Allg. Chem.* **199**, 91 (1931).
801. Brügel, W., T. Ankel, et al., *Z. Elektrochem.* **64**, 1121 (1960); *C. A.* **55**, 10075 (1961).
802. Brühl, J. W., *Ann. Chem.* **200**, 139 (1880); *J. Chem. Soc.* **1880A**, 295.
803. Brühl, J. W., *Ann. Chem.* **203**, 1 (1880); *J. Chem. Soc.* **1880A**, 781.
804. Brühl, J. W., *Z. Phys. Chem.* (Leipzig) **16**, 193 (1895).
805. Brühl, J. W., *Z. Phys. Chem.* (Leipzig) **22**, 373 (1897).
806. Brühl, J. W. and H. Schröder, *Z. Phys. Chem.* (Leipzig) **51**, 1 (1905).
807. Brummer, S. B., *J. Chem. Phys.* **42**, 1636 (1965).
808. Brun, *Bull. Soc. Chim. Fr.* **12**, 452 (1945); *C. A.* **40**, 847 (1946).
809. Brunel, R. F., *Ber. Deut. Chem. Gesell.* **44**, 1000 (1911).
810. Brunel, R. F., *J. Amer. Chem. Soc.* **45**, 1334 (1923).
811. Brunel, R. F., J. L. Crenshaw, et al., *J. Amer. Chem. Soc.* **43**, 561 (1921).
812. Bruner, L., *Compt. Rend.* **120**, 912 (1895).
813. Bruni, G. and M. Amadori, *Gazz. Chim. Ital.* **40, II**, 1 (1910) [1909]; *Chem. Zentr.*
     1910, II, 1437.
814. Bruni, G. and P. Berti, *Atti Real Acad. Lincei* **9** (V), 393 (1900); *J. Chem. Soc.*
     1900, Aii, 592.
815. Bruni, G. and B. Sala, *Gazz. Chim. Ital.* **34, II**, 479 (1905); *Chem. Zentr.* **1905, I**,
     673; *J. Chem. Soc.* **1905, Aii**, 146.
816. Bruni, G. and A. Trovanelli, *Atti R. Accad. Lincei Roma* **13, II**, 176 (1904); *Chem.
     Zentr.* **1904, II**, 944; from [3942].
817. Brunjes, A. S. and C. C. Furnas, *Ind. Eng. Chem.* **27**, 396 (1935).
818. Bruson, H. A. and R. N. Washburne, U.S. Patent 2,184,934, Dec. 26, 1939; *Official
     Gaz. U.S. Patent Office*, **509**.
818a. Bruun, J. H., *Ind. Eng. Chem. Anal. Ed.* **8**, 224 (1936).
819. Bruun, J. H., *Ind. Eng. Chem. Anal. Ed.* **1**, 212 (1929).
819a. Bruun, J. H., and W. B. M. Faulconer, *Ind. Eng. Chem. Anal. Ed.* **9**, 192 (1937).
820. Bruun, J. H., M. M. Hicks-Bruun, et al., *J. Amer. Chem. Soc.* **59**, 2355 (1937).
821. Bruus, B. P., *Z. Anorg. Allg. Chem.* **163**, 126 (1927).
822. Bruylants, P., *Annual Tables of Physical Constants*, Sec. 301(c), 514(c), 921(c), 1941.
823. Bruylants, P. and A. Castille, *Bull. Soc. Chim. Belges* **34**, 261 (1925); *Chem. Zentr.*
     1926, I, 1962.
824. Bruyne, I. M. A., R. M. Davis, et al., *Phys. Z.* **33**, 719 (1932); *C. A.* **27**, 1248 (1933).
825. Bryant, P. J. R. and A. W. H. Wardrop, *J. Chem. Soc.* **1957**, 895.
826. Brzostowski, W. and S. Warycha, *Bull. Acad. Pol. Sci., Ser. Sci. Chim.* **11**, 539
     (1963); *C. A.* **60**, 4868 (1964).

827. Burrows, G., *J. Soc. Chem. Ind.* (London) **65**, 360 (1946).
828. Bursey, M. M. and F. W. McLafferty, *J. Amer. Chem. Soc.* **88**, 529 (1966).
829. Butskus, P. F., *Zh. Obshch. Khim.* **30**, 1816 (1960); C. A. **55**, 7408 (1961).
829a. Buck, F., R., K. F. Coles, et al., *J. Chem. Soc.* **1949**, 2377.
830. Buck, F. R., B. B. Elsner, et al., *J. Inst. Petrol.* **34**, 339 (1948); C. A. **42**, 7011 (1948).
831. Buckingham, A. D., J. Y. H. Chau, et al., *J. Chem. Soc.* **1956**, 1405.
832. Buckingham, A. D., B. Harris, et al., *J. Chem. Soc.* **1953**, 1626.
833. Buckingham, A. D. and R. J. W. Le Fèvre, *J. Chem. Soc.* **1953**, 4169.
834. Buckley, A., *J. Chem. Educ.* **42**, 674 (1965).
835. Buckley, E. and E. F. G. Herrington, *Trans. Faraday Soc.* **61**, 1618 (1965).
836. Buckley, P. and P. A. Giguère, *Can. J. Chem.* **45**, 397 (1967).
837. Buczkowski, Z. and T. Urbanski, *Spectrochim. Acta* **18**, 1187 (1962); C. A. **58**, 5166 (1963).
838. Budde, T., *Pharm. Zentralh. Deut.* **54**, 1054 (1913); *Apoth. Ztg.* **28**, 709 (1913).
839. Budzikiewicz, H., C. Djerassi, et al., *J. Chem. Soc.* **1964**, 1949.
840. Buehler, C. A., T. S. Gardner, et al., *J. Org. Chem.* **2**, 167 (1937–1938).
841. Buess, C. M., J. V. Karabinos, et al., *Natl. Advisory Comm. Aeronaut., Tech. Note No.* **1021** (1946); C. A. **41**, 4113 (1947).
842. Bugarszky, S., *Z. Phys. Chem.* (Leipzig) **71**, 705 (1910).
843. Buhmann, H., *Arch. Pharm.* (Weinheim) **266**, 123 (1928); C. A. **22**, 4717 (1928).
844. Buist, G. J. and H. J. Lucas, *J. Amer. Chem. Soc.* **79**, 6157 (1957).
845. Bukala, M., B. Burczyk, et al., *Chem. Stosow.* **3**, 497 (1959); C. A. **54**, 14568 (1960); ibid. **4**, 129 (1960); C. A. **54**, 22318 (1960).
846. Bulkley, R., *J. Res. Nat. Bur. Stand.* **6**, 89 (1931).
847. Bunton, C. A., G. J. Minkoff, et al., *J. Chem. Soc.* **1947**, 1416.
848. Burawoy, A and I. Markowitsch-Burawoy, *J. Chem. Soc.* **1936**, 36.
849. Burawoy, A. and A. R. Thompson, *J. Chem. Soc.* **1956**, 4314.
850. Burdette, J. L. and M. T. Rogers, *J. Amer. Chem. Soc.* **86**, 2105 (1964).
851. Burdun, G. D. and P. B. Kantor, *Dokl. Akad. Nauk SSSR* **67**, 985 (1949); C. A. **43**, 8769 (1949).
852. Bureau International des Etalons, *Annual Tables of Physical Constants*, Sec. 301(c), 1941.
853. Bureau International des Etalons, *Annual Tables of Physical Constants*, Sec. 381(c), 1941.
854. Bureau International des Etalons, *Annual Tables of Physical Constants*, Sec. 505(c), 1941.
855. Bureau International des Etalons, *Annual Tables of Physical Constants*, Sec. 514(c), 1941.
856. Bureau International des Etalons, *Annual Tables of Physical Constants*, Sec. 921(c), 1941.
857. Burger, L. L., *J. Chem. Eng. Data* **9**, 112 (1964).
858. Burger, L. L. and R. M. Wagner, *Ind. Eng. Chem.*, Chem. Eng. Data Series **3**, 310 (1958).
859. Burke, J. J. and P. C. Lauterbur, *J. Amer. Chem. Soc.* **86**, 1870 (1964).
860. Burkhard, O. and L. Kahovec, *Monatsh. Chem. Wien* **71**, 333 (1938); *Chem. Zentr.* **1939, I**, 79; Beil **EIII3**, 473.
861. Burlew, J. S., *J. Amer. Chem. Soc.* **62**, 681, 690 (1940).
862. Burnette, L. W., J. B. Johns, et al., *Ind. Eng. Chem.* **40**, 502 (1948).
863. Burriel-Marti, F., *Bull. Soc. Chim. Belges* **39**, 590 (1930); *Chem. Zentr.* **193, 1I**, 3442.
864. Burwell, R. L. and C. H. Langford, *J. Amer. Chem. Soc.* **81**, 3799 (1959).
865. Buss, G., *Z. Phys.* **82**, 445 (1933); C. A. **27**, 4228 (1933).

866. Buswell, A. M., J. R. Downing, et al., *J. Amer. Chem. Soc.* **62,** 2759.
867. Buswell, A. M., E. C. Dunlop, et al., *J. Amer. Chem. Soc.* **62,** 325 (1940).
868. Buswell, A. M., W. H. Rodebush, et al., *J. Amer. Chem. Soc.* **60,** 2239 (1938).
869. Buswell, A. M., W. H. Rodebush, et al., *J. Amer. Chem. Soc.* **60,** 2244 (1938).
870. Buswell, A. M., W. H. Rodebush, et al., *J. Amer. Chem. Soc.* **69,** 770 (1947).
871. Butler, J. A. V., C. N. Ramchandani, et al., *J. Chem. Soc.* **1935,** 280.
872. Butler, J. A. V., D. W. Thomson, et al., *J. Chem. Soc.* **1933,** 674.
873. Butler, J. B. and J. Lielmezs, *J. Chem. Eng. Data* **13,** 122 (1968).
874. Butler, J. C. and W. P. Webb, *Ind. Eng. Chem.*, Chem. Eng. Data Series **2,** 42 (1957).
875. Buu-Hoï and J. Janicaud, *Bull. Soc. Chim. Fr.* **12,** 640 (1945); *C. A.* **40,** 3733 (1946).
876. Buxton, L. O. and R. Kapp, *J. Amer. Chem. Soc.* **62,** 986 (1940).
877. Byers, W. H., *J. Chem. Phys.* **7,** 175 (1939).
878. Bylewski, T., *Rocz. Chem.* **12,** 311 (1932).
879. Bywater, S., *J. Polym. Sci.* **9,** 417 (1953); *C. A.* **47,** 2004 (1953).
879a. Cailletet, L. and E. Mathias, *Compt. Rend.* **102,** 1202 (1886); *J. Chem. Soc.* **1886A,** 758.
880. Calas, R., M. L. Josien, et al., *Compt. Rend.* **247,** 2008 (1958); *C. A.* **53,** 14020 (1959)
881. Calcott, W. S., F. L. English, et al., *Ind. Eng. Chem.* **17,** 942 (1925).
882. Calderazzo, B., *Inorg. Chem.* **4,** 293 (1965).
883. Calderbank, K. E. and R. J. W. Le Fèvre, *J. Chem. Soc.* **1949,** 1462.
884. Caldirola, P. and L. Giulotto, *Nuovo Cimento* **18,** 45 (1941); *C. A.* **36,** 5092 (1942).
885. Call, F., *J. Sci. Food Agr.* **8,** 81 (1957).
886. Calus, H. and B. Zyczynska, *Zeszyty Nauk Politech. Warszaw-Chem.* No. **2,** 61 (1957); *C. A.* **53,** 20984 (1959).
887. Cameron, A. E. and E. Wichers, *J. Amer. Chem. Soc.* **84,** 4175 (1962).
888. Camin, D. L. and F. D. Rossini, *J. Phys. Chem.* **59,** 1173 (1955).
889. Camin, D. L. and F. D. Rossini, *J. Phys. Chem.* **60,** 1446 (1956).
890. Campbell, A. W., Previously unpublished observations.
891. Campbell, A. W., U.S. Patent 3,015,674 (Cl. 260–561) Jan. 2, 1962.
892. Campbell, A. W. and P. F. Tryon, *Ind. Eng. Chem.* **45,** 125 (1953).
893. Campbell, D. H., U.S. Patent 2,998,461, Aug. 29, 1961.
894. Campbell, H. J. and J. T. Edwards, *Can. J. Chem.* **38,** 2109 (1960).
895. Campbell, K. N. and L. T. Eby, *J. Amer. Chem. Soc.* **63,** 2683 (1941).
896. Campbell, T. W., S. Linden, et al., *J. Amer. Chem. Soc.* **69,** 880 (1947).
897. Canals, E., M. Mousseron, et al., *Bull. Soc. Chim. Fr.* **4,** 2048 (1937); *C. A.* **32,** 8269 (1938).
898. Cannon, M. R. and M. R. Fenske, *Ind. Eng. Chem. Anal. Ed.* **10,** 297 (1938).
899. Cannon, M. R. and M. R. Fenske, *Oil Gas J.* **33,** 52 (1935).
900. Cannon, M. R. and M. R. Fenske, *Oil Gas J.* **34,** 45 (1936).
901. Cantacuzene, J., *Bull. Soc. Chim. Fr.* **1962,** 741; *C. A.* **57,** 5787 (1962).
902. Cantoni, A. and J. Feldman, *Ind. Eng. Chem.* **45,** 2580 (1953).
903. Capalbi, A., C. Franconi, et al., *SIPS, Sci Tec.* **5**(3–4), 125 (1961); *C. A.* **61,** 5108 (1964).
904. Capinjola, J. V., *J. Amer. Chem. Soc.* **67,** 1615 (1945).
905. Carbide and Carbon Chemicals Corp., *Synthetic Organic Chemicals*, 12th ed., New York, 1945.
906. Carius, L., *Ann. Chem.* **110,** 210 (1859).
907. Carlisle, P. J. and A. A. Levine, *Ind. Eng. Chem.* **24,** 1164 (1932).
908. Carney, G. E. and J. K. Sanford, *Anal. Chem.* **25,** 1417 (1953).
909. Carothers, W. H., *Organic Syntheses*, Vol. 13, Wiley, New York, 1933.

910. Carothers, W. H. and R. Adams, *J. Amer. Chem. Soc.* **46**, 1675 (1924).
911. Carpenter, C. P., C. S. Weil, et al., *Arch. Ind. Hyg. Occupational Med.* **8**, 219(1953); *C. A.* **48**, 272 (1954)
912. Carpenter, G. B. and J. Donohue, *J. Amer. Chem. Soc.* **72**, 2315 (1950).
913. Carpenter, W., A. M. Duffield, et al., *J. Amer Chem. Soc.* **89**, 6167 (1967).
914. Carr, C. and J. A. Riddick, *Ind. Eng. Chem.* **43**, 692 (1951).
915. Carr, E. P. and H. Stücklen, *J. Chem. Phys.* **4**, 760 (1936).
916. Carr, E. P. and H. Stücklen, *Z. Phys Chem.* (Leipzig) **23B**, 57 (1934).
917. Carr, E. P. and G. F. Walker, *J. Chem. Phys.* **4**, 756.
918. Carrington, R. A. G., *Anal. Chem.* **31**, 1117 (1959).
919. Carroll, B. H., G. K. Rollefson, et al., *J. Amer. Chem. Soc.* **47**, 1791 (1925).
920. Carson, A. S., W. Carter, et al., *Proc. Roy. Soc.* **A260**, 550 (1961); *C. A.* **55**, 15101 (1961).
921. Carson, B. B. and V. N. Ipatieff, *J. Amer. Chem. Soc.* **59**, 645 (1937).
922. Carswell, T. S., *Ind. Eng. Chem.* **20**, 728 (1928).
923. Carswell, T. S. and H. L. Morrill, *Ind. Eng. Chem.* **29**, 1247 (1937).
924. Carswell, T. S. and C. E. Pfeifer, *J. Amer. Chem. Soc.* **50**, 1765 (1928).
925. Cartwright, C. H., *Phys. Rev.* **35**, 415 (1930).
926. Carvajal, C., K. J. Tölle, et al., *J. Amer. Chem. Soc.* **87**, 5548 (1965).
927. Casnati, G., M. Crisafulli, et al., *Tetrahedron Lett.* **1965**, 243; *C. A.* **62**, 9049 (1965).
928. Cass, O. W., U.S. Patent 2,489,265, Nov. 29, 1949; *C. A.* **44**, 1543 (1950).
929. Cass, R. C., S. E. Fletcher, et al., *J. Chem. Soc.* **1958**, 958.
930. Cass, R. C., S. E. Fletcher, et al., *J. Chem. Soc.* **1958**, 1406.
931. Castellano, S. and R. Kostelnik, *J. Chem. Phys.* **46**, 327 (1967).
932. Castellano, S. and J. S. Waugh, *J. Chem. Phys.* **34**, 295 (1961).
933. Castille, A. and V. Henri, *Bull. Soc. Chim. Biol.* **6**, 299 (1924); *C. A.* **18**, 3165 (1924).
934. Catterall, R., L. P. Stodulski, et al., *J. Chem. Soc.* **1968A** 437.
935. Catterall, W. E., U.S. Patent 2,787,586, Apr. 2, 1957.
936. Cauwood, J. D. and W. E. S. Turner, *J. Chem. Soc.* **1915**, 276.
937. Cavanaugh, J. R. and B. P. Dailey, *J. Chem. Phys.* **34**, 1094, 1099 (1961).
938. Celiano, A. V., P. S. Gentile, et al., *J. Chem. Eng. Data* **7**, 391 (1962).
939. Centnerszwer, M., *Z. Phys. Chem.* (Leipzig) **49**, 199 (1904).
940. Čepelák, J., *Plast. Kautschuk* **2**, 159 (1955); *C. A.* **51**, 7052.
941. Ceuterick, P., *Bull. Soc. Chim. Belges* **45**, 545 (1936).
942. Chadwell, H. M., *J. Amer. Chem. Soc.* **48**, 1912 (1926).
943. Chaikin, S. W. and W. G. Brown, *J. Amer. Chem. Soc.* **71**, 122 (1949).
944. Chalmers, W., *Can. J. Research* **7**, 464 (1932); *C. A.* **27**, 701 (1933).
945. Chamberlain, N. F., *Anal. Chem.* **31**, 56 (1959).
946. Chan, S. C. and J. P. Valleau, *Can. J. Chem.* **46**, 853 (1968).
947. Chandra, A. K. and A. B. Sannigrahi, *J. Phys. Chem.* **69**, 2494 (1965).
948. Chang, Cheng-Ch'i, *Predel'no Dopustimye Kontsentratsii Atm. Zagraznenii Sb.* **1961**, No. 5, 94; *C. A.* **57**, 17011 (1962).
949. Chao, J. and F. D. Rossini, *J. Chem. Eng. Data* **10**, 374 (1965).
950. Chapman, D. and J. F. Nacey, *Analyst* **83**, 377 (1958).
951. Chapman, O. L. and R. W. King, *J. Amer. Chem. Soc.* **86**, 1258 (1964); see also [4823].
952. Charnley, T., H. A. Skinner, et al., *J. Chem. Soc.* **1952**, 2288.
953. Charon, E., *Bull. Soc. Chim. Fr.* **15**, 390 (1896); *Chem. Zentr.* **1896**, I, 992.
954. Charton-Koechlin, M. and M. A. Leroy, *J. Chim. Phys.* **56**, 850 (1959); *C. A.* **55**, 21808 (1961).

955. Chavanne, G., *Annual Tables of Physical Constants*, Sec. 514(c), 1941.
956. Chavanne, G. and G. Tock. *Bull. Soc. Chim. Belges* **41**, 630 (1932).
956a. *C & EN*, **44**, 18, 1966.
957. Chemische Fabrik Griesheim-Elektron, German Patent 271,381, Mar. 13, 1914; *Chem. Zentr.* **1914, I**, 1316.
957a. *C & EN*, **43**, 62 (1965); *Chem. Zentr.* **1911**, 1, 1292–93.
958. Chen, C-Y, R. J. W. Le Fèvre, et al., *J. Chem. Soc.* **1965**, 553.
959. Chen, D. T. Y. and L. J. Laidler, *Trans. Faraday Soc.* **58**, 480 (1962).
960. Chenevier, A., *Z. Anal. Chem.* **31**, 68 (1892).
961. Cheney, H. A. and S. H. McAllister, U.S. Patent 2,441,095, May 4, 1948; *C. A.* **42**, 6849 (1948).
962. Cheng, D. C-H., *Chem. Eng. Sci.* **18**, 715 (1963).
963. Cheng, D. C-H. and J. C. McCoubrey, *J. Chem. Soc.* **1963**, 4993.
964. Cheng, D. C-H., J. C. McCoubrey, et al., *Trans. Faraday Soc.* **58**, 224 (1962).
965. Cheng, Hua-Chih, *Z. Phys. Chem.* (Leipzig) **24B**, 293 (1934); *Chem. Zentr.* **1934, I**, 3440.
966. Chenon, M. T. and N. Lumbroso-Bader, *J. Chim. Phys.* **62**, 1075 (1965); *C. A.* **64**, 4908 (1966).
967. Cherkasova, L. M. and L. K. Gorin, *Koks Khim.* **1963**, No. 1, 44; *C. A.* **58**, 10020 (1963).
968. Cherrier, C., *Compt. Rend.* **225**, 930 (1947); *C. A.* **42**, 2519 (1948).
969. Cherrier, C., *Compt. Rend.* **228**, 379 (1949); *C. A.* **43**, 4955 (1949).
970. Chiang, Y. and E. B. Whipple, *J. Amer. Chem. Soc.* **85**, 2763 (1963).
971. Chiao, T.-T. and A. R. Thompson, *J. Chem. Eng. Data* **6**, 192 (1961).
972. Child, W., Thesis, University of Wisconsin, 1955.
973. Chiorboli, P. and P. Mirone, *Ann. Chim.* (Rome) **48**, 363 (1958); *C. A.* **52**, 15247 (1958).
974. Chiurodoglu, G. and W. Masschelein, *Bull. Soc. Chim. Belges* **68**, 484 (1959); *C. A.* **54**, 8284 (1960).
975. Chivate, M. R. and S. M. Shah, *Trans. Indian Inst. Chem. Engrs.* **6**, 170 (1953–1954); *C. A.* **50**, 1382 (1950).
976. Choudhuri, B. K., *Indian J. Physics* **11**, 203 (1937); *C. A.* **31**, 8376 (1937).
977. Christopher, P. M., *J. Chem. Eng. Data* **5**, 568 (1960).
978. Christopher, P. M. and A. Schilman, *J. Chem. Eng. Data* **12**, 333 (1967).
979. Chu, B., *J. Chem. Phys.* **41**, 226 (1964).
980. Chu, K-Yu and A. R. Thompson, *J. Chem. Eng. Data* **5**, 147 (1960).
981. Chueh, P. L. and S. W. Briggs, *J. Chem. Eng. Data* **9**, 207 (1964).
982. Ciamician, G., *Ber. Deut. Chem. Gesell.* **37**, 4200 (1904).
983. Ciccone, A. *Nuovo Cimento* **5**, 489 (1948); *C. A.* **43**, 4573 (1949).
984. Cislak, F. E. and M. M. Otto, U.S. Patent 2,432,062, Dec. 2, 1947; 2,432,063, Dec. 2, 1947; *C. A.* **42**, 1967 (1948); 2,456,581, Dec. 14, 1948; *C. A.* **43**, 3459 (1949).
985. Cislan, I. and A. Cornilescu, *Rev. Chim.* (Bucharest) **16**, 98 (1965); *C. A.* **63**, 6842 (1965).
986. Citroni, M., *Arch. Ital. Sci. Farmacol.* **1**, 284 (1951); *C. A.* **50**, 1213 (1956).
987. Claisen, L., *Ber. Deut. Chem. Gesell.* **20**, 646 (1887).
988. Clark, L. B. and W. T. Simpson, *J. Chem. Phys.* **43**, 3666 (1965).
989. Clarke, A. J. and L. Crombie, *Chem. Ind.* (London) **1957**, 143.
990. Clarke, G. A. and S. Sandler, *Chemist-Analyst* **50**, 76 (1961).
991. Clarke, H. T., *Organic Syntheses*, Vol. 3, Wiley, New York, 1923.
992. Clarke, H. T., *Organic Syntheses*, Vol. 10, Wiley, New York, 1930.

993. Clarke, H. T. and E. R. Taylor, *J. Amer. Chem. Soc.* **45,** 830 (1923).
994. Clarke, J., R. Robinson, et al., *J. Chem. Soc.* **1927,** 2647.
995. Clarke, J. T. and E. R. Blout, *J. Polym. Sci.* **1,** 419 (1946).
996. Claverie, N. and C. Garrigou-Lagrange, *J. Chim. Phys.* **61,** 889 (1964); *C. A.* **61,** 11480 (1964).
997. Clavier, A., *Bull. Soc. Chim. Fr.* **1954,** 646; *C. A.* **48,** 10540 (1954).
997a. Claxton, G. and W. H. Hoffert, *J. Soc. Chem. Ind.* **65,** 333, 341 (1946).
998. Clay, J., A. J. Dekker, et al., *Physica* **10,** 768 (1943); *C. A.* **38,** 5120 (1944).
999. Clayton, J. W., H. Sherman, et al., *Amer. Ind. Hyg. Assoc. J.* **27,** 332 (1966).
1000. Clayton, W. R. and E. E. Reid, *J. Amer. Chem. Soc.* **64,** 908 (1942).
1001. Cleaves, A. P. and M. E. Sherrick, *Natl. Advisory Comm. Aeronaut.*, *ARR* No. ESF27 (1945); *C. A.* **42,** 7960 (1948).
1002. Clemett, C. and M. Davies, *Trans. Faraday Soc.* **58,** 1718 (1962).
1003. Clendenning, K. A., F. G. MacDonald, et al., *Can. J. Research* **28B,** 608 (1950); *C. A.* **45,** 7007 (1951).
1004. Clerbaux, Th. and P. Huyskens, *Ann. Soc. Sci. Bruxelles Ser.* I, **80,** 130 (1966); *C. A.* **65,** 13517 (1966).
1005. Cleveland, F. F., *J. Phys. Chem.* **11,** 1 (1943); *C. A.* **37,** 831 (1943).
1006. Cleveland, F. F. and D. E. Lee, *Phys. Rev.* **65,** 350 (1944).
1007. Cleveland, F. F. and M. J. Murray, *J. Chem. Phys.* **5,** 752 (1937).
1008. Cleveland, F. F., M. J. Murray, et al., *J. Chem. Phys.* **8,** 153 (1940).
1009. Cleveland, F. F., M. J. Murray, et al., *J. Chem. Phys.* **10,** 18 (1942).
1010. Clever, H. L. and C. C. Snead, *J. Phys. Chem.* **67,** 918 (1963).
1011. Closson, W. D., S. F. Brady, et al., *J. Org. Chem.* **30,** 4026 (1965).
1012. Cluett, M. L., *Anal. Chem.* **34,** 1491 (1962).
1013. Cluzet, J. and T. Kofman, *Compt. Rend. Soc. Biol.* **103,** 783(1930); *Chem. Zentr.* **1930, II,** 12.
1014. Cocivera, M., *J. Amer. Chem. Soc.* **88,** 672 (1966).
1015. Cockerille, F. O., U.S. Patent 2,352,253, June 27, 1944; *C. A.* **38,** 5508 (1944).
1016. Coe, J. R. and T. B. Godfrey, *J. Applied Phys.* **15,** 625 (1944).
1017. Coetzee, J. F., *Pure Appl. Chem.* **13,** 429 (1966).
1018. Coetzee, J. F. and G. P. Cunningham, *J. Amer. Chem. Soc.* **87,** 2529 (1965).
1019. Coetzee, J. F., G. P. Cunningham, et al., *Anal. Chem.* **34,** 1139 (1962).
1020. Coetzee, J. F. and I. M. Kolthoff, *J. Amer. Chem. Soc.* **79,** 6110 (1957).
1021. Coetzee, J. F. and D K. McGuire, *J. Phys. Chem.* **67,** 1810 (1963).
1022. Coetzee, J. F. and R. Mei-Shun Lok, *J. Phys. Chem.* **69,** 2690 (1965).
1023. Coetzee, J. F. and G. R. Padmanabhan, *J. Phys. Chem.* **66,** 1708 (1962).
1024. Coffey, S. and C. F. Ward, *J. Chem. Soc.* **1921,** 1301.
1025. Coggeshall, N. D. and A. S. Glessner, *J. Amer. Chem. Soc.* **71,** 3150 (1949).
1026. Cogrossi, C., *Ann. Chim.* (Rome) **56,** 270 (1966); *C. A.* **64,** 19368 (1966).
1027. Cohen, E. and L C. J. te Boekhorst, *Z. Phys. Chem.* (Leipzig) **24B,** 241 (1934).
1028. Cohen, E. and J. S. Buij, *Z. Phys. Chem.* (Leipzig) **35B,** 270 (1937); *C. A.* **31,** 4174 (1937).
1029. Cohen, E. and A. L. T. Moesveld, *Z. Phys. Chem.* (Leipzig) **100,** 151 (1922).
1030. Cohen, H. and J. D. Mier, *Chem. Ind.* (London) **1965,** 349.
1031. Cohen, I., *J. Amer. Chem. Soc.* **52,** 2827 (1930).
1032. Cole, A. R. H., P. R. Jefferies, et al., *J. Chem. Soc.* **1959,** 1222.
1033. Cole, J. R., J. O. Knox, et al., *Chemist-Analyst* **48,** 38 (1959).
1034. Cole, P. J., *J. Chem. Eng. Data* **5,** 367 (1960).
1035. Cole, P. J., U.S. Patent 2,393,888, Jan. 29, 1946; *C. A.* **40,** 2468 (1946).

1036. Cole, P. J. and G. W. Burtt, U.S. Patent 2,348,329; *C. A.* **39,** 532 (1945).
1037. Coleman, C. F., *J. Phys. Chem.* **72,** 365 (1968).
1038. Coleman, C. F. and T. De Vries, *J. Amer. Chem. Soc.* **71,** 2839 (1949).
1039. Coleman, D. J. and G. Pilcher, *Trans. Faraday Soc.* **62,** 821 (1966).
1040. Coleman, G. H. and A. M. Alvarado, *Organic Syntheses,* Vol. 3, Wiley, New York, 1923.
1041. Coles, H. W. and W. E. Tournay, *Ind. Eng. Chem., Anal. Ed.* **14,** 20 (1942).
1042. Coles, H. W. and W. E. Tournay, *Anal. Chem.* **19,** 936 (1947).
1043. Collerson, R. R., J. F. Counsell, et al., *J. Chem. Soc.* **1965,** 3697.
1044. Collin, J., *Bull. Soc. Chim. Belges* **69,** 575 (1960); *C. A.* **55,** 14048 (1961).
1045. Collin, J., *Bull. Soc. Roy. Sci. Liege* **25,** 520 (1956); *C. A.* **51,** 17418 (1957).
1046. Collin, J. E. and M. J. Franklin, *Bull. Roy. Soc. Liege* **35,** 285 (1966); *C. A.* **65,** 12983 (1966).
1047. Collin, P. J. and S. Sternhell, *Aust. J. Chem.* **19,** 317 (1966); *C. A.* **64,** 16867 (1966).
1048. Collins, B. T., C. F. Coleman, et al., *J. Amer. Chem. Soc.* **71,** 2929 (1949).
1049. Colomina, M., C. Latorre, et al., *Pure Appl. Chem.* **2,** 133 (1961).
1050. Colomina, M., A. S. Pell, et al., *Trans. Faraday Soc.* **61,** 2641 (1965).
1051. Colon, A. A. and H. A. Frediani, *J. Amer. Pharm. Assoc.,* Sci. Ed. **40,** 607 (1951).
1052. Commercial Solvents Corp., British Patent 898,388, June 6, 1962.
1053. Commercial Solvents Corp., "The Nitroparaffins."
1054. Commercial Solvents Corp., "Acetone," Technical Data Sheet No. 13, IC Series, New York, 1962.
1055. Commercial Solvents Corp., "Butyl Stearate," Technical Data Sheet No. 27, IC Series, New York, 1963.
1056. Commercial Solvents Corp., "Dibutyl Phthalate," Technical Data Sheet No. 14, IC Series, New York, 1962.
1057. Commercial Solvents Corp., "Ethyl Acetate," Technical Data Sheet No. 20, IC Series, New York, 1962.
1058. Commercial Solvents Corp., "Storage and Handling of Nitromethane," NP Technical Data Sheet, TDS No. 2, New York, 1961.
1059. Commercial Solvents Corp., "Tributyl Phosphate," Technical Data Sheet No. 7, IC Series, New York, 1965.
1060. Conant, J. B., *Organic Syntheses,* Vol, 2, Wiley, New York, 1922.
1061. Conant, J. B., *Organic Syntheses,* Vol. 9, Wiley, New York, 1929.
1062. Conant, J. B. and G. H. Carlson, *J. Amer. Chem. Soc.* **51,** 3464 (1929).
1063. Conant, J. B., C. N. Webb, et al., *J. Amer. Chem. Soc.* **51,** 1249 (1929).
1064. Connolly, J. F., *J. Chem. Eng. Data* **11,** 13 (1966).
1065. Connolly, J. F. and G. A. Kandalic, *J. Chem. Eng. Data* **7,** 137 (1962).
1066. Connor, W. P. and C. P. Smyth, *J. Amer. Chem. Soc.* **65,** 382 (1943).
1067. Conrad, M. and W. R. Hodgkinson, *Ann. Chem.* **193,** 320 (1885); beil **I, VI,** 435.
1068. Conrad-Bilbroth, H., *Z. Phys. Chem.* (Leipzig) **23B,** 315 (1933); *C. A.* **28,** 948 (1934).
1069. Constable, F. H. and S. Tegul, *Nature* **157,** 735 (1946).
1070. Cook, A. N., *J. Amer. Chem. Soc.* **26,** 303 (1904).
1071. Cook, A. N. and A. L. Haines, *Proc. Iowa Acad. Sci.* **9,** 86 (1901); *J. Amer. Chem. Soc.* **25R,** 559 (1903).
1072. Cooke, R. G., *Chem. Ind.* **1955,** 142.
1073. Cookson, R. C., *J. Chem. Soc.* **1954,** 282.
1074. Cookson, R. C. and S. H. Dandegaonker, *J. Chem. Soc.* **1955,** 1651.
1075. Coolidge, A. S., *J. Amer. Chem. Soc.* **50,** 2166 (1928).
1076. Coolidge, A. S., *J. Amer. Chem. Soc.* **52,** 1874 (1930).

1077. Coomber, D. I. and J. R. Partington, *J. Chem. Soc.* **1938**, 1444.
1078. Coomber, D. I. and B. A. Rose, *J. Pharm. Pharmacol.* **11**, 670 (1959); *C. A.* **54**, 9204 (1960).
1079. Coon, E. D. and F. Daniels, *J. Phys. Chem.* **37**, 1 (1933).
1080. Coop, I. E. and L. E. Sutton, *J. Chem. Soc.* **1938**, 1869.
1081. Cooper, J. R., M. Kini, et al., *J. Neurochem.* **9**, 119 (1962); *C. A.* **57**, 8844 (1962).
1082. Coops, J., D. Mulder, et al., *Rec. Trav. Chim. Pays-Bas* **65**, 128 (1946); *C. A.* **40** 3972 (1946).
1083. Copeland, C. S. and M. W. Rigg, *J. Amer. Chem. Soc.* **73**, 3584 (1951).
1084. Copelin, H. B., U.S. Patent 2,904,600, Sept. 15, 1959; *C. A.* **55**, 851 (1961).
1085. Copley, M. J., E. Ginsberg, et al., *J. Amer. Chem. Soc.* **63**, 254 (1941).
1086. Copp, J. L. and D. H. Everett, *Discussion Faraday Soc.* No. 15, 174 (1953); from [4773].
1087. Copp, J. L. and T. J. V. Findlay, *Trans. Faraday Soc.* **56**, 13 (1960).
1088. Cordes, H. F. and C. W. Tait, *Anal. Chem.* **29**, 485 (1957).
1089. Corey, C. J. and M. Chaykovsky, *J. Amer. Chem. Soc.* **86**, 1639 (1964).
1090. Corin, C., *Bull. Soc. Roy. Sci. Liege* **10**, 99 (1941); *C. A.* **39**, 4548 (1945).
1091. Corin, C., *J. Chim. Phys.* **33**, 448 (1936); *C. A.* **30**, 7038 (1936).
1092. Corin, C. and G. B. B. M. Sutherland, *Proc. Roy. Soc.* **165A**, 43 (1938).
1093. Corish, P. J. and D. Chapman, *J. Chem. Soc.* **1957**, 1746.
1094. Cornubert, R., G. Barraud, et al., *Bull. Soc. Chim. Fr.* **1955**, 400; *C. A.* **50**, 2449.
1095. Cornubert, R., V. Kondrachov, et al., *Bull. Soc. Chim. Fr.* **1959**, 385; *C. A.* **53**, 16016 (1959); Beil EIII6, 69, 73.
1096. Cornubert, R., M. Lafont-Lemoine, *Compt. Rend.* **237**, 469 (1953); *C. A.* **48**, 13643 (1954).
1097. Corruccini, R. J. and D. C. Ginnings, *J. Amer. Chem. Soc.* **69**, 2291 (1947).
1098. Corson, B. B. and V. N. Ipatieff, *J. Amer. Chem. Soc.* **59**, 645 (1937).
1099. Cosner, J. L., J. E. Gagliardo, et al., *J. Chem. Eng. Data* **6**, 360 (1961).
1100. Costello, J. M. and S. T. Bowden, *Chem. Ind.* **1956**, 1041.
1101. Costello, J. M. and S. T. Bowden, *Rec. Trav. Chim. Pays-Bas* **78**, 391 (1949); from [4773].
1102. Cotton, F. A. and R. Francis, *J. Amer. Chem. Soc.* **82**, 2986 (1960).
1103. Cottrell, F. G., *J. Amer. Chem. Soc.* **41**, 721 (1919).
1104. Coujolle, F. and G. Roux, *Compt. Rend.* **239**, 680 (1954); *C. A.* **49**, 4181 (1955).
1105. Coulson, E. A. and J. B. Ditcham, *J. Appl. Chem.* (London) **2**, 236 (1952).
1106. Coulson, E. A., J. L. Hales, et al., *J. Chem. Soc.* **1951**, 2125.
1107. Counsell, J. F., D. H. Everett, et al., *Pure Appl. Chem.* **2**, 335 (1961).
1108. Counsell, J. F., J. H. S. Green, et al., *Trans. Faraday Soc.* **61**, 212 (1965).
1109. Counsell, J. F., J. L. Hales, et al., *Trans. Faraday Soc.* **61**, 1869 (1965).
1110. Counsell, J. F., E. B. Lees, et al., *J. Chem. Soc.* **1968A**, 1819.
1111. Courtot, C., *Ann. Soc. Sci. Bruxelles, Ser. I, Sci. Math. Phys. Astron.* **60**, 122 (1946); *C. A.* **41**, 4717 (1947).
1112. Courtoy, C., *Ann. Soc. Sci. Bruxelles, Ser. I*, **61**, 139 (1947); *C. A.* **43**, 4140 (1949).
1113. Cowan, D. M., G. H. Jeffery, et al., *J. Chem. Soc.* **1940**, 171.
1114. Cowan, D. M. and A. I. Vogel, *J. Chem. Soc.* **1940**, 1528.
1115. Cowley, E. G., *J. Chem. Soc.* **1952**, 3557.
1116. Cowley, E. G. and J. R. Partington, *J. Chem. Soc.* **1933**, 1257.
1117. Cowley, E. G. and J. R. Partington, *J. Chem. Soc.* **1935**, 604.
1118. Cowley, E. G. and J. R. Partington, *J. Chem. Soc.* **1936**, 1184.
1119. Cowley, E. G. and J. R. Partington, *J. Chem. Soc.* **1938**, 977.

1120. Cox, H. L. and L. H. Cretcher, *J. Amer. Chem. Soc.* **48,** 451 (1926).
1121. Cox, J. D., *Pure Appl. Chem.* **2,** 125 (1961).
1122. Cox, J. D., A. R. Challoner, et al., *J. Chem. Soc.* **1954,** 265.
1123. Crable, G. F., G. L. Kearns, et al., *Anal. Chem.* **32,** 13 (1960).
1124. Crafts, J. M., *Ber. Deut. Chem. Gesell.* **20,** 709 (1887).
1125. Craig, B. M., *Can. J. Chem.* **31,** 499 (1953).
1126. Cramer, J. S. N., *Rec. Trav. Chim. Pays-Bas* **62,** 606 (1943); *C. A.* **38,** 6148 (1944).
1127. Cramer, P. L. and V. A. Miller, *J. Amer. Chem. Soc.* **62,** 1452 (1940).
1128. Cramer, P. L. and M. J. Mulligan, *J. Amer. Chem. Soc.* **58,** 373 (1936).
1129. Crawford, J. W. C. and J. McGrath, U.S. Patent 2,140,469, Dec. 13, 1938.
1130. Creamer, R. M. and D. H. Chambers, *J. Electrochem. Soc.* **101,** 162 (1954).
1131. Cretcher, L. H. and W. H. Pittenger, *J. Amer. Chem. Soc.* **46,** 1503 (1924).
1132. Cretcher, L. H. and W. H. Pittenger, *J. Amer. Chem. Soc.* **47,** 163 (1925).
1133. Crisler, R. O. and A. M. Burrill, *Anal. Chem.* **31,** 2055 (1959).
1134. Crisp, S., S. R. C. Hughes, et al., *J. Chem. Soc.* **1968A,** 603.
1135. Critchfield, F. E., J. A. Gibson, et al., *J. Amer. Chem. Soc.* **75,** 1991 (1953).
1136. Crompton, T. R. and D. Buckley, *Analyst,* **90,** 76 (1965).
1137. Cropper, F. R. and S. Kaminsky, *Anal. Chem.* **35,** 735 (1963).
1138. Crowder, G. A. and B. R. Cook, *J. Phys. Chem.* **71,** 914 (1967).
1139. Crowe, R. W. and C. P. Smyth, *J. Amer. Chem. Soc.* **72,** 4009 (1950).
1140. Crowe, R. W. and C. P. Smyth, *J. Amer. Chem. Soc.* **73,** 5406 (1951).
1141. Crown Zellerbach Corp., Chemical Products Division, "Dimethyl Sulfoxide Technical Bulletin" and supplement, Camas, Washington, 1963 and 1966 ed.
1142. Crown Zellerbach Corp., Chemical Products Division, "Properties and Suggested Uses of Dimethyl Sulfide (DMS)," Camas, Washington, 1957; Loose Leaf Revision, 1965.
1143. Croxall, W. J. and H. T. Neher, U.S. Patent 2,448,660 Sept. 7, 1948; *C. A.* **43,** 1433 (1949).
1144. Cryder, D. S. and J. O. Maloney, *Trans. Inst. Chem. Eng.* **37,** 827 (1941).
1145. Cuisa, *Gazz. Chim. Ital.* **41a,** 688 (1911); from [3942], p. 607.
1146. Culbertson, G. and R. Pettit, *J. Amer. Chem. Soc.* **85,** 741 (1963).
1147. Cullinane, N. M. and W. C. Davies, British Patent 600,837, Apr. 20, 1948; *C. A.* **42,** 7334 (1948).
1148. Cummins, E. G. and J. E. Page, *J. Chem. Soc.* **1957,** 3847.
1149. Cumper, C. W. N., J. F. Read, et al., *J. Chem. Soc.* **1965,** 5323.
1150. Cumper, C. W. N., J. F. Read, et al., *J. Chem. Soc.* **1966A,** 239.
1151. Cumper, C. W. N. and A. Singleton, *J. Chem. Soc.* **1968B,** 645.
1152. Cumper, C. W. N. and A. I. Vogel, *J. Chem. Soc.* **1959,** 3521.
1153. Cumper, C. W. N. and A. I. Vogel, *J. Chem. Soc.* **1960,** 4723.
1154. Cumper, C. W. N. and A. I. Vogel, et al., *J. Chem. Soc.* **1957,** 3640.
1155. Cundiff, R. H. and P. C. Markunas, *Anal. Chem.* **30,** 1450 (1958); **33,** 1028 (1961).
1156. Curl, R. F., *J. Chem. Phys.* **30,** 1529 (1959).
1157. Curme, G. O. and F. Johnston, eds., *Glycols,* Reinhold, New York, 1952.
1158. Curran, C., *J. Amer. Chem. Soc.* **67,** 1835 (1945).
1159. Cutler, J. A., *J. Chem. Phys.* **16,** 136 (1948).
1160. Cuypers, R., *Annual Tables of Physical Constants,* Sec. 301(c), 1941.
1161. Czapska, W., *Compt. Rend.* **189,** 32 (1929); *C. A.* **23,** 5108 (1929)
1162. Czerny, M. and P. Mollet, *Z. Phys.* **108,** 85 (1937); *C. A.* **32,** 4432 (1938).

D

1163. Daasch, L. W., *J. Chem. Phys.* **22,** 1293 (1954).
1164. Daasch, L. W., *J. Phys. Chem.* **69,** 3196 (1965).
1165. D'Adams, A. F. and R. H. Kienle, *J. Amer. Chem. Soc.* **77,** 4408 (1955).
1166. Dadieu, A., *Monatsh. Chem.* **57,** 437 (1931); *C. A.* **25,** 4794 (1931).
1167. Dadieu, A. and K. W. F. Kohlrausch, *Ber. Deut. Chem. Gesell.* **63,** 251 (1930).
1168. Dadieu, A. and K. W. F. Kohlrausch, *Monatsh. Chem.* **52,** 220 (1929); *Chem. Zentr.* **1929,** II, 697, 1776.
1169. Dadieu, A. and K. W. F. Kohlrausch, *Monatsh. Chem.* **52,** 379 (1929); *Chem. Zentr.* **1930,** I, 2218.
1170. Dadieu, A. and K. W. F. Kohlrausch, *Monatsh. Chem.* **55,** 58 (1930); *C. A.* **24,** 2105 (1930); ibid. **53/54,** 282 (1929); *C. A.* **24,** 552 (1930).
1171. Dadieu, A. and K. Kohlrausch, *Monatsh. Chem.* **55,** 201 (1930); *C. A.* **24,** 3437 (1930).
1172. Dadieu, A. and K. W. F. Kohlrausch, *Monatsh. Chem.* **55,** 379 (1930); *Chem. Zentr.* **1930,** II, 1340.
1173. Dadieu, A. and K. W. F. Kohlrausch, *Monatsh. Chem.* **56,** 461 (1930); *Chem. Zentr.* **1931,** I, 2170.
1174. Dadieu, A. and K. W. F. Kohlrausch, *Monatsh. Chem.* **57,** 225 (1930); *Chem. Zentr.* **1931,** I, 2170.
1175. Dadieu, A. and K. W. F. Kohlrausch, *Naturwissenschaften* **17,** 366 (1929); *C. A.* **23.** 5108 (1929).
1176. Dadieu, A. and K. W. F. Kohlrausch, *Phys. Z.* **30,** 384; *Chem. Zentr.* **1929,** II, 970.
1177. Dadieu, A. and K. W. F. Kohlrausch, *Sitzb. Akad. Wiss. Wien*, Abt. IIa, **139,** 459, (1930); *Monatsh. Chem.* **57,** 225 (1931); *C. A.* **25,** 2365 (1931).
1178. Dadieu, A., A. Pongratz, et al., *Monatsh. Chem.* **61,** 409, 426; *Sitzb. Akad. Wiss. Wien, Math.-Naturw. Klasse*, Abt. II, **141,** 477; *C. A.* **27,** 4172 (1933).
1179. Dahlgard, M. and R. Q. Brewster, *J. Amer. Chem. Soc.* **80,** 5861 (1958).
1180. Dakshinamurty, P., G. J. Rao, et al., *J. Appl. Chem.* (London) **11,** 226 (1961).
1181. Dale, J., *Acta Chem. Scand.* **11,** 971 (1957); *C. A.* **52,** 14325 (1958).
1182. D'Alelio, G. F. and E. E. Reid, *J. Amer. Chem. Soc.* **59,** 109 (1937).
1183. Dalton, F., G. Meakins, et al., *J. Chem. Soc.* **1962,** 1566.
1184. Danforth, W. E., *J. Phys. Rev.* **38,** 1224 (1931); *Chem. Zentr.* **1931,** II, 3582.
1185. Dangyan, M. T., *Bull. Armenian Branch Acad. Sci. USSR* **1942,** No. 9/10 (23/24), 53; *C. A.* **40,** 3399 (1946).
1186. Daniels, F., J. H. Mathews, et al., *Experimental Physical Chemistry*, 3rd ed., McGraw-Hill, New York, 1941.
1187. Danner, P. S. and J. H. Hildebrand, *J. Amer. Chem. Soc.* **44,** 2824 (1922).
1188. Dannhauser, W. and L. W. Bahe, *J. Chem. Phys.* **40,** 3058 (1964).
1189. Dannhauser, W., L. W. Bahe, et al., *Chem. Phys.* **43,** 257 (1965).
1190. Dannhauser, W. and R. H. Cole, *J. Chem. Phys.* **23,** 1762 (1955).
1191. Dannahuser, W. and A. F. Flueckinger, *J. Phys. Chem.* **68,** 1814 (1964).
1192. Duragan, B., *Bull. Soc. Chim. Belges* **44,** 597 (1935).
1193. DaSilva, D. J., *Rec. Trav. Chim. Pays-Bas* **53,** 1097 (1934).
1194. Dasler, W. and C. D. Bauer, *Ind. Eng. Chem., Anal. Ed.* **18,** 52 (1946).
1195. Datin, P., *Ann. Phys.* **5,** 218 (1916).
1196. Daub, L. and J. M. Vandenbelt, *J. Amer. Chem. Soc.* **69,** 2714 (1947).
1197. Dauben, W. G., G. J. Fonken, et al., *J. Amer. Chem. Soc.* **78,** 2579 (1956).
1198. Davanathan, T., *Proc. Indian Acad. Sci.* Sect. A**57,** No. 2, 109 (1963); *C. A.* **59,** 2250 (1963).

1199. David, J. G. and H. E. Hallam, *Trans. Faraday Soc.* **60,** 2013 (1964).
1200. Davidson, J. G., *Ind. Eng. Chem.* **18,** 669 (1926).
1201. Davies, C. W., *Phil. Mag.* **4,** 244 (1927).
1202. Davies, D. G., *J. Chem. Soc.* **1935,** 1166.
1203. Davies, G. F. and E. C. Gilbert, *J. Amer. Chem. Soc.* **63,** 1585 (1941).
1204. Davies, G. F. and E. C. Gilbert, *J. Amer. Chem. Soc.* **63,** 2730 (1941).
1205. Davies, M., *J. Chem. Phys.* **16,** 267 (1948).
1206. Davies, M., A. H. Jones, et al., *Trans. Faraday Soc.* **55,** 1100 (1959).
1207. Davies, M. and B. Kybett, *Trans. Faraday Soc.* **61,** 1608 (1965).
1208. Davies, M. and D. K. Thomas, *J. Phys. Chem.* **60,** 767 (1956).
1209. Davies, M. M., *J. Chem. Phys.* **8,** 577 (1940).
1210. Davies, R. M., *Phil. Mag.* **21,** 1; 1008 (1936).
1211. Davies, R. R. and H. H. Hodgson, *J. Chem. Soc.* **1943,** 84.
1212. Davies, R. R. and H. H. Hodgson, *J. Soc. Chem. Ind. Trans.* (London) **62,** 128 (1943).
1213. Davies, W. and J. B. Jagger, et al., *J. Soc. Chem. Ind.* (London) **68,** 26 (1949).
1214. Davies, W. H. and L. L. McGee, *J. Chem. Soc.* **1950,** 678.
1215. Davis, D. S., *Ind. Eng. Chem.* **33,** 401 (1941).
1216. Davis, H. S. and O. F. Wiedeman, *Ind. Eng. Chem.* **37,** 482 (1945).
1217. Davis, J. C., *J. Chem. Educ.* **43,** 611 (1966).
1218. Davis, M. M., *Acid-Base Behavior in Aprotic Organic Solvents*, Washington, D.C., U.S. Government Printing Office, 1968.
1219. Davis, M. M. and P. J. Schumann, *J. Res. Nat. Bur. Stand.* **39,** 221 (1947).
1220. Davis, R., H. S. Bridge, et al., *J. Amer. Chem. Soc.* **65,** 857 (1943).
1221. Davison, J. A., *J. Amer. Chem. Soc.* **67,** 228 (1945).
1222. Davison, R. R., W. H. Smith, et al., *J. Chem. Eng. Data* **5,** 420 (1960).
1223. Davison, W. H. T. and G. R. Bates, *J. Chem. Soc.* **1953,** 2607.
1224. Davy, L. G. and N. V. Sidgwick, *J. Chem. Soc.* **1933,** 281.
1225. Dawson, L. R., M. Golben, et al., *J. Electrochem. Soc.* **99,** 28 (1952).
1226. Dawson, L. R., R. H. Graves, et al., *J. Amer. Chem. Soc.* **79,** 298 (1957).
1227. Dawson, L. R., T. M. Newell, et al., *J. Amer. Chem. Soc.* **76,** 6024 (1954).
1228. Dawson, L. R., P. G. Sears, et al., *J. Amer. Chem. Soc.* **77,** 1986 (1955).
1229. Dawson, L. R., J. W. Vaughn, et al., *J. Phys. Chem.* **66,** 2684 (1962).
1230. Dawson, L. R., E. D. Wilhoit, et al., *J. Amer. Chem. Soc.* **78,** 1569 (1956).
1231. Dawson, L. R., E. D. Wilhoit, et al., *J. Amer. Chem. Soc.* **79,** 3004 (1957).
1232. Day, H. O. and W. A. Felsing, *J. Amer. Chem. Soc.* **74,** 1951 (1952).
1233. Day, H. O., D. E. Nicholson, et al., *J. Amer. Chem. Soc.* **70,** 1784 (1948).
1234. Day, J. H. and A. Joachim, *J. Org. Chem.* **30,** 4107 (1965).
1235. Day, R. A., A. E. Robinson, et al., *J. Amer. Chem. Soc.* **72,** 1379 (1950).
1236. Deanesly, R. M. and L. T. Carleton, *J. Phys. Chem.* **45,** 1104 (1941).
1237. Dearden, J. C. and W. F. Forbes, *Can. J. Chem.* **37,** 1294 (1959).
1238. Dearden, J. C. and W. F. Forbes, *Can. J. Chem.* **37,** 1305 (1959).
1239. Deb, A. R., *Indian J. Phys.* **27,** 457 (1953); *Sci. Abstr.* **57A,** 698 (1954); *C. A.* **51,** 7856 (1957).
1240. Deb, K. K., *Indian J. Phys.* **35,** 535 (1961); *C. A.* **57,** 300 (1962).
1241. Deb, K. K., *Indian J. Phys.* **36,** 557 (1962); *C. A.* **58,** 12084 (1963).
1242. de Beule, P., *Bull. Soc. Chim. Belges* **40,** 195 (1931).
1243. de Brouckère, L. and A. Prigogine, *Bull. Soc. Chim. Belges* **47,** 382 (1938); *C. A.* **32,** 8888 (1938).
1244. de Brouckère, L. and A. Prigogine, *Bull. Soc. Chim. Belges* **47,** 399 (1938); *C. A.* **32,** 8888 (1938).

1245. De Bruyne, J. M. A., R. M. Davis, et al., *J. Amer. Chem. Soc.* **55**, 3936 (1933).
1246. Deckenbrock, W., *Deut. Lebensm.-Rundschau* **49**, 30 (1953); *C. A.* **49**, 7804 (1955).
1247. Decker, R. and H. Halz, German Patent 837,995, May 5, 1952; *C. A.* **51**, 15548 (1957).
1248. Deffet, L., *Bull. Soc. Chim. Belges.* **40**, 385 (1931); *C. A.* **26**, 352 (1932).
1249. Deffet, L., *Bull. Soc. Chim. Belges.* **44**, 41, 97 (1935); *Chem. Zentr.* **1935, II**, 33.
1250. Deffet, L., *Bull. Soc. Chim. Belges.* **47**, 446 (1938); *Chem. Zentr.* **1939, I**, 3149.
1251. Deffet, L., *Bull. Soc. Chim. Belges.* **49**, 223 (1940); *C. A.* **36**, 322 (1942).
1252. de Forcrand, R., *Compt. Rend.* **114**, 1062; *J. Chem. Soc.* **1892A**, 1066.
1253. de Forcrand, R., *Compt. Rend.* **130**, 1622 (1900); *Chem. Zentr.* **1900, II**, 160.
1254. de Forcrand, R., *Compt. Rend.* **132**, 569 (1901); *Chem. Zentr.* **1901, I**, 818.
1255. de Forcrand, R., *Compt. Rend.* **136**, 1037 (1903); *J. Chem. Soc.* **1903Ai**, 455.
1256. de Forcrand, R., *Compt. Rend.* **154**, 1327 (1912); *J. Chem. Soc.* **1912Ai**, 548.
1257. de Forcrand, R., *Compt. Rend.* **154**, 1767 (1912); *J. Chem. Soc.* **1912Aii**, 735.
1258. De Francesco, F., *Riv. Ital. Sostanze Grasse* **41**, 20 (1964); *C. A.* **61**, 2070 (1964).
1259. De Graaf, D. E. and G. B. B. M. Sutherland, *J. Chem. Phys.* **26**, 716 (1957).
1260. Dehn, W. M. and K. E. Jackson, *J. Amer. Chem. Soc.* **55**, 4284 (1933).
1261. Deichmann, W. B. and S. Witherup, *J. Pharmacol.* **80**, 233 (1944); *C. A.* **38**, 2389 (1944).
1262. Deizenrot, I. V., V. B. Kogan, et al., *Fridman, Zh. Prikl. Khim.* **39**, 1880 (1966); *C. A.* **65**, 16119 (1966).
1263. de Kolossowski, N. and A. Alemov, *Bull. Soc. Chim. Fr.* **1**, 877 (1934); Beil EIII3, 5.
1264. de Kolossowski, N. and W. W. Udowenko, *Compt. Rend.* **197**, 519 (1933); *Chem. Zentr.* **1933, II**, 2375; Beil EIII2, 1610.
1265. de Kowalewski, D. G., *J. Phys. Radium* **23**, 255 (1962); *C. A.* **57**, 11096 (1962).
1266. de Kowalewski, D. G. and V. J. Kowalewski, *Anales. Asoc. Quim. Arg.* **48**, 157 (1960); *C. A.* **55**, 18311 (1961).
1267. Delaby, R. and J. Lecomte, *Bull. Soc. Chim. Fr.* **4**, 738 (1937).
1268. Delbouille, L., *Bull. Classe Sci. Acad. Roy. Belg.* **44**, 791 (1958); *C. A.* **53**, 19571 (1959).
1269. Delépine, M., *Compt. Rend.* **124**, 292 (1897); *J. Chem. Soc.* **1897Ai**, 394.
1270. Delépine, M., *Compt. Rend.* **126**, 964 (1898); *Chem. Zentr.* **1898, I**, 989.
1271. Delépine, M., *Compt. Rend.* **126**, 1794 (1898); *J. Chem. Soc.* **1898Aii**, 559.
1272. Delorme, P., V. Lorenzelli, et al., *Compt. Rend.* **261**, (Groupe 6), 3331 (1965); *C. A.* **64**, 6450 (1966).
1273. Deluzarche, A., A. Maillard, J. Marie, et al., *Bull. Soc. Chim. Fr.* **1963**, 89; *C. A.* **58**, 13463 (1963); Ref. [5181].
1274. Delvaux, E., *J. Pharm. Belg.* **18**, 101, 131, 153 (1936); *Chem. Zentr.* **1936, I**, 3769.
1275. de Maine, P. A. D., L. H. Daly, et al., *Can. J. Chem.* **38**, 1921 (1960).
1276. de Melle, U., French Patent 871,565, Ap. 30, 1942; *C. A.* **43**, 5032 (1949); British Patent 614,165, Dec. 10, 1948; *C. A.* **43**, 4682 (1949).
1277. Demidenkova, I. V., *Tr. Gos. Inst. Prikl. Khim.* No. **52**, 157 (1964); *C. A.* **63**, 14230 (1965).
1278. de Montmollin, M. and P. Matile, *Helv. Chim. Acta* **7**, 106 (1924).
1279. Denison, R. B. and B. D. Steele, *J. Chem. Soc.* **1906**, 999, 1386.
1280. Deno, N., T. Edwards, et al., *J. Amer. Chem. Soc.* **79**, 2108 (1957).
1281. Deno, N. C., R. W. Gaugler, et al., *J. Org. Chem.* **31**, 1967 (1966).
1282. Deno, N. C., R. W. Gaugler, et al., *J. Org. Chem.* **31**, 1968 (1966)
1283. Deno, N. C. and J. O. Turner, *J. Org. Chem.* **31**, 1969 (1966).
1284. Deno, N. C. and M. J. Wisotsky, *J. Amer. Chem. Soc.* **85**, 1735 (1963).

1285. Denyer, R. L., F. A. Fidler, et al., *Ind. Eng. Chem.* **41**, 2727 (1949).
1286. de Pauw, F. and G. E. Limido, *Compt. Rend.* 27^e Congr. Intern. Chim. Ind. Brussels **1954**, 2; *Ind. Chim. Belge* **20**, Spec. No., 178–81 (1955); *C. A.* **50**, 14573 (1956).
1287. Du Puy, C. H. and B. W. Ponder, *J. Amer. Chem. Soc.* **81**, 4629 (1959).
1288. Dereppe, J. M. and M. van Meerssche, *Bull. Soc. Chim. Belges* **69**, 466 (1960); from [4773].
1289. Deslandres, H., *Compt. Rend.* **212**, 832 (1941); *C. A.* **38**, 5455 (1944).
1290. Deslandres, H., *Compt. Rend.* **213**, 749 (1941); *C. A.* **37**, 4013 (1943).
1290a. Desreux, V., *Bull. Soc. Chim. Belges.* **44**, 11, 249 (1935).
1291. Desty, D. H. and F. A. Fidler, *Ind. Eng. Chem.* **43**, 905 (1951).
1292. Dever, D. F., A. Finch, et al., *J. Phys. Chem.* **59**, 668 (1955).
1293. Devoto, G., *Gazz. Chim. Ital.* **63**, 495 (1933); *Chem. Zentr.* **1933**, II, 3394.
1294. Devoto, G. and R. Di Nola, *Gazz. Chim. Ital.* **63**, 495 (1933); *C. A.* **28**, 744 (1934).
1295. De Vries, T. and H. Soffer, *J. Phys. Colloid Chem.* **55**, 406 (1951).
1296. Dewar, M. J. S., R. C. Fahey, et al., *Tetrahedron Lett.* **1963**, 343; *C. A.* **59**, 2604 (1963).
1297. Deyanova, E. V., *Novoe v Oblasti Sanit.-Khim. Analiza* (*Raboty po Prom.-Sanit. Khim*) **1962**, 120; *C. A.* **59**, 2094 (1963).
1298. Dharmatti, S. S., G. Govil, et al., *Proc. Indian Acad. Sci.* **56A**, 86 (1962); *C. A.* **58**, 4069 (1963).
1299. Dial, W., U.S. Patent 3,025,331, Mar. 13, 1962; *C. A.* **57**, 1182 (1962).
1300. D'Ianni, J. and H. Adkins, *J. Amer. Chem. Soc.* **61**, 1675 (1939).
1301. Dibeler, V. H., *J. Research Nat. Bur. Stand.* **49**, 235 (1952).
1302. Dibeler, V. H. and R. M. Reese, *J. Research Nat. Bur. Stand.* **54**, 127 (1955).
1303. Dibeler, V. H., R. M. Reese, et al., *J. Chem. Phys.* **26**, 304 (1957).
1304. Di Bello, L. M., H. M. McDevitt, et al., *J. Phys. Chem.* **72**, 1405 (1968).
1305. Di Ciommox, *Nuovo Cimento* **3**, 97 (1903).
1306. Dickinson, J. D. and C. Eaborn, *Chem. Ind.* (London) **1956**, 959.
1307. Dieckmann, W., *Ber. Deut. Chem. Gesell.* **55**, 2470 (1922).
1308. Diehl, P. and G. Svegliado, *Helv. Chim. Acta* **46**, 461 (1963).
1309. Dieke, S. H., G. S. Allen, et al., *J. Pharmacol. Exp. Ther.* **90**, 260 (1947).
1310. Dietrich, M. W., J. S. Nash, et al., *Anal. Chem.* **38**, 1479 (1966).
1311. Di Macco, G., *Scientia Med. Ital.* **4**, 100 (1955); *C. A.* **50**, 14111 (1956).
1312. Dimroth, O., *Angew. Chem.* **46**, 571 (1933).
1313. Dippy, J. F. J., *Chem. Rev.* **25**, 151 (1939).
1314. Dippy, J. F. J., *J. Chem. Soc.* **1938**, 1222.
1315. Dirksen, H.-W., Diploma Thesis, Universität Freiberg, 1959; from [3042].
1316. Dirksen, H.-W., Ph.D. Thesis, Universität Freiberg, 1961; from [3042].
1317. Dischler, B., *Z. Naturforsch.* **20A**, 888 (1965); *C. A.* **63**, 17,343 (1965).
1318. Distillers Co. Ltd., British Patent 883,746, Dec. 6, 1961; *C. A.* **57**, 13684 (1962).
1319. Divorken, A. and M. Guillamin, *J. Chim. Phys.* **63**, 53, 57 (1966); *C. A.* **64**, 18517 (1966).
1320. Dixon, R. L., R. H. Adamson, et al., *Arch. Int. Pharmacodyn.* **160**, 333 (1966).
1321. Djerassi, C., *J. Amer. Chem. Soc.* **71**, 1003 (1949).
1322. Djerassi, C. and C. Fenselau, *J. Amer. Chem. Soc.* **87**, 5756 (1965).
1323. Djerassi, C., J. Osiecki, et al., *J. Amer. Chem. Soc.* **83**, 4433 (1961).
1324. Dobrinskaya, A. A., M. B. Neĭman, et al., *Dokl. Akad. Nauk SSSR* **63**, 549 (1948) *C. A.* **43**, 2865 (1949).
1325. Dobrosserdow, D., *J. Russ. Phys. Chem. Soc.* **43**, 73 (1911).
1326. Dobrosserdow, D., *Sapiski Kasaner Univ.* **1909**, 1; *Chem. Zentr.* **1911**, I, 953.

1327. Dobry, A. and R. Keller, *J. Phys. Chem.* **61,** 1148 (1957).
1328. Doerffel, K. and K. Dathe, *Wies. Z. Tech. Hochsch. Chem. Launa-Merseburg* **7,** 306 (1965); *C. A.* **64,** 17385 (1966).
1329. Doldi, S., *Ann. Chim.* (Rome) *Applicata* **28,** 454 (1938); *C. A.* **33,** 5250 (1939).
1330. Dolique, R., *Ann. Chim.* (Paris) **15,** 425 (1931); *C. A.* **26,** 2791 (1932).
1331. Dolique, R., *Bull. Sci. Pharmacol.* **39,** 129 (1932).
1332. Dolique, R., *Compt. Rend.* **194,** 289 (1932).
1333. Dolliver, M. A., T. L. Gresham, et al., *J. Amer. Chem. Soc.* **60,** 440 (1938).
1334. Donaldson, R. E. and O. R. Quayle, *J. Amer. Chem. Soc.* **72,** 35 (1950).
1335. Donle, H. L. and K. A. Gehrckens, *Z. Phys. Chem.* (Leipzig) **18B,** 316 (1932).
1336. Donle, H. L. and G. Volkert, *Z. Phys. Chem.* (Leipzig) **8B,** 60 (1930).
1337. Donoghue, J. T. and R. S. Drago, *Inorg. Chem.* **1,** 866 (1962); **2,** 572 (1963).
1338. Doolittle, A. K., *Ind. Eng. Chem.* **27,** 1169 (1935).
1339. Doolittle, A. K., *The Technology of Solvents and Plasticizers,* Wiley, New York, 1954.
1340. Doolittle, A. K. and R. H. Peterson, *J. Amer. Chem. Soc.* **73,** 2145 (1951).
1341. D'Or, L., J. Momigny and P. Natalis, *Advan. Mass. Spectrometry,* Proc. Conf., 2d, Oxford, 1961 **2,** 370 (Pub. 1963); *C. A.* **59,** 6225 (1963).
1342. Dorinson, A., M. R. McCorkle, et al., *J. Amer. Chem. Soc.* **64,** 2739 (1942).
1343. Dorinson, A. and A. W. Ralston, *J. Amer. Chem. Soc.* **66,** 361 (1944).
1344. Dorman, F. H., *J. Chem. Phys.* **44,** 3856 (1966).
1345. Dornte, R. W. and C. P. Smyth, *J. Amer. Chem. Soc.* **52,** 3546 (1930).
1346. Dorough, G. L., H. B. Glass, et al., *J. Amer. Chem. Soc.* **63,** 3100 (1941).
1347. Doub, L. and J. M. Vandenbelt, *J. Amer. Chem. Soc.* **69,** 2714 (1947).
1348. Douglas, A. W. and J. H. Goldstein, *J. Mol. Spectrosc.,* **16,** 1 (1965); *C. A.* **63,** 1374 (1965).
1349. Douglas, T. B., *J. Amer. Chem. Soc.* **70,** 2001 (1948).
1350. Douglas, T. B., G. T. Furukawa, et al., *J. Res. Nat. Bur. Stand.* **53,** 139 (1954).
1351. Douslin, D. R. and H. M. Huffman, *J. Amer. Chem. Soc.* **68,** 1704 (1946).
1352. Douslin, D. R., R. T. Moore, et al., *J. Amer. Chem. Soc.* **80,** 2031 (1958).
1353. Dow Chemical Co., British Patent 916,129, Jan. 23, 1963; *C. A.* **58,** 13792 (1963).
1354. Dow Chemical Co., "Chlorothene NU," Midland, 1962.
1355. Dow Chemical Co., "DOWANOL, Glycerol-Ether-Solvents," Midland, 1958.
1356. Dow Chemical Co., "Dow Products 1964–1965," Midland.
1357. Dow Chemical Co., "Ethanolamines," Midland, 1954.
1358. Dow Chemical Co., "Ethylenimine," Midland.
1359. Dow Chemical Co., data sheet, "Physical Properties of the Dow Alkene Oxides," Midland.
1360. Dow Chemical Co., "Vinylidene Chloride Monomer," Midland, 1966.
1361. Drago, R. S., R. L. Carlson, et al., *Inorg. Chem.* **4,** 15 (1965).
1362. Drago, R. S., V. A. Mode, et al., *J. Amer. Chem. Soc.* **87,** 5010 (1965).
1363. Drahowzal, F. and D. Klamann, *Monatsh. Chem.* **82,** 588 (1951); *C. A.* **46,** 8036 (1952).
1364. Draper, O. J. and A. L. Pollard, *Science* **109,** 448 (1949).
1365. Drefahl, G. and H. Schick, *Z. Chem.* **4,** 347 (1964); *C. A.* **61,** 13178 (1964).
1366. Dreisbach, R. R., *Ind. Eng. Chem. Anal. Ed.* **12,** 160 (1940).
1367. Dreisbach, R. R., *Ind. Eng. Chem.* **40,** 2269 (1948).
1368. Dreisbach, R. R., *J. Amer. Chem. Soc.* **73,** 3147 (1951).
1369. Dreisbach, R. R., *Physical Properties of Chemical Compounds,* American Chemical Society, Washington, D.C., 1955.
1370. Dreisbach, R. R., *Physical Properties of Chemical Compounds II,* American Chemical Society, Washington, D.C., 1959.

1371. Dreisbach, R. R., *Physical Properties of Chemical Compounds III*, American Chemical Society, Washington, D. C., 1961.
1372. Dreisbach, R. R. and R. A. Martin, *Ind. Eng. Chem.* **41**, 2875 (1949).
1373. Dreisbach, R. R. and S. A. Shrader, *Ind. Eng. Chem.* **41**, 2879 (1949).
1374. Dreisbach, R. R. and R. S. Spencer, *Ind. Eng. Chem.* **41**, 176 (1949).
1375. Dreisbach, R. R. and R. S. Spencer, *Ind. Eng. Chem.* **41**, 1363 (1949).
1376. Drenth, W., G. L. Hekkert, et al., *Rec. Trav. Chim. Pays-Bas* **79**, 1056 (1960); *C. A.* **51**, 6361 (1961).
1377. Drinker, P. and W. A. Cook, *J. Ind. Hyg. Toxicol.* **31**, 51 (1949); *C. A.* **43**, 4400 (1949).
1378. Drucker, C., *Ostwald-Luther Handund Hilfobuch zur Ausführung physikochemischer Messungen*, Dover, New York, 1943.
1379. Drucker, C., *Z. Phys. Chem.* (Leipzig) **52**, 641 (1905).
1380. Drucker, C. and H. Weissbach, *Z. Phys. Chem.* (Leipzig) **117**, 209 (1925).
1381. Drude, P., *Z. Phys. Chem.* (Leipzig) **23**, 267 (1897).
1382. Dubois, J. E., *Compt. Rend.* **244**, 1234 (1947); *C. A.* **41**, 6525 (1947).
1383. Dubois, J. E. and H. Viellard, *J. Chim. Phys.* **62**, 699 (1965); *C. A.* **63**, 14178 (1965).
1384. Duchesne, J., *Nature* **157**, 733 (1946).
1385. Duchesne, J., *Physica* **8**, 144 (1941); *C. A.* **35**, 7832 (1941).
1386. Duclaux, J. and A. Lanzenberg, *Bull. Soc. Chim. Fr.*, **27**, 779 (1920); *Chem. Zentr* **1921**, I, 277.
1387. Ducom, J. and M. H. Normant, *Compt. Rend.* **264**, 722 (1967).
1388. Duffield, A. M., H. Budzikiewicz, et al., *J. Amer. Chem. Soc.* **86**, 5536 (1964).
1389. Duffield, A. M., H. Budzikiewicz, et al., *J. Amer. Chem. Soc.* **87**, 810 (1965).
1390. Duffield, A. M., H. Budzikiewicz, et al., *J. Amer. Chem. Soc.* **87**, 2920 (1965).
1391. Dugacheva, G. M. and A. G. Anikin, *Zh. Fiz. Khim.* **38** (1), 208 (1964); *C. A.* **60**, 10523 (1964).
1392. Dugan, P. R., *Anal. Chem.* **33**, 1630 (1961).
1393. Duggan, R. J. and L. O. Winstrom, French Patent 1,316,057, Jan. 25, 1963; *C. A.* **59**, 2719 (1963).
1394. Duke, F. R. and G. F. Smith, *Ind. Eng. Chem., Anal. Ed.* **12**, 201 (1940).
1395. Duncan, N. E. and G. J. Janz, *J. Chem. Phys.* **23**, 434 (1955).
1396. Duncan, W. A., J. P. Sheridan, et al., *Trans. Faraday Soc.* **62**, 1090 (1966).
1397. Duncan, W. A. and F. L. Swinton, *Trans. Faraday Soc.* **62**, 1082 (1966).
1398. Dunlap, M. K., J. K. Dodama, et al., *A. M. A. Arch. Ind. Health* **18**, 303 (1958); *C. A.* **53**, 1591 (1959).
1399. Dunstan, A. E., *J. Chem. Soc.* **1904**, 817.
1400. Dunstan, A. E., T. P. Hilditch, et al., *J. Chem. Soc.* **1913**, 133.
1401. Dunstan, A. E. and A. G. Mussell, *J. Chem. Soc.* **1910**, 1935.
1402. Dunstan, A. E. and J. A. Stubbs, *J. Chem. Soc.* **1908**, 1919.
1403. Dunstan, A. E. and F. B. Thole, *J. Chem. Soc.* **1908**, 1815.
1404. Dunstan, A. E. and F. B. Thole, *J. Chem. Soc.* **1913**, 127.
1405. Dunstan, A. E., F. B. Thole, et al., *J. Chem. Soc.* **1914**, 782.
1406. Dupont, G., R. Dulou, et al., *Bull. Soc. Chim. Fr.*, **3**, 1639 (1936); *C. A.* **30**, 8026 (1936).
1407. du Pont de Nemours and Co., E. I., Patent British 575,362, Feb. 14, 1946; *C. A.* **41**, 5551 (1947).
1408. du Pont de Nemours & Co., E. I., *du Pont DMAC*, Wilmington, Delaware.
1409. du Pont de Nemours & Co., E. I., *du Pont DMF*, Wilmington, Delaware, 1967.
1410. du Pont de Nemours & Co., E. I., "Freon," Technical Bulletin B-1, Wilmington, Delaware, 1965.

1411. du Pont de Nemours & Co., E. I., "Freon," Technical Bulletin B-2, Wilmington, Delaware, 1966.
1412. du Pont de Nemours & Co., E. I., "Freon," Technical Bulletin B-4B, Wilmington, Delaware, 1964.
1413. du Pont de Nemours & Co., E. I., French Patent 1,311,076, Dec. 7, 1962; *C. A.* **58,** 13916 (1963).
1414. du Pont de Nemours & Co., E. I., *Ind. Eng. Chem.* **28,** 1160 (1936).
1415. du Pont de Nemours & Co., E. I., "Properties and Uses of THF du Pont Tetrahydrofuran," Wilmington, Delaware, 1961.
1416. du Pont de Nemours & Co., E. I., "A Review of Catalytic and Synthetic Applications for UMF and UMAC," Wilmington, Delaware.
1417. Durand, J. F., *Bull. Soc. Chim. Fr.* **4,** 67 (1937); *C. A.* **31,** 4185 (1937).
1418. Durand, J. F. and E. Rouge, *Bull. Soc. Chim. Fr.* **37,** 697 (1925); *C. A.* **19,** 2903 (1925).
1419. Durocher, G. and C. Sandorfy, *J. Mol. Spectrosc.* **15,** 22 (1965); *C. A.* **62,** 4787 (1965).
1420. Dutt, I. S., *Quart. J. Chem. Soc.* **1,** 297 (1925); *C. A.* **19,** 2475 (1925).
1421. Dutt, P. K., *J. Chem. Soc.* **1923,** 2714.
1422. Duyckaerts, G. and G. Michel, *Anal. Chim. Acta* **2,** 750 (1948); *C. A.* **43,** 7858 (1949).
1423. Dyke, M. D., P. G. Sears, et al., *J. Phys. Chem.* **71,** 4140 (1967).
1424. Dykstra, H. B., J. F. Lewis, et al., *J. Amer. Chem. Soc.* **52,** 3396 (1930).
1425. Dykyj, J., M. Seprakova, et al., *Chem. Zvesti* **15,** 465 (1961); *C. A.* **56,** 10930 (1962).
1426. Dyrssen, D., *Acta Chem. Scand.* **11,** 1771 (1957); from [2956].
1427. Dzung, L. S., *Brown Boveri Rev.* **33,** 158 (1946); *C. A.* **41,** 4372 (1947).

E

1428. Earle, J. C., *J. Soc. Chem. Ind.* (London) **37,** 2749 (1918).
1429. Earle, R. B. and H. L. Jackson, *J. Amer. Chem. Soc.* **28,** 104 (1906).
1430. Eastman Chemical Products Co. Inc., "Isobutyronitrile, Properties, Applications," TDR No. N-102, Kingsport, 1957.
1431. Eastman Chemical Products Co. Inc., Technical Data, "n-Butyronitrile," Kingsport.
1432. Easton, B. C., M. K. Hargreaves, et al., *J. Appl. Chem.* (London) **7,** 198 (1957).
1433. Easton, B. C. and M. K. Hargreaves, *J. Chem. Soc.* **1959,** 1413.
1434. Eberhardt, H. and O. Osberghaus, *Z. Naturforsch.* **13a,** 16 (1958); *C. A.* **52,** 8750 (1958).
1435. Eberius, E. and W. Kowalski, *Chem.-Ztg.* **81,** 75 (1957); *C. A.* **51,** 9417 (1957).
1436. Ebersole, E. R., *Natl. Advisory Comm. Aeronaut. Tech. Notes* **1164,** 6 pp. (1946); *C. A.* **41,** 2010 (1947).
1437. Edgar, G., *Ind. Eng. Chem.* **19,** 145 (1927).
1438. Edgar, G. and G. Calingaert, *J. Amer. Chem. Soc.* **51,** 1483 (1929).
1439. Edgar, G. and G. Calingaert, *J. Amer. Chem. Soc.* **51,** 1540 (1929).
1440. Edgell, W. F. and G. Glockler, *J. Chem. Phys.* **9,** 484 (1941).
1441. Edgell, W. F., M. T. Yang, et al., *J. Amer. Chem. Soc.* **87,** 3080 (1965).
1442. Edmonds, W. J., U.S. Patent 1,550,792 (Cl. 260-156), Aug. 25, 1925.
1443. Edsall, J. T., *J. Amer. Chem. Soc.* **65,** 1767 (1943).
1444. Edsall, J. T., *J. Chem. Phys.* **5,** 225 (1937).
1445. Edwards, D. A. and C. F. Bonilla, *Ind. Eng. Chem.* **36,** 1038 (1944).
1446. Edward, J. T., *Chem. Ind.* (London) **1963,** 489.
1447. Edward, J. T., J. B. Leane, et al., *Can. J. Chem.* **40,** 1521 (1962).
1448. Edward, J. T. and I. C. Wang, *Can. J. Chem.* **40,** 966 (1962).
1449. Edwards, P. R., *J. Chem. Soc.* **1925,** 744.
1450. Eeckelaers, R., Belgian Patent 505,125 (1952); *C. A.* **49,** 15950 (1955).

1451. Eftring, E., *Annual Tables of Physical Constants*, Frick Chemical Laboratory, Princeton, New Jersey, 1941.
1452. Eggenberger, D. N., F. K. Broome, et al., *J. Org. Chem.* **14**, 1108 (1949).
1453. Egloff, G., *Physical Constants of Hydrocarbons*, Reinhold, New York, 1946.
1454. Ehrhardt, E. and O. Osberghaus, *Z. Naturforsch.* **15a**, 575 (1960); *C. A.* **54**, 23719 (1960).
1455. Ehrlich, F., *Ber. Deut. Chem. Gesell.* **40**, 2538 (1907).
1456. Eichmann, O., *Z. Phys.* **82**, 470 (1933); *C. A.* **27**, 4228 (1933).
1457. Eidinoff, M. L., *J. Amer. Chem. Soc.* **67**, 2072 (1945).
1458. Eigenberger, E., *J. Prakt. Chem.* **130**, 75 (1931).
1458a. Eijkman, J. F., see Eykman, J. F.
1459. Eilingsfeld, H., M. Seefelder, et al., *Angew. Chem.* **72**, 836 (1960); from [3042].
1460. Eliassaf, J., R. M. Fuoss, et al., *J. Phys. Chem.* **67**, 1724 (1963).
1461. Eliel, E. L. and R. G. Haber, *J. Org. Chem.* **23**, 2041 (1958).
1462. Eliel, E. L. and C. A. Lukach, *J. Amer. Chem. Soc.* **79**, 5986 (1957).
1463. Eliel, E. L., C. C. Price, et al., *Spectrochim. Acta* **10**, 423 (1958); *C. A.* **55**, 25466 (1961).
1464. Elizarova, O. N., *Uch. Zap., Mosk. Nauchn.-Issled. Inst. Gigieny* No. **9**, 105 (1961); *C. A.* **61**, 3605 (1964).
1465. Elkarim, A. A., *Rev. Sci. Instrum.* **19**, 833 (1948).
1466. Elkins, H. B., *Ind. Med.* **8**, 426 (1939); *C. A.* **34**, 5196 (1940).
1467. Elliott, J. J. and S. F. Mason, *J. Chem. Soc.* **1959**, 1275.
1468. Elliott, M. A., A. R. Jones, et al., *Anal. Chem.* **19**, 10 (1947).
1469. Ellis, J. W., *J. Amer. Chem. Soc.* **51**, 1384 (1929).
1470. Ellis, J. W., *Phys. Rev.* **22**, 200 (1923).
1471. Ellis, J. W., *Phys. Rev.* **23**, 48 (1924); *C. A.* **18**, 1086 (1924).
1472. Ellis, L. M. and E. E. Reid, *J. Amer. Chem. Soc.* **54**, 1674 (1932).
1473. Emblem, H. G. and C. A. McDowell, *J. Chem. Soc.* **1946**, 641.
1474. Emde, H. and T. Hornemann, *Helv. Chem. Acta* **14**, 892 (1931).
1475. Emery, A. G. and F. G. Benedict, *Am. J. Physiol.* **28**, 301; *C. A.* **6**, 234 (1912).
1476. Emery, E. M., *Anal. Chem.* **32**, 1495 (1960).
1477. Emschwiller, G. and J. Lecomte, *J. Phys. Radium* **8**, 130 (1937); *C. A.* **31**, 5680 (1937).
1478. Emschwiller, M. G., *Compt. Rend.* **191**, 208 (1930).
1479. Engel, K. H., U.S. Patent 2,339,388, Jan. 18, 1944; *C. A.* **38**, 3672 (1944).
1480. Engel, K. H., U.S. Patent 2,382,142, Aug. 14, 1945; *C. A.* **39**, 4893 (1945).
1481. Engel, W. F., Dutch Patent 59,179, Apr. 15, 1947; *C. A.* **41**, 5893 (1947).
1482. Engelhard, H., H. Schilfarth, et al., *Ann. Chem.* **563**, 239 (1949).
1483. Englin, B. A., A. F. Plate, et al., *Khim. Tekhnol. Topliv Masel* **10**, 42 (1965); *C. A.* **63**, 14608 (1965).
1484. English, S. and W. E. S. Turner, *J. Chem. Soc.* **1914**, 1656.
1485. Enjay Chemical Co., "Isopropyl Alcohol," New York, 1961.
1486. Ens, A. and F. E. Murray, *Can. J. Chem.* **35**, 170 (1957).
1487. Eötvös, R., *Ann. Phys.* (Leipzig) **27**, 448 (1886).
1488. Erdös, E., L. Jäger, et al., *Chem. Listy* **46**, 770 (1952); *C. A.* **47**, 4183 (1953).
1489. Erić, B., E. U. Goode, et al., *J. Chem. Soc.* **1960**, 55.
1490. Erlandsson, G. and H. Selen, *Arkiv. Fysik.* **14**, 61 (1958); *C. A.* **53**, 5785 (1959).
1491. Ernst, R. C., E. E. Litkenhous, et al., *J. Phys. Chem.* **36**, 842 (1932).
1492. Ernst, R. C., C. H. Watkins, et al., *J. Phys. Chem.* **40**, 627 (1936).
1493. Errera, J., *Phys. Z.* **27**, 764 (1926); *Chem. Zentr.* **1927**, I, 1928.
1494. Errera, J. and M. L. Sherrill, *J. Amer. Chem. Soc.* **52**, 1993 (1930).
1495. Esso Research and Engineering Co., British Patent 797,242, 1958,

1496. Estermann, J., Z. Phys. Chem. (Leipzig) Abt. B 1, 134 (1928).
1497. Estok, G. K. and J. S. Dehn, J. Amer. Chem. Soc. 77, 4769 (1955).
1498. Estok, G. K. and J. H. Sikes, J. Amer. Chem. Soc. 75, 2745 (1953).
1499. Estok, G. K. and W. W. Wendlandt, J. Amer. Chem. Soc. 77, 4767 (1955).
1500. Ethyl Corp., British Patent 935,315, Aug. 28, 1963; C. A. 59, 14171 (1963).
1501. Ethyl Corp., Technical Bulletin, "Vinylidine Chloride," New York, 1965.
1502. Ettel, V. and A. Kohlik, Collect. Trav. Chim. Tchecoslovaquie 3, 585 (1931); Chem. Zentr. 1932, I, 1513.
1503. Eubank, P. T. and J. M. Smith, J. Chem. Eng. Data 7, 75 (1962).
1504. Evans, A. G. and S. D. Hamann, Trans. Faraday Soc. 47, 34 (1951); R. G. Bates and H. B. Hetzer, J. Phys. Chem. 65, 667 (1961).
1505. Evans, A. G. and E. Tyrrall, J. Polym. Sci. 2, 387 (1947).
1506. Evans, D. F., J. Chem. Soc. 1959, 877.
1507. Evans, D. P., W. C. Davies, et al., J. Chem. Soc. 1930, 1310.
1508. Evans, E. B., J. Inst. Petrol. Technol. 24, 537 (1938); Beil E III 5, 1257.
1509. Evans, F. D. and P. F. Tiley, J. Chem. Soc. 1966B, 134.
1510. Evans, F. W., D. M. Fairbrother, et al., Trans. Faraday Soc. 55, 399 (1959).
1511. Evans, F. W. and H. A. Skinner, Trans. Faraday Soc. 55, 255 (1959).
1512. Evans, F. W. and H. A. Skinner, Trans. Faraday Soc. 55, 260 (1959).
1513. Evans, H. B., A. R. Tarpley, et al., J. Phys. Chem. 72, 2552 (1968).
1514. Evans, J. C., J. Chem. Phys. 22, 1228 (1954).
1515. Evans, J. C., J. Chem. Phys. 30, 934 (1959).
1516. Evans, J. C., Spectrochim. Acta 16, 1382 (1960); C. A. 55, 12039 (1961).
1517. Evans, J. C. and H. J. Bernstein, Can. J. Chem. 33, 1746 (1955).
1518. Evans, J. C. and H. J. Bernstein, Can. J. Chem. 34, 1083 (1956).
1519. Evans, J. C. and G. Y.-S. Lo, J. Amer. Chem. Soc. 88, 2118 (1966).
1520. Evans, T., U.S. Patent 2,154,930, Apr. 18, 1939; C. A. 33, 5411 (1939).
1521. Evans, T. and W. C. Fetsch, J. Amer. Chem. Soc. 26, 1158 (1904).
1522. Everard, K. B., L. Kumar, et al., J. Chem. Soc. 1951, 2807.
1523. Everett, D. H., Unpublished result; R. G. Bates and H. B. Hetzer, J. Phys. Chem. 65, 667 (1961).
1524. Everett, D. H. and B. R. W. Pinsent, Proc. Roy. Soc. 215A, 416 (1952); R. G. Bates and H. B. Hetzer, J. Phys. Chem. 65, 667 (1961).
1525. Evers, E. C. and A. G. Knox, J. Amer. Chem. Soc. 73, 1739 (1951).
1526. Ewell, R. H. and J. F. Bourland, J. Chem. Phys. 8, 635 (1940).
1527. Ewert, M., Bull. Soc. Chim. Belges 45, 509 (1936).
1528. Ewert, M., Bull. Soc. Chim. Belges 46, 90 (1937).
1529. Ewins, A. J., J. Chem. Soc. 1914, 350.
1530. Eykmann, J. F., Rec. Trav. Chim. Pays-Bas 14, 185 (1895); J. Chem. Soc. 1896Aii, 133.
1531. Eykman, J. F., Z. Phys. Chem. (Leipzig) 3, 113 (1889).
1532. Eykman, J. F., Z. Phys. Chem. (Leipzig) 3, 203 (1899).
1533. Eykman, J. F., Z. Phys. Chem. (Leipzig) 4, 497 (1889).
1534. Eykman, J. F., F. Bergema, et al., Chem. Weekbl. 2, 59; Chem. Zentr. 1905, I, 814.
1535. Eyster, E. H., J. Chem. Phys. 6, 576 (1938).

F

1536. Fainberg, A. H. and W. T. Miller, J. Org. Chem. 30, 864 (1965).
1537. Fairbairn, A. W., H. A. Cheney, et al., Chem. Eng. Progr. 43, No. 6, Trans. A.I.C.H.E. 280 (1947); C. A. 41, 5090 (1947).
1538. Fairbrother, F., J. Chem. Soc. 1934, 1846.

1539. Fairbrother, F., *Proc. Roy. Soc.* **142A,** 173 (1933).

1540. Falk, K. G., *J. Amer. Chem. Soc.* **31,** 806 (1909).

1541. Fanelli, R., *J. Amer. Chem. Soc.* **70,** 1792 (1948).

1542. Farkas, A. and A. F. Stribley, U.S. Patent 2,410,642, Nov. 5, 1946; *C. A.* **41,** 1833 (1947).

1543. Favini, G., S. Carra, et al., *Nuovo Cimento* **8,** 60 (1958); *C. A.* **52,** 16031 (1958).

1544. Favini, G. and M. Simonetta, *Gazz. Chim. Ital.* **89,** 2111 (1959); *C. A.* **55,** 8036 (1961).

1545. Favret, A. G. and R. Meister, *J. Chem. Phys.* **41,** 1011 (1964).

1546. Fawcett, F. S., *Ind. Eng. Chem.* **38,** 338 (1946).

1547. Fawcett, F. S. and H. E. Rasmussen, *J. Amer. Chem. Soc.* **67,** 1705 (1945).

1548. Fawsitt, C. E., *J. Chem. Soc.* **1919,** 790.

1549. Feairheller, W. R. and J. E. Katon, *Spectrochim. Acta* **23A,** 2225 (1967); *Chem. Zentr.* **1968,** 0834.

1550. Fedin, E. I., P. V. Petrovskii, et al., *Neftekhimiya* **2,** 270 (1962); *C. A.* **58,** 2042 (1963).

1551. Fedosow, J., *Soc. Sci. Fennica Commentationes Phys.-Math.* **14,** No. 1,1 (1948); *C. A.* **42,** 6672 (1948).

1552. Feeny, J., A. Ledwith, et al., *J. Chem. Soc.* **1962,** 2021.

1553. Fehnel, E. A. and M. Carmack, *J. Amer. Chem. Soc.* **71,** 84 (1949).

1554. Fehnel, E. A. and M. Carmack, *J. Amer. Chem. Soc.* **71,** 2932 (1949).

1555. Feigl, F., *Spot Tests in Inorganic Analysis*, 5th ed., Elsevier, New York, 1958.

1556. Fenske, M. R., W. G. Braun, et al., *Ind. Eng. Ch m.*, *Anal. Ed.* **19,** 700 (1947).

1557. Fenske, M. R. and H. S. Myers, *Ind. Eng. Chem.* **42,** 649 (1950).

1558. Ferguson, A. and J. T. Miller, *Proc. Phys. Soc.* **45,** 194 (1933); *C. A.* **27,** 2372 (1933).

1558a. Fenske, M. R. and D. Quiggle, et al., *Ind. Eng. Chem.* **24,** 408 (1932).

1558b. Fenske, M. R., D. Quiggle, et al., *Ind. Eng. Chem.* **24,** 542 (1932).

1559. Ferguson, E. E., R. L. Hudson, et al., *J. Chem. Phys.* **21,** 1736 (1953).

1560. Ferguson, J., *J. Chem. Soc.* **1954,** 304.

1561. Ferguson, J. and R. L. Werner, *J. Chem. Soc.* **1954,** 3645.

1562. Ferguson, J. B., M. Freed, et al., *J. Phys. Chem.* **37,** 87 (1933).

1563. Ferguson, J. W., *Proc. Indiana Acad. Sci.* **63,** 131 (1953); *C. A.* **49,** 8094 (1955).

1564. Ferrier, R., *Compt. Rend.* **220,** 460 (1945); *C. A* **40,** 1475 (1946).

1565. Few, A. V. and J. W. Smith, *J. Chem. Soc.* **1949,** 753.

1566. Few, A. V. and J. W. Smith, *J. Chem. Soc.* **1949,** 2663.

1567. Fialkovskaya, O. V., *Izv. Akad. Nauk SSSR, Ser. Fiz.* **22,** 1093 (1958); *C. A.* **53,** 862 (1959).

1568. Field, F. H., M. S. B. Munson, et al., *Advan. Chem. Ser.* No. **58,** 167 (1966); *C. A.* **65,** 17868 (1966).

1569 Fierz-David, H. E., *Chimia* **1,** 246 (1947); *C. A.* **42,** 2228 (1948).

1570. Fife, H. R. and E. W. Reid, *Ind. Eng. Chem.* **22,** 513 (1930).

1571. Filar, L. J., U.S. Patent 2,948,758, Aug. 9, 1960; *C. A.* **55,** 3521 (1961).

1572. Finch, J. N. and E. R. Lippincott, *J. Phys. Chem.* **61,** 894 (1957).

1573. Finger, G. C. and F. H. Reed, *Trans. Illinois State Acad. Sci.* **29,** No. 2, 89 (1936); *C. A.* **31,** 2999 (1937).

1574. Finke, H. L., M. E. Gross, et al., *J. Amer. Chem. Soc.* **76,** 333 (1954).

1575. Fiock, E. F., D. C. Ginnings, et al., *J. Res. Nat. Bur. Stand.* **6,** 881 (1931).

1576. Fischer, E., *J. Chem. Soc.* **1955,** 1382.

1577. Fischer, E. and F. Rogowski, *Phys. Z.* **40,** 331 (1939); *C. A.* **33,** 7633 (1939).

1578. Fischer, E. and A. Speier, *Ber. Beut. Chem. Gesell.* **28,** 3252 (1895).

1579. Fischer, W. and W. Klemm, *Z. Phys. Chem.* (Leipzig) **147A,** 275 (1930).

1580. Fishbein, L. and J. A. Gallaghan, *J. Amer. Chem. Soc.* **78**, 1218 (1956).
1581. Fisher, F. R. and R. A. Baxter, *Mines Mag.* (Colorado School of Mines) **30**, 447 (1940); *C. A.* **34**, 8111 (1940).
1582. Fittig, R. and W. B. Chanlarow, *Ann. Chem., Justus Liebig* **226**, 331 (1884).
1583. Fitzgerald, W. E. and G. J. Janz, *Mol. Spectrosc.* **1**, 49 (1957); *C. A.* **51**, 11856 (1957).
1584. Flanagan, P. W. and H. F. Smith, *Anal. Chem.* **37**, 1699 (1965).
1585. Flannery, J. B. and G. J. Janz, *J. Chem. Eng. Data* **10**, 387 (1965).
1586. Fletcher, W. H., *J. Amer. Chem. Soc.* **68**, 2726 (1946).
1587. Flett, M. St. C., *J. Chem. Soc.* **1951**, 962.
1588. Flett, M. St. C., *Trans. Faraday Soc.* **44**, 767 (1948); *C. A.* **43**, 3715 (1949).
1589. Foerster, D. R., R. Miller, et al., "10th Annual Report on Research, Petroleum Research Fund," p. 15, American Chemical Society, Washington, D.C. 1966.
1590. Fogg, E. T., A. N. Hixson, et al., *Anal. Chem.* **27**, 1609 (1955).
1591. Foley, R. A., W.-M. Lee, et al., *Anal. Chem.* **36**, 1100 (1964).
1592. Fonteyne, R., *Natuurwetensch. Tijdschr.* **24**, 161 (1942); *Chem. Zentr.* **1943**, I, 265.
1593. Forbes, G. S. and A. S. Coolidge, *J. Amer. Chem. Soc.* **41**, 150 (1919).
1594. Forbes, J. W., *Anal. Chem.* **34**, 1125 (1962).
1595. Forcier, G. A. and J. W. Olver, *Anal. Chem.* **37**, 1447 (1965).
1596. Fordham, S., *Research* (London) **1**, Research Suppl. 336 (1948); *C. A.* **48**, 5292 (1948).
1597. Fordyce, C. R. and L. W. A. Meyer, *Ind. Eng. Chem.* **32**, 1053 (1940).
1598. Forel, M. T., A. Lafaix, et al., *J. Chim. Phys.* **60**, 875 (1963); *C. A.* **59**, 10894 (1963).
1599. Foresti, B., *Ann. Chim. Applicata* **27**, 359 (1937); *C. A.* **32**, 2089 (1938).
1600. Forman, E. J. and D. N. Hume, *J. Phys. Chem.* **63**, 1949 (1959).
1601. Forster, E. O., *J. Chem. Phys.* **37**, 1021 (1962).
1602. Forster, E. O., *J. Chem. Phys.* **40**, 86 (1964).
1603. Forster, E. O., *J. Chem. Phys.* **40**, 91 (1964).
1604. Forster, G., R. Skrabal, et al., *Z. Electrochem.* **43**, 290 (1937); *C. A.* **31**, 5273 (1937).
1605. Fort, R. J. and W. R. Moore, *Trans. Faraday Soc.* **62**, 1112 (1966).
1606. Forziati, A. F., *J. Res. Nat. Bur. Stand.* **44**, 373 (1950).
1607. Forziati, A. F., D. L. Camin, and F. D. Raesini, *J. Res. Nat. Bur. Stand.* **45**, 406 (1950).
1608. Forziati, A. F., A. R. Glasgow, et al., *J. Res. Nat. Bur. Stand.* **36**, 129 (1946).
1609. Forziati, A. F., B. J. Mair, et al., *J. Res. Nat. Bur. Stand.* **35**, 513 (1945).
1609a. Fourth Congress on Calorimetry, *Chem. Eng. News* **27**, 2772 (1949).
1610. Fougue, G. and M. Cabanac, *Bull. Soc. Chim. Fr.* **39**, 1184 (1926); *C. A.* **20**, 3687 (1926).
1611. Fowler, L., W. N. Trump, et al., *J. Chem. Eng. Data* **13**, 209 (1968).
1612. Fox, J. J. and M. F. Barker, *J. Soc. Chem. Ind.* **37**, 268T (1918).
1613. Fox, J. J. and A. E. Martin, *Proc. Roy. Soc.* **162A**, 419 (1937); *C. A.* **31**, 8374 (1937).
1614. Foxton, A. A., G. H. Jeffery, et al., *J. Chem. Soc.* **1966 A**, 249.
1615. Fraenkel, G., *Proc. Nat. Acad. Sci. U.S.* **44**, 688 (1958); *C. A.* **53**, 52 (1959).
1616. Franchimont, A. P. N., *Rec. Trav. Chim. Pays-Bas* **3**, 226 (1884); from [3042].
1617. Francis, A. W., *Critical Solution Temperatures*, American Chemical Society, Washington, D.C., 1961.
1618. Francis, C. V., *Anal. Chem.* **21**, 1238 (1949).
1619. Francis, F. and F. J. E. Collins, *J. Chem. Soc.* **1936**, 137.
1620. Frandsen, B. S., *J. Res. Nat. Bur. Stand.* **7**, 477 (1931).
1621. Franeau, J. and G. Birnbaum, *J. Appl. Phys.* **20**, 817 (1949); *C. A.* **43**, 8769 (1949).
1622. Franke, E., *Z. Phys. Chem.* (Leipzig) **16**, 463 (1895).

1623. Frankforter, G. B. and L. Cohen, *J. Amer. Chem. Soc.* **36**, 1103 (1914).

1624. Frankiss, S. G., *J. Phys. Chem.* **67**, 752 (1963).

1625. Franzen, V. and H.-E. Driesen, *Chem. Ber.* **96**, 1881 (1963).

1626. Freeman, R., *J. Chem. Phys.* **40**, 3571 (1964).

1627. Freeman, R., N. S. Bhacca, et al., *J. Chem. Phys.* **38**, 293 (1963).

1628. Freiser, H., M. E. Hobbs, et al., *J. Amer. Chem. Soc.* **71**, 111 (1949).

1629. Frejacques, J. L. M., U.S. Patent 2,461,048, Feb. 8, 1949; *C. A.* **43**, 7275 (1949).

1630. French, C. M., *Trans. Faraday Soc.* **43**, 357 (1947).

1631. French, C. M. and K. H. Glover, *Trans. Faraday Soc.* **51**, 1418 (1955).

1632. French, C. M. and K. H. Glover, *Trans. Faraday Soc.* **51**, 1427 (1955).

1633. French, C. M. and I. G. Roe, *Trans. Faraday Soc.* **49**, 314, 791 (1953).

1634. French, C. M. and U. C. Trew, *Trans. Faraday Soc.* **47**, 365 (1951); *C. A.* **45**, 8827 (1951).

1635. French, H. E. and G. G. Wrightsman, *J. Amer. Chem. Soc.* **60**, 50 (1938).

1636. French, K. H. V., Chairman, *Anal. Chem.* **36**, 339 (1964).

1637. French, K. H. V. and G. Claxton, *J. Soc. Chem. Ind.* **65**, 344 (1946).

1638. French Patent 784,656, July 22, 1935; *C. A.* **30**, 108 (1936).

1639. French Patent 1,045,726, Dec. 1, 1953; *C. A.* **52**, 11888 (1958).

1640. Frenzel, C. A., D. W. Scott, et al., *U.S. Bur. Mines, Rept. Invest.* **5658**, 17 pp. (1960); *C. A.* **55**, 2277 (1961).

1641. Freri, M., *Gazz. Chim. Ital.* **78**, 286 (1948); *C. A.* **42**, 8647 (1948).

1642. Freudenberg, K. J., German Patent 281,473, Jan. 8, 1915; *Chem. Zentr.* **1915, I,** 230.

1643. Frey, P. R. and E. C. Gilbert, *J. Amer. Chem. Soc.* **59**, 1344 (1937).

1644. Freyer, F. A., *Mfg. Chemist* **33**, 100 (1962); *C. A.* **57**, 1206 (1962); Ref. [2123].

1645. Freymann, M., *Ann. Chim.* (Paris) **11**, 11 (1939); *C. A.* **33**, 3694 (1939).

1646. Freymann, M. and R. Freymann, *Compt. Rend.* **219**, 515 (1944); *C. A.* **40**, 1393 (1946).

1647. Freymann, M. and R. Freymann, *Compt. Rend.* **248**, 677 (1959); *C. A.* **53**, 14694 (1959).

1648. Freymann, M. and R. Freymann, *Groupment Franc. Dévelop. Recherches Aeronaut.* Note Tech. No. 11 (1943); *C. A.* **43**, 3715 (1949).

1649. Freymann, M., R. Freymann, et al., *Arch. Sci.* **13**, 506 (1960); *C. A.* **58**, 3289 (1963).

1650. Freymann, R., *Compt. Rend.* **194**, 1471 (1932); *C. A.* **26**, 3991 (1932).

1651. Freymann, R., *Compt. Rend.* **202**, 952 (1936); *C. A.* **30**, 3691 (1936).

1652. Freymann, R., *Compt. Rend.* **261** (Groupe 7), 2637 (1965); *C. A.* **64**, 5943 (1966).

1653. Freymann, R., *J. Phys. Radium* **10**, 1 (1939); *C. A.* **33**, 3694 (1939).

1654. Freymann, R. and P. Duhamel, *Bull. Soc. Chim. Fr.* **1961**, 1238.

1655. Freymann, R., M. Freymann, et al., *Arch. Sci.* (Geneva) **12**, Fasc. Spec. 207 (1959); *C. A.* **54**, 11705 (1960).

1656. Fried, V., J. Pick, et al., *Chem. Listy* **48**, 774 (1954); *C. A.* **48**, 10398 (1954).

1657. Friedel, R. H., *J. Amer. Chem. Soc.* **73**, 2881 (1951).

1658. Friedel, R. A. and D. S. McKinney, *J. Amer. Chem. Soc.* **69**, 604 (1947).

1659. Friedel, R. A., J. L. Schultz, et al., *Anal. Chem.* **28**, 926 (1956).

1660. Friedel, R. A. and A. G. Sharkey, *Anal. Chem.* **28**, 940 (1956).

1661. Friedman, I. and H. Stechter, *J. Org. Chem.* **25**, 877 (1960).

1662. Friedman, L. and F. A. Long, *J. Amer. Chem. Soc.* **75**, 2832 (1953).

1663. Friedman, L., F. A. Long, et al., *J. Chem. Phys.* **27**, 613 (1957).

1644. Friend, J. N. and W. D. Hargreaves, *Phil. Mag.* **34**, 643 (1943); Beil **EIII2**, 39.

1665. Friend, J. N. and W. D. Hargreaves, *Phil. Mag.* **35**, 619 (1944); *C. A.* **39**, 1098 (1945).

1666. Friend, J. N. and W. D. Hargreaves, *Phil. Mag.* **37**, 120 (1946).
1667. Friend, J. N. and W. D. Hargreaves, *Phil. Mag.* **37**, 201 (1946).
1668. Fritz, J. F. and L. W. Marple, *Anal. Chem.* **34**, 921 (1962).
1669. Frolov, A. F., M. A. Loginova, et al., *Zh. Fiz. Khim.* **36**, 2282 (1962); *C. A.* **58**, 3957 (1963).
1670. Frolov, Y. L., A. V. Kalabina, et al., *Zh. Strukt. Khim.* **6**, 397 (1965); *C. A.* **63**, 9780 (1965).
1671. Frost, A. A. and D. R. Kalkwarf, *J. Chem. Phys.* **21**, 264 (1953).
1672. Fryer, F. A., *Mfg. Chemist* **33**, 100 (1962); *C. A.* **57**, 1206 (1962); Ref. [2123].
1673. Fuchs, L., *Spectrochim. Acta* **2**, 243 (1942); *C. A.* **37**, 5313 (1943).
1674. Fuchs, O., *Z. Phys.* **63**, 824 (1930).
1675. Fuge, E. T. J., S. T. Bowden, et al., *J. Phys. Chem.* **56**, 1013 (1952).
1676. Fuguitt, R. E., W. D. Stallcup, et al., *J. Amer. Chem. Soc.* **64**, 2978 (1942).
1677. Fühner, H., *Ber. Deut. Chem. Gesell.* **57**, 510 (1924).
1678. Fujimura, T. and K. Kamiyoshi, *Sci. Rept. Res. Inst. Tohoku Univ.* Ser. A, **13**, 320 (1961), *C. A.* **56**, 12411 (1962).
1679. Fukumoto, Y., *Sci. Repts. Tohoku Imp. Univ.*, First Ser. **25**, 1162 (1937); *C. A.* **31**, 5273 (1937).
1680. Furukawa, G. T., D. C. Ginnings, et al., *J. Res. Nat. Bur. Stand.* **46**, 195 (1951).
1681. Fuson, R. C., *Organic Syntheses*, Vol. 18, Wiley, New York, 1938.

G

1682. Gabriel, C. L., *Ind. Eng. Chem.* **32**, 887 (1940).
1683. Gabriel, S. and R. Stelzner, *Ber. Deut. Chem. Gesell.* **28**, 2929 (1895).
1684. Gakhokidse, A., *J. Gen. Chem.* (USSR) **17**, 1329 (1947); Beil **EIII2**, 1278.
1685. Gălăteanu, D. B., *Bull. Soc. Roumaine Phys.* **43**, 5 (1942); *Chem. Zentr.* **1943**, I, 2189; *C. A.* **38**, 4512 (1944).
1686. Gal'chenko, G. L., M. M. Ammar, et al., *Vestn. Mosk. Univ. Ser. II, Khim.* **20**, 3 (1965); *C. A.* **63**, 7703 (1965).
1687. Gallaugher, A. F. and H. Hibbert, *J. Amer. Chem. Soc.* **58**, 813 (1936).
1688. Gallaugher, A. F., and H. Hibbert, *J. Amer. Chem. Soc.* **59**, 2514 (1937).
1689. Gallaugher, A. F. and H. Hibbert, *J. Amer. Chem. Soc.* **59**, 2521 (1937).
1690. Gallegos, E. J. and R. W. Kiser, *J. Amer. Chem. Soc.* **83**, 773 (1961).
1691. Gallegos, E. J. and R. W. Kiser, *J. Phys. Chem.* **66**, 136 (1962).
1692. Gal'pern, G. D., M. M. Kusakov, et al., *Fiz. Sbornik L'vov. Univ.* **1957**, No. 3, 334; *C. A.* **56**, 3378 (1962).
1693. Gand, E., *Ann. Faculté Sci. Marseille* (2) **12**, 134 (1939); *C. A.* **35**, 681 (1941).
1694. Gand, E., *Ann. Faculté Sci. Marseille* **15**, 29 (1941); *C. A.* **40**, 3967 (1946).
1695. Gand, E., *Bull. Soc. Chim. Fr.* **10**, 465 (1943); *C. A.* **38**, 3951 (1944).
1696. Gand, E., *Compt. Rend. Trav. Faculté Sci. Marseille* **1**, 129 (1943); *C. A.* **40**, 3967 (1946).
1697. Gaponova, N. E., M. P. Lisitsa, et al., *Opt. Spektrosk.* **8**, 465 (1960); *C. A.* **57**, 298 (1962).
1698. Garach, J., *Compt. Rend.* **236**, 1414 (1953); *C. A.* **47**, 9152 (1953).
1699. Garach, J., *Groupment Franc. Dévelop. Recherches Aeronaut.* Note Tech. No. **15**, (1944); *C. A.* **43**, 2156 (1949).
1700. Garach, J., *Groupment Franc. Dévelop. Recherches Aeronaut.* Note Tech. No. **61**, (1946); *C. A.* **43**, 2175 (1949).
1701. Garach, J. and J. Lecomte, *Compt. Rend.* **222**, 74 (1946); *C. A.* **41**, 1933 (1947).
1702. Garcia, M. O. G., *Rev. Fac. Cienc. Univ. Oviedo* **5**, 143 (1964); *C. A.* **62**, 7547 (1965).

1703. Gardner, G. S. and J. E. Brewer, *Ind. Eng. Chem.* **29,** 179 (1937).
1704. Gardner, J. H. and P. Borgstrom, *J. Amer. Chem. Soc.* **51,** 3375 (1929).
1705. Garelli, F. and C. Montanari, *Gazz. Chim. Ital.* **246,** 229 (1894).
1706. Garg, S. K. and P. K. Kadaba, *J. Phys. Chem.* **68,** 737 (1964).
1707. Garg, S. K. and C. P. Smyth, *J. Chem. Phys.* **42,** 1397 (1965).
1708. Garg, S. K. and C. P. Smyth, *J. Phys. Chem.* **69,** 1294 (1965).
1709. Garland, C. E. and E. E. Reid, *J. Amer. Chem. Soc.* **47,** 2333 (1925).
1710. Garner, J. B., B. Saxton, et al., *Amer. Chem. J.* **46,** 236 (1911).
1711. Garner, W. E. and C. L. Abernethy, *Proc. Roy. Soc.* **99A,** 213 (1921); *Chem. Zentr.* **1921, III,** 866; Beil **EII5,** 172.
1712. Garner, W. E., F. C. Madden, et al., *J. Chem. Soc.* **1926,** 2491.
1713. Garner, W. E. and F. C. Randall, *J. Chem. Soc.* **1924,** 881.
1714. Garnsey, R. and J. E. Prue, *Trans. Faraday Soc.* **64,** 1206 (1968).
1715. Garrick, F. J., *Trans. Faraday Soc.* **23,** 560 (1927).
1716. Garrigou-Lagrange, C., N. Claverie, et al., *J. Chim. Phys.* **58,** 559 (1961); *C. A.* **55,** 23047 (1961).
1717. Gartenmeister, R., *Z. Phys. Chem.* (Leipzig) **6,** 524 (1890).
1718. Gary, J. T. and L. W. Pickett, *J. Chem. Phys.* **22,** 1266 (1954).
1719. Gasopoulos, I., *Ber. Deut. Chem. Gesell.* **59,** 2184 (1926).
1720. Gasparič, J. and J. Borecky, *J. Chromatogr.* **5,** 466 (1961); *C. A.* **56,** 9393 (1962).
1721. Gasparič, J., J. Petranek, et al., *J. Chromatogr.* **5,** 408 (1961); *C. A.* **56,** 9393 (1962).
1722. Gast, J. H. and F. L. Estes, *Anal. Chem.* **32,** 1712 (1960).
1723. Gates, C. B., *J. Phys. Chem.* **15,** 97 (1911).
1724. Gates, D. M., *J. Chem. Phys.* **17,** 393 (1949).
1725. Gaumann, T., *Helv. Chim. Acta* **41,** 1956 (1958).
1726. Gause, E. M. and F. M. Ernsberger, *Ind. Eng. Chem., Chem. Eng. Data Series* **2,** 28 (1957).
1727. Geddes, A. L., *J. Phys. Chem.* **58,** 1062 (1954).
1728. Geddes, J. A. and E. C. Bingham, *J. Amer. Chem. Soc.* **56,** 2625 (1934).
1729. Geiseler, G., J. Fruwert, et al., *Chem. Ber.* **99,** 1594 (1966).
1730. Geiseler, G. and H. Kessler, *Ber. Bunsenges. Physik. Chem.* **68,** 571 (1964); *C. A.* **61,** 15540 (1964).
1731. Geiseler, G. and E. Pilz, *Chem. Ber.* **95,** 96 (1962).
1732. Geist, J. M. and M. R. Cannon, *Ind. Eng. Chem., Anal. Ed.* **18,** 611 (1946).
1733. Geldof, H. and J. P. Wibaut, *Rec. Trav. Chim. Pays-Bas* **67,** 105 (1948); *C. A.* **42,** 5409 (1948).
1734. Geller, B. E., *Zh. Fiz. Khim.* **35,** 2210 (1961); *C. A.* **56,** 14,940 (1962).
1735. Gelles, E. and K. S. Pitzer, *J. Amer. Chem. Soc.* **75,** 5259 (1953).
1736. General Aniline & Film Corp., "Butyrolactone," New York, 1964.
1737. General Aniline & Film Corp., "Methylpyrrolidone," New York, 1961.
1738. Gensler, W. J. and C. M. Samour, *J. Org. Chem.* **18,** 9 (1953).
1739. Gent, W. L. G., *J. Chem. Soc.* **1957,** 58.
1740. George, M. V. and G. F. Wright, *Can. J. Chem.* **36,** 189 (1958).
1741. Gerding, H. and J. Lecompte, *Rec. Trav. Chim. Pays-Bas* **58,** 614 (1939); *C. A.* **33,** 9141 (1939).
1742. Gerding, H. and G. W. A. Rijnders, *Rec. Trav. Chim. Pays-Bas* **65,** 143 (1946); *C. A.* **40,** 5644 (1946).
1743. Germann, F. E. and O. S. Knight, *J. Amer. Chem. Soc.* **55,** 4150 (1933).
1744. Gero, A., *J. Org. Chem.* **19,** 469, 1960 (1954).
1745. Gero, A., *J. Org. Chem.* **26,** 3156 (1961).

1746. Gero, A. and J. J. Markham, *J. Org. Chem.* **16,** 1835 (1951).
1747. Gershteĭn, N. A. and M. F. Shostakovskiĭ, *J. Gen. Chem.* (USSR) **18,** 451 (1948) *C. A.* **42,** 7243 (1948).
1748. Getman, F. H., *J. Amer. Chem. Soc.* **62,** 2179 (1940).
1749. Getman, F. H., *Rec. Trav. Chim. Pays-Bas* **55,** 231 (1936); *C. A.* **30,** 4743 (1936).
1750. Ghazanfar, S. A. S., J. T. Edsall, et al., *J. Amer. Chem. Soc.* **86,** 559 (1964).
1751. Gheorghiu, A. and T. Gheorghiu, *Ann. Combustibles Liquides,* **8,** 451 (1933); *C. A.* **29,** 2448 (1935).
1752. Ghiringhelli, L., *Med. Lavoro* **47,** 192 (1956); *C. A.* **51,** 1461 (1957).
1753. Ghosh, J. C., S. K. Bhattacharyya, et al., *Curr. Sci.* **16,** 88 (1947); *C. A.* **41,** 4725 (1947).
1754. Giauque, W. F. and R. Wiebe, *J. Amer. Chem. Soc.* **50,** 101 (1928).
1755. Gibbons, L. C., J. F. Thompson, et al., *J. Amer. Chem. Soc.* **68,** 1130 (1946).
1756. Gibbons, W. A. and V. M. S. Gil, *Mol. Phys.* **9,** 163 (1965); *C. A.* **63,** 9783 (1965).
1757. Gibbs, H. D., *J. Amer. Chem. Soc.* **49,** 839 (1927).
1758. Gibbs, J. H. and C. P. Smyth, *J. Amer. Chem. Soc.* **73,** 5115 (1951).
1759. Gibby, C. W. and J. Hall, *J. Chem. Soc.* **1931,** 691.
1760. Gibson, C. S. and J. B. A. Johnson, *J. Chem. Soc.* **1931,** 266.
1761. Gibson, G. E. and W. F. Giauque, *J. Amer. Chem. Soc.* **45,** 93 (1923).
1762. Gibson, R. E. and J. F. Kincaid, *J. Amer. Chem. Soc.* **59,** 579 (1937).
1763. Gibson, R. E. and J. F. Kincaid, *J. Amer. Chem. Soc.* **60,** 511 (1938).
1764. Gibson, R. E. and O. H. Loeffler, *J. Amer. Chem. Soc.* **61,** 2515 (1939).
1765. Gibson, S. G. and J. D. A. Johnson, *J. Chem. Soc.* **1930,** 2525.
1766. Giertz, H. W. and J. MacPhersson, *Norsk Skogsin.* **10,** 348 (1956).
1767. Giessner-Prettre, C., *Compt. Rend.* **252,** 3238 (1961); *C. A.* **55,** 21,808 (1961).
1768. Giessner-Prettre, C., *Compt. Rend.* **254,** 4165 (1962); *C. A.* **59,** 149 (1963).
1769. Gifford, A. P., S. M. Rock, et al., *Anal. Chem.* **21,** 1026 (1949).
1770. Gil, V. M. S. and J. N. Murrell, *Trans. Faraday Soc.* **60,** 248 (1964).
1771. Gil'denblat, I. A., A. S. Furmanov, et al., *Zh. Prikl. Khim.* **33,** 246 (1960); *C. A.* **54,** 9410 (1960).
1772. Gill, A. H. and F. P. Dexter, *Ind. Eng. Chem.* **26,** 881 (1934).
1773. Giller, E. B. and H. G. Drickamer, *Ind. Eng. Chem.* **41,** 2067 (1949).
1774. Gillespie, D. T. C., A. K. Macbeth, et al., *J. Chem. Soc.* **1940,** 280.
1775. Gillo, L., *Ann. Chim.* **12,** 281 (1939); *C. A.* **34,** 1530 (1940).
1776. Gillo, L., *Bull. Soc. Chim. Belges* **48,** 341 (1939); *C. A.* **34,** 2221 (1940).
1777. Gilman, H., *Organic Syntheses,* Vol. 6, Wiley, New York, 1926.
1778. Gilman, H. and A. H. Blatt, *Organic Syntheses,* Collective Vol. 1, 2d ed., Wiley, New York, 1941.
1779. Gilmann, H. H. and P. Gross, *J. Amer. Chem. Soc.* **60,** 1525 (1938).
1780. Gilpin, J. A., *Anal. Chem.* **31,** 935 (1959).
1781. Gilson, L. E., *J. Amer. Chem. Soc.* **54,** 1445 (1932).
1782. Ginnings, D. C. and G. T. Furukawa, *J. Amer. Chem. Soc.* **75,** 522 (1953).
1783. Ginnings, P. M. and R. Baum, *J. Amer. Chem. Soc.* **59,** 1111 (1937).
1784. Ginnings, P. M., D. Plonk, et al., *J. Amer. Chem. Soc.* **62,** 1923 (1940).
1785. Ginnings, P. M. and R. Webb, *J. Amer. Chem. Soc.* **60,** 1388 (1938).
1786. Gisseleire, J. and R. Bilterys, *Bull. Soc. Chim. Belges* **44,** 567 (1935); *Chem. Zentr.* **1936, I,** 3999.
1787. Giua, M. and G. Guastalla, *Chim. Ind.* **29,** 268 (1933).
1788. Givaudan-Delawanna, Inc., *The Givaudan Index,* 2d ed., New York, 1961.
1789. Gladstone, J. H., *J. Chem. Soc.* **1884,** 241.

1790. Glagolena, A. A., *Zh. Obshch. Khim.* (*J. Gen. Chem.*) **18**, 1005 (1948); *C. A.* **43**, 6900 (1949).

1791. Glaser, F. and H. Ruland, *Chem.-Ing.-Tech.* **29**, 772 (1957).

1792. Glasgow, A. R., *J. Res. Nat. Bur. Stand.* **24**, 509 (1940).

1793. Glasgow, A. R., N. C. Krouskop, et al., *Anal. Chem.* **20**, 410 (1948).

1794. Glasgow, A. R., E. T. Murphy, et al., *J. Res. Nat. Bur. Stand.* **37**, 141 (1946).

1795. Glasgow, A. R., A. J. Streiff, et al., *J. Res. Nat. Bur. Stand.* **35**, 355 (1945).

1796. Glass, H. M. and W. M. Madgin, *J. Chem. Soc.* **1934**, 1292.

1797. Glasstone, S., *Text Book of Physical Chemistry*, Van Nostrand, Princeton, New Jersey, 1940.

1798. Glasstone, S. and A. F. Schram, *J. Amer. Chem. Soc.* **69**, 1213 (1947).

1799. Glockler, G. and W. F. Edgell, *Ind. Eng. Chem.* **34**, 532 (1942).

1800. Glockler, G. and W. F. Edgell, *J. Chem. Phys.* **9**, 527 (1941).

1801. Glockler, G. and C. Sage, *J. Chem. Phys.* **9**, 387 (1941).

1802. Gloor, W. E., *Ind. Eng. Chem.* **39**, 1125 (1947).

1803. Glover, C. A. and C. P. Hill, *Anal. Chem.* **25**, 1379 (1953).

1804. Gluzman, L. D., V. A. Ivanushkina, et al., *Org. Chem. Ind.* (USSR) **5**, 682 (1938); *C. A.* **33**, 5158 (1939).

1805. *Gmelins Handbuch der Anorganischen Chemie*, 8th ed., Pt. 5, Weinheim/Bergstr., Verlag Chemie, 1963.

1806. Godard, J., *J. Chim. Phys.* **34**, 70 (1937); *C. A.* **31** 4595 (1937).

1807. Godchot, M. and G. Cauquil, *Compt. Rend.* **198**, 663 (1934); *C. A.* **28**, 2686 (1934).

1808. Godnev, I. N., A. S. Sverdlin, et al., *Zh. Fiz. Khim.* **24**, 807 (1950); *C. A.* **45**, 4128 (1951).

1809. Goebel, H. L. and H. H. Wenzke, *J. Amer. Chem. Soc.* **59**, 2301 (1937).

1810. Goering, H. L. and F. H. McCarron, *J. Amer. Chem. Soc* **80**, 2287 (1958).

1811. Goering, H. L. and C. Serres, *J. Amer. Chem. Soc.* **74**, 5908 (1952).

1812. Goetz, H., F. Nerdel, et al., *Ann. Chem.* **681**, 1 (1965); *C. A.* **62**, 11,666 (1965).

1813. Gohlke, R. S. and F. W. McLafferty, *Anal. Chem.* **34**, 1281 (1962).

1814. Gokhale, S. D., N. L. Phalnikar, et al., *J. Univ. Bombay* **11A**, Pt. 5, 56 (1943) *C. A.* **37**, 5293 (1943).

1815. Goldblum, K. B., R. W. Martin, et al., *Ind. Eng. Chem.* **39**, 1474 (1947).

1816. Goldfarb, A. R., A. Mele, et al., *J. Amer. Chem. Soc.* **77**, 6194 (1955).

1817. Goldschmidt, H., *Z. Phys. Chem.* (Leipzig) **89**, 130 (1915).

1818. Goldschmidt, H., *Z. Phys. Chem.* (Leipzig) **124**, 23 (1926).

1819. Goldschmidt, H. and E. Mathiesen, *Z. Phys. Chem.* (Leipzig) **121**, 153 (1926).

1820. Goldschmidt, H. and L. Oslan, *Ber. Deut. Chem. Gesell.* **33**, 1140 (1900).

1821. Goldschmidt, H. and R. Salcher, *Z. Phys. Chem.* (Leipzig) **29**, 89 (1899).

1822. Goldschmidt, H. and L. Thomas, *Z. Phys. Chem.* (Leipzig) **126**, 24 (1927).

1823. Goldschmidt, H. and A. Thuesen, *Z. Phys. Chem.* (Leipzig) **81**, 30 (1913).

1824. Goldschmid, O., *J. Amer. Chem. Soc.* **75**, 3780 (1953).

1825. Gol'dshtein, I. P, E. N. Gur'yanova, et al., *Zh. Strukt. Khim.* **7**, 222 (1966); *C. A.* **65**, 16,840 (1966).

1826. Gol'dshtein, I. P., Y. M. Kessler, et al., *Zh. Strukt. Khim.* **4**, 445 (1963); *C. A.* **59**, 5894 (1963).

1827. Golendeev, V. P. and I. S. Okrokova, *Tr. Khim. Khim. Tekhnol.* **1964**, 531; *C. A.* **64**, 19,401 (1966).

1828. Golik, A. Z. and S. D. Ravikovich, *Zh. Fiz. Chim.* **23**, 86 (1949). *C. A.* **43**, 4064 (1949).

1829. Golumbic, C., *J. Amer. Chem. Soc.* **71**, 2627 (1949).

1830. Gomberg, M. and C. C. Buchler, *J. Amer. Chem. Soc.* **42**, 2059 (1920).
1831. Gomel, M. and H. Lumbroso, *Bull. Soc. Chim. Fr.* **1962**, 2200; *C. A.* **58**, 13,762 (1963).
1832. Gomel, M., H. Lumbroso, et al., *Bull. Soc. Chim. Fr.* **1959**, 1908; *C. A.* **54**, 14,171 (1960)
1833. Gonzalez R. E. *Ciencia* (Mex.) **8**, 175 (1947); *C. A.* **43**, 127 (1949).
1834. Good, W. D., J. L. Lacina, et al., *J. Phys. Chem.* **65**, 2229 (1961).
1835. Good, W. D., D. W. Scott, et al., *J. Phys. Chem.* **60**, 1080 (1956).
1836. Goode, E. V. and D. A. Ibbitson, *J. Chem. Soc.* **1960**, 4265.
1837. Goodeve, J. W., *Trans. Faraday Soc.* **30**, 501 (1934); *Chem. Zentr.* **1934, II**, 3232.
1838. Goodeve, J. W., *Trans. Faraday Soc.* **30**, 504 (1934); *Chem. Zentr.* **1934, II**, 3228.
1839. Goodeve, J. W., *Trans. Faraday Soc.* **34**, 1239 (1938); *Brit. Chem. Abstr.* **1938AI**, 598.
1840. Gopal, R. and S. A. Rizvi, *J. Indian Chem. Soc.* **43**, 179 (1966); *C. A.* **65**, 3739 (1966).
1841. Gordon, G. S. and R. L. Burwell, *J. Amer. Chem. Soc.* **71**, 2355 (1949); **70**, 3128 (1948).
1842. Gordon, J. and W. F. Giauque, *J. Amer. Chem. Soc.* **70**, 1506 (1948).
1843. Gordon, J. S., *J. Chem. Phys.* **29**, 889 (1958).
1844. Gordon, M., J. G. Miller, et al., *J. Amer. Chem. Soc.* **71**, 1245 (1949).
1845. Gordy, W., *J. Chem. Phys.* **7**, 93 (1939).
1846. Gordy, W., and S. C. Stanford, *J. Amer. Chem. Soc.* **62**, 497 (1940).
1847. Gordy, W. and S. C. Stanford, *J. Chem. Phys.* **9**, 204 (1941).
1848. Gordy, W. and D. Williams, *J. Chem. Phys.* **4**, 85 (1936).
1849. Gornowski, E. J., E. H. Amick, et al., *Ind. Eng. Chem.* **39**, 1348 (1947).
1850. Gorodetskiĭ, I. Y. A. and V. M. Dlevskiĭ, *Vestn. Leningrad Univ.* **15**, No. 16, *Ser. Fiz. i. Khim.* No. 3, 102 (1960); R. T. Leslie and E. C. Kuehner, *Anal. Chem.* **34**, 50R (1962).
1851. Goss, F. R., *J. Chem. Soc.* **1940**, 888.
1852. Goto, K., *Sci. Light* (Tokyo) **11**, 119 (1962); *C. A.* **61**, 183 (1964).
1853. Götz, I. D., *Z. Phys. Chem.* (Leipzig) **94**, 181 (1920).
1854. Goubeau, J., *Reichsamt Wirtschaftsausbau, Prüf-Nr.* **43**, (PB 52003), 129 (1940); *C. A.* **41**, 6491 (1947).
1855. Goubeau, J. and V. von Schneider, *Angew. Chem.* **53**, 531 (1940); *C. A.* **35**, 1355 (1941).
1856. Gough, G. A. C., H. Hunter, et al., *J. Chem. Soc.* **1926**, 2052.
1857. Gouravelev, D. J., *J. Phys. Chem.* (USSR) **9**, 877 (1937).
1858. Gouw, T. H. and J. C. Vlugter, *J. Amer. Oil Chem. Soc.* **41**, 142, 426, 675 (1964).
1859. Goy, C. A. and H. O. Pritchard, *J. Phys. Chem.* **69**, 3040 (1965).
1860. Grammaticakis, I. P., *Bull. Soc. Chim. Fr.* **1953**, 821; *C. A.* **48**, 1148 (1954).
1861. Grammer, G. N. and P. W. Trotter, U.S. Patent 3,238,137 (Cl. 252-171), Mar. 1, 1966.
1862. Granier, J., *Compt. Rend.* **223**, 893 (1946); *C. A.* **41**, 1511 (1947).
1863. Grant, D. M. and R. C. Hirst, *J. Chem. Phys.* **38**, 470 (1963).
1864. Grasshof, H., *Chem. Ber.* **84**, 916 (1951).
1865. Grau, A. and H. Lumbroso, *Bull. Soc. Chim. Fr.* **1961**, 1860; *C. A.* **57**, 5404 (1962).
1866. Graul, R. J. and J. V. Karabinos, *Science*, **104**, 557 (1946).
1867. Gray, H. LeB. and G. O. Gutekunst, *J. Amer. Chem. Soc.* **42**, 856 (1920).
1868. Gray, P. and A. Jones, *Trans. Faraday Soc.* **61**, 2161 (1965).
1869. Greathause, L. H., H. J. Janssen, et al., *Anal. Chem.* **28**, 357 (1956).
1870. Grédy, B. and T. Piaux, *Bull. Soc. Chim. Pays-Bas* **1**, 1481 (1934); *C. A.* **29**, 2449 (1935).

1871. Grédy, B. and L. Piaux, *Compt. Rend.* **198**, 1235 (1934); *C. A.* **28**, 3379 (1934).
1872. Green, C., J. Marsden, et al., *Can. J. Res.* **9**, 396 (1933); Beil **EIII2**, 269; 1042.
1873. Green, J. H. S., *Chem. Ind.* (London) **1960**, 1215.
1874. Green, J. H. S., *Chem. Ind.* (London) **1961**, 369; 1218.
1875. Green, J. H. S., *Chem. Ind.* (London) **1962**, 1575.
1876. Green, J. H. S., *J. Appl. Chem.* (London) **11**, 397 (1961).
1877. Green, J. H. S., *J. Chem. Soc.* **1961**, 2236.
1878. Green, J. H. S., *J. Chem. Soc.* **1961**, 2241.
1879. Green, J. H. S., *Spectrochim. Acta* **18**, 39 (1962); *C. A.* **57**, 6765 (1962).
1880. Green, J. H. S., *Trans. Faraday Soc.* **57**, 2132 (1961).
1881. Green, J. H. S., *Trans. Faraday Soc.* **59**, 1559 (1963).
1882. Green, J. H. S. and D. J. Holden, *J. Chem. Soc.* **1962**, 1794.
1883. Green, J. H. and K. R. Ryan, *Proc. Roy. Soc.* Ser. A **286** (1405), 178 (1965), *C. A.* **63**, 2511 (1965).
1884. Greenburg, R. B., U.S. Patent 2,398,526, Apr. 16, 1946; *C. A.* **40**, 3774 (1946).
1885. Greenwald, W. C., U.S. Patent 2,456,184, Dec. 14, 1948; *C. A.* **43**, 3437 (1949).
1886. Greenwood, F. L., F. C. Whitmore, et al., *J. Amer. Chem. Soc.* **60**, 2028 (1938).
1887. Gregory, M. D., H. E. Affsprung, et al., *J. Phys. Chem.* **72**, 1748 (1968).
1888. Greinacher, H., *Helv. Phys. Acta* **21**, 261 (1948); *C. A.* **42**, 8554 (1948).
1889. Grekov, A. P., USSR Patent 168,296 (Cl. C 07d), Feb. 18, 1965; *C. A.* **62**, 16,068 (1965).
1890. Gresham, W. F., U.S. Patent 2,439,426 (Cl 260-465.2), Apr. 13, 1948; *C. A.* **42**, 6841 (1948).
1891. Griffin, D. M., *J. Amer. Chem. Soc.* **71**, 1423 (1949).
1892. Griffing, V., M. A. Cargyl, et al., *J. Phys. Chem.* **58**, 1054 (1954).
1893. Griffiths, E. H., *Phil. Trans.* **182A**, 43 (1891); from [4772].
1894. Griffiths, P. H., W. A. Walkey, et al., *J. Chem. Soc.* **1934**, 631.
1895. Griffiths, V. S., *J. Chem. Soc.* **1954**, 860.
1896. Grignard, V., *Compt. Rend.* **130**, 1322 (1900); *Chem. Zentr.* **1901, II**, 622.
1897. Grignard, V., *Compt. Rend.* **132**, 336 (1901); *Chem. Zentr.* **1901, I**, 612.
1898. Grignard, V. and G. Vegnon, *Compt. Rend.* **144**, 1358 (1907); *Chem. Zentr.* **1907, II**, 681; *C. A.* **1**, 2553 (1907).
1899. Grimm, F. V. and W. A. Patrick, *J. Amer. Chem. Soc.* **45**, 2794 (1923).
1900. Grimm, H. G., *Z. Phys. Chem.* (Leipzig) **140A**, 321 (1929).
1901. Griswold, A. A. and P. S. Starcher, *J. Org. Chem.* **30**, 1687 (1965).
1902. Grodde, K.-H., *Phys. Z.* **39**, 772 777 (1938); Beil **EIII5**, 915.
1903. Groggins, P. H. and R. H. Nagel, *Ind. Eng. Chem.* **26**, 1313 (1934).
1904. Groggins, P. H., R. H. Nagel, et al., *Ind. Eng. Chem.* **26**, 1317 (1934).
1905. Grolle, H. P. A. and M. W. Tamele, U.S. Patent 2,010,358, Aug. 6, 1935; Official Gaz. U.S. Patent Office, No. 457.
1906. Gronowitz, S., A. B. Hornfeldt, et al., *Ark. Kemi* **18**, 133 (1961); *C. A.* **56**, 10,075 (1962).
1907. Gronowitz, S., G. Sorlin, et al., *Arkiv. Kemi* **19**, 483 (1962); *C. A.* **58**, 2041 (1963).
1908. Gross, B. and M.-T. Forel, *J. Chim. Phys.* **62**, 1163 (1965); *C. A.* **64**, 9089 (1966).
1909. Gross, P., *J. Amer. Chem. Soc.* **51**, 2362 (1929).
1910. Gross, P., *Z. Phys.* **32**, 587 (1931); *Chem. Zentr.* **102**, 2700 (1931).
1911. Gross, P., J. C. Rintelen, et al., *J. Phys. Chem.* **43**, 197 (1939).
1912. Gross, P. M. and J. H. Saylor, *J. Amer. Chem. Soc.* **53**, 1744 (1931).
1913. Gross, P. M., J. H. Saylor, et al., *J. Amer. Chem. Soc.* **55**, 650 (1933).

1914. Grosse, A. V., E. J. Rosenbaum, et al., *Ind. Eng. Chem.*, *Anal. Ed.* **12**, 191 (1940).
1915. Grosse, A. V., R. C. Wacker, et al., *J. Phys. Chem.* **44**, 275 (1940).
1916. Groves, C. E., *Ann. Chem.*, *Justus Liebigs* **174**, 376 (1874).
1917. Groves, L. G., *J. Chem. Soc.* **1938**, 1195.
1918. Groves, L. G. and S. Sudgen, *J. Chem. Soc.* **1934**, 1094.
1919. Groves, L. G. and S. Sudgen, *J. Chem. Soc.* **1937**, 158.
1920. Groves, L. G. and S. Sudgen, *J. Chem. Soc.* **1937**, 1779.
1921. Groves, L. G. and S. Sudgen, *J. Chem. Soc.* **1937**, 1782.
1922. Groves, L. G. and S. Sudgen, *J. Chem. Soc.* **1937**, 1992.
1923. Grude, K., J. Haupt, et al., *Z. Naturforsch.* **21A**, 1231 (1966); *C. A.* **65**, 17,936 (1966).
1924. Grünzweig, C., *Ann. Chem.*, *Justus Liebigs* **162**, 213 (1871 or 1872).
1925. Grzeskowiak, R., G. H. Jeffery, et al., *J. Chem. Soc.* **1960**, 4719.
1926. Grzeskowiak, R., G. H. Jeffery, et al., *J. Chem. Soc.* **1960**, 4728.
1927. Gudzinowicz, B. J., R. H. Campbell, et al, *J. Chem. Eng. Data* **8**, 201 (1963).
1928. Guenther, E., *The Essential Oils*, Vol. 1, Van Nostrand, Princeton, New Jersey, 1948.
1929. Guenther, E., *The Essential Oils*, Vol. 2, Van Nostrand, Princeton, New Jersey, 1949.
1930. Guérin, H. and J. Adam-Gironne, *Bull. Soc. Chim. Fr.* **1949**, 607; *C. A.* **44**, 1788. (1950).
1931. Guerin, J., *Bull. Soc. Pharm. Lille* **1956**, No. 2, 71; *C. A.* **51**, 3504 (1957).
1932. Guinchant, J., *Ann. Chim.* **9**, 49 (1918); *Chem. Zentr.* **1918**, II, 609.
1933. Guinchant, J., *Ann. Chim.* **10**, 30 (1918); *Chem. Zentr.* **1919**, I, 532.
1934. Guldberg, C. M., *Z. Phys. Chem.* (Leipzig) **5**, 374 (1890).
1935. Gundry, H. A., A. J. Head, et al., *Trans. Faraday Soc.* **58**, 1309 (1962).
1936. Gunn, S. R., *Anal. Chem.* **34**, 1292 (1962).
1937. Günther, P., A. D. von der Horst, et al., *Z. Elektrochem.* **34**, 616 (1928).
1938. Gutbier, H., *Z. Naturforsch.* **9A**, 348 (1954); *C. A.* **48**, 9806 (1954).
1939. Guthrie, G. B. and H. M. Huffman, *J. Amer. Chem. Soc.* **65**, 1139 (1943).
1940. Guthrie, G B., D. W. Scott, et al., *J. Amer. Chem. Soc.* **74**, 4662 (1952).
1941. Gutmann, F. and L. M. Simmons, *J. Chem. Phys.* **18**, 696 (1950).
1942. Gutmann, V., M. Michlmayr, et al., *Anal. Chem.* **40**, 619 (1968).
1943. Gutmann, V. and G. Schober, *Z. Anal. Chem.* **171**, 339 (1959); S. Wawzonek, *Anal. Chem.* **34**, 182R (1962).
1944. Gutowsky, H. S., *Pure Appl. Chem.* **7**, 93 (1963).
1945. Gutowsky, H. S., R. L. Rutledge, et al., *J. Amer. Chem. Soc.* **76**, 4242 (1954).
1946. Gutt, J., *Ber. Deut. Chem. Gesell.* **40**, 2061 (1907).
1947. Guttmann, L. F., *J. Amer. Chem. Soc.* **29**, 345 (1907).
1948. Guye, P. A. and E. Mallet, *Arch. Sci. Phys. Nat. Geneve* **13**, 274 (1902); *Chem. Zentr.* **1902, 1**, 1314.
1949. Guye, P. A. and E. Mallet, *Compt. Rend.* **133**, 1287 (1901); *J. Chem. Soc.* **1902Aii**, 195.
1950. Guye, P. A. and E. Mallet, *Compt. Rend.* **134**, 168 (1902); *J. Chem. Soc.* **1902Aii**, 243.
1951. Gwinn, W. D. and K. S. Pitzer, *J. Chem. Phys.* **16**, 303 (1948).

H

1952. Haarer, E. and K. Rühl, German Patent 1,002,305, Feb. 14, 1957; *C. A.* **54**, 1305 (1960).
1953. Hack, C. W. and M. Van Winkle, *Ind. Eng. Chem.* **46**, 2392 (1954).
1954. Hackel, W., *Phys. Z.* **38**, 195 (1937).
1955. Haefele, J. W. and R. W. Broge, *Kosmetik-Parfum-Drogen Rundschau* **8**, 1 (1961) *C. A.* **56**, 4591 (1962).

1956. Haensel, V. and V. N. Ipatieff, *Ind. Eng. Chem.* **39**, 853 (1947).
1957. Hagemeyer, H. J. and G. C. DeCroes, "The Chemistry of Isobutyraldehyde and Its Derivatives," Eastman, Kingsport, Tennessee, 1953.
1958. Haggenmacher, J. E., *Ind. Eng. Chem.* **40**, 436 (1948).
1959. Haggenmacher, J. E., *J. Amer. Chem. Soc.* **68**, 1633 (1946).
1960. Haggerty, C. J. and J. F. Weiler, *J. Amer. Chem. Soc.* **51**, 1623 (1929).
1961. Hagglund, E., B. Lindberg, et al., *Acta Chem. Scand.* **10**, 1160 (1956).
1962. Hahn, F. L. and R. Klockmann, *Z. Phys. Chem.* (Leipzig) **146**, 373 (1930).
1963. Haider, S. Z., M. H. Khundkar, et al., *J. Appl. Chem.* (London) **4**, 93 (1954).
1964. Haines, W. E., R. V. Helm, et al., *J. Phys. Chem.* **58**, 270 (1954).
1965. Haines, W. E., R. V. Helm, et al., *J. Phys. Chem.* **60**, 549 (1956).
1965a. Hakala, R. W., *Chem. Eng. News*, March 16, 1959, p. 43.
1966. Hála, E., J. Pick, et al., *Vapor-Liquid Equilibrium*, Pergamon Press, New York, 1958.
1967. Hale, W. J. and G. H. Cheney, U.S. Patent 1,729,775 (Cl. 260-130.5), Oct. 1, 1929.
1968. Hales, J., J. Cox and E. Lees, *Trans. Faraday Soc.* **59**, 1544 (1963).
1969. Halford, J. O., L. C. Anderson, et al., *J. Chem. Phys.* **5**, 927 (1937).
1970. Halford, R. S. and O. A. Schaeffer, *J. Chem. Phys.* **14**, 141 (1946).
1971. Hall, E. E. and A. R. Payne, *Phys. Rev.* **20**, 249 (1922); *Chem. Zentr.* **1923**, III, 1380.
1972. Hall, H. K., *J. Amer. Chem. Soc.* **79**, 5441 (1957).
1973. Hall, H. K., *J. Amer. Chem. Soc.* **79**, 5444 (1957).
1974. Hall, H. K., *J. Phys. Chem.* **60**, 63 (1956).
1975. Hall, H. K. and R. Zbinden, *J. Amer. Chem. Soc.* **80**, 6428 (1958).
1976. Hall, N. F., *J. Amer. Chem. Soc.* **52**, 5115 (1930).
1977. Hall, N. F. and M. R. Sprinkle, *J. Amer. Chem. Soc.* **54**, 3469 (1932).
1978. Hall, N. F. and H. H. Voge, *J. Amer. Chem. Soc.* **55**, 239 (1933).
1979. Haller, A., *Compt. Rend.* **114**, 1326 (1892).
1980. Haller, W. and H. C. Duecker, *J. Res. Nat. Bur. Stand.* **64A**, 527 (1960).
1981. Hallgren, B., R. Ryhage, et al., *Acta Chem. Scand.* **13**, 845 (1959); *C. A.* **55**, 12,024 (1961).
1982. Halpern, B. and J. W. Westley, *Aust. J. Chem.* **19**, 1533 (1966); *C. A.* **65**, 19,996 (1966).
1983. Halverson, F., R. F. Stamm, et al., *J. Chem. Phys.* **16**, 808 (1948).
1984. Halverstadt, I. F. and W. D. Kumler, *J. Amer. Chem. Soc.* **64**, 1982 (1942).
1985. Halverstadt, I. F. and W. D. Kumler, *J. Amer. Chem. Soc.* **64**, 2988 (1942).
1986. Hamamoto, K. and H. Tone, *Ann. Repts. Shionogi Research Lab.* **1**, 107 (1952); *C. A.* **51**, 8004 (1957).
1987. Hamelin, R., *Bull. Soc. Chim. Fr.* **1961**, 926; *C. A.* **55**, 27,027 (1961).
1988. Hamer, J., L. Placek, et al., *Nitro Compounds.*, *Proc. Intern. Symp.*, *Warsaw* **1963** 395 (Pub. 1964); *C. A.* **64**, 162 (1966).
1989. Hammaker, R. M., *Can. J. Chem.* **43**, 2916 (1965).
1990. Hammett, L. P. and N. Dietz, *J. Amer. Chem. Soc.* **52**, 4795 (1930).
1991. Hammick, D. L. and J. Howard, *J. Chem. Soc.* **1932**, 2915.
1992. Hammick, D. L., A. Norris, et al., *J. Chem. Soc.* **1938**, 1755.
1993. Hammick, D. L. and M. Roberts, *J. Chem. Soc.* **1948**, 73.
1994. Hammond, B. R. and R. H. Stokes, *Trans. Faraday Soc.* **51**, 1641 (1955).
1995. Hammond, J. A. S., U.S. Patent 2,356,758, Aug. 15, 1944; *C. A.* **39**, 90 (1945).
1996. Hampton, J. and J. A. Riddick, Unpublished data from the Research and Development Division, Commercial Solvents Corp.
1997. Hampton, R. R., *Anal. Chem.* **21**, 923 (1949).
1998. Hamstead, A. C. and L. S. Van Delinder, *J. Chem. Eng. Data* **5**, 383 (1960).

1999. Hanai, S., *J. Chem. Soc. Jap.* **62,** 1208 (1941); *C. A.* **41,** 3436 (1947).

2000. Hancock, C. K., G. M. Watson, et al., *J. Phys. Chem.* **58,** 127 (1954).

2001. Handley, R., *Ind. Chem. Mfg.* **31,** 535 (1955); Ref. [2123].

2002. Hands, C. H. G. and W. S. Norman, *Ind. Chemist* **21,** 307 (1945); *C. A.* **39,** 4273 (1945).

2003. Hanford, W. E. and D. L. Fuller, *Ind. Eng. Chem.* **40,** 1171 (1948).

2004. Hannan, M. C. and M. C. Markham, *J. Amer. Chem. Soc.* **71,** 1120 (1949).

2005. Hannay, N. B. and C. P. Smyth, *J. Amer. Chem. Soc.* **68,** 1005 (1946).

2006. Hannay, N. B. and C. P. Smyth, *J. Amer. Chem. Soc.* **68,** 1357 (1946).

2007. Hannotte, T., *Bull. Soc. Chim. Belges* **35,** 86 (1926); *C. A.* **20,** 2657 (1926); *Chem. Zentr.* **1926, II,** 741.

2008. Hanson, M. W. and J. B. Bouck, *J. Amer. Chem. Soc.* **79,** 5631 (1957).

2009. Hantzsch, A., *Ber. Deut. Chem. Gesell.* **39,** 1084 (1906).

2010. Hantzsch, A. and H. Freese, *Ber. Deut. Chem. Gesell.* **27,** 2529, 2966 (1894).

2011. Hantzsch, A. and O. K. Hofmann, *Ber. Deut. Chem. Gesell.* **44,** 1776 (1911).

2012. Hantzsch, A. and E. Scharf, *Ber. Deut. Chem. Gesell.* **46,** 3570 (1913).

2013. Happ, G. P. and D. W. Stewart, *J. Amer. Chem. Soc.* **74,** 4404 (1952).

2014. Hardies, D. E., Belgian Patent 620,486, Aug. 14, 1962.

2015. Hardies, D. E. and B. O. Pray, U.S. Patent 3,070,634 (Cl. 260-652.5), Dec. 25, 1962; *C. A.* **58,** 10,075 (1963).

2016. Hardy, C. J., D. Fairhurst, et al., *Trans. Faraday Soc.* **60,** 1626 (1964).

2017. Hardy, D. V. N., *J. Soc. Chem. Ind.* (London) **67,** 81 (1948).

2018. Haresnap, D., F. A. Fidler, et al., *Ind. Eng. Chem.* **41,** 2691 (1949).

2019. Hargreaves, M. K., *J. Chem. Soc.* **1956,** 3679.

2020. Hariharan, T. A., *J. Indian Inst. Sci.* **36,** 189 (1954); *C. A.* **49,** 724 (1955).

2021. Hariharan, T. A., *J. Indian Inst. Sci.* **36A,** 215 (1954); *C. A.* **49,** 2871 (1955).

2022. Harkins, W. D., *Z. Phys. Chem.* (Leipzig) **139A,** 647 (1928).

2023. Harkins, W. D., F. E. Brown, et al., *J. Amer. Chem. Soc.* **39,** 354 (1917).

2024. Harkins, W. D., G. L. Clark, et al., *J. Amer. Chem. Soc.* **42,** 700 (1920).

2025. Harkins, W. D. and R. W. Wampler, *J. Amer. Chem. Soc.* **53,** 850 (1931).

2026. Harley-Mason, J., et al., *J. Chem. Soc.* **1966B,** 396.

2027. Harlow, G. A., C. M. Noble, et al., *Anal. Chem.* **28,** 787 (1956).

2028. Harned, H. S. and R. W. Ehlers, *J. Amer. Chem. Soc.* **54,** 1350 (1932).

2029. Harned, H. S. and R. W. Ehlers, *J. Amer. Chem. Soc.* **55,** 652 (1933).

2030. Harned, H. S. and R. W. Ehlers, *J. Amer. Chem. Soc.* **55,** 2379 (1933).

2031. Harned, H. S. and N. D. Embree, *J. Amer. Chem. Soc.* **56,** 1042 (1934).

2032. Harned, H. S. and J. O. Morrison, *J. Amer. Chem. Soc.* **58,** 1908 (1936).

2033. Harned, H. S. and B. B. Owen, *J. Amer. Chem. Soc.* **52,** 5079 (1930).

2034. Harned, H. S. and R. A. Robinson, *Trans. Faraday Soc.* **36,** 973 (1940); from A. K. Covington, R. A. Robinson and R. G. Bates, *J. Phys. Chem.* **70,** 3820 (1966).

2035. Harrand, M. and H. Martin, *Bull. Soc. Chim. Fr.* **1956,** 1383; *C. A.* **51,** 4143 (1957).

2036. Harris, B., R. J. W. Le Fèvre, et al., *J. Chem. Soc.* **1953,** 1622.

2037. Harris, F. E. and C. T. O'Konski, *J. Amer. Chem. Soc.* **76,** 4317 (1954).

2038. Harris, H., *J. Chem. Soc.* **1925,** 1049.

2039. Harris, J. F. and L. L. Zoch, *Anal. Chem.* **34,** 201 (1962).

2040. Harris, R. K. and N. Sheppard, *Trans. Faraday Soc.* **59,** 606 (1963).

2041. Harris, R. K. and R. A. Spragg, *Chem. Commun.* **1966,** 314.

2042. Harrison, A. J., B. J. Cederholm, et al., *J. Chem. Phys.* **30,** 355 (1959).

2043. Harrison, D. and E. A. Moelwyn-Hughes, *Proc. Roy. Soc.* **239A,** 230 (1957); from [4773].

2044. Harrison, G. R., R. C. Lord, et al., *Practical Spectroscopy*, Prentice-Hall, Englewood Cliffs, N.J. 1948.

2045. Harrison, H. R. and E. J. Eisenbraun, *J. Org. Chem.* **31**, 1294 (1966).

2046. Hartley H. and H. R. Raikes, *J. Chem. Soc.* **1925**, 524.

2047. Hartley, W. N. and J. J. Dobbie, *J. Chem. Soc.* **1900**, 846.

2048. Hartmann, H., A. Neumann, et al., *Z. Phys. Chem.* (Frankfurt) **44**, 204 (1965).

2049. Hartsuch, P. J., *J. Amer. Chem. Soc.* **61**, 1142 (1939).

2050. Hartwell, E. J., R. E. Richards, et al., *J. Chem. Soc.* **1948**, 1436.

2051. Hasabe, N., *J. Chem. Soc. Jap.* **64**, 1041 (1943); *C. A.* **41**, 3586 (1947).

2052. Hashimoto, T., H. Shina, et al. *Kogyo Kagaku Zasshi* **68**, 1434 (1965); *C. A.* **63**, 17,826 (1965).

2053. Hass, H. B., E. B. Hodge, et al., *Ind. Eng. Chem.* **28**, 339 (1936).

2054. Hass, H. B. and J. A. Patterson, *Ind. Eng. Chem.* **30**, 67 (1938).

2055. Hass, H. B, and J. A. Patterson, *Ind. Eng. Chem.* **33**, 615 (1941).

2056. Hassel, O. and E. Naeshagen, *Tids. Kjemoig Bergvesen* **10**, 81 (1930); *C. A.* **25**, 2698 (1931).

2057. Hassel, O. and E. Naeshagen, *Z. Phys. Chem.* (Leipzig) **12B**, 79 (1931).

2058. Hassion, F. X. and R. H. Cole, *J. Chem. Phys.* **23**, 1756 (1955).

2059. Hastings, S. H. and D. E. Nicholson, *J. Phys. Chem.* **61**, 730 (1957).

2060. Haszeldine, R. N., *J. Chem. Soc.* **1953**, 1764.

2061. Haszeldine, R. N., *J. Chem. Soc.* **1953**, 2525.

2062. Haszeldine, R. N., *J. Chem. Soc.* **1953**, 2622.

2063. Hatch, G. B. and H. Adkins, *J. Amer. Chem. Soc.* **59**, 1694 (1937).

2064. Hatch, L. F. and T. W. Evans, U.S. Patent 2,434,110, Jan. 6, 1948; *C. A.* **42**, 2267 (1948).

2065. Hatch, L. F. and A. C. Moore, *J. Amer. Chem. Soc.* **66**, 285 (1944).

2066. Hatch, L. F. and S. S. Nesbitt, *J. Amer. Chem. Soc.* **72**, 727 (1950).

2067. Hatschek, E., *The Viscosity of Liquids*, Van Nostrand, Princeton, N.J., 1928.

2068. Hatton, J. V. and R. E. Richards, *Trans. Faraday Soc.* **56**, 315 (1960).

2069. Hatton, W. E., D. L. Hildenbrand, et al., *J. Chem. Eng. Data* **7**, 229 (1962).

2070. Hauptman, E., *Acta Phys. Polon.* **7**, 86 (1938); *C. A.* **33**, 1594 (1938).

2071. Hauser, C. R., W. J. Humphlett, et al., *J. Amer. Chem. Soc.* **70**, 426 (1948).

2072. Hausser, K. W., R. Kuhn, et al., *Z. Phys. Chem.* (Leipzig) **29B**, 371 (1935); G. N. Lewis and M. Calvin, *Chem. Rev.* **25**, 273 (1939); *Chem. Zentr.* **1936**, I, 2326.

2073. Hausser, K. W., R. Kuhn, et al., *Z. Phys. Chem.* (Leipzig) **29B**, 378 (1935); *Chem. Zentr.* **1936**, I, 2327.

2074. Hawkins, J. E. and G. T. Armstrong, *J. Amer. Chem. Soc.* **76**, 3756 (1954).

2075. Hawkins, J. E. and W. T. Eriksen, *J. Amer. Chem. Soc.* **76**, 2669 (1954).

2076. Haworth, W. N., *J. Chem. Soc.* **1913**, 1242.

2077. Hayden, A. L., O. R. Sammul, et al., *J. Assoc. Offic. Agr. Chemist* **45**, 797 (1962).

2078. Hazlet, S. E. and R. B. Callison, *J. Amer. Soc. Chem.* **66**, 1248 (1944).

2079. Heap, J. G., W. J. Jones, et al., *J. Amer. Chem. Soc.* **43**, 1936 (1921).

2080. Heard, J. R., U.S. Patent 2,417,685 (Cl. 260-348.6), Mar. 18, 1947; *C. A.* **41**, 3481 (1947).

2081. Hearne, G., M. Tamele, et al., *Ind. Eng. Chem.* **33**, 805 (1941).

2082. Hebeisen, J., *Ann. Phys.* (Leipzig) (4) **77**, 216 (1925); *J. Chem. Soc.* **1925Aii**, 763.

2083. Hecht, K. T. and D. L. Wood, *Proc. Roy. Soc.* **235A**, 174 (1956).

2084. Hedlund, A. I., Swedish Patent 119,077, June 25, 1947; *C. A.* **42**, 1752 (1948).

2085. Heel, H. and W. Zeil, *Z. Elektrochem.* **64**, 962 (1960); *C. A.* **55**, 115 (1961).

2086. Hehlgans, F., *Phys. Z.* **30**, 942 (1929).

2087. Heigl, J. J., J. F. Black, et al., *Anal. Chem.* **21**, 556 (1949).
2088. Heilbron, Ian, ed., *Dictionary of Organic Compounds*, 4th ed., Oxford University Press, New York, 1965.
2089. Heilmann, R., J. M. Bonnier, et al., *Compt. Rend.* **244**, 1787 (1957); *C. A.* **51**, 14538 (1957).
2090. Heim, G., *Bull. Soc. Chim. Belges* **42**, 461 (1933).
2091. Heinemann, H., U.S. Patent 2,451,949, Oct. 19, 1948; *C. A.* **43**, 2222 (1949).
2092. Heinert, D. and A. E. Martell, *J. Amer. Chem. Soc.* **85**, 183 (1963).
2093. Heinze, A. J., *Hydrocarbon Process. Petrol. Refiner* **41**, No. 3, 187 (1962); *C. A.* **58**, 1281 (1963).
2094. Held, R. P. and C. M. Criss, *J. Phys. Chem.* **69**, 2611 (1965).
2095. Helferich, B. and J. Hausen, *Ber. Deut. Chem. Gesell.* **57**, 795 (1924).
2096. Heller, H. E., *J. Amer. Chem. Soc.* **74**, 4858 (1952).
2097. Helm, R. V., W. J. Lanum, et al., *J. Phys. Chem.* **62**, 858 (1958).
2098. Hemptinne, M. de, *Contrib. Étude Structure Mol, Vol. Commem. Victor Henri* **1947/48**, 151; *C. A.* **43**, 2865 (1949).
2099. Hende, A. van den, and R. Fonteyne, *Natuurw. Tijdschr.* (Ghent) **25**, 24 (1943); *Chem. Zentr.* **1943**, I, 1463.
2100. Hendra, P. J. and D. B. Powell, *J. Chem. Soc.* **1960**, 5105.
2101. Hendricks, S. B., and M. E. Jefferson, *J. Opt. Soc. Amer.* **23**, 299 (1933); *C. A.* **27**, 5594 (1933).
2102. Hennaut-Roland, M. and M. Lek, *Bull. Soc. Chim. Belges* **40**, 177 (1931); *C. A.* **25**, 5322 (1931).
2103. Henne, A. L. and T. Alderson, *J. Amer. Chem. Soc.* **67**, 918 (1945).
2104. Henne, A. L. and C. J. Fox, *J. Amer. Chem. Soc.* **73**, 2323 (1951).
2105. Henne, A. L. and K. W. Greenlee, *J. Amer. Chem. Soc.* **67**, 484 (1945).
2106. Henne, A. L. and D. M. Hubbard, *J. Amer. Chem. Soc.* **58**, 404 (1936).
2107. Henne, A. L. and P. Trott, *J. Amer. Chem. Soc.* **69**, 1820 (1947).
2108. Hennelly, E. J., W. M. Heston, et al., *J. Amer. Chem. Soc.* **70**, 4102 (1948).
2109. Hennion, G. F. and W. S. Murray, *J. Amer. Chem. Soc.* **64**, 1220 (1942).
2110. Henri, M. V., *Compt. Rend.* **178**, 846 (1924).
2111. Henri, V. and P. Angenot, *J. Chim. Phys.* **33**, 641 (1936); *C. A.* **31**, 3786 (1937).
2112. Henry, L., *Ber. Deut. Chem. Gesell.* **5**, 569 (1872).
2113. Henry, L., *Ber. Deut. Chem. Gesell.* **6**, 728 (1873).
2114. Henry, P., *Z. Phys. Chem.* (Leipzig) **10**, 96 (1892).
2115. Hentz, R. R., *J. Phys. Chem.* **66**, 1622 (1962).
2116. Heppel, L. A., P. A. Neal, et al., *J. Ind. Hyg. Toxicol.* **28**, 113 (1946); *C. A.* **40**, 6170 (1946).
2117. Hepworth, H., *J. Chem. Soc.* **1921**, 1249.
2118. Herail, F., *Compt. Rend., Ser.* **262C**, 22 (1966); *C. A.* **64**, 18,706 (1966).
2119. Herald, A. E. and H. A. Price, *Oil Gas J.* **46**, No. 9, 94 (1947); *C. A.* **41**, 5818 (1947).
2120. Herb, S. F., *J. Amer. Oil Chem. Soc.* **32**, 153 (1955).
2121. Herbison-Evans, D. and R. E. Richards, *Trans. Faraday Soc.* **58**, 845 (1962).
2122. Herington, E. F. G., *J. Amer. Chem. Soc.* **73**, 5883 (1951).
2123. Herington, E. F. G., *Zone Melting of Organic Compounds*, Wiley, New York, 1963.
2124. Herington, E. F. G., A. B. Densham, et al., *J. Chem. Soc.* **1954**, 2643.
2125. Herington, E. F. G., R. Handley, et al., *Chem. Ind.* (London) **1956**, 292.
2126. Herington, E. F. G. and W. Kynaston, *Trans. Faraday Soc.* **53**, 138 (1957).
2127. Herington, E. F. G. and J. F. Martin, *Trans. Faraday Soc.* **49**, 154 (1953); *C. A.* **47**, 8442 (1953).

2128. Herlan, A., *Brennst-Chem.* **46**, 264 (1965); *C. A.* **64**, 10,541 (1966).
2129. Herman, R. C., *J. Chem. Phys.* **8**, 252 (1940).
2130. Hernandez, G. J., *J. Chem. Phys.* **38**, 2233 (1963).
2131. Hernandez, G. J., *J. Chem. Phys.* **39**, 1355 (1963).
2132. Hernandez, G. J. and A. B. F. Duncan, *J. Chem. Phys.* **36**, 1504 (1962).
2133. Herold, W., *Z. Phys. Chem.* (Leipzig) **18B**, 265 (1932); *C. A.* **26**, 5257 (1932); *Chem. Zentr.* **1932, II**, 2807.
2134. Herold, W. and K. L. Wolf, *Z. Phys. Chem.* (Leipzig) **12B**, 194 (1931); *C. A.* **25**, 2921 (1931).
2135. Hershberg, E. B., *J. Org. Chem.* **13**, 542 (1948).
2136. Herz, E., *Monatsh. Chem.* **74**, 160 (1943); *C. A.* **38**, 1428 (1944).
2137. Herz, E., *Monatsh. Chem.* **76**, 1 (1946); *C. A.* **41**, 1558 (1947).
2138. Herz, E., L. Kahovec, et al., *Monatsh. Chem.* **76**, 100 (1946); *C. A.* **41**, 3370 (1947).
2139. Herz, E. and K. W. F. Kohlrausch, *Monatsh. Chem.* **74**, 175 (1943); *C. A.* **38**, 1428 (1944).
2140. Herz, E., K. W. F. Kohlrausch, et al., *Monatsh. Chem.* **76**, 112 (1946); *C. A.* **41**, 3370 (1947).
2141. Herz, E., K. W. F. Kohlrausch, et al., *Monatsh. Chem.* **76**, 200, 231 (1947); *C. A.* **41**, 6153 (1947).
2142. Herz, W., *Z. Elektrochem.* **23**, 24 (1917); *C. A.* **11**, 2631 (1917).
2143. Herz, W. and W. Bloch, *Z. Phys. Chem.* (Leipzig) **110**, 23 (1924).
2144. Herz, W. and E. Lorentz, *Z. Phys. Chem.* (Leipzig) **140**, 406 (1929).
2145. Herz, W. and E. Neukirch, *Z. Phys. Chem.* (Leipzig) **104**, 433 (1923).
2146. Herz, W. and W. Rathmann, *Chem.-Ztg.* **36**, 1417 (1912); *J. Chem. Soc.* **1913Aii**, 26.
2147. Herz, W. and W. Rathmann, *Chem.-Ztg.* **37**, 621 (1913); *C. A.* **7**, 3866 (1913).
2148. Herz, W. and P. Schuftan, *Z. Phys. Chem.* (Leipzig) **101**, 269 (1922).
2149. Herzberg, G., *Infrared and Raman Spectra of Polyatomic Molecules*, Van Nostrand, Princeton, New Jersey, 1945.
2150. Herzberg, G. and R. Kolsch, *Z. Elektrochem.* **39**, 572 (1933); *C. A.* **27**, 4734 (1933).
2151. Herzog, R., *Ind. Eng. Chem.* **36**, 997 (1944).
2152. Hess, K., *Ber. Deut. Chem. Gesell.* **63**, 518 (1930).
2153. Hess, K. and H. Frahm, *Ber. Deut. Chem. Gesell.* **71B**, 2627 (1938).
2154. Hess, K. and H. Haber, *Ber. Deut. Chem. Gesell.* **70**, 2205 (1937).
2155. Hesse, G. and H. Schildknecht, *Angew. Chem.* **67**, 737 (1955); *C. A.* **50**, 3675 (1956).
2156. Hesse, G. and R. Schrödel, *Angew. Chem.* **68**, 438 (1956); *C. A.* **51**, 11,341 (1957).
2157. Heston, W. M., E. J. Hennelly, et al., *J. Amer. Chem. Soc.* **70**, 4093 (1948).
2158. Heston, W. M., E. J. Hennelly, et al., *J. Amer. Chem. Soc.* **72**, 2071 (1950).
2159. Hetzer, H. B., R. G. Bates, et al., *J. Phys. Chem.* **70**, 2869 (1966).
2160. Hetzer, H. B., R. A. Robinson, et al., *J. Phys. Chem.* **72**, 2081 (1968).
2161. Heumann, W. R., O. Dionne, et al., *Can. J. Chem.* **43**, 1096 (1965).
2162. Heuser, R. V., U.S. Patent 2,409,124 (Cl. 260-464), Oct. 8, 1946; *C. A.* **41**, 1235 (1947).
2163. Heydweiller, A., *Ann. Phys.* (Leipzig) **59**, 193 (1896).
2164. Heyns, K., R. Stute, et al., *Tetrahedron* **22**, 2223 (1966).
2165. Hibben, J. H., *J. Chem. Phys.* **5**, 706 (1932).
2166. Hibbert, H., U.S. Patent 1,008,333, Nov. 14, 1911; *Official Gaz. U.S. Patent Office* **172**, 280 (1911).
2167. Hibbert, H. and J. S. Allen, *J. Amer. Chem. Soc.* **54**, 4115 (1932).
2168. Hickman, K. C. D., *J. Phys. Chem.* **34**, 627 (1930).
2169. Hickman, K. C. D., J. C. Hecker, et al., *Ind. Eng. Chem. Anal. Ed.* **9**, 264 (1937).

2170. Hickman, K. C. D., and D. J. Trevoy, *Ind. Eng. Chem.* **44**, 1882 (1952).
2171. Hicks, J. F. G., J. G. Hooley, et al., *J. Amer. Chem. Soc.* **66**, 1064 (1944).
2172. Hicks-Brunn, M. M. and J. H. Brunn, *J. Res. Nat. Bur. Stand.* **8**, 525 (1932).
2173. Hidalgo, A. and M. T. Sardiña, *An. Real Soc. Espan. Fís y Quim* (Madrid) **52B**, 627 (1956); *C. A.* **54**, 4153 (1960).
2174. Hieber, W. and A. Woerner, *Z. Elektrochem.* **40**, 252, 256 (1934); *Chem. Zentr.* **1934, II**, 920.
2175. Higashi, K., *Bull. Inst. Phys.-Chem. Research* (Tokyo) **12**, 22 (1933); *C. A.* **27**, 2358 (1933).
2176. Higashi, K., *Bull. Inst. Phys.-Chem. Res.* (Tokyo) **12**, 771 (1933); *C. A.* **28**, 386 (1934).
2177. Higashi, K., *Bull. Inst. Phys.-Chem. Res.* (Tokyo) **15**, 766 (1936); *C. A.* **31**, 4863 (1937).
2178. Higashi, K., *Sci. Papers Inst. Phys.-Chem. Res.* (Tokyo) **31**, 311 (1937); *C. A.* **31**, 6559 (1937).
2179. Higgins, C. E., W. H. Baldwin, et al., *J. Phys. Chem.* **63**, 112 (1959).
2180. Higgins, E. R. and J. Lielmezo, *J. Chem. Eng. Data* **10**, 178 (1965).
2181. High, M. E., *Phys. Rev.* **38**, 1837 (1931); *C. A.* **26**, 920 (1932).
2182. Highet, R. J. and W. C. Wildman, *J. Amer. Chem. Soc.* **77**, 4399 (1955).
2183. Higuchi, T., C. H. Barnstein, et al., *Anal. Chem.* **34**, 400 (1962).
2184. Hildenbrand, D. L. and R. A. McDonald, *J. Phys. Chem.* **63**, 1521 (1959).
2185. Hildenbrand, D. L., R. A. McDonald, et al., *J. Chem. Phys.* **30**, 930 (1959).
2186. Hilditch, T. P., *The Chemical Constitution of Natural Fats*, 2d ed., p. 373, Wiley, New York, 1940; from Markley [3151].
2187. Hilditch, T. P., *The Chemical Constitution of Natural Fats*, 3rd ed., p. 575, Chapman & Hall, London, 1956; from Markley [3151].
2188. Hilditch, T. P., *The Industrial Chemistry of Fats and Waxes*, 2d ed., p. 88, Ballière, Tindall & Cox, London, 1941; from Markley [3151].
2189. Hill, A. E., *J. Amer. Chem. Soc.* **45**, 1143 (1923).
2190. Hill, A. E. and T. B. Fitzgerald, *J. Amer. Chem. Soc.* **57**, 250 (1935).
2191. Hill, A. E. and W. N. Malisoff, *J. Amer. Chem. Soc.* **48**, 918 (1926).
2192. Hill, A. V., *Proc. Roy. Soc.* **127A**, 9 (1930).
2193. Hill, F. N. and A. Brown, *Anal. Chem.* **22**, 562 (1950).
2194. Hill, R. D. and G. D. Meakins, *J. Chem. Soc.* **1958**, 760.
2195. Hinegardner, W. S., U.S. Patent 2,525,410, Oct. 10, 1950; British Patent 642,969, Dec. 19, 1946; *C. A.* **45**, 2025 (1951).
2196. Hipple, J. A. and M. Shepherd, *Anal. Chem.* **21**, 32 (1949).
2197. Hippel, A. v. and L. G. Wesson, *Ind. Eng. Chem.* **38**, 1121 (1946).
2198. Hirakawa, A. Y. and M. Tsuboi, *Proc. Intern. Symp. Mol. Struct. Spectry. Tokyo* **1962** (A107); *C. A.* **61**, 1396 (1964).
2199. Hirano, K., *Bull. Chem. Soc. Jap.* **38**, 842 (1965).
2200. Hirota, K., K. Nagoshi, et al., *Bull. Soc. Chem. Jap.* **34**, 226 (1961); *C. A.* **55**, 15,105 (1961).
2201. Hirota, K., J. Takezaki, et al., *Nippon Hoshasan Kobunshi Kenkyu Kyokai Nenpo* **6**, 247 (1964–1965); *C. A.* **64**, 18,636 (1966).
2202. Hirschler, A. E. and W. B. M. Faulconer, *J. Amer. Chem. Soc.* **68**, 210 (1946).
2203. Hirsjärvi, P., *Suomen Kemistilehti* **29B**, 200 (1956); *C. A.* **51**, 4820 (1957).
2204. Hirt, R. C. and J. P. Howe, *J. Chem. Phys.* **16**, 480 (1948).
2205. Hissel, J., *Bull. Soc. Roy. Sci. Liege* **21**, 457 (1952); *C. A.* **47**, 11,961 (1953).
2206. Hitch, E. F. and H. N. Gilbert, *J. Amer. Chem. Soc.* **35**, 1780 (1913).
2207. Hite, H. B., *Amer. Chem. J.* **17**, 507 (1895).

2208. Hobbs, M. E. and W. W. Bates, *J. Amer. Chem. Soc.* **74**, 746 (1952).
2209. Hobson, R. W., R. J. Hartman, et al., *J. Amer. Chem. Soc.* **63**, 2094 (1941).
2210. Hochstrasser, G., *Helv. Phys. Acta* **34**, 189 (1961); *C. A.* **55**, 23,059 (1961).
2211. Hodge, E. B., *Ind. Eng. Chem.* **32**, 748 (1940).
2212. Hodge, H. C., *Proc. Soc. Exp. Biol. Med.* **53**, 20 (1943); *C. A.* **37**, 4472 (1943).
2213. Hodgman, C. D., ed., *Handbook of Chemistry and Physics*, 44th ed., Chemical Rubber Publishing, Cleveland, 1962.
2214. Hodgson, H. H., *J. Chem. Soc.* **1943**, 380.
2215. Hoepe, G. and W. D. Treadwell, *Helv. Chim. Acta* **25**, 353 (1942).
2216. Hoerr, C. W. and H. J. Harwood, *J. Phys. Chem.* **56**, 1068 (1952).
2217. Hoerr, C. W., M. R. McCorkle, et al., *J. Amer. Chem. Soc.* **65**, 328 (1943).
2218. Hoerr, C. W. and R. S. Sedgwick, *J. Org. Chem.* **11**, 603 (1946).
2219. Höfermann, H., *Chem.-Ing.-Tech.* **36**, 422 (1964).
2220. Hoffmann, C. de and E. Barbier, *Bull. Soc. Chim. Belges* **45**, 565 (1936); *Chem. Zentr.* **1937, I**, 2590; Beil **EIII2**, 655.
2221. Hoffman, H. T., G. E. Evans, et al., *J. Amer. Chem. Soc.* **73**, 3028 (1951).
2222. Hoffman, J. and C. E. Boord, *J. Amer. Chem. Soc.* **77**, 3139 (1955).
2223. Hoffman, R. A., B. Gestblom, et al., *J. Chem. Phys.* **39**, 486 (1963).
2224. Hoffmann, E. C., *Z. Phys. Chem.* (Leipzig) **53B**, 179 (1943); *C. A.* **38**, 300 (1944).
2225. Hoffmann, U., German Patent 618,972, Sept. 19, 1935; *C. A.* **30**, 1066 (1936).
2226. Hofman, W., L. Stefaniak, et al., *J. Amer. Chem. Soc.* **86**, 554 (1964).
2227. Hofmann and H. Oettel, *Naunyn-Schmiedebergs Arch. Pharmakol. Exp. Pathol.* **222**, 233 (1954); *C. A.* **48**, 7788 (1954).
2228. Hofmann, H. E. and E. W. Reid, *Ind. Eng. Chem.* **21**, 955 (1929).
2229. Hogler, R. and L. Kohovec, *Monatsh. Chem.* **76**, 27 (1946); *C. A.* **41**, 1558 (1947).
2230. Hogness, T. R., A. E. Sidwell, et al., *J. Biol. Chem.* **120**, 239 (1937).
2231. Hoigné, J. and T. Gäumann, *Helv. Chim. Acta* **41**, 1933 (1958).
2232. Højendahl, K., *Beret. 18th Skand. Naturforskedmode Copenhagen* **1929**, 344; *C. A.* **24**, 2649 (1930).
2233. Højendahl, K., Dissertation, Copenhagen, 1928; from R. N. Kerr, *J. Chem. Soc.* **1926**, 2798.
2234. Højendahl, K., *Kgl. Danske Videnskab. Selskab. Mat.-Fys. Medd.* **24**, No. 2, 11 pp. (1946); *C. A.* **41**, 4985 (1947).
2235. Højendahl, K., *Nord. Kemikermode, Forh.* **5**, 209 (1939); *C. A.* **38**, 2249 (1944).
2236. Højendahl, K., *Phys. Z.* **30**, 394; Beil **EII6**, 209; *Chem. Zentr.* **1929, II**, 1898.
2237. Holcomb, D. E. and C. L. Dorsey, *Ind. Eng. Chem.* **41**, 2788 (1949).
2238. Hollmann, R., *Z. Phys. Chem.* (Leipzig) **43**, 129 (1903).
2239. Holman, R. T. and P. R. Edmondson, *Anal. Chem.* **28**, 1533 (1956).
2240. Homer, J. and L. F. Thomas, *J. Chem. Soc.* **1966B**, 141.
2241. Hood, G. C., O. Redlich, et al., *J. Chem. Phys.* **23**, 2229 (1955).
2242. Hood, H. E., U.S. Patent 2,971,893, Feb. 14, 1961; *Official Gaz. U.S. Patent Office* **763**, 454 (1961).
2243. Hooper, G. S. and C. A. Kraus, *J. Amer. Chem. Soc.* **56**, 2265 (1934).
2244. Hoosier Solvents and Chemical Corporation, "Physical Properties of Modern Organic Solvents," Indianapolis, Indiana.
2245. Hoover, T. B., *J. Phys. Chem.* **68**, 876 (1964).
2246. Hoover, T. B., *J. Phys. Chem.* **68**, 3003 (1964).
2247. Hopke, E. R. and G. W. Sears, *J. Amer. Chem. Soc.* **70**, 3801 (1948).
2248. Hopton, F. J., A. J. Rest, et al., *J. Chem. Soc.* **1966A**, 1326.
2249. Horák, M. and O. Exner, *Chem. Listy* **52**, 1451 (1958); *C. A.* **52**, 19,894 (1958).

2250. Horn, H. J., *Toxicol. Appl. Pharmacol.* **3**, 12 (1961).

2251. Horning, E. C., ed., *Organic Synthesis, Collective Vol. 3*, Wiley, New York, 1955.

2252. Hornstein, I. and P. F. Crowe, *Anal. Chem.* **34**, 1037 (1962).

2253. Horsley, G. F. and Imperial Chemical Industries, British Patent 313,466, February 6, 1928; *C. A.* **24**, 1125 (1930).

2254. Horsley, L. H., *Azeotropic Data*, American Chemical Society, Washington, D.C., 1952.

2255. Horsley, L. H., *Azeotropic Data II*, American Chemical Society, Washington, D.C., 1962.

2256. Horton, C. A. and J. C. White, *Talanta* **7**, 215 (1961); *C. A.* **55**, 23,045 (1961).

2257. Hosoya, H., J. Tanaka, et al., *J. Mol. Spectrosc.* **8**, 257 (1962); *C. A.* **57**, 2990 (1962).

2258. Hougen, O. A. and K. M. Watson, *Chemical Process Principles*, Wiley, New York, 1943.

2259. Hough, E. W., D. M. Mason, et al., *J. Amer. Chem. Soc.* **72**, 5775 (1950).

2260. Houston, D. F., *J. Amer. Chem. Soc.* **55**, 4131 (1933).

2261. Houtman, J. P. W., J. van Steenis, et al., *Rec. Trav. Chim. Pays-Bas* **65**, 781 (1946); *C. A.* **41**, 3049 (1947).

2262. Houtman, T., T. E. Zurawic, et al., U.S. Patent 2,779,721 (Cl. 202-42), Jan. 29, 1957; *C. A.* **51**, 7073 (1957); *Official Gaz. U.S. Patent Office* **714**, 1018 (1957).

2263. Hovermale, R. A., P. G. Sears, et al., *J. Chem. Eng. Data* **8**, 490 (1963).

2264. Hovermale, R. A. and P. G. Sears, *J. Phys. Chem.* **60**, 1579 (1956).

2265. Hovorka, F. and D. Dreisbach, *J. Amer. Chem. Soc.* **56**, 1664 (1934).

2266. Hovorka, F. and F. E. Geiger, *J. Amer. Chem. Soc.* **55**, 4759 (1933).

2267. Hovorka, F. and H. P. Lankelma, et al., *J. Amer. Chem. Soc.* **62**, 2372 (1940).

2268. Hovorka, F., H. P. Lankelma, et al., *J. Amer. Chem. Soc.* **60**, 820 (1938).

2269. Howard, F. L., T. W. Mears, et al., *J. Res. Nat. Bur. Stand.* **38**, 365 (1947).

2270. Howell, O. W., *Proc. Roy. Soc.* **137A**, 418 (1932).

2271. Howlett, K. E., *J. Chem. Soc.* **1951**, 1409.

2272. Howlett, K. E., *J. Chem. Soc.* **1955**, 1784.

2273. Howlett, L. E., *Can. J. Res.* **4**, 79 (1931); *Chem. Zentr.* **1931**, I, 3437.

2274. Hoyer, H., *Z. Phys. Chem.* (Leipzig) **45B**, 389 (1940); *C. A.* **34**, 4990 (1940).

2275. Hoyer, H., *Z. Elektrochem.* **47**, 451 (1941); *C. A.* **36**, 4415 (1942).

2276. Hoyt, C. W. and C. K. Fink, *J. Phys. Chem.* **41**, 453 (1937).

2277. Hoyt, H. E., U.S. Patent 2,381,055, Aug. 7, 1945; *C. A.* **39**, 5106 (1945).

2278. Hruska, F., E. Bock, et al., *Can. J. Chem.* **41**, 3034 (1963).

2279. Hrynakowski, K. and A. Smoczkiewiczowa, *Rocz. Chem.* **17**, 140 (1937); *C. A.* **31**, 4883 (1937).

2280. Hrynakowski, K. and A. Smoczkiewiczowa, *Rocz. Chem.* **17**, 165 (1937); *C. A.* **31**, 6096 (1937).

2281. Hrynakowski, K. and M. Szmyt, *Z. Phys. Chem.* (Leipzig) **182A**, 405 (1938).

2282. Hsia, A. W., *Z. Tech. Phys.* **12**, 550 (1931).

2283. Huang, T. C. and K. P. Sung, *Science Repts. Natl. Tsinghua Univ.* (A) **2**, 303 (1934); Beil **EIII6**, 1454.

2284. Hubbard, W. N., F. R. Frow, et al., *J. Phys. Chem.* **65**, 1326 (1961).

2285. Hubbard, W. N., W. O. Good, et al., *J. Phys. Chem.* **62**, 614 (1958).

2286. Hubbard, W. N., C. Katz, et al., *J. Phys. Chem.* **58**, 142 (1954).

2287. Hubbard, W. N., J. W. Knowlton, et al., *J. Phys. Chem.* **58**, 396 (1954).

2288. Hubbard, W. N., D. W. Scott, et al., *J. Amer. Chem. Soc.* **77**, 5857 (1955).

2289. Hückel, W. and P. Ackermann, *J. Pr. Chem.* **136**, 15 (1933); *Brit. Chem. Abs.* **1933A**, 372.

2290. Hückel, W., J. Datow, et al., *Z. Phys. Chem.* (Leipzig) **A186**, 129 (1940); *C. A.* **35**, 1688 (1941).

2291. Hückel, W. and K. Hagenguth, *Ber. Deut. Chem. Gesell.* **64B**, 2892 (1931).

2292. Hückel, W. and H. Harder, *Chem. Ber.* **80**, 357 (1947).

2293. Hückel, W. and A. Hubele, *Ann. Chem.* **613**, 27 (1958); *Chem. Zentr.* **1959**, 6411.

2294. Hückel, W., K. Kumetat, et al., *Ann. Chem.* **517**, 184 (1935); *C. A.* **29**, 6126 (1935).

2295. Hückel, W. and J. Kurz, *Chem. Ber.* **91**, 1290 (1958).

2296. Hudson, G. H., J. C. McCoubrey, et al., *Trans. Faraday Soc.* **56**, 1144 (1960).

2297. Huff, W. J., *U.S. Bur. Mines, Rept. Invest.* **3794**, (1945); *C. A.* **39**, 1291 (1945).

2298. Huff, W. J., *U.S. Bur. Mines, Rept. Invest.* **4031**, (1946); *C. A.* **41**, 6719 (1947).

2299. Huffman, H. M., M. E. Gross, et al., *J. Phys. Chem.* **65**, 495 (1961).

2300. Huffman, H. M., G. S. Parks, et al., *J. Amer. Chem. Soc.* **52**, 1547 (1930).

2301. Huffman, H. M., G. S. Parks, et al., *J. Amer. Chem. Soc.* **52**, 3241 (1930).

2302. Huffman, H. M., G. S. Parks, et al., *J. Amer. Chem. Soc.* **53**, 3876 (1931).

2303. Huffman, H. M., S. S. Todd, et al., *J. Amer. Chem. Soc.* **71**, 584 (1949).

2304. Huggins, C. M., G. C. Pimentel, et al., *J. Phys. Chem.* **60**, 1311 (1956).

2305. Hughes, E. C. and J. R. Johnson, *J. Amer. Chem. Soc.* **53**, 737 (1931).

2306. Hughes, E. D. and U. G. Shapiro, *J. Chem. Soc.* **1937**, 1177.

2307. Hughes, O. L. and H. Hartley, *Phil. Mag.* **15**, 610 (1933).

2308. Hughes, S. R. C., *J. Chem. Soc.* **1957**, 634.

2309. Huisgen, R. and H. Brade, *Chem. Ber.* **90**, 1432 (1957).

2310. Huisgen, R. and H. Walz, *Chem. Ber.* **89**, 2616 (1956).

2311. Huisgen, R., H. Brade, et al., *Chem. Ber.* **90**, 1437 (1957).

2312. Huisman, J. and B. H. Sage, *J. Chem. Eng. Data* **9**, 223 (1964).

2313. Hukumoto, Y., *J. Chem. Phys.* **3**, 165 (1935).

2314. Hukumoto, Y., *Sci. Rep. Tohoku Imp. Univ.* (1) **22**, 23 (1933); *Chem. Zentr.* **1933**, II, 1968.

2315. Hukumoto, Y., *Sci. Rep. Tohoku Imp. Univ. Ser. I* **25**, 1162 (1937); *Chem. Zentr.* **1937**, II, 1548.

2316. Hull, D. C. and A. H. Agett, U.S. Patent 2,422,016, June 10, 1947; *C. A.* **41**, 6275 (1947).

2317. Hull, G. F., *J. Chem. Phys.* **3**, 534 (1935).

2318. Humphries, C. M., A. D. Walsh, et al., *Trans. Faraday Soc.* **63**, 513 (1967).

2319. Hunecke, H., *Ber. Deut. Chem. Gesell.* **60**, 1451 (1927).

2320. Hünig, S., *Ann. Chem.* **569**, 198 (1950); *C. A.* **45**, 7951 (1951).

2321. Hunt, H. and H. T. Briscoe, *J. Phys. Chem.* **33**, 1495 (1929).

2322. Hunt, H. D. and W. T. Simpson, *J. Amer. Chem. Soc.* **75**, 4540 (1953).

2323. Hunter, E. C. E. and J. R. Partington, *J. Chem. Soc.* **1932**, 2812.

2324. Hunter, E. C. E., and J. R. Partington, *J. Chem. Soc.* **1933**, 309.

2325. Hunter, W., British Patent 606,608, Aug. 17, 1948; *C. A.* **43**, 4692 (1949).

2326. Hurd, A. D. and D. G. Botteron, *J. Amer. Chem. Soc.* **68**, 1200 (1946).

2327. Hurd, C. D. and J. S. Strong, *Anal. Chem.* **23**, 542 (1951).

2328. Hurdis, E. C. and C. P. Smyth, *J. Amer. Chem. Soc.* **64**, 2212 (1942).

2329. Hurdis, E. C. and C. P. Smyth, *J. Amer. Chem. Soc.* **64**, 2829 (1942).

2330. Hurdis, E. C. and C. P. Smyth, *J. Amer. Chem. Soc.* **65**, 89 (1943).

2331. Huston, R. C. and C. O. Bastwick, *J. Org. Chem.* **13**, 331 (1948).

2332. Huston, R. C., W. B. Fox, et al., *J. Org. Chem.* **3**, 251 (1938–1939).

2333. Huston, R. C. and I. A. Kaye, *J. Amer. Chem. Soc.* **64**, 1576 (1942).

2334. Hutchinson, A. W. and G. C. Chandler, *J. Amer. Chem. Soc.* **53**, 2881 (1931).

2335. Hutchinson, M. H. and L. E. Sutton, *J. Chem. Soc.* **1958**, 4382.

2336. Hüttner, K. and G. Tammann, *Z. Anorg. Chem.* **43**, 218 (1905).
2337. Hygienic Guide Series, *Am. Ind. Hyg. Assoc. J.* **26**, 196 (1965).
2338. Hygienic Guide Series, *Am. Ind. Hyg. Assoc. J.* **27**, 571 (1966).

I

2339. Ievskaya, N. M. and R. M. Umarkhodzhaev, *Paramagnitn. Rezonans, Kazansk. Univ., Sbornik* **1960** 137; *C. A.* **56**, 8197 (1962).
2340. Imanishi, S. and Y. Kanda, *J. Sci. Research Inst.* (Tokyo) **43**, 1 (1949); *C. A.* **43**, 8885 (1949).
2341. Imanov, L. M., *Dokl. Akad. Nauk Azerbaĭdzhan SSR* **12**, 531 (1956); *C. A.* **51**, 2345 (1957).
2342. Imbaud, J. P. and G. Berthet, *Compt. Rend.* **261**, 953 (1965); *C. A.* **64**, 190 (1966).
2343. Indest, H., H. Massat, et al., German Patent 1,111,193 (Cl. 12p) Dec. 20, 1955; *C. A.* **56**, 2338 (1962).
2344. Indest, H., H. Massat, et al., German Patent 1,139,122 (Cl. 12p), Nov. 8, 1962; *C. A.* **58**, 5840 (1963).
2345. Ingersoll, L. R., *J. Opt. Soc. Amer.* **6**, 663 (1922).
2346. Ingle, J. D. and H. P. Cady, *J. Phys. Chem.* **42**, 397 (1938).
2347. Ingold, C. K. and W. J. Powell, *J. Chem. Soc.* **1921**, 1222.
2348. Ingold, E. H. and C. K. Ingold, *J. Chem. Soc.* **1932**, 756.
2349. *International Critical Tables*, McGraw-Hill, New York, 1930.
2350. Ioffe, B. V., *Zh. Obsch. Khim.* **25**, 902 (1955); *C. A.* **49**, 13,717 (1955).
2351. Ioffe, I. I. and E. S. Yampol'skaya, *J. Appl. Chem.* (USSR) **17**, 527 (1944); *C. A.* **39**, 3986 (1945).
2352. Ipatieff, V. N. and B. B. Corson, *Ind. Eng. Chem.* **30**, 1039 (1938).
2353. Ipatieff, V. and A. V. Grosse, U.S. Patent 2,104,424 (Cl 260-168), Jan. 4, 1938.
2354. Ipatiew, V., B. Dolgow, et al., *Ber. Deut. Chem. Gesell.* **63**, 3072 (1930).
2355. Ipatiew, W., *Ber. Deut. Chem. Gesell.* **40**, 1281 (1907).
2356. Iredale, T. and A. G. Mills, *Nature* **126**, 604 (1930); *Chem. Zentr.* **1931**, I, 23.
2357. Ishida, K., *Bunko Kenkyu* **9**, 27 (1960); *C. A.* **57**, 15,808 (1962).
2358. Ishihara, M. and R. Oda, *J. Soc. Chem. Ind. Jap.* **46**, 1268 (1943); *C. A.* **42**, 6333 (1948).
2359. Issoire, J., *Mém. Poudres* **42**, 333 (1960); *C. A.* **55**, 14421 (1961).
2360. Itakura, T., *J. Chem. Soc. Jap.* **63**, 1400 (1942); *C. A.* **41**, 3041 (1947).
2361. Ito, M., *J. Mol. Spectrosc.* **4**, 125 (1960); *C. A.* **55**, 13,048 (1961).
2362. Ito, S. and I. Miura, *Bull. Chem. Soc. Jap.* **38**, 2197 (1965).
2363. Ivanchev, S. S., A. I. Yurzhenko, et al., *Zh. Fiz. Khim.* **39**, 1900 (1965); *C. A.* **64**, 166 (1966).
2364. Ivanov, V. A., *Tr. Voronezhsk Gos. Med. Inst.* **29**, 23 (1957).
2365. Ivanskii, V. I. and B. N. Dolgov, *Kinetika i Kataliz* **4**, 165 (1963); *C. A.* **59**, 538 (1963).
2366. Iwai, I., H. Shindo, et al., *Yakugaku Zasshi* **80**, 1588 (1960); cf. *Chem. Pharm. Bull.* **8**, 815 (1960); *C. A.* **55**, 10,063 (1961).
2367. Izergin, A. P., *Izv. Vyssh. Ucheb. Zaved. Fiz.* **1958**, No. 5, 115; *C. A.* **53**, 6719 (1959); Ref. [5181].

J

2368. Jackman, L. M., A. K. Macbeth, et al., *J. Chem. Soc.* **1949**, 1717.
2369. Jackson, L. C., *Phil. Mag.* **43**, 481 (1922).
2370. Jacobs, C. J. and G. S. Parks, *J. Amer. Chem. Soc.* **56**, 1513 (1934).

2371. Jacobs, T. L., J. D. Roberts, et al., *J. Amer. Chem. Soc.* **66**, 656 (1944).
2372. Jaeger, F. M., *Z. Anorg. Chem.* **101**, 1 (1917); *Chem. Zentr.* **1918**, I, 324; Beil **EI6**, 169.
2373. Jaffé, G., *Ann. Phys.* (Leipzig) **28**, 326 (1909).
2374. Jakobsen, R. J., *U.S. Dept. Commerce, Office Tech. Serv. PB Rept.* **171,342** (1960); *C. A.* **58**, 4055 (1963).
2375. Jander, G. and G. Winkler, *J. Inorg. Nucl. Chem.* **9**, 24, 32, 39 (1959); *C. A.* **53**, 8775 (1959).
2376. Janz, G. J. and S. S. Danyluk, *J. Amer. Chem. Soc.* **81**, 3846 (1959).
2377. Jasper, J. J. and E. R. Kerr, *J. Amer. Chem. Soc.* **76**, 2659 (1954).
2378. Jasper, J. J., R. E. Kerr, et al., *J. Amer. Chem. Soc.* **75**, 5252 (1953).
2379. Jasper, J. J. and E. V. Kring, *J. Phys. Chem.* **59**, 1019 (1955).
2380. Jasper, J. J. and H. L. Wedlick, *J. Chem. Eng. Data* **9**, 446 (1964).
2381. Jatkar, S. K. K. and C. M. Deshpande, *J. Indian Chem. Soc.* **37**, 1 (1960); *C. A.* **54**, 18,000 (1960).
2382. Jatkar, S. K. K. and C. M. Deshpande, *J. Indian Chem. Soc.* **37**, 15 (1960); *C. A.* **54**, 18,001 (1960).
2383. Jatkar, S. K. K. and C. M. Deshpande, *J. Univ. Poona Sci. and Technol.* **1954**, 48; *C. A.* **49**, 15,316 (1955).
2384. Jatkar, S. K. K. and D. Lakshminarayanan, *J. Indian Inst. Sci.* **28A**, 1 (1946); *C. A.* **41**, 1901 (1947).
2385. Jatkar, S. K. K. and R. Padmanabhan, *Indian J. Phys.* **10**, 55 (1936); *C. A.* **30**, 6284 (1936).
2385a. *J. Chem. Soc.* **1964**, 2991.
2386. Jeffery, G. H. and A. I. Vogel, *J. Chem. Soc.* **1948**, 658.
2387. Jeffery, G. H. and A. I. Vogel, *J. Chem. Soc.* **1948**, 674.
2388. Jelatis, J. G., *J. Appl. Phys.* **19**, 419 (1948).
2389. Jen, T.-Ko., *Union Ind. Res. Inst. Rept.* No. **17** (Taiwan), (1956); *C. A.* **55**, 22,116 (1961).
2390. Jen, T.-Ko, *Union Ind. Res. Inst. Rept.* No. **35** (Taiwan), (1958).
2391. Jenkin, C. F., *Trans. Faraday Soc.* **18**, 197 (1922).
2392. Jenner, P. M., E C. Hagan, et al., *Food Cosmet. Toxicol.* **2**, 327 (1964); *C. A.* **62**, 4516 (1965).
2393. Jennings, A. L. and J. E. Boggs, *J. Org. Chem.* **29**, 2065 (1964).
2394. Jenny, R., *Compt. Rend.* **246**, 3477 (1958); *C. A.* **53**, 2122 (1959).
2395. Jensen, F. R., D. S. Noyce, et al., *J. Amer. Chem. Soc.* **84**, 386 (1962).
2396. Jensen, O. and R. A. Gartner, *J. Phys. Chem.* **36**, 3138 (1932).
2397. Jesse, R. H., *J. Amer. Chem. Soc.* **34**, 1337 (1912).
2398. Jessup, R. S., *J. Res. Nat. Bur. Stand.* **18**, 115 (1937).
2399. Jhon, M. S., E. R. Van Artsdalen, et al., *J. Chem. Phys.* **47**, 2231 (1967).
2400. Jockmann, W., *Arch. Toxikol.* **18**, 698 (1961); *C. A.* **56**, 10,497 (1962).
2401. Joglekar, M. S. and V. N. Thatte, *Z. Phys.* **98**, 692 (1936); *Chem. Zentr.* **1937**, I, 568.
2402. Johns, I. B., E. A. McElhill, et al., *J. Chem. Eng. Data* **7**, 277 (1962).
2403. Johnsen, S. E. J., *Anal. Chem.* **19**, 305 (1947).
2404. Johnson, A. W., *J. Chem. Soc.* **1946**, 1014.
2405. Johnson, E. L., K. H. Pool, et al., *Anal. Chem.* **38**, 183 (1966).
2406. Johnson, H. E., U.S. Patent 2,459,432, Jan. 18, 1949; *C. A.* **43**, 2476 (1949).
2407. Johnson, J. F. and R. L. LeTourneau, *J. Amer. Chem. Soc.* **75**, 1743 (1953).
2408. Johnson, J. R., *Organic Syntheses*, Vol. 16, Wiley, New York, 1936.
2409. Johnson, M. D., *J. Chem. Soc.* **1965**, 805.

2410. Johnson, R. D., R. J. Myers, et al., *J. Chem. Phys.* **21**, 1425 (1953).
2411. Johnson, W. H., E. J. Prosen, et al., *J. Res. Nat. Bur. Stand.* **37**, 51 (1946).
2412. Johnson, W. H., E. J. Prosen, et al., *J. Res. Nat. Bur. Stand.* **38**, 419 (1947).
2413. Johnson, W. H., E. J. Prosen, et al., *J. Res. Nat. Bur. Stand.* **39**, 49 (1947).
2414. Johnstone, R. T., *A. M. A. Arch. Ind. Health* **20**, 445 (1959); *C. A.* **54**, 9130 (1960).
2415. Jonathan, N., S. Gordon, et al., *J. Chem. Phys.* **36**, 2443 (1962).
2416. Joncich, M. J. and D. R. Bailey, *Anal. Chem.* **32**, 1578 (1960).
2417. Jones, D. C., *J. Chem. Soc.* **1929**, 799.
2418. Jones, D. C. and H. F. Betts, *J. Chem. Soc.* **1928**, 1177.
2419. Jones, D. G., British Patent 699,079, Oct. 28, 1953; *C. A.* **49**, 3262 (1955).
2420. Jones, E. C. S. and J. Kenner, *J. Chem. Soc.* **1932**, 711.
2421. Jones, E. R., *J. Phys. Chem.* **31**, 1316 (1927).
2422. Jones, G. and S. M. Christian, *J. Amer. Chem. Soc.* **61**, 82 (1939).
2423. Jones, G. and H. J. Fornwalt, *J. Amer. Chem. Soc.* **60**, 1683 (1938).
2424. Jones, G. T., M. Randic, et al., *Croat. Chem. Acta* **36**, 111 (1964); *C. A.* **62**, 8541 (1965).
2425. Jones, G. W. and G. S. Scott, *U.S. Bur. Mines, Rept. Invest.* **3881**, 5 pp. (1946); *C. A.* **40**, 5919 (1946).
2426. Jones, H. C., *Z. Phys. Chem.* (Leipzig) **31**, 114 (1899).
2427. Jones, J. R. and C. B. Monk, *J. Chem. Soc.* **1963**, 2633.
2428. Jones, L. C. and L. W. Taylor, *Anal. Chem.* **27**, 228 (1955).
2429. Jones, L. C. and R. A. Friedel, *Ind. Eng. Chem., Anal. ed.* **17**, 349 (1945).
2430. Jones, L. R. and J. A. Riddick, *Anal. Chem.* **23**, 349 (1951); **24**, 1533 (1952); **28**, 1137, 1493 (1956).
2431. Jones, M. and A. Lapworth, *J. Chem. Soc.* **1914**, 1804.
2432. Jones, R. A. Y., A. R. Katritzky, et al., *J. Chem. Soc.* **1962**, 2576.
2433. Jones, R. L. *J. Mol. Spectrosc.* **11**, 411 (1963); *C. A.* **60**, 3620 (1964).
2434. Jones, R. L., *Spectrochim. Acta* **22**, 1555 (1966); *C. A.* **65**, 16,827 (1966).
2435. Jones, R. N., *Nat. Res. Council Bull.* (Canada) No. **5** (1957); *C. A.* **52**, 3516 (1958).
2436. Jones, R. N., J. B. DiGiorgio, et al., *J. Org. Chem.* **30**, 1822 (1965).
2437. Jones, R. N., A. F. McKay, et al., *J. Amer. Chem. Soc.* **74**, 2575 (1952).
2438. Jones, R. N. and K. Noack, *Can. J. Chem.* **39**, 2214 (1961).
2439. Jones, R. N. and G. D. Thorn, *Can. J. Res.* **27B**, 580 (1949); *C. A.* **43**, 8273 (1949).
2440. Jones, W. J., S. T. Bowden, et al., *J. Phys. Colloid Chem.* **52**, 753 (1948).
2441. Jones, W. J. and K. J. P. Orton, *J. Chem. Soc.* **1909**, 1056.
2442. Jones, W. M. and W. F. Giauque, *J. Amer. Chem. Soc.* **69**, 983 (1947).
2443. Joop, N. and H. Zimmerman, *Z. Elektrochem.* **66**, 440 (1962); *C. A.* **57**, 10,681 (1962).
2444. Jordan, T. E., *Vapor Pressure of Organic Compounds*, Interscience, New York, 1950.
2445. Joris, G. G. and H. S. Taylor, *J. Chem. Phys.* **16**, 45 (1948).
2446. Joris, G. G. and J. Vitrone, U.S. Patent 2,829,166, Apr. 23, 1956.
2447. Josefowicz, E., *Rocz. Chem.* **18**, 577 (1938).
2448. Joshi, D. V. and J. R. Merchant, *J. Sci. Ind. Res.* (India) **14B**, 482 (1955); *C. A.* **50**, 11,272 (1956).
2449. Joshi, G. J., *Indian J. Pure Appl. Phys.* **4**, 40 (1966); *C. A.* **64**, 16,843 (1966).
2450. Josien, M. L., *Compt. Rend.* **237**, 175 (1953); *C. A.* **48**, 4977 (1954).
2451. Josien, M. L. and J. M. Lebas, *Bull. Soc. Chim. Fr.* **1956**, 53; *C. A.* **50**, 6920 (1956).
2452. Jost, F. and M. Harrand, *J. Phys. Radium* **23**, 308 (1962); *C. A.* **57**, 8085 (1962).
2453. Joukovsky, I., *Bull. Soc. Chim. Belges* **43**, 397 (1934); *Chem. Zentr.* **1935**, I, 3126.
2454. Jouve, P., *Ann. Phys.* (Paris) **1**, 127 (1966); *C. A.* **65**, 1633 (1966).
2454a. *J. Pharm. Chem.* **23**, 230 (1906).

2455. Ju, T. Y., C. E. Wood, et al., *J. Inst. Petrol.* **28**, 159 (1942); *C. A.* **37**, 4886 (1943).
2456. Judson, C. M. and M. Kilpatrick, *J. Amer. Chem. Soc.* **71**, 3110 (1949).
2457. Jullander, I. and K. Brune, *Acta Chem. Scand.* **2**, 204 (1948); *C. A.* **43**, 4184 (1949).
2458. Junell, R., *Ark. Kem., Mineral. Geol., Ser.* **B11**, No. 30, 6/3 (1934); *Chem. Zentr.* **1934, I,** 3727.
2459. Junell, R., *Svensk Kem. Tidskr.* **46**, 125 (1934); Dissertation, Univ. of Uppsala, 1935.
2460. Junell, R., *Z. Phys. Chem.* (Leipzig) **141A,** 71 (1929).
2461. Jungnickel, J. L. and J. W. Forbes, *Anal. Chem.* **35**, 938 (1963).

K

2462. Kablukov, I. A. and V. T. Malischeva, *J. Amer. Chem. Soc.* **47**, 1553 (1925).
2463. Kablukow, I. A. and F. M. Perelman, *Dokl. Akad. Nauk SSSR* **1930**, 519; *Chem. Zentr.* **1931, II,** 23.
2464. Kabo, G. Y. and D. N. Andreevskii, *Izv. Vyssh. Ucheb. Zaved., Khim. Khim. Tekhnol.* **8**, 574 (1965); *C. A.* **64**, 4343 (1966).
2465. Kagarise, R. E. and L. W. Daasch, *J. Chem. Phys.* **23**, 113 (1955).
2466. Kahane, E., *Compt. Rend.* **227**, 841 (1948); *C. A.* **43**, 1982i (1949).
2467. Kahlbaum, G. W. A., *Ber. Deut. Chem. Gesell.* **13**, 2348 (1880).
2468. Kahlbaum, G. W. A., *Ber. Deut. Chem. Gesell.* **18**, 2108 (1885).
2469. Kahlbaum, G. W. A., *Z. Phys. Chem.* (Leipzig) **26**, 577 (1898).
2470. Kahlenberg, L., *J. Phys. Chem.* **5**, 215, 284 (1901); *J. Chem. Soc.* **1901Aii**, 492 (1902), **Aii,** 195.
2471. Kahlenberg, L., *J. Phys. Chem.* **6**, 45 (1902); *J. Chem. Soc.* **1902Aii**, 310.
2472. Kahlenberg, L. and A. T. Lincoln, *J. Phys. Chem.* **3**, 12 (1899); *I. C. T.* **6**, 144.
2473. Kahovec, L., *Z. Phys. Chem.* (Leipzig) **40B,** 135 (1938); *C. A.* **32**, 6549 (1938).
2474. Kahovec, L., *Z. Phys. Chem.* (Leipzig) **43B,** 109 (1939); *C. A.* **33**, 6718 (1939).
2475. Kahovec, L. and K. W. F. Kohlrausch, *Monatsh. Chem.* **68**, 359 (1936); *C. A.* **31,** 36 (1937).
2476. Kahovec, L. and K. W. F. Kohlrausch, *Z. Phys. Chem.* (Leipzig) **35B,** 29 (1937); *C. A.* **31**, 2931 (1937).
2477. Kahovec, L. and K. W. F. Kohlrausch, *Z. Phys. Chem.* (Leipzig) **46B,** 165 (1940); *C. A.* **35**, 977 (1941).
2478. Kahovec, L. and A. W. Reitz, *Monatsh. Chem.* **69**, 363 (1936); *C. A.* **31**, 2515 (1937).
2479. Kahovec, L. and J. Wagner, *Z. Phys. Chem.* (Leipzig) **47B,** 48 (1940); *C. A.* **35**, 977 (1941).
2480. Kahovec, L. and H. Wassmuth, *Z. Phys. Chem.* (Leipzig) **48B,** 70 (1940); *C. A.* **35**, 3172 (1941).
2481. Kailan, A. and K. Melkus, *Monatsh. Chem.* **48**, 9 (1927).
2482. Kakac, B. and M. Hudlichy, *Collect. Czech. Chem. Commun.* **30**, 745 (1965); *C. A.* **63**, 1343 (1965).
2483. Kambara, S., *J. Soc. Chem. Ind. Jap.* **47**, 518 (1944); *C. A.* **43**, 1594 (1949).
2484. Kamei, H., *Bull. Chem. Soc. Jap.* **38**(7), 1212 (1965); *C. A.* **63**, 12,539 (1965).
2485. Kamei, H., *Jap. J. Appl. Phys.* **4**, 212 (1965); *C. A.* **63**, 2543 (1965).
2486. Kameo, T., T. Hirashima, et al., *Kagaku To Kogyo* (Osaka) **40**, 135 (1966); *C. A.* **65,** 10,466 (1966).
2487. Kamiyoshi, K. and T. Fujimura, *J. Phys. Radium* **23**, 311 (1962); *C. A.* **57**, 9322 (1962).
2488. Kamm, O., *Qualitative Organic Analysis*, 2d ed., Wiley, New York, 1932.
2489. Kamm, O. and C. S. Marvel, *J. Amer. Chem. Soc.* **42**, 299 (1920).
2490. Kamm, O. and J. H. Waldo, *J. Amer. Chem. Soc.* **43**, 2225 (1941).

2491. Kamm, W. F. and A. O. Mathews, *J. Amer. Pharm. Assoc.* **11**, 599 (1922).
2492. Kanbayashi, U. and K. Nukada, *Nippon Kagaku Zasshi* **84**, 297 (1963).
2493. Kanerevskaya, A. A., *Prom. Toksikol. Moscow Sb.* **1960**, 15; *C. A.* **57**, 6270 (1962).
2494. Kanomata, I., *Bull. Chem. Soc. Jap.* **34**, 1596 (1961); *C. A.* **57**, 239 (1962).
2495. Kantor, S. W. and C. R. Hauser, *J. Amer. Chem. Soc.* **75**, 1744 (1953).
2496. Kapff, S. F. and R. B. Jacobs, *Rev. Sci. Instrum.* **18**, 581 (1947).
2497. Kapustin, A. P., *J. Exptl. Theoret. Phys.* (USSR) **17**, 30 (1947); *C. A.* **42**, 1092 (1948).
2498. Karabinos, J. V., *Science Counselor* **17**, 3 (1954); *C. A.* **49**, 4508 (1955).
2499. Karabinos, J. V. and K. T. Serijan, *J. Amer. Chem. Soc.* **67**, 1856 (1945).
2499a. Karash, N., *Organic Sulfur Compounds*, Pergamon, New York, 1961.
2500. Karnatz, F. A. and F. C. Whitmore, *J. Amer. Chem. Soc.* **60**, 3082 (1938).
2501. Karpov, B. D., *Farmakol. Toksikol.* **17**, No. 1, 49 (1954); *C. A.* **48**, 13,984 (1954).
2502. Karpov, B. D., *Gig. Sanit.* **1955**, No. 8, 19; *C. A.* **50**, 2199 (1956).
2503. Karr, A. E., W. M. Bowes, et al., *Anal. Chem.* **23**, 459 (1951).
2504. Karvonen, A., *Ann. Acad. Sci. Fennicae* **20**, No. 9 (1924); *C. A.* **18**, 1981 (1924).
2505. Karyakin, A. V. and A. F. Petrov, *Zh. Anal. Khim.* **19**, 1234 (1964); *C. A.* **62**, 3405 (1965).
2506. Kashin, L. M., *Gig. Sanit.* **30**, 23 (1965); *C. A.* **63**, 7558 (1965).
2507. Kassel, L. S., *J. Amer. Chem. Soc.* **58**, 670 (1936).
2508. Katagiri, S., *Sci. Repts. Tohoku Univ. First Ser.* **44**, 165 (1960); *C. A.* **55**, 25,406 (1961).
2509. Katasura, S., *J. Chem. Soc. Jap.* **63**, 1483 (1942); *C. A.* **41**, 3449 (1947).
2510. Katayama, M. and Y. Morino, *Repts. Radiation Chem. Res. Inst. Tokyo Univ.* No. 4, 1 (1949).
2510a. Katayama, M., *Sci. Repts Tohoku Imp. Univ.* **4**, 373 (1916); *C. A.* **10**, 994 (1916).
2511. Kato, J. and M. Sakuma, *Denki Kagaku* **25**, 126 (1957); *C. A.* **52**, 4468 (1958).
2512. Kato, J., M. Sakuma, et al., *Denki Kagaku* **25**, 331 (1957); *C. A.* **52**, 4469 (1958).
2513. Kato, S. and F. Someno, *Sci. Papers Inst. Phys. Chem. Research* (Tokyo) **34**, 905 (1938); *C. A.* **32**, 8267 (1938).
2514. Kato, Y., *J. Chem. Phys.* **44**, 2824 (1966).
2515. Katon, J. E., W. R. Feairheller, et al., *Anal. Chem.* **36**, 2126 (1964).
2516. Katon, J. E., W. R. Feairheller, et al., *J. Mol. Spectrosc.* **13**, 72 (1964); *C. A.* **60**, 15316 (1964).
2517. Katritzky, A. R. and N. A. Coats, *J. Chem. Soc.* **1959**, 2062.
2518. Katritzky, A. R. and R. A. Jones, *J. Chem. Soc.* **1959**, 3670.
2519. Katritzky, A. R. and J. M. Lagowski, *J. Chem. Soc.* **1958**, 4155.
2520. Katritzky, A. R. and P. Simmons, *J. Chem. Soc.* **1959**, 2051.
2521. Katritzky, A. R. and P. Simmons, *J. Chem. Soc.* **1959**, 2058.
2522. Katsuno, M., *J. Soc. Chem. Ind. Jap.* **41**, 75 (1938); *C. A.* **32**, 5372 (1938).
2523. Katsuno, M., *J. Soc. Chem. Ind. Jap.* **44**, 898 (1941); *C. A.* **42**, 2575 (1948).
2524. Katsuno, M., *J. Soc. Chem. Ind. Jap.* **44**, 903 (1941); *C. A.* **42**, 2575 (1948).
2525. Katsuno, M., *J. Soc. Chem. Ind. Jap.* **46**, 1028 (1943); *C. A.* **42**, 7285 (1948).
2526. Katti, P. K. and M. M. Chaudhri, *J. Chem. Eng. Data* **9**, 128 (1964).
2527. Katti, P. K. and M. M. Chaudhri, *J. Chem. Eng. Data* **9**, 442 (1964).
2528. Kauck, E. A. and A. R. Diesslin, *Ind. Eng. Chem.* **43**, 2332 (1951).
2529. Kauer, E., L. Grote, et al., *Z. Phys. Chem.* (Leipzig) **232**, 356 (1966); *Wiss. Z. Tech. Hochsch. Chem., Leuna-Merseberg* **7**, (2) 73 (1965); *C. A.* **63**, 17,222 (1965).
2530. Kaufmann, W. E. and R. Adams, *J. Amer. Chem. Soc.* **45**, 3029 (1923).
2531. Kawazoe, Y. and M. Ohnishi, *Chem. Pharm. Bull.* (Tokyo) **11**, 243 (1963); *C. A.* **59**, 14,774 (1963).

2532. Kay, R. L., G. A. Vidulich, et al., *J. Chem. Phys.* **47**, 866 (1967).
2533. Kay, W. B., *J. Amer. Chem. Soc.* **68**, 1336 (1946).
2534. Kay, W. B., *J. Amer. Chem. Soc.* **69**, 1273 (1947).
2535. Kay, W. B. and F. M. Warzel, *Ind. Eng. Chem.* **43**, 1150 (1951).
2536. Kaye, W., *Spectrochim. Acta* **6**, 257 (1954); *C. A.* **49**, 769 (1955).
2537. Kay-Fries Chemicals, Inc., "Technical Data," New York.
2538. Kazanskiĭ, B. A., A. L. Liberman, et al., *J. Gen. Chem.* (USSR) **17**, 1503 (1947); *C. A.* **42**, 2225 (1948).
2539. Keading, W. W., R. O. Lindblom, et al., *Ind. Eng. Chem.* **53**, 805 (1961).
2540. Kecki, Z., *Spectrochim. Acta* **18**, 1155 (1962); *C. A.* **58**, 5162 (1963).
2541. Kecki, Z. and H. Wincel, *Nukleonika* **8** (2), 117 (1963); *C. A.* **60**, 3595 (1964).
2542. Keffler, L., *Bull. Soc. Chim. Belges* **44**, 429 (1935); *C. A.* **30**, 719 (1936).
2543. Keffler, L. and J. H. McLean, *J. Chem. Soc. Ind.* (*London*) **54**, 178T (1935).
2544. Keffler, L. J. P., *J. Phys. Chem.* **34**, 1319 (1930).)
2545. Keicher, G. and O. Klopfer, German Patent 943,866, Jan. 1, 1956; *Chem. Zentr.* **1956**, 13,257.
2546. Keller, H. and H. v. Halban, *Helv. Chim. Acta* **27**, 1439 (1944).
2547. Kelley, K. K., *J. Amer. Chem. Soc.* **51**, 180 (1929).
2548. Kelley, K. K., *J. Amer. Chem. Soc.* **51**, 779 (1929).
2549. Kelley, K. K., *J. Amer. Chem. Soc.* **51**, 1145 (1929).
2550. Kelley, K. K., *J. Amer. Chem. Soc.* **51**, 1400 (1929).
2551. Kelso, E. A. and W. A. Felsing, *J. Amer. Chem. Soc.* **62**, 3132 (1940).
2552. Kempa, R. and W. H. Lee, *J. Chem. Soc.* **1958**, 1936.
2553. Kempa, R. F. and W. H. Lee, *Talanta* **9**, 325 (1962); *C. A.* **56**, 13,630 (1962).
2554. Kempter, H., *Z. Phys.* **116**, 1 (1940); *C. A.* **35**, 3170 (1941).
2555. Kendall, J. and J. J. Beaver, *J. Amer. Chem. Soc.* **43**, 1853 (1921).
2556. Kendall, J. and J. E. Booge, *J. Amer. Chem. Soc.* **38**, 1712 (1916).
2557. Kendall, J. and P. M. Gross, *J. Amer. Chem. Soc.* **43**, 1426 (1921).
2558. Kendall, J. and A. H. Wright, *J. Amer. Chem. Soc.* **42**, 1776 (1920).
2559. Kennedy, H. and J. Lielmezs, *Ind. Eng. Chem., Fundamentals* **6**, 310 (1967).
2560. Kenyon, J., H Phillips, et al., *J. Chem. Soc.* **1935**, 1072.
2561. Kenyon, J. and H. E. Strauss, *J. Chem. Soc.* **1949**, 2153.
2562. Kepner, R. E. and L. J. Andrews, *J. Org. Chem.* **13**, 208 (1948).
2563. Keswani, R. and H. Freiser, *J. Amer. Chem. Soc.* **71**, 218 (1949).
2564. Ketelaar, J. A. A., P. F. Velden, et al., *Rec. Trav. Chim. Pays-Bas* **66**, 721 (1947); *C. A.* **42**, 3631 (1948); Beil **EIII1**, 651 ff.
2565. Ketelaar, J. A. A., L. de Vries, et al., *Rec. Trav. Chim.* **66**, 733 (1947); *C. A.* **42**, 3632 (1948); Beil **EIII1**, 651 ff.
2566. Ketteler, E., *Ann. Phys. Chem.* (2), **33**, 353; 506 (1888); *J. Chem. Soc.* **1888A**, 541.
2567. Kettering, C. F. and W. W. Sleator, *Physics* **4**, 39 (1933); *C. A.* **27**, 2095 (1933).
2568. Keyes, D. B. and J. H. Hildebrand, *J. Amer. Chem. Soc.* **39**, 2126 (1917).
2569. Keyes, D. B., S. Swann, et al., *Trans. Amer. Electrochem. Soc.* **54** (preprint) 7 pp (1928).
2570. Keyes, F. G. and W. J. Winninghoff, *J. Amer. Chem. Soc.* **38**, 1178 (1916).
2571. Keyworth, D. A., *Talanta* **8**, 461 (1961).
2572. Kezdy, F. and A. Bruylants, *Bull. Soc. Chim. Fr.* **1959**, 947; *C. A.* **54**, 4370 (1960).
2573. Khalilov, A., K., *Izv. Akad. Nauk SSSR, Ser. Fiz.* **17**, 586 (1953); *C. A.* **48**, 5652 (1954).
2574. Khalilov, A. K. and S. Z. Rzaeva, *Dokl. Akad. Nauk Azerbaidzhan SSR* **12**, 441 (1956); *C. A.* **51**, 855 (1957).

2575. Khalilov, A. K. and P. P. Shorygin, *Dokl. Akad. Nauk SSSR* **78**, 1177 (1951); *C. A.* **46**, 4368 (1952).

2576. Khalilov, K., *J. Exptl. Theoret. Phys.* (USSR) **9**, 335 (1939); *C. A.* **33**, 8069 (1939).

2577. Kharasch, M. S., *J. Res. Nat. Bur. Stand.* **2**, 359 (1929).

2578. Kharasch, M. S. and C. F. Fuchs, *J. Org. Chem.* **10**, 159 (1945).

2579. Kharasch, M. S. and W. H. Urry, *J. Org. Chem.* **13**, 101 (1948).

2580. Kheilo, G. I. and S. N. Kremneva, *Gigiena Truda Prof. Zabolevaniya* **10**, 13 (1966); *C. A.* **65**, 1281 (1966).

2581. Khol'kin, Y. I., *Izv. Vyssh. Ucheb. Zaved. Lesnoĭ Zhur.* **3**, No. 4, 131 (1960); *C. A.* **55**, 14,421 (1961).

2582. Khol'kin, Y. I. and G. N. Chernyaeva, *Gidrolizn. Lesokhim. Prom.* **16**, 6 (1963); *C. A.* **60**, 1674 (1964).

2583. Khol'kin, Y. I., L. S. Solov'er, et al., *Izv. Sib. Odt. Akad. Nauk. SSSR, Ser. Khim. Nauk* **1964**, 105; *C. A.* **63**, 2873 (1965).

2584. Khotinsky, E., *Ber. Deut. Chem. Gesell.* **42**, 2506 (1909).

2585. Khramchenkov, V. A. and E. N. Kokoreva, *Zavod. Lab.* **28**, 1355 (1962); *C. A.* **59**, 4990 (1963).

2586. Kieffer, R., J. C. Marie, et al., *Bull. Soc. Chim. Fr.* **1965**, 3271; *C. A.* **64**, 6473 (1966).

2587. Kienitz, H., *Z. Elektrochem.* **59**, 168 (1955); *C. A.* **49**, 15341 (1955).

2588. Kierstead, H. A. and J. Turkevich, *J. Chem. Phys.* **12**, 24 (1944).

2589. Kilpatrick, J. E., E. J. Prosen, et al., *J. Research Nat. Bur. Stand.* **36**, 559 (1946).

2590. Kim, H., R. Keller, et al., *J. Chem. Phys.* **37**, 2748 (1962).

2591. Kimbrough, R. and T. B. Gaines, *Nature* **211**, 146 (1966).

2592. Kimura, C., K. Kashiwaya, et al., *Yuki Gosei Kagaku Kyokai Shi.* **33** (3), 254 (1965); *C. A.* **62**, 1304 (1965).

2593. Kimura, K. and R. Fujishiro, *Bull. Chem. Soc. Jap.* **39**, 608 (1966).

2594. Kimura, K., Y. Toshiyasu, et al., *Bull. Chem. Soc. Jap.* **39**, 1681 (1966).

2595. Kimura, O., *J. Chem. Soc. Jap.* **63**, 98 (1942); *C. A.* **41**, 2951 (1947).

2596. Kimura, O., *J. Chem. Soc. Jap.* **63**, 423; 426; 429; 814 (1942); *C. A.* **41**, 2951 (1947).

2597. King, H. and K. J. P. Orton, *J. Chem. Soc.* **1911**, 1377.

2598. King, H. and E. V. Wright, *J. Chem. Soc.* **1939**, 1168.

2599. King, J. F. and B. Vig, *Can. J. Chem.* **40**, 1023 (1962).

2600. Kini, M. M. and J. R. Cooper, *Biochim. et Biophys. Acta* **44**, 599 (1960); *C. A.* **55**, 13,670 (1961).

2601. Kini, M. M. and J. R. Cooper, *Biochem. J.* **82**, 164 (1962).

2602. Kinney, C. R., *Ind. Eng. Chem.* **32**, 559 (1940).

2603. Kinney, C. R., *Ind. Eng. Chem.* **33**, 791 (1941).

2604. Kinney, C. R., *J. Amer. Chem. Soc.* **60**, 3032 (1938).

2605. Kinsey, E. L. and J. W. Ellis, *J. Chem. Phys.* **5**, 399 (1937).

2606. Kirchner, H. H., *Z. Phys. Chem.* (Frankfurt) **39**, 273 (1963); *C. A.* **60**, 14013 (1964).

2607. Kirejew, W. A. and W. A. Nikiforowa, *Z. Obshch. Khim.* **6**, 75 (1936); *Chem. Zentr.* **1937**, I, 4219; Beil **EIII1**, 1344.

2608. Kirk, R. E. and D. F. Othmer, *Encyclopedia of Chemical Technology*, Interscience, New York, 1947–1951.

2609. Kirk-Othmer, *Encyclopedia of Chemical Technology*, 2d ed., Interscience-Wiley, New York, 1963 ff.

2610. Kirrmann, A., *Bull. Soc. Chim. Fr.* **5**, 915 (1938); *C. A.* **32**, 6619 (1938).

2611. Kirrmann, A., *Bull. Soc. Chim. Fr.* **39**, 988 (1926); *C. A.* **20**, 3443 (1926).

2612. Kirrmann, A., *Compt. Rend.* **308**, 353 (1939); *C. A.* **33**, 2413 (1939).

2613. Kirrmann, A. and N. Hamaide, *Bull. Soc. Chim. Fr.* **1957**, 789; *C. A.* **51**, 16,496 (1957).

2614. Kirsanov, A. V. and Y. M. Zolotov, *Zh. Obshch. Khim.* **19**, 220 (1949); *C. A.* **44**, 4446 (1950).

2615. Kirsanov, A. V. and Y. M. Zolotov, *Zh. Obshch. Khim.* **20**, 284 (1950); *C. A.* **44**, 6384 (1950).

2616. Kirsanova, R. P. and S. S. Byk, *Zh. Prikl. Khim.* **34**, 1373 (1961); *Anal. Chem.* **36**, 56R (1964), ref. 22F.

2617. Kiss, Á., J. Molnár, et al., *Compt. Rend.* **227**, 724 (1948); *C. A.* **43**, 2512 (1949).

2618. Kiss, Á. I. and B. R. Muth, *Magy. Tudományoo Akad. Központi Fiz. Kutató Intézetének Közleményei* **7**, 147 (1959); *C. A.* **54**, 14,932 (1960).

2619. Kistiakowsky, G. B., J. R. Ruhoff, et al., *J. Amer. Chem. Soc.* **58**, 137 (1936).

2620. Kiyama, R., K. Suzuki, et al., *Rev. Phys. Chem. Jap.* **21**, 50 (1951); *C. A.* **46**, 3818 (1952).

2621. Klaboe, P., *Acta Chem. Scand.* **17**, 1179 (1963); *C. A.* **60**, 10,062 (1964).

2622. Klaboe, P. and J. R. Nielsen, *J. Mol. Spectrosc.* **6**, 379 (1961); *C. A.* **55**, 19,479 (1961).

2623. Kland-English, M. J., R. K. Summerbell, et al., *J. Amer. Chem. Soc.* **75**, 3709 (1953).

2624. Klein, A. K., *J. Assoc. Offic. Agr. Chemists* **32**, 349 (1949).

2625. Kleman, B, *Can. J. Phys.* **41**, 2034 (1963); *C. A.* **60**, 3626 (1964).

2626. Klemenc, A. and M. Low, *Rec. Trav. Chim. Pays-Bas* **49**, 629 (1930); *Chem. Zentr.* **1930, II**, 3815; Beil **EIII5**, 545.

2627. Klemm, L., W. Klemm, et al., *Z. Phys. Chem.* (Leipzig) **165A**, 379 (1933); *Chem. Zentr.* **1933, II**, 1965; Beil **EII5**, 223, 1570.

2628. Klemperer, W., M. W. Cronyn, et al., *J. Amer. Chem. Soc.* **76**, 5846 (1954).

2629. Klesment, I. R. and K. M. Soo, *Khim. Tekhnol. Goryuch. Slantsev Produktov Pererabotki* **1962**, No. 11, 307; *C. A.* **58**, 13,681 (1963).

2630. Kletz, T. A. and W. C. Price, *J. Chem. Soc.* **1947**, 644.

2631. Klevens, H. B., *J. Phys. Colloid Chem.* **54**, 283 (1950).

2632. Klevens, H. B. and J. R. Platt, *J. Amer. Chem. Soc.* **69**, 3055 (1947).

2633. Kliment, V., V. Fried, et al., *Collect. Czech. Chem. Commun.* **29**, 2008 (1964); *C. A.* **61**, 12,689 (1964).

2634. Kling, A. and P. Roy, *Compt. Rend.* **144**, 111; *C. A.* **1**, 2233 (1907).

2635. Klingstedt, F. W., *Compt. Rend.* **176**, 248 (1923).

2636. Knecht, L. A. and I. M. Kolthoff, *Inorg. Chem.* **1**, 195 (1962).

2637. Knight, H. M. and J. T. Kelly, *Ind. Eng. Chem.* **51**, 1355 (1959).

2638. Knight, H. S. and F. T. Weiss, *Anal. Chem.* **34**, 749 (1962).

2639. Knoevenagel, E., *Ber. Deut. Chem. Gesell.* **40**, 508 (1907).

2640. Knorr, L., *Ber. Deut. Chem. Gesell.* **30**, 915 (1897).

2641. Knorr, L., *Ber. Deut. Chem. Gesell.* **30**, 918 (1897).

2642. Knorr, L., *Ber. Deut. Chem. Gesell.* **30**, 1492 (1897).

2643. Knorr, L. O. Rothe, et al., *Ber. Deut. Chem. Gesell.* **44**, 1138 (1911).

2644. Knowles, C. L., *Ind. Eng. Chem.* **12**, 881 (1920).

2645. Knowlton, J. W. and F. D. Rossini, *J. Res. Nat. Bur. Stand.* **22**, 415 (1939).

2646. Knox, J. H. and J. Riddick, *Trans. Faraday Soc.* **62**, 1190 (1966).

2647. Kobayashi, R., Y. Fukula, et al., *J. Chem. Soc. Jap.*, Ind. Chem. Sect. **56**, 102 (1953); *C. A.* **48**, 8581 (1954).

2648. Kobayashi, R., T. Saito, et al., *J. Chem. Soc. Jap.*, Ind. Chem. Sect. **56**, 169 (1953); *C. A.* **48**, 9099 (1954).

2649. Kobayashi, U., *J. Soc. Org. Synthet. Chem. Jap.* **12**, 63, 112, 319, 358, 503 (1954); *C. A.* **51**, 952 (1957).

2650. Kobe, K. A., H. R. Crawford, et al., *Ind. Eng. Chem.* **47**, 1767 (1955).
2651. Kobe, K. A. and R. E. Lynn, *Chem. Rev.* **52**, 117 (1953).
2652. Kobe, K. A., T. S. Okabe, et al., *J. Amer. Chem. Soc.* **63**, 3251 (1941).
2653. Kobe, K. A., A. E. Ravicz, et al., *Ind. Eng. Chem.*, *Chem. Eng. Data Series* **1**, 50 (1956).
2654. Koch, F. K. V., *J. Chem. Soc.* **1927**, 647.
2655. Koch, F. K. V., *J. Chem. Soc.* **1928**, 269.
2656. Koch, H. P., *J. Chem. Soc.* **1949**, 387.
2657. Koefoid, J. and J. V. Villadsen, *Acta Chim. Scand.* **12**, 1124 (1958); from [4773].
2658. Kofod, H., L. E. Sutton, et al., *J. Chem. Soc.* **1952**, 1467.
2659. Kohlrausch, K. W. F., *Monatsh. Chem.* **68**, 349 (1936); *C. A.* **31**, 36 (1937).
2660. Kohlrausch, K. W. F., *Monatsh. Chem.* **76**, 231 (1947); *C. A.* **41**, 6153 (1947).
2661. Kohlrausch, K. W. F. and F. Köppl, *Monatsh. Chem.* **63**, 255 (1933); *Chem. Zentr.* **1934, I**, 3440.
2662. Kohlrausch, K. W. F., and F. Köppl, *Monatsh. Chem.* **65**, 185 (1935); *C. A.* **29**, 2850 (1935).
2663. Kohlrausch, K. W. F. and F. Köppl, *Z. Phys. Chem.* (Leipzig) **24B**, 370 (1934); *Chem. Zentr.* **1934, I**, 3440.
2664. Kohlrausch, K. W. F. and F. Köppl, *Z. Phys. Chem.* (Leipzig) **26B**, 209 (1934); *C. A.* **28**, 6065 (1934).
2665. Kohlrausch, K. W. F., F. Köppl, et al., *Z. Phys. Chem.* (Leipzig) **21B**, 242 (1933); *Chem. Zentr.* **1933, II**, 336.
2666. Kohlrausch, K. W. F., F. Köppl, et al., *Z. Phys. Chem.* (Leipzig) **22B**, 359 (1933); *Chem. Zentr.* **1933, II**, 3666.
2667. Kohlrausch, K. W. F. and O. Paulsen, *Monatsh. Chem.* **72**, 268 (1939); *C. A.* **33**, 8117 (1939).
2668. Kohlrausch, K. W. F. and A. Pongratz, *Ber. Deut. Chem. Gesell.* **66**, 1357 (1933).
2669. Kohlrausch, K. W. F. and A. Pongratz, *Ber. Deut. Chem. Gesell.* **67**, 976, 1465 (1934).
2670. Kohlrausch, K. W. F. and A. Pongratz, *Monatsh. Chem.* **63**, 427 (1934); *Chem. Zentr.* **1934, I**, 3441.
2671. Kohlrausch, K. W. F. and A. Pongratz, *Monatsh. Chem.* **65**, 6 (1934); *C. A.* **29**, 1325 (1935).
2672. Kohlrausch, K. W. F. and A. Pongratz, *Monatsh. Chem.* **70**, 226 (1937); *C. A.* **31**, 4902 (1937); Beil **EIII4**, 145.
2673. Kohlrausch, K. W. F. and A. Pongratz, *Z. Phys. Chem.* (Leipzig) **27B**, 176 (1934); *C. A.* **29**, 1325 (1935); *Chem. Zentr.* **1935, I**, 2798.
2674. Kohlrausch, K. W. F., A. Pongratz, et al., *Ber. Deut. Chem. Gesell.* **66**, 1 (1933).
2675. Kohlrausch, K. W. F. and H. W. Reitz, *Z. Phys. Chem.* (Leipzig) **45B**, 249 (1939); *C. A.* **34**, 4990 (1940).
2676. Kohlrausch, K. W. F. and A. Reitz, *Z. Phys. Chem.* (Leipzig) **45B**, 269 (1940); *C. A.* **34**, 4990 (1940).
2677. Kohlrausch, K. W. F. and R. Seka, *Ber. Deut. Chem. Gesell.* **71**, 985 (1938).
2678. Kohlrausch, K. W. F. and R. Skrabl, *Monatsh. Chem.* **70**, 377 (1937); *C. A.* **31**, 7758 (1937).
2679. Kohlrausch, K. W. F. and J. Wagner, *Z. Phys. Chem.* (Leipzig) **45B**, 93 (1939); *C. A.* **34**, 4990 (1940).
2680. Kohlrausch, K. W. F. and H. Wittek, *Monatsh. Chem.* **74**, 1 (1941); *C. A.* **36**, 6083 (1942).
2681. Kohlrausch, K. W. F. and G. P. Ypsilanti, *Z. Phys. Chem.* (Leipzig) **29B**, 274 (1935); *Chem. Zentr.* **1936, I**, 2065; Beil **222**, 298.

2682. Kohlrausch, K. W. F. and G. P. Ypsilanti, *Z. Phys. Chem.* (Leipzig) **32B**, 407 (1936); *C. A.* **30**, 5127 (1936).

2683. Kohltoff, I. M. and T. B. Reddy, *Inorg. Chem.* **1**, 189 (1962).

2684. Kohn, C. A. and W. Trantom, *J. Chem. Soc.* **1899**, 1155.

2685. Koizumi, N., *J. Chem. Phys.* **27**, 625 (1957).

2686. Koizumi, N. and T. Hanai, *Bull. Inst. Chem. Res., Kyoto Univ.* **33**, 14 (1955); *C. A.* **50**, 634 (1956).

2687. Koizumi, N. and T. Hanai, *J. Phys. Chem.* **60**, 1496 (1956).

2688. Kolling, O. W., K. K. O'Hare, et al., *Trans. Kansas Acad. Sci.* **66**, No. 3, 435 (1963).

2688a. Kolosovskiĭ, N., see also N. de Kolossowski.

2689. Kolosovskiĭ, N. A. and V. V. Udovenko, *J. Gen. Chem.* (USSR) **4**, 1027 (1934); *C. A.* **29**, 3588 (1935).

2690. Kolossowski, N. de and W. W. Udowenko, *Compt. Rend.* **197**, 519 (1933); *Chem. Zentr.* **1933**, II, 2375; Beil **EIII2**, 593.

2691. Kolthoff, I. M., S. Bruckenstein, et al., *J. Amer. Chem. Soc.* **83**, 3927 (1961).

2692. Kolthoff, I. M. and J. F. Coetzee, *J. Amer. Chem. Soc.* **79**, 870 (1957).

2693. Kolthoff, I. M. and P. J. Elving, *Treatise on Analytical Chemistry*, Part I, Vol. 1, Interscience-Wiley, New York, 1959.

2694. Kolthoff, I. M. and P. J. Elving, eds., *Treatise on Analytical Chemistry*, Part 1, Vol. 8, Interscience-Wiley, New York, 1968.

2695. Kolthoff, I. M. and T. B. Reddy, *Inorg. Chem.* **1**, 189 (1962).

2696. Kolthoff, I. M. and T. B. Reddy, *J. Electrochem. Soc.* **108**, 980 (1961).

2697. Kolthoff, J. M., *Biochem. Z.* **162**, 289 (1925).

2698. Komandin, A. V. and A. K. Bonetskaya, *Zh. Fiz. Khim.* **28**, 1113 (1954); *C. A.* **49**, 7905 (1955).

2699. Komandin, A. V. and B. D. Shimit, *Zh. Fiz. Khim.* **37** (3) 510 (1963); *C. A.* **59**, 4623 (1963).

2700. Komarewsky, V. I., S. C. Uhlick, et al., *J. Amer. Chem. Soc.* **67**, 557 (1945).

2701. Komarowsky, A., *Chem.-Ztg.* **27**, 807, 1086 (1903); *Chem. Zentr.* **1903**, II, 724, 1396.

2702. Komers, R., K. Kochloefl, et al., *Chem. Ind.* (London) **1958**, 1405.

2703. Kondilenko, I. I., P. A. Korotkov, et al., *Opt. Spektrosk.* **11**, 169 (1961); *C. A.* **56**, 3038 (1962).

2704. Kondilenko, I. I., and V. E. Pogorelov, *Opt. Spektrosk.* **19**, 41 (1965); *C. A.* **64**, 188 (1966).

2705. Kondo, S., Japanese Patent 5312, Oct. 16, 1953; *C. A.* **49**, 6987 (1955).

2706. Konek, v. Fr., *Ber. Deut. Chem. Gesell.* **39**, 2263 (1906).

2707. Kononenko, O. K. and K. M. Herstein, *Ind. Eng. Chem., Chem. Eng. Data Series* **1**, 87 (1956).

2708. Kopf, R., A. Loeser, et al., *Arch. Exptl. Path. Pharmakol.* **210**, 346 (1950); *C. A.* **45**, 5308 (1951).

2709. Kopper, H., R. Seka, et al., *Monatsh. Chem.* **61**, 397; *Sitzber. Akad. Wiss. Wien. Math.-Naturw. Klasse*, Abt. IIa, **141**, 465 (1932); *C. A.* **27**, 4172 (1933).

2710. Koptyug, V. A., A. G. Khmel'nitskii, et al., *Izv. Sib. Otd. Akad. Nauk SSSR, Ser. Khim. Nauk* **1964**, 116; *C. A.* **61**, 11,813 (1964).

2711. Korbakova, A. I., *Vestn. Akad. Nauk SSSR* **19**, 17 (1964); *C. A.* **61**, 16,694 (1964).

2712. Kordes, E., *Z. Elektrochem.* **58**, 424 (1954); *C. A.* **49**, 2140 (1955).

2713. Kornblum, N. and D. C. Iffland, *J. Amer. Chem. Soc.* **77**, 6653 (1955).

2714. Kornblum, N., H. O. Larson, et al., *J. Amer. Chem. Soc.* **78**, 1497 (1956).

2715. Kornev, K. A., A. P. Grekov, et al., *Khim. Prom.* **1963** (1), 16; *C. A.* **59**, 6252 (1963).

2716. Korschun, T. W. and K. W. Roll, *J. Russ. Phys.-Chem. Gesell.* **49**, 153 (1917); *Chem. Zentr.* **1923, III**, 775.
2717. Korshunov, A. V. and P. S. Sarapkin, *Trudy Sibir. Tekhnol. Inst.* **1959**, No. 24, 13; *C. A.* **56**, 3038 (1962).
2718. Kortuem, G. and E. Faltusz, *Chem.-Ing.-Tech.* **33**, 599 (1961); *C. A.* **56**, 283 (1962).
2719. Kortüm, G., *Z. Phys. Chem.* (Leipzig) **42B**, 39 (1939); *C. A.* **33**, 3261 (1939).
2720. Kosakewitsch, P. P., *Z. Phys. Chem.* (Leipzig) **133A**, 1 (1928).
2721. Koskikallio, J. and S. Syrjaepalo, *Suomen Kemistilehti* **37B**, 120 (1964); *C. A.* **63**, 14,119 (1965).
2722. Kostyuk, N. G., S. V. L'vov, et al., *Zh. Prikl. Khim.* **35**, 698 (1962); *C. A.* **57**, 2068 (1962).
2723. Kotake, M., Japanese Patent 133,250, Nov. 14, 1939; *C. A.* **35**, 4042 (1941).
2724. Kotera, A., S. Nishimura, et al., *J. Chem. Soc. Jap.* **65**, 527 (1944); *C. A.* **41**, 3335 (1947).
2725. Kotera, A., S. Shibata, et al., *J. Amer. Chem. Soc.* **77**, 6183 (1955).
2726. Koteswaram, P., *Z. Phys.* **110**, 118 (1938); *C. A.* **32**, 8938 (1938).
2727. Kotlyarevskiĭ, I. L., A. N. Volkov, et al., *Izv. Sib. Otd. Akad. Nauk SSSR* **1959**, No. 3, 62; *C. A.* **53**, 21,605 (1959).
2728. Kováts, E., *Helv. Chim. Acta* **41I**, 1915 (1958).
2729. Kowalewski, V. J. and D. G. de Kowalewski, *Ark. Kemi.* **16**, 373 (1960).
2730. Kowalewski, V. J. and D. G. de Kowalewski, *J. Chem. Phys.* **32**, 1272 (1960).
2731. Kowalewski, V. J. and D. G. de Kowalewski, *J. Chem. Phys.* **33**, 1794 (1960).
2732. Kozima, K. and S. Mizushima, *Sci. Papers Inst. Phys. Chem. Research* (Tokyo) **31**, No. 697, 296 (1937); *C. A.* **31**, 6558 (1937).
2733. Kozina, M. P. and S. M. Skuratov, *Dokl. Akad. Nauk SSSR* **127**, 561 (1959); *C. A.* **54**, 284 (1960).
2734. Kozlov, N. W. and S. Y. Chumakov, *Zh. Prikl. Khim.* **30**, 318 (1957); *C. A.* **51**, 12,816 (1957).
2735. Kraemer, G. and M. Grodzki, *Chem. Ber.* **9**, 1928 (1876).
2736. Krafft, F., *Ber. Deut. Chem. Gesell.* **15**, 1687 (1882).
2737. Krafft, F. and H. Noerdlinger, *Ber. Deut. Chem. Gesell.* **22**, 816 (1889).
2738. Kranzfelder, A. L., and F. J. Sowa, *J. Amer. Chem. Soc.* **59**, 1490 (1937).
2739. Kranzfelder, A. L. and R. R. Vogt, *J. Amer. Chem. Soc.* **60**, 1714 (1938).
2740. Krase, N. W. and J. B. Goodman, *Ind. Eng. Chem.* **22**, 13 (1930).
2741. Kraus, C. A. and J. E. Bishop, *J. Amer. Chem. Soc.* **43**, 1568 (1921).
2742. Kraus, C. A. and J. E. Bishop, *J. Amer. Chem. Soc.* **44**, 2206 (1922).
2743. Kraus, C. A. and C. C. Callis, *J. Amer. Chem. Soc.* **45**, 2624 (1923).
2744. Kraus, C. A. and R. M. Fuoss, *J. Amer. Chem. Soc.* **55**, 21 (1933).
2745. Kraus, C. A. and R. A. Vingee, *J. Amer. Chem. Soc.* **56**, 511 (1934).
2746. Krausz, F., *Ann. Chim.* **4**, 811 (1949); *C. A.* **44**, 5811 (1950); See also Beil **78**, 603.
2747. Krchma, I. J. and J. W. Williams, *J. Amer. Chem. Soc.* **49**, 2408 (1927).
2748. Kreevoy, M. M., B. E. Eichinger, et al., *J. Org. Chem.* **29**, 1641 (1964).
2749. Kreglewski, A., *Bull. Acad. Pol. Sci. Ser. Sci. Chim.* **10**, 629 (1962); *C. A.* **59**, 2180 (1963).
2750. Kreimeier, O. R., U.S. Patent 2,106,181, Jan. 25, 1938; *Official Gaz. U.S. Patent Office* **486**, 726 (1938).
2751. Kremann, R., F. Gugl, et al., *Monatsh. Chem.* **35**, 1365 (1914).
2752. Kretow, A. E., *J. Russ. Phys. Chem. Soc.* **61**, 2350 (1929); *Chem. Zentr.* **1930, II**, 370.
2753. Kretschmer, C. B., *J. Phys. Colloid Chem.* **55**, 1351 (1951).
2754. Kretschmer, C. B. and R. Wiebe, *J. Amer. Chem. Soc.* **74**, 1276 (1952).

2755. Kreutzberger, A. and P. A. Kalter, *J. Phys. Chem.* **65**, 624 (1961).
2756. Kreuzer, J. and R. Mecke, *Z. Phys. Chem.* (Leipzig) **49B**, 309 (1941); *C. A.* **36**, 6898 (1942).
2757. Krishna, B. and A. N. Srivastava, *Aust. J. Chem.* **19**, 1847 (1966); *C. A.* **65**, 18,475 (1966).
2758. Krishna, B. and K. K. Srivastava, *J. Chem. Phys.* **32**, 663 (1960).
2759. Krishna, K. V. G., *Indian J. Phys.* **31**, 283 (1957); *C. A.* **51**, 17,293 (1957).
2760. Krishna, K. V. G., *Trans. Faraday Soc.* **53**, 767 (1957).
2761. Krishnan, K., *Proc. Indian. Acad. Sci.* **53A**, 151 (1961); *C. A.* **55**, 19,478 (1961).
2762. Kronberger, H. and J. Weiss, *J. Chem. Soc.* **1944**, 464,
2763. Krongauz, V. A., *Dokl. Akad. Nauk SSSR* **162**, 1300 (1965); *C. A.* **63**, 9780 (1965).
2764. Kross, R. D., V. A. Fassel, et al., *J. Amer. Chem. Soc.* **78**, 1332 (1956).
2765. Krueger, J. H. and W. A. Johnson, *Inorg. Chem.* **7**, 679 (1968).
2766. Krueger, P. J. and H. D. Mettee, *Can. J. Chem.* **43**, 2970 (1965).
2767. Krueger, P. J. and H. D. Mettee, *J. Mol. Spectrosc.* **18**, 131 (1965); *C. A.* **64**, 4906 (1966).
2768. Kubo, M., *Sci. Papers Inst. Phys. Chem. Res.* (Tokyo) **27**, 65 (1935); *C. A.* **29**, 6481 (1935).
2769. Kubo, M., *Sci. Papers Inst. Phys. Chem. Res.* (Tokyo), **30**, 238 (1936); *C. A.* **31**, 1669 (1937).
2770. Kubo, V. M., *Sci. Papers Inst. Phys. Chem. Res.* (Tokyo) **29**, 181 (1936); *C. A.* **30**, 7002 (1936); Beil **EIII1**, 2570.
2771. Kuchinskaya, K., *Sbornik Trudov Optynogo Zavoda im. Akad. S. V. Lebedeva* **1938**, 27; *Khim. Referat. Zh.* **2**, No. 4, 23 (1939); *C. A.* **34**, 3147 (1940).
2771a. Kudchadker, A. P., G. H. Alani, et al., *Chem. Rev.* **68**, 659 (1968).
2772. Kudchadker, A. P. and J. J. McKetta, *A.I.C.H.E. Journal* **7**, 707 (1961).
2773. Kudryashov, I. V. and R. I. Savachenko, *Izv. Vyssh. Ucheb. Zaved., Khim. Khim. Tekhnol* **8**, 602 (1965); *C. A.* **64**, 18,438 (1966).
2774. Kuivila, H. G. and O. F. Beumel, *J. Amer. Chem. Soc.* **83**, 1246 (1961).
2775. Kulagina, N. K. and T. A. Kochetkova, *Toksikol. Novykh Prom. Khim. Veshchestv* No. **7**, 56 (1965); *C. A.* **63**, 7548 (1965).
2776. Kumler, W. D., *J. Amer. Chem. Soc.* **74**, 261 (1952).
2777. Kumler, W. D., *J. Amer. Chem. Soc.* **83**, 4983 (1961).
2778. Kumler, W. D. and I. F. Halverstadt, *J. Amer. Chem. Soc.* **63**, 2182 (1941).
2779. Kumler, W. D. and C. W. Porter, *J. Amer. Chem. Soc.* **56**, 2549 (1934).
2780. Kurbatow, V. F. J., *Comm. Trav. Scient. Techn. Rep. Leningrad* **17**, 24 (1925); *Annual Tables of Physical Constants*, Sec. 514 (c), 1941.
2781. Kurbatow, V. Y., *J. Gen. Chem.* (USSR) **17**, 1999 (1947); *C. A.* **42**, 4829 (1948).
2782. Kurbatow, V. Y., *Zh. Obshch. Khim.* (*J. Gen. Chem.*) **18**, 372 (1948); *C. A.* **43**, 30 (1949).
2783. Kurbatow, W. A., *J. Russ. Phys. Chem. Soc.* **34**, 766 (1902); *J. Chem. Soc.* **1903Aii**, 246.
2784. Kurbatow, W. A., *J. Russ. Phys. Chem. Soc.* **35**, 319 (1903); *Chem. Zentr.* **1903, II**, 323; *J. Chem. Soc.* **1903Aii**, 710.
2785. Kurbatow, W. A., *J. Russ. Phys. Chem. Soc.* **40**, 1471, 1474, 1480; *Chem. Zentr.* **1909**, 635, 636; *J. Chem. Soc.* **1909Aii**, 119.
2786. Kurnakow, N. and S. Zemcuzny, *Z. Phys. Chem.* (Leipzig) **83**, 481 (1913).
2787. Kuroda, Y. and M. Kubo, *J. Polym. Sci.* **26**, 323 (1957); *C. A.* **52**, 10,715 (1958).
2788. Kurtz, S. S., S. Amon, et al., *Ind. Eng. Chem.* **42**, 174 (1950).
2789. Kurtz, S. S. and M. R. Lipkin, *J. Amer. Chem. Soc.* **63**, 2158 (1941).

2790. Kusama, T. and D. Koike, *J. Chem. Soc. Jap.*, Pure Chem. Sect. **72**, 229 (1951); *C. A.* **47**, 2155 (1953).

2791. Kusano, K., *Nippon Kagaku Zasshi* **78**, 614 (1958); *C. A.* **53**, 3786 (1959).

2792. Kushner, L. M., R. W. Crowe, et al., *J. Amer. Chem. Soc.* **72**, 1091 (1950).

2793. Kuskov, V. K. and V. A. Zhukova, *Izv. Akad. Nauk SSSR, Otd. Khim. Nauk* **1956**, 733; *C. A.* **51**, 1877 (1957).

2794. Kuss, E., *Z. Angew. Phys.* **7**, 372 (1955); *C. A.* **50**, 3028 (1956).

2795. Kuss, E., J. Moos, et al., *Naturwissemschaften* **48**, 73 (1961); *C. A.* **55**, 19,748 (1961).

2796. Kutsenko, A. and V. Lyubomilov, *Zh. Prikl. Khim.* **31**, 1419 (1958); *C. A.* **53**, 2070 (1959).

2797. Kuz'micheva, M. N., *Gig. Sanit.* **27**, 41 (1962); *C. A.* **58**, 9546 (1963).

L

2798. Labbauf, A., J. R. Greenshields, et al., *J. Chem. Eng. Data* **6**, 261 (1961).

2798a. *Laboratory, The*, 35 (4), 106 (1967); *C & EN* "Concentrates," May 1, 1967, p. 45.

2799. Lacher, J. R., L. E. Hummel, et al., *J. Amer. Chem. Soc.* **72**, 5486 (1950).

2800. Lacher, J. R., R. E. Scruby, et al., *J. Amer. Chem. Soc.* **71**, 1797 (1949).

2801. Lacher, J. R. and C. H. Walden, et al., *J. Amer. Chem. Soc.* **71**, 3026 (1949).

2802. Lafontaine, I., *Bull. Soc. Chim. Belges* **67**, 153 (1958); *C. A.* **53**, 791 (1959); see also [4773].

2803. Lagemann, R. T., D. R. McMillan, et al., *J. Chem. Phys.* **17**, 369 (1949).

2804. Lagemann, R. T. and H. N. Nielsen, *J. Chem. Phys.* **10**, 668 (1942).

2805. Lake, J. S. and A. J. Harrison, *J. Chem. Phys.* **30**, 361 (1959).

2806. Lalande, A., *J. Chim. Phys.* **31**, 583 (1934); *C. A.* **29**, 5339 (1935).

2807. Lambert, P. and J. Lecomte, *Ann. Phys. Paris* **10**, 503 (1938); *C. A.* **33**, 4127 (1939).

2808. Lambert, P. and J. Lecomte, *Ann. Phys. Paris* **18**, 329 (1932); *C. A.* **27**, 1825 (1933).

2809. Lambert, P. and J. Lecomte, *Compt. Rend.* **194**, 77, 960 (1932).

2810. Lambert, P. and J. Lecomte, *Compt. Rend.* **208**, 740 (1939); *C. A.* **33**, 3693 (1939.

2811. Lamond, J., *Analyst* **74**, 560 (1949).

2812. Lamprecht, W., *Fette, Seifen, Anstrichmittle* **61**, 96 (1959); *Chem. Zentr.* **1960**, 5263.

2813. Landee, F. A. and I. B. Johns, *J. Amer. Chem. Soc.* **63**, 2891 (1941).

2814. Landolt, H., *Ann. Chem.* **189**, 333 (1877).

2815. Landsberg, G. S. and S. A. Ukholin, *Compt. Rend. Acad. Sci. URSS* **16**, 391 (1937).

2816. Langseth, A. and B. Bak, *Kgl. Danske Videnskab. Selskab, Mat.-Fys. Medd.* **24**, No. 3, 16 pp. (1947); *C. A.* **42**, 459 (1948).

2817. Lane, M. R., J. W. Linnett, et al., *Proc. Roy. Soc.* **216A**, 316 (1953).

2818. Lane, W. H., *Ind. Eng. Chem. Anal. Ed.* **18**, 295 (1946).

2819. Lang, H. R., *Proc. Roy. Soc.* **118A**, 138 (1928).

2820. Lange, J., *Z. Phys. Chem.* (Leipzig) **161A**, 77 (1932); *Chem. Zentr.* **1932**, II, 2848.

2821. Lange, N. A., ed., *Handbook of Chemistry*, 11th ed., McGraw-Hill, New York, 1967.

2822. Langseth, L. H., *Nature* **124**, 92 (1929); *C. A.* **23**, 5108 (1929).

2823. Lannung, A., *Z. Phys. Chem.* (Leipzig) **161A**, 255 (1932).

2824. Lantz, V., *J. Amer. Chem. Soc.* **62**, 3260 (1940).

2825. LaPaglia, S. R., *Trans. Faraday Soc.* **60**, 1210 (1964).

2826. Lapik, A. S., *Izv. Sib. Odt. Akad. Nauk SSSR, Ser. Biol.-Med. Nauk* **1965**, 91; *C. A.* **64**, 20,497 (1966).

2827. La Planche, L. A. and M. T. Rogers, *J. Amer. Chem. Soc.* **86**, 337 (1964).

2828. La Planche, L. A., H. B. Thompson, et al., *J. Phys. Chem.* **69**, 1482 (1965).

2829. Lappert, M. F., *Chem. Rev.* **56**, 959 (1956).

2830. Lardicci, L., R. Rossi, et al., *Chim. Ind.* (Milan) **44**, 1002 (1962); *C. A.* **60**, 13,138 (1964).

2831. Lardicci, L., P. Salvadori, et al., *Ann. Chim.* (Rome) **52**, 652 (1962); *C. A.* **58**, 5494 (1963).
2832. LaRochelle, J. H. and A. A. Vernon, *J. Amer. Chem. Soc.* **72**, 3293 (1950).
2833. Larson, R. C. and R. T. Iwamoto, *Inorg. Chem.* **1**, 316 (1962).
2834. Larson, R. C. and R. T. Iwamoto, *J. Amer. Chem. Soc.* **82**, 3239 (1960).
2835. Larson, R. G. and H. Hunt, *J. Phys. Chem.* **43**, 417 (1939).
2836. Larsson, E., *Z. Phys. Chem.* (Leipzig) **159A**, 315 (1932); *Chem. Zeit.* **1932, II**, 678.
2837. Lascombe, J. and M. L. Josien, *Bull. Soc. Chim. Fr.* **1955**, 1227; *C. A.* **50**, 2296 (1956).
2838. Lassar-Cohn, *Ann. Chem. Justus Liebigs* **284**, 226 (1895).
2839. Lastovskii, R. P. and V. Y. Temkina, *Khim. Prom.* **1958**, 219; *C. A.* **53**, 1101 (1959).
2840. Laszlo, H. de, *J. Amer. Chem. Soc.* **49**, 2106 (1927).
2841. Lauer, K., *Ber. Deut. Chem. Gesell.* **70**, 1127, 1288 (1937).
2842. Laughlin, K. C. and F. C. Whitmore, *J. Amer. Chem. Soc.* **55**, 2607 (1933).
2843. Lauterbur, P. C., *J. Chem. Phys.* **43**, 360 (1965).
2844. Lawrie, J. W., "Glycerol and the Glycols," Chemical Catalog Company, Inc., New York, 1928.
2845. Lazlo, P., *Bull. Soc. Chim. Fr.* **1964**, 2658; *C. A.* **62**, 8540 (1965).
2846. Lazniewski, M., *Rocz. Chem.* **14**, 560 (1934); *Chem. Zentr.* **1936, I**, 986.
2847. Leader, G. R., *J. Amer. Chem. Soc.* **73**, 856 (1951).
2848. Leader, G. R. and G. F. Gormley, *J. Amer. Chem. Soc.* **73**, 5731 (1951).
2849. Le Bel, R. G. and D. A. I. Goring, *J. Chem. Eng. Data* **7**, 100 (1962).
2850. LeBlanc, R. B., *Anal. Chem.* **30**, 1797 (1958).
2851. Lebo, R. B., *J. Amer. Chem. Soc.* **43**, 1005 (1921).
2852. Lecat, M., *Mem. Acad. Belges* **23**, No. 3, 34 (1949).
2853. Lecky, H. S. and R. H. Ewell, *Ind. Eng. Chem. Anal. Ed.* **12**, 544 (1940).
2854. Leclercq, M., *Mem. Poudres* **43**, 369 (1961); *C. A.* **57**, 6766 (1962).
2855. Lecomte, J., *Bull. Soc. Chim. Fr.* **1946**, 415; *C. A.* **41**, 1933 (1947).
2856. Lecomte, J., *Compt. Rend.* **178**, 1530 (1924).
2857. Lecomte, J., *Compt. Rend.* **178**, 1698 (1924).
2858. Lecomte, J., *Compt. Rend.* **178**, 2073 (1924).
2859. Lecomte, J., *Compt. Rend.* **207**, 395 (1938); *C. A.* **32**, 7821 (1938).
2860. Lecomte, J., *J. Phys. Radium* **3**, 193 (1942); *C. A.* **38**, 3549 (1944).
2861. Lecomte, J., *J. Phys. Radium* **6**, 127 (1945); *C. A.* **40**, 1394 (1946).
2862. Lecomte, J., *J. Phys. Radium* **8**, 489 (1937); *C. A.* **32**, 2833 (1938).
2863. Lecomte, J., *J. Phys. Radium* **9**, 13 (1938); *C. A.* **32**, 2833 (1938).
2864. Lecomte, J. and R. Freymann, *Bull. Soc. Chim. Fr.* **8**, 601 (1941); *C. A.* **36**, 2477 (1942); Beil **159**, 175.
2865. Lecomte, J., E. Gray, et al., *Bull. Soc. Chim. Fr.* **1947**, 774; *C. A.* **42**, 1129 (1948).
2866. Ledaal, T., *Tetrahedron Lett.* **1966**, 1653; *C. A.* **64**, 18736 (1966).
2866a. Lederer, E. L., *Allgem. Öl-u. Fettzig* **27**, 237 (1930); *C. A.* **25**, 2013 (1931).
2867. Lederer, E. L., *Seifensieder-Ztg.* **57**, 67 (1930); Ref. [3150]. pp. 508 ff; *C. A.* **24**, 2625 (1930).
2868. Lederer, E. L., *Seifensieder-Ztg.* **57**, 68, 329 (1930); *Chem. Zentr.* **1932, II**, 1095; Beil **EIII2**, 1388.
2869. Lederer, E. L. and O. Hartleb, *Seifensieder-Ztg.* **56**, 345 (1929); *C. A.* **24**, 741–2 (1929); from [3150].
2870. Lee, C. M. and W. D. Kumler, *J. Amer. Chem. Soc.* **83**, 4593 (1961).
2871. Lee, S., *J. Soc. Chem. Ind. Jap.* **43**, Suppl. binding 190 (1940); *C. A.* **34**, 7677 (1940).
2872. Leech, J. W., *Proc. Phys. Soc.* **62B**, 390 (1949); *Chem. Zentr.* **1950, I**, 2339.

2873. Le Fèvre, C. G. and R. J. W. Le Fèvre, *J. Chem. Soc.* **1936**, 487.
2874. Le Fèvre, C. G. and R. J. W. Le Lèvre, *J. Chem. Soc.* **1936**, 1130.
2875. Le Fèvre, C. G. and R. J. W. Le Fèvre, *J. Chem. Soc.* **1950**, 1829.
2876. Le Fèvre, C. G. and R. J. W. Le Fèvre, *J. Chem. Soc.* **1954**, 1577.
2877. Le Fèvre, C. G., R. J. W. Le Fèvre, et al., *Aust. J. Chem.* **10**, 218 (1957); *C. A.* **52**, 819 (1958).
2878. Le Fèvre, C. G., R. J. W. Le Fèvre, et al., *J. Chem. Soc.* **1935**, 480.
2879. Le Fèvre, C. G., R. J. W. Le Fèvre, et al., *J. Chem. Soc.* **1959**, 1188.
2880. Le Fèvre, C. G., R. J. W. Le Fèvre, et al., *J. Chem. Soc.* **1960**, 123.
2881. Le Fèvre, C. G., R. J. W. Le Fèvre, et al., *J. Chem. Soc.* **1960**, 1814.
2882. Le Fèvre, R. J. W. and F. Maramba, *J. Chem. Soc.* **1952**, 235.
2883. Le Fèvre, R. J. W. and B. J. Orr, *J. Chem. Soc.* **1965**, 5349.
2884. Le Fèvre, R. J. W. and B. J. Orr, *J. Chem. Soc.* **1966B**, 37.
2885. Le Fèvre, R. J. W. and B. J. Orr, et al., *J. Chem. Soc.* **1965**, 2499.
2886. Le Fèvre, R. J. W., B. J. Orr, et al., *J. Chem. Soc.* **1965**, 3619.
2887. Le Fèvre, R. J. W., R. K. Pierens, et al., *Aust. J. Chem.* **19**, 1325 (1966); *C. A.* **65**, 14,572 (1966).
2888. Le Fèvre, R. J. W., W. P. H. Roberts, et al., *J. Chem. Soc.* **1949**, 902.
2889. Le Fèvre, R. J. W., R. Roper, et al., *Aust. J. Chem.* **12**, 743 (1959); *C. A.* **54**, 5246 (1960).
2890. Le Fèvre, R. J. W. and J. W. Smith, *J. Chem. Soc.* **1932**, 2810.
2891. Le Fèvre, R. J. W. and K. M. S. Sundaram, *J. Chem. Soc.* **1963**, 1880.
2892. Le Fèvre, R. J. W. and A. J. Williams, *J. Chem. Soc.* **1960**, 108.
2893. Le Fèvre, R. J. W. and A. J. Williams, *J. Chem. Soc.* **1960**, 115.
2894. Le Fèvre, R. J. W. and A. J. Williams, *J. Chem. Soc.* **1960**, 1825.
2895. Le Fèvre, R. J. W. and A. J. Williams, *J. Chem. Soc.* **1965**, 4185.
2896. Lěibush, A. G. and E. D. Shorina, *J. Appl. Chem.* (USSR) **20**, 69 (1947); *C. A.* **41**, 5446 (1947); Beil **EIII4**, 696.
2897. Leighton, P. A., R. W. Crary, et al., *J. Amer. Chem. Soc.* **53**, 3017 (1931).
2898. Leis, D. G. and B. C. Currans, *J. Amer. Chem. Soc.* **67**, 79 (1945).
2899. Lek, L., *Annual Tables of Physical Constants*, Sec. 700 (c), 1941.
2900. Lemaire, H. and H. J. Lucas, *J. Amer. Chem. Soc.* **73**, 5198 (1951).
2901. Lemons, J. F. and W. A. Felsing, *J. Amer. Chem. Soc.* **65**, 46 (1943).
2902. Lemoult, P., *Compt. Rend.* **143**, 746 (1906).
2903. Lenormant, H., *Bull. Soc. Chim. Fr.* **1948**, 33; *C. A.* **42**, 4060 (1948).
2904. Lenormant, H. and P. L. Clement, *Bull. Soc. Chim. Fr.* **1946**, 559; *C. A.* **41**, 2642 (1947).
2905. Lenz, W., *Arch. Pharm.* **249**, 289 (1911).
2906. Leonard, N. J. and L. E. Sutton, *J. Amer. Chem. Soc.* **70**, 1564 (1948).
2907. Lenormant, H., *Compt. Rend.* **228**, 1861 (1949); *C. A.* **43**, 8274 (1949).
2908. Leonard, L. J., *Organic Syntheses*, Vol. 36, Wiley, New York, 1956.
2909. Leroy, Y. and E. Constant, *Compt. Rend., Ser. A, B* **262B**, 1391 (1966); *C. A.* **65**, 8198 (1966).
2910. Lespieau, R., *Bull. Soc. Chim. Fr.* (4) **3**, 638 (1908); *Chem. Zentr.* **1909**, **II**, 151.
2911. Lespieau, R., *Compt. Rend.* **169**, 31 (1919).
2912. Lespieau, R. and B. Gredy, *Compt. Rend.* **196**, 399 (1933); *C. A.* **27**, 2673 (1933).
2913. Letaw, H. and A. H. Gropp, *J. Phys. Chem.* **57**, 964 (1953).
2914. Levene, P. A. and L. H. Cretcher, *J. Biol. Chem.* **33**, 505 (1918).
2915. Levene, P. A. and H. L. Haller, *J. Biol. Chem.* **69**, 569 (1926).
2916. Levene, P. A., A. Rothen, et al., *J. Biol. Chem.* **115**, 401 (1936).

2917. Levene, P. A. and A. Rothen, *J. Biol. Chem.* **115,** 415 (1936).
2918. Levene, P. A. and A. Rothen, *J. Biol. Chem.* **116,** 209 (1936).
2919. Levene, P. A. and F. A. Taylor, *J. Biol. Chem.* **54,** 351 (1922).
2920. Levene, P. A. and A. Walti, *J. Biol. Chem.* **68,** 415 (1926).
2921. Levene, P. A. and A. Walti, *J. Biol. Chem.* **94,** 361 (1931).
2922. Levene, P. A. and A. Walti, *J. Biol. Chem.* **94,** 367 (1931).
2923. Levene, P. A., A. Walti, et al., *J. Biol. Chem.* **71,** 465 (1926–1927).
2924. Levi, G., *Compt. Rend* **261** (Groupe 6), 4007 (1965); *C. A.* **64,** 4905 (1966).
2925. Levi, L., L. G. Chatten, et al., *J. Amer. Pharm. Assoc., Sci. Ed.* **44,** 61 (1955).
2926. Levin, B. Y., *Zh. Fiz. Khim.* **28,** 1399 (1954); *C. A.* **49,** 7980 (1955).
2927. Levy, E. J. and W. A. Stahl, *Anal. Chem.* **33,** 707 (1961).
2928. Lewin, A. H. and S. Winstein, *J. Amer. Chem. Soc.* **84,** 2464 (1962).
2929. Lewin, S. Z. and J. E. Vance, *J. Amer. Chem. Soc.* **74,** 1433 (1952).
2930. Lewis, D. T., *J. Chem. Soc.* **1938,** 1056.
2931. Lewis, D. T., *J. Chem. Soc.* **1940,** 32.
2932. Lewis, E., *J. Soc. Chem. Ind.* **41,** 97T (1922).
2933. Lewis, G. L. and C. P. Smyth, *J. Amer. Chem. Soc.* **61,** 3067 (1939).
2934. Lewis, G. L., and C. P. Smyth, *J. Chem. Phys.* **7,** 1085 (1939).
2935. Lewis, G. N., "Valence and Structure of Atoms and Molecules," Chemical Catalog Company, New York, 1923.
2936. Lewis, G. N., *J. Franklin Inst.* **226,** 293 (1938).
2937. Lewis, H. F., R. Hendricks, et al., *J. Amer. Chem. Soc.* **50,** 1993 (1928).
2938. Lewis, J., J. R. Miller, et al., *J. Chem. Soc.* **1965,** 5850.
2939. Lewis, J. R., *J. Amer. Chem. Soc.* **47,** 626 (1925).
2940. Ley, H. and B. Arends, *Z. Phys. Chem.* (Leipzig) **15B,** 311 (1932); *Chem. Zentr.* **1932, I,** 2549.
2941. Ley, H. and B. Arends, *Z. Phys. Chem.* (Leipzig) **17B,** 177 (1932); *Chem. Zentr.* **1932, II,** 835.
2941a. Ley, H. and H. Hunecke, *Ber. Deut. Chem. Gesell.* **59,** 510 (1926).
2942. Li, C.-C., *Hua Hsueh Tung Pao* **1960,** No. 3, 128; *C. A.* **55,** 24,167 (1961).
2943. Li, J. C. M. and K. S. Pitzer, *J. Amer. Chem. Soc.* **78,** 1077 (1956).
2944. Li, J. C. M. and F. D. Rossini, *J. Chem. Eng. Data* **6,** 268 (1961).
2945. Li, K., *J. Phys. Chem* **61,** 782 (1957).
2946. Li, N. C. C., *J. Chem. Phys.* **7,** 1068 (1939).
2947. Li, N. C. C. and P. C. Hsü, *J. Chinese Chem. Soc.* **13,** 11 (1946); *C. A.* **41,** 3672 (1947).
2948. Li, N. C. C. and T. D. Terry, *J. Amer. Chem. Soc.* **70,** 344 (1948).
2949. Liberman, A. L., Z. N. Parnes, et al., *Bull. Acad. Sci. URSS,* Classe Sci. Chim. **1948,** 101; *C. A.* **42,** 5291 (1948).
2950. Liddell, U., *Ann. N.Y. Acad. Sci.* **69,** 70 (1957).
2951. Liddel, U. and C. Kasper, *J. Res. Nat. Bur. Stand.* **11,** 599 (1933); *C. A.* **28,** 1601 (1934).
2952. Liebaert, R., A. Lebrun, et al., *Compt. Rend.* **253,** 2496 (1961); *C. A.* **56,** 8070 (1962)
2953. Lieben, A. and S. Zeisel, *Ber. Deut. Chem. Gesell.* **14,** 515 (1881).
2954. Liebermann, L. N., *Phys. Rev.* **60,** 496 (1941).
2955. Liebush, A. G. and E. D. Shorina, *J. Applied Chem.* (USSR) **20,** 69 (1947); *C. A.* **41,** 5447 (1947).
2956. Liem, H. D., *Acta Chem. Scand.* **22,** 753 (1968).
2957. Liler, M. and D. Kosanović, *J. Chem. Soc.* **1958,** 1084.
2958. Lin, R.-Y. and W. Dannhauser, *J. Phys. Chem.* **67,** 1805 (1963).

2959. Lin, Wei-Chuwan and Fu-Tai Tuan, *J. Chinese Chem. Soc.* **5**, 33 (1958); *C. A.* **53**, 10,872 (1959).

2960. Linard, J., *Bull. Soc. Chim. Belges* **34**, 363 (1925).

2961. Linch, A. L., *Am. Ind. Hyg. Assoc. J.* **26**, 95 (1965).

2962. Lincoln, A. T., *J. Phys. Chem.* **3**, 457 (1899).

2963. Lind, E. L. and T. F. Young, *J. Chem. Phys.* **1**, 266 (1933).

2964. Lindberg, J. J., J. Kenttama, et al., *Suomen Kemistilehti* **34B**, 98 (1961), in English; *C. A.* **58**, 1946 (1963).

2965. Lindenberg, A. B., *Compt. Rend. Ser. C* **262**, 1504 (1966); *C. A.* **65**, 6450 (1966).

2966. Linder, E. G., *J. Phys. Chem.* **35**, 531 (1931).

2967. Ling, A. C. and J. E. Willard, *J. Phys. Chem.* **72**, 3349 (1968).

2968. Ling, T. D. and M. Van Winkle, *Ind. Eng. Chem., Chem. Eng. Data Series* **3**, 88 (1958).

2969. Linnell, R. H., *J. Org. Chem.* **25**, 290 (1960).

2970. Lipp, A., *Ann. Chem.* **205**, 1 (1880); *J. Chem. Soc.* **1881A**.

2971. Lippert, E. and H. Prigge, *Ann. Chem.* **659**, 81 (1962); *C. A.* **58**, 8878 (1963).

2972. Lippincott, E. R., E. R. Myers, et al., *Spectrochim. Acta* **22**, 1493 (1966); *C. A.* **65**, 9963 (1966).

2973. Lippincott, S. B., U.S. Patent 2,423,783, July 8, 1947; *C. A.* **41**, 6275 (1947).

2974. Lippincott, S. B. and H. B. Hass, *Ind. Eng. Chem.* **31**, 118 (1939).

2975. Lippmaa, E., A. Olivson, et al., *Eesti NSV Tead. Akad. Toim., Fuusik.-Mat. ja Tehnikateaduste Seer* **14**, 473 (1965); *C. A.* **64**, 7552 (1966).

2976. Lister, M. W., *J. Amer. Chem. Soc.* **63**, 143 (1941).

2977. Litovitz, T. A., R. Higgs, et al., *J. Chem. Phys.* **22**, 1281 (1954).

2978. Little, M. H. and A. E. Martell, *J. Phys. Colloid Chem.* **53**, 472 (1949).

2979. Little, R. C. and C. R. Sengleterry, *J. Phys. Chem.* **68**, 2709 (1964).

2980. Liu, K. F. and W. T. Ziegler, *J. Chem. Eng. Data* **11**, 187 (1966).

2981. Livingston, J., R. Morgan, et al., *J. Amer. Chem. Soc.* **33**, 1713 (1911).

2982. Livingston, J., R. Morgan, et al., *J. Amer. Chem. Soc.* **35**, 1505 (1913).

2983. Livingston, J., R. Morgan, et al., *J. Amer. Chem. Soc.* **35**, 1821 (1913).

2984. Livingston, J., R. Morgan, et al., *J. Amer. Chem. Soc.* **46**, 881 (1924).

2985. Livingston, R., *J. Amer. Chem. Soc.* **69**, 1220 (1947).

2986. Lochte, H. L., *Ind. Eng. Chem.* **16**, 956 (1924).

2987. Locke, E. G., W. R. Brode, et al., *J. Amer. Chem. Soc.* **56**, 1726 (1934).

2988. Locquin, M., *Bull. Mens. Soc. Linnéenne Lyon* **11**, 95 (1942); *C. A.* **40**, 427 (1946).

2989. Locquin, R., *Ann. Chim.* **19**, 32 (1923); *C. A.* **17**, 2559 (1923).

2990. Loder, D. J., U.S. Patent 2,204,371, June 11, 1940; *Official Gaz. U.S. Patent Office* **515**, 494 (1940).

2991. Loeser, A., *Pharmazie* **4**, 263 (1949); *C. A.* **43**, 8558 (1949).

2992. Loewenherz, R., *Z. Phys. Chem.* (Leipzig) **6**, 552 (1890).

2993. Lombard, R., *Bull. Soc. Chim. Fr.* **1946**, 598; *C. A.* **41**, 2875 (1947).

2994. Lomonova, G. V., *Prom. Toksikol. Klinika Prof. Zabolevanii Khim. Etiol. Sb.* **1962** 160; *C. A.* **61**, 3600 (1964).

2995. Long, F. A. and P. Ballinger, *Electrolytes, Proc. Intern. Symp., Trieste, Yugoslavia* **1959**, 152 (Pub. 1962); *C. A.* **61**, 10,098 (1964).

2996. Long, T. V. and R. A. Plane, *J. Chem. Phys.* **43**, 457 (1965).

2997. Longinow, W. and A. Prjaschnikow, *Trans. Inst. Pure Chem. Reagents* (Moscow) **1931**, 48.

2998. Longster, G. F. and E. E. Walker, *Trans. Faraday Soc.* **49**, 228 (1953); *C. A.* **47**, 9697 (1953).

2999. Lord, R. C. and F. A. Miller, *J. Chem. Phys.* **10**, 328 (1942).

3000. Louder, E. A., T. R. Briggs, et al., *Ind. Eng. Chem.* **16**, 932 (1924).
3000a. Louguinine, see also Luginin.
3001. Louguinine, W., *Ann. Chim. Phys.* **13**, 289 (1898); *Chem. Zentr.* **1898, I**, 824.
3002. Louguinine, W., *J. Chim. Phys.* **2**, 1 (1904); *Chem. Zentr.* **1904, II**, 436.
3003. Louguinine, W. and G. Dupont, *Bull. Soc. Chim. Fr.* **9**, 219 (1911); *C. A.* **5**, 1857 (1911).
3004. Louguinine, W. E., *J. Chim. Phys.* **2**, 1 (1904); *Chem. Zentr.* **1904, II**, 436.
3005. Lovering, E. G. and K. J. Laidler, *Can. J. Chem.* **38**, 2367 (1960).
3006. Lovering, E. G. and O. bin. M. Nor, *Can. J. Chem.* **40**, 199 (1962).
3007. Löwe, K. F., *Ann. Phys.* (Leipzig) **66**, 390 (1898).
3008. Löwenherz, R., *Z. Phys. Chem.* (Leipzig) **25**, 385 (1898).
3009. Lowry, T. M., *J. Chem. Soc.* **1914**, 81.
3010. Lowry, T. M., *Trans. Faraday Soc.* **20**, 13 (1924).
3011. Lowry, T. M. and C. B. Allsopp, *Proc. Roy. Soc.* **113A**, 26 (1931).
3012. Lozac'h, N., *Bull. Soc. Chim. Fr.* **1949**, 286; *C. A.* **43**, 6162 (1949).
3013. Lucatu, E., *Compt. Rend.* **207**, 1403 (1938); *C. A.* **33**, 1563 (1939).
3014. Luder, W. F. and S. Zuffanti, *The Electronic Theory of Acids and Bases*, Wiley, New York, 1946.
3014a. Luginin, see also Louguinine.
3015. Luginin, W., *Arch. Sci. Phys. Nat. Gèneve* **9**, 5; *J. Chem. Soc.* **1900Aii**, 334.
3016. Luginin, W. F., *Ann. Chim. Phys.* (Paris) **26**, 288 (1902); *J. Chem. Soc.* **1902Aii**, 547.
3017. Luginin, W. F., *Ann. Chim. Phys.* (Paris) **27**, 105 (1902); *J. Chem. Soc.* **1903Aii**, 7.
3018. Luginin, W. F., *Arch. Sci. Phys. Nat.* **9**, 5 (1900).
3019. Luginin, W. F., *Compt. Rend.* **128**, 366 (1899); *J. Chem. Soc.* **1899Aii**, 354.
3020. Lumbroso, H., *Ann. Fac. Sci. Univ. Toulouse Sci.-Math et Sci. Phys.* **14**, 21, 35, 56, 108 (1950); *C. A.* **48**, 5781 (1954).
3021. Lumbroso, H., *Compt. Rend.* **225**, 1003 (1947); *C. A.* **42**, 2827 (1948).
3022. Lumbroso, H., *Compt. Rend.* **228**, 77 (1949); *C. A.* **43**, 4061 (1949).
3023. Lumbroso, H., D. M. Bertin, et al., *Bull. Soc. Chim. Fr.* **1966**, 540; *C. A.* **64**, 16,752 (1966).
3024. Lumbroso, H. and P. Rumpf, *Bull. Soc. Chim. Fr.* **1950**, 371; *C. A.* **44**, 7106 (1950).
3025. Lumbroso, N., T. K. Wu, et al., *J. Phys. Chem.* **67**, 2469 (1963).
3026. Lumsden, J. S., *J. Chem. Soc.* **1905**, 90.
3027. Lumsden, J. S., *J. Chem. Soc.* **1919**, 1366.
3028. Lunazzi, L. and F. Taddei, *Bull. Soc. Fac. Chim. Ind. Bologna* **22**, 91 (1964); *C. A.* **63**, 12,538 (1965).
3029. Lund, H. and J. Bjerrum, *Ber. Deut. Chem. Gesell.* **64**, 210 (1931).
3030. Lundin, H. G., M. L. Crowder, *Proc. Sci. Sec. Toilet Goods Assoc.* **18**, 1 (1952); *C. A.* **47**, 3698 (1953).
3031. Lundin, N. P., *Farmatsiya* **9**, No. 2, 39 (1946); *C. A.* **41**, 4445 (1947).
3032. Lunelli, B. and C. Pecile, *Gazz. Chim. Ital.* **96**, (1966); *C. A.* **65**, 9946 (1966).
3033. Lunge, G., *Ber. Deut. Chem. Gesell.* **14**, 1755 (1881).
3034. Lunt, R. W. and M. A. G. Rau, *Proc. Roy. Soc.* **126A**, 213 (1930).
3035. Luther, H., *Z. Elektrochem.* **52**, 210 (1948); *C. A.* **43**, 8888 (1949).
3036. Luther, H. and E. Lohrengel, *Brennst-Chem.* **35**, 338 (1954); *C. A.* **49**, 3736 (1955).
3037. Luther, R. and F. Weigert, *Z. Phys. Chem.* (Leipzig) **51**, 297 (1905).
3038. Lüthy, A., *Z. Phys. Chem.* (Leipzig) **107**, 285 (1923).
3039. Lutskiĭ, A. E., *Zh. Obshch. Khim.* **24**, 440 (1954); *C. A.* **48**, 8609 (1954).
3040. Lutskiĭ, A. E. and S. A. Mikhailenko, *Zh. Strukt. Khim.* **3**, 523 (1962); *C. A.* **59**, 4622 (1963).
3041. Lutskiĭ, A. E. and A. N. Panova, *Zh. Fiz. Khim.* **33**, 970 (1959); *C. A.* **54**, 8261 (1960).

3042. Lüttringhaus, A. and H. W. Dirksen, *Angew. Chem. Intern. Ed. Engl.* **3**, 260 (1964).
3043. Lüttringhaus, A. and J. Grohmann, *Z. Naturforsch.* **10B**, 367 (1955); from [3042].
3044. Lynch, B. M., B. C. Macdonald, et al., *Tetrahedron* **24**, 3595 (1968).
3045. Lynch, E. J. and C. R. Wilke, *J. Chem. Eng. Data* **5**, 300 (1960).
3046. Lynn, E. V., *J. Amer. Chem. Soc.* **41**, 361 (1919).
3047. Lynn, J. W., *J. Org. Chem.* **21**, 578 (1956).
3048. Lyons, L. E., *J. Proc. Roy. Soc. N.S. Wales* **83**, 75 (1949); *C. A.* **45**, 7430 (1951).
3049. Lyubomilov, V. I. and N. M. Merkula, *Zh. Obshch. Khim.* **33**, 22 (1963); *C. A.* **59**, 425 (1963).

M

3050. McAllan, D. T., T. V. Cullum, et al., *J. Amer. Chem. Soc.* **73**, 3626 (1951).
3051. McAlpine, K. B. and C. P. Smyth, *J. Chem. Phys.* **3**, 55 (1935).
3052. McArdle, E. H. and D. M. Mason, U.S. Patent 2,435,792, Feb. 10, 1948; *C. A.* **42**, 2988 (1948).
3053. McAuliffe, C., *J. Phys. Chem.* **70**, 1267 (1966).
3054. McBee, E. T., V. V. Lindgren, et al., *Ind. Eng. Chem.* **39**, 378 (1947).
3055. McCaulay, D. A., B. H. Shoemaker, et al., *Ind. Eng. Chem.* **42**, 2103 (1950).
3056. McClellan, A. L., *Tables of Experimental Dipole Momenti*, W. H. Freeman, San Francisco, 1963.
3057. McClellan, A. L. and S. W. Nicksic, *J. Phys. Chem.* **69**, 446 (1965).
3057a. MacCleod, D. B., *Trans. Faraday Soc.* **19**, 38 (1923); *C. A.* **17**, 2072 (1923).
3058. McClure, H. B., *Chem. Eng. News* **22**, 421 (1944).
3059. McClure, J. H., T. M. Roder, et al., *Anal. Chem.* **27**, 1599 (1955).
3060. McCombie, H., B. C. Saunders, et al., *J. Chem. Soc.* **1944**, 24.
3061. McCullough, J. P., H. L. Finke, et al., *J. Phys. Chem.* **61**, 289 (1957).
3062. McCullough, J. P., H. L. Finke, et al., *J. Phys. Chem.* **61**, 1105 (1957).
3063. McCullough, J. P. and W. D. Good, *J. Phys. Chem.* **65**, 1430 (1961).
3064. McCullough, J. P., W. N. Hubbard, et al., *J. Amer. Chem. Soc.* **79**, 561 (1957).
3065. McCullough, J. P., R. E. Pennington, et al., *J. Amer. Chem. Soc.* **81**, 5880 (1959).
3066. McCullough, J. P., D. W. Scott, et al., *J. Amer. Chem. Soc.* **76**, 4791 (1954).
3067. McCurdy, K. G. and K. J. Laidler, *Can. J. Chem.* **41**, 1867 (1963).
3068. McDevitt, N. T., A. L. Rozek, et al., *J. Chem. Phys.* **42**, 1173 (1965).
3069. McDonald, H. J., *J. Chem. Phys.* **48**, 47 (1944).
3070. McDonald, R. A., S. A. Shrader, et al., *J. Chem. Eng. Data* **4**, 311 (1959).
3071. McDougall, F. H., *J. Amer. Chem. Soc.* **58**, 2585 (1936).
3072. McDougall, L. A. and J. E. Kilpatrick, *J. Chem. Phys.* **42**, 2307 (1965).
3073. McEachern, D. M. and J. E. Kilpatrick, *J. Chem. Phys.* **41**, 3127 (1964).
3074. McEwen, B., *J. Chem. Soc.* **1923**, 2284.
3075. McEwen, W. K., *J. Amer. Chem. Soc.* **58**, 1124 (1936).
3076. McFadden, W. H., E. A. Day, et al., *Anal. Chem.* **37**, 89 (1965).
3077. McFadden, W. H., J. Wasserman, et al., *Anal. Chem.* **36**, 1031 (1964).
3078. McGovern, E. W., *Ind. Eng. Chem.* **35**, 1230 (1943).
3079. McKay, R. A. and B. N. Sage, *J. Chem. Eng. Data* **5**, 21 (1960).
3080. McKelvy, E. C. and D. H. Simpson, *J. Amer. Chem. Soc.* **44**, 105 (1922).
3081. McKenna, F. E., H. V. Tartar, et al., *J. Amer. Chem. Soc.* **71**, 729 (1949).
3082. McKinley, C., British Patent 705,670, Mar. 17, 1954; *C. A.* **49**, 9026 (1955).
3083. McKinley, C. and J. P. McKinley, *J. Amer. Chem. Soc.* **72**, 5331 (1950).
3084. McKinney, D. S., C. E. Leberknight, et al., *J. Amer. Chem. Soc.* **59**, 481 (1937).
3085. McLafferty, F. W., *Anal. Chem.* **28**, 306 (1956).

3086. McLafferty, F. W., *Anal. Chem.* **29**, 1782 (1957).
3087. McLafferty, F. W., *Anal. Chem.* **34**, 2, 16, 26 (1962).
3088. McLafferty, F. W., *Atlas of Mass Spectra*, Wiley-Interscience, New York, 1968.
3089. McMillan, D. R., *Phys. Rev.* **57**, 941 (1939).
3090. McMullan, R. K. and J. D. Corbett, *J. Chem. Educ.* **33**, 313 (1956).
3091. McMurry, H. L., *J. Chem. Phys.* **9**, 231 (1941).
3092. McMurry, H. L., *J. Chem. Phys.* **9**, 241 (1941).
3093. McNeight, S. A. and C. P. Smyth, *J. Amer. Chem. Soc.* **58**, 1718 (1936).
3094. McQuillin, F. J. and W. O. Ord, *J. Chem. Soc.* **1959**, 3169.
3095. Macbeth, A. K. and J. A. Mills, *J. Chem. Soc.* **1945**, 709.
3096. Macbeth, A. K. and J. A. Mills, *J. Chem. Soc.* **1947**, 205.
3097. Maccoll, A. and P. J. Thomas, *J. Chem. Soc.* **1955**, 979.
3098. Maciel, G. E. and G. B. Savitsky, *J. Phys. Chem.* **69**, 3925 (1965).
3099. MacInnes, D. A. and T. Shedlovsky, *J. Amer. Chem. Soc.* **54**, 1429 (1932).
3100. Mackinney, G. and O. Temmer, *J. Amer. Chem. Soc.* **70**, 3586 (1948).
3101. Mackle, H. and R. T. B. McClean, *Trans. Faraday Soc.* **60**, 817 (1964).
3102. Mackle, H. and P. A. G. O'Hare, *Tetrahedron* **19**, 961 (1963).
3103. Mackle, H. and P. A. G. O'Hare, *Trans. Faraday Soc.* **58**, 1912 (1962).
3104. Mackor, E. L., A. Hofstra, et al., *Trans. Faraday Soc.* **54**, 186 (1958).
3105. Maclean, M. E., P. J. Jencks, et al., *J. Res. Nat. Bur. Stand.* **34**, 271 (1945).
3106. Maesen, F. van der, *Physica* **15**, 481 (1949); *C. A.* **44**, 3319 (1950).
3107. Magill, P. L., *Ind. Eng. Chem.* **26**, 611 (1934).
3108. Magnusson, L. B., C. Postmus, et al., *J. Amer. Chem. Soc.* **85**, 1711 (1963).
3109. Mahanti, P. C., *J. Indian Chem. Soc.* **5**, 673 (1928).
3110. Mahanti, P. C., *J. Indian Chem. Soc.* **6**, 743 (1929); *C. A.* **24**, 1001 (1930).
3111. Mahanti, P. C. and R. N. Das-Gupta, *J. Indian Chem. Soc.* **6**, 411 (1929); *C. A.* **24**, 1001 (1930).
3112. Mahanti, P. C. and D. N. Sen-Gupta, *J. Indian Chem. Soc.* **5**, 673 (1928); *C. A.* **23**, 2615 (1929).
3113. Maibaum, B. K., *J. Exptl. Theoret. Phys.* (URSS) **9**, 1383 (1939); *C. A.* **35**, 5006 (1941).
3114. Maier, W. and K. R. Fröhner, *Spectrochim. Acta* **1959**, 977; *C. A.* **54**, 7341 (1960).
3115. Mair, B. J., *J. Res. Nat. Bur. Stand.* **9**, 457 (1932).
3115a. Mair, B. J., *J. Res. Nat. Bur. Stand.* **34**, 435 (1945).
3116. Mair, B. J. and A. F. Forziati, *J. Research Nat. Bur. Stand.* **32**, 151 (1944).
3116a. Mair, B. J. and A. F. Forziati, *J. Res. Nat. Bur. Stand.* **32**, 165 (1944).
3117. Mair, B. J., A. R. Glasgow, et al., *J. Res. Nat. Bur. Stand.* **26**, 591 (1941).
3118. Mair, B. J. and A. J. Streiff, *J. Res. Nat. Bur. Stand.* **24**, 395 (1940).
3119. Mair, B. J. and A. J. Streiff, *J. Res. Nat. Bur. Stand.* **27**, 343 (1941).
3120. Mair, B. J., D. J. Termini, et al., *J. Res. Nat. Bur. Stand.* **37**, 229 (1946).
3121. Majewska, J. and S. Urbanowicz, *Polimery* **9** (1), 18 (1964); *C. A.* **61**, 8409 (1964).
3122. Makarov, Y. A., L. N. Voronina, et al., USSR Patent 162,835, May 27, 1964; *C. A.* **61**, 8485 (1964).
3123. Malanowski, S., *Bull. Acad. Pol. Sci. Ser. Sci. Chim.* **9**, No. 2, 71 (1961).
3124. Malecki, J., *Acta Phys. Pol.* **21**, 13 (1962), in English; *C. A.* **58**, 3985 (1963).
3125. Malherbe, F. E. and H. J. Bernstein, *J. Amer. Chem. Soc.* **74**, 4408 (1952).
3126. Malinowski, E. R., T. Vladimiroff, et al., *J. Phys. Chem.* **70**, 2046 (1966).
3127. Malinovskiĭ, M. S., E. E. Volkova, et al., *Zh. Obshch. Khim.* (*J. Gen. Chem.*) **19**, 114 (1949); *C. A.* **43**, 6155 (1949).

3128. Malinowski, S., H. Jedrzejewska, et al., *Rocz. Chem.* **31**, 71 (1957); *C. A.* **51**, 14,557 (1957).

3129. Malmberg, C. G. and A. A. Maryott, *J. Research Nat. Bur. Stand.* **56**, 1 (1956).

3130. Malyshev, V. I., *Bull. Acad. Sci. URSS, Ser. Phys.* **5**, No. 1, 13 (1941); *Khim. Referat. Zhur.* **4**, No. 9, 6 (1941); *C. A.* **38**, 679 (1944).

3131. Malyshev, V. I. and V. N. Murzin, *Izv. Akad. Nauk SSSR, Ser. Fiz.* **22**, 1107 (1958); *C. A.* **53**, 789 (1959).

3132. Malyshev, V. I. and M. V. Shishkina, *Dokl. Akad. Nauk SSSR* **66**, 833 (1949); *C. A.* **43**, 7345 (1949).

3133. Maman, A., *Pub. Sci. Tech. Ministère Air* (France) No. **66** (1935); *C. A.* **30**, 7095 (1936).

3134. Mandric, G., *Igiena* (Bucharest) **9**, 271 (1961); *C. A.* **57**, 8847 (1962).

3135. Mandric, G., *Rev. Chim.* (Bucharest) **12**, 503 (1961); *C. A.* **56**, 6317 (1962).

3136. Mangini, A., *Gazz. Chim. Ital.* **88**, 1063 (1958); *C. A.* **53**, 18,624 (1959).

3137. Manley, R. St. J., *Svensk Papperstidn.* **61**, 96 (1958).

3138. Manly, D. G., U.S. Patent 3,021,342, Feb. 13, 1962; *C. A.* **56**, 15,489 (1962); *Official Gaz. U.S. Patent Office* **775**, 558 (1962).

3139. Manly, D. G. and E. D. Amstutz, *J. Org. Chem.* **22**, 323 (1957).

3140. Manufacturing Chemists Association Data Sheet SD-29, *J. Chem. Ed.* **41**, A578 (1964).

3141. Manzoni-Ansidei, R., *Boll. Sci. Fac. Chim. Ind. Bologna* **1940**, 137; *C. A.* **36**, 6904 (1942).

3142. Manzoni-Ansidei, R., *Ric. Sci.* **10**, 328 (1939); *Chem. Zentr.* **1940**, I, 35.

3143. Manzoni-Ansidei, R. and M. Rolla, *Atti. Accad. Naz. Lincei, Rend Cl. Sci. Fis. Mat. Nat.* **27**, 410 (1938); *C. A.* **32**, 8933 (1938).

3144. Marckwald, W. and A. McKenzie, *Ber. Deut. Chem. Gesell.* **34**, 485 (1901).

3145. Marcom, K. W. and D. N. Traverse, *Trans. Faraday Soc.* **61**, 230 (1965).

3146. Marcus, S. H., W. F. Reynolds, et al., *J. Org. Chem.* **31**, 1872 (1966).

3147. Mardles, E. W. J., *Trans. Faraday Soc.* **29**, 476 (1933); *C. A.* **27**, 4150 (1933).

3148. Marenzi, A. D. and F. Vilallonga, *Rev. Soc. Argentina Biol.* **17**, 232 (1941); *C. A.* **36**, 500 (1942).

3149. Marinangeli, A., *Ann. Chim.* (Rome) **44**, 211, 219 (1954); *C. A.* **49**, 2158 (1955).

3150. Markley, K. S., Ed., *Fatty Acids*, 2d ed., pt. 1, Interscience, New York, 1960.

3151. Markley, K. S., Ed., *Fatty Acids*, 2d ed., pt. 2, Interscience, New York, 1961.

3152. Markley, K. S., Ed., *Fatty Acids*, 2d ed., pt. 3, Interscience, New York, 1964.

3153. Markova, S. V., P. A. Bazhulin, et al., *Opt. Spectrosk.* **1**, No. 1, 41 (1956); *C. A.* **50**, 13,413 (1956).

3154. Markownikow, W. and W. Tscherdynzew, *J. Russ. Phys. Chem. Ges.* **32**, 302; *Chem. Zentr.* **1900**, II, 630.

3155. Markunas, P. C. and J. A. Riddick, *Anal. Chem.* **23**, 337 (1951).

3156. Maroni, P., *Ann. Chim.* (Paris) **2**, 757 (1957); *C. A.* **52**, 8728 (1958).

3157. Marquardt, R. P. and E. N. Luce, *Ind. Eng. Chem. Anal. Ed.* **16**, 751 (1944).

3158. Marrison, L. W., *J. Chem. Soc.* **1951**, 1614.

3159. Marschalko, B. and J. Barna, *Acta Tech. Acad. Sci. Hung.* **19**, 85 (1957); *C. A.* **52**, 15,994 (1958).

3160. Marsden, C. and S. Mann, *Solvents Guide* 2d ed., Interscience, New York, 1963.

3161. Marsden, J. and A. C. Cuthbertson, *Can. J. Res.* **9**, 419 (1933); *C. A.* **28**, 696 (1934); and ref. [4568].

3162. Marsden, R. J. B. and L. E. Sutton, *J. Chem. Soc.* **139**, 1383 (1936).

3163. Marshall, M. J., *Ind. Eng. Chem.* **20**, 1379 (1928).

3164. Martelli, D., *Med. Lavoro* **51**, 123 (1960); *C. A.* **54**, 23,008 (1960).

3165. Martens, T. F., Belgian Patent 613,121, July 25, 1962; *C. A.* **58**, 8902 (1963).

3166. Martin, A. R., *J. Chem. Soc.* **1928**, 3270.

3167. Martin, A. R. and B. Collie, *J. Chem. Soc.* **1932**, 2658.

3168. Martin, A. R. and C. M. George, *J. Chem. Soc.* **1933**, 413.

3169. Martin, D., A. Weise, et al., *Angew. Chem. Intern. Ed.* **6**, 318 (1967).

3170. Martin, D. R., *J. Phys. Colloid Chem.* **51**, 1400 (1947).

3171. Martin, G. J. and M. L. Martin, *Bull. Soc. Chim. Fr.* **1966**, 1636; *C. A.* **65**, 13,511 (1966).

3172. Martin, J. and B. P. Dailey, *J. Chem. Phys.* **37**, 2594 (1962).

3173. Martin, M., *Ann. Phys.* (Paris) **7**, 35 (1962); *C. A.* **59**, 148 (1963).

3174. Martin, M., *Bull. Soc. Roy. Sci. Liege* **31**, 434 (1962); *C. A.* **57**, 6772 (1962).

3175. Martin, M., *J. Chim. Phys.* **59**, 736 (1962); *C. A.* **58**, 158 (1963).

3176. Martin, M. and F. Herail, *Compt. Rend.* **248**, 1994 (1959); *C. A.* **56**, 10,983 (1962).

3177. Martin, M. and M. Quilbeuf, *Compt. Rend.* **252**, 4151 (1961); *C. A.* **56**, 3046 (1962).

3178. Martin, R., *Rev. Inst. Franc. Pétrole* **2**, 323 (1947); *C. A.* **42**, 4497 (1948).

3179. Maruta, Y. and I. Matubora, *Nippon Nogei Kagaku Kaishi* **28**, 125 (1954); *C. A.* **51**, 9416 (1957).

3180. Maruyama, M., *Polarography* **3**, 22 (1955); *C. A.* **50**, 7348 (1956).

3181. Marvel, C. S., *Organic Syntheses*, Vol. 5, Wiley, New York, 1925.

3182. Marvel, C. S., *Organic Syntheses*, Vol. 11, Wiley, New York, 1931.

3183. Marvel, C. S., J. Harkema, et al., *J. Amer. Chem. Soc.* **63**, 1609 (1941).

3184. Maryott, A. A., *J. Amer. Chem. Soc.* **63**, 3079 (1941).

3185. Maryott, A. A., M. E. Hobbs, et al., *J. Amer. Chem. Soc.* **63**, 659 (1941).

3186. Maryott, A. A., M. E. Hobbs, et al., *J. Amer. Chem. Soc.* **71**, 1671 (1949).

3187. Mascarelli, C. and I. Musatty, *Gazz. Chim. Ital.* **41a**, 80 (1911); from 3942 p. 607.

3188. Mascarelli, L., *Atti Accud. Naz. Lincei Rend., Cl. Sci. Fis. Mat. Nat.* **16** (V), 924 (1907); *J. Chem. Soc.* **1907Aii**, 602.

3189. Mascarelli, L., *Atti Accad. Naz. Lincei Rend., Cl. Sci. Fis. Mat. Nat.* **17** (V), 494 (1908); *Chem. Zentr.* **1909**, I, 169.

3190. Mascarelli, L. and F. Benati, *Gazz. Chim. Ital.* **39II**, 642 (1909); *Chem. Zentr.* **1910**, I, 1024.

3191. Mascarelli, L. and L. Vecchiotti, *Atti Accad. Naz. Lincei, Rend., Cl. Sci. Fis. Mat. Nat.* **1911**, 410 (1910); *Chem. Zentr.* **1911**, I, 75.

3192. Mashiko, Y., *Nippon Kagaku Zasshi* **80**, 593 (1959); *C. A.* **53**, 13,778 (1959).

3193. Maslov, P. G. and Y. P. Maslov, *Khim. Tekhnol. Topl. Masel* **3**, No. 10, 50 (1958); *C. A.* **53**, 1910 (1959).

3194. Mason, M. E., B. Johnson, et al., *Anal. Chem.* **37**, 760 (1965).

3195. Massart, L., *Bull. Soc. Chim. Belges* **45**, 76 (1936); *C. A.* **30**, 4062 (1936).

3196. Massol, G. and A. Faucon, *Compt. Rend.* **149**, 345 (1909); *J. Chem. Soc.* **1909Aii**, 791.

3197. Massol, G. and A. Faucon, *Compt. Rend.* **157**, 386 (1913).

3198. Massol, G. and A. Faucon, *Compt. Rend.* **159**, 314 (1914); *C. A.* **9**, 10 (1915).

3199. Massol, G. and A. Faucon, *Compt. Rend.* **163**, 92 (1916); *Chem. Zentr.* **1916II**, 725.

3200. Massol, G. and A. Faucon, *Compt. Rend.* **164**, 308 (1917); *C. A.* **11**, 2560 (1917).

3201. Masson, I., *Nature* **128**, 726 (1931).

3202. Massy, N. B., F. L. Warren, et al., *J. Chem. Soc.* **1932**, 91.

3203. Mastrangelo, S. V. R., *Anal. Chem.* **29**, 841 (1957).

3204. Mastrangelo, S. R. V. and J. G. Aston, *Anal. Chem.* **26**, 764 (1954).

3205. Matheson, M., E. Auer, et al., *J. Amer. Chem. Soc.* **71**, 2610 (1949).

3206. Mathews, J. F. and J. J. McKetta, *J. Phys. Chem.* **65**, 758 (1961).
3207. Mathews, J. H., *J. Amer. Chem. Soc.* **39**, 1125 (1917).
3208. Mathews, J. H., *J. Amer. Chem. Soc.* **48**, 562 (1926).
3209. Mathews, J. H. and K. E. Faville, *J. Phys. Chem.* **22**, 1 (1918).
3210. Mathews, J. H. and P. R. Fehlandt, *J. Amer. Chem. Soc.* **53**, 3212 (1931).
3211. Mathias, A., *Tetrahedron* **22**, 217 (1966).
3212. Mathias, S., *J. Amer. Chem. Soc.* **72**, 1897 (1950).
3213. Mathias, S. and R. G. Cecchini, *Anales Real Soc. Espan. Fis. Quim.* Ser. B **60**, 241 (1964); *C. A.* **62**, 1149 (1965).
3214. Mathias, S. and E. de C. Filho, *J. Phys. Chem.* **62**, 1427 (1958).
3215. Mathias, S., E. de C. Filho, et al., *J. Phys. Chem.* **65**, 425 (1961).
3216. Mathieu, J. P. and D. Massignon, *Ann. Phys.* **16**, 5 (1941); *Chem. Zentr.* **1942**, II, 24; *C. A.* **37**, 4304 (1943).
3217. Mathieu, M. P., *Acad. Roy. Belg., Classe Sci. Mém. Collect.* in-8, **28**, No. 2, 224 pp. (1953); *C. A.* **47**, 10,288 (1953).
3218. Mathieu, M. P., *Bull. Soc. Chim. Belges* **61**, 683 (1952); *C. A.* **48**, 7365 (1954).
3219. Mathis, A., *Anal. Chim. Acta* **31**, 598 (1964).
3220. Mathur, R., E. D. Becker, et al., *J. Phys. Chem.* **67**, 2190 (1963).
3221. Mathur, R., S. M. Wang, et al., *J. Phys. Chem.* **68**, 2140 (1964).
3222. Matignon, C., H. Moureu, et al., *Bull. Soc. Chim. Fr.* **1**, 1308 (1934); *Chem. Zentr.* **1935**, I, 1855.
3223. Matossi, F., *Phys. Z.* **45**, 304 (1944); *C. A.* **40**, 6993 (1946).
3224. Matsen, F. A., N. Ginsburg, et al., *J. Chem. Phys.* **13**, 309 (1945).
3225. Matsen, F. A., W. W. Robertson, et al., *Chem. Rev.* **41**, 273 (1947).
3226. Matsen, J. M. and E. F. Johnson, *J. Chem. Eng. Data* **5**, 531 (1960).
3227. Matsui, H. and S. Ishimoto, *Tetrahedron Lett.* **1966**, 1827; *C. A.* **65**, 593 (1966).
3228. Matsui, K., *J. Chem. Soc. Jap.* **64**, 1417 (1943); *C. A.* **41**, 3753 (1947).
3229. Matsuno, K. and K. Han, *Bull. Chem. Soc. Jap.* **11**, 321 (1936); *C. A.* **30**, 5887 (1936).
3230. Matsuoka, S., S. Hattori, et al., *J. Phys. Soc. Jap.* **20**, 1212 (1965); *C. A.* **63**, 7799 (1965).
3231. Matterson, A. H. S. and L. A. Woodward, *Proc. Roy. Soc.* **231A**, 514 (1955); *C. A.* **50**, 676 (1956).
3232. Matthews, J. B., J. F. Sumner, et al., *Trans. Faraday Soc.* **46**, 797 (1950).
3233. Mattox, W. J., U.S. Patent 2,436,932, Mar. 2, 1948; *Official Gaz. U.S. Patent Office*, **608**, 119 (1948).
3234. Matuszko, A. J. and M. S. Chang, *Chem. Ind.* (London) **1963**, 822.
3235. Mavel, G., *Compt. Rend.* **250**, 1477 (1960).
3236. Maxim, N. and R. Mavrodineanu, *Bull. Soc. Chim. Fr.* **2**, 591 (1935); *C. A.* **29**, 4328 (1935).
3237. Mayer-Pitsch, E. and H. Duftschmid-Hinrichs, et al., *Z. Elektrochem.* **49**, 368 (1943); *C. A.* **38**, 21 (1944).
3238. Mayland, B. J., U.S. Patent 2,412,823 (Cl. 260-347), Dec. 17, 1946; *C. A.* **41**, 1711 (1947).
3239. Mayo, F. R. and F. M. Lewis, *J. Amer. Chem. Soc.* **66**, 1596 (1944).
3240. Mazonski, T., D. Gasztych, et al., *Przem. Chem.* **41**, 251 (1962); *C. A.* **58**, 11,202 (1963).
3241. Mazumder, M., *Indian J. Phys.* **27**, 406 (1953); *C. A.* **51**, 7862 (1957); **30**, 384 (1956); *C. A.* **51**, 9319 (1957).
3242. Mazur, I. J., *Acta Phys. Polon.* **7**, 285 (1938); *C. A.* **33**, 8068 (1939).
3243. Mears, T. W., A. Fookson, et al., *J. Res. Nat. Bur. Stand.* **44**, 299 (1950).

3244. Mears, T. W., C. L. Stanley, et al., *J. Res. Nat. Bur. Stand.* **67A,** 475 (1963).

3245. Mecka, R. and A. Reuter, *Z. Naturforsch.* **4A,** 368 (1949); *C. A.* **44,** 8182 (1950).

3246. Mecke, R., *Z. Phys. Chem.* (Leipzig) **36B,** 347 (1937); *C. A.* **31,** 8376 (1937).

3247. Mecke, R. and R. Joeckle, *Z. Elektrochem.* **66,** 255 (1962); *C. A.* **57,** 9322 (1962).

3248. Mecke, Ro., Rei. Mecke, et al., *Z. Naturforsch.* **10B,** 367 (1955); from [3042].

3249. Mecke, R. and G. Rossmy, *Z. Elektrochem.* **59,** 866 (1955); *Chem. Zentr.* **1956,** 5506.

3250. Mecke, R. and H. Specht, *Z. Elektrochem.* **62,** 500 (1958); *C. A.* **52,** 16,819 (1958).

3251. Meda, E. and A. Bertino, *Minerva Farm.* **11,** 45 (1962); *C. A.* **57,** 11,313 (1962).

3252. Médard, L., *J. Chim. Phys.* **33,** 626 (1936); *Chem. Zentr.* **1936, II,** 3531.

3253. Médard, L. and F. Déguillon, *Compt. Rend.* **203,** 1518 (1936); *C. A.* **31,** 1295 (1937).

3254. Medoks, G. V. and L. E. Ozerskaya, *Zh. Obshch. Khim.* **30,** 1643 (1960); *C. A.* **55,** 1410 (1961).

3255. Meeker, R. L., F. E. Critchfield, et al., *Anal. Chem.* **34,** 1510 (1962).

3256. Meerwein, H., B. J. Bock, et al., *J. Prakt. Chem.* **147,** 211 (1936); *C. A.* **31,** 656 (1937).

3257. Megaloikonomos, J. G., *Praktika Akad. Athenon* **12,** 226 (1937); *C. A.* **33,** 5321 (1939).

3258. Mehl, W., *Chem. Fabrik.* **7,** 240 (1934); *Chem. Zentr.* **1934, II,** 1958.

3259. Mehl, W., *Z. Ges. Kalte-Ind.* **41,** 152 (1934); *Chem. Zentr.* **1934, II,** 3736.

3260. Meighan, R. M. and R. H. Cole, *J. Phys. Chem.* **68,** 503 (1964).

3261. Meisenheimer, J. and O. Dorner, *Ann. Chem. Justus Liebigs* **482,** 130 (1930).

3262. Meissner, H. P. and A. S. Michaels, *Ind. Eng. Chem.* **41,** 2782 (1949).

3263. Meissner, H. P. and E. M. Redding, *Ind. Eng. Chem.* **34,** 521 (1942).

3264. Meissner, W., *Z. Angew. Phys.* **1,** 75 (1948); *C. A.* **43,** 4915 (1949).

3265. Mekhiev, S. D., R. G. Rizaev, ct al., *Azerb. Khim. Zh.* **1965,** 70; *C. A.* **64,** 10,539 (1966).

3266. Meldrum, W. B., L. P. Saxer, et al., *J. Amer. Chem. Soc.* **65,** 2023 (1943).

3267. Melia, T. P., *Polymer* **3,** 317 (1962), Part 8.

3268. Mellan, I., *Polyhedric Alcohols*, Spartan Books, Washington, D.C., 1962.

3269. Melton, C. E. and P. S. Rudolph, *J. Chem. Phys.* **31,** 1485 (1959).

3270. Menashi, J., W. L. Reynolds, et al., *Inorg. Chem.* **4,** 299 (1965).

3271. Menczel, S., *Z. Phys. Chem.* (Leipzig) **125,** 161 (1927).

3272. Menschutkin, B. N., *J. Russ. Phys. Chem. Soc.* **39,** 121 (1907); *C. A.* **1,** 1547 (1907); ibid. **40,** 1415 (1909); *C. A.* **4,** 2262 (1910)

3273. Menschutkin, N., *J. Chem. Soc.* **89,** 1532 (1906).

3274. Menzies, A. W. C. and S. L. Wright, *J. Amer. Chem. Soc.* **43,** 2314 (1921).

3275. Merckx, R., J. Verhulst, et al., *Bull. Soc. Chim. Belges* **42,** 177 (1933); *Chem. Zentr.* **1933, II,** 1173.

3276. Meredith, C. C. and G. F. Wright, *Can. J. Chem.* **38,** 1177 (1960).

3277. Merz, V. and W. Weith. *Ber. Deut. Chem. Gesell.* **14,** 187 (1881).

3278. Merzlin, R. W., *J. Russ. Phys. Chem. Soc.* **67,** 161 (1935).

3278a. Messerly, G. H. and J. G. Aston, *J. Amer. Chem. Soc.* **62,** 886 (1940).

3279. Messerly, G. H. and R. M. Kennedy, *J. Amer. Chem. Soc.* **62,** 2988 (1940).

3280. Messerly, J. F., G. B. Guthrie, et al., *J. Chem. Eng. Data* **12,** 338 (1967).

3281. Messerly, J. F., S. S. Todd, et al., *J. Phys. Chem.* **69,** 4304 (1965).

3282. Metcalf, L. D. and A. A. Schmitz, *Anal. Chem.* **33,** 363 (1961).

3283. Métra, M., L. Lesage, et al., *Compt. Rend.* **206,** 1026 (1938); *C. A.* **32,** 4908 (1938).

3284. Metz, D. J. and A. Glines, *J. Phys. Chem.* **71,** 1158 (1967).

3285. Meyer, F. and A. G. Harrison, *Can. J. Chem.* **42,** 2008 (1964).

3286. Meyer, H., *Analyse und Konstitutionsermittlung Organischer Verbindungen*, Ann. Arbor, Edwards Brothers, Inc., reproduced 1943 (1st German ed. 1938).

3287. Meyer, H., *Lehrb. der Organ.-Chem. Methodik*, 4th ed., Vol. 1, Berlin (1922).
3288. Meyer, K. H. and R. Luhdemann, *Helv. Chim. Acta* **18**, 307 (1935).
3289. Meyer, K. H. and F. G. Willson, *Ber. Deut. Chem. Gesell.* **47**, 837 (1914).
3290. Meyer, L. W. A. and W. M. Gearhart, *Ind. Eng. Chem.* **40**, 1478 (1948).
3291. Meyer, R. and P. Jaeger, *Ber. Deut. Chem. Gesell.* **36**, 1555 (1903).
3292. Meyer, V., *Ber. Deut. Chem. Gesell.* **19**, 3259 (1886).
3293. Meyer, V. and coworkers, *Ber. Deut. Chem. Gesell.* **5**, 203, 399, 514, 1029, 1034 (1872).
3294. Meyers, C. Y., *J. Org. Chem.* **26**, 1046 (1961).
3295. Michael, A., E. Scharf, et al., *J. Amer. Chem. Soc.* **38**, 653 (1916).
3296. Michaelis, L., *Ber. Deut. Chem. Gesell.* **46**, 3683 (1913).
3297. Michaelson, J. B. and D. J. Huntsman, *J. Med. Chem.* **7**, 378 (1964).
3298. Michalczyk, J., *Rocz. Chem.* **38**, 697 (1964); *C. A.* **62**, 1152 (1965).
3299. Michalski, H., S. Michalowski, et al., *Zeszyty Nauk. Politech. Lodz. Chem.* **12**, 73 (1962); *C. A.* **61**, 6445 (1964).
3300. Michel, G., *Bull. Soc. Chim. Belges* **68**, 643 (1959); *C. A.* **54**, 16,178 (1960).
3301. Michel, J., *Bull. Soc. Chim. Belges* **48**, 105 (1939); *C. A.* **33**, 7650 (1939).
3302. Michler, W. and C. Escherich, *Ber. Deut. Chem. Gesell.* **12**, 1162 (1879).
3303. *Microfilm Abstract* **7**, No. 2, 42–4 (1947).
3304. Middelhoek, J. and C. J. F. Bottcher, *Chem. Soc.* (London), *Spec. Publ.* **20**, 69 (1966); *C. A.* **65**, 8130 (1966).
3305. Middleton, B. A. and J. R. Partington, *Nature* **141**, 516 (1938); *C. A.* **32**, 4399 (1938).
3306. Midzushima, S., Y. Morina, et al., *J. Chem. Soc. Jap.* **65**, 127 (1944).
3307. Mikawa, Y., *Bull. Chem. Soc. Jap.* **29**, 110 (1956); *C. A.* **51**, 7143 (1957).
3308. Mikhailets, G. A., *Toksikol. Serasgan. Soedin. Ufa., Sb.* **1964**, 4; *C. A.* **63**, 7550 (1965).
3309. Mikhailov, G. P. and B. I. Sazhin, *Zh. Tekh. Fiz.* **25**, 1696 (1955); *C. A.* **50**, 634 (1956).
3310. Mikhailov, L., *Khim. Ind.* (Sofia) **33**, No. 4, 117 (1961); *C. A.* **57**, 8525 (1962).
3311. Mikhailov, V. A., S. K. Kharchenko, et al., *Izv. Sib. Otd. Akad. Nauk SSSR* **1962**, No. 7, 50; *C. A.* **58**, 976 (1963).
3312. Mikhant'ev, B. I. and E. A. Pyrakhina, *Zh. Obshch. Khim.* **30**, 958 (1960); *C. A.* **55**, 369 (1961).
3313. Mikulak, R. and O. Runquist, *J. Chem. Ed.* **38**, 557 (1961).
3314. Milas, N. A., *J. Amer. Chem. Soc.* **53**, 221 (1931).
3315. Milas, N. A., U.S. Patent 2,414,385, Jan. 14, 1947; *C. A.* **41**, 5894 (1947).
3316. Milas, N. A. and L. H. Perry, *J. Amer. Chem. Soc.* **68**, 1938 (1946).
3317. Milazzo, G., *Boll. Sci. Fac. Chim. Ind. Bologna* **1941**, 94; *Chem. Zentr.* **1942**, I, 1995.
3318. Milazzo, G., *Rend. Ist. Super. Sanità* **11**, 372 (1948); *C. A.* **43**, 2094 (1949).
3319. Milazzo, G., *Rend. Ist. Super. Sanità* **19**, 322, 342 (1956); *C. A.* **51**, 7791 (1957).
3320. Milburn, A. H. and E. V. Truter, *J. Chem. Soc.* **1954**, 3344.
3321. Miles, C. B. and H. Hunt, *J. Phys. Chem.* **45**, 1346 (1941).
3322. Millard, B. J. and D. F. Shaw, *J. Chem. Soc.* **1966**, Sec. B, 664.
3323. Miller, C. C., *Proc. Roy. Soc.* **106A**, 724 (1924).
3324. Miller, G. A., *J. Chem. Eng. Data* **8**, 69 (1963).
3325. Miller, G. H., U.S. Patent 2,428,120 (Cl. 260–347); Sept. 30, 1947; *C. A.* **42**, 610 (1948).
3326. Miller, K. J., *Ind. Eng. Chem., Chem. Eng. Data Series* **3**, 239 (1958).
3327. Miller, O., *Bull. Soc. Chim. Belges* **41**, 217 (1932).
3328. Miller, R. E. and G. E. Bennett, *Ind. Eng. Chem.* **53**, 33 (1961).

3329. Millero, F. J., *J. Phys. Chem.* **72**, 3209 (1968).

3330. Millikan, A. F., U.S. Patent 3,031,383, Apr. 24, 1962; *C. A.* **57**, 3722 (1962).

3331. Mills, J. E., *J. Amer. Chem. Soc.* **31**, 1099 (1909).

3332. Milone, M., *IX Congr. Intern. Quim. Pura Aplicada* **2**, 191 (1934); *C. A.* **29**, 7139 (1935).

3333. Minami, T. and T. Ando, (Osaka Ind. Res. Inst.) *Osaka Kôgyô Gijyutsu Shikenjyo Kihô* **8**, 51-4 (1957); *C. A.* **53**, 13,649.

3334. Miner, C. S. and N. N. Dalton, *Glycerol*, Reinhold, New York, 1953.

3335. Miner, C. S., J. P. Trickey, et al., *Chem. Met. Eng.* **27**, 299; 362 (1922).

3336. Mingaleva, K. S., K. G. Golodova, et al., *Zh. Org. Khim.* **1**, 2078 (1965); *C. A.* **64**, 9570 (1966).

3337. Minkov, P., *Mashinostroene* **11**, 19 (1962); *C. A.* **57**, 9295 (1962).

3338. Mirone, P., *Atti Accad. Naz. Lincei, Rend., Cl. Sci. Fis., Mat. Nat.* **16**, 483 (1954); *C. A.* **49**, 5967 (1955).

3339. Mirone, P. and G. F. Fabbri, *Gazz. Chim. Ital.* **84**, 187 (1954); *C. A.* **49**, 10,741 (1955).

3340. Mirov, N. T., *J. Forestry* **45**, 659 (1947); *C. A.* **42**, 386 (1948).

3341. Miskidzh'yan, S. P. and N. A. Trifonov, *J. Gen. Chem.* (USSR) **17**, 1231 (1947); *C. A.* **43**, 922 (1949).

3342. Mita, I., *J. Chem. Soc. Jap.* **63**, 760 (1942); *C. A.* **41**, 3038 (1947).

3343. Mitchell, J. and D. M. Smith, *Aquametry*, Interscience, New York, 1948.

3344. Mitchell, J. H. and H. R. Kraybill, *J. Amer. Chem. Soc.* **64**, 988 (1942).

3345. Mitra, B. C., P. Ghosh, et al., *Anal. Chem.* **36**, 673 (1964).

3346. Mitsukuri, S. and A. Nakatsuchi, *Sci. Rep. Tohoku Univ.* **15**, 45 (1926); *Chem. Zentr.* **1926, II**, 545; *C. A.* **20**, 1020 (1926).

3347. Mitsutani, A. and T. Kominami, *Nippon Kagaku Zasshi* **80**, 895 (1959); *C. A.* **55**, 4349 (1961).

3348. Mitzner, B. M., E. T. Theimer, et al., *Appl. Spectrosc.* **19**, 169 (1965); *C. A.* **64**, 4443 (1966).

3349. Miyake, A., *J. Amer. Chem. Soc.* **82**, 3040 (1960).

3350. Miyazawa, T. and K. S. Pitzer, *J. Amer. Chem. Soc.* **80**, 60 (1958).

3351. Miyazawa, T., T. Shimanouchi, et al., *J. Chem. Phys.* **24**, 408 (1955).

3352. Mizuhara, S. and W. F. Sever, *J. Amer. Chem. Soc.* **75**, 3274 (1953).

3353. Mizushima, S. and Y. Morino, *Proc. Indian Acad. Sci.* **8A**, 315 (1938); *C. A.* **33**, 6158 (1938).

3354. Mizushima, S. and Y. Morino, et al., *Phys. Z.* **38**, 459 (1937); *C. A.* **31**, 6111 (1937).

3355. Mizushima, S., Y. Morino, et al., *Sci. Paper Inst. Phys. Chem. Res.* (*Tokyo*) **29**, 111, June 1936; *Chem. Zentr.* **1937, I**, 4489; Beil **EIII1**, 161.

3356. Mizushima, S., Y. Morino, et al., *Sci. Papers Inst. Phys. Chem. Res.* (*Tokyo*) **36**, 281 (1939); *C. A.* **34**, 2707 (1940).

3357. Mizushima, S.-I., T. Simanouti, et al., *J. Amer. Chem. Soc.* **72**, 3490 (1950).

3358. Mizushima, S., Y. Uehara, et al., *Bull. Chem. Soc. Jap.* **12**, 132 (1937).

3359. Mizutani, M., *Z. Phys. Chem.* (Leipzig) **116**, 350 (1925).

3360. Mizutani, M., *Z. Phys. Chem.* (Leipzig) **118**, 327 (1925).

3361. Moffat, J. B., *J. Chem. Eng. Data* **13**, 36 (1968).

3362. Moffett, R. B. and B. D. Aspergen, *Chem. Eng. News* **32**, 4328 (1954).

3363. Mohler, F. L., L. Williamson, et al., *J. Res. Nat. Bur. Stand.* **44**, 291 (1950).

3364. Mohler, H., *Helv. Chim. Acta* **20**, 1188 (1937).

3365. Mohler, H. and H. Lohr, *Helv. Chim. Acta* **21**, 485 (1938).

3366. Mohler, H. and J. Polya, *Helv. Chim. Acta* **20**, 96 (1937).

3367. Mohler, H. and J. Sorge, *Helv. Chim. Acta* **23**, 119 (1940).

3368. Mohler, H. and J. Sorge, *Helv. Chim. Acta* **23**, 1200 (1940).
3369. Mohr, O., *Mikrochemie* **2**, 154 (1930); *C. A.* **24**, 4242 (1930).
3370. Moldavskiĭ, B. L. and L. E. Turetskaya, *J. Gen. Chem.* (USSR) **16**, 445 (1946); *C. A.* **41**, 951 (1947).
3371. Mole, M. F., W. S. Holmes, et al., *J. Chem. Soc.* **1964**, 5144.
3372. Moles, E., *Z. Phys. Chem.* (Leipzig) **80**, 531 (1912).
3373. Molinari, J. G. D., *Ind. Chemist* **37**, 323 (1961); Ref. [5340].
3374. Moliński, S., F. Nowotny, et al., *Przem. Chem.* **23**, 30 (1939); *C. A.* **33**, 4593 (1939).
3375. Moll, W. L. H., *Kolloid-Beihefte* **49**, 1 (1939); *C. A.* **33**, 2018 (1939); Beil **EIII3**, 474; Beil **EIII2**, 305.
3376. Möller, R., *Phys. Z.* **32**, 697 (1931).
3377. Momigny, J., *Bull. Soc. Roy. Sci. Liege* **22**, 541 (1953); *C. A.* **48**, 9806 (1954).
3378. Momigny, J., *Bull. Soc. Roy. Sci. Liege* **24**, 111 (1955); *C. A.* **50**, 1455 (1956).
3379. Monahan, J. E. and H. E. Stanton, *J. Chem. Phys.* **37**, 2654 (1962).
3380. Monica, M. D., L. Jannelli, et al., *J. Phys. Chem.* **72**, 1068 (1968).
3381. Monica, M. D., U. Lamanna, et al., *J. Chem. Phys.* **72**, 2124 (1968).
3382. Moniz, W. B. and J. A. Dixon, *J. Amer. Chem. Soc.* **83**, 1671 (1961).
3383. Monroe, K. P., *Ind. Eng. Chem.* **13**, 133 (1921).
3384. Monsanto Chemical Company, "Dimethylacetamide (DMAC)", Technical Data Sheet, St. Louis, Missouri 1960.
3385. "Montecatini" Soc. Gen. per l'industria Mineraria et Agricola, French Patent 808,057, Jan. 28, 1937; *C. A.* **31**, 6262 (1937).
3386. Moor, V. G., Kanep, E. K., et al., *Trans. Exptl. Research Lab. Khemgas, Materials on Cracking and Chemical Treatment of Cracking Products*, USSR **3**, 320 (1937); *C. A.* **31**, 6072 (1937).
3387. Moore, D. W., L. A. Burkardt, et al., *J. Chem. Phys.* **25**, 1235 (1956).
3388. Moore, E. M. and M. E. Hobbs, *J. Amer. Chem. Soc.* **71**, 411 (1949).
3389. Moore, F. J. and I. B. Johns, *J. Amer. Chem. Soc.* **63**, 3336 (1941).
3390. Moore, G. E. and G. S. Parks, *J. Amer. Chem. Soc.* **61**, 2561 (1939).
3391. Moore, G. E., M. L. Renquist, et al., *J. Amer. Chem. Soc.* **62**, 1505 (1940).
3392. Mopsik, F. I., *J. Res. Nat. Bur. Stand.* **71A**, 287 (1967).
3393. Morávek, J., *Chem. Prum.* **7**, 49 (1957); *C. A.* **51**, 14,351 (1957).
3394. Morcillo, J. and J. M. Orza, *Anal. Real Soc. Espan. Fis. Quim* **56B**, 231, 253 (1960); *C. A.* **54**, 19,126 (1960).
3395. Morcom, K. W. and D. N. Travers, *Trans. Faraday Soc.* **61**, 230 (1965).
3396. Moreno, J. M. M., *Anal. Real Soc. Espan. Fis. quim.* **43**, 261 (1947); *C. A.* **41**, 6420 (1947).
3397. Morino, Y., I. Watanabe, et al., *Sci. Papers Inst. Phys. Chem. Research* (Tokyo) **39**, 396 (1942); *C. A.* **41**, 6154 (1947).
3398. Morgan, J. L. R. and A. McD. McAfee, *J. Amer. Chem. Soc.* **33**, 1275 (1911).
3399. Morgan, S. O. and H. H. Lowry, *J. Phys. Chem.* **34**, 2385 (1930).
3400. Morgan, S. O. and W. A. Yager, *Ind. Eng. Chem.* **32**, 1519 (1940).
3401. Mori, N., S. Omura, et al., *Bull. Chem. Soc. Jap.* **38**, 2149 (1965).
3402. Mori, N., S. Omura, et al., *Bull. Chem. Soc. Jap.* **38**, 2199 (1965).
3403. Moriarty, R. M. and J. M. Kliegman, *J. Org. Chem.* **31**, 3007 (1966).
3404. Morikawa, T. and K. Yoshida, *Kagaku to Kogyo* (Osaka) **37**, 107 (1963); *C. A.* **60**, 3161 (1964).
3405. Morín, M. G., C. C. de Pérez, et al., *Rev. Col. Quim. Puerto Rico* **16**, 19 (1959); *C. A.* **54**, 1996 (1960).

3406. Morín, M. G., C. C. de Pérez, et al., *Rev. Col. Quim. Puerto Rico* **16**, No. 19, 21 (1959); *C. A.* **57**, 1612 (1962).

3407. Morino, Y., *Inst. Phys. Chem. Res.* (Tokyo) *Sci. Pap.* **23**, 49 (1933).

3408. Morino, Y. and I. Watanabe, *J. Chem. Soc. Jap.* **63**, 377 (1942); *C. A.* **41**, 2995 (1947).

3409. Morino, Y., S. Yamaguchi, et al., *Sci. Papers Inst. Phys. Chem. Res.* (Tokyo) **42**, Chemistry, 1 (1944); *C. A.* **41**, 6089 (1947).

3410. Morman, D. H. and G. A. Harlow, *Anal. Chem.* **39**, 1869 (1967).

3411. Morris, H. J., A. A. Nelson, et al., *J. Pharmacol.* **74**, 266 (1942).

3412. Morris, J. C., *J. Chem. Phys.* **11**, 230 (1943).

3413. Morris, J. C., W. J. Lanum, et al., *J. Chem. Eng. Data* **5**, 112 (1960).

3414. Morrison, G. O. and T. P. G. Shaw, *Trans. Electrochem. Soc.* **63**, 443 (1933); *Chem. Zentr.* **1933, II**, 131; Beil **EIII2**, 269.

3415. Mortimer, F. S. and R. B. Blodgett, et al., *J. Am. Chem. Soc.* **69**, 822 (1947).

3416. Morton, A. A. and J. G. Mark, *Ind. Eng. Chem., Anal. Ed.* **6**, 151 (1934).

3417. Morton, R. A. and A. J. A. de Gouveia, *J. Chem. Soc.* **1934**, 916.

3418. Morton, R. A. and A. L. Stubbs, *J. Chem. Soc.* **1940**, 1347.

3419. Mosher, W. A., *J. Amer. Chem. Soc.* **62**, 552 (1940).

3420. Moskalev, V. V. and F. I. Skripov, *Fiz. Probl. Spektroakopii, Akad. Nauk SSSR, Materialy 13-go (Trinadtsatogo) Soveshoh.* Leningrad, 1960 **2**, 148; *C. A.* **59**, 13,504 (1963).

3421. Moskalyk, R. E., L. G. Chatten, et al., *J. Pharm. Sci.* **50**, 179 (1961).

3422. Motodo, T. and Y. Yoshie, *Kogyo Kagaku Zasshi* **68**, 1669 (1965); *C. A.* **64**, 195 (1966).

3423. Motoyama, I. and C. H. Jarboe, *J. Phys. Chem.* **70**, 3226 (1966).

3424. Mountcastle, W. R., D. F. Smith, et al., *J. Phys. Chem.* **64**, 1342 (1960).

3425. Mouradoff, L. and E. Darmois, *Bull. Soc. Chim. Fr.* **1949**, 446D; *C. A.* **44**, 2301 (1950).

3425a. Moureu, C., *Ann. Chim. Phys.* [7] **2**, 187, 191 (1894).

3426. Moureu, C., *Bull. Soc. Chim. Fr.* **9**, 386; *J. Chem. Soc.* **1893Ai**, 548.

3427. Moureu, C., *Bull. Soc. Chim. Fr.* **9**, 424; *J. Chem. Soc.* **1893Ai**, 682.

3428. Moureu, C., A. Boutaric, et al., *J. Chim. Phys.* **18**, 333 (1921); *C. A.* **15**, 2417 (1921); Beil **EII1**, 783.

3429. Moureu, C. and A. Boutaric, *J. Chim. Phys.* **18**, 348 (1920); *C. A.* **15**, 2220 (1921).

3430. Moureu, C., M. Murat, et al., *Compt. Rend.* **172**, 1267 (1921); *C. A.* **16**, 55 (1922); Beil **EII2**, 386.

3431. Moureu, H. and M. Dode, *Bull. Soc. Chim. Fr.* **4**, 637 (1937); *C. A.* **31**, 4884 (1937).

3432. Mousseron, M., R. Granger, et al., *Bull. Soc. Chim. Fr.* **1947**, 459; *C. A.* **42**, 1895 (1948).

3433. Mukerji, S. K., *Phil. Mag.* **19**, 1079 (1935); *C. A.* **29**, 6505 (1935).

3434. Mukherjee, D. K., *Indian J. Phys.* **34**, 402 (1960); *C. A.* **55**, 7041 (1961).

3435. Mukherjee, D. K., P. K. Bishui, et al., *Indian J. Phys.* **39**, 537 (1965); *C. A.* **64**, 16,838 (1966).

3436. Mukherjee, L. M. and S. Bruckenstein, *Pure. Appl. Chem.* **13**, 421 (1966).

3437. Mukhitov, B. M., *Gig. Sanit.* **27**, No. 6, 16 (1962); *C. A.* **57**, 12,830 (1962).

3438. Müller, A., *Fette Seifen* **49**, 572 (1942); *C. A.* **37**, 6510 (1943).

3439. Müller, A., *Fette Seifen* **53**, 462 (1951); *C. A.* **46**, 2757 (1952).

3440. Müller, A. H., *Z. Phys. Chem.* (Leipzig) **86**, 177 (1914).

3441. Müller, E. and K. Ehrmann, *Ber. Deut. Chem. Gesell.* **69**, 2207 (1936).

3342. Müller, F., Landolt Bornstein, *Physikalisch Chemische Tabellen*, 5 ed., Suppl. 2, pt. 2, p. 1196, Berlin, Julius Springer, 1931.

3443. Müller, F. H., *Phys. Z.* **38**, 283 (1937); *C. A.* **31**, 4551 (1937).

3444. Müller, H., *Phys. Z.* **35**, 346 (1934).

3445. Müller, H. and H. Sack, *Phys. Z.* **31**, 815 (1930); *Chem. Zentr.* **1930, II, 3374.**

3446. Müller, N., *J. Chem. Phys.* **42**, 4309 (1965).

3447. Müller, R. and H. Brenneis, *Z. Electrochem.* **38**, 450 (1932); *Chem. Zentr.* **1932, II,** 1420.

3448. Müller, R., F. Griengl, et al., *Monatsh. Chem.* **47**, 88 (1926).

3449. Müller, R., V. Raschka, et al., *Monatsh. Chem.* **48**, 659 (1927).

3450. Mulliken, R. S., *J. Chem. Phys.* **7**, 339 (1939).

3451. Mulliken, R. S. and E. Teller, *Phys. Rev.* **61**, 283 (1942).

3452. Mumford, S. A. and J. W. C. Phillips, *J. Chem. Soc.* **1950**, 75.

3453. Munch, J. C., *J. Amer. Chem. Soc.* **48**, 994 (1926).

3454. Muncke, *Berz. Jb.* **27**, 453.

3455. Muney, W. S. and J. F. Coetzee, *J. Phys. Chem.* **66**, 89 (1962).

3456. Munson, M. S. B. and F. H. Field, *J. Amer. Chem. Soc.* **88**, 4337 (1966).

3457. Murayama, K. and K. Nukada, *Bull. Chem. Soc. Jap.* **36**, 1223 (1963); *C. A.* **60,** 1249 (1964).

3458. Murkerjee, D. K., P. K. Bishui, et al., *Indian J. Phys.* **39**, 537 (1965); *C. A.* **64,** 16,838 (1966).

3459. Muroi, K., K. Ogana, et al., *Bull. Soc. Chem. Jap.* **38**, 1176 (1965).

3460. Murphy, R. A. and J. C. Davis, *J. Phys. Chem.* **72**, 3111 (1968).

3461. Murray, M. J. and F. F. Cleveland, *J. Chem. Phys.* **9**, 129 (1941).

3462. Murray-Rust, D. M., H. J. Hadow, et al., *J. Chem. Soc.* **1931**, 215.

3463. Murrell, J. N. and V. M. S. Gill, *Trans. Faraday Soc.* **61,** 402 (1965).

3464. Murty, C. R. K., *J. Sci. Ind. Res* (India) **18B**, 268 (1959).

3465. Murty, C. R. and D. V. G. L. N. Rao, *J. Sci. Ind. Res.* (India) **15B**, 350 (1956); *C. A.* **51**, 5484 (1957).

3466. Murty, G. V. L. N. and T. R. Seshadri, *Proc. Indian Acad. Sci.* **8A**, 519 (1938); *C. A.* **33**, 4874 (1939).

3467. Murty, G. V. L. N. and T. R. Seshadri, *Proc. Indian Acad. Sci.* **10A**, 307 (1939); *C. A.* **34**, 2707 (1940).

3468. Murty, G. V. L. N. and T. R. Seshadri, *Proc. Indian Acad. Sci.* **11A**, 32 (1940); *C. A.* **34**, 6881 (1940).

3469. Murty, G. V. L. N. and T. R. Seshadri, *Proc. Indian Acad. Sci.* **11A**, 424 (1940); *C. A.* **34**, 7744 (1940).

3470. Murty, G. V. L. N. and T. R. Seshadri, *Proc. Indian Acad. Sci.* **14A**, 593 (1941); *C. A.* **36**, 4026 (1942).

3471. Murty, G. V. L. N. and T. R. Seshadri, *Proc. Indian Acad. Sci.* **16A**, 264 (1942); *C. A.* **37**, 3669 (1943).

3472. Musher, J. I. and R. G. Gordon, *J. Chem. Phys.* **36**, 3097 (1962).

3473. Musher, J. and R. E. Richards, *Proc. Chem. Soc.* **1958**, 230.

3474. Mussell, A. G., F. B. Thole, et al., *J. Chem. Soc.* **101**, 1008 (1912).

3475. Musser, D. M. and H. Adkins, *J. Amer. Chem. Soc.* **60**, 664 (1938).

3476. Myers, R. T., *J. Phys. Chem.* **69**, 700 (1965).

3477. Maryott, A. A. and E. A. Smith, *Table of Dielectric Constants of Pure Liquids*, NBS Circular 514, Aug. 10, 1951.

N

3478. Nadeschdin, *J. Russ. Phys. Chem. Soc.* (2) **14**, 538; Beil **EIII1,** p. 1873.

3479. Naegeli, C., L. Gruntuch, et al., *Helv. Chim. Acta* **12**, 227 (1929).

3480. Nagai, H., *Kumamoto J. Sci.*, Ser. **A2**, 100 (1954); *C. A.* **50**, 13,649 (1956).
3481. Nagakura, S. and A. Kuboyama, *Rept. Inst. Sci. Technol.*, *Univ. Tokyo* **5**, 27 (1951); *C. A.* **46**, 1315 (1952).
3482. Nagornow, N. and L. Rotinjanz, *Ann. Inst. Anal. Physio-Chim. Leningrad* **3**, 162 (1926); *Chem. Zentr.* **1927**, I, 2648.
3483. Nakagawa, I., *Nippon Kagaku Zasshi* **79**, 1353 (1958); *C. A.* **53**, 7699 (1959).
3484. Nakamura, K., *Nippon Kagaku Zasshi* **78**, 1164 (1957); *C. A.* **52**, 1048 (1958).
3485. Nakamura, S., *J. Chem. Soc. Jap.* **60**, 1010 (1939); *C. A.* **34**, 1563 (1940).
3486. Nakata, N., *Ber. Deut. Chem. Gesell.* **64**, 2059 (1931).
3487. Nametkin, S. S., *J. Russ. Phys. Chem. Soc.* **55**, 75 (1923–1924).
3488. Nametkin, S. S. and E. S. Pokrovskaya, *Z. Obsc. Chim.* **7**, 962 (1937); Beil **EIII5,** 1256.
3489. Naqvi, N. and Q. Fernando, *J. Org. Chem.* **25**, 551 (1960).
3490. Narasimhan, P. T., *J. Indian Inst. Sci.* **37A**, 30 (1955); *C. A.* **49**, 7906 (1955).
3491. Narayanaswamy, P. K., *Proc. Indian Acad. Sci.* **26A**, 121 (1947); *C. A.* **42**, 2519 (1948).
3492. Narayanaswamy, P. K., *Proc. Indian Acad. Sci.* **27A**, 336 (1948); *C. A.* **42**, 6665 (1948).
3493. Narsimham, G., *Chem. Process Eng.* **46**, 498 (1965); *C. A.* **63**, 16,196 (1965).
3494. Natelson, S. and J. E. Bonas, *Microchem. J.* **9**, 68 (1965); *C. A.* **63**, 4920 (1965).
3495. National Aniline Division, *Aniline*, Allied Chemical Corp., New York, 1964.
3496. National Bureau of Standards, Circular No. 19, *Standard Density and Volumetric Tables*, 6th ed., Government Printing Office, Washington, D.C., 1924.
3497. National Formulary, The, 12th ed., American Pharmaceutical Association, Washington, D.C., 1965.
3498. Natradze, A. G. and K. E. Novikova, *Med. Prom.* SSSR **12**, No. 4, 33 (1958); *C. A.* **53**, 11,714 (1959).
3499. Naves, Y. R., *Mfg. Chemist* **17**, 187 (1946); *C. A.* **42**, 6056 (1948).
3500. Nazarov, I. N., L. D. Bergel'son, et al., *Izv. Akad. Nauk SSSR, Otd. Khim. Nauk* **1953**, 889; *C. A.* **49**, 1082 (1955).
3501. Nazarov, I. N., S. M. Makin, et al., *Zh. Obshch. Khim.* **29**, 111 (1959); *C. A.* **54**, 255 (1960).
3502. Nazarova, I. M., Y. K. Syrkin, *Izv. Akad. Nauk SSSR, Otd. Khim. Nauk* **1949**, 35; *C. A.* **43**, 4913 (1949).
3503. Nechai, F., *Zh. Tekh. Fiz.* **26**, 436 (1956); *C. A.* **51**, 29 (1957).
3504. Neelakantan, P., *Proc. Indian Akad. Sci.* Sect. **A57**, 94 (1963); *C. A.* **58**, 13,317 (1963).
3505. Neelakantan, P., *Proc. Indian Acad. Sci.*, Sect. **A60**, 422 (1964); *C. A.* **62**, 15,608 (1965).
3506. Neikam, W. C. and B. P. Dailey, *J. Chem. Phys.* **38**, 445 (1963).
3507. Nekrasov, A. S. and B. A. Krentsel, *Zh. Obshch. Khim.* **19**, 948 (1949); *C. A.* **44**, 1006 (1950).
3508. Nelson, K. W., J. F. Ege, et al., *J. Ind. Hyg. Toxicol.* **25**, 282 (1943); *C. A.* **39**, 5001 (1945).
3509. Nel'son, K. V. and Z. D. Stepanova, *Kolebatel'nye Spektry i Molekul. Protsessy v Kauchukakh, Vses. Nauchn.-Issled. Inst. Sintekich. Kauchuka* **1965**, 127; *C. A.* **64**, 9476 (1966).
3510. Nelson, O. A., *Ind. Eng. Chem.* **20**, 1382 (1928).
3511. Nelson, O. A., *Ind. Eng. Chem.* **22**, 971 (1930).
3512. Nelson, R. A. and R. S. Jessup, *J. Research Natl. Bur. Stand.* **48**, 206 (1952).
3513. Nelson, R. D., C. A. Billings, et al., *J. Phys. Chem.* **71**, 2742 (1967).

3514. Nemes, L. and W. J. Orville-Thomas, *Trans. Faraday Soc.* **61**, 1839 (1965).
3515. Nemirowsky, J., *J. Prak. Chem.* (2) **28**, 439; *J. Chem. Soc.* **1884A**, 419.
3516. Nernst, W., *Z. Phys. Chem.* (Leipzig) **14**, 622 (1894).
3517. Neu, J. T., A. Ottenberg, et al., *J. Chem. Phys.* **16**, 1004 (1948).
3518. Neuberg, C. and E. Kerb, *Biochem. Z.* **92**, 96 (1918).
3519. Neuman, R. C. and M. L. Rahm, *J. Org. Chem.* **31**, 1857 (1966).
3520. Neuman, R. C., W. Snider, et al., *J. Phys. Chem.* **72**, 2469 (1968).
3521. Neunhoeffer, O., *J. Prakt. Chem.* **133**, 95 (1932); *C. A.* **26**, 2435 (1932).
3522. Nevgi, G. V. and S. K. K. Jatkar, *J. Indian Inst. Sci.* **17A**, 175 (1934); *C. A.* **29**, 6505 (1935).
3523. Nevgi, G. V. and S. K. K. Jatkar, *J. Indian Inst. Sci.* **17A**, 189 (1934); *C. A.* **29**, 6505 (1935).
3524. Newitt, D. M., R. P. Linstead, et al., *J. Chem. Soc.* **1937**, 876.
3525. Newitt, D. M. and P. S. Mene, *J. Chem. Soc.* **1946**, 97.
3526. Newman, M., C. B. Hayworth, et al., *Ind. Eng. Chem.* **41**, 2039 (1949).
3527. Newmark, R., AEC Accession No. **5934**, Rept. No. **UCRL-11649**, Avail. OTS, 194 pp. (1964); *C. A.* **63**, 163 (1965).
3528. Newmark, R. A. and C. H. Sederholm, *J. Chem. Phys* **43**, 602 (1965).
3529. Nicholls, D., C. Sutjohen, et al., *J. Phys. Chem.* **72**, 1021 (1968).
3530. Nicholson, D. E., *Anal. Chem.* **32**, 1372 (1960).
3531. Nicholson, D. E., *Anal. Chem.* **32**, 1634 (1960).
3532. Nicholson, D. E., *Anal. Chem.* **34**, 370 (1966).
3533. Nickerson, J. D. and R. McIntosh, *Can. J. Chem.* **35**, 1325 (1957).
3534. Nickerson, J. K., K. A. Kobe, et al., *J. Phys. Chem.* **65**, 1037 (1961).
3535. Nicolescu, I. V. and O. Serban, *Analele Univ. Bucuresti, Ser. Stiint. Nat.* **12**, 165 (1963); *C. A.* **64**, 8012 (1966).
3536. Nicolini, E., *Ann. Chim.* (*Paris*) **6**, 582 (1951); *C. A.* **46**, 796 (1952).
3537. Nicolini, E. and P. Laffitte, *Compt. Rend.* **229**, 757 (1949); *C. A.* **44**, 4409 (1950).
3538. Nielsen, J. R. and D. C. Smith, *Ind. Eng. Chem., Anal. Ed.* **15**, 609 (1943).
3539. Nielsen, J. R. and N. E. Ward, *J. Chem. Phys.* **10**, 81 (1942).
3540. Niini, A., *Suomen Kemistilehti* **11A**, 19 (1938); *C. A.* **32**, 4861 (1938).
3541. Nikitina, A. N., V. A. Petukhov, et al., *Opt. Spektrosk.* **16**, 976 (1964); *C. A.* **61**, 12,803 (1964).
3542. Nikolaev, A. V., Yu. A. Afanas'ev, et al., *Dokl. Akad. Nauk SSSR* **168**, 351 (1966); *C. A.* **65**, 9826 (1966).
3543. Nikuradse, A., *Z. Phys. Chem.* (Leipzig) **155A**, 59 (1931).
3543a. Noller, C. R., Ed., *Organic Syntheses*, Vol. 15, Wiley, New York, 1935.
3544. Normant, H., *Angew. Chem. Intern. Ed. Engl.* **6**, 1046 (1967).
3545. Normant, H., *Bull. Soc. Chim. Fr.*, **1968**, 791.
3546. Normant, H., *Compt. Rend.* **239**, 1510 (1954); *C. A.* **50**, 228 (1956).
3547. Norris, F. A. and D. E. Terry, *Oil & Soap* **22**, 41 (1945).
3548. Norris, J. F. and F. Cortese, *J. Amer. Chem. Soc.* **49**, 2640 (1927).
3549. Norris, J. F. and J. N. Ingraham, *J. Amer. Chem. Soc.* **60**, 1421 (1938).
3550. Norris, J. F. and G. W. Rigby, *J. Amer. Chem. Soc.* **54**, 2088 (1932).
3551. Norris, J. F. and B. M. Sturgis, *J. Amer. Chem. Soc.* **61**, 1413 (1939).
3552. Norton, F. H. and H. B. Hass, *J. Amer. Chem. Soc.* **58**, 2147 (1936).
3553. Notley, J. M. and M. Spiro, *J. Chem. Soc.* **1966B**, 362.
3554. Novák, J., J. Matouš, et al., *Collect. Czech. Chem. Commun.* **25**, 583 (1960); *C. A.* **54**, 16,067 (1960) and [4773].
3555. Noyce, D. S. and D. B. Denney, *J. Amer. Chem. Soc.* **72**, 5743 (1950).
3556. Noyce, D. S. and D. B. Denney, *J. Amer. Chem. Soc.* **74**, 5912 (1952).

3557. Noyes, W. A., *J. Amer. Chem. Soc.* **45**, 857 (1923).
3558. Nozaki, K. and P. D. Bartlett, *J. Amer. Chem. Soc.* **68**, 2377 (1946).
3559. Nukada, K., *Bull. Chem. Soc. Jap.* **33**, 1606 (1960); *C. A.* **55**, 13,064 (1961).
3560. Nukada, K., *Nippon Kagaku Zasshi* **80**, 976 (1959); *C. A.* **53**, 21,156 (1959).
3561. Nukada, K., *Nippon Kagaku Zasshi* **80**, 1112 (1959); *C. A.* **54**, 1071 (1960).
3562. Nukada, K., *Spectrochim. Acta* **18**, 745 (1962); *C. A.* **57**, 8085 (1962).
3563. Nukada, K. and U. Maeda, *Bull. Chem. Soc. Jap.* **32**, 6556 (1959); *C. A.* **54**, 6316 (1960).
3564. Nystrom, R. F. and W. G. Brown, *J. Amer. Chem. Soc.* **69**, 1197 (1947).
3565. Nystrom, R. F. and W. G. Brown, *J. Amer. Chem. Soc.* **69**, 2548 (1947).

O

3566. Obach, E., *J. Prakt. Chem.* **26**, 299 (1882).
3567. Obolentsev, R. D., N. G. Marina, et al., *Khim. Seraorgan. Soedin., Soderzhashch. v Neft. i Nefteprod., Akad. Nauk SSSR, Bashkirsk. Filial* **6**, 220 (1962); *C. A.* **61**, 9369 (1964).
3568. O'Brien, K. G., *Aust. J. Chem.* **10**, 91 (1957); *C. A.* **51**, 10,367 (1957).
3569. O'Connor, J. G. and M. S. Norris, *Anal. Chem.* **36**, 1391 (1964).
3570. O'Connor, R. T., *J. Amer. Oil Chem. Soc.* **32**, 624 (1955).
3571. O'Connor, R. T., E. T. Field, et al., *J. Amer. Oil Chemist's Soc.* **28**, 154 (1951); *C. A.* **45**, 4558 (1951).
3572. Odan, M., S. Midzushima and Y. Morino, *Sci. Papers. Inst. Phys. Chem. Res.* (Tokyo) **42**, Chemistry 27 (1944); *C. A.* **41**, 6089 (1947).
3573. Oddo, B. and F. Tognacchini, *Gazz. Chim. Ital.* **52II**, 347 (1922); *Chem. Zentr.* **1923**, III, 923.
3574. O'Donnell, J. F., J. T. Ayres, et al., *Anal. Chem.* **37**, 1161 (1965).
3575. Oehme, F., *Chem.-Ztg.* **82**, 33 (1958); *C. A.* **52**, 8419 (1958).
3576. Oehme, F., *Farbe Lack* **64**, 183 (1958); *C. A.* **53**, 790 (1957).
3577. Oehme, F., *Fette, Seifen, Anstrichmittel* **62**, 910 (1960); *C. A.* **55**, 20,460 (1961).
3578. Oetting, F. L., *J. Chem. Phys.* **41**, 149 (1964).
3579. Oetting, F. L., *J. Phys. Chem.* **67**, 2757 (1963).
3580. Oglukian, R. L., Ph.D. Dissertation, Tulane University, 1968.
3581. Oguri, S., S. Hinonishi, et al., *Mem. Faculty Sci. & Eng., Waseda Univ., Tokyo* **1937**, 99; *C. A.* **32**, 6520 (1938).
3582. Oka, E., Y. Numata, et al., Japanese Patent 5769, Aug. 18, 1955; *C. A.* **51**, 15,549 (1957).
3583. Okazaki, H., *J. Chem. Soc. Jap.* **63**, 1314, 1500 (1942); *C. A.* **41**, 3370 (1947).
3584. Okazaki, H., *J. Chem. Soc. Jap.* **64**, 1255 (1943); *C. A.* **41**, 3740 (1947).
3585. O'Kelly, A. A., J. Kellett, et al., *Ind. Eng. Chem.* **39**, 154 (1947).
3586. Oki, M. and H. Iwamura, *Bull. Chem. Soc. Jap.* **33**, 1632 (1960); *C. A.* **55**, 23,055 (1961).
3587. Oki, M. and H. Iwamura, *Bull. Chem. Soc. Jap.* **32**, 567 (1959); *C. A.* **54**, 6307 (1960).
3588. Oliver, G. D., M. Eaton, et al., *J. Amer. Chem. Soc.* **70**, 1502 (1948).
3589. Olivier, S. C. J., *Rec. Trav. Chim. Pays-Bas* **55**, 1027 (1936); *C. A.* **31**, 1759 (1937).
3590. Olsen, A. L. and E. R. Washburn, *J. Amer. Chem. oSc.* **57**, 303 (1935).
3591. Olsen, W. T., H. F. Hipsher, et al., *J. Amer. Chem. Soc.* **69**, 2451 (1947).
3592. Olsson, H., *Z. Phys. Chem.* (Leipzig) **125**, 243 (1927).
3593. O'Malley, J. A., C. Owens, et al., *J. Phys. Chem.* **72**, 3584 (1968).
3594. Omel'chenko, F. S., *Izv. Vyssh. Ucheb. Zaved., Pishch. Tekhnol.* **1962**, No. 3, 97; *C. A.* **57**, 11,920 (1962).

3595. Oosterhout, J. C. D. and T. C. Roddy, U.S. Patent 2,419,499 (Cl. 260–677), Apr. 22, 1947; *C. A.* **41**, 5151 (1947).
3596. O'Rourke, C. E., L. B. Clapp, et al., *J. Amer. Chem. Soc.* **78**, 2159 (1956).
3597. Orton, K. J. P. and D. C. Jones, *J. Chem. Soc.* **1919**, 1055; 1194.
3598. Orton, K. J. P. and M. Jones, *J. Chem. Soc.* **1912**, 1708.
3599. Osaka, H., *Bull. Inst. Phys. Chem. Res.* (Tokyo) **7**, 873 (1928); English ed., I, 80 (1928); *C. A.* **23**, 20 (1929).
3600. Osborn, A. G. and D. R. Douslin, *J. Chem. Eng. Data* **13**, 534 (1968).
3601. Osborne, N. S. and D. C. Ginnings, *J. Research Nat. Bur. Stand.* **39**, 453 (1947).
3602. Osborne, N. S., E. C. McKelvey, et al., *Bull. Nat. Bur. Stand.* **9**, 327 (1913); *J. Wash. Acad. Sci.* **2**, 95; *C. A.* **6**, 1085 (1912); **7**, 2889 (1913).
3603. Osborne, N. S., H. F. Stimson, et al., *J. Research Nat. Bur. Stand.* **23**, 197 (1939).
3604. Osina, T. M., *Trudy Nauch. Sessii Leningrad. Nauch.-Issledovatel. Inst. Gigieny Truda Profzabolev.* **1958**, 256; *C. A.* **56**, 2681 (1962).
3605. Orchin, M., *J. Assoc. Official Agr. Chem.* **25**, 839 (1942).
3606. Orndorff, W. R. and F. K. Cameron, *Am. Chem. J.* **17**, 517 (1895).
3607. Orton, K. J. P. and A. E. Bradfield, *J. Chem. Soc.* **1924**, 960.
3608. Orton, K. J. P. and A. E. Bradfield, *J. Chem. Soc.* **1927**, 983.
3609. Orton, K. J. P., M. G. Edwards, et al., *J. Chem. Soc.* **1911**, 1178.
3610. Orton, K. J. P. and M. Jones, *J. Chem. Soc.* **1912**, 1720.
3611. Orville-Thomas, W. J., A. E. Parsons, et al., *J. Chem. Soc.* **1958**, 1047.
3612. Osborn, A. G. and D. R. Douslin, *J. Chem. Eng. Data* **11**, 502 (1966).
3613. Osborne, N. S. and C. H. Meyers, *J. Research Nat. Bur. Stand.* **13**, 1 (1934).
3614. Ostwald, W., *Z. Phys. Chem.* (Leipzig) **3**, 369 (1889).
3615. Othmer, D. F., S. A. Savitt, et al., *Ind. Eng. Chem.* **41**, 572 (1949).
3616. Othmer, D. F. and T. O. Wentworth, *Ind. Eng. Chem.* **32**, 1588 (1940).
3617. Othmer, D. F. and R. E. White, U.S. Patent 2,275,802, March 10, 1942; *Official Gaz. U.S. Patent Office* **536**, 359 (1942).
3618. Otto, M. M., *J. Amer. Chem. Soc.* **57**, 1476 (1935).
3619. Ouellette, R. J., *Can. J. Chem.* **43**, 707 (1965).
3620. Owen, K., O. R. Quayle, et al., *J. Amer. Chem. Soc.* **61**, 900 (1939).
3621. Owen, K., O. R. Quayle, et al., *J. Amer. Chem. Soc.* **64**, 1294 (1942).
3622. Oxford, A. E., *Biochem. J.* **28**, 1325 (1934).
3623. Ozeki, T. and T. Nishida, Japanese Patent 7060, June 14, 1960; *C. A.* **55**, 7354 (1961).
3624. Ozol, R. J. and C. R. Masterson, U.S. Patent 2,356,689, Aug. 22, 1944; *C. A.* **39**, 86 (1945).

P

3625. Paal, C., *Ber. Deut. Chem. Gesell.* **25**, 1202 (1892).
3626. Packendorff, K., *Ber. Deut. Chem. Gesell.* **67**, 905 (1934).
3627. Page, J. M., C. C. Buchler, et al., *Ind. Eng. Chem.* **25**, 418 (1933).
3628. Pahlavouni, E., *Bull. Soc. Chim. Belges.* **36**, 533 (1927); *Chem. Zentr.* **1928**, I, 477.
3629. Pajeau, R., *Bull. Soc. Chim. Fr.* **1946**, 544; *C. A.* **41**, 2643 (1947).
3630. Pajeau, R., *Bull. Soc. Chim. Fr.* **1948**, 59; *C. A.* **42**, 4958 (1948).
3631. Palfray, L., *Bull. Soc. Chim. Fr.* **7**, 401 (1940); *C. A.* **36**, 2837 (1942).
3632. Palma, F. E. and E. A. Piotrowski, et al., *J. Mol. Spectrosc.* **13**, 119 (1964); *C. A.* **61**, 196 (1964).
3633. Palomaa, M. H., *Ber. Deut. Chem. Gesell.* **42**, 3873 (1909).
3634. Palomaa, M. H. and V. Aalto, *Ber. Deut. Chem. Gesell.* **66**, 471 (1933).
3635. Palomaa, M. H. and T. Honkanen, *Ber. Deut. Chem. Gesell.* **70**, 2199 (1937).
3636. Palomaa, M. and A. Juvala, *Ber. Deut. Chem. Gesell.* **61**, 1770 (1928).

3637. Palomaa, M. H. and I. Mikkila, *Ber. Deut. Chem. Gesell.* **75**, 1659 (1942).
3638. Palomaa, M. H. and T. A. Siitonen, *Ber. Deut. Chem. Gesell.* **63**, 3117 (1930).
3639. Palyi, G. and F. Peter, *Magy. Kim. Lapja* **17**, 354 (1962); *C. A.* **58**, 6820 (1963).
3640. Panov, N. and I. Dudnikov, *Khim. Referat. Zh.* **2**, No. 4, 24 (1939); *C. A.* **34**, 1514 (1940).
3641. Papini, G. and S. Cuomo, *Antincendio* **8**, 338 (1956); *C. A.* **51**, 4118 (1957).
3642. Papoušek, D., *Collect. Czech. Chem. Commun.* **24**, 2666 (1959); *C. A.* **54**, 1005 (1960)
3643. Paranjpe, G. R. and D. J. Davar, *Indian J. Phys.* **12**, 283 (1938); *C. A.* **33**, 3222 (1939).
3644. Pardee, W. A. and W. Weinrich, *Ind. Eng. Chem.* **36**, 595 (1944).
3645. Parkinson, A. E. and E. C. Wagner, *Ind. Eng. Chem., Anal. Ed.* **6**, 434 (1934).
3646. Parks, G. S., *J. Amer. Chem. Soc.* **47**, 338 (1925).
3647. Parks, G. S. and C. T. Anderson, *J. Amer. Chem. Soc.* **48**, 1506 (1926).
3648. Parks, G. S. and B. Barton, *J. Amer. Chem. Soc.* **50**, 24 (1928).
3649. Parks, G. S. and J. A. Hatton, *J. Amer. Chem. Soc.* **71**, 2773 (1949).
3650. Parks, G. S. and H. M. Huffman, *J. Amer. Chem. Soc.* **48**, 2788 (1926).
3651. Parks, G. S. and H. M. Huffman, *J. Amer. Chem. Soc.* **52**, 4381 (1930).
3652. Parks, G. S., H. M. Huffman, et al., *J. Amer. Chem. Soc.* **52**, 1032 (1930).
3653. Parks, G. S., H. M. Huffman, et al., *J. Amer. Chem. Soc.* **55**, 2733 (1933).
3654. Parks, G. S. and K. K. Kelly, *J. Amer. Chem. Soc.* **47**, 2089 (1925).
3655. Parks, G. S., W. D. Kennedy, et al., *J. Amer. Chem. Soc.* **78**, 56 (1956).
3656. Parks, G. S., K. E. Manchester, et al., *J. Chem. Phys.* **22**, 2089 (1954).
3657. Parks, G. S. and G. E. Moore, *J. Chem. Phys.* **7**, 1066 (1939).
3658. Parks, G. S., J. R. Mosley, et al., *J. Chem. Phys.* **18**, 152 (1950).
3659. Parks, G. S. and W. K. Nelson, *J. Phys. Chem.* **32**, 61 (1928).
3660. Parks, G. S., S. S. Todd, et al., *J. Amer. Chem. Soc.* **58**, 398 (1936).
3661. Parks, G. S., S. S. Todd, et al., *J. Amer. Chem. Soc.* **58**, 2505 (1936).
3662. Parks, G. S., G. E. Warren, et al., *J. Amer. Chem. Soc.* **57**, 616 (1935).
3663. Parks, G. S., T. J. West, et al., *J. Amer. Chem. Soc.* **68**, 2524 (1946).
3664. Parry, H. L., U.S. Patent 2,653,172, Sept. 22, 1953; *C. A.* **48**, 12,169 (1954).
3665. Parshall, G. W., *J. Org. Chem.* **27**, 4649 (1962).
3666. Parti, Y. P. and R. Samuel, *Proc. Roy. Soc.* **49**, 568 (1937); *C. A.* **31**, 8371 (1937).
3667. Partington, J. R., *Trans. Faraday Soc.* **17**, 734 (1922).
3668. Parts, A., *Z. Phys. Chem.* (Leipzig) **7B**, 327 (1930).
3669. Parts, A., *Z. Phys. Chem.* (Leipzig) **10B**, 264 (1930).
3670. Pascual, O. S. and E. Almeda, *Philippine At. Energy Comm.* (Rept) **PAEC(D)CH-634**, App. (1963); *C. A.* **60**, 10,521 (1964).
3671. Passino, H. J., U.S. Patent 2,396,966, Mar. 19, 1966; *C. A.* **40**, 3595 (1946).
3672. Paternò, E., *Gazz. Chem. Ital.* **19**, 640 (1899); from [3942].
3673. Paternò, E. and C. Montemartini, *Gazz. Chem. Ital.* **24b**, 197 (1894); *J. Chem. Soc.* **1895Aii**, 207.
3674. Paterson, W. G. and N. R. Tipman, *Can. J. Chem.* **40**, 2122 (1962).
3675. Patrick, C. R. and G. S. Prosser, *Trans. Faraday Soc.* **60**, 700 (1964).
3676. Patten, H. E., *J. Phys. Chem.* **6**, 554 (1902).
3677. Patterson, T. S. and W. C. Forsyth, *J. Chem. Soc.* **1913**, 2263.
3678. Patterson, T. S. and D. Thompson, *J. Chem. Soc.* **1908**, 355.
3679. Patterson, W. A., *Anal. Chem.* **26**, 823 (1954).
3680. Patty, F. A., *Industrial Hygiene and Toxicology*, 2d rev. ed., Vol. 1, Interscience, New York, 1958.
3681. Patty, F. A., *Industrial Hygiene and Toxicology*, 2d rev. ed., Vol. 2, Interscience, New York, 1963.

3682. Paul, D. E., D. Lipkin, et al., *J. Amer. Chem. Soc.* **78,** 116 (1956).
3683. Paul, R. and S. Tchelitcheff, *Bull. Soc. Chim. Fr.* **1947,** 453; *Brit. Chem. Abs.* **1948AII** 335.
3684. Paulopoulos, T. and M. A. El-Sayed, *J. Chem. Phys.* **41,** 1082 (1964).
3685. Paulsen, O., *Monatsh. Chem.* **72,** 244 (1939); *C. A.* **33,** 9138 (1939).
3686. Pavlic, A. A. and H. Adkins, *J. Amer. Chem. Soc.* **68,** 1471 (1946).
3687. Pawlewski, Br., *Ber. Deut. Chem. Gesell.* **15,** 2460 3034 (1882).
3688. Pawlewski, Br., *Ber. Deut. Chem. Gesell.* **21,** 2141 (1888).
3689. Peach, M. E. and T. C. Waddington, *J. Chem. Soc.* **1962,** 2680.
3690. Pearce, J. N., *J. Phys. Chem.* **19,** 14 (1915).
3691. Pearce, J. N. and L. F. Berhenke, *J. Phys. Chem.* **39,** 1005 (1935).
3692. Pearce, J. N. and F. S. Mortimer, *J. Amer. Chem. Soc.* **40,** 509 (1918).
3693. Pearson, B. D. and J. E. Ollerenshaw, *Chem. Ind.* **1966,** 370.
3694. Pearson, R. G., *J. Amer. Chem. Soc.* **70,** 204 (1948).
3695. Pearson, R. G. and R. L. Dillon, *J. Amer. Chem. Soc.* **75,** 2439 (1953).
3696. Pearson, R. G. and J. M. Mills, *J. Amer. Chem. Soc.* **72,** 1692 (1950).
3697. Pearson, R. G. and D. C. Vogelsong, *J. Amer. Chem. Soc.* **80,** 1038 (1958).
3698. Pearson, R. G. and F. V. Williams, *J. Amer. Chem. Soc.* **75,** 3073 (1953).
3699. Pecchiai, L. and U. Saffotti, *Med. Lavoro* **48,** 247 (1957); *C. A.* **51,** 15,040 (1957)
3700. Pechalin, L. I. and G. M. Panchenkov, *Vestn. Mosk. Univ. Ser. II. Khim.* **21,** 30 (1966); *C. A.* **65,** 11,374 (1966).
3701. Pechalin, L. I. and G. M. Panchenkov, *Zh. Fiz. Khim.* **40,** 1140 (1966); *C. A.* **65,** 6328 (1966).
3702. Peddle, C. J., and W. E. S. Turner, *J. Chem. Soc.* **1913,** 1202.
3703. Pell, A. S. and G. Pilcher, *Trans. Faraday Soc.* **61,** 71 (1965).
3704. Peltonen, R. J., P. Neuenschwander, et al., *Z. Lebensm.-Unters. Forsch.* **102,** 245 (1955); *C. A.* **50,** 1259 (1956).
3705. Pendl, E., A. W. Reitz, et al., *Proc. Indian Acad. Sci.* **8A,** 508 (1938).
3706. Pennington, R. E. and K. A. Kobe, *J. Amer. Chem. Soc.* **79,** 300 (1957).
3707. Peppiatt, E. G. and R. J. Wicker, *J. Chem. Soc.* **1955,** 3122.
3708. Perchard, J. P., M. T. Forel, et al., *J. Chim. Phys.* **61,** 645 (1964); *C. A.* **61,** 10,188 (1964).
3709. Peregudov, G. V., S. V. Markova, et al., *Izv. Akad. Nauk SSSR, Otd. Khim. Nauk* **1957,** 37; *C. A.* **51,** 12,657 (1957).
3710. Perichon, J. and R. Buvet, *Bull. Soc. Chim. Fr.* **1968,** 1279, 1282.
3711. Perkin, W. H., *J. Chem. Soc.* **1889,** 680.
3712. Perkin, W. H., *J. Chem. Soc.* **1892,** 287.
3713. Perkin, W. H., *J. Chem. Soc.* **1894,** 815.
3714. Perkin, W. H., *J. Chem. Soc.* **1896,** 1025.
3715. Perkin, W. H. and C. Weizmann, *J. Chem. Soc.* **1906,** 1649.
3716. Perkins, R. P., U.S. Patent 2,190,607 (Cl. 260–600), Feb. 13, 1940; *C. A.* **34,** 4081 (1940).
3717. Perlick, A., *Bull. Inst, Intern. Froid* **1,** 4 (1937).
3718. Perrin, D. D., *Aust. J. Chem.* **17,** 484, (1964).
3719. Perry, E. S. and W. H. Weber, *J. Amer. Chem. Soc.* **71,** 3726 (1949).
3720. Perry, J. H., *J. Phys. Chem.* **31,** 1737 (1927).
3721. Perry, R. E. and G. Thodos, *Ind. Eng. Chem.* **44,** 1649 (1952).
3722. Person, W. B. and G. C. Pimentel, *J. Amer. Chem. Soc.* **75,** 532 (1953).
3723. Pesce, B., *Gazz. Chim. Ital.* **70,** 710 (1940); *C. A.* **35,** 5091 (1941).
3724. Pesez, M., *Ann. Pharm. Fr.* **6,** 279 (1948); *C. A.* **43,** 5341 (1949).

3725. Pestemer, M. and A. Alslev-Klinker, *Z. Electrochem. Agnew. Phys. Chem.* **53**, 387 (1949); *Chem. Zentr.* **1950**, II, 515.

3726. Pestemer, M. and H. Flaschka, *Monatsh. Chem.* **71**, 325 (1938); *C. A.* **32**, 8266 (1938).

3727. Peters, L. M., K. E. Marple, et al., *Ind. Eng. Chem.* **40**, 2046 (1948).

3728. Peterson, B. H., *Proc. Iowa Acad. Sci.* **50**, 253 (1943).

3729. Peterson, P. E., *J. Org. Chem.* **31**, 439 (1966).

3730. Petrescu, V., *Ann. Sci. Univ. Jassy, Sect. I*, **26**, 233 (1940); *C. A.* **34**, 4626 (1940).

3731. Petrikaln, A., *Z. Phys. Chem.* (Leipzig) **3B**, 362 (1929).

3732. Petrikaln, A. and J. Hochberg, *Z. Phys. Chem.* (Leipzig) Abt. **B.4**, 299 (1929); *C. A.* **23**, 4620 (1929).

3733. Petro, A. J. and C. P. Smyth, *J. Amer. Chem. Soc.* **80**, 73 (1958).

3734. Petrov, A. A., *J. Gen. Chem.* (USSR) **11**, 713 (1941); *C.A.* **36**, 404 (1942).

3735. Petrov, A. A. and G. I. Semenov, *Zh. Obshch. Khim.* **27**, 2941 (1957); *C. A.* **52**, 8102 (1958).

3736. Petrov, K. D. and M. I. Itkina, *J. Gen. Chem.* (USSR) **17**, 220 (1947); *C. A.* **42**, 516 (1948).

3737. Petrov, K. D. and E. S. Lagucheva, *Zh. Obshch. Khim.* **18**, 1150 (1948); *C. A.* **43**, 1742 (1949).

3738. Pettit, L. D. and S. Bruckenstein, *J. Inorg. Nucl. Chem.* **24**, 1478 (1962).

3739. Pfann, W., *Zone Melting*, 2d ed., Wiley, New York, 1966.

3740. Pfeiffer, A., *Ber. Deut. Chem. Gesell.* **5**, 699 (1872).

3741. Pfenninger, H., *Helv. Chim. Acta* **45**, 460 (1962).

3742. Phadke, R. P., *J. Indian Inst. Sci.* **34**, 189 (1953); *C. A.* **47**, 3060 (1953).

3743. Phadke, S. R., S. D. Gokhale, et al., *J. Indian Chem. Soc.* **22**, 235 (1945); *C. A.* **40**, 5310 (1946); Beil E**III**2, 306.

3744. Philip, J. C. and S. C. Waterton, *J. Chem. Soc.* **1930**, 2783.

3745. Philip, N. M., *Proc. Indian Acad. Sci.* **9A**, 109 (1939); *C. A.* **33**, 4484 (1939).

3746. Philippe, R. and A. M. Piette, *Bull. Soc. Chim. Belges* **64**, 600 (1955); *C. A.* **50**, 15,151 (1956); [4773].

3747. Phillips Petroleum Co., Chemical Dept., Bulletin 522, "Phillips Hydrocarbons," 6th ed., Bartlesville, Oklahoma, 1964.

3748. Phillips, W. D., H. C. Miller, et al., *J. Amer. Chem. Soc.* **81**, 4496 (1959).

3749. Philpot, J. P., E. C. Rhodes, et al., *J. Chem. Soc.* **1940**, 84.

3750. Philpotts, A. R., W. Thain, et al., *Anal. Chem.* **23**, 268 (1951).

3751. Piatti, L., *Z. Phys. Chem.* (Leipzig) **152A**, 36 (1931).

3752. Pick, J., V. Fried, et al., *Chem. Listy* **49**, 1720 (1955); *C. A.* **50**, 636 (1956).

3753. Pickard, R. H., and J. Kenyon, *J. Chem. Soc.* **1911**, 45.

3754. Pickard, R. H. and J. Kenyon, *J. Chem. Soc.* **1912**, 620.

3755. Pickard, R. H. and J. Kenyon, *J. Chem. Soc.* **1913**, 1923.

3756. Pickard, R. H. and J. Kenyon, *J. Chem. Soc.* **1914**, 830.

3757. Pickering, R. A. and C. C. Price, *J. Amer. Chem. Soc.* **80**, 4931 (1958).

3758. Pickett, L. W., *J. Chem. Phys.* **10**, 660 (1942).

3759. Pickett, L. W., N. J. Hoeflich, et al., *J. Amer. Chem. Soc.* **73**, 4865 (1951).

3760. Pickett, L. W., M. Muntz, et al., *J. Amer. Chem. Soc.* **73**, 4862 (1951).

3761. Pickett, O. A. and J. M. Peterson, *Ind. Eng. Chem.* **21**, 325 (1929).

3762. Pictet, A. and A. Geleznoff, *Ber. Deut. Chem. Gesell.* **36**, 2219 (1903).

3763. Pickara, A. and B. Pickara, *Compt. Rend.* **198**, 1018 (1934); *C. A.* **28**, 2961 (1934).

3764. Pierce, L. and M. Hayashi, *J. Chem. Phys.* **35**, 479 (1961).

3765. Pierre, I. and E. Puchot, *Ann. Chem.* **163**, 283 (1872).

3766. Pierson, R. H., A. N. Fletcher, et al., *Anal. Chem.* **28**, 1218 (1956).

3767. Pilcher, G. and J. D. M. Chadwick, *Trans. Faraday Soc.* **63**, 2357 (1967).
3768. Pilcher, G., H. A. Skinner, et al., *Trans. Faraday Soc.* **59**, 316 (1963).
3769. Pilpel, N., *J. Amer. Chem. Soc.* **77**, 2949 (1955).
3769a. Pines, H., U.S. Patent 2,422,798, June 24, 1947; *C. A.* **41**, 6270 (1947).
3770. Pipparelli, E., A. Balducci, et al., *Ann. Chim.* (Rome) **46**, 112 (1956); *C. A.* **50**, 13,862 (1956).
3771. Pirlot, G., *Bull. Soc. Chim. Belges* **58**, 28 (1949); *C. A.* **43**, 7366 (1949).
3772. Pirlot, G., *Bull. Soc. Roy. Sci. Liége* **18**, 115 (1949); *C. A.* **43**, 7817 (1949).
3773. Pittsburgh Industrial Chemicals, *PX-138 Dioctyl Phthalate*, Pittsburgh, Pennsylvania,
3774. Pitzer, K. S., *J. Amer. Chem. Soc.* **62**, 331 (1940).
3775. Pitzer, K. S., *J. Amer. Chem. Soc.* **62**, 1224 (1940).
3776. Pitzer, K. S., *J. Amer. Chem. Soc.* **63**, 2413 (1941).
3777. Pitzer, K. S., L. Guttman, et al., *J. Amer. Chem. Soc.* **68**, 2209 (1946).
3778. Pitzer, K. S. and W. D. Gwinn, *J. Amer. Chem. Soc.* **63**, 3313 (1941).
3779. Pitzer, K. S. and J. L. Hollenberg, *J. Amer. Chem. Soc.* **76**, 1493 (1954).
3780. Pitzer, K. S. and W. Weltner, *J. Amer. Chem. Soc.* **71**, 2842 (1949).
3781. Plamondon, J. E., R. J. Buenker, et al., *Proc. Iowa Acad. Sci.* **70**, 163 (1963); *C. A.* **61**, 11,424 (1964).
3782. Plate, A. F. and G. A. Tarasova, *J. Appl. Chem.* (USSR) **17**, 576 (1944); *C. A.* **40**, 2318 (1946).
3782a. Plato, C. and A. R. Glasgow, *Anal. Chem.* **41**, 330 (1969).
3783. Platt, J. R. and H. B. Klevens, *J. Chem. Phys.* **16**, 832 (1948).
3784. Platt, J. R., H. B. Klevens, et al., *J. Chem. Phys.* **17**, 466 (1949).
3785. Platt, J. R., I. Rusoff, et al., *J. Chem. Phys.* **11**, 535 (1943).
3786. Plebański, T., *Bull. Acad. Pol. Sci. Ser. Sci. Chim.* **8**, 239 (1960); *C. A.* **55**, 20,553 (1961).
3787. Pliva, J. and V. Herout, *Collect. Czech. Chem. Commun.* **15**, 160 (1950); *C. A.* **45**, 8481 (1951).
3788. Plum, W. B., *J. Chem. Phys.* **5**, 172 (1937).
3789. Plyler, E. K., *J. Chem. Phys.* **16**, 1008 (1948).
3790. Plyler, E. K., *J. Chem. Phys.* **17**, 218 (1949).
3791. Plyler, E. K., *J. Opt. Soc. Amer.* **37**, 746 (1947).
3792. Plyler, E. K. and N. Acquista, *J. Res. Nat. Bur. Stand.* **43**, 37 (1949).
3793. Plyler, E. K., H. C. Allen, et al., *J. Res. Nat. Bur. Stand.* **58**, 255 (1957).
3794. Plyler, E. K. and W. S. Benedict, *J. Res. Nat. Bur. Stand.* **47**, 202 (1951).
3795. Plyler, E. K. and C. J. Humphreys, *J. Res. Nat. Bur. Stand.* **39**, 59 (1947).
3796. Plyler, E. K., R. Stair, et al., *J. Res. Nat. Bur. Stand.* **38**, 211 (1947).
3797. Pohl, H. A., M. E. Hobbs, et al., *J. Chem. Phys.* **9**, 408 (1941).
3798. Pohland, E. and W. Mehl, *Z. Phys. Chem.* (Leipzig) **164A**, 48 (1933).
3799. Pohoryles, L. A., S. Sarel, et al., *J. Org. Chem.* **24**, 1878 (1959).
3800. Polak, J. and I. Mertl, *Collect. Czech. Chem. Commun.* **30**, 3526 (1965); *C. A.* **63**, 14,072 (1965).
3801. Polak, L. S., R. A. Khmel'nitskiĭ, et al., *Petrol. Chem. USSR* **2** (1), 1 (1963); *C. A.* · **61**, 13,166 (1964).
3802. Poley, J. P., *Appl. Sci. Res.* **4B**, 337 (1955); *C. A.* **49**, 15,316 (1955).
3803. Polonskiĭ, T. M., *Khim. Referat. Zh.* **4**, No. 3, 6 (1941); *C. A.* **37**, 3668 (1943).
3804. Polyakova, A. A., T. I. Popova, et al., *Neftekhimiya* **6**, 150 (1966); *C. A.* **64**, 18,635 (1966).
3805. Pomerantz, P., A. Fookson, et al., *J. Res. Nat. Bur. Stand.* **52**, 59 (1954).
3806. Pool, W. O. and A. W. Ralston, *Ind. Eng. Chem.* **34**, 1104 (1942).
3807. Pople, J. A. and T. Schaefer, *Mol. Phys.* **3**, 547 (1960); *C. A.* **55**, 8045 (1961).

3808. Pople, J. A., W. G. Schneider, et al., *Can. J. Chem.* **35,** 1060 (1957).
3809. Popov, M. M. and P. K. Shirokich, *Z. Phys. Chem.* (Leipzig) **167A,** 183 (1933); *C. A.* **28,** 2573 (1934).
3810. Porcaro, P. J. and V. D. Johnston, *Anal. Chem.* **33,** 361 (1961).
3811. Porret, D. and C. F. Goodeve, *Trans. Faraday Soc.* **33,** 690 (1937); *C. A.* **31,** 7337 (1937).
3812. Porte, A. L. and H. S. Gutowsky, *J. Org. Chem.* **28,** 3216 (1963).
3813. Porter, F. and J. N. Cosby, U.S. Patent 2,507,048, May 9, 1950; *C. A.* **44,** 7879 (1950).
3814. Potter, A. E. and H. L. Ritter, *J. Phys. Chem.* **58,** 1040 (1954).
3815. Potter, J. C. and J. H. Saylor, *J. Amer. Chem. Soc.* **73,** 90 (1951).
3816. Potts, W. A., *Anal. Chem.* **27,** 1027 (1955).
3817. Potts, W. J., *J. Chem. Phys.* **20,** 809 (1952).
3818. Potts, W. J., *J. Chem. Phys.* **23,** 65 (1955).
3819. Pound, J. R., *J. Phys. Chem.* **31,** 547 (1927).
3820. Powers, R. M., M. T. Tetenbaum, et al., *Anal. Chem.* **34,** 1132 (1962).
3821. Powles, J. G. and A. Hartland, *Arch. Sci.* (Geneva) **13,** Spec. No. 475 (1960).
3822. Pozzani, U. C., E. R. Kinkead, *et al.*, *Am. Ind. Hyg. Assoc. J.* **29,** 202 (1968).
3823. Prakash, S., *J. Sci. Res. Banaras Hindu Univ.* **13** (1), 144 (1962–63); *C. A.* **59,** 2296 (1963).
3824. Prakash, S. and N. L. Singh, *J. Sci. Ind. Res.* **21B,** 512 (1962); *C. A.* **58,** 5155 (1963).
3825. Pratt, H. R. C., *Ind. Chemist* **23,** 658, 727 (1947); *C. A.* **42,** 3233 (1948).
3826. Preston, S. T., *J. Gas Chromatog.* **1** (3), 8 (1963).
3827. Prévost, C., *Compt. Rend.* **183,** 1292 (1926).
3828. Prévost, C. and A. Valette, *Compt. Rend.* **222,** 326 (1946); *C. A.* **40,** 3719 (1946).
3829. Prey, V. and W. Unger, *Ann. Chem.* **651,** 154 (1962); *C. A.* **57,** 3540 (1962).
3830. Price, C. C. and M. Osgan, *J. Amer. Chem. Soc.* **78,** 4787 (1957).
3831. Price, W. C. and D. M. Simpson, *Proc. Roy. Soc.* **165A,** 272 (1938).
3832. Price, W. C. and A. D. Walsh, *Proc. Roy. Soc.* **179A,** 201 (1941); *C. A.* **36,** 336 (1942).
3833. Price, W. C. and A. D. Walsh, *Proc. Roy. Soc.* **197A,** 22 (1947); *C. A.* **42,** 1504 (1948).
3834. Prichard, W. H. and W. J. Orville-Thomas, *Trans. Faraday Soc.* **61,** 1549 (1965).
3835. Prideaux, E. B. R. and R. N. Coleman, *J. Chem. Soc.* **1936,** 1346.
3836. Private communication to the authors.
3837. Privett, O. S., E. Breault, et al., *J. Amer. Oil Chem. Soc.* **35,** 366 (1958).
3838. Produits Chimiques Pechiney-Saint-Gobain, *French Addn.* **83,** 361 (Cl. C 07c), July 31, 1964; *C. A.* **61,** 15974 (1964).
3839. Prosen, E. J., W. H. Johnson, et al., *J. Res. Nat. Bur. Stand.* **39,** 173 (1947).
3840. Prosen, E. J. and F. D. Rossini, *J. Res. Nat. Bur. Stand.* **33,** 255 (1944).
3841. Prosen, E. J. and F. D. Rossini, *J. Res. Nat. Bur. Stand.* **34,** 59 (1945).
3842. Prout, F. S. and J. E. Spikner, *J. Org. Chem.* **27,** 1488 (1962).
3843. Prud'homme, M., *Bull. Soc. Chim. Fr.* **31,** 295 (1922); *Chem. Zentr.* **1922,** III, 1160.
3844. Prue, J. E. and P. J. Sherrington, *Trans. Faraday Soc.* **57,** 1795 (1961).
3845. Pryor, W. A. and R. E. Jentoft, *J. Chem. Eng. Data* **6,** 36 (1961).
3846. Puck, T. T. and H. Wise, *J. Phys. Chem.* **50,** 329 (1946).
3847. Pudowik, A. N., *Z. Obshch. Chim.* **19,** 1187 (1949); *C. A.* **44,** 1005 (1950).
3848. Pugacheva, A. E., *Uchenye Zapiski Moskov. Oblast. Pedagog. Inst.* **93,** 43 (1960); *C. A.* **56,** 2073 (1962).
3849. Pukirev, A. B., *Trans. Inst. Pure Chem. Reagents* (Moscow) **15,** 45 (1937); *C. A.* **32,** 5378 (1938).
3850. Pulfrich, C., *Neues Jahar, Beilageband,* **V,** 167 (1887); from [4772].

3851. Pummer, W. J., *J. Chem. Eng. Data* **6**, 76 (1961).
3852. Puranik, P. G., *J. Chem. Phys.* **23**, 761 (1955).
3853. Puranik, P. G., *Proc. Indian Acad. Sci.* **42A**, 326 (1955); *C. A.* **50**, 7595 (1956).
3854. Puranik, P. G. and K. V. Ramiah, *J. Mol. Spectrosc.* **3**, 486 (1959); *C. A.* **55**, 10,064 (1961).
3855. Puranik, P. G. and K. V. Ramiah, *Proc. Indian Acad. Sci.* **54**, 69 (1961); *C. A.* **56**, 9592 (1962).
3856. Purcell, J. M., S. G. Morris, et al., *Anal. Chem.* **38**, 588 (1966).
3857. Purcell, J. M. and H. Susi, *Appl. Spectrosc.* **19**, 105 (1965); *C. A.* **63**, 13,032 (1965).
3858. Purvis, J. E., *J. Chem. Soc.* **1910**, 1546.
3859. Purvis, J. E., *J. Chem. Soc.* **1912**, 1315.
3860. Purvis, J. E., *J. Chem. Soc.* **1924**, 406.
3861. Purvis, J. E., *Proc. Cambridge Phil. Soc.* **23**, 588 (1925); *Chem. Zentr.* **1927**, **II**, 379.
3862. Purvis, J. E. and N. P. McCleland, *J. Chem. Soc.* **1913**, 1088.
3863. Puschin, N. A. and A. A. Glagoleva, *J. Chem. Soc.* **1922**, 2813.
3864. Puschin, N. A. and P. G. Malavulj, *Z. Phys. Chem.* (Leipzig) **158A**, 161, 162, 290, 341, 415 (1932).
3865. Puschin, N. A. and T. Pinter, *Z. Phys. Chem.* (Leipzig) **142A**, 211 (1929).
3866. Putnam, W. E., D. M. McEachern, et al., *J. Chem. Phys.* **42**, 749 (1965).
3867. Pyryalova, P. S. and E. N. Zil'berman, *Tr. Khim. Khim. Tekhnol.* **1963** (2), 353; *C. A.* **61**, 4211 (1964).

Q

3868. Quaker Oats Company, The, "QO Furfural" Bulletin 203-A, Chicago.
3869. Quaker Oats Company, Chemicals Division, "Furfuryl Alcohol," Bulletin 205-A, Chicago, 1963.
3870. Quaker Oats Company, "QO Tetrahydrofurfuryl Alcohol," Bulletin 206-A, Chicago, 1960.
3871. Quaker Oats Company, Tech. Bulletin No. 148, "QO Tetrahydrofuran," Chicago, 1964.
3872. Quaker Oats Company, Tech. Bulletin No. 149, "QO Furan," Chicago, 1965.
3872a. Quayle, O. R., *Chem. Rev.* **53**, 439 (1953).
3873. Quayle, O. R., R. A. Day, *J. Amer. Chem. Soc.* **66**, 938 (1944).
3874. Quiggle, D., C. O. Tongberg, et al., *Ind. Eng. Chem. Anal. Ed.* **6**, 466 (1934).

R

3875. Rabinovitch, I. B., V. I. Tel'noi, et al., *Dokl. Akad. Nauk SSSR* **143**, 133 (1962); *C. A.* **57**, 4557 (1962).
3876. Rabinowitsch, M., *Z. Phys. Chem.* (Leipzig) **119**, 59, 70 (1926).
3877. Rabjohn, N., *Organic Syntheses, Collective Vol. iv*, 1st ed., Wiley, New York, 1963.
3878. Radinger, G. and H. Wittek, *Z. Phys. Chem.* (Leipzig) **45B**, 329 (1939); *C. A.* **34**, 4991 (1940).
3879. Radulescu, D. and M. Alexa, *Bull. Soc. Chim. Romania* **20A**, 89 (1938); *C. A.* **34**, 934 (1940).
3880. Radzitzky, P. de, and J. Hanotier, *Ind. Chim. Belge* **27**, 125 (1962); *C. A.* **58**, 4464 (1963).
3881. Raffauf, R. F., *J. Amer. Chem. Soc.* **72**, 753 (1950).
3882. Rahman, S. M. F., A. M. Roy, et al., *J. Nat. Sci. Math.* **5**, 203 (1965); *C. A.* **65**, 129 (1966).
3883. Railing, W. E., *J. Amer. Chem. Soc.* **61**, 3349 (1939).

3884. Rall, H. T. and H. M. Smith, *Ind. Eng. Chem., Anal. Ed.* **8**, 324 (1936).

3885. Ralph, E. K. and W. A. Gilkerson, *J. Amer. Chem. Soc.* **86**, 4783 (1964).

3886. Ralston, A. W. and C. W. Hoerr, *J. Org. Chem.* **7**, 546 (1942).

3887. Ramart-Lucas, Mme., Mlle. Bequard, et al., *Compt. Rend.* **190**, 1196 (1930).

3888. Ramart-Lucas, de Mmo, and M. F. Salmon-Legagneur, *Compt. Rend.* **190**, 492 (1930).

3889. Rambidi, N. G., *Dokl. Akad. Nauk SSSR* **102**, 747 (1955); *C. A.* **49**, 15,485 (1955).

3890. Rambraud, R. and D. Besserre, *Bull. Soc. Chim. Fr.* **1955**, 45; *C. A.* **50**, 11,940 (1956).

3891. Ramiah, K. V. and V. V. Chalapathi, *Proc. Indian Acad. Sci., Sect. A*, **60**, 242 (1964); *C. A.* **63**, 155 (1965).

3892. Ramiah, K. V., V. V. Chalapathi, et al., *Curr. Sci.* **35**, 350 (1966); *C. A.* **65**, 13,038 (1966).

3893. Ramsay, W. and J. Shields, *Z. Phys. Chem.* (Leipzig) **12**, 433 (1893).

3894. Ramsay, W. and B. D. Steele, *Z. Phys. Chem.* (Leipzig) **44**, 348 (1903).

3895. Ramsey, B. G., *J. Amer. Chem. Soc.* **88**, 5358 (1966).

3896. Ramsey, D. A., *Proc. Roy. Soc.* **190A**, 562 (1947); *C. A.* **42**, 4059 (1948).

3897. Ramsey, J. B. and F. T. Aldridge, *J. Amer. Chem. Soc.* **77**, 2561 (1955).

3898. Randall, H. M., R. G. Fowler, et al., *Infrared Determination of Organic Structures*, Van Nostrand, Princeton, N.J., 1949.

3899. Rank, D. H., *J. Opt. Soc. Amer.* **37**, 798 (1947).

3900. Rank, D. H. and E. R. Bordner, *J. Chem. Phys.* **3**, 248 (1935).

3901. Rank, D. H., R. W. Scott, et al., *Ind. Eng. Chem., Anal. Ed.* **14**, 816 (1942).

3902. Rank, D. H., N. Sheppard, et al., *J. Chem. Phys.* **16**, 698 (1948).

3903. Rao, A. L. S., *Indian J. Phys.* **14**, Pt. 5, 365 (1941); *C. A.* **35**, 6187 (1941).

3904. Rao, A. L. S., *J. Indian Chem. Soc.* **18**, 337 (1941); *C. A.* **36**, 967 (1942).

3905. Rao, D. V. G. L. N., *Curr. Sci.* **25**, 217 (1956); *C. A.* **51**, 2348 (1957).

3906. Rao, I. A., *J. Sci. Ind. Res.* **20B**, 145 (1961); *C. A.* **55**, 24,228 (1961).

3907. Rao, M. R. A. and B. S. Rao, *J. Indian Chem. Soc.* **12**, 322 (1935); *Chem. Zentr.* **1936**, I, 1389; *C. A.* **29**, 7148 (1935).

3908. Rao, M. V., *Proc. Indian Acad. Sci.* **24A**, 510 (1946); *C. A.* **41**, 3697 (1947).

3909. Rao, P. R., *Proc. Indian Acad. Sci.* Sect. **55A**, 232 (1962); *C. A.* **57**, 6755 (1962).

3910. Rao, P. R. and N. L. Singh, *Proc. Indian Acad. Sci.* **55A**, 178 (1962); *C. A.* **57**, 5475 (1962).

3911. Raoult, F. M., *Ann. Chim. Phys.* **2**, 66 (1884); *J. Chem. Soc.* **1884A**, 952.

3912. Rapean, J. C., D. L. Pearson, et al., *Ind. Eng. Chem.* **51**, No. 2, 77A (1959).

3913. Raphael, R. A. and F. Sondheimer, *J. Chem. Soc.* **1951**, 2693.

3914. Raposo, M. R., C. Estevens, et al., *Rev. Port. Farm.* **11**, No. 2, 79 (1961); *C. A.* **56**, 5217 (1962).

3915. Rapp, D. E., U.S. Patent 2,923,747, Feb. 2, 1960.

3916. Raskin, S. S. and F. I. Skripov, *Zh. Eksp. Teor. Fiz.* **26**, 479 (1954); *C. A.* **49**, 15,484 (1955).

3917. Rasmussen, R. S., *J. Chem. Phys.* **16**, 712 (1948).

3918. Rasmussen, R. S. and R. R. Brattain, *J. Amer. Chem. Soc.* **71**, 1073 (1949).

3919. Rasmussen, R. S., R. R. Brattain, et al., *J. Chem. Phys.* **15**, 135 (1947).

3920. Rasmussen, R. S., D. D. Tunnieliff, et al., *J. Amer. Chem. Soc.* **71**, 1068 (1949).

3921. Rast, K., *Ber. Deut. Chem. Gesell.* **55**, 1051 (1922).

3922. Rastogi, R. P., R. K. Nigam, et al., *J. Phys. Chem.* **39**, 3042 (1963).

3923. Ratchford, W., C. Rehberg, et al., *J. Amer. Chem. Soc.* **66**, 1864 (1944).

3924. Ravikovich, S. D. and V. P. Solomko, *Ukr. Khim. Zh.* **24**, 7 (1958); *C. A.* **52**, 15,173 (1958).

3925. Rea, D. G., *Anal. Chem.* **32**, 1638 (1960).
3926. Read, R. R., L. S. Foster, et al., *Org. Syntheses* **25**, 11 (1945); *C. A.* **40**, 320 (1946).
3927. Reay, W. H. and G. T. Merrall, British Patent 891,823, Mar. 21, 1962; *C. A.* **57**, 4515 (1962).
3928. Rechenberg, C. v., *Einfache und Fraktionierte Destillation in Theorie und Praxis*, Selbstverlag von Schimmel and Company, Miltitz bei Leipzig, 1923.
3929. Reddy, G. S., *Z. Naturforsch.* **A21**, 609 (1966); *C. A.* **65**, 13,505 (1966).
3930. Reddy, G. S. and J. H. Goldstein, *J. Amer. Chem. Soc.* **83**, 5020 (1961).
3931. Reddy, G. S. and J. H. Goldstein, *J. Amer. Chem. Soc.* **84**, 583 (1962).
3932. Reddy, G. S., J. H. Goldstein, et al., *J. Amer. Chem. Soc.* **83**, 1300 (1961).
3933. Redlich, O. and W. Stricks, *Monatsh. Chem.* **68**, 47 (1936); *Chem. Zentr.* **1936, II**, 2521.
3934. Rees, H. L. and D. H. Anderson, *Anal. Chem.* **21**, 989 (1949).
3935. Reeves, L. W. and W. G. Schneider, *Trans. Faraday Soc.* **54**, 314 (1958).
3936. Rehberg, C. E. and C. H. Fisher, *Ind. Eng. Chem.* **40**, 1429 (1948).
3937. Rehberg, C. E. and C. H. Fisher, *J. Amer. Chem. Soc.* **66**, 1203 (1944).
3938. Reichel, L. and O. Strasser, *Ber. Deut. Chem. Gesell.* **64**, 1997 (1931).
3939. Reichert, J. S., J. H. Bailey, et al., *J. Amer. Chem. Soc.* **45**, 1552 (1923).
3940. Reid, E. E., U.S. Patent 1,821,667, Sept. 1, 1931; *Official Gaz. U.S. Patent Office* **410**, 221 (1931).
3941. Reilley, C. A. and J. D. Swalen, *J. Chem. Phys.* **35**, 1522 (1961).
3942. Reilly, J. and W. N. Rae, *Physico-Chemical Methods*, 3rd ed., Van Nostrand, Princeton, New Jersey, 1939.
3943. Reilly, J., R. Wolter, et al., *Sci. Proc. Roy. Dublin Soc.* **19**, 467 (1930); *Chem. Zentr.* **1931, I**, 1899.
3944. Reinhardt, R. C., *Ind. Eng. Chem.* **35**, 422 (1943).
3945. Reinisch, L., *J. Chem. Phys.* **51**, 113 (1954); *C. A.* **49**, 2801 (1955).
3946. Reinke, R. C. and E. N. Luce, *Ind. Eng. Chem., Anal. Ed.* **18**, 244 (1946).
3947. Reisse, J. and R. Ottinger, *Spectrochim. Acta* **22**, 969 (1966); *C. A.* **65**, 1935 (1966).
3948. Reith, H. and H. Eckardt, *Freiberger Forschungsh.* **A250**, 39 (1962); *C. A.* **59**, 11,158 (1963).
3949. Reitmeier, R. E., V. Sivertz, et al., *J. Amer. Chem. Soc.* **62**, 1943 (1940).
3950. Reitz, A. W., *Z. Phys. Chem.* (Leipzig) **38B**, 179 (1937); *Chem. Zentr.* **1937, II**, 367.
3951. Reitz, A. W., *Z. Phys. Chem.* (Leipzig) **38B**, 275 (1937); *C. A.* **32**, 2837 (1938).
3952. Reitz, A. W. and R. Sabathy, *Monatsh. Chem.* **71**, 100 (1937); *C. A.* **32**, 2027 (1938).
3953. Reitz, A. W. and R. Sabathy, *Monatsh. Chem.* **71**, 131 (1938); *C. A.* **32**, 4076 (1938).
3954. Reitz, A. W. and R. Skrabal, *Monatsh. Chem.* **70**, 398 (1937); *C. A.* **31**, 7758 (1937).
3955. Remiz, E. K. and A. V. Frost, *J. Appl. Chem.* (USSR) **13**, 210 (1940); *C. A.* **35**, 433 (1941).
3956. Renault, L., French Patent 868,182, Dec. 23, 1941; *C. A.* **43**, 5034 (1949).
3957. Reppe, B. W., *Chemie und Technik der Acetylene-Druckvaktionen*, Weinheim, 1951.
3958. Reppe, J. W., *Acetylene Chemistry*, Charles A. Meyer, New York, 1949.
3959. Reppe, J. W., *Ann. Chem. Justus Liebigs* **596**, 25 (1955); *C. A.* **50**, 16,774 (1956).
3960. Reuther, H. and G. Reichel, *Plast Kautschuk* **10**, 720 (1963); *C. A.* **60**, 11,389 (1964).
3961. Rey-Lafon, M., J. Lascombe, et al., *Ann. Chim.* (Paris) **8**, 493 (1963); *C. A.* **60**, 10,065 (1964).
3962. Reynolds, P. W. and D. M. Grudgings, British Patent 622, 937, May 10, 1949; *C. A.* **43**, 8397 (1949).
3963. Reynolds, W. L. and R. H. Weiss, *J. Amer. Chem. Soc.* **81**, 1790 (1959).
3964. Rhodes, R. E., M. Barber, et al., *Anal. Chem.* **38**, 48 (1966).
3965. Rice, F. O., *Proc. Roy. Soc.* **91A**, 76 (1914).

3966. Richards, R. E. and T. Schaefer, *Trans. Faraday Soc.* **54,** 1447 (1958).
3967. Richards, R. E. and H. W. Thompson, *Proc. Roy. Soc.* **195A,** 1 (1948); *C. A.* **43,** 4140 (1949).
3968. Richards, T. W. and F. Barry, *J. Amer. Chem. Soc.* **37,** 993 (1915).
3969. Richards, T. W. and L. B. Coombs, *J. Amer. Chem. Soc.* **37,** 1656 (1915).
3970. Richards, T. W. and J. H. Mathews, *J. Amer. Chem. Soc.* **30,** 8 (1908).
3971. Richards, T. W. and J. W. Shipley, *J. Amer. Chem. Soc.* **41,** 2002 (1919).
3972. Richards, W. T. and J. H. Wallace, *J. Amer. Chem. Soc.* **54,** 2705 (1932).
3973. Richardson, G. M. and P. W. Robertson, *J. Chem. Soc.* **1928,** 1775.
3974. Riddick, J. A., *Anal. Chem.* **24,** 41 (1952); **26,** 77 (1954); **28,** 679 (1956); **30,** 793 (1958); **32,** 172R (1960).
3975. Riddick, J. A., Unpublished data from Research and Development Dept., Commercial Solvents Corporation.
3976. Riddick, J. A. and W. B. Bunger, Calculated value.
3977. Riddick, J. A. and J. Hampton, Unpublished data from Research and Development Division, Commercial Solvents Corporation.
3978. Riddle, E., "Monomeric Acrylic Esters," New York, 1954.
3979. Rieche, A., *Z. Angew. Chem.* **44,** 896 (1931).
3980. Riedel, L., *Z. Ges. Kalti-Ind.* **45,** 221 (1938); *C. A.* **33,** 5735 (1939); *Chem. Zentr.* **1939, I,** 3349.
3981. Rigamonti, R. and G. Schiavina, *Chim. Ind.* (Milan) **36,** 611 (1954); *C. A.* **49, 3510** (1955).
3982. Riiber, C. N., *Z. Elektrochem.* **29,** 334 (1923); *C. A.* **17,** 3434 (1923).
3983. Riiber, C. N., T. Sorensen, et al., *Ber. Deut. Chem. Gesell.* **58,** 964 (1925).
3984. Rikovski, I. and V. Carich, *Glas. Khem. Drus., Beograd* **22,** 87 (1957); *C. A.* **54,** 2859 (1960).
3985. Riley, F. T. and K. C. Bailey, *Proc. Roy. Irish Acad.* **38B,** 450 (1929); *British Abstr.* **1929A,** 1137.
3986. Rinkenbach, W. H., *Ind. Eng. Chem.* **19,** 474 (1927).
3987. Rinkenbach, W. H. and H. A. Aaronson, *Ind. Eng. Chem.* **23,** 160 (1931).
3988. Rintelen, J. C., J. H. Saylor, et al., *J. Amer. Chem. Soc.* **59,** 1129 (1937).
3989. Risseghem, H. Van, *Bull, Soc. Chim. Belges.* **47,** 194; 221 (1938).
3990. Risseghem, H. Van, *Bull. Soc. Chim. Fr.* **1959,** 1661, *C. A.* **55,** 27,248 (1961).
3991. Robert, L., *Chim. Ind., Genie Chim.* **97,** 337 (1967).
3992. Roberti, D. M. and C. P. Smyth, *J. Amer. Chem. Soc.* **82,** 2106 (1960).
3993. Roberts, D. E., W. W. Walton, et al., *J. Res. Nat. Bur. Stand.* **38,** 627 (1947).
3994. Roberts, G. N., *J. Amer. Chem. Soc.* **75,** 2264 (1953).
3995. Roberts, H. M. and C. R. Bury, *J. Chem. Soc.* **1923,** 2037.
3996. Roberts, J. D. and V. C. Chambers, *J. Amer. Chem. Soc.* **73,** 5030 (1951).
3997. Roberts, R. H. and S. E. J. Johnsen, *Anal. Chem.* **20,** 690; 1225 (1948).
3998. Robertson, A. and D. G. Jones, British Patent 698,677, Oct. 21, 1953; *C. A.* **48,** 5878 (1954).
3999. Robertson, N. C., U.S. Patent 2,477,087, July 26, 1949; *C. A.* **43,** 9079 (1949).
4000. Robertson, P. W., N. T. Clare, et al., *J. Chem. Soc.* **1937,** 335.
4001. Robertson, W. W., N. Ginsburg, et al., *Ind. Eng. Chem., Anal. Ed.* **18,** 746 (1946).
4002. Robertson, W. W. and F. A. Matsen, *J. Amer. Chem. Soc.* **72,** 5248 (1950).
4003. Robertson, W. W. and F. A. Matsen, *J. Amer. Chem. Soc.* **72,** 5250 (1950).
4004. Robertson, W. W. and F. A. Matsen, *J. Amer. Chem. Soc.* **72,** 5252 (1950).
4005. Robertson, W. W., F. A. Matsen, et al., *Phys. Rev.* **73,** 652 (1947); *C. A.* **43,** 4094 (1949).
4006. Robertson, W. W., J. F. Music, et al., *J. Amer. Chem. Soc.* **72,** 5260 (1950).

4007. Robertson, W. W., A. J. Seriff, et al., *J. Amer. Chem. Soc.* **72**, 1539 (1950).
4008. Robinson, R. A. and A. K. Kiang, *Trans. Faraday Soc.* **52**, 327 (1956).
4009. Robles, H. de V., *Rec. Trav. Chem. Pays-Bas* **58**, 111 (1939); *C. A.* **33**, 2785 (1939).
4010. Röck, H., *Naturwissenschaften* **43**, 81 (1956); Ref. [2123], [5181].
4011. Rockenfeller, J. D. and F. D. Rossini, *J. Phys. Chem.* **65**, 267 (1961).
4012. Rodionov, V., *Bull. Soc. Chem. Fr.* **30**, 305 (1926) *C.A.* **20**, 1795 (1926).
4013. Rodziewicz, W., *Roz. Chem.* **39**, 539 (1965); *C. A.* **63**, 8169 (1965).
4014. Rogers, M. T., *J. Amer. Chem. Soc.* **69**, 457 (1947).
4015. Rogers, M. T., *J. Amer. Chem. Soc.* **69**, 1243 (1947).
4016. Rogers, M. T., *J. Amer. Chem. Soc.* **69**, 2544 (1947).
4017. Rogers, M. T., *J. Amer. Chem. Soc.* **77**, 3681 (1955).
4018. Rogers, M. T. and M. B. Panish, *J. Amer. Chem. Soc.* **77**, 3684 (1955).
4019. Rogers, M. T. and M. B. Panish, *J. Amer. Chem. Soc.* **77**, 4230 (1955).
4020. Röhler, H., *Z. Elektrochem.* **16**, 419; *C. A.* **4**, 2597 (1910).
4021. Rohm & Haas Company, "Acrylic and Methacrylic Acids and Esters," Philadelphia, 1963.
4022. Rohm & Haas Company, Special Products Department, *Glacial Methacrylic Acid—Glacial Acrylic Acid*, Philadelphia, 1960.
4023. Roland, M., *Bull. Soc. Chim. Belges.* **37**, 117 (1928).
4024. Rolla, M., *Atti Accad. Italia, Rend. Cl. Sci. Fis., Mat. Nat.* (7), **1**, 756 (1940); *C. A.* **37**, 832 (1943).
4025. Rollett, A. and A. Leimüller, *Monatsh. Chem.* **73**, 197 (1940); *C. A.* **35**, 4347 (1941).
4026. Rollier, M. A., *Gazz. Chim. Ital.* **77**, 366 (1947); *C. A.* **42**, 2826 (1948).
4027. Román, U. D. and A. M. Górriz, *An. Real Soc. Espan. Fis. Quim.* **54B**, 559 (1958); *C. A.* **53**, 16,628 (1959).
4028. Romans, J. B. and C. R. Singleterry, *J. Chem. Eng. Data* **6**, 56 (1961).
4029. Roper, E. E., *J. Amer. Chem. Soc.* **60**, 1693 (1938).
4030. Ropp, G. A. and C. E. Melton, *J. Amer. Chem. Soc.* **80**, 3509 (1958).
4031. Rosado-Lojo, O., C. K. Hancock, et al., *J. Org. Chem.* **31**, 1899 (1966).
4032. Rosanoff, M. A. and R. A. Dunphy, *J. Amer. Chem. Soc.* **36**, 1411 (1914).
4033. Roschdestwensky, M. S., A. Pukirew, et al., *Trans. Inst. Pure Chem. Reagents* (USSR) **14**, 58 (1935); *C. A.* **30**, 597 (1936).
4034. Rose, A., J. A. Acciarri, et al., *Ind. Eng. Chem.* **49**, 104 (1957).
4035. Rose, A., J. A. Acciarri, et al., *Ind. Eng. Chem., Chem. Eng. Data Series* **3**, 210 (1958).
4036. Rose, A., B. T. Papahornis, et al., *Ind. Eng. Chem., Chem. Eng. Data Series* **3**, 216 (1958).
4037. Rose, A. and E. Rose, *The Condensed Chemical Dictionary*, 6th ed., Reinhold, New York, 1961.
4038. Rose, A. and W. R. Suoina, *J. Chem. Eng. Data* **6**, 173 (1961).
4039. Rose, F. W., *J. Res. Nat. Bur. Stand.* **19**, 143 (1937).
4040. Rosenbaum, E. J., A. V. Grosse, et al., *J. Amer. Chem. Soc.* **61**, 689 (1939).
4041. Rosenbaum, J. and M. C. R. Symons, *J. Chem. Soc.* **1961**, 1.
4042. Rosin, J., *Reagent Chemicals and Standards*, 5th ed., Van Nostrand, Princeton, New Jersey, 1967.
4043. Ross, G. R. and W. J. Heideger, *J. Chem. Eng. Data* **7**, 505 (1962).
4044. Ross, G. S. and H. D. Dixon, *J. Res. Nat. Bur. Stand.* **67A**, 247 (1963).
4045. Ross, G. S. and L. J. Frolen, *J. Res. Nat. Bur. Stand.* **67A**, 607 (1963).
4046. Ross, G. S. and A. R. Glasgow, *Anal. Chem.* **36**, 700 (1964).
4047. Ross, J. D. M. and I. C. Somerville, *J. Chem. Soc.* **1926**, 2770.

4048. Ross, S., *J. Amer. Chem. Soc.* **68**, 1484 (1946).
4049. Ross, S. D. and M. M. Labes, *J. Amer. Chem. Soc.* **79**, 4155 (1957).
4050. Rossini, F. D., *Anal. Chem.* **20**, 110 (1948).
4051. Rossini, F. D., *J. Res. Nat. Bur. Stand.* **11**, 553 (1933); *C. A.* **28**, 961 (1934).
4052. Rossini, F. D., *J. Res. Nat. Bur. Stand.* **12**, 735 (1934).
4053. Rossini, F. D., *J. Res. Nat. Bur. Stand.* **13**, 189 (1934).
4054. Rossini, F. D., *Pure Appl. Chem.* **9**, 453 (1964).
4055. Roth, R., *Diplomarbeiten*, 1956, from [4124].
4056. Roth, W. A. and I. Meyer, *Z. Elektrochem.* **39**, 35 (1933); *Chem. Zentr.* **1933, I**, 1418.
4057. Roth, W. A. and F. Müller, *Ber. Deut. Chem. Gesell.* **62**, 1188 (1929).
4058. Rothenbach, E., *Wochschr. Brau.* **56**, 273 (1939); *C. A.* **34**, 6076 (1940).
4059. Rotinjantz, L. and N. Nagornow, *Z. Phys. Chem.* (Leipzig) **169A**, 20 (1934); *C. A.* **28**, 5743 (1934).
4060. Rousselot, M. M. and M. Martin, *Compt. Rend., Ser. C*, **262**, 1445 (1966); *C. A.* **65**, 6545 (1966).
4061. Rowe, J. J. M., K. B. Gibney, et al., *J. Amer. Chem. Soc.* **90**, 1924 (1968).
4062. Rowley, H. H. and W. R. Reed, *J. Amer. Chem. Soc.* **73**, 2960 (1951).
4063. Rozental, D., *Bull. Soc. Chim. Belges* **45**, 585 (1936); *C. A.* **31**, 1686 (1937).
4064. Rubin, H. and R. E. Swarbrick, *Anal. Chem.* **33**, 217 (1961).
4065. Rubin, T. R., B. H. Levedahl, et al., *J. Amer. Chem. Soc.* **66**, 279 (1944).
4066. Ruby, W. R. and R. P. Loveland, *J. Phys. Chem.* **50**, 345 (1946).
4067. Ruch, J. E. and F. E. Critchfield, *Anal. Chem.* **33**, 1569 (1961).
4068. Rudakov, G. A. and S. Y. Korotov, *J. Appl. Chem.* (USSR) **10**, 312 (1937); *C. A.* **31**, 4554 (1937).
4069. Rudenko, V. A., A. Y. Yakubovich, et al., *J. Gen. Chem.* (USSR) **17**, 2256 (1947); *C. A.* **42**, 4918 (1948).
4070. Rudolph, H. D. and H. Seiler, *Z. Naturforsch.* **20A**, 1682 (1965); *C. A.* **64**, 13,573 (1966).
4071. Rueggeberg, W. H. C., M. L. Cushing, et al., *J. Amer. Chem. Soc.* **68**, 191 (1946).
4072. Ruehrwein, R. A. and H. M. Huffmann, *J. Amer. Chem. Soc.* **65**, 1620 (1943).
4073. Ruff, O. and O. Bretschneider, *Z. Anorg. Allg. Chem.* **210**, 173 (1933); *Chem. Z.* **1933I**, 1925.
4074. Ruff, O. and H. Golla, *Z. Anorg. Allge Chem.* **138**, 17 (1921).
4075. Ruhoff, J. R. and E. E. Reid, *J. Amer. Chem. Soc.* **59**, 401 (1937).
4076. Rumpf, K. and R. Mecke, *Z. Phys. Chem.* (Leipzig) **44B**, 299 (1939); *C. A.* **34**, 676 (1940).
4077. Ruppol, E., *Bull. Acad. Roy. Belges Cl. Sci.* **21**, 236 (1935); *Chem. Zentr.* **1935, II**, 2659; *Bull. Soc. Chim. Belges* **44**, 551 (1935); *Chem. Zentr.* **1936, I**, 3316.
4078. Rusoff, I. I., J. R. Platt, et al., *J. Amer. Chem. Soc.* **67**, 673 (1945).
4079. Rutledge, G. P. and W. T. Smith, *J. Amer. Chem. Soc.* **75**, 5762 (1953).
4080. Ruzicka, L., *Helv. Chim. Acta* **2**, 144 (1919); *C. A.* **13**, 1320 (1919).
4081. Ryland, G., *Amer. Chem. J.* **22**, 384 (1899).

S

4082. Sabatier, P. and A. Mailhe, *Ann. Chim. Phys.* (8), **10**, 527; *C. A.* **2**, 793 (1908); *Chem. Zentr.* **1907, I**, 1695.
4083. Sabatier, P. and A. Mailhe, *Compt. Rend.* **138**, 1321 (1904); *J. Chem. Soc.* **1904Ai**, 666.
4084. Sabatier, P. and M. Murat, *Compt. Rend.* **154**, 1390 (1912); *C. A.* **6**, 2082 (1912).
4085. Sabatier, P. and J. B. Senderens, *Compt. Rend.* **137**, 302 (1903); *Chem. Zentr.* **1903, II**, 708.

4086. Sabatier, P. and J. B. Senderens, *Compt. Rend.* **138,** 457 (1904).

4087. Sachs, F., *Ber. Deut. Chem. Gesell.* **34,** 494 (1901).

4088. Sah, P. P. T. and S.-L. Chien, *J. Amer. Chem. Soc.* **53,** 3901 (1931).

4089. Sah, P. P. T. and H. H. Lei, *Sci. Repts. Nat. Ysing Hua Univ. Ser. A,* **1,** 193 (1932) Beil E**III6,** 1454.

4090. Saidel, L. J., *Arch. Biochem. Biophys.* **54,** 184 (1955).

4091. Saidel, L. J., *Nature* **172,** 955 (1953).

4092. Saier, E. L., L. R. Cousins, et al., *Anal. Chem.* **34,** 824 (1962).

4093. Saint-Gilles, L. P., de, *Ann. Chim. (Paris)* **55,** 396 (1859).

4094. Sakai, S. and N. Tsunakawa, Japanese Patent **6757**('63), May 22; *C. A.* **59,** 11,435 (1963).

4095. Sakiadis, B. C. and J. Coates, *A.I.C.H.E. J.* **1,** 275 (1955).

4096. Saksena, B. D., *Proc. Indian Acad. Sci.* **10A,** 333 (1939); *C. A.* **34,** 3173 (1940).

4097. Saksena, B. D., *Proc. Indian Acad. Sci.* **12A,** 312 (1940); *C. A.* **35,** 3172 (1941).

4098. Saksena, B. D., *Proc. Indian Acad. Sci.* **12A,** 321 (1940); *C. A.* **35,** 3173 (1941).

4099. Saksena, B. D., *Proc. Indian Acad. Sci.* **12A,** 416 (1940); *C. A.* **35,** 3173 (1941).

4100. Salimov, M. A., L. A. Erivanskaya, et al., *Azerbaĭdzhan. Khim. Zhur.* **1961,** No. 3, 123; No. 4, 93; *C. A.* **56,** 8184 (1962).

4101. Salimov, M. A., E. A. Viktorova, et al., *Azerb. Khim. Zh.* **1960,** 99; *C. A.* **55,** 19,481 (1961).

4102. Salmi, E. J. and E. Väihkönen, *Suomen Kemistilehti* **19B,** 132 (1946); *C. A.* **41,** 5481 (1947).

4103. Salmon, M., E. Cortes, et al., *Bol. Isnst. Quim. Univ. Nacl. Auton. Mex.* **17,** 151 (1965) *C. A.* **65,** 13,506 (1966).

4104. Salonen, A. K., *Ann. Acad. Sci. Fennicae Ser.* **A6,** No. 67 (1961); *C. A.* **55,** 20,624 (1961).

4105. Samitov, Y. Y. and R. M. Aminova, *Zh. Strukt. Khim.* **5,** 538 (1964); *C. A.* **61,** 15,555 (1964).

4106. Sammis, J. L., *J. Phys. Chem.* **10,** 593 (1906).

4107. Sammul, O. R., W. L. Brannon, et al., *J. Assoc. Offic. Agr. Chem.* **47,** 918 (1964).

4108. Samsanova, N. I., N. S. Kalyazina, et al., *Tr. Vses. Nauchn.-Issled. Inst. Khim. Reaktivov* No. **25,** 434 (1963); *C. A.* **61,** 2984 (1964).

4109. Sanger, R., *Phys. Z.* **27,** 556 (1926).

4110. Sangewald, R. and A. Weissberger, *Phys. Z.* **30,** 268 (1929).

4111. Sanyal, S. B., *Indian J. Phys.* **24,** 151 (1950); *C. A.* **65,** 450 (1951).

4112. Sapgir, S., *Bull. Soc. Chim. Belges* **38,** 392 (1929).

4113. Sappenfield, J. W., *Phys. Rev.* **33,** 37 (1929); *Chem. Zentr.* **1929,** I, 1419.

4114. Sarancha, E. T. and A. P. Dubovikova, *Zavodsk. Lab.* **27,** 398 (1961); *C. A.* **57,** 6588 (1962).

4115. Sasa, T., *J. Soc. Org. Synthetic Chem. (Japan)* **11,** 463 (1953); *C. A.* **49,** 919 (1955).

4116. Sastry, C. R. and K. R. Rao, *Indian J. Phys.* **21,** 313 (1947); *C. A.* **42,** 6237 (1948).

4117. Saunders, R. H., M. J. Murray, et al., *J. Amer. Chem. Soc.* **65,** 1309 (1943).

4118. Savard, J., *Ann. Chim.* (Paris) **11,** 287 (1929).

4119. Savedoff, L. G., *J. Amer. Chem. Soc.* **88,** 664 (1966).

4120. Sawyer, D. T. and J. R. Brannan, *Anal. Chem.* **38,** 192 (1966).

4121. Sawyer, R. A., *Experimental Spectroscopy,* Prentice-Hall, Englewood Cliffs, New Jersey, 1944.

4122. Saxton, B. and G. Waters, *J. Amer. Chem. Soc.* **59,** 1048 (1937).

4123. Scatchard, G. and C. L. Raymond, *J. Amer. Chem. Soc.* **60,** 1278 (1938).

4124. Scatchard, G. and G. M. Wilson, *J. Amer. Chem. Soc.* **86,** 133 (1964).

4125. Scatchard, G. and S. E. Wood, *J. Phys. Chem.* **43**, 119 (1939).

4126. Scattergood, A., W. H. Miller, et al., *J. Amer. Chem. Soc.* **67**, 2150 (1945).

4127. Schaaffs, W., *Z. Phys. Chem.* (Leipzig) **194**, 170 (1944); *C. A.* **42**, 8550 (1948).

4128. Schaap, W. B., R. E. Bayer, et al., *Record Chem. Progr.* **22**, 197 (1961).

4129. Schaefer, T. and W. G. Schneider, *Can. J. Chem.* **37**, 2078 (1959); *J. Chem. Phys.* **32**, 1218 (1960).

4130. Schaefer, T. and W. G. Schneider, *J. Chem. Phys.* **32**, 1224 (1960).

4131. Schafer, Sr. M. and C. Curran, *Inorg. Chem.* **4**, 623 (1965).

4132. Schafer, Sr. M. and C. Curran, *Inorg. Chem.* **5**, 265 (1966).

4133. Schall, C. and C. Thieme-Wiedtmarckter, *Z. Elektrochem.* **35**, 337 (1929).

4134. Schatzberg, P., *J. Phys. Chem.* **67**, 776 (1963).

4135. Schauffle, R. F. and L. Goodman, *Trans. Faraday Soc.* **61**, 597 (1965).

4136. Scheibe, G. and H. Grieneisen, *Z. Phys. Chem.* (Leipzig) **25B**, 52 (1934).

4137. Scheibe, G., F. May, et al., *Ber. Deut. Chem. Gesell.* **57**, 1330 (1924).

4138. Scheibler, H. and J. Magasanik, *Ber. Deut. Chem. Gesell.* **48**, 1810 (1915).

4139. Scherer, J. R. and J. C. Evans, *Spectrochim. Acta* **19**, 1739 (1963); *C. A.* **59**, 13,482 (1963).

4140. Schiemann, G., *Z. Phys. Chem.* (Leipzig) **156A**, 397 (1931).

4141. Schierholtz, O. J. and M. L. Staples, *J. Amer. Chem. Soc.* **57**, 2709 (1935).

4142. Schiff, R., *Ann. Chim.* (Paris) **220**, 278 (1884); *J. Chem. Soc.* **1884A**, 386; Beil **I**, 291.

4143. Schiff, R., *Ann. Chim.* (Paris) **234**, 300 (1886); *J. Chem. Soc.* **1887A**, 6.

4144. Schiff, R., *Z. Phys. Chem.* (Leipzig) **1**, 376 (1887).

4145. Schildknecht, C. E., *Vinyl and Related Polymers*, Wiley, New York, 1952.

4146. Schildknecht, H., *Chimia* (Aarua) **17**, 145 (1963); *Chem. Zentr.* **1965**, No. 10, 1738; Ref. [5181].

4147. Schildknecht, H., *Z. Anal. Chem.* **181**, 254 (1961); Ref. [2123].

4148. Schildknecht, H., *Zonenschmelzen*, Verlag Chemie, Weiheim/Bergstr., 1964.

4149. Schildknecht, H. and V. Hopf, *Z. Anal. Chem.* **193**, 401 (1963); *C. A.* **59**, 6324 (1963).

4150. Schildknecht, H. and H. Vetter, *Angew. Chem.* **73**, 240 (1961); *C. A.* **55**, 22,934 (1961).

4151. Schildknecht, C. E., A. O. Zoss, et al., *Ind. Eng. Chem.* **39**, 180 (1947).

4152. Schiller, A., *Z. Phys.* **105**, 175 (1937); *C. A.* **31**, 5271 (1937).

4153. Shimozawa, T. and M. K. Wilson, *Spectrochim. Acta* **22**, 1599, 1609 (1966); *C. A.* **65**, 16,261, 16,262 (1966).

4154. Schindlbauer, H. and G. Hajek, *Chem. Ber.* **96**, 2601 (1963).

4155. Schioper, I. M., O. Bot, et al., *Bul. Mst. Politeh. Iasi* **7**, 115 (1961); *C. A.* **58**, 10,793 (1963).

4156. Schjänberg, E., *Z. Phys. Chem.* (Leipzig) **172A**, 197 (1935).

4157. Schläfer, H. L. and W. Schaffernicht, *Angew. Chem.* **72**, 618 (1960); *Chem. Zentr.* **1961**, 2543; *C. A.* **55**, 14,036 (1961).

4158. Schlegel, H., *J. Chim. Phys.* **31**, 517 (1934).

4159. Schlesinger, H. I. and A. W. Martin, *J. Amer. Chem. Soc.* **36**, 1589 (1914).

4160. Schlundt, H., *J. Phys. Chem.* **5**, 157 (1901); *Chem. Zentr.* **1901**, I, 1135.

4161. Schlundt, H., *J. Phys. Chem.* **5**, 503 (1901); *Chem. Zentr.* **1902**, I, 3.

4162. Schmerling, L., U.S. Patent 2,385,303, Sept. 18, 1945; *C. A.* **40**, 601 (1946).

4163. Schmidt, A. W., V. Schoeller, et al., *Ber. Deut. Chem. Gesell.* **74B**, 1313 (1941).

4164. Schmidt, E., *Ber. Deut. Chem. Gesell.* **52**, 400 (1919).

4165. Schmidt, G. C., *Z. Phys. Chem.* (Leipzig) **121**, 221 (1926).

4166. Schmidt, M., Thesis, Paris (1934), *Annual Table of Physical Constants*, Sec. 514(c), 1941; 513(c), 1941.

4167. Schmidt, M. R. and H. C. Jones, *Amer. Chem. J.* **42**, 37 (1909).

4168. Schmidt, O., *Z. Phys. Chem.* (Leipzig) **58**, 513 (1907).
4169. Schmidt, W. and K. Seydel, German Patent 872,342, Mar. 30, 1953.
4170. Schmitz-Dumont, W., *Chem.-Ztg.* **21**, 511 (1897).
4171. Schmulbach, C. D. and R. S. Drago, *J. Amer. Chem. Soc.* **82**, 4484 (1960).
4172. Schnaible, H. E. and J. M. Smith, *Chem. Eng. Progr. Symp. Ser.* **49**, No. 7, 159 (1953).
4173. Schneider, G. and G. Wilhelm, *Z. Phys. Chem.* (Frankfurt) **20**, 219 (1959); *Chem. Zentr.* **1960**, 1453.
4174. Schnurmann, R. and W. J. Morris, *Research* (London) **2**, Suppl. 394 (1949); *C. A.* **44**, 3759 (1950).
4175. Schnurmann, R. and S. Whincup, *Petroleum* (London) **8**, 122, 142 (1945); *C. A.* **40**, 4000 (1946).
4176. Schöber, G. and V. Gutmann, *Monatsh. Chem.* **89**, 649 (1958); *C. A.* **53**, 7826 (1959); Ref. [3436].
4177. Scholl, R. and H. Berblinger, *Ber. Deut. Chem. Gesell.* **36**, 3427 (1903).
4178. Scholte, Th. G., *Physica* **15**, 450 (1949); *C. A.* **44**, 3319 (1950).
4179. Schorger, A. W., *J. Ind. Eng. Chem.* **6**, 631 (1914).
4180. Schorigin, P., *Ber. Deut. Chem. Gesell.* **56**, 186 (1923).
4181. Schorigin, P. and Makaroff-Semljanski, *Ber. Deut. Chem. Gesell.* **65**, 1293 (1932).
4182. Schott, H., *J. Chem. Eng. Data* **6**, 19 (1961).
4183. Schouteden, F. and C. J. Deveux, *Natuurwetensch. Tijds.* **18**, 242 (1936).
4184. Schreiner, L., *Ber. Deut. Chem. Gesell.* **12**, 179 (1879).
4185. Schroeder, J. and H. Steiner, *J. Prakt. Chem.* **79**, 49 (1909); *C. A.* **3**, 1374 (1909).
4186. Schröer, E., *Z. Phys. Chem.* (Leipzig) **140**, 381 (1929).
4187. Schrauth, W. and K. Gorig, *Ber. Deut. Chem. Gesell.* **56B**, 1900 (1923).
4188. Schreiber, K. C., *Anal. Chem.* **21**, 1168 (1949).
4189. Schreiner, E. and O. E. Frivold, *Z. Phys. Chem.* **124**, 1 (1926).
4190. Schröer, E., *Z. Phys. Chem.* (Leipzig) **49B**, 271 (1941); *C. A.* **36**, 6389 (1942).
4191. Schroeter, G., *Ann. Chem.* **426**, 1 (1922).
4192. Schulek, E., E. Pungor, et al., *Mikrochim. Acta* **1958**, 52; *C. A.* **53**, 977 (1959).
4193. Schumann, S. C., Univ. Microfilms (Ann Arbor, Michigan) Pub. No. 241, (1941); *C. A.* **35**, 4666 (1941).
4194. Schumann, S. C., J. G. Aston, et al., *J. Amer. Chem. Soc.* **64**, 1039 (1942).
4195. Schupp, R. L. and R. Mecke, *Z. Elektrochem.* **52**, 54 (1948); *C. A.* **43**, 6480 (1949).
4196. Schurz, J. and E. Kienzl, *Monatsh. Chem.* **88**, 78 (1957); *C. A.* **51**, 11,852 (1957).
4197. Schurz, J., H. Zah, et al., *Z. Phys. Chem.* (Frankfurt) **21**, 185 (1959).
4198. Schutz, G., *Z. Phys. Chem.* (Leipzig) **40B**, 156 (1938).
4199. Schwab, F. W. and E. Wiehers, *J. Res. Nat. Bur. Stand.* **25**, 747 (1940).
4200. Schwable, C., *Z. Farben Textilechemie* **3**, 461 (1904); *Chem. Zentr.* **1905**, 360.
4201. Schwartz, D. P. and O. W. Parks, *Anal. Chem.* **33**, 1396 (1961).
4202. Schwers, F., *J. Phys. Chem.* **9**, 15 (1911); *I. C. T.* **3**, 29, 33.
4203. Scott, D. W., W. T. Berg, et al., *J. Phys. Chem.* **64**, 906 (1960).
4204. Scott, D. W., W. T. Berg, et al., *J. Phys. Chem.* **71**, 2263 (1967).
4205. Scott, D. W., H. L. Finke, et al., *J. Amer. Chem. Soc.* **74**, 4656 (1952).
4206. Scott, D. W., H. L. Finke, et al., *J. Amer. Chem. Soc.* **79**, 1062 (1957).
4207. Scott, D. W., P. J. McCullough, et al., *J. Amer. Chem. Soc.* **73**, 1707 (1951).
4208. Scott, D. W., J. P. McCullough, et al., *J. Amer. Chem. Soc.* **78**, 5457 (1956).
4209. Scott, D. W., J. P. McCullough, et al., *J. Amer. Chem. Soc.* **78**, 5463 (1956).
4210. Scott, N. D., U.S. Patent 2,004,350, June 11, 1935; *Official Gaz. U.S. Patent Office* **455**, 363 (1935).

4211. Scott, P. T., D. E. Pearson, et al., *Org. Chem. J.* **19**, 1815 (1954).
4212. Scott, R. B. and F. G. Brickwedde, *J. Res. Nat. Bur. Stand.* **35**, 501 (1945).
4213. Scott, R. B. and J. W. Mellors, *J. Res. Nat. Bur. Stand.* **34**, 243 (1945).
4214. Scott, T. A., D. Macmillan, et al., *Ind. Eng. Chem.* **44**, 172 (1952).
4215. Scudder, H., from Miner and Dalton [3334], ref. 145, p. 333).
4216. Searles, S. and C. F. Butler, *J. Amer. Chem. Soc.* **76**, 56 (1954).
4217. Searles, S., M. Tamres, et al., *J. Amer. Chem. Soc.* **78**, 4917 (1956).
4218. Sears, G. W. and E. R. Hopke, *J. Amer. Chem. Soc.* **71**, 1632 (1949).
4219. Sears, P. G., W. H. Fortune, et al., *J. Chem. Eng. Data* **11**, 406 (1966).
4220. Sears, P. G., W. Siegfried, et al., *J. Chem. Eng. Data* **9**, 261 (1964).
4221. Sears, P. G., E. D. Wilhoit, et al., *J. Phys. Chem.* **59**, 373 (1955).
4222. Sechkarev, A. V. and N. J. Dvorovenko, *Izv. Vyssh. Ucheb. Zaved. Fiz.* **9**, 111 (1966); *C. A.* **65**, 6539 (1966).
4223. Segal, L. and F. V. Eggerton, *Appl. Spectrosc.* **15**, 116 (1961); *C. A.* **56**, 4264 (1962).
4224. Segur, J. B. and H. E. Oberstar, *Ind. Eng. Chem.* **43**, 2117 (1951).
4225. Seidell, A., *Solubilities of Organic Compounds*, Vol. 2, 3rd ed., Van Nostrand, Princeton, New Jersey, 1941.
4226. Seidman, J., *Anal. Chem.* **23**, 559 (1951).
4227. Seikel, M. K., *Ind. Eng. Chem., Anal. Ed.* **13**, 388 (1941).
4228. Sekino, M., *Repts. Research Lab., Asahi Glass Co.* **1**, 96 (1950); *C. A.* **50**, 1584 (1956).
4229. Seliverstov, V. M. and M. P. Sergievskaya, *Tr. Leningr. Inst. Vodn. Transp.* No. **75**, 47 (1964); *C. A.* **64**, 5828 (1966).
4230. Sellers, P. and S. Sunner, *Acta Chem. Scund.* **16**, 46 (1962); *C. A.* **56**, 15,000 (1962).
4231. Semeria, G. B. and G. Ribotti-Lissone, *Gazz. Chim. Ital.* **60**, 862 (1930); *C. A.* **25**, 1800 (1931).
4232. Sen, S. K., *Indian J. Phys.* **30**, 321 (1956); *C. A.* **51**, 9312, (1957).
4233. Sen, S. K., *Indian J. Phys.* **34**, 237 (1960); *C. A.* **54**, 23,728 (1960).
4234. Senderens, J. B., *Compt. Rend.* **176**, 813; **177**, 15, 1183 (1923).
4235. Senderens, J. B. and J. Aboulenc, *Compt. Rend.* **152**, 1671 (1911); **155**, 1012, 1254 (1913); **156**, 1620 (1913); from Markley [3151].
4236. Senkus, M., U.S. Patent 2,406,713, Aug. 27, 1946; *C. A.* **41**, 150 (1947).
4237. Seprakova, M., J. Paulech, et al., *Chem. Zvesti* **13**, 313 (1959); *Chem. Zentr.* **133**, 12,605 (1962).
4238. Serijan, K. T., I. A. Goodman, et al., *Natl. Advisory Comm.* Aeronaut. Tech. Note No. 2557 (1951); *C. A.* **46**, 3406 (1952).
4239. Serpinskii, V. V., S. A. Voïtkevich, et al., *Zh. Fiz. Khim.* **28**, 810 (1954); *C. A.* **49**, 6677 (1955).
4240. Serpinsky, V. V., S. A. Voïtkevich, et al., *Zh. Fiz. Khim.* **31**, 1278 (1957); *C. A.* **52**, 3270 (1958).
4241. Sevryugova, N. N., V. A. Sokol'skii, et al., *Khim. Prom.* **1963**, 572; *C. A.* **60**, 1580 (1964).
4242. Seward, R. P. and E. C. Vierira, *J. Phys. Chem.* **62**, 127 (1958).
4243. Seyer, W. F., *J. Amer. Chem. Soc.* **75**, 616 (1953).
4244. Seyer, W. F. and G. M. Barrow, *J. Amer. Chem. Soc.* **70**, 802 (1948).
4245. Seyer, W. F. and C. H. Davenport, *J. Amer. Chem. Soc.* **63**, 2425 (1941).
4246. Seyer, W. F. and J. D. Leslie, *J. Amer. Chem. Soc.* **64**, 1912 (1942).
4247. Seyer, W. F. and C. W. Mann, *J. Amer. Chem. Soc.* **67**, 328 (1945).
4248. Seyer, W. F. and R. D. Walker, *J. Amer. Chem. Soc.* **60**, 2125 (1938).
4249. Seyer, W. F., M. M. Wright, et al., *Ind. Eng. Chem.* **31**, 759 (1939).

4250. Shablygin, M. V., D. N. Shigorin, et al., *Zh. Prikl. Spektrosk. Akad. Nauk Belorussk. SSR* **3**, 56 (1965); *C. A.* **64**, 5939 (1966).

4251. Shafer, P. R., D. R. Davis, et al., *Proc. Nat. Acad. Sci. U.S.* **47**, 49 (1961); *C. A.* **55**, 12,044 (1961).

4252. Shakhparonov, M. I. and N. G. Shlenkina, *Dokl. Akad. Nauk SSSR* **96**, 55 (1954); *C. A.* **50**, 62 (1956).

4253. Shamonin, Y. Y. and K. A. Gold'gammer, *Dokl. Akad. Nauk SSSR* **140**, 1136 (1961); *C. A.* **56**, 12,449 (1962).

4254. Shanley, E. S. and F. P. Greenspan, *Ind. Eng. Chem.* **39**, 1536 (1947).

4255. Shannon, T. W., T. E. Mead, et al., *Anal. Chem.* **39**, 1748 (1967).

4256. Shapet'ko, N. N., D. N. Shigorin, et al., *Opt. Spectrosk.* **17**, 459 (1964); *C. A.* **62**, 145 (1965).

4257. Sharkey, A. G., J. L. Schultz, et al., *Anal. Chem.* **28**, 934 (1956).

4258. Sharkey, A. G., J. L. Schultz, et al., *Anal. Chem.* **31**, 87 (1959).

4259. Sharpe, A. N. and S. Walker, *J. Chem. Soc.* **1961**, 2974.

4260. Sharpless, N. E. and M. Flavin, *Biochim.* **5**, 2963 (1966).

4261. Shchekin, V. V., *Trudy Inst. Nefti, Akad. Nauk SSSR* **1**, No. 2, 33 (1952); *C. A.* **49**, 832 (1955).

4262. Sheinker, Y. N., E. M. Peresleni, et al., *Zh. Fiz. Khim.* **29**, 518 (1955).

4263. Shelepin, I. V., G. M. Dugacheva, et al., *Plasticheskie Massy* **1963** (9), 50; *C. A.* **59**, 15,176 (1963).

4264. Shell Chemical Corp., "Allyl Alcohol," New York, 1946.

4265. Shell Chemical Corp., "Organic Chemicals," 2nd ed., San Francisco, 1942.

4266. Shell Chemical Corp., "Technical Booklet," SC-49-30, 1949.

4267. Shell Chemical Corp., Industrial Chemicals Division, Technical Bulletin No. IC:63-13R, "Sulfalane," New York.

4268. Shell Development Co., "Acrolein," Emeryville, California, 1947.

4269. Shell Development Co., Emeryville, California, Unpublished data, Private Communications.

4270. Shepard, A. F. and A. L. Henne, *Ind. Eng. Chem.* **22**, 356 (1930).

4271. Shepard, A. F., A. L. Henne, et al., *J. Amer. Chem. Soc.* **53**, 1948 (1931).

4272. Shepherd, E. J. and J. A. Kitchener, *J. Chem. Soc.* **1956**, 2448.

4273. Sheppard, N., *J. Chem. Phys.* **16**, 690 (1948).

4274. Sheppard, N., *J. Chem. Phys.* **17**, 79 (1949).

4275. Sheppard, N. and G. B. B. M. Southerland, *Proc. Roy. Soc.* **196A**, 195 (1949); *C. A.* **44**, 1806 (1950).

4276. Sherman, E. O. and D. C. Olson, *Anal. Chem.* **40**, 1174 (1968).

4277. Sherrill, M. L., *J. Amer. Chem. Soc.* **52**, 1982 (1930).

4278. Sherrill, M. L., K. E. Mayer, et al., *J. Amer. Chem. Soc.* **56**, 926 (1934).

4279. Sherrill, M. L. and P. Mollet, *J. Chim. Phys.* **33**, 701 (1936); *C. A.* **31**, 3784 (1937).

4280. Sherrill, M. L. and G. F. Walters, *J. Amer. Chem. Soc.* **58**, 742 (1936).

4281. Shidlovskaya, A. N. and Y. K. Syrkin, *Zh. Fiz. Khim.* **22**, 913 (1948); *C. A.* **43**, 454 (1949).

4282. Shieh, N. and C. C. Price, *J. Org. Chem.* **24**, 1169 (1959).

4283. Shiga, K. and T. Tsuruta, Japanese Patent 223, Jan. 21, 1953; *C. A.* **48**, 8812 (1954).

4284. Shigeru, T., A. Hisashi, et al., *Technol. Repts. Osaka Univ.* **6**, 163 (1956).

4285. Shigorin, D. N., *Zh. Fiz. Khim.* **27**, 689 (1953); *C. A.* **48**, 13,633 (1954).

4286. Shimanko, N. A., M. V. Shishkina, et al., *Izv. Anal. Nauk SSSR, Ser. Fiz.* **26**, 1252 (1962); *C. A.* **58**, 7511 (1963).

4287. Shimanouchi, T., *J. Chem. Soc. Jap.* **62**, 1264 (1941); *C. A.* **41**, 2994 (1947).

4288. Shimanskii, Yu. I., *Nauk Povidomlennya Kiiv Univ.* **1956**, No. 1, 44; *C. A.* **54**, 9411 (1960).

4289. Shimizu, H. and S. Fujiwara. *Chem. Pharm. Bull.* (Tokyo) **8**, 272 (1960); *C. A.* **55**, 9048 (1961).

4290. Shimizu, T., T. Tsuda, et al., *Kogyo Kagaku Zasshi* **68**, 2473 (1965); *C. A.* **65**, 5343 (1966).

4291. Shinoda, K., T. Yamanaka, et al., *J. Phys. Chem.* **63**, 648 (1959).

4292. Shipsey, K. and E. A. Werner, *J. Chem. Soc.* **1913**, 1255.

4293. Shishkina, M. V., *Neftekhimiya* **1**, 255 (1961); *C. A.* **57**, 6765 (1962).

4294. Shmulyakovskii, Ya. E., *Vestn. Leningrad Univ.* **1950**, No. 3, 65; *C. A.* **48**, 12,609 (1954).

4295. Shokal, E. C. and T. W. Evans, U.S. Patent 2,428,590, Oct. 7, 1947; *Official Gaz. U.S. Patent Office* **603**, 124 (1947); British Patent 577,992, June 11, 1946.

4296. Shorygin, P. P., T. N. Shkurina, et al., *Izv. Akad. Nauk SSSR, Otd. Khim. Nauk* **1959**, 2208; *C. A.* **54**, 10,515 (1960).

4297. Shorygin, P. P., A. A. Simanovskaya, et al., *J. Appl. Chem.* (USSR) **9**, 1442 (1936) *C. A.* **31**, 2590 (1937).

4298. Shostakovskiĭ, M. F., *J. Appl. Chem.* (USSR) **16**, 66 (1943); *C. A.* **38**, 3020 (1944).

4299. Shostakovskiĭ, M. F. and N. A. Gershtein, *Akad. Nauk SSSR, Inst. Org. Khim., Sintezy Org. Saedinenii, Sbornik* **2**, 154 (1952); *C. A.* **48**, 569 (1954).

4300. Shostakovskiĭ, M. F. and N. A. Gershtein, *J. Gen. Chem.* (USSR) **16**, 937 (1946); *C. A.* **41**, 1999 (1947).

4301. Shostakovskiĭ, M. F. and E. N. Prilezhaeva, *J. Gen. Chem.* (USSR) **17**, 1129 (1947); *C. A.* **42**, 3633 (1948).

4302. Shott-L'vova, E. A., Y. K. Syrkin, *J. Phys. Chem.* (USSR) **12**, 479 (1938); *C. A.* **33**, 4839 (1939).

4303. Shreve, O. D., M. R. Heether, et al., *Anal. Chem.* **22**, 1261, 1498 (1950).

4304. Shreve, O. D., M. R. Heether, et al., *Anal. Chem.* **23**, 277 (1951).

4305. Shreve, O. D., M. R. Heether, et al., *Anal. Chem.* **23**, 282 (1951).

4306. Shubin, A. A., *Izv. Akad. Nauk SSSR, Ser. Fiz.* **14**, 422 (1950); *C. A.* **45**, 3245 (1951).

4307. Shuler, W. E. and R. C. Axtmann, *U.S. Atomic Energy Comm.* **DP474** (1960); *C. A.* **55**, 115 (1961).

4308. Siddigi, K., *Curr. Sci.* **12**, 253 (1943); *C. A.* **38**, 1169 (1944).

4309. Sideri, C. N. and A. Osol, *J. Amer. Chem. Assoc., Sci. Ed.* **42**, 586 (1953).

4310. Sidgwick, N. V., *Proc. Chem. Soc.* (London) **26**, 60; *Chem. Zentr.* **1910**, I, 1828.

4311. Sidgwick, N. V. and E. N. Allott, *J. Chem. Soc.* **1923**, 2819.

4312. Sidgwick, N. V. and N. S. Bayliss, *J. Chem. Soc.* **1930**, 2027.

4313. Sidgwick, N. V. and E. K. Ewbank, *J. Chem. Soc.* **1924**, 2268.

4314. Sidgwick, N. V. and H. E. Rubie, *J. Chem. Soc.* **1921**, 1013.

4315. Sidgwick, N. V., W. J. Spurrell, et al., *J. Chem. Soc.* **107**, 1202 (1915).

4316. Sidgwick, N. V. and E. E. Sutton, *J. Chem. Soc.* **1930**, 1323.

4317. Sidorova, N. G. and I. A. Tuchinskaya, *Zh. Obshch. Khim.* **27**, 1763 (1957); *C. A.* **52**, 4568 (1958).

4318. Siegel, S., *J. Amer. Chem. Soc.* **75**, 1317 (1953).

4319. Signaigo, F. K. and P. L. Cramer, *J. Amer. Chem. Soc.* **55**, 3326 (1933).

4320. Silbermann, G. B., S. T. Raschewskaja, *Z. Prikl. Chem.* **9**, 1832 (1936); *Chem. Zentr.* **1937**, I, 4786.

4321. Simanouti, T., H. Turuta, et al., *Sci. Papers Inst. Phys. Chem. Res.* (Tokyo) **42**, 51 (1946); *C. A.* **42**, 459 (1948).

4322. Simon, A. and G. Heintz, *Chem. Ber.* **95**, 2333 (1962).

4323. Simon, A. and J. Huter, *Z. Elektrochem.* **41**, 28 (1935); *C. A.* **29**, 5002 (1935).
4324. Simon, I., *Bull. Soc. Chim. Belges.* **38**, 47 (1929); *Chem. Zentr.* **1929, I**, 2519.
4325. Simon, M., *Annual Tables of Physical Constants*, Sec. 514 (c), 1941.
4326. Simonetta, M., *Chim. Ind.* (Milan) **29**, 37 (1947); *C. A.* **41**, 7165 (1947).
4327. Simonetta, M. and E. Mugnaini, *Chim. Ind.* (Milan) **30**, 73 (1948); *Brit. Chem. Abstr.* **1948A**, 338.
4328. Simons, J. H. and S. Archer, *J. Amer. Chem. Soc.* **60**, 2952 (1938).
4329. Simons, J. H. and H. Hart, *J. Amer. Chem. Soc.* **66**, 1309 (1944).
4330. Simons, J. H. and H. Hart, *J. Amer. Chem. Soc.* **69**, 979 (1947).
4331. Simons, J. H. and K. E. Lorentzen, *J. Amer. Chem. Soc.* **72**, 1426 (1950).
4332. Simonsen, D. R. and E. R. Washburn, *J. Amer. Chem. Soc.* **68**, 235 (1946).
4333. Sims, L. L., U.S. Patent 3,159,582 (Cl. 252-153), Dec. 1, 1964.
4334. Sinclair, R. G., A. F. McKay, et al., *J. Amer. Chem. Soc.* **74**, 2578 (1952).
4335. Singh, B. K. and B. K. K. Nayar, *Proc. Indian Acad. Sci.* **27A**, 61 (1948); *C. A.* **43**, 610 (1949).
4336. Singh, R. and L. W. Shemil, *J. Chem. Phys.* **23**, 1370 (1955).
4337. Singh, V. B., R. N. Singh, et al., *Spectrochim. Acta* **22**, 927 (1966); *C. A.* **65**, 1600 (1966).
4338. Sinke, G. C., *J. Phys. Chem.* **63**, 2063 (1959).
4339. Sinke, G. C. and D. L. Hildenbrand, *J. Chem. Eng. Data* **7**, 74 (1962).
4340. Sinke, G. C. and F. L. Oetting, *J. Phys. Chem.* **68**, 1354 (1964).
4341. Sinke, G. C. and D. R. Stull, *J. Phys. Chem.* **62**, 397 (1958).
4342. Sinyakov, Yu. I., A. I. Gorbanev, et al., *Izv. Akad. Nauk SSSR, Otd. Khim. Nauk* **1961**, 1514; *C. A.* **56**, 952 (1962).
4343. Sirkar, S. C. and B. M. Bishui, *Indian J. Phys.* **20**, 35 (1946); *C. A.* **41**, 4718 (1947).
4344. Sirkar, S. C. and B. M. Bishui, *Indian J. Phys.* **20**, 111 (1946); *C. A.* **41**, 4719 (1947).
4345. Sirkar, S. C. and B. M. Bishui, *Proc. Nat. Inst. Sci. India* **9**, 287 (1943); *C. A.* **42**, 8651 (1948).
4346. Sirkar, S. C. and B. M. Bishui, *Sci. Cult.* (Calcutta) **11**, 273 (1945); *C. A.* **40**, 2073 (1946).
4347. Sirkar, S. C., D. K. Mukherjee, et al., *Indian J. Phys.* **38**, 610 (1964); *C. A.* **62**, 14,045 (1965).
4348. Sirkar, S. C. and S. B. Sanyal, *Indian J. Phys.* **17**, 309 (1944); *C. A.* **39**, 868 (1945).
4349. Sisler, H. H., H. H. Batey, et al., *J. Amer. Chem. Soc.* **70**, 3821 (1948).
4350. Sjogren, H., *Ark. Fysik* **31**, 159 (1966); *C. A.* **65**, 3718 (1966).
4351. Skau, E. L., *J. Phys. Chem.* **37**, 609 (1933).
4352. Skau, E. L. and R. McCullough, *J. Amer. Chem. Soc.* **57**, 2439 (1935).
4353. Skeeters, M. J., U.S. Patent 3,076,040 (Cl. 260–652.5), Apr. 4, 1960; *C. A.* **58**, 13,792 (1963).
4354. Skeeters, M. J. and W. J. Esselstyn, U.S. Patent 2,567,621 (Cl. 260–652.5), Sept. 11, 1951.
4355. Skerlak, T. and B. Ninkov, *Glas. Drust. Hem. Teknol. SR Bosne Hercogovine* No. **11**, 43 (1962); *C. A.* **61**, 2496 (1964).
4356. Skinner, G. S., *J. Amer. Chem. Soc.* **55**, 2036 (1933).
4357. Skinner, H. A. and A. Snelson, *Trans. Faraday Soc.* **56**, 1776 (1960).
4358. Skinner, J. F., E. L. Cussler, et al., *J. Phys. Chem.* **72**, 1057 (1968).
4359. Skita, A., *Ber. Deut. Chem. Gesell.* **53**, 1792 (1920).
4360. Skita, A. and W. Faust, *Ber. Deut. Chem. Gesell.* **64**, 2878 (1931).
4361. Skita, A. and W. Faust, *Ber. Deut. Chem. Gesell.* **72B**, 1127 (1939).
4362. Skita, A., H. Hauber, et al., *Ann. Chem.* **431**, 1 (1923); *Chem. Zentr.* **1923, I**, 1320.

4363. Skita, A. and A. Schneck, *Ber. Deut. Chem. Gesell.* **55**, 144 (1922).
4364. Sklyarenko, S. I., B. I. Markin, et al., *Zh. Fiz. Khim.* **32**, 1916 (1958); *C. A.* **53**, 4848 (1959).
4365. Skulski, T., *Bull. Acad. Pol. Sci., Ser. Sci. Chim.* **14**, 23 (1966); *C. A.* **65**, 1605 (1966).
4366. Skuratov, S. M. and M. P. Kozina, *Dokl. Akad. Nauk SSSR* **122**, 109 (1958); *C. A.* **54**, 23,701 (1960).
4367. Skuratov, S. M., A. A. Strepikheev, et al., *Dokl. Akad. Nauk SSSR* **117**, 452 (1957); *C. A.* **52**, 14,598 (1958).
4368. Sladkov, A. M. and L. K. Luneva, *Tr. Nauchn.-Issled. Inst. Sintetich. Spirtov i Org. Produktov* **1960** (2) 253.
4369. Small, P. A., K. W. Small, et al., *Trans. Faraday Soc.* **44**, 810 (1948); *C. A.* **43**, 3677 (1949).
4370. Smeets, A. and E. Ruppol, *Bull. Soc. Chim. Biol.* **16**, 865 (1934); *Chem. Zentr.* **1935**, I, 2619.
4371. Smeets, F. and J. Verhulet, *Bull. Soc. Chim. Belges* **61**, 694 (1952); *C. A.* **48**, 6989 (1954).
4372. Smets, G., *Trav. Lab. Chim. Gen., Univ. Louvain*, **1942–1947**, 66 pp., *Acad. Roy. Belg., Classe Sci. Mem.* Collection in **21**, 3 (1947); *C. A.* **44**, 8315 (1950).
4373. Smiley, H. M. and P. G. Sears, *Trans. Kentucky Acad. Sci.* **18** (2–3), 40 (1957).
4374. Smit, W. M., *Purity Control by Thermal Analysis*, Elsevier, New York, 1957.
4375. Smith, A. W. and C. E. Boord, *J. Amer. Chem. Soc.* **48**, 1512 (1926).
4376. Smith, A. W., C. E. Boord, et al., *J. Amer. Chem. Soc.* **49**, 1335 (1927).
4377. Smith, D. C., C.-Y. Pan, et al., *J. Chem. Phys.* **18**, 706 (1950).
4378. Smith, D. M. and W. M. D. Bryant, *J. Amer. Chem. Soc.* **57**, 61 (1935).
4379. Smith, D. M. and W. M. D. Bryant, *J. Amer. Chem. Soc.* **57**, 841 (1935).
4380. Smith, E. R., *J. Res. Nat. Bur. Stand.* **24**, 229 (1940).
4381. Smith, E. R., *J. Res. Nat. Bur. Stand.* **26**, 129 (1941).
4382. Smith, E. R. and H. Matheson, *J. Res. Nat. Bur. Stand.* **20**, 641 (1938).
4383. Smith, E. R. and M. Wojciechowski, *J. Res. Nat. Bur. Stand.* **18**, 461 (1937).
4384. Smith, F. and L. M. Turton, *J. Chem. Soc.* **1951**, 1701.
4385. Smith, F. A. and E. C. Creitz, *J. Res. Nat. Bur. Stand.* **46**, 145 (1951).
4386. Smith, G. E. and Wm. Hunter, Brit. Patent 606,607, Aug. 17, 1948; *C. A.* **43**, 3839 (1949).
4387. Smith, G. F., *J. Chem. Soc.* **1931**, 3257.
4388. Smith, G. F., *J. Chem. Soc.* **1940**, 869.
4389. Smith, G. W. and L. V. Sorg, *J. Phys. Chem.* **45**, 671 (1941).
4390. Smith, J. C., *J. Chem. Soc.* **1931**, 802.
4391. Smith, J. C., *J. Chem. Soc.* **1939**, 974.
4392. Smith, J. C., N. J. Foecking, et al., *Ind. Eng. Chem.* **41**, 2289 (1949).
4393. Smith, J. M., *Chem. Eng. Progr.* **44**, 521 (1948).
4394. Smith, J. M., *Trans. A.I.C.H.E.* **42**, 983 (1946).
4395. Smith, J. W. and S. M. Walshaw, *J. Chem. Soc.* **1957**, 3217.
4396. Smith, L. and L. Bjellerup, *Acta Chem. Scand.* **1**, 566 (1947); *C. A.* **42**, 3248 (1948).
4397. Smith, L. and S. Sunner, *The Svedberg* (Mem. Vol.) **1944**, 352; *C. A.* **39**, 1100 (1945).
4398. Smith, L. I., Ed., *Organic Synthesis*, Vol. 23, Wiley, New York, 1943.
4399. Smith, L. T. and H. V. Claborn, *Ind. Eng. Chem.* **32**, 692 (1940).
4400. Smith, M. S., P. A. D. de Maine, et al., *J. Miss. Acad. Sci.* **8**, 244 (1962); *C. A.* **58**, 2028 (1963).
4401. Smith, N. K. and W. D. Good, *J. Chem. Eng. Data* **12**, 572 (1967).

4402. Smith, S. G., A. H. Fainberg et al., *J. Amer. Chem. Soc.* **83**, 618 (1961).

4403. Smith, T. E. and R. F. Bonner, *Anal. Chem.* **24**, 517 (1952).

4404. Smith, T. E. and R. F. Bonner, *Ind. Eng. Chem.* **43**, 1169 (1951).

4405. Smith, W. M., K. C. Eberly, et al., *J. Amer. Chem. Soc.* **78**, 626 (1956).

4406. Smith, W. T., S. Greenbaum, et al., *J. Phys. Chem.* **58**, 443 (1954).

4407. Smittenberg, J., H. Hoog, et al., *J. Amer. Chem. Soc.* **60**, 17 (1938).

4408. Smyrl, W. H. and C. W. Tobias, *J. Electrochem. Soc.* **115**, 33 (1968).

4409. Smyth, C. P., "Dielectric Constant and Molecular Structure," Chemical Catalog Company, New York, 1931.

4410. Smyth, C. P., *J. Amer. Chem. Soc.* **47**, 1894 (1925).

4411. Smyth, C. P., *J. Amer. Chem. Soc.* **63**, 57 (1941).

4412. Smyth, C. P., R. W. Dornte, et al., *J. Amer. Chem. Soc.* **53**, 4242 (1931).

4413. Smyth, C. P. and E. W. Engel, *J. Amer. Chem. Soc.* **51**, 2646 (1929).

4414. Smyth, C. P. and S. E. Kamerling, *J. Amer. Chem. Soc.* **53**, 2988 (1931).

4415. Smyth, C. P. and K. B. McAlpine, *J. Amer. Chem. Soc.* **56**, 1697 (1934).

4416. Smyth, C. P. and K. B. McAlpine, *J. Chem. Phys.* **1**, 190 (1933).

4417. Smyth, C. P. and K. B. McAlpine, *J. Chem. Phys.* **2**, 499 (1934).

4418. Smyth, C. P. and K. B. McAlpine, *J. Chem. Phys.* **3**, 347 (1935).

4419. Smyth, C. P. and S. A. McNeight, *J. Amer. Chem. Soc.* **58**, 1597 (1936).

4420. Smyth, C. P. and S. O. Morgan, *J. Amer. Chem. Soc.* **49**, 1030 (1927).

4421. Smyth, C. P. and S. O. Morgan, *J. Amer. Chem. Soc.* **50**, 1547 (1928).

4422. Smyth, C. P. and H. E. Rogers, *J. Amer. Chem. Soc.* **52**, 1824 (1930).

4423. Smyth, C. P. and H. E. Rogers, *J. Amer. Chem. Soc.* **52**, 2227 (1930).

4424. Smyth, C. P. and W. N. Stoops, *J. Amer. Chem. Soc.* **50**, 1883 (1928).

4425. Smyth, C. P. and W. N. Stoops, *J. Amer. Chem. Soc.* **51**, 3312, 3330 (1929).

4426. Smyth, C. P. and W. S. Walls, *J. Amer. Chem. Soc.* **53**, 527, 2115 (1931).

4427. Smyth, C. P. and W. S. Walls, *J. Amer. Chem. Soc.* **54**, 1854 (1932).

4428. Smyth, C. P. and W. S. Walls, *J. Amer. Chem. Soc.* **54**, 2261 (1932).

4429. Smyth, C. P. and W. S. Walls, *J. Amer. Chem. Soc.* **54**, 3230 (1932).

4430. Smyth, C. P. and W. S. Walls, *J. Chem. Phys.* **3**, 557 (1935).

4431. Smyth, H. F., C. P. Carpenter, et al., *Arch. Ind. Hyg. Occupational Med.* **4**, 119 (1951); *C. A.* **45**, 9710 (1951).

4432. Smyth, H. F., C. P. Carpenter, et al., *Arch. Ind. Hyg. Occupational Med.* **10**, 61 (1954); *C. A.* **48**, 13,951 (1954).

4433. Snead, C. C. and H. L. Clever, *J. Chem. Eng. Data* **7**, 393 (1962).

4434. Sneddon, R., *Petroleum Engr.* **16**, No. 13, 148 (1945); *C. A.* **40**, 4568 (1946).

4435. Snelson, A. and H. A. Skinner, *Trans. Faraday Soc.* **57**, 2125 (1961).

4436. Snethlage, H. C. S., *Z. Phys. Chem.* (Leipzig) **90**, 1 (1915).

4437. Snider, O. E., R. H. Belden, et al., U.S. Patent 3,021,326 (Cl. 260–239.3), Feb. 13, 1962.

4438. Snyder, E. I., *J. Amer. Chem. Soc.* **85**, 2624 (1963).

4439. Snyder, L. R. and B. E. Buell, *J. Chem. Eng. Data* **11**, 545 (1966).

4440. Sobotka, H. and J. Kahn, *J. Amer. Chem. Soc.* **53**, 2935 (1931).

4441. Soc. Anon. Distilleries des Deux-Sevres, Brit. Patent 274, 488, July 14, 1926; *C. A.* **22**, 2171 (1928).

4442. Solvay & Cie, Belgian Patent 625,490, May 29, 1963; *C. A.* **60**, 12,243 (1964).

4443. Solvay & Cie, Netherlands Appl. 289,737, May 11, 1964; *C. A.* **62**, 2707 (1965).

4444. Sommer, L. H., H. D. Blankman, et al., *J. Amer. Chem. Soc.* **76**, 803 (1954).

4445. Sommer, S. and G. Tauberger, *Arzneim.-Forsch.* **14**, 1050 (1964); *C. A.* **62**, 989 (1965).

4446. Somsen, G. and J. Coops, *Rec. Trav. Chim. Pays-Bas* **84**, 985 (1965); *C. A.* **63** 17,222 (1965).

4447. Sordes, J., *Compt. Rend.* **195,** 247 (1932); *C. A.* **26,** 5008 (1932).
4448. Sorensen, P., *Chem. Ind.* (London) **1959,** 1593.
4449. Sorendararajan, S., *Indian J. Chem.* **1,** 503 (1963).
4450. Southard, J. C. and D. H. Andrews, *J. Franklin Inst.* **209,** 349 (1930); *C. A.* **24,** 2332 (1930).
4451. Spaght, M. E., S. B. Thomas, et al., *J. Phys. Chem.* **36,** 882 (1932).
4452. Spasovski, M., *Khigiena* (Sofia) **7,** 38 (1964); *C. A.* **62,** 9684 (1965).
4453. Spassow, A., *Ber. Deut. Chem. Gesell.* **70,** 1926 (1937); *C. A.* **31,** 8503 (1937).
4454. Späth, E., *Monatsh. Chem.* **35,** 324 (1914).
4455. Späth, E. and H. Bretschneider, *Ber. Deut. Chem. Gesell.* **61,** 327 (1928).
4456. Spells, K. E., *Trans. Faraday Soc.* **32,** 530 (1936).
4457. Spencer, H. M., *J. Amer. Chem. Soc.* **67,** 1859 (1945).
4458. Spencer, H. M. and G. N. Flannagan, *J. Amer. Chem. Soc.* **64,** 2511 (1942).
4459. Speros, D. M. and F. D. Rossini, *J. Phys. Chem.* **64,** 1723 (1960).
4460. Spiers, C. W. F. and J. P. Wibaut, *Rec. Trav. Chim. Pays-Bas* **56,** 573 (1937); *C. A.* **31,** 5272 (1937).
4461. Spinner, E. and A. Burawoy, *Spectrochim. Acta* **17,** 558 (1961); *C. A.* **55,** 19,473 (1961).
4462. Spitzer, R. and H. M. Huffman, *J. Amer. Chem. Soc.* **69,** 211 (1947).
4463. Spitzer, R. and K. S. Pitzer, *J. Amer. Chem. Soc.* **68,** 2537 (1946).
4464. Sponer, H., *Chem. Rev.* **41,** 281 (1947).
4465. Sponer, H., *Rev. Mod. Phys.* **14,** 224 (1942); *C. A.* **37,** 1651 (1943).
4466. Sponer, H. and M. B. Hall, *Contrib. Étude Structure Mol.*, Vol. Commen. Victor Henri **1947/48,** 222; *C. A.* **43,** 2864 (1949).
4467. Sponer, H. and J. S. Kirby-Smith, *J. Chem. Phys.* **9,** 667 (1941).
4468. Sponer, H. and H. Stuklen, *J. Chem. Phys.* **14,** 101 (1946).
4469. Springer, C. S. and D. W. Meek, *J. Phys. Chem.* **70,** 481 (1966).
4470. Srivastava, G. P. and Y. P. Varshni, *Z. Phys. Chem.* (Leipzig) **213,** 30 (1960).
4471. Srivastava, N. K. and J. B. Lal, *J. Sci. Soc., Harcourt Butler Technol. Inst. and Indian Inst. Sugor Technol., Kanpur, India* **4,** 9 (1955); *C. A.* **54,** 12,051 (1960).
4472. Staff, *Chemical Processing,* Feb. 1962, p. 49.
4473. Stafford, O. F., *J. Amer. Chem. Soc.* **55,** 3987 (1933).
4474. Stair, R. and W. W. Coblentz, *J. Res. Nat. Bur. Stand.* **15,** 295 (1935).
4475. Stanford, S. C. and W. Gordy, *J. Amer. Chem. Soc.* **62,** 1247 (1940).
4476. Stasenkova, K. P. and T. A. Kochetkova, *Toksikol. Novykh Prom. Khim. Veshchestv.* **1962,** No. 4, 19; *C. A.* **58,** 9543 (1963).
4477. Stasenkova, K. P. and T. A. Kochetkova, *Toksikol. Novykh Prom. Khim. Veshchestv.* **1962,** No. 4, 29; *C. A.* **58,** 10,655 (1963).
4478. Stasenkova, K. P. and T. A. Kochetkova, *Toksikol. Novykh Prom. Khim. Veshchestv.* No. **7,** 27–38 (1965); *C. A.* **63,** 8947 (1965).
4479. Staudhammer, P. and W. F. Sayer, *J. Amer. Chem. Soc.* **80,** 6491 (1958).
4480. Staudinger, H., *Z. Angew. Chem.* **35,** 658 (1922).
4481. Staudinger, H. and A. Schwalbach, *Ann. Chim.* **488,** 8 (1931); *C. A.* **25,** 5138 (1931).
4482. Stauffer, "Chlorinated Hydrocarbons," Stauffer Chemical Co., New York, 1965–1966.
4483. Staveley, L. A. K. and A. K. Gupta, *Trans. Faraday Soc.* **45,** 50 (1949); *C. A.* **43,** 5276 (1949).
4484. Staveley, L. A. K., W. I. Tupman, et al., *Trans. Faraday Soc.* **51,** 323 (1955).
4485. Staverman, A. J., *Rec. Trav. Chim. Pays-Bas* **60,** 836 (1941); Beil **EIII1,** 154.
4486. Stearn, A. E. and C. P. Smyth, *J. Amer. Chem. Soc.* **56,** 1667 (1934).
4487. Stehling, F. C., *Anal. Chem.* **35,** 773 (1963).

4488. Stein, E. and O. Bayer, German Patent 888,689, Nov. 17, 1951; from [3042].
4489. Steiner, E. C. and J. M. Gilbert, *J. Amer. Chem. Soc.* **87**, 382 (1965).
4490. Steiner, L. A., *Ind. Eng. Chem. Anal. Ed.* **10**, 582 (1938).
4491. Steiner, L. E. and J. Johnston, *J. Phys. Chem.* **32**, 912 (1928).
4492. Stepanenko, N., *J. Exptl. Theoret. Phys.* (USSR) **14**, 163 (1944); *C. A.* **39**, 1337 (1945).
4493. Stepanenko, N. N., B. A. Agranat, et al., *J. Phys. Chem.* (USSR) **21**, 893 (1947); *C. A.* **42**, 2485 (1948).
4494. Stephen, H. and T. Stephen, *Solubilities of Inorganic and Organic Compounds*, Vol. 1, Binary Systems, Pts. 1 and 2, Macmillan, New York, 1963.
4495. Stephens, O. C., *J. Soc. Chem. Ind.* **43**, 313T, 327T (1924).
4496. Stern, A. and K. Thalmayer, *Z. Phys. Chem.* (Leipzig) **31B**, 403 (1936); *Chem. Zentr.* **1936**, **II**, 1148.
4497. Stevens, D. R., *Ind. Eng. Chem.* **35**, 655 (1943).
4498. Stevenson, D. P. and J. Y. Beach, *J. Chem. Phys.* **6**, 25 (1938).
4499. Stewart, J. E., *J. Chem. Phys.* **23**, 986 (1955).
4500. Stewart, J. E., R. O. Brace, et al., *Nature* **186**, 628 (1960).
4501. Stewart, P. B. and E. S. Starkman, *Chem. Eng. Progr.* **53**, 41J (1957).
4502. Stewart, R. and J. P. O'Donnell, *Can. J. Chem.* **42**, 168 (1964).
4503. Stewart, R. and K. Yates, *J. Amer. Chem. Soc.* **80**, 6355 (1958).
4504. Stewart, T. D. and B. J. Fontana, *J. Amer. Chem. Soc.* **62**, 3281 (1940).
4505. Stich, K., G. Rotzler, et al., *Helv. Chim. Acta* **42**, 1480 (1959).
4506. Stock, A. and P. Seelig, *Ber. Deut. Chem. Gesell.* **52**, 672 (1919).
4507. Stockhardt, J. S. and C. M. Hull, *Ind. Eng. Chem.* **23**, 1438 (1931).
4508. Stokes, S., *J. Chem. Phys.* **39**, 2309 (1963).
4509. Stokinger, H. E., Chairman, *Documentation of Threshold Limit Values*, Revised ed., American Conference of Governmental Industrial Hygienists, Cincinnati, Ohio, 1966.
4510. Storey, W. H., *J. Amer. Oil Chem. Soc.* **37**, 676 (1960).
4511. Stork, G. and W. N. White, *J. Amer. Chem. Soc.* **78**, 4609 (1956).
4512. Stothers, J. B. and P. C. Lauterbur, *Can. J. Chem.* **42**, 1563 (1964).
4513. Stotz, E., W. W. Westerfield, et al., *J. Biol. Chem.* **152**, 41 (1944).
4514. Stoughton, R. W. and P. D. Lamson, *J. Pharmacol.* **58**, 74 (1936); *C. A.* **31**, 465 (1937).
4515. Stout, A. W. and H. A. Schuette, *Ind. Eng. Chem., Anal. Ed.* **5**, 100 (1933).
4516. Stout, J. W. and L. H. Fisher, *J. Chem. Phys.* **9**, 163 (1941).
4517. Stranathan, J. D., *J. Chem. Phys.* **5**, 828 (1937).
4518. Stranathan, J. D., *J. Chem. Phys.* **6**, 395 (1938).
4519. Strecher, P. G., ed., *The Merck Index of Chemicals and Drugs*, 7th ed., Merck, Rahway, N.J., 1960.
4520. Strecker, W. and R. Spitaler, *Ber. Deut. Chem. Gesell.* **59**, 1754 (1926).
4521. Streiff, A. J., A. R. Hulme, et al., *Anal. Chem.* **27**, 411 (1955).
4522. Streiff, A. J., E. T. Murphy, et al., *J. Res. Nat. Bur. Stand.* **37**, 331 (1946).
4523. Streiff, A. J., E. T. Murphy, et al., *J. Res. Nat. Bur. Stand.* **38**, 53 (1947).
4524. Streiff, A. J., E. T. Murphy, et al., *J. Res. Nat. Bur. Stand.* **39**, 321 (1947).
4525. Streiff, A. J. and F. D. Rossini, *J. Res. Nat. Bur. Stand.* **32**, 185 (1944).
4526. Streiff, A. J., L. H. Schultz, et al., *Anal. Chem.* **29**, 361 (1957).
4527. Streiff, A. J., L. F. Soule, et al., *J. Res. Nat. Bur. Stand.* **45**, 173 (1950).
4528. Streiff, A. J., J. C. Zimmerman, et al., *J. Res. Nat. Bur. Stand.* **41**, 323 (1948).
4529. Streim, H. G., E. A. Boyce, et al., *Anal. Chem.* **33**, 85 (1961).
4530. Strepikheev, A. A., S. M. Skuratov, et al., *Dokl. Akad. Nauk SSSR* **102**, 105 (1955); *C. A.* **50**, 4903 (1956).

4531. Stroh, H. and G. Westphal, *Chem. Ber.* **96**, 184 (1963).
4532. Stuart, A. H., *Petroleum* (London) **10**, 74 (1947) *C. A.* **41**, 5293 (1947).
4533. Stuart, A. V. and G. B. B. M. Southerland, *J. Chem. Phys.* **24**, 559 (1956).
4534. Stuckey, J. M. and J. H. Saylor, *J. Amer. Chem. Soc.* **62**, 2922 (1940).
4535. Stull, D. R., *Ind. Eng. Chem.* **39**, 517 (1947).
4536. Stull, D. R., *J. Amer. Chem. Soc.* **59**, 2726 (1937).
4537. Su, Fu-Ti and Yü-L. P'an, *Union Ind. Research Inst. Rept.* No. 24 (1957); *C. A.* **54**, 22,460 (1960).
4538. Sucharda, E. and H. Kuczyński, *Rocz. Chem.* **14**, 1182 (1934); *C. A.* **29**, 6214 (1935).
4539. Sudgen, S., *J. Chem. Soc.* **1924**, 1167.
4540. Sudgen, S., *J. Chem. Soc.* **1933**, 768.
4541. Sudgen, S. and H. Whittaker, *J. Chem. Soc.* **1925**, 1868.
4542. Sudgen, S. and H. Wilkins, *J. Chem. Soc.* **1927**, 139.
4543. Sue, P., J. Pauly, et al., *Bull. Soc. Chim. Fr.* **5**, 593 (1958); *C. A.* **52**, 15,177 (1958); Ref. [5181].
4544. Suetaka, W., *Gazz. Chim. Ital.* **86**, 783 (1956); *C. A.* **52**, 5974 (1958).
4545. Suga, H. and S. Seki, *Bull. Chem. Soc. Jap.* **35**, 1905 (1962); *C. A.* **58**, 3027 (1963).
4546. Sugarman, B., *Proc. Phys. Soc.* **55**, 429 (1943); *C. A.* **38**, 1427 (1944).
4547. Suhrmann, R. and P. Klein, *Z. Phys. Chem.* (Leipzig) **50B**, 23 (1941); *Chem. Zentr.* **1942I**, 2251; *C. A.* **37**, 3343 (1943).
4548. Sukhodol, V. F. and P. A. Chatskii, *Spirt. Prom.* **28**, No. 3, 35 (1962); *C. A.* **57**, 5124 (1962).
4549. Sultanov, G. A. and M. M. Musaev, *Izv. Akad. Nauk Azerb. SSR* **1957**, No. 1, 31; *C. A.* **51**, 11,072 (1957).
4550. Sundara Rao, A. L., *J. Indian Chem. Soc.* **22**, 260 (1945); *C. A.* **40**, 4292 (1946).
4551. Sunier, A. A., *J. Phys. Chem.* **35**, 1756 (1931).
4552. Sunner, S., *Acta Chem. Scand.* **9**, 837, 847 (1955); *C. A.* **50**, 2269, 2270 (1956).
4553. Surinder, P. V. and K. A. Kobe, *J. Chem. Eng. Data* **4**, 329 (1959).
4554. Sus, A. N., *Compt. Rend. Acad. Sci. URSS* **33**, 310 (1941); *C. A.* **37**, 1308.
4555. Susz, B. and E. Briner, *Helv. Chim. Acta* **22**, 117 (1939).
4556. Sutton, L. E., *Proc. Roy. Soc.*, Ser. **A133**, 668 (1931); *Chem. Zentr.* **1932, I**, 3265.
4557. Sutton, L. E., R. G. A. New, et al., *J. Chem. Soc.* **1933**, 652.
4558. Svyentoslavskii, V., *J. Russ. Phys. Chem. Soc.* **41**, 920; *Chem. Zentr.* **1909, II**, 2144.
4559. Suyver, J. F. and J. P. Wibaut, *Rec. Trav. Chim. Pays-Bas* **64**, 65 (1945); *C. A.* **40**, 3428 (1946); Beil **EIII5**, 1580.
4560. Suzuki, I., *Bull. Chem. Soc. Jap.* **35**, 540 (1962); from [2827].
4561. Suzuki, K., *Rev. Phys. Chem. Jap.* **23**, 57 (1953); *C. A.* **49**, 3007 (1955).
4562. Suzuki, S., *Kagaku* **23**, 535 (1953); *C. A.* **47**, 11,892 (1953).
4563. Svirbely, W. J., J. E. Ablard, et al., *J. Amer. Chem. Soc.* **57**, 652 (1935).
4564. Svirbely, W. J., W. M. Eareckson, et al., *J. Amer. Chem. Soc.* **71**, 507 (1949).
4565. Swalen, J. D. and D. R. Herschbach, *J. Chem. Phys.* **27**, 100 (1957).
4566. Swallen, L. C. and C. E. Boord, *J. Amer. Chem. Soc.* **52**, 651 (1930)
4567. Swallow, J. C. and R. O. Gibson, *J. Chem. Soc.* **1934**, 440.
4568. Swamy, P. A. and M. Van Winkle, *J. Chem. Eng. Data* **10**, 214 (1965).
4569. Swarts, F., *Bull. Acad. Roy. Belg. Classe Sci.* **1909**, 43; *Chem. Zentr.* **1909, I**, 1989.
4570. Swarts, F., *Bull. Sci. Acad. Roy. Belg.* **8**, 343 (1922); *C. A.* **17**, 769 (1923).
4571. Swarts, F., *Bull. Soc. Chim. Belges* **38**, 99 (1929).
4572. Swarts, F., *J. Chim. Phys.* **28**, 622 (1931); *Beil* **EIII5**, 676.
4573. Swarts, F., *J. Pharm. Chim.* **28**, 622 (1931).
4574. Swarts, F., *Rec. Trav. Chim. Pays-Bas* **33**, 281 (1914); *Bull. Acad. Roy. Belg.* **1914**, 18; *C. A.* **9**, 445 (1915).

4575. Swensen, R. F. and D. A. Keyworth, *Anal. Chem.* **35**, 863 (1963).
4576. Swern, D., ed., *Bailey's Industrial Oil and Fat Products*, 3rd ed., Interscience, New York, 1964.
4577. Swietoslawski, W., *Ebulliometric Measurements*, Reinhold, New York, 1945.
4578. Swietoslawski, W., *J. Phys. Chem.* **31**, 1169 (1934).
4579. Swietoslawski, W., *Przeglad Chem.* **6**, 249 (1948); *C. A.* **43**, 8977 (1949).
4580. Swietoslawski, W., *Z. Phys. Chem.* (Leipzig) **65**, 513 (1909).
4581. Swietoslawski, W., *Z. Phys. Chem.* (Leipzig) **72**, 49 (1910).
4582. Swietoslawski, W. and E. Bartoszewicz, *Rocz. Chem.* **11**, 78 (1931); *Chem. Zentr.* **1931**, I, 2787.
4583. Swietoslawski, W. and M. Popow, *J. Chim. Phys.* **22**, 395 (1925); *Chem. Zentr.* **1926**, I, 600.
4584. Swietoslawski, W., S. Rybicka, et al., *Rocz. Chem.* **11**, 35 (1931).
4585. Swift, E., *J. Amer. Chem. Soc.* **64**, 115 (1942).
4586. Swift, E. and C. R. Calkins, *J. Amer. Chem. Soc.* **65**, 2415 (1943).
4587. Swindells, J. F., J. R. Coe, et al., *J. Res. Nat. Bur. Stand.* **48**, 1 (1952).
4588. Sydow, E. v., *Acta Chem. Scand.* **18**, 1099 (1964); *C. A.* **61**, 15,504 (1964).
4589. Szasz, G. J., J. H. Morrison, et al., *J. Chem. Phys.* **15**, 562 (1947).
4590. Szekely, A., *Acta Chim. Acad. Sci. Hung.* **5**, 317 (1955); *C. A.* **49**, 15,426 (1955).
4591. Szobel, L., *Compt. Rend.* **218**, 347 (1944); *C. A.* **40**, 2740 (1946).

T

4592. Tabuteau, J. and H. Gautier, French Patent 860,484, Jan. 16, 1941; *C. A.* **42**, 6841 (1948).
4593. Tadayon, J., A. H. Nisson, et al., *Anal. Chem.* **21**, 1532 (1949).
4594. Tafel, J. and G. Frederichs, *Ber. Deut. Chem. Gesell* **37**, 3187 (1904).
4595. Tafel, J. and O. Wassmuth, *Ber. Deut. Chem. Gesell* **40**, 2831 (1907).
4596. Takao, M. and M. Kumamoto, Japanese Patent 6566, Oct. 15, 1954; *C. A.* **50**, 4194 (1956).
4597. Takamitsu, N. and H. Sumida, *J. Chem. Soc. Jap., Ind. Chem. Sect.* **70**, 2138 (1967).
4598. Takada, M. and H. S. Gutowsky, *J. Chem. Phys.* **26**, 577 (1957).
4599. Takeda, M., *J. Chem. Soc. Jap.* **62**, 896 (1941); *C. A.* **41**, 5390 (1947).
4600. Tallman, R. C., *J. Amer. Chem. Soc.* **56**, 126 (1934).
4601. Tammann, G., *Z. Phys. Chem.* (Leipzig) **85**, 273 (1913).
4602. Tamris, M., S. Searles, et al., *J. Amer. Chem. Soc.* **76**, 3983 (1954).
4603. Tamura, Z. and S. Kawai, *Bunsaki Kagaku* **15**, 64 (1955); *C. A.* **65**, 6522 (1966).
4604. Tanaka, C., *Mem. Coll. Sci., Kyota Imp. Univ.* Ser. A, **13**, 239 (1930); *Chem. Zentr.* **1930**, II, 1523.
4605. Tanaka, C., *Nippon Kagaku Zasshi* **83**, 521 (1962); *C. A.* **58**, 10,882 (1963).
4606. Tanaka, N., *J. Chem. Soc. Jap.* **63**, 134 (1942); *C. A.* **41**, 3063 (1947).
4607. Tanaka, T. and M. Hashimoto, *Sci. Repts. Saitama Univ.* Ser. A, **2**, 117 (1956); *C. A.* **51**, 6334 (1957).
4608. Tananka, S., *Japan Analyst* **4**, 109 (1955); *C. A.* **50**, 5465 (1956).
4609. Tarasenkov, D. N. and E. N. Polozhintzwa, *Chem. Ber.* **65B**, 184 (1932).
4610. Tavernier, P. and M. Lamouroux, *Mem. Poudres.* **38**, 65 (1956); *C. A.* **51**, 14403 (1957).
4611. Taylor, C. A. and W. H. Rinkenbach, *Ind. Eng. Chem.* **18**, 676 (1926).
4612. Taylor, C. A. and W. H. Rinkenbach, *J. Amer. Chem. Soc.* **48**, 1305 (1926).
4613. Taylor, G. B. and M. B. Hall, *Anal. Chem.* **23**, 947 (1951).
4614. Taylor, H. A. and T. W. Davis, *J. Phys. Chem.* **32**, 1467 (1928).

4615. Taylor, J. M., P. M. Jenner, et al., *Toxicol. and Appl. Pharmacol.* **6**, 378 (1964).
4616. Taylor, M. D. and L. R. Grant, *J. Org. Chem.* **25**, 678 (1960).
4617. Taylor, M. D. and M. B. Templeman, *J. Amer. Chem. Soc.* **78**, 2950 (1956).
4618. Taylor, R. C., R. A. Brown, et al., *Anal. Chem.* **20**, 396 (1948).
4619. Taylor, R. K., *J. Amer. Chem. Soc.* **75**, 2521 (1953).
4620. Taylor, W. J. and F. D. Rossini, *J. Res. Nat. Bur. Stand.* **32**, 197 (1944).
4621. Taylor, W. J., D. D. Wagman, et al., *J. Res. Nat. Bur. Stand.* **37**, 95 (1946).
4622. Tedder, J. M., *J. Chem. Soc.* **1954**, 2646.
4623. Telang, M. S., *J. Phys. Chem.* **50**, 373 (1946).
4624. Teramoto, S., M. Ishikawa, et al., *J. Fermentation Technol.* (Japan) **33**, 307 (1955);
      C. A. **49**, 16,323 (1955).
4625. Terry, T. D., R. E. Kepner, et al., *J. Chem. Eng. Data* **5**, 403 (1960).
4626. Teter, J. W. and W. J. Merwin, U.S. Patent 2,411,346, (Cl. 202–42), Nov. 19, 1946;
      C. A. **41**, 983 (1947).
4627. Tézé, M. and R. Schaal, *Bull. Soc. Chim. Fr.* **1962**, 1372; *Chem. Zentr.* **1963**, 7875.
4628. Thatte, V. N. and A. S. Ganesan, *Phil. Mag.* **12**, 823 (1931); *C. A.* **26**, 655 (1932).
4629. Thatte, V. N. and M. S. Joglekar, *Phil. Mag.* **19**, 1116 (1935); *C. A.* **29**, 6505 (1935).

  * References 4630–4704: Thermodynamics Research Center Data Project (formerly
Manufacturing Chemists Association Research Project), Thermodynamics Research Center,
College Station, Texas A & M University.

4630.* MCA Catalog of Infrared Spectra.
4631. Thermodynamic Research Center Data Project, Catalog of Mass Spectra.
4632. MCA Catalog of Nuclear Magnetic Resonance Spectra.
4633. MCA Catalog of Raman Spectra.
4634. MCA Catalog of Ultraviolet Spectra.
4635. Table 2-1-a, Dec. 31, 1962.
4636. Table 2-1-i, June 30, 1959.
4637. Table 2-1-(1.02)-fa, Dec. 31, 1966.
4638. Table 23-2-1-(1.1000)-a, Dec. 31, 1966.
4639. Table 23-2-1-(1.1000)-k, June 30, 1965.
4640. Table 23-2-1-(1.1000)-m, Dec. 31, 1965.
4641. Table 23-2-1 (1.1000) p, Dec. 31, 1965.
4642. Table 23-2-1(1.1001)-a, June 30, 1963.
4643. Table 23-2-1-(1.1020)-a, Dec. 31, 1967.
4644. Table 23-2-1-(1.1020)-d, June 30, 1966.
4645. Table 23-2-1-(1.1020)-i, Dec. 31, 1966.
4646. Table 23-2-1-(1.1020)-k, June 30, 1965.
4647. Table 23-2-1-(1.1020)-m, June 30, 1965 and Dec. 31, 1965.
4648. Table 23-2-1-(1.1020)-p, June 30, 1965.
4649. Table 23-2-1-(1.1020)-v. Dec. 31, 1965.
4650. Table 23-2-1-(1.1021)-a, Dec. 31, 1967.
4651. Table 23-2-1-(1.1021)-d, June 30, 1966.
4652. Table 23-2-1-(1.1021)-k, June 30, 1965.
4653. Table 23-2-1-(1.1036)-a, June 30, 1966.
4654. Table 23-2-1-(1.1036)-k, June 30, 1966.
4655. Table 23-2-1-(1.1100)-k, Dec. 31, 1960.
4656. Table 23-2-1-(1.1105)-a, Dec. 31, 1964.
4657. Table 23-2-1-(1.1130)-a, Dec. 31, 1965.
4658. Table 23-2-1-(1.1130)-i, June 30, 1966.

4659. Table 23-2-1-(1.1130)-k, Dec. 31, 1965.
4660. Table 23-2-1-(1.1200)-a, June 30, 1961.
4661. Table 23-2-1-(1.1200)-k, Dec. 31, 1960.
4662. Table 23-2-1-(1.1210)-i, Dec. 31, 1966.
4663. Table 23-2-1-(1.1300)-i, Dec. 31, 1966.
4664. Table 23-2-1-(1.1320)-i, Dec. 31, 1966.
4665. Table 23-2-1-(1.1340)-i, Dec. 31, 1966.
4666. Table 23-2-1-(1.2115)-a, Dec. 31, 1965.
4667. Table 23-2-1-(1.2115)-m, Dec. 31, 1965.
4668. Table 23-2-1-(1.2115)-p, Dec. 31, 1965.
4669. Table 23-2-1-(1.2120)-k, June 30, 1963.
4670. Table 23-2-1-(1.2121)-a, June 30, 1963.
4671. Table 23-2-1-(1.2121)-m, Dec. 31, 1965.
4672. Table 23-2-1-(1.2121)-p, Dec. 31. 1965.
4673. Table 23-2-1-(33.1000)-a. June 30, 1962.
4675. Table 23-2-1-(33.1000)-d, Dec. 31, 1962.
4674. Table 23-2-1-(33.1000)-i, Dec. 31, 1963.
4676. Table 23-2-1-(33.1000)-k, June 30, 1962.
4677. Table 23-2-1-(33.7000)-m, Dec. 31, 1965.
4678. Table 23-2-1-(33.7000)-p, Dec. 31, 1965.
4679. Table 23-10-2-(1.0111)-k, June 30, 1956.
4680. Table 23-10-2-(1.013)-a, June 30, 1959.
4681. Table 23-10-2-(1.013)-i, Dec. 31, 1966.
4682. Table 23-10-2-(2.011)-a, Dec. 31, 1964.
4683. Table 23-10-2-(2.0111)-k, Dec. 31, 1957.
4684. Table 23-10-2-(2.013)-a, Dec. 31, 1964.
4685. Table 23-10-2-(2.013)-i, Dec. 31, 1966.
4686. Table 23-10-2-(3.011)-i, Dec. 31, 1966.
4687. Table 23-10-2-(4.011)-i, Dec. 31, 1966.
4688. Table 23-10-2-(10.011)-v, Dec. 31, 1967.
4689. Table 23-10-2-(10.011)-w, Dec. 31, 1967.
4690. Table 23-10-9-2-(10.011)-w, Dec. 31, 1967.
4691. Table 23-11-2-(1.0110)-i, Dec. 31, 1966.
4692. Table 23-11-2-(1.013)-a, Dec. 31, 1958.
4693. Table 23-11-2-(2.013)-a, Dec. 31, 1964.
4694. Table 23-11-2-(2.013)-i, Dec. 31, 1966.
4695. Table 23-12-2-(1.0111)-k, June 30, 1956.
4696. Table 23-12-2-(1.013)-a, Dec. 31, 1964.
4697. Table 23-12-2-(2.011)-a, Dec. 31, 1964.
4698. Table 23-12-2-(2.0111)-k, Dec. 31, 1963.
4699. Table 23-18-2-(1.01111)-k, Dec. 31, 1956.
4700. Table 23-18-2-(1.0113)-a, Dec. 31, 1961.
4701. Table 23-18-2-(1.0125)-a, Dec. 31, 1964.
4702. Table 23-18-2-(1.0211)-a, June 30, 1960.
4703 Table 23-18-2-(1.0133-a.)
4704. Table 23-18-2-(1.0213)-a, June 30, 1960.
4705. Thiessen, P. A. and C. Stüber, *Ber. Deut. Chem. Gesell.* **71B**, 2103 (1938).
4706. Thirion, P. and E. C. Craven, *J. Appl. Chem.* (London) **2**, 210 (1952).
4707. Thizy, A., *Compt. Rend. Congr. Intern. Chim.* **31**ᵉ, *Liège*, **1958**, (Pub. as *Ind. Chim. Belg. Suppl.*) **1**, 619; *C. A.* **54**, 3274 (1960).

4708. Thodos, G., *Amer. Inst. Chim. Eng. Journal* **1**, 168 (1955).
4709. Thole, F. B., *J. Chem. Soc.* **1910**, 2596.
4710. Thole, F. B., *J. Chem. Soc.* **1913**, 317.
4711. Thole, F. B., *J. Chem. Soc.* **1914**, 2004.
4712. Thole, F. B., *Z. Phys. Chem.* (Leipzig) **74**, 683 (1910).
4713. Thomas, A. B. and E. G. Rochow, *J. Amer. Chem. Soc.* **79**, 1843 (1957).
4714. Thomas, A. F. and B. Willhalm, *Helv. Chim. Acta* **47**, 475 (1964).
4715. Thomas, B. W. and W. D. Seyfried, *Anal. Chem.* **21**, 1022 (1949).
4716. Thomas, G. A. and J. E. Hawkins, *J. Amer. Chem. Soc.* **76**, 4856 (1954).
4717. Thomas, G. L. and S. Young, *J. Chem. Soc.* **1895**, 1071.
4718. Thomas, J. R. and W. D. Gwinn, *J. Amer. Chem. Soc.* **71**, 2785 (1949).
4719. Thomas, L. and E. Maxum, *Z. Phys. Chem.* (Leipzig) **143A**, 191 (1929).
4720. Thomas, L. H., *J. Chem. Soc.* **1960**, 4960.
4721. Thomas, L. H., *Nature* **158**, 622 (1946).
4722. Thomas, L. H. and R. Meatyard, *J. Chem. Soc.* **1963**, 1986.
4723. Thomas, R. M. and G. K. Weisse, U.S. Patent 3,020,316, February 6, 1962; *Official Gaz. U.S. Patent Office* **775**, 245 (1962).
4724. Thomé, L. G., *Ber. Deut. Chem. Gesell.* **36**, 582 (1903).
4725. Thompson, C. J., H. J. Coleman, et al., *J. Amer. Chem. Soc.* **76**, 3445 (1954).
4726. Thompson, H. B. and L. A. LaPlanche, *J. Phys. Chem.* **67**, 2230 (1963).
4727. Thompson, H. W., *Endeavor* **4**, 154 (1945); *C. A.* **40**, 2729 (1946).
4728. Thompson, H. W., *J. Amer. Chem. Soc.* **61**, 1396 (1939).
4729. Thompson, H. W., *J. Chem. Phys.* **7**, 441 (1939).
4730. Thompson, H. W., *J. Chem. Phys.* **7**, 453 (1939).
4731. Thompson, H. W. and W. T. Cave, *Trans. Faraday Soc.* **47**, 951 (1951); *C. A.* **46**, 2908 (1952).
4732. Thompson, H. W. and G. P. Harris, *Trans. Faraday Soc.* **38**, 37 (1942); *C. A.* **36**, 5091 (1942).
4733. Thompson, H. W. and J. W. Linnett, *Trans. Faraday Soc.* **32**, 681 (1936); *C. A.* **30**, 4371 (1936).
4734. Thompson, H. W. and C. H. Purkis, *Trans. Faraday Soc.* **32**, 674 (1936); *C. A.* **30**, 4402 (1936).
4735. Thompson, H. W. and R. B. Temple, *J. Chem. Soc.* **1948**, 1432.
4736. Thompson, H. W. and R. B. Temple, *Trans. Faraday Soc.* **41**, 27 (1945); *C. A.* **39**, 2454 (1945).
4737. Thompson, H. W. and P. Torkington, *J. Chem. Soc.* **1944**, 597.
4738. Thompson, H. E. and P. Torkington, *J. Chem. Soc.* **1945**, 640.
4739. Thompson, H. W. and P. Torkington, *Proc. Roy. Soc.* **184A**, 3 (1945); *C. A.* **40**, 276 (1946).
4740. Thompson, H. W. and P. Torkington, *Proc. Roy. Soc.* **184A**, 21 (1945); *C. A.* **40**, 277 (1946).
4741. Thompson, H. W. and P. Torkington, *Trans. Faraday Soc.* **42**, 432 (1946); *C. A.* **40**, 5641 (1946).
4742. Thompson, S. D., D. G. Carroll, et al., *J. Chem. Phys.* **45**, 1367 (1966).
4743. Thomsen, J., *Z. Phys. Chem.* (Leipzig) **52**, 343 (1905).
4744. Thomson, G., *J. Chem. Soc.* **1937**, 1051.
4745. Thomson, G., *J. Chem. Soc.* **1939**, 1118.
4746. Thomson, G. W., *Chem. Rev.* **38**, 1 (1946).
4747. Thorpe, T. E. and J. W. Rodger, *J. Chem. Soc.* **1897**, 360.
4748. Thorpe, T. E. and J. W. Rodger, *Phil. Trans.* **185A**, 397 (1894).

4749. Thosar, B. V., *Z. Phys.* **107**, 780 (1937); *C. A.* **32**, 2027 (1938).
4750. Thosar, B. V. and B. K. Singh, *Proc. Indian Acad. Sci.* **6A**, 105 (1937); *C. A.* **32**, 48 (1938).
4751. Thurber, F. H. and R. C. Thielke, *J. Amer. Chem. Soc.* **53**, 1030 (1931).
4752. Thwing, C. B., *Z. Phys. Chem.* (Leipzig) **14**, 286 (1894).
4753. Tiers, G. V. D., *J. Phys. Chem.* **67**, 929 (1963).
4754. Tiganik, L., *Z. Phys. Chem.* (Leipzig) **13B**, 425 (1931); *C. A.* **25**, 5805 (1931).
4755. Tiganik, L., *Z. Phys. Chem.* (Leipzig) **14B**, 135 (1931).
4756. Tilicheev, M. D. and N. V. Milovidova, *Zh. Prikl. Khim.* **22**, 611 (1949); *C. A.* **43**, 8124 (1949).
4757. Tilicheev, M. D., V. P. Peshkov, et al., *Zh. Obshch. Khim.* **21**, 1229 (1951); *C. A.* **46**, 1953 (1952).
4758. Tilitschejew, M., *J. Russ. Phys. Chem. Soc.* **58**, 447 (1926); *Chem. Zentr.* **1927**, I, 440.
4759. Tilton, L. W. and J. K. Taylor, *J. Res. Nat. Bur. Stand.* **20**, 419 (1938).
4760. Timm, B. and R. Mecke, *Z. Phys.* **97**, 221 (1935); *C. A.* **30**, 383 (1936).
4761. Timmermans, J., *Bull. Soc. Chim. Belges.* **24**, 244 (1910); *C. A.* **4**, 2896 (1910).
4762. Timmermans, J., *Bull. Soc. Chim. Belges.* **25**, 300 (1911); *Chem. Zentr.* **1911**, II, 1015.
4763. Timmermans, J., *Bull. Soc. Chim. Belges.* **27**, 334 (1914); *Chem. Zentr.* **1914**, I, 618.
4764. Timmermans, J., *Bull. Soc. Chim. Belges.* **30**, 62 (1921); *Chem. Zentr.* **1921**, III, 287.
4765. Timmermans, J., *Bull. Soc. Chim. Belges.* **30**, 213 (1921).
4766. Timmermans, J., *Bull. Soc. Chim. Belges.* **31**, 389 (1923); *Chem. Zentr.* **1923**, III, 1137.
4767. Timmermans, J., *Bull. Soc. Chim. Belges.* **36**, 502 (1927); *C. A.* **22**, 56 (1928).
4768. Timmermans, J., *Bull. Soc. Chim. Belges.* **43**, 626 (1934); *C. A.* **29**, 1699 (1935).
4769. Timmermans, J., *Bull. Soc. Chim. Belges.* **44**, 17 (1935); *C. A.* **29**, 2433 (1935).
4770. Timmermans, J., *Bull. Soc. Chim. Belges.* **61**, 393 (1952); *C. A.* **48**, 540 (1954).
4771. Timmermans, J., *Comm. Phys. Lab. Univ. Leiden*, Suppl. No. **64**, 3 (1929); *C. A.* **23**, 4390 (1929).
4772. Timmermans, J., *Physio-Chemical Constants of Pure Organic Compounds*, Elsevier, New York, 1950.
4773. Timmermans, J., *Physico-Chemical Constants of Pure Organic Compounds*, Vol. 2, Elsevier, New York, 1965.
4774. Timmermans, J., Previous publication or unpublished "Annual Tables of Physical Constants," Sec. 505 (c), 1941.
4775. Timmermans, J. and Y. Delcourt, *J. Chim. Phys.* **31**, 85 (1934); *Chem. Zentr.* **1934**, II, 1277.
4776. Timmermans, J. and L. Gillo, *Rocz. Chem.* **18**, 812 (1938); *C. A.* **33**, 4582 (1939).
4777. Timmermans, J. and Mme. Hennaut-Roland, *An. Real Soc. Espan. Fis. Quim.* **27**, 460 (1929); *Chem. Zentr.* **1930**, I, 1612.
4778. Timmermans, J. and Mme. Hennaut-Roland, *J. Chim. Phys.* **27**, 401 (1930); *C. A.* **25**, 2038 (1931); *Chem. Zentr.* **1931**, I, 236.
4779. Timmermans, J. and Mme. Hennaut-Roland, *J. Chim. Phys.* **29**, 529 (1932); *Chem. Zentr.* **1933**, I, 2226.
4780. Timmermans, J. and Mme. Hennaut-Roland, *J. Chim. Phys.* **32**, 501, 589 (1935); *C. A.* **30**, 2072 (1936).
4781. Timmermans, J. and Mme. Hennaut-Roland, *J. Chim. Phys.* **34**, 693 (1937).
4782. Timmermans, J. and Mme. Hennaut-Roland, *J. Chim. Phys.* **56**, 984 (1959); from [4773].
4783. Timmermans, J. and F. Martin, *J. Chim. Phys.* **23**, 747 (1926); *Chem. Zentr.* **1927**, I, 836.

4784. Timmermans, J. and F. Martin, *J. Chim. Phys.* **25**, 411 (1928); *C. A.* **22**, 4024 (1928).
4785. Timmermans, J. and R. E. Oesper, *Chemical Species*, Chemical Publishing Company, New York, 1940.
4786. Timmermans, J., A. M. Piette, et al., *Bull. Soc. Chim. Belges* **64**, 5 (1955); *C. A.* **49**, 13,711 (1955).
4787. Timmermans, J. and H. Van der Horst, et al., *Arch. Neerland.* **6**, IIIA, 180 (1923).
4788. Timmermans, J. and H. Van der Horst, et al., *Compt. Rend.* **174**, 365 (1922).
4789. Timofejew, G., *Isv. Kieff Polytechn. Inst.* **1905**, 1.
4790. Tintea H. *Bull. Sect. Sci Acad. Roumaine* **22**, 16 (1939); *C.A.* **34**, 5753 (1940).
4791. Tintea, H., *Bull. Soc. Roumaine Phys.* **43**, 9 (1942); *Chem. Zentr.* **1943**, I, 2188; *C. A* **38**, 4512 (1944).
4792. Tissier, *Ann. Chim. Phys.* (6) **29**, 340; Beil I, 406.
4793. Titani, T., *Bull. Inst. Res. Jap.* **1927**, 671; *Bull. Chem. Soc. Jap.* **2**, 95, 161, 196, 225 (1927).
4794. Tjebbes, J., *Acta Chem. Scand.* **16**, 953 (1962); *C. A.* **57**, 9298 (1962).
4795. Tjebbes, J., *Pure Appl. Chem.* **2**, 129 (1961).
4796. Tkachev, P. G., *Gig. Sanit.* **28** (4), 3 (1963); *C. A.* **59**, 4471 (1963).
4797. Tobin, M. C., *J. Phys. Chem.* **61**, 1392 (1957).
4798. Todd, S. S., G. D. Oliver, et al., *J. Amer. Chem. Soc.* **69**, 1519 (1947).
4799. Torkelson, T. R., M. A. Wolf, et al., *Amer. Ind. Hyg. Assoc. J.* **20**, 224 (1959); *C. A.* **54**, 6967 (1960).
4800. Torkington, P., *Trans. Faraday Soc.* **45**, 445 (1949); *C. A.* **43**, 7342 (1949).
4801. Tokuhiro, T., *J. Chem. Phys.* **41**, 438 (1964).
4802. Tolkmith, H., *J. Amer. Chem. Soc.* **75**, 5273 (1953).
4803. Tollens, B., *Ber. Deut. Chem. Gesell.* **5**, 68 (1872).
4804. Tomecko, J. W. and W. H. Hatcher, *Trans. Roy. Soc. Can.* **45**, 39 (1951); *C. A.* **46**, 9362 (1952).
4805. Tong, L. K. J. and W. O. Kenyon, *J. Amer. Chem. Soc.* **71**, 1925 (1949).
4806. Tongberg, C. O. and M. R. Fenske, *Ind. Eng. Chem.* **24**, 814 (1932).
4807. Tongberg, C. O. and F. Johnston, *Ind. Eng. Chem.* **25**, 733 (1933).
4808. Tonomura, T., *Sci. Repts. Tohoker Imp. Univ.* 1st series **22**, 104 (1933).
4809. Tooke, J. W., U.S. Patent 2,368,050, Jan. 23, 1945; *C. A.* **39**, 4091 (1945).
4810. Toops, E. E., *J. Phys. Chem.* **60**, 304 (1956).
4811. Toops, E. E., Previously unpublished data.
4812. Toriyama, K. and T. Shimanouchi, *Proc. Intern. Symp. Mol. Struct. Spectry. Tokyo* **1962** (A103); *C. A.* **61**, 1397 (1964).
4813. Tornoe, H., *Ber. Deut. Chem. Gesell.* **24**, 2670 (1891).
4814. Toropov, A. P. and A. M. Yakubov, *Zh. Fiz. Khim.* **30**, 1702 (1956); *C. A.* **51**, 7193 (1957).
4815. Tourky, A. R., H. A. Rizk, et al., *Z. Phys. Chem.* (Frankfurt) **31**, 161 (1962); *C. A.* **57**, 201 (1962).
4816. Toussaint, W. J.. U.S. Patent 2,060,267, Nov. 10, 1936; *C. A.* **31**, 420 (1937).
4817. Trabert, E. and K. Schaum, *Z. Wiss. Phot.* **35**, 153 (1936); *C. A.* **31**, 319 (1937).
4818. Trachuk, S. V., *Khim. Prom.* **1963** (2), 14; *C. A.* **59**, 8374 (1963).
4819. Tranchant, J., *Bull. Soc. Chim. Fr.* **1968**, 2216.
4820. Transue, L. F., E. R. Washburn, et al., *J. Amer. Chem. Soc.* **64**, 274 (1942).
4821. Traube, J., *Ber. Deut. Chem. Gesell.* **29**, 1715 (1896).
4822. Traynard, P., *Compt. Rend.* **223**, 991 (1946); *C. A.* **41**, 5021 (1947).
4823. Traynham, J. G. and G. A. Knesel, *J. Amer. Chem. Soc.* **87**, 4220 (1965); see also [951].

4824. Triebs, W. and H. Röhnert, *Chem. Ber.* **84,** 433 (1951).

4825. Treon, J. F., *Ind. Hyg. Toxicol.* **25,** 199 (1943).

4826. Treon, J. F., W. E. Crutchfield, et al., *J. Ind. Hyg. Toxicol.* **25,** 323 (1943).

4827. Treszczanowicz, E. and S. Bakowski, *Przem. Chem.* **3,** 429 (1947); *C. A.* **42,** 2920 (1948).

4828. Trevoy, D. J. and W. A. Torpey, *Anal. Chem.* **26,** 492 (1954).

4829. Trew, V. C. G. and G. M. C. Watkins, *Trans. Faraday Soc.* **29,** 1310 (1933); *C. A.* **28,** 3954 (1934).

4830. Trickey, J. P., *Ind. Eng. Chem.* **19,** 643 (1927).

4831. Trieschmann, H. G., *Z. Phys. Chem.* (Leipzig) **29B,** 328 (1935).

4832. Trimble, F., *Ind. Eng. Chem.* **33,** 660 (1941).

4832a. Trimble, H. M. and W. Potts, *Ind. Eng. Chem.* **27,** 66 (1935).

4833. Trinh, Ng.-Q., *Compt. Rend.* **227,** 394 (1948); Beil **EIII1,** 2256.

4834. Tromp, S. T. J., *Rec. Trav. Chim. Pays-Bas* **41,** 278 (1922); *Chem. Zentr.* **1922, III,** 755.

4835. Troshina, M. M., *Toksikol. Novykh Prom. Khim. Veshchestv.* No. **6,** 45 (1964); *C. A.* **63,** 17,017 (1965).

4836. Trotter, I. F. and H. W. Thompson, *J. Chem. Soc.* **1946,** 481.

4837. Trunel, P., *Ann. Chim.* **12,** 93 (1939); *C. A.* **33,** 8066 (1939).

4838. Trusell, F. and H. Diehl, *Anal. Chem.* **35,** 674 (1963).

4839. Tschamler, H., *Monatsh. Chem.* **79,** 162 (1948); *C. A.* **42,** 4609 (1948); Beil **EIII1,** 1349.

4840. Tschamler, H. and H. Voetter, *Monatsh. Chem.* **83,** 302 (1952); *C. A.* **46,** 6935 (1952).

4841. Tschamler, H., F. Wettig, et al., *Monatsh. Chem.* **80,** 572 (1949); *C. A.* **44,** 421 (1950); Beil **EIII1,** 1350.

4842. Tschelinzew, W. and B. Pawlow, *J. Russ. Phys. Chem. Soc.* **45,** 298 (1913).

4843. Tschelinzew, W. and B. Tronow, *J. Russ. Phys. Chem. Gesell.* **46,** 1876; *Chem. Zentr.* **1915, II,** 470; Beil **EI20,** 36.

4844. Tschugaeff, L., *Ber. Deut. Chem. Gesell.* **32,** 3332 (1899).

4845. Tschugaeff, L., *Ber. Deut. Chem. Gesell.* **33,** 735, 3118 (1900; **34,** 2276 (1901); **37,** 1481 (1904); **42,** 4631 (1909).

4846. Tseng, Yu-M. and A. R. Thompson, *J. Chem. Eng. Data* **7,** 483 (1962).

4847. Tseng, Yu-M. and A. R. Thompson, *J. Chem. Eng. Data* **9,** 264 (1964).

4848. Tsirlin, Yu. A. and G. S. Muromtseva, *Sb. Tr., Gos. Nauch.-Issled. Inst. Gidrolizn. Sul'fitno-Spirt. Prom.* **9,** 205 (1961); *C. A.* **60,** 11,965 (1964).

4849. Tsubomura, H., K. Kimura, et al., *Bull. Chem. Soc. Jap.* **37,** 417 (1964); *C. A.* **61,** 184 (1964).

4850. Tsuchiya, M., S. Matsuhira, et al., *Bunseki Kagaku* **14,** 465 (1965); *C. A.* **63,** 5499 (1965).

4851. Tsuda, K., R. Hayatsu, et al., *Chem. Ind.* (London) **1959,** 1411.

4852. Tsuzuki, Y., K. Tanabe, et al., *Bull. Chem. Soc. Jap.* **39,** 1389 (1966); *C. A.* **65,** 13,508 (1966).

4853. Tuck, D. G., *Trans. Faraday Soc.* **57,** 1297 (1961).

4854. Tucker, S. W. and S. Walker, *J. Chem. Phys.* **45,** 1302 (1966).

4855. Tucker, W. C. and J. E. Hawkins, *Ind. Eng. Chem.* **46,** 2387 (1954).

4856. Tuot, M. and J. Lecomte, *Bull. Soc. Chim. Fr.* **10,** 542 (1943); *C. A.* **38,** 5146 (1944).

4857. Turkevich, A. and C. P. Smyth, *J. Amer. Chem. Soc.* **62,** 2468 (1940).

4858. Turkevich, J. and P. C. Stevenson, *J. Chem. Phys.* **11,** 328 (1943).

4859. Turnbull, D. and S. H. Maron, *J. Amer. Chem. Soc.* **65,** 212 (1943).

4860. Turner, D. W., *J. Chem. Soc.,* **1959,** 30.

4861. Turner, D. W., *J. Chem. Soc.* **1962**, 847.
4862. Turner, W. E. S., *J. Chem. Soc.*, **1910**, 1184.
4863. Turner, W. E. S. and E. W. Merry, *J. Chem. Soc.* **1910**, 2069.
4864. Turner, W. E. S. and C. T. Pollard, *J. Chem. Soc.* **1914**, 1751.
4865. Turova-Pollak, M. B. and M. A. Maslova, *Zh. Obshch. Khim.* **26**, 2185 (1956); *C. A.* **51**, 4972 (1957).
4866. Tyrer, D., *J. Chem. Soc.* **1911**, 1633.
4867. Tyrer, D., *J. Chem. Soc.* **1912**, 81.

U

4868. Udovenko, V. V. and R. P. Aïrapetova, *J. Gen. Chem.* (USSR) **17**, 665 (1947); *C. A.* **42**, 3650 (1948).
4869. Udovenko, V. V. and L. P. Aleksanorova, *Zh. Fiz. Khim.* **34**, 1366 (1960); *Anal. Chem.* **36**, 56R (1964); ref. 70F.
4870. Udovenko, V. V. and Ts. B. Frid, *Zh. Fiz. Khim.* **22**, 1126 (1948); *C. A.* **43**, 476 (1949).
4871. Udránsky, L. v., *Z. Phys. Chem.* (Leipzig) **12**, 385 (1888); **11**, 537; **12**, 33.
4872. Udránsky, L. v., *Z. Phys. Chem.* (Leipzig) **13**, 251 (1889).
4873. Uenaka, F. and T. Shono, *Kagaku To Kôgyô* (Osaka) **35**, 175 (1961); *C. A.* **55**, 25,223 (1961).
4874. Ueno, S. and S. Komori, *J. Soc. Chem. Ind. Jap.* **49**, 161 (1946); *C. A.* **42**, 6738 (1948).
4875. *Ullmanns Encyklopädie der technischen Chemie*, W. Foerst, ed., München-Berlin, Urban & Schwarzenberger.
4876. Ul'yanova, O. D. and V. M. Tatevskii, *Vestn. Moskov Univ.* **6**, No. 5, *Ser. Fiz.-Mat.*, *Estest. Nauk*, No. 3, 87 (1951); *C. A.* **46**, 343 (1952).
4877. Underwood, H. W. and G. C. Toone, *J. Amer. Chem. Soc.* **52**, 391 (1930).
4878. Ungnade, H. E., *J. Amer. Chem. Soc.* **75**, 432 (1953).
4879. Ungnade, H. E., V. Kerr, et al., *Science* **113**, 601 (1951).
4880. Ungnade, H. E. and D. V. Nightingale, *J. Amer. Chem. Soc.* **66**, 1218 (1944).
4881. Union Carbide Chemicals Co., "Acids and Anhydrides," New York, 1960.
4882. Union Carbide Chemicals Co., "Acrylonitrile," New York, 1960.
4883. Union Carbide Chemicals Co., "Alcohols," New York, 1961.
4884. Union Carbide Chemicals Co., "Aldehydes," New York, 1960.
4885. Union Carbide Chemicals Co., "Alkanolamines and Morpholines," New York, 1960.
4886. Union Carbide Chemicals Co., "Alkylene Oxides," New York, 1961.
4887. Union Carbide Chemicals Co., "Esters," New York, 1962.
4888. Union Carbide Chemicals Co., "Ethers and Oxides," New York, 1957.
4889. Union Carbide Chemicals Co., "Flexol Plasticizer DOP," Technical Information, New York, 1957.
4890. Union Carbide Chemicals Co., "Glycols," New York, 1958.
4891. Union Carbide Chemicals Co., "Organic Chlorine Compounds," New York, 1960.
4892. Union Carbide Corp., "Isopropanol," Product Bulletin, New York, 1965.
4893. Union Carbide Corp., Chemicals Division, "Physical Properties Synthetic Organic Chemicals," 1965 ed., New York.
4894. Unni, A. K. R., L. Elias, et al., *J. Phys. Chem.* **67**, 1216 (1963).
4895. Upadhya, K. N., *J. Sci. Res. Banaras Hindu Univ.* **14**, 173 (1963–1964); *C. A.* **61**, 11,485 (1964).
4896. Urbański, T., *Tetrahedron* **6**, 1 (1959).
4897. Urazovskii, S. S. and P. A. Chernyavskii, *Trudy Khar'kov. Politekh. Inst. im. V. I. Lenina* **26**, *Ser. Khim.-Tekhnol.* No. 6, 29 (1959); *C. A.* **56**, 10,930 (1962).

4898. Urazovskiǐ, S. S. and I. I. Ezhik, *Zh. Fiz. Khim.* **36**, 951 (1962); *C. A.* **58**, 5123 (1963).
4899. Urbancová, L., *Chem. Zvesti* **13**, 224 (1959); *C. A.* **53**, 19,500 (1959).
4900. Urone, P., J. E. Smith, et al., *Anal. Chem.* **34**, 476 (1962).
4901. Ushakov, S. and J. Feinstein, *Ind. Eng. Chem.* **26**, 561 (1934).
4902. Usines de Melle, British Patent 586,470, Mar. 20, 1947; *C. A.* **41**, 6894 (1947).
4903. Utermark, W. and W. Schicke, *Melting Point Tables of Organic Compounds*, 2d revised and suppl. ed., Interscience-Wiley, New York, 1963.

V

4904. Valette, A., *Ann. Chim.* **3**, 644 (1948); *C. A.* **43**, 2577 (1949).
4905. Valyashko, N. A. and Y. S. Rozum, *J. Gen. Chem.* (USSR) **16**, 593 (1946); *C. A.* **41**, 1152 (1947).
4906. Van Arkel, A. E., P. Meerburg, et al., *Rec. Trav. Chim. Pays-Bas* **61**, 767 (1942); *C. A.* **38**, 3174 (1944).
4907. Van Arkel, A. E. and J. L. Snoek, *Z. Phys. Chem.* (Leipzig) **18B**, 159 (1932).
4908. Van Arkel, A. E. and S. E. Vles, *Rec. Trav. Chim. Pays-Bas* **55**, 407 (1936); *Chem. Zentr.* **1936, II**, 1329; Beil **EIII1**, 154.
4909. van den Hende, A., *Bull. Soc. Chim. Belges.* **54**, 88 (1945); *C. A.* **40**, 6271 (1946).
4910. van den Hende, A., *Bull. Soc. Chim. Belges.* **56**, 328 (1947); *C. A.* **42**, 6667 (1948).
4911. van der Kelen, G. P., *Bull. Soc. Chim. Belges* **71**, 421 (1962); *C. A.* **58**, 2934 (1963).
4912. van der Vlies, C., *Rec. Trav. Chim. Pays-Bas* **84**, 1289 (1965); *C. A.* **64**, 4474 (1966).
4913. van de Vloed, A., *Bull. Soc. Chim. Belges.* **48**, 229 (1939); *Chem. Zentr.* **1939, II**, 3973
4914. van Os, D., *Chem. Weekblad* **22**, 18 (1925); *Chem. Zentr.* **1925, I**, 1076.
4915. van Pelt, A. J. and J. P. Wibaut, *Rec. Trav. Chim. Pays-Bas* **60**, 55 (1941); *C. A.* **35**, 5089 (1941).
4916. Van Reis, M. A., *Ann. Phys.* (Leipzig) **13**, 447 (1881).
4917. van Risseghem, H., *Bull. Soc. Chim. Belges* **35**, 328 (1926); *Chem. Zentr.* **1927, I**, 53.
4918. Van Tussenbroeck, M. J., Thesis, Delft, 1929; "Annual Tables of Physical Constants," Sec. 921 (c), 1941.
4919. Van Wazer, J. R., C. F. Callis, *J. Amer. Chem. Soc.* **78**, 5715 (1956).
4920. Van Winkle, W., *J. Pharmacol.* **72**, 227 (1941); *C. A.* **35**, 5989 (1941).
4921. Varshni, Y. P., *Z. Phys. Chem.* (Leipzig) **203**, 247 (1954).
4922. Varshni, Y. P. and S. S. Mitra, *Z. Phys. Chem.* (Leipzig) **203**, 380 (1954).
4923. Vaughan, W. C., *Phil. Mag.* **27**, 669 (1939); *C. A.* **33**, 7162 (1939).
4924. Vaughn, J. W. and C. F. Hawkins, *J. Chem. Eng. Data* **9**, 140 (1964).
4925. Vaughn, W. E., S. B. W. Roeder, et al., *J. Chem. Phys.* **39**, 701 (1963).
4926. Veatch, F., J. D. Idol, et al., *Hydrocarbon Process. Petrol. Refiner* **43**, No. 4, 177 (1964).
4927. Veibel, S., F. Lundqvist, et al., *Bull. Soc. Chim. Pays-Bas* **6**, 990 (1939); *C. A.* **33**, 7273 (1939).
4928. Veksler, M. A., *J. Exptl. Theoret. Phys.* (USSR) **9**, 616 (1939); *C. A.* **33**, 9072 (1939).
4929. Veksler, R. I., *Zh. Anal. Khim.* **4**, 14 (1949); *C. A.* **44**, 484 (1950).
4930. Velasco-Durantez, M., *An. Real Soc. Espan. Fis. Quim.* **25**, 252 (1927); *Chem. Zentr.* **1927, II**, 2649.
4931. Velsicol Chemical Corp., "Technical Data Bulletin," Sales Service Bulletin No. 8163, Chattanooga, 1965.
4932. Vencov, St., *Bull. Sect. Sci. Acad. Roumaine* **26**, 89 (1943); *C. A.* **39**, 2696 (1945).
4933. Veneraki, I. E. and N. K. Bolotin, *Gaz. Prom.* **9** (3), 30 (1964); *C. A.* **60**, 1545T (1964).
4934. Venkareswarlu, K. and G. Thyagarajan, *Z. Phys.* **154**, 70 (1959); *C. A.* **53**, 5871 (1959).

4935. Venkatesetty, H. V. and G. H. Brown, *J. Phys. Chem.* **66**, 2075 (1962).
4936. Venkateswaran, C. S. and N. S. Pandya, *Proc. Indian Acad. Sci.* **15A**, 390 (1942); *C. A.* **36**, 6904 (1942).
4937. Venkateswaran, S., *Nature* **122**, 506 (1928); *Chem. Zentr.* **1928, II**, 2705.
4938. Venkateswaran, S., *Phil. Mag.* **7**, 597 (1929); *C. A.* **23**, 5108 (1929).
4939. Venkateswaran, S. and S. Bhagavantam, *Indian J. Phys.* **7**, 585 (1933); *C. A.* **27**, 4481 (1933).
4940. Venkateswaran, V. and S. Bhagavantum, *Proc. Roy. Soc.* **128A**, 252 (1930); *C. A.* **24**, 5627 (1930).
4941. Venkateswarlu, K., *Curr. Sci.* **16**, 21 (1947); *C. A.* **41**, 3697 (1947).
4942. Venkateswarlu, K., *Proc. Indian Acad. Sci.* **19A**, 111 (1944); *C. A.* **38**, 4867 (1944).
4943. Venkateswarlu, K. and S. Mariam, *Z. Phys.* **168**, 195 (1962); *C. A.* **57**, 4202 (1962).
4944. Venkateswarlu, K., M. G. Pillai, et al., *Proc. Indian Acad. Sci.* **53A**, 195 (1961); *C. A.* **55**, 24,232 (1961).
4945. Vereinigte Glanzstoff-Fabriken Akt.-Ges., British Patent 803,375, Oct. 22, 1958; *C. A.* **53**, 4822 (1959).
4946. Vergnoux, A. M. and R. Dadillon, *Compt. Rend.* **213**, 166 (1941); *C. A.* **36**, 5705 (1942).
4947. Verhoek, F. H., *J. Amer. Chem. Soc.* **58**, 2577 (1936).
4948. Verkade, P. E. and J. Coops, *Rec. Trav. Chim. Pays-Bas* **46**, 903 (1927); *C. A.* **22**, 571 (1928).
4949. Vermillion, G. and M. A. Hill, *J. Amer. Chem. Soc.* **67**, 2209 (1945).
4950. Vermillion, H. E., B. Werbel, et al., *J. Amer. Chem. Soc.* **63**, 1346 (1941).
4951. Vernon, A. A. and E. V. Kring, *J. Amer. Chem. Soc.* **71**, 1888 (1949).
4952. Vespignani, G. B., *Gazz. Chim. Ital.* **33**, 73 (1903); *J. Chem. Soc.* **1903Ai**, 545.
4953. Vidulich, G. A., D. F. Evans, et al., *J. Phys. Chem.* **71**, 656 (1967).
4954. Vilallonga, F., *Pubs. Centro Invest. Tisiol.* (Buenos Aires) **10**, 59 (1946); *C. A.* **41**, 5020 (1947).
4955. Vinokurov, D. M., *Izv. Vyssh. Ucheb. Zaved., Lesnaĭ Zhur.* **2**, No. 2, 155 (1959); *C. A.* **54**, 266 (1960).
4956. Vinokurov, D. M., *Nauch. Dokl. Vyssheĭ Shkoly. Lessinzhener. Delo* **1958**, No. 4, 193; *C. A.* **53**, 21,628 (1959).
4957. Vinson, C. G. and I. I. Martin, *I. Chem. Eng. Data* **8**, 74 (1963).
4958. Virginia Chemicals, Inc., *Virginia Amines*, West Norfolk, Virginia 1967.
4959. Virin, L. I., Ya. A. Safin, et al., *Zh. Fiz. Khim.* **39**, 2824 (1965); *C. A.* **64**, 9054 (1966).
4960. Virobyants, R. A., M. A. Nechaeva, et al., *Khim. Sera-organ. Soedin., Soderzhashch. Neft. Nefteprod., Akad. Nauk SSSR, Bashkirsk. Filial* **6**, 193 (1964); *C. A.* **61**, 5592 (1964).
4961. Vistron Corp., "Methacrylonitrile," Technical Bulletin, Cleveland, 1967.
4962. Viswanath, D. S. and N. R. Kuloor, *J. Chem. Eng. Data* **11**, 69 (1966).
4963. Viswanath, D. S. and N. R. Kuloor, *J. Chem. Eng. Data* **11**, 544 (1966).
4964. Vogel, A. I., *J. Chem. Soc.* **1934**, 333.
4965. Vogel, A. I., *J. Chem. Soc.* **1938**, 1323.
4966. Vogel, A. I., *J. Chem. Soc.* **1943**, 636.
4967. Vogel, A. I., *J. Chem. Soc.* **1946**, 133.
4968. Vogel, A. I., *J. Chem. Soc.* **1948**, 607.
4969. Vogel, A. I., *J. Chem. Soc.* **1948**, 610.
4970. Vogel, A. I., *J. Chem. Soc.* **1948**, 616.
4971. Vogel, A. I., *J. Chem. Soc.* **1948**, 624.
4972. Vogel, A. I., *J. Chem. Soc.* **1948**, 644.
4973. Vogel, A. I., *J. Chem. Soc.* **1948**, 654.

4974. Vogel, A. I., *J. Chem. Soc.* **1948,** 1809.
4975. Vogel, A. I., *J. Chem. Soc.* **1948,** 1814.
4976. Vogel, A. I., *J. Chem. Soc.* **1948,** 1820.
4977. Vogel, A. I., *J. Chem. Soc.* **1948,** 1825.
4978. Vogel, A. I., *J. Chem. Soc.* **1948,** 1833.
4979. Vogel, A. I. and D. M. Cowan, *J. Chem. Soc.* **1943,** 16.
4980. Vogel, A. I., W. T. Cresswell, et al., *J. Chem. Soc.* **1952,** 514.
4981. Vogel-Högler, R., *Acta Phys. Austriaca* **1,** 311 (1948); *C. A.* **42,** 6663 (1948).
4982. Vohra, S. P. and K. A. Kobe, *J. Chem. Eng. Data* **4,** 329 (1959).
4983. Volarovich, M. P. and N. N. Stepanenko, *J. Exptl. Theoret. Phys.* (USSR) **10,** 817 (1940); *C. A.* **35,** 5365 (1941).
4984. Volarovich, M. and N. Stepanenko, *J. Exptl. Theoret. Phys.* (USSR) **14,** 313 (1944); *C. A.* **39,** 1338 (1944).
4985. Vold, R. D., *J. Amer. Chem. Soc.* **59,** 1515 (1937).
4986. Vol'kenshtein, M. V., *Acta Physicochim. URSS* **16,** 120 (1942); *C. A.* **37,** 2267 (1943).
4987. Vol'kenshtein, M. V. and M. A. Eliashevich, *Compt. Rend. Acad. Sci. URSS* **43,** 51 (1944); *C. A.* **39,** 2030 (1945).
4988. Volmer, M. and M. Marder, *Z. Phys. Chem.* (Leipzig) **154A,** 971 (1931).
4989. Volyak, L. D., *Zh. Fiz. Khim.* **30** 2244 (1956); *C. A.* **51,** 9284 (1957).
4989a. Vondráček, R. and J. Dostál, *Chem. Listy* **31,** 15 (1937); *C. A.* **31,** 3675 (1937).
4990. von Euler, H. and A. Ölander, *Z. Phys. Chem.* (Leipzig) **131,** 107 (1927).
4991. von Mikusch, J. D., *Farbe Lack* **55,** 241 (1949); *C. A.* **43,** 8700 (1949).
4992. Vorländer, D., *Ann. Chim.* (*Paris*) **280,** 167 (1894); *J. Chem. Soc.* **1895A,** 17.
4993. Vorländer, D., *Ann. Chem., Justus Liebig* **280,** 167 (1894).
4994. Vorländer, D. and R. Walter, *Z. Phys. Chem.* (Leipzig) **118,** 1 (1925).
4995. Voronkov, M. G., *Zh. Fiz. Khim.* **23,** 1311 (1949); *C. A.* **44,** 3759 (1950).
4996. Voronkov, M. G., *Zh. Fiz. Khim.* **26,** 813 (1952); *C. A.* **46,** 11,125.
4997. Voronkov, M. G., *Zh. Obshcheĭ Khim.* **20,** 2060 (1950); *C. A.* **45,** 5607 (1951); *Chem. Zentr.* **1951,** II, 661.
4998. Voroshin, E. M., *Izv. Akad. Nauk SSSR, Ser. Fiz.* **17,** 717 (1953); *C. A.* **48,** 6826 (1954).
4999. Vorozhtsov, N. N., V. E. Platonov, et al., *Izv. Akad. Nauk SSSR, Ser. Khim.* **1963,** 1524; *C. A.* **59,** 13,846 (1963).
5000. Vriens, G. N. and A. G. Hill, *Ind. Eng. Chem.* **44,** 2732 (1952).
5001. Vuks, M., *Acta Physico-Chim.* (URSS) **20,** 851 (1945); *C. A.* **40,** 4293 (1946).
5002. Vvedenskiĭ, A. A., P. Y. Ivannikov, et al., *Zh. Obshch. Khim.* (J. Gen. Chem) **19,** 1094 (1949); *C. A.* **43,** 8247 (1949).
5003. Vvedenskiĭ, A. A. and D. M. Maĭorov, *Zh. Obshch. Khim.* **27,** 2052 (1957); *C. A.* **52,** 4968 (1958).
5004. Vvedenskiĭ, A. A. and Y. A. Timin, *Novosti Neft. Gaz. Tekhn. Neftepererabolka Neftekhim.* **1961,** No. 11, 31; *C. A.* **58,** 3962 (1963).
5005. Vvedenskiĭ, A. A., M. I. Yakushkin, et al., *Khim. Prom.* **1962,** 11; *C. A.* **58,** 2366 (1963).

W

5006. Wacker-Chemie G.m.b.H., British Patent 842,731, July 27, 1960; *C. A.* **55,** 3432 (1961).
5007. Waddington, G. and D. R. Douslin, *J. Amer. Chem. Soc.* **69,** 2275 (1947).
5008. Waddington, G., J. W. Knowlton, et al., *J. Amer. Chem. Soc.* **71,** 797 (1949).
5009. Waddington, G., J. C. Smith, et al., *J. Phys. Chem.* **66,** 1074 (1962).

5010. Waddington, G., S. S. Todd, *J. Amer. Chem. Soc.* **69**, 22 (1947).
5011. Wade, J. and H. Finnemore, *J. Chem. Soc.* **1909**, 1842.
5012. Wadsö, I., *Acta Chem. Scand.* **20**, 544 (1966).
5013. Wagman, D. D., J. E. Kilpatric, et al., *J. Res. Nat. Bur. Stand.* **34**, 143 (1945).
5014. Wagner, C. D., R. H. Smith, et al., *Anal. Chem.* **19**, 976 (1947).
5015. Wagner, E. C., *J. Chem. Educ.* **7**, 1135 (1930).
5016. Wagner, J., *Z. Phys. Chem.* (Leipzig) **40B**, 36 (1938); *C. A.* **32**, 6549 (1938).
5017. Wagner, J., *Z. Phys. Chem.* (Leipzig) **40B**, 439 (1938); *C. A.* **32**, 8937 (1938).
5018. Wagner, J., *Z. Phys. Chem.* (Leipzig) **45B**, 69 (1939); *C. A.* **34**, 940 (1940).
5019. Waksmundzki, A., *Rocz. Chem.* **18**, 865 (1938); *C. A.* **33**, 6688 (1939).
5020. Walden, P., *Ber. Deut. Chem. Gesell.* **40**, 3214 (1907).
5021. Walden, P., *Bull. St. Petersburg Acad. Sci.* **1911**, 1055; *Nature* **1912**, 387; *Pharm. J.* **88**, 125; *C. A.* **6**, 1394 (1912); *Chem. Zentr.* **1912, I,** 122.
5022. Walden, P., *Bull. Acad. St. Petersburg* **1913**, 559; *Chem. Zentr.* **1913, II,** 331.
5023. Walden, P., *Bull. Acad. St. Petersburg* **1915**, 1485; *Chem. Zentr.* **1925, I,** 1676.
5024. Walden, P., *Z. Anorg. Chem.* (Leipzig) **68**, 307 (1910).
5025. Walden, P., *Z. Phys. Chem.* (Leipzig) **43**, 385 (1903).
5026. Walden, P., *Z. Phys. Chem.* (Leipzig) **46**, 103 (1903).
5027. Walden, P., *Z. Phys. Chem.* (Leipzig) **54**, 129 (1906).
5028. Walden, P., *Z. Phys. Chem.* (Leipzig) **55**, 207 (1906).
5029. Walden, P., *Z. Phys. Chem.* (Leipzig) **55**, 281 (1906).
5030. Walden, P., *Z. Phys. Chem.* (Leipzig) **58**, 479 (1907).
5031. Walden, P., *Z. Phys. Chem.* (Leipzig) **59**, 385 (1907).
5032. Walden, P., *Z. Phys. Chem.* (Leipzig) **65**, 129 (1909).
5033. Walden, P., *Z. Phys. Chem.* (Leipzig) **66**, 385 (1909).
5034. Walden, P., *Z. Phys. Chem.* (Leipzig) **70**, 569 (1910).
5035. Walden, P., *Z. Phys. Chem.* (Leipzig) **73**, 257 (1910).
5036. Walden, P., *Z. Phys. Chem.* (Leipzig) **78**, 257 (1912).
5037. Walden, P., *Z. Phys. Chem.* (Leipzig) **147A**, 1, (1930).
5038. Walden, P., *Z. Phys. Chem.* (Leipzig) **162A**, 1 (1932); *Chem. Zentr.* **1933, I,** 1252.
5039. Walden, P., L. F. Audrieth, et al., *Z. Phys. Chem.* (Leipzig) **160A**, 337 (1932).
5040. Walden, P. and E. J. Birr, *Z. Phys. Chem.* (Leipzig) **144A**, 269 (1929).
5041. Walden, P. and E. J. Birr, *Z. Phys. Chem.* (Leipzig) **153A**, 1 (1931).
5042. Walden, P. and E. J. Birr, *Z. Phys. Chem.* (Leipzig) **163A**, 281 (1933).
5043. Walden, P. and E. J. Birr, *Z. Phys. Chem.* (Leipzig) **165A**, 26 (1933).
5044. Walden, P. and E. J. Birr, *Z. Phys. Chem.* (Leipzig) **168A**, 107 (1934).
5045. Walden, P. and G. Busch, *Z. Phys. Chem.* (Leipzig) **140A**, 89 (1929).
5046. Walden, P. and M. Centnerszwer, *Z. Phys. Chem.* (Leipzig) **55**, 321 (1906).
5047. Walden, P. and H. Gloy, *Z. Phys. Chem.* (Leipzig) **144A**, 395 (1929).
5048. Walden, P. and R. Swinne, *Z. Phys. Chem.* (Leipzig) **79**, 700 (1912).
5049. Walden, P., H. Ulich, et al., *Z. Phys. Chem.* (Leipzig) **114**, 275 (1925).
5050. Walden, P., H. Ulich, et al., *Z. Phys. Chem.* (Leipzig) **116**, 261 (1925).
5051. Walden, P. and O. Werner, *Z. Phys. Chem.* (Leipzig) **111**, 465 (1924).
5052. Walden, P. and O. Werner, *Z. Phys. Chem.* (Leipzig) **124**, 405 (1926).
5053. Walden, P., S. Zastrow, et al., *Ann. Acad. Sci. Fennicae*, **29A**, No. 23 (1929).
5054. Wladislaw, B., A. Giora, et al., *J. Chem. Soc.* **1966B**, 586.
5055. Waldron, J. W., U.S. Patent 2,454,019 (Cl. 260–290), Nov. 16, 1948; *C. A.* **43**, 1810 (1949).
5056. Walisch, W. and H. A. Ruppersberg, *Chem. Ber.* **92**, 2622 (1959).
5057. Walker, J., *Z. Phys. Chem.* (Leipzig) **4**, 319 (1889).

5058. Walker, J. W. and F. M. G. Johnson, *J. Chem. Soc.* **1905**, 1597.
5059. Wall, F. T. and W. F. Claussen, *J. Amer. Chem. Soc.* **61**, 2679 (1939).
5060. Wall, F. T. and W. F. Claussen, *J. Amer. Chem. Soc.* **61**, 2812 (1939).
5061. Wallace, T. A. and W. R. G. Atkins, *J. Chem. Soc.* **1912**, 1179.
5062. Wallace, T. A. and W. R. G. Atkins, *J. Chem. Soc.* **1912**, 1958.
5063. Wallace, W. J. and A. L. Mathews, *J. Chem. Eng. Data* **8**, 496 (1963).
5064. Wallace, W. J. and A. L. Mathews, *J. Chem. Eng. Data* **9**, 267 (1964).
5065. Wallach, O., *Ann. Chem., Justus Liebigs* **275**, 195 (1888).
5066. Wallach, O., *Ann. Chem., Justus Liebig* **359**, 287 (1908); *C. A.* **2**, 1831 (1908).
5067. Wallerstein, R., T. Alba, et al., U.S. Patent 2,366,990, Jan. 9, 1945; *C. A.* **39**, 2378 (1945).
5068. Walls, W. S. and C. P. Smyth, *J. Chem. Phys.* **1**, 337 (1933).
5069. Walsh, A. D., *Trans. Faraday Soc.* **42**, 62 (1946); *C. A.* **40**, 4957 (1946).
5070. Walsh, A. D. and P. A. Warsop, *Trans. Faraday Soc.* **63**, 524 (1967).
5071. Walsh, A. D. and P. A. Warsop, *Trans. Faraday Soc.* **64**, 1418 (1968).
5072. Walsh, A. D. and P. A. Warsop, *Trans. Faraday Soc.* **64**, 1425 (1968).
5073. Walsh, A. D., P. A. Warsop, et al., *Trans. Faraday Soc.* **64**, 1432 (1968).
5074. Walsh, P. N. and N. O. Smith, *Chem. Eng. Data* **6**, 33 (1961).
5075. Walther, R., Arch. *Gewerbepath. Gewerbehyg.* **11**, 326 (1942); *C. A.* **37**, 5492 (1943).
5076. Walton, J. H. and L. L. Withrow, *J. Amer. Chem. Soc.* **45**, 2689 (1923).
5077. Wan, S. W., *J. Phys. Chem.* **45**, 903 (1941).
5078. Wang, T. S., *J. Org. Chem.* **28**, 245 (1963).
5079. Ward, A. L. and S. S. Kurtz, *Ind. Eng. Chem., Anal. Ed.* **10**, 559 (1938).
5080. Ward, H. L., *J. Phys. Chem.* **38**, 761 (1934).
5081. Ware, D. R. and B. A. Dunell, *Trans. Faraday Soc.* **61**, 834 (1965).
5082. Waring, W., *Chem. Rev.* **51**, 171 (1952).
5083. Warshowsky, B. and P. J. Elving, *Ind. Eng. Chem., Anal. Ed.*, **18**, 253 (1946).
5084. Washburn, E. R. and C. H. Shildneck, *J. Amer. Chem. Soc.* **55**, 2354 (1933).
5085. Washburn, E. W. and J. W. Read, *J. Amer. Chem. Soc.* **41**, 729 (1919).
5086. Washburn, R. M., E. Levens, et al., *Metallic-Organic Compounds*, pp. 129 ff., Advances in Chemistry series, Washington, American Chemical Society, 1959.
5087. Washburn, W. H., *Anal. Chem.* **30**, 156 (1958).
5088. Wassermann, A., *Z. Phys. Chem.* (Leipzig) **146**, 418 (1930).
5089. Waterman, H. I. and S. H. Bertram, *Rec. Trav. Chim. Pays-Bas* **46**, 699 (1927); *C. A.* **22**, 218 (1928).
5090. Waterman, H. I. and W. J. C. de Kok, *Rec. Trav. Chim. Pays-Bas* **52**, 251 (1933); *C. A.* **27**, 2931 (1933).
5091. Waterman, H. I. and W. J. C. de Kok, *Rec. Trav. Chim. Pays-Bas* **52**, 298 (1933); *C. A.* **27**, 2932 (1933).
5092. Waterman, H. I. and W. J. C. de Kok, *Rec. Trav. Chim. Pays-Bas* **53**, 725 (1934); *C. A.* **28**, 5397 (1934).
5093. Waterman, H. I., P. vant Spijker, et al., *Rec. Trav. Chim. Pays-Bas* **48**, 1191 (1929); *C. A.* **24**, 824 (1930).
5094. Waterman, H. I. and H. A. Van Westen, *Rec. Trav. Chim. Pays-Bas* **48**, 637 (1929); *C. A.* **23**, 4455 (1929).
5095. Watt, G. W. and J. B. Otto, *J. Amer. Chem. Soc.* **69**, 836 (1947).
5096. Wawzonek, S. and M. E. Runner, *J. Electrochem. Soc.* **99**, 457 (1952).
5097. Way, K. R. and M. E. Russell, *J. Phys. Chem.* **69**, 4420 (1965).
5098. Wayland, B. B. and R. S. Drago, *J. Amer. Chem. Soc.* **87**, 2372 (1965).
5099. Wearn, R. B., W. M. Murray, et al., *Anal. Chem.* **20**, 922 (1938).

5100. Webb, J. L. A. and J. D. Webb, *J. Amer. Chem. Soc.* **71,** 2285 (1949).
5101. Webb, T. J. and C. H. Lindsley, *J. Amer. Chem. Soc.* **56,** 874 (1934).
5102. Webber, H., *Res. Develop. Ind.*, No. 10, **47** (1962); *C. A.* **58,** 3608 (1963).
5103. Weber, L. A. and J. E. Kilpatrick, *J. Chem. Phys.* **36,** 829 (1926).
5104. Wedekind, E. and T. Goost, *Ber. Deut. Chem. Gesell.* **49,** 942 (1916).
5105. Wehle, H., *Z. Anal. Chem.* **169,** 241 (1959); *Chem. Zentr.* **1960,** 1279.
5106. Wehman, T. C., Private communication.
5107. Weidmann, H. and H. K. Zimmerman, *Texas J. Sci.* **11,** 212 (1959); *C. A.* **54,** 1987 (1960).
5108. Weigert, F., *Optische Methoden der Chemie*, Leipzig, Akademische Verlagsgesellschaft, 1927.
5109. Weinberg, D. S. and C. Djerassi, *J. Org. Chem.* **31,** 115 (1966).
5110. Weissberger, A., *Physical Methods of Organic Chemistry*, 3rd ed., Vol. I, Pt. I, Interscience, New York, 1959.
5111. Weissberger, A., *Physical Methods of Organic Chemistry*, 2d ed., Vol. I, Pt. II, Interscience, New York, 1950.
5112. Weissberger, A., *Physical Methods of Organic Chemistry*, 3rd ed., Vol. I, Pt. IV, Interscience, New York, 1960.
5113. Weissberger, A. and K. Fasold, *Z. Phys. Chem.* (Leipzig) **157A,** 65 (1931).
5114. Weissberger, A. and R. Sangewald, *Ber. Deut. Chem. Gesell.* **65,** 701 (1932).
5115. Weissberger, A. and E. Strasser, *J. Prakt. Chem.* **135,** 209 (1933).
5116. Weissenberger, G., R. Henke, et al., *Monatsh. Chem.* **46,** 483 (1925); *C. A.* **20,** 2851 (1926).
5117. Weissenberger, G., F. Schuster, et al., *Monatsh. Chem.* **46,** 281 (1926); *C. A.* **20,** 1785 (1926); *Chem. Zentr.* **1926, I,** 2454.
5118. Weissgerber, R. and O. Kruber, *Ber. Deut. Chem. Gesell.* **53,** 1551 (1920).
5119. Weissler, A., *J. Amer. Chem. Soc.* **70,** 1634 (1948).
5120. Weitz, E., A. Roth, et al., *Ann. Chem.* **425,** 176 (1921).
5121. Weizmann, C., U.S. Patent 2,456,549, Dec. 14, 1948; *C. A.* **43,** 2632 (1949).
5122. Wells, A. J. and E. B. Wilson, *J. Chem. Phys.* **9,** 314 (1941).
5123. Weltner, W. and K. S. Pitzner, *J. Amer. Chem. Soc.* **73,** 2606 (1951).
5124. Weniger, W., *Phys. Rev.* **31,** 388 (1910).
5125. Wentworth, T. O., British Patent 567,288, Feb. 7, 1945; *C. A.* **41,** 1805 (1947).
5126. Werner, A., P. Ferchland, *Z. Anorg. Chem.* **15,** 1 (1897); *Chem. Zentr.* **1897, II,** 461.
5127. Werner, E. A., *Analyst* **58,** 335 (1933).
5128. Werner, M., *Mitt. Ver. Grosskesselbesitzer* **41,** 119 (1956); *C. A.* **50,** 12,606 (1956).
5129. Werner, R. L. and K. G. O'Brien, *Aust. J. Chem.* **8,** 355 (1955); *C. A.* **49,** 15,486 (1955).
5130. Wertyporoch, E. and B. Adamus, *Z. Phys. Chem.* (Leipzig) **168A,** 31 (1934).
5131. Wertporoch, E. and W. Firla, *Z. Phys. Chem.* (Leipzig) **162A,** 398 (1932).
5132. Wheeler, D. H. and R. W. Riemeneschneider, *Oil & Soap* **16,** 207 (1939); *C. A.* **34,** 278 (1940); Ref. [3152], pp. 2088.
5133. Wheeler, O. H. and J. L. Mateos, *Can. J. Chem.* **36,** 1431 (1958).
5134. Wheeler, O. H. and J. L. Mateos, *J. Org. Chem.* **21,** 1110 (1956).
5135. Wheland, G. W. and J. Farr, *J. Amer. Chem. Soc.* **65,** 1433 (1943).
5136. Whetsel, K., W. E. Roberson, et al., *Anal. Chem.* **30,** 1594, 1598 (1958).
5137. Whiffen, D. H. and H. W. Thompson, *J. Chem. Soc.* **1945,** 268.
5138. Whipple, E. B., W. E. Stewart, et al., *J. Chem. Phys.* **34,** 2136 (1961).
5139. Whitby, G. S., *J. Chem. Soc.* **1926,** 1458.
5140. White, A. H. and S. O. Morgan, *J. Chem. Phys.* **5,** 655 (1937).

5141. White, G. F. and H. C. Jones, *Amer. Chem. J.* **44**, 157 (1910).
5142. White, G. F., A. B. Morrison, et al., *J. Amer. Chem. Soc.* **46**, 961 (1924).
5143. White, H. F., *Anal. Chem.* **36**, 1291 (1964).
5144. White, J. D. and F. W. Rose, *J. Res. Nat. Bur. Stand.* **7**, 907 (1931).
5145. White, J. D. and F. W. Rose, *J. Res. Nat. Bur. Stand.* **9**, 711 (1932).
5146. White, J. R., *J. Amer. Chem. Soc.* **72**, 1859 (1950).
5147. White, P. T., D. G. Barnard-Smith, et al., *Ind. Eng. Chem.* **44**, 1430 (1952).
5148. White, T., *J. Chem. Soc.* **1943**, 238.
5149. White, W. P., *J. Phys. Chem.* **24**, 393 (1920).
5150. Whitehead, E. V., R. A. Dean, et al., *J. Amer. Chem. Soc.* **73**, 3632 (1951).
5151. Whitford, E. L., *J. Amer. Chem. Soc.* **47**, 2934 (1925).
5152. Whitman, G. M., U.S. Patent 2,418,441, Apr. 1, 1947; *C. A.* **41**, 4169 (1947).
5153. Whitman, J. L. and D. M. Hurt, *J. Amer. Chem. Soc.* **52**, 4762 (1930).
5154. Whitmore, F. C., *Organic Chemistry*, Van Nostrand, Princeton, New Jersey, 1937.
5155. Whitmore, F. C., *Organic Syntheses*, Vol. 7, Wiley, New York, 1927.
5156. Whitmore, F. C., *Organic Syntheses*, Vol. 12, Wiley, New York, 1932.
5157. Whitmore, F. C. and G. H. Fleming, *J. Amer. Chem. Soc.* **55**, 3803 (1933).
5158. Whitmore, F. C. and J. M. Herndon, *J. Amer. Chem. Soc.* **55**, 3428 (1933).
5159. Whitmore, F. C. and F. A. Karnatz, *J. Amer. Chem. Soc.* **60**, 2533 (1938).
5160. Whitmore, F. C. and J. H. Olewine, *J. Amer. Chem. Soc.* **60**, 2569 (1938).
5161. Whitmore, F. C. and H. S. Rothrock, *J. Amer. Chem. Soc.* **54**, 3431 (1932).
5162. Whitmore, F. C. and C. T. Simpson, *J. Amer. Chem. Soc.* **55**, 3809 (1933).
5163. Whitmore, F. C. and J. D. Surmatis, *J. Amer. Chem. Soc.* **62**, 995 (1940).
5164. Whitmore, F. C., J. S. Whitaker, et al., *J. Amer. Chem. Soc.* **63**, 643 (1941).
5165. Whitmore, W. F. and E. Lieber, *Ind. Eng. Chem., Anal. Ed.* **7**, 127 (1935).
5166. Whitnack, G. C. and R. W. Moshier, *Ind. Eng. Chem., Anal. Ed.* **16**, 496 (1944).
5167. Whittaker, A. G. and S. Siegel, *J. Chem. Phys.* **42**, 3320 (1965).
5168. Wibaut, J. P. and H. Geldof, *Rec. Trav. Chim. Pays-Bas* **65**, 125 (1946); *C. A.* **40**, 3952 (1946).
5169. Wibaut, J. P., H. Hoog, et al., *Rec. Trav. Chim. Pays-Bas* **58**, 329 (1939); *C. A.* **33**, 6231 (1939).
5170. Wibaut, J. P. and S. L. Langedijk, *Rec. Trav. Chim. Pays-Bas* **59**, 1220 (1940); *C. A.* **35**, 5089 (1941).
5171. Wibaut, J. P. and A. J. van Pelt, *Rec. Trav. Chim. Pays-Bas* **57**, 1055 (1938); *C. A.* **33**, 1263 (1939).
5172. Wicker, R. J., *J. Chem. Soc.* **1956**, 2165.
5173. Wicker, R. J., *J. Chem. Soc.* **1957**, 3299.
5174. Widiger, A. H., U.S. Patent 3,087,969, Apr. 30, 1963. *Official Gaz. U.S. Patent Office* **789** (1963).
5175. Widmark, G., *Acta Chem. Scand.* **9**, 925 (1955); *C. A.* **50**, 4503 (1956).
5176. Wiemann, J., O. Convert, et al., *Bull. Soc. Chim. Fr.* **1966**, 1760; *C. A.* **65**, 12,087 (1966).
5177. Wienhaus, H., *Ber. Deut. Chem. Gesell.* **53**, 1656 (1920).
5178. Wiezevich, P. J. and P. K. Frolich, *Ind. Eng. Chem.* **26**, 267 (1934).
5179. Wightmann, E. P. and H. C. Jones, *Amer. Chem. J.* **46**, 56 (1911).
5180. Wikoff, H. L., B. R. Cohen, et al., *Ind. Eng. Chem. Anal. Ed.* **12**, 92 (1940).
5181. Wilcox, W. R., R. Friedenberg, et al., *Chem. Rev.* **64**, 187 (1964).
5182. Wilde, J. H. de, *Z. Anorg. Allg. Chem.* **233**, 411 (1937); *C. A.* **32**, 13 (1938).
5183. Wilkinson, R., *J. Chem. Soc.* **1931**, 3057.
5184. Wille, H. and L. Rappen, German Patent 1,099,546, Feb. 16, 1961; *C. A.* **55**, 27,860 (1961).

5185. Wille, H. and L. Rappen. German Patent 1,124,046, Feb. 22, 1962; *C. A.* **57**, 7174 (1962).
5186. Williams, D., *Phys. Rev.* **54**, 504 (1938).
5187. Williams, D., T. Gatica, et al., *J. Phys. Chem.* **41**, 645 (1937).
5188. Williams, F. T., P. K. W. Flanagan, et al., *J. Org. Chem.* **30**, 2674 (1965).
5189. Williams, J. W., *J. Amer. Chem. Soc.* **52**, 1831 (1930).
5190. Williams, J. W., *Phys. Z.* **29**, 174 (1928); *C. A.* **22**, 1897 (1928).
5191. Williams, J. W., *Phys. Z.* **29**, 204 (1928).
5192. Williams, J. W., *Phys. Z.* **29**, 271 (1928); *C. A.* **22**, 3089 (1928).
5193. Williams, J. W., *Z. Phys. Chem.* (Leipzig) **138A**, 75 (1928).
5194. Williams, J. W. and F. Daniels, *J. Amer. Chem. Soc.* **46**, 903 (1924).
5195. Williams, J. W. and F. Daniels, *J. Amer. Chem. Soc.* **47**, 1490 (1925).
5196. Williams, J. W. and J. M. Fogelberg, *J. Amer. Chem. Soc.* **53**, 2096 (1931).
5197. Williams, J. W. and I. J. Krchma, *J. Amer. Chem. Soc.* **48**, 1888 (1926).
5198. Williams, J. W. and E. F. Ogg, *J. Amer. Chem. Soc.* **50**, 94 (1928).
5199. Williams, R. B., S. H. Hastings, et al., *Anal. Chem.* **24**, 1911 (1952).
5200. Williams, T. and J J. Sudborough, *J. Chem. Soc.* **1912**, 414.
5201. Williamson, I., *Nature* **167**, 316 (1951).
5202. Williamson, K D. and R. H. Harrison, *J. Chem. Phys.* **26**, 1409 (1957).
5203. Willingham, C. B. and F. D. Rossini, *J. Res. Nat. Bur. Stand.* **37**, 15 (1946).
5204. Willingham, C. B., W. J. Taylor, et al., *J. Res. Nat. Bur. Stand.* **35**, 219 (1945).
5205. Willson, J. E., D. E Brown, et al., *Toxicol. Appl. Pharmacol.* **7**, 104 (1965).
5206. Willstätter, R. and V. L. King, *Ber. Deut. Chem. Gesell.* **46**, 527 (1913).
5207. Willstätter, R. and F. Seitz, *Ber. Deut. Chem. Gesell.* **56**, 1388 (1923).
5208. Willstätter, R. and F. Seitz, *Ber. Deut. Chem. Gesell.* **57**, 683 (1924).
5209. Wilmshurst, J. K., *Can. J. Chem.* **36**, 285 (1958).
5210. Wilmshurst, J. K. and H. J. Bernstein, *Can. J. Chem.* **35**, 911 (1957).
5211. Wilson, A. L., *Ind. Eng. Chem.* **27**, 867 (1935).
5212. Wilson, C. L., British Patent 569,625, July 1, 1945; *C. A.* **41**, 6275 (1947).
2513. Wilson, H. N. and A. E. Heron, *J. Soc. Chem. Ind.* **60**, 168 (1941).
5214. Wilson, H. N. and W. C. Hughes, *J. Soc. Chem. Ind.* **58**, 74 (1939).
5215. Wilson, J. M., *Experientia* **16**, 403 (1960); *C. A.* **55**, 7043 (1961).
5216. Winccl, II. and Z. Kecki, *Nukleonika* **10**, 567 (1965); *C. A.* **64**, 17,390 (1966).
5217. Windwer, S. and B. R. Sundheim, *J. Phys. Chem.* **66**, 1254 (1962).
5218. Wing, J. and W. H. Johnston, *J. Amer. Chem. Soc.* **79**, 864 (1957).
5219. Winkler, L. W., *Ber. Deut. Chem. Gesell.* **38**, 3612 (1905).
5220. Winters, G. and J. Shelmerdine, British Patent 777,961, July 3, 1957; *C. A.* **51**, 15,098 (1957).
5221. Wirth, H. E. and P. I. Slick, *J. Phys. Chem.* **66**, 2277 (1962).
5222. Wislicenus, J., *Ann. Chem. Justus Liebig* **92**, 106 (1878); *J. Chem. Soc.* **1878A**, 776.
5223. Wislicenus, J., *Ber. Deut. Chem. Gesell.* **7**, 893 (1874).
5224. Wislicenus, J. and L. Kaufmann, *Ber. Deut. Chem. Gesell.* **28**, 1323 (1895); *J. Prakt. Chem.* **54** (2), 54 (1896).
5225. Wiswall, R. H. and C. P. Smyth, *J. Chem. Phys.* **9**, 356 (1941).
5226. Witanowski, M., T. Urbański, et al., *J. Amer. Chem. Soc.* **86**, 2569 (1964).
5227. Witschonke, C. R., *Anal. Chem.* **26**, 562 (1954).
5228. Wittek, H., *Monatsh. Chem.* **73**, 231 (1941); *C. A.* **36**, 5092 (1942).
5229. Wittek, H., *Z. Phys. Chem.* (Leipzig) **48B**, 1 (1940); *C. A.* **35**, 3172 (1941).
5230. Wittek, H., *Z. Phys. Chem.* (Leipzig) **51B**, 187 (1942); *C. A.* **37**, 4013 (1943).
5231. Wittek, H., *Z. Phys. Chem.* (Leipzig) **52B**, 315, 330 (1942); Beil **EIII6**, 799.
5232. Wladislaw, B., A. Giora, et al., *J. Chem. Soc.* **1966B**, 586.

5233. Wöhler, F., *Ann. Chem.*, *Justus Liebigs* **81**, 376 (1852).
5234. Wojciechowski, M., *J. Res. Nat. Bur. Stand.* **17**, 453 (1936).
5235. Wojciechowski, M., *J. Res. Nat. Bur. Stand.* **17**, 459 (1936).
5236. Wojcieshowski, M., *J. Res. Nat. Bur. Stand.* **17**, 721 (1936).
5237. Wojciechowski, M., *J. Res. Nat. Bur. Stand.* **19**, 347 (1937).
5238. Wojcieshowski, M., *Ind. Chim.* (Warsaw) **5**, 129 (1939).
5239. Wojciechowski, M. and E. R. Smith, *J. Res. Nat. Bur. Stand.* **18**, 499 (1937).
5240. Wojciechowski, M. and E. R. Smith, *Rocz. Chem.* **17**, 118 (1937); *Chem. Zentr.* **1937, II,** 1554.
5241. Wolf, H. C. and H. P. Deutsch, *Naturwissenschaften*, **41**, 425 (1954); *Chem. Zentr.* **1955**, 10,708.
5242. Wolf, K. L., *Phys. Z.* **31**, 227 (1930); *Chem. Zentr.* **1930, I,** 2521.
5243. Wolf, K. L. and W. J. Gross, *Z. Phys. Chem.* (Leipzig) **14B**, 305 (1931); *Chem. Zentr.* **1931, II,** 3581.
5244. Wolf, K. L. and W. Herold, *Z. Phys. Chem.* (Leipzig) **13B**, 201 (1931).
5245. Wolf, K. L., H. Pahlke, et al., *Z. Phys. Chem.* (Leipzig) **28B**, 1 (1935).
5246. Wolf, K. L. and O. Strasser, *Z. Phys. Chem.* (Leipzig) **21B**, 389 (1933).
5247. Wolf, K. L. and H. Weghofer, *Z. Phys. Chem.* (Leipzig) **39B**, 194 (1938); *Chem. Zentr.* **1938, II,** 2722.
5248. Wolfe, W. C., *Anal. Chem.* **34**, 1328 (1962).
5249. Wolff, H. and D. Staschewski, *Z. Elektrochem.* **65**, 840 (1961); *C. A.* **57**, 6758 (1962).
5250. Wollman, S. H., *J. Chem. Phys.* **14**, 123 (1946).
5251. Wood, C. E., J. E. Such, et al., *J. Chem. Soc.* **1923**, 600.
5252. Wood, S. E., *J. Amer. Chem. Soc.* **59**, 1510 (1937).
5253. Wood, S. E. and J. P. Brusie, *J. Amer. Chem. Soc.* **65**, 1891 (1943).
5254. Wood, S. E. and J. A. Gray, *J. Amer. Chem. Soc.* **74**, 3729 (1952).
5255. Woodman, A. L., W. J. Murbach et al., *J. Phys. Chem.* **64**, 658 (1960).
5256. Worden, E. F., *U.S. Atomic Energy Comm.* **UCRL-8508;** *C. A.* **53**, 13,772 (1959).
5257. Woringer, B., *Z. Phys. Chem.* (Leipzig) **34**, 257 (1900).
5258. Woronkow, M. G., *J. Phys. Chem.* (Leningrad) **22,** 975 (1948); *Chem. Zentr.* **1948, I,** E193.
5259. Worsham, J. E. and M. E. Hobbs, *J. Amer. Chem. Soc.* **76**, 206 (1954).
5260. Wotiz, J. H. and F. A. Miller, *J. Amer. Chem. Soc.* **71**, 3441 (1949).
5261. Wotiz, J. H., F. A. Miller, et al., *J. Amer. Chem. Soc.* **72**, 5055 (1950).
5262. Wowzonek, S. and M. E. Runner, *J. Electrochem. Soc.* **99**, 457 (1952).
5263. Wrewsky, M. S. and A. A. Glagoleva, *Z. Phys. Chem.* (Leipzig) **133A**, 370 (1928).
5264. Wrewsky, M. S., K. P. Miscenko, et al., *Z. Phys. Chem.* (Leipzig) **133A**, 362 (1928).
5265. Wright, C. P., M. Murray-Rust, et al., *J. Chem. Soc.* **1931**, 199.
5266. Wright, F. J., *J. Chem. Eng. Data* **6**, 454 (1961).
5267. Wright, O. L., and L. E. Mura, *J. Chem. Educ.* **43**, 150 (1966).
5268. Wright, R. H. and R. S. E. Serenius, *J. Appl. Chem.* (London) **4**, 615 (1954).
5269. Wrobel, J. and K. Galuszko, *Tetrahedron Lett.* **1965**, 4381; *C. A.* **64**, 9550 (1966).
5270. Wu, T. Y., *J. Chem. Phys.* **7**, 965 (1939).
5271. Wulf, O. R., E. J. Jones, et al., *J. Chem. Phys.* **8**, 753 (1940).
5272. Wulff, C. A. and E. F. Westrum, *J. Phys. Chem.* **67**, 2376 (1963).
5273. Wuyfs, H. and R. Billeux, *Bull. Acad. Roy. Belges.* **29**, 55 (1920); *Chem. Zentr.* **1920, I,** 817; *C. A.* **16**, 2111 (1922).
5274. Wuyts, H. and A. Lacourt, *Bull. Soc. Chim. Belges.* **39**, 157 (1930).
5275. Wyman, J., *Phys. Rev.* **35**, 623 (1930).
5276. Wymore, C. E., *Ind. Eng. Chem.*, *Prod. Res. Develop.* **1**, 173 (1962).

5277. Wynn, E. A., *Microchem. J.* **5**, 175 (1961); *Chem. Zentr.* **1962**, 16,789.
5278. Wyss, H. R., R. D. Werder, et al., *Spectrochim. Acta* **20**, 573 (1964); *C. A.* **61**, 5095 (1964).

Y

5279. Yabroff, D. L., *Ind. Eng. Chem.* **32**, 257 (1940).
5280. Yada, S., T. Yamauchi, et al., *Shokubai* (Tokyo) **5** (1), 2 (1963); *C. A.* **59**, 8582 (1963).
5281. Yagudaev, M. R., E. M. Popov, et al., *Izv. Akad. Nauk SSSR, Ser. Khim.* **1964**, 1189 *C. A.* **64**, 18,705 (1966).
5282. Yamaguchi, I., *Bull. Chem. Soc. Jap.* **34**, 353 (1961); *C. A.* **55**, 22,210 (1961).
5283. Yamomoto, K., Y' Hirosawa, et al., *Bull. Chem. Soc. Jap.* **27**, 386 (1954); *C. A.* **50**, 159 (1956).
5284. Yamamoto, M. and S. Ohara, *J. Pharm. Soc. Jap.* **60**, 535 (1940); *C. A.* **35**, 1773 (1941).
5285. Yamashita, M. and T. Matsumura, *J. Chem. Soc. Jap.* **64**, 506 (1943); *C. A.* **41**, 3753 (1947).
5286. Yaroslavskii, N. G., *J. Phys. Chem.* (USSR) **22**, 265 (1948); *C. A.* **42**, 5772 (1948).
5287. Yates, K. and R. Stewart, *Can. J. Chem.* **37**, 664 (1959).
5288. Yates, W. R. and J. L. Ihrig, *J. Amer. Chem. Soc.* **87**, 710 (1965).
5289. Yeh, P.-Y., C.-T. Chen, et al., *Formosan Sci.* **7**, 27 (1953); *C. A.* **49**, 3033 (1955).
5290. Yerger, E. A., Thesis, Northwastern University, 1956.
5291. Yokoto, K. and Y. Ishii, *Kogyo Kagaku Zasshi* **69**, 1083 (1966); *C. A.* **65**, 12,303 (1966).
5292. Yonezawa, T., I. Morishima, et al., *Bull. Chem. Soc. Jap.* **39**, 1398 (1966).
5293. Yonezawa, T., H. Saito, *Bull. Chem. Soc. Jap.* **38**, 1431 (1965).
5294. Yoroslavskii, N. G., and A. N. Terenin, *Dokl. Akad. Nauk SSSR* **66**, 885 (1949); *C. A.* **43**, 7343 (1949).
5295. Yoshida, Zen-ichi, E. Osawa, et al., *J. Phys. Chem.* **68**, 2895 (1964).
5296. Yoshino, T. and H. J. Bernstein, *Can. J. Chem.* **35**, 339 (1957).
5297. Young, S., *J. Chem. Soc.* **1889**, 483.
5298. Young, S., *J. Chem. Soc.* **1897**, 446.
5299. Young, S., *J. Chem. Soc.* **1898**, 675.
5299a. Young, S., *J. Chem. Soc.* **1899**, 172.
5300. Young, S., *J. Chem. Soc.* **1902**, 707.
5301. Young, S., *Sci. Proc. Roy. Dublin Soc.* **12**, 374 (1909–1910); *C. A.* **5**, 406 (1911).
5302. Young, S., *Z. Phys. Chem.* (Leipzig) **29**, 193 (1899).
5303. Young, S., *Z. Phys. Chem.* (Leipzig) **70**, 620 (1910).
5304. Young, S. and E. C. Fortey, *J. Chem. Soc.* **1899**, 873.
5305. Young, S. and E. Fortney, *J. Chem. Soc.* **1903**, 45.
5306. Young, S. and G. L. Thomas, *J. Chem. Soc.* **1893**, 1191.
5307. Young, W., *J. Soc. Chem. Ind.* **52**, 449 (1933).
5308. Young, W. G. and L. J. Andrews, *J. Amer. Chem. Soc.* **66**, 421 (1944).
5309. Young, W. G., W. H. Hartung, et al., *J. Amer. Chem. Soc.* **58**, 100 (1936).
5310. Yuan, H-S. and Tseng, C-Y., *Chlorine and Alkali News* (Formosa) No. **12**, 42 (1953); *C. A.* **49**, 6819 (1955).
5311. Yura, S. and R. Oda, *J. Soc. Chem. Ind. Jap.* **46**, 82 (1943); *C. A.* **42**, 6339 (1948).
5312. Yvernault, T., *Aleagineux* **1**, 189 (1946); *C. A.* **41**, 3697 (1947).

Z

5313. Zabetakis, M. G., *Flammable Characteristics of Combustible Gases and Vapors*, Bulletin 627, Washington, Bureau of Mines, 1965.

5314. Zabetakis, M. G., G. S. Scott, et al., *Ind. Eng. Chem.* **43**, 2120 (1951).
5315. Zagradnik, R., M. Chvapil, et al., *Farmakol. Toksikol.* **25**, 618 (1962); *C. A.* **58**, 11,886 (1963).
5316. Zahn, C. T., *Phys. Rev.* **37**, 1516 (1931); *C. A.* **25**, 5320 (1931).
5317. Zahn, C. T., *Phys. Z.* **33**, 525 (1932); *C. A.* **26**, 5803 (1932).
5318. Zahn, C. T., *Phys. Z.* **34**, 570 (1933); *Chem. Zentr.* **1933**, II, 2647.
5319. Zakhakin, L. I. and I. M. Khorlina, *Dokl. Akad. Nauk SSSR* **116**, 422 (1957); *C. A.* **52**, 8040 (1958).
5320. Zal'kind, Y. S. and I. F. Markov, *J. Appl. Chem.* (USSR) **12**, 437 (1939); *C. A.* **33**, 6809 (1939).
5321. Zaoral, M., *Chem. Listy* **47**, 1872 (1953); *C. A.* **49**, 966 (1955).
5322. Zapletálek, A., *Collect. Czech. Chem. Commun.* **11**, 28 (1939); *C. A.* **33**, 4550 (1939).
5323. Zappi, E. V. and H. Degiorgi, *Anal. Asoc. Quim Argentina* **18**, 210 (1930); *Chem. Zentr.* **1931**, II, 1121.
5324. Zuagg, H. E., B. W. Horrom, et al., *J. Amer. Chem. Soc.* **82**, 2895 (1960).
5325. Zaugg, H. E. and A. D. Schaefer, *Anal. Chem.* **36**, 2121 (1964).
5326. Zaugg, H. E. and A. D. Schaefer, *J. Amer. Chem. Soc.* **87**, 1857 (1965).
5327. Zecherl, M. K., *Mikrochim. Acta* **33**, 387 (1948); *C. A.* **42**, 6537 (1948).
5328. Zeiss, H. H. and M. Tsutsui, *J. Amer. Chem. Soc.* **75**, 897 (1953).
5329. Zelinsky, N., *Ber. Deut. Chem. Gesell.* **34**, 2877 (1901).
5330. Zelinsky, N. and N. Rosanoff, *Z. Phys. Chem.* (Leipzig) **78**, 629 (1912).
5331. Zelinsky, N. D. and M. B. Turowa-Pollak, *Ber. Deut. Chem. Gesell.* **65**, 1299 (1932).
5332. Zellner, R. J. and C. A. R. Johnson, U.S. Patent 3,031,511, Apr. 24, 1962; *C. A.* **57**, 13,728 (1962).
5333. Zepalova-Mikhailova, L. A., *Trans. Inst. Pure Chem. Reagents* (USSR), No. **16**, 51 (1939); *Khim. Referat. Zhur.* **1939**, No. 6, 70; *C. A.* **32**, 5288 (1938); **34**, 3960 (1940).
5334. Zerewitinoff, T., *Z. Anal. Chem.* **50**, 683 (1911); *Ber. Deut. Chem. Gesell.* **47**, 2417 (1914).
5335. Zerweck, W., H. Ritter, et al., German Patent 855,107, Oct. 11, 1952; *Chem. Zentr.* **1953**, I, 2353.
5336. Zetzsche, F. and P. Zala, *Helv. Chim. Acta* **9**, 288 (1926); *C. A.* **20**, 2996 (1926).
5337. Zhdanov, A. K., *J. Gen. Chem.* (USSR) **11**, No. 7, 471 (1941); *C. A.* **35**, 7275 (1941).
5338. Zhuravlev, D. I., *J. Phys. Chem.* (USSR) **9**, 875 (1937); *C. A.* **31**, 8287 (1937).
5339. Ziebell, G., German (East) Patent 26,288 (Cl. C 07c 3), Oct. 26, 1963; *C. A.* **60**, 15,734 (1964).
5340. Zief, M. and W. R. Wilcox, eds., *Fractional Solidification*, Vol. 1. Dekker, New York, 1967.
5341. Ziegler, R., *Z. Phys.* **116**, 716 (1940); *C. A.* **35**, 4682 (1941).
5342. Zietlow, J. P., F. F. Cleveland, et al., *J. Chem. Phys.* **18**, 1076 (1950).
5343. Zilberman, E. N., *Zh. Prikl. Khim.* **24**, 776 (1951); *C. A.* **47**, 4172 (1953).
5344. Zil'berman-Granovskaya, A. A., *J. Phys. Chem.* (USSR) **14**, 759 (1940); *C. A.* **35**, 3867 (1941).
5345. Zil'berman-Granovskaya, A. A., *J. Phys. Chem.* (USSR) **14**, 768 (1940); *C. A.* **35**, 3867 (1941).
5346. Zilberman-Granovskaya, A. A., *J. Phys. Chem.* (USSR) **14**, 1004 (1940); *C. A.* **35**, 3867 (1941).
5347. Zimakov, P. V. and V. A. Sokolova, *Zh. Fiz. Khim.* **27**, 1079 (1953); *C. A.* **49**, 3597 (1955).
5348. Zimm, B. H., *J. Phys. Colloid Chem.* **54**, 1306 (1950).
5349. Zirnitis, V. A. and M. M. Suschinskii, *Opt. Spektrosk. Akad. Nauk SSSR, Otd. Fiz.-Mat. Nauk, Sb. Statei* **2**, 153 (1963); *C. A.* **59**, 13,500 (1963).

5350. Zmaczyński, A., *J. Chim. Phys.* **27**, 503 (1930); *Brit. Abstr.* **1931A**, 156.
5351. Zmaczyński, M. A., *Svensk Kem. Tidskr.* **48**, 268 (1936); *Chem. Zentr.* **1937, I**, 2762, 4994; *Rocz. Chem.* **16**, 486 (1936).
5352. Zobel, R. and A. B. F. Duncan, *J. Amer. Chem. Soc.* **77**, 2611 (1955).
5353. Zozulya, A. P. and E. V. Novikova, *Zh. Analit. Khim.* **18**, 1105 (1963); *C. A.* **60**, 1119 (1964).
5354. Zumwalt, L. R. and R. M. Badger, *J. Amer. Chem. Soc.* **62**, 305 (1940).
5355. Zweig, A., J. E. Lehnsen, et al., *J. Amer. Chem. Soc.* **85**, 3933 (1963).
5356. Zweig, A., J. E. Lehnsen, et al., *J. Amer. Chem. Soc.* **85**, 3940 (1963).
5357. Zwolenik, J. J. and R. M. Fuoss, *J. Phys. Chem.* **68**, 903 (1964).
5358. Zwolinski, B. J. and R. C. Wilhoit, "Heats of Formation and Heats of Combustion," Thermodynamics Research Center, College Station, Texas, 1968.

INDEX

The number immediately following the name of the solvent is its code number and is in (parentheses). The code number is followed by a colon. The first number following the colon refers to the page number of the table of physical properties of the subject solvent in Chapter III. The second number refers to methods of preparation, purification, criteria of purity and safety in Chapter V. Other pertinent information is listed as subentries. Example:

3-Methylpentane, (9): 81, 596

density equation, 30

special grades available, 580

Synonyms listed in Chapter I for the solvents are included in the index to facilitate the use of the book. They are followed, in order, by the abbreviation "syn.," the name of the solvent, the page number of the table of physical properties, and the page number of the methods of preparation, purification, etc. Example: Allyl alcohol, syn., 2-propen-1-ol, 192, 683. Complete indexing will be found under the name of the solvent.